D1290342

WIN the complete library of RSMeans Online data!

Register your book below to receive your FREE quarterly updates, PLUS you will be entered into a quarterly drawing to win the COMPLETE RSMeans Online library of 2015 data!

Be sure to keep up-to-date in 2015!

Fill out the card below for RSMeans' free quarterly updates, as well as a chance to win the complete RSMeans Online library. Please provide your name, address, and email below and return this card by mail, or to register online:

http://info.thegordiangroup.com/RSMeans.html

Name _____

email_____ Title_____

Company _____

Street _____

City/Town _____ State/Prov. _____ Zip/Postal Code _____

RSMeans

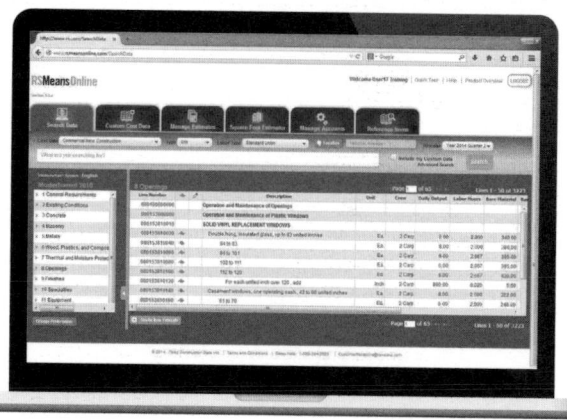

WIN the complete library of RSMeans Online 2015 data!

See other side for details.

RSMeans Plumbing Cost Data

2015   38th annual edition

Melville J. Mossman, PE, Senior Editor

Engineering Director
Bob Mewis, CCC (*1, 14, 41*)

Contributing Editors
Christopher Babbitt
Adrian C. Charest, PE
(*26, 27, 28, 48*)
Gary W. Christensen
Cheryl Elsmore
David Fiske
Robert Fortier, PE
(*2, 31, 32, 33, 34, 35, 44, 46*)
Robert J. Kuchta (*8*)

*Numbers in italics are the divisional responsibilities
for each editor. Please contact the designated editor
directly with any questions.*

Thomas Lane (*6, 7*)
Robert C. McNichols
Genevieve Medeiros
Melville J. Mossman, PE (*21, 22, 23*)
Jeannene D. Murphy
Marilyn Phelan, AIA (*9, 10, 11, 12*)
Stephen C. Plotner (*3, 5*)
Siobhan Soucie (*4, 13*)

Vice President Data & Engineering
Chris Anderson

Cover Design
Sara Rutan

Product Manager
Andrea Sillah

Production Manager
Debbie Panarelli

Production
Jill Goodman
Jonathan Forgit
Sara Rutan
Mary Lou Geary

Technical Support
Gary L. Hoitt
Kathryn S. Rodriguez

RSMeans
Construction Publishers & Consultants
700 Longwater Drive
Norwell, MA 02061
United States of America
1-877-709-6968
www.rsmeans.com

Copyright 2014 by RSMeans
All rights reserved.
Cover photo ©
KuLouKu/iStock/Thinkstock

Printed in the United States of America
ISSN 1537-8411
ISBN 978-1-940238-64-7

0215 $206.95 per copy (in United States)
Price is subject to change without prior notice.

Related RSMeans Products and Services

The engineers at RSMeans suggest the following products and services as companion information resources to *RSMeans Plumbing Cost Data*

Construction Cost Data Books
Assemblies Cost Data 2015
Square Foot Costs 2015

Reference Books
Estimating Building Costs
RSMeans Estimating Handbook
Green Building: Project Planning & Estimating
How to Estimate with RSMeans Data
Plan Reading & Material Takeoff
Project Scheduling & Management for Construction
Universal Design Ideas for Style, Comfort & Safety

Seminars and In-House Training
Unit Price Estimating
RSMeans Online® Training
Practical Project Management for Construction Professionals
Scheduling with MSProject for Construction Professionals
Mechanical & Electrical Estimating

RSMeans Online Bookstore
Visit RSMeans at **www.rsmeans.com** for a one-stop portal for the most reliable and current resources available on the market. Here you can learn about our more than 20 annual cost data books and eBooks, also available on CD and online versions, as well as our libary of reference books and RSMeans seminars.

RSMeans Electronic Data
Get the information found in RSMeans cost books electronically at **RSMeansOnline.com**.

RSMeans Business Solutions
Cost engineers and analysts conduct facility life cycle and benchmark research studies and predictive cost modeling, as well as offer consultation services for conceptual estimating and real property management. Research studies are designed with the application of proprietary cost and project data from RSMeans' extensive North American databases. Analysts offer building product manufacturers qualitative and quantitative market research, as well as time/motion studies for new products, and Web-based dashboard for market opportunity analysis. Clients are from both the public and private sectors, including federal, state, and municipal agencies; corporations; institutions; construction management firms; hospitals and associations.

Construction Costs for Software Applications
More than 25 unit price and assemblies cost databases are available through a number of leading estimating and facilities management software partners. For more information see the "Other RSMeans Products and Services" pages at the back of this publication.

RSMeans data is also available to federal, state, and local government agencies as multi-year, multi-seat licenses.

For information on our current partners, visit:
www.rsmeans.com/partners or call 877.709.6968.

Table of Contents

Foreword

Our Mission

Since 1942, RSMeans has been actively engaged in construction cost publishing and consulting throughout North America.

Today, more than 70 years after RSMeans began, our primary objective remains the same: to provide you, the construction and facilities professional, with the most current and comprehensive construction cost data possible.

Whether you are a contractor, owner, architect, engineer, facilities manager, or anyone else who needs a reliable construction cost estimate, you'll find this publication to be a highly useful and necessary tool.

With the constant flow of new construction methods and materials today, it's difficult to find the time to look at and evaluate all the different construction cost possibilities. In addition, because labor and material costs keep changing, last year's cost information is not a reliable basis for today's estimate or budget.

That's why so many construction professionals turn to RSMeans. We keep track of the costs for you, along with a wide range of other key information, from city cost indexes . . . to productivity rates . . . to crew composition . . . to contractor's overhead and profit rates.

RSMeans performs these functions by collecting data from all facets of the industry and organizing it in a format that is instantly accessible to you. From the preliminary budget to the detailed unit price estimate, you'll find the data in this book useful for all phases of construction cost determination.

The Staff, the Organization, and Our Services

When you purchase one of RSMeans' publications, you are, in effect, hiring the services of a full-time staff of construction and engineering professionals.

Our thoroughly experienced and highly qualified staff works daily at collecting, analyzing, and disseminating comprehensive cost information for your needs. These staff members have years of practical construction experience and engineering training prior to joining the firm. As a result, you can count on them not only for accurate cost figures, but also for additional background reference information that will help you create a realistic estimate.

The RSMeans organization is always prepared to help you solve construction problems through its variety of data solutions, including online, CD, and print book formats, as well as cost estimating expertise available via our business solutions, training, and seminars.

Besides a full array of construction cost estimating books, RSMeans also publishes a number of other reference works for the construction industry. Subjects include construction estimating and project and business management, green building, and a library of facility management references.

In addition, you can access all of our construction cost data electronically in convenient CD format or on the web. Visit **RSMeansOnline.com** for more information on our 24/7 online cost data.

What's more, you can increase your knowledge and improve your construction estimating and management performance with an RSMeans construction seminar or in-house training program. These two-day seminar programs offer unparalleled opportunities for everyone in your organization to become updated on a wide variety of construction-related issues.

RSMeans is also a worldwide provider of construction cost management and analysis services for commercial and government owners.

In short, RSMeans can provide you with the tools and expertise for constructing accurate and dependable construction estimates and budgets in a variety of ways.

Robert Snow Means Established a Tradition of Quality That Continues Today

Robert Snow Means spent years building RSMeans, making certain he always delivered a quality product.

Today, at RSMeans, we do more than talk about the quality of our data and the usefulness of our books. We stand behind all of our data, from historical cost indexes to construction materials and techniques to current costs.

If you have any questions about our products or services, please call us toll-free at 1-877-709-6968. Our customer service representatives will be happy to assist you. You can also visit our website at **www.rsmeans.com**

How the Book Is Built: An Overview

The Construction Specifications Institute (CSI) and Construction Specifications Canada (CSC) have produced the 2014 edition of MasterFormat®, a system of titles and numbers used extensively to organize construction information.

All unit price data in the RSMeans cost data books is now arranged in the 50-division MasterFormat® 2014 system.

Powerful Construction Tool

You have in your hands one of the most powerful construction tools available today. A successful project is built on the foundation of an accurate and dependable estimate. This book will enable you to construct just such an estimate.

For the casual user the book is designed to be:

- quickly and easily understood so you can get right to your estimate.
- filled with valuable information so you can understand the necessary factors that go into the cost estimate.

For the regular user, the book is designed to be:

- a handy desk reference that can be quickly referred to for key costs.
- a comprehensive, fully reliable source of current construction costs and productivity rates so you'll be prepared to estimate any project.
- a source book for preliminary project cost, product selections, and alternate materials and methods.

To meet all of these requirements, we have organized the book into the following clearly defined sections.

Quick Start

See our "Quick Start" instructions on the following page to get started right away.

Estimating with RSMeans Unit Price Cost Data

Please refer to these steps for guidance on completing an estimate using RSMeans unit price cost data.

How to Use the Book: The Details

This section contains an in-depth explanation of how the book is arranged . . . and how you can use it to determine a reliable construction cost estimate. It includes information about how we develop our cost figures and how to completely prepare your estimate.

Unit Price Section

All cost data has been divided into the 50 divisions according to the MasterFormat system of classification and numbering. For a listing of these divisions and an outline of their subdivisions, see the Unit Price Section Table of Contents. *Estimating tips are included at the beginning of each division.*

Assemblies Section

The cost data in this section has been organized in an "Assemblies" format. These assemblies are the functional elements of a building and are arranged according to the 7 elements of the UNIFORMAT II classification system. For a complete explanation of a typical "Assemblies" page, see "How RSMeans Assemblies Data Works."

Reference Section

This section includes information on Equipment Rental Costs, Crew Listings, Historical Cost Indexes, City Cost Indexes, Location Factors, Reference Tables, Change Orders, Square Foot Costs, and a listing of Abbreviations.

Equipment Rental Costs: This section contains the average costs to rent and operate hundreds of pieces of construction equipment.

Crew Listings: This section lists all of the crews referenced in the book. For the purposes of this book, a crew is composed of more than one trade classification and/or the addition of power equipment to any trade classification. Power equipment is included in the cost of the crew. Costs are shown both with bare labor rates and with the installing contractor's overhead and profit added. For each, the total crew cost per eight-hour day and the composite cost per labor-hour are listed.

Historical Cost Indexes: These indexes provide you with data to adjust construction costs over time.

City Cost Indexes: All costs in this book are U.S. national averages. Costs vary because of the regional economy. You can adjust costs by CSI Division to over 700 locations throughout the U.S. and Canada by using the data in this section.

Location Factors: You can adjust total project costs to over 900 locations throughout the U.S. and Canada by using the data in this section.

Reference Tables: At the beginning of selected major classifications in the Unit Price section are reference numbers shown in a shaded box. These numbers refer you to related information in the Reference Section. In this section, you'll find reference tables, explanations, estimating information that support how we develop the unit price data, technical data, and estimating procedures.

Change Orders: This section includes information on the factors that influence the pricing of change orders.

Square Foot Costs: This section contains costs for 59 different building types that allow you to make a rough estimate for the overall cost of a project or its major components.

Abbreviations: A listing of abbreviations used throughout this book, along with the terms they represent, is included in this section.

Index

A comprehensive listing of all terms and subjects in this book will help you quickly find what you need when you are not sure where it occurs in MasterFormat.

The Scope of This Book

This book is designed to be as comprehensive and as easy to use as possible. To that end we have made certain assumptions and limited its scope in two key ways:

1. We have established material prices based on a national average.
2. We have computed labor costs based on a 30-city national average of union wage rates.

For a more detailed explanation of how the cost data is developed, see "How To Use the Book: The Details."

Project Size/Type

The material prices in RSMeans cost data books are "contractor's prices." They are the prices that contractors can expect to pay at the lumberyards, suppliers'/distributors' warehouses, etc. Small orders of speciality items would be higher than the costs shown, while very large orders, such as truckload lots, would be less. The variation would depend on the size, timing, and negotiating power of the contractor. The labor costs are primarily for new construction or major renovation rather than repairs or minor alterations. *With reasonable exercise of judgment, the figures can be used for any building work.*

Absolute Essentials for a Quick Start

If you feel you are ready to use this book and don't think you will need the detailed instructions that begin on the following page, this Absolute Essentials for a Quick Start page is for you. These steps will allow you to get started estimating in a matter of minutes.

1 Scope
Think through the project that you will be estimating, and identify the many individual work tasks that will need to be covered in your estimate.

2 Quantify
Determine the number of units that will be required for each work task that you identified.

3 Pricing
Locate individual Unit Price line items that match the work tasks you identified. The Unit Price Section Table of Contents that begins on page 1 and the Index in the back of the book will help you find these line items.

4 Multiply
Multiply the Total Incl O&P cost for a Unit Price line item in the book by your quantity for that item. The price you calculate will be an estimate for a completed item of work performed by a subcontractor. Keep adding line items in this manner to build your estimate.

5 Project Overhead
Include project overhead items in your estimate. These items are needed to make the job run and are typically, but not always, provided by the General Contractor. They can be found in Division 1. An alternate method of estimating project overhead costs is to apply a percentage of the total project cost.

Include rented tools not included in crews, waste, rubbish handling, and cleanup.

6 Estimate Summary
Include General Contractor's markup on subcontractors, General Contractor's office overhead and profit, and sales tax on materials and equipment.

Adjust your estimate to the project's location by using the City Cost Indexes or Location Factors found in the Reference Section.

Editors' Note: We urge you to spend time reading and understanding the supporting material in the front of this book. An accurate estimate requires experience, knowledge, and careful calculation. The more you know about how we at RSMeans developed the data, the more accurate your estimate will be. In addition, it is important to take into consideration the reference material in the back of the book such as Equipment Listings, Crew Listings, City Cost Indexes, Location Factors, and Reference Tables.

Estimating with RSMeans Unit Price Cost Data

[fol]lowing these steps will allow you to [com]plete an accurate estimate using [RS]Means Unit Price cost data.

Scope Out the Project

Think through the project and identify those CSI Divisions needed in your estimate.

Identify the individual work tasks that will need to be covered in your estimate.

The Unit Price data in this book has been divided into 50 Divisions according to the CSI MasterFormat® 2014—their titles are listed on the back cover of your book.

The Unit Price Section Table of Contents on page 1 may also be helpful when scoping out your project.

Experienced estimators find it helpful to begin with Division 2 and continue through completion. Division 1 can be estimated after the full project scope is known.

Quantify

Determine the number of units required for each work task that you identified.

Experienced estimators include an allowance for waste in their quantities. (Waste is not included in RSMeans Unit Price line items unless so stated.)

Price the Quantities

Use the Unit Price Table of Contents, and the Index, to locate individual Unit Price line items for your estimate.

Reference Numbers indicated within a Unit Price section refer to additional information that you may find useful.

The crew indicates who is performing the work for that task. Crew codes are expanded in the Crew Listings in the Reference Section to include all trades and equipment that comprise the crew.

- The Daily Output is the amount of work the crew is expected to do in one day.

- The Labor-Hours value is the amount of time it will take for the crew to install one unit of work.

- The abbreviated Unit designation indicates the unit of measure upon which the crew, productivity, and prices are based.

- Bare Costs are shown for materials, labor, and equipment needed to complete the Unit Price line item. Bare costs do not include waste, project overhead, payroll insurance, payroll taxes, main office overhead, or profit.

- The Total Incl O&P cost is the billing rate or invoice amount of the installing contractor or subcontractor who performs the work for the Unit Price line item.

4 Multiply

- Multiply the total number of units needed for your project by the Total Incl O&P cost for each Unit Price line item.

- Be careful that your take off unit of measure matches the unit of measure in the Unit column.

- The price you calculate is an estimate for a completed item of work.

- Keep scoping individual tasks, determining the number of units required for those tasks, matching each task with individual Unit Price line items in the book, and multiply quantities by Total Incl O&P costs.

- An estimate completed in this manner is priced as if a subcontractor, or set of subcontractors, is performing the work. The estimate does not yet include Project Overhead or Estimate Summary components such as General Contractor markups on subcontracted work, General Contractor office overhead and profit, contingency, and location factor.

5 Project Overhead

- Include project overhead items from Division 1 – General Requirements.

- These items are needed to make the job run. They are typically, but not always, provided by the General Contractor. Items include, but are not limited to, field personnel, insurance, performance bond, permits, testing, temporary utilities, field office and storage facilities, temporary scaffolding and platforms, equipment mobilization and demobilization, temporary roads and sidewalks, winter protection, temporary barricades and fencing, temporary security, temporary signs, field engineering and layout, final cleaning and commissioning.

- Each item should be quantified, and matched to individual Unit Price line items in Division 1, and then priced and added to your estimate.

- An alternate method of estimating project overhead costs is to apply a percentage of the total project cost, usually 5% to 15% with an average of 10% (see General Conditions, page ix).

- Include other project related expenses in your estimate such as:
  - Rented equipment not itemized in the Crew Listings
  - Rubbish handling throughout the project (see 02 41 19.19)

6 Estimate Summary

- Include sales tax as required by laws of your state or county.
- Include the General Contractor's markup on self-performed work, usually 5% to 15% with an average of 10%.
- Include the General Contractor's markup on subcontracted work, usually 5% to 15% with an average of 10%.
- Include General Contractor's main office overhead and profit:
  - □ RSMeans gives general guidelines on the General Contractor's main office overhead (see section 01 31 13.60 and Reference Number R013113-50).
  - □ RSMeans gives no guidance on the General Contractor's profit.

  - □ Markups will depend on the size of the General Contractor's operations, his projected annual revenue, the level of risk he is taking on, and on the level of competition in the local area and for this project in particular.
- Include a contingency, usually 3% to 5%, if appropriate.
- Adjust your estimate to the project's location by using the City Cost Indexes or the Location Factors in the Reference Section:
  - □ Look at the rules on the pages for "How to Use the City Cost Indexes" to see how to apply the Indexes for your location.
  - □ When the proper Index or Factor has been identified for the project's location, convert it to a multiplier by

dividing it by 100, and then multiply that multiplier by your estimated total cost. The original estimated total cost will now be adjusted up or down from the national average to a total that is appropriate for your location.

Editors' Notes:
We urge you to spend time reading and understanding the supporting material in the front of this book. An accurate estimate requires experience, knowledge, and careful calculation. The more you know about how we at RSMeans developed the data, the more accurate your estimate will be. In addition, it is important to take into consideration the reference material in the back of the book such as Equipment Listings, Crew Listings, City Cost Indexes, Location Factors, and Reference Tables.

How to Use the Book: The Details

What's Behind the Numbers? The Development of Cost Data

The staff at RSMeans continually monitors developments in the construction industry in order to ensure reliable, thorough, and up-to-date cost information. While overall construction costs may vary relative to general economic conditions, price fluctuations within the industry are dependent upon many factors. Individual price variations may, in fact, be opposite to overall economic trends. Therefore, costs are constantly tracked and complete updates are published yearly. Also, new items are frequently added in response to changes in materials and methods.

Costs—$ (U.S.)

All costs represent U.S. national averages and are given in U.S. dollars. The RSMeans City Cost Indexes can be used to adjust costs to a particular location. The City Cost Indexes for Canada can be used to adjust U.S. national averages to local costs in Canadian dollars. No exchange rate conversion is necessary.

G The processes or products identified by the green symbol in our publications have been determined to be environmentally responsible and/or resource-efficient solely by the RSMeans engineering staff. The inclusion of the green symbol does not represent compliance with any specific industry association or standard.

Material Costs

The RSMeans staff contacts manufacturers, dealers, distributors, and contractors all across the U.S. and Canada to determine national average material costs. If you have access to current material costs for your specific location, you may wish to make adjustments to reflect differences from the national average. Included within material costs are fasteners for a normal installation. RSMeans engineers use manufacturers' recommendations, written specifications, and/or standard construction practice for size and spacing of fasteners. Adjustments to material costs may be required for your specific application or location. The manufacturer's warranty is assumed. Extended warranties are not included in the material costs. Material costs do not include sales tax.

Labor Costs

Labor costs are based on the average of wage rates from 30 major U.S. cities. Rates are determined from labor union agreements or prevailing wages for construction trades for the current year. Rates, along with overhead and profit markups, are listed on the inside back cover of this book.

- If wage rates in your area vary from those used in this book, or if rate increases are expected within a given year, labor costs should be adjusted accordingly.

Labor costs reflect productivity based on actual working conditions. In addition to actual installation, these figures include time spent during a normal weekday on tasks such as, material receiving and handling, mobilization at site, site movement, breaks, and cleanup.

Productivity data is developed over an extended period so as not to be influenced by abnormal variations and reflects a typical average.

Equipment Costs

Equipment costs include not only rental, but also operating costs for equipment under normal use. The operating costs include parts and labor for routine servicing such as repair and replacement of pumps, filters, and worn lines. Normal operating expendables, such as fuel, lubricants, tires, and electricity (where applicable), are also included. Extraordinary operating expendables with highly variable wear patterns, such as diamond bits and blades, are excluded. These costs are included under materials. Equipment rental rates are obtained from industry sources throughout North America—contractors, suppliers, dealers, manufacturers, and distributors.

Rental rates can also be treated as reimbursement costs for contractor-owned equipment. Owned equipment costs include depreciation, loan payments, interest, taxes, insurance, storage and major repairs.

Equipment costs do not include operators' wages; nor do they include the cost to move equipment to a job site (mobilization) or from a job site (demobilization).

Equipment Cost/Day—The cost of power equipment required for each crew is included in the Crew Listings in the Reference Section (small tools that are considered as essential everyday tools are not listed out separately). The Crew Listings itemize specialized tools and heavy equipment along with labor trades. The daily cost of itemized equipment included in a crew is based on dividing the weekly bare rental rate by 5 (number of working days per week) and then adding the hourly operating cost times 8 (the number of hours per day). This Equipment Cost/Day is shown in the last column of the Equipment Rental Cost pages in the Reference Section.

Mobilization/Demobilization—The cost to move construction equipment from an equipment yard or rental company to the job site and back again is not included in equipment costs. Mobilization (to the site) and demobilization (from the site) costs can be found in the Unit Price Section. If a piece of equipment is already at the job site, it is not appropriate to utilize mob./demob. costs again in an estimate.

Overhead and Profit

Total Cost including O&P for the *Installing Contractor* is shown in the last column on the Unit Price and/or the Assemblies pages of this book. This figure is the sum of the bare material cost plus 10% for profit, the bare labor cost plus total overhead and profit, and the bare equipment cost plus 10% for profit. Details for the calculation of Overhead and Profit on labor are shown on the inside back cover and in the Reference Section of this book. (See "How RSMeans Unit Price Data Works" for an example of this calculation.)

General Conditions

Cost data in this book is presented in two ways: Bare Costs and Total Cost including O&P (Overhead and Profit). General Conditions, or General Requirements, of the contract should also be added to the Total Cost including O&P when applicable. Costs for General Conditions are listed in Division 1 of the Unit Price Section and the Reference

Section of this book. General Conditions for the *Installing Contractor* may range from 0% to 10% of the Total Cost including O&P. For the *General* or *Prime Contractor*, costs for General Conditions may range from 5% to 15% of the Total Cost including O&P, with a figure of 10% as the most typical allowance. If applicable, the Assemblies and Models sections of this book use costs that include the installing contractor's overhead and profit (O&P).

Factors Affecting Costs

Costs can vary depending upon a number of variables. Here's how we have handled the main factors affecting costs.

Quality—The prices for materials and the workmanship upon which productivity is based represent sound construction work. They are also in line with U.S. government specifications.

Overtime—We have made no allowance for overtime. If you anticipate premium time or work beyond normal working hours, be sure to make an appropriate adjustment to your labor costs.

Productivity—The productivity, daily output, and labor-hour figures for each line item are based on working an eight-hour day in daylight hours in moderate temperatures. For work that extends beyond normal work hours or is performed under adverse conditions, productivity may decrease. (See "How RSMeans Unit Price Data Works" for more on productivity.)

Size of Project—The size, scope of work, and type of construction project will have a significant impact on cost. Economies of scale can reduce costs for large projects. Unit costs can often run higher for small projects.

Location—Material prices in this book are for metropolitan areas. However, in dense urban areas, traffic and site storage limitations may increase costs. Beyond a 20-mile radius of large cities, extra trucking or transportation charges may also increase the material costs slightly. On the other hand, lower wage rates may be in effect. Be sure to consider both of these factors when preparing an estimate, particularly if the job site is located in a central city or remote rural location. In addition, highly specialized subcontract items may require travel and per-diem expenses for mechanics.

Other Factors——
- season of year
- contractor management
- weather conditions
- local union restrictions
- building code requirements
- availability of:
  - adequate energy
  - skilled labor
  - building materials
- owner's special requirements/restrictions
- safety requirements
- environmental considerations

Unpredictable Factors—General business conditions influence "in-place" costs of all items. Substitute materials and construction methods may have to be employed. These may affect the installed cost and/or life cycle costs. Such factors may be difficult to evaluate and cannot necessarily be predicted on the basis of the job's location in a particular section of the country. Thus, where these factors apply, you may find significant but unavoidable cost variations for which you will have to apply a measure of judgment to your estimate.

Rounding of Costs

In general, all unit prices in excess of $5.0 have been rounded to make them easier to use and still maintain adequate precision of the results. The rounding rules we have chosen are in the following table.

Prices from . . .	Rounded to the nearest . . .
$.01 to $5.00	$.0
$5.01 to $20.00	$.0
$20.01 to $100.00	$.5
$100.01 to $300.00	$1.0
$300.01 to $1,000.00	$5.0
$1,000.01 to $10,000.00	$25.0
$10,000.01 to $50,000.00	$100.0
$50,000.01 and above	$500.0

How Subcontracted Items Affect Costs

A considerable portion of all large construction jobs is usually subcontracted. I fact, the percentage done by subcontractors is constantly increasing and may run over 90%. Since the workers employed by these companies do nothing else but install their particular product, they soon become expert in that line. The result is, installation by these firms is accomplished so efficiently that the total in-place cost, even adding the general contractor's overhead and profit, is no more, and often less, than if the principa contractor had handled the installation himself/herself. Companies that deal with construction specialties are anxious to have their product perform well and, consequentl the installation will be the best possible.

ntingencies

e allowance for contingencies generally
vides for unforeseen construction
ficulties. On alterations or repair jobs,
% is not too much. If drawings are
al and only field contingencies are
ing considered, 2% or 3% is probably
fficient, and often nothing need be
ded. Contractually, changes in plans
l be covered by extras. The contractor
uld consider inflationary price trends
d possible material shortages during
e course of the job. These escalation
ctors are dependent upon both economic
nditions and the anticipated time between
e estimate and actual construction. If
awings are not complete or approved, or a
dget cost is wanted, it is wise to add 5%
10%. Contingencies, then, are a matter
judgment. Additional allowances for
ntingencies are shown in Division 1.

nportant Estimating Considerations

e productivity or daily output of each
aftsman or crew assumes a well-managed
o where tradesmen with the proper tools
d equipment, along with the appropriate
nstruction materials, are present. Included
e daily mobilization and cleanup time,
eak time and plan layout time. Unless
herwise indicated, time for material
ovement on site (for items that can be
ansported by hand) of up to 200' into the
ilding and to the first or second floor is also
cluded. If material has to be transported
other means, over greater distances, or to
gher floors, an additional allowance should
considered by the estimator.

hile horizontal movement is typically
sole function of distances, vertical
ansport introduces other variables that
n significantly impact productivity. In
occupied building, the use of elevators
ssuming access, size, and required
otective measures are acceptable) must be
derstood at the time of the estimate. For
ew construction, hoist wait and cycle times
n easily be 15 minutes and may result
scheduled access extending beyond the
ormal work day. Finally, all vertical transport
ill impose strict weight limits likely to

preclude the use of any motorized material
handling.

The productivity, or daily output, also assumes
installation that meets manufacturer/designer/
standard specifications. A time allowance for
quality control checks, minor adjustments
and any task required to ensure the proper
function or operation is also included. For
items that require connections to services,
time is included for positioning, leveling,
securing the unit and for making all the
necessary connections (and start up where
applicable) ensuring a complete installation.
Estimating of the services themselves
(electrical, plumbing, water, steam,
hydraulics, dust collection, etc.) is separate.

In some cases, the estimator must consider
the use of a crane and an appropriate crew
for the installation of large or heavy items.
For those situations where a crane is not
included in the assigned crew and as part of
the line item cost, equipment rental costs,
mobilization and demobilization costs, and
operator and support personnel costs must
be considered.

Labor-Hours

The labor-hours expressed in this publication
are derived by dividing the total daily labor
hours for the crew by the daily output.
Based on average installation time and the
assumptions listed above, the labor-hours
include: direct labor, indirect labor, and
nonproductive time. A typical day for a
craftsman might include, but not be
limited to:

- Direct Work
  □ Measuring and layout
  □ Preparing materials
  □ Actual installation
  □ Quality assurance/quality control
- Indirect Work
  □ Reading plans or specifications
  □ Preparing space
  □ Receiving materials
  □ Material movement
  □ Giving or receiving instruction
  □ Miscellaneous

- Non-Work
  □ Chatting
  □ Personal issues
  □ Breaks
  □ Interruptions (i.e., sickness, weather, material or equipment shortages, etc.)

If any of the items for a typical day do
not apply to the particular work or project
situation, the estimator should make any
necessary adjustments.

Final Checklist

Estimating can be a straightforward process
provided you remember the basics. Here's
a checklist of some of the steps you should
remember to complete before finalizing
your estimate.

Did you remember to . . .

- factor in the City Cost Index for your locale?
- take into consideration which items have been marked up and by how much?
- mark up the entire estimate sufficiently for your purposes?
- read the background information on techniques and technical matters that could impact your project time span and cost?
- include all components of your project in the final estimate?
- double check your figures for accuracy?
- call RSMeans if you have any questions about your estimate or the data you've found in our publications? Remember, RSMeans stands behind its publications. If you have any questions about your estimate . . . about the costs you've used from our books . . . or even about the technical aspects of the job that may affect your estimate, feel free to call the RSMeans editors at 1-877-709-6968.

Free Quarterly Updates

Stay up-to-date throughout 2015 with
RSMeans' free cost data updates four times
a year. Sign up online to make sure you have
access to the newest data. Every quarter
we provide the city cost adjustment factors
for hundreds of cities and key materials.
Register at:
http://info.thegordiangroup.com/RSMeans.html.

Unit Price Section

Table of Contents

1

How RSMeans Unit Price Data Works

All RSMeans unit price data is organized in the same way.

It is important to understand the structure, so that you can find information easily and use it correctly.

RSMeans **Line Numbers** consist of 12 characters, which identify a unique location in the database for each task. The first 6 or 8 digits conform to the Construction Specifications Institute MasterFormat® 2014. The remainder of the digits are a further breakdown by RSMeans in order to arrange items in understandable groups of similar tasks. Line numbers are consistent across all RSMeans publications, so a line number in any RSMeans product will always refer to the same unit of work.

RSMeans engineers have created **reference** information to assist you in your estimate. If there is information that applies to a section, it will be indicated at the start of the section. In this case, R033105-10 provides information on the proportionate quantities of formwork, reinforcing, and concrete used in cast-in-place concrete items such as footings, slabs, beams, and columns. The Reference Section is located in the back of the book on the pages with a gray edge.

RSMeans **Descriptions** are shown in a hierarchical structure to make them readable. In order to read a complete description, read up through the indents to the top of the section. Include everything that is above and to the left that is not contradicted by information below. For instance, the complete description for line 03 30 53.40 3550 is "Concrete in place, including forms (4 uses), Grade 60 rebar, concrete (Portland cement Type 1), placement and finishing unless otherwise indicated; Equipment pad (3000 psi), 4' x 4' x 6" thick."

03 30 Cast-In-Place Concrete

03 30 53 – Miscellaneous Cast-In-Place Concrete

03 30 53.40 Concrete In Place

Line	Description	Reference
0010	**CONCRETE IN PLACE**	R033105-10
0020	Including forms (4 uses), Grade 60 rebar, concrete (Portland cement	R033105-20
0050	Type I), placement and finishing unless otherwise indicated	R033105-50
0300	Beams (3500 psi), 5 kip per L.F., 10' span	R033105-65
0350	25' span	R033105-70
0500	Chimney foundations (5000 psi), over 5 C.Y.	R033105-85
0510	(3500 psi), under 5 C.Y.	
0700	Columns, square (4000 psi), 12" x 12", less than 2% reinforcing	
3450	Over 10,000 S.F.	
3500	Add per floor for 3 to 6 stories high	
3520	For 7 to 20 stories high	
3540	Equipment pad (3000 psi), 3' x 3' x 6" thick	
3550	4' x 4' x 6" thick	
3560	5' x 5' x 8" thick	
3570	6' x 6' x 8" thick	

When using **RSMeans data**, it is important to read through an entire section to ensure that you use the data that most closely matches your work. Note that sometimes there is additional information shown in the section that may improve your price. There are frequently lines that further describe, add to, or adjust data for specific situations.

The data published in RSMeans print books represents a "national average" cost. This data should be modified to the project location using the **City Cost Indexes** or **Location Factors** tables found in the Reference Section (see pages 602–650). Use the location factors to adjust estimate totals if the project covers multiple trades. Use the city cost indexes (CCI) for single trade projects or projects where a more detailed analysis is required. All figures in the two tables are derived from the same research. The last row of data in the CCI, the weighted average, is the same as the numbers reported for each location in the location factor table.

Crews include labor or labor and equipment necessary to accomplish each task. In this case, Crew C-14H is used. RSMeans selects a crew to represent the workers and equipment that are typically used for that task. In this case, Crew C-14H consists of one carpenter foreman (outside), two carpenters, one rodman, one laborer, one cement finisher, and one gas engine vibrator. Details of all crews can be found in the reference section.

Crews

Crew No.	Bare Costs		Incl. Subs O&P		Cost Per Labor-Hour	
Crew C-14H	Hr.	Daily	Hr.	Daily	Bare Costs	Incl. O&P
1 Carpenter Foreman (outside)	$48.95	$391.60	$75.35	$602.80	$46.33	$71.08
2 Carpenters	46.95	751.20	72.25	1156.00		
1 Rodman (reinf.)	52.55	420.40	82.10	656.80		
1 Laborer	37.60	300.80	57.85	462.80		
1 Cement Finisher	45.00	360.00	66.65	533.20		
1 Gas Engine Vibrator		31.20		34.32	.65	.71
48 L.H., Daily Totals		$2255.20		$3445.92	$46.98	$71.79

The **Daily Output** is the amount of work that the crew can do in a normal 8-hour workday, including mobilization, layout, movement of materials, and cleanup. In this case, crew C-14H can install thirty 4' x 4' x 6" thick concrete pads in a day. Daily output is variable, based on many factors, including the size of the job, location, and environmental conditions. RSMeans data represents work done in daylight (or adequate lighting) and temperate conditions.

Bare Costs are the costs of materials, labor, and equipment that the installing contractor pays. They represent the cost, in U.S. dollars, for one unit of work. They do not include any markups for profit or labor burden.

Crew	Daily Output	Labor-Hours	Unit	Material	2015 Bare Costs Labor	Equipment	Total	Total Incl O&P
C-14A	15.62	12.804	C.Y.	320	605	48	973	1,325
"	18.55	10.782		335	510	40.50	885.50	1,200
C-14C	32.22	3.476		150	157	.97	307.97	405
"	23.71	4.724		177	213	1.32	391.32	525
C-14A	11.96	16.722		365	790	63	1,218	1,700
	2200	.025		.94	1.07	.33	2.34	3.02
	31800	.002			.07	.02	.09	.13
	21200	.003			.11	.03	.14	.21
C-14H	45	1.067	Ea.	47	49.50	.69	97.19	128
	30	1.600		69.50	74	1.04	144.54	191
	18	2.667		122	124	1.73	247.73	325
	14	3.429		164	159	2.23	325.23	425

The **Total Incl O&P column** is the total cost, including overhead and profit, that the installing contractor will charge the customer. This represents the cost of materials plus 10% profit, the cost of labor plus labor burden and 10% profit, and the cost of equipment plus 10% profit. It does not include the general contractor's overhead and profit. Note: See the inside back cover for details of how RSMeans calculates labor burden.

The **Total column** represents the total bare cost for the installing contractor, in U.S. dollars. In this case, the sum of $69.50 for material + $74.00 for labor + $1.04 for equipment is $144.54.

The figure in the **Labor Hours** column is the amount of labor required to perform one unit of work—in this case the amount of labor required to construct one 4' x 4' equipment pad. This figure is calculated by dividing the number of hours of labor in the crew by the daily output (48 labor hours divided by 30 pads = 1.6 hours of labor per pad). Multiply 1.600 times 60 to see the value in minutes: 60 x 1.6 = 96 minutes. Note: the labor hour figure is not dependent on the crew size. A change in crew size will result in a corresponding change in daily output, but the labor hours per unit of work will not change.

All RSMeans unit cost data includes the typical **Unit of Measure** used for estimating that item. For concrete-in-place the typical unit is cubic yards (C.Y.) or each (Ea.). For installing broadloom carpet it is square yard, and for gypsum board it is square foot. The estimator needs to take special care that the unit in the data matches the unit in the take-off. Unit conversions may be found in the Reference Section.

Sample Estimate

This sample demonstrates the elements of an estimate, including a tally of the RSMeans data lines, and a summary of the markups on a contractor's work to arrive at a total cost to the owner. The RSMeans Location Factor is added at the bottom of the estimate to adjust the cost of the work to a specific location.

Work Performed: The body of the estimate shows the RSMeans data selected, including line number, a brief description of each item, its take-off unit and quantity, and the bare costs of materials, labor, and equipment. This estimate also includes a column titled "SubContract." This data is taken from the RSMeans column "Total Incl O&P," and represents the total that a subcontractor would charge a general contractor for the work, including the sub's markup for overhead and profit.

Division 1, General Requirements: This is the first division numerically, but the last division estimated. Division 1 includes project-wide needs provided by the general contractor. These requirements vary by project, but may include temporary facilities and utilities, security, testing, project cleanup, etc. For small projects a percentage can be used, typically between 5% and 15% of project cost. For large projects the costs may be itemized and priced individually.

Bonds: Bond costs should be added to the estimate. The figures here represent a typical performance bond, ensuring the owner that if the general contractor does not complete the obligations in the construction contract the bonding company will pay the cost for completion of the work.

Location Adjustment: RSMeans published data is based on national average costs. If necessary, adjust the total cost of the project using a location factor from the "Location Factor" table or the "City Cost Index" table. Use location factors if the work is general, covering multiple trades. If the work is by a single trade (e.g., masonry) use the more specific data found in the "City Cost Indexes."

This estimate is based on an interactive spreadsheet. A copy of this spreadsheet is located on the RSMeans website at **http://www.reedconstructiondata.com/rsmeans/extras/546011.** You are free to download it and adjust it to your methodology.

Project Name: Pre-Engineered Steel Building				Architect: As Shown	
Location:	**Anywhere, USA**				
Line Number	**Description**		**Qty**	**Unit**	**Mate**
03 30 53.40 3940	Strip footing, 12" x 24", reinforced		34	C.Y.	$4,726
03 30 53.40 3950	Strip footing, 12" x 36", reinforced		15	C.Y.	$1,995
03 11 13.65 3000	Concrete slab edge forms		500	L.F.	$175
03 22 11.10 0200	Welded wire fabric reinforcing		150	C.S.F.	$2,580
03 31 13.35 0300	Ready mix concrete, 4000 psi for slab on grade		278	C.Y.	$29,746
03 31 13.70 4300	Place, strike off & consolidate concrete slab		278	C.Y.	$0
03 35 13.30 0250	Machine float & trowel concrete slab		15,000	S.F.	$0
03 15 16.20 0140	Cut control joints in concrete slab		950	L.F.	$57
03 39 23.13 0300	Sprayed concrete curing membrane		150	C.S.F.	$1,657
Division 03	**Subtotal**				**$40,936**
08 36 13.10 2650	Manual 10' x 10' steel sectional overhead door		8	Ea.	$8,800
08 36 13.10 2860	Insulation and steel back panel for OH door		800	S.F.	$3,800
Division 08	**Subtotal**				**$12,600**
13 34 19.50 1100	Pre-Engineered Steel Building, 100' x 150' x 24'		15,000	SF Flr.	$0
13 34 19.50 6050	Framing for PESB door opening, 3' x 7'		4	Opng.	$0
13 34 19.50 6100	Framing for PESB door opening, 10' x 10'		8	Opng.	$0
13 34 19.50 6200	Framing for PESB window opening, 4' x 3'		6	Opng.	$0
13 34 19.50 5750	PESB door, 3' x 7', single leaf		4	Opng.	$2,340
13 34 19.50 7750	PESB sliding window, 4' x 3' with screen		6	Opng.	$2,280
13 34 19.50 6550	PESB gutter, eave type, 26 ga., painted		300	L.F.	$2,085
13 34 19.50 8650	PESB roof vent, 12" wide x 10' long		15	Ea.	$540
13 34 19.50 6900	PESB insulation, vinyl faced, 4" thick		27,400	S.F.	$11,508
Division 13	**Subtotal**				**$18,753.**
	Subtotal				**$72,289.**
Division 01	**General Requirements @ 7%**				**5,060.**
	Estimate Subtotal				**$77,349.**
	Sales Tax @ 5%				**3,867.**
	Subtotal A				**81,217.**
	GC O & P				**8,121.**
	Subtotal B				**89,338.**
	Contingency @ 5%				
	Subtotal C				
	Bond @ $12/1000 +10% O&P				
	Subtotal D				
	Location Adjustment Factor				
	Grand Total				

4

s example shows the cost to construct a pre-engineered steel building. The foundation, doors, windows, and insulation will be
talled by the general contractor. A subcontractor will install the structural steel, roofing, and siding.

		01/01/15	STD	
Labor	Equipment	SubContract	Estimate Total	
570.00	$22.10	$0.00		
260.00	$7.80	$0.00		
,190.00	$0.00	$0.00		
050.00	$0.00	$0.00		
$0.00	$0.00	$0.00		
753.80	$158.46	$0.00		
000.00	$450.00	$0.00		
,380.00	$85.50	$0.00		
952.50	$0.00	$0.00		
156.30	$723.86	$0.00	$66,816.66	Division 03
,320.00	$0.00	$0.00		
$0.00	$0.00	$0.00		
320.00	$0.00	$0.00	$15,920.00	Division 08
$0.00	$0.00	$352,500.00		
$0.00	$0.00	$2,280.00		
$0.00	$0.00	$9,200.00		
$0.00	$0.00	$3,330.00		
$672.00	$0.00	$0.00		
$576.00	$67.20	$0.00		
$789.00	$0.00	$0.00		
,165.00	$0.00	$0.00		
,042.00	$0.00	$0.00		
,244.00	$67.20	$367,310.00	$400,374.20	Division 13
,720.30	$791.06	$367,310.00	$483,110.86	Subtotal
990.42	55.37	25,711.70		Gen. Requirements
,710.72	$846.43	$393,021.70	$483,110.86	Estimate Subtotal
	42.32	9,825.54		Sales tax
,710.72	888.76	402,847.24		Subtotal
,912.34	88.88	40,284.72		GC O&P
,623.06	977.63	443,131.97	$604,071.64	Subtotal
			30,203.58	Contingency
			$634,275.22	Subtotal
			8,372.43	Bond
			$642,647.66	Subtotal
102.30			14,780.90	Location Adjustment
			$657,428.55	Grand Total

Sales Tax: If the work is subject to state or local sales taxes, the amount must be added to the estimate. Sales tax may be added to material costs, equipment costs, and subcontracted work. In this case, sales tax was added in all three categories. It was assumed that approximately half the subcontracted work would be material cost, so the tax was applied to 50% of the subcontract total.

GC O&P: This entry represents the general contractor's markup on material, labor, equipment, and subcontractor costs. RSMeans' standard markup on materials, equipment, and subcontracted work is 10%. In this estimate, the markup on the labor performed by the GC's workers uses "Skilled Workers Average" shown in Column F on the table "Installing Contractor's Overhead & Profit," which can be found on the inside-back cover of the book.

Contingency: A factor for contingency may be added to any estimate to represent the cost of unknowns that may occur between the time that the estimate is performed and the time the project is constructed. The amount of the allowance will depend on the stage of design at which the estimate is done, and the contractor's assessment of the risk involved. Refer to section 01 21 16.50 for contigency allowances.

Estimating Tips

01 20 00 Price and Payment Procedures

- Allowances that should be added to estimates to cover contingencies and job conditions that are not included in the national average material and labor costs are shown in section 01 21.

- When estimating historic preservation projects (depending on the condition of the existing structure and the owner's requirements), a 15%–20% contingency or allowance is recommended, regardless of the stage of the drawings.

01 30 00 Administrative Requirements

- Before determining a final cost estimate, it is a good practice to review all the items listed in Subdivisions 01 31 and 01 32 to make final adjustments for items that may need customizing to specific job conditions.

- Requirements for initial and periodic submittals can represent a significant cost to the General Requirements of a job. Thoroughly check the submittal specifications when estimating a project to determine any costs that should be included.

01 40 00 Quality Requirements

- All projects will require some degree of Quality Control. This cost is not included in the unit cost of construction listed in each division. Depending upon the terms of the contract, the various costs of inspection and testing can be the responsibility of either the owner or the contractor. Be sure to include the required costs in your estimate.

01 50 00 Temporary Facilities and Controls

- Barricades, access roads, safety nets, scaffolding, security, and many more requirements for the execution of a safe project are elements of direct cost. These costs can easily be overlooked when preparing an estimate. When looking through the major classifications of this subdivision, determine which items apply to each division in your estimate.

- Construction Equipment Rental Costs can be found in the Reference Section in section 01 54 33. Operators' wages are not included in equipment rental costs.

- Equipment mobilization and demobilization costs are not included in equipment rental costs and must be considered separately in section 01 54 36.50.

- The cost of small tools provided by the installing contractor for his workers is covered in the "Overhead" column on the "Installing Contractor's Overhead and Profit" table that lists labor trades, base rates and markups and, therefore, is included in the "Total Incl. O&P" cost of any Unit Price line item.

01 70 00 Execution and Closeout Requirements

- When preparing an estimate, thoroughly read the specifications to determine the requirements for Contract Closeout. Final cleaning, record documentation, operation and maintenance data, warranties and bonds, and spare parts and maintenance materials can all be elements of cost for the completion of a contract. Do not overlook these in your estimate.

Reference Numbers

Reference numbers are shown in shaded boxes at the beginning of some major classifications. These numbers refer to related items in the Reference Section. The reference information may be an estimating procedure, an alternate pricing method, or technical information.

Note: Not all subdivisions listed here necessarily appear in this publication. ■

01 11 Summary of Work

01 11 31 – Professional Consultants

01 11 31.10 Architectural Fees	Crew	Daily Output	Labor-Hours	Unit	Material	2015 Bare Costs Labor	Equipment	Total	Total Incl O&P
0010 **ARCHITECTURAL FEES**									
0020 For new construction									
0060 Minimum				Project				4.90%	4.90%
0090 Maximum								16%	16%
0100 For alteration work, to $500,000, add to new construction fee								50%	50%
0150 Over $500,000, add to new construction fee								25%	25%

01 11 31.20 Construction Management Fees									
0010 **CONSTRUCTION MANAGEMENT FEES**									
0020 $1,000,000 job, minimum				Project				4.50%	4.50%
0050 Maximum								7.50%	7.50%
0300 $50,000,000 job, minimum								2.50%	2.50%
0350 Maximum								4%	4%

01 11 31.30 Engineering Fees									
0010 **ENGINEERING FEES** R011110-30									
0020 Educational planning consultant, minimum				Project				.50%	.50%
0100 Maximum				"				2.50%	2.50%
0200 Electrical, minimum				Contrct				4.10%	4.10%
0300 Maximum								10.10%	10.10%
0400 Elevator & conveying systems, minimum								2.50%	2.50%
0500 Maximum								5%	5%
0600 Food service & kitchen equipment, minimum								8%	8%
0700 Maximum								12%	12%
0800 Landscaping & site development, minimum								2.50%	2.50%
0900 Maximum								6%	6%
1000 Mechanical (plumbing & HVAC), minimum								4.10%	4.10%
1100 Maximum								10.10%	10.10%
1200 Structural, minimum				Project				1%	1%
1300 Maximum				"				2.50%	2.50%

01 21 Allowances

01 21 16 – Contingency Allowances

01 21 16.50 Contingencies

	Crew	Daily Output	Labor-Hours	Unit	Material	Labor	Equipment	Total	Total Incl O&P
0010 **CONTINGENCIES**, Add to estimate									
0020 Conceptual stage				Project				20%	20%
0050 Schematic stage								15%	15%
0100 Preliminary working drawing stage (Design Dev.)								10%	10%
0150 Final working drawing stage								3%	3%

01 21 53 – Factors Allowance

01 21 53.50 Factors

	Crew	Daily Output	Labor-Hours	Unit	Material	Labor	Equipment	Total	Total Incl O&P
0010 **FACTORS** Cost adjustments R012153-10									
0100 Add to construction costs for particular job requirements									
0500 Cut & patch to match existing construction, add, minimum				Costs	2%	3%			
0550 Maximum					5%	9%			
0800 Dust protection, add, minimum					1%	2%			
0850 Maximum					4%	11%			
1100 Equipment usage curtailment, add, minimum					1%	1%			
1150 Maximum					3%	10%			
1400 Material handling & storage limitation, add, minimum					1%	1%			
1450 Maximum					6%	7%			
1700 Protection of existing work, add, minimum					2%	2%			

01 21 53 – Factors Allowance

01 21 53.50 Factors		Crew	Daily Output	Labor-Hours	Unit	Material	2015 Bare Costs Labor	2015 Bare Costs Equipment	Total	Total Incl O&P
1750	Maximum				Costs	5%	7%			
2000	Shift work requirements, add, minimum						5%			
2050	Maximum						30%			
2300	Temporary shoring and bracing, add, minimum					2%	5%			
2350	Maximum					5%	12%			
2400	Work inside prisons and high security areas, add, minimum						30%			
2450	Maximum						50%			

01 21 55 – Job Conditions Allowance

01 21 55.50 Job Conditions

		Crew	Daily Output	Labor-Hours	Unit	Material	2015 Bare Costs Labor	2015 Bare Costs Equipment	Total	Total Incl O&P
0010	**JOB CONDITIONS** Modifications to applicable									
0020	cost summaries									
0100	Economic conditions, favorable, deduct				Project				2%	2%
0200	Unfavorable, add								5%	5%
0300	Hoisting conditions, favorable, deduct								2%	2%
0400	Unfavorable, add								5%	5%
0700	Labor availability, surplus, deduct								1%	1%
0800	Shortage, add								10%	10%
0900	Material storage area, available, deduct								1%	1%
1000	Not available, add								2%	2%
1100	Subcontractor availability, surplus, deduct								5%	5%
1200	Shortage, add								12%	12%
1300	Work space, available, deduct								2%	2%
1400	Not available, add								5%	5%

01 21 57 – Overtime Allowance

01 21 57.50 Overtime

		Crew	Daily Output	Labor-Hours	Unit	Material	2015 Bare Costs Labor	2015 Bare Costs Equipment	Total	Total Incl O&P
0010	**OVERTIME** for early completion of projects or where R012909-90									
0020	labor shortages exist, add to usual labor, up to				Costs		100%			

01 21 61 – Cost Indexes

01 21 61.10 Construction Cost Index

		Crew	Daily Output	Labor-Hours	Unit	Material	2015 Bare Costs Labor	2015 Bare Costs Equipment	Total	Total Incl O&P
0010	**CONSTRUCTION COST INDEX** (Reference) over 930 zip code locations in									
0020	the U.S. and Canada, total bldg. cost, min. (Fayetteville, AR)				%				75.40%	75.40%
0050	Average								100%	100%
0100	Maximum (New York, NY)								131.80%	131.80%

01 21 61.20 Historical Cost Indexes

		Crew	Daily Output	Labor-Hours	Unit	Material	2015 Bare Costs Labor	2015 Bare Costs Equipment	Total	Total Incl O&P
0010	**HISTORICAL COST INDEXES** (See Reference Section)									

01 21 61.30 Labor Index

		Crew	Daily Output	Labor-Hours	Unit	Material	2015 Bare Costs Labor	2015 Bare Costs Equipment	Total	Total Incl O&P
0010	**LABOR INDEX** (Reference) For over 930 zip code locations in									
0020	the U.S. and Canada, minimum (Beaufort, SC)				%		44.90%			
0050	Average						100%			
0100	Maximum (New York, NY)						168.90%			

01 21 61.50 Material Index

		Crew	Daily Output	Labor-Hours	Unit	Material	2015 Bare Costs Labor	2015 Bare Costs Equipment	Total	Total Incl O&P
0010	**MATERIAL INDEX** (Reference) For over 930 zip code locations in									
0020	the U.S. and Canada, minimum (Oil Clty, PA)				%	92.40%				
0040	Average					100%				
0060	Maximum (Ketchikan, AK)					132%				

01 21 63 – Taxes

01 21 63.10 Taxes

		Crew	Daily Output	Labor-Hours	Unit	Material	2015 Bare Costs Labor	2015 Bare Costs Equipment	Total	Total Incl O&P
0010	**TAXES** R012909-80									
0020	Sales tax, State, average				%	5.04%				
0050	Maximum R012909-85					7.50%				

01 21 Allowances

01 21 63 – Taxes

01 21 63.10 Taxes		Crew	Daily Output	Labor-Hours	Unit	Material	2015 Bare Costs			Total
							Labor	Equipment	Total	Incl O&P
0200	Social Security, on first $117,000 of wages				%		7.65%			
0300	Unemployment, combined Federal and State, minimum						.60%			
0350	Average						7.80%			
0400	Maximum				↓		12.87%			

01 31 Project Management and Coordination

01 31 13 – Project Coordination

01 31 13.20 Field Personnel

		Crew	Daily Output	Labor-Hours	Unit	Material	Labor	Equipment	Total	Incl O&P
0010	**FIELD PERSONNEL**									
0020	Clerk, average				Week		450		450	695
0100	Field engineer, minimum						1,075		1,075	1,650
0120	Average						1,400		1,400	2,150
0140	Maximum						1,575		1,575	2,450
0160	General purpose laborer, average						1,500		1,500	2,325
0180	Project manager, minimum						1,975		1,975	3,075
0200	Average						2,275		2,275	3,525
0220	Maximum						2,600		2,600	4,025
0240	Superintendent, minimum						1,925		1,925	2,975
0260	Average						2,125		2,125	3,275
0280	Maximum						2,400		2,400	3,725
0290	Timekeeper, average				↓		1,225		1,225	1,900

01 31 13.30 Insurance

			Crew	Daily Output	Labor-Hours	Unit	Material	Labor	Equipment	Total	Incl O&P
0010	**INSURANCE**	R013113-40									
0020	Builders risk, standard, minimum					Job				.24%	.24%
0050	Maximum	R013113-60								.64%	.64%
0200	All-risk type, minimum									.25%	.25%
0250	Maximum					↓				.62%	.62%
0400	Contractor's equipment floater, minimum					Value				.50%	.50%
0450	Maximum					"				1.50%	1.50%
0600	Public liability, average					Job				2.02%	2.02%
0800	Workers' compensation & employer's liability, average										
0850	by trade, carpentry, general					Payroll		14.93%			
1000	Electrical							5.76%			
1150	Insulation							11.66%			
1450	Plumbing							6.98%			
1550	Sheet metal work (HVAC)					↓		8.82%			

01 31 13.50 General Contractor's Mark-Up

		Crew	Daily Output	Labor-Hours	Unit	Material	Labor	Equipment	Total	Incl O&P
0010	**GENERAL CONTRACTOR'S MARK-UP** on Change Orders									
0200	Extra work, by subcontractors, add				%				10%	10%
0250	By General Contractor, add								15%	15%
0400	Omitted work, by subcontractors, deduct all but								5%	5%
0450	By General Contractor, deduct all but								7.50%	7.50%
0600	Overtime work, by subcontractors, add								15%	15%
0650	By General Contractor, add				↓				10%	10%

01 31 13.60 Installing Contractor's Main Office Overhead

		Crew	Daily Output	Labor-Hours	Unit	Material	Labor	Equipment	Total	Incl O&P
0010	**INSTALLING CONTRACTOR'S MAIN OFFICE OVERHEAD**									
0020	As percent of direct costs, minimum				%				5%	
0050	Average								13%	
0100	Maximum								30%	

10

01 31 Project Management and Coordination

01 31 13 – Project Coordination

01 31 13.80 Overhead and Profit

		Crew	Daily Output	Labor-Hours	Unit	Material	2015 Bare Costs Labor	Equipment	Total	Total Incl O&P
0010	**OVERHEAD & PROFIT** Allowance to add to items in this									
0020	book that do not include Subs O&P, average				%				25%	
0100	Allowance to add to items in this book that									
0110	do include Subs O&P, minimum				%				5%	5%
0150	Average								10%	10%
0200	Maximum								15%	15%
0300	Typical, by size of project, under $100,000								30%	
0350	$500,000 project								25%	
0400	$2,000,000 project								20%	
0450	Over $10,000,000 project								15%	

01 31 13.90 Performance Bond

		Crew	Daily Output	Labor-Hours	Unit	Material	2015 Bare Costs Labor	Equipment	Total	Total Incl O&P
0010	**PERFORMANCE BOND** R013113-80									
0020	For buildings, minimum				Job				.60%	.60%
0100	Maximum				"				2.50%	2.50%

01 32 Construction Progress Documentation

01 32 33 – Photographic Documentation

01 32 33.50 Photographs

		Crew	Daily Output	Labor-Hours	Unit	Material	2015 Bare Costs Labor	Equipment	Total	Total Incl O&P
0010	**PHOTOGRAPHS**									
0020	8" x 10", 4 shots, 2 prints ea., std. mounting				Set	475			475	525
0100	Hinged linen mounts					530			530	580
0200	8" x 10", 4 shots, 2 prints each, in color					425			425	470
0300	For I.D. slugs, add to all above					5.30			5.30	5.85
1500	Time lapse equipment, camera and projector, buy				Ea.	2,650			2,650	2,925
1550	Rent per month				"	1,600			1,600	1,750
1700	Cameraman and film, including processing, B.&W.				Day	1,225			1,225	1,350
1720	Color				"	1,425			1,425	1,550

01 41 Regulatory Requirements

01 41 26 – Permit Requirements

01 41 26.50 Permits

		Crew	Daily Output	Labor-Hours	Unit	Material	2015 Bare Costs Labor	Equipment	Total	Total Incl O&P
0010	**PERMITS**									
0020	Rule of thumb, most cities, minimum				Job				.50%	.50%
0100	Maximum				"				2%	2%

01 51 Temporary Utilities

01 51 13 – Temporary Electricity

01 51 13.80 Temporary Utilities

		Crew	Daily Output	Labor-Hours	Unit	Material	2015 Bare Costs Labor	Equipment	Total	Total Incl O&P
0010	**TEMPORARY UTILITIES**									
0350	Lighting, lamps, wiring, outlets, 40,000 SF building, 8 strings	1 Elec	34	.235	CSF Flr	4.91	12.85		17.76	24.50
0360	16 strings	"	17	.471		9.80	25.50		35.30	49.50
0400	Power for temp lighting only, 6.6 KWH, per month								.92	1.01
0430	11.8 KWH, per month								1.65	1.82
0450	23.6 KWH, per month								3.30	3.63
0600	Power for job duration incl. elevator, etc., minimum								47	51.50
0650	Maximum								110	121
1000	Toilet, portable, see Equip. Rental 01 54 33 in Reference Section									

11

01 52 Construction Facilities

01 52 13 – Field Offices and Sheds

01 52 13.20 Office and Storage Space	Crew	Daily Output	Labor-Hours	Unit	Material	2015 Bare Costs Labor	Equipment	Total	Total Incl O&P
0010 **OFFICE AND STORAGE SPACE**									
0020 Office trailer, furnished, no hookups, 20' x 8', buy	2 Skwk	1	16	Ea.	10,400	780		11,180	12,600
0250 Rent per month					188			188	206
0300 32' x 8', buy	2 Skwk	.70	22.857		15,500	1,100		16,600	18,800
0350 Rent per month					239			239	262
0400 50' x 10', buy	2 Skwk	.60	26.667		24,100	1,300		25,400	28,500
0450 Rent per month					340			340	375
0500 50' x 12', buy	2 Skwk	.50	32		29,400	1,550		30,950	34,700
0550 Rent per month					410			410	455
0700 For air conditioning, rent per month, add					48.50			48.50	53.50
0800 For delivery, add per mile				Mile	11			11	12.10
0890 Delivery each way				Ea.	200			200	220
0900 Bunk house trailer, 8' x 40' duplex dorm with kitchen, no hookups, buy	2 Carp	1	16		37,300	750		38,050	42,200
0910 9 man with kitchen and bath, no hookups, buy		1	16		38,000	750		38,750	43,000
0920 18 man sleeper with bath, no hookups, buy		1	16		49,000	750		49,750	55,000
1000 Portable buildings, prefab, on skids, economy, 8' x 8'		265	.060	S.F.	25	2.83		27.83	32
1100 Deluxe, 8' x 12'		150	.107	"	20	5		25	29.50
1200 Storage boxes, 20' x 8', buy	2 Skwk	1.80	8.889	Ea.	2,775	430		3,205	3,725
1250 Rent per month					82.50			82.50	91
1300 40' x 8', buy	2 Skwk	1.40	11.429		3,850	555		4,405	5,075
1350 Rent per month					101			101	111

01 54 Construction Aids

01 54 16 – Temporary Hoists

01 54 16.50 Weekly Forklift Crew

	Crew	Daily Output	Labor-Hours	Unit	Material	Labor	Equipment	Total	Total Incl O&P
0010 **WEEKLY FORKLIFT CREW**									
0100 All-terrain forklift, 45' lift, 35' reach, 9000 lb. capacity	A-3P	.20	40	Week		1,950	2,625	4,575	5,850

01 54 19 – Temporary Cranes

01 54 19.50 Daily Crane Crews

	Crew	Daily Output	Labor-Hours	Unit	Material	Labor	Equipment	Total	Total Incl O&P
0010 **DAILY CRANE CREWS** for small jobs, portal to portal									
0100 12-ton truck-mounted hydraulic crane	A-3H	1	8	Day		415	860	1,275	1,575
0200 25-ton	A-3I	1	8			415	990	1,405	1,725
0300 40-ton	A-3J	1	8			415	1,225	1,640	1,975
0400 55-ton	A-3K	1	16			775	1,625	2,400	2,975
0500 80-ton	A-3L	1	16			775	2,350	3,125	3,750
0600 100-ton	A-3M	1	16			775	2,325	3,100	3,750
0900 If crane is needed on a Saturday, Sunday or Holiday									
0910 At time-and-a-half, add				Day		50%			
0920 At double time, add				"		100%			

01 54 19.60 Monthly Tower Crane Crew

	Crew	Daily Output	Labor-Hours	Unit	Material	Labor	Equipment	Total	Total Incl O&P
0010 **MONTHLY TOWER CRANE CREW**, excludes concrete footing									
0100 Static tower crane, 130' high, 106' jib, 6200 lb. capacity	A-3N	.05	176	Month		9,100	24,800	33,900	41,100

01 54 23 – Temporary Scaffolding and Platforms

01 54 23.70 Scaffolding

	Crew	Daily Output	Labor-Hours	Unit	Material	Labor	Equipment	Total	Total Incl O&P
0010 **SCAFFOLDING** R015423-10									
0015 Steel tube, regular, no plank, labor only to erect & dismantle									
0090 Building exterior, wall face, 1 to 5 stories, 6'-4" x 5' frames	3 Carp	8	3	C.S.F.		141		141	217
0200 6 to 12 stories	4 Carp	8	4			188		188	289
0301 13 to 20 stories	5 Clab	8	5			188		188	289
0460 Building interior, wall face area, up to 16' high	3 Carp	12	2			94		94	145

01 54 23 – Temporary Scaffolding and Platforms

01 54 23.70 Scaffolding		Crew	Daily Output	Labor-Hours	Unit	Material	2015 Bare Costs Labor	Equipment	Total	Total Incl O&P
0560	16' to 40' high	3 Carp	10	2.400	C.S.F.		113		113	173
0800	Building interior floor area, up to 30' high	↓	150	.160	C.C.F.		7.50		7.50	11.55
0900	Over 30' high	4 Carp	160	.200	"		9.40		9.40	14.45
0906	Complete system for face of walls, no plank, material only rent/mo				C.S.F.	35.50			35.50	39.50
0908	Interior spaces, no plank, material only rent/mo				C.C.F.	4.33			4.33	4.76
0910	Steel tubular, heavy duty shoring, buy									
0920	Frames 5' high 2' wide				Ea.	81.50			81.50	90
0925	5' high 4' wide					93			93	102
0930	6' high 2' wide					93.50			93.50	103
0935	6' high 4' wide				↓	109			109	120
0940	Accessories									
0945	Cross braces				Ea.	15.50			15.50	17.05
0950	U-head, 8" x 8"					19.10			19.10	21
0955	J-head, 4" x 8"					13.90			13.90	15.30
0960	Base plate, 8" x 8"					15.50			15.50	17.05
0965	Leveling jack				↓	33.50			33.50	36.50
1000	Steel tubular, regular, buy									
1100	Frames 3' high 5' wide				Ea.	75			75	82.50
1150	5' high 5' wide					86.50			86.50	95
1200	6'-4" high 5' wide					118			118	129
1350	7'-6" high 6' wide					158			158	173
1500	Accessories cross braces					15.50			15.50	17.05
1550	Guardrail post					16.40			16.40	18.05
1600	Guardrail 7' section					8.70			8.70	9.60
1650	Screw jacks & plates					21.50			21.50	24
1700	Sidearm brackets					30.50			30.50	33.50
1750	8" casters					31			31	34
1800	Plank 2" x 10" x 16'-0"					57.50			57.50	63.50
1900	Stairway section					286			286	315
1910	Stairway starter bar					32			32	35
1920	Stairway inside handrail					58.50			58.50	64
1930	Stairway outside handrail					86.50			86.50	95
1940	Walk-thru frame guardrail				↓	41.50			41.50	46
2000	Steel tubular, regular, rent/mo.									
2100	Frames 3' high 5' wide				Ea.	5			5	5.50
2150	5' high 5' wide					5			5	5.50
2200	6'-4" high 5' wide					5.50			5.50	6.05
2250	7'-6" high 6' wide					10			10	11
2500	Accessories, cross braces					1			1	1.10
2550	Guardrail post					1			1	1.10
2600	Guardrail 7' section					1			1	1.10
2650	Screw jacks & plates					2			2	2.20
2700	Sidearm brackets					2			2	2.20
2750	8" casters					8			8	8.80
2800	Outrigger for rolling tower					3			3	3.30
2850	Plank 2" x 10" x 16'-0"					10			10	11
2900	Stairway section					35			35	38.50
2940	Walk-thru frame guardrail				↓	2.50			2.50	2.75
3000	Steel tubular, heavy duty shoring, rent/mo.									
3250	5' high 2' & 4' wide				Ea.	8.50			8.50	9.35
3300	6' high 2' & 4' wide					8.50			8.50	9.35
3500	Accessories, cross braces					1			1	1.10
3600	U - head, 8" x 8"				↓	2.50			2.50	2.75

01 54 23.70 Scaffolding	Crew	Daily Output	Labor-Hours	Unit	Material	Labor	Equipment	Total	Total Incl O&P
3650	J - head, 4" x 8"				Ea.	2.50			2.50	2.7
3700	Base plate, 8" x 8"					1			1	1.1
3750	Leveling jack					2.50			2.50	2.7
5700	Planks, 2" x 10" x 16'-0", labor only to erect & remove to 50' H	3 Carp	72	.333			15.65		15.65	24
5800	Over 50' high	4 Carp	80	.400			18.80		18.80	29
6000	Heavy duty shoring for elevated slab forms to 8'-2" high, floor area									
6100	Labor only to erect & dismantle	4 Carp	16	2	C.S.F.		94		94	145
6110	Materials only, rent.mo				"	44.50			44.50	49
6500	To 14'-8" high									
6600	Labor only to erect & dismantle	4 Carp	10	3.200	C.S.F.		150		150	231
6610	Materials only, rent/mo				"	65			65	71.50

01 54 23.75 Scaffolding Specialties

		Crew	Daily Output	Labor-Hours	Unit	Material	Labor	Equipment	Total	Total Incl O&P
0010	**SCAFFOLDING SPECIALTIES**									
1200	Sidewalk bridge, heavy duty steel posts & beams, including									
1210	parapet protection & waterproofing (material cost is rent/month)									
1220	8' to 10' wide, 2 posts	3 Carp	15	1.600	L.F.	45	75		120	166
1230	3 posts	"	10	2.400	"	69.50	113		182.50	250
1500	Sidewalk bridge using tubular steel scaffold frames including									
1510	planking (material cost is rent/month)	3 Carp	45	.533	L.F.	8.55	25		33.55	48
1600	For 2 uses per month, deduct from all above					50%				
1700	For 1 use every 2 months, add to all above					100%				
1900	Catwalks, 20" wide, no guardrails, 7' span, buy				Ea.	166			166	183
2000	10' span, buy					233			233	256
3720	Putlog, standard, 8' span, with hangers, buy					162			162	178
3730	Rent per month					20			20	22
3750	12' span, buy					203			203	223
3755	Rent per month					25			25	27.50
3760	Trussed type, 16' span, buy					365			365	400
3770	Rent per month					30			30	33
3790	22' span, buy					420			420	465
3795	Rent per month					40			40	44
3800	Rolling ladders with handrails, 30" wide, buy, 2 step					267			267	294
4000	7 step					760			760	835
4050	10 step					1,075			1,075	1,175
4100	Rolling towers, buy, 5' wide, 7' long, 10' high					1,200			1,200	1,325
4200	For 5' high added sections, to buy, add					204			204	224
4300	Complete incl. wheels, railings, outriggers,									
4350	21' high, to buy				Ea.	2,025			2,025	2,225
4400	Rent/month = 5% of purchase cost				"	200			200	220

01 54 36.50 Mobilization

		Crew	Daily Output	Labor-Hours	Unit	Material	Labor	Equipment	Total	Total Incl O&P
0010	**MOBILIZATION** (Use line item again for demobilization) R015436-50									
0015	Up to 25 mi. haul dist. (50 mi. RT for mob/demob crew)									
1200	Small equipment, placed in rear of, or towed by pickup truck	A-3A	4	2	Ea.		97	39	136	191
1300	Equipment hauled on 3-ton capacity towed trailer	A-3Q	2.67	3			146	67	213	295
1400	20-ton capacity	B-34U	2	8			355	237	592	795
1500	40-ton capacity	B-34N	2	8			365	375	740	965
1600	50-ton capacity	B-34V	1	24			1,125	1,050	2,175	2,850
1700	Crane, truck-mounted, up to 75 ton (driver only)	1 Eqhv	4	2			103		103	157
1800	Over 75 ton (with chase vehicle)	A-3E	2.50	6.400			294	62.50	356.50	515
2400	Crane, large lattice boom, requiring assembly	B-34W	.50	144			6,275	7,500	13,775	17,800
2500	For each additional 5 miles haul distance, add						10%	10%		

14

For customer support on your Plumbing Cost Data, call 877.709.6968.

01 54 Construction Aids

01 54 36 – Equipment Mobilization

01 54 36.50 Mobilization	Crew	Daily Output	Labor-Hours	Unit	Material	2015 Bare Costs Labor	Equipment	Total	Total Incl O&P
3000	For large pieces of equipment, allow for assembly/knockdown								
3100	For mob/demob of micro-tunneling equip, see Section 33 05 23.19								

01 55 Vehicular Access and Parking

01 55 23 – Temporary Roads

01 55 23.50 Roads and Sidewalks

	01 55 23.50 Roads and Sidewalks	Crew	Daily Output	Labor-Hours	Unit	Material	Labor	Equipment	Total	Total Incl O&P
0010	**ROADS AND SIDEWALKS** Temporary									
0050	Roads, gravel fill, no surfacing, 4" gravel depth	B-14	715	.067	S.Y.	4.04	2.67	.51	7.22	9.10
0100	8" gravel depth	"	615	.078	"	8.10	3.10	.59	11.79	14.30
1000	Ramp, 3/4" plywood on 2" x 6" joists, 16" O.C.	2 Carp	300	.053	S.F.	1.58	2.50		4.08	5.60
1100	On 2" x 10" joists, 16" O.C.	"	275	.058	"	2.24	2.73		4.97	6.65

01 56 Temporary Barriers and Enclosures

01 56 13 – Temporary Air Barriers

01 56 13.60 Tarpaulins

	01 56 13.60 Tarpaulins	Crew	Daily Output	Labor-Hours	Unit	Material	Labor	Equipment	Total	Total Incl O&P
0010	**TARPAULINS**									
0020	Cotton duck, 10 oz. to 13.13 oz. per S.Y., 6'x8'				S.F.	.83			.83	.91
0050	30'x30'					.59			.59	.65
0100	Polyvinyl coated nylon, 14 oz. to 18 oz., minimum					1.36			1.36	1.50
0150	Maximum					1.36			1.36	1.50
0200	Reinforced polyethylene 3 mils thick, white					.04			.04	.04
0300	4 mils thick, white, clear or black					.09			.09	.10
0400	5.5 mils thick, clear					.18			.18	.20
0500	White, fire retardant					.44			.44	.48
0600	12 mils, oil resistant, fire retardant					.39			.39	.43
0700	8.5 mils, black					.57			.57	.63
0710	Woven polyethylene, 6 mils thick					.18			.18	.20
0730	Polyester reinforced w/integral fastening system 11 mils thick					.21			.21	.23
0740	Mylar polyester, non-reinforced, 7 mils thick					1.17			1.17	1.29

01 56 13.90 Winter Protection

	01 56 13.90 Winter Protection	Crew	Daily Output	Labor-Hours	Unit	Material	Labor	Equipment	Total	Total Incl O&P
0010	**WINTER PROTECTION**									
0100	Framing to close openings	2 Clab	500	.032	S.F.	.46	1.20		1.66	2.35
0200	Tarpaulins hung over scaffolding, 8 uses, not incl. scaffolding		1500	.011		.24	.40		.64	.88
0250	Tarpaulin polyester reinf. w/integral fastening system 11 mils thick		1600	.010		.21	.38		.59	.81
0300	Prefab fiberglass panels, steel frame, 8 uses		1200	.013		2.40	.50		2.90	3.41

01 56 16 – Temporary Dust Barriers

01 56 16.10 Dust Barriers, Temporary

	01 56 16.10 Dust Barriers, Temporary	Crew	Daily Output	Labor-Hours	Unit	Material	Labor	Equipment	Total	Total Incl O&P
0010	**DUST BARRIERS, TEMPORARY**									
0020	Spring loaded telescoping pole & head, to 12', erect and dismantle	1 Clab	240	.033	Ea.		1.25		1.25	1.93
0025	Cost per day (based upon 250 days)				Day	.26			.26	.29
0030	To 21', erect and dismantle	1 Clab	240	.033	Ea.		1.25		1.25	1.93
0035	Cost per day (based upon 250 days)				Day	.44			.44	.48
0040	Accessories, caution tape reel, erect and dismantle	1 Clab	480	.017	Ea.		.63		.63	.96
0045	Cost per day (based upon 250 days)				Day	.36			.36	.40
0060	Foam rail and connector, erect and dismantle	1 Clab	240	.033	Ea.		1.25		1.25	1.93
0065	Cost per day (based upon 250 days)				Day	.10			.10	.11
0070	Caution tape	1 Clab	384	.021	C.L.F.	2.70	.78		3.48	4.18
0080	Zipper, standard duty		60	.133	Ea.	8	5		13	16.50

For customer support on your Plumbing Cost Data, call 877.709.6968.

15

01 56 Temporary Barriers and Enclosures

01 56 16 – Temporary Dust Barriers

01 56 16.10 Dust Barriers, Temporary	Crew	Daily Output	Labor-Hours	Unit	Material	2015 Bare Costs Labor	Equipment	Total	Total Incl O&P
0090	Heavy duty	1 Clab	48	.167	Ea.	9.50	6.25		15.75	20
0100	Polyethylene sheet, 4 mil		37	.216	Sq.	2.90	8.15		11.05	15.70
0110	6 mil		37	.216	"	4.02	8.15		12.17	16.90
1000	Dust partition, 6 mil polyethylene, 1" x 3" frame	2 Carp	2000	.008	S.F.	.30	.38		.68	.90
1080	2" x 4" frame	"	2000	.008	"	.35	.38		.73	.97

01 56 23 – Temporary Barricades

01 56 23.10 Barricades

		Crew	Daily Output	Labor-Hours	Unit	Material	2015 Bare Costs Labor	Equipment	Total	Total Incl O&P
0010	**BARRICADES**									
0020	5' high, 3 rail @ 2" x 8", fixed	2 Carp	20	.800	L.F.	6	37.50		43.50	64.50
0150	Movable		30	.533		5	25		30	44
1000	Guardrail, wooden, 3' high, 1" x 6", on 2" x 4" posts		200	.080		1.27	3.76		5.03	7.20
1100	2" x 6", on 4" x 4" posts		165	.097		2.46	4.55		7.01	9.70
1200	Portable metal with base pads, buy					15.55			15.55	17.10
1250	Typical installation, assume 10 reuses	2 Carp	600	.027		2.55	1.25		3.80	4.74
1300	Barricade tape, polyethylene, 7 mil, 3" wide x 500' long roll				Ea.	25			25	27.50
3000	Detour signs, set up and remove									
3010	Reflective aluminum, MUTCD, 24" x 24", post mounted	1 Clab	20	.400	Ea.	2.46	15.05		17.51	25.50
4000	Roof edge portable barrier stands and warning flags, 50 uses	1 Rohe	9100	.001	L.F.	.07	.03		.10	.12
4010	100 uses	"	9100	.001	"	.03	.03		.06	.09

01 56 26 – Temporary Fencing

01 56 26.50 Temporary Fencing

		Crew	Daily Output	Labor-Hours	Unit	Material	2015 Bare Costs Labor	Equipment	Total	Total Incl O&P
0010	**TEMPORARY FENCING**									
0020	Chain link, 11 ga., 4' high	2 Clab	400	.040	L.F.	2.95	1.50		4.45	5.55
0100	6' high		300	.053		2.95	2.01		4.96	6.35
0200	Rented chain link, 6' high, to 1000' (up to 12 mo.)		400	.040		4.29	1.50		5.79	7.05
0250	Over 1000' (up to 12 mo.)		300	.053		4.29	2.01		6.30	7.80
0350	Plywood, painted, 2" x 4" frame, 4' high	A-4	135	.178		6.20	7.95		14.15	19
0400	4" x 4" frame, 8' high	"	110	.218		12	9.75		21.75	28
0500	Wire mesh on 4" x 4" posts, 4' high	2 Carp	100	.160		9.85	7.50		17.35	22.50
0550	8' high	"	80	.200		14.90	9.40		24.30	31

01 56 29 – Temporary Protective Walkways

01 56 29.50 Protection

		Crew	Daily Output	Labor-Hours	Unit	Material	2015 Bare Costs Labor	Equipment	Total	Total Incl O&P
0010	**PROTECTION**									
0020	Stair tread, 2" x 12" planks, 1 use	1 Carp	75	.107	Tread	5.10	5		10.10	13.30
0100	Exterior plywood, 1/2" thick, 1 use		65	.123		1.97	5.80		7.77	11.05
0200	3/4" thick, 1 use		60	.133		2.78	6.25		9.03	12.70
2200	Sidewalks, 2" x 12" planks, 2 uses		350	.023	S.F.	.85	1.07		1.92	2.59
2300	Exterior plywood, 2 uses, 1/2" thick		750	.011		.33	.50		.83	1.13
2400	5/8" thick		650	.012		.39	.58		.97	1.32
2500	3/4" thick		600	.013		.46	.63		1.09	1.47

01 56 32 – Temporary Security

01 56 32.50 Watchman

		Crew	Daily Output	Labor-Hours	Unit	Material	2015 Bare Costs Labor	Equipment	Total	Total Incl O&P
0010	**WATCHMAN**									
0020	Service, monthly basis, uniformed person, minimum				Hr.				25	27.50
0100	Maximum								45.50	50
0200	Person and command dog, minimum								31	34
0300	Maximum								54.50	60
0500	Sentry dog, leased, with job patrol (yard dog), 1 dog				Week				290	320
0600	2 dogs				"				390	430
0800	Purchase, trained sentry dog, minimum				Ea.				1,375	1,500

01 56 Temporary Barriers and Enclosures

01 56 32 - Temporary Security

01 56 32.50 Watchman	Crew	Daily Output	Labor-Hours	Unit	Material	2015 Bare Costs Labor	Equipment	Total	Total Incl O&P
0900 Maximum				Ea.				2,725	3,000

01 58 Project Identification

01 58 13 - Temporary Project Signage

01 58 13.50 Signs

	Crew	Daily Output	Labor-Hours	Unit	Material	2015 Bare Costs Labor	Equipment	Total	Total Incl O&P
0010 **SIGNS**									
0020 High intensity reflectorized, no posts, buy				Ea.	25			25	27.50

01 74 Cleaning and Waste Management

01 74 13 - Progress Cleaning

01 74 13.20 Cleaning Up

	Crew	Daily Output	Labor-Hours	Unit	Material	2015 Bare Costs Labor	Equipment	Total	Total Incl O&P
0010 **CLEANING UP**									
0020 After job completion, allow, minimum				Job				.30%	.30%
0040 Maximum				"				1%	1%
0050 Cleanup of floor area, continuous, per day, during const.	A-5	24	.750	M.S.F.	2.18	28.50	2.56	33.24	48.50
0100 Final by GC at end of job	"	11.50	1.565	"	2.31	59	5.35	66.66	99.50

01 91 Commissioning

01 91 13 - General Commissioning Requirements

01 91 13.50 Building Commissioning

	Crew	Daily Output	Labor-Hours	Unit	Material	2015 Bare Costs Labor	Equipment	Total	Total Incl O&P
0010 **BUILDING COMMISSIONING**									
0100 Basic building commissioning, minimum				%				.25%	.25%
0150 Maximum								.50%	.50%
0200 Enhanced building commissioning, minimum								.50%	.50%
0250 Maximum								1%	1%

01 93 Facility Maintenance

01 93 13 - Facility Maintenance Procedures

01 93 13.15 Mechanical Facilities Maintenance

	Crew	Daily Output	Labor-Hours	Unit	Material	2015 Bare Costs Labor	Equipment	Total	Total Incl O&P
0010 **MECHANICAL FACILITIES MAINTENANCE**									
0100 Air conditioning system maintenance									
0130 Belt, replace	1 Stpi	15	.533	Ea.		32		32	48
0170 Fan, clean		16	.500			30		30	45
0180 Filter, remove, clean, replace		12	.667			40		40	60
0190 Flexible coupling alignment, inspect		40	.200			11.95		11.95	18.05
0200 Gas leak locate and repair		4	2			120		120	180
0250 Pump packing gland, remove and replace		11	.727			43.50		43.50	65.50
0270 Tighten		32	.250			14.95		14.95	22.50
0290 Pump disassemble and assemble		4	2			120		120	180
0300 Air pressure regulator, disassemble, clean, assemble	1 Skwk	4	2			97.50		97.50	150
0310 Repair or replace part	"	6	1.333			65		65	100
0320 Purging system	1 Stpi	16	.500			30		30	45
0400 Compressor, air, remove or install fan wheel	1 Skwk	20	.400			19.45		19.45	30
0410 Disassemble or assemble 2 cylinder, 2 stage		4	2			97.50		97.50	150
0420 4 cylinder, 4 stage		1	8			390		390	600
0430 Repair or replace part		2	4			195		195	300

01 93 13.15 Mechanical Facilities Maintenance		Crew	Daily Output	Labor-Hours	Unit	Material	2015 Bare Costs Labor	Equipment	Total	Total Incl O&P
0700	Demolition, for mech. demolition see Section 23 05 05.10 or 22 05 05.10									
0800	Ductwork, clean									
0810	Rectangular									
0820	6"	G	1 Shee	187.50	.043	L.F.		2.39		2.39	3.6
0830	8"	G		140.63	.057			3.18		3.18	4.8
0840	10"	G		112.50	.071			3.98		3.98	6.1
0850	12"	G		93.75	.085			4.77		4.77	7.3
0860	14"	G		80.36	.100			5.55		5.55	8.5
0870	16"	G		70.31	.114			6.35		6.35	9.7
0900	Round									
0910	4"	G	1 Shee	358.10	.022	L.F.		1.25		1.25	1.9
0920	6"	G		238.73	.034			1.87		1.87	2.8
0930	8"	G		179.05	.045			2.50		2.50	3.8
0940	10"	G		143.24	.056			3.12		3.12	4.7
0950	12"	G		119.37	.067			3.75		3.75	5.7
0960	16"	G		89.52	.089			5		5	7.6
1000	Expansion joint, not screwed, install or remove		1 Stpi	3	2.667	Ea.		159		159	241
1010	Repack		"	6	1.333	"		79.50		79.50	120
1200	Fire protection equipment									
1220	Fire hydrant, replace		Q-1	3	5.333	Ea.		282		282	425
1230	Service, lubricate, inspect, flush, clean		1 Plum	7	1.143			67		67	101
1240	Test		"	11	.727			42.50		42.50	64.5
1310	Inspect valves, pressure, nozzle,		1 Spri	4	2	System		112		112	170
1500	Instrumentation preventive maintenance, rule of thumb									
1510	Annual cost per instrument for calibration & trouble shooting									
1520	Includes flow meters, analytical transmitters, flow control									
1530	chemical feed systems, recorders and distrib. control sys.						Ea.			325	325
1800	Plumbing fixtures, for installation see Section 22 41 00									
1801	Plumbing fixtures									
1840	Open drain with snake		1 Plum	13	.615	Ea.		36		36	54.50
1850	Open drain with toilet auger			16	.500			29.50		29.50	44.50
1860	Grease trap, clean			2	4			235		235	355
1870	Plaster trap, clean			6	1.333			78.50		78.50	118
1881	Clean commode		1 Clab	20	.400		.37	15.05		15.42	23.50
1882	Clean commode seat			48	.167		.18	6.25		6.43	9.85
1883	Clean single sink			28	.286		.24	10.75		10.99	16.80
1884	Clean double sink			20	.400		.37	15.05		15.42	23.50
1885	Clean lavatory			28	.286		.24	10.75		10.99	16.80
1886	Clean bathtub			12	.667		.73	25		25.73	39.50
1887	Clean fiberglass shower			12	.667		.73	25		25.73	39.50
1888	Clean fiberglass tub/shower			8	1		.92	37.50		38.42	59
1889	Clean faucet set			32	.250		.15	9.40		9.55	14.60
1891	Clean shower head			80	.100		.02	3.76		3.78	5.80
1893	Clean water heater			16	.500		1.28	18.80		20.08	30.50
1900	Relief valve, test and adjust		1 Stpi	20	.400			24		24	36
1910	Clean pump, heater, or motor for whirlpool		1 Clab	16	.500		.18	18.80		18.98	29
1920	Clean thermal cover for whirlpool		"	16	.500		1.83	18.80		20.63	31
2000	Repair or replace-steam trap		1 Stpi	8	1			60		60	90
2020	Y-type or bell strainer			6	1.333			79.50		79.50	120
2040	Water trap or vacuum breaker, screwed joints			13	.615			37		37	55.50
2100	Steam specialties, clean									
2120	Air separator with automatic trap, 1" fittings		1 Stpi	12	.667	Ea.		40		40	60
2130	Bucket trap, 2" pipe			7	1.143			68.50		68.50	103

For customer support on your Plumbing Cost Data, call 877.709.6968.

01 93 13 – Facility Maintenance Procedures

01 93 13.15 Mechanical Facilities Maintenance		Crew	Daily Output	Labor-Hours	Unit	Material	2015 Bare Costs Labor	Equipment	Total	Total Incl O&P
2150	Drip leg, 2" fitting	1 Stpi	45	.178	Ea.		10.60		10.60	16.05
2200	Thermodynamic trap, 1" fittings		50	.160			9.55		9.55	14.45
2210	Thermostatic		65	.123			7.35		7.35	11.10
2240	Screen and seat in y-type strainer, plug type.		25	.320			19.10		19.10	29
2242	Screen and seat in y-type strainer, flange type.		12	.667			40		40	60
2500	Valve, replace broken handwheel		24	.333			19.90		19.90	30
3000	Valve, overhaul, regulator, relief, flushometer, mixing									
3040	Cold water, gas	1 Stpi	5	1.600	Ea.		95.50		95.50	144
3050	Hot water, steam		3	2.667			159		159	241
3080	Globe, gate, check up to 4" cold water, gas		10	.800			48		48	72
3090	Hot water, steam		5	1.600			95.50		95.50	144
3100	Over 4" ID hot or cold line		1.40	5.714			340		340	515
3120	Remove and replace, gate, globe or check up to 4"	Q-5	6	2.667			143		143	217
3130	Over 4"	"	2	8			430		430	650
3150	Repack up to 4"	1 Stpi	13	.615			37		37	55.50
3160	Over 4"	"	4	2			120		120	180

Division Notes

		CREW	DAILY OUTPUT	LABOR-HOURS	UNIT	BARE COSTS				TOTAL INCL O&P
						MAT.	LABOR	EQUIP.	TOTAL	

Estimating Tips

02 30 00 Subsurface Investigation

In preparing estimates on structures involving earthwork or foundations, all information concerning soil characteristics should be obtained. Look particularly for hazardous waste, evidence of prior dumping of debris, and previous stream beds.

02 40 00 Demolition and Structure Moving

The costs shown for selective demolition do not include rubbish handling or disposal. These items should be estimated separately using RSMeans data or other sources.

- Historic preservation often requires that the contractor remove materials from the existing structure, rehab them, and replace them. The estimator must be aware of any related measures and precautions that must be taken when doing selective demolition and cutting and patching. Requirements may include special handling and storage, as well as security.

- In addition to Subdivision 02 41 00, you can find selective demolition items in each division. Example: Roofing demolition is in Division 7.

02 40 00 Building Deconstruction

This section provides costs for the careful dismantling and recycling of most of low-rise building materials.

02 50 00 Containment of Hazardous Waste

This section addresses on-site hazardous waste disposal costs.

02 80 00 Hazardous Material Disposal/ Remediation

This subdivision includes information on hazardous waste handling, asbestos remediation, lead remediation, and mold remediation. See reference R028213-20 and R028319-60 for further guidance in using these unit price lines.

02 90 00 Monitoring Chemical Sampling, Testing Analysis

This section provides costs for on-site sampling and testing hazardous waste.

Reference Numbers

Reference numbers are shown in shaded boxes at the beginning of some major classifications. These numbers refer to related items in the Reference Section. The reference information may be an estimating procedure, an alternate pricing method, or technical information.

Note: Not all subdivisions listed here necessarily appear in this publication. ∎

Did you know?

RSMeans Online gives you the same access to RSMeans' data with 24/7 access:

- Quickly locate costs in the searchable database.
- Build cost lists, estimates, and reports in minutes.
- Adjust costs to any location in the U.S. and Canada with the click of a button.

Start your free trial today at **www.rsmeansonline.com**

RSMeansOnline

02 21 Surveys

02 21 13 – Site Surveys

02 21 13.09 Topographical Surveys		Crew	Daily Output	Labor-Hours	Unit	Material	2015 Bare Costs Labor	2015 Bare Costs Equipment	Total	Total Incl O&P
0010	**TOPOGRAPHICAL SURVEYS**									
0020	Topographical surveying, conventional, minimum	A-7	3.30	7.273	Acre	20	375	16.60	411.60	615
0100	Maximum	A-8	.60	53.333	"	60	2,675	91	2,826	4,250

02 21 13.13 Boundary and Survey Markers

		Crew	Daily Output	Labor-Hours	Unit	Material	Labor	Equipment	Total	Total Incl O&P
0010	**BOUNDARY AND SURVEY MARKERS**									
0300	Lot location and lines, large quantities, minimum	A-7	2	12	Acre	35	620	27.50	682.50	1,025
0320	Average	"	1.25	19.200		55	990	44	1,089	1,625
0400	Small quantities, maximum	A-8	1	32		75	1,600	54.50	1,729.50	2,600
0600	Monuments, 3' long	A-7	10	2.400	Ea.	40	124	5.45	169.45	240
0800	Property lines, perimeter, cleared land	"	1000	.024	L.F.	.05	1.24	.05	1.34	2.02
0900	Wooded land	A-8	875	.037	"	.07	1.83	.06	1.96	2.9

02 21 13.16 Aerial Surveys

		Crew	Daily Output	Labor-Hours	Unit	Material	Labor	Equipment	Total	Total Incl O&P
0010	**AERIAL SURVEYS**									
1500	Aerial surveying, including ground control, minimum fee, 10 acres				Total				4,700	4,700
1510	100 acres								9,400	9,400
1550	From existing photography, deduct								1,625	1,625
1600	2' contours, 10 acres				Acre				470	470
1850	100 acres								94	94
2000	1000 acres								90	90
2050	10,000 acres								85	85

02 41 Demolition

02 41 13 – Selective Site Demolition

02 41 13.17 Demolish, Remove Pavement and Curb

		Crew	Daily Output	Labor-Hours	Unit	Material	Labor	Equipment	Total	Total Incl O&P
0010	**DEMOLISH, REMOVE PAVEMENT AND CURB** R024119-10									
5010	Pavement removal, bituminous roads, up to 3" thick	B-38	690	.058	S.Y.		2.48	1.85	4.33	5.80
5050	4" to 6" thick	"	420	.095	"		4.08	3.04	7.12	9.60

02 41 13.23 Utility Line Removal

		Crew	Daily Output	Labor-Hours	Unit	Material	Labor	Equipment	Total	Total Incl O&P
0010	**UTILITY LINE REMOVAL**									
0015	No hauling, abandon catch basin or manhole	B-6	7	3.429	Ea.		142	52	194	274
0020	Remove existing catch basin or manhole, masonry		4	6			248	91	339	480
0030	Catch basin or manhole frames and covers, stored		13	1.846			76	28	104	148
0040	Remove and reset		7	3.429			142	52	194	274
0900	Hydrants, fire, remove only	B-21A	5	8			375	94.50	469.50	675
0950	Remove and reset	"	2	20			940	237	1,177	1,675
2900	Pipe removal, sewer/water, no excavation, 12" diameter	B-6	175	.137	L.F.		5.65	2.08	7.73	10.95
2930	15"-18" diameter	B-12Z	150	.160			6.75	10.70	17.45	22
2960	21"-24" diameter		120	.200			8.45	13.40	21.85	27.50
3000	27"-36" diameter		90	.267			11.30	17.85	29.15	37
3200	Steel, welded connections, 4" diameter	B-6	160	.150			6.20	2.28	8.48	11.95
3300	10" diameter	"	80	.300			12.40	4.55	16.95	24

02 41 13.30 Minor Site Demolition

		Crew	Daily Output	Labor-Hours	Unit	Material	Labor	Equipment	Total	Total Incl O&P
0010	**MINOR SITE DEMOLITION** R024119-10									
4000	Sidewalk removal, bituminous, 2" thick	B-6	350	.069	S.Y.		2.83	1.04	3.87	5.45
4010	2-1/2" thick		325	.074			3.05	1.12	4.17	5.90
4100	Concrete, plain, 4"		160	.150			6.20	2.28	8.48	11.95
4110	Plain, 5"		140	.171			7.05	2.60	9.65	13.70
4120	Plain, 6"		120	.200			8.25	3.04	11.29	16

02 41 Demolition

02 41 19 – Selective Demolition

02 41 19.19 Selective Demolition	Crew	Daily Output	Labor-Hours	Unit	Material	2015 Bare Costs Labor	Equipment	Total	Total Incl O&P
0010 **SELECTIVE DEMOLITION,** Rubbish Handling R024119-10									
0020 The following are to be added to the demolition prices									
0600 Dumpster, weekly rental, 1 dump/week, 6 C.Y. capacity (2 Tons)				Week	415			415	455
0700 10 C.Y. capacity (3 Tons)					480			480	530
0725 20 C.Y. capacity (5 Tons) R024119-20					565			565	625
0800 30 C.Y. capacity (7 Tons)					730			730	800
0840 40 C.Y. capacity (10 Tons)					775			775	850
2000 Load, haul, dump and return, 0 – 50' haul, hand carried	2 Clab	24	.667	C.Y.		25		25	38.50
2005 Wheeled		37	.432			16.25		16.25	25
2040 0 – 100' haul, hand carried		16.50	.970			36.50		36.50	56
2045 Wheeled		25	.640			24		24	37
2080 Haul and return, add per each extra 100' haul, hand carried		35.50	.451			16.95		16.95	26
2085 Wheeled		54	.296			11.15		11.15	17.15
2120 For travel in elevators, up to 10 floors, add		140	.114			4.30		4.30	6.60
2130 0 – 50' haul, incl. up to 5 riser stair, hand carried		23	.696			26		26	40
2135 Wheeled		35	.457			17.20		17.20	26.50
2140 6 – 10 riser stairs, hand carried		22	.727			27.50		27.50	42
2145 Wheeled		34	.471			17.70		17.70	27
2150 11 – 20 riser stairs, hand carried		20	.800			30		30	46.50
2155 Wheeled		31	.516			19.40		19.40	30
2160 21 – 40 riser stairs, hand carried		16	1			37.50		37.50	58
2165 Wheeled		24	.667			25		25	38.50
2170 0 – 100' haul, incl. 5 riser stair, hand carried		15	1.067			40		40	61.50
2175 Wheeled		23	.696			26		26	40
2180 6 – 10 riser stair, hand carried		14	1.143			43		43	66
2185 Wheeled		21	.762			28.50		28.50	44
2190 11 – 20 riser stair, hand carried		12	1.333			50		50	77
2195 Wheeled		18	.889			33.50		33.50	51.50
2200 21 – 40 riser stair, hand carried		8	2			75		75	116
2205 Wheeled		12	1.333			50		50	77
2210 Haul and return, add per each extra 100' haul, hand carried		35.50	.451			16.95		16.95	26
2215 Wheeled		54	.296			11.15		11.15	17.15
2220 For each additional flight of stairs, up to 5 risers, add		550	.029	Flight		1.09		1.09	1.68
2225 6 – 10 risers, add		275	.058			2.19		2.19	3.37
2230 11 – 20 risers, add		138	.116			4.36		4.36	6.70
2235 21 – 40 risers, add		69	.232			8.70		8.70	13.40
3000 Loading & trucking, including 2 mile haul, chute loaded	B-16	45	.711	C.Y.		27.50	15.35	42.85	59
3040 Hand loading truck, 50' haul	"	48	.667			26	14.40	40.40	55.50
3080 Machine loading truck	B-17	120	.267			10.90	6.45	17.35	24
5000 Haul, per mile, up to 8 C.Y. truck	B-34B	1165	.007			.28	.59	.87	1.07
5100 Over 8 C.Y. truck	"	1550	.005			.21	.45	.66	.80

02 41 19.20 Selective Demolition, Dump Charges

	Crew	Daily Output	Labor-Hours	Unit	Material	2015 Bare Costs Labor	Equipment	Total	Total Incl O&P
0010 **SELECTIVE DEMOLITION, DUMP CHARGES** R024119-10									
0100 Building construction materials				Ton	74			74	81
0300 Rubbish only					63			63	69.50
0500 Reclamation station, usual charge					74			74	81

02 41 19.27 Selective Demolition, Torch Cutting

	Crew	Daily Output	Labor-Hours	Unit	Material	2015 Bare Costs Labor	Equipment	Total	Total Incl O&P
0010 **SELECTIVE DEMOLITION, TORCH CUTTING** R024119-10									
0020 Steel, 1" thick plate	E-25	333	.024	L.F.	.84	1.31	.03	2.18	3.24
0040 1" diameter bar	"	600	.013	Ea.	.14	.73	.02	.89	1.44
1000 Oxygen lance cutting, reinforced concrete walls									
1040 12" to 16" thick walls	1 Clab	10	.800	L.F.		30		30	46.50

For customer support on your Plumbing Cost Data, call 877.709.6968.

23

02 41 19 - Selective Demolition

02 41 19.27 Selective Demolition, Torch Cutting		Crew	Daily Output	Labor-Hours	Unit	Material	2015 Bare Costs		Total	Total Incl O&P
							Labor	Equipment		
1080	24" thick walls	1 Clab	6	1.333	L.F.		50		50	7

02 65 Underground Storage Tank Removal

02 65 10 - Underground Tank and Contaminated Soil Removal

02 65 10.30 Removal of Underground Storage Tanks

			Crew	Daily Output	Labor-Hours	Unit	Material	Labor	Equipment	Total	Total Incl O&P
0010	**REMOVAL OF UNDERGROUND STORAGE TANKS**	R026510-20									
0011	Petroleum storage tanks, non-leaking										
0100	Excavate & load onto trailer										
0110	3000 gal. to 5000 gal. tank	G	B-14	4	12	Ea.		475	91	566	830
0120	6000 gal. to 8000 gal. tank	G	B-3A	3	13.333			535	345	880	1,200
0130	9000 gal. to 12000 gal. tank	G	"	2	20			805	515	1,320	1,800
0190	Known leaking tank, add					%				100%	100%
0200	Remove sludge, water and remaining product from tank bottom										
0201	of tank with vacuum truck										
0300	3000 gal. to 5000 gal. tank	G	A-13	5	1.600	Ea.		78	153	231	286
0310	6000 gal. to 8000 gal. tank	G		4	2			97	191	288	360
0320	9000 gal. to 12000 gal. tank	G		3	2.667			130	254	384	475
0390	Dispose of sludge off-site, average					Gal.				6.25	6.8
0400	Insert inert solid CO_2 "dry ice" into tank										
0401	For cleaning/transporting tanks (1.5 lb./100 gal. cap)	G	1 Clab	500	.016	Lb.	1.17	.60		1.77	2.2
1020	Haul tank to certified salvage dump, 100 miles round trip										
1023	3000 gal. to 5000 gal. tank					Ea.				760	830
1026	6000 gal. to 8000 gal. tank									880	960
1029	9,000 gal. to 12,000 gal. tank									1,050	1,150
1100	Disposal of contaminated soil to landfill										
1110	Minimum					C.Y.				145	160
1111	Maximum					"				400	440
1120	Disposal of contaminated soil to										
1121	bituminous concrete batch plant										
1130	Minimum					C.Y.				80	88
1131	Maximum					"				115	125
2010	Decontamination of soil on site incl poly tarp on top/bottom										
2011	Soil containment berm, and chemical treatment										
2020	Minimum	G	B-11C	100	.160	C.Y.	7.80	7.05	3.64	18.49	23.5
2021	Maximum	G	"	100	.160		10.10	7.05	3.64	20.79	26
2050	Disposal of decontaminated soil, minimum									135	150
2055	Maximum									400	440

02 81 Transportation and Disposal of Hazardous Materials

02 81 20 - Hazardous Waste Handling

02 81 20.10 Hazardous Waste Cleanup/Pickup/Disposal

			Crew	Daily Output	Labor-Hours	Unit	Material	Labor	Equipment	Total	Total Incl O&P
0010	**HAZARDOUS WASTE CLEANUP/PICKUP/DISPOSAL**										
0100	For contractor rental equipment, i.e., Dozer,										
0110	Front end loader, Dump truck, etc., see 01 54 33 Reference Section										
1000	Solid pickup										
1100	55 gal. drums					Ea.				240	265
1120	Bulk material, minimum					Ton				190	210
1130	Maximum					"				595	655
1200	Transportation to disposal site										

02 81 Transportation and Disposal of Hazardous Materials

02 81 20 – Hazardous Waste Handling

02 81 20.10 Hazardous Waste Cleanup/Pickup/Disposal	Crew	Daily Output	Labor-Hours	Unit	Material	2015 Bare Costs Labor	Equipment	Total	Total Incl O&P
1220	Truckload = 80 drums or 25 C.Y. or 18 tons									
1260	Minimum				Mile				3.95	4.45
1270	Maximum				"				7.25	7.35
3000	Liquid pickup, vacuum truck, stainless steel tank									
3100	Minimum charge, 4 hours									
3110	1 compartment, 2200 gallon				Hr.				140	155
3120	2 compartment, 5000 gallon				"				200	225
3400	Transportation in 6900 gallon bulk truck				Mile				7.95	8.75
3410	In teflon lined truck				"				10.20	11.25
5000	Heavy sludge or dry vacuumable material				Hr.				140	160
6000	Dumpsite disposal charge, minimum				Ton				140	155
6020	Maximum				"				415	455

02 82 Asbestos Remediation

02 82 13 – Asbestos Abatement

02 82 13.39 Asbestos Remediation Plans and Methods

		Crew	Daily Output	Labor-Hours	Unit	Material	Labor	Equipment	Total	Total Incl O&P
0010	**ASBESTOS REMEDIATION PLANS AND METHODS**									
0100	Building Survey-Commercial Building				Ea.				2,200	2,400
0200	Asbestos Abatement Remediation Plan				"				1,350	1,475

02 82 13.41 Asbestos Abatement Equipment

		Crew	Daily Output	Labor-Hours	Unit	Material	Labor	Equipment	Total	Total Incl O&P
0010	**ASBESTOS ABATEMENT EQUIPMENT** R028213-20									
0011	Equipment and supplies, buy									
0200	Air filtration device, 2000 CFM				Ea.	960			960	1,050
0250	Large volume air sampling pump, minimum					345			345	380
0260	Maximum					335			335	365
0300	Airless sprayer unit, 2 gun					4,500			4,500	4,950
0350	Light stand, 500 watt					49			49	54
0400	Personal respirators									
0410	Negative pressure, 1/2 face, dual operation, min.				Ea.	26.50			26.50	29
0420	Maximum					29.50			29.50	32.50
0450	P.A.P.R., full face, minimum					124			124	137
0460	Maximum					165			165	182
0470	Supplied air, full face, incl. air line, minimum					168			168	185
0480	Maximum					405			405	445
0500	Personnel sampling pump					226			226	249
1500	Power panel, 20 unit, incl. GFI					600			600	660
1600	Shower unit, including pump and filters					1,275			1,275	1,425
1700	Supplied air system (type C)					3,425			3,425	3,750
1750	Vacuum cleaner, HEPA, 16 gal., stainless steel, wet/dry					435			435	475
1760	55 gallon					560			560	615
1800	Vacuum loader, 9 – 18 ton/hr.					94,500			94,500	104,000
1900	Water atomizer unit, including 55 gal. drum					281			281	310
2000	Worker protection, whole body, foot, head cover & gloves, plastic					8.65			8.65	9.55
2500	Respirator, single use					25			25	27.50
2550	Cartridge for respirator					5.75			5.75	6.30
2570	Glove bag, 7 mil, 50" x 64"					9.20			9.20	10.10
2580	10 mil, 44" x 60"					5.75			5.75	6.35
2590	6 mil, 44" x 60"					5.75			5.75	6.35
3000	HEPA vacuum for work area, minimum					305			305	335
3050	Maximum					815			815	900
6000	Disposable polyethylene bags, 6 mil, 3 C.F.					.82			.82	.90

For customer support on your Plumbing Cost Data, call 877.709.6968.

25

02 82 13 – Asbestos Abatement

02 82 13.41 Asbestos Abatement Equipment	Crew	Daily Output	Labor-Hours	Unit	Material	2015 Bare Costs Labor	Equipment	Total	Total Incl O&P
6300	Disposable fiber drums, 3 C.F.				Ea.	18.20			18.20	20
6400	Pressure sensitive caution labels, 3" x 5"					3.47			3.47	3.82
6450	11" x 17"					7.35			7.35	8.05
6500	Negative air machine, 1800 CFM			▼	845			845	930

02 82 13.42 Preparation of Asbestos Containment Area

		Crew	Daily Output	Labor-Hours	Unit	Material	Labor	Equipment	Total	Total Incl O&P
0010	**PREPARATION OF ASBESTOS CONTAINMENT AREA** R028213-20									
0100	Pre-cleaning, HEPA vacuum and wet wipe, flat surfaces	A-9	12000	.005	S.F.	.02	.28		.30	.45
0200	Protect carpeted area, 2 layers 6 mil poly on 3/4" plywood	"	1000	.064		2.04	3.35		5.39	7.45
0300	Separation barrier, 2" x 4" @ 16", 1/2" plywood ea. side, 8' high	2 Carp	400	.040		3.32	1.88		5.20	6.55
0310	12' high		320	.050		3.32	2.35		5.67	7.25
0320	16' high		200	.080		2.30	3.76		6.06	8.35
0400	Personnel decontam. chamber, 2" x 4" @ 16", 3/4" ply ea. side		280	.057		4.34	2.68		7.02	8.90
0450	Waste decontam. chamber, 2" x 4" studs @ 16", 3/4" ply ea. side		360	.044	▼	4.34	2.09		6.43	8
0500	Cover surfaces with polyethylene sheeting									
0501	Including glue and tape									
0550	Floors, each layer, 6 mil	A-9	8000	.008	S.F.	.04	.42		.46	.70
0551	4 mil		9000	.007		.03	.37		.40	.61
0560	Walls, each layer, 6 mil		6000	.011		.04	.56		.60	.92
0561	4 mil	▼	7000	.009	▼	.03	.48		.51	.78
0570	For heights above 12', add						20%			
0575	For heights above 20', add						30%			
0580	For fire retardant poly, add					100%				
0590	For large open areas, deduct					10%	20%			
0600	Seal floor penetrations with foam firestop to 36 sq. in.	2 Carp	200	.080	Ea.	7.75	3.76		11.51	14.30
0610	36 sq. in. to 72 sq. in.		125	.128		15.50	6		21.50	26.50
0615	72 sq. in. to 144 sq. in.		80	.200		31	9.40		40.40	48.50
0620	Wall penetrations, to 36 square inches		180	.089		7.75	4.17		11.92	14.90
0630	36 sq. in. to 72 sq. in.		100	.160		15.50	7.50		23	28.50
0640	72 sq. in. to 144 sq. in.	▼	60	.267	▼	31	12.50		43.50	53.50
0800	Caulk seams with latex	1 Carp	230	.035	L.F.	.17	1.63		1.80	2.70
0900	Set up neg. air machine, 1-2k CFM/25 M.C.F. volume	1 Asbe	4.30	1.860	Ea.		97.50		97.50	152
0950	Set up and remove portable shower unit	2 Asbe	4	4	"		209		209	325

02 82 13.43 Bulk Asbestos Removal

		Crew	Daily Output	Labor-Hours	Unit	Material	Labor	Equipment	Total	Total Incl O&P
0010	**BULK ASBESTOS REMOVAL**									
0020	Includes disposable tools and 2 suits and 1 respirator filter/day/worker									
0200	Boiler insulation	A-9	480	.133	S.F.	.45	7		7.45	11.40
0210	With metal lath, add				%		50%			50%
0300	Boiler breeching or flue insulation	A-9	520	.123	S.F.	.36	6.45		6.81	10.45
0310	For active boiler, add				%		100%			100%
0400	Duct or AHU insulation	A-10B	440	.073	S.F.	.21	3.82		4.03	6.20
0500	Duct vibration isolation joints, up to 24 sq. in. duct	A-9	56	1.143	Ea.	3.30	60		63.30	97
0520	25 sq. in. to 48 sq. in. duct		48	1.333		3.85	70		73.85	113
0530	49 sq. in. to 76 sq. in. duct		40	1.600	▼	4.62	84		88.62	136
0600	Pipe insulation, air cell type, up to 4" diameter pipe		900	.071	L.F.	.21	3.73		3.94	6.05
0610	4" to 8" diameter pipe		800	.080		.23	4.19		4.42	6.80
0620	10" to 12" diameter pipe		700	.091		.26	4.79		5.05	7.75
0630	14" to 16" diameter pipe		550	.116	▼	.34	6.10		6.44	9.85
0650	Over 16" diameter pipe		650	.098	S.F.	.28	5.15		5.43	8.35
0700	With glove bag up to 3" diameter pipe		200	.320	L.F.	9.15	16.75		25.90	36
1000	Pipe fitting insulation up to 4" diameter pipe		320	.200	Ea.	.58	10.50		11.08	16.95
1100	6" to 8" diameter pipe		304	.211		.61	11.05		11.66	17.85
1110	10" to 12" diameter pipe		192	.333	▼	.96	17.45		18.41	28

26

For customer support on your Plumbing Cost Data, call 877.709.6968.

02 82 Asbestos Remediation

02 82 13 - Asbestos Abatement

02 82 13.43 Bulk Asbestos Removal

		Crew	Daily Output	Labor-Hours	Unit	Material	2015 Bare Costs Labor	Equipment	Total	Total Incl O&P
1120	14" to 16" diameter pipe	A-9	128	.500	Ea.	1.44	26		27.44	42.50
1130	Over 16" diameter pipe		176	.364	S.F.	1.05	19.05		20.10	30.50
1200	With glove bag, up to 8" diameter pipe		75	.853	L.F.	6.25	44.50		50.75	76.50
2000	Scrape foam fireproofing from flat surface		2400	.027	S.F.	.08	1.40		1.48	2.26
2100	Irregular surfaces		1200	.053		.15	2.80		2.95	4.52
3000	Remove cementitious material from flat surface		1800	.036		.10	1.86		1.96	3.01
3100	Irregular surface		1000	.064		.13	3.35		3.48	5.35
6000	Remove contaminated soil from crawl space by hand		400	.160	C.F.	.46	8.40		8.86	13.55
6100	With large production vacuum loader	A-12	700	.091	"	.26	4.79	1.09	6.14	8.95
7000	Radiator backing, not including radiator removal	A-9	1200	.053	S.F.	.15	2.80		2.95	4.52
9000	For type B (supplied air) respirator equipment, add				%		10%		10%	

02 82 13.44 Demolition In Asbestos Contaminated Area

		Crew	Daily Output	Labor-Hours	Unit	Material	2015 Bare Costs Labor	Equipment	Total	Total Incl O&P
0010	**DEMOLITION IN ASBESTOS CONTAMINATED AREA**									
0200	Ceiling, including suspension system, plaster and lath	A-9	2100	.030	S.F.	.09	1.60		1.69	2.59
0210	Finished plaster, leaving wire lath		585	.109		.32	5.75		6.07	9.30
0220	Suspended acoustical tile		3500	.018		.05	.96		1.01	1.55
0230	Concealed tile grid system		3000	.021		.06	1.12		1.18	1.81
0240	Metal pan grid system		1500	.043		.12	2.24		2.36	3.62
0250	Gypsum board		2500	.026		.07	1.34		1.41	2.17
0260	Lighting fixtures up to 2' x 4'		72	.889	Ea.	2.56	46.50		49.06	75.50
0400	Partitions, non load bearing									
0410	Plaster, lath, and studs	A-9	690	.093	S.F.	.88	4.86		5.74	8.50
0450	Gypsum board and studs	"	1390	.046	"	.13	2.41		2.54	3.91
9000	For type B (supplied air) respirator equipment, add				%				10%	10%

02 82 13.45 OSHA Testing

		Crew	Daily Output	Labor-Hours	Unit	Material	2015 Bare Costs Labor	Equipment	Total	Total Incl O&P
0010	**OSHA TESTING**									
0100	Certified technician, minimum				Day				200	220
0110	Maximum								300	330
0120	Industrial hygienist, minimum								250	250
0130	Maximum								400	440
0200	Asbestos sampling and PCM analysis, NIOSH 7400, minimum	1 Asbe	8	1	Ea.	2.96	52.50		55.46	85
0210	Maximum		4	2		3.26	105		108.26	167
1000	Cleaned area samples		8	1		2.81	52.50		55.31	84.50
1100	PCM air sample analysis, NIOSH 7400, minimum		8	1		31.50	52.50		84	117
1110	Maximum		4	2		3.42	105		108.42	167
1200	TEM air sample analysis, NIOSH 7402, minimum								80	106
1210	Maximum								360	450

02 82 13.46 Decontamination of Asbestos Containment Area

		Crew	Daily Output	Labor-Hours	Unit	Material	2015 Bare Costs Labor	Equipment	Total	Total Incl O&P
0010	**DECONTAMINATION OF ASBESTOS CONTAINMENT AREA**									
0100	Spray exposed substrate with surfactant (bridging)									
0200	Flat surfaces	A-9	6000	.011	S.F.	.36	.56		.92	1.27
0250	Irregular surfaces		4000	.016	"	.31	.84		1.15	1.65
0300	Pipes, beams, and columns		2000	.032	L.F.	.56	1.68		2.24	3.23
1000	Spray encapsulate polyethylene sheeting		8000	.008	S.F.	.34	.42		.76	1.02
1100	Roll down polyethylene sheeting		8000	.008	"		.42		.42	.65
1500	Bag polyethylene sheeting		400	.160	Ea.	.81	8.40		9.21	13.95
2000	Fine clean exposed substrate, with nylon brush		2400	.027	S.F.		1.40		1.40	2.18
2500	Wet wipe substrate		4800	.013			.70		.70	1.09
2600	Vacuum surfaces, fine brush		6400	.010			.52		.52	.82
3000	Structural demolition									
3100	Wood stud walls	A-9	2800	.023	S.F.		1.20		1.20	1.87
3500	Window manifolds, not incl. window replacement		4200	.015			.80		.80	1.24

For customer support on your Plumbing Cost Data, call 877.709.6968.

27

02 82 Asbestos Remediation

02 82 13 – Asbestos Abatement

02 82 13.46 Decontamination of Asbestos Containment Area	Crew	Daily Output	Labor-Hours	Unit	Material	2015 Bare Costs Labor	Equipment	Total	Total Incl O&P
3600	Plywood carpet protection	A-9	2000	.032	S.F.		1.68		1.68	2.61
4000	Remove custom decontamination facility	A-10A	8	3	Ea.	15.40	158		173.40	262
4100	Remove portable decontamination facility	3 Asbe	12	2	"	13.05	105		118.05	177
5000	HEPA vacuum, shampoo carpeting	A-9	4800	.013	S.F.	.07	.70		.77	1.17
9000	Final cleaning of protected surfaces	A-10A	8000	.003	"		.16		.16	.25

02 82 13.47 Asbestos Waste Pkg., Handling, and Disp.

		Crew	Daily Output	Labor-Hours	Unit	Material	Labor	Equipment	Total	Total Incl O&P
0010	**ASBESTOS WASTE PACKAGING, HANDLING, AND DISPOSAL**									
0100	Collect and bag bulk material, 3 C.F. bags, by hand	A-9	400	.160	Ea.	.82	8.40		9.22	13.95
0200	Large production vacuum loader	A-12	880	.073		.86	3.81	.87	5.54	7.85
1000	Double bag and decontaminate	A-9	960	.067		.82	3.49		4.31	6.35
2000	Containerize bagged material in drums, per 3 C.F. drum	"	800	.080		18.20	4.19		22.39	26.50
3000	Cart bags 50' to dumpster	2 Asbe	400	.040			2.09		2.09	3.26
5000	Disposal charges, not including haul, minimum				C.Y.				61	67
5020	Maximum				"				355	395
9000	For type B (supplied air) respirator equipment, add				%				10%	10%

02 82 13.48 Asbestos Encapsulation With Sealants

		Crew	Daily Output	Labor-Hours	Unit	Material	Labor	Equipment	Total	Total Incl O&P
0010	**ASBESTOS ENCAPSULATION WITH SEALANTS**									
0100	Ceilings and walls, minimum	A-9	21000	.003	S.F.	.31	.16		.47	.59
0110	Maximum		10600	.006	"	.45	.32		.77	.99
0300	Pipes to 12" diameter including minor repairs, minimum		800	.080	L.F.	.43	4.19		4.62	7
0310	Maximum		400	.160	"	1.12	8.40		9.52	14.30

02 85 Mold Remediation

02 85 16 – Mold Remediation Preparation and Containment

02 85 16.50 Preparation of Mold Containment Area

		Crew	Daily Output	Labor-Hours	Unit	Material	Labor	Equipment	Total	Total Incl O&P
0010	**PREPARATION OF MOLD CONTAINMENT AREA**									
0100	Pre-cleaning, HEPA vacuum and wet wipe, flat surfaces	A-9	12000	.005	S.F.	.02	.28		.30	.45
0300	Separation barrier, 2" x 4" @ 16", 1/2" plywood ea. side, 8' high	2 Carp	400	.040		3.32	1.88		5.20	6.55
0310	12' high		320	.050		3.32	2.35		5.67	7.25
0320	16' high		200	.080		2.30	3.76		6.06	8.35
0400	Personnel decontam. chamber, 2" x 4" @ 16", 3/4" ply ea. side		280	.057		4.34	2.68		7.02	8.90
0450	Waste decontam. chamber, 2" x 4" studs @ 16", 3/4" ply each side		360	.044		4.34	2.09		6.43	8
0500	Cover surfaces with polyethylene sheeting									
0501	Including glue and tape									
0550	Floors, each layer, 6 mil	A-9	8000	.008	S.F.	.04	.42		.46	.70
0551	4 mil		9000	.007		.03	.37		.40	.61
0560	Walls, each layer, 6 mil		6000	.011		.04	.56		.60	.92
0561	4 mil		7000	.009		.03	.48		.51	.78
0570	For heights above 12', add						20%			
0575	For heights above 20', add						30%			
0580	For fire retardant poly, add					100%				
0590	For large open areas, deduct					10%	20%			
0600	Seal floor penetrations with foam firestop to 36 sq. in.	2 Carp	200	.080	Ea.	7.75	3.76		11.51	14.30
0610	36 sq. in. to 72 sq. in.		125	.128		15.50	6		21.50	26.50
0615	72 sq. in. to 144 sq. in.		80	.200		31	9.40		40.40	48.50
0620	Wall penetrations, to 36 square inches		180	.089		7.75	4.17		11.92	14.90
0630	36 sq. in. to 72 sq. in.		100	.160		15.50	7.50		23	28.50
0640	72 sq. in. to 144 sq. in.		60	.267		31	12.50		43.50	53.50
0800	Caulk seams with latex caulk	1 Carp	230	.035	L.F.	.17	1.63		1.80	2.70
0900	Set up neg. air machine, 1-2k CFM/25 M.C.F. volume	1 Asbe	4.30	1.860	Ea.		97.50		97.50	152

02 85 33 – Removal and Disposal of Materials with Mold

02 85 33.50 Demolition in Mold Contaminated Area	Crew	Daily Output	Labor-Hours	Unit	Material	2015 Bare Costs Labor	Equipment	Total	Total Incl O&P
0010	**DEMOLITION IN MOLD CONTAMINATED AREA**									
0200	Ceiling, including suspension system, plaster and lath	A-9	2100	.030	S.F.	.09	1.60		1.69	2.59
0210	Finished plaster, leaving wire lath		585	.109		.32	5.75		6.07	9.30
0220	Suspended acoustical tile		3500	.018		.05	.96		1.01	1.55
0230	Concealed tile grid system		3000	.021		.06	1.12		1.18	1.81
0240	Metal pan grid system		1500	.043		.12	2.24		2.36	3.62
0250	Gypsum board		2500	.026		.07	1.34		1.41	2.17
0255	Plywood		2500	.026		.07	1.34		1.41	2.17
0260	Lighting fixtures up to 2' x 4'		72	.889	Ea.	2.56	46.50		49.06	75.50
0400	Partitions, non load bearing									
0410	Plaster, lath, and studs	A-9	690	.093	S.F.	.88	4.86		5.74	8.50
0450	Gypsum board and studs		1390	.046		.13	2.41		2.54	3.91
0465	Carpet & pad		1390	.046		.13	2.41		2.54	3.91
0600	Pipe insulation, air cell type, up to 4" diameter pipe		900	.071	L.F.	.21	3.73		3.94	6.05
0610	4" to 8" diameter pipe		800	.080		.23	4.19		4.42	6.80
0620	10" to 12" diameter pipe		700	.091		.26	4.79		5.05	7.75
0630	14" to 16" diameter pipe		550	.116		.34	6.10		6.44	9.85
0650	Over 16" diameter pipe		650	.098	S.F.	.28	5.15		5.43	8.35
9000	For type B (supplied air) respirator equipment, add				%				10%	10%

For customer support on your Plumbing Cost Data, call 877.709.6968.

29

Division Notes

	CREW	DAILY OUTPUT	LABOR-HOURS	UNIT	BARE COSTS				TOTAL INCL O&P
					MAT.	LABOR	EQUIP.	TOTAL	

Estimating Tips
General
- Carefully check all the plans and specifications. Concrete often appears on drawings other than structural drawings, including mechanical and electrical drawings for equipment pads. The cost of cutting and patching is often difficult to estimate. See Subdivision 03 81 for Concrete Cutting, Subdivision 02 41 19.16 for Cutout Demolition, Subdivision 03 05 05.10 for Concrete Demolition, and Subdivision 02 41 19.19 for Rubbish Handling (handling, loading and hauling of debris).
- Always obtain concrete prices from suppliers near the job site. A volume discount can often be negotiated, depending upon competition in the area. Remember to add for waste, particularly for slabs and footings on grade.

03 10 00 Concrete Forming and Accessories
- A primary cost for concrete construction is forming. Most jobs today are constructed with prefabricated forms. The selection of the forms best suited for the job and the total square feet of forms required for efficient concrete forming and placing are key elements in estimating concrete construction. Enough forms must be available for erection to make efficient use of the concrete placing equipment and crew.
- Concrete accessories for forming and placing depend upon the systems used. Study the plans and specifications to ensure that all special accessory requirements have been included in the cost estimate, such as anchor bolts, inserts, and hangers.
- Included within costs for forms-in-place are all necessary bracing and shoring.

03 20 00 Concrete Reinforcing
- Ascertain that the reinforcing steel supplier has included all accessories, cutting, bending, and an allowance for lapping, splicing, and waste. A good rule of thumb is 10% for lapping, splicing, and waste. Also, 10% waste should be allowed for welded wire fabric.
- The unit price items in the subdivisions for Reinforcing In Place, Glass Fiber Reinforcing, and Welded Wire Fabric include the labor to install accessories such as beam and slab bolsters, high chairs, and bar ties and tie wire. The material cost for these accessories is not included; they may be obtained from the Accessories Division.

03 30 00 Cast-In-Place Concrete
- When estimating structural concrete, pay particular attention to requirements for concrete additives, curing methods, and surface treatments. Special consideration for climate, hot or cold, must be included in your estimate. Be sure to include requirements for concrete placing equipment, and concrete finishing.
- For accurate concrete estimating, the estimator must consider each of the following major components individually: forms, reinforcing steel, ready-mix concrete, placement of the concrete, and finishing of the top surface. For faster estimating, Subdivision 03 30 53.40 for Concrete-In-Place can be used; here, various items of concrete work are presented that include the costs of all five major components (unless specifically stated otherwise).

03 40 00 Precast Concrete
03 50 00 Cast Decks and Underlayment
- The cost of hauling precast concrete structural members is often an important factor. For this reason, it is important to get a quote from the nearest supplier. It may become economically feasible to set up precasting beds on the site if the hauling costs are prohibitive.

Reference Numbers
Reference numbers are shown in shaded boxes at the beginning of some major classifications. These numbers refer to related items in the Reference Section. The reference information may be an estimating procedure, an alternate pricing method, or technical information.

Note: Not all subdivisions listed here necessarily appear in this publication. ■

03 11 Concrete Forming

03 11 13 – Structural Cast-In-Place Concrete Forming

03 11 13.40 Forms In Place, Equipment Foundations		Crew	Daily Output	Labor-Hours	Unit	Material	2015 Bare Costs Labor	Equipment	Total	Total Incl O&P
0010	**FORMS IN PLACE, EQUIPMENT FOUNDATIONS**									
0020	1 use	C-2	160	.300	SFCA	3.45	13.70		17.15	25
0050	2 use		190	.253		1.90	11.55		13.45	19.90
0100	3 use		200	.240		1.38	11		12.38	18.40
0150	4 use		205	.234		1.12	10.70		11.82	17.75

03 11 13.45 Forms In Place, Footings										
0010	**FORMS IN PLACE, FOOTINGS**									
0020	Continuous wall, plywood, 1 use	C-1	375	.085	SFCA	6.80	3.81		10.61	13.35
0050	2 use		440	.073		3.74	3.24		6.98	9.10
0100	3 use		470	.068		2.72	3.04		5.76	7.65
0150	4 use		485	.066		2.22	2.94		5.16	6.95
5000	Spread footings, job-built lumber, 1 use		305	.105		2.28	4.68		6.96	9.70
5050	2 use		371	.086		1.27	3.85		5.12	7.30
5100	3 use		401	.080		.91	3.56		4.47	6.50
5150	4 use		414	.077		.74	3.45		4.19	6.10

03 11 13.65 Forms In Place, Slab On Grade										
0010	**FORMS IN PLACE, SLAB ON GRADE**									
3000	Edge forms, wood, 4 use, on grade, to 6" high	C-1	600	.053	L.F.	.35	2.38		2.73	4.04
6000	Trench forms in floor, wood, 1 use		160	.200	SFCA	1.94	8.90		10.84	15.90
6050	2 use		175	.183		1.06	8.15		9.21	13.70
6100	3 use		180	.178		.77	7.95		8.72	13.05
6150	4 use		185	.173		.63	7.70		8.33	12.55

03 15 Concrete Accessories

03 15 05 – Concrete Forming Accessories

03 15 05.75 Sleeves and Chases

03 15 05.75 Sleeves and Chases		Crew	Daily Output	Labor-Hours	Unit	Material	2015 Bare Costs Labor	Equipment	Total	Total Incl O&P
0010	**SLEEVES AND CHASES**									
0100	Plastic, 1 use, 12" long, 2" diameter	1 Carp	100	.080	Ea.	2.15	3.76		5.91	8.15
0150	4" diameter		90	.089		6.05	4.17		10.22	13.05
0200	6" diameter		75	.107		10.65	5		15.65	19.40
0250	12" diameter		60	.133		31.50	6.25		37.75	44
5000	Sheet metal, 2" diameter	G	100	.080		1.36	3.76		5.12	7.30
5100	4" diameter	G	90	.089		1.70	4.17		5.87	8.25
5150	6" diameter	G	75	.107		1.82	5		6.82	9.70
5200	12" diameter	G	60	.133		3.63	6.25		9.88	13.65
6000	Steel pipe, 2" diameter	G	100	.080		4.55	3.76		8.31	10.80
6100	4" diameter	G	90	.089		16.30	4.17		20.47	24.50
6150	6" diameter	G	75	.107		36.50	5		41.50	47.50
6200	12" diameter	G	60	.133		86.50	6.25		92.75	105

03 15 16 – Concrete Construction Joints

03 15 16.20 Control Joints, Saw Cut

03 15 16.20 Control Joints, Saw Cut		Crew	Daily Output	Labor-Hours	Unit	Material	2015 Bare Costs Labor	Equipment	Total	Total Incl O&P
0010	**CONTROL JOINTS, SAW CUT**									
0100	Sawcut control joints in green concrete									
0120	1" depth	C-27	2000	.008	L.F.	.04	.36	.08	.48	.66
0140	1-1/2" depth		1800	.009		.06	.40	.09	.55	.75
0160	2" depth		1600	.010		.08	.45	.10	.63	.88
0180	Sawcut joint reservoir in cured concrete									
0182	3/8" wide x 3/4" deep, with single saw blade	C-27	1000	.016	L.F.	.06	.72	.17	.95	1.31
0184	1/2" wide x 1" deep, with double saw blades		900	.018		.11	.80	.19	1.10	1.52
0186	3/4" wide x 1-1/2" deep, with double saw blades		800	.020		.23	.90	.21	1.34	1.82

03 15 Concrete Accessories

03 15 16 – Concrete Construction Joints

	03 15 16.20 Control Joints, Saw Cut		Crew	Daily Output	Labor-Hours	Unit	Material	2015 Bare Costs Labor	Equipment	Total	Total Incl O&P
0190	Water blast joint to wash away laitance, 2 passes		C-29	2500	.003	L.F.		.12	.03	.15	.22
0200	Air blast joint to blow out debris and air dry, 2 passes		C-28	2000	.004	▼		.18	.01	.19	.28
0300	For backer rod, see Section 07 91 23.10										
0342	For joint sealant, see Section 07 92 13.20										

03 15 19 – Cast-In Concrete Anchors

03 15 19.10 Anchor Bolts

			Crew	Daily Output	Labor-Hours	Unit	Material	Labor	Equipment	Total	Total Incl O&P
0010	**ANCHOR BOLTS**										
0015	Made from recycled materials										
0025	Single bolts installed in fresh concrete, no templates										
0030	Hooked w/nut and washer, 1/2" diameter, 8" long	G	1 Carp	132	.061	Ea.	1.33	2.85		4.18	5.85
0040	12" long	G		131	.061		1.48	2.87		4.35	6.05
0050	5/8" diameter, 8" long	G		129	.062		2.98	2.91		5.89	7.75
0060	12" long	G		127	.063		3.67	2.96		6.63	8.60
0070	3/4" diameter, 8" long	G		127	.063		3.67	2.96		6.63	8.60
0080	12" long	G	▼	125	.064	▼	4.59	3		7.59	9.65
0090	2-bolt pattern, including job-built 2-hole template, per set										
0100	J-type, incl. hex nut & washer, 1/2" diameter x 6" long	G	1 Carp	21	.381	Set	5.20	17.90		23.10	33.50
0110	12" long	G		21	.381		5.80	17.90		23.70	34
0120	18" long	G		21	.381		6.70	17.90		24.60	35
0130	3/4" diameter x 8" long	G		20	.400		10.20	18.80		29	40
0140	12" long	G		20	.400		12	18.80		30.80	42
0150	18" long	G		20	.400		14.75	18.80		33.55	45.50
0160	1" diameter x 12" long	G		19	.421		20.50	19.75		40.25	53
0170	18" long	G		19	.421		24	19.75		43.75	57
0180	24" long	G		19	.421		28.50	19.75		48.25	62
0190	36" long	G		18	.444		38	21		59	74
0200	1-1/2" diameter x 18" long	G		17	.471		37.50	22		59.50	75.50
0210	24" long	G		16	.500		44	23.50		67.50	84.50
0300	L-type, incl. hex nut & washer, 3/4" diameter x 12" long	G		20	.400		12.90	18.80		31.70	43
0310	18" long	G		20	.400		15.75	18.80		34.55	46.50
0320	24" long	G		20	.400		18.60	18.80		37.40	49.50
0330	30" long	G		20	.400		23	18.80		41.80	54
0340	36" long	G		20	.400		25.50	18.80		44.30	57.50
0350	1" diameter x 12" long	G		19	.421		19.55	19.75		39.30	52
0360	18" long	G		19	.421		23.50	19.75		43.25	56.50
0370	24" long	G		19	.421		28.50	19.75		48.25	62
0380	30" long	G		19	.421		33	19.75		52.75	67
0390	36" long	G		18	.444		37.50	21		58.50	73.50
0400	42" long	G		18	.444		45	21		66	81.50
0410	48" long	G		18	.444		50	21		71	87
0420	1-1/4" diameter x 18" long	G		18	.444		30	21		51	65
0430	24" long	G		18	.444		35	21		56	70.50
0440	30" long	G		17	.471		40	22		62	78
0450	36" long	G	▼	17	.471		45	22		67	83.50
0460	42" long	G	2 Carp	32	.500		50.50	23.50		74	91.50
0470	48" long	G		32	.500		57	23.50		80.50	99
0480	54" long	G		31	.516		67	24		91	111
0490	60" long	G		31	.516		73	24		97	118
0500	1-1/2" diameter x 18" long	G		33	.485		51	23		74	91
0510	24" long	G		32	.500		59	23.50		82.50	101
0520	30" long	G		31	.516		66	24		90	110
0530	36" long	G	▼	30	.533		75.50	25		100.50	122

33

03 15 19 – Cast-In Concrete Anchors

03 15 19.10 Anchor Bolts		Crew	Daily Output	Labor-Hours	Unit	Material	2015 Bare Costs Labor	Equipment	Total	Total Incl O&P
0540	42" long	G	2 Carp	30	.533	Set	85.50	25		110.50	133
0550	48" long	G		29	.552		96	26		122	145
0560	54" long	G		28	.571		116	27		143	170
0570	60" long	G		28	.571		127	27		154	182
0580	1-3/4" diameter x 18" long	G		31	.516		74.50	24		98.50	120
0590	24" long	G		30	.533		87	25		112	134
0600	30" long	G		29	.552		101	26		127	151
0610	36" long	G		28	.571		114	27		141	168
0620	42" long	G		27	.593		128	28		156	184
0630	48" long	G		26	.615		141	29		170	200
0640	54" long	G		26	.615		174	29		203	236
0650	60" long	G		25	.640		187	30		217	252
0660	2" diameter x 24" long	G		27	.593		111	28		139	166
0670	30" long	G		27	.593		125	28		153	181
0680	36" long	G		26	.615		137	29		166	196
0690	42" long	G		25	.640		153	30		183	214
0700	48" long	G		24	.667		175	31.50		206.50	241
0710	54" long	G		23	.696		208	32.50		240.50	280
0720	60" long	G		23	.696		223	32.50		255.50	297
0730	66" long	G		22	.727		239	34		273	315
0740	72" long	G		21	.762		261	36		297	340
1000	4-bolt pattern, including job-built 4-hole template, per set										
1100	J-type, incl. hex nut & washer, 1/2" diameter x 6" long	G	1 Carp	19	.421	Set	7.55	19.75		27.30	39
1110	12" long	G		19	.421		8.75	19.75		28.50	40
1120	18" long	G		18	.444		10.55	21		31.55	43.50
1130	3/4" diameter x 8" long	G		17	.471		17.55	22		39.55	53.50
1140	12" long	G		17	.471		21	22		43	57.50
1150	18" long	G		17	.471		26.50	22		48.50	63.50
1160	1" diameter x 12" long	G		16	.500		38	23.50		61.50	77.50
1170	18" long	G		15	.533		45	25		70	88
1180	24" long	G		15	.533		54	25		79	98
1190	36" long	G		15	.533		73	25		98	119
1200	1-1/2" diameter x 18" long	G		13	.615		72	29		101	124
1210	24" long	G		12	.667		85.50	31.50		117	142
1300	L-type, incl. hex nut & washer, 3/4" diameter x 12" long	G		17	.471		23	22		45	59.50
1310	18" long	G		17	.471		28.50	22		50.50	65.50
1320	24" long	G		17	.471		34.50	22		56.50	72
1330	30" long	G		16	.500		43	23.50		66.50	83
1340	36" long	G		16	.500		48.50	23.50		72	89.50
1350	1" diameter x 12" long	G		16	.500		36	23.50		59.50	76
1360	18" long	G		15	.533		44.50	25		69.50	87.50
1370	24" long	G		15	.533		54	25		79	98
1380	30" long	G		15	.533		63.50	25		88.50	109
1390	36" long	G		15	.533		72	25		97	118
1400	42" long	G		14	.571		87	27		114	138
1410	48" long	G		14	.571		97.50	27		124.50	149
1420	1-1/4" diameter x 18" long	G		14	.571		57	27		84	104
1430	24" long	G		14	.571		67	27		94	115
1440	30" long	G		13	.615		77	29		106	129
1450	36" long	G		13	.615		87	29		116	141
1460	42" long	G	2 Carp	25	.640		98	30		128	154
1470	48" long	G		24	.667		112	31.50		143.50	171
1480	54" long	G		23	.696		131	32.50		163.50	195

03 15 Concrete Accessories

03 15 19 – Cast-In Concrete Anchors

03 15 19.10 Anchor Bolts		Crew	Daily Output	Labor-Hours	Unit	Material	2015 Bare Costs Labor	Equipment	Total	Total Incl O&P
1490	60" long	G 2 Carp	23	.696	Set	144	32.50		176.50	209
1500	1-1/2" diameter x 18" long	G	25	.640		99	30		129	155
1510	24" long	G	24	.667		115	31.50		146.50	174
1520	30" long	G	23	.696		129	32.50		161.50	193
1530	36" long	G	22	.727		148	34		182	216
1540	42" long	G	22	.727		169	34		203	238
1550	48" long	G	21	.762		189	36		225	263
1560	54" long	G	20	.800		230	37.50		267.50	310
1570	60" long	G	20	.800		251	37.50		288.50	335
1580	1-3/4" diameter x 18" long	G	22	.727		146	34		180	214
1590	24" long	G	21	.762		171	36		207	243
1600	30" long	G	21	.762		199	36		235	273
1610	36" long	G	20	.800		226	37.50		263.50	305
1620	42" long	G	19	.842		254	39.50		293.50	340
1630	48" long	G	18	.889		278	41.50		319.50	370
1640	54" long	G	18	.889		345	41.50		386.50	445
1650	60" long	G	17	.941		370	44		414	480
1660	2" diameter x 24" long	G	19	.842		220	39.50		259.50	305
1670	30" long	G	18	.889		247	41.50		288.50	335
1680	36" long	G	18	.889		272	41.50		313.50	365
1690	42" long	G	17	.941		305	44		349	405
1700	48" long	G	16	1		345	47		392	455
1710	54" long	G	15	1.067		415	50		465	530
1720	60" long	G	15	1.067		445	50		495	565
1730	66" long	G	14	1.143		475	53.50		528.50	605
1740	72" long	G	14	1.143		520	53.50		573.50	655
1990	For galvanized, add				Ea.	75%				

03 15 19.45 Machinery Anchors

		Crew	Daily Output	Labor-Hours	Unit	Material	Labor	Equipment	Total	Total Incl O&P
0010	**MACHINERY ANCHORS**, heavy duty, incl. sleeve, floating base nut,									
0020	lower stud & coupling nut, fiber plug, connecting stud, washer & nut.									
0030	For flush mounted embedment in poured concrete heavy equip. pads.									
0200	Stud & bolt, 1/2" diameter	G E-16	40	.400	Ea.	72.50	21.50	3.65	97.65	122
0300	5/8" diameter	G	35	.457		80.50	24.50	4.17	109.17	136
0500	3/4" diameter	G	30	.533		93	28.50	4.86	126.36	157
0600	7/8" diameter	G	25	.640		101	34.50	5.85	141.35	177
0800	1" diameter	G	20	.800		117	43	7.30	167.30	212
0900	1-1/4" diameter	G	15	1.067		141	57	9.75	207.75	265

03 21 Reinforcement Bars

03 21 11 – Plain Steel Reinforcement Bars

03 21 11.60 Reinforcing In Place

		Crew	Daily Output	Labor-Hours	Unit	Material	Labor	Equipment	Total	Total Incl O&P
0010	**REINFORCING IN PLACE**, 50-60 ton lots, A615 Grade 60									
0020	Includes labor, but not material cost, to install accessories									
0030	Made from recycled materials									
0502	Footings, #4 to #7	G 4 Rodm	4200	.008	Lb.	.48	.40		.88	1.16
0552	#8 to #18	G	7200	.004		.48	.23		.71	.89
0602	Slab on grade, #3 to #7	G	4200	.008		.48	.40		.88	1.16
0900	For other than 50 – 60 ton lots									
1000	Under 10 ton job, #3 to #7, add					25%	10%			
1010	#8 to #18, add					20%	10%			
1050	10 – 50 ton job, #3 to #7, add					10%				

For customer support on your Plumbing Cost Data, call 877.709.6968.

35

03 21 Reinforcement Bars

03 21 11 – Plain Steel Reinforcement Bars

03 21 11.60 Reinforcing In Place	Crew	Daily Output	Labor-Hours	Unit	Material	2015 Bare Costs Labor	Equipment	Total	Total Incl O&P
1060	#8 to #18, add					5%				
1100	60 – 100 ton job, #3 to #7, deduct					5%				
1110	#8 to #18, deduct					10%				
1150	Over 100 ton job, #3 to #7, deduct					10%				
1160	#8 to #18, deduct					15%				

03 22 Fabric and Grid Reinforcing

03 22 11 – Plain Welded Wire Fabric Reinforcing

03 22 11.10 Plain Welded Wire Fabric

		Crew	Daily Output	Labor-Hours	Unit	Material	Labor	Equipment	Total	Total Incl O&P
0010	**PLAIN WELDED WIRE FABRIC** ASTM A185									
0020	Includes labor, but not material cost, to install accessories									
0030	Made from recycled materials									
0050	Sheets									
0100	6 x 6 - W1.4 x W1.4 (10 x 10) 21 lb. per C.S.F. **G**	2 Rodm	35	.457	C.S.F.	14.50	24		38.50	53.50

03 22 13 – Galvanized Welded Wire Fabric Reinforcing

03 22 13.10 Galvanized Welded Wire Fabric

		Crew	Daily Output	Labor-Hours	Unit	Material	Labor	Equipment	Total	Total Incl O&P
0010	**GALVANIZED WELDED WIRE FABRIC**									
0100	Add to plain welded wire pricing for galvanized welded wire				Lb.	.23			.23	.25

03 22 16 – Epoxy-Coated Welded Wire Fabric Reinforcing

03 22 16.10 Epoxy-Coated Welded Wire Fabric

		Crew	Daily Output	Labor-Hours	Unit	Material	Labor	Equipment	Total	Total Incl O&P
0010	**EPOXY-COATED WELDED WIRE FABRIC**									
0100	Add to plain welded wire pricing for epoxy-coated welded wire				Lb.	.21			.21	.23

03 30 Cast-In-Place Concrete

03 30 53 – Miscellaneous Cast-In-Place Concrete

03 30 53.40 Concrete In Place

		Crew	Daily Output	Labor-Hours	Unit	Material	Labor	Equipment	Total	Total Incl O&P
0010	**CONCRETE IN PLACE**									
0020	Including forms (4 uses), Grade 60 rebar, concrete (Portland cement									
0050	Type I), placement and finishing unless otherwise indicated									
3540	Equipment pad (3000 psi), 3' x 3' x 6" thick	C-14H	45	1.067	Ea.	47	49.50	.69	97.19	128
3550	4' x 4' x 6" thick		30	1.600		69.50	74	1.04	144.54	191
3560	5' x 5' x 8" thick		18	2.667		122	124	1.73	247.73	325
3570	6' x 6' x 8" thick		14	3.429		164	159	2.23	325.23	425
3580	8' x 8' x 10" thick		8	6		350	278	3.90	631.90	815
3590	10' x 10' x 12" thick		5	9.600		595	445	6.25	1,046.25	1,350
3800	Footings (3000 psi), spread under 1 C.Y.	C-14C	28	4	C.Y.	166	180	1.12	347.12	460
3825	1 C.Y. to 5 C.Y.		43	2.605		201	117	.73	318.73	405
3850	Over 5 C.Y. R033053-60		75	1.493		185	67.50	.42	252.92	310
3900	Footings, strip (3000 psi), 18" x 9", unreinforced	C-14L	40	2.400		125	105	.79	230.79	300
3920	18" x 9", reinforced	C-14C	35	3.200		148	144	.90	292.90	385
3925	20" x 10", unreinforced	C-14L	45	2.133		122	93.50	.70	216.20	278
3930	20" x 10", reinforced	C-14C	40	2.800		140	126	.78	266.78	350
3935	24" x 12", unreinforced	C-14L	55	1.745		120	76.50	.58	197.08	250
3940	24" x 12", reinforced	C-14C	48	2.333		139	105	.65	244.65	315
3945	36" x 12", unreinforced	C-14L	70	1.371		116	60	.45	176.45	219
3950	36" x 12", reinforced	C-14C	60	1.867		133	84	.52	217.52	276
4000	Foundation mat (3000 psi), under 10 C.Y.		38.67	2.896		204	131	.81	335.81	425
4050	Over 20 C.Y.		56.40	1.986		178	89.50	.56	268.06	335

03 30 Cast-In-Place Concrete

03 30 53 – Miscellaneous Cast-In-Place Concrete

03 30 53.40 Concrete In Place

03 30 53.40 Concrete In Place		Crew	Daily Output	Labor-Hours	Unit	Material	2015 Bare Costs Labor	Equipment	Total	Total Incl O&P
4650	Slab on grade (3500 psi), not including finish, 4" thick	C-14E	60.75	1.449	C.Y.	124	67.50	.51	192.01	242
4700	6" thick	"	92	.957	↓	119	44.50	.33	163.83	200
4701	Thickened slab edge (3500 psi), for slab on grade poured									
4702	monolithically with slab; depth is in addition to slab thickness;									
4703	formed vertical outside edge, earthen bottom and inside slope									
4705	8" deep x 8" wide bottom, unreinforced	C-14L	2190	.044	L.F.	3.47	1.92	.01	5.40	6.80
4710	8" x 8", reinforced	C-14C	1670	.067		5.75	3.02	.02	8.79	10.95
4715	12" deep x 12" wide bottom, unreinforced	C-14L	1800	.053		7.05	2.34	.02	9.41	11.35
4720	12" x 12", reinforced	C-14C	1310	.086		11.20	3.85	.02	15.07	18.30
4725	16" deep x 16" wide bottom, unreinforced	C-14L	1440	.067		11.85	2.92	.02	14.79	17.50
4730	16" x 16", reinforced	C-14C	1120	.100		16.80	4.51	.03	21.34	25.50
4735	20" deep x 20" wide bottom, unreinforced	C-14L	1150	.083		17.90	3.66	.03	21.59	25.50
4740	20" x 20", reinforced	C-14C	920	.122		24	5.50	.03	29.53	35
4745	24" deep x 24" wide bottom, unreinforced	C-14L	930	.103		25	4.53	.03	29.56	35
4750	24" x 24", reinforced	C-14C	740	.151	↓	33.50	6.80	.04	40.34	47

03 31 Structural Concrete

03 31 13 – Heavyweight Structural Concrete

03 31 13.35 Heavyweight Concrete, Ready Mix

		Crew	Daily Output	Labor-Hours	Unit	Material	2015 Bare Costs Labor	Equipment	Total	Total Incl O&P
0010	**HEAVYWEIGHT CONCRETE, READY MIX**, delivered									
0012	Includes local aggregate, sand, Portland cement (Type I) and water									
0015	Excludes all additives and treatments									
0020	2000 psi				C.Y.	97			97	107
0100	2500 psi					99.50			99.50	109
0150	3000 psi					102			102	112
0200	3500 psi					104			104	115
0300	4000 psi					107			107	118
1000	For high early strength (Portland cement Type III), add					10%				
1300	For winter concrete (hot water), add					4.05			4.05	4.46
1410	For mid-range water reducer, add					3.31			3.31	3.64
1420	For high-range water reducer/superplasticizer, add					5.65			5.65	6.20
1430	For retarder, add					3.23			3.23	3.55
1440	For non-Chloride accelerator, add					5.50			5.50	6.05
1450	For Chloride accelerator, per 1%, add					2.90			2.90	3.19
1460	For fiber reinforcing, synthetic (1 lb./C.Y.), add					6.65			6.65	7.30
1500	For Saturday delivery, add				↓	10.80			10.80	11.90
1510	For truck holding/waiting time past 1st hour per load, add				Hr.	94			94	103
1520	For short load (less than 4 C.Y.), add per load				Ea.	60.50			60.50	66.50
2000	For all lightweight aggregate, add				C.Y.	45%				

03 31 13.70 Placing Concrete

		Crew	Daily Output	Labor-Hours	Unit	Material	2015 Bare Costs Labor	Equipment	Total	Total Incl O&P
0010	**PLACING CONCRETE**									
0020	Includes labor and equipment to place, level (strike off) and consolidate									
1900	Footings, continuous, shallow, direct chute	C-6	120	.400	C.Y.		15.65	.52	16.17	24.50
1950	Pumped	C-20	150	.427			17.25	5.20	22.45	32.50
2000	With crane and bucket	C-7	90	.800			32.50	13.50	46	65
2100	Footings, continuous, deep, direct chute	C-6	140	.343			13.45	.45	13.90	21
2150	Pumped	C-20	160	.400			16.15	4.89	21.04	30
2200	With crane and bucket	C-7	110	.655			27	11.05	38.05	53
2400	Footings, spread, under 1 C.Y., direct chute	C-6	55	.873			34	1.13	35.13	53.50
2450	Pumped	C-20	65	.985			40	12.05	52.05	74.50
2500	With crane and bucket	C-7	45	1.600	↓		65.50	27	92.50	130

For customer support on your Plumbing Cost Data, call 877.709.6968.

37

03 31 Structural Concrete

03 31 13 – Heavyweight Structural Concrete

03 31 13.70 Placing Concrete		Crew	Daily Output	Labor-Hours	Unit	Material	2015 Bare Costs		Total	Total Incl O&P
							Labor	Equipment		
2600	Over 5 C.Y., direct chute	C-6	120	.400	C.Y.		15.65	.52	16.17	24.5
2650	Pumped	C-20	150	.427			17.25	5.20	22.45	32.5
2700	With crane and bucket	C-7	100	.720			29.50	12.15	41.65	58.5
2900	Foundation mats, over 20 C.Y., direct chute	C-6	350	.137			5.35	.18	5.53	8.4
2950	Pumped	C-20	400	.160			6.45	1.96	8.41	12
3000	With crane and bucket	C-7	300	.240	↓		9.80	4.05	13.85	19.4

03 35 Concrete Finishing

03 35 13 – High-Tolerance Concrete Floor Finishing

03 35 13.30 Finishing Floors, High Tolerance

		Crew	Daily Output	Labor-Hours	Unit	Material	2015 Bare Costs		Total	Total Incl O&P
							Labor	Equipment		
0010	**FINISHING FLOORS, HIGH TOLERANCE**									
0012	Finishing of fresh concrete flatwork requires that concrete									
0013	first be placed, struck off & consolidated									
0015	Basic finishing for various unspecified flatwork									
0100	Bull float only	C-10	4000	.006	S.F.		.26		.26	.3
0125	Bull float & manual float		2000	.012			.51		.51	.7
0150	Bull float, manual float, & broom finish, w/edging & joints		1850	.013			.55		.55	.8
0200	Bull float, manual float & manual steel trowel	↓	1265	.019	↓		.81		.81	1.2
0210	For specified Random Access Floors in ACI Classes 1, 2, 3 and 4 to achieve									
0215	Composite Overall Floor Flatness and Levelness values up to FF35/FL25									
0250	Bull float, machine float & machine trowel (walk-behind)	C-10C	1715	.014	S.F.		.60	.03	.63	.9
0300	Power screed, bull float, machine float & trowel (walk-behind)	C-10D	2400	.010			.43	.05	.48	.6
0350	Power screed, bull float, machine float & trowel (ride-on)	C-10E	4000	.006	↓		.26	.06	.32	.4
0352	For specified Random Access Floors in ACI Classes 5, 6, 7 and 8 to achieve									
0354	Composite Overall Floor Flatness and Levelness values up to FF50/FL50									
0356	Add for two-dimensional restraightening after power float	C-10	6000	.004	S.F.		.17		.17	.2
0358	For specified Random or Defined Access Floors in ACI Class 9 to achieve									
0360	Composite Overall Floor Flatness and Levelness values up to FF100/FL100									
0362	Add for two-dimensional restraightening after bull float & power float	C-10	3000	.008	S.F.		.34		.34	.5
0364	For specified Superflat Defined Access Floors in ACI Class 9 to achieve									
0366	Minimum Floor Flatness and Levelness values of FF100/FL100									
0368	Add for 2-dim'l restraightening after bull float, power float, power trowel	C-10	2000	.012	S.F.		.51		.51	.76

03 63 Epoxy Grouting

03 63 05 – Grouting of Dowels and Fasteners

03 63 05.10 Epoxy Only

		Crew	Daily Output	Labor-Hours	Unit	Material	2015 Bare Costs		Total	Total Incl O&P
							Labor	Equipment		
0010	**EPOXY ONLY**									
1500	Chemical anchoring, epoxy cartridge, excludes layout, drilling, fastener									
1530	For fastener 3/4" diam. x 6" embedment	2 Skwk	72	.222	Ea.	5.65	10.80		16.45	23
1535	1" diam. x 8" embedment		66	.242		8.45	11.80		20.25	27.50
1540	1-1/4" diam. x 10" embedment		60	.267		16.90	12.95		29.85	38.50
1545	1-3/4" diam. x 12" embedment		54	.296		28	14.40		42.40	53.50
1550	14" embedment		48	.333		34	16.20		50.20	62
1555	2" diam. x 12" embedment		42	.381		45	18.55		63.55	78
1560	18" embedment	↓	32	.500	↓	56.50	24.50		81	99.50

03 82 Concrete Boring

03 82 13 – Concrete Core Drilling

03 82 13.10 Core Drilling

	Crew	Daily Output	Labor-Hours	Unit	Material	2015 Bare Costs Labor	Equipment	Total	Total Incl O&P
0010 **CORE DRILLING**									
0015　Includes bit cost, layout and set-up time									
0020　Reinforced concrete slab, up to 6" thick									
0100　　1" diameter core	B-89A	17	.941	Ea.	.18	40.50	6.80	47.48	70
0150　　　For each additional inch of slab thickness in same hole, add		1440	.011		.03	.48	.08	.59	.86
0200　　2" diameter core		16.50	.970		.28	42	7	49.28	72.50
0250　　　For each additional inch of slab thickness in same hole, add		1080	.015		.05	.64	.11	.80	1.15
0300　　3" diameter core		16	1		.38	43	7.25	50.63	75
0350　　　For each additional inch of slab thickness in same hole, add		720	.022		.06	.96	.16	1.18	1.73
0500　　4" diameter core		15	1.067		.49	46	7.70	54.19	80
0550　　　For each additional inch of slab thickness in same hole, add		480	.033		.08	1.44	.24	1.76	2.58
0700　　6" diameter core		14	1.143		.76	49.50	8.25	58.51	86
0750　　　For each additional inch of slab thickness in same hole, add		360	.044		.13	1.92	.32	2.37	3.45
0900　　8" diameter core		13	1.231		1.05	53	8.90	62.95	93
0950　　　For each additional inch of slab thickness in same hole, add		288	.056		.17	2.40	.40	2.97	4.32
1100　　10" diameter core		12	1.333		1.48	57.50	9.65	68.63	101
1150　　　For each additional inch of slab thickness in same hole, add		240	.067		.25	2.88	.48	3.61	5.25
1300　　12" diameter core		11	1.455		1.78	62.50	10.50	74.78	110
1350　　　For each additional inch of slab thickness in same hole, add		206	.078		.30	3.35	.56	4.21	6.10
1500　　14" diameter core		10	1.600		2.08	69	11.55	82.63	121
1550　　　For each additional inch of slab thickness in same hole, add		180	.089		.35	3.83	.64	4.82	7
1700　　18" diameter core		9	1.778		2.82	76.50	12.85	92.17	135
1750　　　For each additional inch of slab thickness in same hole, add		144	.111		.47	4.79	.80	6.06	8.80
1754　　24" diameter core		8	2		4.02	86.50	14.45	104.97	153
1756　　　For each additional inch of slab thickness in same hole, add		120	.133		.67	5.75	.96	7.38	10.65
1760　　For horizontal holes, add to above						20%	20%		
1770　Prestressed hollow core plank, 8" thick									
1780　　1" diameter core	B-89A	17.50	.914	Ea.	.24	39.50	6.60	46.34	68.50
1790　　　For each additional inch of plank thickness in same hole, add		3840	.004		.03	.18	.03	.24	.34
1794　　2" diameter core		17.25	.928		.38	40	6.70	47.08	69.50
1796　　　For each additional inch of plank thickness in same hole, add		2880	.006		.05	.24	.04	.33	.46
1800　　3" diameter core		17	.941		.51	40.50	6.80	47.81	70.50
1810　　　For each additional inch of plank thickness in same hole, add		1920	.008		.06	.36	.06	.48	.69
1820　　4" diameter core		16.50	.970		.66	42	7	49.66	73
1830　　　For each additional inch of plank thickness in same hole, add		1280	.013		.08	.54	.09	.71	1.02
1840　　6" diameter core		15.50	1.032		1.01	44.50	7.45	52.96	78
1850　　　For each additional inch of plank thickness in same hole, add		960	.017		.13	.72	.12	.97	1.38
1860　　8" diameter core		15	1.067		1.40	46	7.70	55.10	81
1870　　　For each additional inch of plank thickness in same hole, add		768	.021		.17	.90	.15	1.22	1.75
1880　　10" diameter core		14	1.143		1.98	49.50	8.25	59.73	87.50
1890　　　For each additional inch of plank thickness in same hole, add		640	.025		.25	1.08	.18	1.51	2.13
1900　　12" diameter core		13.50	1.185		2.37	51	8.55	61.92	91
1910　　　For each additional inch of plank thickness in same hole, add		548	.029		.30	1.26	.21	1.77	2.50

03 82 16 – Concrete Drilling

03 82 16.10 Concrete Impact Drilling

	Crew	Daily Output	Labor-Hours	Unit	Material	2015 Bare Costs Labor	Equipment	Total	Total Incl O&P
0010 **CONCRETE IMPACT DRILLING**									
0020　Includes bit cost, layout and set-up time, no anchors									
0050　Up to 4" deep in concrete/brick floors/walls									
0100　　Holes, 1/4" diameter	1 Carp	75	.107	Ea.	.07	5		5.07	7.75
0150　　　For each additional inch of depth in same hole, add		430	.019		.02	.87		.89	1.36
0200　　3/8" diameter		63	.127		.06	5.95		6.01	9.20
0250　　　For each additional inch of depth in same hole, add		340	.024		.01	1.10		1.11	1.72

For customer support on your Plumbing Cost Data, call 877.709.6968.

39

03 82 Concrete Boring

03 82 16 - Concrete Drilling

03 82 16.10 Concrete Impact Drilling	Crew	Daily Output	Labor-Hours	Unit	Material	2015 Bare Costs Labor	Equipment	Total	Total Incl O&P
0300	1/2" diameter	1 Carp	50	.160	Ea.	.06	7.50		7.56	11.6
0350	For each additional inch of depth in same hole, add		250	.032		.01	1.50		1.51	2.3
0400	5/8" diameter		48	.167		.09	7.85		7.94	12.1
0450	For each additional inch of depth in same hole, add		240	.033		.02	1.56		1.58	2.4
0500	3/4" diameter		45	.178		.12	8.35		8.47	13
0550	For each additional inch of depth in same hole, add		220	.036		.03	1.71		1.74	2.6
0600	7/8" diameter		43	.186		.16	8.75		8.91	13.6
0650	For each additional inch of depth in same hole, add		210	.038		.04	1.79		1.83	2.7
0700	1" diameter		40	.200		.17	9.40		9.57	14.6
0750	For each additional inch of depth in same hole, add		190	.042		.04	1.98		2.02	3.0
0800	1-1/4" diameter		38	.211		.26	9.90		10.16	15.5
0850	For each additional inch of depth in same hole, add		180	.044		.06	2.09		2.15	3.2
0900	1-1/2" diameter		35	.229		.39	10.75		11.14	16.9
0950	For each additional inch of depth in same hole, add	▼	165	.048	▼	.10	2.28		2.38	3.6
1000	For ceiling installations, add						40%			

Estimating Tips

05 05 00 Common Work Results for Metals

- Nuts, bolts, washers, connection angles, and plates can add a significant amount to both the tonnage of a structural steel job and the estimated cost. As a rule of thumb, add 10% to the total weight to account for these accessories.

- Type 2 steel construction, commonly referred to as "simple construction," consists generally of field-bolted connections with lateral bracing supplied by other elements of the building, such as masonry walls or x-bracing. The estimator should be aware, however, that shop connections may be accomplished by welding or bolting. The method may be particular to the fabrication shop and may have an impact on the estimated cost.

05 10 00 Structural Steel

- Steel items can be obtained from two sources: a fabrication shop or a metals service center. Fabrication shops can fabricate items under more controlled conditions than can crews in the field. They are also more efficient and can produce items more economically. Metal service centers serve as a source of long mill shapes to both fabrication shops and contractors.

- Most line items in this structural steel subdivision, and most items in 05 50 00 Metal Fabrications, are indicated as being shop fabricated. The bare material cost for these shop fabricated items is the "Invoice Cost" from the shop and includes the mill base price of steel plus mill extras, transportation to the shop, shop drawings and detailing where warranted, shop fabrication and handling, sandblasting and a shop coat of primer paint, all necessary structural bolts, and delivery to the job site. The bare labor cost and bare equipment cost for these shop fabricated items is for field installation or erection.

- Line items in Subdivision 05 12 23.40 Lightweight Framing, and other items scattered in Division 5, are indicated as being field fabricated. The bare material cost for these field fabricated items is the "Invoice Cost" from the metals service center and includes the mill base price of steel plus mill extras, transportation to the metals service center, material handling, and delivery of long lengths of mill shapes to the job site. Material costs for structural bolts and welding rods should be added to the estimate. The bare labor cost and bare equipment cost for these items is for both field fabrication and field installation or erection, and include time for cutting, welding and drilling in the fabricated metal items. Drilling into concrete and fasteners to fasten field fabricated items to other work are not included and should be added to the estimate.

05 20 00 Steel Joist Framing

- In any given project the total weight of open web steel joists is determined by the loads to be supported and the design. However, economies can be realized in minimizing the amount of labor used to place the joists. This is done by maximizing the joist spacing, and therefore minimizing the number of joists required to be installed on the job. Certain spacings and locations may be required by the design, but in other cases maximizing the spacing and keeping it as uniform as possible will keep the costs down.

05 30 00 Steel Decking

- The takeoff and estimating of metal deck involves more than simply the area of the floor or roof and the type of deck specified or shown on the drawings. Many different sizes and types of openings may exist. Small openings for individual pipes or conduits may be drilled after the floor/roof is installed, but larger openings may require special deck lengths as well as reinforcing or structural support. The estimator should determine who will be supplying this reinforcing. Additionally, some deck terminations are part of the deck package, such as screed angles and pour stops, and others will be part of the steel contract, such as angles attached to structural members and cast-in-place angles and plates. The estimator must ensure that all pieces are accounted for in the complete estimate.

05 50 00 Metal Fabrications

- The most economical steel stairs are those that use common materials, standard details, and most importantly, a uniform and relatively simple method of field assembly. Commonly available A36 channels and plates are very good choices for the main stringers of the stairs, as are angles and tees for the carrier members. Risers and treads are usually made by specialty shops, and it is most economical to use a typical detail in as many places as possible. The stairs should be pre-assembled and shipped directly to the site. The field connections should be simple and straightforward to be accomplished efficiently, and with minimum equipment and labor.

Reference Numbers

Reference numbers are shown in shaded boxes at the beginning of some major classifications. These numbers refer to related items in the Reference Section. The reference information may be an estimating procedure, an alternate pricing method, or technical information.

Note: Not all subdivisions listed here necessarily appear in this publication. ■

05 05 19 – Post-Installed Concrete Anchors

05 05 19.10 Chemical Anchors	Crew	Daily Output	Labor-Hours	Unit	Material	2015 Bare Costs Labor	Equipment	Total	Total Incl O&P
0010 **CHEMICAL ANCHORS**									
0020 Includes layout & drilling									
1430 Chemical anchor, w/rod & epoxy cartridge, 3/4" diam. x 9-1/2" long	B-89A	27	.593	Ea.	9.80	25.50	4.28	39.58	55
1435 1" diameter x 11-3/4" long		24	.667		17.55	29	4.82	51.37	69
1440 1-1/4" diameter x 14" long		21	.762		36.50	33	5.50	75	96.5
1445 1-3/4" diameter x 15" long		20	.800		65	34.50	5.80	105.30	131
1450 18" long		17	.941		78	40.50	6.80	125.30	156
1455 2" diameter x 18" long		16	1		103	43	7.25	153.25	187
1460 24" long		15	1.067		134	46	7.70	187.70	227

05 05 19.20 Expansion Anchors		Crew	Daily Output	Labor-Hours	Unit	Material	Labor	Equipment	Total	Total Incl O&P
0010 **EXPANSION ANCHORS**										
0100 Anchors for concrete, brick or stone, no layout and drilling										
0200 Expansion shields, zinc, 1/4" diameter, 1-5/16" long, single	G	1 Carp	90	.089	Ea.	.43	4.17		4.60	6.8
0300 1-3/8" long, double	G		85	.094		.54	4.42		4.96	7.4
0400 3/8" diameter, 1-1/2" long, single	G		85	.094		.64	4.42		5.06	7.5
0500 2" long, double	G		80	.100		1.18	4.70		5.88	8.5
0600 1/2" diameter, 2-1/16" long, single	G		80	.100		1.18	4.70		5.88	8.5
0700 2-1/2" long, double	G		75	.107		1.91	5		6.91	9.80
0800 5/8" diameter, 2-5/8" long, single	G		75	.107		2.05	5		7.05	9.95
0900 2-3/4" long, double	G		70	.114		2.70	5.35		8.05	11.20
1000 3/4" diameter, 2-3/4" long, single	G		70	.114		3.09	5.35		8.44	11.65
1100 3-15/16" long, double	G		65	.123		5	5.80		10.80	14.40
2100 Hollow wall anchors for gypsum wall board, plaster or tile										
2300 1/8" diameter, short	G	1 Carp	160	.050	Ea.	.23	2.35		2.58	3.86
2400 Long	G		150	.053		.23	2.50		2.73	4.10
2500 3/16" diameter, short	G		150	.053		.40	2.50		2.90	4.29
2600 Long	G		140	.057		.59	2.68		3.27	4.78
2700 1/4" diameter, short	G		140	.057		.59	2.68		3.27	4.78
2800 Long	G		130	.062		.65	2.89		3.54	5.15
3000 Toggle bolts, bright steel, 1/8" diameter, 2" long	G		85	.094		.19	4.42		4.61	7
3100 4" long	G		80	.100		.25	4.70		4.95	7.55
3200 3/16" diameter, 3" long	G		80	.100		.26	4.70		4.96	7.55
3300 6" long	G		75	.107		.36	5		5.36	8.10
3400 1/4" diameter, 3" long	G		75	.107		.34	5		5.34	8.05
3500 6" long	G		70	.114		.51	5.35		5.86	8.80
3600 3/8" diameter, 3" long	G		70	.114		.84	5.35		6.19	9.15
3700 6" long	G		60	.133		1.36	6.25		7.61	11.15
3800 1/2" diameter, 4" long	G		60	.133		1.88	6.25		8.13	11.70
3900 6" long	G		50	.160		2.45	7.50		9.95	14.25
4000 Nailing anchors										
4100 Nylon nailing anchor, 1/4" diameter, 1" long		1 Carp	3.20	2.500	C	21	117		138	204
4200 1-1/2" long			2.80	2.857		26	134		160	235
4300 2" long			2.40	3.333		30.50	157		187.50	275
4400 Metal nailing anchor, 1/4" diameter, 1" long	G		3.20	2.500		17.85	117		134.85	201
4500 1-1/2" long	G		2.80	2.857		23.50	134		157.50	232
4600 2" long	G		2.40	3.333		27.50	157		184.50	272
5000 Screw anchors for concrete, masonry,										
5100 stone & tile, no layout or drilling included										
5700 Lag screw shields, 1/4" diameter, short	G	1 Carp	90	.089	Ea.	.34	4.17		4.51	6.75
5800 Long	G		85	.094		.38	4.42		4.80	7.20
5900 3/8" diameter, short	G		85	.094		.67	4.42		5.09	7.55
6000 Long	G		80	.100		.87	4.70		5.57	8.20

05 05 19 – Post-Installed Concrete Anchors

05 05 19.20 Expansion Anchors		Crew	Daily Output	Labor-Hours	Unit	Material	2015 Bare Costs Labor	Equipment	Total	Total Incl O&P
6100	1/2" diameter, short	G 1 Carp	80	.100	Ea.	.95	4.70		5.65	8.30
6200	Long	G	75	.107		1.22	5		6.22	9.05
6300	5/8" diameter, short	G	70	.114		1.45	5.35		6.80	9.85
6400	Long	G	65	.123		1.86	5.80		7.66	10.95
6600	Lead, #6 & #8, 3/4" long	G	260	.031		.18	1.44		1.62	2.42
6700	#10 - #14, 1-1/2" long	G	200	.040		.37	1.88		2.25	3.30
6800	#16 & #18, 1-1/2" long	G	160	.050		.41	2.35		2.76	4.06
6900	Plastic, #6 & #8, 3/4" long		260	.031		.04	1.44		1.48	2.26
7000	#8 & #10, 7/8" long		240	.033		.04	1.56		1.60	2.45
7100	#10 & #12, 1" long		220	.036		.05	1.71		1.76	2.69
7200	#14 & #16, 1-1/2" long	↓	160	.050	↓	.07	2.35		2.42	3.69
8000	Wedge anchors, not including layout or drilling									
8050	Carbon steel, 1/4" diameter, 1-3/4" long	G 1 Carp	150	.053	Ea.	.41	2.50		2.91	4.30
8100	3-1/4" long	G	140	.057		.54	2.68		3.22	4.72
8150	3/8" diameter, 2-1/4" long	G	145	.055		.50	2.59		3.09	4.54
8200	5" long	G	140	.057		.88	2.68		3.56	5.10
8250	1/2" diameter, 2-3/4" long	G	140	.057		.99	2.68		3.67	5.20
8300	7" long	G	125	.064		1.70	3		4.70	6.50
8350	5/8" diameter, 3-1/2" long	G	130	.062		1.86	2.89		4.75	6.50
8400	8-1/2" long	G	115	.070		3.95	3.27		7.22	9.40
8450	3/4" diameter, 4-1/4" long	G	115	.070		2.90	3.27		6.17	8.25
8500	10" long	G	95	.084		6.60	3.95		10.55	13.35
8550	1" diameter, 6" long	G	100	.080		9.30	3.76		13.06	16.05
8575	9" long	G	85	.094		12.10	4.42		16.52	20
8600	12" long	G	75	.107		13.10	5		18.10	22
8650	1-1/4" diameter, 9" long	G	70	.114		24.50	5.35		29.85	35.50
8700	12" long	G ↓	60	.133	↓	31.50	6.25		37.75	44
8750	For type 303 stainless steel, add					350%				
8800	For type 316 stainless steel, add					450%				
8950	Self-drilling concrete screw, hex washer head, 3/16" diam. x 1-3/4" long	G 1 Carp	300	.027	Ea.	.20	1.25		1.45	2.15
8960	2-1/4" long	G	250	.032		.24	1.50		1.74	2.57
8970	Phillips flat head, 3/16" diam. x 1-3/4" long	G	300	.027		.21	1.25		1.46	2.16
8980	2-1/4" long	G ↓	250	.032	↓	.23	1.50		1.73	2.56

05 05 21 – Fastening Methods for Metal

05 05 21.15 Drilling Steel

		Crew	Daily Output	Labor-Hours	Unit	Material	2015 Bare Costs Labor	Equipment	Total	Total Incl O&P
0010	**DRILLING STEEL**									
1910	Drilling & layout for steel, up to 1/4" deep, no anchor									
1920	Holes, 1/4" diameter	1 Sswk	112	.071	Ea.	.09	3.76		3.85	6.65
1925	For each additional 1/4" depth, add		336	.024		.09	1.25		1.34	2.28
1930	3/8" diameter		104	.077		.09	4.05		4.14	7.15
1935	For each additional 1/4" depth, add		312	.026		.09	1.35		1.44	2.44
1940	1/2" diameter		96	.083		.10	4.39		4.49	7.70
1945	For each additional 1/4" depth, add		288	.028		.10	1.46		1.56	2.65
1950	5/8" diameter		88	.091		.14	4.79		4.93	8.45
1955	For each additional 1/4" depth, add		264	.030		.14	1.60		1.74	2.93
1960	3/4" diameter		80	.100		.18	5.25		5.43	9.35
1965	For each additional 1/4" depth, add		240	.033		.18	1.75		1.93	3.25
1970	7/8" diameter		72	.111		.23	5.85		6.08	10.40
1975	For each additional 1/4" depth, add		216	.037		.23	1.95		2.18	3.65
1980	1" diameter		64	.125		.24	6.60		6.84	11.70
1985	For each additional 1/4" depth, add	↓	192	.042		.24	2.19		2.43	4.07
1990	For drilling up, add				↓		40%			

For customer support on your Plumbing Cost Data, call 877.709.6968.

43

05 05 21 – Fastening Methods for Metal

05 05 21.90 Welding Steel		Crew	Daily Output	Labor-Hours	Unit	Material	2015 Bare Costs Labor	Equipment	Total	Total Incl O&P
0010	**WELDING STEEL**, Structural	R050521-20								
0020	Field welding, 1/8" E6011, cost per welder, no operating engineer	E-14	8	1	Hr.	4.33	54.50	18.25	77.08	120
0200	With 1/2 operating engineer	E-13	8	1.500		4.33	79	18.25	101.58	157
0300	With 1 operating engineer	E-12	8	2		4.33	103	18.25	125.58	194
0500	With no operating engineer, 2# weld rod per ton	E-14	8	1	Ton	4.33	54.50	18.25	77.08	120
0600	8# E6011 per ton	"	2	4		17.30	219	73	309.30	480
0800	With one operating engineer per welder, 2# E6011 per ton	E-12	8	2		4.33	103	18.25	125.58	194
0900	8# E6011 per ton	"	2	8		17.30	415	73	505.30	775
1200	Continuous fillet, down welding									
1300	Single pass, 1/8" thick, 0.1#/L.F.	E-14	150	.053	L.F.	.22	2.91	.97	4.10	6.3
1400	3/16" thick, 0.2#/L.F.		75	.107		.43	5.85	1.94	8.22	12.7
1500	1/4" thick, 0.3#/L.F.		50	.160		.65	8.75	2.92	12.32	19.1
1610	5/16" thick, 0.4#/L.F.		38	.211		.87	11.50	3.84	16.21	25
1800	3 passes, 3/8" thick, 0.5#/L.F.		30	.267		1.08	14.55	4.86	20.49	32
2010	4 passes, 1/2" thick, 0.7#/L.F.		22	.364		1.52	19.85	6.65	28.02	43.5
2200	5 to 6 passes, 3/4" thick, 1.3#/L.F.		12	.667		2.81	36.50	12.15	51.46	80
2400	8 to 11 passes, 1" thick, 2.4#/L.F.		6	1.333		5.20	73	24.50	102.70	159
2600	For vertical joint welding, add						20%			
2700	Overhead joint welding, add						300%			
2900	For semi-automatic welding, obstructed joints, deduct						5%			
3000	Exposed joints, deduct						15%			
4000	Cleaning and welding plates, bars, or rods									
4010	to existing beams, columns, or trusses	E-14	12	.667	L.F.	1.08	36.50	12.15	49.73	78

05 05 23 – Metal Fastenings

05 05 23.30 Lag Screws

			Crew	Daily Output	Labor-Hours	Unit	Material	Labor	Equipment	Total	Total Incl O&P
0010	**LAG SCREWS**										
0020	Steel, 1/4" diameter, 2" long	G	1 Carp	200	.040	Ea.	.09	1.88		1.97	2.9
0100	3/8" diameter, 3" long	G		150	.053		.30	2.50		2.80	4.18
0200	1/2" diameter, 3" long	G		130	.062		.64	2.89		3.53	5.15
0300	5/8" diameter, 3" long	G		120	.067		1.15	3.13		4.28	6.10

05 05 23.35 Machine Screws

			Crew	Daily Output	Labor-Hours	Unit	Material	Labor	Equipment	Total	Total Incl O&P
0010	**MACHINE SCREWS**										
0020	Steel, round head, #8 x 1" long	G	1 Carp	4.80	1.667	C	3.93	78.50		82.43	124
0110	#8 x 2" long	G		2.40	3.333		6.20	157		163.20	248
0200	#10 x 1" long	G		4	2		5.20	94		99.20	151
0300	#10 x 2" long	G		2	4		8.65	188		196.65	299

05 05 23.50 Powder Actuated Tools and Fasteners

			Crew	Daily Output	Labor-Hours	Unit	Material	Labor	Equipment	Total	Total Incl O&P
0010	**POWDER ACTUATED TOOLS & FASTENERS**										
0020	Stud driver, .22 caliber, single shot					Ea.	148			148	163
0100	.27 caliber, semi automatic, strip					"	440			440	485
0300	Powder load, single shot, .22 cal, power level 2, brown					C	5.35			5.35	5.90
0400	Strip, .27 cal, power level 4, red						7.70			7.70	8.50
0600	Drive pin, .300 x 3/4" long	G	1 Carp	4.80	1.667		4.11	78.50		82.61	125
0700	.300 x 3" long with washer	G	"	4	2		12.65	94		106.65	159

05 05 23.55 Rivets

			Crew	Daily Output	Labor-Hours	Unit	Material	Labor	Equipment	Total	Total Incl O&P
0010	**RIVETS**										
0100	Aluminum rivet & mandrel, 1/2" grip length x 1/8" diameter	G	1 Carp	4.80	1.667	C	7.65	78.50		86.15	128
0200	3/16" diameter	G		4	2		11.40	94		105.40	158
0300	Aluminum rivet, steel mandrel, 1/8" diameter	G		4.80	1.667		10.25	78.50		88.75	131
0400	3/16" diameter	G		4	2		16.45	94		110.45	163
0500	Copper rivet, steel mandrel, 1/8" diameter	G		4.80	1.667		9.50	78.50		88	130

05 05 23 – Metal Fastenings

05 05 23.55 Rivets

		Crew	Daily Output	Labor-Hours	Unit	Material	2015 Bare Costs Labor	Equipment	Total	Total Incl O&P
0800	Stainless rivet & mandrel, 1/8" diameter	G 1 Carp	4.80	1.667	C	24.50	78.50		103	147
0900	3/16" diameter	G	4	2		39	94		133	188
1000	Stainless rivet, steel mandrel, 1/8" diameter	G	4.80	1.667		15.70	78.50		94.20	137
1100	3/16" diameter	G	4	2		25.50	94		119.50	173
1200	Steel rivet and mandrel, 1/8" diameter	G	4.80	1.667		7.40	78.50		85.90	128
1300	3/16" diameter	G	4	2		12	94		106	158
1400	Hand riveting tool, standard				Ea.	70.50			70.50	77.50
1500	Deluxe					365			365	400
1600	Power riveting tool, standard					500			500	555
1700	Deluxe					1,650			1,650	1,825

05 05 23.70 Structural Blind Bolts

		Crew	Daily Output	Labor-Hours	Unit	Material	2015 Bare Costs Labor	Equipment	Total	Total Incl O&P
0010	**STRUCTURAL BLIND BOLTS**									
0100	1/4" diameter x 1/4" grip	G 1 Sswk	240	.033	Ea.	1.24	1.75		2.99	4.41
0150	1/2" grip	G	216	.037		1.33	1.95		3.28	4.85
0200	3/8" diameter x 1/2" grip	G	232	.034		1.75	1.82		3.57	5.10
0250	3/4" grip	G	208	.038		1.84	2.02		3.86	5.55
0300	1/2" diameter x 1/2" grip	G	224	.036		3.99	1.88		5.87	7.65
0350	3/4" grip	G	200	.040		5.60	2.11		7.71	9.80
0400	5/8" diameter x 3/4" grip	G	216	.037		8.25	1.95		10.20	12.50
0450	1" grip	G	192	.042		9.50	2.19		11.69	14.25

05 05 23.90 Welding Rod

		Crew	Daily Output	Labor-Hours	Unit	Material	2015 Bare Costs Labor	Equipment	Total	Total Incl O&P
0010	**WELDING ROD**									
0020	Steel, type 6011, 1/8" diam., less than 500#				Lb.	2.16			2.16	2.38
0100	500# to 2,000#					1.95			1.95	2.15
0200	2,000# to 5,000#					1.83			1.83	2.02
0300	5/32" diameter, less than 500#					2.20			2.20	2.42
0310	500# to 2,000#					1.98			1.98	2.18
0320	2,000# to 5,000#					1.86			1.86	2.05
0400	3/16" diam., less than 500#					2.46			2.46	2.71
0500	500# to 2,000#					2.22			2.22	2.44
0600	2,000# to 5,000#					2.09			2.09	2.30
0620	Steel, type 6010, 1/8" diam., less than 500#					2.23			2.23	2.45
0630	500# to 2,000#					2.01			2.01	2.21
0640	2,000# to 5,000#					1.89			1.89	2.08
0650	Steel, type 7018 Low Hydrogen, 1/8" diam., less than 500#					2.16			2.16	2.38
0660	500# to 2,000#					1.95			1.95	2.15
0670	2,000# to 5,000#					1.83			1.83	2.02
0700	Steel, type 7024 Jet Weld, 1/8" diam., less than 500#					2.50			2.50	2.75
0710	500# to 2,000#					2.25			2.25	2.48
0720	2,000# to 5,000#					2.12			2.12	2.33
1550	Aluminum, type 4043 TIG, 1/8" diam., less than 10#					5.10			5.10	5.65
1560	10# to 60#					4.61			4.61	5.05
1570	Over 60#					4.33			4.33	4.77
1600	Aluminum, type 5356 TIG, 1/8" diam., less than 10#					5.45			5.45	6
1610	10# to 60#					4.90			4.90	5.40
1620	Over 60#					4.61			4.61	5.05
1900	Cast iron, type 8 Nickel, 1/8" diam., less than 500#					22			22	24
1910	500# to 1,000#					19.75			19.75	21.50
1920	Over 1,000#					18.55			18.55	20.50
2000	Stainless steel, type 316/316L, 1/8" diam., less than 500#					7.10			7.10	7.85
2100	500# to 1000#					6.40			6.40	7.05
2220	Over 1000#					6.05			6.05	6.65

For customer support on your Plumbing Cost Data, call 877.709.6968.

45

05 12 Structural Steel Framing

05 12 23 – Structural Steel for Buildings

05 12 23.40 Lightweight Framing

05 12 23.40 Lightweight Framing		Crew	Daily Output	Labor-Hours	Unit	Material	2015 Bare Costs Labor	Equipment	Total	Total Incl O&P
0010	**LIGHTWEIGHT FRAMING**									
0015	Made from recycled materials									
0400	Angle framing, field fabricated, 4" and larger	G	E-3	440	.055	Lb.	.77	2.91	.33	4.01	6.2
0450	Less than 4" angles	G		265	.091		.80	4.83	.55	6.18	9.9
0600	Channel framing, field fabricated, 8" and larger	G		500	.048		.80	2.56	.29	3.65	5.6
0650	Less than 8" channels	G		335	.072		.80	3.82	.44	5.06	8
1000	Continuous slotted channel framing system, shop fab, simple framing	G	2 Sswk	2400	.007		4.11	.35		4.46	5.1
1200	Complex framing	G	"	1600	.010		4.64	.53		5.17	6
1250	Plate & bar stock for reinforcing beams and trusses	G					1.46			1.46	1.6
1300	Cross bracing, rods, shop fabricated, 3/4" diameter	G	E-3	700	.034		1.59	1.83	.21	3.63	5.1
1310	7/8" diameter	G		850	.028		1.59	1.51	.17	3.27	4.5
1320	1" diameter	G		1000	.024		1.59	1.28	.15	3.02	4.1
1330	Angle, 5" x 5" x 3/8"	G		2800	.009		1.59	.46	.05	2.10	2.6
1350	Hanging lintels, shop fabricated	G		850	.028		1.59	1.51	.17	3.27	4.5
1380	Roof frames, shop fabricated, 3'-0" square, 5' span	G	E-2	4200	.013		1.59	.69	.36	2.64	3.3
1400	Tie rod, not upset, 1-1/2" to 4" diameter, with turnbuckle	G	2 Sswk	800	.020		1.72	1.05		2.77	3.72
1420	No turnbuckle	G		700	.023		1.66	1.20		2.86	3.9
1500	Upset, 1-3/4" to 4" diameter, with turnbuckle	G		800	.020		1.72	1.05		2.77	3.72
1520	No turnbuckle	G		700	.023		1.66	1.20		2.86	3.9

05 12 23.60 Pipe Support Framing

05 12 23.60 Pipe Support Framing		Crew	Daily Output	Labor-Hours	Unit	Material	Labor	Equipment	Total	Total Incl O&P
0010	**PIPE SUPPORT FRAMING**									
0020	Under 10#/L.F., shop fabricated	G	E-4	3900	.008	Lb.	1.78	.44	.04	2.26	2.75
0200	10.1 to 15#/L.F.	G		4300	.007		1.75	.40	.03	2.18	2.65
0400	15.1 to 20#/L.F.	G		4800	.007		1.72	.35	.03	2.10	2.54
0600	Over 20#/L.F.	G		5400	.006		1.70	.32	.03	2.05	2.45

05 54 Metal Floor Plates

05 54 13 – Floor Plates

05 54 13.20 Checkered Plates

05 54 13.20 Checkered Plates		Crew	Daily Output	Labor-Hours	Unit	Material	Labor	Equipment	Total	Total Incl O&P
0010	**CHECKERED PLATES**, steel, field fabricated									
0015	Made from recycled materials									
0020	1/4" & 3/8", 2000 to 5000 S.F., bolted	G	E-4	2900	.011	Lb.	.88	.59	.05	1.52	2.05
0100	Welded	G		4400	.007	"	.84	.39	.03	1.26	1.63
0300	Pit or trench cover and frame, 1/4" plate, 2' to 3' wide	G		100	.320	S.F.	11.35	17	1.46	29.81	43.50
0400	For galvanizing, add	G				Lb.	.29			.29	.32
0500	Platforms, 1/4" plate, no handrails included, rectangular	G	E-4	4200	.008		3.21	.41	.03	3.65	4.27
0600	Circular	G	"	2500	.013		4.01	.68	.06	4.75	5.65

Estimating Tips
06 05 00 Common Work Results for Wood, Plastics, and Composites

- Common to any wood-framed structure are the accessory connector items such as screws, nails, adhesives, hangers, connector plates, straps, angles, and hold-downs. For typical wood-framed buildings, such as residential projects, the aggregate total for these items can be significant, especially in areas where seismic loading is a concern. For floor and wall framing, the material cost is based on 10 to 25 lbs. per MBF. Hold-downs, hangers, and other connectors should be taken off by the piece.

  Included with material costs are fasteners for a normal installation. RSMeans engineers use manufacturer's recommendations, written specifications, and/or standard construction practice for size and spacing of fasteners. Prices for various fasteners are shown for informational purposes only. Adjustments should be made if unusual fastening conditions exist.

06 10 00 Carpentry

- Lumber is a traded commodity and therefore sensitive to supply and demand in the marketplace. Even in "budgetary" estimating of wood-framed projects, it is advisable to call local suppliers for the latest market pricing.

- Common quantity units for wood-framed projects are "thousand board feet" (MBF). A board foot is a volume of wood, 1" x 1' x 1', or 144 cubic inches. Board-foot quantities are generally calculated using nominal material dimensions— dressed sizes are ignored. Board foot per lineal foot of any stick of lumber can be calculated by dividing the nominal cross-sectional area by 12. As an example, 2,000 lineal feet of 2 x 12 equates to 4 MBF by dividing the nominal area, 2 x 12, by 12, which equals 2, and multiplying by 2,000 to give 4,000 board feet. This simple rule applies to all nominal dimensioned lumber.

- Waste is an issue of concern at the quantity takeoff for any area of construction. Framing lumber is sold in even foot lengths, i.e., 10', 12', 14', 16' and, depending on spans, wall heights, and the grade of lumber, waste is inevitable. A rule of thumb for lumber waste is 5%–10% depending on material quality and the complexity of the framing.

- Wood in various forms and shapes is used in many projects, even where the main structural framing is steel, concrete, or masonry. Plywood as a back-up partition material and 2x boards used as blocking and cant strips around roof edges are two common examples. The estimator should ensure that the costs of all wood materials are included in the final estimate.

06 20 00 Finish Carpentry

- It is necessary to consider the grade of workmanship when estimating labor costs for erecting millwork and interior finish. In practice, there are three grades: premium, custom, and economy. The RSMeans daily output for base and case moldings is in the range of 200 to 250 L.F. per carpenter per day. This is appropriate for most average custom-grade projects. For premium projects, an adjustment to productivity of 25%–50% should be made, depending on the complexity of the job.

Reference Numbers

Reference numbers are shown in shaded boxes at the beginning of some major classifications. These numbers refer to related items in the Reference Section. The reference information may be an estimating procedure, an alternate pricing method, or technical information.

Note: Not all subdivisions listed here necessarily appear in this publication. ■

06 11 Wood Framing

06 11 10 – Framing with Dimensional, Engineered or Composite Lumber

06 11 10.24 Miscellaneous Framing	Crew	Daily Output	Labor-Hours	Unit	Material	2015 Bare Costs Labor	Equipment	Total	Total Incl O&P
0010 **MISCELLANEOUS FRAMING**									
8500 Firestops, 2" x 4"	2 Carp	.51	31.373	M.B.F.	635	1,475		2,110	2,97!
8505 Pneumatic nailed		.62	25.806		640	1,200		1,840	2,57!
8520 2" x 6"		.60	26.667		665	1,250		1,915	2,65(
8525 Pneumatic nailed		.73	21.858		670	1,025		1,695	2,32!
8540 2" x 8"		.60	26.667		690	1,250		1,940	2,67!
8560 2" x 12"		.70	22.857		860	1,075		1,935	2,60(
8600 Nailers, treated, wood construction, 2" x 4"		.53	30.189		765	1,425		2,190	3,025
8605 Pneumatic nailed		.64	25.157		770	1,175		1,945	2,675
8620 2" x 6"		.75	21.333		780	1,000		1,780	2,40(
8625 Pneumatic nailed	▼	.90	17.778	▼	785	835		1,620	2,15(

06 16 Sheathing

06 16 36 – Wood Panel Product Sheathing

06 16 36.10 Sheathing	Crew	Daily Output	Labor-Hours	Unit	Material	2015 Bare Costs Labor	Equipment	Total	Total Incl O&P
0010 **SHEATHING**									
0012 Plywood on roofs, CDX									
0050 3/8" thick	2 Carp	1525	.010	S.F.	.60	.49		1.09	1.42
0055 Pneumatic nailed		1860	.009		.60	.40		1	1.28
0200 5/8" thick		1300	.012		.78	.58		1.36	1.75
0205 Pneumatic nailed		1586	.010		.78	.47		1.25	1.59
0300 3/4" thick		1200	.013		.93	.63		1.56	1.98
0305 Pneumatic nailed		1464	.011		.93	.51		1.44	1.81
0500 Plywood on walls, with exterior CDX, 3/8" thick		1200	.013		.60	.63		1.23	1.62
0505 Pneumatic nailed		1488	.011		.60	.50		1.10	1.44
0700 5/8" thick		1050	.015		.78	.72		1.50	1.96
0705 Pneumatic nailed		1302	.012		.78	.58		1.36	1.75
0800 3/4" thick		975	.016		.93	.77		1.70	2.21
0805 Pneumatic nailed	▼	1209	.013	▼	.93	.62		1.55	1.98

Estimating Tips

07 10 00 Dampproofing and Waterproofing

- Be sure of the job specifications before pricing this subdivision. The difference in cost between waterproofing and dampproofing can be great. Waterproofing will hold back standing water. Dampproofing prevents the transmission of water vapor. Also included in this section are vapor retarding membranes.

07 20 00 Thermal Protection

- Insulation and fireproofing products are measured by area, thickness, volume or R-value. Specifications may give only what the specific R-value should be in a certain situation. The estimator may need to choose the type of insulation to meet that R-value.

07 30 00 Steep Slope Roofing
07 40 00 Roofing and Siding Panels

- Many roofing and siding products are bought and sold by the square. One square is equal to an area that measures 100 square feet.

  This simple change in unit of measure could create a large error if the estimator is not observant. Accessories necessary for a complete installation must be figured into any calculations for both material and labor.

07 50 00 Membrane Roofing
07 60 00 Flashing and Sheet Metal
07 70 00 Roofing and Wall Specialties and Accessories

- The items in these subdivisions compose a roofing system. No one component completes the installation, and all must be estimated. Built-up or single-ply membrane roofing systems are made up of many products and installation trades. Wood blocking at roof perimeters or penetrations, parapet coverings, reglets, roof drains, gutters, downspouts, sheet metal flashing, skylights, smoke vents, and roof hatches all need to be considered along with the roofing material. Several different installation trades will need to work together on the roofing system. Inherent difficulties in the scheduling and coordination of various trades must be accounted for when estimating labor costs.

07 90 00 Joint Protection

- To complete the weather-tight shell, the sealants and caulkings must be estimated. Where different materials meet—at expansion joints, at flashing penetrations, and at hundreds of other locations throughout a construction project—they provide another line of defense against water penetration. Often, an entire system is based on the proper location and placement of caulking or sealants. The detailed drawings that are included as part of a set of architectural plans show typical locations for these materials. When caulking or sealants are shown at typical locations, this means the estimator must include them for all the locations where this detail is applicable. Be careful to keep different types of sealants separate, and remember to consider backer rods and primers if necessary.

Reference Numbers

Reference numbers are shown in shaded boxes at the beginning of some major classifications. These numbers refer to related items in the Reference Section. The reference information may be an estimating procedure, an alternate pricing method, or technical information.

Note: Not all subdivisions listed here necessarily appear in this publication. ■

07 65 10 – Sheet Metal Flashing

07 65 10.10 Sheet Metal Flashing and Counter Flashing	Crew	Daily Output	Labor-Hours	Unit	Material	2015 Bare Costs Labor	Equipment	Total	Total Incl O&P
0010 **SHEET METAL FLASHING AND COUNTER FLASHING**									
0011 Including up to 4 bends									
0020 Aluminum, mill finish, .013" thick	1 Rofc	145	.055	S.F.	.79	2.21		3	4.6
0030 .016" thick		145	.055		.91	2.21		3.12	4.7
0060 .019" thick		145	.055		1.15	2.21		3.36	5.0
0100 .032" thick		145	.055		1.35	2.21		3.56	5.2
0200 .040" thick		145	.055		2.31	2.21		4.52	6.3
0300 .050" thick		145	.055		2.43	2.21		4.64	6.4
0325 Mill finish 5" x 7" step flashing, .016" thick		1920	.004	Ea.	.15	.17		.32	.4
0350 Mill finish 12" x 12" step flashing, .016" thick		1600	.005	"	.50	.20		.70	.8
0400 Painted finish, add				S.F.	.29			.29	.3
1000 Mastic-coated 2 sides, .005" thick	1 Rofc	330	.024		1.82	.97		2.79	3.6
1100 .016" thick		330	.024		2.01	.97		2.98	3.8
1600 Copper, 16 oz., sheets, under 1000 lb.		115	.070		7.65	2.79		10.44	13.2
1700 Over 4000 lb.		155	.052		7.65	2.07		9.72	12
1900 20 oz. sheets, under 1000 lb.		110	.073		9.40	2.92		12.32	15.3
2000 Over 4000 lb.		145	.055		8.95	2.21		11.16	13.6
2200 24 oz. sheets, under 1000 lb.		105	.076		13.30	3.06		16.36	19.8
2300 Over 4000 lb.		135	.059		12.65	2.38		15.03	17.9
2500 32 oz. sheets, under 1000 lb.		100	.080		17.80	3.21		21.01	25
2600 Over 4000 lb.		130	.062		16.90	2.47		19.37	23
5800 Lead, 2.5 lb. per S.F., up to 12" wide		135	.059		4.99	2.38		7.37	9.5
5900 Over 12" wide		135	.059		3.99	2.38		6.37	8.4
8900 Stainless steel sheets, 32 ga.,		155	.052		3	2.07		5.07	6.85
9000 28 ga.		155	.052		3.99	2.07		6.06	7.90
9100 26 ga.		155	.052		4.25	2.07		6.32	8.20
9200 24 ga.		155	.052		4.75	2.07		6.82	8.80
9400 Terne coated stainless steel, .015" thick, 28 ga		155	.052		7.35	2.07		9.42	11.65
9500 .018" thick, 26 ga		155	.052		8.30	2.07		10.37	12.65
9600 Zinc and copper alloy (brass), .020" thick		155	.052		9.30	2.07		11.37	13.80
9700 .027" thick		155	.052		11	2.07		13.07	15.65
9800 .032" thick		155	.052		14	2.07		16.07	18.95
9900 .040" thick		155	.052		18.50	2.07		20.57	24

07 65 13 – Laminated Sheet Flashing

07 65 13.10 Laminated Sheet Flashing

	Crew	Daily Output	Labor-Hours	Unit	Material	Labor	Equipment	Total	Total Incl O&P
0010 **LAMINATED SHEET FLASHING**, Including up to 4 bends									
0500 Aluminum, fabric-backed 2 sides, mill finish, .004" thick	1 Rofc	330	.024	S.F.	1.54	.97		2.51	3.35
0700 .005" thick		330	.024		1.82	.97		2.79	3.66
0750 Mastic-backed, self adhesive		460	.017		3.51	.70		4.21	5.05
0800 Mastic-coated 2 sides, .004" thick		330	.024		1.54	.97		2.51	3.35
2800 Copper, paperbacked 1 side, 2 oz.		330	.024		1.84	.97		2.81	3.68
2900 3 oz.		330	.024		2.15	.97		3.12	4.03
3100 Paperbacked 2 sides, 2 oz.		330	.024		1.97	.97		2.94	3.83
3150 3 oz.		330	.024		2.25	.97		3.22	4.14
3200 5 oz.		330	.024		3.03	.97		4	4.99
3250 7 oz.		330	.024		5.30	.97		6.27	7.50
3400 Mastic-backed 2 sides, copper, 2 oz.		330	.024		1.82	.97		2.79	3.66
3500 3 oz.		330	.024		2.20	.97		3.17	4.08
3700 5 oz.		330	.024		3.30	.97		4.27	5.30
3800 Fabric-backed 2 sides, copper, 2 oz.		330	.024		2.10	.97		3.07	3.97
4000 3 oz.		330	.024		2.40	.97		3.37	4.30
4100 5 oz.		330	.024		3.40	.97		4.37	5.40

For customer support on your Plumbing Cost Data, call 877.709.6968.

07 65 Flexible Flashing

07 65 13 – Laminated Sheet Flashing

07 65 13.10 Laminated Sheet Flashing	Crew	Daily Output	Labor-Hours	Unit	Material	2015 Bare Costs Labor	Equipment	Total	Total Incl O&P
4300	Copper-clad stainless steel, .015" thick, under 500 lb.	1 Rofc	115	.070	S.F.	6.40	2.79		9.19	11.80
4400	Over 2000 lb.		155	.052		6.10	2.07		8.17	10.25
4600	.018" thick, under 500 lb.		100	.080		7.20	3.21		10.41	13.40
4700	Over 2000 lb.		145	.055		6.80	2.21		9.01	11.30
8550	Shower pan, 3 ply copper and fabric, 3 oz.		155	.052		3.50	2.07		5.57	7.40
8600	7 oz.		155	.052		4.20	2.07		6.27	8.15
9300	Stainless steel, paperbacked 2 sides, .005" thick	▼	330	.024	▼	3.62	.97		4.59	5.65

07 65 19 – Plastic Sheet Flashing

07 65 19.10 Plastic Sheet Flashing and Counter Flashing	Crew	Daily Output	Labor-Hours	Unit	Material	2015 Bare Costs Labor	Equipment	Total	Total Incl O&P
0010	PLASTIC SHEET FLASHING AND COUNTER FLASHING									
7300	Polyvinyl chloride, black, 10 mil	1 Rofc	285	.028	S.F.	.21	1.13		1.34	2.15
7400	20 mil		285	.028		.28	1.13		1.41	2.23
7600	30 mil		285	.028		.35	1.13		1.48	2.31
7700	60 mil		285	.028		.85	1.13		1.98	2.86
7900	Black or white for exposed roofs, 60 mil	▼	285	.028	▼	1.80	1.13		2.93	3.90
8060	PVC tape, 5" x 45 mils, for joint covers, 100 L.F./roll				Ea.	170			170	187
8850	Polyvinyl chloride, 30 mil	1 Rofc	160	.050	S.F.	1.30	2.01		3.31	4.85

07 72 Roof Accessories

07 72 53 – Snow Guards

07 72 53.10 Snow Guard Options	Crew	Daily Output	Labor-Hours	Unit	Material	2015 Bare Costs Labor	Equipment	Total	Total Incl O&P
0010	SNOW GUARD OPTIONS									
0100	Slate & asphalt shingle roofs, fastened with nails	1 Rofc	160	.050	Ea.	12.55	2.01		14.56	17.20
0200	Standing seam metal roofs, fastened with set screws		48	.167		16.95	6.70		23.65	30
0300	Surface mount for metal roofs, fastened with solder		48	.167	▼	6.05	6.70		12.75	18.05
0400	Double rail pipe type, including pipe	▼	130	.062	L.F.	32	2.47		34.47	39

07 72 73 – Pitch Pockets

07 72 73.10 Pitch Pockets, Variable Sizes	Crew	Daily Output	Labor-Hours	Unit	Material	2015 Bare Costs Labor	Equipment	Total	Total Incl O&P
0010	PITCH POCKETS, VARIABLE SIZES									
0100	Adjustable, 4" to 7", welded corners, 4" deep	1 Rofc	48	.167	Ea.	16.25	6.70		22.95	29.50
0200	Side extenders, 6"	"	240	.033	"	2.50	1.34		3.84	5.05

07 76 Roof Pavers

07 76 16 – Roof Decking Pavers

07 76 16.10 Roof Pavers and Supports	Crew	Daily Output	Labor-Hours	Unit	Material	2015 Bare Costs Labor	Equipment	Total	Total Incl O&P
0010	ROOF PAVERS AND SUPPORTS									
1000	Roof decking pavers, concrete blocks, 2" thick, natural	1 Clab	115	.070	S.F.	3.37	2.62		5.99	7.75
1100	Colors		115	.070	"	3.73	2.62		6.35	8.10
1200	Support pedestal, bottom cap		960	.008	Ea.	3	.31		3.31	3.78
1300	Top cap		960	.008		4.80	.31		5.11	5.80
1400	Leveling shims, 1/16"		1920	.004		1.20	.16		1.36	1.56
1500	1/8"		1920	.004		1.20	.16		1.36	1.56
1600	Buffer pad		960	.008	▼	2.50	.31		2.81	3.23
1700	PVC legs (4" SDR 35)		2880	.003	Inch	.12	.10		.22	.29
2000	Alternate pricing method, system in place		101	.079	S.F.	7	2.98		9.98	12.30

For customer support on your Plumbing Cost Data, call 877.709.6968.

51

07 84 13 – Penetration Firestopping

07 84 13.10 Firestopping	Crew	Daily Output	Labor-Hours	Unit	Material	2015 Bare Costs Labor	Equipment	Total	Total Incl O&P
0010 **FIRESTOPPING**									
0100 Metallic piping, non insulated									
0110 Through walls, 2" diameter	1 Carp	16	.500	Ea.	17.20	23.50		40.70	55
0120 4" diameter		14	.571		26	27		53	70.50
0130 6" diameter		12	.667		35.50	31.50		67	87
0140 12" diameter		10	.800		62.50	37.50		100	127
0150 Through floors, 2" diameter		32	.250		9.80	11.75		21.55	29
0160 4" diameter		28	.286		14.20	13.40		27.60	36
0170 6" diameter		24	.333		18.50	15.65		34.15	44.50
0180 12" diameter		20	.400		31.50	18.80		50.30	63.50
0190 Metallic piping, insulated									
0200 Through walls, 2" diameter	1 Carp	16	.500	Ea.	24	23.50		47.50	62.50
0210 4" diameter		14	.571		33	27		60	77.50
0220 6" diameter		12	.667		42	31.50		73.50	94
0230 12" diameter		10	.800		68.50	37.50		106	134
0240 Through floors, 2" diameter		32	.250		16.60	11.75		28.35	36.50
0250 4" diameter		28	.286		21	13.40		34.40	43.50
0260 6" diameter		24	.333		25.50	15.65		41.15	52
0270 12" diameter		20	.400		31.50	18.80		50.30	63.50
0280 Non metallic piping, non insulated									
0290 Through walls, 2" diameter	1 Carp	12	.667	Ea.	68.50	31.50		100	124
0300 4" diameter		10	.800		86	37.50		123.50	153
0310 6" diameter		8	1		120	47		167	205
0330 Through floors, 2" diameter		16	.500		54	23.50		77.50	95.50
0340 4" diameter		6	1.333		66.50	62.50		129	170
0350 6" diameter		6	1.333		80.50	62.50		143	185
0370 Ductwork, insulated & non insulated, round									
0380 Through walls, 6" diameter	1 Carp	12	.667	Ea.	35	31.50		66.50	86.50
0390 12" diameter		10	.800		69.50	37.50		107	135
0400 18" diameter		8	1		113	47		160	198
0410 Through floors, 6" diameter		16	.500		18.60	23.50		42.10	56.50
0420 12" diameter		14	.571		34	27		61	79
0430 18" diameter		12	.667		59	31.50		90.50	113
0440 Ductwork, insulated & non insulated, rectangular									
0450 With stiffener/closure angle, through walls, 6" x 12"	1 Carp	8	1	Ea.	28	47		75	104
0460 12" x 24"		6	1.333		37.50	62.50		100	138
0470 24" x 48"		4	2		107	94		201	262
0480 With stiffener/closure angle, through floors, 6" x 12"		10	.800		15.40	37.50		52.90	75
0490 12" x 24"		8	1		28	47		75	103
0500 24" x 48"		6	1.333		54	62.50		116.50	156
0510 Multi trade openings									
0520 Through walls, 6" x 12"	1 Carp	2	4	Ea.	59	188		247	355
0530 12" x 24"	"	1	8		238	375		613	840
0540 24" x 48"	2 Carp	1	16		950	750		1,700	2,200
0550 48" x 96"	"	.75	21.333		3,700	1,000		4,700	5,625
0560 Through floors, 6" x 12"	1 Carp	2	4		39	188		227	330
0570 12" x 24"	"	1	8		157	375		532	755
0580 24" x 48"	2 Carp	.75	21.333		625	1,000		1,625	2,250
0590 48" x 96"	"	.50	32		2,450	1,500		3,950	4,975
0600 Structural penetrations, through walls									
0610 Steel beams, W8 x 10	1 Carp	8	1	Ea.	37.50	47		84.50	114
0620 W12 x 14		6	1.333		59	62.50		121.50	162
0630 W21 x 44		5	1.600		118	75		193	246

For customer support on your Plumbing Cost Data, call 877.709.6968.

07 84 Firestopping

07 84 13 – Penetration Firestopping

07 84 13.10 Firestopping

		Crew	Daily Output	Labor-Hours	Unit	Material	2015 Bare Costs Labor	Equipment	Total	Total Incl O&P
0640	W36 x 135	1 Carp	3	2.667	Ea.	287	125		412	510
0650	Bar joists, 18" deep		6	1.333		54.50	62.50		117	156
0660	24" deep		6	1.333		67.50	62.50		130	171
0670	36" deep		5	1.600		101	75		176	228
0680	48" deep		4	2		118	94		212	275
0690	Construction joints, floor slab at exterior wall									
0700	Precast, brick, block or drywall exterior									
0710	2" wide joint	1 Carp	125	.064	L.F.	8.50	3		11.50	13.95
0720	4" wide joint	"	75	.107	"	17	5		22	26.50
0730	Metal panel, glass or curtain wall exterior									
0740	2" wide joint	1 Carp	40	.200	L.F.	20	9.40		29.40	36.50
0750	4" wide joint	"	25	.320	"	28	15		43	53.50
0760	Floor slab to drywall partition									
0770	Flat joint	1 Carp	100	.080	L.F.	8.40	3.76		12.16	15.05
0780	Fluted joint		50	.160		17.40	7.50		24.90	30.50
0790	Etched fluted joint		75	.107		11.10	5		16.10	19.90
0800	Floor slab to concrete/masonry partition									
0810	Flat joint	1 Carp	75	.107	L.F.	18.75	5		23.75	28
0820	Fluted joint	"	50	.160	"	22.50	7.50		30	36
0830	Concrete/CMU wall joints									
0840	1" wide	1 Carp	100	.080	L.F.	10.25	3.76		14.01	17.10
0850	2" wide		75	.107		18.75	5		23.75	28
0860	4" wide		50	.160		38	7.50		45.50	53
0870	Concrete/CMU floor joints									
0880	1" wide	1 Carp	200	.040	L.F.	5.15	1.88		7.03	8.55
0890	2" wide		150	.053		9.40	2.50		11.90	14.20
0900	4" wide		100	.080		17.90	3.76		21.66	25.50

07 91 Preformed Joint Seals

07 91 13 – Compression Seals

07 91 13.10 Compression Seals

		Crew	Daily Output	Labor-Hours	Unit	Material	2015 Bare Costs Labor	Equipment	Total	Total Incl O&P
0010	**COMPRESSION SEALS**									
4900	O-ring type cord, 1/4"	1 Bric	472	.017	L.F.	.37	.78		1.15	1.60
4910	1/2"		440	.018		.95	.84		1.79	2.33
4920	3/4"		424	.019		1.80	.87		2.67	3.31
4930	1"		408	.020		3.40	.91		4.31	5.10
4940	1-1/4"		384	.021		6.30	.96		7.26	8.40
4950	1-1/2"		368	.022		7.95	1		8.95	10.30
4960	1-3/4"		352	.023		13.25	1.05		14.30	16.20
4970	2"		344	.023		18.50	1.07		19.57	22

07 91 16 – Joint Gaskets

07 91 16.10 Joint Gaskets

		Crew	Daily Output	Labor-Hours	Unit	Material	2015 Bare Costs Labor	Equipment	Total	Total Incl O&P
0010	**JOINT GASKETS**									
4400	Joint gaskets, neoprene, closed cell w/adh, 1/8" x 3/8"	1 Bric	240	.033	L.F.	.28	1.54		1.82	2.66
4500	1/4" x 3/4"		215	.037		.59	1.72		2.31	3.27
4700	1/2" x 1"		200	.040		1.30	1.85		3.15	4.25
4800	3/4" x 1-1/2"		165	.048		1.45	2.24		3.69	5

07 91 Preformed Joint Seals

07 91 23 – Backer Rods

07 91 23.10 Backer Rods	Crew	Daily Output	Labor-Hours	Unit	Material	2015 Bare Costs Labor	Equipment	Total	Total Incl O&P
0010 **BACKER RODS**									
0030 Backer rod, polyethylene, 1/4" diameter	1 Bric	4.60	1.739	C.L.F.	2.13	80.50		82.63	125
0050 1/2" diameter		4.60	1.739		3.40	80.50		83.90	127
0070 3/4" diameter		4.60	1.739		5.55	80.50		86.05	129
0090 1" diameter		4.60	1.739		8.85	80.50		89.35	133

07 91 26 – Joint Fillers

07 91 26.10 Joint Fillers

	Crew	Daily Output	Labor-Hours	Unit	Material	2015 Bare Costs Labor	Equipment	Total	Total Incl O&P
0010 **JOINT FILLERS**									
4360 Butyl rubber filler, 1/4" x 1/4"	1 Bric	290	.028	L.F.	.22	1.27		1.49	2.19
4365 1/2" x 1/2"		250	.032		.89	1.48		2.37	3.23
4370 1/2" x 3/4"		210	.038		1.34	1.76		3.10	4.15
4375 3/4" x 3/4"		230	.035		2.01	1.61		3.62	4.66
4380 1" x 1"		180	.044		2.68	2.05		4.73	6.05
4390 For coloring, add					12%				
4980 Polyethylene joint backing, 1/4" x 2"	1 Bric	2.08	3.846	C.L.F.	12	178		190	284
4990 1/4" x 6"		1.28	6.250	"	28	288		316	470
5600 Silicone, room temp vulcanizing foam seal, 1/4" x 1/2"		1312	.006	L.F.	.33	.28		.61	.80
5610 1/2" x 1/2"		656	.012		.67	.56		1.23	1.59
5620 1/2" x 3/4"		442	.018		1	.84		1.84	2.38
5630 3/4" x 3/4"		328	.024		1.50	1.13		2.63	3.37
5640 1/8" x 1"		1312	.006		.33	.28		.61	.80
5650 1/8" x 3"		442	.018		1	.84		1.84	2.38
5670 1/4" x 3"		295	.027		2	1.25		3.25	4.11
5680 1/4" x 6"		148	.054		3.99	2.49		6.48	8.20
5690 1/2" x 6"		82	.098		8	4.50		12.50	15.65
5700 1/2" x 9"		52.50	.152		12	7.05		19.05	24
5710 1/2" x 12"		33	.242		15.95	11.20		27.15	34.50

07 92 Joint Sealants

07 92 13 – Elastomeric Joint Sealants

07 92 13.20 Caulking and Sealant Options

	Crew	Daily Output	Labor-Hours	Unit	Material	2015 Bare Costs Labor	Equipment	Total	Total Incl O&P
0010 **CAULKING AND SEALANT OPTIONS**									
0050 Latex acrylic based, bulk				Gal.	26.50			26.50	29.50
0055 Bulk in place 1/4" x 1/4" bead	1 Bric	300	.027	L.F.	.08	1.23		1.31	1.97
0060 1/4" x 3/8"		294	.027		.14	1.26		1.40	2.07
0065 1/4" x 1/2"		288	.028		.19	1.28		1.47	2.17
0075 3/8" x 3/8"		284	.028		.21	1.30		1.51	2.21
0080 3/8" x 1/2"		280	.029		.28	1.32		1.60	2.32
0085 3/8" x 5/8"		276	.029		.35	1.34		1.69	2.43
0095 3/8" x 3/4"		272	.029		.42	1.36		1.78	2.53
0100 1/2" x 1/2"		275	.029		.37	1.34		1.71	2.46
0105 1/2" x 5/8"		269	.030		.47	1.37		1.84	2.61
0110 1/2" x 3/4"		263	.030		.56	1.40		1.96	2.76
0115 1/2" x 7/8"		256	.031		.65	1.44		2.09	2.92
0120 1/2" x 1"		250	.032		.75	1.48		2.23	3.07
0125 3/4" x 3/4"		244	.033		.84	1.51		2.35	3.24
0130 3/4" x 1"		225	.036		1.12	1.64		2.76	3.74
0135 1" x 1"		200	.040		1.50	1.85		3.35	4.47
0190 Cartridges				Gal.	32			32	35.50
0200 11 fl. oz. cartridge				Ea.	2.76			2.76	3.04

07 92 Joint Sealants

07 92 13 – Elastomeric Joint Sealants

07 92 13.20 Caulking and Sealant Options		Crew	Daily Output	Labor-Hours	Unit	Material	2015 Bare Costs Labor	Equipment	Total	Total Incl O&P
0500	1/4" x 1/2"	1 Bric	288	.028	L.F.	.23	1.28		1.51	2.21
0600	1/2" x 1/2"		275	.029		.45	1.34		1.79	2.55
0800	3/4" x 3/4"		244	.033		1.01	1.51		2.52	3.43
0900	3/4" x 1"		225	.036		1.35	1.64		2.99	4
1000	1" x 1"		200	.040		1.69	1.85		3.54	4.68
1400	Butyl based, bulk				Gal.	35			35	38.50
1500	Cartridges				"	38.50			38.50	42
1700	1/4" x 1/2", 154 L.F./gal.	1 Bric	288	.028	L.F.	.23	1.28		1.51	2.21
1800	1/2" x 1/2", 77 L.F./gal.	"	275	.029	"	.45	1.34		1.79	2.55
2300	Polysulfide compounds, 1 component, bulk				Gal.	72			72	79
2600	1 or 2 component, in place, 1/4" x 1/4", 308 L.F./gal.	1 Bric	300	.027	L.F.	.23	1.23		1.46	2.14
2700	1/2" x 1/4", 154 L.F./gal.		288	.028		.47	1.28		1.75	2.47
2900	3/4" x 3/8", 68 L.F./gal.		272	.029		1.06	1.36		2.42	3.23
3000	1" x 1/2", 38 L.F./gal.		250	.032		1.89	1.48		3.37	4.33
3200	Polyurethane, 1 or 2 component				Gal.	49			49	54
3300	Cartridges				"	64			64	70.50
3500	Bulk, in place, 1/4" x 1/4"	1 Bric	300	.027	L.F.	.16	1.23		1.39	2.06
3655	1/2" x 1/4"		288	.028		.32	1.28		1.60	2.31
3800	3/4" x 3/8"		272	.029		.72	1.36		2.08	2.87
3900	1" x 1/2"		250	.032		1.28	1.48		2.76	3.66
4100	Silicone rubber, bulk				Gal.	49			49	54
4200	Cartridges				"	51			51	56

07 92 16 – Rigid Joint Sealants

07 92 16.10 Rigid Joint Sealants		Crew	Daily Output	Labor-Hours	Unit	Material	Labor	Equipment	Total	Total Incl O&P
0010	**RIGID JOINT SEALANTS**									
5800	Tapes, sealant, PVC foam adhesive, 1/16" x 1/4"				C.L.F.	10			10	11
5900	1/16" x 1/2"					10			10	11
5950	1/16" x 1"					16			16	17.60
6000	1/8" x 1/2"					9.50			9.50	10.45

07 92 19 – Acoustical Joint Sealants

07 92 19.10 Acoustical Sealant		Crew	Daily Output	Labor-Hours	Unit	Material	Labor	Equipment	Total	Total Incl O&P
0010	**ACOUSTICAL SEALANT**									
0020	Acoustical sealant, elastomeric, cartridges				Ea.	8.50			8.50	9.35
0025	In place, 1/4" x 1/4"	1 Bric	300	.027	L.F.	.35	1.23		1.58	2.26
0030	1/4" x 1/2"		288	.028		.69	1.28		1.97	2.72
0035	1/2" x 1/2"		275	.029		1.39	1.34		2.73	3.57
0040	1/2" x 3/4"		263	.030		2.08	1.40		3.48	4.43
0045	3/4" x 3/4"		244	.033		3.12	1.51		4.63	5.75
0050	1" x 1"		200	.040		5.55	1.85		7.40	8.90

For customer support on your Plumbing Cost Data, call 877.709.6968.

55

Division Notes

	CREW	DAILY OUTPUT	LABOR-HOURS	UNIT	BARE COSTS				TOTAL INCL O&P
					MAT.	LABOR	EQUIP.	TOTAL	

Estimating Tips
General
- Room Finish Schedule: A complete set of plans should contain a room finish schedule. If one is not available, it would be well worth the time and effort to obtain one.

09 20 00 Plaster and Gypsum Board
- Lath is estimated by the square yard plus a 5% allowance for waste. Furring, channels, and accessories are measured by the linear foot. An extra foot should be allowed for each accessory miter or stop.

- Plaster is also estimated by the square yard. Deductions for openings vary by preference, from zero deduction to 50% of all openings over 2 feet in width. The estimator should allow one extra square foot for each linear foot of horizontal interior or exterior angle located below the ceiling level. Also, double the areas of small radius work.

- Drywall accessories, studs, track, and acoustical caulking are all measured by the linear foot. Drywall taping is figured by the square foot. Gypsum wallboard is estimated by the square foot. No material deductions should be made for door or window openings under 32 S.F.

09 60 00 Flooring
- Tile and terrazzo areas are taken off on a square foot basis. Trim and base materials are measured by the linear foot. Accent tiles are listed per each. Two basic methods of installation are used. Mud set is approximately 30% more expensive than thin set. In terrazzo work, be sure to include the linear footage of embedded decorative strips, grounds, machine rubbing, and power cleanup.

- Wood flooring is available in strip, parquet, or block configuration. The latter two types are set in adhesives with quantities estimated by the square foot. The laying pattern will influence labor costs and material waste. In addition to the material and labor for laying wood floors, the estimator must make allowances for sanding and finishing these areas, unless the flooring is prefinished.

- Sheet flooring is measured by the square yard. Roll widths vary, so consideration should be given to use the most economical width, as waste must be figured into the total quantity. Consider also the installation methods available, direct glue down or stretched.

09 70 00 Wall Finishes
- Wall coverings are estimated by the square foot. The area to be covered is measured, length by height of wall above baseboards, to calculate the square footage of each wall. This figure is divided by the number of square feet in the single roll which is being used. Deduct, in full, the areas of openings such as doors and windows. Where a pattern match is required allow 25%–30% waste.

09 80 00 Acoustic Treatment
- Acoustical systems fall into several categories. The takeoff of these materials should be by the square foot of area with a 5% allowance for waste. Do not forget about scaffolding, if applicable, when estimating these systems.

09 90 00 Painting and Coating
- A major portion of the work in painting involves surface preparation. Be sure to include cleaning, sanding, filling, and masking costs in the estimate.

- Protection of adjacent surfaces is not included in painting costs. When considering the method of paint application, an important factor is the amount of protection and masking required. These must be estimated separately and may be the determining factor in choosing the method of application.

Reference Numbers
Reference numbers are shown in shaded boxes at the beginning of some major classifications. These numbers refer to related items in the Reference Section. The reference information may be an estimating procedure, an alternate pricing method, or technical information.

Note: Not all subdivisions listed here necessarily appear in this publication. ■

09 22 Supports for Plaster and Gypsum Board

09 22 03 – Fastening Methods for Finishes

09 22 03.20 Drilling Plaster/Drywall		Crew	Daily Output	Labor-Hours	Unit	Material	2015 Bare Costs Labor	Equipment	Total	Total Incl O&P
0010	**DRILLING PLASTER / DRYWALL**									
1100	Drilling & layout for drywall/plaster walls, up to 1" deep, no anchor									
1200	Holes, 1/4" diameter	1 Carp	150	.053	Ea.	.01	2.50		2.51	3.8
1300	3/8" diameter		140	.057		.01	2.68		2.69	4.1
1400	1/2" diameter		130	.062		.01	2.89		2.90	4.4
1500	3/4" diameter		120	.067		.01	3.13		3.14	4.8
1600	1" diameter		110	.073		.02	3.41		3.43	5.2
1700	1-1/4" diameter		100	.080		.03	3.76		3.79	5.8
1800	1-1/2" diameter		90	.089		.05	4.17		4.22	6.4
1900	For ceiling installations, add						40%			

09 91 Painting

09 91 23 – Interior Painting

09 91 23.52 Miscellaneous, Interior		Crew	Daily Output	Labor-Hours	Unit	Material	2015 Bare Costs Labor	Equipment	Total	Total Incl O&P
0010	**MISCELLANEOUS, INTERIOR**									
3800	Grilles, per side, oil base, primer coat, brushwork	1 Pord	520	.015	S.F.	.14	.62		.76	1.0
3850	Spray		1140	.007		.15	.28		.43	.5
3880	Paint 1 coat, brushwork		520	.015		.22	.62		.84	1.1
3900	Spray		1140	.007		.24	.28		.52	.70
3920	Paint 2 coats, brushwork		325	.025		.42	.99		1.41	1.97
3940	Spray		650	.012		.48	.50		.98	1.28
3950	Prime & paint 1 coat		325	.025		.36	.99		1.35	1.8
3960	Prime & paint 2 coats		270	.030		.35	1.20		1.55	2.19
4500	Louvers, one side, primer, brushwork		524	.015		.09	.62		.71	1.03
4520	Paint one coat, brushwork		520	.015		.10	.62		.72	1.04
4530	Spray		1140	.007		.11	.28		.39	.5
4540	Paint two coats, brushwork		325	.025		.18	.99		1.17	1.70
4550	Spray		650	.012		.20	.50		.70	.97
4560	Paint three coats, brushwork		270	.030		.27	1.20		1.47	2.10
4570	Spray		500	.016		.30	.65		.95	1.30
5000	Pipe, 1" - 4" diameter, primer or sealer coat, oil base, brushwork	2 Pord	1250	.013	L.F.	.10	.52		.62	.89
5100	Spray		2165	.007		.09	.30		.39	.55
5200	Paint 1 coat, brushwork		1250	.013		.11	.52		.63	.90
5300	Spray		2165	.007		.10	.30		.40	.56
5350	Paint 2 coats, brushwork		775	.021		.19	.83		1.02	1.47
5400	Spray		1240	.013		.22	.52		.74	1.02
5420	Paint 3 coats, brushwork		775	.021		.29	.83		1.12	1.58
5450	5" - 8" diameter, primer or sealer coat, brushwork		620	.026		.20	1.04		1.24	1.79
5500	Spray		1085	.015		.32	.60		.92	1.25
5550	Paint 1 coat, brushwork		620	.026		.29	1.04		1.33	1.89
5600	Spray		1085	.015		.33	.60		.93	1.26
5650	Paint 2 coats, brushwork		385	.042		.39	1.68		2.07	2.96
5700	Spray		620	.026		.43	1.04		1.47	2.04
5720	Paint 3 coats, brushwork		385	.042		.58	1.68		2.26	3.16
5750	9" - 12" diameter, primer or sealer coat, brushwork		415	.039		.29	1.56		1.85	2.66
5800	Spray		725	.022		.35	.89		1.24	1.73
5850	Paint 1 coat, brushwork		415	.039		.30	1.56		1.86	2.67
6000	Spray		725	.022		.33	.89		1.22	1.70
6200	Paint 2 coats, brushwork		260	.062		.58	2.48		3.06	4.38
6250	Spray		415	.039		.64	1.56		2.20	3.05
6270	Paint 3 coats, brushwork		260	.062		.86	2.48		3.34	4.69

09 91 23 - Interior Painting

09 91 23.52 Miscellaneous, Interior	Crew	Daily Output	Labor-Hours	Unit	Material	2015 Bare Costs Labor	Equipment	Total	Total Incl O&P
6300	13" - 16" diameter, primer or sealer coat, brushwork	2 Pord	310	.052	L.F.	.39	2.08		2.47	3.57
6350	Spray		540	.030		.44	1.20		1.64	2.28
6400	Paint 1 coat, brushwork		310	.052		.40	2.08		2.48	3.58
6450	Spray		540	.030		.44	1.20		1.64	2.29
6500	Paint 2 coats, brushwork		195	.082		.78	3.31		4.09	5.85
6550	Spray		310	.052		.86	2.08		2.94	4.09
6600	Radiators, per side, primer, brushwork	1 Pord	520	.015	S.F.	.09	.62		.71	1.04
6620	Paint, one coat		520	.015		.09	.62		.71	1.04
6640	Two coats		340	.024		.18	.95		1.13	1.63
6660	Three coats		283	.028		.27	1.14		1.41	2.02

Division Notes

	CREW	DAILY OUTPUT	LABOR-HOURS	UNIT	BARE COSTS				TOTAL INCL O&P
					MAT.	LABOR	EQUIP.	TOTAL	

Estimating Tips
General

- The items in this division are usually priced per square foot or each.

- Many items in Division 10 require some type of support system or special anchors that are not usually furnished with the item. The required anchors must be added to the estimate in the appropriate division.

- Some items in Division 10, such as lockers, may require assembly before installation. Verify the amount of assembly required. Assembly can often exceed installation time.

10 20 00 Interior Specialties

- Support angles and blocking are not included in the installation of toilet compartments, shower/dressing compartments, or cubicles. Appropriate line items from Divisions 5 or 6 may need to be added to support the installations.

- Toilet partitions are priced by the stall. A stall consists of a side wall, pilaster, and door with hardware. Toilet tissue holders and grab bars are extra.

- The required acoustical rating of a folding partition can have a significant impact on costs. Verify the sound transmission coefficient rating of the panel priced to the specification requirements.

- Grab bar installation does not include supplemental blocking or backing to support the required load. When grab bars are installed at an existing facility, provisions must be made to attach the grab bars to solid structure.

Reference Numbers

Reference numbers are shown in shaded boxes at the beginning of some major classifications. These numbers refer to related items in the Reference Section. The reference information may be an estimating procedure, an alternate pricing method, or technical information.

Note: Not all subdivisions listed here necessarily appear in this publication. ■

10 21 13 – Toilet Compartments

10 21 13.13 Metal Toilet Compartments

		Crew	Daily Output	Labor-Hours	Unit	Material	2015 Bare Costs Labor	2015 Bare Costs Equipment	Total	Total Incl O&P
0010	**METAL TOILET COMPARTMENTS**									
0110	Cubicles, ceiling hung									
0200	Powder coated steel	2 Carp	4	4	Ea.	525	188		713	865
0500	Stainless steel	"	4	4		1,075	188		1,263	1,475
0600	For handicap units, incl. 52" grab bars, add				↓	450			450	495
0900	Floor and ceiling anchored									
1000	Powder coated steel	2 Carp	5	3.200	Ea.	590	150		740	875
1300	Stainless steel	"	5	3.200		1,275	150		1,425	1,625
1400	For handicap units, incl. 52" grab bars, add				↓	315			315	345
1610	Floor anchored									
1700	Powder coated steel	2 Carp	7	2.286	Ea.	590	107		697	815
2000	Stainless steel	"	7	2.286		1,400	107		1,507	1,725
2100	For handicap units, incl. 52" grab bars, add					310			310	345
2200	For juvenile units, deduct				↓	41.50			41.50	45.50
2450	Floor anchored, headrail braced									
2500	Powder coated steel	2 Carp	6	2.667	Ea.	390	125		515	625
2804	Stainless steel	"	4.60	3.478		1,025	163		1,188	1,375
2900	For handicap units, incl. 52" grab bars, add					370			370	410
3000	Wall hung partitions, powder coated steel	2 Carp	7	2.286		635	107		742	865
3300	Stainless steel	"	7	2.286		1,650	107		1,757	2,000
3400	For handicap units, incl. 52" grab bars, add				↓	370			370	410
4000	Screens, entrance, floor mounted, 58" high, 48" wide									
4200	Powder coated steel	2 Carp	15	1.067	Ea.	242	50		292	345
4500	Stainless steel	"	15	1.067	"	910	50		960	1,075
4650	Urinal screen, 18" wide									
4704	Powder coated steel	2 Carp	6.15	2.602	Ea.	211	122		333	420
5004	Stainless steel	"	6.15	2.602	"	580	122		702	830
5100	Floor mounted, head rail braced									
5300	Powder coated steel	2 Carp	8	2	Ea.	230	94		324	400
5600	Stainless steel	"	8	2	"	570	94		664	770
5750	Pilaster, flush									
5800	Powder coated steel	2 Carp	10	1.600	Ea.	278	75		353	420
6100	Stainless steel		10	1.600		625	75		700	805
6300	Post braced, powder coated steel		10	1.600		163	75		238	295
6600	Stainless steel		10	1.600		450	75		525	610
6700	Wall hung, bracket supported									
6800	Powder coated steel	2 Carp	10	1.600	Ea.	163	75		238	295
7100	Stainless steel		10	1.600		278	75		353	420
7400	Flange supported, powder coated steel		10	1.600		106	75		181	232
7700	Stainless steel		10	1.600		310	75		385	455
7800	Wedge type, powder coated steel		10	1.600		134	75		209	264
8100	Stainless steel		10	1.600		575	75		650	750

10 21 13.16 Plastic-Laminate-Clad Toilet Compartments

		Crew	Daily Output	Labor-Hours	Unit	Material	2015 Bare Costs Labor	2015 Bare Costs Equipment	Total	Total Incl O&P
0010	**PLASTIC-LAMINATE-CLAD TOILET COMPARTMENTS**									
0110	Cubicles, ceiling hung									
0300	Plastic laminate on particle board	2 Carp	4	4	Ea.	520	188		708	860
0600	For handicap units, incl. 52" grab bars, add				"	450			450	495
0900	Floor and ceiling anchored									
1100	Plastic laminate on particle board	2 Carp	5	3.200	Ea.	790	150		940	1,100
1400	For handicap units, incl. 52" grab bars, add				"	315			315	345
1610	Floor mounted									
1800	Plastic laminate on particle board	2 Carp	7	2.286	Ea.	535	107		642	750

10 21 Compartments and Cubicles

10 21 13 – Toilet Compartments

10 21 13.16 Plastic-Laminate-Clad Toilet Compartments

		Crew	Daily Output	Labor-Hours	Unit	Material	2015 Bare Costs Labor	Equipment	Total	Total Incl O&P
2450	Floor mounted, headrail braced									
2600	Plastic laminate on particle board	2 Carp	6	2.667	Ea.	750	125		875	1,025
3400	For handicap units, incl. 52" grab bars, add					370			370	410
4300	Entrance screen, floor mtd., plas. lam., 58" high, 48" wide	2 Carp	15	1.067		610	50		660	745
4800	Urinal screen, 18" wide, ceiling braced, plastic laminate		8	2		194	94		288	360
5400	Floor mounted, headrail braced		8	2		200	94		294	365
5900	Pilaster, flush, plastic laminate		10	1.600		505	75		580	675
6400	Post braced, plastic laminate		10	1.600		305	75		380	450
6700	Wall hung, bracket supported									
6900	Plastic laminate on particle board	2 Carp	10	1.600	Ea.	94.50	75		169.50	220
7450	Flange supported									
7500	Plastic laminate on particle board	2 Carp	10	1.600	Ea.	230	75		305	370

10 21 13.19 Plastic Toilet Compartments

		Crew	Daily Output	Labor-Hours	Unit	Material	2015 Bare Costs Labor	Equipment	Total	Total Incl O&P
0010	**PLASTIC TOILET COMPARTMENTS**									
0110	Cubicles, ceiling hung									
0250	Phenolic	2 Carp	4	4	Ea.	865	188		1,053	1,250
0600	For handicap units, incl. 52" grab bars, add				"	450			450	495
0900	Floor and ceiling anchored									
1050	Phenolic	2 Carp	5	3.200	Ea.	810	150		960	1,125
1400	For handicap units, incl. 52" grab bars, add				"	315			315	345
1610	Floor mounted									
1750	Phenolic	2 Carp	7	2.286	Ea.	750	107		857	990
2100	For handicap units, incl. 52" grab bars, add					310			310	345
2200	For juvenile units, deduct					41.50			41.50	45.50
2450	Floor mounted, headrail braced									
2550	Phenolic	2 Carp	6	2.667	Ea.	750	125		875	1,025

10 21 13.40 Stone Toilet Compartments

		Crew	Daily Output	Labor-Hours	Unit	Material	2015 Bare Costs Labor	Equipment	Total	Total Incl O&P
0010	**STONE TOILET COMPARTMENTS**									
0100	Cubicles, ceiling hung, marble	2 Marb	2	8	Ea.	1,800	350		2,150	2,500
0600	For handicap units, incl. 52" grab bars, add					450			450	495
0800	Floor & ceiling anchored, marble	2 Marb	2.50	6.400		1,975	278		2,253	2,600
1400	For handicap units, incl. 52" grab bars, add					315			315	345
1600	Floor mounted, marble	2 Marb	3	5.333		1,225	232		1,457	1,700
2400	Floor mounted, headrail braced, marble	"	3	5.333		1,150	232		1,382	1,625
2900	For handicap units, incl. 52" grab bars, add					370			370	410
4100	Entrance screen, floor mounted marble, 58" high, 48" wide	2 Marb	9	1.778		795	77.50		872.50	990
4600	Urinal screen, 18" wide, ceiling braced, marble	D-1	6	2.667		755	112		867	1,000
5100	Floor mounted, head rail braced									
5200	Marble	D-1	6	2.667	Ea.	645	112		757	880
5700	Pilaster, flush, marble		9	1.778		840	75		915	1,050
6200	Post braced, marble		9	1.778		825	75		900	1,025

For customer support on your Plumbing Cost Data, call 877.709.6968.

63

10 28 Toilet, Bath, and Laundry Accessories

10 28 13 – Toilet Accessories

10 28 13.13 Commercial Toilet Accessories	Crew	Daily Output	Labor-Hours	Unit	Material	2015 Bare Costs Labor	Equipment	Total	Total Incl O&P
0010 COMMERCIAL TOILET ACCESSORIES									
0200 Curtain rod, stainless steel, 5' long, 1" diameter	1 Carp	13	.615	Ea.	28.50	29		57.50	75.50
0300 1-1/4" diameter	"	13	.615	"	29	29		58	76
0500 Dispenser units, combined soap & towel dispensers,									
0510 mirror and shelf, flush mounted	1 Carp	10	.800	Ea.	310	37.50		347.50	400
0600 Towel dispenser and waste receptacle,									
0610 18 gallon capacity	1 Carp	10	.800	Ea.	300	37.50		337.50	390
0800 Grab bar, straight, 1-1/4" diameter, stainless steel, 18" long		24	.333		29	15.65		44.65	56
0900 24" long		23	.348		29	16.35		45.35	56.50
1000 30" long		22	.364		31.50	17.05		48.55	61
1100 36" long		20	.400		38.50	18.80		57.30	71.50
1105 42" long		20	.400		46	18.80		64.80	79.50
1120 Corner, 36" long		20	.400		85.50	18.80		104.30	123
1200 1-1/2" diameter, 24" long		23	.348		31	16.35		47.35	59
1300 36" long		20	.400		33.50	18.80		52.30	65.50
1310 42" long		18	.444		38	21		59	73.50
1500 Tub bar, 1-1/4" diameter, 24" x 36"		14	.571		92.50	27		119.50	144
1600 Plus vertical arm		12	.667		97.50	31.50		129	155
1900 End tub bar, 1" diameter, 90° angle, 16" x 32"		12	.667		109	31.50		140.50	168
2300 Hand dryer, surface mounted, electric, 115 volt, 20 amp		4	2		445	94		539	630
2400 230 volt, 10 amp		4	2		745	94		839	965
2450 Hand dryer, touch free, 1400 watt, 81,000 rpm		4	2		1,025	94		1,119	1,275
2600 Hat and coat strip, stainless steel, 4 hook, 36" long		24	.333		68	15.65		83.65	99
2700 6 hook, 60" long		20	.400		124	18.80		142.80	166
3000 Mirror, with stainless steel 3/4" square frame, 18" x 24"		20	.400		49	18.80		67.80	83
3100 36" x 24"		15	.533		109	25		134	159
3200 48" x 24"		10	.800		150	37.50		187.50	223
3300 72" x 24"		6	1.333		289	62.50		351.50	415
3500 With 5" stainless steel shelf, 18" x 24"		20	.400		194	18.80		212.80	242
3600 36" x 24"		15	.533		236	25		261	298
3700 48" x 24"		10	.800		245	37.50		282.50	325
3800 72" x 24"		6	1.333		300	62.50		362.50	425
4100 Mop holder strip, stainless steel, 5 holders, 48" long		20	.400		89	18.80		107.80	127
4200 Napkin/tampon dispenser, recessed		15	.533		590	25		615	690
4220 Semi-recessed		6.50	1.231		310	58		368	430
4250 Napkin receptacle, recessed		6.50	1.231		165	58		223	271
4300 Robe hook, single, regular		36	.222		18.95	10.45		29.40	37
4400 Heavy duty, concealed mounting		36	.222		19.60	10.45		30.05	37.50
4600 Soap dispenser, chrome, surface mounted, liquid		20	.400		46.50	18.80		65.30	80
4700 Powder		20	.400		56.50	18.80		75.30	91.50
5000 Recessed stainless steel, liquid		10	.800		154	37.50		191.50	227
5600 Shelf, stainless steel, 5" wide, 18 ga., 24" long		24	.333		80.50	15.65		96.15	113
5700 48" long		16	.500		157	23.50		180.50	208
5800 8" wide shelf, 18 ga., 24" long		22	.364		69	17.05		86.05	103
5900 48" long		14	.571		121	27		148	175
6000 Toilet seat cover dispenser, stainless steel, recessed		20	.400		167	18.80		185.80	212
6050 Surface mounted		15	.533		34.50	25		59.50	76.50
6100 Toilet tissue dispenser, surface mounted, SS, single roll		30	.267		17.80	12.50		30.30	39
6200 Double roll		24	.333		23.50	15.65		39.15	50
6240 Plastic, twin/jumbo dbl. roll		24	.333		28.50	15.65		44.15	55
6400 Towel bar, stainless steel, 18" long		23	.348		41.50	16.35		57.85	71
6500 30" long		21	.381		111	17.90		128.90	151
6700 Towel dispenser, stainless steel, surface mounted		16	.500		44.50	23.50		68	85

64

For customer support on your Plumbing Cost Data, call 877.709.6968.

10 28 Toilet, Bath, and Laundry Accessories

10 28 13 – Toilet Accessories

10 28 13.13 Commercial Toilet Accessories		Crew	Daily Output	Labor-Hours	Unit	Material	2015 Bare Costs Labor	Equipment	Total	Total Incl O&P
6800	Flush mounted, recessed	1 Carp	10	.800	Ea.	257	37.50		294.50	340
6900	Plastic, touchless, battery operated		16	.500		88	23.50		111.50	133
7000	Towel holder, hotel type, 2 guest size		20	.400		52	18.80		70.80	86.50
7200	Towel shelf, stainless steel, 24" long, 8" wide		20	.400		60	18.80		78.80	95
7400	Tumbler holder, for tumbler only		30	.267		17.80	12.50		30.30	39
7410	Tumbler holder, recessed		20	.400		9.80	18.80		28.60	40
7500	Soap, tumbler & toothbrush		30	.267		19.60	12.50		32.10	41
7510	Tumbler & toothbrush holder		20	.400		13.60	18.80		32.40	44
7700	Wall urn ash receiver, surface mount, 11" long		12	.667		95	31.50		126.50	152
7800	7-1/2", long		18	.444		102	21		123	144
8000	Waste receptacles, stainless steel, with top, 13 gallon		10	.800		295	37.50		332.50	385
8100	36 gallon		8	1		405	47		452	520

10 28 16 – Bath Accessories

10 28 16.20 Medicine Cabinets

		Crew	Daily Output	Labor-Hours	Unit	Material	2015 Bare Costs Labor	Equipment	Total	Total Incl O&P
0010	**MEDICINE CABINETS**									
0020	With mirror, sst frame, 16" x 22", unlighted	1 Carp	14	.571	Ea.	98	27		125	150
0100	Wood frame		14	.571		128	27		155	183
0300	Sliding mirror doors, 20" x 16" x 4-3/4", unlighted		7	1.143		124	53.50		177.50	219
0400	24" x 19" x 8-1/2", lighted		5	1.600		179	75		254	315
0600	Triple door, 30" x 32", unlighted, plywood body		7	1.143		325	53.50		378.50	445
0700	Steel body		7	1.143		375	53.50		428.50	495
0900	Oak door, wood body, beveled mirror, single door		7	1.143		199	53.50		252.50	300
1000	Double door		6	1.333		380	62.50		442.50	515
1200	Hotel cabinets, stainless, with lower shelf, unlighted		10	.800		200	37.50		237.50	277
1300	Lighted		5	1.600		305	75		380	450

10 28 19 – Tub and Shower Enclosures

10 28 19.10 Partitions, Shower

		Crew	Daily Output	Labor-Hours	Unit	Material	2015 Bare Costs Labor	Equipment	Total	Total Incl O&P
0010	**PARTITIONS, SHOWER** floor mounted, no plumbing									
0400	Cabinet, one piece, fiberglass, 32" x 32"	2 Carp	5	3.200	Ea.	530	150		680	815
0420	36" x 36"		5	3.200		570	150		720	855
0440	36" x 48"		5	3.200		1,375	150		1,525	1,725
0460	Acrylic, 32" x 32"		5	3.200		310	150		460	570
0480	36" x 36"		5	3.200		1,025	150		1,175	1,350
0500	36" x 48"		5	3.200		1,500	150		1,650	1,900
0520	Shower door for above, clear plastic, 24" wide	1 Carp	8	1		186	47		233	277
0540	28" wide		8	1		206	47		253	300
0560	Tempered glass, 24" wide		8	1		200	47		247	293
0580	28" wide		8	1		226	47		273	320
2400	Glass stalls, with doors, no receptors, chrome on brass	2 Shee	3	5.333		1,650	298		1,948	2,275
2700	Anodized aluminum	"	4	4		1,150	224		1,374	1,625
2900	Marble shower stall, stock design, with shower door	2 Marb	1.20	13.333		2,475	580		3,055	3,600
3000	With curtain		1.30	12.308		2,200	535		2,735	3,225
3200	Receptors, precast terrazzo, 32" x 32"		14	1.143		360	49.50		409.50	470
3300	48" x 34"		9.50	1.684		465	73.50		538.50	620
3500	Plastic, simulated terrazzo receptor, 32" x 32"		14	1.143		148	49.50		197.50	239
3600	32" x 48"		12	1.333		215	58		273	325
3800	Precast concrete, colors, 32" x 32"		14	1.143		186	49.50		235.50	281
3900	48" x 48"		8	2		251	87		338	410
4100	Shower doors, economy plastic, 24" wide	1 Shee	9	.889		144	49.50		193.50	234
4200	Tempered glass door, economy		8	1		258	56		314	370
4400	Folding, tempered glass, aluminum frame		6	1.333		390	74.50		464.50	545
4500	Sliding, tempered glass, 48" opening		6	1.333		540	74.50		614.50	705

10 28 19 – Tub and Shower Enclosures

10 28 19.10 Partitions, Shower	Crew	Daily Output	Labor-Hours	Unit	Material	2015 Bare Costs Labor	Equipment	Total	Total Incl O&P
4700	Deluxe, tempered glass, chrome on brass frame, 42" to 44"	1 Shee	8	1	Ea.	385	56		441	505
4800	39" to 48" wide		1	8		625	450		1,075	1,375
4850	On anodized aluminum frame, obscure glass		2	4		540	224		764	930
4900	Clear glass	↓	1	8	↓	625	450		1,075	1,375
5100	Shower enclosure, tempered glass, anodized alum. frame									
5120	2 panel & door, corner unit, 32" x 32"	1 Shee	2	4	Ea.	1,025	224		1,249	1,500
5140	Neo-angle corner unit, 16" x 24" x 16"	"	2	4		1,075	224		1,299	1,550
5200	Shower surround, 3 wall, polypropylene, 32" x 32"	1 Carp	4	2		455	94		549	645
5220	PVC, 32" x 32"		4	2		380	94		474	565
5240	Fiberglass		4	2		420	94		514	605
5250	2 wall, polypropylene, 32" x 32"		4	2		315	94		409	495
5270	PVC		4	2		395	94		489	580
5290	Fiberglass	↓	4	2		400	94		494	585
5300	Tub doors, tempered glass & frame, obscure glass	1 Shee	8	1		229	56		285	340
5400	Clear glass		6	1.333		535	74.50		609.50	705
5600	Chrome plated, brass frame, obscure glass		8	1		305	56		361	420
5700	Clear glass		6	1.333		745	74.50		819.50	930
5900	Tub/shower enclosure, temp. glass, alum. frame, obscure glass		2	4		410	224		634	790
6200	Clear glass		1.50	5.333		855	298		1,153	1,400
6500	On chrome-plated brass frame, obscure glass		2	4		565	224		789	965
6600	Clear glass	↓	1.50	5.333		1,200	298		1,498	1,775
6800	Tub surround, 3 wall, polypropylene	1 Carp	4	2		256	94		350	425
6900	PVC		4	2		390	94		484	575
7000	Fiberglass, obscure glass		4	2		400	94		494	585
7100	Clear glass	↓	3	2.667	↓	680	125		805	945

10 44 Fire Protection Specialties

10 44 13 – Fire Protection Cabinets

10 44 13.53 Fire Equipment Cabinets

10 44 13.53		Crew	Daily Output	Labor-Hours	Unit	Material	2015 Bare Costs Labor	Equipment	Total	Total Incl O&P
0010	**FIRE EQUIPMENT CABINETS**, not equipped, 20 ga. steel box	D4020-310								
0040	recessed, D.S. glass in door, box size given									
1000	Portable extinguisher, single, 8" x 12" x 27", alum. door & frame	Q-12	8	2	Ea.	155	101		256	325
1100	Steel door and frame		8	2		116	101		217	281
1200	Stainless steel door and frame		8	2		205	101		306	380
2000	Portable extinguisher, large, 8" x 12" x 36", alum. door & frame		8	2		258	101		359	435
2100	Steel door and frame	D4020-330	8	2		163	101		264	330
2200	Stainless steel door and frame		8	2		270	101		371	450
2500	8" x 16" x 38", aluminum door & frame		8	2		236	101		337	410
2600	Steel door and frame		8	2		224	101		325	400
2700	Fire blanket & extinguisher cab, inc blanket, rec stl., 14" x 40" x 8"		7	2.286		188	116		304	380
2800	Fire blanket cab, inc blanket, surf mtd, stl, 15"x10"x5", w/pwdr coat fin		8	2		94	101		195	257
3000	Hose rack assy., 1-1/2" valve & 100' hose, 24" x 40" x 5-1/2"									
3100	Aluminum door and frame	Q-12	6	2.667	Ea.	370	135		505	610
3200	Steel door and frame	D4020-410	6	2.667		246	135		381	475
3300	Stainless steel door and frame		6	2.667		455	135		590	710
4000	Hose rack assy., 2-1/2" x 1-1/2" valve, 100' hose, 24" x 40" x 8"									
4100	Aluminum door and frame	Q-12	6	2.667	Ea.	375	135		510	615
4200	Steel door and frame	R211226-10	6	2.667		253	135		388	480
4300	Stainless steel door and frame		6	2.667		495	135		630	750
5000	Hose rack assy., 2-1/2" x 1-1/2" valve, 100' hose									
5010	and extinguisher, 30" x 40" x 8"									

10 44 Fire Protection Specialties

10 44 13 – Fire Protection Cabinets

10 44 13.53 Fire Equipment Cabinets

		Crew	Daily Output	Labor-Hours	Unit	Material	2015 Bare Costs Labor	Equipment	Total	Total Incl O&P
5100	Aluminum door and frame	Q-12	5	3.200	Ea.	475	162		637	770
5200	Steel door and frame		5	3.200		264	162		426	535
5300	Stainless steel door and frame		5	3.200		535	162		697	835
6000	Hose rack assy., 1-1/2" valve, 100' hose									
6010	and 2-1/2" FD valve, 24" x 44" x 8"									
6100	Aluminum door and frame	Q-12	5	3.200	Ea.	405	162		567	690
6200	Steel door and frame		5	3.200		258	162		420	530
6300	Stainless steel door and frame		5	3.200		560	162		722	865
7000	Hose rack assy., 1-1/2" valve & 100' hose, 2-1/2" FD valve R211226-20									
7010	and extinguisher, 30" x 44" x 8"									
7100	Aluminum door and frame	Q-12	5	3.200	Ea.	500	162		662	795
7200	Steel door and frame		5	3.200		315	162		477	590
7300	Stainless steel door and frame		5	3.200		665	162		827	980
8000	Valve cabinet for 2-1/2" FD angle valve, 18" x 18" x 8"									
8100	Aluminum door and frame	Q-12	12	1.333	Ea.	164	67.50		231.50	282
8200	Steel door and frame		12	1.333		136	67.50		203.50	251
8300	Stainless steel door and frame		12	1.333		221	67.50		288.50	345

10 44 16 – Fire Extinguishers

10 44 16.13 Portable Fire Extinguishers

		Crew	Daily Output	Labor-Hours	Unit	Material	2015 Bare Costs Labor	Equipment	Total	Total Incl O&P
0010	**PORTABLE FIRE EXTINGUISHERS**									
0140	CO_2, with hose and "H" horn, 10 lb.				Ea.	275			275	305
0160	15 lb.					355			355	390
0180	20 lb.					395			395	435
1000	Dry chemical, pressurized									
1040	Standard type, portable, painted, 2-1/2 lb.				Ea.	37.50			37.50	41
1060	5 lb.					51			51	56.50
1080	10 lb.					81			81	89.50
1100	20 lb.					136			136	150
1120	30 lb.					425			425	470
1300	Standard type, wheeled, 150 lb.					2,425			2,425	2,650
2000	ABC all purpose type, portable, 2-1/2 lb.					21			21	23.50
2060	5 lb.					26.50			26.50	29
2080	9-1/2 lb.					44			44	48
2100	20 lb.					79.50			79.50	87.50
5000	Pressurized water, 2-1/2 gallon, stainless steel					102			102	112
5060	With anti-freeze					106			106	116
9400	Installation of extinguishers, 12 or more, on nailable surface	1 Carp	30	.267			12.50		12.50	19.25
9420	On masonry or concrete	"	15	.533			25		25	38.50

10 44 16.16 Wheeled Fire Extinguisher Units

		Crew	Daily Output	Labor-Hours	Unit	Material	2015 Bare Costs Labor	Equipment	Total	Total Incl O&P
0010	**WHEELED FIRE EXTINGUISHER UNITS**									
0350	CO_2, portable, with swivel horn									
0360	Wheeled type, cart mounted, 50 lb.				Ea.	1,125			1,125	1,250
0400	100 lb.				"	3,850			3,850	4,250
2200	ABC all purpose type									
2300	Wheeled, 45 lb.				Ea.	735			735	810
2360	150 lb.				"	1,850			1,850	2,025

Division Notes

	CREW	DAILY OUTPUT	LABOR-HOURS	UNIT	BARE COSTS				TOTAL INCL O&P
					MAT.	LABOR	EQUIP.	TOTAL	

Estimating Tips
General

- The items in this division are usually priced per square foot or each. Many of these items are purchased by the owner for installation by the contractor. Check the specifications for responsibilities and include time for receiving, storage, installation, and mechanical and electrical hookups in the appropriate divisions.

- Many items in Division 11 require some type of support system that is not usually furnished with the item. Examples of these systems include blocking for the attachment of casework and support angles for ceiling-hung projection screens. The required blocking or supports must be added to the estimate in the appropriate division.

- Some items in Division 11 may require assembly or electrical hookups. Verify the amount of assembly required or the need for a hard electrical connection and add the appropriate costs.

Reference Numbers

Reference numbers are shown in shaded boxes at the beginning of some major classifications. These numbers refer to related items in the Reference Section. The reference information may be an estimating procedure, an alternate pricing method, or technical information.

Note: Not all subdivisions listed here necessarily appear in this publication. ■

11 11 Vehicle Service Equipment

11 11 13 – Compressed-Air Vehicle Service Equipment

11 11 13.10 Compressed Air Equipment	Crew	Daily Output	Labor-Hours	Unit	Material	2015 Bare Costs Labor	2015 Bare Costs Equipment	Total	Total Incl O&P
0010 **COMPRESSED AIR EQUIPMENT**									
0030 Compressors, electric, 1-1/2 H.P., standard controls	L-4	1.50	16	Ea.	455	710		1,165	1,600
0550 Dual controls		1.50	16		810	710		1,520	2,000
0600 5 H.P., 115/230 volt, standard controls		1	24		2,575	1,050		3,625	4,475
0650 Dual controls		1	24		3,400	1,050		4,450	5,375

11 11 19 – Vehicle Lubrication Equipment

11 11 19.10 Lubrication Equipment

	Crew	Daily Output	Labor-Hours	Unit	Material	Labor	Equipment	Total	Total Incl O&P
0010 **LUBRICATION EQUIPMENT**									
3000 Lube equipment, 3 reel type, with pumps, not including piping	L-4	.50	48	Set	8,800	2,125		10,925	13,000
3700 Pump lubrication, pneumatic, not incl. air compressor									
3710 Oil/gear lube	Q-1	9.60	1.667	Ea.	860	88		948	1,075
3720 Grease	"	9.60	1.667	"	1,075	88		1,163	1,300

11 11 33 – Vehicle Spray Painting Equipment

11 11 33.10 Spray Painting Equipment

	Crew	Daily Output	Labor-Hours	Unit	Material	Labor	Equipment	Total	Total Incl O&P
0010 **SPRAY PAINTING EQUIPMENT**									
4000 Spray painting booth, 26' long, complete	L-4	.40	60	Ea.	16,400	2,650		19,050	22,100

11 21 Retail and Service Equipment

11 21 53 – Barber and Beauty Shop Equipment

11 21 53.10 Barber Equipment

	Crew	Daily Output	Labor-Hours	Unit	Material	Labor	Equipment	Total	Total Incl O&P
0010 **BARBER EQUIPMENT**									
0020 Chair, hydraulic, movable, minimum	1 Carp	24	.333	Ea.	560	15.65		575.65	640
0050 Maximum	"	16	.500		3,475	23.50		3,498.50	3,850
0500 Sink, hair washing basin, rough plumbing not incl.	1 Plum	8	1		495	58.50		553.50	635
1000 Sterilizer, liquid solution for tools					161			161	177

11 21 73 – Commercial Laundry and Dry Cleaning Equipment

11 21 73.13 Dry Cleaning Equipment

0010 **DRY CLEANING EQUIPMENT**									

11 21 73.16 Drying and Conditioning Equipment

	Crew	Daily Output	Labor-Hours	Unit	Material	Labor	Equipment	Total	Total Incl O&P
0010 **DRYING AND CONDITIONING EQUIPMENT**									
0100 Dryers, Not including rough-in									
1500 Industrial, 30 lb. capacity	1 Plum	2	4	Ea.	3,175	235		3,410	3,825
1600 50 lb. capacity	"	1.70	4.706	"	3,400	276		3,676	4,175

11 21 73.26 Commercial Washers and Extractors

	Crew	Daily Output	Labor-Hours	Unit	Material	Labor	Equipment	Total	Total Incl O&P
0010 **COMMERCIAL WASHERS AND EXTRACTORS**, not including rough-in									
6000 Combination washer/extractor, 20 lb. capacity	L-6	1.50	8	Ea.	5,950	460		6,410	7,225
6100 30 lb. capacity		.80	15		9,275	860		10,135	11,500
6200 50 lb. capacity		.68	17.647		10,900	1,000		11,900	13,500
6300 75 lb. capacity		.30	40		20,400	2,300		22,700	26,000
6350 125 lb. capacity		.16	75		27,700	4,300		32,000	36,900

11 21 73.33 Coin-Operated Laundry Equipment

	Crew	Daily Output	Labor-Hours	Unit	Material	Labor	Equipment	Total	Total Incl O&P
0010 **COIN-OPERATED LAUNDRY EQUIPMENT**									
0990 Dryer, gas fired									
1000 Commercial, 30 lb. capacity, coin operated, single	1 Plum	3	2.667	Ea.	3,300	157		3,457	3,850
1100 Double stacked	"	2	4	"	7,275	235		7,510	8,350
5290 Clothes washer									
5300 Commercial, coin operated, average	1 Plum	3	2.667	Ea.	1,250	157		1,407	1,600

11 21 Retail and Service Equipment

11 21 83 – Photo Processing Equipment

11 21 83.13 Darkroom Equipment	Crew	Daily Output	Labor-Hours	Unit	Material	2015 Bare Costs Labor	2015 Bare Costs Equipment	Total	Total Incl O&P
0010 **DARKROOM EQUIPMENT**									
0020 Developing sink, 5" deep, 24" x 48"	Q-1	2	8	Ea.	4,150	425		4,575	5,225
0050 48" x 52"		1.70	9.412		4,425	495		4,920	5,625
0200 10" deep, 24" x 48"		1.70	9.412		1,550	495		2,045	2,450
0250 24" x 108"		1.50	10.667		3,650	565		4,215	4,875
3500 Washers, round, minimum sheet 11" x 14"		2	8		3,275	425		3,700	4,250
3550 Maximum sheet 20" x 24"		1	16		3,600	845		4,445	5,250
3800 Square, minimum sheet 20" x 24"		1	16		3,150	845		3,995	4,750
3900 Maximum sheet 50" x 56"	▼	.80	20	▼	4,900	1,050		5,950	7,000
4500 Combination tank sink, tray sink, washers, with									
4510 Dry side tables, average	Q-1	.45	35.556	Ea.	10,800	1,875		12,675	14,600

11 30 Residential Equipment

11 30 13 – Residential Appliances

11 30 13.15 Cooking Equipment

	Crew	Daily Output	Labor-Hours	Unit	Material	Labor	Equipment	Total	Total Incl O&P
0010 **COOKING EQUIPMENT**									
0020 Cooking range, 30" free standing, 1 oven, minimum	2 Clab	10	1.600	Ea.	440	60		500	580
0050 Maximum		4	4		2,025	150		2,175	2,450
0150 2 oven, minimum		10	1.600		1,225	60		1,285	1,425
0200 Maximum	▼	10	1.600	▼	2,550	60		2,610	2,900

11 30 13.16 Refrigeration Equipment

	Crew	Daily Output	Labor-Hours	Unit	Material	Labor	Equipment	Total	Total Incl O&P
0010 **REFRIGERATION EQUIPMENT**									
5200 Icemaker, automatic, 20 lb. per day	1 Plum	7	1.143	Ea.	895	67		962	1,075
5350 51 lb. per day	"	2	4	"	1,575	235		1,810	2,100

11 30 13.17 Kitchen Cleaning Equipment

	Crew	Daily Output	Labor-Hours	Unit	Material	Labor	Equipment	Total	Total Incl O&P
0010 **KITCHEN CLEANING EQUIPMENT**									
2750 Dishwasher, built-in, 2 cycles, minimum	L-1	4	4	Ea.	238	227		465	600
2800 Maximum		2	8		435	455		890	1,150
2950 4 or more cycles, minimum		4	4		375	227		602	755
2960 Average		4	4		500	227		727	890
3000 Maximum	▼	2	8	▼	1,100	455		1,555	1,875

11 30 13.18 Waste Disposal Equipment

	Crew	Daily Output	Labor-Hours	Unit	Material	Labor	Equipment	Total	Total Incl O&P
0010 **WASTE DISPOSAL EQUIPMENT**									
3300 Garbage disposal, sink type, minimum	L-1	10	1.600	Ea.	87.50	90.50		178	232
3350 Maximum	"	10	1.600	"	220	90.50		310.50	380

11 30 13.19 Kitchen Ventilation Equipment

	Crew	Daily Output	Labor-Hours	Unit	Material	Labor	Equipment	Total	Total Incl O&P
0010 **KITCHEN VENTILATION EQUIPMENT**									
4150 Hood for range, 2 speed, vented, 30" wide, minimum	L-3	5	3.200	Ea.	70.50	164		234.50	330
4200 Maximum		3	5.333		780	273		1,053	1,275
4300 42" wide, minimum		5	3.200		168	164		332	435
4330 Custom		5	3.200		1,650	164		1,814	2,075
4350 Maximum	▼	3	5.333		2,025	273		2,298	2,650
4500 For ventless hood, 2 speed, add					18.65			18.65	20.50
4650 For vented 1 speed, deduct from maximum				▼	50			50	55

11 30 13.24 Washers

	Crew	Daily Output	Labor-Hours	Unit	Material	Labor	Equipment	Total	Total Incl O&P
0010 **WASHERS**									
5000 Residential, 4 cycle, average	1 Plum	3	2.667	Ea.	875	157		1,032	1,200
6650 Washing machine, automatic, minimum		3	2.667		485	157		642	765
6700 Maximum	▼	1	8	▼	1,525	470		1,995	2,375

For customer support on your Plumbing Cost Data, call 877.709.6968.

71

11 30 Residential Equipment

11 30 13 – Residential Appliances

11 30 13.25 Dryers	Crew	Daily Output	Labor-Hours	Unit	Material	2015 Bare Costs Labor	2015 Bare Costs Equipment	Total	Total Incl O&P
0010 **DRYERS**									
0500 Gas fired residential, 16 lb. capacity, average	1 Plum	3	2.667	Ea.	675	157		832	975
7450 Vent kits for dryers	1 Carp	10	.800	"	39	37.50		76.50	101

11 30 15 – Miscellaneous Residential Appliances

11 30 15.13 Sump Pumps

	Crew	Daily Output	Labor-Hours	Unit	Material	Labor	Equipment	Total	Total Incl O&P
0010 **SUMP PUMPS**									
6400 Cellar drainer, pedestal, 1/3 H.P., molded PVC base	1 Plum	3	2.667	Ea.	135	157		292	385
6450 Solid brass	"	2	4	"	289	235		524	675
6460 Sump pump, see also Section 22 14 29.16									

11 30 15.23 Water Heaters

	Crew	Daily Output	Labor-Hours	Unit	Material	Labor	Equipment	Total	Total Incl O&P
0010 **WATER HEATERS**									
6900 Electric, glass lined, 30 gallon, minimum	L-1	5	3.200	Ea.	430	181		611	745
6950 Maximum		3	5.333		595	300		895	1,100
7100 80 gallon, minimum		2	8		1,225	455		1,680	2,025
7150 Maximum		1	16		1,700	905		2,605	3,225
7180 Gas, glass lined, 30 gallon, minimum	2 Plum	5	3.200		805	188		993	1,175
7220 Maximum		3	5.333		1,125	315		1,440	1,700
7260 50 gallon, minimum		2.50	6.400		845	375		1,220	1,500
7300 Maximum		1.50	10.667		1,175	625		1,800	2,225
7310 Water heater, see also Section 22 33 30.13									

11 30 15.43 Air Quality

	Crew	Daily Output	Labor-Hours	Unit	Material	Labor	Equipment	Total	Total Incl O&P
0010 **AIR QUALITY**									
2450 Dehumidifier, portable, automatic, 15 pint	1 Elec	4	2	Ea.	152	109		261	330
2550 40 pint		3.75	2.133		209	117		326	405
3550 Heater, electric, built-in, 1250 watt, ceiling type, minimum		4	2		107	109		216	282
3600 Maximum		3	2.667		175	146		321	410
3700 Wall type, minimum		4	2		172	109		281	355
3750 Maximum		3	2.667		185	146		331	420
3900 1500 watt wall type, with blower		4	2		172	109		281	355
3950 3000 watt		3	2.667		350	146		496	605
4850 Humidifier, portable, 8 gallons per day					133			133	146
5000 15 gallons per day					211			211	232

11 32 Unit Kitchens

11 32 13 – Metal Unit Kitchens

11 32 13.10 Commercial Unit Kitchens

	Crew	Daily Output	Labor-Hours	Unit	Material	Labor	Equipment	Total	Total Incl O&P
0010 **COMMERCIAL UNIT KITCHENS**									
1500 Combination range, refrigerator and sink, 30" wide, minimum	L-1	2	8	Ea.	1,100	455		1,555	1,900
1550 Maximum		1	16		1,475	905		2,380	3,000
1570 60" wide, average		1.40	11.429		1,525	650		2,175	2,675
1590 72" wide, average		1.20	13.333		1,625	755		2,380	2,900
1600 Office model, 48" wide		2	8		2,025	455		2,480	2,925
1620 Refrigerator and sink only		2.40	6.667		2,525	380		2,905	3,350
1640 Combination range, refrigerator, sink, microwave									
1660 Oven and ice maker	L-1	.80	20	Ea.	4,550	1,125		5,675	6,700

11 41 Foodservice Storage Equipment

11 41 13 – Refrigerated Food Storage Cases

11 41 13.20 Refrigerated Food Storage Equipment	Crew	Daily Output	Labor-Hours	Unit	Material	2015 Bare Costs Labor	Equipment	Total	Total Incl O&P
0010 **REFRIGERATED FOOD STORAGE EQUIPMENT**									
2350 Cooler, reach-in, beverage, 6' long	Q-1	6	2.667	Ea.	3,600	141		3,741	4,175
4300 Freezers, reach-in, 44 C.F.		4	4		4,425	211		4,636	5,175
4500 68 C.F.		3	5.333		4,825	282		5,107	5,725

11 44 Food Cooking Equipment

11 44 13 – Commercial Ranges

11 44 13.10 Cooking Equipment

	Crew	Daily Output	Labor-Hours	Unit	Material	2015 Bare Costs Labor	Equipment	Total	Total Incl O&P
0010 **COOKING EQUIPMENT**									
0020 Bake oven, gas, one section	Q-1	8	2	Ea.	5,450	106		5,556	6,150
0300 Two sections		7	2.286		9,075	121		9,196	10,200
0600 Three sections		6	2.667		11,200	141		11,341	12,500
0900 Electric convection, single deck	L-7	4	7		6,225	320		6,545	7,325
6350 Kettle, w/steam jacket, tilting, w/positive lock, SS, 20 gallons		7	4		8,225	182		8,407	9,325
6600 60 gallons		6	4.667		11,000	212		11,212	12,400

11 46 Food Dispensing Equipment

11 46 83 – Ice Machines

11 46 83.10 Commercial Ice Equipment

	Crew	Daily Output	Labor-Hours	Unit	Material	2015 Bare Costs Labor	Equipment	Total	Total Incl O&P
0010 **COMMERCIAL ICE EQUIPMENT**									
5800 Ice cube maker, 50 pounds per day	Q-1	6	2.667	Ea.	1,600	141		1,741	2,000
6050 500 pounds per day	"	4	4	"	2,775	211		2,986	3,375

11 48 Foodservice Cleaning and Disposal Equipment

11 48 13 – Commercial Dishwashers

11 48 13.10 Dishwashers

		Crew	Daily Output	Labor-Hours	Unit	Material	2015 Bare Costs Labor	Equipment	Total	Total Incl O&P
0010 **DISHWASHERS**										
2700 Dishwasher, commercial, rack type										
2720 10 to 12 racks per hour		Q-1	3.20	5	Ea.	3,525	264		3,789	4,275
2730 Energy star rated, 35 to 40 racks/hour	G		1.30	12.308		4,700	650		5,350	6,150
2740 50 to 60 racks/hour	G		1.30	12.308		10,200	650		10,850	12,200
2800 Automatic, 190 to 230 racks per hour		L-6	.35	34.286		13,900	1,975		15,875	18,300
2820 235 to 275 racks per hour			.25	48		31,500	2,750		34,250	38,800
2840 8,750 to 12,500 dishes per hour			.10	120		55,500	6,875		62,375	71,500

11 53 Laboratory Equipment

11 53 13 – Laboratory Fume Hoods

11 53 13.13 Recirculating Laboratory Fume Hoods

	Crew	Daily Output	Labor-Hours	Unit	Material	2015 Bare Costs Labor	Equipment	Total	Total Incl O&P
0010 **RECIRCULATING LABORATORY FUME HOODS**									
0670 Service fixtures, average				Ea.	240			240	264
0680 For sink assembly with hot and cold water, add	1 Plum	1.40	5.714	"	735	335		1,070	1,325

For customer support on your Plumbing Cost Data, call 877.709.6968.

73

11 53 Laboratory Equipment

11 53 19 – Laboratory Sterilizers

11 53 19.13 Sterilizers	Crew	Daily Output	Labor-Hours	Unit	Material	2015 Bare Costs Labor	Equipment	Total	Total Incl O&P
0010 **STERILIZERS**									
0700 Glassware washer, undercounter, minimum	L-1	1.80	8.889	Ea.	6,325	505		6,830	7,700
0710 Maximum	"	1	16		13,300	905		14,205	16,000
1850 Utensil washer-sanitizer	1 Plum	2	4		11,300	235		11,535	12,800

11 53 23 – Laboratory Refrigerators

11 53 23.13 Refrigerators

	Crew	Daily Output	Labor-Hours	Unit	Material	Labor	Equipment	Total	Total Incl O&P
0010 **REFRIGERATORS**									
1200 Blood bank, 28.6 C.F. emergency signal				Ea.	9,650			9,650	10,600
1210 Reach-in, 16.9 C.F.				"	8,525			8,525	9,375

11 53 33 – Emergency Safety Appliances

11 53 33.13 Emergency Equipment

	Crew	Daily Output	Labor-Hours	Unit	Material	Labor	Equipment	Total	Total Incl O&P
0010 **EMERGENCY EQUIPMENT**									
1400 Safety equipment, eye wash, hand held				Ea.	410			410	450
1450 Deluge shower				"	770			770	850

11 53 43 – Service Fittings and Accessories

11 53 43.13 Fittings

	Crew	Daily Output	Labor-Hours	Unit	Material	Labor	Equipment	Total	Total Incl O&P
0010 **FITTINGS**									
1600 Sink, one piece plastic, flask wash, hose, free standing	1 Plum	1.60	5	Ea.	1,950	294		2,244	2,600
1610 Epoxy resin sink, 25" x 16" x 10"	"	2	4	"	221	235		456	600
8000 Alternate pricing method: as percent of lab furniture									
8050 Installation, not incl. plumbing & duct work				% Furn.				22%	22%
8100 Plumbing, final connections, simple system								10%	10%
8110 Moderately complex system								15%	15%
8120 Complex system								20%	20%
8150 Electrical, simple system								10%	10%
8160 Moderately complex system								20%	20%
8170 Complex system								35%	35%

11 71 Medical Sterilizing Equipment

11 71 10 – Medical Sterilizers & Distillers

11 71 10.10 Sterilizers and Distillers

	Crew	Daily Output	Labor-Hours	Unit	Material	Labor	Equipment	Total	Total Incl O&P
0010 **STERILIZERS AND DISTILLERS**									
0700 Distiller, water, steam heated, 50 gal. capacity	1 Plum	1.40	5.714	Ea.	19,200	335		19,535	21,700
5600 Sterilizers, floor loading, 26" x 62" x 42", single door, steam					122,000			122,000	134,500
5650 Double door, steam					206,000			206,000	227,000
5800 General purpose, 20" x 20" x 38", single door					13,100			13,100	14,400
6000 Portable, counter top, steam, minimum					3,550			3,550	3,900
6020 Maximum					4,325			4,325	4,750
6050 Portable, counter top, gas, 17" x 15" x 32-1/2"					39,800			39,800	43,700
6150 Manual washer/sterilizer, 16" x 16" x 26"	1 Plum	2	4		54,500	235		54,735	60,500
6200 Steam generators, electric 10 kW to 180 kW, freestanding									
6250 Minimum	1 Elec	3	2.667	Ea.	8,775	146		8,921	9,875
6300 Maximum	"	.70	11.429		29,200	625		29,825	33,000
8200 Bed pan washer-sanitizer	1 Plum	2	4		7,500	235		7,735	8,600

11 73 Patient Care Equipment

11 73 10 – Patient Treatment Equipment

11 73 10.10 Treatment Equipment

		Crew	Daily Output	Labor-Hours	Unit	Material	2015 Bare Costs Labor	Equipment	Total	Total Incl O&P
0010	**TREATMENT EQUIPMENT**									
1800	Heat therapy unit, humidified, 26" x 78" x 28"				Ea.	3,600			3,600	3,975
2100	Hubbard tank with accessories, stainless steel,									
2110	125 GPM at 45 psi water pressure				Ea.	27,200			27,200	29,900
2150	For electric overhead hoist, add					2,975			2,975	3,250
3600	Paraffin bath, 126°F, auto controlled					1,100			1,100	1,200
4600	Station, dietary, medium, with ice					16,700			16,700	18,400
8400	Whirlpool bath, mobile, sst, 18" x 24" x 60"					4,900			4,900	5,400
8450	Fixed, incl. mixing valves	1 Plum	2	4		9,700	235		9,935	11,100

11 74 Dental Equipment

11 74 10 – Dental Office Equipment

11 74 10.10 Diagnostic and Treatment Equipment

		Crew	Daily Output	Labor-Hours	Unit	Material	2015 Bare Costs Labor	Equipment	Total	Total Incl O&P
0010	**DIAGNOSTIC AND TREATMENT EQUIPMENT**									
0020	Central suction system, minimum	1 Plum	1.20	6.667	Ea.	1,575	390		1,965	2,325
0100	Maximum	"	.90	8.889		4,450	520		4,970	5,700
0300	Air compressor, minimum	1 Skwk	.80	10		3,025	485		3,510	4,075
0400	Maximum		.50	16		8,950	780		9,730	11,100
0600	Chair, electric or hydraulic, minimum		.50	16		2,250	780		3,030	3,675
0700	Maximum		.25	32		7,825	1,550		9,375	11,000
2000	Light, ceiling mounted, minimum		8	1		1,175	48.50		1,223.50	1,350
2100	Maximum		8	1		2,025	48.50		2,073.50	2,300
2200	Unit light, minimum	2 Skwk	5.33	3.002		735	146		881	1,025
2210	Maximum		5.33	3.002		1,575	146		1,721	1,950
2220	Track light, minimum		3.20	5		1,575	243		1,818	2,100
2230	Maximum		3.20	5		2,650	243		2,893	3,300
2300	Sterilizers, steam portable, minimum					1,300			1,300	1,425
2350	Maximum					10,500			10,500	11,600
2600	Steam, institutional					3,275			3,275	3,600
2650	Dry heat, electric, portable, 3 trays					1,225			1,225	1,350

11 76 Operating Room Equipment

11 76 10 – Operating Room Equipment

11 76 10.10 Surgical Equipment

		Crew	Daily Output	Labor-Hours	Unit	Material	2015 Bare Costs Labor	Equipment	Total	Total Incl O&P
0010	**SURGICAL EQUIPMENT**									
5000	Scrub, surgical, stainless steel, single station, minimum	1 Plum	3	2.667	Ea.	4,300	157		4,457	4,950
5100	Maximum				"	6,975			6,975	7,675

For customer support on your Plumbing Cost Data, call 877.709.6968.

75

11 78 Mortuary Equipment

11 78 13 – Mortuary Refrigerators

11 78 13.10 Mortuary and Autopsy Equipment	Crew	Daily Output	Labor-Hours	Unit	Material	2015 Bare Costs Labor	2015 Bare Costs Equipment	Total	Total Incl O&P
0010 **MORTUARY AND AUTOPSY EQUIPMENT**									
0015 Autopsy table, standard	1 Plum	1	8	Ea.	9,750	470		10,220	11,400
0020 Deluxe	"	.60	13.333	"	15,100	785		15,885	17,800

11 81 Facility Maintenance Equipment

11 81 19 – Vacuum Cleaning Systems

11 81 19.10 Vacuum Cleaning

	Crew	Daily Output	Labor-Hours	Unit	Material	Labor	Equipment	Total	Total Incl O&P
0010 **VACUUM CLEANING**									
0020 Central, 3 inlet, residential	1 Skwk	.90	8.889	Total	1,075	430		1,505	1,850
0200 Commercial		.70	11.429		1,225	555		1,780	2,200
0400 5 inlet system, residential		.50	16		1,500	780		2,280	2,850
0600 7 inlet system, commercial		.40	20		1,700	975		2,675	3,375
0800 9 inlet system, residential		.30	26.667		3,750	1,300		5,050	6,125
4010 Rule of thumb: First 1200 S.F., installed								1,425	1,575
4020 For each additional S.F., add				S.F.				.26	.26

11 91 Religious Equipment

11 91 13 – Baptisteries

11 91 13.10 Baptistry

	Crew	Daily Output	Labor-Hours	Unit	Material	Labor	Equipment	Total	Total Incl O&P
0010 **BAPTISTRY**									
0150 Fiberglass, 3'-6" deep, x 13'-7" long,									
0160 steps at both ends, incl. plumbing, minimum	L-8	1	20	Ea.	5,725	985		6,710	7,800
0200 Maximum	"	.70	28.571		9,325	1,400		10,725	12,500
0250 Add for filter, heater and lights					1,850			1,850	2,050

11 91 23 – Sanctuary Equipment

11 91 23.10 Sanctuary Furnishings

	Crew	Daily Output	Labor-Hours	Unit	Material	Labor	Equipment	Total	Total Incl O&P
0010 **SANCTUARY FURNISHINGS**									
0020 Altar, wood, custom design, plain	1 Carp	1.40	5.714	Ea.	2,550	268		2,818	3,250
0050 Deluxe	"	.20	40	"	12,400	1,875		14,275	16,500

11 98 Detention Equipment

11 98 30 – Detention Cell Equipment

11 98 30.10 Cell Equipment

	Crew	Daily Output	Labor-Hours	Unit	Material	Labor	Equipment	Total	Total Incl O&P
0010 **CELL EQUIPMENT**									
3000 Toilet apparatus including wash basin, average	L-8	1.50	13.333	Ea.	3,400	655		4,055	4,750

Estimating Tips

General

- The items in this division are usually priced per square foot or each. Most of these items are purchased by the owner and installed by the contractor. Do not assume the items in Division 12 will be purchased and installed by the contractor. Check the specifications for responsibilities and include receiving, storage, installation, and mechanical and electrical hookups in the appropriate divisions.

- Some items in this division require some type of support system that is not usually furnished with the item. Examples of these systems include blocking for the attachment of casework and heavy drapery rods. The required blocking must be added to the estimate in the appropriate division.

Reference Numbers

Reference numbers are shown in shaded boxes at the beginning of some major classifications. These numbers refer to related items in the Reference Section. The reference information may be an estimating procedure, an alternate pricing method, or technical information.

Note: Not all subdivisions listed here necessarily appear in this publication. ■

12 32 23 – Hardwood Casework

12 32 23.10 Manufactured Wood Casework, Stock Units	Crew	Daily Output	Labor-Hours	Unit	Material	2015 Bare Costs Labor	Equipment	Total	Total Incl O&P
0010 **MANUFACTURED WOOD CASEWORK, STOCK UNITS**									
0700 Kitchen base cabinets, hardwood, not incl. counter tops,									
0710 24" deep, 35" high, prefinished									
0800 One top drawer, one door below, 12" wide	2 Carp	24.80	.645	Ea.	265	30.50		295.50	340
0840 18" wide		23.30	.687		300	32		332	380
0880 24" wide		22.30	.717		365	33.50		398.50	450
1000 Four drawers, 12" wide		24.80	.645		279	30.50		309.50	350
1040 18" wide		23.30	.687		310	32		342	395
1060 24" wide		22.30	.717		345	33.50		378.50	430
1200 Two top drawers, two doors below, 27" wide		22	.727		390	34		424	485
1260 36" wide		20.30	.788		455	37		492	555
1300 48" wide		18.90	.847		515	40		555	630
1500 Range or sink base, two doors below, 30" wide		21.40	.748		350	35		385	440
1540 36" wide		20.30	.788		395	37		432	490
1580 48" wide		18.90	.847		435	40		475	540
1800 For sink front units, deduct					161			161	177
9000 For deluxe models of all cabinets, add					40%				
9500 For custom built in place, add					25%	10%			
9558 Rule of thumb, kitchen cabinets not including									
9560 appliances & counter top, minimum	2 Carp	30	.533	L.F.	176	25		201	232
9600 Maximum	"	25	.640	"	395	30		425	480

12 32 23.30 Manufactured Wood Casework Vanities

	Crew	Daily Output	Labor-Hours	Unit	Material	2015 Bare Costs Labor	Equipment	Total	Total Incl O&P
0010 **MANUFACTURED WOOD CASEWORK VANITIES**									
8000 Vanity bases, 2 doors, 30" high, 21" deep, 24" wide	2 Carp	20	.800	Ea.	310	37.50		347.50	400
8050 30" wide		16	1		370	47		417	480
8100 36" wide		13.33	1.200		360	56.50		416.50	480
8150 48" wide		11.43	1.400		470	65.50		535.50	620
9000 For deluxe models of all vanities, add to above					40%				
9500 For custom built in place, add to above					25%	10%			

Estimating Tips
General

- The items and systems in this division are usually estimated, purchased, supplied, and installed as a unit by one or more subcontractors. The estimator must ensure that all parties are operating from the same set of specifications and assumptions, and that all necessary items are estimated and will be provided. Many times the complex items and systems are covered, but the more common ones, such as excavation or a crane, are overlooked for the very reason that everyone assumes nobody could miss them. The estimator should be the central focus and be able to ensure that all systems are complete.

- Another area where problems can develop in this division is at the interface between systems. The estimator must ensure, for instance, that anchor bolts, nuts, and washers are estimated and included for the air-supported structures and pre-engineered buildings to be bolted to their foundations. Utility supply is a common area where essential items or pieces of equipment can be missed or overlooked, because each subcontractor may feel it is another's responsibility. The estimator should also be aware of certain items which may be supplied as part of a package but installed by others, and ensure that the installing contractor's estimate includes the cost of installation. Conversely, the estimator must also ensure that items are not costed by two different subcontractors, resulting in an inflated overall estimate.

13 30 00 Special Structures

- The foundations and floor slab, as well as rough mechanical and electrical, should be estimated, as this work is required for the assembly and erection of the structure. Generally, as noted in the book, the pre-engineered building comes as a shell. Pricing is based on the size and structural design parameters stated in the reference section. Additional features, such as windows and doors with their related structural framing, must also be included by the estimator. Here again, the estimator must have a clear understanding of the scope of each portion of the work and all the necessary interfaces.

Reference Numbers

Reference numbers are shown in shaded boxes at the beginning of some major classifications. These numbers refer to related items in the Reference Section. The reference information may be an estimating procedure, an alternate pricing method, or technical information.

Note: Not all subdivisions listed here necessarily appear in this publication. ■

13 11 Swimming Pools

13 11 13 – Below-Grade Swimming Pools

13 11 13.50 Swimming Pools

		Crew	Daily Output	Labor-Hours	Unit	Material	2015 Bare Costs Labor	Equipment	Total	Total Incl O&P
0010	**SWIMMING POOLS** Residential in-ground, vinyl lined, concrete									
0020	Swimming pools,resi in-ground,vyl lined,conc sides,W/ equip ,sand bot	B-52	300	.187	SF Surf	23.50	8.15	1.98	33.63	40
0100	Metal or polystyrene sides	B-14	410	.117		19.55	4.66	.89	25.10	29.50
0200	Add for vermiculite bottom					1.49			1.49	1.64
0500	Gunite bottom and sides, white plaster finish									
0600	12' x 30' pool	B-52	145	.386	SF Surf	43.50	16.85	4.10	64.45	78.50
0720	16' x 32' pool		155	.361		39	15.75	3.83	58.58	71
0750	20' x 40' pool		250	.224		35	9.75	2.38	47.13	56
0810	Concrete bottom and sides, tile finish									
0820	12' x 30' pool	B-52	80	.700	SF Surf	44	30.50	7.45	81.95	103
0830	16' x 32' pool		95	.589		36.50	25.50	6.25	68.25	86.50
0840	20' x 40' pool		130	.431		29	18.80	4.57	52.37	66
1100	Motel, gunite with plaster finish, incl. medium									
1150	capacity filtration & chlorination	B-52	115	.487	SF Surf	53.50	21	5.15	79.65	97
1200	Municipal, gunite with plaster finish, incl. high									
1250	capacity filtration & chlorination	B-52	100	.560	SF Surf	69.50	24.50	5.95	99.95	120
1350	Add for formed gutters				L.F.	102			102	112
1360	Add for stainless steel gutters				"	300			300	330
1600	For water heating system, see Section 23 52 28.10									
1700	Filtration and deck equipment only, as % of total				Total				20%	20%
1800	Deck equipment, rule of thumb, 20' x 40' pool				SF Pool				1.18	1.30
1900	5000 S.F. pool				"				1.73	1.90
3000	Painting pools, preparation + 3 coats, 20' x 40' pool, epoxy	2 Pord	.33	48.485	Total	1,775	1,950		3,725	4,900
3100	Rubber base paint, 18 gallons	"	.33	48.485		1,225	1,950		3,175	4,300
3500	42' x 82' pool, 75 gallons, epoxy paint	3 Pord	.14	171		7,500	6,925		14,425	18,700
3600	Rubber base paint	"	.14	171		5,075	6,925		12,000	16,000

R131113-20

13 11 46 – Swimming Pool Accessories

13 11 46.50 Swimming Pool Equipment

		Crew	Daily Output	Labor-Hours	Unit	Material	2015 Bare Costs Labor	Equipment	Total	Total Incl O&P
0010	**SWIMMING POOL EQUIPMENT**									
0020	Diving stand, stainless steel, 3 meter	2 Carp	.40	40	Ea.	15,100	1,875		16,975	19,500
0300	1 meter	"	2.70	5.926		9,175	278		9,453	10,500
2100	Lights, underwater, 12 volt, with transformer, 300 watt	1 Elec	1	8		330	440		770	1,025
2200	110 volt, 500 watt, standard		1	8		294	440		734	980
2400	Low water cutoff type		1	8		300	440		740	985
2800	Heaters, see Section 23 52 28.10									

13 17 Tubs and Pools

13 17 13 – Hot Tubs

13 17 13.10 Redwood Hot Tub System

		Crew	Daily Output	Labor-Hours	Unit	Material	2015 Bare Costs Labor	Equipment	Total	Total Incl O&P
0010	**REDWOOD HOT TUB SYSTEM**									
7050	4' diameter x 4' deep	Q-1	1	16	Ea.	3,225	845		4,070	4,825
7100	5' diameter x 4' deep		1	16		4,100	845		4,945	5,775
7150	6' diameter x 4' deep		.80	20		4,950	1,050		6,000	7,025
7200	8' diameter x 4' deep		.80	20		7,250	1,050		8,300	9,575

13 17 33 – Whirlpool Tubs

13 17 33.10 Whirlpool Bath

		Crew	Daily Output	Labor-Hours	Unit	Material	2015 Bare Costs Labor	Equipment	Total	Total Incl O&P
0010	**WHIRLPOOL BATH**									
6000	Whirlpool, bath with vented overflow, molded fiberglass									
6100	66" x 36" x 24"	Q-1	1	16	Ea.	3,475	845		4,320	5,100
6400	72" x 36" x 21"		1	16		1,800	845		2,645	3,275

For customer support on your Plumbing Cost Data, call 877.709.6968.

13 17 Tubs and Pools

13 17 33 – Whirlpool Tubs

13 17 33.10 Whirlpool Bath		Crew	Daily Output	Labor-Hours	Unit	Material	2015 Bare Costs Labor	Equipment	Total	Total Incl O&P
6500	60" x 34" x 21"	Q-1	1	16	Ea.	1,800	845		2,645	3,250
6600	72" x 42" x 23"	↓	1	16		2,175	845		3,020	3,675
6710	For color add					10%				
6711	For designer colors and trim add				↓	25%				

13 24 Special Activity Rooms

13 24 16 – Saunas

13 24 16.50 Saunas and Heaters

		Crew	Daily Output	Labor-Hours	Unit	Material	2015 Bare Costs Labor	Equipment	Total	Total Incl O&P
0010	**SAUNAS AND HEATERS**									
0020	Prefabricated, incl. heater & controls, 7' high, 6' x 4', C/C	L-7	2.20	12.727	Ea.	5,175	580		5,755	6,575
0050	6' x 4', C/P		2	14		4,700	635		5,335	6,125
0400	6' x 5', C/C		2	14		5,875	635		6,510	7,450
0450	6' x 5', C/P		2	14		5,325	635		5,960	6,825
0600	6' x 6', C/C		1.80	15.556		6,225	705		6,930	7,925
0650	6' x 6', C/P		1.80	15.556		5,675	705		6,380	7,300
0800	6' x 9', C/C		1.60	17.500		7,975	795		8,770	10,000
0850	6' x 9', C/P		1.60	17.500		7,225	795		8,020	9,150
1000	8' x 12', C/C		1.10	25.455		11,700	1,150		12,850	14,700
1050	8' x 12', C/P		1.10	25.455		10,500	1,150		11,650	13,400
1200	8' x 8', C/C		1.40	20		9,175	910		10,085	11,500
1250	8' x 8', C/P		1.40	20		8,450	910		9,360	10,700
1400	8' x 10', C/C		1.20	23.333		10,200	1,050		11,250	12,800
1450	8' x 10', C/P		1.20	23.333		9,250	1,050		10,300	11,800
1600	10' x 12', C/C		1	28		12,200	1,275		13,475	15,500
1650	10' x 12', C/P	↓	1	28		11,000	1,275		12,275	14,100
2500	Heaters only (incl. above), wall mounted, to 200 C.F.					685			685	755
2750	To 300 C.F.					930			930	1,025
3000	Floor standing, to 720 C.F., 10,000 watts, w/controls	1 Elec	3	2.667		2,950	146		3,096	3,475
3250	To 1,000 C.F., 16,000 watts	"	3	2.667	↓	3,825	146		3,971	4,425

13 24 26 – Steam Baths

13 24 26.50 Steam Baths and Components

		Crew	Daily Output	Labor-Hours	Unit	Material	2015 Bare Costs Labor	Equipment	Total	Total Incl O&P
0010	**STEAM BATHS AND COMPONENTS**									
0020	Heater, timer & head, single, to 140 C.F.	1 Plum	1.20	6.667	Ea.	2,200	390		2,590	3,000
0500	To 300 C.F.		1.10	7.273		2,425	425		2,850	3,325
1000	Commercial size, with blow-down assembly, to 800 C.F.		.90	8.889		5,725	520		6,245	7,100
1500	To 2500 C.F.	↓	.80	10		7,650	585		8,235	9,300
2000	Multiple, motels, apts., 2 baths, w/blow-down assm., 500 C.F.	Q-1	1.30	12.308		6,325	650		6,975	7,950
2500	4 baths	"	.70	22.857		10,200	1,200		11,400	13,000
2700	Conversion unit for residential tub, including door				↓	3,550			3,550	3,925

13 34 Fabricated Engineered Structures

13 34 23 – Fabricated Structures

13 34 23.10 Comfort Stations	Crew	Daily Output	Labor-Hours	Unit	Material	2015 Bare Costs Labor	Equipment	Total	Total Incl O&P
0010 **COMFORT STATIONS** Prefab., stock, w/doors, windows & fixt.									
0100 Not incl. interior finish or electrical									
0300 Mobile, on steel frame, 2 unit				S.F.	163			163	180
0350 7 unit					271			271	298
0400 Permanent, including concrete slab, 2 unit	B-12J	50	.320		249	14.30	17.65	280.95	315
0500 6 unit	"	43	.372	↓	191	16.60	20.50	228.10	259
0600 Alternate pricing method, mobile, 2 fixture				Fixture	5,650			5,650	6,225
0650 7 fixture					9,750			9,750	10,700
0700 Permanent, 2 unit	B-12J	.70	22.857		20,500	1,025	1,250	22,775	25,500
0750 6 unit	"	.50	32	↓	17,400	1,425	1,775	20,600	23,300

13 42 Building Modules

13 42 63 – Detention Cell Modules

13 42 63.16 Steel Detention Cell Modules

	Crew	Daily Output	Labor-Hours	Unit	Material	Labor	Equipment	Total	Total Incl O&P
0010 **STEEL DETENTION CELL MODULES**									
2000 Cells, prefab., 5' to 6' wide, 7' to 8' high, 7' to 8' deep,									
2010 bar front, cot, not incl. plumbing	E-4	1.50	21.333	Ea.	9,775	1,125	97.50	10,997.50	12,800

13 47 Facility Protection

13 47 13 – Cathodic Protection

13 47 13.16 Cathodic Prot. for Underground Storage Tanks

	Crew	Daily Output	Labor-Hours	Unit	Material	Labor	Equipment	Total	Total Incl O&P
0010 **CATHODIC PROTECTION FOR UNDERGROUND STORAGE TANKS**									
1000 Anodes, magnesium type, 9 #	R-15	18.50	2.595	Ea.	41.50	140	16.30	197.80	273
1010 17 #		13	3.692		71.50	199	23	293.50	400
1020 32 #		10	4.800		124	258	30	412	555
1030 48 #	↓	7.20	6.667		164	360	42	566	765
1100 Graphite type w/epoxy cap, 3" x 60" (32 #)	R-22	8.40	4.438		128	206		334	455
1110 4" x 80" (68 #)		6	6.213		241	289		530	705
1120 6" x 72" (80 #)		5.20	7.169		1,500	335		1,835	2,150
1130 6" x 36" (45 #)	↓	9.60	3.883		760	181		941	1,100
2000 Rectifiers, silicon type, air cooled, 28 V/10 A	R-19	3.50	5.714		2,300	315		2,615	3,025
2010 20 V/20 A		3.50	5.714		2,375	315		2,690	3,075
2100 Oil immersed, 28 V/10 A		3	6.667		3,175	365		3,540	4,025
2110 20 V/20 A	↓	3	6.667	↓	3,175	365		3,540	4,050
3000 Anode backfill, coke breeze	R-22	3850	.010	Lb.	.22	.45		.67	.92
4000 Cable, HMWPE, No. 8		2.40	15.533	M.L.F.	500	720		1,220	1,650
4010 No. 6		2.40	15.533		745	720		1,465	1,925
4020 No. 4		2.40	15.533		1,125	720		1,845	2,350
4030 No. 2		2.40	15.533		1,750	720		2,470	3,025
4040 No. 1		2.20	16.945		2,400	790		3,190	3,850
4050 No. 1/0		2.20	16.945		2,975	790		3,765	4,475
4060 No. 2/0		2.20	16.945		4,625	790		5,415	6,300
4070 No. 4/0	↓	2	18.640	↓	6,000	865		6,865	7,925
5000 Test station, 7 terminal box, flush curb type w/lockable cover	R-19	12	1.667	Ea.	71.50	91.50		163	216
5010 Reference cell, 2" dia PVC conduit, cplg., plug, set flush	"	4.80	4.167	"	151	228		379	505

Estimating Tips
General

■ Many products in Division 14 will require some type of support or blocking for installation not included with the item itself. Examples are supports for conveyors or tube systems, attachment points for lifts, and footings for hoists or cranes. Add these supports in the appropriate division.

14 10 00 Dumbwaiters
14 20 00 Elevators

■ Dumbwaiters and elevators are estimated and purchased in a method similar to buying a car. The manufacturer has a base unit with standard features. Added to this base unit price will be whatever options the owner or specifications require. Increased load capacity, additional vertical travel, additional stops, higher speed, and cab finish options are items to be considered. When developing an estimate for dumbwaiters and elevators, remember that some items needed by the installers may have to be included as part of the general contract.

Examples are:

□ shaftway

□ rail support brackets

□ machine room

□ electrical supply

□ sill angles

□ electrical connections

□ pits

□ roof penthouses

□ pit ladders

Check the job specifications and drawings before pricing.

■ Installation of elevators and handicapped lifts in historic structures can require significant additional costs. The associated structural requirements may involve cutting into and repairing finishes, moldings, flooring, etc. The estimator must account for these special conditions.

14 30 00 Escalators and Moving Walks

■ Escalators and moving walks are specialty items installed by specialty contractors. There are numerous options associated with these items. For specific options, contact a manufacturer or contractor. In a method similar to estimating

dumbwaiters and elevators, you should verify the extent of general contract work and add items as necessary.

14 40 00 Lifts
14 90 00 Other Conveying Equipment

■ Products such as correspondence lifts, chutes, and pneumatic tube systems, as well as other items specified in this subdivision, may require trained installers. The general contractor might not have any choice as to who will perform the installation, or when it will be performed. Long lead times are often required for these products, making early decisions in scheduling necessary.

Reference Numbers

Reference numbers are shown in shaded boxes at the beginning of some major classifications. These numbers refer to related items in the Reference Section. The reference information may be an estimating procedure, an alternate pricing method, or technical information.

Note: Not all subdivisions listed here necessarily appear in this publication. ■

14 45 10.10 Hydraulic Lifts	Crew	Daily Output	Labor-Hours	Unit	Material	2015 Bare Costs Labor	Equipment	Total	Total Incl O&P
0010 **HYDRAULIC LIFTS**									
2200 Single post, 8000 lb. capacity	L-4	.40	60	Ea.	5,550	2,650		8,200	10,200
2810 Double post, 6000 lb. capacity		2.67	8.989		7,825	400		8,225	9,250
2815 9000 lb. capacity		2.29	10.480		18,600	465		19,065	21,200
2820 15,000 lb. capacity		2	12		21,000	530		21,530	23,900
2822 Four post, 26,000 lb. capacity		1.80	13.333		14,200	590		14,790	16,500
2825 30,000 lb. capacity		1.60	15		46,400	665		47,065	52,000
2830 Ramp style, 4 post, 25,000 lb. capacity		2	12		18,200	530		18,730	20,800
2835 35,000 lb. capacity		1	24		84,500	1,050		85,550	94,500
2840 50,000 lb. capacity		1	24		94,500	1,050		95,550	105,500
2845 75,000 lb. capacity		1	24		110,000	1,050		111,050	122,500
2850 For drive thru tracks, add, minimum					1,175			1,175	1,275
2855 Maximum					2,000			2,000	2,200
2860 Ramp extensions, 3' (set of 2)					960			960	1,050
2865 Rolling jack platform					3,325			3,325	3,675
2870 Electric/hydraulic jacking beam					8,925			8,925	9,825
2880 Scissor lift, portable, 6000 lb. capacity					8,750			8,750	9,625

Estimating Tips

Pipe for fire protection and all uses is located in Subdivisions 21 11 13 and 22 11 13.

The labor adjustment factors listed in Subdivision 22 01 02.20 also apply to Division 21.

Many, but not all, areas in the U.S. require backflow protection in the fire system. It is advisable to check local building codes for specific requirements.

For your reference, the following is a list of the most applicable Fire Codes and Standards which may be purchased from the NFPA, 1 Batterymarch Park, Quincy, MA 02169-7471.

- NFPA 1: Uniform Fire Code
- NFPA 10: Portable Fire Extinguishers
- NFPA 11: Low-, Medium-, and High-Expansion Foam
- NFPA 12: Carbon Dioxide Extinguishing Systems (Also companion 12A)
- NFPA 13: Installation of Sprinkler Systems (Also companion 13D, 13E, and 13R)
- NFPA 14: Installation of Standpipe and Hose Systems
- NFPA 15: Water Spray Fixed Systems for Fire Protection
- NFPA 16: Installation of Foam-Water Sprinkler and Foam-Water Spray Systems
- NFPA 17: Dry Chemical Extinguishing Systems (Also companion 17A)
- NFPA 18: Wetting Agents
- NFPA 20: Installation of Stationary Pumps for Fire Protection
- NFPA 22: Water Tanks for Private Fire Protection
- NFPA 24: Installation of Private Fire Service Mains and their Appurtenances
- NFPA 25: Inspection, Testing and Maintenance of Water-Based Fire Protection

Reference Numbers

Reference numbers are shown in shaded boxes at the beginning of some major classifications. These numbers refer to related items in the Reference Section. The reference information may be an estimating procedure, an alternate pricing method, or technical information.

Note: Not all subdivisions listed here necessarily appear in this publication. ■

21 05 Common Work Results for Fire Suppression

21 05 23 – General-Duty Valves for Water-Based Fire-Suppression Piping

21 05 23.50 General-Duty Valves		Crew	Daily Output	Labor-Hours	Unit	Material	2015 Bare Costs Labor	Equipment	Total	Total Incl O&P
0010	**GENERAL-DUTY VALVES**, for water-based fire suppression									
6200	Valves and components									
6210	Alarm, includes									
6220	retard chamber, trim, gauges, alarm line strainer									
6260	3" size	Q-12	3	5.333	Ea.	1,625	269		1,894	2,200
6280	4" size	"	2	8		1,700	405		2,105	2,475
6300	6" size	Q-13	4	8		1,900	430		2,330	2,750
6320	8" size	"	3	10.667	↓	2,125	570		2,695	3,225
6500	Check, swing, C.I. body, brass fittings, auto. ball drip									
6520	4" size	Q-12	3	5.333	Ea.	350	269		619	795
6540	6" size	Q-13	4	8		650	430		1,080	1,350
6580	8" size	"	3	10.667	↓	1,225	570		1,795	2,225
6800	Check, wafer, butterfly type, C.I. body, bronze fittings									
6820	4" size	Q-12	4	4	Ea.	1,000	202		1,202	1,400
6840	6" size	Q-13	5.50	5.818		1,625	310		1,935	2,250
6860	8" size		5	6.400		1,850	340		2,190	2,575
6880	10" size	↓	4.50	7.111		2,250	380		2,630	3,050
8800	Flow control valve, includes trim and gauges, 2" size	Q-12	2	8		4,700	405		5,105	5,775
8820	3" size	"	1.50	10.667		5,175	540		5,715	6,500
8840	4" size	Q-13	2.80	11.429		5,775	610		6,385	7,275
8860	6" size	"	2	16	↓	6,675	855		7,530	8,650
9200	Pressure operated relief valve, brass body	1 Spri	18	.444	↓	580	25		605	675
9600	Waterflow indicator, with recycling retard and									
9610	two single pole retard switches, 2" thru 6" pipe size	1 Spri	8	1	Ea.	137	56		193	236

21 11 Facility Fire-Suppression Water-Service Piping

21 11 13 – Facility Water Distribution Piping

21 11 13.16 Pipe, Plastic

		Crew	Daily Output	Labor-Hours	Unit	Material	Labor	Equipment	Total	Total Incl O&P
0010	**PIPE, PLASTIC**									
0020	CPVC, fire suppression, (C-UL-S. FM, NFPA 13, 13D & 13R)									
0030	Socket joint, no couplings or hangers									
0100	SDR 13.5, (ASTM F442)									
0120	3/4" diameter	Q-12	420	.038	L.F.	1.67	1.93		3.60	4.75
0130	1" diameter		340	.047		2.58	2.38		4.96	6.45
0140	1-1/4" diameter		260	.062		4.10	3.11		7.21	9.20
0150	1-1/2" diameter		190	.084		5.65	4.26		9.91	12.70
0160	2" diameter		140	.114		9	5.80		14.80	18.65
0170	2-1/2" diameter		130	.123		16.65	6.20		22.85	28
0180	3" diameter	↓	120	.133	↓	25.50	6.75		32.25	38

21 11 13.18 Pipe Fittings, Plastic

		Crew	Daily Output	Labor-Hours	Unit	Material	Labor	Equipment	Total	Total Incl O&P
0010	**PIPE FITTINGS, PLASTIC**									
0020	CPVC, fire suppression, (C-UL-S. FM, NFPA 13, 13D & 13R)									
0030	Socket joint									
0100	90° Elbow									
0120	3/4"	1 Plum	26	.308	Ea.	2.10	18.05		20.15	30
0130	1"		22.70	.352	↓	4.62	20.50		25.12	36
0140	1-1/4"		20.20	.396		5.85	23.50		29.35	41.50
0150	1-1/2"	↓	18.20	.440		8.30	26		34.30	48
0160	2"	Q-1	33.10	.483		10.30	25.50		35.80	50
0170	2-1/2"		24.20	.661		19.80	35		54.80	75
0180	3"	↓	20.80	.769	↓	27	40.50		67.50	91

21 11 13 – Facility Water Distribution Piping

21 11 13.18 Pipe Fittings, Plastic	Crew	Daily Output	Labor-Hours	Unit	Material	2015 Bare Costs Labor	Equipment	Total	Total Incl O&P
0200	45° Elbow								
0210	3/4"	1 Plum	26	.308	Ea.	2.89	18.05		20.94	30.50
0220	1"		22.70	.352		3.39	20.50		23.89	34.50
0230	1-1/4"		20.20	.396		4.90	23.50		28.40	40.50
0240	1-1/2"		18.20	.440		6.85	26		32.85	46.50
0250	2"	Q-1	33.10	.483		8.50	25.50		34	48
0260	2-1/2"		24.20	.661		15.30	35		50.30	70
0270	3"		20.80	.769		22	40.50		62.50	85.50
0300	Tee								
0310	3/4"	1 Plum	17.30	.462	Ea.	2.89	27		29.89	44
0320	1"		15.20	.526		5.70	31		36.70	53
0330	1-1/4"		13.50	.593		8.55	35		43.55	62
0340	1-1/2"		12.10	.661		12.60	39		51.60	72.50
0350	2"	Q-1	20	.800		18.65	42.50		61.15	84.50
0360	2-1/2"		16.20	.988		30.50	52		82.50	113
0370	3"		13.90	1.151		47.50	61		108.50	144
0400	Tee, reducing x any size								
0420	1"	1 Plum	15.20	.526	Ea.	4.84	31		35.84	52
0430	1-1/4"		13.50	.593		8.85	35		43.85	62.50
0440	1-1/2"		12.10	.661		10.75	39		49.75	70.50
0450	2"	Q-1	20	.800		19.85	42.50		62.35	86
0460	2-1/2"		16.20	.988		23.50	52		75.50	105
0470	3"		13.90	1.151		27	61		88	122
0500	Coupling								
0510	3/4"	1 Plum	26	.308	Ea.	2.02	18.05		20.07	29.50
0520	1"		22.70	.352		2.67	20.50		23.17	34
0530	1-1/4"		20.20	.396		3.89	23.50		27.39	39.50
0540	1-1/2"		18.20	.440		5.55	26		31.55	45
0550	2"	Q-1	33.10	.483		7.50	25.50		33	47
0560	2-1/2"		24.20	.661		11.45	35		46.45	65.50
0570	3"		20.80	.769		14.90	40.50		55.40	78
0600	Coupling, reducing								
0610	1" x 3/4"	1 Plum	22.70	.352	Ea.	2.67	20.50		23.17	34
0620	1-1/4" x 1"		20.20	.396		4.04	23.50		27.54	39.50
0630	1-1/2" x 3/4"		18.20	.440		6.05	26		32.05	45.50
0640	1-1/2" x 1"		18.20	.440		5.85	26		31.85	45.50
0650	1-1/2" x 1-1/4"		18.20	.440		5.55	26		31.55	45
0660	2" x 1"	Q-1	33.10	.483		7.80	25.50		33.30	47
0670	2" x 1-1/2"	"	33.10	.483		7.50	25.50		33	47
0700	Cross								
0720	3/4"	1 Plum	13	.615	Ea.	4.54	36		40.54	59.50
0730	1"		11.30	.708		5.70	41.50		47.20	69.50
0740	1-1/4"		10.10	.792		7.85	46.50		54.35	78.50
0750	1-1/2"		9.10	.879		10.90	51.50		62.40	90
0760	2"	Q-1	16.60	.964		17.75	51		68.75	96.50
0770	2-1/2"	"	12.10	1.322		39	70		109	148
0800	Cap								
0820	3/4"	1 Plum	52	.154	Ea.	1.12	9.05		10.17	14.90
0830	1"		45	.178		1.74	10.45		12.19	17.65
0840	1-1/4"		40	.200		2.81	11.75		14.56	21
0850	1-1/2"		36.40	.220		3.89	12.90		16.79	24
0860	2"	Q-1	66	.242		5.85	12.80		18.65	26
0870	2-1/2"		48.40	.331		8.45	17.45		25.90	36

For customer support on your Plumbing Cost Data, call 877.709.6968.

87

21 11 13 – Facility Water Distribution Piping

21 11 13.18 Pipe Fittings, Plastic

		Crew	Daily Output	Labor-Hours	Unit	Material	2015 Bare Costs Labor	Equipment	Total	Total Incl O&P
0880	3"	Q-1	41.60	.385	Ea.	13.60	20.50		34.10	45.50
0900	Adapter, sprinkler head, female w/metal thd. insert, (s x FNPT)									
0920	3/4" x 1/2"	1 Plum	52	.154	Ea.	5.45	9.05		14.50	19.65
0930	1" x 1/2"		45	.178		5.75	10.45		16.20	22
0940	1" x 3/4"	↓	45	.178	↓	9.05	10.45		19.50	25.50

21 11 16 – Facility Fire Hydrants

21 11 16.50 Fire Hydrants for Buildings

		Crew	Daily Output	Labor-Hours	Unit	Material	2015 Bare Costs Labor	Equipment	Total	Total Incl O&P
0010	**FIRE HYDRANTS FOR BUILDINGS**									
3750	Hydrants, wall, w/caps, single, flush, polished brass									
3800	2-1/2" x 2-1/2"	Q-12	5	3.200	Ea.	213	162		375	480
3840	2-1/2" x 3"		5	3.200		430	162		592	715
3860	3" x 3"	↓	4.80	3.333		350	168		518	640
3900	For polished chrome, add				↓	20%				
3950	Double, flush, polished brass									
4000	2-1/2" x 2-1/2" x 4"	Q-12	5	3.200	Ea.	570	162		732	875
4040	2-1/2" x 2-1/2" x 6"		4.60	3.478		825	176		1,001	1,175
4080	3" x 3" x 4"		4.90	3.265		1,225	165		1,390	1,600
4120	3" x 3" x 6"	↓	4.50	3.556		1,250	180		1,430	1,650
4200	For polished chrome, add				↓	10%				
4350	Double, projecting, polished brass									
4400	2-1/2" x 2-1/2" x 4"	Q-12	5	3.200	Ea.	254	162		416	525
4450	2-1/2" x 2-1/2" x 6"	"	4.60	3.478	"	520	176		696	835
4460	Valve control, dbl. flush/projecting hydrant, cap &									
4470	chain, extension rod & cplg., escutcheon, polished brass	Q-12	8	2	Ea.	300	101		401	485
4480	Four-way square, flush, polished brass									
4540	2-1/2"(4) x 6"	Q-12	3.60	4.444	Ea.	3,075	225		3,300	3,725

21 11 19 – Fire-Department Connections

21 11 19.50 Connections for the Fire-Department

		Crew	Daily Output	Labor-Hours	Unit	Material	2015 Bare Costs Labor	Equipment	Total	Total Incl O&P
0010	**CONNECTIONS FOR THE FIRE-DEPARTMENT**									
6000	Roof manifold, horiz., brass, without valves & caps									
6040	2-1/2" x 2-1/2" x 4"	Q-12	4.80	3.333	Ea.	168	168		336	440
6060	2-1/2" x 2-1/2" x 6"		4.60	3.478		184	176		360	470
6080	2-1/2" x 2-1/2" x 2-1/2" x 4"		4.60	3.478		271	176		447	565
6090	2-1/2" x 2-1/2" x 2-1/2" x 6"	↓	4.60	3.478	↓	280	176		456	575
7000	Sprinkler line tester, cast brass					24			24	26.50
7140	Standpipe connections, wall, w/plugs & chains									
7160	Single, flush, brass, 2-1/2" x 2-1/2"	Q-12	5	3.200	Ea.	159	162		321	420
7180	2-1/2" x 3"	"	5	3.200	"	164	162		326	425
7240	For polished chrome, add					15%				
7280	Double, flush, polished brass									
7300	2-1/2" x 2-1/2" x 4"	Q-12	5	3.200	Ea.	520	162		682	815
7330	2-1/2" x 2-1/2" x 6"		4.60	3.478		725	176		901	1,050
7340	3" x 3" x 4"		4.90	3.265		945	165		1,110	1,300
7370	3" x 3" x 6"	↓	4.50	3.556	↓	1,125	180		1,305	1,525
7400	For polished chrome, add					15%				
7440	For sill cock combination, add				Ea.	90.50			90.50	99.50
7580	Double projecting, polished brass									
7600	2-1/2" x 2-1/2" x 4"	Q-12	5	3.200	Ea.	475	162		637	765
7630	2-1/2" x 2-1/2" x 6"	"	4.60	3.478	"	800	176		976	1,150
7680	For polished chrome, add					15%				
7900	Three way, flush, polished brass									
7920	2-1/2" (3) x 4"	Q-12	4.80	3.333	Ea.	1,650	168		1,818	2,075

21 11 19 – Fire-Department Connections

21 11 19.50 Connections for the Fire-Department	Crew	Daily Output	Labor-Hours	Unit	Material	2015 Bare Costs Labor	Equipment	Total	Total Incl O&P
7930 2-1/2" (3) x 6"	Q-12	4.80	3.333	Ea.	1,650	168		1,818	2,075
8000 For polished chrome, add			↓		9%				
8020 Three way, projecting, polished brass									
8040 2-1/2"(3) x 4"	Q-12	4.80	3.333	Ea.	1,575	168		1,743	1,975
8070 2-1/2" (3) x 6"	"	4.60	3.478		1,575	176		1,751	2,000
8100 For polished chrome, add			↓		12%				
8200 Four way, square, flush, polished brass,									
8240 2-1/2"(4) x 6"	Q-12	3.60	4.444	Ea.	1,250	225		1,475	1,725
8300 For polished chrome, add				"	10%				
8550 Wall, vertical, flush, cast brass									
8600 Two way, 2-1/2" x 2-1/2" x 4"	Q-12	5	3.200	Ea.	395	162		557	680
8660 Four way, 2-1/2"(4) x 6"		3.80	4.211		1,300	213		1,513	1,750
8680 Six way, 2-1/2"(6) x 6"	↓	3.40	4.706	↓	1,550	238		1,788	2,075
8700 For polished chrome, add				↓	10%				
8800 Sidewalk siamese unit, polished brass, two way									
8820 2-1/2" x 2-1/2" x 4"	Q-12	2.50	6.400	Ea.	635	325		960	1,175
8850 2-1/2" x 2-1/2" x 6"		2	8		780	405		1,185	1,475
8860 3" x 3" x 4"		2.50	6.400		900	325		1,225	1,475
8890 3" x 3" x 6"	↓	2	8	↓	1,225	405		1,630	1,950
8940 For polished chrome, add					12%				
9100 Sidewalk siamese unit, polished brass, three way									
9120 2-1/2" x 2-1/2" x 2-1/2" x 6"	Q-12	2	8	Ea.	1,000	405		1,405	1,700
9160 For polished chrome, add				"	15%				

21 12 Fire-Suppression Standpipes

21 12 13 – Fire-Suppression Hoses and Nozzles

21 12 13.50 Fire Hoses and Nozzles

				Unit	Material	Labor	Equipment	Total	Total Incl O&P
0010 **FIRE HOSES AND NOZZLES**									
0200 Adapters, rough brass, straight hose threads R211226-10									
0220 One piece, female to male, rocker lugs									
0240 1" x 1" R211226-20				Ea.	48			48	53
0260 1-1/2" x 1"					48			48	52.50
0280 1-1/2" x 1-1/2"					12.90			12.90	14.20
0300 2" x 1-1/2"					66			66	72.50
0320 2" x 2"					42			42	46
0340 2-1/2" x 1-1/2"					14.70			14.70	16.15
0360 2-1/2" x 2"					38			38	41.50
0380 2-1/2" x 2-1/2"					19			19	21
0400 3" x 2-1/2"					125			125	137
0420 3" x 3"					83.50			83.50	92
0500 For polished brass, add					50%				
0520 For polished chrome, add				↓	75%				
0700 One piece, female to male, hexagon									
0740 1-1/2" x 3/4"				Ea.	40.50			40.50	44.50
0760 2" x 1-1/2"					84			84	92.50
0780 2-1/2" x 1"					161			161	178
0800 2-1/2" x 1-1/2"					55.50			55.50	61
0820 2-1/2" x 2"					38			38	41.50
0840 3" x 2-1/2"					70			70	77
0900 For polished chrome, add				↓	75%				
1100 Swivel, female to female, pin lugs									

21 12 Fire-Suppression Standpipes

21 12 13 – Fire-Suppression Hoses and Nozzles

21 12 13.50 Fire Hoses and Nozzles	Crew	Daily Output	Labor-Hours	Unit	Material	2015 Bare Costs Labor	Equipment	Total	Total Incl O&P
1120　1-1/2" x 1-1/2"				Ea.	61.50			61.50	67.50
1200　2-1/2" x 2-1/2"					120			120	132
1260　For polished brass, add					50%				
1280　For polished chrome, add					75%				
1400　Couplings, sngl. & dbl. jacket, pin lug or rocker lug, cast brass									
1410　1-1/2"				Ea.	53.50			53.50	58.50
1420　2-1/2"				"	70			70	77
1500　For polished brass, add					20%				
1520　For polished chrome, add					40%				
1580　Reducing, F x M, interior installation, cast brass									
1590　2" x 1-1/2"				Ea.	66			66	72.50
1600　2-1/2" x 1-1/2"					14.70			14.70	16.15
1680　For polished brass, add					50%				
1720　For polished chrome, add					75%				
2200　Hose, less couplings									
2260　Synthetic jacket, lined, 300 lb. test, 1-1/2" diameter	Q-12	2600	.006	L.F.	3.24	.31		3.55	4.03
2280　2-1/2" diameter		2200	.007		5.60	.37		5.97	6.70
2360　High strength, 500 lb. test, 1-1/2" diameter		2600	.006		3.35	.31		3.66	4.16
2380　2-1/2" diameter		2200	.007		5.90	.37		6.27	7.05
5000　Nipples, straight hose to tapered iron pipe, brass									
5060　Female to female, 1-1/2" x 1-1/2"				Ea.	19.85			19.85	22
5100　2-1/2" x 2-1/2"					37			37	40.50
5190　For polished chrome, add					75%				
5200　Double male or male to female, 1" x 1"					42			42	46
5220　1-1/2" x 1"					62			62	68
5230　1-1/2" x 1-1/2"					14.35			14.35	15.75
5260　2" x 1-1/2"					67.50			67.50	74
5270　2" x 2"					84			84	92.50
5280　2-1/2" x 1-1/2"					55.50			55.50	61
5300　2-1/2" x 2"					45.50			45.50	50
5310　2-1/2" x 2-1/2"					25.50			25.50	28.50
5340　For polished chrome, add					75%				
5600　Nozzles, brass									
5620　Adjustable fog, 3/4" booster line				Ea.	111			111	122
5630　1" booster line					137			137	150
5640　1-1/2" leader line					101			101	112
5660　2-1/2" direct connection					151			151	166
5680　2-1/2" playpipe nozzle					223			223	246
5780　For chrome plated, add					8%				
5850　Electrical fire, adjustable fog, no shock									
5900　1-1/2"				Ea.	415			415	455
5920　2-1/2"					560			560	615
5980　For polished chrome, add					6%				
6200　Heavy duty, comb. adj. fog and str. stream, with handle									
6210　1" booster line				Ea.	375			375	410
6240　1-1/2"					450			450	495
6260　2-1/2", for playpipe					750			750	825
6280　2-1/2" direct connection					560			560	615
6300　2-1/2" playpipe combination					560			560	615
6480　For polished chrome, add					7%				
6500　Plain fog, polished brass, 1-1/2"					118			118	129
6540　Chrome plated, 1-1/2"					117			117	128
6700　Plain stream, polished brass, 1-1/2" x 10"					48			48	52.50

90

For customer support on your Plumbing Cost Data, call 877.709.6968.

21 12 13 – Fire-Suppression Hoses and Nozzles

21 12 13.50 Fire Hoses and Nozzles	Crew	Daily Output	Labor-Hours	Unit	Material	2015 Bare Costs Labor	Equipment	Total	Total Incl O&P
6760 2-1/2" x 15" x 7/8" or 1-1/2"				Ea.	100			100	110
6860 For polished chrome, add					20%				
7000 Underwriters playpipe, 2-1/2" x 30" with 1-1/8" tip				Ea.	239			239	263
9200 Storage house, hose only, primed steel					855			855	940
9220 Aluminum					1,875			1,875	2,050
9280 Hose and hydrant house, primed steel					1,150			1,150	1,275
9300 Aluminum					1,675			1,675	1,825
9340 Tools; crowbar and brackets	1 Carp	12	.667		70	31.50		101.50	125
9360 Combination hydrant wrench and spanner					31.50			31.50	35
9380 Fire axe and brackets									
9400 6 lb.	1 Carp	12	.667	Ea.	116	31.50		147.50	175

21 12 16 – Fire-Suppression Hose Reels

21 12 16.50 Fire-Suppression Hose Reels	Crew	Daily Output	Labor-Hours	Unit	Material	Labor	Equipment	Total	Total Incl O&P
0010 **FIRE-SUPPRESSION HOSE REELS**									
2990 Hose reel, swinging, for 1-1/2" polyester neoprene lined hose									
3000 50' long	Q-12	14	1.143	Ea.	131	58		189	232
3020 100' long		14	1.143		184	58		242	290
3060 For 2-1/2" cotton rubber hose, 75' long		14	1.143		212	58		270	320
3100 150' long		14	1.143		248	58		306	360

21 12 19 – Fire-Suppression Hose Racks

21 12 19.50 Fire Hose Racks	Crew	Daily Output	Labor-Hours	Unit	Material	Labor	Equipment	Total	Total Incl O&P
0010 **FIRE HOSE RACKS**									
2600 Hose rack, swinging, for 1-1/2" diameter hose,									
2620 Enameled steel, 50' & 75' lengths of hose	Q-12	20	.800	Ea.	61	40.50		101.50	128
2640 100' and 125' lengths of hose		20	.800		61	40.50		101.50	128
2680 Chrome plated, 50' and 75' lengths of hose		20	.800		98.50	40.50		139	169
2700 100' and 125' lengths of hose		20	.800		104	40.50		144.50	176
2780 For hose rack nipple, 1-1/2" polished brass, add					27			27	29.50
2820 2-1/2" polished brass, add					49			49	54
2840 1-1/2" polished chrome, add					39.50			39.50	43.50
2860 2-1/2" polished chrome, add					52			52	57.50

21 12 23 – Fire-Suppression Hose Valves

21 12 23.70 Fire Hose Valves	Crew	Daily Output	Labor-Hours	Unit	Material	Labor	Equipment	Total	Total Incl O&P
0010 **FIRE HOSE VALVES**									
0020 Angle, combination pressure adjust/restricting, rough brass R211226-10									
0030 1-1/2"	1 Spri	12	.667	Ea.	85.50	37.50		123	151
0040 2-1/2" R211226-20	"	7	1.143	"	175	64		239	289
0042 Nonpressure adjustable/restricting, rough brass									
0044 1-1/2"	1 Spri	12	.667	Ea.	53.50	37.50		91	115
0046 2-1/2"	"	7	1.143	"	90	64		154	196
0050 For polished brass, add					30%				
0060 For polished chrome, add					40%				
0080 Wheel handle, 300 lb., 1-1/2"	1 Spri	12	.667	Ea.	95	37.50		132.50	161
0090 2-1/2"	"	7	1.143	"	175	64		239	289
0100 For polished brass, add					35%				
0110 For polished chrome, add					50%				
1000 Ball drip, automatic, rough brass, 1/2"	1 Spri	20	.400	Ea.	16.15	22.50		38.65	52
1010 3/4"	"	20	.400	"	18	22.50		40.50	54
1100 Ball, 175 lb., sprinkler system, FM/UL, threaded, bronze									
1120 Slow close									
1150 1" size	1 Spri	19	.421	Ea.	241	23.50		264.50	300

21 12 23.70 Fire Hose Valves		Crew	Daily Output	Labor-Hours	Unit	Material	2015 Bare Costs Labor	Equipment	Total	Total Incl O&P
1160	1-1/4" size	1 Spri	15	.533	Ea.	260	30		290	330
1170	1-1/2" size		13	.615		440	34.50		474.50	540
1180	2" size	↓	11	.727		415	41		456	520
1190	2-1/2" size	Q-12	15	1.067	↓	560	54		614	700
1230	For supervisory switch kit, all sizes									
1240	One circuit, add	1 Spri	48	.167	Ea.	165	9.35		174.35	196
1280	Quarter turn for trim									
1300	1/2" size	1 Spri	22	.364	Ea.	32	20.50		52.50	66
1310	3/4" size		20	.400		34.50	22.50		57	71.50
1320	1" size		19	.421		38	23.50		61.50	78
1330	1-1/4" size		15	.533		62.50	30		92.50	114
1340	1-1/2" size		13	.615		78.50	34.50		113	139
1350	2" size	↓	11	.727		93.50	41		134.50	165
1400	Caps, polished brass with chain, 3/4"					42			42	46
1420	1"					53.50			53.50	59
1440	1-1/2"					14.35			14.35	15.75
1460	2-1/2"					21.50			21.50	24
1480	3"					33.50			33.50	37
1900	Escutcheon plate, for angle valves, polished brass, 1-1/2"					15.20			15.20	16.70
1920	2-1/2"					24.50			24.50	27
1940	3"					31			31	34
1980	For polished chrome, add					15%				
2000	Foam, control valve, 3"	1 Spri	6	1.333		1,825	75		1,900	2,150
2020	Supply valve, 2-1/2"		7	1.143		154	64		218	266
2040	Proportioner, 8"	↓	2	4		3,000	225		3,225	3,650
2060	Oscillating foam monitor with electric remote control	Q-12	5.33	3.002		11,300	152		11,452	12,600
3000	Gate, hose, wheel handle, N.R.S., rough brass, 1-1/2"	1 Spri	12	.667		135	37.50		172.50	205
3040	2-1/2", 300 lb.	"	7	1.143		189	64		253	305
3080	For polished brass, add					40%				
3090	For polished chrome, add					50%				
3800	Hydrant, screw type, crank handle, brass									
3840	2-1/2" size	Q-12	11	1.455	Ea.	345	73.50		418.50	490
3880	For chrome, same price									
4200	Hydrolator, vent and draining, rough brass, 1-1/2"	1 Spri	12	.667	Ea.	84	37.50		121.50	149
4280	For polished brass, add					50%				
4290	For polished chrome, add					90%				
5000	Pressure restricting, adjustable rough brass, 1-1/2"	1 Spri	12	.667		107	37.50		144.50	174
5020	2-1/2"	"	7	1.143		160	64		224	273
5080	For polished brass, add					30%				
5090	For polished chrome, add					45%				
8000	Wye, leader line, ball type, swivel female x male x male									
8040	2-1/2" x 1-1/2" x 1-1/2" polished brass				Ea.	262			262	288
8060	2-1/2" x 1-1/2" x 1-1/2" polished chrome				"	245			245	270

21 13 13 – Wet-Pipe Sprinkler Systems

21 13 13.50 Wet-Pipe Sprinkler System Components	Crew	Daily Output	Labor-Hours	Unit	Material	2015 Bare Costs Labor	Equipment	Total	Total Incl O&P
0010	**WET-PIPE SPRINKLER SYSTEM COMPONENTS**								
1100	Alarm, electric pressure switch (circuit closer) R211313-10	1 Spri	26	.308	Ea.	88.50	17.30		105.80	124
1140	For explosion proof, max 20 PSI, contacts close or open		26	.308		535	17.30		552.30	610
1220	Water motor, complete with gong		4	2		415	112		527	625
1800	Firecycle system, controls, includes panel,									
1820	batteries, solenoid valves and pressure switches	Q-13	1	32	Ea.	20,900	1,700		22,600	25,600
1860	Detector	1 Spri	16	.500	"	715	28		743	830
1900	Flexible sprinkler head connectors									
1910	Braided stainless steel hose with mounting bracket									
1920	1/2" and 3/4" outlet size									
1940	40" length	1 Spri	30	.267	Ea.	70.50	14.95		85.45	100
1960	60" length	"	22	.364	"	81	20.50		101.50	120
1982	May replace hard-pipe armovers									
1984	For wet, pre-action, deluge or dry pipe systems									
2000	Release, emergency, manual, for hydraulic or pneumatic system	1 Spri	12	.667	Ea.	218	37.50		255.50	296
2060	Release, thermostatic, for hydraulic or pneumatic release line		20	.400		675	22.50		697.50	775
2200	Sprinkler cabinets, 6 head capacity		16	.500		79	28		107	129
2260	12 head capacity		16	.500		83.50	28		111.50	135
2340	Sprinkler head escutcheons, standard, brass tone, 1" size		40	.200		2.85	11.25		14.10	20
2360	Chrome, 1" size		40	.200		3.04	11.25		14.29	20.50
2400	Recessed type, bright brass		40	.200		10.10	11.25		21.35	28
2440	Chrome or white enamel		40	.200		3.35	11.25		14.60	20.50
2600	Sprinkler heads, not including supply piping									
3700	Standard spray, pendent or upright, brass, 135°F to 286°F									
3720	1/2" NPT, 3/8" orifice	1 Spri	16	.500	Ea.	15.40	28		43.40	59.50
3730	1/2" NPT, 7/16" orifice		16	.500		15.20	28		43.20	59.50
3740	1/2" NPT, 1/2" orifice		16	.500		9.90	28		37.90	53.50
3760	1/2" NPT, 17/32" orifice		16	.500		12.90	28		40.90	56.50
3780	3/4" NPT, 17/32" orifice		16	.500		11.90	28		39.90	55.50
3800	For open sprinklers, deduct					15%				
3840	For chrome, add				Ea.	3.38			3.38	3.72
3860	For wax and lead coating, add					35			35	38
3880	For wax coating, add					21			21	23
3900	For lead coating, add					22.50			22.50	25
3920	For 360°F, same cost									
3930	For 400°F				Ea.	93			93	102
3940	For 500°F				"	93			93	102
4200	Sidewall, vertical brass, 135°F to 286°F									
4240	1/2" NPT, 1/2" orifice	1 Spri	16	.500	Ea.	25	28		53	70
4280	3/4" NPT, 17/32" orifice	"	16	.500		70	28		98	119
4360	For satin chrome, add					2.75			2.75	3.03
4400	For 360°F, same cost									
4500	Sidewall, horizontal, brass, 135°F to 286°F									
4520	1/2" NPT, 1/2" orifice	1 Spri	16	.500	Ea.	25	28		53	70
4540	For 360°F, same cost									
4800	Recessed pendent, brass, 135°F to 286°F									
4820	1/2" NPT, 3/8" orifice	1 Spri	10	.800	Ea.	42.50	45		87.50	115
4830	1/2" NPT, 7/16" orifice		10	.800		18.40	45		63.40	88
4840	1/2" NPT, 1/2" orifice		10	.800		14.25	45		59.25	83.50
4860	1/2" NPT, 17/32" orifice		10	.800		42.50	45		87.50	115
4900	For satin chrome, add					2.75			2.75	3.03
5000	Recessed-vertical sidewall, brass, 135°F to 286°F									
5020	1/2" NPT, 3/8" orifice	1 Spri	10	.800	Ea.	31	45		76	102

21 13 13 – Wet-Pipe Sprinkler Systems

21 13 13.50 Wet-Pipe Sprinkler System Components	Crew	Daily Output	Labor-Hours	Unit	Material	2015 Bare Costs Labor	Equipment	Total	Total Incl O&P
5030	1/2" NPT, 7/16" orifice	1 Spri	10	.800	Ea.	31	45		76	102
5040	1/2" NPT, 1/2" orifice	↓	10	.800	↓	31	45		76	102
5100	For bright nickel, same cost									
5600	Concealed, complete with cover plate									
5620	1/2" NPT, 1/2" orifice, 135°F to 212°F	1 Spri	9	.889	Ea.	24	50		74	102
5800	Window, brass, 1/2" NPT, 1/4" orifice		16	.500		34	28		62	79.50
5810	1/2" NPT, 5/16" orifice		16	.500		34	28		62	79.50
5820	1/2" NPT, 3/8" orifice		16	.500		34	28		62	79.50
5830	1/2" NPT, 7/16" orifice		16	.500		34	28		62	79.50
5840	1/2" NPT, 1/2" orifice	↓	16	.500		36	28		64	82
5860	For polished chrome, add					4.55			4.55	5
5880	3/4" NPT, 5/8" orifice	1 Spri	16	.500		37.50	28		65.50	83.50
5890	3/4 NPT, 3/4" orifice	"	16	.500		37.50	28		65.50	83.50
6000	Sprinkler head guards, bright zinc, 1/2" NPT					6			6	6.60
6020	Bright zinc, 3/4" NPT				↓	6			6	6.60
6025	Residential sprinkler components, (one and two family)									
6026	Water motor alarm, with strainer	1 Spri	4	2	Ea.	415	112		527	625
6027	Fast response, glass bulb, 135°F to 155°F									
6028	1/2" NPT, pendent, brass	1 Spri	16	.500	Ea.	27	28		55	72.50
6029	1/2" NPT, sidewall, brass		16	.500		27	28		55	72.50
6030	1/2" NPT, pendent, brass, extended coverage		16	.500		24	28		52	68.50
6031	1/2" NPT, sidewall, brass, extended coverage		16	.500		22	28		50	66.50
6032	3/4" NPT sidewall, brass, extended coverage	↓	16	.500	↓	24	28		52	69
6033	For chrome, add					15%				
6034	For polyester/teflon coating add					20%				
6100	Sprinkler head wrenches, standard head				Ea.	24.50			24.50	26.50
6120	Recessed head					37.50			37.50	41
6160	Tamper switch, (valve supervisory switch)	1 Spri	16	.500	↓	102	28		130	155

21 13 16 – Dry-Pipe Sprinkler Systems

21 13 16.50 Dry-Pipe Sprinkler System Components	Crew	Daily Output	Labor-Hours	Unit	Material	2015 Bare Costs Labor	Equipment	Total	Total Incl O&P
0010	**DRY-PIPE SPRINKLER SYSTEM COMPONENTS**									
0600	Accelerator	1 Spri	8	1	Ea.	755	56		811	915
0800	Air compressor for dry pipe system, automatic, complete									
0820	30 gal. system capacity, 3/4 HP	1 Spri	1.30	6.154	Ea.	765	345		1,110	1,375
0860	30 gal. system capacity, 1 HP		1.30	6.154		790	345		1,135	1,400
0910	30 gal. system capacity, 1-1/2 HP		1.30	6.154		825	345		1,170	1,425
0920	30 gal. system capacity, 2 HP		1.30	6.154		875	345		1,220	1,475
0960	Air pressure maintenance control		24	.333		345	18.70		363.70	410
1600	Dehydrator package, incl. valves and nipples R211313-20	↓	12	.667	↓	710	37.50		747.50	840
2600	Sprinkler heads, not including supply piping									
2640	Dry, pendent, 1/2" orifice, 3/4" or 1" NPT									
2660	1/2" to 6" length	1 Spri	14	.571	Ea.	124	32		156	185
2670	6-1/4" to 8" length		14	.571		129	32		161	191
2680	8-1/4" to 12" length		14	.571		135	32		167	197
2690	12-1/4" to 15" length		14	.571		140	32		172	203
2700	15-1/4" to 18" length		14	.571		145	32		177	209
2710	18-1/4" to 21" length		13	.615		151	34.50		185.50	219
2720	21-1/4" to 24" length		13	.615		156	34.50		190.50	225
2730	24-1/4" to 27" length		13	.615		162	34.50		196.50	231
2740	27-1/4" to 30" length		13	.615		168	34.50		202.50	237
2750	30-1/4" to 33" length		13	.615		173	34.50		207.50	244
2760	33-1/4" to 36" length	↓	13	.615	↓	179	34.50		213.50	250

21 13 16 – Dry-Pipe Sprinkler Systems

21 13 16.50 Dry-Pipe Sprinkler System Components	Crew	Daily Output	Labor-Hours	Unit	Material	2015 Bare Costs Labor	Equipment	Total	Total Incl O&P
2780	36-1/4" to 39" length	1 Spri	12	.667	Ea.	185	37.50		222.50	260
2790	39-1/4" to 42" length	↓	12	.667		190	37.50		227.50	265
2800	For each inch or fraction, add				↓	2.84			2.84	3.12
6330	Valves and components									
6340	Alarm test/shut off valve, 1/2"	1 Spri	20	.400	Ea.	21.50	22.50		44	58
8000	Dry pipe air check valve, 3" size	Q-12	2	8		1,775	405		2,180	2,550
8200	Dry pipe valve, incl. trim and gauges, 3" size		2	8		2,575	405		2,980	3,450
8220	4" size	↓	1	16		2,850	810		3,660	4,375
8240	6" size	Q-13	2	16		3,350	855		4,205	4,975
8280	For accelerator trim with gauges, add	1 Spri	8	1	↓	239	56		295	350

21 13 20 – Fire-Cycle Sprinkler Systems

21 13 20.50 Firecycle Fire-Suppression Sprinkler Systems

		Crew	Daily Output	Labor-Hours	Unit	Material	Labor	Equipment	Total	Total Incl O&P
0010	**FIRECYCLE FIRE-SUPPRESSION SPRINKLER SYSTEMS**									
8400	Firecycle package, includes swing check									
8420	and flow control valves with required trim									
8440	2" size	Q-12	2	8	Ea.	4,625	405		5,030	5,675
8460	3" size		1.50	10.667		5,075	540		5,615	6,400
8480	4" size	↓	1	16		5,650	810		6,460	7,450
8500	6" size	Q-13	1.40	22.857	↓	6,525	1,225		7,750	9,025

21 13 26 – Deluge Fire-Suppression Sprinkler Systems

21 13 26.50 Deluge Fire-Suppression Sprinkler Sys. Comp.

		Crew	Daily Output	Labor-Hours	Unit	Material	Labor	Equipment	Total	Total Incl O&P
0010	**DELUGE FIRE-SUPPRESSION SPRINKLER SYSTEM COMPONENTS**									
1400	Deluge system, monitoring panel w/deluge valve & trim	1 Spri	18	.444	Ea.	10,800	25		10,825	11,900
6200	Valves and components									
7000	Deluge, assembly, incl. trim, pressure									
7020	operated relief, emergency release, gauges									
7040	2" size	Q-12	2	8	Ea.	3,675	405		4,080	4,625
7060	3" size		1.50	10.667		4,100	540		4,640	5,350
7080	4" size	↓	1	16		4,700	810		5,510	6,375
7100	6" size	Q-13	1.80	17.778	↓	5,525	950		6,475	7,525
7800	Pneumatic actuator, bronze, required on all									
7820	pneumatic release systems, any size deluge	1 Spri	18	.444	Ea.	415	25		440	495

21 13 39 – Foam-Water Systems

21 13 39.50 Foam-Water System Components

		Crew	Daily Output	Labor-Hours	Unit	Material	Labor	Equipment	Total	Total Incl O&P
0010	**FOAM-WATER SYSTEM COMPONENTS**									
2600	Sprinkler heads, not including supply piping									
3600	Foam-water, pendent or upright, 1/2" NPT	1 Spri	12	.667	Ea.	210	37.50		247.50	287

21 21 Carbon-Dioxide Fire-Extinguishing Systems

21 21 16 – Carbon-Dioxide Fire-Extinguishing Equipment

21 21 16.50 CO2 Fire Extinguishing System

		Crew	Daily Output	Labor-Hours	Unit	Material	Labor	Equipment	Total	Total Incl O&P
0010	**CO_2 FIRE EXTINGUISHING SYSTEM**									
0042	For detectors and control stations, see Section 28 31 23.50									
0100	Control panel, single zone with batteries (2 zones det., 1 suppr.)	1 Elec	1	8	Ea.	1,725	440		2,165	2,550
0150	Multizone (4) with batteries (8 zones det., 4 suppr.)	"	.50	16		3,275	875		4,150	4,900
1000	Dispersion nozzle, CO_2, 3" x 5"	1 Plum	18	.444		67	26		93	113
2000	Extinguisher, CO_2 system, high pressure, 75 lb. cylinder	Q-1	6	2.667		1,275	141		1,416	1,625
2100	100 lb. cylinder	"	5	3.200		1,300	169		1,469	1,700
3000	Electro/mechanical release	L-1	4	4	↓	167	227		394	525

21 21 Carbon-Dioxide Fire-Extinguishing Systems

21 21 16 – Carbon-Dioxide Fire-Extinguishing Equipment

21 21 16.50 CO2 Fire Extinguishing System	Crew	Daily Output	Labor-Hours	Unit	Material	2015 Bare Costs Labor	Equipment	Total	Total Incl O&P
3400	Manual pull station	1 Plum	6	1.333	Ea.	60.50	78.50		139	185
4000	Pneumatic damper release	"	8	1	↓	223	58.50		281.50	335

21 22 Clean-Agent Fire-Extinguishing Systems

21 22 16 – Clean-Agent Fire-Extinguishing Equipment

21 22 16.50 FM200 Fire Extinguishing System

		Crew	Daily Output	Labor-Hours	Unit	Material	2015 Bare Costs Labor	Equipment	Total	Total Incl O&P
0010	**FM200 FIRE EXTINGUISHING SYSTEM**									
1100	Dispersion nozzle FM200, 1-1/2"	1 Plum	14	.571	Ea.	67	33.50		100.50	124
2400	Extinguisher, FM200 system, filled, with mounting bracket									
2460	26 lb. container	Q-1	8	2	Ea.	2,300	106		2,406	2,675
2480	44 lb. container		7	2.286		3,050	121		3,171	3,550
2500	63 lb. container		6	2.667		3,575	141		3,716	4,150
2520	101 lb. container		5	3.200		4,775	169		4,944	5,500
2540	196 lb. container	↓	4	4	↓	7,775	211		7,986	8,875
6000	FM200 system, simple nozzle layout, with broad dispersion				C.F.	1.76			1.76	1.94
6020	Complex nozzle layout and/or including underfloor dispersion				"	3.50			3.50	3.85

21 31 Centrifugal Fire Pumps

21 31 13 – Electric-Drive, Centrifugal Fire Pumps

21 31 13.50 Electric-Drive Fire Pumps

		Crew	Daily Output	Labor-Hours	Unit	Material	2015 Bare Costs Labor	Equipment	Total	Total Incl O&P
0010	**ELECTRIC-DRIVE FIRE PUMPS** Including controller, fittings and relief valve									
3100	250 GPM, 55 psi, 15 HP, 3550 RPM, 2" pump	Q-13	.70	45.714	Ea.	14,500	2,450		16,950	19,600
3200	500 GPM, 50 psi, 27 HP, 1770 RPM, 4" pump		.68	47.059		14,900	2,525		17,425	20,200
3250	500 GPM, 100 psi, 47 HP, 3550 RPM, 3" pump		.66	48.485		16,100	2,600		18,700	21,600
3300	500 GPM, 125 psi, 64 HP, 3550 RPM, 3" pump		.62	51.613		17,700	2,750		20,450	23,600
3350	750 GPM, 50 psi, 44 HP, 1770 RPM, 5" pump		.64	50		15,500	2,675		18,175	21,200
3400	750 GPM, 100 psi, 66 HP, 3550 RPM, 4" pump		.58	55.172		17,900	2,950		20,850	24,200
3450	750 GPM, 165 psi, 120 HP, 3550 RPM, 4" pump		.56	57.143		24,400	3,050		27,450	31,400
3500	1000 GPM, 50 psi, 48 HP 1770 RPM, 5" pump		.60	53.333		16,600	2,850		19,450	22,500
3550	1000 GPM, 100 psi, 86 HP, 3550 RPM, 5" pump		.54	59.259		22,600	3,175		25,775	29,600
3600	1000 GPM, 150 psi, 142 HP, 3550 RPM, 5" pump		.50	64		26,400	3,425		29,825	34,300
3650	1000 GPM, 200 psi, 245 HP, 1770 RPM, 6" pump		.36	88.889		39,500	4,750		44,250	50,500
3660	1250 GPM, 75 psi, 75 HP, 1770 RPM, 5" pump		.55	58.182		21,500	3,100		24,600	28,300
3700	1500 GPM, 50 psi, 66 HP, 1770 RPM, 6" pump		.50	64		20,900	3,425		24,325	28,100
3750	1500 GPM, 100 psi, 139 HP, 1770 RPM, 6" pump		.46	69.565		26,100	3,725		29,825	34,300
3800	1500 GPM, 150 psi, 200 HP, 1770 RPM, 6" pump		.36	88.889		41,800	4,750		46,550	53,000
3850	1500 GPM, 200 psi, 279 HP, 1770 RPM, 6" pump		.32	100		44,700	5,350		50,050	57,000
3900	2000 GPM, 100 psi, 167 HP, 1770 RPM, 6" pump		.34	94.118		28,800	5,025		33,825	39,300
3950	2000 GPM, 150 psi, 292 HP, 1770 RPM, 6" pump		.28	114		42,300	6,100		48,400	56,000
4000	2500 GPM, 100 psi, 213 HP, 1770 RPM, 8" pump		.30	106		35,800	5,700		41,500	48,000
4040	2500 GPM, 135 psi, 339 HP, 1770 RPM, 8" pump		.26	123		50,500	6,575		57,075	65,500
4100	3000 GPM, 100 psi, 250 HP, 1770 RPM, 8" pump		.28	114		54,000	6,100		60,100	69,000
4150	3000 GPM, 140 psi, 428 HP, 1770 RPM, 10" pump		.24	133		59,500	7,125		66,625	76,500
4200	3500 GPM, 100 psi, 300 HP, 1770 RPM, 10" pump		.26	123		54,500	6,575		61,075	69,500
4250	3500 GPM, 140 psi, 450 HP, 1770 RPM, 10" pump		.24	133		69,000	7,125		76,125	87,000
5000	For jockey pump 1", 3 HP, with control, add	Q-12	2	8	↓	2,600	405		3,005	3,450

21 31 16 – Diesel-Drive, Centrifugal Fire Pumps

21 31 16.50 Diesel-Drive Fire Pumps	Crew	Daily Output	Labor-Hours	Unit	Material	2015 Bare Costs Labor	Equipment	Total	Total Incl O&P
0010 **DIESEL-DRIVE FIRE PUMPS** Including controller, fittings and relief valve									
0050 500 GPM, 50 psi, 27 HP, 4" pump	Q-13	.64	50	Ea.	33,800	2,675		36,475	41,200
0100 500 GPM, 100 psi, 62 HP, 4" pump		.60	53.333		39,600	2,850		42,450	47,900
0150 500 GPM, 125 psi, 78 HP, 4" pump		.56	57.143		42,100	3,050		45,150	51,000
0200 750 GPM, 50 psi, 44 HP, 5" pump		.60	53.333		34,800	2,850		37,650	42,500
0250 750 GPM, 100 psi, 80 HP, 4" pump		.56	57.143		39,100	3,050		42,150	47,600
0300 750 GPM, 165 psi, 203 HP, 5" pump		.52	61.538		49,700	3,300		53,000	59,500
0350 1000 GPM, 50 psi, 48 HP, 5" pump		.58	55.172		36,800	2,950		39,750	45,000
0400 1000 GPM, 100 psi, 89 HP, 4" pump		.56	57.143		39,400	3,050		42,450	47,900
0450 1000 GPM, 150 psi, 148 HP, 4" pump		.48	66.667		51,000	3,575		54,575	61,500
0470 1000 GPM, 200 psi, 280 HP, 5" pump		.40	80		59,500	4,275		63,775	72,000
0480 1250 GPM, 75 psi, 75 HP, 5" pump		.54	59.259		39,900	3,175		43,075	48,700
0500 1500 GPM, 50 psi, 66 HP, 6" pump		.50	64		39,100	3,425		42,525	48,200
0550 1500 GPM, 100 psi, 140 HP, 6" pump		.46	69.565		50,000	3,725		53,725	60,500
0600 1500 GPM, 150 psi, 228 HP, 6" pump		.42	76.190		56,000	4,075		60,075	68,000
0650 1500 GPM, 200 psi, 279 HP, 6" pump		.38	84.211		70,500	4,500		75,000	84,500
0700 2000 GPM, 100 psi, 167 HP, 6" pump		.34	94.118		49,900	5,025		54,925	62,500
0750 2000 GPM, 150 psi, 284 HP, 6" pump		.30	106		64,000	5,700		69,700	79,000
0800 2500 GPM, 100 psi, 213 HP, 8" pump		.32	100		54,500	5,350		59,850	67,500
0820 2500 GPM, 150 psi, 365 HP, 8" pump		.26	123		69,500	6,575		76,075	86,500
0850 3000 GPM, 100 psi, 250 HP, 8" pump		.28	114		71,000	6,100		77,100	88,000
0900 3000 GPM, 150 psi, 384 HP, 10" pump		.20	160		92,500	8,550		101,050	115,000
0950 3500 GPM, 100 psi, 300 HP, 10" pump		.24	133		71,000	7,125		78,125	89,000
1000 3500 GPM, 150 psi, 518 HP, 10" pump		.20	160		100,000	8,550		108,550	123,000

For customer support on your Plumbing Cost Data, call 877.709.6968.

97

Division Notes

	CREW	DAILY OUTPUT	LABOR-HOURS	UNIT	BARE COSTS				TOTAL INCL O&P
					MAT.	LABOR	EQUIP.	TOTAL	

Estimating Tips

22 10 00 Plumbing Piping and Pumps

This subdivision is primarily basic pipe and related materials. The pipe may be used by any of the mechanical disciplines, i.e., plumbing, fire protection, heating, and air conditioning.

Note: CPVC plastic piping approved for fire protection is located in 21 11 13.

- The labor adjustment factors listed in Subdivision 22 01 02.20 apply throughout Divisions 21, 22, and 23. CAUTION: the correct percentage may vary for the same items. For example, the percentage add for the basic pipe installation should be based on the maximum height that the craftsman must install for that particular section. If the pipe is to be located 14' above the floor but it is suspended on threaded rod from beams, the bottom flange of which is 18' high (4' rods), then the height is actually 18' and the add is 20%. The pipe coverer, however, does not have to go above the 14', and so the add should be 10%.

- Most pipe is priced first as straight pipe with a joint (coupling, weld, etc.) every 10' and a hanger usually every 10'. There are exceptions with hanger spacing such as for cast iron pipe (5') and plastic pipe (3 per 10'). Following each type of pipe there are several lines listing sizes and the amount to be subtracted to delete couplings and hangers. This is for pipe that is to be buried or supported together on trapeze hangers. The reason that the couplings are deleted is that these runs are usually long, and frequently longer lengths of pipe are used. By deleting the couplings, the estimator is expected to look up and add back the correct reduced number of couplings.

- When preparing an estimate, it may be necessary to approximate the fittings. Fittings usually run between 25% and 50% of the cost of the pipe. The lower percentage is for simpler runs, and the higher number is for complex areas, such as mechanical rooms.

- For historic restoration projects, the systems must be as invisible as possible, and pathways must be sought for pipes, conduit, and ductwork. While installations in accessible spaces (such as basements and attics) are relatively straightforward to estimate, labor costs may be more difficult to determine when delivery systems must be concealed.

22 40 00 Plumbing Fixtures

- Plumbing fixture costs usually require two lines: the fixture itself and its "rough-in, supply, and waste."

- In the Assemblies Section (Plumbing D2010) for the desired fixture, the System Components Group at the center of the page shows the fixture on the first line. The rest of the list (fittings, pipe, tubing, etc.) will total up to what we refer to in the Unit Price section as "Rough-in, supply, waste, and vent." Note that for most fixtures we allow a nominal 5' of tubing to reach from the fixture to a main or riser.

- Remember that gas- and oil-fired units need venting.

Reference Numbers

Reference numbers are shown in shaded boxes at the beginning of some major classifications. These numbers refer to related items in the Reference Section. The reference information may be an estimating procedure, an alternate pricing method, or technical information.

Note: Not all subdivisions listed here necessarily appear in this publication. ■

22 01 02 – Labor Adjustments

	22 01 02.10 Boilers, General		Crew	Daily Output	Labor-Hours	Unit	Material	2015 Bare Costs Labor	Equipment	Total	Total Incl O&P
0010	**BOILERS, GENERAL,** Prices do not include flue piping, elec. wiring,										
0020	gas or oil piping, boiler base, pad, or tankless unless noted										
0100	Boiler H.P.: 10 KW = 34 lb./steam/hr. = 33,475 BTU/hr.										
0150	To convert SFR to BTU rating: Hot water, 150 x SFR;										
0160	Forced hot water, 180 x SFR; steam, 240 x SFR										

22 01 02.20 Labor Adjustment Factors

			Crew	Daily Output	Labor-Hours	Unit	Material	2015 Bare Costs Labor	Equipment	Total	Total Incl O&P
0010	**LABOR ADJUSTMENT FACTORS,** (For Div. 21, 22 and 23)	R220102-20									
0100	Labor factors, The below are reasonable suggestions, however										
0110	each project must be evaluated for its own peculiarities, and										
0120	the adjustments be increased or decreased depending on the										
0130	severity of the special conditions.										
1000	Add to labor for elevated installation (Above floor level)										
1080	10' to 14.5' high	R221113-70						10%			
1100	15' to 19.5' high							20%			
1120	20' to 24.5' high							25%			
1140	25' to 29.5' high							35%			
1160	30' to 34.5' high							40%			
1180	35' to 39.5' high							50%			
1200	40' and higher							55%			
2000	Add to labor for crawl space										
2100	3' high							40%			
2140	4' high							30%			
3000	Add to labor for multi-story building										
3100	Add per floor for floors 3 thru 19							2%			
3140	Add per floor for floors 20 and up							4%			
4000	Add to labor for working in existing occupied buildings										
4100	Hospital							35%			
4140	Office building							25%			
4180	School							20%			
4220	Factory or warehouse							15%			
4260	Multi dwelling							15%			
5000	Add to labor, miscellaneous										
5100	Cramped shaft							35%			
5140	Congested area							15%			
5180	Excessive heat or cold							30%			
9000	Labor factors, The above are reasonable suggestions, however										
9010	each project should be evaluated for its own peculiarities.										
9100	Other factors to be considered are:										
9140	Movement of material and equipment through finished areas										
9180	Equipment room										
9220	Attic space										
9260	No service road										
9300	Poor unloading/storage area										
9340	Congested site area/heavy traffic										

100

For customer support on your Plumbing Cost Data, call 877.709.6968.

22 05 05 – Selective Demolition for Plumbing

22 05 05.10 Plumbing Demolition		Crew	Daily Output	Labor-Hours	Unit	Material	2015 Bare Costs Labor	2015 Bare Costs Equipment	Total	Total Incl O&P
0010	**PLUMBING DEMOLITION**	R220105-10								
0400	Air compressor, up thru 2 H.P.	Q-1	10	1.600	Ea.		84.50		84.50	128
0410	3 H.P. thru 7-1/2 H.P.	R024119-10	5.60	2.857			151		151	228
0420	10 H.P. thru 15 H.P.		1.40	11.429			605		605	910
0430	20 H.P. thru 30 H.P.	Q-2	1.30	18.462			1,000		1,000	1,525
0500	Backflow preventer, up thru 2" diameter	1 Plum	17	.471			27.50		27.50	41.50
0510	2-1/2" thru 3" diameter	Q-1	10	1.600			84.50		84.50	128
0520	4" thru 6" diameter	"	5	3.200			169		169	255
0530	8" thru 10" diameter	Q-2	3	8			440		440	660
0700	Carriers and supports									
0710	Fountains, sinks, lavatories and urinals	1 Plum	14	.571	Ea.		33.50		33.50	50.50
0720	Water closets	"	12	.667	"		39		39	59
0730	Grinder pump or sewage ejector system									
0732	Simplex	Q-1	7	2.286	Ea.		121		121	182
0734	Duplex	"	2.80	5.714			300		300	455
0738	Hot water dispenser	1 Plum	36	.222			13.05		13.05	19.70
0740	Hydrant, wall		26	.308			18.05		18.05	27.50
0744	Ground		12	.667			39		39	59
0760	Cleanouts and drains, up thru 4" pipe diameter		10	.800			47		47	71
0764	5" thru 8" pipe diameter	Q-1	10	1.600			84.50		84.50	128
0780	Industrial safety fixtures	1 Plum	8	1			58.50		58.50	88.50
1020	Fixtures, including 10' piping									
1100	Bathtubs, cast iron	1 Plum	4	2	Ea.		117		117	177
1120	Fiberglass		6	1.333			78.50		78.50	118
1140	Steel		5	1.600			94		94	142
1150	Bidet	Q-1	7	2.286			121		121	182
1200	Lavatory, wall hung	1 Plum	10	.800			47		47	71
1220	Counter top		8	1			58.50		58.50	88.50
1300	Sink, single compartment		8	1			58.50		58.50	88.50
1320	Double compartment		7	1.143			67		67	101
1340	Shower, stall and receptor	Q-1	6	2.667			141		141	213
1350	Group	"	7	2.286			121		121	182
1400	Water closet, floor mounted	1 Plum	8	1			58.50		58.50	88.50
1420	Wall mounted	"	7	1.143			67		67	101
1440	Wash fountain, 36" diameter	Q-2	8	3			164		164	248
1442	54" diameter	"	7	3.429			188		188	284
1500	Urinal, floor mounted	1 Plum	4	2			117		117	177
1520	Wall mounted	"	7	1.143			67		67	101
1590	Whirl pool or hot tub	Q-1	2.60	6.154			325		325	490
1600	Water fountains, free standing	1 Plum	8	1			58.50		58.50	88.50
1620	Wall or deck mounted		6	1.333			78.50		78.50	118
1800	Medical gas specialties		8	1			58.50		58.50	88.50
1810	Plumbing demo, Floor drain, Remove		12	.667			39		39	59
1820	Plumbing demo, Roof drain, Remove		9	.889			52		52	79
1900	Piping fittings, single connection, up thru 1-1/2" diameter		30	.267			15.65		15.65	23.50
1910	2" thru 4" diameter		14	.571			33.50		33.50	50.50
1980	Pipe hanger/support removal		80	.100			5.85		5.85	8.85
1990	Glass pipe with fittings, 1" thru 3" diameter		200	.040	L.F.		2.35		2.35	3.55
1992	4" thru 6" diameter		150	.053			3.13		3.13	4.73
2000	Piping, metal, up thru 1-1/2" diameter		200	.040			2.35		2.35	3.55
2050	2" thru 3-1/2" diameter		150	.053			3.13		3.13	4.73
2100	4" thru 6" diameter	2 Plum	100	.160			9.40		9.40	14.20
2150	8" thru 14" diameter	"	60	.267			15.65		15.65	23.50

22 05 05.10 Plumbing Demolition	Crew	Daily Output	Labor-Hours	Unit	Material	2015 Bare Costs Labor	2015 Bare Costs Equipment	Total	Total Incl O&P
2153	16" thru 20" diameter	Q-18	70	.343	L.F.		19.10	.82	19.92	30
2155	24" thru 26" diameter		55	.436			24.50	1.05	25.55	37.50
2156	30" thru 36" diameter	↓	40	.600			33.50	1.44	34.94	52
2160	Plastic pipe with fittings, up thru 1-1/2" diameter	1 Plum	250	.032			1.88		1.88	2.84
2162	2" thru 3" diameter	"	200	.040			2.35		2.35	3.55
2164	4" thru 6" diameter	Q-1	200	.080			4.23		4.23	6.40
2166	8" thru 14" diameter		150	.107			5.65		5.65	8.50
2168	16" diameter		100	.160	↓		8.45		8.45	12.75
2170	Prison fixtures, lavatory or sink		18	.889	Ea.		47		47	71
2172	Shower		5.60	2.857			151		151	228
2174	Urinal or water closet		13	1.231			65		65	98
2180	Pumps, all fractional horse-power		12	1.333			70.50		70.50	106
2184	1 H.P. thru 5 H.P.		6	2.667			141		141	213
2186	7-1/2 H.P. thru 15 H.P.	↓	2.50	6.400			340		340	510
2188	20 H.P. thru 25 H.P.	Q-2	4	6			330		330	495
2190	30 H.P. thru 60 H.P.		.80	30			1,650		1,650	2,475
2192	75 H.P. thru 100 H.P.		.60	40			2,200		2,200	3,300
2194	150 H.P.	↓	.50	48			2,625		2,625	3,975
2198	Pump, sump or submersible	1 Plum	12	.667			39		39	59
2200	Receptors and interceptors, up thru 20 GPM	"	8	1			58.50		58.50	88.50
2204	25 thru 100 GPM	Q-1	6	2.667			141		141	213
2208	125 thru 300 GPM	"	2.40	6.667			350		350	530
2211	325 thru 500 GPM	Q-2	2.60	9.231	↓		505		505	765
2212	Deduct for salvage, aluminum scrap				Ton				700	770
2214	Brass scrap								2,450	2,675
2216	Copper scrap								3,200	3,525
2218	Lead scrap								520	570
2220	Steel scrap				↓				180	200
2230	Temperature maintenance cable	1 Plum	1200	.007	L.F.		.39		.39	.59
2240	Toilet partitions, see Section 10 21 13									
2250	Water heater, 40 gal.	1 Plum	6	1.333	Ea.		78.50		78.50	118
3100	Tanks, water heaters and liquid containers									
3110	Up thru 45 gallons	Q-1	22	.727	Ea.		38.50		38.50	58
3120	50 thru 120 gallons		14	1.143			60.50		60.50	91
3130	130 thru 240 gallons		7.60	2.105			111		111	168
3140	250 thru 500 gallons	↓	5.40	2.963			157		157	236
3150	600 thru 1000 gallons	Q-2	1.60	15			820		820	1,250
3160	1100 thru 2000 gallons		.70	34.286			1,875		1,875	2,825
3170	2100 thru 4000 gallons	↓	.50	48			2,625		2,625	3,975
6000	Remove and reset fixtures, easy access	1 Plum	6	1.333			78.50		78.50	118
6100	Difficult access	"	4	2	↓		117		117	177
6500	Solar heating system									
6510	Solar panel, hot water system	1 Plum	304	.026	S.F.		1.55		1.55	2.33
7910	Solar panel for pool	"	300	.027	"		1.57		1.57	2.36
8000	Sprinkler system									
8100	Exposed wet/dry	1 Plum	2500	.003	SF Flr.		.19		.19	.28
8200	Concealed wet/dry	"	1250	.006	"		.38		.38	.57
9100	Valve, metal valves or strainers and similar, up thru 1-1/2" diameter	1 Stpi	28	.286	Ea.		17.05		17.05	26
9110	2" thru 3" diameter	Q-1	11	1.455			77		77	116
9120	4" thru 6" diameter	"	8	2			106		106	160
9130	8" thru 14" diameter	Q-2	8	3			164		164	248
9140	16" thru 20" diameter		2	12			655		655	995
9150	24" diameter	↓	1.20	20	↓		1,100		1,100	1,650

22 05 05 – Selective Demolition for Plumbing

22 05 05.10 Plumbing Demolition	Crew	Daily Output	Labor-Hours	Unit	Material	2015 Bare Costs Labor	Equipment	Total	Total Incl O&P
9200	Valve, plastic, up thru 1-1/2" diameter	1 Plum	42	.190	Ea.		11.20		11.20	16.90
9210	2" thru 3" diameter		15	.533			31.50		31.50	47.50
9220	4" thru 6" diameter		12	.667			39		39	59
9300	Vent flashing and caps		55	.145			8.55		8.55	12.90
9350	Water filter, commercial, 1" thru 1-1/2"	Q-1	2	8			425		425	640
9360	2" thru 2-1/2"	"	1.60	10			530		530	800
9400	Water heaters									
9410	Up thru 245 GPH	Q-1	2.40	6.667	Ea.		350		350	530
9420	250 thru 756 GPH		1.60	10			530		530	800
9430	775 thru 1640 GPH		.80	20			1,050		1,050	1,600
9440	1650 thru 4000 GPH	Q-2	.50	48			2,625		2,625	3,975
9470	Water softener	Q-1	2	8			425		425	640

22 05 23 – General-Duty Valves for Plumbing Piping

22 05 23.10 Valves, Brass

		Crew	Daily Output	Labor-Hours	Unit	Material	Labor	Equipment	Total	Total Incl O&P
0010	**VALVES, BRASS**									
0032	For motorized valves, see Section 23 09 53.10									
0500	Gas cocks, threaded									
0510	1/4"	1 Plum	26	.308	Ea.	14.70	18.05		32.75	43.50
0520	3/8"		24	.333		14.70	19.55		34.25	45.50
0530	1/2"		24	.333		13.15	19.55		32.70	44
0540	3/4"		22	.364		15.50	21.50		37	49
0550	1"		19	.421		31	24.50		55.50	71.50
0560	1-1/4"		15	.533		44.50	31.50		76	96.50
0570	1-1/2"		13	.615		61	36		97	122
0580	2"		11	.727		103	42.50		145.50	179
0672	For larger sizes use lubricated plug valve, Section 23 05 23.70									

22 05 23.20 Valves, Bronze

			Crew	Daily Output	Labor-Hours	Unit	Material	Labor	Equipment	Total	Total Incl O&P
0010	**VALVES, BRONZE**	R220523-80									
1020	Angle, 150 lb., rising stem, threaded										
1030	1/8"	R220523-90	1 Plum	24	.333	Ea.	128	19.55		147.55	170
1040	1/4"			24	.333		128	19.55		147.55	170
1050	3/8"			24	.333		130	19.55		149.55	173
1060	1/2"			22	.364		142	21.50		163.50	188
1070	3/4"			20	.400		193	23.50		216.50	249
1080	1"			19	.421		279	24.50		303.50	345
1090	1-1/4"			15	.533		360	31.50		391.50	445
1100	1-1/2"			13	.615		470	36		506	570
1110	2"			11	.727		755	42.50		797.50	895
1300	Ball										
1304	Soldered										
1312	3/8"		1 Plum	21	.381	Ea.	14.95	22.50		37.45	50.50
1316	1/2"			18	.444		14.95	26		40.95	56
1320	3/4"			17	.471		25	27.50		52.50	68.50
1324	1"			15	.533		31	31.50		62.50	82
1328	1-1/4"			13	.615		52	36		88	112
1332	1-1/2"			11	.727		66.50	42.50		109	138
1336	2"			9	.889		83.50	52		135.50	171
1340	2-1/2"			7	1.143		395	67		462	535
1344	3"			5	1.600		455	94		549	640
1350	Single union end										
1358	3/8"		1 Plum	21	.381	Ea.	21.50	22.50		44	57.50
1362	1/2"			18	.444		22.50	26		48.50	64.50

For customer support on your Plumbing Cost Data, call 877.709.6968.

103

22 05 23 – General-Duty Valves for Plumbing Piping

22 05 23.20 Valves, Bronze		Crew	Daily Output	Labor-Hours	Unit	Material	2015 Bare Costs Labor	Equipment	Total	Total Incl O&P
1366	3/4"	1 Plum	17	.471	Ea.	43	27.50		70.50	89
1370	1"		15	.533		52	31.50		83.50	105
1374	1-1/4"		13	.615		88.50	36		124.50	152
1378	1-1/2"		11	.727		111	42.50		153.50	187
1382	2"		9	.889		173	52		225	269
1398	Threaded, 150 psi									
1400	1/4"	1 Plum	24	.333	Ea.	14.15	19.55		33.70	45
1430	3/8"		24	.333		14.15	19.55		33.70	45
1450	1/2"		22	.364		14.15	21.50		35.65	47.50
1460	3/4"		20	.400		23.50	23.50		47	61
1470	1"		19	.421		33.50	24.50		58	74.50
1480	1-1/4"		15	.533		58.50	31.50		90	112
1490	1-1/2"		13	.615		76.50	36		112.50	139
1500	2"		11	.727		93	42.50		135.50	167
1510	2-1/2"		9	.889		310	52		362	420
1520	3"		8	1		470	58.50		528.50	610
1600	Butterfly, 175 psi, full port, solder or threaded ends									
1610	Stainless steel disc and stem									
1620	1/4"	1 Plum	24	.333	Ea.	17.35	19.55		36.90	48.50
1630	3/8"		24	.333		17.35	19.55		36.90	48.50
1640	1/2"		22	.364		18.65	21.50		40.15	52.50
1650	3/4"		20	.400		29.50	23.50		53	68
1660	1"		19	.421		36	24.50		60.50	77
1670	1-1/4"		15	.533		58	31.50		89.50	112
1680	1-1/2"		13	.615		75	36		111	137
1690	2"		11	.727		94	42.50		136.50	168
1750	Check, swing, class 150, regrinding disc, threaded									
1800	1/8"	1 Plum	24	.333	Ea.	66.50	19.55		86.05	103
1830	1/4"		24	.333		66.50	19.55		86.05	103
1840	3/8"		24	.333		70.50	19.55		90.05	107
1850	1/2"		24	.333		75.50	19.55		95.05	113
1860	3/4"		20	.400		100	23.50		123.50	146
1870	1"		19	.421		144	24.50		168.50	196
1880	1-1/4"		15	.533		208	31.50		239.50	277
1890	1-1/2"		13	.615		242	36		278	320
1900	2"		11	.727		355	42.50		397.50	455
1910	2-1/2"	Q-1	15	1.067		800	56.50		856.50	965
1920	3"	"	13	1.231		1,075	65		1,140	1,275
2000	For 200 lb., add					5%	10%			
2040	For 300 lb., add					15%	15%			
2060	Check swing, 300#, sweat, 3/8" size	1 Plum	24	.333	Ea.	63	19.55		82.55	98.50
2070	1/2"		24	.333		63	19.55		82.55	98.50
2080	3/4"		20	.400		79.50	23.50		103	123
2090	1"		19	.421		114	24.50		138.50	163
2100	1-1/4"		15	.533		159	31.50		190.50	223
2110	1-1/2"		13	.615		194	36		230	268
2120	2"		11	.727		296	42.50		338.50	390
2130	2-1/2"	Q-1	15	1.067		650	56.50		706.50	800
2140	3"	"	13	1.231		870	65		935	1,050
2850	Gate, N.R.S., soldered, 125 psi									
2900	3/8"	1 Plum	24	.333	Ea.	61	19.55		80.55	96.50
2920	1/2"		24	.333		54	19.55		73.55	89
2940	3/4"		20	.400		63.50	23.50		87	106

For customer support on your Plumbing Cost Data, call 877.709.6968.

22 05 Common Work Results for Plumbing

22 05 23 – General-Duty Valves for Plumbing Piping

22 05 23.20 Valves, Bronze	Crew	Daily Output	Labor-Hours	Unit	Material	2015 Bare Costs Labor	Equipment	Total	Total Incl O&P
2950 1"	1 Plum	19	.421	Ea.	76.50	24.50		101	122
2960 1-1/4"		15	.533		126	31.50		157.50	187
2970 1-1/2"		13	.615		142	36		178	212
2980 2"		11	.727		185	42.50		227.50	269
2990 2-1/2"	Q-1	15	1.067		450	56.50		506.50	580
3000 3"	"	13	1.231		560	65		625	715
3350 Threaded, class 150									
3410 1/4"	1 Plum	24	.333	Ea.	75.50	19.55		95.05	113
3420 3/8"		24	.333		75.50	19.55		95.05	113
3430 1/2"		24	.333		71	19.55		90.55	108
3440 3/4"		20	.400		81.50	23.50		105	125
3450 1"		19	.421		102	24.50		126.50	150
3460 1-1/4" size		15	.533		138	31.50		169.50	200
3470 1-1/2"		13	.615		198	36		234	273
3480 2"		11	.727		237	42.50		279.50	325
3490 2-1/2"	Q-1	15	1.067		605	56.50		661.50	750
3500 3" size	"	13	1.231		855	65		920	1,050
3600 Gate, flanged, 150 lb.									
3610 1"	1 Plum	7	1.143	Ea.	1,150	67		1,217	1,350
3620 1-1/2"		6	1.333		1,425	78.50		1,503.50	1,700
3630 2"		5	1.600		2,100	94		2,194	2,450
3634 2-1/2"	Q-1	5	3.200		3,900	169		4,069	4,525
3640 3"	"	4.50	3.556		3,775	188		3,963	4,425
3850 Rising stem, soldered, 300 psi									
3900 3/8"	1 Plum	24	.333	Ea.	141	19.55		160.55	185
3920 1/2"		24	.333		116	19.55		135.55	157
3940 3/4"		20	.400		132	23.50		155.50	181
3950 1"		19	.421		179	24.50		203.50	235
3960 1-1/4"		15	.533		247	31.50		278.50	320
3970 1-1/2"		13	.615		300	36		336	385
3980 2"		11	.727		480	42.50		522.50	595
3990 2-1/2"	Q-1	15	1.067		1,025	56.50		1,081.50	1,200
4000 3"	"	13	1.231		1,575	65		1,640	1,850
4250 Threaded, class 150									
4310 1/4"	1 Plum	24	.333	Ea.	68.50	19.55		88.05	105
4320 3/8"		24	.333		68.50	19.55		88.05	105
4330 1/2"		24	.333		63	19.55		82.55	98.50
4340 3/4"		20	.400		73.50	23.50		97	116
4350 1"		19	.421		98.50	24.50		123	146
4360 1-1/4"		15	.533		134	31.50		165.50	195
4370 1-1/2"		13	.615		169	36		205	240
4380 2"		11	.727		227	42.50		269.50	315
4390 2-1/2"	Q-1	15	1.067		530	56.50		586.50	670
4400 3"	"	13	1.231		740	65		805	910
4500 For 300 psi, threaded, add					100%	15%			
4540 For chain operated type, add					15%				
4850 Globe, class 150, rising stem, threaded									
4920 1/4"	1 Plum	24	.333	Ea.	97	19.55		116.55	136
4940 3/8"		24	.333		95.50	19.55		115.05	135
4950 1/2"		24	.333		95.50	19.55		115.05	135
4960 3/4"		20	.400		99	23.50		122.50	145
4970 1"		19	.421		154	24.50		178.50	207
4980 1-1/4"		15	.533		245	31.50		276.50	315

For customer support on your Plumbing Cost Data, call 877.709.6968.

105

22 05 23 – General-Duty Valves for Plumbing Piping

22 05 23.20 Valves, Bronze		Crew	Daily Output	Labor-Hours	Unit	Material	2015 Bare Costs Labor	2015 Bare Costs Equipment	Total	Total Incl O&P
4990	1-1/2"	1 Plum	13	.615	Ea.	320	36		356	410
5000	2"	↓	11	.727		465	42.50		507.50	580
5010	2-1/2"	Q-1	15	1.067		1,175	56.50		1,231.50	1,350
5020	3"	"	13	1.231	↓	1,675	65		1,740	1,925
5120	For 300 lb. threaded, add					50%	15%			
5600	Relief, pressure & temperature, self-closing, ASME, threaded									
5640	3/4"	1 Plum	28	.286	Ea.	253	16.75		269.75	305
5650	1"		24	.333		405	19.55		424.55	475
5660	1-1/4"		20	.400		895	23.50		918.50	1,025
5670	1-1/2"	↓	18	.444		1,250	26		1,276	1,425
5680	2"	↓	16	.500	↓	1,350	29.50		1,379.50	1,550
5950	Pressure, poppet type, threaded									
6000	1/2"	1 Plum	30	.267	Ea.	78.50	15.65		94.15	110
6040	3/4"	"	28	.286	"	73.50	16.75	*	90.25	106
6400	Pressure, water, ASME, threaded									
6440	3/4"	1 Plum	28	.286	Ea.	116	16.75		132.75	153
6450	1"		24	.333		260	19.55		279.55	315
6460	1-1/4"		20	.400		390	23.50		413.50	465
6470	1-1/2"		18	.444		570	26		596	665
6480	2"		16	.500		820	29.50		849.50	950
6490	2-1/2"	↓	15	.533	↓	3,150	31.50		3,181.50	3,525
6900	Reducing, water pressure									
6920	300 psi to 25-75 psi, threaded or sweat									
6940	1/2"	1 Plum	24	.333	Ea.	395	19.55		414.55	465
6950	3/4"		20	.400		405	23.50		428.50	485
6960	1"		19	.421		630	24.50		654.50	730
6970	1-1/4"		15	.533		1,100	31.50		1,131.50	1,250
6980	1-1/2"		13	.615		1,650	36		1,686	1,875
6990	2"	↓	11	.727		2,450	42.50		2,492.50	2,775
7100	For built-in by-pass or 10-35 psi, add				↓	39			39	42.50
7700	High capacity, 250 psi to 25-75 psi, threaded									
7740	1/2"	1 Plum	24	.333	Ea.	540	19.55		559.55	625
7780	3/4"		20	.400		540	23.50		563.50	630
7790	1"		19	.421		775	24.50		799.50	895
7800	1-1/4"		15	.533		1,350	31.50		1,381.50	1,525
7810	1-1/2"		13	.615		1,975	36		2,011	2,225
7820	2"		11	.727		2,875	42.50		2,917.50	3,225
7830	2-1/2"		9	.889		4,100	52		4,152	4,600
7840	3"	↓	8	1		4,925	58.50		4,983.50	5,500
7850	3" flanged (iron body)	Q-1	10	1.600		4,850	84.50		4,934.50	5,450
7860	4" flanged (iron body)	"	8	2	↓	6,300	106		6,406	7,075
7920	For higher pressure, add					25%				
8000	Silent check, bronze trim									
8010	Compact wafer type, for 125 or 150 lb. flanges									
8020	1-1/2"	1 Plum	11	.727	Ea.	335	42.50		377.50	430
8021	2"	"	9	.889		355	52		407	470
8022	2-1/2"	Q-1	9	1.778		370	94		464	550
8023	3"		8	2		415	106		521	615
8024	4"	↓	5	3.200		710	169		879	1,025
8025	5"	Q-2	6	4		940	219		1,159	1,350
8026	6"	"	5	4.800	↓	1,100	263		1,363	1,625
8050	For 250 or 300 lb. flanges, thru 6" no change									
8060	Full flange wafer type, 150 lb.									

22 05 23 – General-Duty Valves for Plumbing Piping

22 05 23.20 Valves, Bronze		Crew	Daily Output	Labor-Hours	Unit	Material	2015 Bare Costs Labor	Equipment	Total	Total Incl O&P
8063	1-1/2"	1 Plum	11	.727	Ea.	415	42.50		457.50	525
8064	2"	"	9	.889		560	52		612	695
8065	2-1/2"	Q-1	9	1.778		630	94		724	835
8066	3"		8	2		700	106		806	930
8067	4"	↓	5	3.200		1,025	169		1,194	1,375
8068	5"	Q-2	6	4		1,325	219		1,544	1,800
8069	6"	"	5	4.800		1,750	263		2,013	2,325
8080	For 300 lb., add				↓	40%	10%			
8100	Globe type, 150 lb.									
8110	2"	1 Plum	9	.889	Ea.	690	52		742	840
8111	2-1/2"	Q-1	9	1.778		855	94		949	1,075
8112	3"		8	2		1,025	106		1,131	1,275
8113	4"	↓	5	3.200		1,425	169		1,594	1,800
8114	5"	Q-2	6	4		1,725	219		1,944	2,225
8115	6"	"	5	4.800	↓	2,400	263		2,663	3,050
8130	For 300 lb., add					20%	10%			
8140	Screwed end type, 250 lb.									
8141	1/2"	1 Plum	24	.333	Ea.	62.50	19.55		82.05	98
8142	3/4"		20	.400		62.50	23.50		86	104
8143	1"		19	.421		71	24.50		95.50	116
8144	1-1/4"		15	.533		98.50	31.50		130	156
8145	1-1/2"		13	.615		110	36		146	176
8146	2"	↓	11	.727	↓	149	42.50		191.50	229
8350	Tempering, water, sweat connections									
8400	1/2"	1 Plum	24	.333	Ea.	98	19.55		117.55	138
8440	3/4"	"	20	.400	"	126	23.50		149.50	175
8650	Threaded connections									
8700	1/2"	1 Plum	24	.333	Ea.	126	19.55		145.55	169
8740	3/4"		20	.400		770	23.50		793.50	880
8750	1"		19	.421		865	24.50		889.50	995
8760	1-1/4"		15	.533		1,350	31.50		1,381.50	1,525
8770	1-1/2"		13	.615		1,475	36		1,511	1,650
8780	2"	↓	11	.727	↓	2,200	42.50		2,242.50	2,500

22 05 23.40 Valves, Lined, Corrosion Resistant/High Purity

		Crew	Daily Output	Labor-Hours	Unit	Material	Labor	Equipment	Total	Total Incl O&P
0010	**VALVES, LINED, CORROSION RESISTANT/HIGH PURITY**									
3500	Check lift, 125 lb., cast iron flanged									
3510	Horizontal PPL or SL lined									
3530	1"	1 Plum	14	.571	Ea.	635	33.50		668.50	750
3540	1-1/2"		11	.727		770	42.50		812.50	910
3550	2"	↓	8	1		895	58.50		953.50	1,075
3560	2-1/2"	Q-1	5	3.200		1,175	169		1,344	1,550
3570	3"		4.50	3.556		1,450	188		1,638	1,875
3590	4"	↓	3	5.333		1,925	282		2,207	2,550
3610	6"	Q-2	3	8		3,275	440		3,715	4,250
3620	8"	"	2.50	9.600	↓	7,225	525		7,750	8,750
4250	Vertical PPL or SL lined									
4270	1"	1 Plum	14	.571	Ea.	570	33.50		603.50	680
4290	1-1/2"		11	.727		705	42.50		747.50	840
4300	2"	↓	8	1		820	58.50		878.50	995
4310	2-1/2"	Q-1	5	3.200		1,250	169		1,419	1,625
4320	3"		4.50	3.556		1,250	188		1,438	1,625
4340	4"	↓	3	5.333		1,625	282		1,907	2,225

107

For customer support on your Plumbing Cost Data, call 877.709.6968.

22 05 23.40 Valves, Lined, Corrosion Resistant/High Purity		Crew	Daily Output	Labor-Hours	Unit	Material	2015 Bare Costs Labor	Equipment	Total	Total Incl O&P
4360	6"	Q-2	3	8	Ea.	2,750	440		3,190	3,675
4370	8"	"	2.50	9.600	↓	4,975	525		5,500	6,250

22 05 23.60 Valves, Plastic

		Crew	Daily Output	Labor-Hours	Unit	Material	Labor	Equipment	Total	Total Incl O&P
0010	**VALVES, PLASTIC** R220523-90									
1100	Angle, PVC, threaded									
1110	1/4"	1 Plum	26	.308	Ea.	39.50	18.05		57.55	71
1120	1/2"		26	.308		56.50	18.05		74.55	89.50
1130	3/4"		25	.320		67	18.80		85.80	103
1140	1"	↓	23	.348	↓	81.50	20.50		102	121
1150	Ball, PVC, socket or threaded, true union									
1230	1/2"	1 Plum	26	.308	Ea.	40.50	18.05		58.55	72
1240	3/4"		25	.320		40.50	18.80		59.30	73
1250	1"		23	.348		48	20.50		68.50	84
1260	1-1/4"		21	.381		83.50	22.50		106	126
1270	1-1/2"		20	.400		83.50	23.50		107	128
1280	2"	↓	17	.471		110	27.50		137.50	163
1290	2-1/2"	Q-1	26	.615		188	32.50		220.50	256
1300	3"		24	.667		282	35		317	365
1310	4"	↓	20	.800		450	42.50		492.50	560
1360	For PVC, flanged, add					100%	15%			
1450	Double union 1/2"	1 Plum	26	.308		32.50	18.05		50.55	63.50
1460	3/4"		25	.320		42.50	18.80		61.30	75
1470	1"		23	.348		50.50	20.50		71	86.50
1480	1-1/4"		21	.381		69.50	22.50		92	111
1490	1-1/2"		20	.400		84	23.50		107.50	128
1500	2"	↓	17	.471	↓	111	27.50		138.50	164
1650	CPVC, socket or threaded, single union									
1700	1/2"	1 Plum	26	.308	Ea.	58	18.05		76.05	91
1720	3/4"		25	.320		77	18.80		95.80	113
1730	1"		23	.348		87.50	20.50		108	127
1750	1-1/4"		21	.381		140	22.50		162.50	188
1760	1-1/2"		20	.400		140	23.50		163.50	190
1770	2"	↓	17	.471		192	27.50		219.50	254
1780	3"	Q-1	24	.667		760	35		795	890
1840	For CPVC, flanged, add					65%	15%			
1880	For true union, socket or threaded, add				↓	50%	5%			
2050	Polypropylene, threaded									
2100	1/4"	1 Plum	26	.308	Ea.	45	18.05		63.05	77
2120	3/8"		26	.308		45	18.05		63.05	77
2130	1/2"		26	.308		45	18.05		63.05	77
2140	3/4"		25	.320		54.50	18.80		73.30	88.50
2150	1"		23	.348		62.50	20.50		83	99.50
2160	1-1/4"		21	.381		84	22.50		106.50	127
2170	1-1/2"		20	.400		103	23.50		126.50	150
2180	2"	↓	17	.471		138	27.50		165.50	194
2190	3"	Q-1	24	.667		315	35		350	400
2200	4"	"	20	.800	↓	550	42.50		592.50	670
2550	PVC, three way, socket or threaded									
2600	1/2"	1 Plum	26	.308	Ea.	62.50	18.05		80.55	96
2640	3/4"		25	.320	.	71	18.80		89.80	107
2650	1"		23	.348		76.50	20.50		97	115
2660	1-1/2"	↓	20	.400	↓	181	23.50		204.50	235

22 05 23 – General-Duty Valves for Plumbing Piping

22 05 23.60 Valves, Plastic		Crew	Daily Output	Labor-Hours	Unit	Material	2015 Bare Costs Labor	Equipment	Total	Total Incl O&P
2670	2"	1 Plum	17	.471	Ea.	206	27.50		233.50	269
2680	3"	Q-1	24	.667		430	35		465	530
2740	For flanged, add					60%	15%			
3150	Ball check, PVC, socket or threaded									
3200	1/4"	1 Plum	26	.308	Ea.	45.50	18.05		63.55	78
3220	3/8"		26	.308		45.50	18.05		63.55	78
3240	1/2"		26	.308		45.50	18.05		63.55	78
3250	3/4"		25	.320		51	18.80		69.80	84.50
3260	1"		23	.348		64	20.50		84.50	102
3270	1-1/4"		21	.381		107	22.50		129.50	152
3280	1-1/2"		20	.400		107	23.50		130.50	154
3290	2"		17	.471		146	27.50		173.50	203
3310	3"	Q-1	24	.667		405	35		440	500
3320	4"	"	20	.800		575	42.50		617.50	700
3360	For PVC, flanged, add					50%	15%			
3750	CPVC, socket or threaded									
3800	1/2"	1 Plum	26	.308	Ea.	70.50	18.05		88.55	105
3840	3/4"		25	.320		83.50	18.80		102.30	121
3850	1"		23	.348		99.50	20.50		120	140
3860	1-1/2"		20	.400		171	23.50		194.50	224
3870	2"		17	.471		230	27.50		257.50	295
3880	3"	Q-1	24	.667		620	35		655	735
3920	4"	"	20	.800		840	42.50		882.50	990
3930	For CPVC, flanged, add					40%	15%			
4340	Polypropylene, threaded									
4360	1/2"	1 Plum	26	.308	Ea.	44	18.05		62.05	76
4400	3/4"		25	.320		61	18.80		79.80	95.50
4440	1"		23	.348		65	20.50		85.50	103
4450	1-1/2"		20	.400		125	23.50		148.50	173
4460	2"		17	.471		158	27.50		185.50	216
4500	For polypropylene flanged, add					200%	15%			
4850	Foot valve, PVC, socket or threaded									
4900	1/2"	1 Plum	34	.235	Ea.	73	13.80		86.80	101
4930	3/4"		32	.250		83	14.70		97.70	114
4940	1"		28	.286		107	16.75		123.75	144
4950	1-1/4"		27	.296		206	17.40		223.40	254
4960	1-1/2"		26	.308		206	18.05		224.05	255
4970	2"		24	.333		365	19.55		384.55	430
4980	3"		20	.400		835	23.50		858.50	955
4990	4"		18	.444		1,275	26		1,301	1,475
5000	For flanged, add					25%	10%			
5050	CPVC, socket or threaded									
5060	1/2"	1 Plum	34	.235	Ea.	72	13.80		85.80	101
5070	3/4"		32	.250		83	14.70		97.70	114
5080	1"		28	.286		102	16.75		118.75	139
5090	1-1/4"		27	.296		164	17.40		181.40	207
5100	1-1/2"		26	.308		164	18.05		182.05	208
5110	2"		24	.333		211	19.55		230.55	262
5120	3"		20	.400		430	23.50		453.50	505
5130	4"		18	.444		780	26		806	895
5140	For flanged, add					25%	10%			
5280	Needle valve, PVC, threaded									
5300	1/4"	1 Plum	26	.308	Ea.	55.50	18.05		73.55	88.50

For customer support on your Plumbing Cost Data, call 877.709.6968.

109

22 05 23 – General-Duty Valves for Plumbing Piping

22 05 23.60 Valves, Plastic		Crew	Daily Output	Labor-Hours	Unit	Material	2015 Bare Costs Labor	Equipment	Total	Total Incl O&P
5340	3/8"	1 Plum	26	.308	Ea.	64	18.05		82.05	98
5360	1/2"	↓	26	.308		64	18.05		82.05	98
5380	For polypropylene, add				↓	10%				
5800	Y check, PVC, socket or threaded									
5820	1/2"	1 Plum	26	.308	Ea.	55	18.05		73.05	88
5840	3/4"		25	.320		89	18.80		107.80	126
5850	1"		23	.348		96.50	20.50		117	137
5860	1-1/4"		21	.381		151	22.50		173.50	201
5870	1-1/2"		20	.400		165	23.50		188.50	218
5880	2"		17	.471		205	27.50		232.50	268
5890	2-1/2"	↓	15	.533		435	31.50		466.50	525
5900	3"	Q-1	24	.667		435	35	·	470	530
5910	4"	"	20	.800		710	42.50		752.50	845
5960	For PVC flanged, add				↓	45%	15%			
6350	Y sediment strainer, PVC, socket or threaded									
6400	1/2"	1 Plum	26	.308	Ea.	55	18.05		73.05	88
6440	3/4"		24	.333		58.50	19.55		78.05	93.50
6450	1"		23	.348		69.50	20.50		90	108
6460	1-1/4"		21	.381		117	22.50		139.50	162
6470	1-1/2"		20	.400		117	23.50		140.50	164
6480	2"		17	.471		144	27.50		171.50	200
6490	2-1/2"		15	.533		350	31.50		381.50	435
6500	3"	Q-1	24	.667		350	35		385	440
6510	4"	"	20	.800		585	42.50		627.50	705
6560	For PVC, flanged, add				↓	55%	15%			

22 05 29 – Hangers and Supports for Plumbing Piping and Equipment

22 05 29.10 Hangers & Supp. for Plumb'g/HVAC Pipe/Equip.

		Crew	Daily Output	Labor-Hours	Unit	Material	2015 Bare Costs Labor	Equipment	Total	Total Incl O&P
0010	**HANGERS AND SUPPORTS FOR PLUMB'G/HVAC PIPE/EQUIP.**									
0011	TYPE numbers per MSS-SP58									
0050	Brackets									
0060	Beam side or wall, malleable iron, TYPE 34									
0070	3/8" threaded rod size	1 Plum	48	.167	Ea.	3.11	9.80		12.91	18.20
0080	1/2" threaded rod size		48	.167		4.39	9.80		14.19	19.65
0090	5/8" threaded rod size		48	.167		8.20	9.80		18	24
0100	3/4" threaded rod size		48	.167		11.30	9.80		21.10	27
0110	7/8" threaded rod size	↓	48	.167		11.70	9.80		21.50	27.50
0120	For concrete installation, add				↓		30%			
0150	Wall, welded steel, medium, TYPE 32									
0160	0 size, 12" wide, 18" deep	1 Plum	34	.235	Ea.	172	13.80		185.80	210
0170	1 size, 18" wide, 24" deep		34	.235		211	13.80		224.80	253
0180	2 size, 24" wide, 30" deep	↓	34	.235	↓	290	13.80		303.80	340
0200	Beam attachment, welded, TYPE 22									
0202	3/8"	Q-15	80	.200	Ea.	7.15	10.55	.72	18.42	24.50
0203	1/2"		76	.211		7.25	11.10	.76	19.11	25.50
0204	5/8"		72	.222		8.45	11.75	.80	21	28
0205	3/4"		68	.235		12.30	12.45	.85	25.60	33
0206	7/8"		64	.250		15.75	13.20	.90	29.85	38.50
0207	1"	↓	56	.286	↓	31.50	15.10	1.03	47.63	58.50
0300	Clamps									
0310	C-clamp, for mounting on steel beam flange, w/locknut, TYPE 23									
0320	3/8" threaded rod size	1 Plum	160	.050	Ea.	2.05	2.94		4.99	6.70
0330	1/2" threaded rod size		160	.050	↓	2.20	2.94		5.14	6.85

22 05 29.10 Hangers & Supp. for Plumb'g/HVAC Pipe/Equip.	Crew	Daily Output	Labor-Hours	Unit	Material	2015 Bare Costs Labor	Equipment	Total	Total Incl O&P
0340	5/8" threaded rod size	1 Plum	160	.050	Ea.	3.70	2.94		6.64	8.50
0350	3/4" threaded rod size		160	.050		4.55	2.94		7.49	9.45
0352	7/8" threaded rod size		140	.057		15	3.35		18.35	21.50
0400	High temperature to 1050°F, alloy steel									
0410	4" pipe size	Q-1	106	.151	Ea.	26.50	7.95		34.45	41.50
0420	6" pipe size		106	.151		45.50	7.95		53.45	62
0430	8" pipe size		97	.165		50.50	8.70		59.20	68.50
0440	10" pipe size		84	.190		74	10.05		84.05	96.50
0450	12" pipe size		72	.222		83	11.75		94.75	109
0460	14" pipe size		64	.250		209	13.20		222.20	250
0470	16" pipe size		56	.286		230	15.10		245.10	275
0480	Beam clamp, flange type, TYPE 25									
0482	For 3/8" bolt	1 Plum	48	.167	Ea.	3.42	9.80		13.22	18.55
0483	For 1/2" bolt		44	.182		4.39	10.65		15.04	21
0484	For 5/8" bolt		40	.200		4.98	11.75		16.73	23.50
0485	For 3/4" bolt		36	.222		4.98	13.05		18.03	25
0486	For 1" bolt		32	.250		4.98	14.70		19.68	27.50
0500	I-beam, for mounting on bottom flange, strap iron, TYPE 21									
0510	2" flange size	1 Plum	96	.083	Ea.	17.65	4.89		22.54	27
0520	3" flange size		95	.084		21	4.94		25.94	30.50
0530	4" flange size		93	.086		5.75	5.05		10.80	13.95
0540	5" flange size		92	.087		6.30	5.10		11.40	14.65
0550	6" flange size		90	.089		7.15	5.20		12.35	15.75
0560	7" flange size		88	.091		7.95	5.35		13.30	16.80
0570	8" flange size		86	.093		8.55	5.45		14	17.65
0600	One hole, vertical mounting, malleable iron									
0610	1/2" pipe size	1 Plum	160	.050	Ea.	1.08	2.94		4.02	5.60
0620	3/4" pipe size		145	.055		1.17	3.24		4.41	6.20
0630	1" pipe size		136	.059		1.24	3.45		4.69	6.55
0640	1-1/4" pipe size		128	.063		2.19	3.67		5.86	7.95
0650	1-1/2" pipe size		120	.067		2.52	3.91		6.43	8.65
0660	2" pipe size		112	.071		3.64	4.19		7.83	10.35
0670	2-1/2" pipe size		104	.077		7.05	4.52		11.57	14.60
0680	3" pipe size		96	.083		8.55	4.89		13.44	16.80
0690	3-1/2" pipe size		90	.089		10.45	5.20		15.65	19.40
0700	4" pipe size		84	.095		11.90	5.60		17.50	21.50
0750	Riser or extension pipe, carbon steel, TYPE 8									
0756	1/2" pipe size	1 Plum	52	.154	Ea.	2.52	9.05		11.57	16.40
0760	3/4" pipe size		48	.167		2.52	9.80		12.32	17.55
0770	1" pipe size		47	.170		2.64	10		12.64	18
0780	1-1/4" pipe size		46	.174		3.21	10.20		13.41	18.95
0790	1-1/2" pipe size		45	.178		3.42	10.45		13.87	19.50
0800	2" pipe size		43	.186		3.54	10.90		14.44	20.50
0810	2-1/2" pipe size		41	.195		3.68	11.45		15.13	21.50
0820	3" pipe size		40	.200		4	11.75		15.75	22
0830	3-1/2" pipe size		39	.205		4.50	12.05		16.55	23
0840	4" pipe size		38	.211		5.45	12.35		17.80	24.50
0850	5" pipe size		37	.216		6.80	12.70		19.50	26.50
0860	6" pipe size		36	.222		8.35	13.05		21.40	29
0870	8" pipe size		34	.235		14.60	13.80		28.40	37
0880	10" pipe size		32	.250		22.50	14.70		37.20	47
0890	12" pipe size		28	.286		30	16.75		46.75	58.50
0900	For plastic coating 3/4" to 4", add					190%				

22 05 29.10 Hangers & Supp. for Plumb'g/HVAC Pipe/Equip.	Crew	Daily Output	Labor-Hours	Unit	Material	2015 Bare Costs Labor	Equipment	Total	Total Incl O&P
0910	For copper plating 3/4" to 4", add					58%				
0950	Two piece, complete, carbon steel, medium weight, TYPE 4									
0960	1/2" pipe size	Q-1	137	.117	Ea.	2.29	6.15		8.44	11.80
0970	3/4" pipe size		134	.119		2.29	6.30		8.59	12.05
0980	1" pipe size		132	.121		2.59	6.40		8.99	12.50
0990	1-1/4" pipe size		130	.123		2.80	6.50		9.30	12.90
1000	1-1/2" pipe size		126	.127		3.64	6.70		10.34	14.15
1010	2" pipe size		124	.129		4.48	6.80		11.28	15.25
1020	2-1/2" pipe size		120	.133		4.84	7.05		11.89	15.95
1030	3" pipe size		117	.137		5.35	7.20		12.55	16.80
1040	3-1/2" pipe size		114	.140		10	7.40		17.40	22
1050	4" pipe size		110	.145		7.10	7.70		14.80	19.40
1060	5" pipe size		106	.151		17.40	7.95		25.35	31
1070	6" pipe size		104	.154		19.20	8.15		27.35	33.50
1080	8" pipe size		100	.160		23.50	8.45		31.95	38.50
1090	10" pipe size		96	.167		46	8.80		54.80	64
1100	12" pipe size		89	.180		55.50	9.50		65	76
1110	14" pipe size		82	.195		73.50	10.30		83.80	96.50
1120	16" pipe size	▼	68	.235		95	12.45		107.45	124
1130	For galvanized, add				▼	45%				
1150	Insert, concrete									
1160	Wedge type, carbon steel body, malleable iron nut, galvanized									
1170	1/4" threaded rod size	1 Plum	96	.083	Ea.	9.55	4.89		14.44	17.90
1180	3/8" threaded rod size		96	.083		10.30	4.89		15.19	18.75
1190	1/2" threaded rod size		96	.083		10.65	4.89		15.54	19.15
1200	5/8" threaded rod size		96	.083		11.25	4.89		16.14	19.80
1210	3/4" threaded rod size		96	.083		11.95	4.89		16.84	20.50
1220	7/8" threaded rod size	▼	96	.083	▼	12.35	4.89		17.24	21
1250	Pipe guide sized for insulation									
1260	No. 1, 1" pipe size, 1" thick insulation	1 Stpi	26	.308	Ea.	124	18.40		142.40	164
1270	No. 2, 1-1/4"-2" pipe size, 1" thick insulation		23	.348		151	21		172	198
1280	No. 3, 1-1/4"-2" pipe size, 1-1/2" thick insulation		21	.381		151	23		174	201
1290	No. 4, 2-1/2"-3-1/2" pipe size, 1-1/2" thick insulation		18	.444		151	26.50		177.50	206
1300	No. 5, 4"-5" pipe size, 1-1/2" thick insulation	▼	16	.500		171	30		201	234
1310	No. 6, 5"-6" pipe size, 2" thick insulation	Q-5	21	.762		199	41		240	281
1320	No. 7, 8" pipe size, 2" thick insulation		16	1		251	54		305	355
1330	No. 8, 10" pipe size, 2" thick insulation	▼	12	1.333		370	71.50		441.50	520
1340	No. 9, 12" pipe size, 2" thick insulation	Q-6	17	1.412		370	78.50		448.50	530
1350	No. 10, 12"-14" pipe size, 2-1/2" thick insulation		16	1.500		435	83.50		518.50	605
1360	No. 11, 16" pipe size, 2-1/2" thick insulation		10.50	2.286		435	127		562	670
1370	No. 12, 16"-18" pipe size, 3" thick insulation		9	2.667		640	149		789	930
1380	No. 13, 20" pipe size, 3" thick insulation		7.50	3.200		640	178		818	975
1390	No. 14, 24" pipe size, 3" thick insulation	▼	7	3.429	▼	845	191		1,036	1,225
1400	Bands									
1410	Adjustable band, carbon steel, for non-insulated pipe, TYPE 7									
1420	1/2" pipe size	Q-1	142	.113	Ea.	.30	5.95		6.25	9.35
1430	3/4" pipe size		140	.114		.30	6.05		6.35	9.45
1440	1" pipe size		137	.117		.30	6.15		6.45	9.65
1450	1-1/4" pipe size		134	.119		.33	6.30		6.63	9.90
1460	1-1/2" pipe size		131	.122		.33	6.45		6.78	10.10
1470	2" pipe size		129	.124		.33	6.55		6.88	10.25
1480	2-1/2" pipe size		125	.128		.60	6.75		7.35	10.85
1490	3" pipe size		122	.131		.64	6.95		7.59	11.15

For customer support on your Plumbing Cost Data, call 877.709.6968.

22 05 29.10 Hangers & Supp. for Plumb'g/HVAC Pipe/Equip.	Crew	Daily Output	Labor-Hours	Unit	Material	2015 Bare Costs Labor	Equipment	Total	Total Incl O&P
1500	3-1/2" pipe size	Q-1	119	.134	Ea.	1.04	7.10		8.14	11.90
1510	4" pipe size		114	.140		1.04	7.40		8.44	12.35
1520	5" pipe size		110	.145		1.66	7.70		9.36	13.45
1530	6" pipe size		108	.148		1.94	7.85		9.79	13.95
1540	8" pipe size		104	.154		2.93	8.15		11.08	15.45
1550	For copper plated, add					50%				
1560	For galvanized, add					30%				
1570	For plastic coating, add					30%				
1600	Adjusting nut malleable iron, steel band, TYPE 9									
1610	1/2" pipe size, galvanized band	Q-1	137	.117	Ea.	4.88	6.15		11.03	14.65
1620	3/4" pipe size, galvanized band		135	.119		4.88	6.25		11.13	14.80
1630	1" pipe size, galvanized band		132	.121		5.05	6.40		11.45	15.20
1640	1-1/4" pipe size, galvanized band		129	.124		5.05	6.55		11.60	15.45
1650	1-1/2" pipe size, galvanized band		126	.127		5.20	6.70		11.90	15.90
1660	2" pipe size, galvanized band		124	.129		5.45	6.80		12.25	16.30
1670	2-1/2" pipe size, galvanized band		120	.133		8.60	7.05		15.65	20
1680	3" pipe size, galvanized band		117	.137		9	7.20		16.20	21
1690	3-1/2" pipe size, galvanized band		114	.140		35.50	7.40		42.90	50
1700	4" pipe size, cadmium plated band		110	.145		23	7.70		30.70	36.50
1740	For plastic coated band, add					35%				
1750	For completely copper coated, add					45%				
1800	Clevis, adjustable, carbon steel, for non-insulated pipe, TYPE 1									
1810	1/2" pipe size	Q-1	137	.117	Ea.	.93	6.15		7.08	10.30
1820	3/4" pipe size		135	.119		.93	6.25		7.18	10.45
1830	1" pipe size		132	.121		.98	6.40		7.38	10.75
1840	1-1/4" pipe size		129	.124		1.11	6.55		7.66	11.10
1850	1-1/2" pipe size		126	.127		1.22	6.70		7.92	11.50
1860	2" pipe size		124	.129		1.43	6.80		8.23	11.85
1870	2-1/2" pipe size		120	.133		2.39	7.05		9.44	13.30
1880	3" pipe size		117	.137		2.58	7.20		9.78	13.75
1890	3-1/2" pipe size		114	.140		3	7.40		10.40	14.50
1900	4" pipe size		110	.145		3.24	7.70		10.94	15.15
1910	5" pipe size		106	.151		4.48	7.95		12.43	17
1920	6" pipe size		104	.154		5.45	8.15		13.60	18.20
1930	8" pipe size		100	.160		7.95	8.45		16.40	21.50
1940	10" pipe size		96	.167		14.20	8.80		23	29
1950	12" pipe size		89	.180		18.40	9.50		27.90	34.50
1960	14" pipe size		82	.195		31	10.30		41.30	49.50
1970	16" pipe size		68	.235		49.50	12.45		61.95	73.50
1971	18" pipe size		54	.296		59.50	15.65		75.15	89
1972	20" pipe size		38	.421		123	22		145	170
1980	For galvanized, add					66%				
1990	For copper plated 1/2" to 4", add					77%				
2000	For light weight 1/2" to 4", deduct					13%				
2010	Insulated pipe type, 3/4" to 12" pipe, add					180%				
2020	Insulated pipe type, chrome-moly U-strap, add					530%				
2250	Split ring, malleable iron, for non-insulated pipe, TYPE 11									
2260	1/2" pipe size	Q-1	137	.117	Ea.	4.39	6.15		10.54	14.15
2270	3/4" pipe size		135	.119		4.43	6.25		10.68	14.30
2280	1" pipe size		132	.121		4.70	6.40		11.10	14.80
2290	1-1/4" pipe size		129	.124		5.80	6.55		12.35	16.30
2300	1-1/2" pipe size		126	.127		7.05	6.70		13.75	17.95
2310	2" pipe size		124	.129		8	6.80		14.80	19.10

For customer support on your Plumbing Cost Data, call 877.709.6968.

113

22 05 29.10 Hangers & Supp. for Plumb'g/HVAC Pipe/Equip.	Crew	Daily Output	Labor-Hours	Unit	Material	2015 Bare Costs Labor	Equipment	Total	Total Incl O&P
2320	2-1/2" pipe size	Q-1	120	.133	Ea.	11.55	7.05		18.60	23.50
2330	3" pipe size		117	.137		14.45	7.20		21.65	27
2340	3-1/2" pipe size		114	.140		15.05	7.40		22.45	28
2350	4" pipe size		110	.145		15.05	7.70		22.75	28
2360	5" pipe size		106	.151		20	7.95		27.95	34
2370	6" pipe size		104	.154		44.50	8.15		52.65	61.50
2380	8" pipe size		100	.160		65.50	8.45		73.95	85
2390	For copper plated, add					8%				
2400	Channels, steel, 3/4" x 1-1/2"	1 Plum	80	.100	L.F.	4.10	5.85		9.95	13.35
2404	1-1/2" x 1-1/2"		70	.114		5.40	6.70		12.10	16.10
2408	1-7/8" x 1-1/2"		60	.133		23	7.85		30.85	37.50
2412	3" x 1-1/2"		50	.160		39.50	9.40		48.90	57.50
2416	Hangers, trapeze channel support, 12 ga. 1-1/2" x 1/1/2", 12" wide, steel		8.80	.909	Ea.	29	53.50		82.50	112
2418	18" wide, steel		8.30	.964		31.50	56.50		88	120
2430	Spring nuts, long, 1/4"		120	.067		1.17	3.91		5.08	7.20
2432	3/8"		100	.080		1.23	4.70		5.93	8.45
2434	1/2"		80	.100		1.33	5.85		7.18	10.30
2436	5/8"		80	.100		3.62	5.85		9.47	12.85
2438	3/4"		75	.107		6.15	6.25		12.40	16.20
2440	Spring nuts, short, 1/4"		120	.067		1.16	3.91		5.07	7.20
2442	3/8"		100	.080		1.20	4.70		5.90	8.40
2444	1/2"		80	.100		1.67	5.85		7.52	10.70
2500	Washer, flat steel									
2502	3/8"	1 Plum	240	.033	Ea.	.09	1.96		2.05	3.05
2503	1/2"		220	.036		.19	2.13		2.32	3.43
2504	5/8"		200	.040		.22	2.35		2.57	3.79
2505	3/4"		180	.044		.54	2.61		3.15	4.53
2506	7/8"		160	.050		.66	2.94		3.60	5.15
2507	1"		140	.057		.94	3.35		4.29	6.10
2508	1-1/4"		120	.067		1.17	3.91		5.08	7.20
2520	Nut, steel, hex									
2522	3/8"	1 Plum	200	.040	Ea.	.14	2.35		2.49	3.70
2523	1/2"		180	.044		.35	2.61		2.96	4.33
2524	5/8"		160	.050		.56	2.94		3.50	5.05
2525	3/4"		140	.057		.94	3.35		4.29	6.10
2526	7/8"		120	.067		1.51	3.91		5.42	7.55
2527	1"		100	.080		2.07	4.70		6.77	9.40
2528	1-1/4"		80	.100		5.20	5.85		11.05	14.55
2532	Turnbuckle, TYPE 13									
2534	3/8"	1 Plum	80	.100	Ea.	4.82	5.85		10.67	14.15
2535	1/2"		72	.111		5.45	6.50		11.95	15.85
2536	5/8"		64	.125		9.45	7.35		16.80	21.50
2537	3/4"		56	.143		12.40	8.40		20.80	26.50
2538	7/8"		48	.167		29.50	9.80		39.30	47.50
2539	1"		40	.200		42	11.75		53.75	64
2540	1-1/4"		32	.250		78	14.70		92.70	108
2650	Rods, carbon steel									
2660	Continuous thread									
2670	1/4" thread size	1 Plum	144	.056	L.F.	1.70	3.26		4.96	6.80
2680	3/8" thread size		144	.056		1.81	3.26		5.07	6.90
2690	1/2" thread size		144	.056		2.86	3.26		6.12	8.10
2700	5/8" thread size		144	.056		4.05	3.26		7.31	9.40
2710	3/4" thread size		144	.056		7.15	3.26		10.41	12.80

For customer support on your Plumbing Cost Data, call 877.709.6968.

22 05 Common Work Results for Plumbing

22 05 29 – Hangers and Supports for Plumbing Piping and Equipment

22 05 29.10 Hangers & Supp. for Plumb'g/HVAC Pipe/Equip.		Crew	Daily Output	Labor-Hours	Unit	Material	2015 Bare Costs Labor	Equipment	Total	Total Incl O&P
2720	7/8" thread size	1 Plum	144	.056	L.F.	8.95	3.26		12.21	14.80
2721	1" thread size	Q-1	160	.100		15.25	5.30		20.55	25
2722	1-1/8" thread size	"	120	.133		22.50	7.05		29.55	35
2725	1/4" thread size, bright finish	1 Plum	144	.056		1.24	3.26		4.50	6.30
2726	1/2" thread size, bright finish	"	144	.056		4.23	3.26		7.49	9.60
2730	For galvanized, add					40%				
2820	Rod couplings									
2821	3/8"	1 Plum	60	.133	Ea.	1.43	7.85		9.28	13.35
2822	1/2"		54	.148		1.99	8.70		10.69	15.35
2823	5/8"		48	.167		2.50	9.80		12.30	17.55
2824	3/4"		44	.182		3.51	10.65		14.16	19.95
2825	7/8"		40	.200		5.95	11.75		17.70	24.50
2826	1"		34	.235		7.85	13.80		21.65	29.50
2827	1-1/8"		30	.267		15.55	15.65		31.20	40.50
2860	Pipe hanger assy, adj. clevis, saddle, rod, clamp, insul. allowance									
2864	1/2" pipe size	Q-5	35	.457	Ea.	30	24.50		54.50	70
2866	3/4" pipe size		34.80	.460		31	24.50		55.50	72
2868	1" pipe size		34.60	.462		18.50	25		43.50	58
2869	1-1/4" pipe size		34.30	.466		18.90	25		43.90	59
2870	1-1/2" pipe size		33.90	.472		19.20	25.50		44.70	59.50
2872	2" pipe size		33.30	.480		21.50	26		47.50	62.50
2874	2-1/2" pipe size		32.30	.495		25	26.50		51.50	67.50
2876	3" pipe size		31.20	.513		30	27.50		57.50	74.50
2880	4" pipe size		30.70	.521		32.50	28		60.50	78
2884	6" pipe size		29.80	.537		41.50	29		70.50	89
2888	8" pipe size		28	.571		52.50	30.50		83	104
2892	10" pipe size		25.20	.635		59.50	34		93.50	117
2896	12" pipe size		23.20	.690		88.50	37		125.50	153
2900	Rolls									
2910	Adjustable yoke, carbon steel with CI roll, TYPE 43									
2918	2" pipe size	Q-1	140	.114	Ea.	10.60	6.05		16.65	21
2920	2-1/2" pipe size		137	.117		10.60	6.15		16.75	21
2930	3" pipe size		131	.122		11.25	6.45		17.70	22
2940	3-1/2" pipe size		124	.129		15.80	6.80		22.60	27.50
2950	4" pipe size		117	.137		15.80	7.20		23	28.50
2960	5" pipe size		110	.145		23	7.70		30.70	36.50
2970	6" pipe size		104	.154		30.50	8.15		38.65	46
2980	8" pipe size		96	.167		43.50	8.80		52.30	61.50
2990	10" pipe size		80	.200		53.50	10.55		64.05	75
3000	12" pipe size		68	.235		85.50	12.45		97.95	113
3010	14" pipe size		56	.286		162	15.10		177.10	201
3020	16" pipe size		48	.333		195	17.60		212.60	241
3050	Chair, carbon steel with CI roll									
3060	2" pipe size	1 Plum	68	.118	Ea.	11.80	6.90		18.70	23.50
3070	2-1/2" pipe size		65	.123		12.80	7.20		20	25
3080	3" pipe size		62	.129		13.80	7.55		21.35	26.50
3090	3-1/2" pipe size		60	.133		17.80	7.85		25.65	31.50
3100	4" pipe size		58	.138		16.25	8.10		24.35	30
3110	5" pipe size		56	.143		19.10	8.40		27.50	33.50
3120	6" pipe size		53	.151		25.50	8.85		34.35	41.50
3130	8" pipe size		50	.160		34.50	9.40		43.90	52
3140	10" pipe size		48	.167		47	9.80		56.80	67
3150	12" pipe size		46	.174		72	10.20		82.20	95

For customer support on your Plumbing Cost Data, call 877.709.6968.

115

22 05 29.10 Hangers & Supp. for Plumb'g/HVAC Pipe/Equip.		Crew	Daily Output	Labor-Hours	Unit	Material	2015 Bare Costs Labor	Equipment	Total	Total Incl O&P
3170	Single pipe roll, (see line 2650 for rods), TYPE 41, 1" pipe size	Q-1	137	.117	Ea.	7.90	6.15		14.05	18
3180	1-1/4" pipe size		131	.122		8.15	6.45		14.60	18.75
3190	1-1/2" pipe size		129	.124		8.45	6.55		15	19.20
3200	2" pipe size		124	.129		8.75	6.80		15.55	19.90
3210	2-1/2" pipe size		118	.136		9.10	7.15		16.25	21
3220	3" pipe size		115	.139		9.40	7.35		16.75	21.50
3230	3-1/2" pipe size		113	.142		10.30	7.50		17.80	22.50
3240	4" pipe size		112	.143		10.60	7.55		18.15	23
3250	5" pipe size		110	.145		12.10	7.70		19.80	25
3260	6" pipe size		101	.158		26	8.35		34.35	41
3270	8" pipe size		90	.178		26.50	9.40		35.90	43.50
3280	10" pipe size		80	.200		35	10.55		45.55	54.50
3290	12" pipe size		68	.235		57.50	12.45		69.95	82
3300	Saddles (add vertical pipe riser, usually 3" diameter)									
3310	Pipe support, complete, adjust., CI saddle, TYPE 36									
3320	2-1/2" pipe size	1 Plum	96	.083	Ea.	92.50	4.89		97.39	109
3330	3" pipe size		88	.091		94	5.35		99.35	112
3340	3-1/2" pipe size		79	.101		95.50	5.95		101.45	114
3350	4" pipe size		68	.118		139	6.90		145.90	163
3360	5" pipe size		64	.125		142	7.35		149.35	167
3370	6" pipe size		59	.136		142	7.95		149.95	168
3380	8" pipe size		53	.151		147	8.85		155.85	175
3390	10" pipe size		50	.160		171	9.40		180.40	202
3400	12" pipe size		48	.167		181	9.80		190.80	214
3450	For standard pipe support, one piece, CI deduct					34%				
3460	For stanchion support, CI with steel yoke, deduct					60%				
3550	Insulation shield 1" thick, 1/2" pipe size, TYPE 40	1 Asbe	100	.080	Ea.	2.74	4.19		6.93	9.50
3560	3/4" pipe size		100	.080		3.06	4.19		7.25	9.85
3570	1" pipe size		98	.082		3.30	4.27		7.57	10.30
3580	1-1/4" pipe size		98	.082		3.52	4.27		7.79	10.50
3590	1-1/2" pipe size		96	.083		3.45	4.36		7.81	10.60
3600	2" pipe size		96	.083		3.65	4.36		8.01	10.80
3610	2-1/2" pipe size		94	.085		3.77	4.46		8.23	11.10
3620	3" pipe size		94	.085		4.53	4.46		8.99	11.95
3630	2" thick, 3-1/2" pipe size		92	.087		4.84	4.55		9.39	12.40
3640	4" pipe size		92	.087		6.75	4.55		11.30	14.55
3650	5" pipe size		90	.089		9.15	4.65		13.80	17.30
3660	6" pipe size		90	.089		10.45	4.65		15.10	18.70
3670	8" pipe size		88	.091		17	4.76		21.76	26
3680	10" pipe size		88	.091		21.50	4.76		26.26	31
3690	12" pipe size		86	.093		24.50	4.87		29.37	34.50
3700	14" pipe size		86	.093		40.50	4.87		45.37	52
3710	16" pipe size		84	.095		43.50	4.99		48.49	55.50
3720	18" pipe size		84	.095		48	4.99		52.99	61
3730	20" pipe size		82	.098		53	5.10		58.10	66
3732	24" pipe size		80	.100		62	5.25		67.25	76.50
3750	Covering protection saddle, TYPE 39									
3760	1" covering size									
3770	3/4" pipe size	1 Plum	68	.118	Ea.	4.61	6.90		11.51	15.50
3780	1" pipe size		68	.118		4.61	6.90		11.51	15.50
3790	1-1/4" pipe size		68	.118		4.61	6.90		11.51	15.50
3800	1-1/2" pipe size		66	.121		4.92	7.10		12.02	16.15
3810	2" pipe size		66	.121		4.92	7.10		12.02	16.15

116

22 05 Common Work Results for Plumbing

22 05 29 – Hangers and Supports for Plumbing Piping and Equipment

22 05 29.10 Hangers & Supp. for Plumb'g/HVAC Pipe/Equip.	Crew	Daily Output	Labor-Hours	Unit	Material	2015 Bare Costs Labor	Equipment	Total	Total Incl O&P
3820	2-1/2" pipe size	1 Plum	64	.125	Ea.	4.92	7.35		12.27	16.50
3830	3" pipe size		64	.125		6.60	7.35		13.95	18.35
3840	3-1/2" pipe size		62	.129		7.30	7.55		14.85	19.50
3850	4" pipe size		62	.129		7.30	7.55		14.85	19.50
3860	5" pipe size		60	.133		7.30	7.85		15.15	19.85
3870	6" pipe size		60	.133		8.75	7.85		16.60	21.50
3900	1-1/2" covering size									
3910	3/4" pipe size	1 Plum	68	.118	Ea.	5.50	6.90		12.40	16.50
3920	1" pipe size		68	.118		5.50	6.90		12.40	16.50
3930	1-1/4" pipe size		68	.118		5.50	6.90		12.40	16.50
3940	1-1/2" pipe size		66	.121		5.50	7.10		12.60	16.80
3950	2" pipe size		66	.121		5.95	7.10		13.05	17.30
3960	2-1/2" pipe size		64	.125		7.20	7.35		14.55	19
3970	3" pipe size		64	.125		7.20	7.35		14.55	19
3980	3-1/2" pipe size		62	.129		7.60	7.55		15.15	19.80
3990	4" pipe size		62	.129		7.60	7.55		15.15	19.80
4000	5" pipe size		60	.133		7.60	7.85		15.45	20
4010	6" pipe size		60	.133		10.25	7.85		18.10	23
4020	8" pipe size		58	.138		12.85	8.10		20.95	26.50
4022	10" pipe size		56	.143		12.85	8.40		21.25	27
4024	12" pipe size		54	.148		31.50	8.70		40.20	47.50
4028	2" covering size									
4029	2-1/2" pipe size	1 Plum	62	.129	Ea.	7.25	7.55		14.80	19.45
4032	3" pipe size		60	.133		8.55	7.85		16.40	21
4033	4" pipe size		58	.138		8.55	8.10		16.65	21.50
4034	6" pipe size		56	.143		12.25	8.40		20.65	26
4035	8" pipe size		54	.148		14.40	8.70		23.10	29
4080	10" pipe size		58	.138		15.35	8.10		23.45	29
4090	12" pipe size		56	.143		35	8.40		43.40	51
4100	14" pipe size		56	.143		35	8.40		43.40	51
4110	16" pipe size		54	.148		45.50	8.70		54.20	63
4120	18" pipe size		54	.148		45.50	8.70		54.20	63
4130	20" pipe size		52	.154		50.50	9.05		59.55	69.50
4150	24" pipe size		50	.160		59.50	9.40		68.90	79
4160	30" pipe size		48	.167		65	9.80		74.80	86.50
4180	36" pipe size		45	.178		75.50	10.45		85.95	99
4186	2-1/2" covering size									
4187	3" pipe size	1 Plum	58	.138	Ea.	8.65	8.10		16.75	22
4188	4" pipe size		56	.143		9.40	8.40		17.80	23
4189	6" pipe size		52	.154		12.80	9.05		21.85	28
4190	8" pipe size		48	.167		15.85	9.80		25.65	32.50
4191	10" pipe size		44	.182		17.40	10.65		28.05	35.50
4192	12" pipe size		40	.200		37.50	11.75		49.25	59
4193	14" pipe size		36	.222		37.50	13.05		50.55	60.50
4194	16" pipe size		32	.250		48	14.70		62.70	74.50
4195	18" pipe size		28	.286		53.50	16.75		70.25	84.50
4200	Sockets									
4210	Rod end, malleable iron, TYPE 16									
4220	1/4" thread size	1 Plum	240	.033	Ea.	1.42	1.96		3.38	4.51
4230	3/8" thread size		240	.033		1.42	1.96		3.38	4.51
4240	1/2" thread size		230	.035		1.87	2.04		3.91	5.15
4250	5/8" thread size		225	.036		3.50	2.09		5.59	7
4260	3/4" thread size		220	.036		4.91	2.13		7.04	8.60

For customer support on your Plumbing Cost Data, call 877.709.6968.

117

22 05 29.10 Hangers & Supp. for Plumb'g/HVAC Pipe/Equip.	Crew	Daily Output	Labor-Hours	Unit	Material	2015 Bare Costs Labor	Equipment	Total	Total Incl O&P
4270	7/8" thread size	1 Plum	210	.038	Ea.	6.90	2.24		9.14	11
4290	Strap, 1/2" pipe size, TYPE 26	Q-1	142	.113		1.54	5.95		7.49	10.70
4300	3/4" pipe size		140	.114		1.61	6.05		7.66	10.85
4310	1" pipe size		137	.117		2.37	6.15		8.52	11.90
4320	1-1/4" pipe size		134	.119		2.41	6.30		8.71	12.20
4330	1-1/2" pipe size		131	.122		2.57	6.45		9.02	12.60
4340	2" pipe size		129	.124		2.59	6.55		9.14	12.75
4350	2-1/2" pipe size		125	.128		3.97	6.75		10.72	14.55
4360	3" pipe size		122	.131		5.20	6.95		12.15	16.20
4370	3-1/2" pipe size		119	.134		5.40	7.10		12.50	16.70
4380	4" pipe size		114	.140		5.70	7.40		13.10	17.45
4400	U-bolt, carbon steel									
4410	Standard, with nuts, TYPE 42									
4420	1/2" pipe size	1 Plum	160	.050	Ea.	.98	2.94		3.92	5.50
4430	3/4" pipe size		158	.051		1.03	2.97		4	5.60
4450	1" pipe size		152	.053		1.07	3.09		4.16	5.85
4460	1-1/4" pipe size		148	.054		1.35	3.17		4.52	6.30
4470	1-1/2" pipe size		143	.056		1.43	3.28		4.71	6.55
4480	2" pipe size		139	.058		1.55	3.38		4.93	6.80
4490	2-1/2" pipe size		134	.060		2.53	3.50		6.03	8.10
4500	3" pipe size		128	.063		2.86	3.67		6.53	8.70
4510	3-1/2" pipe size		122	.066		3.08	3.85		6.93	9.20
4520	4" pipe size		117	.068		3.12	4.01		7.13	9.50
4530	5" pipe size		114	.070		3.66	4.12		7.78	10.25
4540	6" pipe size		111	.072		6.95	4.23		11.18	14.05
4550	8" pipe size		109	.073		7.60	4.31		11.91	14.85
4560	10" pipe size		107	.075		13.20	4.39		17.59	21
4570	12" pipe size		104	.077		19.15	4.52		23.67	28
4580	For plastic coating on 1/2" thru 6" size, add					150%				
4700	U-hook, carbon steel, requires mounting screws or bolts									
4710	3/4" thru 2" pipe size									
4720	6" long	1 Plum	96	.083	Ea.	.73	4.89		5.62	8.20
4730	8" long		96	.083		1.06	4.89		5.95	8.55
4740	10" long		96	.083		1.64	4.89		6.53	9.20
4750	12" long		96	.083		2.24	4.89		7.13	9.85
4760	For copper plated, add					50%				
7000	Roof supports									
7006	Duct									
7010	Rectangular, open, 12" off roof									
7020	To 18" wide	Q-9	26	.615	Ea.	150	31		181	213
7030	To 24" wide	"	22	.727		175	36.50		211.50	249
7040	To 36" wide	Q-10	30	.800		200	42		242	284
7050	To 48" wide		28	.857		225	45		270	315
7060	To 60" wide		24	1		250	52		302	355
7100	Equipment									
7120	Equipment support	Q-5	20	.800	Ea.	65	43		108	137
7300	Pipe									
7310	Roller type									
7320	Up to 2-1/2" diam. pipe									
7324	3-1/2" off roof	Q-5	24	.667	Ea.	16	36		52	71.50
7326	Up to 10" off roof	"	20	.800	"	21	43		64	88
7340	2-1/2" to 3-1/2" diam. pipe									
7342	Up to 16" off roof	Q-5	18	.889	Ea.	45.50	48		93.50	122

22 05 29 – Hangers and Supports for Plumbing Piping and Equipment

22 05 29.10 Hangers & Supp. for Plumb'g/HVAC Pipe/Equip.	Crew	Daily Output	Labor-Hours	Unit	Material	2015 Bare Costs Labor	Equipment	Total	Total Incl O&P
7360	4" to 5" diam. pipe									
7362	Up to 12" off roof	Q-5	16	1	Ea.	65	54		119	153
7400	Strut/channel type									
7410	Up to 2-1/2" diam. pipe									
7424	3-1/2" off roof	Q-5	24	.667	Ea.	14	36		50	69.50
7426	Up to 10" off roof	"	20	.800	"	19	43		62	86
7440	Strut and roller type									
7452	2-1/2" to 3-1/2" diam. pipe									
7454	Up to 16" off roof	Q-5	18	.889	Ea.	65.50	48		113.50	144
7460	Strut and hanger type									
7470	Up to 3" diam. pipe									
7474	Up to 8" off roof	Q-5	19	.842	Ea.	41.50	45.50		87	115
8000	Pipe clamp, plastic, 1/2" CTS	1 Plum	80	.100		.23	5.85		6.08	9.10
8010	3/4" CTS		73	.110		.24	6.45		6.69	9.95
8020	1" CTS		68	.118		.54	6.90		7.44	11.05
8080	Economy clamp, 1/4" CTS		175	.046		.05	2.68		2.73	4.11
8090	3/8" CTS		168	.048		.07	2.80		2.87	4.30
8100	1/2" CTS		160	.050		.07	2.94		3.01	4.51
8110	3/4" CTS		145	.055		.15	3.24		3.39	5.05
8200	Half clamp, 1/2" CTS		80	.100		.08	5.85		5.93	8.95
8210	3/4" CTS		73	.110		.12	6.45		6.57	9.85
8300	Suspension clamp, 1/2" CTS		80	.100		.23	5.85		6.08	9.10
8310	3/4" CTS		73	.110		.24	6.45		6.69	9.95
8320	1" CTS		68	.118		.54	6.90		7.44	11.05
8400	Insulator, 1/2" CTS		80	.100		.37	5.85		6.22	9.25
8410	3/4" CTS		73	.110		.39	6.45		6.84	10.15
8420	1" CTS		68	.118		.40	6.90		7.30	10.90
8500	J hook clamp with nail 1/2" CTS		240	.033		.13	1.96		2.09	3.09
8501	3/4" CTS	▼	240	.033	▼	.13	1.96		2.09	3.09
8800	Wire cable support system									
8810	Cable with hook terminal and locking device									
8830	2 mm, (.079") dia cable, (100 lb. cap.)									
8840	1 m, (3.3') length, with hook	1 Shee	96	.083	Ea.	3.03	4.66		7.69	10.45
8850	2 m, (6.6') length, with hook		84	.095		3.50	5.35		8.85	12
8860	3 m, (9.9') length, with hook	▼	72	.111		3.97	6.20		10.17	13.85
8870	5 m, (16.4') length, with hook	Q-9	60	.267		4.97	13.45		18.42	26
8880	10 m, (32.8') length, with hook	"	30	.533	▼	7.20	27		34.20	49
8900	3mm, (.118") dia cable, (200 lb. cap.)									
8910	1 m, (3.3') length, with hook	1 Shee	96	.083	Ea.	3.88	4.66		8.54	11.35
8920	2 m, (6.6') length, with hook		84	.095		4.36	5.35		9.71	12.95
8930	3 m, (9.9') length, with hook		72	.111		4.81	6.20		11.01	14.80
8940	5 m, (16.4') length, with hook	Q-9	60	.267		5.95	13.45		19.40	27
8950	10 m, (32.8') length, with hook	"	30	.533	▼	8.55	27		35.55	50.50
9000	Cable system accessories									
9010	Anchor bolt, 3/8", with nut	1 Shee	140	.057	Ea.	1.16	3.20		4.36	6.15
9020	Air duct corner protector		160	.050		.52	2.80		3.32	4.85
9030	Air duct support attachment	▼	140	.057	▼	.99	3.20		4.19	6
9040	Flange clip, hammer-on style									
9044	For flange thickness 3/32" - 9/64", 160 lb. cap.	1 Shee	180	.044	Ea.	.29	2.49		2.78	4.12
9048	For flange thickness 1/8" - 1/4", 200 lb. cap.		160	.050		.29	2.80		3.09	4.60
9052	For flange thickness 5/16" - 1/2", 200 lb. cap.		150	.053		.56	2.98		3.54	5.20
9056	For flange thickness 9/16" - 3/4", 200 lb. cap.	▼	140	.057	▼	.75	3.20		3.95	5.70
9060	Wire insulation protection tube	▼	180	.044	L.F.	.34	2.49		2.83	4.17

For customer support on your Plumbing Cost Data, call 877.709.6968.

119

22 05 29 – Hangers and Supports for Plumbing Piping and Equipment

22 05 29.10 Hangers & Supp. for Plumb'g/HVAC Pipe/Equip.	Crew	Daily Output	Labor-Hours	Unit	Material	2015 Bare Costs Labor	Equipment	Total	Total Incl O&P
9070	Wire cutter				Ea.	41.50			41.50	45.50

22 05 33 – Heat Tracing for Plumbing Piping

22 05 33.20 Temperature Maintenance Cable

		Crew	Daily Output	Labor-Hours	Unit	Material	Labor	Equipment	Total	Total Incl O&P
0010	**TEMPERATURE MAINTENANCE CABLE**									
0040	Components									
0080	Heating cable									
0100	208 V									
0150	140°F	Q-1	1060.80	.015	L.F.	9.20	.80		10	11.35
0200	120 V									
0220	125°F	Q-1	1060.80	.015	L.F.	8.10	.80		8.90	10.10
0300	Power kit w/1 end seal	1 Elec	48.80	.164	Ea.	76.50	8.95		85.45	97.50
0310	Splice kit		35.50	.225		90.50	12.35		102.85	118
0320	End seal		160	.050		8.80	2.74		11.54	13.80
0330	Tee kit w/1 end seal		26.80	.299		96	16.35		112.35	130
0340	Powered splice w/2 end seals		20	.400		102	22		124	145
0350	Powered tee kit w/3 end seals		18	.444		131	24.50		155.50	181
0360	Cross kit w/2 end seals		18.60	.430		135	23.50		158.50	185
0500	Recommended thickness of fiberglass insulation									
0510	Pipe size									
0520	1/2" - 1" use 1" insulation									
0530	1-1/4" - 2" use 1-1/2" insulation									
0540	2-1/2" - 6" use 2" insulation									
0560	NOTE: For pipe sizes 1-1/4" and smaller use 1/4" larger diameter									
0570	insulation to allow room for installation over cable.									

22 05 48 – Vibration and Seismic Controls for Plumbing Piping and Equipment

22 05 48.40 Vibration Absorbers

		Crew	Daily Output	Labor-Hours	Unit	Material	Labor	Equipment	Total	Total Incl O&P
0010	**VIBRATION ABSORBERS**									
0100	Hangers, neoprene flex									
0200	10 – 120 lb. capacity				Ea.	26			26	28.50
0220	75 – 550 lb. capacity					37			37	40.50
0240	250 – 1100 lb. capacity					72			72	79
0260	1000 – 4000 lb. capacity					134			134	148
0500	Spring flex, 60 lb. capacity					36			36	39.50
0520	450 lb. capacity					80.50			80.50	88.50
0540	900 lb. capacity					72			72	79
0560	1100 – 1300 lb. capacity					72			72	79
0600	Rubber in shear									
0610	45 – 340 lb., up to 1/2" rod size	1 Stpi	22	.364	Ea.	14.30	21.50		35.80	49
0620	130 – 700 lb., up to 3/4" rod size		20	.400		31	24		55	70
0630	50 – 1000 lb., up to 3/4" rod size		18	.444		36.50	26.50		63	80
1000	Mounts, neoprene, 45 – 380 lb. capacity					17.25			17.25	19
1020	250 – 1100 lb. capacity					49.50			49.50	54
1040	1000 – 4000 lb. capacity					111			111	122
1100	Spring flex, 60 lb. capacity					70.50			70.50	78
1120	165 lb. capacity					72			72	79
1140	260 lb. capacity					74.50			74.50	82
1160	450 lb. capacity					96.50			96.50	106
1180	600 lb. capacity					99.50			99.50	110
1200	750 lb. capacity					110			110	121
1220	900 lb. capacity					108			108	119
1240	1100 lb. capacity					117			117	129
1260	1300 lb. capacity					117			117	129

22 05 Common Work Results for Plumbing

22 05 48 – Vibration and Seismic Controls for Plumbing Piping and Equipment

22 05 48.40 Vibration Absorbers		Crew	Daily Output	Labor-Hours	Unit	Material	2015 Bare Costs Labor	Equipment	Total	Total Incl O&P
1280	1500 lb. capacity				Ea.	176			176	194
1300	1800 lb. capacity					189			189	208
1320	2200 lb. capacity					189			189	208
1340	2600 lb. capacity				↓	197			197	216
1399	Spring type									
1400	2 Piece									
1410	50 – 1000 lb.	1 Stpi	12	.667	Ea.	130	40		170	203
1420	1100 – 1600 lb.	"	12	.667	"	141	40		181	215
1500	Double spring open									
1510	150 – 450 lb.	1 Stpi	24	.333	Ea.	89	19.90		108.90	128
1520	500 – 1000 lb.		24	.333		89	19.90		108.90	128
1530	1100 – 1600 lb.		24	.333		89	19.90		108.90	128
1540	1700 – 2400 lb.		24	.333		97	19.90		116.90	136
1550	2500 – 3400 lb.	↓	24	.333		165	19.90		184.90	212
2000	Pads, cork rib, 18" x 18" x 1", 10-50 psi					179			179	197
2020	18" x 36" x 1", 10-50 psi					350			350	385
2100	Shear flexible pads, 18" x 18" x 3/8", 20-70 psi					78			78	86
2120	18" x 36" x 3/8", 20-70 psi				↓	179			179	197
2150	Laminated neoprene and cork									
2160	1" thick	1 Stpi	16	.500	S.F.	88.50	30		118.50	143
2200	Neoprene elastomer isolation bearing pad									
2230	464 PSI, 5/8" x 19-11/16" x 39-3/8"	1 Stpi	16	.500	Ea.	295	30		325	370
3000	Note overlap in capacities due to deflections									

22 05 53 – Identification for Plumbing Piping and Equipment

22 05 53.10 Piping System Identification Labels

		Crew	Daily Output	Labor-Hours	Unit	Material	2015 Bare Costs Labor	Equipment	Total	Total Incl O&P
0010	**PIPING SYSTEM IDENTIFICATION LABELS**									
0100	Indicate contents and flow direction									
0106	Pipe markers									
0110	Plastic snap around									
0114	1/2" pipe	1 Plum	80	.100	Ea.	6.50	5.85		12.35	16
0116	3/4" pipe		80	.100		6.50	5.85		12.35	16
0118	1" pipe		80	.100		6.50	5.85		12.35	16
0120	2" pipe		75	.107		8.10	6.25		14.35	18.35
0122	3" pipe		70	.114		12.40	6.70		19.10	24
0124	4" pipe		60	.133		12.40	7.85		20.25	25.50
0126	6" pipe		60	.133		13.20	7.85		21.05	26.50
0128	8" pipe		56	.143		17.40	8.40		25.80	32
0130	10" pipe		56	.143		17.40	8.40		25.80	32
0200	Over 10" pipe size	↓	50	.160	↓	21	9.40		30.40	37
1110	Self adhesive									
1114	1" pipe	1 Plum	80	.100	Ea.	3.20	5.85		9.05	12.35
1116	2" pipe		75	.107		3.20	6.25		9.45	12.95
1118	3" pipe		70	.114		5.10	6.70		11.80	15.75
1120	4" pipe		60	.133		5	7.85		12.85	17.30
1122	6" pipe		60	.133		5	7.85		12.85	17.30
1124	8" pipe		56	.143		7.20	8.40		15.60	20.50
1126	10" pipe		56	.143		7.20	8.40		15.60	20.50
1200	Over 10" pipe size	↓	50	.160	↓	10.20	9.40		19.60	25.50
2000	Valve tags									
2010	Numbered plus identifying legend									
2100	Brass, 2" diameter	1 Plum	40	.200	Ea.	3	11.75		14.75	21
2200	Plastic, 1-1/2" diam.	"	40	.200	"	2.70	11.75		14.45	20.50

22 05 76.10 Cleanouts

		Crew	Daily Output	Labor-Hours	Unit	Material	2015 Bare Costs Labor	Equipment	Total	Total Incl O&P
0010	**CLEANOUTS**									
0060	Floor type									
0080	Round or square, scoriated nickel bronze top									
0100	2" pipe size	1 Plum	10	.800	Ea.	197	47		244	288
0120	3" pipe size		8	1		295	58.50		353.50	415
0140	4" pipe size		6	1.333		295	78.50		373.50	445
0160	5" pipe size		4	2		370	117		487	585
0180	6" pipe size	Q-1	6	2.667		375	141		516	630
0200	8" pipe size	"	4	4		660	211		871	1,050
0340	Recessed for tile, same price									
0980	Round top, recessed for terrazzo									
1000	2" pipe size	1 Plum	9	.889	Ea.	197	52		249	296
1080	3" pipe size		6	1.333		295	78.50		373.50	445
1100	4" pipe size		4	2		295	117		412	500
1120	5" pipe size	Q-1	6	2.667		375	141		516	630
1140	6" pipe size		5	3.200		375	169		544	670
1160	8" pipe size		4	4		660	211		871	1,050
2000	Round scoriated nickel bronze top, extra heavy duty									
2060	2" pipe size	1 Plum	9	.889	Ea.	270	52		322	375
2080	3" pipe size		6	1.333		370	78.50		448.50	525
2100	4" pipe size		4	2		370	117		487	580
2120	5" pipe size	Q-1	6	2.667		440	141		581	700
2140	6" pipe size		5	3.200		440	169		609	740
2160	8" pipe size		4	4		735	211		946	1,125
4000	Wall type, square smooth cover, over wall frame									
4060	2" pipe size	1 Plum	14	.571	Ea.	320	33.50		353.50	405
4080	3" pipe size		12	.667		345	39		384	440
4100	4" pipe size		10	.800		370	47		417	480
4120	5" pipe size		9	.889		540	52		592	675
4140	6" pipe size		8	1		595	58.50		653.50	745
4160	8" pipe size	Q-1	11	1.455		795	77		872	990
5000	Extension, C.I.; bronze countersunk plug, 8" long									
5040	2" pipe size	1 Plum	16	.500	Ea.	169	29.50		198.50	231
5060	3" pipe size		14	.571		196	33.50		229.50	267
5080	4" pipe size		13	.615		200	36		236	274
5100	5" pipe size		12	.667		310	39		349	405
5120	6" pipe size		11	.727		380	42.50		422.50	480

22 05 76.20 Cleanout Tees

		Crew	Daily Output	Labor-Hours	Unit	Material	2015 Bare Costs Labor	Equipment	Total	Total Incl O&P
0010	**CLEANOUT TEES**									
0100	Cast iron, B&S, with countersunk plug									
0200	2" pipe size	1 Plum	4	2	Ea.	267	117		384	470
0220	3" pipe size		3.60	2.222		292	130		422	515
0240	4" pipe size		3.30	2.424		365	142		507	615
0260	5" pipe size	Q-1	5.50	2.909		790	154		944	1,100
0280	6" pipe size	"	5	3.200		980	169		1,149	1,325
0300	8" pipe size	Q-3	5	6.400		1,125	360		1,485	1,775
0500	For round smooth access cover, same price									
0600	For round scoriated access cover, same price									
0700	For square smooth access cover, add				Ea.	60%				
2000	Cast iron, no hub									
2010	Cleanout tee, with 2 couplings									
2012	2"	Q-1	22	.727	Ea.	43.50	38.50		82	106

22 05 76.20 Cleanout Tees	Crew	Daily Output	Labor-Hours	Unit	Material	2015 Bare Costs Labor	Equipment	Total	Total Incl O&P
2014	3"	Q-1	19	.842	Ea.	55.50	44.50		100	128
2016	4"	↓	16.50	.970		78	51		129	164
2018	6"	Q-2	20	1.200	↓	176	65.50		241.50	294
2040	Cleanout plug, no hub, with 1 coupling									
2042	2"	Q-1	44	.364	Ea.	12.10	19.20		31.30	42.50
2046	3"		38	.421		18.80	22		40.80	54
2048	4"	↓	33	.485		32.50	25.50		58	74
2050	6"	Q-2	40	.600		112	33		145	173
2052	8"	"	33	.727	↓	155	40		195	231
4000	Plastic, tees and adapters. Add plugs									
4010	ABS, DWV									
4020	Cleanout tee, 1-1/2" pipe size	1 Plum	15	.533	Ea.	23	31.50		54.50	72.50
4030	2" pipe size	Q-1	27	.593		25	31.50		56.50	75
4040	3" pipe size		21	.762		47	40.50		87.50	113
4050	4" pipe size	↓	16	1		101	53		154	191
4100	Cleanout plug, 1-1/2" pipe size	1 Plum	32	.250		3.94	14.70		18.64	26.50
4110	2" pipe size	Q-1	56	.286		5.20	15.10		20.30	28.50
4120	3" pipe size		36	.444		8.35	23.50		31.85	44.50
4130	4" pipe size	↓	30	.533		14.65	28		42.65	58.50
4180	Cleanout adapter fitting, 1-1/2" pipe size	1 Plum	32	.250		6.25	14.70		20.95	29
4190	2" pipe size	Q-1	56	.286		9.60	15.10		24.70	33.50
4200	3" pipe size		36	.444		24.50	23.50		48	62.50
4210	4" pipe size	↓	30	.533	↓	45	28		73	92
5000	PVC, DWV									
5010	Cleanout tee, 1-1/2" pipe size	1 Plum	15	.533	Ea.	17.10	31.50		48.60	66.50
5020	2" pipe size	Q-1	27	.593		19.95	31.50		51.45	69.50
5030	3" pipe size		21	.762		35.50	40.50		76	100
5040	4" pipe size	↓	16	1		69.50	53		122.50	157
5090	Cleanout plug, 1-1/2" pipe size	1 Plum	32	.250		4.08	14.70		18.78	26.50
5100	2" pipe size	Q-1	56	.286		4.53	15.10		19.63	28
5110	3" pipe size		36	.444		8.10	23.50		31.60	44.50
5120	4" pipe size		30	.533		11.95	28		39.95	55.50
5130	6" pipe size		24	.667		39	35		74	96
5170	Cleanout adapter fitting, 1-1/2" pipe size	1 Plum	32	.250		5.45	14.70		20.15	28
5180	2" pipe size	Q-1	56	.286		7	15.10		22.10	30.50
5190	3" pipe size		36	.444		19.75	23.50		43.25	57
5200	4" pipe size		30	.533		32.50	28		60.50	78
5210	6" pipe size	↓	24	.667	↓	94.50	35		129.50	157
5300	Cleanout tee, with plug									
5310	4"	Q-1	14	1.143	Ea.	74	60.50		134.50	172
5320	6"	"	8	2		63	106		169	229
5330	8"	Q-2	11	2.182		132	120		252	325
5340	12"	"	10	2.400	↓	320	131		451	555
5400	Polypropylene, Schedule 40									
5410	Cleanout tee with plug									
5420	1-1/2"	1 Plum	10	.800	Ea.	36.50	47		83.50	112
5430	2"	Q-1	17	.941		39	49.50		88.50	118
5440	3"		11	1.455		83	77		160	208
5450	4"	↓	9	1.778	↓	113	94		207	266

For customer support on your Plumbing Cost Data, call 877.709.6968.

123

22 07 Plumbing Insulation

22 07 16 – Plumbing Equipment Insulation

22 07 16.10 Insulation for Plumbing Equipment		Crew	Daily Output	Labor-Hours	Unit	Material	2015 Bare Costs Labor	Equipment	Total	Total Incl O&P
0010	**INSULATION FOR PLUMBING EQUIPMENT**									
2900	Domestic water heater wrap kit									
2920	1-1/2" with vinyl jacket, 20-60 gal.	G	1 Plum	8	1	Ea.	18.60	58.50		77.10	109
2925	50 to 80 gallons	G	"	8	1	"	23.50	58.50		82	115

22 07 19 – Plumbing Piping Insulation

22 07 19.10 Piping Insulation

22 07 19.10 Piping Insulation		Crew	Daily Output	Labor-Hours	Unit	Material	2015 Bare Costs Labor	Equipment	Total	Total Incl O&P
0010	**PIPING INSULATION**									
0100	Rule of thumb, as a percentage of total mechanical costs				Job				10%	10%
0110	Insulation req'd. is based on the surface size/area to be covered									
0230	Insulated protectors, (ADA)									
0235	For exposed piping under sinks or lavatories									
0240	Vinyl coated foam, velcro tabs									
0245	P Trap, 1-1/4" or 1-1/2"	1 Plum	32	.250	Ea.	22.50	14.70		37.20	46.50
0260	Valve and supply cover									
0265	1/2", 3/8", and 7/16" pipe size	1 Plum	32	.250	Ea.	22	14.70		36.70	46
0280	Tailpiece offset (wheelchair)									
0285	1-1/4" pipe size	1 Plum	32	.250	Ea.	13.65	14.70		28.35	37
0600	Pipe covering (price copper tube one size less than IPS)									
1000	Mineral wool									
1010	Preformed, 1200°F, plain									
1014	1" wall									
1016	1/2" iron pipe size	G	Q-14	230	.070	L.F.	1.65	3.28		4.93	6.90
1018	3/4" iron pipe size	G		220	.073		1.72	3.43		5.15	7.25
1022	1" iron pipe size	G		210	.076		1.77	3.59		5.36	7.55
1024	1-1/4" iron pipe size	G		205	.078		1.82	3.68		5.50	7.75
1026	1-1/2" iron pipe size	G		205	.078		1.90	3.68		5.58	7.85
1028	2" iron pipe size	G		200	.080		2.32	3.77		6.09	8.40
1030	2-1/2" iron pipe size	G		190	.084		2.46	3.97		6.43	8.90
1032	3" iron pipe size	G		180	.089		2.55	4.19		6.74	9.30
1034	4" iron pipe size	G		150	.107		3.14	5.05		8.19	11.30
1036	5" iron pipe size	G		140	.114		3.55	5.40		8.95	12.30
1038	6" iron pipe size	G		120	.133		3.79	6.30		10.09	13.95
1040	7" iron pipe size	G		110	.145		4.43	6.85		11.28	15.50
1042	8" iron pipe size	G		100	.160		6.10	7.55		13.65	18.45
1044	9" iron pipe size	G		90	.178		6.65	8.40		15.05	20.50
1046	10" iron pipe size	G	▼	90	.178	▼	7.50	8.40		15.90	21.50
1050	1-1/2" wall									
1052	1/2" iron pipe size	G	Q-14	225	.071	L.F.	2.82	3.35		6.17	8.30
1054	3/4" iron pipe size	G		215	.074		2.89	3.51		6.40	8.65
1056	1" iron pipe size	G		205	.078		2.93	3.68		6.61	8.95
1058	1-1/4" iron pipe size	G		200	.080		2.99	3.77		6.76	9.15
1060	1-1/2" iron pipe size	G		200	.080		3.14	3.77		6.91	9.30
1062	2" iron pipe size	G		190	.084		3.32	3.97		7.29	9.85
1064	2-1/2" iron pipe size	G		180	.089		3.62	4.19		7.81	10.50
1066	3" iron pipe size	G		165	.097		3.77	4.57		8.34	11.25
1068	4" iron pipe size	G		140	.114		4.53	5.40		9.93	13.40
1070	5" iron pipe size	G		130	.123		4.84	5.80		10.64	14.35
1072	6" iron pipe size	G		110	.145		4.99	6.85		11.84	16.15
1074	7" iron pipe size	G		100	.160		5.90	7.55		13.45	18.25
1076	8" iron pipe size	G		90	.178		6.55	8.40		14.95	20.50
1078	9" iron pipe size	G		85	.188		7.55	8.85		16.40	22
1080	10" iron pipe size	G	▼	80	.200	▼	8.30	9.45		17.75	24

For customer support on your Plumbing Cost Data, call 877.709.6968.

22 07 19 – Plumbing Piping Insulation

22 07 19.10 Piping Insulation		Crew	Daily Output	Labor-Hours	Unit	Material	2015 Bare Costs Labor	Equipment	Total	Total Incl O&P
1082	12" iron pipe size	G Q-14	75	.213	L.F.	9.45	10.05		19.50	26
1084	14" iron pipe size	G	70	.229		10.80	10.75		21.55	28.50
1086	16" iron pipe size	G	65	.246		12.30	11.60		23.90	31.50
1088	18" iron pipe size	G	60	.267		13.85	12.55		26.40	35
1090	20" iron pipe size	G	55	.291		15.25	13.70		28.95	38.50
1092	22" iron pipe size	G	50	.320		17.45	15.10		32.55	42.50
1094	24" iron pipe size	G	45	.356		18.65	16.75		35.40	46.50
1100	2" wall									
1102	1/2" iron pipe size	G Q-14	220	.073	L.F.	3.65	3.43		7.08	9.35
1104	3/4" iron pipe size	G	210	.076		3.78	3.59		7.37	9.75
1106	1" iron pipe size	G	200	.080		4.01	3.77		7.78	10.25
1108	1-1/4" iron pipe size	G	190	.084		4.30	3.97		8.27	10.95
1110	1-1/2" iron pipe size	G	190	.084		4.48	3.97		8.45	11.15
1112	2" iron pipe size	G	180	.089		4.70	4.19		8.89	11.65
1114	2-1/2" iron pipe size	G	170	.094		5.30	4.44		9.74	12.75
1116	3" iron pipe size	G	160	.100		5.70	4.71		10.41	13.60
1118	4" iron pipe size	G	130	.123		6.60	5.80		12.40	16.35
1120	5" iron pipe size	G	120	.133		7.35	6.30		13.65	17.90
1122	6" iron pipe size	G	100	.160		7.65	7.55		15.20	20
1124	8" iron pipe size	G	80	.200		9.25	9.45		18.70	25
1126	10" iron pipe size	G	70	.229		11.35	10.75		22.10	29
1128	12" iron pipe size	G	65	.246		12.55	11.60		24.15	32
1130	14" iron pipe size	G	60	.267		14.05	12.55		26.60	35
1132	16" iron pipe size	G	55	.291		15.90	13.70		29.60	39
1134	18" iron pipe size	G	50	.320		17.20	15.10		32.30	42.50
1136	20" iron pipe size	G	45	.356		19.80	16.75		36.55	48
1138	22" iron pipe size	G	45	.356		21.50	16.75		38.25	49.50
1140	24" iron pipe size	G	40	.400		22.50	18.85		41.35	54.50
1150	4" wall									
1152	1-1/4" iron pipe size	G Q-14	170	.094	L.F.	11.65	4.44		16.09	19.70
1154	1-1/2" iron pipe size	G	165	.097		12.10	4.57		16.67	20.50
1156	2" iron pipe size	G	155	.103		12.35	4.87		17.22	21
1158	4" iron pipe size	G	105	.152		15.60	7.20		22.80	28.50
1160	6" iron pipe size	G	75	.213		18.40	10.05		28.45	35.50
1162	8" iron pipe size	G	60	.267		21	12.55		33.55	42.50
1164	10" iron pipe size	G	50	.320		25	15.10		40.10	51
1166	12" iron pipe size	G	45	.356		27.50	16.75		44.25	56.50
1168	14" iron pipe size	G	40	.400		29.50	18.85		48.35	62
1170	16" iron pipe size	G	35	.457		33.50	21.50		55	70
1172	18" iron pipe size	G	32	.500		36	23.50		59.50	76
1174	20" iron pipe size	G	30	.533		38.50	25		63.50	81.50
1176	22" iron pipe size	G	28	.571		43.50	27		70.50	89.50
1178	24" iron pipe size	G	26	.615		46	29		75	95.50
4280	Cellular glass, closed cell foam, all service jacket, sealant,									
4281	working temp. (-450°F to +900°F), 0 water vapor transmission									
4284	1" wall									
4286	1/2" iron pipe size	G Q-14	120	.133	L.F.	6.80	6.30		13.10	17.25
4300	1-1/2" wall,									
4301	1" iron pipe size	G Q-14	105	.152	L.F.	10.40	7.20		17.60	22.50
4304	2-1/2" iron pipe size	G	90	.178		14.20	8.40		22.60	28.50
4306	3" iron pipe size	G	85	.188		17.60	8.85		26.45	33
4308	4" iron pipe size	G	70	.229		21.50	10.75		32.25	40.50
4310	5" iron pipe size	G	65	.246		22	11.60		33.60	42

22 07 19.10 Piping Insulation		Crew	Daily Output	Labor-Hours	Unit	Material	2015 Bare Costs Labor	Equipment	Total	Total Incl O&P
4320	2" wall,									
4322	1" iron pipe size	G Q-14	100	.160	L.F.	13.85	7.55		21.40	27
4324	2-1/2" iron pipe size	G	85	.188		21.50	8.85		30.35	37.50
4326	3" iron pipe size	G	80	.200		21.50	9.45		30.95	38
4328	4" iron pipe size	G	65	.246		26	11.60		37.60	46.50
4330	5" iron pipe size	G	60	.267		27	12.55		39.55	49
4332	6" iron pipe size	G	50	.320		31	15.10		46.10	58
4336	8" iron pipe size	G	40	.400		36.50	18.85		55.35	70
4338	10" iron pipe size	G	35	.457		45.50	21.50		67	83.50
4350	2-1/2" wall,									
4360	12" iron pipe size	G Q-14	32	.500	L.F.	60.50	23.50		84	103
4362	14" iron pipe size	G "	28	.571	"	67	27		94	116
4370	3" wall,									
4378	6" iron pipe size	G Q-14	48	.333	L.F.	41.50	15.70		57.20	70
4380	8" iron pipe size	G	38	.421		52	19.85		71.85	88
4382	10" iron pipe size	G	33	.485		60	23		83	102
4384	16" iron pipe size	G	25	.640		84.50	30		114.50	140
4386	18" iron pipe size	G	22	.727		108	34.50		142.50	172
4388	20" iron pipe size	G	20	.800		117	37.50		154.50	188
4400	3-1/2" wall,									
4412	12" iron pipe size	G Q-14	27	.593	L.F.	71.50	28		99.50	122
4414	14" iron pipe size	G "	25	.640	"	74	30		104	129
4430	4" wall,									
4446	16" iron pipe size	G Q-14	22	.727	L.F.	96	34.50		130.50	160
4448	18" iron pipe size	G	20	.800		105	37.50		142.50	174
4450	20" iron pipe size	G	18	.889		118	42		160	195
4480	Fittings, average with fabric and mastic									
4484	1" wall,									
4486	1/2" iron pipe size	G 1 Asbe	40	.200	Ea.	5.85	10.45		16.30	23
4500	1-1/2" wall,									
4502	1" iron pipe size	G 1 Asbe	38	.211	Ea.	7.80	11		18.80	26
4504	2-1/2" iron pipe size	G	32	.250		14.15	13.10		27.25	36
4506	3" iron pipe size	G	30	.267		14.15	13.95		28.10	37
4508	4" iron pipe size	G	28	.286		17.85	14.95		32.80	43
4510	5" iron pipe size	G	24	.333		22	17.45		39.45	51
4520	2" wall,									
4522	1" iron pipe size	G 1 Asbe	36	.222	Ea.	11.30	11.65		22.95	30.50
4524	2-1/2" iron pipe size	G	30	.267		17.85	13.95		31.80	41
4526	3" iron pipe size	G	28	.286		17.85	14.95		32.80	43
4528	4" iron pipe size	G	24	.333		22	17.45		39.45	51
4530	5" iron pipe size	G	22	.364		32.50	19.05		51.55	65
4532	6" iron pipe size	G	20	.400		45.50	21		66.50	82.50
4536	8" iron pipe size	G	12	.667		76	35		111	139
4538	10" iron pipe size	G	8	1		88.50	52.50		141	179
4550	2-1/2" wall,									
4560	12" iron pipe size	G 1 Asbe	6	1.333	Ea.	162	70		232	287
4562	14" iron pipe size	G "	4	2	"	214	105		319	400
4570	3" wall,									
4578	6" iron pipe size	G 1 Asbe	16	.500	Ea.	65.50	26		91.50	114
4580	8" iron pipe size	G	10	.800		98	42		140	173
4582	10" iron pipe size	G	6	1.333		133	70		203	255
4900	Calcium silicate, with 8 oz. canvas cover									
5100	1" wall, 1/2" iron pipe size	G Q-14	170	.094	L.F.	4.07	4.44		8.51	11.40

126

22 07 19 – Plumbing Piping Insulation

22 07 19.10 Piping Insulation		Crew	Daily Output	Labor-Hours	Unit	Material	2015 Bare Costs Labor	Equipment	Total	Total Incl O&P
5130	3/4" iron pipe size	G	Q-14	170	.094	L.F.	4.10	4.44		8.54	11.40
5140	1" iron pipe size	G		170	.094		3.98	4.44		8.42	11.30
5150	1-1/4" iron pipe size	G		165	.097		4.07	4.57		8.64	11.60
5160	1-1/2" iron pipe size	G		165	.097		4.13	4.57		8.70	11.65
5170	2" iron pipe size	G		160	.100		4.74	4.71		9.45	12.55
5180	2-1/2" iron pipe size	G		160	.100		5.05	4.71		9.76	12.90
5190	3" iron pipe size	G		150	.107		5.45	5.05		10.50	13.85
5200	4" iron pipe size	G		140	.114		6.75	5.40		12.15	15.85
5210	5" iron pipe size	G		135	.119		7.10	5.60		12.70	16.55
5220	6" iron pipe size	G		130	.123		7.55	5.80		13.35	17.40
5280	1-1/2" wall, 1/2" iron pipe size	G		150	.107		4.40	5.05		9.45	12.70
5310	3/4" iron pipe size	G		150	.107		4.48	5.05		9.53	12.80
5320	1" iron pipe size	G		150	.107		4.88	5.05		9.93	13.20
5330	1-1/4" iron pipe size	G		145	.110		5.25	5.20		10.45	13.90
5340	1-1/2" iron pipe size	G		145	.110		5.65	5.20		10.85	14.30
5350	2" iron pipe size	G		140	.114		6.20	5.40		11.60	15.20
5360	2-1/2" iron pipe size	G		140	.114		6.75	5.40		12.15	15.85
5370	3" iron pipe size	G		135	.119		7.10	5.60		12.70	16.50
5380	4" iron pipe size	G		125	.128		8.20	6.05		14.25	18.45
5390	5" iron pipe size	G		120	.133		9.25	6.30		15.55	20
5400	6" iron pipe size	G		110	.145		9.55	6.85		16.40	21
5460	2" wall, 1/2" iron pipe size	G		135	.119		6.80	5.60		12.40	16.15
5490	3/4" iron pipe size	G		135	.119		7.15	5.60		12.75	16.55
5500	1" iron pipe size	G		135	.119		7.50	5.60		13.10	16.95
5510	1-1/4" iron pipe size	G		130	.123		8.05	5.80		13.85	17.90
5520	1-1/2" iron pipe size	G		130	.123		8.40	5.80		14.20	18.30
5530	2" iron pipe size	G		125	.128		8.90	6.05		14.95	19.20
5540	2-1/2" iron pipe size	G		125	.128		10.55	6.05		16.60	21
5550	3" iron pipe size	G		120	.133		10.65	6.30		16.95	21.50
5560	4" iron pipe size	G		115	.139		12.30	6.55		18.85	23.50
5570	5" iron pipe size	G		110	.145		14	6.85		20.85	26
5580	6" iron pipe size	G		105	.152		15.30	7.20		22.50	28
5600	Calcium silicate, no cover										
5720	1" wall, 1/2" iron pipe size	G	Q-14	180	.089	L.F.	3.70	4.19		7.89	10.55
5740	3/4" iron pipe size	G		180	.089		3.70	4.19		7.89	10.55
5750	1" iron pipe size	G		180	.089		3.55	4.19		7.74	10.40
5760	1-1/4" iron pipe size	G		175	.091		3.61	4.31		7.92	10.65
5770	1-1/2" iron pipe size	G		175	.091		3.64	4.31		7.95	10.70
5780	2" iron pipe size	G		170	.094		4.20	4.44		8.64	11.50
5790	2-1/2" iron pipe size	G		170	.094		4.44	4.44		8.88	11.80
5800	3" iron pipe size	G		160	.100		4.77	4.71		9.48	12.60
5810	4" iron pipe size	G		150	.107		5.95	5.05		11	14.40
5820	5" iron pipe size	G		145	.110		6.20	5.20		11.40	14.90
5830	6" iron pipe size	G		140	.114		6.50	5.40		11.90	15.55
5900	1-1/2" wall, 1/2" iron pipe size	G		160	.100		3.91	4.71		8.62	11.65
5920	3/4" iron pipe size	G		160	.100		3.98	4.71		8.69	11.75
5930	1" iron pipe size	G		160	.100		4.34	4.71		9.05	12.10
5940	1-1/4" iron pipe size	G		155	.103		4.67	4.87		9.54	12.75
5950	1-1/2" iron pipe size	G		155	.103		5	4.87		9.87	13.10
5960	2" iron pipe size	G		150	.107		5.55	5.05		10.60	13.95
5970	2-1/2" iron pipe size	G		150	.107		6.05	5.05		11.10	14.50
5980	3" iron pipe size	G		145	.110		6.30	5.20		11.50	15.05
5990	4" iron pipe size	G		135	.119		7.30	5.60		12.90	16.75

22 07 19.10 Piping Insulation		Crew	Daily Output	Labor-Hours	Unit	Material	2015 Bare Costs Labor	Equipment	Total	Total Incl O&P
6000	5" iron pipe size	G	Q-14	130	.123	L.F.	8.25	5.80		14.05	18.10
6010	6" iron pipe size	G		120	.133		8.40	6.30		14.70	19.05
6020	7" iron pipe size	G		115	.139		9.85	6.55		16.40	21
6030	8" iron pipe size	G		105	.152		11.15	7.20		18.35	23.50
6040	9" iron pipe size	G		100	.160		13.35	7.55		20.90	26.50
6050	10" iron pipe size	G		95	.168		14.75	7.95		22.70	28.50
6060	12" iron pipe size	G		90	.178		17.50	8.40		25.90	32.50
6070	14" iron pipe size	G		85	.188		19.80	8.85		28.65	36
6080	16" iron pipe size	G		80	.200		22	9.45		31.45	39
6090	18" iron pipe size	G		75	.213		24.50	10.05		34.55	42.50
6120	2" wall, 1/2" iron pipe size	G		145	.110		6.20	5.20		11.40	14.90
6140	3/4" iron pipe size	G		145	.110		6.50	5.20		11.70	15.25
6150	1" iron pipe size	G		145	.110		6.85	5.20		12.05	15.65
6160	1-1/4" iron pipe size	G		140	.114		7.35	5.40		12.75	16.50
6170	1-1/2" iron pipe size	G		140	.114		7.70	5.40		13.10	16.85
6180	2" iron pipe size	G		135	.119		8.05	5.60		13.65	17.55
6190	2-1/2" iron pipe size	G		135	.119		9.70	5.60		15.30	19.35
6200	3" iron pipe size	G		130	.123		9.75	5.80		15.55	19.75
6210	4" iron pipe size	G		125	.128		11.25	6.05		17.30	22
6220	5" iron pipe size	G		120	.133		12.85	6.30		19.15	24
6230	6" iron pipe size	G		115	.139		14.05	6.55		20.60	25.50
6240	7" iron pipe size	G		110	.145		15.20	6.85		22.05	27.50
6250	8" iron pipe size	G		105	.152		16.75	7.20		23.95	29.50
6260	9" iron pipe size	G		100	.160		18.90	7.55		26.45	33
6270	10" iron pipe size	G		95	.168		21	7.95		28.95	35.50
6280	12" iron pipe size	G		90	.178		23.50	8.40		31.90	38.50
6290	14" iron pipe size	G		85	.188		25.50	8.85		34.35	42
6300	16" iron pipe size	G		80	.200		28.50	9.45		37.95	45.50
6310	18" iron pipe size	G		75	.213		31	10.05		41.05	49.50
6320	20" iron pipe size	G		65	.246		38	11.60		49.60	60
6330	22" iron pipe size	G		60	.267		42	12.55		54.55	66
6340	24" iron pipe size	G		55	.291		43	13.70		56.70	69
6360	3" wall, 1/2" iron pipe size	G		115	.139		11.20	6.55		17.75	22.50
6380	3/4" iron pipe size	G		115	.139		11.30	6.55		17.85	22.50
6390	1" iron pipe size	G		115	.139		11.35	6.55		17.90	22.50
6400	1-1/4" iron pipe size	G		110	.145		11.55	6.85		18.40	23.50
6410	1-1/2" iron pipe size	G		110	.145		11.60	6.85		18.45	23.50
6420	2" iron pipe size	G		105	.152		12	7.20		19.20	24.50
6430	2-1/2" iron pipe size	G		105	.152		14.10	7.20		21.30	26.50
6440	3" iron pipe size	G		100	.160		14.20	7.55		21.75	27.50
6450	4" iron pipe size	G		95	.168		18.15	7.95		26.10	32.50
6460	5" iron pipe size	G		90	.178		19.70	8.40		28.10	34.50
6470	6" iron pipe size	G		90	.178		22	8.40		30.40	37.50
6480	7" iron pipe size	G		85	.188		24.50	8.85		33.35	41
6490	8" iron pipe size	G		85	.188		26.50	8.85		35.35	43
6500	9" iron pipe size	G		80	.200		30	9.45		39.45	47.50
6510	10" iron pipe size	G		75	.213		32	10.05		42.05	50.50
6520	12" iron pipe size	G		70	.229		35	10.75		45.75	56
6530	14" iron pipe size	G		65	.246		39.50	11.60		51.10	61.50
6540	16" iron pipe size	G		60	.267		43.50	12.55		56.05	67.50
6550	18" iron pipe size	G		55	.291		48	13.70		61.70	74
6560	20" iron pipe size	G		50	.320		57	15.10		72.10	86
6570	22" iron pipe size	G		45	.356		62	16.75		78.75	94

22 07 Plumbing Insulation

22 07 19 – Plumbing Piping Insulation

22 07 19.10 Piping Insulation		Crew	Daily Output	Labor-Hours	Unit	Material	2015 Bare Costs Labor	Equipment	Total	Total Incl O&P
6580	24" iron pipe size	G	Q-14	40	.400	L.F.	66.50	18.85		85.35	103
6600	Fiberglass, with all service jacket										
6640	1/2" wall, 1/2" iron pipe size	G	Q-14	250	.064	L.F.	.68	3.02		3.70	5.45
6660	3/4" iron pipe size	G		240	.067		.77	3.14		3.91	5.75
6670	1" iron pipe size	G		230	.070		.80	3.28		4.08	6
6680	1-1/4" iron pipe size	G		220	.073		.85	3.43		4.28	6.30
6690	1-1/2" iron pipe size	G		220	.073		.98	3.43		4.41	6.45
6700	2" iron pipe size	G		210	.076		1.05	3.59		4.64	6.75
6710	2-1/2" iron pipe size	G		200	.080		1.09	3.77		4.86	7.05
6840	1" wall, 1/2" iron pipe size	G		240	.067		.83	3.14		3.97	5.80
6860	3/4" iron pipe size	G		230	.070		.90	3.28		4.18	6.10
6870	1" iron pipe size	G		220	.073		.97	3.43		4.40	6.40
6880	1-1/4" iron pipe size	G		210	.076		1.05	3.59		4.64	6.75
6890	1-1/2" iron pipe size	G		210	.076		1.13	3.59		4.72	6.85
6900	2" iron pipe size	G		200	.080		1.22	3.77		4.99	7.20
6910	2-1/2" iron pipe size	G		190	.084		1.39	3.97		5.36	7.75
6920	3" iron pipe size	G		180	.089		1.49	4.19		5.68	8.15
6930	3-1/2" iron pipe size	G		170	.094		1.61	4.44		6.05	8.65
6940	4" iron pipe size	G		150	.107		1.97	5.05		7.02	10
6950	5" iron pipe size	G		140	.114		2.22	5.40		7.62	10.85
6960	6" iron pipe size	G		120	.133		2.36	6.30		8.66	12.40
6970	7" iron pipe size	G		110	.145		2.81	6.85		9.66	13.75
6980	8" iron pipe size	G		100	.160		3.85	7.55		11.40	16
6990	9" iron pipe size	G		90	.178		4.06	8.40		12.46	17.50
7000	10" iron pipe size	G		90	.178		4.09	8.40		12.49	17.55
7010	12" iron pipe size	G		80	.200		4.45	9.45		13.90	19.60
7020	14" iron pipe size	G		80	.200		5.35	9.45		14.80	20.50
7030	16" iron pipe size	G		70	.229		6.80	10.75		17.55	24
7040	18" iron pipe size	G		70	.229		7.55	10.75		18.30	25
7050	20" iron pipe size	G		60	.267		8.40	12.55		20.95	29
7060	24" iron pipe size	G		60	.267		10.25	12.55		22.80	31
7080	1-1/2" wall, 1/2" iron pipe size	G		230	.070		1.57	3.28		4.85	6.85
7100	3/4" iron pipe size	G		220	.073		1.57	3.43		5	7.10
7110	1" iron pipe size	G		210	.076		1.68	3.59		5.27	7.45
7120	1-1/4" iron pipe size	G		200	.080		1.83	3.77		5.60	7.85
7130	1-1/2" iron pipe size	G		200	.080		1.92	3.77		5.69	7.95
7140	2" iron pipe size	G		190	.084		2.11	3.97		6.08	8.50
7150	2-1/2" iron pipe size	G		180	.089		2.27	4.19		6.46	9
7160	3" iron pipe size	G		170	.094		2.37	4.44		6.81	9.50
7170	3-1/2" iron pipe size	G		160	.100		2.60	4.71		7.31	10.20
7180	4" iron pipe size	G		140	.114		2.70	5.40		8.10	11.35
7190	5" iron pipe size	G		130	.123		3.02	5.80		8.82	12.35
7200	6" iron pipe size	G		110	.145		3.19	6.85		10.04	14.15
7210	7" iron pipe size	G		100	.160		3.54	7.55		11.09	15.65
7220	8" iron pipe size	G		90	.178		4.47	8.40		12.87	17.95
7230	9" iron pipe size	G		85	.188		4.66	8.85		13.51	18.95
7240	10" iron pipe size	G		80	.200		4.82	9.45		14.27	20
7250	12" iron pipe size	G		75	.213		5.45	10.05		15.50	21.50
7260	14" iron pipe size	G		70	.229		6.55	10.75		17.30	24
7270	16" iron pipe size	G		65	.246		8.50	11.60		20.10	27.50
7280	18" iron pipe size	G		60	.267		9.55	12.55		22.10	30
7290	20" iron pipe size	G		55	.291		9.75	13.70		23.45	32.50
7300	24" iron pipe size	G		50	.320		11.95	15.10		27.05	36.50

For customer support on your Plumbing Cost Data, call 877.709.6968.

22 07 19.10 Piping Insulation		Crew	Daily Output	Labor-Hours	Unit	Material	2015 Bare Costs Labor	Equipment	Total	Total Incl O&P
7320	2" wall, 1/2" iron pipe size	G	Q-14	220	.073	L.F.	2.43	3.43		5.86	8
7340	3/4" iron pipe size	G		210	.076		2.51	3.59		6.10	8.35
7350	1" iron pipe size	G		200	.080		2.67	3.77		6.44	8.80
7360	1-1/4" iron pipe size	G		190	.084		2.82	3.97		6.79	9.30
7370	1-1/2" iron pipe size	G		190	.084		2.94	3.97		6.91	9.45
7380	2" iron pipe size	G		180	.089		3.10	4.19		7.29	9.90
7390	2-1/2" iron pipe size	G		170	.094		3.34	4.44		7.78	10.55
7400	3" iron pipe size	G		160	.100		3.54	4.71		8.25	11.25
7410	3-1/2" iron pipe size	G		150	.107		3.84	5.05		8.89	12.05
7420	4" iron pipe size	G		130	.123		4.13	5.80		9.93	13.60
7430	5" iron pipe size	G		120	.133		4.73	6.30		11.03	15
7440	6" iron pipe size	G		100	.160		4.88	7.55		12.43	17.10
7450	7" iron pipe size	G		90	.178		5.55	8.40		13.95	19.15
7460	8" iron pipe size	G		80	.200		5.95	9.45		15.40	21
7470	9" iron pipe size	G		75	.213		6.50	10.05		16.55	23
7480	10" iron pipe size	G		70	.229		7.10	10.75		17.85	24.50
7490	12" iron pipe size	G		65	.246		7.95	11.60		19.55	27
7500	14" iron pipe size	G		60	.267		10.25	12.55		22.80	31
7510	16" iron pipe size	G		55	.291		11.25	13.70		24.95	34
7520	18" iron pipe size	G		50	.320		12.55	15.10		27.65	37.50
7530	20" iron pipe size	G		45	.356		14.25	16.75		31	41.50
7540	24" iron pipe size	G		40	.400		15.30	18.85		34.15	46.50
7560	2-1/2" wall, 1/2" iron pipe size	G		210	.076		2.88	3.59		6.47	8.75
7562	3/4" iron pipe size	G		200	.080		3.01	3.77		6.78	9.15
7564	1" iron pipe size	G		190	.084		3.14	3.97		7.11	9.65
7566	1-1/4" iron pipe size	G		185	.086		3.25	4.08		7.33	9.95
7568	1-1/2" iron pipe size	G		180	.089		3.42	4.19		7.61	10.25
7570	2" iron pipe size	G		170	.094		3.59	4.44		8.03	10.85
7572	2-1/2" iron pipe size	G		160	.100		4.13	4.71		8.84	11.90
7574	3" iron pipe size	G		150	.107		4.35	5.05		9.40	12.65
7576	3-1/2" iron pipe size	G		140	.114		4.73	5.40		10.13	13.60
7578	4" iron pipe size	G		120	.133		4.99	6.30		11.29	15.30
7580	5" iron pipe size	G		110	.145		5.95	6.85		12.80	17.20
7582	6" iron pipe size	G		90	.178		7.15	8.40		15.55	21
7584	7" iron pipe size	G		80	.200		7.20	9.45		16.65	22.50
7586	8" iron pipe size	G		70	.229		7.65	10.75		18.40	25
7588	9" iron pipe size	G		65	.246		8.35	11.60		19.95	27.50
7590	10" iron pipe size	G		60	.267		9.10	12.55		21.65	29.50
7592	12" iron pipe size	G		55	.291		11.05	13.70		24.75	33.50
7594	14" iron pipe size	G		50	.320		12.90	15.10		28	37.50
7596	16" iron pipe size	G		45	.356		14.80	16.75		31.55	42.50
7598	18" iron pipe size	G		40	.400		16.05	18.85		34.90	47
7602	24" iron pipe size	G		30	.533		21	25		46	62.50
7620	3" wall, 1/2" iron pipe size	G		200	.080		3.69	3.77		7.46	9.90
7622	3/4" iron pipe size	G		190	.084		3.90	3.97		7.87	10.50
7624	1" iron pipe size	G		180	.089		4.13	4.19		8.32	11.05
7626	1-1/4" iron pipe size	G		175	.091		4.21	4.31		8.52	11.35
7628	1-1/2" iron pipe size	G		170	.094		4.40	4.44		8.84	11.75
7630	2" iron pipe size	G		160	.100		4.76	4.71		9.47	12.60
7632	2-1/2" iron pipe size	G		150	.107		4.95	5.05		10	13.30
7634	3" iron pipe size	G		140	.114		5.25	5.40		10.65	14.20
7636	3-1/2" iron pipe size	G		130	.123		5.80	5.80		11.60	15.45
7638	4" iron pipe size	G		110	.145		6.25	6.85		13.10	17.55

22 07 Plumbing Insulation

22 07 19 – Plumbing Piping Insulation

22 07 19.10 Piping Insulation		Crew	Daily Output	Labor-Hours	Unit	Material	2015 Bare Costs Labor	Equipment	Total	Total Incl O&P
7640	5" iron pipe size **G**	Q-14	100	.160	L.F.	7.05	7.55		14.60	19.55
7642	6" iron pipe size **G**		80	.200		7.55	9.45		17	23
7644	7" iron pipe size **G**		70	.229		8.65	10.75		19.40	26.50
7646	8" iron pipe size **G**		60	.267		9.40	12.55		21.95	30
7648	9" iron pipe size **G**		55	.291		10.15	13.70		23.85	32.50
7650	10" iron pipe size **G**		50	.320		10.90	15.10		26	35.50
7652	12" iron pipe size **G**		45	.356		13.55	16.75		30.30	41
7654	14" iron pipe size **G**		40	.400		15.65	18.85		34.50	46.50
7656	16" iron pipe size **G**		35	.457		17.75	21.50		39.25	53
7658	18" iron pipe size **G**		32	.500		18.95	23.50		42.45	57.50
7660	20" iron pipe size **G**		30	.533		20.50	25		45.50	61.50
7662	24" iron pipe size **G**		28	.571		26	27		53	71
7664	26" iron pipe size **G**		26	.615		29	29		58	76.50
7666	30" iron pipe size **G**		24	.667		34	31.50		65.50	86.50
7800	For fiberglass with standard canvas jacket, deduct					5%				
7802	For fittings, add 3 L.F. for each fitting									
7804	plus 4 L.F. for each flange of the fitting									
7810	Finishes									
7814	For single layer of felt, add					10%	10%			
7816	For roofing paper, 45 lb. to 55 lb., add					25%	10%			
7820	Polyethylene tubing flexible closed cell foam, UV resistant									
7828	Standard temperature (-90°F to +212°F)									
7830	3/8" wall, 1/8" iron pipe size **G**	1 Asbe	130	.062	L.F.	.22	3.22		3.44	5.25
7831	1/4" iron pipe size **G**		130	.062		.23	3.22		3.45	5.25
7832	3/8" iron pipe size **G**		130	.062		.26	3.22		3.48	5.30
7833	1/2" iron pipe size **G**		126	.063		.29	3.32		3.61	5.45
7834	3/4" iron pipe size **G**		122	.066		.34	3.43		3.77	5.70
7835	1" iron pipe size **G**		120	.067		.39	3.49		3.88	5.90
7836	1-1/4" iron pipe size **G**		118	.068		.47	3.55		4.02	6.05
7837	1-1/2" iron pipe size **G**		118	.068		.60	3.55		4.15	6.20
7838	2" iron pipe size **G**		116	.069		.79	3.61		4.40	6.45
7839	2-1/2" iron pipe size **G**		114	.070		1.01	3.67		4.68	6.80
7840	3" iron pipe size **G**		112	.071		1.56	3.74		5.30	7.50
7842	1/2" wall, 1/8" iron pipe size **G**		120	.067		.33	3.49		3.82	5.80
7843	1/4" iron pipe size **G**		120	.067		.35	3.49		3.84	5.85
7844	3/8" iron pipe size **G**		120	.067		.39	3.49		3.88	5.90
7845	1/2" iron pipe size **G**		118	.068		.43	3.55		3.98	6
7846	3/4" iron pipe size **G**		116	.069		.49	3.61		4.10	6.15
7847	1" iron pipe size **G**		114	.070		.55	3.67		4.22	6.30
7848	1-1/4" iron pipe size **G**		112	.071		.65	3.74		4.39	6.50
7849	1-1/2" iron pipe size **G**		110	.073		.77	3.81		4.58	6.80
7850	2" iron pipe size **G**		108	.074		1.11	3.88		4.99	7.25
7851	2-1/2" iron pipe size **G**		106	.075		1.37	3.95		5.32	7.65
7852	3" iron pipe size **G**		104	.077		1.89	4.03		5.92	8.35
7853	3-1/2" iron pipe size **G**		102	.078		2.14	4.11		6.25	8.75
7854	4" iron pipe size **G**		100	.080		2.44	4.19		6.63	9.20
7855	3/4" wall, 1/8" iron pipe size **G**		110	.073		.50	3.81		4.31	6.50
7856	1/4" iron pipe size **G**		110	.073		.53	3.81		4.34	6.55
7857	3/8" iron pipe size **G**		108	.074		.61	3.88		4.49	6.70
7858	1/2" iron pipe size **G**		106	.075		.73	3.95		4.68	6.95
7859	3/4" iron pipe size **G**		104	.077		.88	4.03		4.91	7.20
7860	1" iron pipe size **G**		102	.078		1.01	4.11		5.12	7.50
7861	1-1/4" iron pipe size **G**		100	.080		1.38	4.19		5.57	8

For customer support on your Plumbing Cost Data, call 877.709.6968.

131

22 07 19.10 Piping Insulation		Crew	Daily Output	Labor-Hours	Unit	Material	2015 Bare Costs Labor	Equipment	Total	Total Incl O&P
7862	1-1/2" iron pipe size	G	1 Asbe	100	.080	L.F.	1.56	4.19		5.75	8.20
7863	2" iron pipe size	G		98	.082		1.98	4.27		6.25	8.85
7864	2-1/2" iron pipe size	G		96	.083		2.39	4.36		6.75	9.45
7865	3" iron pipe size	G		94	.085		3.15	4.46		7.61	10.40
7866	3-1/2" iron pipe size	G		92	.087		3.51	4.55		8.06	10.95
7867	4" iron pipe size	G		90	.089		3.92	4.65		8.57	11.55
7868	1" wall, 1/4" iron pipe size	G		100	.080		1.06	4.19		5.25	7.65
7869	3/8" iron pipe size	G		98	.082		1.08	4.27		5.35	7.85
7870	1/2" iron pipe size	G		96	.083		1.35	4.36		5.71	8.30
7871	3/4" iron pipe size	G		94	.085		1.57	4.46		6.03	8.70
7872	1" iron pipe size	G		92	.087		1.91	4.55		6.46	9.20
7873	1-1/4" iron pipe size	G		90	.089		2.17	4.65		6.82	9.65
7874	1-1/2" iron pipe size	G		90	.089		2.48	4.65		7.13	10
7875	2" iron pipe size	G		88	.091		3.23	4.76		7.99	10.95
7876	2-1/2" iron pipe size	G		86	.093		4.22	4.87		9.09	12.25
7877	3" iron pipe size	G		84	.095		4.74	4.99		9.73	12.95
7878	Contact cement, quart can	G				Ea.	10.50			10.50	11.55
7879	Rubber tubing, flexible closed cell foam										
7880	3/8" wall, 1/4" iron pipe size	G	1 Asbe	120	.067	L.F.	.33	3.49		3.82	5.80
7900	3/8" iron pipe size	G		120	.067		.36	3.49		3.85	5.85
7910	1/2" iron pipe size	G		115	.070		.41	3.64		4.05	6.10
7920	3/4" iron pipe size	G		115	.070		.46	3.64		4.10	6.15
7930	1" iron pipe size	G		110	.073		.52	3.81		4.33	6.50
7940	1-1/4" iron pipe size	G		110	.073		.59	3.81		4.40	6.60
7950	1-1/2" iron pipe size	G		110	.073		.73	3.81		4.54	6.75
8100	1/2" wall, 1/4" iron pipe size	G		90	.089		.54	4.65		5.19	7.85
8120	3/8" iron pipe size	G		90	.089		.60	4.65		5.25	7.90
8130	1/2" iron pipe size	G		89	.090		.67	4.71		5.38	8.10
8140	3/4" iron pipe size	G		89	.090		.75	4.71		5.46	8.20
8150	1" iron pipe size	G		88	.091		.82	4.76		5.58	8.30
8160	1-1/4" iron pipe size	G		87	.092		.95	4.81		5.76	8.55
8170	1-1/2" iron pipe size	G		87	.092		1.15	4.81		5.96	8.75
8180	2" iron pipe size	G		86	.093		1.47	4.87		6.34	9.20
8190	2-1/2" iron pipe size	G		86	.093		1.85	4.87		6.72	9.65
8200	3" iron pipe size	G		85	.094		2.06	4.93		6.99	9.90
8210	3-1/2" iron pipe size	G		85	.094		2.86	4.93		7.79	10.80
8220	4" iron pipe size	G		80	.100		3.06	5.25		8.31	11.50
8230	5" iron pipe size	G		80	.100		4.17	5.25		9.42	12.75
8240	6" iron pipe size	G		75	.107		4.17	5.60		9.77	13.30
8300	3/4" wall, 1/4" iron pipe size	G		90	.089		.85	4.65		5.50	8.20
8320	3/8" iron pipe size	G		90	.089		.92	4.65		5.57	8.25
8330	1/2" iron pipe size	G		89	.090		1.10	4.71		5.81	8.55
8340	3/4" iron pipe size	G		89	.090		1.35	4.71		6.06	8.85
8350	1" iron pipe size	G		88	.091		1.54	4.76		6.30	9.10
8360	1-1/4" iron pipe size	G		87	.092		2.05	4.81		6.86	9.75
8370	1-1/2" iron pipe size	G		87	.092		2.32	4.81		7.13	10.05
8380	2" iron pipe size	G		86	.093		2.68	4.87		7.55	10.55
8390	2-1/2" iron pipe size	G		86	.093		3.58	4.87		8.45	11.55
8400	3" iron pipe size	G		85	.094		4.08	4.93		9.01	12.15
8410	3-1/2" iron pipe size	G		85	.094		4.81	4.93		9.74	12.95
8420	4" iron pipe size	G		80	.100		5.10	5.25		10.35	13.80
8430	5" iron pipe size	G		80	.100		6.15	5.25		11.40	14.95
8440	6" iron pipe size	G		80	.100		7.45	5.25		12.70	16.35

22 07 Plumbing Insulation

22 07 19 – Plumbing Piping Insulation

22 07 19.10 Piping Insulation		Crew	Daily Output	Labor-Hours	Unit	Material	2015 Bare Costs Labor	Equipment	Total	Total Incl O&P
8444	1" wall, 1/2" iron pipe size	G	1 Asbe	86	.093	L.F.	2.05	4.87		6.92	9.85
8445	3/4" iron pipe size	G		84	.095		2.48	4.99		7.47	10.50
8446	1" iron pipe size	G		84	.095		2.89	4.99		7.88	10.95
8447	1-1/4" iron pipe size	G		82	.098		3.23	5.10		8.33	11.50
8448	1-1/2" iron pipe size	G		82	.098		3.76	5.10		8.86	12.10
8449	2" iron pipe size	G		80	.100		4.95	5.25		10.20	13.60
8450	2-1/2" iron pipe size	G		80	.100		6.45	5.25		11.70	15.25
8456	Rubber insulation tape, 1/8" x 2" x 30'	G				Ea.	12.05			12.05	13.25
8460	Polyolefin tubing, flexible closed cell foam, UV stabilized, work										
8462	temp.-165°F to +210°F, 0 water vapor transmission										
8464	3/8" wall, 1/8" iron pipe size	G	1 Asbe	140	.057	L.F.	.48	2.99		3.47	5.20
8466	1/4" iron pipe size	G		140	.057		.50	2.99		3.49	5.20
8468	3/8" iron pipe size	G		140	.057		.55	2.99		3.54	5.25
8470	1/2" iron pipe size	G		136	.059		.61	3.08		3.69	5.45
8472	3/4" iron pipe size	G		132	.061		.62	3.17		3.79	5.60
8474	1" iron pipe size	G		130	.062		.83	3.22		4.05	5.90
8476	1-1/4" iron pipe size	G		128	.063		1.02	3.27		4.29	6.20
8478	1-1/2" iron pipe size	G		128	.063		1.25	3.27		4.52	6.50
8480	2" iron pipe size	G		126	.063		1.50	3.32		4.82	6.80
8482	2-1/2" iron pipe size	G		123	.065		2.17	3.40		5.57	7.70
8484	3" iron pipe size	G		121	.066		2.42	3.46		5.88	8.05
8486	4" iron pipe size	G		118	.068		4.43	3.55		7.98	10.40
8500	1/2" wall, 1/8" iron pipe size	G		130	.062		.71	3.22		3.93	5.80
8502	1/4" iron pipe size	G		130	.062		.76	3.22		3.98	5.85
8504	3/8" iron pipe size	G		130	.062		.81	3.22		4.03	5.90
8506	1/2" iron pipe size	G		128	.063		.89	3.27		4.16	6.10
8508	3/4" iron pipe size	G		126	.063		1.02	3.32		4.34	6.25
8510	1" iron pipe size	G		123	.065		1.18	3.40		4.58	6.60
8512	1-1/4" iron pipe size	G		121	.066		1.39	3.46		4.85	6.95
8514	1-1/2" iron pipe size	G		119	.067		1.67	3.52		5.19	7.35
8516	2" iron pipe size	G		117	.068		2.11	3.58		5.69	7.85
8518	2-1/2" iron pipe size	G		114	.070		2.85	3.67		6.52	8.85
8520	3" iron pipe size	G		112	.071		3.71	3.74		7.45	9.90
8522	4" iron pipe size	G		110	.073		5.15	3.81		8.96	11.60
8534	3/4" wall, 1/8" iron pipe size	G		120	.067		1.08	3.49		4.57	6.65
8536	1/4" iron pipe size	G		120	.067		1.14	3.49		4.63	6.70
8538	3/8" iron pipe size	G		117	.068		1.30	3.58		4.88	7
8540	1/2" iron pipe size	G		114	.070		1.45	3.67		5.12	7.30
8542	3/4" iron pipe size	G		112	.071		1.56	3.74		5.30	7.50
8544	1" iron pipe size	G		110	.073		2.20	3.81		6.01	8.35
8546	1-1/4" iron pipe size	G		108	.074		2.94	3.88		6.82	9.30
8548	1-1/2" iron pipe size	G		108	.074		3.43	3.88		7.31	9.80
8550	2" iron pipe size	G		106	.075		4.34	3.95		8.29	10.90
8552	2-1/2" iron pipe size	G		104	.077		5.10	4.03		9.13	11.85
8554	3" iron pipe size	G		102	.078		6.80	4.11		10.91	13.90
8556	4" iron pipe size	G		100	.080		9	4.19		13.19	16.40
8570	1" wall, 1/8" iron pipe size	G		110	.073		1.95	3.81		5.76	8.10
8572	1/4" iron pipe size	G		108	.074		2	3.88		5.88	8.25
8574	3/8" iron pipe size	G		106	.075		2.04	3.95		5.99	8.40
8576	1/2" iron pipe size	G		104	.077		2.16	4.03		6.19	8.65
8578	3/4" iron pipe size	G		102	.078		2.83	4.11		6.94	9.50
8580	1" iron pipe size	G		100	.080		3.27	4.19		7.46	10.10
8582	1-1/4" iron pipe size	G		97	.082		3.72	4.32		8.04	10.80

22 07 19 – Plumbing Piping Insulation

22 07 19.10 **Piping Insulation**		Crew	Daily Output	Labor-Hours	Unit	Material	2015 Bare Costs Labor	Equipment	Total	Total Incl O&P
8584	1-1/2" iron pipe size [G]	1 Asbe	97	.082	L.F.	4.25	4.32		8.57	11.40
8586	2" iron pipe size [G]		95	.084		6.75	4.41		11.16	14.25
8588	2-1/2" iron pipe size [G]		93	.086		7.85	4.50		12.35	15.60
8590	3" iron pipe size [G]		91	.088		9.55	4.60		14.15	17.65
8606	Contact adhesive (R-320) [G]				Qt.	28			28	30.50
8608	Contact adhesive (R-320) [G]				Gal.	113			113	124
8610	NOTE: Preslit/preglued vs unslit, same price									

22 07 19.30 Piping Insulation Protective Jacketing, PVC

		Crew	Daily Output	Labor-Hours	Unit	Material	2015 Bare Costs Labor	Equipment	Total	Total Incl O&P
0010	**PIPING INSULATION PROTECTIVE JACKETING, PVC**									
0100	PVC, white, 48" lengths cut from roll goods									
0120	20 mil thick									
0140	Size based on OD of insulation									
0150	1-1/2" ID	Q-14	270	.059	L.F.	.20	2.79		2.99	4.57
0152	2" ID		260	.062		.27	2.90		3.17	4.82
0154	2-1/2" ID		250	.064		.32	3.02		3.34	5.05
0156	3" ID		240	.067		.39	3.14		3.53	5.30
0158	3-1/2" ID		230	.070		.44	3.28		3.72	5.60
0160	4" ID		220	.073		.50	3.43		3.93	5.90
0162	4-1/2" ID		210	.076		.56	3.59		4.15	6.20
0164	5" ID		200	.080		.61	3.77		4.38	6.50
0166	5-1/2" ID		190	.084		.67	3.97		4.64	6.95
0168	6" ID		180	.089		.73	4.19		4.92	7.30
0170	6-1/2" ID		175	.091		.79	4.31		5.10	7.55
0172	7" ID		170	.094		.84	4.44		5.28	7.80
0174	7-1/2" ID		164	.098		.91	4.60		5.51	8.15
0176	8" ID		161	.099		.96	4.68		5.64	8.35
0178	8-1/2" ID		158	.101		1.02	4.77		5.79	8.55
0180	9" ID		155	.103		1.08	4.87		5.95	8.80
0182	9-1/2" ID		152	.105		1.14	4.96		6.10	8.95
0184	10" ID		149	.107		1.19	5.05		6.24	9.20
0186	10-1/2" ID		146	.110		1.25	5.15		6.40	9.45
0188	11" ID		143	.112		1.31	5.25		6.56	9.65
0190	11-1/2" ID		140	.114		1.37	5.40		6.77	9.90
0192	12" ID		137	.117		1.42	5.50		6.92	10.10
0194	12-1/2" ID		134	.119		1.49	5.65		7.14	10.40
0195	13" ID		132	.121		1.54	5.70		7.24	10.60
0196	13-1/2" ID		132	.121		1.60	5.70		7.30	10.65
0198	14" ID		130	.123		1.66	5.80		7.46	10.90
0200	15" ID		128	.125		1.78	5.90		7.68	11.10
0202	16" ID		126	.127		1.89	6		7.89	11.40
0204	17" ID		124	.129		2.01	6.10		8.11	11.65
0206	18" ID		122	.131		2.13	6.20		8.33	11.95
0208	19" ID		120	.133		2.24	6.30		8.54	12.25
0210	20" ID		118	.136		2.35	6.40		8.75	12.55
0212	21" ID		116	.138		2.47	6.50		8.97	12.80
0214	22" ID		114	.140		2.58	6.60		9.18	13.15
0216	23" ID		112	.143		2.69	6.75		9.44	13.45
0218	24" ID		110	.145		2.81	6.85		9.66	13.75
0220	25" ID		108	.148		2.93	7		9.93	14.05
0222	26" ID		106	.151		3.04	7.10		10.14	14.45
0224	27" ID		104	.154		3.16	7.25		10.41	14.80
0226	28" ID		102	.157		3.27	7.40		10.67	15.10

134

For customer support on your Plumbing Cost Data, call 877.709.6968.

22 07 Plumbing Insulation

22 07 19 – Plumbing Piping Insulation

22 07 19.30 Piping Insulation Protective Jacketing, PVC	Crew	Daily Output	Labor-Hours	Unit	Material	2015 Bare Costs Labor	Equipment	Total	Total Incl O&P
0228	29" ID	Q-14	100	.160	L.F.	3.39	7.55		10.94	15.50
0230	30" ID	↓	98	.163	↓	3.51	7.70		11.21	15.85
0300	For colors, add				Ea.	10%				
1000	30 mil thick									
1010	Size based on OD of insulation									
1020	2" ID	Q-14	260	.062	L.F.	.40	2.90		3.30	4.96
1022	2-1/2" ID		250	.064		.48	3.02		3.50	5.25
1024	3" ID		240	.067		.58	3.14		3.72	5.55
1026	3-1/2" ID		230	.070		.65	3.28		3.93	5.80
1028	4" ID		220	.073		.74	3.43		4.17	6.15
1030	4-1/2" ID		210	.076		.82	3.59		4.41	6.50
1032	5" ID		200	.080		.91	3.77		4.68	6.85
1034	5-1/2" ID		190	.084		.99	3.97		4.96	7.30
1036	6" ID		180	.089		1.08	4.19		5.27	7.70
1038	6-1/2" ID		175	.091		1.16	4.31		5.47	8
1040	7" ID		170	.094		1.25	4.44		5.69	8.30
1042	7-1/2" ID		164	.098		1.34	4.60		5.94	8.60
1044	8" ID		161	.099		1.43	4.68		6.11	8.85
1046	8-1/2" ID		158	.101		1.51	4.77		6.28	9.10
1048	9" ID		155	.103		1.60	4.87		6.47	9.35
1050	9-1/2" ID		152	.105		1.68	4.96		6.64	9.55
1052	10" ID		149	.107		1.77	5.05		6.82	9.85
1054	10-1/2" ID		146	.110		1.85	5.15		7	10.10
1056	11" ID		143	.112		1.94	5.25		7.19	10.35
1058	11-1/2" ID		140	.114		2.02	5.40		7.42	10.60
1060	12" ID		137	.117		2.11	5.50		7.61	10.85
1062	12-1/2" ID		134	.119		2.20	5.65		7.85	11.15
1063	13" ID		132	.121		2.29	5.70		7.99	11.40
1064	13-1/2" ID		132	.121		2.37	5.70		8.07	11.50
1066	14" ID		130	.123		2.46	5.80		8.26	11.75
1068	15" ID		128	.125		2.63	5.90		8.53	12.05
1070	16" ID		126	.127		2.80	6		8.80	12.40
1072	17" ID		124	.129		2.97	6.10		9.07	12.70
1074	18" ID		122	.131		3.15	6.20		9.35	13.05
1076	19" ID		120	.133		3.32	6.30		9.62	13.45
1078	20" ID		118	.136		3.49	6.40		9.89	13.80
1080	21" ID		116	.138		3.66	6.50		10.16	14.15
1082	22" ID		114	.140		3.83	6.60		10.43	14.50
1084	23" ID		112	.143		4	6.75		10.75	14.90
1086	24" ID		110	.145		4.17	6.85		11.02	15.25
1088	25" ID		108	.148		4.34	7		11.34	15.60
1090	26" ID		106	.151		4.51	7.10		11.61	16.05
1092	27" ID		104	.154		4.68	7.25		11.93	16.45
1094	28" ID		102	.157		4.85	7.40		12.25	16.85
1096	29" ID		100	.160		5	7.55		12.55	17.25
1098	30" ID	↓	98	.163	↓	5.20	7.70		12.90	17.70
1300	For colors, add				Ea.	10%				
2000	PVC, white, fitting covers									
2020	Fiberglass insulation inserts included with sizes 1-3/4" thru 9-3/4"									
2030	Size is based on OD of insulation									
2040	90° Elbow fitting									
2060	1-3/4"	Q-14	135	.119	Ea.	.47	5.60		6.07	9.20
2062	2"	↓	130	.123	↓	.58	5.80		6.38	9.70

For customer support on your Plumbing Cost Data, call 877.709.6968.

135

22 07 19.30 Piping Insulation Protective Jacketing, PVC	Crew	Daily Output	Labor-Hours	Unit	Material	Labor	Equipment	Total	Total Incl O&P
2064	2-1/4"	Q-14	128	.125	Ea.	.65	5.90		6.55	9.85
2068	2-1/2"		126	.127		.71	6		6.71	10.10
2070	2-3/4"		123	.130		.77	6.15		6.92	10.40
2072	3"		120	.133		.82	6.30		7.12	10.70
2074	3-3/8"		116	.138		.93	6.50		7.43	11.10
2076	3-3/4"		113	.142		1.04	6.65		7.69	11.55
2078	4-1/8"		110	.145		1.31	6.85		8.16	12.10
2080	4-3/4"		105	.152		1.58	7.20		8.78	12.95
2082	5-1/4"		100	.160		1.84	7.55		9.39	13.75
2084	5-3/4"		95	.168		2.43	7.95		10.38	15
2086	6-1/4"		90	.178		4.05	8.40		12.45	17.50
2088	6-3/4"		87	.184		4.27	8.65		12.92	18.20
2090	7-1/4"		85	.188		5.35	8.85		14.20	19.70
2092	7-3/4"		83	.193		5.60	9.10		14.70	20.50
2094	8-3/4"		80	.200		7.20	9.45		16.65	22.50
2096	9-3/4"		77	.208		9.65	9.80		19.45	26
2098	10-7/8"		74	.216		10.75	10.20		20.95	27.50
2100	11-7/8"		71	.225		12.35	10.60		22.95	30
2102	12-7/8"		68	.235		15.55	11.10		26.65	34.50
2104	14-1/8"		66	.242		17.80	11.45		29.25	37.50
2106	15-1/8"		64	.250		19.15	11.80		30.95	39.50
2108	16-1/8"		63	.254		21	11.95		32.95	41.50
2110	17-1/8"		62	.258		23	12.15		35.15	44.50
2112	18-1/8"		61	.262		30	12.35		42.35	52.50
2114	19-1/8"		60	.267		40.50	12.55		53.05	64
2116	20-1/8"		59	.271		52	12.80		64.80	77.50
2200	45° Elbow fitting									
2220	1-3/4" thru 9-3/4" same price as 90° Elbow fitting									
2320	10-7/8"	Q-14	74	.216	Ea.	10.75	10.20		20.95	27.50
2322	11-7/8"		71	.225		11.85	10.60		22.45	29.50
2324	12-7/8"		68	.235		13.25	11.10		24.35	32
2326	14-1/8"		66	.242		15.10	11.45		26.55	34.50
2328	15-1/8"		64	.250		16.25	11.80		28.05	36
2330	16-1/8"		63	.254		18.55	11.95		30.50	39
2332	17-1/8"		62	.258		21	12.15		33.15	42
2334	18-1/8"		61	.262		25.50	12.35		37.85	47.50
2336	19-1/8"		60	.267		35	12.55		47.55	58
2338	20-1/8"		59	.271		39.50	12.80		52.30	63.50
2400	Tee fitting									
2410	1-3/4"	Q-14	96	.167	Ea.	.88	7.85		8.73	13.20
2412	2"		94	.170		1	8		9	13.60
2414	2-1/4"		91	.176		1.09	8.30		9.39	14.10
2416	2-1/2"		88	.182		1.19	8.55		9.74	14.65
2418	2-3/4"		85	.188		1.31	8.85		10.16	15.25
2420	3"		82	.195		1.42	9.20		10.62	15.85
2422	3-3/8"		79	.203		1.63	9.55		11.18	16.65
2424	3-3/4"		76	.211		1.85	9.90		11.75	17.50
2426	4-1/8"		73	.219		2.19	10.35		12.54	18.50
2428	4-3/4"		70	.229		2.73	10.75		13.48	19.75
2430	5-1/4"		67	.239		3.28	11.25		14.53	21
2432	5-3/4"		63	.254		4.36	11.95		16.31	23.50
2434	6-1/4"		60	.267		5.75	12.55		18.30	26
2436	6-3/4"		59	.271		7.10	12.80		19.90	27.50

For customer support on your Plumbing Cost Data, call 877.709.6968.

22 07 Plumbing Insulation

22 07 19 – Plumbing Piping Insulation

22 07 19.30 Piping Insulation Protective Jacketing, PVC		Crew	Daily Output	Labor-Hours	Unit	Material	2015 Bare Costs Labor	Equipment	Total	Total Incl O&P
2438	7-1/4"	Q-14	57	.281	Ea.	11.45	13.25		24.70	33
2440	7-3/4"		54	.296		12.55	13.95		26.50	35.50
2442	8-3/4"		52	.308		15.30	14.50		29.80	39.50
2444	9-3/4"		50	.320		18	15.10		33.10	43.50
2446	10-7/8"		48	.333		18.25	15.70		33.95	44.50
2448	11-7/8"		47	.340		20.50	16.05		36.55	47.50
2450	12-7/8"		46	.348		22.50	16.40		38.90	50
2452	14-1/8"		45	.356		24.50	16.75		41.25	53
2454	15-1/8"		44	.364		26.50	17.15		43.65	55.50
2456	16-1/8"		43	.372		28.50	17.55		46.05	58.50
2458	17-1/8"		42	.381		30.50	17.95		48.45	61.50
2460	18-1/8"		41	.390		33.50	18.40		51.90	65.50
2462	19-1/8"		40	.400		36.50	18.85		55.35	69.50
2464	20-1/8"		39	.410		40	19.35		59.35	74
4000	Mechanical grooved fitting cover, including insert									
4020	90° Elbow fitting									
4030	3/4" & 1"	Q-14	140	.114	Ea.	3.70	5.40		9.10	12.45
4040	1-1/4" & 1-1/2"		135	.119		4.68	5.60		10.28	13.85
4042	2"		130	.123		7.40	5.80		13.20	17.20
4044	2-1/2"		125	.128		8.20	6.05		14.25	18.40
4046	3"		120	.133		9.20	6.30		15.50	19.90
4048	3-1/2"		115	.139		10.65	6.55		17.20	22
4050	4"		110	.145		11.85	6.85		18.70	23.50
4052	5"		100	.160		14.75	7.55		22.30	28
4054	6"		90	.178		22	8.40		30.40	37
4056	8"		80	.200		23.50	9.45		32.95	40.50
4058	10"		75	.213		30.50	10.05		40.55	49
4060	12"		68	.235		43.50	11.10		54.60	65.50
4062	14"		65	.246		43.50	11.60		55.10	66
4064	16"		63	.254		59.50	11.95		71.45	84
4066	18"		61	.262		81.50	12.35		93.85	109
4100	45° Elbow fitting									
4120	3/4" & 1"	Q-14	140	.114	Ea.	3.29	5.40		8.69	12
4130	1-1/4" & 1-1/2"		135	.119		4.19	5.60		9.79	13.30
4140	2"		130	.123		6.60	5.80		12.40	16.30
4142	2-1/2"		125	.128		7.30	6.05		13.35	17.45
4144	3"		120	.133		8.20	6.30		14.50	18.80
4146	3-1/2"		115	.139		8.65	6.55		15.20	19.70
4148	4"		110	.145		10.55	6.85		17.40	22.50
4150	5"		100	.160		13.10	7.55		20.65	26
4152	6"		90	.178		19.35	8.40		27.75	34.50
4154	8"		80	.200		21	9.45		30.45	37.50
4156	10"		75	.213		27	10.05		37.05	45
4158	12"		68	.235		39	11.10		50.10	60
4160	14"		65	.246		50.50	11.60		62.10	73.50
4162	16"		63	.254		61	11.95		72.95	86
4164	18"		61	.262		72	12.35		84.35	99
4200	Tee fitting									
4220	3/4" & 1"	Q-14	93	.172	Ea.	4.82	8.10		12.92	17.90
4230	1-1/4" & 1-1/2"		90	.178		6.10	8.40		14.50	19.75
4240	2"		87	.184		9.60	8.65		18.25	24
4242	2-1/2"		84	.190		10.70	9		19.70	26
4244	3"		80	.200		11.95	9.45		21.40	28

For customer support on your Plumbing Cost Data, call 877.709.6968.

137

22 07 19.30 Piping Insulation Protective Jacketing, PVC		Crew	Daily Output	Labor-Hours	Unit	Material	2015 Bare Costs Labor	Equipment	Total	Total Incl O&P
4246	3-1/2"	Q-14	77	.208	Ea.	12.65	9.80		22.45	29
4248	4"		73	.219		15.45	10.35		25.80	33
4250	5"		67	.239		17.05	11.25		28.30	36.50
4252	6"		60	.267		25	12.55		37.55	47
4254	8"		54	.296		27.50	13.95		41.45	51.50
4256	10"		50	.320		39.50	15.10		54.60	67
4258	12"		46	.348		56.50	16.40		72.90	88
4260	14"		43	.372		68.50	17.55		86.05	103
4262	16"		42	.381		82	17.95		99.95	119
4264	18"		41	.390		90.50	18.40		108.90	128

22 07 19.40 Pipe Insulation Protective Jacketing, Aluminum

		Crew	Daily Output	Labor-Hours	Unit	Material	2015 Bare Costs Labor	Equipment	Total	Total Incl O&P
0010	**PIPE INSULATION PROTECTIVE JACKETING, ALUMINUM**									
0100	Metal roll jacketing									
0120	Aluminum with polykraft moisture barrier									
0140	Smooth, based on OD of insulation, .016" thick									
0180	1/2" ID	Q-14	220	.073	L.F.	.20	3.43		3.63	5.55
0190	3/4" ID		215	.074		.26	3.51		3.77	5.75
0200	1" ID		210	.076		.32	3.59		3.91	5.95
0210	1-1/4" ID		205	.078		.38	3.68		4.06	6.15
0220	1-1/2" ID		202	.079		.44	3.73		4.17	6.30
0230	1-3/4" ID		199	.080		.50	3.79		4.29	6.45
0240	2" ID		195	.082		.56	3.87		4.43	6.60
0250	2-1/4" ID		191	.084		.62	3.95		4.57	6.85
0260	2-1/2" ID		187	.086		.69	4.03		4.72	7.05
0270	2-3/4" ID		184	.087		.75	4.10		4.85	7.25
0280	3" ID		180	.089		.82	4.19		5.01	7.40
0290	3-1/4" ID		176	.091		.88	4.28		5.16	7.60
0300	3-1/2" ID		172	.093		.94	4.38		5.32	7.90
0310	3-3/4" ID		169	.095		1	4.46		5.46	8.05
0320	4" ID		165	.097		1.06	4.57		5.63	8.25
0330	4-1/4" ID		161	.099		1.12	4.68		5.80	8.55
0340	4-1/2" ID		157	.102		1.19	4.80		5.99	8.80
0350	4-3/4" ID		154	.104		1.25	4.90		6.15	9
0360	5" ID		150	.107		1.31	5.05		6.36	9.30
0370	5-1/4" ID		146	.110		1.37	5.15		6.52	9.55
0380	5-1/2" ID		143	.112		1.43	5.25		6.68	9.75
0390	5-3/4" ID		139	.115		1.49	5.45		6.94	10.10
0400	6" ID		135	.119		1.55	5.60		7.15	10.40
0410	6-1/4" ID		133	.120		1.61	5.65		7.26	10.60
0420	6-1/2" ID		131	.122		1.67	5.75		7.42	10.80
0430	7" ID		128	.125		1.79	5.90		7.69	11.10
0440	7-1/4" ID		125	.128		1.85	6.05		7.90	11.45
0450	7-1/2" ID		123	.130		1.91	6.15		8.06	11.65
0460	8" ID		121	.132		2.03	6.25		8.28	11.95
0470	8-1/2" ID		119	.134		2.15	6.35		8.50	12.20
0480	9" ID		116	.138		2.27	6.50		8.77	12.60
0490	9-1/2" ID		114	.140		2.39	6.60		8.99	12.95
0500	10" ID		112	.143		2.51	6.75		9.26	13.25
0510	10-1/2" ID		110	.145		2.63	6.85		9.48	13.55
0520	11" ID		107	.150		2.75	7.05		9.80	14
0530	11-1/2" ID		105	.152		2.88	7.20		10.08	14.35
0540	12" ID		103	.155		3	7.30		10.30	14.70

22 07 Plumbing Insulation

22 07 19 – Plumbing Piping Insulation

22 07 19.40 Pipe Insulation Protective Jacketing, Aluminum	Crew	Daily Output	Labor-Hours	Unit	Material	2015 Bare Costs Labor	Equipment	Total	Total Incl O&P
0550	12-1/2" ID	Q-14	100	.160	L.F.	3.12	7.55		10.67	15.20
0560	13" ID		99	.162		3.24	7.60		10.84	15.40
0570	14" ID		98	.163		3.49	7.70		11.19	15.85
0580	15" ID		96	.167		3.73	7.85		11.58	16.35
0590	16" ID		95	.168		3.98	7.95		11.93	16.75
0600	17" ID		93	.172		4.22	8.10		12.32	17.25
0610	18" ID		92	.174		4.38	8.20		12.58	17.55
0620	19" ID		90	.178		4.70	8.40		13.10	18.20
0630	20" ID		89	.180		4.95	8.45		13.40	18.65
0640	21" ID		87	.184		5.20	8.65		13.85	19.20
0650	22" ID		86	.186		5.45	8.75		14.20	19.60
0660	23" ID		84	.190		5.70	9		14.70	20.50
0670	24" ID		83	.193		5.90	9.10		15	20.50
0710	For smooth .020" thick, add					27%	10%			
0720	For smooth .024" thick, add					52%	20%			
0730	For smooth .032" thick, add					104%	33%			
0800	For stucco embossed, add					1%				
0820	For corrugated, add					2.50%				
0900	White aluminum with polysurlyn moisture barrier									
0910	Smooth, % is an add to polykraft lines of same thickness									
0940	For smooth .016" thick, add				L.F.	35%				
0960	For smooth .024" thick, add				"	22%				
1000	Aluminum fitting covers									
1010	Size is based on OD of insulation									
1020	90° LR elbow, 2 piece									
1100	1-1/2"	Q-14	140	.114	Ea.	3.45	5.40		8.85	12.20
1110	1-3/4"		135	.119		3.45	5.60		9.05	12.50
1120	2"		130	.123		4.34	5.80		10.14	13.80
1130	2-1/4"		128	.125		4.34	5.90		10.24	13.90
1140	2-1/2"		126	.127		4.34	6		10.34	14.05
1150	2-3/4"		123	.130		4.34	6.15		10.49	14.30
1160	3"		120	.133		4.71	6.30		11.01	15
1170	3-1/4"		117	.137		4.71	6.45		11.16	15.25
1180	3-1/2"		115	.139		5.55	6.55		12.10	16.35
1190	3-3/4"		113	.142		5.70	6.65		12.35	16.70
1200	4"		110	.145		6	6.85		12.85	17.25
1210	4-1/4"		108	.148		6.15	7		13.15	17.60
1220	4-1/2"		106	.151		6.15	7.10		13.25	17.85
1230	4-3/4"		104	.154		6.15	7.25		13.40	18.05
1240	5"		102	.157		6.95	7.40		14.35	19.10
1250	5-1/4"		100	.160		8	7.55		15.55	20.50
1260	5-1/2"		97	.165		9.95	7.75		17.70	23
1270	5-3/4"		95	.168		9.95	7.95		17.90	23.50
1280	6"		92	.174		8.50	8.20		16.70	22
1290	6-1/4"		90	.178		8.50	8.40		16.90	22.50
1300	6-1/2"		87	.184		10.25	8.65		18.90	25
1310	7"		85	.188		12.40	8.85		21.25	27.50
1320	7-1/4"		84	.190		12.40	9		21.40	27.50
1330	7-1/2"		83	.193		18.25	9.10		27.35	34
1340	8"		82	.195		13.60	9.20		22.80	29.50
1350	8-1/2"		80	.200		25	9.45		34.45	42
1360	9"		78	.205		25	9.65		34.65	42.50
1370	9-1/2"		77	.208		18.50	9.80		28.30	36

For customer support on your Plumbing Cost Data, call 877.709.6968.

139

22 07 19.40 Pipe Insulation Protective Jacketing, Aluminum		Crew	Daily Output	Labor-Hours	Unit	Material	2015 Bare Costs Labor	Equipment	Total	Total Incl O&P
1380	10"	Q-14	76	.211	Ea.	18.50	9.90		28.40	36
1390	10-1/2"		75	.213		18.90	10.05		28.95	36.50
1400	11"		74	.216		18.90	10.20		29.10	37
1410	11-1/2"		72	.222		21.50	10.45		31.95	40
1420	12"		71	.225		21.50	10.60		32.10	40
1430	12-1/2"		69	.232		38	10.95		48.95	59
1440	13"		68	.235		38	11.10		49.10	59.50
1450	14"		66	.242		51.50	11.45		62.95	74.50
1460	15"		64	.250		53.50	11.80		65.30	77.50
1470	16"		63	.254		58.50	11.95		70.45	83
2000	45° Elbow, 2 piece									
2010	2-1/2"	Q-14	126	.127	Ea.	3.59	6		9.59	13.25
2020	2-3/4"		123	.130		3.59	6.15		9.74	13.50
2030	3"		120	.133		4.05	6.30		10.35	14.25
2040	3-1/4"		117	.137		4.05	6.45		10.50	14.50
2050	3-1/2"		115	.139		4.62	6.55		11.17	15.30
2060	3-3/4"		113	.142		4.62	6.65		11.27	15.50
2070	4"		110	.145		4.70	6.85		11.55	15.80
2080	4-1/4"		108	.148		5.35	7		12.35	16.75
2090	4-1/2"		106	.151		5.35	7.10		12.45	17
2100	4-3/4"		104	.154		5.35	7.25		12.60	17.20
2110	5"		102	.157		6.30	7.40		13.70	18.45
2120	5-1/4"		100	.160		6.30	7.55		13.85	18.70
2130	5-1/2"		97	.165		6.60	7.75		14.35	19.35
2140	6"		92	.174		6.60	8.20		14.80	20
2150	6-1/2"		87	.184		9.10	8.65		17.75	23.50
2160	7"		85	.188		9.10	8.85		17.95	24
2170	7-1/2"		83	.193		9.20	9.10		18.30	24.50
2180	8"		82	.195		9.20	9.20		18.40	24.50
2190	8-1/2"		80	.200		11.45	9.45		20.90	27.50
2200	9"		78	.205		11.45	9.65		21.10	27.50
2210	9-1/2"		77	.208		15.90	9.80		25.70	33
2220	10"		76	.211		15.90	9.90		25.80	33
2230	10-1/2"		75	.213		14.85	10.05		24.90	32
2240	11"		74	.216		14.85	10.20		25.05	32
2250	11-1/2"		72	.222		17.85	10.45		28.30	36
2260	12"		71	.225		17.85	10.60		28.45	36
2270	13"		68	.235		21	11.10		32.10	40.50
2280	14"		66	.242		25.50	11.45		36.95	46.50
2290	15"		64	.250		39.50	11.80		51.30	62
2300	16"		63	.254		43	11.95		54.95	66
2310	17"		62	.258		42	12.15		54.15	65
2320	18"		61	.262		48	12.35		60.35	72
2330	19"		60	.267		58.50	12.55		71.05	84
2340	20"		59	.271		56.50	12.80		69.30	82
2350	21"		58	.276		61	13		74	87
3000	Tee, 4 piece									
3010	2-1/2"	Q-14	88	.182	Ea.	29	8.55		37.55	45.50
3020	2-3/4"		86	.186		29	8.75		37.75	45.50
3030	3"		84	.190		32.50	9		41.50	50
3040	3-1/4"		82	.195		32.50	9.20		41.70	50.50
3050	3-1/2"		80	.200		34.50	9.45		43.95	52
3060	4"		78	.205		34.50	9.65		44.15	53

22 07 Plumbing Insulation

22 07 19 – Plumbing Piping Insulation

22 07 19.40 Pipe Insulation Protective Jacketing, Aluminum		Crew	Daily Output	Labor-Hours	Unit	Material	2015 Bare Costs Labor	Equipment	Total	Total Incl O&P
3070	4-1/4"	Q-14	76	.211	Ea.	36.50	9.90		46.40	55.50
3080	4-1/2"		74	.216		36.50	10.20		46.70	56
3090	4-3/4"		72	.222		36.50	10.45		46.95	56.50
3100	5"		70	.229		37.50	10.75		48.25	58.50
3110	5-1/4"		68	.235		37.50	11.10		48.60	59
3120	5-1/2"		66	.242		39.50	11.45		50.95	61.50
3130	6"		64	.250		39.50	11.80		51.30	62
3140	6-1/2"		60	.267		43.50	12.55		56.05	67
3150	7"		58	.276		43.50	13		56.50	67.50
3160	7-1/2"		56	.286		48	13.45		61.45	74
3170	8"		54	.296		48	13.95		61.95	74.50
3180	8-1/2"		52	.308		49.50	14.50		64	77
3190	9"		50	.320		49.50	15.10		64.60	78
3200	9-1/2"		49	.327		36.50	15.40		51.90	64.50
3210	10"		48	.333		36.50	15.70		52.20	65
3220	10-1/2"		47	.340		39	16.05		55.05	68
3230	11"		46	.348		39	16.40		55.40	68.50
3240	11-1/2"		45	.356		41	16.75		57.75	71.50
3250	12"		44	.364		41	17.15		58.15	72
3260	13"		43	.372		43	17.55		60.55	75
3270	14"		42	.381		45.50	17.95		63.45	78
3280	15"		41	.390		49	18.40		67.40	82.50
3290	16"		40	.400		50.50	18.85		69.35	85
3300	17"		39	.410		58	19.35		77.35	93.50
3310	18"		38	.421		60	19.85		79.85	97
3320	19"		37	.432		66.50	20.50		87	105
3330	20"		36	.444		68	21		89	107
3340	22"		35	.457		86.50	21.50		108	129
3350	23"		34	.471		89	22		111	132
3360	24"		31	.516		91	24.50		115.50	138

22 07 19.50 Pipe Insulation Protective Jacketing, St. Stl.

22 07 19.50		Crew	Daily Output	Labor-Hours	Unit	Material	2015 Bare Costs Labor	Equipment	Total	Total Incl O&P
0010	**PIPE INSULATION PROTECTIVE JACKETING, STAINLESS STEEL**									
0100	Metal roll jacketing									
0120	Type 304 with moisture barrier									
0140	Smooth, based on OD of insulation, .010" thick									
0260	2-1/2" ID	Q-14	250	.064	L.F.	2.12	3.02		5.14	7.05
0270	2-3/4" ID		245	.065		2.31	3.08		5.39	7.35
0280	3" ID		240	.067		2.50	3.14		5.64	7.65
0290	3-1/4" ID		235	.068		2.69	3.21		5.90	7.95
0300	3-1/2" ID		230	.070		2.88	3.28		6.16	8.25
0310	3-3/4" ID		225	.071		3.07	3.35		6.42	8.60
0320	4" ID		220	.073		3.26	3.43		6.69	8.95
0330	4-1/4" ID		215	.074		3.44	3.51		6.95	9.25
0340	4-1/2" ID		210	.076		3.63	3.59		7.22	9.60
0350	5" ID		200	.080		4.01	3.77		7.78	10.25
0360	5-1/2" ID		190	.084		4.39	3.97		8.36	11.05
0370	6" ID		180	.089		4.76	4.19		8.95	11.75
0380	6-1/2" ID		175	.091		5.15	4.31		9.46	12.35
0390	7" ID		170	.094		5.50	4.44		9.94	12.95
0400	7-1/2" ID		164	.098		5.90	4.60		10.50	13.65
0410	8" ID		161	.099		6.25	4.68		10.93	14.20
0420	8-1/2" ID		158	.101		6.65	4.77		11.42	14.75

22 07 19.50 Pipe Insulation Protective Jacketing, St. Stl.		Crew	Daily Output	Labor-Hours	Unit	Material	2015 Bare Costs Labor	Equipment	Total	Total Incl O&P
0430	9" ID	Q-14	155	.103	L.F.	7.05	4.87		11.92	15.35
0440	9-1/2" ID		152	.105		7.40	4.96		12.36	15.85
0450	10" ID		149	.107		7.80	5.05		12.85	16.45
0460	10-1/2" ID		146	.110		8.15	5.15		13.30	17.05
0470	11" ID		143	.112		8.55	5.25		13.80	17.60
0480	12" ID		137	.117		9.30	5.50		14.80	18.75
0490	13" ID		132	.121		10.05	5.70		15.75	19.95
0500	14" ID	Q-14	130	.123		10.80	5.80		16.60	21
0700	For smooth .016" thick, add					45%	33%			
1000	Stainless steel, Type 316, fitting covers									
1010	Size is based on OD of insulation									
1020	90° LR Elbow, 2 piece									
1100	1-1/2"	Q-14	126	.127	Ea.	12.50	6		18.50	23
1110	2-3/4"		123	.130		12.50	6.15		18.65	23.50
1120	3"		120	.133		13.05	6.30		19.35	24
1130	3-1/4"		117	.137		13.05	6.45		19.50	24.50
1140	3-1/2"		115	.139		13.75	6.55		20.30	25.50
1150	3-3/4"		113	.142		14.50	6.65		21.15	26.50
1160	4"		110	.145		15.95	6.85		22.80	28
1170	4-1/4"		108	.148		21	7		28	34.50
1180	4-1/2"		106	.151		21	7.10		28.10	34.50
1190	5"		102	.157		21.50	7.40		28.90	35.50
1200	5-1/2"		97	.165		32.50	7.75		40.25	47.50
1210	6"		92	.174		35.50	8.20		43.70	52
1220	6-1/2"		87	.184		49	8.65		57.65	67
1230	7"		85	.188		49	8.85		57.85	67.50
1240	7-1/2"		83	.193		57	9.10		66.10	76.50
1250	8"		80	.200		57	9.45		66.45	77
1260	8-1/2"		80	.200		59	9.45		68.45	79.50
1270	9"		78	.205		89.50	9.65		99.15	114
1280	9-1/2"		77	.208		88	9.80		97.80	112
1290	10"		76	.211		88	9.90		97.90	112
1300	10-1/2"		75	.213		105	10.05		115.05	132
1310	11"		74	.216		101	10.20		111.20	127
1320	12"		71	.225		114	10.60		124.60	143
1330	13"		68	.235		158	11.10		169.10	191
1340	14"		66	.242		159	11.45		170.45	193
2000	45° Elbow, 2 piece									
2010	2-1/2"	Q-14	126	.127	Ea.	10.40	6		16.40	21
2020	2-3/4"		123	.130		10.40	6.15		16.55	21
2030	3"		120	.133		11.20	6.30		17.50	22
2040	3-1/4"		117	.137		11.20	6.45		17.65	22.50
2050	3-1/2"		115	.139		11.35	6.55		17.90	22.50
2060	3-3/4"		113	.142		11.35	6.65		18	23
2070	4"		110	.145		14.60	6.85		21.45	26.50
2080	4-1/4"		108	.148		20.50	7		27.50	33.50
2090	4-1/2"		106	.151		20.50	7.10		27.60	33.50
2100	4-3/4"		104	.154		20.50	7.25		27.75	34
2110	5"		102	.157		21	7.40		28.40	34.50
2120	5-1/2"		97	.165		21	7.75		28.75	35
2130	6"		92	.174		24.50	8.20		32.70	40
2140	6-1/2"		87	.184		24.50	8.65		33.15	40.50
2150	7"		85	.188		42	8.85		50.85	60

22 07 Plumbing Insulation

22 07 19 – Plumbing Piping Insulation

22 07 19.50 Pipe Insulation Protective Jacketing, St. Stl.	Crew	Daily Output	Labor-Hours	Unit	Material	2015 Bare Costs Labor	Equipment	Total	Total Incl O&P
2160	7-1/2"	Q-14	83	.193	Ea.	42.50	9.10		51.60	61
2170	8"		82	.195		42.50	9.20		51.70	61.50
2180	8-1/2"		80	.200		50	9.45		59.45	69.50
2190	9"		78	.205		50	9.65		59.65	70
2200	9-1/2"		77	.208		58.50	9.80		68.30	80
2210	10"		76	.211		58.50	9.90		68.40	80
2220	10-1/2"		75	.213		70.50	10.05		80.55	93
2230	11"		74	.216		70.50	10.20		80.70	93.50
2240	12"		71	.225		76.50	10.60		87.10	101
2250	13"		68	.235		90	11.10		101.10	116

22 11 Facility Water Distribution

22 11 13 – Facility Water Distribution Piping

22 11 13.14 Pipe, Brass

		Crew	Daily Output	Labor-Hours	Unit	Material	Labor	Equipment	Total	Total Incl O&P
0010	**PIPE, BRASS**, Plain end									
0900	Field threaded, coupling & clevis hanger assembly 10' O.C.									
0920	Regular weight									
1120	1/2" diameter	1 Plum	48	.167	L.F.	7.15	9.80		16.95	22.50
1140	3/4" diameter		46	.174		9.30	10.20		19.50	25.50
1160	1" diameter		43	.186		13.30	10.90		24.20	31
1180	1-1/4" diameter	Q-1	72	.222		19.90	11.75		31.65	40
1200	1-1/2" diameter		65	.246		23.50	13		36.50	45.50
1220	2" diameter		53	.302		33	15.95		48.95	60.50
1240	2-1/2" diameter		41	.390		51	20.50		71.50	87.50
1260	3" diameter		31	.516		71.50	27.50		99	120
1300	4" diameter	Q-2	37	.649		126	35.50		161.50	192
1930	To delete coupling & hanger, subtract									
1940	1/2" diam.					14%	46%			
1950	3/4" diam. to 1-1/2" diam.					8%	47%			
1960	2" diam. to 4" diam.					10%	37%			

22 11 13.16 Pipe Fittings, Brass

		Crew	Daily Output	Labor-Hours	Unit	Material	Labor	Equipment	Total	Total Incl O&P
0010	**PIPE FITTINGS, BRASS**, Rough bronze, threaded.									
1000	Standard wt., 90° Elbow									
1040	1/8"	1 Plum	13	.615	Ea.	18	36		54	74.50
1060	1/4"		13	.615		18	36		54	74.50
1080	3/8"		13	.615		18	36		54	74.50
1100	1/2"		12	.667		18	39		57	79
1120	3/4"		11	.727		24	42.50		66.50	91
1140	1"		10	.800		39	47		86	114
1160	1-1/4"	Q-1	17	.941		63.50	49.50		113	145
1180	1-1/2"		16	1		78.50	53		131.50	166
1200	2"		14	1.143		126	60.50		186.50	230
1220	2-1/2"		11	1.455		305	77		382	450
1240	3"		8	2		465	106		571	675
1260	4"	Q-2	11	2.182		945	120		1,065	1,225
1280	5"		8	3		2,625	164		2,789	3,125
1300	6"		7	3.429		3,850	188		4,038	4,525
1500	45° Elbow, 1/8"	1 Plum	13	.615		22	36		58	79
1540	1/4"		13	.615		22	36		58	79
1560	3/8"		13	.615		22	36		58	79
1580	1/2"		12	.667		22	39		61	83.50

For customer support on your Plumbing Cost Data, call 877.709.6968.

143

22 11 13.16 Pipe Fittings, Brass		Crew	Daily Output	Labor-Hours	Unit	Material	2015 Bare Costs Labor	Equipment	Total	Total Incl O&P
1600	3/4"	1 Plum	11	.727	Ea.	31.50	42.50		74	99
1620	1"	↓	10	.800		53.50	47		100.50	130
1640	1-1/4"	Q-1	17	.941		85	49.50		134.50	169
1660	1-1/2"		16	1		107	53		160	198
1680	2"		14	1.143		173	60.50		233.50	282
1700	2-1/2"		11	1.455		330	77		407	480
1720	3"	↓	8	2		510	106		616	720
1740	4"	Q-2	11	2.182		1,125	120		1,245	1,425
1760	5"		8	3		2,125	164		2,289	2,600
1780	6"	↓	7	3.429		2,950	188		3,138	3,525
2000	Tee, 1/8"	1 Plum	9	.889		21	52		73	103
2040	1/4"		9	.889		21	52		73	103
2060	3/8"		9	.889		21	52		73	103
2080	1/2"		8	1		21	58.50		79.50	112
2100	3/4"		7	1.143		30	67		97	134
2120	1"		6	1.333		54.50	78.50		133	178
2140	1-1/4"	Q-1	10	1.600		93.50	84.50		178	231
2160	1-1/2"		9	1.778		105	94		199	258
2180	2"		8	2		175	106		281	350
2200	2-1/2"		7	2.286		415	121		536	640
2220	3"	↓	5	3.200		635	169		804	955
2240	4"	Q-2	7	3.429		1,575	188		1,763	2,000
2260	5"		5	4.800		3,275	263		3,538	4,000
2280	6"	↓	4	6		5,150	330		5,480	6,150
2500	Coupling, 1/8"	1 Plum	26	.308		15.05	18.05		33.10	44
2540	1/4"		22	.364		15.05	21.50		36.55	48.50
2560	3/8"		18	.444		15.05	26		41.05	56
2580	1/2"		15	.533		15.05	31.50		46.55	64
2600	3/4"		14	.571		21	33.50		54.50	74
2620	1"		13	.615		36	36		72	94
2640	1-1/4"	Q-1	22	.727		60	38.50		98.50	125
2660	1-1/2"		20	.800		78.50	42.50		121	150
2680	2"		18	.889		130	47		177	213
2700	2-1/2"		14	1.143		221	60.50		281.50	335
2720	3"	↓	10	1.600		305	84.50		389.50	465
2740	4"	Q-2	12	2		630	110		740	860
2760	5"		10	2.400		1,175	131		1,306	1,500
2780	6"	↓	9	2.667	↓	1,675	146		1,821	2,075
3000	Union, 125 lb.									
3020	1/8"	1 Plum	12	.667	Ea.	51.50	39		90.50	116
3040	1/4"		12	.667		51.50	39		90.50	116
3060	3/8"		12	.667		51.50	39		90.50	116
3080	1/2"		11	.727		51.50	42.50		94	121
3100	3/4"		10	.800		70.50	47		117.50	149
3120	1"		9	.889		106	52		158	196
3140	1-1/4"	Q-1	16	1		154	53		207	249
3160	1-1/2"		15	1.067		183	56.50		239.50	286
3180	2"		13	1.231		247	65		312	370
3200	2-1/2"		10	1.600		675	84.50		759.50	870
3220	3"	↓	7	2.286		1,050	121		1,171	1,325
3240	4"	Q-2	10	2.400	↓	3,250	131		3,381	3,775

22 11 13 – Facility Water Distribution Piping

22 11 13.23 Pipe/Tube, Copper		Crew	Daily Output	Labor-Hours	Unit	Material	2015 Bare Costs Labor	Equipment	Total	Total Incl O&P
0010	**PIPE/TUBE, COPPER**, Solder joints	R221113-50								
0100	Solder									
0120	Solder, lead free, roll				Lb.	9.35			9.35	10.30
1000	Type K tubing, couplings & clevis hanger assemblies 10' O.C.									
1100	1/4" diameter	R221113-70	1 Plum	84	.095	L.F.	3.71	5.60		9.31	12.55
1120	3/8" diameter			82	.098		4.31	5.75		10.06	13.40
1140	1/2" diameter			78	.103		4.78	6		10.78	14.35
1160	5/8" diameter			77	.104		5.90	6.10		12	15.65
1180	3/4" diameter			74	.108		7.80	6.35		14.15	18.15
1200	1" diameter			66	.121		10.35	7.10		17.45	22
1220	1-1/4" diameter			56	.143		12.20	8.40		20.60	26
1240	1-1/2" diameter			50	.160		15.60	9.40		25	31.50
1260	2" diameter			40	.200		23.50	11.75		35.25	43.50
1280	2-1/2" diameter		Q-1	60	.267		35	14.10		49.10	60
1300	3" diameter			54	.296		48	15.65		63.65	76
1320	3-1/2" diameter			42	.381		64	20		84	101
1330	4" diameter			38	.421		80	22		102	122
1340	5" diameter			32	.500		152	26.50		178.50	207
1360	6" diameter		Q-2	38	.632		222	34.50		256.50	297
1380	8" diameter		"	34	.706		385	38.50		423.50	485
1390	For other than full hard temper, add						13%				
1440	For silver solder, add							15%			
1800	For medical clean, (oxygen class), add						12%				
1950	To delete cplgs. & hngrs., 1/4"-1" pipe, subtract						27%	60%			
1960	1-1/4"-3" pipe, subtract						14%	52%			
1970	3-1/2"-5" pipe, subtract						10%	60%			
1980	6"-8" pipe, subtract						19%	53%			
2000	Type L tubing, couplings & clevis hanger assemblies 10' O.C.										
2100	1/4" diameter		1 Plum	88	.091	L.F.	2.77	5.35		8.12	11.10
2120	3/8" diameter			84	.095		3.50	5.60		9.10	12.30
2140	1/2" diameter			81	.099		3.70	5.80		9.50	12.80
2160	5/8" diameter			79	.101		5.35	5.95		11.30	14.85
2180	3/4" diameter			76	.105		5.20	6.20		11.40	15.05
2200	1" diameter			68	.118		7.85	6.90		14.75	19.05
2220	1-1/4" diameter			58	.138		10.35	8.10		18.45	23.50
2240	1-1/2" diameter			52	.154		12.95	9.05		22	28
2260	2" diameter			42	.190		18.80	11.20		30	37.50
2280	2-1/2" diameter		Q-1	62	.258		29.50	13.65		43.15	53
2300	3" diameter			56	.286		37	15.10		52.10	63.50
2320	3-1/2" diameter			43	.372		53	19.65		72.65	87.50
2340	4" diameter			39	.410		65.50	21.50		87	105
2360	5" diameter			34	.471		120	25		145	170
2380	6" diameter		Q-2	40	.600		166	33		199	232
2400	8" diameter		"	36	.667		285	36.50		321.50	370
2410	For other than full hard temper, add						21%				
2590	For silver solder, add							15%			
2900	For medical clean, (oxygen class), add						12%				
2940	To delete cplgs. & hngrs., 1/4"-1" pipe, subtract						37%	63%			
2960	1-1/4"-3" pipe, subtract						12%	53%			
2970	3-1/2"-5" pipe, subtract						12%	63%			
2980	6"-8" pipe, subtract						24%	55%			
3000	Type M tubing, couplings & clevis hanger assemblies 10' O.C.										

For customer support on your Plumbing Cost Data, call 877.709.6968.

145

22 11 13.23 Pipe/Tube, Copper		Crew	Daily Output	Labor-Hours	Unit	Material	2015 Bare Costs Labor	Equipment	Total	Total Incl O&P
3100	1/4" diameter	1 Plum	90	.089	L.F.	3	5.20		8.20	11.20
3120	3/8" diameter		87	.092		3.05	5.40		8.45	11.50
3140	1/2" diameter		84	.095		3.05	5.60		8.65	11.80
3160	5/8" diameter		81	.099		4.49	5.80		10.29	13.70
3180	3/4" diameter		78	.103		4.13	6		10.13	13.65
3200	1" diameter		70	.114		6.60	6.70		13.30	17.40
3220	1-1/4" diameter		60	.133		8.80	7.85		16.65	21.50
3240	1-1/2" diameter		54	.148		11.55	8.70		20.25	26
3260	2" diameter	↓	44	.182		17.35	10.65		28	35
3280	2-1/2" diameter	Q-1	64	.250		25.50	13.20		38.70	48
3300	3" diameter		58	.276		34	14.55		48.55	59.50
3320	3-1/2" diameter		45	.356		47.50	18.80		66.30	80.50
3340	4" diameter		40	.400		60	21		81	98
3360	5" diameter	↓	36	.444		120	23.50		143.50	168
3370	6" diameter	Q-2	42	.571		166	31.50		197.50	230
3380	8" diameter	"	38	.632	↓	285	34.50		319.50	370
3440	For silver solder, add						15%			
3960	To delete cplgs. & hngrs., 1/4"-1" pipe, subtract					35%	65%			
3970	1-1/4"-3" pipe, subtract					19%	56%			
3980	3-1/2"-5" pipe, subtract					13%	65%			
3990	6"-8" pipe, subtract					28%	58%			
4000	Type DWV tubing, couplings & clevis hanger assemblies 10' O.C.									
4100	1-1/4" diameter	1 Plum	60	.133	L.F.	9.15	7.85		17	22
4120	1-1/2" diameter		54	.148		11.15	8.70		19.85	25.50
4140	2" diameter	↓	44	.182		14.60	10.65		25.25	32
4160	3" diameter	Q-1	58	.276		26.50	14.55		41.05	51
4180	4" diameter		40	.400		44	21		65	80
4200	5" diameter	↓	36	.444		107	23.50		130.50	154
4220	6" diameter	Q-2	42	.571		153	31.50		184.50	216
4240	8" diameter	"	38	.632	↓	310	34.50		344.50	395
4730	To delete cplgs. & hngrs., 1-1/4"-2" pipe, subtract					16%	53%			
4740	3"-4" pipe, subtract					13%	60%			
4750	5"-8" pipe, subtract					23%	58%			
5200	ACR tubing, type L, hard temper, cleaned and									
5220	capped, no couplings or hangers									
5240	3/8" OD				L.F.	1.51			1.51	1.66
5250	1/2" OD					2.30			2.30	2.53
5260	5/8" OD					2.81			2.81	3.09
5270	3/4" OD					3.97			3.97	4.37
5280	7/8" OD					4.41			4.41	4.85
5290	1-1/8" OD					6.40			6.40	7.05
5300	1-3/8" OD					8.60			8.60	9.45
5310	1-5/8" OD					11			11	12.10
5320	2-1/8" OD					17.30			17.30	19.05
5330	2-5/8" OD					26			26	28.50
5340	3-1/8" OD					34.50			34.50	38
5350	3-5/8" OD					45			45	49.50
5360	4-1/8" OD				↓	57.50			57.50	63
5380	ACR tubing, type L, hard, cleaned and capped									
5381	No couplings or hangers									
5384	3/8"	1 Stpi	160	.050	L.F.	1.51	2.99		4.50	6.15
5385	1/2"		160	.050		2.30	2.99		5.29	7.05
5386	5/8"	↓	160	.050	↓	2.81	2.99		5.80	7.60

22 11 13 – Facility Water Distribution Piping

22 11 13.23 Pipe/Tube, Copper

		Crew	Daily Output	Labor-Hours	Unit	Material	2015 Bare Costs Labor	Equipment	Total	Total Incl O&P
5387	3/4"	1 Stpi	130	.062	L.F.	3.97	3.68		7.65	9.90
5388	7/8"		130	.062		4.41	3.68		8.09	10.40
5389	1-1/8"		115	.070		6.40	4.16		10.56	13.35
5390	1-3/8"		100	.080		8.60	4.78		13.38	16.65
5391	1-5/8"		90	.089		11	5.30		16.30	20
5392	2-1/8"		80	.100		17.30	6		23.30	28
5393	2-5/8"	Q-5	125	.128		26	6.90		32.90	39
5394	3-1/8"		105	.152		34.50	8.20		42.70	50.50
5395	4-1/8"		95	.168		57.50	9.05		66.55	76.50
5800	Refrigeration tubing, dryseal, 50' coils									
5840	1/8" OD				Coil	32			32	35.50
5850	3/16" OD					37.50			37.50	41.50
5860	1/4" OD					43.50			43.50	47.50
5870	5/16" OD					58			58	64
5880	3/8" OD					60			60	66
5890	1/2" OD					87			87	96
5900	5/8" OD					112			112	124
5910	3/4" OD					133			133	147
5920	7/8" OD					199			199	219
5930	1-1/8" OD					300			300	330
5940	1-3/8" OD					520			520	575
5950	1-5/8" OD					660			660	730

22 11 13.25 Pipe/Tube Fittings, Copper

		Crew	Daily Output	Labor-Hours	Unit	Material	2015 Bare Costs Labor	Equipment	Total	Total Incl O&P
0010	**PIPE/TUBE FITTINGS, COPPER**, Wrought unless otherwise noted									
0020	For silver solder, add						15%			
0040	Solder joints, copper x copper									
0070	90° elbow, 1/4"	1 Plum	22	.364	Ea.	3.28	21.50		24.78	35.50
0090	3/8"		22	.364		3.11	21.50		24.61	35.50
0100	1/2"		20	.400		1.10	23.50		24.60	36.50
0110	5/8"		19	.421		2.46	24.50		26.96	40
0120	3/4"		19	.421		2.46	24.50		26.96	40
0130	1"		16	.500		6.05	29.50		35.55	51
0140	1-1/4"		15	.533		9	31.50		40.50	57.50
0150	1-1/2"		13	.615		14.05	36		50.05	70
0160	2"		11	.727		25.50	42.50		68	92.50
0170	2-1/2"	Q-1	13	1.231		51.50	65		116.50	155
0180	3"		11	1.455		64.50	77		141.50	187
0190	3-1/2"		10	1.600		225	84.50		309.50	375
0200	4"		9	1.778		152	94		246	310
0210	5"		6	2.667		645	141		786	925
0220	6"	Q-2	9	2.667		865	146		1,011	1,175
0230	8"	"	8	3		3,475	164		3,639	4,075
0250	45° elbow, 1/4"	1 Plum	22	.364		5.90	21.50		27.40	38.50
0270	3/8"		22	.364		5	21.50		26.50	37.50
0280	1/2"		20	.400		2.01	23.50		25.51	37.50
0290	5/8"		19	.421		9.55	24.50		34.05	48
0300	3/4"		19	.421		3.52	24.50		28.02	41.50
0310	1"		16	.500		8.85	29.50		38.35	54
0320	1-1/4"		15	.533		11.85	31.50		43.35	60.50
0330	1-1/2"		13	.615		14.25	36		50.25	70
0340	2"		11	.727		24	42.50		66.50	90.50
0350	2-1/2"	Q-1	13	1.231		50.50	65		115.50	154

For customer support on your Plumbing Cost Data, call 877.709.6968.

147

22 11 13 – Facility Water Distribution Piping

22 11 13.25 Pipe/Tube Fittings, Copper		Crew	Daily Output	Labor-Hours	Unit	Material	2015 Bare Costs Labor	Equipment	Total	Total Incl O&P
0360	3"	Q-1	13	1.231	Ea.	70.50	65		135.50	176
0370	3-1/2"		10	1.600		125	84.50		209.50	266
0380	4"		9	1.778		151	94		245	310
0390	5"		6	2.667		545	141		686	815
0400	6"	Q-2	9	2.667		860	146		1,006	1,175
0410	8"	"	8	3		3,400	164		3,564	3,975
0450	Tee, 1/4"	1 Plum	14	.571		6.55	33.50		40.05	57.50
0470	3/8"		14	.571		5.25	33.50		38.75	56.50
0480	1/2"		13	.615		1.87	36		37.87	56.50
0490	5/8"		12	.667		12.05	39		51.05	72.50
0500	3/4"		12	.667		4.51	39		43.51	64
0510	1"		10	.800		13.95	47		60.95	86.50
0520	1-1/4"		9	.889		18.85	52		70.85	99.50
0530	1-1/2"		8	1		29	58.50		87.50	121
0540	2"		7	1.143		45.50	67		112.50	151
0550	2-1/2"	Q-1	8	2		91.50	106		197.50	261
0560	3"		7	2.286		131	121		252	325
0570	3-1/2"		6	2.667		380	141		521	635
0580	4"		5	3.200		320	169		489	605
0590	5"		4	4		975	211		1,186	1,400
0600	6"	Q-2	6	4		1,325	219		1,544	1,800
0610	8"	"	5	4.800		5,125	263		5,388	6,050
0612	Tee, reducing on the outlet, 1/4"	1 Plum	15	.533		16.45	31.50		47.95	65.50
0613	3/8"		15	.533		15.40	31.50		46.90	64.50
0614	1/2"		14	.571		10.05	33.50		43.55	61.50
0615	5/8"		13	.615		32	36		68	90
0616	3/4"		12	.667		6.45	39		45.45	66
0617	1"		11	.727		18.50	42.50		61	85
0618	1-1/4"		10	.800		20.50	47		67.50	93.50
0619	1-1/2"		9	.889		21.50	52		73.50	103
0620	2"		8	1		39	58.50		97.50	132
0621	2-1/2"	Q-1	9	1.778		107	94		201	259
0622	3"		8	2		110	106		216	281
0623	4"		6	2.667		203	141		344	435
0624	5"		5	3.200		1,125	169		1,294	1,500
0625	6"	Q-2	7	3.429		1,550	188		1,738	1,975
0626	8"	"	6	4		5,950	219		6,169	6,875
0630	Tee, reducing on the run, 1/4"	1 Plum	15	.533		13.90	31.50		45.40	63
0631	3/8"		15	.533		19.60	31.50		51.10	69
0632	1/2"		14	.571		11.35	33.50		44.85	63
0633	5/8"		13	.615		19.40	36		55.40	76
0634	3/4"		12	.667		15.15	39		54.15	75.50
0635	1"		11	.727		16.65	42.50		59.15	83
0636	1-1/4"		10	.800		26	47		73	100
0637	1-1/2"		9	.889		46	52		98	130
0638	2"		8	1		59	58.50		117.50	154
0639	2-1/2"	Q-1	9	1.778		152	94		246	310
0640	3"		8	2		202	106		308	380
0641	4"		6	2.667		425	141		566	680
0642	5"		5	3.200		1,075	169		1,244	1,425
0643	6"	Q-2	7	3.429		1,625	188		1,813	2,075
0644	8"	"	6	4		5,950	219		6,169	6,875
0650	Coupling, 1/4"	1 Plum	24	.333		.75	19.55		20.30	30.50

For customer support on your Plumbing Cost Data, call 877.709.6968.

22 11 13.25 Pipe/Tube Fittings, Copper		Crew	Daily Output	Labor-Hours	Unit	Material	2015 Bare Costs Labor	Equipment	Total	Total Incl O&P
0670	3/8"	1 Plum	24	.333	Ea.	1	19.55		20.55	30.50
0680	1/2"		22	.364		.83	21.50		22.33	33
0690	5/8"		21	.381		2.50	22.50		25	37
0700	3/4"		21	.381		1.67	22.50		24.17	36
0710	1"		18	.444		5.55	26		31.55	45.50
0715	1-1/4"		17	.471		5.85	27.50		33.35	48
0716	1-1/2"		15	.533		7.70	31.50		39.20	56
0718	2"		13	.615		12.90	36		48.90	68.50
0721	2-1/2"	Q-1	15	1.067		27.50	56.50		84	115
0722	3"		13	1.231		38.50	65		103.50	141
0724	3-1/2"		8	2		74	106		180	242
0726	4"		7	2.286		81.50	121		202.50	272
0728	5"		6	2.667		187	141		328	420
0731	6"	Q-2	8	3		310	164		474	590
0732	8"	"	7	3.429		1,000	188		1,188	1,375
0741	Coupling, reducing, concentric									
0743	1/2"	1 Plum	23	.348	Ea.	2.16	20.50		22.66	33.50
0745	3/4"		21.50	.372		4.70	22		26.70	38
0747	1"		19.50	.410		5.95	24		29.95	43
0748	1-1/4"		18	.444		8.30	26		34.30	48.50
0749	1-1/2"		16	.500		10.80	29.50		40.30	56.50
0751	2"		14	.571		23	33.50		56.50	75.50
0752	2-1/2"		13	.615		47	36		83	107
0753	3"	Q-1	14	1.143		53.50	60.50		114	150
0755	4"	"	8	2		113	106		219	284
0757	5"	Q-2	7.50	3.200		585	175		760	905
0759	6"		7	3.429		1,075	188		1,263	1,475
0761	8"		6.50	3.692		2,150	202		2,352	2,650
0771	Cap, sweat									
0773	1/2"	1 Plum	40	.200	Ea.	.80	11.75		12.55	18.65
0775	3/4"		38	.211		1.48	12.35		13.83	20.50
0777	1"		32	.250		3.52	14.70		18.22	26
0778	1-1/4"		29	.276		4.66	16.20		20.86	29.50
0779	1-1/2"		26	.308		6.80	18.05		24.85	35
0781	2"		22	.364		12.30	21.50		33.80	45.50
0791	Flange, sweat									
0793	3"	Q-1	22	.727	Ea.	174	38.50		212.50	249
0795	4"		18	.889		243	47		290	340
0797	5"		12	1.333		455	70.50		525.50	605
0799	6"	Q-2	18	1.333		475	73		548	630
0801	8"	"	16	1.500		1,000	82		1,082	1,225
0850	Unions, 1/4"	1 Plum	21	.381		26	22.50		48.50	63
0870	3/8"		21	.381		26.50	22.50		49	63
0880	1/2"		19	.421		14.05	24.50		38.55	53
0890	5/8"		18	.444		60	26		86	106
0900	3/4"		18	.444		17.60	26		43.60	59
0910	1"		15	.533		30.50	31.50		62	81
0920	1-1/4"		14	.571		50	33.50		83.50	105
0930	1-1/2"		12	.667		65.50	39		104.50	131
0940	2"		10	.800		113	47		160	195
0950	2-1/2"	Q-1	12	1.333		245	70.50		315.50	375
0960	3"	"	10	1.600		635	84.50		719.50	825
0980	Adapter, copper x male IPS, 1/4"	1 Plum	20	.400		10.85	23.50		34.35	47.50

For customer support on your Plumbing Cost Data, call 877.709.6968.

149

22 11 13 – Facility Water Distribution Piping

22 11 13.25 Pipe/Tube Fittings, Copper	Crew	Daily Output	Labor-Hours	Unit	Material	2015 Bare Costs Labor	Equipment	Total	Total Incl O&P
0990	3/8"	1 Plum	20	.400	Ea.	5.40	23.50		28.90	41.50
1000	1/2"		18	.444		2.29	26		28.29	42
1010	3/4"		17	.471		3.84	27.50		31.34	45.50
1020	1"		15	.533		9.95	31.50		41.45	58.50
1030	1-1/4"		13	.615		14.55	36		50.55	70.50
1040	1-1/2"		12	.667		16.70	39		55.70	77.50
1050	2"		11	.727		28.50	42.50		71	95.50
1060	2-1/2"	Q-1	10.50	1.524		108	80.50		188.50	241
1070	3"		10	1.600		135	84.50		219.50	276
1080	3-1/2"		9	1.778		161	94		255	320
1090	4"		8	2		182	106		288	360
1200	5", cast		6	2.667		4,600	141		4,741	5,275
1210	6", cast	Q-2	8.50	2.824		5,125	155		5,280	5,850
1214	Adapter, copper x female IPS									
1216	1/2"	1 Plum	18	.444	Ea.	3.64	26		29.64	43.50
1218	3/4"		17	.471		4.99	27.50		32.49	47
1220	1"		15	.533		11.55	31.50		43.05	60
1221	1-1/4"		13	.615		16.70	36		52.70	73
1222	1-1/2"		12	.667		26	39		65	88
1224	2"		11	.727		35.50	42.50		78	104
1250	Cross, 1/2"		10	.800		18	47		65	91
1260	3/4"		9.50	.842		30.50	49.50		80	108
1270	1"		8	1		38	58.50		96.50	131
1280	1-1/4"		7.50	1.067		85.50	62.50		148	189
1290	1-1/2"		6.50	1.231		122	72.50		194.50	243
1300	2"		5.50	1.455		237	85.50		322.50	390
1310	2-1/2"	Q-1	6.50	2.462		530	130		660	780
1320	3"	"	5.50	2.909		620	154		774	915
1500	Tee fitting, mechanically formed,(Type 1, 'branch sizes up to 2 in.')									
1520	1/2" run size, 3/8" to 1/2" branch size	1 Plum	80	.100	Ea.		5.85		5.85	8.85
1530	3/4" run size, 3/8" to 3/4" branch size		60	.133			7.85		7.85	11.80
1540	1" run size, 3/8" to 1" branch size		54	.148			8.70		8.70	13.15
1550	1-1/4" run size, 3/8" to 1-1/4" branch size		48	.167			9.80		9.80	14.80
1560	1-1/2" run size, 3/8" to 1-1/2" branch size		40	.200			11.75		11.75	17.75
1570	2" run size, 3/8" to 2" branch size		35	.229			13.40		13.40	20.50
1580	2-1/2" run size, 1/2" to 2" branch size		32	.250			14.70		14.70	22
1590	3" run size, 1" to 2" branch size		26	.308			18.05		18.05	27.50
1600	4" run size, 1" to 2" branch size		24	.333			19.55		19.55	29.50
1640	Tee fitting, mechanically formed, (Type 2, branches 2-1/2" thru 4")									
1650	2-1/2" run size, 2-1/2" branch size	1 Plum	12.50	.640	Ea.		37.50		37.50	56.50
1660	3" run size, 2-1/2" to 3" branch size		12	.667			39		39	59
1670	3-1/2" run size, 2-1/2" to 3-1/2" branch size		11	.727			42.50		42.50	64.50
1680	4" run size, 2-1/2" to 4" branch size		10.50	.762			44.50		44.50	67.50
1698	5" run size, 2" to 4" branch size		9.50	.842			49.50		49.50	74.50
1700	6" run size, 2" to 4" branch size		8.50	.941			55.50		55.50	83.50
1710	8" run size, 2" to 4" branch size		7	1.143			67		67	101
1800	ACR fittings, OD size									
1802	Tee, straight									
1808	5/8"	1 Stpi	12	.667	Ea.	1.87	40		41.87	62
1810	3/4"		12	.667		12.05	40		52.05	73.50
1812	7/8"		10	.800		4.51	48		52.51	77
1813	1"		10	.800		33	48		81	108
1814	1-1/8"		10	.800		13.95	48		61.95	87.50

22 11 13 – Facility Water Distribution Piping

22 11 13.25 Pipe/Tube Fittings, Copper		Crew	Daily Output	Labor-Hours	Unit	Material	2015 Bare Costs Labor	Equipment	Total	Total Incl O&P
1816	1-3/8"	1 Stpi	9	.889	Ea.	18.85	53		71.85	101
1818	1-5/8"		8	1		29	60		89	122
1820	2-1/8"		7	1.143		45.50	68.50		114	153
1822	2-5/8"	Q-5	8	2		91.50	108		199.50	263
1824	3-1/8"		7	2.286		131	123		254	330
1826	4-1/8"		5	3.200		320	172		492	610
1830	90° elbow									
1836	5/8"	1 Stpi	19	.421	Ea.	3.67	25		28.67	42
1838	3/4"		19	.421		6.65	25		31.65	45.50
1840	7/8"		16	.500		6.60	30		36.60	52.50
1842	1-1/8"		16	.500		8.85	30		38.85	55
1844	1-3/8"		15	.533		9	32		41	58
1846	1-5/8"		13	.615		14.05	37		51.05	71
1848	2-1/8"		11	.727		25.50	43.50		69	93.50
1850	2-5/8"	Q-5	13	1.231		51.50	66		117.50	157
1852	3-1/8"		11	1.455		64.50	78		142.50	189
1854	4-1/8"		9	1.778		170	95.50		265.50	330
1860	Coupling									
1866	5/8"	1 Stpi	21	.381	Ea.	1.15	23		24.15	36
1868	3/4"		21	.381		2.19	23		25.19	37
1870	7/8"		18	.444		1.67	26.50		28.17	42
1871	1"		18	.444		5.70	26.50		32.20	46.50
1872	1-1/8"		18	.444		3.33	26.50		29.83	43.50
1874	1-3/8"		17	.471		5.85	28		33.85	49
1876	1-5/8"		15	.533		7.70	32		39.70	56.50
1878	2-1/8"		13	.615		12.90	37		49.90	69.50
1880	2-5/8"	Q-5	15	1.067		29	57.50		86.50	119
1882	3-1/8"		13	1.231		39.50	66		105.50	144
1884	4-1/8"		7	2.286		90	123		213	285
2000	DWV, solder joints, copper x copper									
2030	90° Elbow, 1-1/4"	1 Plum	13	.615	Ea.	10.45	36		46.45	66
2050	1-1/2"		12	.667		14.55	39		53.55	75
2070	2"		10	.800		28	47		75	102
2090	3"	Q-1	10	1.600		56.50	84.50		141	191
2100	4"	"	9	1.778		335	94		429	505
2150	45° Elbow, 1-1/4"	1 Plum	13	.615		9.65	36		45.65	65
2170	1-1/2"		12	.667		8	39		47	68
2180	2"		10	.800		18.35	47		65.35	91
2190	3"	Q-1	10	1.600		36.50	84.50		121	169
2200	4"	"	9	1.778		184	94		278	345
2250	Tee, Sanitary, 1-1/4"	1 Plum	9	.889		20.50	52		72.50	102
2270	1-1/2"		8	1		26	58.50		84.50	117
2290	2"		7	1.143		30	67		97	134
2310	3"	Q-1	7	2.286		138	121		259	335
2330	4"	"	6	2.667		330	141		471	575
2400	Coupling, 1-1/4"	1 Plum	14	.571		4.91	33.50		38.41	56
2420	1-1/2"		13	.615		6.10	36		42.10	61
2440	2"		11	.727		8.45	42.50		50.95	74
2460	3"	Q-1	11	1.455		15.40	77		92.40	133
2480	4"	"	10	1.600		49	84.50		133.50	182
2602	Traps, see Section 22 13 16.60									
3500	Compression joint fittings									
3510	As used for plumbing and oil burner work									

151

For customer support on your Plumbing Cost Data, call 877.709.6968.

22 11 Facility Water Distribution

22 11 13 – Facility Water Distribution Piping

22 11 13.25 Pipe/Tube Fittings, Copper

		Crew	Daily Output	Labor-Hours	Unit	Material	2015 Bare Costs Labor	2015 Bare Costs Equipment	Total	Total Incl O&P
3520	Fitting price includes nuts and sleeves									
3540	Sleeve, 1/8"				Ea.	.22			.22	.24
3550	3/16"					.22			.22	.24
3560	1/4"					.08			.08	.09
3570	5/16"					.23			.23	.25
3580	3/8"					.46			.46	.51
3600	1/2"					.61			.61	.67
3620	Nut, 1/8"					.32			.32	.35
3630	3/16"					.34			.34	.37
3640	1/4"					.33			.33	.36
3650	5/16"					.49			.49	.54
3660	3/8"					.54			.54	.59
3670	1/2"					.95			.95	1.05
3710	Union, 1/8"	1 Plum	26	.308		2	18.05		20.05	29.50
3720	3/16"		24	.333		1.92	19.55		21.47	31.50
3730	1/4"		24	.333		2.01	19.55		21.56	31.50
3740	5/16"		23	.348		2.58	20.50		23.08	34
3750	3/8"		22	.364		2.60	21.50		24.10	35
3760	1/2"		22	.364		4	21.50		25.50	36.50
3780	5/8"		21	.381		4.43	22.50		26.93	39
3820	Union tee, 1/8"		17	.471		7.60	27.50		35.10	50
3830	3/16"		16	.500		4.42	29.50		33.92	49.50
3840	1/4"		15	.533		4	31.50		35.50	52
3850	5/16"		15	.533		5.65	31.50		37.15	53.50
3860	3/8"		15	.533		5.70	31.50		37.20	54
3870	1/2"		15	.533		9.65	31.50		41.15	58
3910	Union elbow, 1/4"		24	.333		2.47	19.55		22.02	32
3920	5/16"		23	.348		4.34	20.50		24.84	36
3930	3/8"		22	.364		4.19	21.50		25.69	36.50
3940	1/2"		22	.364		6.85	21.50		28.35	39.50
3980	Female connector, 1/8"		26	.308		1.55	18.05		19.60	29
4000	3/16" x 1/8"		24	.333		1.98	19.55		21.53	31.50
4010	1/4" x 1/8"		24	.333		1.75	19.55		21.30	31.50
4020	1/4"		24	.333		2.35	19.55		21.90	32
4030	3/8" x 1/4"		22	.364		2.35	21.50		23.85	34.50
4040	1/2" x 3/8"		22	.364		3.75	21.50		25.25	36
4050	5/8" x 1/2"		21	.381		5.30	22.50		27.80	40
4090	Male connector, 1/8"		26	.308		1.25	18.05		19.30	29
4100	3/16" x 1/8"		24	.333		1.29	19.55		20.84	31
4110	1/4" x 1/8"		24	.333		1.34	19.55		20.89	31
4120	1/4"		24	.333		1.34	19.55		20.89	31
4130	5/16" x 1/8"		23	.348		1.58	20.50		22.08	32.50
4140	5/16" x 1/4"		23	.348		1.79	20.50		22.29	33
4150	3/8" x 1/8"		22	.364		1.87	21.50		23.37	34
4160	3/8" x 1/4"		22	.364		2.15	21.50		23.65	34.50
4170	3/8"		22	.364		2.15	21.50		23.65	34.50
4180	3/8" x 1/2"		22	.364		2.95	21.50		24.45	35.50
4190	1/2" x 3/8"		22	.364		2.86	21.50		24.36	35
4200	1/2"		22	.364		3.21	21.50		24.71	35.50
4210	5/8" x 1/2"		21	.381		3.71	22.50		26.21	38
4240	Male elbow, 1/8"		26	.308		2.58	18.05		20.63	30.50
4250	3/16" x 1/8"		24	.333		2.85	19.55		22.40	32.50
4260	1/4" x 1/8"		24	.333		2.15	19.55		21.70	32

152

22 11 13 – Facility Water Distribution Piping

22 11 13.25 Pipe/Tube Fittings, Copper		Crew	Daily Output	Labor-Hours	Unit	Material	2015 Bare Costs Labor	Equipment	Total	Total Incl O&P
4270	1/4"	1 Plum	24	.333	Ea.	2.45	19.55		22	32
4280	3/8" x 1/4"		22	.364		3.11	21.50		24.61	35.50
4290	3/8"		22	.364		5.35	21.50		26.85	38
4300	1/2" x 1/4"		22	.364		5.15	21.50		26.65	37.50
4310	1/2" x 3/8"		22	.364		8.10	21.50		29.60	41
4340	Female elbow, 1/8"		26	.308		5.55	18.05		23.60	33.50
4350	1/4" x 1/8"		24	.333		3.15	19.55		22.70	33
4360	1/4"		24	.333		4.13	19.55		23.68	34
4370	3/8" x 1/4"		22	.364		4.07	21.50		25.57	36.50
4380	1/2" x 3/8"		22	.364		6.95	21.50		28.45	39.50
4390	1/2"		22	.364		9.35	21.50		30.85	42.50
4420	Male run tee, 1/4" x 1/8"		15	.533		3.86	31.50		35.36	52
4430	5/16" x 1/8"		15	.533		6.65	31.50		38.15	55
4440	3/8" x 1/4"		15	.533		5.95	31.50		37.45	54
4480	Male branch tee, 1/4" x 1/8"		15	.533		4.47	31.50		35.97	52.50
4490	1/4"		15	.533		4.81	31.50		36.31	53
4500	3/8" x 1/4"		15	.533		5.60	31.50		37.10	53.50
4510	1/2" x 3/8"		15	.533		8.70	31.50		40.20	57
4520	1/2"		15	.533		13.95	31.50		45.45	63
4800	Flare joint fittings									
4810	Refrigeration fittings									
4820	Flare joint nuts and labor not incl. in price. Add 1 nut per jnt.									
4830	90° Elbow, 1/4"				Ea.	2.32			2.32	2.55
4840	3/8"					3.32			3.32	3.65
4850	1/2"					6.20			6.20	6.85
4860	5/8"					9.10			9.10	10
4870	3/4"					13.80			13.80	15.15
5030	Tee, 1/4"					2.99			2.99	3.29
5040	5/16"					3.58			3.58	3.94
5050	3/8"					3.77			3.77	4.15
5060	1/2"					5.85			5.85	6.45
5070	5/8"					8.05			8.05	8.85
5080	3/4"					8.50			8.50	9.35
5140	Union, 3/16"					1.37			1.37	1.51
5150	1/4"					.94			.94	1.03
5160	5/16"					1.70			1.70	1.87
5170	3/8"					1.58			1.58	1.74
5180	1/2"					2.35			2.35	2.59
5190	5/8"					3.78			3.78	4.16
5200	3/4"					10.55			10.55	11.60
5260	Long flare nut, 3/16"	1 Stpi	42	.190		1.74	11.40		13:14	19.10
5270	1/4"		41	.195		.96	11.65		12.61	18.65
5280	5/16"		40	.200		1.74	11.95		13.69	19.95
5290	3/8"		39	.205		1.66	12.25		13.91	20.50
5300	1/2"		38	.211		2.61	12.60		15.21	22
5310	5/8"		37	.216		5.85	12.90		18.75	26
5320	3/4"		34	.235		11.65	14.05		25.70	34
5380	Short flare nut, 3/16"		42	.190		2.88	11.40		14.28	20.50
5390	1/4"		41	.195		.68	11.65		12.33	18.35
5400	5/16"		40	.200		.81	11.95		12.76	18.95
5410	3/8"		39	.205		1.06	12.25		13.31	19.65
5420	1/2"		38	.211		1.58	12.60		14.18	20.50
5430	5/8"		36	.222		2.20	13.30		15.50	22.50

22 11 13 – Facility Water Distribution Piping

22 11 13.25 Pipe/Tube Fittings, Copper		Crew	Daily Output	Labor-Hours	Unit	Material	2015 Bare Costs Labor	Equipment	Total	Total Incl O&P
5440	3/4"	1 Stpi	34	.235	Ea.	7.35	14.05		21.40	29
5500	90° Elbow flare by MIPS, 1/4"					2.59			2.59	2.85
5510	3/8"					2.82			2.82	3.10
5520	1/2"					4.97			4.97	5.45
5530	5/8"					9.95			9.95	10.95
5540	3/4"					16.65			16.65	18.30
5600	Flare by FIPS, 1/4"					4.73			4.73	5.20
5610	3/8"					5.15			5.15	5.65
5620	1/2"					7.35			7.35	8.10
5670	Flare by sweat, 1/4"					2.64			2.64	2.90
5680	3/8"					3.43			3.43	3.77
5690	1/2"					7			7	7.70
5700	5/8"					22			22	24
5760	Tee flare by IPS, 1/4"					4.73			4.73	5.20
5770	3/8"					6.65			6.65	7.30
5780	1/2"					11.10			11.10	12.25
5790	5/8"					8.25			8.25	9.05
5850	Connector, 1/4"					1.02			1.02	1.12
5860	3/8"					1.53			1.53	1.68
5870	1/2"					2.49			2.49	2.74
5880	5/8"					4.34			4.34	4.77
5890	3/4"					6.50			6.50	7.15
5950	Seal cap, 1/4"					.84			.84	.92
5960	3/8"					1.01			1.01	1.11
5970	1/2"					1.26			1.26	1.39
5980	5/8"					2.62			2.62	2.88
5990	3/4"				▼	7.65			7.65	8.40
6000	Water service fittings									
6010	Flare joints nut and labor are included in the fitting price.									
6020	90° Elbow, C x C, 3/8"	1 Plum	19	.421	Ea.	44.50	24.50		69	86.50
6030	1/2"		18	.444		44.50	26		70.50	88.50
6040	3/4"		16	.500		57	29.50		86.50	107
6050	1"		15	.533		108	31.50		139.50	166
6080	2"		10	.800		390	47		437	500
6090	90° Elbow, C x MPT, 3/8"		19	.421		34	24.50		58.50	75
6100	1/2"		18	.444		34	26		60	77
6110	3/4"		16	.500		39.50	29.50		69	87.50
6120	1"		15	.533		107	31.50		138.50	166
6130	1-1/4"		13	.615		220	36		256	298
6140	1-1/2"		12	.667		220	39		259	300
6150	2"		10	.800		268	47		315	365
6160	90° Elbow, C x FPT, 3/8"		19	.421		36.50	24.50		61	77.50
6170	1/2"		18	.444		36.50	26		62.50	79.50
6180	3/4"		16	.500		45	29.50		74.50	94
6190	1"		15	.533		104	31.50		135.50	162
6200	1-1/4"		13	.615		108	36		144	174
6210	1-1/2"		12	.667		415	39		454	520
6220	2"		10	.800		600	47		647	730
6230	Tee, C x C x C, 3/8"		13	.615		65	36		101	126
6240	1/2"		12	.667		65	39		104	131
6250	3/4"		11	.727		76.50	42.50		119	149
6260	1"		10	.800		135	47		182	220
6330	Tube nut, C x nut seat, 3/8"		40	.200		13.90	11.75		25.65	33

22 11 13 – Facility Water Distribution Piping

22 11 13.25 Pipe/Tube Fittings, Copper		Crew	Daily Output	Labor-Hours	Unit	Material	2015 Bare Costs Labor	Equipment	Total	Total Incl O&P
6340	1/2"	1 Plum	38	.211	Ea.	13.90	12.35		26.25	34
6350	3/4"		34	.235		13.20	13.80		27	35.50
6360	1"		32	.250		24	14.70		38.70	48.50
6380	Coupling, C x C, 3/8"		19	.421		42	24.50		66.50	84
6390	1/2"		18	.444		42	26		68	86
6400	3/4"		16	.500		50.50	29.50		80	100
6410	1"		15	.533		93.50	31.50		125	151
6420	1-1/4"		12	.667		151	39		190	226
6430	1-1/2"		12	.667		229	39		268	310
6440	2"		10	.800		345	47		392	450
6450	Adapter, C x FPT, 3/8"		19	.421		32	24.50		56.50	72.50
6460	1/2"		18	.444		32	26		58	74.50
6470	3/4"		16	.500		36.50	29.50		66	84.50
6480	1"		15	.533		78	31.50		109.50	133
6490	1-1/4"		13	.615		101	36		137	167
6500	1-1/2"		12	.667		166	39		205	241
6510	2"		10	.800		212	47		259	305
6520	Adapter, C x MPT, 3/8"		19	.421		27	24.50		51.50	67
6530	1/2"		18	.444		27	26		53	69
6540	3/4"		16	.500		37.50	29.50		67	85.50
6550	1"		15	.533		66	31.50		97.50	120
6560	1-1/4"		13	.615		164	36		200	236
6570	1-1/2"		12	.667		164	39		203	240
6580	2"		10	.800		223	47		270	315
6992	Tube connector fittings, See Section 22 11 13.76 for plastic ftng.									
7000	Insert type Brass/copper, 100 psi @ 180°F, CTS									
7010	Adapter MPT 3/8" x 3/8" CTS	1 Plum	29	.276	Ea.	2.93	16.20		19.13	27.50
7020	1/2" x 1/2"		26	.308		2.95	18.05		21	31
7030	3/4" x 1/2"		26	.308		3.77	18.05		21.82	31.50
7040	3/4" x 3/4"		25	.320		4.35	18.80		23.15	33.50
7050	Adapter CTS 1/2" x 1/2" sweat		24	.333		3.96	19.55		23.51	34
7060	3/4" x 3/4" sweat		22	.364		1.39	21.50		22.89	33.50
7070	Coupler center set 3/8" CTS		25	.320		1.44	18.80		20.24	30
7080	1/2" CTS		23	.348		3.96	20.50		24.46	35.50
7090	3/4" CTS		22	.364		1.39	21.50		22.89	33.50
7100	Elbow 90°, copper 3/8"		25	.320		3.16	18.80		21.96	32
7110	1/2" CTS		23	.348		2.02	20.50		22.52	33
7120	3/4" CTS		22	.364		2.46	21.50		23.96	34.50
7130	Tee copper 3/8" CTS		17	.471		3.75	27.50		31.25	45.50
7140	1/2" CTS		15	.533		2.58	31.50		34.08	50.50
7150	3/4" CTS		14	.571		3.96	33.50		37.46	55
7160	3/8" x 3/8" x 1/2"		16	.500		2.98	29.50		32.48	48
7170	1/2" x 3/8" x 1/2"		15	.533		1.93	31.50		33.43	49.50
7180	3/4" x 1/2" x 3/4"		14	.571		2.98	33.50		36.48	54

22 11 13.27 Pipe/Tube, Grooved Joint for Copper

		Crew	Daily Output	Labor-Hours	Unit	Material	2015 Bare Costs Labor	Equipment	Total	Total Incl O&P
0010	**PIPE/TUBE, GROOVED JOINT FOR COPPER**									
4000	Fittings: coupling material required at joints not incl. in fitting price.									
4001	Add 1 selected coupling, material only, per joint for installed price.									
4010	Coupling, rigid style									
4018	2" diameter	1 Plum	50	.160	Ea.	21	9.40		30.40	37.50
4020	2-1/2" diameter	Q-1	80	.200		24	10.55		34.55	42.50
4022	3" diameter		67	.239		27	12.60		39.60	48.50

For customer support on your Plumbing Cost Data, call 877.709.6968.

155

22 11 13 – Facility Water Distribution Piping

22 11 13.27 Pipe/Tube, Grooved Joint for Copper

		Crew	Daily Output	Labor-Hours	Unit	Material	2015 Bare Costs Labor	Equipment	Total	Total Incl O&P
4024	4" diameter	Q-1	50	.320	Ea.	40	16.90		56.90	69.50
4026	5" diameter	↓	40	.400		68	21		89	107
4028	6" diameter	Q-2	50	.480	↓	89.50	26.50		116	138
4100	Elbow, 90° or 45°									
4108	2" diameter	1 Plum	25	.320	Ea.	37.50	18.80		56.30	69.50
4110	2-1/2" diameter	Q-1	40	.400		40.50	21		61.50	76.50
4112	3" diameter		33	.485		57	25.50		82.50	101
4114	4" diameter		25	.640		130	34		164	194
4116	5" diameter	↓	20	.800		370	42.50		412.50	470
4118	6" diameter	Q-2	25	.960	↓	590	52.50		642.50	730
4200	Tee									
4208	2" diameter	1 Plum	17	.471	Ea.	61.50	27.50		89	109
4210	2-1/2" diameter	Q-1	27	.593		65	31.50		96.50	119
4212	3" diameter		22	.727		97	38.50		135.50	165
4214	4" diameter		17	.941		213	49.50		262.50	310
4216	5" diameter	↓	13	1.231		595	65		660	755
4218	6" diameter	Q-2	17	1.412	↓	735	77.50		812.50	925
4300	Reducer, concentric									
4310	3" x 2-1/2" diameter	Q-1	35	.457	Ea.	54	24		78	95.50
4312	4" x 2-1/2" diameter		32	.500		110	26.50		136.50	161
4314	4" x 3" diameter		29	.552		110	29		139	165
4316	5" x 3" diameter		25	.640		310	34		344	390
4318	5" x 4" diameter	↓	22	.727		310	38.50		348.50	400
4320	6" x 3" diameter	Q-2	28	.857		335	47		382	440
4322	6" x 4" diameter		26	.923		335	50.50		385.50	445
4324	6" x 5" diameter	↓	24	1	↓	335	55		390	455
4350	Flange, w/groove gasket									
4351	ANSI class 125 and 150									
4355	2" diameter	1 Plum	23	.348	Ea.	162	20.50		182.50	209
4356	2-1/2" diameter	Q-1	37	.432		169	23		192	220
4358	3" diameter		31	.516		177	27.50		204.50	235
4360	4" diameter		23	.696		194	37		231	269
4362	5" diameter	↓	19	.842		260	44.50		304.50	355
4364	6" diameter	Q-2	23	1.043	↓	283	57		340	395

22 11 13.29 Pipe, Fittings and Valves, Copper, Pressed-Joint

		Crew	Daily Output	Labor-Hours	Unit	Material	2015 Bare Costs Labor	Equipment	Total	Total Incl O&P
0010	**PIPE, FITTINGS AND VALVES, COPPER, PRESSED-JOINT**									
0040	Pipe/tube includes coupling & clevis type hanger assy's, 10' O.C.									
0120	Type K									
0130	1/2" diameter	1 Plum	78	.103	L.F.	4.98	6		10.98	14.60
0134	3/4" diameter		74	.108		8.05	6.35		14.40	18.45
0138	1" diameter		66	.121		10.65	7.10		17.75	22.50
0142	1-1/4" diameter		56	.143		12.70	8.40		21.10	26.50
0146	1-1/2" diameter		50	.160		16.80	9.40		26.20	32.50
0150	2" diameter	↓	40	.200		24.50	11.75		36.25	45
0154	2-1/2" diameter	Q-1	60	.267		40	14.10		54.10	65.50
0158	3" diameter		54	.296		54	15.65		69.65	83
0162	4" diameter		38	.421	↓	86	22		108	128
0180	To delete cplgs. & hngrs., 1/2" pipe, subtract					19%	48%			
0184	3/4"-2" pipe, subtract					14%	46%			
0186	2-1/2"-4" pipe, subtract					24%	34%			
0220	Type L									
0230	1/2" diameter	1 Plum	81	.099	L.F.	3.90	5.80		9.70	13.05

22 11 Facility Water Distribution

22 11 13 – Facility Water Distribution Piping

22 11 13.29 Pipe, Fittings and Valves, Copper, Pressed-Joint		Crew	Daily Output	Labor-Hours	Unit	Material	2015 Bare Costs Labor	Equipment	Total	Total Incl O&P
0234	3/4" diameter	1 Plum	76	.105	L.F.	5.45	6.20		11.65	15.35
0238	1" diameter		68	.118		8.10	6.90		15	19.40
0242	1-1/4" diameter		58	.138		10.80	8.10		18.90	24
0246	1-1/2" diameter		52	.154		14.15	9.05		23.20	29.50
0250	2" diameter		42	.190		20	11.20		31.20	39
0254	2-1/2" diameter	Q-1	62	.258		35	13.65		48.65	58.50
0258	3" diameter		56	.286		43.50	15.10		58.60	70.50
0262	4" diameter		39	.410		71.50	21.50		93	112
0280	To delete cplgs. & hngrs., 1/2" - 1"pipe, subtract					21%	52%			
0284	1-1/4"-2" pipe, subtract					17%	46%			
0286	2-1/2"-4" pipe, subtract					23%	35%			
0320	Type M									
0330	1/2" diameter	1 Plum	84	.095	L.F.	3.25	5.60		8.85	12.05
0334	3/4" diameter		78	.103		4.38	6		10.38	13.90
0338	1" diameter		70	.114		6.85	6.70		13.55	17.70
0342	1-1/4" diameter		60	.133		9.25	7.85		17.10	22
0346	1-1/2" diameter		54	.148		12.75	8.70		21.45	27
0350	2" diameter		44	.182		18.55	10.65		29.20	36.50
0354	2-1/2" diameter	Q-1	64	.250		31	13.20		44.20	54
0358	3" diameter		58	.276		40.50	14.55		55.05	66.50
0362	4" diameter		40	.400		66	21		87	105
0380	To delete cplgs. & hngrs., 1/2" pipe, subtract					32%	49%			
0384	3/4"-2" pipe, subtract					21%	46%			
0386	2-1/2"-4" pipe, subtract					25%	36%			
1600	Fittings									
1610	Press joints, copper x copper									
1620	Note: Reducing fittings show most expensive size combination.									
1800	90° elbow, 1/2"	1 Plum	36.60	.219	Ea.	3.11	12.85		15.96	23
1810	3/4"		27.50	.291		5.25	17.10		22.35	32
1820	1"		25.90	.309		10.50	18.15		28.65	39
1830	1-1/4"		20.90	.383		21	22.50		43.50	57.50
1840	1-1/2"		18.30	.437		40	25.50		65.50	83
1850	2"		15.70	.510		55.50	30		85.50	106
1860	2-1/2"	Q-1	25.90	.618		161	32.50		193.50	228
1870	3"		22	.727		203	38.50		241.50	281
1880	4"		16.30	.982		251	52		303	355
2000	45° elbow, 1/2"	1 Plum	36.60	.219		3.61	12.85		16.46	23.50
2010	3/4"		27.50	.291		4.44	17.10		21.54	31
2020	1"		25.90	.309		14.20	18.15		32.35	43
2030	1-1/4"		20.90	.383		20.50	22.50		43	56.50
2040	1-1/2"		18.30	.437		33	25.50		58.50	75
2050	2"		15.70	.510		46	30		76	95.50
2060	2-1/2"	Q-1	25.90	.618		107	32.50		139.50	167
2070	3"		22	.727		152	38.50		190.50	226
2080	4"		16.30	.982		212	52		264	310
2200	Tee, 1/2"	1 Plum	27.50	.291		4.74	17.10		21.84	31
2210	3/4"		20.70	.386		8.35	22.50		30.85	43.50
2220	1"		19.40	.412		15	24		39	53
2230	1-1/4"		15.70	.510		24	30		54	71.50
2240	1-1/2"		13.80	.580		49	34		83	106
2250	2"		11.80	.678		61	40		101	127
2260	2-1/2"	Q-1	19.40	.825		199	43.50		242.50	285
2270	3"		16.50	.970		249	51		300	350

For customer support on your Plumbing Cost Data, call 877.709.6968.

157

22 11 13.29 Pipe, Fittings and Valves, Copper, Pressed-Joint		Crew	Daily Output	Labor-Hours	Unit	Material	2015 Bare Costs Labor	Equipment	Total	Total Incl O&P
2280	4"	Q-1	12.20	1.311	Ea.	350	69.50		419.50	490
2400	Tee, reducing on the outlet									
2410	3/4"	1 Plum	20.70	.386	Ea.	7	22.50		29.50	42
2420	1"		19.40	.412		17.15	24		41.15	55.50
2430	1-1/4"		15.70	.510		24	30		54	71.50
2440	1-1/2"		13.80	.580		52.50	34		86.50	109
2450	2"		11.80	.678		84	40		124	153
2460	2-1/2"	Q-1	19.40	.825		278	43.50		321.50	370
2470	3"		16.50	.970		305	51		356	415
2480	4"		12.20	1.311		420	69.50		489.50	565
2600	Tee, reducing on the run									
2610	3/4"	1 Plum	20.70	.386	Ea.	13.35	22.50		35.85	49
2620	1"		19.40	.412		26	24		50	65
2630	1-1/4"		15.70	.510		50	30		80	100
2640	1-1/2"		13.80	.580		76.50	34		110.50	136
2650	2"		11.80	.678		84	40		124	153
2660	2-1/2"	Q-1	19.40	.825		284	43.50		327.50	380
2670	3"		16.50	.970		340	51		391	455
2680	4"		12.20	1.311		485	69.50		554.50	640
2800	Coupling, 1/2"	1 Plum	36.60	.219		2.80	12.85		15.65	22.50
2810	3/4"		27.50	.291		4.23	17.10		21.33	30.50
2820	1"		25.90	.309		8.55	18.15		26.70	37
2830	1-1/4"		20.90	.383		10.60	22.50		33.10	45.50
2840	1-1/2"		18.30	.437		19.70	25.50		45.20	60.50
2850	2"		15.70	.510		25	30		55	72.50
2860	2-1/2"	Q-1	25.90	.618		80.50	32.50		113	138
2870	3"		22	.727		102	38.50		140.50	171
2880	4"		16.30	.982		142	52		194	235
3000	Union, 1/2"	1 Plum	36.60	.219		23	12.85		35.85	45
3010	3/4"		27.50	.291		30.50	17.10		47.60	59.50
3020	1"		25.90	.309		48.50	18.15		66.65	81
3030	1-1/4"		20.90	.383		69.50	22.50		92	111
3040	1-1/2"		18.30	.437		94.50	25.50		120	143
3050	2"		15.70	.510		154	30		184	214
3200	Adapter, tube to MPT									
3210	1/2"	1 Plum	15.10	.530	Ea.	3.65	31		34.65	51
3220	3/4"		13.60	.588		6.60	34.50		41.10	59.50
3230	1"		11.60	.690		12.05	40.50		52.55	74.50
3240	1-1/4"		9.90	.808		26.50	47.50		74	101
3250	1-1/2"		9	.889		37	52		89	120
3260	2"		7.90	1.013		71.50	59.50		131	169
3270	2-1/2"	Q-1	13.40	1.194		161	63		224	274
3280	3"		10.60	1.509		203	79.50		282.50	345
3290	4"		7.70	2.078		235	110		345	425
3400	Adapter, tube to FPT									
3410	1/2"	1 Plum	15.10	.530	Ea.	4.40	31		35.40	52
3420	3/4"		13.60	.588		7.15	34.50		41.65	60
3430	1"		11.60	.690		13.15	40.50		53.65	75.50
3440	1-1/4"		9.90	.808		29.50	47.50		77	104
3450	1-1/2"		9	.889		43	52		95	127
3460	2"		7.90	1.013		73.50	59.50		133	171
3470	2-1/2"	Q-1	13.40	1.194		216	63		279	335
3480	3"		10.60	1.509		283	79.50		362.50	430

22 11 13.29 Pipe, Fittings and Valves, Copper, Pressed-Joint	Crew	Daily Output	Labor-Hours	Unit	Material	2015 Bare Costs Labor	Equipment	Total	Total Incl O&P
3490	4"	Q-1	7.70	2.078	Ea.	365	110		475	565
3600	Flange									
3620	1"	1 Plum	36.20	.221	Ea.	123	12.95		135.95	155
3630	1-1/4"		29.30	.273		178	16.05		194.05	220
3640	1-1/2"		25.60	.313		197	18.35		215.35	245
3650	2"		22	.364		221	21.50		242.50	275
3660	2-1/2"	Q-1	36.20	.442		247	23.50		270.50	310
3670	3"		30.80	.519		299	27.50		326.50	370
3680	4"		22.80	.702		335	37		372	425
3800	Cap, 1/2"	1 Plum	53.10	.151		5.95	8.85		14.80	19.90
3810	3/4"		39.80	.201		10.05	11.80		21.85	29
3820	1"		37.50	.213		15.50	12.50		28	36
3830	1-1/4"		30.40	.263		17.55	15.45		33	43
3840	1-1/2"		26.60	.301		27	17.65		44.65	56.50
3850	2"		22.80	.351		33	20.50		53.50	67.50
3860	2-1/2"	Q-1	37.50	.427		106	22.50		128.50	150
3870	3"		31.90	.502		134	26.50		160.50	187
3880	4"		23.60	.678		178	36		214	250
4000	Reducer									
4010	3/4"	1 Plum	27.50	.291	Ea.	12.65	17.10		29.75	40
4020	1"		25.90	.309		22	18.15		40.15	52
4030	1-1/4"		20.90	.383		32.50	22.50		55	70
4040	1-1/2"		18.30	.437		39	25.50		64.50	82
4050	2"		15.70	.510		53	30		83	104
4060	2-1/2"	Q-1	25.90	.618		159	32.50		191.50	225
4070	3"		22	.727		206	38.50		244.50	284
4080	4"		16.30	.982		265	52		317	370
4100	Stub out, 1/2"	1 Plum	50	.160		7.85	9.40		17.25	23
4110	3/4"		35	.229		13.90	13.40		27.30	36
4120	1"		30	.267		19.05	15.65		34.70	44.50
6000	Valves									
6200	Ball valve									
6210	1/2"	1 Plum	25.60	.313	Ea.	43.50	18.35		61.85	75
6220	3/4"		19.20	.417		61	24.50		85.50	104
6230	1"		18.10	.442		75	26		101	122
6240	1-1/4"		14.70	.544		122	32		154	183
6250	1-1/2"		12.80	.625		169	36.50		205.50	241
6260	2"		11	.727		269	42.50		311.50	360
6400	Check valve									
6410	1/2"	1 Plum	30	.267	Ea.	32.50	15.65		48.15	59.50
6420	3/4"		22.50	.356		41.50	21		62.50	77.50
6430	1"		21.20	.377		48	22		70	86.50
6440	1-1/4"		17.20	.465		69.50	27.50		97	118
6450	1-1/2"		15	.533		98.50	31.50		130	157
6460	2"		12.90	.620		180	36.50		216.50	253
6600	Butterfly valve, lug type									
6660	2-1/2"	Q-1	9	1.778	Ea.	154	94		248	310
6670	3"		8	2		188	106		294	365
6680	4"		5	3.200		235	169		404	515

For customer support on your Plumbing Cost Data, call 877.709.6968.

159

22 11 Facility Water Distribution

22 11 13 – Facility Water Distribution Piping

22 11 13.44 Pipe, Steel		Crew	Daily Output	Labor-Hours	Unit	Material	2015 Bare Costs Labor	Equipment	Total	Total Incl O&P
0010	**PIPE, STEEL**	R221113-50								
0020	All pipe sizes are to Spec. A-53 unless noted otherwise	R221113-70								
0032	Schedule 10, see Line 22 11 13.48 0500									
0050	Schedule 40, threaded, with couplings, and clevis hanger									
0060	assemblies sized for covering, 10' O.C.									
0540	Black, 1/4" diameter	1 Plum	66	.121	L.F.	6.55	7.10		13.65	17.95
0550	3/8" diameter		65	.123		7.30	7.20		14.50	18.95
0560	1/2" diameter		63	.127		3.32	7.45		10.77	14.90
0570	3/4" diameter		61	.131		3.91	7.70		11.61	15.95
0580	1" diameter		53	.151		5.05	8.85		13.90	18.95
0590	1-1/4" diameter	Q-1	89	.180		6.20	9.50		15.70	21
0600	1-1/2" diameter		80	.200		7.10	10.55		17.65	24
0610	2" diameter		64	.250		9	13.20		22.20	30
0620	2-1/2" diameter		50	.320		14	16.90		30.90	41
0630	3" diameter		43	.372		17.80	19.65		37.45	49
0640	3-1/2" diameter		40	.400		24.50	21		45.50	59
0650	4" diameter		36	.444		27	23.50		50.50	65
0809	A-106, gr. A/B, seamless w/cplgs. & clevis hanger assemblies									
0811	1/4" diameter	1 Plum	66	.121	L.F.	8.60	7.10		15.70	20
0812	3/8" diameter		65	.123		8.55	7.20		15.75	20.50
0813	1/2" diameter		63	.127		9.55	7.45		17	22
0814	3/4" diameter		61	.131		10.95	7.70		18.65	23.50
0815	1" diameter		53	.151		12.60	8.85		21.45	27.50
0816	1-1/4" diameter	Q-1	89	.180		15.10	9.50		24.60	31
0817	1-1/2" diameter		80	.200		21.50	10.55		32.05	40
0819	2" diameter		64	.250		19.05	13.20		32.25	41
0821	2-1/2" diameter		50	.320		23	16.90		39.90	51
0822	3" diameter		43	.372		30	19.65		49.65	62.50
0823	4" diameter		36	.444		46.50	23.50		70	86.50
1220	To delete coupling & hanger, subtract									
1230	1/4" diam. to 3/4" diam.					31%	56%			
1240	1" diam. to 1-1/2" diam.					23%	51%			
1250	2" diam. to 4" diam.					23%	41%			
1280	All pipe sizes are to Spec. A-53 unless noted otherwise									
1281	Schedule 40, threaded, with couplings and clevis hanger									
1282	assemblies sized for covering, 10' O. C.									
1290	Galvanized, 1/4" diameter	1 Plum	66	.121	L.F.	9.10	7.10		16.20	21
1300	3/8" diameter		65	.123		9.95	7.20		17.15	22
1310	1/2" diameter		63	.127		3.69	7.45		11.14	15.30
1320	3/4" diameter		61	.131		4.24	7.70		11.94	16.30
1330	1" diameter		53	.151		5.75	8.85		14.60	19.70
1340	1-1/4" diameter	Q-1	89	.180		7	9.50		16.50	22
1350	1-1/2" diameter		80	.200		8.05	10.55		18.60	25
1360	2" diameter		64	.250		10.35	13.20		23.55	31.50
1370	2-1/2" diameter		50	.320		16.30	16.90		33.20	43.50
1380	3" diameter		43	.372		20.50	19.65		40.15	52.50
1390	3-1/2" diameter		40	.400		26	21		47	61
1400	4" diameter		36	.444		30	23.50		53.50	68.50
1750	To delete coupling & hanger, subtract									
1760	1/4" diam. to 3/4" diam.					31%	56%			
1770	1" diam. to 1-1/2" diam.					23%	51%			
1780	2" diam. to 4" diam.					23%	41%			

160

22 11 13.44 Pipe, Steel		Crew	Daily Output	Labor-Hours	Unit	Material	2015 Bare Costs Labor	Equipment	Total	Total Incl O&P
1900	Pipe nipple std black 2"lg 1/2"	1 Plum	19	.421	Ea.	.97	24.50		25.47	38.50
1910	Pipe nipple std black 2"lg 1" diameter	"	15	.533		1.78	31.50		33.28	49.50
1920	3"lg 2-1/2"	Q-1	16	1	↓	11.10	53		64.10	92.50
2000	Welded, sch. 40, on yoke & roll hanger assy's, sized for covering, 10' O.C.									
2040	Black, 1" diameter	Q-15	93	.172	L.F.	4.72	9.10	.62	14.44	19.65
2050	1-1/4" diameter		84	.190		5.90	10.05	.69	16.64	22.50
2060	1-1/2" diameter		76	.211		6.45	11.10	.76	18.31	24.50
2070	2" diameter		61	.262		7.85	13.85	.95	22.65	30.50
2080	2-1/2" diameter		47	.340		12.75	18	1.23	31.98	42.50
2090	3" diameter		43	.372		15.45	19.65	1.34	36.44	48
2100	3-1/2" diameter		39	.410		18.55	21.50	1.48	41.53	54.50
2110	4" diameter		37	.432		21.50	23	1.56	46.06	60
2120	5" diameter	↓	32	.500		34.50	26.50	1.81	62.81	80
2130	6" diameter	Q-16	36	.667		43	36.50	1.60	81.10	104
2140	8" diameter		29	.828		68.50	45.50	1.99	115.99	146
2150	10" diameter		24	1		89.50	55	2.40	146.90	183
2160	12" diameter		19	1.263		105	69	3.03	177.03	224
2170	14" diameter, (two rod roll type hanger for 14" diam. and up)		15	1.600		98.50	87.50	3.84	189.84	244
2180	16" diameter, (two rod roll type hanger)		13	1.846		148	101	4.43	253.43	320
2190	18" diameter, (two rod roll type hanger)		11	2.182		141	120	5.25	266.25	340
2200	20" diameter, (two rod roll type hanger)		9	2.667		143	146	6.40	295.40	385
2220	24" diameter, (two rod roll type hanger)	↓	8	3	↓	183	164	7.20	354.20	460
2560	To delete hanger, subtract									
2570	1" diam. to 1-1/2" diam.					15%	34%			
2580	2" diam. to 3-1/2" diam.					9%	21%			
2590	4" diam. to 12" diam.					5%	12%			
2596	14" diam. to 24" diam.					3%	10%			
3250	Flanged, 150 lb. weld neck, on yoke & roll hangers									
3260	sized for covering, 10' O.C.									
3290	Black, 1" diameter	Q-15	70	.229	L.F.	9.90	12.10	.83	22.83	30
3300	1-1/4" diameter		64	.250		11.10	13.20	.90	25.20	33
3310	1-1/2" diameter		58	.276		11.65	14.55	1	27.20	36
3320	2" diameter		45	.356		13.90	18.80	1.28	33.98	45
3330	2-1/2" diameter		36	.444		19.50	23.50	1.60	44.60	59
3340	3" diameter		32	.500		23	26.50	1.81	51.31	67
3350	3-1/2" diameter		29	.552		28	29	1.99	58.99	77
3360	4" diameter		26	.615		31	32.50	2.22	65.72	85.50
3370	5" diameter	↓	21	.762		49.50	40.50	2.75	92.75	119
3380	6" diameter	Q-16	25	.960		59	52.50	2.30	113.80	147
3390	8" diameter		19	1.263		96.50	69	3.03	168.53	214
3400	10" diameter		16	1.500		138	82	3.60	223.60	279
3410	12" diameter	↓	14	1.714		165	94	4.11	263.11	330
3470	For 300 lb. flanges, add					63%				
3480	For 600 lb. flanges, add				↓	310%				
3960	To delete flanges & hanger, subtract									
3970	1" diam. to 2" diam.					76%	65%			
3980	2-1/2" diam. to 4" diam.					62%	59%			
3990	5" diam. to 12" diam.					60%	46%			
4750	Schedule 80, threaded, with couplings, and clevis hanger assemblies									
4760	sized for covering, 10' O.C.									
4790	Black, 1/4" diameter	1 Plum	54	.148	L.F.	10.95	8.70		19.65	25
4800	3/8" diameter		53	.151		13.90	8.85		22.75	28.50
4810	1/2" diameter	↓	52	.154	↓	4.77	9.05		13.82	18.90

22 11 13 – Facility Water Distribution Piping

22 11 13.44 Pipe, Steel	Crew	Daily Output	Labor-Hours	Unit	Material	2015 Bare Costs Labor	Equipment	Total	Total Incl O&P
4820	3/4" diameter	1 Plum	50	.160	L.F.	5.45	9.40		14.85	20
4830	1" diameter	↓	45	.178		7.20	10.45		17.65	23.50
4840	1-1/4" diameter	Q-1	75	.213		9.45	11.25		20.70	27.50
4850	1-1/2" diameter		69	.232		10.65	12.25		22.90	30.50
4860	2" diameter		56	.286		14.10	15.10		29.20	38.50
4870	2-1/2" diameter		44	.364		21	19.20		40.20	52
4880	3" diameter		38	.421		27	22		49	63
4890	3-1/2" diameter		35	.457		33	24		57	72.50
4900	4" diameter	↓	32	.500	↓	38.50	26.50		65	82.50
5430	To delete coupling & hanger, subtract									
5440	1/4" diam. to 1/2" diam.					31%	54%			
5450	3/4" diam. to 1-1/2" diam.					28%	49%			
5460	2" diam. to 4" diam.					21%	40%			
5510	Galvanized, 1/4" diameter	1 Plum	54	.148	L.F.	9.90	8.70		18.60	24
5520	3/8" diameter		53	.151		10.85	8.85		19.70	25.50
5530	1/2" diameter		52	.154		5.30	9.05		14.35	19.50
5540	3/4" diameter		50	.160		6.40	9.40		15.80	21.50
5550	1" diameter	↓	45	.178		8.50	10.45		18.95	25
5560	1-1/4" diameter	Q-1	75	.213		11.35	11.25		22.60	29.50
5570	1-1/2" diameter		69	.232		13.15	12.25		25.40	33
5580	2" diameter		56	.286		17.45	15.10		32.55	42
5590	2-1/2" diameter		44	.364		26	19.20		45.20	57.50
5600	3" diameter		38	.421		34	22		56	70.50
5610	3-1/2" diameter		35	.457		53	24		77	95
5620	4" diameter	↓	32	.500	↓	40	26.50		66.50	84
5930	To delete coupling & hanger, subtract									
5940	1/4" diam. to 1/2" diam.					31%	54%			
5950	3/4" diam. to 1-1/2" diam.					28%	49%			
5960	2" diam. to 4" diam.					21%	40%			
6000	Welded, on yoke & roller hangers									
6010	sized for covering, 10' O.C.									
6040	Black, 1" diameter	Q-15	85	.188	L.F.	7.35	9.95	.68	17.98	24
6050	1-1/4" diameter		79	.203		9.50	10.70	.73	20.93	27.50
6060	1-1/2" diameter		72	.222		10.75	11.75	.80	23.30	30.50
6070	2" diameter		57	.281		13.55	14.85	1.01	29.41	38.50
6080	2-1/2" diameter		44	.364		20	19.20	1.31	40.51	52.50
6090	3" diameter		40	.400		25.50	21	1.44	47.94	61.50
6100	3-1/2" diameter		34	.471		31	25	1.70	57.70	73.50
6110	4" diameter		33	.485		36	25.50	1.75	63.25	80
6120	5" diameter, A-106B	↓	26	.615		51.50	32.50	2.22	86.22	108
6130	6" diameter, A-106B	Q-16	30	.800		70.50	44	1.92	116.42	146
6140	8" diameter, A-106B		25	.960		106	52.50	2.30	160.80	199
6150	10" diameter, A-106B		20	1.200		172	65.50	2.88	240.38	292
6160	12" diameter, A-106B	↓	15	1.600	↓	385	87.50	3.84	476.34	560
6540	To delete hanger, subtract									
6550	1" diam. to 1-1/2" diam.					30%	14%			
6560	2" diam. to 3" diam.					23%	9%			
6570	3-1/2" diam. to 5" diam.					12%	6%			
6580	6" diam. to 12" diam.					10%	4%			
7250	Flanged, 300 lb. weld neck, on yoke & roll hangers									
7260	sized for covering, 10' O.C.									
7290	Black, 1" diameter	Q-15	66	.242	L.F.	13.75	12.80	.88	27.43	35.50
7300	1-1/4" diameter	↓	61	.262	↓	15.95	13.85	.95	30.75	39.50

22 11 13 – Facility Water Distribution Piping

22 11 13.44 Pipe, Steel		Crew	Daily Output	Labor-Hours	Unit	Material	2015 Bare Costs Labor	Equipment	Total	Total Incl O&P
7310	1-1/2" diameter	Q-15	54	.296	L.F.	17.20	15.65	1.07	33.92	43.50
7320	2" diameter		42	.381		22	20	1.38	43.38	56
7330	2-1/2" diameter		33	.485		29.50	25.50	1.75	56.75	73
7340	3" diameter		29	.552		35	29	1.99	65.99	84.50
7350	3-1/2" diameter		24	.667		47	35	2.41	84.41	107
7360	4" diameter		23	.696		52	37	2.51	91.51	115
7370	5" diameter		19	.842		75.50	44.50	3.04	123.04	153
7380	6" diameter	Q-16	23	1.043		95	57	2.50	154.50	193
7390	8" diameter		17	1.412		147	77.50	3.39	227.89	283
7400	10" diameter		14	1.714		247	94	4.11	345.11	420
7410	12" diameter		12	2		480	110	4.80	594.80	700
7470	For 600 lb. flanges, add					100%				
7940	To delete flanges & hanger, subtract									
7950	1" diam. to 1-1/2" diam.					75%	66%			
7960	2" diam. to 3" diam.					62%	60%			
7970	3-1/2" diam. to 5" diam.					54%	66%			
7980	6" diam. to 12" diam.					55%	62%			
9000	Threading pipe labor, one end, all schedules through 80									
9010	1/4" through 3/4" pipe size	1 Plum	80	.100	Ea.		5.85		5.85	8.85
9020	1" through 2" pipe size		73	.110			6.45		6.45	9.70
9030	2-1/2" pipe size		53	.151			8.85		8.85	13.40
9040	3" pipe size		50	.160			9.40		9.40	14.20
9050	3-1/2" pipe size	Q-1	89	.180			9.50		9.50	14.35
9060	4" pipe size		73	.219			11.60		11.60	17.50
9070	5" pipe size		53	.302			15.95		15.95	24
9080	6" pipe size		46	.348			18.40		18.40	28
9090	8" pipe size		29	.552			29		29	44
9100	10" pipe size		21	.762			40.50		40.50	61
9110	12" pipe size		13	1.231			65		65	98
9120	Cutting pipe labor, one cut									
9124	Shop fabrication, machine cut									
9126	Schedule 40, straight pipe									
9128	2" pipe size or less	1 Stpi	62	.129	Ea.		7.70		7.70	11.65
9130	2-1/2" pipe size		56	.143			8.55		8.55	12.90
9132	3" pipe size		42	.190			11.40		11.40	17.20
9134	4" pipe size		31	.258			15.40		15.40	23.50
9136	5" pipe size		26	.308			18.40		18.40	28
9138	6" pipe size		19	.421			25		25	38
9140	8" pipe size		14	.571			34		34	51.50
9142	10" pipe size		10	.800			48		48	72
9144	12" pipe size		7	1.143			68.50		68.50	103
9146	14" pipe size	Q-5	10.50	1.524			82		82	124
9148	16" pipe size		8.60	1.860			100		100	151
9150	18" pipe size		7	2.286			123		123	186
9152	20" pipe size		5.80	2.759			148		148	224
9154	24" pipe size		4	4			215		215	325
9160	Schedule 80, straight pipe									
9164	2" pipe size or less	1 Stpi	42	.190	Ea.		11.40		11.40	17.20
9166	2-1/2" pipe size		37	.216			12.90		12.90	19.50
9168	3" pipe size		31	.258			15.40		15.40	23.50
9170	4" pipe size		23	.348			21		21	31.50
9172	5" pipe size		18	.444			26.50		26.50	40
9174	6" pipe size		14.60	.548			32.50		32.50	49.50

22 11 13.44 Pipe, Steel	Crew	Daily Output	Labor-Hours	Unit	Material	2015 Bare Costs		Total	Total Incl O&P
						Labor	Equipment		
9176	8" pipe size	1 Stpi	10	.800	Ea.		48		48	72
9178	10" pipe size	Q-5	14	1.143			61.50		61.50	93
9180	12" pipe size	"	10	1.600	↓		86		86	130
9200	Welding labor per joint									
9210	Schedule 40,									
9230	1/2" pipe size	Q-15	32	.500	Ea.		26.50	1.81	28.31	42
9240	3/4" pipe size		27	.593			31.50	2.14	33.64	50
9250	1" pipe size		23	.696			37	2.51	39.51	58.50
9260	1-1/4" pipe size		20	.800			42.50	2.89	45.39	67
9270	1-1/2" pipe size		19	.842			44.50	3.04	47.54	70.50
9280	2" pipe size		16	1			53	3.61	56.61	84
9290	2-1/2" pipe size		13	1.231			65	4.44	69.44	103
9300	3" pipe size		12	1.333			70.50	4.81	75.31	111
9310	4" pipe size		10	1.600			84.50	5.80	90.30	134
9320	5" pipe size		9	1.778			94	6.40	100.40	149
9330	6" pipe size		8	2			106	7.20	113.20	168
9340	8" pipe size		5	3.200			169	11.55	180.55	268
9350	10" pipe size		4	4			211	14.45	225.45	335
9360	12" pipe size		3	5.333			282	19.25	301.25	445
9370	14" pipe size		2.60	6.154			325	22	347	515
9380	16" pipe size		2.20	7.273			385	26.50	411.50	610
9390	18" pipe size		2	8			425	29	454	670
9400	20" pipe size		1.80	8.889			470	32	502	745
9410	22" pipe size		1.70	9.412			495	34	529	790
9420	24" pipe size	↓	1.50	10.667	↓		565	38.50	603.50	895
9450	Schedule 80,									
9460	1/2" pipe size	Q-15	27	.593	Ea.		31.50	2.14	33.64	50
9470	3/4" pipe size		23	.696			37	2.51	39.51	58.50
9480	1" pipe size		20	.800			42.50	2.89	45.39	67
9490	1-1/4" pipe size		19	.842			44.50	3.04	47.54	70.50
9500	1-1/2" pipe size		18	.889			47	3.21	50.21	74.50
9510	2" pipe size		15	1.067			56.50	3.85	60.35	89
9520	2-1/2" pipe size		12	1.333			70.50	4.81	75.31	111
9530	3" pipe size		11	1.455			77	5.25	82.25	122
9540	4" pipe size		8	2			106	7.20	113.20	168
9550	5" pipe size		6	2.667			141	9.65	150.65	224
9560	6" pipe size		5	3.200			169	11.55	180.55	268
9570	8" pipe size		4	4			211	14.45	225.45	335
9580	10" pipe size		3	5.333			282	19.25	301.25	445
9590	12" pipe size	↓	2	8			425	29	454	670
9600	14" pipe size	Q-16	2.60	9.231			505	22	527	790
9610	16" pipe size		2.30	10.435			570	25	595	895
9620	18" pipe size		2	12			655	29	684	1,025
9630	20" pipe size		1.80	13.333			730	32	762	1,125
9640	22" pipe size		1.60	15			820	36	856	1,300
9650	24" pipe size	↓	1.50	16	↓		875	38.50	913.50	1,375

22 11 13 – Facility Water Distribution Piping

22 11 13.45 Pipe Fittings, Steel, Threaded	Crew	Daily Output	Labor-Hours	Unit	Material	2015 Bare Costs Labor	Equipment	Total	Total Incl O&P
0010 **PIPE FITTINGS, STEEL, THREADED**									
0020 Cast Iron									
0040 Standard weight, black									
0060 90° Elbow, straight									
0070 1/4"	1 Plum	16	.500	Ea.	9.70	29.50		39.20	55
0080 3/8"		16	.500		14.05	29.50		43.55	60
0090 1/2"		15	.533		6.15	31.50		37.65	54.50
0100 3/4"		14	.571		6.40	33.50		39.90	57.50
0110 1"		13	.615		7.60	36		43.60	63
0120 1-1/4"	Q-1	22	.727		10.75	38.50		49.25	70
0130 1-1/2"		20	.800		14.90	42.50		57.40	80.50
0140 2"		18	.889		23.50	47		70.50	96.50
0150 2-1/2"		14	1.143		56	60.50		116.50	153
0160 3"		10	1.600		91.50	84.50		176	229
0170 3-1/2"		8	2		248	106		354	435
0180 4"		6	2.667		169	141		310	400
0250 45° Elbow, straight									
0260 1/4"	1 Plum	16	.500	Ea.	11.95	29.50		41.45	57.50
0270 3/8"		16	.500		12.85	29.50		42.35	58.50
0280 1/2"		15	.533		9.35	31.50		40.85	58
0300 3/4"		14	.571		9.50	33.50		43	61
0320 1"		13	.615		11.10	36		47.10	66.50
0330 1-1/4"	Q-1	22	.727		14.95	38.50		53.45	74.50
0340 1-1/2"		20	.800		24.50	42.50		67	91
0350 2"		18	.889		28.50	47		75.50	103
0360 2-1/2"		14	1.143		74	60.50		134.50	173
0370 3"		10	1.600		118	84.50		202.50	257
0380 3-1/2"		8	2		271	106		377	460
0400 4"		6	2.667		244	141		385	480
0500 Tee, straight									
0510 1/4"	1 Plum	10	.800	Ea.	15.20	47		62.20	87.50
0520 3/8"		10	.800		14.75	47		61.75	87.50
0530 1/2"		9	.889		9.60	52		61.60	89.50
0540 3/4"		9	.889		11.20	52		63.20	91.50
0550 1"		8	1		9.95	58.50		68.45	99.50
0560 1-1/4"	Q-1	14	1.143		18.15	60.50		78.65	111
0570 1-1/2"		13	1.231		23.50	65		88.50	124
0580 2"		11	1.455		33	77		110	152
0590 2-1/2"		9	1.778		85.50	94		179.50	236
0600 3"		6	2.667		131	141		272	355
0610 3-1/2"		5	3.200		264	169		433	545
0620 4"		4	4		264	211		475	610
0660 Tee, reducing, run or outlet									
0661 1/2"	1 Plum	9	.889	Ea.	23.50	52		75.50	105
0662 3/4"		9	.889		20.50	52		72.50	102
0663 1"		8	1		23.50	58.50		82	114
0664 1-1/4"	Q-1	14	1.143		32.50	60.50		93	127
0665 1-1/2"		13	1.231		38.50	65		103.50	141
0666 2"		11	1.455		60	77		137	182
0667 2-1/2"		9	1.778		112	94		206	265
0668 3"		6	2.667		169	141		310	400
0669 3-1/2"		5	3.200		400	169		569	695

For customer support on your Plumbing Cost Data, call 877.709.6968.

165

22 11 13.45 Pipe Fittings, Steel, Threaded		Crew	Daily Output	Labor-Hours	Unit	Material	2015 Bare Costs Labor	2015 Bare Costs Equipment	Total	Total Incl O&P
0670	4"	Q-1	4	4	Ea.	445	211		656	810
0674	Reducer, concentric									
0675	3/4"	1 Plum	18	.444	Ea.	17.20	26		43.20	58.50
0676	1"	"	15	.533		11.65	31.50		43.15	60.50
0677	1-1/4"	Q-1	26	.615		34.50	32.50		67	87
0678	1-1/2"		24	.667		54.50	35		89.50	113
0679	2"		21	.762		62	40.50		102.50	130
0680	2-1/2"		18	.889		88.50	47		135.50	168
0681	3"		14	1.143		139	60.50		199.50	244
0682	3-1/2"		12	1.333		237	70.50		307.50	365
0683	4"		10	1.600		258	84.50		342.50	410
0687	Reducer, eccentric									
0688	3/4"	1 Plum	16	.500	Ea.	36	29.50		65.50	84.50
0689	1"	"	14	.571		39	33.50		72.50	93.50
0690	1-1/4"	Q-1	25	.640		62.50	34		96.50	120
0691	1-1/2"		22	.727		82.50	38.50		121	149
0692	2"		20	.800		119	42.50		161.50	195
0693	2-1/2"		16	1		164	53		217	260
0694	3"		12	1.333		258	70.50		328.50	390
0695	3-1/2"		10	1.600		360	84.50		444.50	530
0696	4"		9	1.778		465	94		559	655
0700	Standard weight, galvanized cast iron									
0720	90° Elbow, straight									
0730	1/4"	1 Plum	16	.500	Ea.	15.95	29.50		45.45	62
0740	3/8"		16	.500		15.95	29.50		45.45	62
0750	1/2"		15	.533		18.20	31.50		49.70	67.50
0760	3/4"		14	.571		17.85	33.50		51.35	70
0770	1"		13	.615		20.50	36		56.50	77
0780	1-1/4"	Q-1	22	.727		32	38.50		70.50	93.50
0790	1-1/2"		20	.800		44	42.50		86.50	113
0800	2"		18	.889		65	47		112	143
0810	2-1/2"		14	1.143		134	60.50		194.50	238
0820	3"		10	1.600		203	84.50		287.50	350
0830	3-1/2"		8	2		325	106		431	515
0840	4"		6	2.667		370	141		511	625
0900	45° Elbow, straight									
0910	1/4"	1 Plum	16	.500	Ea.	18.15	29.50		47.65	64.50
0920	3/8"		16	.500		18.15	29.50		47.65	64.50
0930	1/2"		15	.533		18.15	31.50		49.65	67.50
0940	3/4"		14	.571		21	33.50		54.50	73.50
0950	1"		13	.615		25.50	36		61.50	82.50
0960	1-1/4"	Q-1	22	.727		38	38.50		76.50	100
0970	1-1/2"		20	.800		54	42.50		96.50	124
0980	2"		18	.889		76	47		123	155
0990	2-1/2"		14	1.143		122	60.50		182.50	225
1000	3"		10	1.600		238	84.50		322.50	390
1010	3-1/2"		8	2		435	106		541	640
1020	4"		6	2.667		435	141		576	695
1100	Tee, straight									
1110	1/4"	1 Plum	10	.800	Ea.	16.75	47		63.75	89.50
1120	3/8"		10	.800		19.30	47		66.30	92
1130	1/2"		9	.889		18.60	52		70.60	99.50
1140	3/4"		9	.889		25.50	52		77.50	107

22 11 13 – Facility Water Distribution Piping

22 11 13.45 Pipe Fittings, Steel, Threaded	Crew	Daily Output	Labor-Hours	Unit	Material	2015 Bare Costs Labor	Equipment	Total	Total Incl O&P
1150	1"	1 Plum	8	1	Ea.	28	58.50		86.50	119
1160	1-1/4"	Q-1	14	1.143		49	60.50		109.50	145
1170	1-1/2"		13	1.231		64.50	65		129.50	169
1180	2"		11	1.455		80.50	77		157.50	205
1190	2-1/2"		9	1.778		164	94		258	320
1200	3"		6	2.667		365	141		506	615
1210	3-1/2"		5	3.200		460	169		629	760
1220	4"		4	4		515	211		726	890
1300	Extra heavy weight, black									
1310	Couplings, steel straight									
1320	1/4"	1 Plum	19	.421	Ea.	3.77	24.50		28.27	41.50
1330	3/8"		19	.421		4.10	24.50		28.60	42
1340	1/2"		19	.421		5.55	24.50		30.05	43.50
1350	3/4"		18	.444		5.90	26		31.90	46
1360	1"		15	.533		7.55	31.50		39.05	56
1370	1-1/4"	Q-1	26	.615		12.10	32.50		44.60	62.50
1380	1-1/2"		24	.667		12.10	35		47.10	66.50
1390	2"		21	.762		18.45	40.50		58.95	81.50
1400	2-1/2"		18	.889		27.50	47		74.50	101
1410	3"		14	1.143		32.50	60.50		93	127
1420	3-1/2"		12	1.333		43.50	70.50		114	154
1430	4"		10	1.600		51.50	84.50		136	185
1510	90° Elbow, straight									
1520	1/2"	1 Plum	15	.533	Ea.	31.50	31.50		63	82.50
1530	3/4"		14	.571		32	33.50		65.50	86
1540	1"		13	.615		39	36		75	97.50
1550	1-1/4"	Q-1	22	.727		58	38.50		96.50	122
1560	1-1/2"		20	.800		72	42.50		114.50	143
1580	2"		18	.889		89	47		136	169
1590	2-1/2"		14	1.143		216	60.50		276.50	330
1600	3"		10	1.600		285	84.50		369.50	445
1610	4"		6	2.667		585	141		726	855
1650	45° Elbow, straight									
1660	1/2"	1 Plum	15	.533	Ea.	45	31.50		76.50	97
1670	3/4"		14	.571		43.50	33.50		77	98.50
1680	1"		13	.615		52.50	36		88.50	112
1690	1-1/4"	Q-1	22	.727		86	38.50		124.50	153
1700	1-1/2"		20	.800		95	42.50		137.50	168
1710	2"		18	.889		135	47		182	219
1720	2-1/2"		14	1.143		233	60.50		293.50	345
1800	Tee, straight									
1810	1/2"	1 Plum	9	.889	Ea.	50	52		102	134
1820	3/4"		9	.889		49.50	52		101.50	134
1830	1"		8	1		59.50	58.50		118	154
1840	1-1/4"	Q-1	14	1.143		89	60.50		149.50	189
1850	1-1/2"		13	1.231		115	65		180	224
1860	2"		11	1.455		142	77		219	272
1870	2-1/2"		9	1.778		305	94		399	475
1880	3"		6	2.667		415	141		556	675
1890	4"		4	4		810	211		1,021	1,225
4000	Standard weight, black									
4010	Couplings, steel straight, merchants									
4030	1/4"	1 Plum	19	.421	Ea.	1.01	24.50		25.51	38.50

For customer support on your Plumbing Cost Data, call 877.709.6968.

167

22 11 13.45 Pipe Fittings, Steel, Threaded		Crew	Daily Output	Labor-Hours	Unit	Material	2015 Bare Costs Labor	Equipment	Total	Total Incl O&P
4040	3/8"	1 Plum	19	.421	Ea.	1.22	24.50		25.72	39
4050	1/2"		19	.421		1.30	24.50		25.80	39
4060	3/4"		18	.444		1.64	26		27.64	41.50
4070	1"		15	.533		2.31	31.50		33.81	50
4080	1-1/4"	Q-1	26	.615		2.95	32.50		35.45	52.50
4090	1-1/2"		24	.667		3.73	35		38.73	57
4100	2"		21	.762		5.35	40.50		45.85	67
4110	2-1/2"		18	.889		16.75	47		63.75	89.50
4120	3"		14	1.143		23.50	60.50		84	117
4130	3-1/2"		12	1.333		41.50	70.50		112	152
4140	4"		10	1.600		41.50	84.50		126	174
4166	Plug, 1/4"	1 Plum	38	.211		2.73	12.35		15.08	21.50
4167	3/8"		38	.211		2.73	12.35		15.08	21.50
4168	1/2"		38	.211		2.73	12.35		15.08	21.50
4169	3/4"		32	.250		8.25	14.70		22.95	31
4170	1"		30	.267		8.80	15.65		24.45	33
4171	1-1/4"	Q-1	52	.308		10.10	16.25		26.35	35.50
4172	1-1/2"		48	.333		14.35	17.60		31.95	42.50
4173	2"		42	.381		18.55	20		38.55	51
4176	2-1/2"		36	.444		27.50	23.50		51	65.50
4180	4"		20	.800		47.50	42.50		90	116
4200	Standard weight, galvanized									
4210	Couplings, steel straight, merchants									
4230	1/4"	1 Plum	19	.421	Ea.	1.16	24.50		25.66	39
4240	3/8"		19	.421		1.49	24.50		25.99	39
4250	1/2"		19	.421		1.58	24.50		26.08	39
4260	3/4"		18	.444		1.99	26		27.99	41.50
4270	1"		15	.533		2.78	31.50		34.28	50.50
4280	1-1/4"	Q-1	26	.615		3.55	32.50		36.05	53
4290	1-1/2"		24	.667		4.41	35		39.41	58
4300	2"		21	.762		6.60	40.50		47.10	68.50
4310	2-1/2"		18	.889		20.50	47		67.50	93.50
4320	3"		14	1.143		27.50	60.50		88	121
4330	3-1/2"		12	1.333		48	70.50		118.50	159
4340	4"		10	1.600		48	84.50		132.50	181
4370	Plug, galvanized, square head									
4374	1/2"	1 Plum	38	.211	Ea.	6.60	12.35		18.95	26
4375	3/4"		32	.250		6.60	14.70		21.30	29.50
4376	1"		30	.267		6.60	15.65		22.25	31
4377	1-1/4"	Q-1	52	.308		11	16.25		27.25	36.50
4378	1-1/2"		48	.333		14.90	17.60		32.50	43
4379	2"		42	.381		18.50	20		38.50	51
4380	2-1/2"		36	.444		39	23.50		62.50	78.50
4381	3"		28	.571		50	30		80	101
4382	4"		20	.800		137	42.50		179.50	215
4700	Nipple, black									
4710	1/2" x 4" long	1 Plum	19	.421	Ea.	2.21	24.50		26.71	40
4712	3/4" x 4" long		18	.444		2.67	26		28.67	42.50
4714	1" x 4" long		15	.533		3.70	31.50		35.20	51.50
4716	1-1/4" x 4" long	Q-1	26	.615		4.63	32.50		37.13	54
4718	1-1/2" x 4" long		24	.667		5.45	35		40.45	59
4720	2" x 4" long		21	.762		7.60	40.50		48.10	69.50
4722	2-1/2" x 4" long		18	.889		20.50	47		67.50	94

168

For customer support on your Plumbing Cost Data, call 877.709.6968.

22 11 13.45 Pipe Fittings, Steel, Threaded		Crew	Daily Output	Labor-Hours	Unit	Material	2015 Bare Costs Labor	Equipment	Total	Total Incl O&P
4724	3" x 4" long	Q-1	14	1.143	Ea.	26	60.50		86.50	120
4726	4" x 4" long	↓	10	1.600	↓	35.50	84.50		120	167
4800	Nipple, galvanized									
4810	1/2" x 4" long	1 Plum	19	.421	Ea.	2.71	24.50		27.21	40.50
4812	3/4" x 4" long		18	.444		3.35	26		29.35	43
4814	1" x 4" long	↓	15	.533		4.49	31.50		35.99	52.50
4816	1-1/4" x 4" long	Q-1	26	.615		5.50	32.50		38	55
4818	1-1/2" x 4" long		24	.667		7	35		42	60.50
4820	2" x 4" long		21	.762		8.85	40.50		49.35	71
4822	2-1/2" x 4" long		18	.889		23.50	47		70.50	96.50
4824	3" x 4" long		14	1.143		31	60.50		91.50	125
4826	4" x 4" long	↓	10	1.600	↓	41.50	84.50		126	174
5000	Malleable iron, 150 lb.									
5020	Black									
5040	90° elbow, straight									
5060	1/4"	1 Plum	16	.500	Ea.	4.78	29.50		34.28	50
5070	3/8"		16	.500		4.78	29.50		34.28	50
5080	1/2"		15	.533		3.30	31.50		34.80	51
5090	3/4"		14	.571		3.99	33.50		37.49	55
5100	1"	↓	13	.615		6.95	36		42.95	62
5110	1-1/4"	Q-1	22	.727		11.45	38.50		49.95	70.50
5120	1-1/2"		20	.800		15.05	42.50		57.55	80.50
5130	2"		18	.889		26	47		73	99.50
5140	2-1/2"		14	1.143		58	60.50		118.50	155
5150	3"		10	1.600		84.50	84.50		169	221
5160	3-1/2"		8	2		232	106		338	415
5170	4"	↓	6	2.667	↓	182	141		323	415
5250	45° elbow, straight									
5270	1/4"	1 Plum	16	.500	Ea.	7.20	29.50		36.70	52.50
5280	3/8"		16	.500		7.20	29.50		36.70	52.50
5290	1/2"		15	.533		5.45	31.50		36.95	53.50
5300	3/4"		14	.571		6.75	33.50		40.25	58
5310	1"	↓	13	.615		8.50	36		44.50	64
5320	1-1/4"	Q-1	22	.727		15.05	38.50		53.55	74.50
5330	1-1/2"		20	.800		18.60	42.50		61.10	84.50
5340	2"		18	.889		28	47		75	102
5350	2-1/2"		14	1.143		81.50	60.50		142	181
5360	3"		10	1.600		188	84.50		272.50	335
5370	3-1/2"		8	2		208	106		314	390
5380	4"	↓	6	2.667	↓	208	141		349	440
5450	Tee, straight									
5470	1/4"	1 Plum	10	.800	Ea.	6.95	47		53.95	78.50
5480	3/8"		10	.800		6.95	47		53.95	78.50
5490	1/2"		9	.889		4.42	52		56.42	84
5500	3/4"		9	.889		6.35	52		58.35	86
5510	1"	↓	8	1		10.85	58.50		69.35	100
5520	1-1/4"	Q-1	14	1.143		17.60	60.50		78.10	110
5530	1-1/2"		13	1.231		22	65		87	122
5540	2"		11	1.455		37.50	77		114.50	157
5550	2-1/2"		9	1.778		80.50	94		174.50	231
5560	3"		6	2.667		119	141		260	345
5570	3-1/2"		5	3.200		275	169		444	555
5580	4"	↓	4	4	↓	286	211		497	635

For customer support on your Plumbing Cost Data, call 877.709.6968.

169

22 11 13.45 Pipe Fittings, Steel, Threaded		Crew	Daily Output	Labor-Hours	Unit	Material	2015 Bare Costs Labor	Equipment	Total	Total Incl O&P
5601	Tee, reducing, on outlet									
5602	1/2"	1 Plum	9	.889	Ea.	9.65	52		61.65	89.50
5603	3/4"		9	.889		10.90	52		62.90	91
5604	1"		8	1		17.45	58.50		75.95	108
5605	1-1/4"	Q-1	14	1.143		30.50	60.50		91	125
5606	1-1/2"		13	1.231		34.50	65		99.50	136
5607	2"		11	1.455		48	77		125	169
5608	2-1/2"		9	1.778		121	94		215	275
5609	3"		6	2.667		166	141		307	395
5610	3-1/2"		5	3.200		390	169		559	685
5611	4"		4	4		330	211		541	685
5650	Coupling									
5670	1/4"	1 Plum	19	.421	Ea.	5.95	24.50		30.45	44
5680	3/8"		19	.421		5.95	24.50		30.45	44
5690	1/2"		19	.421		4.58	24.50		29.08	42.50
5700	3/4"		18	.444		5.35	26		31.35	45.50
5710	1"		15	.533		8	31.50		39.50	56.50
5720	1-1/4"	Q-1	26	.615		10.40	32.50		42.90	60.50
5730	1-1/2"		24	.667		14.05	35		49.05	68.50
5740	2"		21	.762		21	40.50		61.50	84
5750	2-1/2"		18	.889		57.50	47		104.50	134
5760	3"		14	1.143		77.50	60.50		138	177
5770	3-1/2"		12	1.333		156	70.50		226.50	278
5780	4"		10	1.600		156	84.50		240.50	300
5840	Reducer, concentric, 1/4"	1 Plum	19	.421		6.25	24.50		30.75	44.50
5850	3/8"		19	.421		8	24.50		32.50	46.50
5860	1/2"		19	.421		6.30	24.50		30.80	44.50
5870	3/4"		16	.500		7.50	29.50		37	53
5880	1"		15	.533		12.60	31.50		44.10	61.50
5890	1-1/4"	Q-1	26	.615		14.10	32.50		46.60	64.50
5900	1-1/2"		24	.667		20	35		55	75.50
5910	2"		21	.762		29	40.50		69.50	93
5911	2-1/2"		18	.889		65	47		112	143
5912	3"		14	1.143		79	60.50		139.50	178
5913	3-1/2"		12	1.333		253	70.50		323.50	385
5914	4"		10	1.600		190	84.50		274.50	335
5981	Bushing, 1/4"	1 Plum	19	.421		1.15	24.50		25.65	39
5982	3/8"		19	.421		6.10	24.50		30.60	44
5983	1/2"		19	.421		6.10	24.50		30.60	44
5984	3/4"		16	.500		6.15	29.50		35.65	51.50
5985	1"		15	.533		8.70	31.50		40.20	57
5986	1-1/4"	Q-1	26	.615		10.85	32.50		43.35	61
5987	1-1/2"		24	.667		9.35	35		44.35	63.50
5988	2"		21	.762		11.65	40.50		52.15	74
5989	Cap, 1/4"	1 Plum	38	.211		4.79	12.35		17.14	24
5991	3/8"		38	.211		5.85	12.35		18.20	25
5992	1/2"		38	.211		3.27	12.35		15.62	22.50
5993	3/4"		32	.250		4.53	14.70		19.23	27
5994	1"		30	.267		5.50	15.65		21.15	29.50
5995	1-1/4"	Q-1	52	.308		7.15	16.25		23.40	32.50
5996	1-1/2"		48	.333		9.95	17.60		27.55	37.50
5997	2"		42	.381		14.50	20		34.50	46.50
6000	For galvanized elbows, tees, and couplings add					20%				

22 11 13 – Facility Water Distribution Piping

22 11 13.45 Pipe Fittings, Steel, Threaded		Crew	Daily Output	Labor-Hours	Unit	Material	2015 Bare Costs Labor	Equipment	Total	Total Incl O&P
6058	For galvanized reducers, caps and bushings add					20%				
6100	90° Elbow, galvanized, 150 lb., reducing									
6110	3/4" x 1/2"	1 Plum	15.40	.519	Ea.	9.50	30.50		40	56.50
6112	1" x 3/4"		14	.571		12.20	33.50		45.70	64
6114	1" x 1/2"		14.50	.552		12.95	32.50		45.45	63.50
6116	1-1/4" x 1"	Q-1	24.20	.661		20	35		55	75
6118	1-1/4" x 3/4"		25.40	.630		24.50	33.50		58	77.50
6120	1-1/4" x 1/2"		26.20	.611		26	32.50		58.50	77
6122	1-1/2" x 1-1/4"		21.60	.741		32.50	39		71.50	95
6124	1-1/2" x 1"		23.50	.681		32.50	36		68.50	90.50
6126	1-1/2" x 3/4"		24.60	.650		32.50	34.50		67	88
6128	2" x 1-1/2"		20.50	.780		38	41		79	105
6130	2" x 1-1/4"		21	.762		44	40.50		84.50	109
6132	2" x 1"		22.80	.702		45.50	37		82.50	106
6134	2" x 3/4"		23.90	.669		46.50	35.50		82	105
6136	2-1/2" x 2"		12.30	1.301		130	68.50		198.50	246
6138	2-1/2" x 1-1/2"		12.50	1.280		144	67.50		211.50	261
6140	3" x 2-1/2"		8.60	1.860		235	98.50		333.50	405
6142	3" x 2"		11.80	1.356		204	71.50		275.50	335
6144	4" x 3"		8.20	1.951		535	103		638	740
6160	90° Elbow, black, 150 lb., reducing									
6170	1" x 3/4"	1 Plum	14	.571	Ea.	8.35	33.50		41.85	59.50
6174	1-1/2" x 1"	Q-1	23.50	.681		19.75	36		55.75	76
6178	1-1/2" x 3/4"		24.60	.650		22.50	34.50		57	77
6182	2" x 1-1/2"		20.50	.780		28.50	41		69.50	94
6186	2" x 1"		22.80	.702		32.50	37		69.50	92
6190	2" x 3/4"		23.90	.669		34.50	35.50		70	91.50
6194	2-1/2" x 2"		12.30	1.301		78.50	68.50		147	191
7000	Union, with brass seat									
7010	1/4"	1 Plum	15	.533	Ea.	25.50	31.50		57	75.50
7020	3/8"		15	.533		16.15	31.50		47.65	65.50
7030	1/2"		14	.571		14.60	33.50		48.10	66.50
7040	3/4"		13	.615		16.80	36		52.80	73
7050	1"		12	.667		22	39		61	83
7060	1-1/4"	Q-1	21	.762		31.50	40.50		72	95.50
7070	1-1/2"		19	.842		39	44.50		83.50	110
7080	2"		17	.941		45.50	49.50		95	125
7090	2-1/2"		13	1.231		136	65		201	247
7100	3"		9	1.778		163	94		257	320
7120	Union, galvanized									
7124	1/2"	1 Plum	14	.571	Ea.	18.65	33.50		52.15	71
7125	3/4"		13	.615		22.50	36		58.50	79
7126	1"		12	.667		29	39		68	91
7127	1-1/4"	Q-1	21	.762		42	40.50		82.50	108
7128	1-1/2"		19	.842		51	44.50		95.50	124
7129	2"		17	.941		59	49.50		108.50	140
7130	2-1/2"		13	1.231		207	65		272	325
7131	3"		9	1.778		286	94		380	455
7500	Malleable iron, 300 lb.									
7520	Black									
7540	90° Elbow, straight, 1/4"	1 Plum	16	.500	Ea.	16.75	29.50		46.25	63
7560	3/8"		16	.500		14.90	29.50		44.40	61
7570	1/2"		15	.533		19.30	31.50		50.80	68.50

For customer support on your Plumbing Cost Data, call 877.709.6968.

171

22 11 13.45 Pipe Fittings, Steel, Threaded		Crew	Daily Output	Labor-Hours	Unit	Material	2015 Bare Costs Labor	Equipment	Total	Total Incl O&P
7580	3/4"	1 Plum	14	.571	Ea.	22	33.50		55.50	74.50
7590	1"		13	.615		28	36		64	85
7600	1-1/4"	Q-1	22	.727		40.50	38.50		79	103
7610	1-1/2"		20	.800		48	42.50		90.50	117
7620	2"		18	.889		69	47		116	147
7630	2-1/2"		14	1.143		179	60.50		239.50	288
7640	3"		10	1.600		205	84.50		289.50	355
7650	4"		6	2.667		480	141		621	745
7700	45° Elbow, straight, 1/4"	1 Plum	16	.500		25	29.50		54.50	72
7720	3/8"		16	.500		28	29.50		57.50	75
7730	1/2"		15	.533		28	31.50		59.50	78
7740	3/4"		14	.571		31.50	33.50		65	85
7750	1"		13	.615		34	36		70	92
7760	1-1/4"	Q-1	22	.727		54.50	38.50		93	118
7770	1-1/2"		20	.800		71.50	42.50		114	143
7780	2"		18	.889		108	47		155	190
7790	2-1/2"		14	1.143		241	60.50		301.50	355
7800	3"		10	1.600		315	84.50		399.50	480
7810	4"		6	2.667		720	141		861	1,000
7850	Tee, straight, 1/4"	1 Plum	10	.800		21	47		68	94
7870	3/8"		10	.800		22.50	47		69.50	95.50
7880	1/2"		9	.889		28.50	52		80.50	111
7890	3/4"		9	.889		30.50	52		82.50	113
7900	1"		8	1		37	58.50		95.50	129
7910	1-1/4"	Q-1	14	1.143		53	60.50		113.50	149
7920	1-1/2"		13	1.231		62	65		127	166
7930	2"		11	1.455		91	77		168	216
7940	2-1/2"		9	1.778		222	94		316	385
7950	3"		6	2.667		315	141		456	560
7960	4"		4	4		900	211		1,111	1,300
8050	Couplings, straight, 1/4"	1 Plum	19	.421		15.10	24.50		39.60	54
8070	3/8"		19	.421		15.10	24.50		39.60	54
8080	1/2"		19	.421		16.85	24.50		41.35	56
8090	3/4"		18	.444		19.40	26		45.40	61
8100	1"		15	.533		22	31.50		53.50	72
8110	1-1/4"	Q-1	26	.615		26.50	32.50		59	78
8120	1-1/2"		24	.667		39.50	35		74.50	96.50
8130	2"		21	.762		57	40.50		97.50	124
8140	2-1/2"		18	.889		105	47		152	186
8150	3"		14	1.143		151	60.50		211.50	257
8160	4"		10	1.600		275	84.50		359.50	430
8200	Galvanized									
8220	90° Elbow, straight, 1/4"	1 Plum	16	.500	Ea.	30	29.50		59.50	77.50
8222	3/8"		16	.500		30.50	29.50		60	78
8224	1/2"		15	.533		36.50	31.50		68	87.50
8226	3/4"		14	.571		41	33.50		74.50	95.50
8228	1"		13	.615		52.50	36		88.50	112
8230	1-1/4"	Q-1	22	.727		82.50	38.50		121	149
8232	1-1/2"		20	.800		89.50	42.50		132	163
8234	2"		18	.889		150	47		197	236
8236	2-1/2"		14	1.143		335	60.50		395.50	455
8238	3"		10	1.600		395	84.50		479.50	565
8240	4"		6	2.667		1,175	141		1,316	1,525

22 11 Facility Water Distribution

22 11 13 – Facility Water Distribution Piping

22 11 13.45 Pipe Fittings, Steel, Threaded	Crew	Daily Output	Labor-Hours	Unit	Material	2015 Bare Costs Labor	Equipment	Total	Total Incl O&P
8280	45° Elbow, straight									
8282	1/2"	1 Plum	15	.533	Ea.	54	31.50		85.50	107
8284	3/4"		14	.571		62.50	33.50		96	120
8286	1"	↓	13	.615		68.50	36		104.50	130
8288	1-1/4"	Q-1	22	.727		106	38.50		144.50	175
8290	1-1/2"		20	.800		137	42.50		179.50	215
8292	2"	↓	18	.889		190	47		237	280
8310	Tee, straight, 1/4"	1 Plum	10	.800		42.50	47		89.50	118
8312	3/8"		10	.800		43.50	47		90.50	119
8314	1/2"		9	.889		54.50	52		106.50	139
8316	3/4"		9	.889		60	52		112	145
8318	1"	↓	8	1		74	58.50		132.50	170
8320	1-1/4"	Q-1	14	1.143		106	60.50		166.50	207
8322	1-1/2"		13	1.231		110	65		175	219
8324	2"		11	1.455		175	77		252	310
8326	2-1/2"		9	1.778		555	94		649	750
8328	3"		6	2.667		630	141		771	910
8330	4"	↓	4	4		1,625	211		1,836	2,100
8380	Couplings, straight, 1/4"	1 Plum	19	.421		19.65	24.50		44.15	59
8382	3/8"		19	.421		28.50	24.50		53	68.50
8384	1/2"		19	.421		29	24.50		53.50	69.50
8386	3/4"		18	.444		33	26		59	75.50
8388	1"	↓	15	.533		43.50	31.50		75	95.50
8390	1-1/4"	Q-1	26	.615		56.50	32.50		89	112
8392	1-1/2"		24	.667		80	35		115	141
8394	2"		21	.762		96.50	40.50		137	167
8396	2-1/2"		18	.889		254	47		301	350
8398	3"		14	1.143		305	60.50		365.50	425
8399	4"	↓	10	1.600	↓	465	84.50		549.50	640
8529	Black									
8530	Reducer, concentric, 1/4"	1 Plum	19	.421	Ea.	20	24.50		44.50	59.50
8531	3/8"		19	.421		20	24.50		44.50	59.50
8532	1/2"		17	.471		24	27.50		51.50	68
8533	3/4"		16	.500		34.50	29.50		64	82
8534	1"	↓	15	.533		41	31.50		72.50	92.50
8535	1-1/4"	Q-1	26	.615		61.50	32.50		94	117
8536	1-1/2"		24	.667		63.50	35		98.50	123
8537	2"	↓	21	.762		92	40.50		132.50	162
8550	Cap, 1/4"	1 Plum	38	.211		15.50	12.35		27.85	35.50
8551	3/8"		38	.211		15.50	12.35		27.85	35.50
8552	1/2"		34	.235		15.35	13.80		29.15	38
8553	3/4"		32	.250		19.40	14.70		34.10	43.50
8554	1"	↓	30	.267		25.50	15.65		41.15	51.50
8555	1-1/4"	Q-1	52	.308		28	16.25		44.25	55.50
8556	1-1/2"		48	.333		41	17.60		58.60	72
8557	2"	↓	42	.381		57	20		77	93.50
8570	Plug, 1/4"	1 Plum	38	.211		3.47	12.35		15.82	22.50
8571	3/8"		38	.211		3.64	12.35		15.99	22.50
8572	1/2"		34	.235		3.82	13.80		17.62	25
8573	3/4"		32	.250		4.76	14.70		19.46	27.50
8574	1"	↓	30	.267		7.30	15.65		22.95	31.50
8575	1-1/4"	Q-1	52	.308		15	16.25		31.25	41
8576	1-1/2"		48	.333		17.15	17.60		34.75	45.50

For customer support on your Plumbing Cost Data, call 877.709.6968.

173

22 11 13.45 Pipe Fittings, Steel, Threaded		Crew	Daily Output	Labor-Hours	Unit	Material	2015 Bare Costs Labor	Equipment	Total	Total Incl O&P
8577	2"	Q-1	42	.381	Ea.	27	20		47	60.50
9500	Union with brass seat, 1/4"	1 Plum	15	.533		34	31.50		65.50	85
9530	3/8"		15	.533		34.50	31.50		66	85.50
9540	1/2"		14	.571		28	33.50		61.50	81.50
9550	3/4"		13	.615		31	36		67	88.50
9560	1"		12	.667		40.50	39		79.50	104
9570	1-1/4"	Q-1	21	.762		66	40.50		106.50	134
9580	1-1/2"		19	.842		69	44.50		113.50	143
9590	2"		17	.941		86	49.50		135.50	170
9600	2-1/2"		13	1.231		276	65		341	405
9610	3"		9	1.778		355	94		449	535
9620	4"		5	3.200		1,125	169		1,294	1,500
9630	Union, all iron, 1/4"	1 Plum	15	.533		53	31.50		84.50	106
9650	3/8"		15	.533		53	31.50		84.50	106
9660	1/2"		14	.571		58.50	33.50		92	115
9670	3/4"		13	.615		60	36		96	121
9680	1"		12	.667		58	39		97	123
9690	1-1/4"	Q-1	21	.762		113	40.50		153.50	185
9700	1-1/2"		19	.842		138	44.50		182.50	219
9710	2"		17	.941		174	49.50		223.50	267
9720	2-1/2"		13	1.231		310	65		375	440
9730	3"		9	1.778		585	94		679	785
9750	For galvanized unions, add					15%				
9757	Forged steel, 3000 lb.									
9758	Black									
9760	90° Elbow, 1/4"	1 Plum	16	.500	Ea.	20.50	29.50		50	67
9761	3/8"		16	.500		20.50	29.50		50	67
9762	1/2"		15	.533		15.70	31.50		47.20	65
9763	3/4"		14	.571		19.70	33.50		53.20	72
9764	1"		13	.615		29.50	36		65.50	86.50
9765	1-1/4"	Q-1	22	.727		55.50	38.50		94	119
9766	1-1/2"		20	.800		72	42.50		114.50	143
9767	2"		18	.889		87	47		134	167
9780	45° Elbow 1/4"	1 Plum	16	.500		26	29.50		55.50	73
9781	3/8"		16	.500		26	29.50		55.50	73
9782	1/2"		15	.533		25.50	31.50		57	75.50
9783	3/4"		14	.571		29.50	33.50		63	83
9784	1"		13	.615		40.50	36		76.50	99
9785	1-1/4"	Q-1	22	.727		56	38.50		94.50	120
9786	1-1/2"		20	.800		80	42.50		122.50	152
9787	2"		18	.889		110	47		157	193
9800	Tee, 1/4"	1 Plum	10	.800		25	47		72	98.50
9801	3/8"		10	.800		25	47		72	98.50
9802	1/2"		9	.889		22.50	52		74.50	104
9803	3/4"		9	.889		30.50	52		82.50	113
9804	1"		8	1		40	58.50		98.50	133
9805	1-1/4"	Q-1	14	1.143		77.50	60.50		138	176
9806	1-1/2"		13	1.231		90.50	65		155.50	198
9807	2"		11	1.455		116	77		193	244
9820	Reducer, concentric, 1/4"	1 Plum	19	.421		12.55	24.50		37.05	51.50
9821	3/8"		19	.421		13.05	24.50		37.55	52
9822	1/2"		17	.471		13.05	27.50		40.55	56
9823	3/4"		16	.500		15.50	29.50		45	61.50

22 11 13 – Facility Water Distribution Piping

22 11 13.45 Pipe Fittings, Steel, Threaded

		Crew	Daily Output	Labor-Hours	Unit	Material	2015 Bare Costs Labor	Equipment	Total	Total Incl O&P
9824	1"	1 Plum	15	.533	Ea.	20	31.50		51.50	69.50
9825	1-1/4"	Q-1	26	.615		35.50	32.50		68	88
9826	1-1/2"		24	.667		38.50	35		73.50	95
9827	2"	▼	21	.762		55.50	40.50		96	122
9840	Cap, 1/4"	1 Plum	38	.211		7.30	12.35		19.65	26.50
9841	3/8"		38	.211		7.70	12.35		20.05	27
9842	1/2"		34	.235		7.50	13.80		21.30	29.50
9843	3/4"		32	.250		10.45	14.70		25.15	33.50
9844	1"		30	.267		16	15.65		31.65	41
9845	1-1/4"	Q-1	52	.308		26	16.25		42.25	53
9846	1-1/2"		48	.333		31	17.60		48.60	60.50
9847	2"	▼	42	.381		44.50	20		64.50	79.50
9860	Plug, 1/4"	1 Plum	38	.211		3.47	12.35		15.82	22.50
9861	3/8"		38	.211		3.64	12.35		15.99	22.50
9862	1/2"		34	.235		3.82	13.80		17.62	25
9863	3/4"		32	.250		4.76	14.70		19.46	27.50
9864	1"	▼	30	.267		7.30	15.65		22.95	31.50
9865	1-1/4"	Q-1	52	.308		15	16.25		31.25	41
9866	1-1/2"		48	.333		17.15	17.60		34.75	45.50
9867	2"	▼	42	.381		27	20		47	60.50
9880	Union, bronze seat, 1/4"	1 Plum	15	.533		57	31.50		88.50	110
9881	3/8"		15	.533		57	31.50		88.50	110
9882	1/2"		14	.571		54.50	33.50		88	111
9883	3/4"		13	.615		73	36		109	135
9884	1"	▼	12	.667		82.50	39		121.50	150
9885	1-1/4"	Q-1	21	.762		160	40.50		200.50	237
9886	1-1/2"		19	.842		165	44.50		209.50	248
9887	2"	▼	17	.941		195	49.50		244.50	290
9900	Coupling, 1/4"	1 Plum	19	.421		8	24.50		32.50	46.50
9901	3/8"		19	.421		8	24.50		32.50	46.50
9902	1/2"		17	.471		6.55	27.50		34.05	48.50
9903	3/4"		16	.500		8.55	29.50		38.05	54
9904	1"	▼	15	.533		14.85	31.50		46.35	64
9905	1-1/4"	Q-1	26	.615		25	32.50		57.50	76.50
9906	1-1/2"		24	.667		32	35		67	88
9907	2"	▼	21	.762	▼	40	40.50		80.50	105

22 11 13.47 Pipe Fittings, Steel

		Crew	Daily Output	Labor-Hours	Unit	Material	2015 Bare Costs Labor	Equipment	Total	Total Incl O&P
0010	**PIPE FITTINGS, STEEL**, Flanged, Welded & Special									
0020	Flanged joints, C.I., standard weight, black. One gasket & bolt									
0040	set, mat'l only, required at each joint, not included (see line 0620)									
0060	90° Elbow, straight, 1-1/2" pipe size	Q-1	14	1.143	Ea.	550	60.50		610.50	695
0080	2" pipe size		13	1.231		325	65		390	455
0090	2-1/2" pipe size		12	1.333		350	70.50		420.50	490
0100	3" pipe size		11	1.455		292	77		369	435
0110	4" pipe size		8	2		355	106		461	550
0120	5" pipe size	▼	7	2.286		855	121		976	1,125
0130	6" pipe size	Q-2	9	2.667		565	146		711	840
0140	8" pipe size		8	3		975	164		1,139	1,325
0150	10" pipe size		7	3.429		2,150	188		2,338	2,625
0160	12" pipe size	▼	6	4	▼	4,300	219		4,519	5,075
0171	90° Elbow, reducing									
0172	2-1/2" by 2" pipe size	Q-1	12	1.333	Ea.	1,075	70.50		1,145.50	1,275

For customer support on your Plumbing Cost Data, call 877.709.6968.

175

22 11 13.47 Pipe Fittings, Steel	Crew	Daily Output	Labor-Hours	Unit	Material	2015 Bare Costs Labor	Equipment	Total	Total Incl O&P
0173	3" by 2-1/2" pipe size	Q-1	11	1.455	Ea.	990	77		1,067	1,225
0174	4" by 3" pipe size		8	2		800	106		906	1,050
0175	5" by 3" pipe size		7	2.286		1,725	121		1,846	2,075
0176	6" by 4" pipe size	Q-2	9	2.667		1,075	146		1,221	1,400
0177	8" by 6" pipe size		8	3		1,450	164		1,614	1,850
0178	10" by 8" pipe size		7	3.429		2,750	188		2,938	3,300
0179	12" by 10" pipe size		6	4		5,300	219		5,519	6,150
0200	45° Elbow, straight, 1-1/2" pipe size	Q-1	14	1.143		665	60.50		725.50	820
0220	2" pipe size		13	1.231		470	65		535	615
0230	2-1/2" pipe size		12	1.333		500	70.50		570.50	655
0240	3" pipe size		11	1.455		485	77		562	650
0250	4" pipe size		8	2		545	106		651	760
0260	5" pipe size		7	2.286		1,300	121		1,421	1,625
0270	6" pipe size	Q-2	9	2.667		895	146		1,041	1,200
0280	8" pipe size		8	3		1,300	164		1,464	1,700
0290	10" pipe size		7	3.429		2,775	188		2,963	3,325
0300	12" pipe size		6	4		4,225	219		4,444	4,975
0310	Cross, straight									
0311	2-1/2" pipe size	Q-1	6	2.667	Ea.	1,025	141		1,166	1,350
0312	3" pipe size		5	3.200		1,075	169		1,244	1,425
0313	4" pipe size		4	4		1,400	211		1,611	1,875
0314	5" pipe size		3	5.333		2,950	282		3,232	3,675
0315	6" pipe size	Q-2	5	4.800		2,950	263		3,213	3,650
0316	8" pipe size		4	6		4,750	330		5,080	5,725
0317	10" pipe size		3	8		5,650	440		6,090	6,850
0318	12" pipe size		2	12		8,600	655		9,255	10,500
0350	Tee, straight, 1-1/2" pipe size	Q-1	10	1.600		635	84.50		719.50	825
0370	2" pipe size		9	1.778		355	94		449	530
0380	2-1/2" pipe size		8	2		515	106		621	725
0390	3" pipe size		7	2.286		360	121		481	580
0400	4" pipe size		5	3.200		550	169		719	860
0410	5" pipe size		4	4		1,475	211		1,686	1,950
0420	6" pipe size	Q-2	6	4		800	219		1,019	1,200
0430	8" pipe size		5	4.800		1,375	263		1,638	1,900
0440	10" pipe size		4	6		3,675	330		4,005	4,525
0450	12" pipe size		3	8		5,775	440		6,215	7,000
0459	Tee, reducing on outlet									
0460	2-1/2" by 2" pipe size	Q-1	8	2	Ea.	1,075	106		1,181	1,325
0461	3" by 2-1/2" pipe size		7	2.286		1,075	121		1,196	1,350
0462	4" by 3" pipe size		5	3.200		1,125	169		1,294	1,500
0463	5" by 4" pipe size		4	4		2,475	211		2,686	3,050
0464	6" by 4" pipe size	Q-2	6	4		1,100	219		1,319	1,525
0465	8" by 6" pipe size		5	4.800		1,700	263		1,963	2,275
0466	10" by 8" pipe size		4	6		4,600	330		4,930	5,550
0467	12" by 10" pipe size		3	8		6,725	440		7,165	8,050
0476	Reducer, concentric									
0477	3" by 2-1/2"	Q-1	12	1.333	Ea.	650	70.50		720.50	820
0478	4" by 3"		9	1.778		730	94		824	940
0479	5" by 4"		8	2		1,125	106		1,231	1,400
0480	6" by 4"	Q-2	10	2.400		975	131		1,106	1,275
0481	8" by 6"		9	2.667		1,225	146		1,371	1,575
0482	10" by 8"		8	3		2,225	164		2,389	2,700
0483	12" by 10"		7	3.429		4,375	188		4,563	5,075

22 11 13 – Facility Water Distribution Piping

22 11 13.47 Pipe Fittings, Steel	Crew	Daily Output	Labor-Hours	Unit	Material	2015 Bare Costs Labor	Equipment	Total	Total Incl O&P
0492 Reducer, eccentric									
0493 4" by 3"	Q-1	8	2	Ea.	1,200	106		1,306	1,475
0494 5" by 4"	"	7	2.286		1,825	121		1,946	2,175
0495 6" by 4"	Q-2	9	2.667		1,125	146		1,271	1,450
0496 8" by 6"		8	3		1,450	164		1,614	1,850
0497 10" by 8"		7	3.429		3,600	188		3,788	4,250
0498 12" by 10"		6	4		4,875	219		5,094	5,700
0500 For galvanized elbows and tees, add					100%				
0520 For extra heavy weight elbows and tees, add					140%				
0620 Gasket and bolt set, 150#, 1/2" pipe size	1 Plum	20	.400		2.75	23.50		26.25	38.50
0622 3/4" pipe size		19	.421		2.90	24.50		27.40	40.50
0624 1" pipe size		18	.444		2.96	26		28.96	43
0626 1-1/4" pipe size		17	.471		4.02	27.50		31.52	46
0628 1-1/2" pipe size		15	.533		4.16	31.50		35.66	52
0630 2" pipe size		13	.615		5.95	36		41.95	61
0640 2-1/2" pipe size		12	.667		6.15	39		45.15	66
0650 3" pipe size		11	.727		6.35	42.50		48.85	71.50
0660 3-1/2" pipe size		9	.889		13.05	52		65.05	93.50
0670 4" pipe size		8	1		14.50	58.50		73	104
0680 5" pipe size		7	1.143		18.85	67		85.85	122
0690 6" pipe size		6	1.333		21.50	78.50		100	142
0700 8" pipe size		5	1.600		21.50	94		115.50	166
0710 10" pipe size		4.50	1.778		45	104		149	208
0720 12" pipe size		4.20	1.905		41.50	112		153.50	215
0730 14" pipe size		4	2		40.50	117		157.50	222
0740 16" pipe size		3	2.667		46	157		203	287
0750 18" pipe size		2.70	2.963		88.50	174		262.50	360
0760 20" pipe size		2.30	3.478		144	204		348	470
0780 24" pipe size		1.90	4.211		180	247		427	575
0790 26" pipe size		1.60	5		244	294		538	715
0810 30" pipe size		1.40	5.714		475	335		810	1,025
0830 36" pipe size		1.10	7.273		890	425		1,315	1,625
0850 For 300 lb. gasket set, add					40%				
2000 Flanged unions, 125 lb., black, 1/2" pipe size	1 Plum	17	.471	Ea.	82.50	27.50		110	133
2040 3/4" pipe size		17	.471		109	27.50		136.50	162
2050 1" pipe size		16	.500		105	29.50		134.50	161
2060 1-1/4" pipe size	Q-1	28	.571		125	30		155	183
2070 1-1/2" pipe size		27	.593		115	31.50		146.50	175
2080 2" pipe size		26	.615		135	32.50		167.50	198
2090 2-1/2" pipe size		24	.667		183	35		218	255
2100 3" pipe size		22	.727		207	38.50		245.50	286
2110 3-1/2" pipe size		18	.889		350	47		397	455
2120 4" pipe size		16	1		282	53		335	390
2130 5" pipe size		14	1.143		660	60.50		720.50	815
2140 6" pipe size	Q-2	19	1.263		615	69		684	785
2150 8" pipe size	"	16	1.500		1,425	82		1,507	1,700
2200 For galvanized unions, add					150%				
2290 Threaded flange									
2300 Cast iron									
2310 Black, 125 lb., per flange									
2320 1" pipe size	1 Plum	27	.296	Ea.	47	17.40		64.40	78.50
2330 1-1/4" pipe size	Q-1	44	.364		56.50	19.20		75.70	91
2340 1-1/2" pipe size		40	.400		52	21		73	89.50

For customer support on your Plumbing Cost Data, call 877.709.6968.

177

22 11 Facility Water Distribution

22 11 13 – Facility Water Distribution Piping

22 11 13.47 Pipe Fittings, Steel		Crew	Daily Output	Labor-Hours	Unit	Material	2015 Bare Costs Labor	Equipment	Total	Total Incl O&P
2350	2" pipe size	Q-1	36	.444	Ea.	52	23.50		75.50	93
2360	2-1/2" pipe size		28	.571		61	30		91	113
2370	3" pipe size		20	.800		78.50	42.50		121	151
2380	3-1/2" pipe size		16	1		110	53		163	201
2390	4" pipe size		12	1.333		106	70.50		176.50	222
2400	5" pipe size		10	1.600		149	84.50		233.50	292
2410	6" pipe size	Q-2	14	1.714		169	94		263	330
2420	8" pipe size		12	2		267	110		377	460
2430	10" pipe size		10	2.400		475	131		606	720
2440	12" pipe size		8	3		1,050	164		1,214	1,400
2460	For galvanized flanges, add					95%				
2490	Blind flange									
2492	Cast iron									
2494	Black, 125 lb., per flange									
2496	1" pipe size	1 Plum	27	.296	Ea.	75	17.40		92.40	109
2500	1-1/2" pipe size	Q-1	40	.400		84.50	21		105.50	125
2502	2" pipe size		36	.444		95	23.50		118.50	141
2504	2-1/2" pipe size		28	.571		102	30		132	158
2506	3" pipe size		20	.800		125	42.50		167.50	202
2508	4" pipe size		12	1.333		161	70.50		231.50	283
2510	5" pipe size		10	1.600		261	84.50		345.50	415
2512	6" pipe size	Q-2	14	1.714		284	94		378	450
2514	8" pipe size		12	2		445	110		555	655
2516	10" pipe size		10	2.400		660	131		791	925
2518	12" pipe size		8	3		1,250	164		1,414	1,625
2520	For galvanized flanges, add					80%				
2570	Threaded flange									
2580	Forged steel,									
2590	Black 150 lb., per flange									
2600	1/2" pipe size	1 Plum	30	.267	Ea.	26.50	15.65		42.15	52.50
2610	3/4" pipe size		28	.286		26.50	16.75		43.25	54.50
2620	1" pipe size		27	.296		26.50	17.40		43.90	55.50
2630	1-1/4" pipe size	Q-1	44	.364		26.50	19.20		45.70	58
2640	1-1/2" pipe size		40	.400		26.50	21		47.50	61
2650	2" pipe size		36	.444		29.50	23.50		53	68
2660	2-1/2" pipe size		28	.571		36	30		66	85.50
2670	3" pipe size		20	.800		36.50	42.50		79	104
2690	4" pipe size		12	1.333		42.50	70.50		113	153
2700	5" pipe size		10	1.600		66.50	84.50		151	202
2710	6" pipe size	Q-2	14	1.714		73	94		167	223
2720	8" pipe size		12	2		126	110		236	305
2730	10" pipe size		10	2.400		228	131		359	450
2860	Black 300 lb., per flange									
2870	1/2" pipe size	1 Plum	30	.267	Ea.	29	15.65		44.65	55.50
2880	3/4" pipe size		28	.286		29	16.75		45.75	57.50
2890	1" pipe size		27	.296		29	17.40		46.40	58.50
2900	1-1/4" pipe size	Q-1	44	.364		29	19.20		48.20	61
2910	1-1/2" pipe size		40	.400		29	21		50	64
2920	2" pipe size		36	.444		34	23.50		57.50	73
2930	2-1/2" pipe size		28	.571		47.50	30		77.50	97.50
2940	3" pipe size		20	.800		50.50	42.50		93	120
2960	4" pipe size		12	1.333		73	70.50		143.50	186
2970	6" pipe size	Q-2	14	1.714		136	94		230	292

22 11 Facility Water Distribution

22 11 13 – Facility Water Distribution Piping

22 11 13.47 Pipe Fittings, Steel	Crew	Daily Output	Labor-Hours	Unit	Material	2015 Bare Costs Labor	Equipment	Total	Total Incl O&P
3000	Weld joint, butt, carbon steel, standard weight									
3040	90° elbow, long radius									
3050	1/2" pipe size	Q-15	16	1	Ea.	45	53	3.61	101.61	133
3060	3/4" pipe size		16	1		45	53	3.61	101.61	133
3070	1" pipe size		16	1		21	53	3.61	77.61	107
3080	1-1/4" pipe size		14	1.143		21	60.50	4.13	85.63	119
3090	1-1/2" pipe size		13	1.231		21	65	4.44	90.44	126
3100	2" pipe size		10	1.600		22.50	84.50	5.80	112.80	159
3110	2-1/2" pipe size		8	2		27.50	106	7.20	140.70	198
3120	3" pipe size		7	2.286		29	121	8.25	158.25	223
3130	4" pipe size		5	3.200		48	169	11.55	228.55	320
3136	5" pipe size		4	4		103	211	14.45	328.45	450
3140	6" pipe size	Q-16	5	4.800		106	263	11.50	380.50	525
3150	8" pipe size		3.75	6.400		200	350	15.35	565.35	765
3160	10" pipe size		3	8		400	440	19.20	859.20	1,125
3170	12" pipe size		2.50	9.600		590	525	23	1,138	1,475
3180	14" pipe size		2	12		970	655	29	1,654	2,100
3190	16" pipe size		1.50	16		1,325	875	38.50	2,238.50	2,850
3191	18" pipe size		1.25	19.200		1,525	1,050	46	2,621	3,325
3192	20" pipe size		1.15	20.870		2,450	1,150	50	3,650	4,475
3194	24" pipe size		1.02	23.529		3,450	1,300	56.50	4,806.50	5,800
3200	45° Elbow, long									
3210	1/2" pipe size	Q-15	16	1	Ea.	62.50	53	3.61	119.11	152
3220	3/4" pipe size		16	1		62.50	53	3.61	119.11	152
3230	1" pipe size		16	1		22	53	3.61	78.61	108
3240	1-1/4" pipe size		14	1.143		22	60.50	4.13	86.63	120
3250	1-1/2" pipe size		13	1.231		22	65	4.44	91.44	127
3260	2" pipe size		10	1.600		22	84.50	5.80	112.30	159
3270	2-1/2" pipe size		8	2		26.50	106	7.20	139.70	197
3280	3" pipe size		7	2.286		27.50	121	8.25	156.75	222
3290	4" pipe size		5	3.200		49.50	169	11.55	230.05	320
3296	5" pipe size		4	4		75	211	14.45	300.45	420
3300	6" pipe size	Q-16	5	4.800		97.50	263	11.50	372	515
3310	8" pipe size		3.75	6.400		163	350	15.35	528.35	725
3320	10" pipe size		3	8		325	440	19.20	784.20	1,025
3330	12" pipe size		2.50	9.600		460	525	23	1,008	1,325
3340	14" pipe size		2	12		615	655	29	1,299	1,700
3341	16" pipe size		1.50	16		1,100	875	38.50	2,013.50	2,600
3342	18" pipe size		1.25	19.200		1,575	1,050	46	2,671	3,375
3343	20" pipe size		1.15	20.870		1,625	1,150	50	2,825	3,550
3345	24" pipe size		1.05	22.857		2,450	1,250	55	3,755	4,650
3346	26" pipe size		.85	28.235		2,700	1,550	68	4,318	5,375
3347	30" pipe size		.45	53.333		2,975	2,925	128	6,028	7,825
3349	36" pipe size		.38	63.158		3,275	3,450	152	6,877	9,025
3350	Tee, straight									
3352	For reducing tees and concentrics see starting line 4600									
3360	1/2" pipe size	Q-15	10	1.600	Ea.	110	84.50	5.80	200.30	255
3370	3/4" pipe size		10	1.600		110	84.50	5.80	200.30	255
3380	1" pipe size		10	1.600		54.50	84.50	5.80	144.80	194
3390	1-1/4" pipe size		9	1.778		68.50	94	6.40	168.90	225
3400	1-1/2" pipe size		8	2		68.50	106	7.20	181.70	243
3410	2" pipe size		6	2.667		54.50	141	9.65	205.15	284
3420	2-1/2" pipe size		5	3.200		75.50	169	11.55	256.05	350

For customer support on your Plumbing Cost Data, call 877.709.6968.

179

22 11 13 – Facility Water Distribution Piping

22 11 13.47 Pipe Fittings, Steel		Crew	Daily Output	Labor-Hours	Unit	Material	2015 Bare Costs Labor	Equipment	Total	Total Incl O&P
3430	3" pipe size	Q-15	4	4	Ea.	84	211	14.45	309.45	430
3440	4" pipe size		3	5.333		118	282	19.25	419.25	575
3446	5" pipe size	↓	2.50	6.400		195	340	23	558	750
3450	6" pipe size	Q-16	3	8		203	440	19.20	662.20	905
3460	8" pipe size		2.50	9.600		355	525	23	903	1,200
3470	10" pipe size		2	12		695	655	29	1,379	1,800
3480	12" pipe size		1.60	15		975	820	36	1,831	2,375
3481	14" pipe size		1.30	18.462		1,700	1,000	44.50	2,744.50	3,450
3482	16" pipe size		1	24		1,900	1,325	57.50	3,282.50	4,150
3483	18" pipe size		.80	30		3,025	1,650	72	4,747	5,875
3484	20" pipe size		.75	32		4,750	1,750	77	6,577	7,950
3486	24" pipe size		.70	34.286		6,150	1,875	82.50	8,107.50	9,675
3487	26" pipe size		.55	43.636		6,775	2,400	105	9,280	11,200
3488	30" pipe size		.30	80		7,425	4,375	192	11,992	15,000
3490	36" pipe size	↓	.25	96		8,200	5,250	230	13,680	17,200
3491	Eccentric reducer, 1-1/2" pipe size	Q-15	14	1.143		63.50	60.50	4.13	128.13	166
3492	2" pipe size		11	1.455		43	77	5.25	125.25	169
3493	2-1/2" pipe size		9	1.778		50	94	6.40	150.40	204
3494	3" pipe size		8	2		59	106	7.20	172.20	233
3495	4" pipe size	↓	6	2.667		75.50	141	9.65	226.15	305
3496	6" pipe size	Q-16	5	4.800		232	263	11.50	506.50	665
3497	8" pipe size		4	6		345	330	14.40	689.40	890
3498	10" pipe size		3	8		435	440	19.20	894.20	1,150
3499	12" pipe size	↓	2.50	9.600		605	525	23	1,153	1,475
3501	Cap, 1-1/2" pipe size	Q-15	28	.571		24.50	30	2.06	56.56	75
3502	2" pipe size		22	.727		22	38.50	2.63	63.13	85
3503	2-1/2" pipe size		18	.889		30	47	3.21	80.21	107
3504	3" pipe size		16	1		33.50	53	3.61	90.11	120
3505	4" pipe size	↓	12	1.333		45	70.50	4.81	120.31	161
3506	6" pipe size	Q-16	10	2.400		92	131	5.75	228.75	305
3507	8" pipe size		8	3		138	164	7.20	309.20	410
3508	10" pipe size		6	4		212	219	9.60	440.60	575
3509	12" pipe size		5	4.800		283	263	11.50	557.50	720
3511	14" pipe size		4	6		293	330	14.40	637.40	835
3512	16" pipe size		4	6		455	330	14.40	799.40	1,000
3513	18" pipe size	↓	3	8	↓	665	440	19.20	1,124.20	1,425
3517	Weld joint, butt, carbon steel, extra strong									
3519	90° elbow, long									
3520	1/2" pipe size	Q-15	13	1.231	Ea.	55.50	65	4.44	124.94	164
3530	3/4" pipe size		12	1.333		55.50	70.50	4.81	130.81	173
3540	1" pipe size		11	1.455		27	77	5.25	109.25	152
3550	1-1/4" pipe size		10	1.600		27	84.50	5.80	117.30	164
3560	1-1/2" pipe size		9	1.778		27	94	6.40	127.40	179
3570	2" pipe size		8	2		27.50	106	7.20	140.70	198
3580	2-1/2" pipe size		7	2.286		38.50	121	8.25	167.75	234
3590	3" pipe size		6	2.667		49.50	141	9.65	200.15	278
3600	4" pipe size		4	4		81.50	211	14.45	306.95	425
3606	5" pipe size	↓	3.50	4.571		195	242	16.50	453.50	595
3610	6" pipe size	Q-16	4.50	5.333		206	292	12.80	510.80	680
3620	8" pipe size		3.50	6.857		390	375	16.45	781.45	1,025
3630	10" pipe size		2.50	9.600		830	525	23	1,378	1,725
3640	12" pipe size	↓	2.25	10.667	↓	1,025	585	25.50	1,635.50	2,025
3650	45° Elbow, long									

For customer support on your Plumbing Cost Data, call 877.709.6968.

22 11 13 – Facility Water Distribution Piping

22 11 13.47 Pipe Fittings, Steel		Crew	Daily Output	Labor-Hours	Unit	Material	2015 Bare Costs Labor	2015 Bare Costs Equipment	Total	Total Incl O&P
3660	1/2" pipe size	Q-15	13	1.231	Ea.	62	65	4.44	131.44	171
3670	3/4" pipe size		12	1.333		62	70.50	4.81	137.31	179
3680	1" pipe size		11	1.455		29	77	5.25	111.25	153
3690	1-1/4" pipe size		10	1.600		29	84.50	5.80	119.30	166
3700	1-1/2" pipe size		9	1.778		29	94	6.40	129.40	181
3710	2" pipe size		8	2		29	106	7.20	142.20	199
3720	2-1/2" pipe size		7	2.286		64	121	8.25	193.25	262
3730	3" pipe size		6	2.667		37	141	9.65	187.65	265
3740	4" pipe size		4	4		58.50	211	14.45	283.95	400
3746	5" pipe size	↓	3.50	4.571		139	242	16.50	397.50	535
3750	6" pipe size	Q-16	4.50	5.333		160	292	12.80	464.80	630
3760	8" pipe size		3.50	6.857		277	375	16.45	668.45	890
3770	10" pipe size		2.50	9.600		535	525	23	1,083	1,400
3780	12" pipe size	↓	2.25	10.667	↓	790	585	25.50	1,400.50	1,775
3800	Tee, straight									
3810	1/2" pipe size	Q-15	9	1.778	Ea.	136	94	6.40	236.40	298
3820	3/4" pipe size		8.50	1.882		131	99.50	6.80	237.30	300
3830	1" pipe size		8	2		52	106	7.20	165.20	225
3840	1-1/4" pipe size		7	2.286		52	121	8.25	181.25	249
3850	1-1/2" pipe size		6	2.667		52	141	9.65	202.65	281
3860	2" pipe size		5	3.200		59.50	169	11.55	240.05	335
3870	2-1/2" pipe size		4	4		99	211	14.45	324.45	445
3880	3" pipe size		3.50	4.571		124	242	16.50	382.50	520
3890	4" pipe size		2.50	6.400		149	340	23	512	700
3896	5" pipe size	↓	2.25	7.111		375	375	25.50	775.50	1,000
3900	6" pipe size	Q-16	2.25	10.667		286	585	25.50	896.50	1,225
3910	8" pipe size		2	12		545	655	29	1,229	1,625
3920	10" pipe size		1.75	13.714		845	750	33	1,628	2,100
3930	12" pipe size	↓	1.50	16		1,225	875	38.50	2,138.50	2,725
4000	Eccentric reducer, 1-1/2" pipe size	Q-15	10	1.600		17.30	84.50	5.80	107.60	153
4010	2" pipe size		9	1.778		47.50	94	6.40	147.90	202
4020	2-1/2" pipe size		8	2		71.50	106	7.20	184.70	246
4030	3" pipe size		7	2.286		59	121	8.25	188.25	256
4040	4" pipe size		5	3.200		95.50	169	11.55	276.05	375
4046	5" pipe size	↓	4.70	3.404		266	180	12.30	458.30	580
4050	6" pipe size	Q-16	4.50	5.333		259	292	12.80	563.80	740
4060	8" pipe size		3.50	6.857		390	375	16.45	781.45	1,025
4070	10" pipe size		2.50	9.600		665	525	23	1,213	1,550
4080	12" pipe size		2.25	10.667		845	585	25.50	1,455.50	1,825
4090	14" pipe size		2.10	11.429		1,575	625	27.50	2,227.50	2,700
4100	16" pipe size		1.90	12.632		2,025	690	30.50	2,745.50	3,300
4151	Cap, 1-1/2" pipe size	Q-15	24	.667		24.50	35	2.41	61.91	82.50
4152	2" pipe size		18	.889		22	47	3.21	72.21	98.50
4153	2-1/2" pipe size		16	1		30	53	3.61	86.61	116
4154	3" pipe size		14	1.143		33.50	60.50	4.13	98.13	132
4155	4" pipe size	↓	10	1.600		45	84.50	5.80	135.30	184
4156	6" pipe size	Q-16	9	2.667		92	146	6.40	244.40	330
4157	8" pipe size		7	3.429		138	188	8.25	334.25	445
4158	10" pipe size		5	4.800		212	263	11.50	486.50	640
4159	12" pipe size	↓	4	6	↓	283	330	14.40	627.40	820
4190	Weld fittings, reducing, standard weight									
4200	Welding ring w/spacer pins, 2" pipe size				Ea.	2.40			2.40	2.64
4210	2-1/2" pipe size				↓	2.45			2.45	2.70

For customer support on your Plumbing Cost Data, call 877.709.6968.

181

22 11 13.47 Pipe Fittings, Steel		Crew	Daily Output	Labor-Hours	Unit	Material	2015 Bare Costs Labor	Equipment	Total	Total Incl O&P
4220	3" pipe size				Ea.	2.50			2.50	2.75
4230	4" pipe size					2.65			2.65	2.92
4236	5" pipe size					3.35			3.35	3.69
4240	6" pipe size					3.60			3.60	3.96
4250	8" pipe size					4.20			4.20	4.62
4260	10" pipe size					4.70			4.70	5.15
4270	12" pipe size					5.40			5.40	5.95
4280	14" pipe size					6.45			6.45	7.10
4290	16" pipe size					7.60			7.60	8.35
4300	18" pipe size					8.60			8.60	9.45
4310	20" pipe size					8.95			8.95	9.85
4330	24" pipe size					11.10			11.10	12.20
4340	26" pipe size					13.70			13.70	15.05
4350	30" pipe size					16.75			16.75	18.45
4370	36" pipe size					20.50			20.50	22.50
4600	Tee, reducing on outlet									
4601	2-1/2" x 2" pipe size	Q-15	5	3.200	Ea.	100	169	11.55	280.55	380
4602	3" x 2-1/2" pipe size		4	4		118	211	14.45	343.45	465
4604	4" x 3" pipe size		3	5.333		124	282	19.25	425.25	585
4605	5" x 4" pipe size		2.50	6.400		268	340	23	631	830
4606	6" x 5" pipe size	Q-16	3	8		370	440	19.20	829.20	1,100
4607	8" x 6" pipe size		2.50	9.600		460	525	23	1,008	1,325
4608	10" x 8" pipe size		2	12		850	655	29	1,534	1,950
4609	12" x 10" pipe size		1.60	15		1,250	820	36	2,106	2,675
4610	16" x 12" pipe size		1.50	16		2,000	875	38.50	2,913.50	3,575
4611	14" x 12" pipe size		1.52	15.789		1,900	865	38	2,803	3,425
4618	Reducer, concentric									
4619	2-1/2" by 2" pipe size	Q-15	10	1.600	Ea.	42.50	84.50	5.80	132.80	181
4620	3" by 2-1/2" pipe size		9	1.778		35	94	6.40	135.40	188
4621	3-1/2" by 3" pipe size		8	2		89.50	106	7.20	202.70	266
4622	4" by 2-1/2" pipe size		7	2.286		54.50	121	8.25	183.75	251
4623	5" by 3" pipe size		7	2.286		114	121	8.25	243.25	315
4624	6" by 4" pipe size	Q-16	6	4		94	219	9.60	322.60	445
4625	8" by 6" pipe size		5	4.800		124	263	11.50	398.50	545
4626	10" by 8" pipe size		4	6		240	330	14.40	584.40	775
4627	12" by 10" pipe size		3	8		296	440	19.20	755.20	1,000
4660	Reducer, eccentric									
4662	3" x 2" pipe size	Q-15	8	2	Ea.	59	106	7.20	172.20	233
4664	4" x 3" pipe size		6	2.667		71.50	141	9.65	222.15	305
4666	4" x 2" pipe size		6	2.667		87.50	141	9.65	238.15	320
4670	6" x 4" pipe size	Q-16	5	4.800		148	263	11.50	422.50	570
4672	6" x 3" pipe size		5	4.800		232	263	11.50	506.50	665
4676	8" x 6" pipe size		4	6		188	330	14.40	532.40	720
4678	8" x 4" pipe size		4	6		345	330	14.40	689.40	890
4682	10" x 8" pipe size		3	8		247	440	19.20	706.20	950
4684	10" x 6" pipe size		3	8		435	440	19.20	894.20	1,150
4688	12" x 10" pipe size		2.50	9.600		385	525	23	933	1,250
4690	12" x 8" pipe size		2.50	9.600		605	525	23	1,153	1,475
4691	14" x 12" pipe size		2.20	10.909		675	600	26	1,301	1,675
4693	16" x 14" pipe size		1.80	13.333		1,025	730	32	1,787	2,250
4694	16" x 12" pipe size		2	12		1,375	655	29	2,059	2,525
4696	18" x 16" pipe size		1.60	15		1,100	820	36	1,956	2,525
5000	Weld joint, socket, forged steel, 3000 lb., schedule 40 pipe									

182

For customer support on your Plumbing Cost Data, call 877.709.6968.

22 11 Facility Water Distribution

22 11 13 – Facility Water Distribution Piping

22 11 13.47 Pipe Fittings, Steel	Crew	Daily Output	Labor-Hours	Unit	Material	2015 Bare Costs Labor	Equipment	Total	Total Incl O&P
5010	90° elbow, straight								
5020	1/4" pipe size	Q-15	22	.727	Ea.	25	38.50	2.63	66.13	88.50
5030	3/8" pipe size		22	.727		25	38.50	2.63	66.13	88.50
5040	1/2" pipe size		20	.800		13.95	42.50	2.89	59.34	82.50
5050	3/4" pipe size		20	.800		14.55	42.50	2.89	59.94	83
5060	1" pipe size		20	.800		18.70	42.50	2.89	64.09	87.50
5070	1-1/4" pipe size		18	.889		36	47	3.21	86.21	114
5080	1-1/2" pipe size		16	1		41.50	53	3.61	98.11	130
5090	2" pipe size		12	1.333		61.50	70.50	4.81	136.81	179
5100	2-1/2" pipe size		10	1.600		174	84.50	5.80	264.30	325
5110	3" pipe size		8	2		300	106	7.20	413.20	500
5120	4" pipe size	▼	6	2.667	▼	795	141	9.65	945.65	1,100
5130	45° Elbow, straight								
5134	1/4" pipe size	Q-15	22	.727	Ea.	25	38.50	2.63	66.13	88.50
5135	3/8" pipe size		22	.727		25	38.50	2.63	66.13	88.50
5136	1/2" pipe size		20	.800		18.60	42.50	2.89	63.99	87.50
5137	3/4" pipe size		20	.800		21.50	42.50	2.89	66.89	91
5140	1" pipe size		20	.800		28.50	42.50	2.89	73.89	98
5150	1-1/4" pipe size		18	.889		39	47	3.21	89.21	117
5160	1-1/2" pipe size		16	1		47.50	53	3.61	104.11	136
5170	2" pipe size		12	1.333		76.50	70.50	4.81	151.81	195
5180	2-1/2" pipe size		10	1.600		201	84.50	5.80	291.30	355
5190	3" pipe size		8	2		335	106	7.20	448.20	535
5200	4" pipe size	▼	6	2.667	▼	660	141	9.65	810.65	950
5250	Tee, straight								
5254	1/4" pipe size	Q-15	15	1.067	Ea.	27.50	56.50	3.85	87.85	120
5255	3/8" pipe size		15	1.067		27.50	56.50	3.85	87.85	120
5256	1/2" pipe size		13	1.231		17.30	65	4.44	86.74	122
5257	3/4" pipe size		13	1.231		21	65	4.44	90.44	126
5260	1" pipe size		13	1.231		29	65	4.44	98.44	134
5270	1-1/4" pipe size		12	1.333		44	70.50	4.81	119.31	159
5280	1-1/2" pipe size		11	1.455		58.50	77	5.25	140.75	186
5290	2" pipe size		8	2		84.50	106	7.20	197.70	261
5300	2-1/2" pipe size		6	2.667		243	141	9.65	393.65	490
5310	3" pipe size		5	3.200		580	169	11.55	760.55	910
5320	4" pipe size	▼	4	4	▼	935	211	14.45	1,160.45	1,350
5350	For reducing sizes, add					60%				
5450	Couplings								
5451	1/4" pipe size	Q-15	23	.696	Ea.	15.90	37	2.51	55.41	76
5452	3/8" pipe size		23	.696		15.90	37	2.51	55.41	76
5453	1/2" pipe size		21	.762		7.35	40.50	2.75	50.60	72
5454	3/4" pipe size		21	.762		9.50	40.50	2.75	52.75	74.50
5460	1" pipe size		20	.800		10.50	42.50	2.89	55.89	78.50
5470	1-1/4" pipe size		20	.800		18.60	42.50	2.89	63.99	87.50
5480	1-1/2" pipe size		18	.889		20.50	47	3.21	70.71	97
5490	2" pipe size		14	1.143		33	60.50	4.13	97.63	132
5500	2-1/2" pipe size		12	1.333		73.50	70.50	4.81	148.81	192
5510	3" pipe size		9	1.778		165	94	6.40	265.40	330
5520	4" pipe size		7	2.286		248	121	8.25	377.25	465
5570	Union, 1/4" pipe size		21	.762		34	40.50	2.75	77.25	102
5571	3/8" pipe size		21	.762		34	40.50	2.75	77.25	102
5572	1/2" pipe size		19	.842		28.50	44.50	3.04	76.04	102
5573	3/4" pipe size	▼	19	.842	▼	33.50	44.50	3.04	81.04	107

22 11 13.47 Pipe Fittings, Steel		Crew	Daily Output	Labor-Hours	Unit	Material	2015 Bare Costs Labor	Equipment	Total	Total Incl O&P
5574	1" pipe size	Q-15	19	.842	Ea.	43	44.50	3.04	90.54	118
5575	1-1/4" pipe size		17	.941		69.50	49.50	3.40	122.40	155
5576	1-1/2" pipe size		15	1.067		75	56.50	3.85	135.35	172
5577	2" pipe size		11	1.455		108	77	5.25	190.25	241
5600	Reducer, 1/4" pipe size		23	.696		40.50	37	2.51	80.01	103
5601	3/8" pipe size		23	.696		43	37	2.51	82.51	105
5602	1/2" pipe size		21	.762		26	40.50	2.75	69.25	92.50
5603	3/4" pipe size		21	.762		26	40.50	2.75	69.25	92.50
5604	1" pipe size		21	.762		36.50	40.50	2.75	79.75	104
5605	1-1/4" pipe size		19	.842		46.50	44.50	3.04	94.04	122
5607	1-1/2" pipe size		17	.941		50	49.50	3.40	102.90	134
5608	2" pipe size		13	1.231		55	65	4.44	124.44	163
5612	Cap, 1/4" pipe size		46	.348		15.15	18.40	1.26	34.81	46
5613	3/8" pipe size		46	.348		15.15	18.40	1.26	34.81	46
5614	1/2" pipe size		42	.381		9.25	20	1.38	30.63	42
5615	3/4" pipe size		42	.381		10.80	20	1.38	32.18	44
5616	1" pipe size		42	.381		16.35	20	1.38	37.73	50
5617	1-1/4" pipe size		38	.421		19.30	22	1.52	42.82	56
5618	1-1/2" pipe size		34	.471		27	25	1.70	53.70	69
5619	2" pipe size		26	.615		41.50	32.50	2.22	76.22	97
5630	T-O-L, 1/4" pipe size, nozzle		23	.696		7.10	37	2.51	46.61	66
5631	3/8" pipe size, nozzle		23	.696		7.20	37	2.51	46.71	66
5632	1/2" pipe size, nozzle		22	.727		7.10	38.50	2.63	48.23	68.50
5633	3/4" pipe size, nozzle		21	.762		8.20	40.50	2.75	51.45	73
5634	1" pipe size, nozzle		20	.800		9.55	42.50	2.89	54.94	77.50
5635	1-1/4" pipe size, nozzle		18	.889		15.10	47	3.21	65.31	91
5636	1-1/2" pipe size, nozzle		16	1		15.10	53	3.61	71.71	101
5637	2" pipe size, nozzle		12	1.333		17.20	70.50	4.81	92.51	130
5638	2-1/2" pipe size, nozzle		10	1.600		57.50	84.50	5.80	147.80	198
5639	4" pipe size, nozzle		6	2.667		131	141	9.65	281.65	370
5640	W-O-L, 1/4" pipe size, nozzle		23	.696		16.80	37	2.51	56.31	76.50
5641	3/8" pipe size, nozzle		23	.696		15.85	37	2.51	55.36	75.50
5642	1/2" pipe size, nozzle		22	.727		14.70	38.50	2.63	55.83	77
5643	3/4" pipe size, nozzle		21	.762		15.50	40.50	2.75	58.75	81
5644	1" pipe size, nozzle		20	.800		16.20	42.50	2.89	61.59	85
5645	1-1/4" pipe size, nozzle		18	.889		19.25	47	3.21	69.46	95.50
5646	1-1/2" pipe size, nozzle		16	1		19.25	53	3.61	75.86	105
5647	2" pipe size, nozzle		12	1.333		19.45	70.50	4.81	94.76	133
5648	2-1/2" pipe size, nozzle		10	1.600		44.50	84.50	5.80	134.80	183
5649	3" pipe size, nozzle		8	2		48.50	106	7.20	161.70	221
5650	4" pipe size, nozzle		6	2.667		61.50	141	9.65	212.15	291
5651	5" pipe size, nozzle		5	3.200		152	169	11.55	332.55	435
5652	6" pipe size, nozzle		4	4		177	211	14.45	402.45	530
5653	8" pipe size, nozzle		3	5.333		340	282	19.25	641.25	820
5654	10" pipe size, nozzle		2.60	6.154		490	325	22	837	1,050
5655	12" pipe size, nozzle		2.20	7.273		930	385	26.50	1,341.50	1,625
5674	S-O-L, 1/4" pipe size, outlet		23	.696		9.45	37	2.51	48.96	68.50
5675	3/8" pipe size, outlet		23	.696		9.45	37	2.51	48.96	68.50
5676	1/2" pipe size, outlet		22	.727		8.80	38.50	2.63	49.93	70.50
5677	3/4" pipe size, outlet		21	.762		8.95	40.50	2.75	52.20	74
5678	1" pipe size, outlet		20	.800		9.90	42.50	2.89	55.29	78
5679	1-1/4" pipe size, outlet		18	.889		16.55	47	3.21	66.76	92.50
5680	1-1/2" pipe size, outlet		16	1		16.55	53	3.61	73.16	102

22 11 Facility Water Distribution

22 11 13 – Facility Water Distribution Piping

22 11 13.47 Pipe Fittings, Steel	Crew	Daily Output	Labor-Hours	Unit	Material	2015 Bare Costs Labor	Equipment	Total	Total Incl O&P
5681 2" pipe size, outlet	Q-15	12	1.333	Ea.	18.80	70.50	4.81	94.11	132
6000 Weld-on flange, forged steel									
6020 Slip-on, 150 lb. flange (welded front and back)									
6050 1/2" pipe size	Q-15	18	.889	Ea.	17.50	47	3.21	67.71	94
6060 3/4" pipe size		18	.889		17.50	47	3.21	67.71	94
6070 1" pipe size		17	.941		17.50	49.50	3.40	70.40	98
6080 1-1/4" pipe size		16	1		17.50	53	3.61	74.11	103
6090 1-1/2" pipe size		15	1.067		17.50	56.50	3.85	77.85	108
6100 2" pipe size		12	1.333		19.45	70.50	4.81	94.76	133
6110 2-1/2" pipe size		10	1.600		25	84.50	5.80	115.30	162
6120 3" pipe size		9	1.778		26.50	94	6.40	126.90	178
6130 3-1/2" pipe size		7	2.286		33	121	8.25	162.25	228
6140 4" pipe size		6	2.667		33	141	9.65	183.65	260
6150 5" pipe size		5	3.200		57.50	169	11.55	238.05	330
6160 6" pipe size	Q-16	6	4		54.50	219	9.60	283.10	400
6170 8" pipe size		5	4.800		82.50	263	11.50	357	500
6180 10" pipe size		4	6		143	330	14.40	487.40	670
6190 12" pipe size		3	8		212	440	19.20	671.20	915
6191 14" pipe size		2.50	9.600		281	525	23	829	1,125
6192 16" pipe size		1.80	13.333		440	730	32	1,202	1,625
6200 300 lb. flange									
6210 1/2" pipe size	Q-15	17	.941	Ea.	23.50	49.50	3.40	76.40	104
6220 3/4" pipe size		17	.941		23.50	49.50	3.40	76.40	104
6230 1" pipe size		16	1		23.50	53	3.61	80.11	109
6240 1-1/4" pipe size		13	1.231		23.50	65	4.44	92.94	128
6250 1-1/2" pipe size		12	1.333		23.50	70.50	4.81	98.81	137
6260 2" pipe size		11	1.455		31	77	5.25	113.25	156
6270 2-1/2" pipe size		9	1.778		35	94	6.40	135.40	188
6280 3" pipe size		7	2.286		38	121	8.25	167.25	233
6290 4" pipe size		6	2.667		55.50	141	9.65	206.15	285
6300 5" pipe size		4	4		95	211	14.45	320.45	440
6310 6" pipe size	Q-16	5	4.800		95.50	263	11.50	370	515
6320 8" pipe size		4	6		163	330	14.40	507.40	690
6330 10" pipe size		3.40	7.059		278	385	16.95	679.95	910
6340 12" pipe size		2.80	8.571		340	470	20.50	830.50	1,100
6400 Welding neck, 150 lb. flange									
6410 1/2" pipe size	Q-15	40	.400	Ea.	25.50	21	1.44	47.94	61.50
6420 3/4" pipe size		36	.444		25.50	23.50	1.60	50.60	65.50
6430 1" pipe size		32	.500		25.50	26.50	1.81	53.81	70
6440 1-1/4" pipe size		29	.552		25.50	29	1.99	56.49	74
6450 1-1/2" pipe size		26	.615		25.50	32.50	2.22	60.22	79.50
6460 2" pipe size		20	.800		29.50	42.50	2.89	74.89	99
6470 2-1/2" pipe size		16	1		32	53	3.61	88.61	119
6480 3" pipe size		14	1.143		35	60.50	4.13	99.63	134
6500 4" pipe size		10	1.600		42.50	84.50	5.80	132.80	181
6510 5" pipe size		8	2		66.50	106	7.20	179.70	241
6520 6" pipe size	Q-16	10	2.400		64	131	5.75	200.75	275
6530 8" pipe size		7	3.429		113	188	8.25	309.25	415
6540 10" pipe size		6	4		180	219	9.60	408.60	540
6550 12" pipe size		5	4.800		262	263	11.50	536.50	695
6551 14" pipe size		4.50	5.333		380	292	12.80	684.80	870
6552 16" pipe size		3	8		570	440	19.20	1,029.20	1,300
6553 18" pipe size		2.50	9.600		800	525	23	1,348	1,700

22 11 13.47 Pipe Fittings, Steel		Crew	Daily Output	Labor-Hours	Unit	Material	2015 Bare Costs Labor	Equipment	Total	Total Incl O&P
6554	20" pipe size	Q-16	2.30	10.435	Ea.	975	570	25	1,570	1,975
6556	24" pipe size		2	12		1,300	655	29	1,984	2,475
6557	26" pipe size		1.70	14.118		1,425	775	34	2,234	2,800
6558	30" pipe size		.90	26.667		1,650	1,450	64	3,164	4,075
6559	36" pipe size		.75	32		1,900	1,750	77	3,727	4,800
6560	300 lb. flange									
6570	1/2" pipe size	Q-15	36	.444	Ea.	31	23.50	1.60	56.10	71.50
6580	3/4" pipe size		34	.471		31	25	1.70	57.70	73.50
6590	1" pipe size		30	.533		31	28	1.93	60.93	78.50
6600	1-1/4" pipe size		28	.571		31	30	2.06	63.06	82
6610	1-1/2" pipe size		24	.667		31	35	2.41	68.41	89.50
6620	2" pipe size		18	.889		38.50	47	3.21	88.71	117
6630	2-1/2" pipe size		14	1.143		44	60.50	4.13	108.63	144
6640	3" pipe size		12	1.333		44.50	70.50	4.81	119.81	160
6650	4" pipe size		8	2		73	106	7.20	186.20	248
6660	5" pipe size		7	2.286		111	121	8.25	240.25	315
6670	6" pipe size	Q-16	9	2.667		113	146	6.40	265.40	355
6680	8" pipe size		6	4		195	219	9.60	423.60	555
6690	10" pipe size		5	4.800		355	263	11.50	629.50	800
6700	12" pipe size		4	6		455	330	14.40	799.40	1,000
7740	Plain ends for plain end pipe, mechanically coupled									
7750	Cplg. & labor required at joints not included, add 1 per									
7760	joint for installed price, see line 9180									
7770	Malleable iron, painted, unless noted otherwise									
7800	90° Elbow 1"				Ea.	112			112	124
7810	1-1/2"					133			133	146
7820	2"					198			198	218
7830	2-1/2"					232			232	255
7840	3"					240			240	264
7860	4"					272			272	300
7870	5" welded steel					315			315	345
7880	6"					395			395	435
7890	8" welded steel					745			745	815
7900	10" welded steel					945			945	1,050
7910	12" welded steel					1,050			1,050	1,150
7970	45° Elbow 1"					67			67	73.50
7980	1-1/2"					100			100	110
7990	2"					210			210	231
8000	2-1/2"					210			210	231
8010	3"					240			240	264
8030	4"					251			251	276
8040	5" welded steel					315			315	345
8050	6"					360			360	395
8060	8"					420			420	460
8070	10" welded steel					435			435	480
8080	12" welded steel					785			785	865
8140	Tee, straight 1"					126			126	138
8150	1-1/2"					162			162	178
8160	2"					162			162	178
8170	2-1/2"					211			211	232
8180	3"					325			325	355
8200	4"					465			465	510
8210	5" welded steel					635			635	695

22 11 Facility Water Distribution

22 11 13 – Facility Water Distribution Piping

22 11 13.47 Pipe Fittings, Steel		Crew	Daily Output	Labor-Hours	Unit	Material	2015 Bare Costs Labor	Equipment	Total	Total Incl O&P
8220	6"				Ea.	550			550	605
8230	8" welded steel					795			795	875
8240	10" welded steel					1,250			1,250	1,375
8250	12" welded steel					1,450			1,450	1,600
8340	Segmentally welded steel, painted									
8390	Wye 2"				Ea.	201			201	221
8400	2-1/2"					201			201	221
8410	3"					225			225	247
8430	4"					335			335	370
8440	5"					410			410	450
8450	6"					560			560	620
8460	8"					730			730	800
8470	10"					1,075			1,075	1,175
8480	12"					1,675			1,675	1,825
8540	Wye, lateral 2"					218			218	240
8550	2-1/2"					251			251	276
8560	3"					298			298	330
8580	4"					410			410	450
8590	5"					695			695	765
8600	6"					720			720	790
8610	8"					1,200			1,200	1,325
8620	10"					1,200			1,200	1,325
8630	12"					2,300			2,300	2,525
8690	Cross, 2"					212			212	234
8700	2-1/2"					212			212	234
8710	3"					253			253	279
8730	4"					345			345	380
8740	5"					490			490	540
8750	6"					650			650	715
8760	8"					830			830	915
8770	10"					1,225			1,225	1,325
8780	12"					1,750			1,750	1,925
8800	Tees, reducing 2" x 1"					138			138	152
8810	2" x 1-1/2"					138			138	152
8820	3" x 1"					138			138	152
8830	3" x 1-1/2"					138			138	152
8840	3" x 2"					139			139	153
8850	4" x 1"					203			203	223
8860	4" x 1-1/2"					203			203	223
8870	4" x 2"					200			200	220
8880	4" x 2-1/2"					206			206	227
8890	4" x 3"					206			206	227
8900	6" x 2"					300			300	330
8910	6" x 3"					325			325	355
8920	6" x 4"					355			355	390
8930	8" x 2"					365			365	405
8940	8" x 3"					375			375	410
8950	8" x 4"					400			400	440
8960	8" x 5"					430			430	475
8970	8" x 6"					465			465	515
8980	10" x 4"					620			620	680
8990	10" x 6"					620			620	680
9000	10" x 8"					655			655	720

22 11 13.47 Pipe Fittings, Steel		Crew	Daily Output	Labor-Hours	Unit	Material	2015 Bare Costs		Total	Total Incl O&P
							Labor	Equipment		
9010	12" x 6"				Ea.	955			955	1,050
9020	12" x 8"					955			955	1,050
9030	12" x 10"				↓	995			995	1,100
9080	Adapter nipples 3" long									
9090	1"				Ea.	19.40			19.40	21.50
9100	1-1/2"					19.40			19.40	21.50
9110	2"					19.40			19.40	21.50
9120	2-1/2"					22.50			22.50	24.50
9130	3"					27			27	30
9140	4"					44.50			44.50	49
9150	6"				↓	117			117	129
9180	Coupling, mechanical, plain end pipe to plain end pipe or fitting									
9190	1"	Q-1	29	.552	Ea.	68	29		97	119
9200	1-1/2"		28	.571		68	30		98	120
9210	2"		27	.593		68	31.50		99.50	122
9220	2-1/2"		26	.615		68	32.50		100.50	124
9230	3"		25	.640		99	34		133	160
9240	3-1/2"		24	.667		114	35		149	179
9250	4"	↓	22	.727		114	38.50		152.50	184
9260	5"	Q-2	28	.857		163	47		210	250
9270	6"		24	1		199	55		254	300
9280	8"		19	1.263		350	69		419	490
9290	10"		16	1.500		455	82		537	625
9300	12"	↓	12	2	↓	570	110		680	795
9310	Outlets for precut holes through pipe wall									
9331	Strapless type, with gasket									
9332	4" to 8" pipe x 1/2"	1 Plum	13	.615	Ea.	74	36		110	136
9333	4" to 8" pipe x 3/4"		13	.615		79	36		115	142
9334	10" pipe and larger x 1/2"		11	.727		74 ·	42.50		116.50	146
9335	10" pipe and larger x 3/4"	↓	11	.727	↓	79	42.50		121.50	152
9341	Thermometer wells with gasket									
9342	4" to 8" pipe, 6" stem	1 Plum	14	.571	Ea.	111	33.50	·	144.50	174
9343	8" pipe and larger, 6" stem	"	13	.615	"	111	36		147	178
9940	For galvanized fittings for plain end pipe, add					20%				

22 11 13.48 Pipe, Fittings and Valves, Steel, Grooved-Joint

22 11 13.48 Pipe, Fittings and Valves, Steel, Grooved-Joint		Crew	Daily Output	Labor-Hours	Unit	Material	Labor	Equipment	Total	Total Incl O&P
0010	**PIPE, FITTINGS AND VALVES, STEEL, GROOVED-JOINT** R221113-70									
0012	Fittings are ductile iron. Steel fittings noted.									
0020	Pipe includes coupling & clevis type hanger assemblies, 10' O.C.									
0500	Schedule 10, black									
0550	2" diameter	1 Plum	43	.186	L.F.	7.15	10.90		18.05	24.50
0560	2-1/2" diameter	Q-1	61	.262		8.85	13.85		22.70	31
0570	3" diameter		55	.291		10.15	15.35		25.50	34
0580	3-1/2" diameter		53	.302		12.80	15.95		28.75	38
0590	4" diameter		49	.327		13.40	17.25		30.65	41
0600	5" diameter	↓	40	.400		16.15	21		37.15	50
0610	6" diameter	Q-2	46	.522		20.50	28.50		49	65.50
0620	8" diameter	"	41	.585	↓	28	32		60	79.50
0700	To delete couplings & hangers, subtract									
0710	2" diam. to 5" diam.					25%	20%			
0720	6" diam. to 8" diam.					27%	15%			
1000	Schedule 40, black									
1040	3/4" diameter	1 Plum	71	.113	L.F.	5.50	6.60		12.10	16

22 11 13 – Facility Water Distribution Piping

22 11 13.48 Pipe, Fittings and Valves, Steel, Grooved-Joint	Crew	Daily Output	Labor-Hours	Unit	Material	2015 Bare Costs Labor	Equipment	Total	Total Incl O&P
1050	1" diameter	1 Plum	63	.127	L.F.	5.30	7.45		12.75	17.05
1060	1-1/4" diameter		58	.138		6.60	8.10		14.70	19.50
1070	1-1/2" diameter		51	.157		7.35	9.20		16.55	22
1080	2" diameter		40	.200		8.70	11.75		20.45	27.50
1090	2-1/2" diameter	Q-1	57	.281		13.65	14.85		28.50	37.50
1100	3" diameter		50	.320		16.75	16.90		33.65	44
1110	4" diameter		45	.356		23.50	18.80		42.30	54.50
1120	5" diameter		37	.432		38	23		61	76.50
1130	6" diameter	Q-2	42	.571		48.50	31.50		80	101
1140	8" diameter		37	.649		78.50	35.50		114	140
1150	10" diameter		31	.774		106	42.50		148.50	181
1160	12" diameter		27	.889		118	48.50		166.50	204
1170	14" diameter		20	1.200		122	65.50		187.50	235
1180	16" diameter		17	1.412		183	77.50		260.50	320
1190	18" diameter		14	1.714		183	94		277	345
1200	20" diameter		12	2		214	110		324	400
1210	24" diameter		10	2.400		241	131		372	465
1740	To delete coupling & hanger, subtract									
1750	3/4" diam. to 2" diam.					65%	27%			
1760	2-1/2" diam. to 5" diam.					41%	18%			
1770	6" diam. to 12" diam.					31%	13%			
1780	14" diam. to 24" diam.					35%	10%			
1800	Galvanized									
1840	3/4" diameter	1 Plum	71	.113	L.F.	5.80	6.60		12.40	16.35
1850	1" diameter		63	.127		7	7.45		14.45	18.95
1860	1-1/4" diameter		58	.138		8.95	8.10		17.05	22
1870	1-1/2" diameter		51	.157		10.10	9.20		19.30	25
1880	2" diameter		40	.200		12.50	11.75		24.25	31.50
1890	2-1/2" diameter	Q-1	57	.281		17.60	14.85		32.45	42
1900	3" diameter		50	.320		22.50	16.90		39.40	50
1910	4" diameter		45	.356		32	18.80		50.80	63.50
1920	5" diameter		37	.432		60	23		83	100
1930	6" diameter	Q-2	42	.571		64.50	31.50		96	119
1940	8" diameter		37	.649		76	35.50		111.50	137
1950	10" diameter		31	.774		122	42.50		164.50	198
1960	12" diameter		27	.889		148	48.50		196.50	237
2540	To delete coupling & hanger, subtract									
2550	3/4" diam. to 2" diam.					36%	27%			
2560	2-1/2" diam. to 5" diam.					19%	18%			
2570	6" diam. to 12" diam.					14%	13%			
2600	Schedule 80, black									
2610	3/4" diameter	1 Plum	65	.123	L.F.	6.60	7.20		13.80	18.15
2650	1" diameter		61	.131		8.10	7.70		15.80	20.50
2660	1-1/4" diameter		55	.145		10.40	8.55		18.95	24.50
2670	1-1/2" diameter		49	.163		11.80	9.60		21.40	27.50
2680	2" diameter		38	.211		14.80	12.35		27.15	35
2690	2-1/2" diameter	Q-1	54	.296		21	15.65		36.65	47
2700	3" diameter		48	.333		27	17.60		44.60	56
2710	4" diameter		44	.364		38	19.20		57.20	71
2720	5" diameter		35	.457		55.50	24		79.50	97.50
2730	6" diameter	Q-2	40	.600		74.50	33		107.50	131
2740	8" diameter		35	.686		113	37.50		150.50	181
2750	10" diameter		29	.828		181	45.50		226.50	268

For customer support on your Plumbing Cost Data, call 877.709.6968.

189

22 11 13.48 Pipe, Fittings and Valves, Steel, Grooved-Joint	Crew	Daily Output	Labor-Hours	Unit	Material	2015 Bare Costs Labor	Equipment	Total	Total Incl O&P
2760	12" diameter	Q-2	24	1	L.F.	395	55		450	520
3240	To delete coupling & hanger, subtract									
3250	3/4" diam. to 2" diam.					30%	25%			
3260	2-1/2" diam. to 5" diam.					14%	17%			
3270	6" diam. to 12" diam.					12%	12%			
3300	Galvanized									
3310	3/4" diameter	1 Plum	65	.123	L.F.	7.55	7.20		14.75	19.20
3350	1" diameter		61	.131		9.35	7.70		17.05	22
3360	1-1/4" diameter		55	.145		12.30	8.55		20.85	26.50
3370	1-1/2" diameter		46	.174		14.30	10.20		24.50	31
3380	2" diameter		38	.211		18.60	12.35		30.95	39
3390	2-1/2" diameter	Q-1	54	.296		26.50	15.65		42.15	53
3400	3" diameter		48	.333		34.50	17.60		52.10	64.50
3410	4" diameter		44	.364		40.50	19.20		59.70	73.50
3420	5" diameter		35	.457		57.50	24		81.50	99.50
3430	6" diameter	Q-2	40	.600		78.50	33		111.50	136
3440	8" diameter		35	.686		147	37.50		184.50	218
3450	10" diameter		29	.828		271	45.50		316.50	365
3460	12" diameter		24	1		530	55		585	670
3920	To delete coupling & hanger, subtract									
3930	3/4" diam. to 2" diam.					30%	25%			
3940	2-1/2" diam. to 5" diam.					15%	17%			
3950	6" diam. to 12" diam.					11%	12%			
3990	Fittings: coupling material required at joints not incl. in fitting price.									
3994	Add 1 selected coupling, material only, per joint for installed price.									
4000	Elbow, 90° or 45°, painted									
4030	3/4" diameter	1 Plum	50	.160	Ea.	60	9.40		69.40	80
4040	1" diameter		50	.160		32	9.40		41.40	49
4050	1-1/4" diameter		40	.200		32	11.75		43.75	53
4060	1-1/2" diameter		33	.242		32	14.25		46.25	56.50
4070	2" diameter		25	.320		32	18.80		50.80	63.50
4080	2-1/2" diameter	Q-1	40	.400		32	21		53	67
4090	3" diameter		33	.485		56.50	25.50		82	101
4100	4" diameter		25	.640		61.50	34		95.50	119
4110	5" diameter		20	.800		146	42.50		188.50	225
4120	6" diameter	Q-2	25	.960		172	52.50		224.50	269
4130	8" diameter		21	1.143		360	62.50		422.50	490
4140	10" diameter		18	1.333		625	73		698	800
4150	12" diameter		15	1.600		1,000	87.50		1,087.50	1,225
4170	14" diameter		12	2		860	110		970	1,100
4180	16" diameter		11	2.182		1,125	120		1,245	1,400
4190	18" diameter	Q-3	15	2.133		1,425	119		1,544	1,725
4200	20" diameter		13	2.462		1,875	138		2,013	2,250
4210	24" diameter		11	2.909		2,700	163		2,863	3,225
4250	For galvanized elbows, add					26%				
4690	Tee, painted									
4700	3/4" diameter	1 Plum	38	.211	Ea.	64.50	12.35		76.85	89.50
4740	1" diameter		33	.242		50	14.25		64.25	76.50
4750	1-1/4" diameter		27	.296		50	17.40		67.40	81.50
4760	1-1/2" diameter		22	.364		50	21.50		71.50	87
4770	2" diameter		17	.471		50	27.50		77.50	96.50
4780	2-1/2" diameter	Q-1	27	.593		50	31.50		81.50	103
4790	3" diameter		22	.727		68	38.50		106.50	133

22 11 13 – Facility Water Distribution Piping

22 11 13.48 Pipe, Fittings and Valves, Steel, Grooved-Joint	Crew	Daily Output	Labor-Hours	Unit	Material	2015 Bare Costs Labor	Equipment	Total	Total Incl O&P
4800	4" diameter	Q-1	17	.941	Ea.	103	49.50		152.50	189
4810	5" diameter	↓	13	1.231		241	65		306	365
4820	6" diameter	Q-2	17	1.412		278	77.50		355.50	420
4830	8" diameter		14	1.714		610	94		704	810
4840	10" diameter		12	2		1,225	110		1,335	1,525
4850	12" diameter		10	2.400		1,700	131		1,831	2,075
4851	14" diameter		9	2.667		1,150	146		1,296	1,500
4852	16" diameter	↓	8	3		1,300	164		1,464	1,675
4853	18" diameter	Q-3	11	2.909		1,625	163		1,788	2,050
4854	20" diameter		10	3.200		2,325	179		2,504	2,850
4855	24" diameter	↓	8	4		3,550	224		3,774	4,250
4900	For galvanized tees, add				↓	24%				
4906	Couplings, rigid style, painted									
4908	1" diameter	1 Plum	100	.080	Ea.	25	4.70		29.70	34.50
4909	1-1/4" diameter		100	.080		25	4.70		29.70	34.50
4910	1-1/2" diameter		67	.119		25	7		32	38
4912	2" diameter	↓	50	.160		31.50	9.40		40.90	48.50
4914	2-1/2" diameter	Q-1	80	.200		36	10.55		46.55	55.50
4916	3" diameter		67	.239		41.50	12.60		54.10	65
4918	4" diameter		50	.320		58	16.90		74.90	89.50
4920	5" diameter	↓	40	.400		75	21		96	115
4922	6" diameter	Q-2	50	.480		99	26.50		125.50	149
4924	8" diameter		42	.571		156	31.50		187.50	219
4926	10" diameter		35	.686		305	37.50		342.50	390
4928	12" diameter		32	.750		340	41		381	430
4930	14" diameter		24	1		274	55		329	385
4931	16" diameter		20	1.200		355	65.50		420.50	490
4932	18" diameter		18	1.333		410	73		483	565
4933	20" diameter		16	1.500		565	82		647	745
4934	24" diameter	↓	13	1.846	↓	720	101		821	950
4940	Flexible, standard, painted									
4950	3/4" diameter	1 Plum	100	.080	Ea.	17.80	4.70		22.50	26.50
4960	1" diameter		100	.080		17.80	4.70		22.50	26.50
4970	1-1/4" diameter		80	.100		23.50	5.85		29.35	34.50
4980	1-1/2" diameter		67	.119		25.50	7		32.50	38.50
4990	2" diameter	↓	50	.160		27	9.40		36.40	44
5000	2-1/2" diameter	Q-1	80	.200		31.50	10.55		42.05	51
5010	3" diameter		67	.239		35	12.60		47.60	57.50
5020	3-1/2" diameter		57	.281		50	14.85		64.85	77.50
5030	4" diameter		50	.320		50.50	16.90		67.40	81
5040	5" diameter	↓	40	.400		76.50	21		97.50	117
5050	6" diameter	Q-2	50	.480		90.50	26.50		117	139
5070	8" diameter		42	.571		147	31.50		178.50	210
5090	10" diameter		35	.686		240	37.50		277.50	320
5110	12" diameter		32	.750		272	41		313	360
5120	14" diameter		24	1		390	55		445	515
5130	16" diameter		20	1.200		510	65.50		575.50	660
5140	18" diameter		18	1.333		595	73		668	765
5150	20" diameter		16	1.500		940	82		1,022	1,150
5160	24" diameter	↓	13	1.846	↓	1,025	101		1,126	1,275
5176	Lightweight style, painted									
5178	1-1/2" diameter	1 Plum	67	.119	Ea.	22.50	7		29.50	35
5180	2" diameter	"	50	.160	↓	27	9.40		36.40	44

For customer support on your Plumbing Cost Data, call 877.709.6968.

191

22 11 13 – Facility Water Distribution Piping

22 11 13.48 Pipe, Fittings and Valves, Steel, Grooved-Joint	Crew	Daily Output	Labor-Hours	Unit	Material	2015 Bare Costs Labor	Equipment	Total	Total Incl O&P
5182	2-1/2" diameter	Q-1	80	.200	Ea.	31.50	10.55		42.05	51
5184	3" diameter		67	.239		35	12.60		47.60	57.50
5186	3-1/2" diameter		57	.281		43	14.85		57.85	70
5188	4" diameter		50	.320		50.50	16.90		67.40	81
5190	5" diameter		40	.400		76.50	21		97.50	117
5192	6" diameter	Q-2	50	.480		90.50	26.50		117	139
5194	8" diameter		42	.571		147	31.50		178.50	210
5196	10" diameter		35	.686		297	37.50		334.50	380
5198	12" diameter		32	.750		335	41		376	425
5200	For galvanized couplings, add					33%				
5220	Tee, reducing, painted									
5225	2" x 1-1/2" diameter	Q-1	38	.421	Ea.	105	22		127	149
5226	2-1/2" x 2" diameter		28	.571		105	30		135	161
5227	3" x 2-1/2" diameter		23	.696		92.50	37		129.50	158
5228	4" x 3" diameter		18	.889		124	47		171	208
5229	5" x 4" diameter		15	1.067		266	56.50		322.50	375
5230	6" x 4" diameter	Q-2	18	1.333		293	73		366	430
5231	8" x 6" diameter		15	1.600		610	87.50		697.50	800
5232	10" x 8" diameter		13	1.846		770	101		871	1,000
5233	12" x 10" diameter		11	2.182		1,175	120		1,295	1,475
5234	14" x 12" diameter		10	2.400		840	131		971	1,125
5235	16" x 12" diameter		9	2.667		1,050	146		1,196	1,375
5236	18" x 12" diameter	Q-3	12	2.667		1,250	149		1,399	1,600
5237	18" x 16" diameter		11	2.909		1,575	163		1,738	1,975
5238	20" x 16" diameter		10	3.200		2,050	179		2,229	2,550
5239	24" x 20" diameter		9	3.556		3,250	199		3,449	3,875
5240	Reducer, concentric, painted									
5241	2-1/2" x 2" diameter	Q-1	43	.372	Ea.	37.50	19.65		57.15	71
5242	3" x 2-1/2" diameter		35	.457		45	24		69	86
5243	4" x 3" diameter		29	.552		54.50	29		83.50	104
5244	5" x 4" diameter		22	.727		75	38.50		113.50	141
5245	6" x 4" diameter	Q-2	26	.923		87	50.50		137.50	172
5246	8" x 6" diameter		23	1.043		224	57		281	335
5247	10" x 8" diameter		20	1.200		440	65.50		505.50	585
5248	12" x 10" diameter		16	1.500		790	82		872	990
5255	Eccentric, painted									
5256	2-1/2" x 2" diameter	Q-1	42	.381	Ea.	78	20		98	117
5257	3" x 2-1/2" diameter		34	.471		89	25		114	136
5258	4" x 3" diameter		28	.571		109	30		139	165
5259	5" x 4" diameter		21	.762		147	40.50		187.50	223
5260	6" x 4" diameter	Q-2	25	.960		172	52.50		224.50	269
5261	8" x 6" diameter		22	1.091		345	60		405	470
5262	10" x 8" diameter		19	1.263		915	69		984	1,100
5263	12" x 10" diameter		15	1.600		1,250	87.50		1,337.50	1,500
5270	Coupling, reducing, painted									
5272	2" x 1-1/2" diameter	1 Plum	52	.154	Ea.	36	9.05		45.05	53
5274	2-1/2" x 2" diameter	Q-1	82	.195		47	10.30		57.30	67
5276	3" x 2" diameter		69	.232		53	12.25		65.25	77
5278	4" x 2" diameter		52	.308		84	16.25		100.25	117
5280	5" x 4" diameter		42	.381		94.50	20		114.50	135
5282	6" x 4" diameter	Q-2	52	.462		142	25.50		167.50	195
5284	8" x 6" diameter	"	44	.545		213	30		243	279
5290	Outlet coupling, painted									

22 11 13 – Facility Water Distribution Piping

22 11 13.48 Pipe, Fittings and Valves, Steel, Grooved-Joint		Crew	Daily Output	Labor-Hours	Unit	Material	2015 Bare Costs		Total	Total Incl O&P
							Labor	Equipment		
5294	1-1/2" x 1" pipe size	1 Plum	65	.123	Ea.	47.50	7.20		54.70	63.50
5296	2" x 1" pipe size	"	48	.167		48.50	9.80		58.30	68.50
5298	2-1/2" x 1" pipe size	Q-1	78	.205		75	10.85		85.85	99
5300	2-1/2" x 1-1/4" pipe size		70	.229		84	12.10		96.10	111
5302	3" x 1" pipe size		65	.246		95.50	13		108.50	125
5304	4" x 3/4" pipe size		48	.333		106	17.60		123.60	143
5306	4" x 1-1/2" pipe size		46	.348		150	18.40		168.40	193
5308	6" x 1-1/2" pipe size	Q-2	44	.545		211	30		241	278
5320	Outlet, strap-on T, painted									
5324	Threaded female branch									
5330	2" pipe x 1-1/2" branch size	1 Plum	50	.160	Ea.	34.50	9.40		43.90	52
5334	2-1/2" pipe x 1-1/2" branch size	Q-1	80	.200		42	10.55		52.55	62.50
5338	3" pipe x 1-1/2" branch size		67	.239		44	12.60		56.60	67
5342	3" pipe x 2" branch size		59	.271		50.50	14.35		64.85	77
5346	4" pipe x 2" branch size		50	.320		53	16.90		69.90	84
5350	4" pipe x 2-1/2" branch size		47	.340		55.50	18		73.50	88
5354	4" pipe x 3" branch size		43	.372		59	19.65		78.65	94.50
5358	5" pipe x 3" branch size		40	.400		73	21		94	112
5362	6" pipe x 2" branch size	Q-2	50	.480		66.50	26.50		93	113
5366	6" pipe x 3" branch size		47	.511		76	28		104	126
5370	6" pipe x 4" branch size		44	.545		84	30		114	138
5374	8" pipe x 2-1/2" branch size		42	.571		116	31.50		147.50	176
5378	8" pipe x 4" branch size		38	.632		126	34.50		160.50	192
5390	Grooved branch									
5394	2" pipe x 1-1/4" branch size	1 Plum	78	.103	Ea.	34.50	6		40.50	47
5398	2" pipe x 1-1/2" branch size	"	50	.160		34.50	9.40		43.90	52
5402	2-1/2" pipe x 1-1/2" branch size	Q-1	80	.200		42	10.55		52.55	62.50
5406	3" pipe x 1-1/2" branch size		67	.239		44	12.60		56.60	67
5410	3" pipe x 2" branch size		59	.271		50.50	14.35		64.85	77
5414	4" pipe x 2" branch size		50	.320		53	16.90		69.90	84
5418	4" pipe x 2-1/2" branch size		47	.340		55.50	18		73.50	88
5422	4" pipe x 3" branch size		43	.372		59	19.65		78.65	94.50
5426	5" pipe x 3" branch size		40	.400		73	21		94	112
5430	6" pipe x 2" branch size	Q-2	50	.480		66.50	26.50		93	113
5434	6" pipe x 3" branch size	Q-1	78	.205		76	10.85		86.85	100
5438	6" pipe x 4" branch size	Q-2	44	.545		84	30		114	138
5442	8" pipe x 2-1/2" branch size		42	.571		116	31.50		147.50	176
5446	8" pipe x 4" branch size		38	.632		126	34.50		160.50	192
5750	Flange, w/groove gasket, black steel									
5754	See Line 22 11 13.47 0620 for gasket & bolt set									
5760	ANSI class 125 and 150, painted									
5780	2" pipe size	1 Plum	23	.348	Ea.	106	20.50		126.50	148
5790	2-1/2" pipe size	Q-1	37	.432		132	23		155	180
5800	3" pipe size		31	.516		142	27.50		169.50	197
5820	4" pipe size		23	.696		189	37		226	264
5830	5" pipe size		19	.842		220	44.50		264.50	310
5840	6" pipe size	Q-2	23	1.043		240	57		297	350
5850	8" pipe size		17	1.412		271	77.50		348.50	415
5860	10" pipe size		14	1.714		430	94		524	610
5870	12" pipe size		12	2		560	110		670	780
5880	14" pipe size		10	2.400		1,100	131		1,231	1,400
5890	16" pipe size		9	2.667		1,275	146		1,421	1,625
5900	18" pipe size		6	4		1,575	219		1,794	2,050

For customer support on your Plumbing Cost Data, call 877.709.6968.

193

22 11 13.48 Pipe, Fittings and Valves, Steel, Grooved-Joint		Crew	Daily Output	Labor-Hours	Unit	Material	2015 Bare Costs Labor	Equipment	Total	Total Incl O&P
5910	20" pipe size	Q-2	5	4.800	Ea.	1,875	263		2,138	2,475
5920	24" pipe size	▼	4.50	5.333	▼	2,400	292		2,692	3,100
5940	ANSI class 350, painted									
5946	2" pipe size	1 Plum	23	.348	Ea.	132	20.50		152.50	177
5948	2-1/2" pipe size	Q-1	37	.432		153	23		176	204
5950	3" pipe size		31	.516		209	27.50		236.50	271
5952	4" pipe size		23	.696		278	37		315	360
5954	5" pipe size	▼	19	.842		315	44.50		359.50	415
5956	6" pipe size	Q-2	23	1.043		370	57		427	490
5958	8" pipe size		17	1.412		420	77.50		497.50	580
5960	10" pipe size	▼	14	1.714		670	94		764	880
5962	12" pipe size	1 Plum	12	.667	▼	720	39		759	850
6100	Cross, painted									
6110	2" diameter	1 Plum	12.50	.640	Ea.	88.50	37.50		126	154
6112	2-1/2" diameter	Q-1	20	.800		88.50	42.50		131	161
6114	3" diameter		15.50	1.032		157	54.50		211.50	256
6116	4" diameter	▼	12.50	1.280		261	67.50		328.50	390
6118	6" diameter	Q-2	12.50	1.920		680	105		785	910
6120	8" diameter		10.50	2.286		900	125		1,025	1,175
6122	10" diameter		9	2.667		1,475	146		1,621	1,850
6124	12" diameter	▼	7.50	3.200	▼	2,125	175		2,300	2,625
7400	Suction diffuser									
7402	Grooved end inlet x flanged outlet									
7410	3" x 3"	Q-1	50	.320	Ea.	860	16.90		876.90	970
7412	4" x 4"		38	.421		1,175	22		1,197	1,300
7414	5" x 5"	▼	30	.533		1,350	28		1,378	1,550
7416	6" x 6"	Q-2	38	.632		1,725	34.50		1,759.50	1,925
7418	8" x 8"		27	.889		3,200	48.50		3,248.50	3,575
7420	10" x 10"		20	1.200		4,350	65.50		4,415.50	4,875
7422	12" x 12"		16	1.500		7,150	82		7,232	7,975
7424	14" x 14"		15	1.600		7,575	87.50		7,662.50	8,475
7426	16" x 14"	▼	14	1.714	▼	7,800	94		7,894	8,725
7500	Strainer, tee type, painted									
7506	2" pipe size	1 Plum	38	.211	Ea.	575	12.35		587.35	650
7508	2-1/2" pipe size	Q-1	62	.258		600	13.65		613.65	680
7510	3" pipe size		50	.320		675	16.90		691.90	770
7512	4" pipe size		38	.421		765	22		787	875
7514	5" pipe size	▼	30	.533		1,100	28		1,128	1,275
7516	6" pipe size	Q-2	38	.632		1,200	34.50		1,234.50	1,375
7518	8" pipe size		27	.889		1,850	48.50		1,898.50	2,100
7520	10" pipe size		20	1.200		2,700	65.50		2,765.50	3,075
7522	12" pipe size		16	1.500		3,475	82		3,557	3,950
7524	14" pipe size		15	1.600		10,200	87.50		10,287.50	11,400
7526	16" pipe size	▼	14	1.714	▼	12,700	94		12,794	14,100
7570	Expansion joint, max. 3" travel									
7572	2" diameter	1 Plum	38	.211	Ea.	720	12.35		732.35	815
7574	3" diameter	Q-1	50	.320		830	16.90		846.90	940
7576	4" diameter	"	38	.421		1,100	22		1,122	1,225
7578	6" diameter	Q-2	38	.632	▼	1,700	34.50		1,734.50	1,925
7800	Ball valve w/handle, carbon steel trim									
7810	1-1/2" pipe size	1 Plum	50	.160	Ea.	144	9.40		153.40	172
7812	2" pipe size	"	38	.211		159	12.35		171.35	194
7814	2-1/2" pipe size	Q-1	62	.258	▼	345	13.65		358.65	400

22 11 13 – Facility Water Distribution Piping

22 11 13.48 Pipe, Fittings and Valves, Steel, Grooved-Joint	Crew	Daily Output	Labor-Hours	Unit	Material	2015 Bare Costs Labor	Equipment	Total	Total Incl O&P
7816　3" pipe size	Q-1	50	.320	Ea.	560	16.90		576.90	640
7818　4" pipe size	↓	38	.421		870	22		892	990
7820　6" pipe size	Q-2	30	.800	↓	2,650	44		2,694	3,000
7830　With gear operator									
7834　2-1/2" pipe size	Q-1	62	.258	Ea.	700	13.65		713.65	790
7836　3" pipe size		50	.320		995	16.90		1,011.90	1,125
7838　4" pipe size	↓	38	.421		1,275	22		1,297	1,425
7840　6" pipe size	Q-2	30	.800	↓	3,000	44		3,044	3,350
7870　Check valve									
7874　2-1/2" pipe size	Q-1	62	.258	Ea.	274	13.65		287.65	320
7876　3" pipe size		50	.320		325	16.90		341.90	380
7878　4" pipe size		38	.421		340	22		362	410
7880　5" pipe size	↓	30	.533		570	28		598	670
7882　6" pipe size	Q-2	38	.632		675	34.50		709.50	795
7884　8" pipe size		27	.889		920	48.50		968.50	1,075
7886　10" pipe size		20	1.200		2,450	65.50		2,515.50	2,800
7888　12" pipe size	↓	16	1.500	↓	2,925	82		3,007	3,325
7900　Plug valve, balancing, w/lever operator									
7906　3" pipe size	Q-1	50	.320	Ea.	680	16.90		696.90	770
7908　4" pipe size	"	38	.421		835	22		857	955
7909　6" pipe size	Q-2	30	.800	↓	1,275	44		1,319	1,475
7916　With gear operator									
7920　3" pipe size	Q-1	50	.320	Ea.	1,275	16.90		1,291.90	1,450
7922　4" pipe size	"	38	.421		1,325	22		1,347	1,500
7924　6" pipe size	Q-2	38	.632		1,800	34.50		1,834.50	2,025
7926　8" pipe size		27	.889		2,775	48.50		2,823.50	3,125
7928　10" pipe size		20	1.200		3,475	65.50		3,540.50	3,925
7930　12" pipe size	↓	16	1.500	↓	5,325	82		5,407	5,975
8000　Butterfly valve, 2 position handle, with standard trim									
8010　1-1/2" pipe size	1 Plum	50	.160	Ea.	257	9.40		266.40	297
8020　2" pipe size	"	38	.211		257	12.35		269.35	300
8030　3" pipe size	Q-1	50	.320		370	16.90		386.90	430
8050　4" pipe size	"	38	.421		405	22		427	480
8070　6" pipe size	Q-2	38	.632		820	34.50		854.50	955
8080　8" pipe size		27	.889		1,100	48.50		1,148.50	1,300
8090　10" pipe size	↓	20	1.200	↓	1,800	65.50		1,865.50	2,100
8200　With stainless steel trim									
8240　1-1/2" pipe size	1 Plum	50	.160	Ea.	325	9.40		334.40	375
8250　2" pipe size	"	38	.211		325	12.35		337.35	380
8270　3" pipe size	Q-1	50	.320		440	16.90		456.90	505
8280　4" pipe size	"	38	.421		475	22		497	560
8300　6" pipe size	Q-2	38	.632		890	34.50		924.50	1,025
8310　8" pipe size		27	.889		2,200	48.50		2,248.50	2,500
8320　10" pipe size		20	1.200		3,675	65.50		3,740.50	4,125
8322　12" pipe size		16	1.500		4,200	82		4,282	4,750
8324　14" pipe size		15	1.600		5,400	87.50		5,487.50	6,050
8326　16" pipe size	↓	14	1.714		7,800	94		7,894	8,725
8328　18" pipe size	Q-3	12	2.667		9,600	149		9,749	10,800
8330　20" pipe size		11	2.909		11,800	163		11,963	13,200
8332　24" pipe size		10	3.200		15,800	179		15,979	17,700
8336　Note: sizes 8" up w/manual gear operator									
9000　Cut one groove, labor									
9010　3/4" pipe size	Q-1	152	.105	Ea.		5.55		5.55	8.40

22 11 Facility Water Distribution

22 11 13 – Facility Water Distribution Piping

22 11 13.48 Pipe, Fittings and Valves, Steel, Grooved-Joint		Crew	Daily Output	Labor-Hours	Unit	Material	2015 Bare Costs Labor	Equipment	Total	Total Incl O&P
9020	1" pipe size	Q-1	140	.114	Ea.		6.05		6.05	9.10
9030	1-1/4" pipe size		124	.129			6.80		6.80	10.30
9040	1-1/2" pipe size		114	.140			7.40		7.40	11.20
9050	2" pipe size		104	.154			8.15		8.15	12.25
9060	2-1/2" pipe size		96	.167			8.80		8.80	13.30
9070	3" pipe size		88	.182			9.60		9.60	14.50
9080	3-1/2" pipe size		83	.193			10.20		10.20	15.40
9090	4" pipe size		78	.205			10.85		10.85	16.35
9100	5" pipe size		72	.222			11.75		11.75	17.75
9110	6" pipe size		70	.229			12.10		12.10	18.25
9120	8" pipe size		54	.296			15.65		15.65	23.50
9130	10" pipe size		38	.421			22		22	33.50
9140	12" pipe size		30	.533			28		28	42.50
9150	14" pipe size		20	.800			42.50		42.50	64
9160	16" pipe size		19	.842			44.50		44.50	67
9170	18" pipe size		18	.889			47		47	71
9180	20" pipe size		17	.941			49.50		49.50	75
9190	24" pipe size	▼	15	1.067	▼		56.50		56.50	85
9210	Roll one groove									
9220	3/4" pipe size	Q-1	266	.060	Ea.		3.18		3.18	4.80
9230	1" pipe size		228	.070			3.71		3.71	5.60
9240	1-1/4" pipe size		200	.080			4.23		4.23	6.40
9250	1-1/2" pipe size		178	.090			4.75		4.75	7.15
9260	2" pipe size		116	.138			7.30		7.30	11
9270	2-1/2" pipe size		110	.145			7.70		7.70	11.60
9280	3" pipe size		100	.160			8.45		8.45	12.75
9290	3-1/2" pipe size		94	.170			9		9	13.60
9300	4" pipe size		86	.186			9.85		9.85	14.85
9310	5" pipe size		84	.190			10.05		10.05	15.20
9320	6" pipe size		80	.200			10.55		10.55	15.95
9330	8" pipe size		66	.242			12.80		12.80	19.35
9340	10" pipe size		58	.276			14.55		14.55	22
9350	12" pipe size		46	.348			18.40		18.40	28
9360	14" pipe size		30	.533			28		28	42.50
9370	16" pipe size		28	.571			30		30	45.50
9380	18" pipe size		27	.593			31.50		31.50	47.50
9390	20" pipe size		25	.640			34		34	51
9400	24" pipe size	▼	23	.696	▼		37		37	55.50

22 11 13.60 Tubing, Stainless Steel

		Crew	Daily Output	Labor-Hours	Unit	Material	2015 Bare Costs Labor	Equipment	Total	Total Incl O&P
0010	**TUBING, STAINLESS STEEL**									
5000	Tubing									
5010	Type 304, no joints, no hangers									
5020	.035 wall									
5021	1/4"	1 Plum	160	.050	L.F.	3.12	2.94		6.06	7.85
5022	3/8"		160	.050		3.94	2.94		6.88	8.75
5023	1/2"		160	.050		5.05	2.94		7.99	10.05
5024	5/8"		160	.050		6.20	2.94		9.14	11.30
5025	3/4"		133	.060		7.50	3.53		11.03	13.60
5026	7/8"		133	.060		7.90	3.53		11.43	14.05
5027	1"	▼	114	.070	▼	8.70	4.12		12.82	15.75
5040	.049 wall									
5041	1/4"	1 Plum	160	.050	L.F.	3.67	2.94		6.61	8.45

22 11 Facility Water Distribution

22 11 13 – Facility Water Distribution Piping

22 11 13.60 Tubing, Stainless Steel		Crew	Daily Output	Labor-Hours	Unit	Material	2015 Bare Costs Labor	Equipment	Total	Total Incl O&P
5042	3/8"	1 Plum	160	.050	L.F.	5.05	2.94		7.99	10.05
5043	1/2"		160	.050		5.60	2.94		8.54	10.65
5044	5/8"		160	.050		7.60	2.94		10.54	12.80
5045	3/4"		133	.060		8.10	3.53		11.63	14.25
5046	7/8"		133	.060		8.45	3.53		11.98	14.60
5047	1"		114	.070		8.95	4.12		13.07	16.05
5060	.065 wall									
5061	1/4"	1 Plum	160	.050	L.F.	4.11	2.94		7.05	8.95
5062	3/8"		160	.050		5.65	2.94		8.59	10.70
5063	1/2"		160	.050		6.85	2.94		9.79	12
5064	5/8"		160	.050		8.30	2.94		11.24	13.55
5065	3/4"		133	.060		8.85	3.53		12.38	15.05
5066	7/8"		133	.060		10.15	3.53		13.68	16.50
5067	1"		114	.070		9.95	4.12		14.07	17.10
5210	Type 316									
5220	.035 wall									
5221	1/4"	1 Plum	160	.050	L.F.	3.96	2.94		6.90	8.80
5222	3/8"		160	.050		5	2.94		7.94	9.95
5223	1/2"		160	.050		7	2.94		9.94	12.15
5224	5/8"		160	.050		9.10	2.94		12.04	14.45
5225	3/4"		133	.060		10.50	3.53		14.03	16.90
5226	7/8"		133	.060		15	3.53		18.53	22
5227	1"		114	.070		11.90	4.12		16.02	19.30
5240	.049 wall									
5241	1/4"	1 Plum	160	.050	L.F.	5.45	2.94		8.39	10.45
5242	3/8"		160	.050		7.15	2.94		10.09	12.30
5243	1/2"		160	.050		7.85	2.94		10.79	13.10
5244	5/8"		160	.050		10.90	2.94		13.84	16.45
5245	3/4"		133	.060		10.50	3.53		14.03	16.90
5246	7/8"		133	.060		13.50	3.53		17.03	20
5247	1"		114	.070		12.30	4.12		16.42	19.75
5260	.065 wall									
5261	1/4"	1 Plum	160	.050	L.F.	5.20	2.94		8.14	10.15
5262	3/8"		160	.050		8.30	2.94		11.24	13.55
5263	1/2"		160	.050		8.90	2.94		11.84	14.25
5264	5/8"		160	.050		12.40	2.94		15.34	18.10
5265	3/4"		133	.060		13.35	3.53		16.88	20
5266	7/8"		133	.060		15.80	3.53		19.33	23
5267	1"		114	.070		15.05	4.12		19.17	23

22 11 13.61 Tubing Fittings, Stainless Steel

22 11 13.61 Tubing Fittings, Stainless Steel		Crew	Daily Output	Labor-Hours	Unit	Material	2015 Bare Costs Labor	Equipment	Total	Total Incl O&P
0010	**TUBING FITTINGS, STAINLESS STEEL**									
8200	Tube fittings, compression type									
8202	Type 316									
8204	90° elbow									
8206	1/4"	1 Plum	24	.333	Ea.	17.60	19.55		37.15	49
8207	3/8"		22	.364		21.50	21.50		43	56
8208	1/2"		22	.364		35.50	21.50		57	71
8209	5/8"		21	.381		40	22.50		62.50	78
8210	3/4"		21	.381		63	22.50		85.50	103
8211	7/8"		20	.400		96.50	23.50		120	142
8212	1"		20	.400		120	23.50		143.50	168
8220	Union tee									

197

22 11 13.61 Tubing Fittings, Stainless Steel

		Crew	Daily Output	Labor-Hours	Unit	Material	2015 Bare Costs Labor	2015 Bare Costs Equipment	Total	Total Incl O&P
8222	1/4"	1 Plum	15	.533	Ea.	25	31.50		56.50	75
8224	3/8"		15	.533		32	31.50		63.50	83
8225	1/2"		15	.533		49.50	31.50		81	102
8226	5/8"		14	.571		55.50	33.50		89	112
8227	3/4"		14	.571		74	33.50		107.50	132
8228	7/8"		13	.615		148	36		184	218
8229	1"		13	.615		159	36		195	230
8234	Union									
8236	1/4"	1 Plum	24	.333	Ea.	12.20	19.55		31.75	43
8237	3/8"		22	.364		17.40	21.50		38.90	51
8238	1/2"		22	.364		26	21.50		47.50	60.50
8239	5/8"		21	.381		34	22.50		56.50	71
8240	3/4"		21	.381		42	22.50		64.50	80
8241	7/8"		20	.400		68.50	23.50		92	111
8242	1"		20	.400		72.50	23.50		96	116
8250	Male connector									
8252	1/4" x 1/4"	1 Plum	24	.333	Ea.	7.80	19.55		27.35	38
8253	3/8" x 3/8"		22	.364		12.20	21.50		33.70	45.50
8254	1/2" x 1/2"		22	.364		18.05	21.50		39.55	52
8256	3/4" x 3/4"		21	.381		27.50	22.50		50	64.50
8258	1" x 1"		20	.400		48.50	23.50		72	88.50

22 11 13.64 Pipe, Stainless Steel

		Crew	Daily Output	Labor-Hours	Unit	Material	2015 Bare Costs Labor	2015 Bare Costs Equipment	Total	Total Incl O&P
0010	**PIPE, STAINLESS STEEL** R221113-70									
0020	Welded, with clevis type hanger assemblies, 10' O.C.									
0500	Schedule 5, type 304									
0540	1/2" diameter	Q-15	128	.125	L.F.	8.25	6.60	.45	15.30	19.55
0550	3/4" diameter		116	.138		10.25	7.30	.50	18.05	23
0560	1" diameter		103	.155		11.35	8.20	.56	20.11	25.50
0570	1-1/4" diameter		93	.172		15.45	9.10	.62	25.17	31.50
0580	1-1/2" diameter		85	.188		18.60	9.95	.68	29.23	36.50
0590	2" diameter		69	.232		31	12.25	.84	44.09	53.50
0600	2-1/2" diameter		53	.302		33.50	15.95	1.09	50.54	62
0610	3" diameter		48	.333		40	17.60	1.20	58.80	72
0620	4" diameter		44	.364		51	19.20	1.31	71.51	86.50
0630	5" diameter		36	.444		103	23.50	1.60	128.10	151
0640	6" diameter	Q-16	42	.571		95.50	31.50	1.37	128.37	154
0650	8" diameter		34	.706		146	38.50	1.69	186.19	221
0660	10" diameter		26	.923		202	50.50	2.22	254.72	300
0670	12" diameter		21	1.143		265	62.50	2.74	330.24	390
0700	To delete hangers, subtract									
0710	1/2" diam. to 1-1/2" diam.					8%	19%			
0720	2" diam. to 5" diam.					4%	9%			
0730	6" diam. to 12" diam.					3%	4%			
0750	For small quantities, add				L.F.	10%				
1250	Schedule 5, type 316									
1290	1/2" diameter	Q-15	128	.125	L.F.	10.30	6.60	.45	17.35	22
1300	3/4" diameter		116	.138		14.75	7.30	.50	22.55	28
1310	1" diameter		103	.155		16.45	8.20	.56	25.21	31
1320	1-1/4" diameter		93	.172		20	9.10	.62	29.72	37
1330	1-1/2" diameter		85	.188		27.50	9.95	.68	38.13	46.50
1340	2" diameter		69	.232		42.50	12.25	.84	55.59	66
1350	2-1/2" diameter		53	.302		54.50	15.95	1.09	71.54	85

198

For customer support on your Plumbing Cost Data, call 877.709.6968.

22 11 13 – Facility Water Distribution Piping

22 11 13.64 Pipe, Stainless Steel		Crew	Daily Output	Labor-Hours	Unit	Material	2015 Bare Costs Labor	Equipment	Total	Total Incl O&P
1360	3" diameter	Q-15	48	.333	L.F.	69	17.60	1.20	87.80	104
1370	4" diameter		44	.364		85.50	19.20	1.31	106.01	124
1380	5" diameter	↓	36	.444		163	23.50	1.60	188.10	216
1390	6" diameter	Q-16	42	.571		159	31.50	1.37	191.87	224
1400	8" diameter		34	.706		239	38.50	1.69	279.19	320
1410	10" diameter		26	.923		340	50.50	2.22	392.72	455
1420	12" diameter	↓	21	1.143		440	62.50	2.74	505.24	585
1490	For small quantities, add				↓	10%				
1940	To delete hanger, subtract									
1950	1/2" diam. to 1-1/2" diam.					5%	19%			
1960	2" diam. to 5" diam.					3%	9%			
1970	6" diam. to 12" diam.					2%	4%			
2000	Schedule 10, type 304									
2040	1/4" diameter	Q-15	131	.122	L.F.	5.40	6.45	.44	12.29	16.20
2050	3/8" diameter		128	.125		6.45	6.60	.45	13.50	17.55
2060	1/2" diameter		125	.128		6.75	6.75	.46	13.96	18.15
2070	3/4" diameter		113	.142		7.80	7.50	.51	15.81	20.50
2080	1" diameter		100	.160		11.15	8.45	.58	20.18	25.50
2090	1-1/4" diameter		91	.176		12.70	9.30	.63	22.63	28.50
2100	1-1/2" diameter		83	.193		14.10	10.20	.70	25	31.50
2110	2" diameter		67	.239		17.10	12.60	.86	30.56	39
2120	2-1/2" diameter		51	.314		22	16.55	1.13	39.68	50.50
2130	3" diameter		46	.348		26.50	18.40	1.26	46.16	58.50
2140	4" diameter		42	.381		32.50	20	1.38	53.88	67.50
2150	5" diameter	↓	35	.457		43	24	1.65	68.65	86
2160	6" diameter	Q-16	40	.600		51	33	1.44	85.44	107
2170	8" diameter		33	.727		88.50	40	1.75	130.25	159
2180	10" diameter		25	.960		122	52.50	2.30	176.80	216
2190	12" diameter	↓	21	1.143		152	62.50	2.74	217.24	265
2250	For small quantities, add				↓	10%				
2650	To delete hanger, subtract									
2660	1/4" diam. to 3/4" diam.					9%	22%			
2670	1" diam. to 2" diam.					4%	15%			
2680	2-1/2" diam. to 5" diam.					3%	8%			
2690	6" diam. to 12" diam.					3%	4%			
2750	Schedule 10, type 316									
2790	1/4" diameter	Q-15	131	.122	L.F.	6.50	6.45	.44	13.39	17.40
2800	3/8" diameter		128	.125		7.60	6.60	.45	14.65	18.80
2810	1/2" diameter		125	.128		8.45	6.75	.46	15.66	20
2820	3/4" diameter		113	.142		9.95	7.50	.51	17.96	23
2830	1" diameter		100	.160		14.10	8.45	.58	23.13	29
2840	1-1/4" diameter		91	.176		17.90	9.30	.63	27.83	34.50
2850	1-1/2" diameter		83	.193		19.75	10.20	.70	30.65	38
2860	2" diameter		67	.239		24	12.60	.86	37.46	46.50
2870	2-1/2" diameter		51	.314		31	16.55	1.13	48.68	60.50
2880	3" diameter		46	.348		36.50	18.40	1.26	56.16	69.50
2890	4" diameter		42	.381		46.50	20	1.38	67.88	83
2900	5" diameter	↓	35	.457		56.50	24	1.65	82.15	100
2910	6" diameter	Q-16	40	.600		68.50	33	1.44	102.94	127
2920	8" diameter		33	.727		130	40	1.75	171.75	205
2930	10" diameter		25	.960		171	52.50	2.30	225.80	270
2940	12" diameter	↓	21	1.143		214	62.50	2.74	279.24	335
2990	For small quantities, add				↓	10%				

22 11 13.64 Pipe, Stainless Steel		Crew	Daily Output	Labor-Hours	Unit	Material	2015 Bare Costs Labor	Equipment	Total	Total Incl O&P
3430	To delete hanger, subtract									
3440	1/4" diam. to 3/4" diam.					6%	22%			
3450	1" diam. to 2" diam.					3%	15%			
3460	2-1/2" diam. to 5" diam.					2%	8%			
3470	6" diam. to 12" diam.					2%	4%			
3500	Threaded, couplings and clevis hanger assemblies, 10' O.C.									
3520	Schedule 40, type 304									
3540	1/4" diameter	1 Plum	54	.148	L.F.	10.85	8.70		19.55	25
3550	3/8" diameter		53	.151		11.05	8.85		19.90	25.50
3560	1/2" diameter		52	.154		12.85	9.05		21.90	28
3570	3/4" diameter		51	.157		13.35	9.20		22.55	28.50
3580	1" diameter		45	.178		19.30	10.45		29.75	37
3590	1-1/4" diameter	Q-1	76	.211		27	11.10		38.10	46.50
3600	1-1/2" diameter		69	.232		29.50	12.25		41.75	50.50
3610	2" diameter		57	.281		45	14.85		59.85	72
3620	2-1/2" diameter		44	.364		77	19.20		96.20	114
3630	3" diameter		38	.421		100	22		122	144
3640	4" diameter	Q-2	51	.471		138	26		164	190
3740	For small quantities, add					10%				
4200	To delete couplings & hangers, subtract									
4210	1/4" diam. to 3/4" diam.					15%	56%			
4220	1" diam. to 2" diam.					18%	49%			
4230	2-1/2" diam. to 4" diam.					34%	40%			
4250	Schedule 40, type 316									
4290	1/4" diameter	1 Plum	54	.148	L.F.	11.65	8.70		20.35	26
4300	3/8" diameter		53	.151		12.70	8.85		21.55	27.50
4310	1/2" diameter		52	.154		15.50	9.05		24.55	30.50
4320	3/4" diameter		51	.157		17.95	9.20		27.15	33.50
4330	1" diameter		45	.178		25	10.45		35.45	43.50
4340	1-1/4" diameter	Q-1	76	.211		34	11.10		45.10	54
4350	1-1/2" diameter		69	.232		38.50	12.25		50.75	61
4360	2" diameter		57	.281		53	14.85		67.85	81
4370	2-1/2" diameter		44	.364		90.50	19.20		109.70	129
4380	3" diameter		38	.421		116	22		138	162
4390	4" diameter	Q-2	51	.471		153	26		179	207
4490	For small quantities, add					10%				
4900	To delete couplings & hangers, subtract									
4910	1/4" diam. to 3/4" diam.					12%	56%			
4920	1" diam. to 2" diam.					14%	49%			
4930	2-1/2" diam. to 4" diam.					27%	40%			
5000	Schedule 80, type 304									
5040	1/4" diameter	1 Plum	53	.151	L.F.	18.10	8.85		26.95	33.50
5050	3/8" diameter		52	.154		20.50	9.05		29.55	36.50
5060	1/2" diameter		51	.157		25	9.20		34.20	41.50
5070	3/4" diameter		48	.167		26.50	9.80		36.30	44
5080	1" diameter		43	.186		33	10.90		43.90	53
5090	1-1/4" diameter	Q-1	73	.219		44.50	11.60		56.10	66.50
5100	1-1/2" diameter		67	.239		56	12.60		68.60	81
5110	2" diameter		54	.296		69.50	15.65		85.15	100
5190	For small quantities, add					10%				
5700	To delete couplings & hangers, subtract									
5710	1/4" diam. to 3/4" diam.					10%	53%			
5720	1" diam. to 2" diam.					14%	47%			

22 11 13 – Facility Water Distribution Piping

22 11 13.64 Pipe, Stainless Steel	Crew	Daily Output	Labor-Hours	Unit	Material	2015 Bare Costs Labor	Equipment	Total	Total Incl O&P
5750	Schedule 80, type 316								
5790	1/4" diameter	1 Plum	53	.151	L.F.	21.50	8.85		30.35	37
5800	3/8" diameter		52	.154		26.50	9.05		35.55	42.50
5810	1/2" diameter		51	.157		31	9.20		40.20	48.50
5820	3/4" diameter		48	.167		34	9.80		43.80	52.50
5830	1" diameter		43	.186		44.50	10.90		55.40	65.50
5840	1-1/4" diameter	Q-1	73	.219		64	11.60		75.60	88
5850	1-1/2" diameter		67	.239		68.50	12.60		81.10	94.50
5860	2" diameter		54	.296		86	15.65		101.65	119
5950	For small quantities, add					10%				
7000	To delete couplings & hangers, subtract									
7010	1/4" diam. to 3/4" diam.					9%	53%			
7020	1" diam. to 2" diam.					14%	47%			
8000	Weld joints with clevis type hanger assemblies, 10' O.C.									
8010	Schedule 40, type 304									
8050	1/8" pipe size	Q-15	126	.127	L.F.	8.70	6.70	.46	15.86	20
8060	1/4" pipe size		125	.128		9.60	6.75	.46	16.81	21.50
8070	3/8" pipe size		122	.131		9.60	6.95	.47	17.02	21.50
8080	1/2" pipe size		118	.136		10.95	7.15	.49	18.59	23.50
8090	3/4" pipe size		109	.147		10.80	7.75	.53	19.08	24
8100	1" pipe size		95	.168		15.15	8.90	.61	24.66	31
8110	1-1/4" pipe size		86	.186		20.50	9.85	.67	31.02	38
8120	1-1/2" pipe size		78	.205		22	10.85	.74	33.59	41
8130	2" pipe size		62	.258		32.50	13.65	.93	47.08	57.50
8140	2-1/2" pipe size		49	.327		48.50	17.25	1.18	66.93	80.50
8150	3" pipe size		44	.364		61	19.20	1.31	81.51	98
8160	3-1/2" pipe size		44	.364		72.50	19.20	1.31	93.01	110
8170	4" pipe size		39	.410		83	21.50	1.48	105.98	125
8180	5" pipe size		32	.500		110	26.50	1.81	138.31	163
8190	6" pipe size	Q-16	37	.649		128	35.50	1.56	165.06	196
8200	8" pipe size		29	.828		179	45.50	1.99	226.49	268
8210	10" pipe size		24	1		305	55	2.40	362.40	420
8220	12" pipe size		20	1.200		490	65.50	2.88	558.38	645
8300	Schedule 40, type 316									
8310	1/8" pipe size	Q-15	126	.127	L.F.	9.25	6.70	.46	16.41	21
8320	1/4" pipe size		125	.128		10.15	6.75	.46	17.36	22
8330	3/8" pipe size		122	.131		10.95	6.95	.47	18.37	23
8340	1/2" pipe size		118	.136		13.15	7.15	.49	20.79	26
8350	3/4" pipe size		109	.147		14.85	7.75	.53	23.13	28.50
8360	1" pipe size		95	.168		20	8.90	.61	29.51	36.50
8370	1-1/4" pipe size		86	.186		26	9.85	.67	36.52	44
8380	1-1/2" pipe size		78	.205		29.50	10.85	.74	41.09	49.50
8390	2" pipe size		62	.258		38.50	13.65	.93	53.08	63.50
8400	2-1/2" pipe size		49	.327		56	17.25	1.18	74.43	89
8410	3" pipe size		44	.364		69	19.20	1.31	89.51	106
8420	3-1/2" pipe size		44	.364		80.50	19.20	1.31	101.01	119
8430	4" pipe size		39	.410		87	21.50	1.48	109.98	130
8440	5" pipe size		32	.500		120	26.50	1.81	148.31	174
8450	6" pipe size	Q-16	37	.649		138	35.50	1.56	175.06	207
8460	8" pipe size		29	.828		184	45.50	1.99	231.49	273
8470	10" pipe size		24	1		245	55	2.40	302.40	355
8480	12" pipe size		20	1.200		330	65.50	2.88	398.38	470
8500	Schedule 80, type 304									

For customer support on your Plumbing Cost Data, call 877.709.6968.

201

22 11 13.64 Pipe, Stainless Steel

		Crew	Daily Output	Labor-Hours	Unit	Material	2015 Bare Costs Labor	Equipment	Total	Total Incl O&P
8510	1/4" pipe size	Q-15	110	.145	L.F.	16.75	7.70	.53	24.98	30.50
8520	3/8" pipe size		109	.147		19.05	7.75	.53	27.33	33.50
8530	1/2" pipe size		106	.151		23	7.95	.54	31.49	38
8540	3/4" pipe size		96	.167		24	8.80	.60	33.40	40.50
8550	1" pipe size		87	.184		28.50	9.70	.66	38.86	47
8560	1-1/4" pipe size		81	.198		34	10.45	.71	45.16	53.50
8570	1-1/2" pipe size		74	.216		43.50	11.40	.78	55.68	66
8580	2" pipe size		58	.276		53	14.55	1	68.55	81.50
8590	2-1/2" pipe size		46	.348		68	18.40	1.26	87.66	104
8600	3" pipe size		41	.390		78.50	20.50	1.41	100.41	119
8610	4" pipe size		33	.485		96.50	25.50	1.75	123.75	146
8630	6" pipe size	Q-16	30	.800		171	44	1.92	216.92	256
8640	Schedule 80, type 316									
8650	1/4" pipe size	Q-15	110	.145	L.F.	19.95	7.70	.53	28.18	34
8660	3/8" pipe size		109	.147		24.50	7.75	.53	32.78	39.50
8670	1/2" pipe size		106	.151		29	7.95	.54	37.49	44
8680	3/4" pipe size		96	.167		30.50	8.80	.60	39.90	47.50
8690	1" pipe size		87	.184		38.50	9.70	.66	48.86	58
8700	1-1/4" pipe size		81	.198		50.50	10.45	.71	61.66	72
8710	1-1/2" pipe size		74	.216		53	11.40	.78	65.18	76
8720	2" pipe size		58	.276		64	14.55	1	79.55	93.50
8730	2-1/2" pipe size		46	.348		80	18.40	1.26	99.66	117
8740	3" pipe size		41	.390		92.50	20.50	1.41	114.41	135
8760	4" pipe size		33	.485		122	25.50	1.75	149.25	174
8770	6" pipe size	Q-16	30	.800		219	44	1.92	264.92	310
9100	Threading pipe labor, sst, one end, schedules 40 & 80									
9110	1/4" through 3/4" pipe size	1 Plum	61.50	.130	Ea.		7.65		7.65	11.55
9120	1" through 2" pipe size		55.90	.143			8.40		8.40	12.70
9130	2-1/2" pipe size		41.50	.193			11.30		11.30	17.10
9140	3" pipe size		38.50	.208			12.20		12.20	18.40
9150	3-1/2" pipe size	Q-1	68.40	.234			12.35		12.35	18.65
9160	4" pipe size		73	.219			11.60		11.60	17.50
9170	5" pipe size		40.70	.393			21		21	31.50
9180	6" pipe size		35.40	.452			24		24	36
9190	8" pipe size		22.30	.717			38		38	57
9200	10" pipe size		16.10	.994			52.50		52.50	79.50
9210	12" pipe size		12.30	1.301			68.50		68.50	104
9250	Welding labor per joint for stainless steel									
9260	Schedule 5 and 10									
9270	1/4" pipe size	Q-15	36	.444	Ea.		23.50	1.60	25.10	37.50
9280	3/8" pipe size		35	.457			24	1.65	25.65	38.50
9290	1/2" pipe size		35	.457			24	1.65	25.65	38.50
9300	3/4" pipe size		28	.571			30	2.06	32.06	48
9310	1" pipe size		25	.640			34	2.31	36.31	53.50
9320	1-1/4" pipe size		22	.727			38.50	2.63	41.13	61
9330	1-1/2" pipe size		21	.762			40.50	2.75	43.25	64
9340	2" pipe size		18	.889			47	3.21	50.21	74.50
9350	2-1/2" pipe size		12	1.333			70.50	4.81	75.31	111
9360	3" pipe size		9.73	1.644			87	5.95	92.95	138
9370	4" pipe size		7.37	2.171			115	7.85	122.85	182
9380	5" pipe size		6.15	2.602			137	9.40	146.40	218
9390	6" pipe size		5.71	2.802			148	10.10	158.10	235
9400	8" pipe size		3.69	4.336			229	15.65	244.65	360

For customer support on your Plumbing Cost Data, call 877.709.6968.

22 11 Facility Water Distribution

22 11 13 – Facility Water Distribution Piping

22 11 13.64 Pipe, Stainless Steel		Crew	Daily Output	Labor-Hours	Unit	Material	2015 Bare Costs Labor	Equipment	Total	Total Incl O&P
9410	10" pipe size	Q-15	2.91	5.498	Ea.		290	19.85	309.85	460
9420	12" pipe size	↓	2.31	6.926	↓		365	25	390	585
9500	Schedule 40									
9510	1/4" pipe size	Q-15	28	.571	Ea.		30	2.06	32.06	48
9520	3/8" pipe size		27	.593			31.50	2.14	33.64	50
9530	1/2" pipe size		25.40	.630			33.50	2.27	35.77	53
9540	3/4" pipe size		22.22	.720			38	2.60	40.60	60.50
9550	1" pipe size		20.25	.790			41.50	2.85	44.35	66
9560	1-1/4" pipe size		18.82	.850			45	3.07	48.07	71.50
9570	1-1/2" pipe size		17.78	.900			47.50	3.25	50.75	75.50
9580	2" pipe size		15.09	1.060			56	3.83	59.83	88.50
9590	2-1/2" pipe size		7.96	2.010			106	7.25	113.25	168
9600	3" pipe size		6.43	2.488			131	9	140	209
9610	4" pipe size		4.88	3.279			173	11.85	184.85	275
9620	5" pipe size		4.26	3.756			198	13.55	211.55	315
9630	6" pipe size		3.77	4.244			224	15.30	239.30	355
9640	8" pipe size		2.44	6.557			345	23.50	368.50	550
9650	10" pipe size		1.92	8.333			440	30	470	700
9660	12" pipe size	↓	1.52	10.526	↓		555	38	593	880
9750	Schedule 80									
9760	1/4" pipe size	Q-15	21.55	.742	Ea.		39	2.68	41.68	62
9770	3/8" pipe size		20.75	.771			40.50	2.78	43.28	64.50
9780	1/2" pipe size		19.54	.819			43.50	2.96	46.46	69
9790	3/4" pipe size		17.09	.936			49.50	3.38	52.88	78
9800	1" pipe size		15.58	1.027			54.50	3.71	58.21	86
9810	1-1/4" pipe size		14.48	1.105			58.50	3.99	62.49	92.50
9820	1-1/2" pipe size		13.68	1.170			62	4.22	66.22	98
9830	2" pipe size		11.61	1.378			73	4.98	77.98	115
9840	2-1/2" pipe size		6.12	2.614			138	9.45	147.45	219
9850	3" pipe size		4.94	3.239			171	11.70	182.70	271
9860	4" pipe size		3.75	4.267			225	15.40	240.40	355
9870	5" pipe size		3.27	4.893			259	17.65	276.65	410
9880	6" pipe size		2.90	5.517			291	19.90	310.90	460
9890	8" pipe size		1.87	8.556			450	31	481	720
9900	10" pipe size		1.48	10.811			570	39	609	905
9910	12" pipe size		1.17	13.675			720	49.50	769.50	1,150
9920	Schedule 160, 1/2" pipe size		17	.941			49.50	3.40	52.90	78.50
9930	3/4" pipe size		14.81	1.080			57	3.90	60.90	90.50
9940	1" pipe size		13.50	1.185			62.50	4.28	66.78	99
9950	1-1/4" pipe size		12.55	1.275			67.50	4.60	72.10	107
9960	1-1/2" pipe size		11.85	1.350			71.50	4.87	76.37	113
9970	2" pipe size		10	1.600			84.50	5.80	90.30	134
9980	3" pipe size		4.28	3.738			198	13.50	211.50	315
9990	4" pipe size	↓	3.25	4.923	↓		260	17.75	277.75	415

22 11 13.66 Pipe Fittings, Stainless Steel

		Crew	Daily Output	Labor-Hours	Unit	Material	Labor	Equipment	Total	Total Incl O&P
0010	**PIPE FITTINGS, STAINLESS STEEL**									
0100	Butt weld joint, schedule 5, type 304									
0120	90° Elbow, long									
0140	1/2"	Q-15	17.50	.914	Ea.	19.80	48.50	3.30	71.60	98.50
0150	3/4"		14	1.143		19.80	60.50	4.13	84.43	118
0160	1"		12.50	1.280		21	67.50	4.62	93.12	130
0170	1-1/4"	↓	11	1.455	↓	28.50	77	5.25	110.75	153

For customer support on your Plumbing Cost Data, call 877.709.6968.

203

22 11 13.66 Pipe Fittings, Stainless Steel		Crew	Daily Output	Labor-Hours	Unit	Material	2015 Bare Costs Labor	Equipment	Total	Total Incl O&P
0180	1-1/2"	Q-15	10.50	1.524	Ea.	24	80.50	5.50	110	155
0190	2"		9	1.778		28.50	94	6.40	128.90	181
0200	2-1/2"		6	2.667		66	141	9.65	216.65	296
0210	3"		4.86	3.292		72.50	174	11.90	258.40	355
0220	3-1/2"		4.27	3.747		182	198	13.55	393.55	515
0230	4"		3.69	4.336		99	229	15.65	343.65	470
0240	5"		3.08	5.195		385	274	18.75	677.75	860
0250	6"	Q-16	4.29	5.594		297	305	13.45	615.45	805
0260	8"		2.76	8.696		625	475	21	1,121	1,425
0270	10"		2.18	11.009		970	605	26.50	1,601.50	2,025
0280	12"		1.73	13.873		1,375	760	33.50	2,168.50	2,700
0320	For schedule 5, type 316, add					30%				
0600	45° Elbow, long									
0620	1/2"	Q-15	17.50	.914	Ea.	19.80	48.50	3.30	71.60	98.50
0630	3/4"		14	1.143		19.80	60.50	4.13	84.43	118
0640	1"		12.50	1.280		21	67.50	4.62	93.12	130
0650	1-1/4"		11	1.455		28.50	77	5.25	110.75	153
0660	1-1/2"		10.50	1.524		24	80.50	5.50	110	155
0670	2"		9	1.778		28.50	94	6.40	128.90	181
0680	2-1/2"		6	2.667		66	141	9.65	216.65	296
0690	3"		4.86	3.292		72.50	174	11.90	258.40	355
0700	3-1/2"		4.27	3.747		182	198	13.55	393.55	515
0710	4"		3.69	4.336		83.50	229	15.65	328.15	455
0720	5"		3.08	5.195		310	274	18.75	602.75	775
0730	6"	Q-16	4.29	5.594		209	305	13.45	527.45	710
0740	8"		2.76	8.696		440	475	21	936	1,225
0750	10"		2.18	11.009		770	605	26.50	1,401.50	1,775
0760	12"		1.73	13.873		965	760	33.50	1,758.50	2,225
0800	For schedule 5, type 316, add					25%				
1100	Tee, straight									
1130	1/2"	Q-15	11.66	1.372	Ea.	59.50	72.50	4.95	136.95	180
1140	3/4"		9.33	1.715		59.50	90.50	6.20	156.20	209
1150	1"		8.33	1.921		62.50	101	6.95	170.45	230
1160	1-1/4"		7.33	2.183		50.50	115	7.90	173.40	238
1170	1-1/2"		7	2.286		49.50	121	8.25	178.75	246
1180	2"		6	2.667		51.50	141	9.65	202.15	281
1190	2-1/2"		4	4		124	211	14.45	349.45	475
1200	3"		3.24	4.938		198	261	17.85	476.85	635
1210	3-1/2"		2.85	5.614		260	297	20.50	577.50	760
1220	4"		2.46	6.504		141	345	23.50	509.50	700
1230	5"		2	8		445	425	29	899	1,150
1240	6"	Q-16	2.85	8.421		350	460	20	830	1,100
1250	8"		1.84	13.043		750	715	31.50	1,496.50	1,925
1260	10"		1.45	16.552		1,200	905	39.50	2,144.50	2,750
1270	12"		1.15	20.870		1,675	1,150	50	2,875	3,625
1320	For schedule 5, type 316, add					25%				
2000	Butt weld joint, schedule 10, type 304									
2020	90° elbow, long									
2040	1/2"	Q-15	17	.941	Ea.	13.50	49.50	3.40	66.40	93.50
2050	3/4"		14	1.143		13.50	60.50	4.13	78.13	110
2060	1"		12.50	1.280		14.25	67.50	4.62	86.37	123
2070	1-1/4"		11	1.455		19.50	77	5.25	101.75	143
2080	1-1/2"		10.50	1.524		16.50	80.50	5.50	102.50	146

For customer support on your Plumbing Cost Data, call 877.709.6968.

22 11 Facility Water Distribution

22 11 13 – Facility Water Distribution Piping

22 11 13.66 Pipe Fittings, Stainless Steel		Crew	Daily Output	Labor-Hours	Unit	Material	2015 Bare Costs Labor	Equipment	Total	Total Incl O&P
2090	2"	Q-15	9	1.778	Ea.	19.50	94	6.40	119.90	171
2100	2-1/2"		6	2.667		45	141	9.65	195.65	273
2110	3"		4.86	3.292		42	174	11.90	227.90	320
2120	3-1/2"		4.27	3.747		124	198	13.55	335.55	450
2130	4"		3.69	4.336		70.50	229	15.65	315.15	440
2140	5"		3.08	5.195		263	274	18.75	555.75	725
2150	6"	Q-16	4.29	5.594		203	305	13.45	521.45	705
2160	8"		2.76	8.696		430	475	21	926	1,225
2170	10"		2.18	11.009		660	605	26.50	1,291.50	1,675
2180	12"		1.73	13.873		940	760	33.50	1,733.50	2,200
2500	45° elbow, long									
2520	1/2"	Q-15	17.50	.914	Ea.	13.50	48.50	3.30	65.30	91.50
2530	3/4"		14	1.143		13.50	60.50	4.13	78.13	110
2540	1"		12.50	1.280		14.25	67.50	4.62	86.37	123
2550	1-1/4"		11	1.455		19.50	77	5.25	101.75	143
2560	1-1/2"		10.50	1.524		16.50	80.50	5.50	102.50	146
2570	2"		9	1.778		19.50	94	6.40	119.90	171
2580	2-1/2"		6	2.667		45	141	9.65	195.65	273
2590	3"		4.86	3.292		34	174	11.90	219.90	315
2600	3-1/2"		4.27	3.747		124	198	13.55	335.55	450
2610	4"		3.69	4.336		57	229	15.65	301.65	425
2620	5"		3.08	5.195		210	274	18.75	502.75	665
2630	6"	Q-16	4.29	5.594		143	305	13.45	461.45	635
2640	8"		2.76	8.696		300	475	21	796	1,075
2650	10"		2.18	11.009		525	605	26.50	1,156.50	1,525
2660	12"		1.73	13.873		655	760	33.50	1,448.50	1,900
2670	Reducer, concentric									
2674	1" x 3/4"	Q-15	13.25	1.208	Ea.	29.50	64	4.36	97.86	133
2676	2" x 1-1/2"	"	9.75	1.641		24	86.50	5.90	116.40	164
2678	6" x 4"	Q-16	4.91	4.888		72	268	11.75	351.75	495
2680	Caps									
2682	1"	Q-15	25	.640	Ea.	30	34	2.31	66.31	86.50
2684	1-1/2"		21	.762		36	40.50	2.75	79.25	104
2685	2"		18	.889		34	47	3.21	84.21	112
2686	4"		7.38	2.168		54	115	7.85	176.85	241
2687	6"	Q-16	8.58	2.797		86.50	153	6.70	246.20	335
3000	Tee, straight									
3030	1/2"	Q-15	11.66	1.372	Ea.	40.50	72.50	4.95	117.95	159
3040	3/4"		9.33	1.715		40.50	90.50	6.20	137.20	188
3050	1"		8.33	1.921		43	101	6.95	150.95	208
3060	1-1/4"		7.33	2.183		49	115	7.90	171.90	236
3070	1-1/2"		7	2.286		49	121	8.25	178.25	245
3080	2"		6	2.667		51	141	9.65	201.65	280
3090	2-1/2"		4	4		35.50	211	14.45	260.95	375
3100	3"		3.24	4.938		67	261	17.85	345.85	490
3110	3-1/2"		2.85	5.614		98.50	297	20.50	416	580
3120	4"		2.46	6.504		96	345	23.50	464.50	650
3130	5"		2	8		305	425	29	759	1,000
3140	6"	Q-16	2.85	8.421		240	460	20	720	980
3150	8"		1.84	13.043		510	715	31.50	1,256.50	1,675
3151	10"		1.45	16.552		825	905	39.50	1,769.50	2,325
3152	12"		1.15	20.870		1,150	1,150	50	2,350	3,025
3154	For schedule 10, type 316, add					25%				

For customer support on your Plumbing Cost Data, call 877.709.6968.

205

22 11 13 – Facility Water Distribution Piping

22 11 13.66 Pipe Fittings, Stainless Steel	Crew	Daily Output	Labor-Hours	Unit	Material	2015 Bare Costs Labor	Equipment	Total	Total Incl O&P
3281	Butt weld joint, schedule 40, type 304								
3284	90° Elbow, long, 1/2"	Q-15	12.70	1.260	Ea.	15.75	66.50	4.55	86.80	123
3288	3/4"		11.10	1.441		15.75	76	5.20	96.95	138
3289	1"		10.13	1.579		16.50	83.50	5.70	105.70	150
3290	1-1/4"		9.40	1.702		22	90	6.15	118.15	167
3300	1-1/2"		8.89	1.800		17.25	95	6.50	118.75	170
3310	2"		7.55	2.119		25	112	7.65	144.65	204
3320	2-1/2"		3.98	4.020		47.50	212	14.50	274	390
3330	3"		3.21	4.984		61	263	18	342	485
3340	3-1/2"		2.83	5.654		225	299	20.50	544.50	720
3350	4"		2.44	6.557		105	345	23.50	473.50	665
3360	5"		2.13	7.512		355	395	27	777	1,025
3370	6"	Q-16	2.83	8.481		310	465	20.50	795.50	1,075
3380	8"		1.83	13.115		615	720	31.50	1,366.50	1,775
3390	10"		1.44	16.667		1,300	915	40	2,255	2,850
3400	12"		1.14	21.053		1,650	1,150	50.50	2,850.50	3,625
3410	For schedule 40, type 316, add					25%				
3460	45° Elbow, long, 1/2"	Q-15	12.70	1.260	Ea.	15.75	66.50	4.55	86.80	123
3470	3/4"		11.10	1.441		15.75	76	5.20	96.95	138
3480	1"		10.13	1.579		16.50	83.50	5.70	105.70	150
3490	1-1/4"		9.40	1.702		22	90	6.15	118.15	167
3500	1-1/2"		8.89	1.800		17.25	95	6.50	118.75	170
3510	2"		7.55	2.119		25	112	7.65	144.65	204
3520	2-1/2"		3.98	4.020		47.50	212	14.50	274	390
3530	3"		3.21	4.984		48	263	18	329	475
3540	3-1/2"		2.83	5.654		225	299	20.50	544.50	720
3550	4"		2.44	6.557		75	345	23.50	443.50	635
3560	5"		2.13	7.512		248	395	27	670	900
3570	6"	Q-16	2.83	8.481		218	465	20.50	703.50	960
3580	8"		1.83	13.115		430	720	31.50	1,181.50	1,575
3590	10"		1.44	16.667		910	915	40	1,865	2,425
3600	12"		1.14	21.053		1,150	1,150	50.50	2,350.50	3,075
3610	For schedule 40, type 316, add					25%				
3660	Tee, straight 1/2"	Q-15	8.46	1.891	Ea.	40.50	100	6.85	147.35	203
3670	3/4"		7.40	2.162		40.50	114	7.80	162.30	226
3680	1"		6.74	2.374		43	125	8.55	176.55	245
3690	1-1/4"		6.27	2.552		100	135	9.20	244.20	325
3700	1-1/2"		5.92	2.703		46	143	9.75	198.75	277
3710	2"		5.03	3.181		68.50	168	11.50	248	340
3720	2-1/2"		2.65	6.038		86.50	320	22	428.50	600
3730	3"		2.14	7.477		87	395	27	509	720
3740	3-1/2"		1.88	8.511		164	450	30.50	644.50	895
3750	4"		1.62	9.877		164	520	35.50	719.50	1,000
3760	5"		1.42	11.268		380	595	40.50	1,015.50	1,350
3770	6"	Q-16	1.88	12.766		365	700	30.50	1,095.50	1,475
3780	8"		1.22	19.672		735	1,075	47	1,857	2,475
3790	10"		.96	25		1,425	1,375	60	2,860	3,725
3800	12"		.76	31.579		1,900	1,725	76	3,701	4,800
3810	For schedule 40, type 316, add					25%				
3820	Tee, reducing on outlet, 3/4" x 1/2"	Q-15	7.73	2.070	Ea.	59.50	109	7.45	175.95	239
3822	1" x 1/2"		7.24	2.210		58.50	117	8	183.50	249
3824	1" x 3/4"		6.96	2.299		54	121	8.30	183.30	252
3826	1-1/4" x 1"		6.43	2.488		132	131	9	272	355

For customer support on your Plumbing Cost Data, call 877.709.6968.

22 11 13 – Facility Water Distribution Piping

22 11 13.66 Pipe Fittings, Stainless Steel		Crew	Daily Output	Labor-Hours	Unit	Material	2015 Bare Costs Labor	Equipment	Total	Total Incl O&P
3828	1-1/2" x 1/2"	Q-15	6.58	2.432	Ea.	79.50	128	8.80	216.30	291
3830	1-1/2" x 3/4"		6.35	2.520		74.50	133	9.10	216.60	293
3832	1-1/2" x 1"		6.18	2.589		57	137	9.35	203.35	280
3834	2" x 1"		5.50	2.909		102	154	10.50	266.50	355
3836	2" x 1-1/2"		5.30	3.019		85.50	159	10.90	255.40	345
3838	2-1/2" x 2"		3.15	5.079		129	268	18.35	415.35	565
3840	3" x 1-1/2"		2.72	5.882		131	310	21	462	640
3842	3" x 2"		2.65	6.038		109	320	22	451	625
3844	4" x 2"		2.10	7.619		272	405	27.50	704.50	940
3846	4" x 3"		1.77	9.040		197	480	32.50	709.50	975
3848	5" x 4"		1.48	10.811		455	570	39	1,064	1,400
3850	6" x 3"	Q-16	2.19	10.959		505	600	26.50	1,131.50	1,500
3852	6" x 4"		2.04	11.765		435	645	28	1,108	1,475
3854	8" x 4"		1.46	16.438		1,025	900	39.50	1,964.50	2,525
3856	10" x 8"		.69	34.783		1,725	1,900	83.50	3,708.50	4,850
3858	12" x 10"		.55	43.636		2,275	2,400	105	4,780	6,250
3950	Reducer, concentric, 3/4" x 1/2"	Q-15	11.85	1.350		30	71.50	4.87	106.37	146
3952	1" x 3/4"		10.60	1.509		31.50	79.50	5.45	116.45	161
3954	1-1/4" x 3/4"		10.19	1.570		77.50	83	5.65	166.15	216
3956	1-1/4" x 1"		9.76	1.639		39	86.50	5.90	131.40	181
3958	1-1/2" x 3/4"		9.88	1.619		65.50	85.50	5.85	156.85	207
3960	1-1/2" x 1"		9.47	1.690		52.50	89.50	6.10	148.10	200
3962	2" x 1"		8.65	1.850		30	97.50	6.70	134.20	188
3964	2" x 1-1/2"		8.16	1.961		26.50	104	7.10	137.60	193
3966	2-1/2" x 1"		5.71	2.802		113	148	10.10	271.10	360
3968	2-1/2" x 2"		5.21	3.071		58	162	11.10	231.10	320
3970	3" x 1"		4.88	3.279		78	173	11.85	262.85	360
3972	3" x 1-1/2"		4.72	3.390		40	179	12.25	231.25	325
3974	3" x 2"		4.51	3.548		37.50	187	12.80	237.30	340
3976	4" x 2"		3.69	4.336		53.50	229	15.65	298.15	420
3978	4" x 3"		2.77	5.776		40	305	21	366	525
3980	5" x 3"		2.56	6.250		237	330	22.50	589.50	785
3982	5" x 4"		2.27	7.048		191	370	25.50	586.50	800
3984	6" x 3"	Q-16	3.57	6.723		122	370	16.15	508.15	705
3986	6" x 4"		3.19	7.524		101	410	18.05	529.05	750
3988	8" x 4"		2.44	9.836		315	540	23.50	878.50	1,175
3990	8" x 6"		2.22	10.811		236	590	26	852	1,175
3992	10" x 6"		1.91	12.565		440	690	30	1,160	1,575
3994	10" x 8"		1.61	14.907		365	815	36	1,216	1,675
3995	12" x 6"		1.63	14.724		675	805	35.50	1,515.50	2,000
3996	12" x 8"		1.41	17.021		620	930	41	1,591	2,125
3997	12" x 10"		1.27	18.898		435	1,025	45.50	1,505.50	2,100
4000	Socket weld joint, 3000 lb., type 304									
4100	90° Elbow									
4140	1/4"	Q-15	13.47	1.188	Ea.	44	63	4.29	111.29	148
4150	3/8"		12.97	1.234		57	65	4.45	126.45	166
4160	1/2"		12.21	1.310		62.50	69	4.73	136.23	179
4170	3/4"		10.68	1.498		72	79	5.40	156.40	205
4180	1"		9.74	1.643		108	87	5.95	200.95	257
4190	1-1/4"		9.05	1.768		189	93.50	6.40	288.90	355
4200	1-1/2"		8.55	1.871		229	99	6.75	334.75	410
4210	2"		7.26	2.204		370	116	7.95	493.95	590
4300	45° Elbow									

22 11 13 – Facility Water Distribution Piping

22 11 13.66 Pipe Fittings, Stainless Steel		Crew	Daily Output	Labor-Hours	Unit	Material	2015 Bare Costs Labor	Equipment	Total	Total Incl O&P
4340	1/4"	Q-15	13.47	1.188	Ea.	83	63	4.29	150.29	191
4350	3/8"		12.97	1.234		83	65	4.45	152.45	194
4360	1/2"		12.21	1.310		83	69	4.73	156.73	201
4370	3/4"		10.68	1.498		94	79	5.40	178.40	229
4380	1"		9.74	1.643		136	87	5.95	228.95	288
4390	1-1/4"		9.05	1.768		224	93.50	6.40	323.90	395
4400	1-1/2"		8.55	1.871		225	99	6.75	330.75	405
4410	2"		7.26	2.204		410	116	7.95	533.95	635
4500	Tee									
4540	1/4"	Q-15	8.97	1.784	Ea.	58	94	6.45	158.45	213
4550	3/8"		8.64	1.852		69	98	6.70	173.70	231
4560	1/2"		8.13	1.968		85	104	7.10	196.10	258
4570	3/4"		7.12	2.247		98	119	8.10	225.10	296
4580	1"		6.48	2.469		132	130	8.90	270.90	350
4590	1-1/4"		6.03	2.653		234	140	9.60	383.60	480
4600	1-1/2"		5.69	2.812		335	149	10.15	494.15	605
4610	2"		4.83	3.313		510	175	11.95	696.95	835
5000	Socket weld joint, 3000 lb., type 316									
5100	90° Elbow									
5140	1/4"	Q-15	13.47	1.188	Ea.	54	63	4.29	121.29	159
5150	3/8"		12.97	1.234		63	65	4.45	132.45	172
5160	1/2"		12.21	1.310		76	69	4.73	149.73	194
5170	3/4"		10.68	1.498		100	79	5.40	184.40	236
5180	1"		9.74	1.643		142	87	5.95	234.95	294
5190	1-1/4"		9.05	1.768		253	93.50	6.40	352.90	425
5200	1-1/2"		8.55	1.871		286	99	6.75	391.75	470
5210	2"		7.26	2.204		485	116	7.95	608.95	720
5300	45° Elbow									
5340	1/4"	Q-15	13.47	1.188	Ea.	104	63	4.29	171.29	214
5350	3/8"		12.97	1.234		104	65	4.45	173.45	217
5360	1/2"		12.21	1.310		104	69	4.73	177.73	224
5370	3/4"		10.68	1.498		121	79	5.40	205.40	259
5380	1"		9.74	1.643		182	87	5.95	274.95	340
5390	1-1/4"		9.05	1.768		258	93.50	6.40	357.90	430
5400	1-1/2"		8.55	1.871		292	99	6.75	397.75	475
5410	2"		7.26	2.204		435	116	7.95	558.95	665
5500	Tee									
5540	1/4"	Q-15	8.97	1.784	Ea.	73	94	6.45	173.45	230
5550	3/8"		8.64	1.852		89.50	98	6.70	194.20	253
5560	1/2"		8.13	1.968		99.50	104	7.10	210.60	274
5570	3/4"		7.12	2.247		123	119	8.10	250.10	325
5580	1"		6.48	2.469		188	130	8.90	326.90	415
5590	1-1/4"		6.03	2.653		299	140	9.60	448.60	555
5600	1-1/2"		5.69	2.812		420	149	10.15	579.15	700
5610	2"		4.83	3.313		670	175	11.95	856.95	1,025
5700	For socket weld joint, 6000 lb., type 304 and 316, add					100%				
6000	Threaded companion flange									
6010	Stainless steel, 150 lb., type 304									
6020	1/2" diam.	1 Plum	30	.267	Ea.	45	15.65		60.65	73
6030	3/4" diam.		28	.286		50	16.75		66.75	80.50
6040	1" diam.		27	.296		55	17.40		72.40	87
6050	1-1/4" diam.	Q-1	44	.364		71	19.20		90.20	107
6060	1-1/2" diam.		40	.400		71	21		92	110

22 11 13.66 Pipe Fittings, Stainless Steel		Crew	Daily Output	Labor-Hours	Unit	Material	2015 Bare Costs Labor	Equipment	Total	Total Incl O&P
6070	2" diam.	Q-1	36	.444	Ea.	93	23.50		116.50	138
6080	2-1/2" diam.		28	.571		130	30		160	189
6090	3" diam.		20	.800		136	42.50		178.50	214
6110	4" diam.		12	1.333		186	70.50		256.50	310
6130	6" diam.	Q-2	14	1.714		335	94		429	505
6140	8" diam.	"	12	2		645	110		755	875
6150	For type 316 add					40%				
6260	Weld flanges, stainless steel, type 304									
6270	Slip on, 150 lb. (welded, front and back)									
6280	1/2" diam.	Q-15	12.70	1.260	Ea.	40	66.50	4.55	111.05	150
6290	3/4" diam.		11.11	1.440		41	76	5.20	122.20	166
6300	1" diam.		10.13	1.579		45	83.50	5.70	134.20	182
6310	1-1/4" diam.		9.41	1.700		61	90	6.15	157.15	210
6320	1-1/2" diam.		8.89	1.800		61	95	6.50	162.50	218
6330	2" diam.		7.55	2.119		79	112	7.65	198.65	264
6340	2-1/2" diam.		3.98	4.020		111	212	14.50	337.50	460
6350	3" diam.		3.21	4.984		119	263	18	400	550
6370	4" diam.		2.44	6.557		162	345	23.50	530.50	730
6390	6" diam.	Q-16	1.89	12.698		247	695	30.50	972.50	1,350
6400	8" diam.	"	1.22	19.672		465	1,075	47	1,587	2,200
6410	For type 316, add					40%				
6530	Weld neck 150 lb.									
6540	1/2" diam.	Q-15	25.40	.630	Ea.	20.50	33.50	2.27	56.27	75.50
6550	3/4" diam.		22.22	.720		25	38	2.60	65.60	87.50
6560	1" diam.		20.25	.790		27	41.50	2.85	71.35	95.50
6570	1-1/4" diam.		18.82	.850		43.50	45	3.07	91.57	119
6580	1-1/2" diam.		17.78	.900		38	47.50	3.25	88.75	117
6590	2" diam.		15.09	1.060		40.50	56	3.83	100.33	133
6600	2-1/2" diam.		7.96	2.010		65	106	7.25	178.25	240
6610	3" diam.		6.43	2.488		66.50	131	9	206.50	282
6630	4" diam.		4.88	3.279		101	173	11.85	285.85	385
6640	5" diam.		4.26	3.756		132	198	13.55	343.55	460
6650	6" diam.	Q-16	5.66	4.240		145	232	10.20	387.20	520
6652	8" diam.		3.65	6.575		261	360	15.80	636.80	850
6654	10" diam.		2.88	8.333		365	455	20	840	1,125
6656	12" diam.		2.28	10.526		705	575	25.50	1,305.50	1,675
6670	For type 316 add					23%				
7000	Threaded joint, 150 lb., type 304									
7030	90° elbow									
7040	1/8"	1 Plum	13	.615	Ea.	25.50	36		61.50	82.50
7050	1/4"		13	.615		25.50	36		61.50	82.50
7070	3/8"		13	.615		29.50	36		65.50	87
7080	1/2"		12	.667		26.50	39		65.50	88
7090	3/4"		11	.727		31.50	42.50		74	99
7100	1"		10	.800		43	47		90	119
7110	1-1/4"	Q-1	17	.941		66.50	49.50		116	149
7120	1-1/2"		16	1		77	53		130	165
7130	2"		14	1.143		111	60.50		171.50	214
7140	2-1/2"		11	1.455		271	77		348	415
7150	3"		8	2		395	106		501	595
7160	4"	Q-2	11	2.182		670	120		790	915
7180	45° elbow									
7190	1/8"	1 Plum	13	.615	Ea.	37	36		73	95.50

209

22 11 13.66 Pipe Fittings, Stainless Steel	Crew	Daily Output	Labor-Hours	Unit	Material	2015 Bare Costs Labor	Equipment	Total	Total Incl O&P
7200	1/4"	1 Plum	13	.615	Ea.	37	36		73	95.50
7210	3/8"		13	.615		37.50	36		73.50	95.50
7220	1/2"		12	.667		38	39		77	101
7230	3/4"		11	.727		41	42.50		83.50	110
7240	1"		10	.800		47.50	47		94.50	123
7250	1-1/4"	Q-1	17	.941		66	49.50		115.50	148
7260	1-1/2"		16	1		85.50	53		138.50	174
7270	2"		14	1.143		121	60.50		181.50	224
7280	2-1/2"		11	1.455		380	77		457	535
7290	3"		8	2		560	106		666	775
7300	4"	Q-2	11	2.182		1,000	120		1,120	1,275
7320	Tee, straight									
7330	1/8"	1 Plum	9	.889	Ea.	39	52		91	122
7340	1/4"		9	.889		39	52		91	122
7350	3/8"		9	.889		41.50	52		93.50	125
7360	1/2"		8	1		39.50	58.50		98	132
7370	3/4"		7	1.143		42	67		109	148
7380	1"		6.50	1.231		54	72.50		126.50	169
7390	1-1/4"	Q-1	11	1.455		91	77		168	216
7400	1-1/2"		10	1.600		116	84.50		200.50	256
7410	2"		9	1.778		145	94		239	300
7420	2-1/2"		7	2.286		390	121		511	610
7430	3"		5	3.200		590	169		759	905
7440	4"	Q-2	7	3.429		1,475	188		1,663	1,875
7460	Coupling, straight									
7470	1/8"	1 Plum	19	.421	Ea.	10.25	24.50		34.75	49
7480	1/4"		19	.421		12.10	24.50		36.60	51
7490	3/8"		19	.421		14.50	24.50		39	53.50
7500	1/2"		19	.421		19.30	24.50		43.80	58.50
7510	3/4"		18	.444		25.50	26		51.50	68
7520	1"		15	.533		41.50	31.50		73	93
7530	1-1/4"	Q-1	26	.615		66.50	32.50		99	123
7540	1-1/2"		24	.667		74.50	35		109.50	135
7550	2"		21	.762		123	40.50		163.50	197
7560	2-1/2"		18	.889		287	47		334	385
7570	3"		14	1.143		390	60.50		450.50	520
7580	4"	Q-2	16	1.500		550	82		632	730
7600	Reducer, concentric, 1/2"	1 Plum	12	.667		20.50	39		59.50	82
7610	3/4"		11	.727		26.50	42.50		69	94
7612	1"		10	.800		44.50	47		91.50	120
7614	1-1/4"	Q-1	17	.941		94.50	49.50		144	179
7616	1-1/2"		16	1		103	53		156	193
7618	2"		14	1.143		161	60.50		221.50	268
7620	2-1/2"		11	1.455		400	77		477	555
7622	3"		8	2		445	106		551	650
7624	4"	Q-2	11	2.182		735	120		855	990
7710	Union									
7720	1/8"	1 Plum	12	.667	Ea.	47.50	39		86.50	112
7730	1/4"		12	.667		47.50	39		86.50	112
7740	3/8"		12	.667		55.50	39		94.50	120
7750	1/2"		11	.727		69.50	42.50		112	141
7760	3/4"		10	.800		96	47		143	176
7770	1"		9	.889		138	52		190	231

22 11 Facility Water Distribution

22 11 13 – Facility Water Distribution Piping

22 11 13.66 Pipe Fittings, Stainless Steel		Crew	Daily Output	Labor-Hours	Unit	Material	2015 Bare Costs Labor	Equipment	Total	Total Incl O&P
7780	1-1/4"	Q-1	16	1	Ea.	340	53		393	455
7790	1-1/2"		15	1.067		370	56.50		426.50	490
7800	2"		13	1.231		465	65		530	615
7810	2-1/2"		10	1.600		880	84.50		964.50	1,100
7820	3"		7	2.286		1,175	121		1,296	1,475
7830	4"	Q-2	10	2.400		1,625	131		1,756	1,975
7838	Caps									
7840	1/2"	1 Plum	24	.333	Ea.	12.90	19.55		32.45	43.50
7841	3/4"		22	.364		18.80	21.50		40.30	52.50
7842	1"		20	.400		30	23.50		53.50	68.50
7843	1-1/2"	Q-1	32	.500		76	26.50		102.50	124
7844	2"	"	28	.571		96	30		126	152
7845	4"	Q-2	22	1.091		380	60		440	505
7850	For 150 lb., type 316, add					25%				

22 11 13.74 Pipe, Plastic

	22 11 13.74 Pipe, Plastic	Crew	Daily Output	Labor-Hours	Unit	Material	2015 Bare Costs Labor	Equipment	Total	Total Incl O&P
0010	**PIPE, PLASTIC** R221113-70									
0020	Fiberglass reinforced, couplings 10' O.C., clevis hanger assy's, 3 per 10'									
0080	General service									
0120	2" diameter	Q-1	59	.271	L.F.	12.55	14.35		26.90	35.50
0140	3" diameter		52	.308		16.10	16.25		32.35	42.50
0150	4" diameter		48	.333		21	17.60		38.60	49.50
0160	6" diameter		39	.410		35.50	21.50		57	72
0170	8" diameter	Q-2	49	.490		55	27		82	101
0180	10" diameter		41	.585		80	32		112	137
0190	12" diameter		36	.667		103	36.50		139.50	168
0600	PVC, high impact/pressure, cplgs. 10' O.C., clevis hanger assy's, 3 per 10'									
1020	Schedule 80									
1070	1/2" diameter	1 Plum	50	.160	L.F.	6.75	9.40		16.15	21.50
1080	3/4" diameter		47	.170		7.80	10		17.80	23.50
1090	1" diameter		43	.186		9.65	10.90		20.55	27
1100	1-1/4" diameter		39	.205		11.85	12.05		23.90	31.50
1110	1-1/2" diameter		34	.235		13.40	13.80		27.20	36
1120	2" diameter	Q-1	55	.291		16.65	15.35		32	41.50
1140	3" diameter		50	.320		31	16.90		47.90	60
1150	4" diameter		46	.348		43.50	18.40		61.90	76
1170	6" diameter		38	.421		94.50	22		116.50	138
1730	To delete coupling & hangers, subtract									
1740	1/2" diam.					62%	80%			
1750	3/4" diam. to 1-1/4" diam.					58%	73%			
1760	1-1/2" diam. to 6" diam.					40%	57%			
1800	PVC, couplings 10' O.C., clevis hanger assemblies, 3 per 10'									
1820	Schedule 40									
1860	1/2" diameter	1 Plum	54	.148	L.F.	4.88	8.70		13.58	18.50
1870	3/4" diameter		51	.157		5.20	9.20		14.40	19.65
1880	1" diameter		46	.174		5.85	10.20		16.05	22
1890	1-1/4" diameter		42	.190		6.55	11.20		17.75	24
1900	1-1/2" diameter		36	.222		6.85	13.05		19.90	27.50
1910	2" diameter	Q-1	59	.271		8.05	14.35		22.40	30.50
1920	2-1/2" diameter		56	.286		10.35	15.10		25.45	34.50
1930	3" diameter		53	.302		12.60	15.95		28.55	38
1940	4" diameter		48	.333		16.10	17.60		33.70	44
1950	5" diameter		43	.372		26.50	19.65		46.15	58.50

211

22 11 13.74 Pipe, Plastic		Crew	Daily Output	Labor-Hours	Unit	Material	2015 Bare Costs Labor	2015 Bare Costs Equipment	Total	Total Incl O&P
1960	6" diameter	Q-1	39	.410	L.F.	27.50	21.50		49	62.50
1970	8" diameter	Q-2	48	.500		36.50	27.50		64	81.50
1980	10" diameter		43	.558		75	30.50		105.50	129
1990	12" diameter		42	.571		90	31.50		121.50	147
2000	14" diameter		31	.774		143	42.50		185.50	222
2010	16" diameter		23	1.043		205	57		262	315
2340	To delete coupling & hangers, subtract									
2360	1/2" diam. to 1-1/4" diam.					65%	74%			
2370	1-1/2" diam. to 6" diam.					44%	57%			
2380	8" diam. to 12" diam.					41%	53%			
2390	14" diam. to 16" diam.					48%	45%			
2420	Schedule 80									
2440	1/4" diameter	1 Plum	58	.138	L.F.	4.35	8.10		12.45	17.05
2450	3/8" diameter		55	.145		4.35	8.55		12.90	17.70
2460	1/2" diameter		50	.160		5	9.40		14.40	19.70
2470	3/4" diameter		47	.170		5.35	10		15.35	21
2480	1" diameter		43	.186		6.10	10.90		17	23
2490	1-1/4" diameter		39	.205		6.85	12.05		18.90	25.50
2500	1-1/2" diameter		34	.235		7.30	13.80		21.10	29
2510	2" diameter	Q-1	55	.291		8.20	15.35		23.55	32
2520	2-1/2" diameter		52	.308		9.85	16.25		26.10	35.50
2530	3" diameter		50	.320		13.95	16.90		30.85	41
2540	4" diameter		46	.348		19.55	18.40		37.95	49.50
2550	5" diameter		42	.381		29.50	20		49.50	62.50
2560	6" diameter		38	.421		35.50	22		57.50	73
2570	8" diameter	Q-2	47	.511		45	28		73	92
2580	10" diameter		42	.571		106	31.50		137.50	164
2590	12" diameter		38	.632		128	34.50		162.50	194
2830	To delete coupling & hangers, subtract									
2840	1/4" diam. to 1/2" diam.					66%	80%			
2850	3/4" diam. to 1-1/4" diam.					61%	73%			
2860	1-1/2" diam. to 6" diam.					41%	57%			
2870	8" diam. to 12" diam.					31%	50%			
2900	Schedule 120									
2910	1/2" diameter	1 Plum	50	.160	L.F.	5.50	9.40		14.90	20.50
2950	3/4" diameter		47	.170		6	10		16	21.50
2960	1" diameter		43	.186		7.05	10.90		17.95	24.50
2970	1-1/4" diameter		39	.205		8.30	12.05		20.35	27.50
2980	1-1/2" diameter		33	.242		9.05	14.25		23.30	31.50
2990	2" diameter	Q-1	54	.296		10.75	15.65		26.40	35.50
3000	2-1/2" diameter		52	.308		15.85	16.25		32.10	42
3010	3" diameter		49	.327		19.10	17.25		36.35	47
3020	4" diameter		45	.356		27.50	18.80		46.30	58.50
3030	6" diameter		37	.432		49.50	23		72.50	89
3240	To delete coupling & hangers, subtract									
3250	1/2" diam. to 1-1/4" diam.					52%	74%			
3260	1-1/2" diam. to 4" diam.					30%	57%			
3270	6" diam.					17%	50%			
3300	PVC, pressure, couplings 10' O.C., clevis hanger assy's, 3 per 10'									
3310	SDR 26, 160 psi									
3350	1-1/4" diameter	1 Plum	42	.190	L.F.	6.20	11.20		17.40	24
3360	1-1/2" diameter	"	36	.222		6.40	13.05		19.45	27
3370	2" diameter	Q-1	59	.271		7.30	14.35		21.65	29.50

For customer support on your Plumbing Cost Data, call 877.709.6968.

22 11 Facility Water Distribution

22 11 13 – Facility Water Distribution Piping

22 11 13.74 Pipe, Plastic

		Crew	Daily Output	Labor-Hours	Unit	Material	2015 Bare Costs Labor	Equipment	Total	Total Incl O&P
3380	2-1/2" diameter	Q-1	56	.286	L.F.	9.80	15.10		24.90	34
3390	3" diameter		53	.302		12.30	15.95		28.25	37.50
3400	4" diameter		48	.333		16.50	17.60		34.10	44.50
3420	6" diameter		39	.410		30.50	21.50		52	66
3430	8" diameter	Q-2	48	.500		45	27.50		72.50	91
3660	To delete coupling & clevis hanger assy's, subtract									
3670	1-1/4" diam.					63%	68%			
3680	1-1/2" diam. to 4" diam.					48%	57%			
3690	6" diam. to 8" diam.					60%	54%			
3720	SDR 21, 200 psi, 1/2" diameter	1 Plum	54	.148	L.F.	4.77	8.70		13.47	18.40
3740	3/4" diameter		51	.157		5.10	9.20		14.30	19.50
3750	1" diameter		46	.174		5.60	10.20		15.80	21.50
3760	1-1/4" diameter		42	.190		6.40	11.20		17.60	24
3770	1-1/2" diameter		36	.222		6.75	13.05		19.80	27
3780	2" diameter	Q-1	59	.271		7.65	14.35		22	30
3790	2-1/2" diameter		56	.286		11.90	15.10		27	36
3800	3" diameter		53	.302		13.10	15.95		29.05	38.50
3810	4" diameter		48	.333		20	17.60		37.60	49
3830	6" diameter		39	.410		34	21.50		55.50	69.50
3840	8" diameter	Q-2	48	.500		50	27.50		77.50	96.50
4000	To delete coupling & hangers, subtract									
4010	1/2" diam. to 3/4" diam.					71%	77%			
4020	1" diam. to 1-1/4" diam.					63%	70%			
4030	1-1/2" diam. to 6" diam.					44%	57%			
4040	8" diam.					46%	54%			
4100	DWV type, schedule 40, couplings 10' O.C., clevis hanger assy's, 3 per 10'									
4210	ABS, schedule 40, foam core type									
4212	Plain end black									
4214	1-1/2" diameter	1 Plum	39	.205	L.F.	5.20	12.05		17.25	24
4216	2" diameter	Q-1	62	.258		5.60	13.65		19.25	26.50
4218	3" diameter		56	.286		8	15.10		23.10	32
4220	4" diameter		51	.314		10.20	16.55		26.75	36
4222	6" diameter		42	.381		18	20		38	50.50
4240	To delete coupling & hangers, subtract									
4244	1-1/2" diam. to 6" diam.					43%	48%			
4400	PVC									
4410	1-1/4" diameter	1 Plum	42	.190	L.F.	5.35	11.20		16.55	23
4420	1-1/2" diameter	"	36	.222		5.30	13.05		18.35	25.50
4460	2" diameter	Q-1	59	.271		5.65	14.35		20	27.50
4470	3" diameter		53	.302		8.05	15.95		24	33
4480	4" diameter		48	.333		10	17.60		27.60	37.50
4490	6" diameter		39	.410		16.40	21.50		37.90	50.50
4500	8" diameter	Q-2	48	.500		21.50	27.50		49	65.50
4510	To delete coupling & hangers, subtract									
4520	1-1/4" diam. to 1-1/2" diam.					48%	60%			
4530	2" diam. to 8" diam.					42%	54%			
4550	PVC, schedule 40, foam core type									
4552	Plain end, white									
4554	1-1/2" diameter	1 Plum	39	.205	L.F.	5	12.05		17.05	23.50
4556	2" diameter	Q-1	62	.258		5.35	13.65		19	26.50
4558	3" diameter		56	.286		7.45	15.10		22.55	31
4560	4" diameter		51	.314		9.10	16.55		25.65	35
4562	6" diameter		42	.381		14.55	20		34.55	46.50

22 11 13 – Facility Water Distribution Piping

22 11 13.74 Pipe, Plastic	Crew	Daily Output	Labor-Hours	Unit	Material	2015 Bare Costs Labor	Equipment	Total	Total Incl O&P
4564	8" diameter	Q-2	51	.471	L.F.	18.75	26		44.75	59.50
4568	10" diameter		48	.500		23.50	27.50		51	67
4570	12" diameter		46	.522		28	28.50		56.50	74
4580	To delete coupling & hangers, subtract									
4582	1-1/2" dia to 2" dia					58%	54%			
4584	3" dia to 12" dia					46%	42%			
4800	PVC, clear pipe, cplgs. 10' O.C., clevis hanger assy's 3 per 10', Sched. 40									
4840	1/4" diameter	1 Plum	59	.136	L.F.	4.91	7.95		12.86	17.40
4850	3/8" diameter		56	.143		5.25	8.40		13.65	18.40
4860	1/2" diameter		54	.148		5.90	8.70		14.60	19.65
4870	3/4" diameter		51	.157		6.60	9.20		15.80	21
4880	1" diameter		46	.174		8.15	10.20		18.35	24.50
4890	1-1/4" diameter		42	.190		9.60	11.20		20.80	27.50
4900	1-1/2" diameter		36	.222		10.70	13.05		23.75	31.50
4910	2" diameter	Q-1	59	.271		13	14.35		27.35	36
4920	2-1/2" diameter		56	.286		19	15.10		34.10	44
4930	3" diameter		53	.302		23.50	15.95		39.45	49.50
4940	3-1/2" diameter		50	.320		32.50	16.90		49.40	61
4950	4" diameter		48	.333		33.50	17.60		51.10	63.50
5250	To delete coupling & hangers, subtract									
5260	1/4" diam. to 3/8" diam.					60%	81%			
5270	1/2" diam. to 3/4" diam.					41%	77%			
5280	1" diam. to 1-1/2" diam.					26%	67%			
5290	2" diam. to 4" diam.					16%	58%			
5300	CPVC, socket joint, couplings 10' O.C., clevis hanger assemblies, 3 per 10'									
5302	Schedule 40									
5304	1/2" diameter	1 Plum	54	.148	L.F.	5.90	8.70		14.60	19.65
5305	3/4" diameter		51	.157		6.75	9.20		15.95	21.50
5306	1" diameter		46	.174		8.10	10.20		18.30	24.50
5307	1-1/4" diameter		42	.190		9.65	11.20		20.85	27.50
5308	1-1/2" diameter		36	.222		10.85	13.05		23.90	31.50
5309	2" diameter	Q-1	59	.271		12.85	14.35		27.20	35.50
5310	2-1/2" diameter		56	.286		19.90	15.10		35	45
5311	3" diameter		53	.302		23.50	15.95		39.45	50
5312	4" diameter		48	.333		31.50	17.60		49.10	61.50
5314	6" diameter		43	.372		57.50	19.65		77.15	92.50
5318	To delete coupling & hangers, subtract									
5319	1/2" diam. to 3/4" diam.					37%	77%			
5320	1" diam. to 1-1/4" diam.					27%	70%			
5321	1-1/2" diam. to 3" diam.					21%	57%			
5322	4" diam. to 6" diam.					16%	57%			
5324	Schedule 80									
5325	1/2" diameter	1 Plum	50	.160	L.F.	6.60	9.40		16	21.50
5326	3/4" diameter		47	.170		7	10		17	23
5327	1" diameter		43	.186		8.70	10.90		19.60	26
5328	1-1/4" diameter		39	.205		10.60	12.05		22.65	30
5329	1-1/2" diameter		34	.235		12.05	13.80		25.85	34.50
5330	2" diameter	Q-1	55	.291		14.70	15.35		30.05	39
5331	2-1/2" diameter		52	.308		22.50	16.25		38.75	49
5332	3" diameter		50	.320		26	16.90		42.90	54
5333	4" diameter		46	.348		35.50	18.40		53.90	67
5334	6" diameter		38	.421		67.50	22		89.50	108
5335	8" diameter	Q-2	47	.511		170	28		198	230

22 11 13 – Facility Water Distribution Piping

22 11 13.74 Pipe, Plastic	Crew	Daily Output	Labor-Hours	Unit	Material	2015 Bare Costs Labor	Equipment	Total	Total Incl O&P
5339	To delete couplings & hangers, subtract								
5340	1/2" diam. to 3/4" diam.					44%	77%		
5341	1" diam. to 1-1/4" diam.					32%	71%		
5342	1-1/2" diam. to 4" diam.					25%	58%		
5343	6" diam. to 8" diam.					20%	53%		
5360	CPVC, threaded, couplings 10' O.C., clevis hanger assemblies, 3 per 10'								
5380	Schedule 40								
5460	1/2" diameter	1 Plum	54	.148	L.F.	6.70	8.70		15.40	20.50
5470	3/4" diameter		51	.157		8.15	9.20		17.35	23
5480	1" diameter		46	.174		9.55	10.20		19.75	26
5490	1-1/4" diameter		42	.190		10.80	11.20		22	29
5500	1-1/2" diameter		36	.222		11.80	13.05		24.85	32.50
5510	2" diameter	Q-1	59	.271		14	14.35		28.35	37
5520	2-1/2" diameter		56	.286		21	15.10		36.10	46
5530	3" diameter		53	.302		25.50	15.95		41.45	52
5540	4" diameter		48	.333		38.50	17.60		56.10	69
5550	6" diameter		43	.372		60.50	19.65		80.15	96.50
5730	To delete coupling & hangers, subtract								
5740	1/2" diam. to 3/4" diam.					37%	77%		
5750	1" diam. to 1-1/4" diam.					27%	70%		
5760	1-1/2" diam. to 3" diam.					21%	57%		
5770	4" diam. to 6" diam.					16%	57%		
5800	Schedule 80								
5860	1/2" diameter	1 Plum	50	.160	L.F.	7.40	9.40		16.80	22.50
5870	3/4" diameter		47	.170		8.40	10		18.40	24.50
5880	1" diameter		43	.186		10.15	10.90		21.05	27.50
5890	1-1/4" diameter		39	.205		11.70	12.05		23.75	31
5900	1-1/2" diameter		34	.235		13	13.80		26.80	35.50
5910	2" diameter	Q-1	55	.291		15.85	15.35		31.20	40.50
5920	2-1/2" diameter		52	.308		23.50	16.25		39.75	50.50
5930	3" diameter		50	.320		27.50	16.90		44.40	56
5940	4" diameter		46	.348		42.50	18.40		60.90	75
5950	6" diameter		38	.421		71	22		93	112
5960	8" diameter	Q-2	47	.511		169	28		197	229
6060	To delete couplings & hangers, subtract								
6070	1/2" diam. to 3/4" diam.					44%	77%		
6080	1" diam. to 1-1/4" diam.					32%	71%		
6090	1-1/2" diam. to 4" diam.					25%	58%		
6100	6" diam. to 8" diam.					20%	53%		
6240	CTS, 1/2" diameter	1 Plum	54	.148	L.F.	5.50	8.70		14.20	19.20
6250	3/4" diameter		51	.157		6.75	9.20		15.95	21.50
6260	1" diameter		46	.174		10.10	10.20		20.30	26.50
6270	1 1/4"		42	.190		14.75	11.20		25.95	33
6280	1 1/2" diameter		36	.222		18.35	13.05		31.40	39.50
6290	2" diameter	Q-1	59	.271		28.50	14.35		42.85	52.50
6370	To delete coupling & hangers, subtract								
6380	1/2" diam.					51%	79%		
6390	3/4" diam.					40%	76%		
6392	1" thru 2" diam.					72%	68%		
6500	Residential installation, plastic pipe								
6510	Couplings 10' O.C., strap hangers 3 per 10'								
6520	PVC, Schedule 40								
6530	1/2" diameter	1 Plum	138	.058	L.F.	1.02	3.40		4.42	6.30

215

22 11 Facility Water Distribution

22 11 13 – Facility Water Distribution Piping

22 11 13.74 Pipe, Plastic		Crew	Daily Output	Labor-Hours	Unit	Material	2015 Bare Costs Labor	Equipment	Total	Total Incl O&P
6540	3/4" diameter	1 Plum	128	.063	L.F.	1.23	3.67		4.90	6.90
6550	1" diameter		119	.067		1.81	3.95		5.76	7.95
6560	1-1/4" diameter		111	.072		2.20	4.23		6.43	8.80
6570	1-1/2" diameter	↓	104	.077		2.44	4.52		6.96	9.50
6580	2" diameter	Q-1	197	.081		3.10	4.29		7.39	9.90
6590	2-1/2" diameter		162	.099		5.10	5.20		10.30	13.50
6600	4" diameter	↓	123	.130	↓	8.60	6.85		15.45	19.85
6700	PVC, DWV, Schedule 40									
6720	1-1/4" diameter	1 Plum	100	.080	L.F.	1.84	4.70		6.54	9.10
6730	1-1/2" diameter	"	94	.085		1.77	5		6.77	9.50
6740	2" diameter	Q-1	178	.090		2.09	4.75		6.84	9.45
6760	4" diameter	"	110	.145	↓	5.70	7.70		13.40	17.90
7280	PEX, flexible, no couplings or hangers									
7282	Note: For labor costs add 25% to the couplings and fittings labor total.									
7285	For fittings see section 23 83 16.10 7000									
7300	Non-barrier type, hot/cold tubing rolls									
7310	1/4" diameter x 100'				L.F.	.49			.49	.54
7350	3/8" diameter x 100'					.55			.55	.61
7360	1/2" diameter x 100'					.61			.61	.67
7370	1/2" diameter x 500'					.61			.61	.67
7380	1/2" diameter x 1000'					.61			.61	.67
7400	3/4" diameter x 100'					1.11			1.11	1.22
7410	3/4" diameter x 500'					1.11			1.11	1.22
7420	3/4" diameter x 1000'					1.11			1.11	1.22
7460	1" diameter x 100'					1.90			1.90	2.09
7470	1" diameter x 300'					1.90			1.90	2.09
7480	1" diameter x 500'					1.90			1.90	2.09
7500	1-1/4" diameter x 100'					3.23			3.23	3.55
7510	1-1/4" diameter x 300'					3.23			3.23	3.55
7540	1-1/2" diameter x 100'					4.39			4.39	4.83
7550	1-1/2" diameter x 300'				↓	4.39			4.39	4.83
7596	Most sizes available in red or blue									
7700	Non-barrier type, hot/cold tubing straight lengths									
7710	1/2" diameter x 20'				L.F.	.60			.60	.66
7750	3/4" diameter x 20'					1.10			1.10	1.21
7760	1" diameter x 20'					1.90			1.90	2.09
7770	1-1/4" diameter x 20'					3.23			3.23	3.55
7780	1-1/2" diameter x 20'					4.39			4.39	4.83
7790	2" diameter				↓	8.60			8.60	9.45
7796	Most sizes available in red or blue									
9000	Polypropylene pipe									
9002	For fusion weld fittings and accessories see line 22 11 13.76 9400									
9004	Note: sizes 1/2" thru 4" use socket fusion									
9005	Sizes 6" thru 10" use butt fusion									
9010	SDR 7.4, (domestic hot water piping)									
9011	Enhanced to minimize thermal expansion and high temperature life									
9016	13' lengths, size is ID, includes joints 13' O.C. and hangers 3 per 10'									
9020	3/8" diameter	1 Plum	53	.151	L.F.	2.35	8.85		11.20	16
9022	1/2" diameter		52	.154		2.80	9.05		11.85	16.75
9024	3/4" diameter		50	.160		3.31	9.40		12.71	17.85
9026	1" diameter		45	.178		4.34	10.45		14.79	20.50
9028	1-1/4" diameter		40	.200		6.20	11.75		17.95	24.50
9030	1-1/2" diameter	↓	35	.229	↓	8.75	13.40		22.15	30

22 11 13.74 Pipe, Plastic

		Crew	Daily Output	Labor-Hours	Unit	Material	2015 Bare Costs Labor	2015 Bare Costs Equipment	Total	Total Incl O&P
9032	2" diameter	Q-1	58	.276	L.F.	12.40	14.55		26.95	35.50
9034	2-1/2" diameter		55	.291		17	15.35		32.35	41.50
9036	3" diameter		52	.308		22.50	16.25		38.75	49
9038	3-1/2" diameter		49	.327		31	17.25		48.25	60.50
9040	4" diameter		46	.348		34	18.40		52.40	65.50
9042	6" diameter		39	.410		46	21.50		67.50	83
9044	8" diameter	Q-2	48	.500		69.50	27.50		97	118
9046	10" diameter	"	43	.558		110	30.50		140.50	167
9050	To delete joint & hangers, subtract									
9052	3/8" diam. to 1" diam.					45%	65%			
9054	1-1/4" diam. to 4" diam.					15%	45%			
9056	6" diam. to 10" diam.					5%	24%			
9060	SDR 11, (domestic cold water piping)									
9062	13' lengths, size is ID, includes joints 13' O.C. and hangers 3 per 10'									
9064	1/2" diameter	1 Plum	57	.140	L.F.	2.48	8.25		10.73	15.20
9066	3/4" diameter		54	.148		2.80	8.70		11.50	16.25
9068	1" diameter		49	.163		3.41	9.60		13.01	18.20
9070	1-1/4" diameter		45	.178		4.83	10.45		15.28	21
9072	1-1/2" diameter		40	.200		6.40	11.75		18.15	25
9074	2" diameter	Q-1	62	.258		9	13.65		22.65	30.50
9076	2-1/2" diameter		59	.271		11.85	14.35		26.20	34.50
9078	3" diameter		56	.286		15.90	15.10		31	40.50
9080	3-1/2" diameter		53	.302		23	15.95		38.95	49.50
9082	4" diameter		50	.320		26	16.90		42.90	54
9084	6" diameter		47	.340		31.50	18		49.50	61.50
9086	8" diameter	Q-2	51	.471		47.50	26		73.50	91.50
9088	10" diameter	"	46	.522		72.50	28.50		101	123
9090	To delete joint & hangers, subtract									
9092	1/2" diam. to 1" diam.					45%	65%			
9094	1-1/4" diam. to 4" diam.					15%	45%			
9096	6" diam. to 10" diam.					5%	24%			

22 11 13.76 Pipe Fittings, Plastic

		Crew	Daily Output	Labor-Hours	Unit	Material	2015 Bare Costs Labor	2015 Bare Costs Equipment	Total	Total Incl O&P
0010	**PIPE FITTINGS, PLASTIC**									
0030	Epoxy resin, fiberglass reinforced, general service									
0100	3"	Q-1	20.80	.769	Ea.	113	40.50		153.50	186
0110	4"		16.50	.970		115	51		166	204
0120	6"		10.10	1.584		225	83.50		308.50	375
0130	8"	Q-2	9.30	2.581		415	141		556	670
0140	10"		8.50	2.824		520	155		675	810
0150	12"		7.60	3.158		750	173		923	1,075
0170	Elbow, 90°, flanged									
0172	2"	Q-1	23	.696	Ea.	197	37		234	273
0173	3"		16	1		226	53		279	330
0174	4"		13	1.231		294	65		359	425
0176	6"		8	2		535	106		641	745
0177	8"	Q-2	9	2.667		965	146		1,111	1,300
0178	10"		7	3.429		1,325	188		1,513	1,725
0179	12"		5	4.800		1,775	263		2,038	2,350
0186	Elbow, 45°, flanged									
0188	2"	Q-1	23	.696	Ea.	197	37		234	273
0189	3"		16	1		226	53		279	330
0190	4"		13	1.231		294	65		359	425

For customer support on your Plumbing Cost Data, call 877.709.6968.

217

22 11 13 – Facility Water Distribution Piping

22 11 13.76 Pipe Fittings, Plastic		Crew	Daily Output	Labor-Hours	Unit	Material	2015 Bare Costs Labor	Equipment	Total	Total Incl O&P
0192	6"	Q-1	8	2	Ea.	535	106		641	745
0193	8"	Q-2	9	2.667		915	146		1,061	1,225
0194	10"		7	3.429		1,150	188		1,338	1,550
0195	12"		5	4.800		1,500	263		1,763	2,050
0352	Tee, flanged									
0354	2"	Q-1	17	.941	Ea.	267	49.50		316.50	370
0355	3"		10	1.600		355	84.50		439.50	520
0356	4"		8	2		395	106		501	595
0358	6"		5	3.200		680	169		849	1,000
0359	8"	Q-2	6	4		1,300	219		1,519	1,750
0360	10"		5	4.800		1,875	263		2,138	2,450
0361	12"		4	6		2,575	330		2,905	3,350
0365	Wye, flanged									
0367	2"	Q-1	17	.941	Ea.	530	49.50		579.50	655
0368	3"		10	1.600		725	84.50		809.50	930
0369	4"		8	2		970	106		1,076	1,225
0371	6"		5	3.200		1,375	169		1,544	1,750
0372	8"	Q-2	6	4		2,250	219		2,469	2,800
0373	10"		5	4.800		3,475	263		3,738	4,225
0374	12"		4	6		5,075	330		5,405	6,075
0380	Couplings									
0410	2"	Q-1	33.10	.483	Ea.	22.50	25.50		48	63.50
0420	3"		20.80	.769		24.50	40.50		65	88.50
0430	4"		16.50	.970		33	51		84	114
0440	6"		10.10	1.584		78.50	83.50		162	212
0450	8"	Q-2	9.30	2.581		133	141		274	360
0460	10"		8.50	2.824		195	155		350	450
0470	12"		7.60	3.158		262	173		435	550
0473	High corrosion resistant couplings, add					30%				
0474	Reducer, concentric, flanged									
0475	2" x 1-1/2"	Q-1	30	.533	Ea.	510	28		538	610
0476	3" x 2"		24	.667		560	35		595	670
0477	4" x 3"		19	.842		625	44.50		669.50	755
0479	6" x 4"		15	1.067		625	56.50		681.50	775
0480	8" x 6"	Q-2	16	1.500		1,025	82		1,107	1,250
0481	10" x 8"		13	1.846		955	101		1,056	1,200
0482	12" x 10"		11	2.182		1,475	120		1,595	1,800
0486	Adapter, bell x male or female									
0488	2"	Q-1	28	.571	Ea.	25.50	30		55.50	73.50
0489	3"		20	.800		38	42.50		80.50	106
0491	4"		17	.941		49.50	49.50		99	130
0492	6"		12	1.333		107	70.50		177.50	224
0493	8"	Q-2	15	1.600		154	87.50		241.50	300
0494	10"	"	11	2.182		221	120		341	425
0528	Flange									
0532	2"	Q-1	46	.348	Ea.	36.50	18.40		54.90	68
0533	3"		32	.500		46.50	26.50		73	91
0534	4"		26	.615		60.50	32.50		93	116
0536	6"		16	1		114	53		167	206
0537	8"	Q-2	18	1.333		175	73		248	305
0538	10"		14	1.714		248	94		342	415
0539	12"		10	2.400		310	131		441	540
2100	PVC schedule 80, socket joint									

22 11 13 – Facility Water Distribution Piping

22 11 13.76 Pipe Fittings, Plastic		Crew	Daily Output	Labor-Hours	Unit	Material	2015 Bare Costs Labor	Equipment	Total	Total Incl O&P
2110	90° elbow, 1/2"	1 Plum	30.30	.264	Ea.	2.48	15.50		17.98	26
2130	3/4"		26	.308		3.17	18.05		21.22	31
2140	1"		22.70	.352		4.95	20.50		25.45	36.50
2150	1-1/4"		20.20	.396		6.85	23.50		30.35	42.50
2160	1-1/2"		18.20	.440		7.30	26		33.30	47
2170	2"	Q-1	33.10	.483		8.85	25.50		34.35	48
2180	3"		20.80	.769		23.50	40.50		64	87
2190	4"		16.50	.970		35.50	51		86.50	117
2200	6"		10.10	1.584		101	83.50		184.50	237
2210	8"	Q-2	9.30	2.581		277	141		418	520
2250	45° elbow, 1/2"	1 Plum	30.30	.264		4.67	15.50		20.17	28.50
2270	3/4"		26	.308		7.15	18.05		25.20	35.50
2280	1"		22.70	.352		10.70	20.50		31.20	43
2290	1-1/4"		20.20	.396		13.60	23.50		37.10	50
2300	1-1/2"		18.20	.440		16.10	26		42.10	56.50
2310	2"	Q-1	33.10	.483		21	25.50		46.50	61.50
2320	3"		20.80	.769		53.50	40.50		94	120
2330	4"		16.50	.970		96	51		147	184
2340	6"		10.10	1.584		121	83.50		204.50	259
2350	8"	Q-2	9.30	2.581		262	141		403	500
2400	Tee, 1/2"	1 Plum	20.20	.396		7	23.50		30.50	42.50
2420	3/4"		17.30	.462		7.35	27		34.35	49
2430	1"		15.20	.526		9.15	31		40.15	56.50
2440	1-1/4"		13.50	.593		25.50	35		60.50	80.50
2450	1-1/2"		12.10	.661		25.50	39		64.50	86.50
2460	2"	Q-1	20	.800		31.50	42.50		74	98.50
2470	3"		13.90	1.151		43	61		104	139
2480	4"		11	1.455		49.50	77		126.50	171
2490	6"		6.70	2.388		169	126		295	375
2500	8"	Q-2	6.20	3.871		395	212		607	750
2510	Flange, socket, 150 lb., 1/2"	1 Plum	55.60	.144		13.60	8.45		22.05	27.50
2514	3/4"		47.60	.168		14.50	9.85		24.35	31
2518	1"		41.70	.192		16.15	11.25		27.40	35
2522	1-1/2"		33.30	.240		16.65	14.10		30.75	40
2526	2"	Q-1	60.60	.264		22.50	13.95		36.45	46
2530	4"		30.30	.528		49	28		77	96
2534	6"		18.50	.865		77	45.50		122.50	154
2538	8"	Q-2	17.10	1.404		138	77		215	267
2550	Coupling, 1/2"	1 Plum	30.30	.264		4.49	15.50		19.99	28.50
2570	3/4"		26	.308		6.05	18.05		24.10	34
2580	1"		22.70	.352		6.25	20.50		26.75	38
2590	1-1/4"		20.20	.396		9.50	23.50		33	45.50
2600	1-1/2"		18.20	.440		10.25	26		36.25	50.50
2610	2"	Q-1	33.10	.483		11	25.50		36.50	50.50
2620	3"		20.80	.769		31	40.50		71.50	95.50
2630	4"		16.50	.970		39	51		90	121
2640	6"		10.10	1.584		84	83.50		167.50	218
2650	8"	Q-2	9.30	2.581		114	141		255	340
2660	10"		8.50	2.824		400	155		555	675
2670	12"		7.60	3.158		450	173		623	755
2700	PVC (white), schedule 40, socket joints									
2760	90° elbow, 1/2"	1 Plum	33.30	.240	Ea.	.48	14.10		14.58	22
2770	3/4"		28.60	.280		.54	16.40		16.94	25.50

For customer support on your Plumbing Cost Data, call 877.709.6968.

219

22 11 13.76 Pipe Fittings, Plastic		Crew	Daily Output	Labor-Hours	Unit	Material	2015 Bare Costs Labor	Equipment	Total	Total Incl O&P
2780	1"	1 Plum	25	.320	Ea.	.97	18.80		19.77	29.50
2790	1-1/4"		22.20	.360		1.71	21		22.71	34
2800	1-1/2"		20	.400		1.85	23.50		25.35	37.50
2810	2"	Q-1	36.40	.440		2.89	23		25.89	38
2820	2-1/2"		26.70	.599		8.80	31.50		40.30	57.50
2830	3"		22.90	.699		10.50	37		47.50	67
2840	4"		18.20	.879		18.80	46.50		65.30	90.50
2850	5"		12.10	1.322		48.50	70		118.50	159
2860	6"		11.10	1.441		60	76		136	181
2870	8"	Q-2	10.30	2.330		154	128		282	365
2980	45° elbow, 1/2"	1 Plum	33.30	.240		.80	14.10		14.90	22.50
2990	3/4"		28.60	.280		1.23	16.40		17.63	26.50
3000	1"		25	.320		1.47	18.80		20.27	30
3010	1-1/4"		22.20	.360		2.03	21		23.03	34
3020	1-1/2"		20	.400		2.60	23.50		26.10	38.50
3030	2"	Q-1	36.40	.440		3.38	23		26.38	38.50
3040	2-1/2"		26.70	.599		8.80	31.50		40.30	57.50
3050	3"		22.90	.699		13.65	37		50.65	70.50
3060	4"		18.20	.879		24.50	46.50		71	97
3070	5"		12.10	1.322		48.50	70		118.50	159
3080	6"		11.10	1.441		60.50	76		136.50	182
3090	8"	Q-2	10.30	2.330		146	128		274	355
3180	Tee, 1/2"	1 Plum	22.20	.360		.60	21		21.60	32.50
3190	3/4"		19	.421		.69	24.50		25.19	38.50
3200	1"		16.70	.479		1.29	28		29.29	44
3210	1-1/4"		14.80	.541		2.01	31.50		33.51	50
3220	1-1/2"		13.30	.602		2.45	35.50		37.95	56
3230	2"	Q-1	24.20	.661		3.56	35		38.56	57
3240	2-1/2"		17.80	.899		11.75	47.50		59.25	84.50
3250	3"		15.20	1.053		15.45	55.50		70.95	101
3260	4"		12.10	1.322		28	70		98	136
3270	5"		8.10	1.975		67.50	104		171.50	232
3280	6"		7.40	2.162		94	114		208	276
3290	8"	Q-2	6.80	3.529		218	193		411	530
3380	Coupling, 1/2"	1 Plum	33.30	.240		.32	14.10		14.42	22
3390	3/4"		28.60	.280		.44	16.40		16.84	25.50
3400	1"		25	.320		.77	18.80		19.57	29.50
3410	1-1/4"		22.20	.360		1.06	21		22.06	33
3420	1-1/2"		20	.400		1.13	23.50		24.63	36.50
3430	2"	Q-1	36.40	.440		1.72	23		24.72	37
3440	2-1/2"		26.70	.599		3.81	31.50		35.31	52
3450	3"		22.90	.699		5.95	37		42.95	62
3460	4"		18.20	.879		8.65	46.50		55.15	79.50
3470	5"		12.10	1.322		15.80	70		85.80	122
3480	6"		11.10	1.441		27.50	76		103.50	145
3490	8"	Q-2	10.30	2.330		51	128		179	249
3600	Cap, schedule 40, PVC socket 1/2"	1 Plum	60.60	.132		.44	7.75		8.19	12.20
3610	3/4"		51.90	.154		.50	9.05		9.55	14.20
3620	1"		45.50	.176		.80	10.30		11.10	16.50
3630	1-1/4"		40.40	.198		1.13	11.60		12.73	18.80
3640	1-1/2"		36.40	.220		1.23	12.90		14.13	21
3650	2"	Q-1	66.10	.242		1.49	12.80		14.29	21
3660	2-1/2"		48.50	.330		4.73	17.45		22.18	31.50

For customer support on your Plumbing Cost Data, call 877.709.6968.

22 11 13 – Facility Water Distribution Piping

22 11 13.76 Pipe Fittings, Plastic	Crew	Daily Output	Labor-Hours	Unit	Material	2015 Bare Costs Labor	Equipment	Total	Total Incl O&P
3670	3"	Q-1	41.60	.385	Ea.	5.10	20.50		25.60	36
3680	4"		33.10	.483		11.75	25.50		37.25	51.50
3690	6"	↓	20.20	.792		28	42		70	94
3700	8"	Q-2	18.60	1.290	↓	71	70.50		141.50	185
3710	Reducing insert, schedule 40, socket weld									
3712	3/4"	1 Plum	31.50	.254	Ea.	.50	14.90		15.40	23
3713	1"		27.50	.291		.92	17.10		18.02	27
3715	1-1/2"	↓	22	.364		1.31	21.50		22.81	33.50
3716	2"	Q-1	40	.400		2.17	21		23.17	34.50
3717	4"		20	.800		11.45	42.50		53.95	76.50
3718	6"	↓	12.20	1.311		28.50	69.50		98	136
3719	8"	Q-2	11.30	2.124	↓	112	116		228	299
3730	Reducing insert, socket weld x female/male thread									
3732	1/2"	1 Plum	38.30	.209	Ea.	2.28	12.25		14.53	21
3733	3/4"		32.90	.243		1.41	14.25		15.66	23
3734	1"		28.80	.278		1.98	16.30		18.28	26.50
3736	1-1/2"	↓	23	.348		3.56	20.50		24.06	35
3737	2"	Q-1	41.90	.382		3.81	20		23.81	34.50
3738	4"	"	20.90	.766	↓	36.50	40.50		77	101
3742	Male adapter, socket weld x male thread									
3744	1/2"	1 Plum	38.30	.209	Ea.	.44	12.25		12.69	19
3745	3/4"		32.90	.243		.48	14.25		14.73	22
3746	1"		28.80	.278		.87	16.30		17.17	25.50
3748	1-1/2"	↓	23	.348		1.41	20.50		21.91	32.50
3749	2"	Q-1	41.90	.382		1.85	20		21.85	32.50
3750	4"	"	20.90	.766	↓	10.20	40.50		50.70	72
3754	Female adapter, socket weld x female thread									
3756	1/2"	1 Plum	38.30	.209	Ea.	.54	12.25		12.79	19.10
3757	3/4"		32.90	.243		.69	14.25		14.94	22.50
3758	1"		28.80	.278		.80	16.30		17.10	25.50
3760	1-1/2"	↓	23	.348		1.41	20.50		21.91	32.50
3761	2"	Q-1	41.90	.382		1.90	20		21.90	32.50
3762	4"	"	20.90	.766	↓	10.70	40.50		51.20	73
3800	PVC, schedule 80, socket joints									
3810	Reducing insert									
3812	3/4"	1 Plum	28.60	.280	Ea.	1.45	16.40		17.85	26.50
3813	1"		25	.320		4.14	18.80		22.94	33
3815	1-1/2"	↓	20	.400		9	23.50		32.50	45.50
3816	2"	Q-1	36.40	.440		12.60	23		35.60	49
3817	4"		18.20	.879		48	46.50		94.50	123
3818	6"	↓	11.10	1.441		67	76		143	189
3819	8"	Q-2	10.20	2.353	↓	385	129		514	620
3830	Reducing insert, socket weld x female/male thread									
3832	1/2"	1 Plum	34.80	.230	Ea.	9.05	13.50		22.55	30.50
3833	3/4"		29.90	.268		5.55	15.70		21.25	29.50
3834	1"		26.10	.307		8.70	18		26.70	36.50
3836	1-1/2"	↓	20.90	.383		10.95	22.50		33.45	46
3837	2"	Q-1	38	.421		16	22		38	51
3838	4"	"	19	.842	↓	77.50	44.50		122	153
3844	Adapter, male socket x male thread									
3846	1/2"	1 Plum	34.80	.230	Ea.	3.50	13.50		17	24.50
3847	3/4"		29.90	.268		3.86	15.70		19.56	28
3848	1"	↓	26.10	.307		6.65	18		24.65	34.50

For customer support on your Plumbing Cost Data, call 877.709.6968.

221

22 11 13.76 Pipe Fittings, Plastic		Crew	Daily Output	Labor-Hours	Unit	Material	2015 Bare Costs Labor	Equipment	Total	Total Incl O&P
3850	1-1/2"	1 Plum	20.90	.383	Ea.	11.20	22.50		33.70	46.50
3851	2"	Q-1	38	.421		16.20	22		38.20	51.50
3852	4"	"	19	.842	↓	36.50	44.50		81	107
3860	Adapter, female socket x female thread									
3862	1/2"	1 Plum	34.80	.230	Ea.	4.20	13.50		17.70	25
3863	3/4"		29.90	.268		6.25	15.70		21.95	30.50
3864	1"		26.10	.307		9.20	18		27.20	37
3866	1-1/2"	↓	20.90	.383		18.25	22.50		40.75	54
3867	2"	Q-1	38	.421		32	22		54	68.50
3868	4"	"	19	.842	↓	97.50	44.50		142	174
3872	Union, socket joints									
3874	1/2"	1 Plum	25.80	.310	Ea.	9.20	18.20		27.40	37.50
3875	3/4"		22.10	.362		11.70	21.50		33.20	45
3876	1"		19.30	.415		13.35	24.50		37.85	51.50
3878	1-1/2"	↓	15.50	.516		30	30.50		60.50	79
3879	2"	Q-1	28.10	.569	↓	40.50	30		70.50	90
3888	Cap									
3890	1/2"	1 Plum	54.50	.147	Ea.	4.39	8.60		12.99	17.85
3891	3/4"		46.70	.171		4.63	10.05		14.68	20.50
3892	1"		41	.195		8.20	11.45		19.65	26.50
3894	1-1/2"	↓	32.80	.244		9.90	14.30		24.20	32.50
3895	2"	Q-1	59.50	.269		26	14.20		40.20	50
3896	4"		30	.533		79	28		107	130
3897	6"	↓	18.20	.879		196	46.50		242.50	286
3898	8"	Q-2	16.70	1.437	↓	252	78.50		330.50	395
4500	DWV, ABS, non pressure, socket joints									
4540	1/4 Bend, 1-1/4"	1 Plum	20.20	.396	Ea.	2.57	23.50		26.07	38
4560	1-1/2"	"	18.20	.440		1.92	26		27.92	41
4570	2"	Q-1	33.10	.483		3.05	25.50		28.55	42
4580	3"		20.80	.769		7.70	40.50		48.20	70
4590	4"		16.50	.970		15.95	51		66.95	95
4600	6"	↓	10.10	1.584	↓	65	83.50		148.50	198
4650	1/8 Bend, same as 1/4 Bend									
4800	Tee, sanitary									
4820	1-1/4"	1 Plum	13.50	.593	Ea.	3.42	35		38.42	56.50
4830	1-1/2"	"	12.10	.661		2.92	39		41.92	61.50
4840	2"	Q-1	20	.800		4.52	42.50		47.02	69
4850	3"		13.90	1.151		12.40	61		73.40	106
4860	4"		11	1.455		22	77		99	140
4862	Tee, sanitary, reducing, 2" x 1-1/2"		22	.727		3.94	38.50		42.44	62.50
4864	3" x 2"		15.30	1.046		9.05	55.50		64.55	93.50
4868	4" x 3"	↓	12.10	1.322	↓	20	70		90	127
4870	Combination Y and 1/8 bend									
4872	1-1/2"	1 Plum	12.10	.661	Ea.	7	39		46	66
4874	2"	Q-1	20	.800		7.85	42.50		50.35	72.50
4876	3"		13.90	1.151		18.50	61		79.50	113
4878	4"		11	1.455		35.50	77		112.50	155
4880	3" x 1-1/2"		15.50	1.032		18.05	54.50		72.55	102
4882	4" x 3"	↓	12.10	1.322		28	70		98	136
4900	Wye, 1-1/4"	1 Plum	13.50	.593		3.94	35		38.94	57
4902	1-1/2"	"	12.10	.661		4.49	39		43.49	63.50
4904	2"	Q-1	20	.800		5.85	42.50		48.35	70.50
4906	3"	↓	13.90	1.151	↓	13.80	61		74.80	107

22 11 13.76 Pipe Fittings, Plastic		Crew	Daily Output	Labor-Hours	Unit	Material	2015 Bare Costs Labor	Equipment	Total	Total Incl O&P
4908	4"	Q-1	11	1.455	Ea.	28	77		105	147
4910	6"		6.70	2.388		81	126		207	280
4918	3" x 1-1/2"		15.50	1.032		11.05	54.50		65.55	94.50
4920	4" x 3"		12.10	1.322		22	70		92	130
4922	6" x 4"		6.90	2.319		66.50	123		189.50	259
4930	Double Wye, 1-1/2"	1 Plum	9.10	.879		13.95	51.50		65.45	93.50
4932	2"	Q-1	16.60	.964		16.65	51		67.65	95.50
4934	3"		10.40	1.538		38	81.50		119.50	165
4936	4"		8.25	1.939		74	102		176	237
4940	2" x 1-1/2"		16.80	.952		14.55	50.50		65.05	92
4942	3" x 2"		10.60	1.509		28	79.50		107.50	151
4944	4" x 3"		8.45	1.893		58.50	100		158.50	216
4946	6" x 4"		7.25	2.207		90	117		207	275
4950	Reducer bushing, 2" x 1-1/2"		36.40	.440		1.56	23		24.56	36.50
4952	3" x 1-1/2"		27.30	.586		6.65	31		37.65	54.50
4954	4" x 2"		18.20	.879		12.70	46.50		59.20	84
4956	6" x 4"		11.10	1.441		34.50	76		110.50	153
4960	Couplings, 1-1/2"	1 Plum	18.20	.440		.95	26		26.95	40
4962	2"	Q-1	33.10	.483		1.28	25.50		26.78	40
4963	3"		20.80	.769		3.58	40.50		44.08	65.50
4964	4"		16.50	.970		6.45	51		57.45	84.50
4966	6"		10.10	1.584		27	83.50		110.50	156
4970	2" x 1-1/2"		33.30	.480		2.75	25.50		28.25	41.50
4972	3" x 1-1/2"		21	.762		7.65	40.50		48.15	69.50
4974	4" x 3"		16.70	.958		12.15	50.50		62.65	90
4978	Closet flange, 4"	1 Plum	32	.250		6.65	14.70		21.35	29.50
4980	4" x 3"	"	34	.235		7.55	13.80		21.35	29.50
5000	DWV, PVC, schedule 40, socket joints									
5040	1/4 bend, 1-1/4"	1 Plum	20.20	.396	Ea.	5.20	23.50		28.70	40.50
5060	1-1/2"	"	18.20	.440		1.48	26		27.48	40.50
5070	2"	Q-1	33.10	.483		2.33	25.50		27.83	41
5080	3"		20.80	.769		6.85	40.50		47.35	69
5090	4"		16.50	.970		13.05	51		64.05	92
5100	6"		10.10	1.584		46	83.50		129.50	177
5105	8"	Q-2	9.30	2.581		65	141		206	286
5106	10"	"	8.50	2.824		290	155		445	555
5110	1/4 bend, long sweep, 1-1/2"	1 Plum	18.20	.440		3.42	26		29.42	43
5112	2"	Q-1	33.10	.483		3.83	25.50		29.33	42.50
5114	3"		20.80	.769		8.75	40.50		49.25	71
5116	4"		16.50	.970		16.65	51		67.65	96
5150	1/8 bend, 1-1/4"	1 Plum	20.20	.396		3.40	23.50		26.90	38.50
5170	1-1/2"	"	18.20	.440		1.44	26		27.44	40.50
5180	2"	Q-1	33.10	.483		2.08	25.50		27.58	41
5190	3"		20.80	.769		6.15	40.50		46.65	68.50
5200	4"		16.50	.970		10.40	51		61.40	89
5210	6"		10.10	1.584		42	83.50		125.50	172
5215	8"	Q-2	9.30	2.581		158	141		299	385
5216	10"		8.50	2.824		206	155		361	460
5217	12"		7.60	3.158		284	173		457	570
5250	Tee, sanitary 1-1/4"	1 Plum	13.50	.593		5.20	35		40.20	58.50
5254	1-1/2"	"	12.10	.661		2.58	39		41.58	61.50
5255	2"	Q-1	20	.800		3.80	42.50		46.30	68
5256	3"		13.90	1.151		10	61		71	103

For customer support on your Plumbing Cost Data, call 877.709.6968.

223

22 11 13.76 Pipe Fittings, Plastic		Crew	Daily Output	Labor-Hours	Unit	Material	2015 Bare Costs Labor	Equipment	Total	Total Incl O&P
5257	4"	Q-1	11	1.455	Ea.	17.75	77		94.75	136
5259	6"	↓	6.70	2.388		74	126		200	273
5261	8"	Q-2	6.20	3.871	↓	173	212		385	510
5276	Tee, sanitary, reducing									
5281	2" x 1-1/2" x 1-1/2"	Q-1	23	.696	Ea.	3.35	37		40.35	59
5282	2" x 1-1/2" x 2"		22	.727		4.05	38.50		42.55	62.50
5283	2" x 2" x 1-1/2"		22	.727		3.35	38.50		41.85	61.50
5284	3" x 3" x 1-1/2"		15.50	1.032		7.05	54.50		61.55	90.50
5285	3" x 3" x 2"		15.30	1.046		7.45	55.50		62.95	91.50
5286	4" x 4" x 1-1/2"		12.30	1.301		19.10	68.50		87.60	125
5287	4" x 4" x 2"		12.20	1.311		15.85	69.50		85.35	122
5288	4" x 4" x 3"		12.10	1.322		21.50	70		91.50	129
5291	6" x 6" x 4"	↓	6.90	2.319	↓	71.50	123		194.50	264
5294	Tee, double sanitary									
5295	1-1/2"	1 Plum	9.10	.879	Ea.	5.75	51.50		57.25	84.50
5296	2"	Q-1	16.60	.964		7.75	51		58.75	85.50
5297	3"		10.40	1.538		21.50	81.50		103	147
5298	4"	↓	8.25	1.939	↓	35	102		137	194
5303	Wye, reducing									
5304	2" x 1-1/2" x 1-1/2"	Q-1	23	.696	Ea.	6.55	37		43.55	62.50
5305	2" x 2" x 1-1/2"		22	.727		5.70	38.50		44.20	64.50
5306	3" x 3" x 2"		15.30	1.046		18.50	55.50		74	104
5307	4" x 4" x 2"		12.20	1.311		13.65	69.50		83.15	120
5309	4" x 4" x 3"	↓	12.10	1.322		18.40	70		88.40	125
5314	Combination Y & 1/8 bend, 1-1/2"	1 Plum	12.10	.661		6.25	39		45.25	65.50
5315	2"	Q-1	20	.800		6.70	42.50		49.20	71.50
5317	3"		13.90	1.151		16.60	61		77.60	110
5318	4"		11	1.455		33	77		110	152
5319	6"	↓	6.70	2.388		140	126		266	345
5320	8"	Q-2	6.20	3.871		202	212		414	540
5321	10"		5.70	4.211		790	231		1,021	1,225
5322	12"	↓	5.10	4.706	↓	930	258		1,188	1,425
5324	Combination Y & 1/8 bend, reducing									
5325	2" x 2" x 1-1/2"	Q-1	22	.727	Ea.	8.80	38.50		47.30	67.50
5327	3" x 3" x 1-1/2"		15.50	1.032		15.65	54.50		70.15	99.50
5328	3" x 3" x 2"		15.30	1.046		11.35	55.50		66.85	96
5329	4" x 4" x 2"	↓	12.20	1.311		17.85	69.50		87.35	125
5331	Wye, 1-1/4"	1 Plum	13.50	.593		6.65	35		41.65	60
5332	1-1/2"	"	12.10	.661		4.87	39		43.87	64
5333	2"	Q-1	20	.800		4.63	42.50		47.13	69
5334	3"		13.90	1.151		12.50	61		73.50	106
5335	4"		11	1.455		22.50	77		99.50	141
5336	6"	↓	6.70	2.388		68.50	126		194.50	267
5337	8"	Q-2	6.20	3.871		124	212		336	455
5338	10"		5.70	4.211		268	231		499	645
5339	12"	↓	5.10	4.706		440	258		698	870
5341	2" x 1-1/2"	Q-1	22	.727		5.70	38.50		44.20	64.50
5342	3" x 1-1/2"		15.50	1.032		8.20	54.50		62.70	91.50
5343	4" x 3"		12.10	1.322		18.40	70		88.40	125
5344	6" x 4"		6.90	2.319		50.50	123		173.50	241
5345	8" x 6"	Q-2	6.40	3.750		93.50	205		298.50	415
5347	Double wye, 1-1/2"	1 Plum	9.10	.879		10.55	51.50		62.05	89.50
5348	2"	Q-1	16.60	.964	↓	12.20	51		63.20	90.50

22 11 13 – Facility Water Distribution Piping

22 11 13.76 Pipe Fittings, Plastic	Crew	Daily Output	Labor-Hours	Unit	Material	2015 Bare Costs Labor	Equipment	Total	Total Incl O&P
5349	3"	Q-1	10.40	1.538	Ea.	25	81.50		106.50	151
5350	4"	↓	8.25	1.939	↓	50.50	102		152.50	211
5353	Double wye, reducing									
5354	2" x 2" x 1-1/2" x 1-1/2"	Q-1	16.80	.952	Ea.	10.75	50.50		61.25	88
5355	3" x 3" x 2" x 2"		10.60	1.509		18.45	79.50		97.95	141
5356	4" x 4" x 3" x 3"		8.45	1.893		40	100		140	195
5357	6" x 6" x 4" x 4"	↓	7.25	2.207		138	117		255	330
5374	Coupling, 1-1/4"	1 Plum	20.20	.396		3.15	23.50		26.65	38.50
5376	1-1/2"	"	18.20	.440		.72	26		26.72	40
5378	2"	Q-1	33.10	.483		.95	25.50		26.45	39.50
5380	3"		20.80	.769		3.29	40.50		43.79	65
5390	4"		16.50	.970		5.65	51		56.65	83.50
5400	6"	↓	10.10	1.584		18.40	83.50		101.90	146
5402	8"	Q-2	9.30	2.581		31.50	141		172.50	249
5404	2" x 1-1/2"	Q-1	33.30	.480		2.13	25.50		27.63	41
5406	3" x 1-1/2"		21	.762		6.25	40.50		46.75	68
5408	4" x 3"		16.70	.958		10.15	50.50		60.65	87.50
5410	Reducer bushing, 2" x 1-1/4"		36.50	.438		2.82	23		25.82	38
5411	2" x 1-1/2"		36.40	.440		1.22	23		24.22	36.50
5412	3" x 1-1/2"		27.30	.586		5.85	31		36.85	53.50
5413	3" x 2"		27.10	.590		3.06	31		34.06	50.50
5414	4" x 2"		18.20	.879		10.20	46.50		56.70	81
5415	4" x 3"		16.70	.958		5.20	50.50		55.70	82
5416	6" x 4"	↓	11.10	1.441		27	76		103	145
5418	8" x 6"	Q-2	10.20	2.353		46	129		175	246
5425	Closet flange 4"	Q-1	32	.500		7.20	26.50		33.70	48
5426	4" x 3"	"	34	.471	↓	5.80	25		30.80	44
5450	Solvent cement for PVC, industrial grade, per quart				Qt.	27.50			27.50	30.50
5500	CPVC, Schedule 80, threaded joints									
5540	90° Elbow, 1/4"	1 Plum	32	.250	Ea.	12.15	14.70		26.85	35.50
5560	1/2"		30.30	.264		7.05	15.50		22.55	31.50
5570	3/4"		26	.308		10.55	18.05		28.60	39
5580	1"		22.70	.352		14.80	20.50		35.30	47.50
5590	1-1/4"		20.20	.396		28.50	23.50		52	66.50
5600	1-1/2"	↓	18.20	.440		30.50	26		56.50	73
5610	2"	Q-1	33.10	.483		41	25.50		66.50	83.50
5620	2-1/2"		24.20	.661		128	35		163	193
5630	3"		20.80	.769		137	40.50		177.50	212
5640	4"		16.50	.970		214	51		265	315
5650	6"	↓	10.10	1.584	↓	252	83.50		335.50	405
5660	45° Elbow same as 90° Elbow									
5700	Tee, 1/4"	1 Plum	22	.364	Ea.	23.50	21.50		45	58
5702	1/2"		20.20	.396		23.50	23.50		47	61
5704	3/4"		17.30	.462		34	27		61	78.50
5706	1"		15.20	.526		36.50	31		67.50	87
5708	1-1/4"		13.50	.593		37	35		72	93
5710	1-1/2"	↓	12.10	.661		38.50	39		77.50	101
5712	2"	Q-1	20	.800		43	42.50		85.50	112
5714	2-1/2"		16.20	.988		211	52		263	310
5716	3"		13.90	1.151		245	61		306	360
5718	4"		11	1.455		580	77		657	750
5720	6"	↓	6.70	2.388		670	126		796	925
5730	Coupling, 1/4"	1 Plum	32	.250		15.45	14.70		30.15	39

For customer support on your Plumbing Cost Data, call 877.709.6968.

225

22 11 13.76 Pipe Fittings, Plastic		Crew	Daily Output	Labor-Hours	Unit	Material	2015 Bare Costs Labor	Equipment	Total	Total Incl O&P
5732	1/2"	1 Plum	30.30	.264	Ea.	12.75	15.50		28.25	37.50
5734	3/4"		26	.308		20.50	18.05		38.55	50
5736	1"		22.70	.352		23.50	20.50		44	57
5738	1-1/4"		20.20	.396		25	23.50		48.50	62
5740	1-1/2"		18.20	.440		26.50	26		52.50	68.50
5742	2"	Q-1	33.10	.483		31.50	25.50		57	73
5744	2-1/2"		24.20	.661		56.50	35		91.50	115
5746	3"		20.80	.769		65.50	40.50		106	134
5748	4"		16.50	.970		133	51		184	225
5750	6"		10.10	1.584		182	83.50		265.50	325
5752	8"	Q-2	9.30	2.581		390	141		531	640
5900	CPVC, Schedule 80, socket joints									
5904	90° Elbow, 1/4"	1 Plum	32	.250	Ea.	11.55	14.70		26.25	34.50
5906	1/2"		30.30	.264		4.53	15.50		20.03	28.50
5908	3/4"		26	.308		5.80	18.05		23.85	34
5910	1"		22.70	.352		9.20	20.50		29.70	41
5912	1-1/4"		20.20	.396		19.85	23.50		43.35	57
5914	1-1/2"		18.20	.440		22	26		48	63.50
5916	2"	Q-1	33.10	.483		26.50	25.50		52	68
5918	2-1/2"		24.20	.661		61.50	35		96.50	121
5920	3"		20.80	.769		69.50	40.50		110	138
5922	4"		16.50	.970		126	51		177	216
5924	6"		10.10	1.584		252	83.50		335.50	405
5926	8"		9.30	1.720		620	91		711	815
5930	45° Elbow, 1/4"	1 Plum	32	.250		17.20	14.70		31.90	41
5932	1/2"		30.30	.264		5.55	15.50		21.05	29.50
5934	3/4"		26	.308		7.95	18.05		26	36.50
5936	1"		22.70	.352		12.75	20.50		33.25	45
5938	1-1/4"		20.20	.396		25	23.50		48.50	62.50
5940	1-1/2"		18.20	.440		25.50	26		51.50	67
5942	2"	Q-1	33.10	.483		29	25.50		54.50	70.50
5944	2-1/2"		24.20	.661		59	35		94	118
5946	3"		20.80	.769		75.50	40.50		116	145
5948	4"		16.50	.970		104	51		155	192
5950	6"		10.10	1.584		305	83.50		388.50	460
5952	8"		9.30	1.720		665	91		756	865
5960	Tee, 1/4"	1 Plum	22	.364		10.60	21.50		32.10	43.50
5962	1/2"		20.20	.396		10.60	23.50		34.10	46.50
5964	3/4"		17.30	.462		10.80	27		37.80	53
5966	1"		15.20	.526		13.25	31		44.25	61
5968	1-1/4"		13.50	.593		28	35		63	83
5970	1-1/2"		12.10	.661		32	39		71	93.50
5972	2"	Q-1	20	.800		36	42.50		78.50	104
5974	2-1/2"		16.20	.988		90.50	52		142.50	179
5976	3"		13.90	1.151		90.50	61		151.50	192
5978	4"		11	1.455		121	77		198	249
5980	6"		6.70	2.388		315	126		441	535
5982	8"	Q-2	6.20	3.871		900	212		1,112	1,300
5990	Coupling, 1/4"	1 Plum	32	.250		12.30	14.70		27	35.50
5992	1/2"		30.30	.264		4.79	15.50		20.29	29
5994	3/4"		26	.308		6.70	18.05		24.75	35
5996	1"		22.70	.352		9	20.50		29.50	41
5998	1-1/4"		20.20	.396		13.50	23.50		37	50

22 11 Facility Water Distribution

22 11 13 – Facility Water Distribution Piping

22 11 13.76 Pipe Fittings, Plastic	Crew	Daily Output	Labor-Hours	Unit	Material	2015 Bare Costs Labor	Equipment	Total	Total Incl O&P
6000	1-1/2"	1 Plum	18.20	.440	Ea.	17	26		43	57.50
6002	2"	Q-1	33.10	.483		19.75	25.50		45.25	60
6004	2-1/2"		24.20	.661		44	35		79	102
6006	3"		20.80	.769		47.50	40.50		88	114
6008	4"		16.50	.970		62.50	51		113.50	147
6010	6"		10.10	1.584		147	83.50		230.50	288
6012	8"	Q-2	9.30	2.581		400	141		541	655
6200	CTS, 100 psi at 180°F, hot and cold water									
6230	90° Elbow, 1/2"	1 Plum	20	.400	Ea.	.48	23.50		23.98	36
6250	3/4"		19	.421		.80	24.50		25.30	38.50
6251	1"		16	.500		2.19	29.50		31.69	47
6252	1-1/4"		15	.533		4.65	31.50		36.15	52.50
6253	1-1/2"		14	.571		8.40	33.50		41.90	60
6254	2"	Q-1	23	.696		16.10	37		53.10	73
6260	45° Elbow, 1/2"	1 Plum	20	.400		.62	23.50		24.12	36
6280	3/4"		19	.421		1.06	24.50		25.56	38.50
6281	1"		16	.500		2.71	29.50		32.21	47.50
6282	1-1/4"		15	.533		5.75	31.50		37.25	54
6283	1-1/2"		14	.571		8.30	33.50		41.80	59.50
6284	2"	Q-1	23	.696		17.40	37		54.40	74.50
6290	Tee, 1/2"	1 Plum	13	.615		.62	36		36.62	55
6310	3/4"		12	.667		1.17	39		40.17	60.50
6311	1"		11	.727		6.10	42.50		48.60	71.50
6312	1-1/4"		10	.800		9.40	47		56.40	81.50
6313	1-1/2"		10	.800		12.25	47		59.25	84.50
6314	2"	Q-1	17	.941		19.80	49.50		69.30	97
6320	Coupling, 1/2"	1 Plum	22	.364		.40	21.50		21.90	32.50
6340	3/4"		21	.381		.53	22.50		23.03	34.50
6341	1"		18	.444		2.38	26		28.38	42
6342	1-1/4"		17	.471		3.07	27.50		30.57	45
6343	1-1/2"		16	.500		4.34	29.50		33.84	49.50
6344	2"	Q-1	28	.571		8.25	30		38.25	54.50
6360	Solvent cement for CPVC, commercial grade, per quart				Qt.	45			45	49.50
7340	PVC flange, slip-on, Sch 80 std., 1/2"	1 Plum	22	.364	Ea.	13.60	21.50		35.10	47
7350	3/4"		21	.381		14.50	22.50		37	50
7360	1"		18	.444		16.15	26		42.15	57.50
7370	1-1/4"		17	.471		16.65	27.50		44.15	60
7380	1-1/2"		16	.500		17	29.50		46.50	63
7390	2"	Q-1	26	.615		22.50	32.50		55	74
7400	2-1/2"		24	.667		35	35		70	91.50
7410	3"		18	.889		38.50	47		85.50	114
7420	4"		15	1.067		49	56.50		105.50	139
7430	6"		10	1.600		77	84.50		161.50	213
7440	8"	Q-2	11	2.182		138	120		258	330
7550	Union, schedule 40, socket joints, 1/2"	1 Plum	19	.421		5.20	24.50		29.70	43.50
7560	3/4"		18	.444		5.35	26		31.35	45.50
7570	1"		15	.533		5.55	31.50		37.05	53.50
7580	1-1/4"		14	.571		16.65	33.50		50.15	69
7590	1-1/2"		13	.615		18.60	36		54.60	75
7600	2"	Q-1	20	.800		25	42.50		67.50	91.50
7992	Polybutyl/polyethyl pipe, for copper fittings see Line 22 11 13.25 7000									
8000	Compression type, PVC, 160 psi cold water									
8010	Coupling, 3/4" CTS	1 Plum	21	.381	Ea.	3.48	22.50		25.98	38

For customer support on your Plumbing Cost Data, call 877.709.6968.

227

22 11 13.76 **Pipe Fittings, Plastic**	Crew	Daily Output	Labor-Hours	Unit	Material	2015 Bare Costs Labor	2015 Bare Costs Equipment	Total	Total Incl O&P
8020	1" CTS	1 Plum	18	.444	Ea.	4.32	26		30.32	44.50
8030	1-1/4" CTS		17	.471		6.10	27.50		33.60	48
8040	1-1/2" CTS		16	.500		8.30	29.50		37.80	53.50
8050	2" CTS		15	.533		11.65	31.50		43.15	60.50
8060	Female adapter, 3/4" FPT x 3/4" CTS		23	.348		5.70	20.50		26.20	37.50
8070	3/4" FPT x 1" CTS		21	.381		6.60	22.50		29.10	41.50
8080	1" FPT x 1" CTS		20	.400		6.60	23.50		30.10	43
8090	1-1/4" FPT x 1-1/4" CTS		18	.444		8.65	26		34.65	49
8100	1-1/2" FPT x 1-1/2" CTS		16	.500		9.90	29.50		39.40	55.50
8110	2" FPT x 2" CTS		13	.615		14.35	36		50.35	70.50
8130	Male adapter, 3/4" MPT x 3/4" CTS		23	.348		4.74	20.50		25.24	36
8140	3/4" MPT x 1" CTS		21	.381		5.65	22.50		28.15	40
8150	1" MPT x 1" CTS		20	.400		5.65	23.50		29.15	41.50
8160	1-1/4" MPT x 1-1/4" CTS		18	.444		7.60	26		33.60	48
8170	1-1/2" MPT x 1-1/2" CTS		16	.500		9.15	29.50		38.65	54.50
8180	2" MPT x 2" CTS		13	.615		11.85	36		47.85	67.50
8200	Spigot adapter, 3/4" IPS x 3/4" CTS		23	.348		2.70	20.50		23.20	34
8210	3/4" IPS x 1" CTS		21	.381		3.30	22.50		25.80	37.50
8220	1" IPS x 1" CTS		20	.400		3.30	23.50		26.80	39
8230	1-1/4" IPS x 1-1/4" CTS		18	.444		4.99	26		30.99	45
8240	1-1/2" IPS x 1-1/2" CTS		16	.500		5.25	29.50		34.75	50.50
8250	2" IPS x 2" CTS		13	.615		6.45	36		42.45	61.50
8270	Price includes insert stiffeners									
8280	250 psi is same price as 160 psi									
8300	Insert type, nylon, 160 & 250 psi, cold water									
8310	Clamp ring stainless steel, 3/4" IPS	1 Plum	115	.070	Ea.	2.62	4.08		6.70	9.05
8320	1" IPS		107	.075		2.67	4.39		7.06	9.60
8330	1-1/4" IPS		101	.079		2.69	4.65		7.34	9.95
8340	1-1/2" IPS		95	.084		3.61	4.94		8.55	11.40
8350	2" IPS		85	.094		4.15	5.50		9.65	12.90
8370	Coupling, 3/4" IPS		22	.364		.89	21.50		22.39	33
8380	1" IPS		19	.421		.92	24.50		25.42	38.50
8390	1-1/4" IPS		18	.444		1.37	26		27.37	41
8400	1-1/2" IPS		17	.471		1.62	27.50		29.12	43.50
8410	2" IPS		16	.500		3.60	29.50		33.10	48.50
8430	Elbow, 90°, 3/4" IPS		22	.364		1.77	21.50		23.27	34
8440	1" IPS		19	.421		1.96	24.50		26.46	39.50
8450	1-1/4" IPS		18	.444		2.19	26		28.19	42
8460	1-1/2" IPS		17	.471		2.58	27.50		30.08	44.50
8470	2" IPS		16	.500		3.60	29.50		33.10	48.50
8490	Male adapter, 3/4" IPS x 3/4" MPT		25	.320		.89	18.80		19.69	29.50
8500	1" IPS x 1" MPT		21	.381		.92	22.50		23.42	35
8510	1-1/4" IPS x 1-1/4" MPT		20	.400		1.45	23.50		24.95	37
8520	1-1/2" IPS x 1-1/2" MPT		18	.444		1.62	26		27.62	41.50
8530	2" IPS x 2" MPT		15	.533		3.11	31.50		34.61	51
8550	Tee, 3/4" IPS		14	.571		1.71	33.50		35.21	52.50
8560	1" IPS		13	.615		2.23	36		38.23	57
8570	1-1/4" IPS		12	.667		3.48	39		42.48	63
8580	1-1/2" IPS		11	.727		3.95	42.50		46.45	69
8590	2" IPS		10	.800		7.80	47		54.80	79.50
8610	Insert type, PVC, 100 psi @ 180°F, hot & cold water									
8620	Coupler, male, 3/8" CTS x 3/8" MPT	1 Plum	29	.276	Ea.	.72	16.20		16.92	25.50
8630	3/8" CTS x 1/2" MPT		28	.286		.72	16.75		17.47	26.50

22 11 13 – Facility Water Distribution Piping

22 11 13.76 Pipe Fittings, Plastic	Crew	Daily Output	Labor-Hours	Unit	Material	2015 Bare Costs Labor	Equipment	Total	Total Incl O&P
8640	1/2" CTS x 1/2" MPT	1 Plum	27	.296	Ea.	.72	17.40		18.12	27.50
8650	1/2" CTS x 3/4" MPT		26	.308		2.29	18.05		20.34	30
8660	3/4" CTS x 1/2" MPT		25	.320		2.07	18.80		20.87	31
8670	3/4" CTS x 3/4" MPT		25	.320		2.46	18.80		21.26	31
8700	Coupling, 3/8" CTS x 1/2" CTS		25	.320		4.58	18.80		23.38	33.50
8710	1/2" CTS		23	.348		5.50	20.50		26	37
8730	3/4" CTS		22	.364		11.60	21.50		33.10	45
8750	Elbow 90°, 3/8" CTS		25	.320		4	18.80		22.80	33
8760	1/2" CTS		23	.348		4.73	20.50		25.23	36
8770	3/4" CTS		22	.364		7.25	21.50		28.75	40
8800	Rings, crimp, copper, 3/8" CTS		120	.067		.22	3.91		4.13	6.15
8810	1/2" CTS		117	.068		.22	4.01		4.23	6.30
8820	3/4" CTS		115	.070		.29	4.08		4.37	6.45
8850	Reducer tee,-bronze, 3/8" x 3/8" x 1/2" CTS		17	.471		4.35	27.50		31.85	46.50
8860	1/2" x 1/2" x 3/4" CTS		15	.533		5.55	31.50		37.05	53.50
8870	3/4" x 1/2" x 1/2" CTS		14	.571		5.55	33.50		39.05	56.50
8890	3/4" x 3/4" x 1/2" CTS		14	.571		5.65	33.50		39.15	57
8900	1" x 1/2" x 1/2" CTS		14	.571		9.35	33.50		42.85	61
8930	Tee, 3/8" CTS		17	.471		3.31	27.50		30.81	45
8940	1/2" CTS		15	.533		2.27	31.50		33.77	50
8950	3/4" CTS		14	.571		3.50	33.50		37	54.50
8960	Copper rings included in fitting price									
9000	Flare type, assembled, acetal, hot & cold water									
9010	Coupling, 1/4" & 3/8" CTS	1 Plum	24	.333	Ea.	3.36	19.55		22.91	33
9020	1/2" CTS		22	.364		3.84	21.50		25.34	36
9030	3/4" CTS		21	.381		5.70	22.50		28.20	40.50
9040	1" CTS		18	.444		7.25	26		33.25	47.50
9050	Elbow 90°, 1/4" CTS		26	.308		3.73	18.05		21.78	31.50
9060	3/8" CTS		24	.333		4	19.55		23.55	34
9070	1/2" CTS		22	.364		4.73	21.50		26.23	37
9080	3/4" CTS		21	.381		7.25	22.50		29.75	42
9090	1" CTS		18	.444		9.15	26		35.15	49.50
9110	Tee, 1/4" CTS		16	.500		4.09	29.50		33.59	49
9114	3/8" CTS		15	.533		4.13	31.50		35.63	52
9120	1/2" CTS		14	.571		5.25	33.50		38.75	56.50
9130	3/4" CTS		13	.615		8.25	36		44.25	63.50
9140	1" CTS		12	.667		11.05	39		50.05	71
9400	Polypropylene, fittings and accessories									
9404	Fittings fusion welded, sizes are I.D.									
9408	Note: sizes 1/2" thru 4" use socket fusion									
9410	Sizes 6" thru 10" use butt fusion									
9416	Coupling									
9420	3/8"	1 Plum	39	.205	Ea.	.88	12.05		12.93	19.15
9422	1/2"		37.40	.214		1.15	12.55		13.70	20
9424	3/4"		35.40	.226		1.28	13.25		14.53	21.50
9426	1"		29.70	.269		1.69	15.80		17.49	26
9428	1-1/4"		27.60	.290		2.02	17		19.02	27.50
9430	1-1/2"		24.80	.323		4.25	18.95		23.20	33
9432	2"	Q-1	43	.372		8.50	19.65		28.15	39
9434	2-1/2"		35.60	.449		9.50	23.50		33	46.50
9436	3"		30.90	.518		21	27.50		48.50	64.50
9438	3-1/2"		27.80	.576		34	30.50		64.50	83.50
9440	4"		25	.640		44.50	34		78.50	100

229

For customer support on your Plumbing Cost Data, call 877.709.6968.

22 11 13 – Facility Water Distribution Piping

22 11 13.76 Pipe Fittings, Plastic		Crew	Daily Output	Labor-Hours	Unit	Material	2015 Bare Costs Labor	2015 Bare Costs Equipment	Total	Total Incl O&P
9442	Reducing coupling, female to female									
9446	2" to 1-1/2"	Q-1	49	.327	Ea.	12.25	17.25		29.50	39.50
9448	2-1/2" to 2"		41.20	.388		13.45	20.50		33.95	46
9450	3" to 2-1/2"	↓	33.10	.483	↓	18.15	25.50		43.65	58.50
9470	Reducing bushing, female to female									
9472	1/2" to 3/8"	1 Plum	38.20	.209	Ea.	1.15	12.30		13.45	19.80
9474	3/4" to 3/8" or 1/2"		36.50	.219		1.28	12.85		14.13	21
9476	1" to 3/4" or 1/2"		33.10	.242		1.71	14.20		15.91	23.50
9478	1-1/4" to 3/4" or 1"		28.70	.279		2.63	16.35		18.98	27.50
9480	1-1/2" to 1/2" thru 1-1/4"		26.20	.305		4.36	17.90		22.26	32
9482	2" to 1/2" thru 1-1/2"	Q-1	43	.372		8.70	19.65		28.35	39
9484	2-1/2" to 1/2" thru 2"		41.20	.388		9.75	20.50		30.25	41.50
9486	3" to 1-1/2" thru 2-1/2"		33.10	.483		21.50	25.50		47	62.50
9488	3-1/2" to 2" thru 3"		29.20	.548		34.50	29		63.50	81.50
9490	4" to 2-1/2" thru 3-1/2"		26.20	.611		54	32.50		86.50	108
9491	6" to 4" SDR 7.4		16.50	.970		65	51		116	149
9492	6" to 4" SDR 11		16.50	.970		65	51		116	149
9493	8" to 6" SDR 7.4		10.10	1.584		95.50	83.50		179	231
9494	8" to 6" SDR 11		10.10	1.584		87.50	83.50		171	223
9495	10" to 8" SDR 7.4		7.80	2.051		131	108		239	310
9496	10" to 8" SDR 11	↓	7.80	2.051	↓	114	108		222	290
9500	90° Elbow									
9504	3/8"	1 Plum	39	.205	Ea.	.97	12.05		13.02	19.25
9506	1/2"		37.40	.214		1.23	12.55		13.78	20.50
9508	3/4"		35.40	.226		1.58	13.25		14.83	21.50
9510	1"		29.70	.269		2.28	15.80		18.08	26.50
9512	1-1/4"		27.60	.290		3.51	17		20.51	29.50
9514	1-1/2"	↓	24.80	.323		7.55	18.95		26.50	37
9516	2"	Q-1	43	.372		11.60	19.65		31.25	42.50
9518	2-1/2"		35.60	.449		26	23.50		49.50	64.50
9520	3"		30.90	.518		43	27.50		70.50	88.50
9522	3-1/2"		27.80	.576		61	30.50		91.50	113
9524	4"		25	.640		93.50	34		127.50	154
9526	6" SDR 7.4		5.55	2.883		107	152		259	345
9528	6" SDR 11	↓	5.55	2.883		83.50	152		235.50	320
9530	8" SDR 7.4	Q-2	8.10	2.963		415	162		577	705
9532	8" SDR 11		8.10	2.963		355	162		517	635
9534	10" SDR 7.4		7.50	3.200		635	175		810	965
9536	10" SDR 11	↓	7.50	3.200	↓	575	175		750	900
9551	45° Elbow									
9554	3/8"	1 Plum	39	.205	Ea.	.94	12.05		12.99	19.25
9556	1/2"		37.40	.214		1.23	12.55		13.78	20.50
9558	3/4"		35.40	.226		1.58	13.25		14.83	21.50
9564	1"		29.70	.269		2.28	15.80		18.08	26.50
9566	1-1/4"		27.60	.290		3.51	17		20.51	29.50
9568	1-1/2"	↓	24.80	.323		7.55	18.95		26.50	37
9570	2"	Q-1	43	.372		11.45	19.65		31.10	42
9572	2-1/2"		35.60	.449		25.50	23.50		49	64
9574	3"		30.90	.518		47	27.50		74.50	93
9576	3-1/2"		27.80	.576		67	30.50		97.50	120
9578	4"		25	.640		103	34		137	164
9580	6" SDR 7.4		5.55	2.883		119	152		271	360
9582	6" SDR 11	↓	5.55	2.883		98	152		250	340

22 11 13 – Facility Water Distribution Piping

22 11 13.76 Pipe Fittings, Plastic		Crew	Daily Output	Labor-Hours	Unit	Material	2015 Bare Costs Labor	Equipment	Total	Total Incl O&P
9584	8" SDR 7.4	Q-2	8.10	2.963	Ea.	325	162		487	605
9586	8" SDR 11		8.10	2.963		284	162		446	555
9588	10" SDR 7.4		7.50	3.200		525	175		700	840
9590	10" SDR 11		7.50	3.200		435	175		610	740
9600	Tee									
9604	3/8"	1 Plum	26	.308	Ea.	1.22	18.05		19.27	29
9606	1/2"		24.90	.321		1.66	18.85		20.51	30.50
9608	3/4"		23.70	.338		2.28	19.80		22.08	32.50
9610	1"		19.90	.402		2.90	23.50		26.40	38.50
9612	1-1/4"		18.50	.432		4.43	25.50		29.93	43.50
9614	1-1/2"		16.60	.482		12.65	28.50		41.15	56.50
9616	2"	Q-1	26.80	.597		18.10	31.50		49.60	67.50
9618	2-1/2"		23.80	.672		30.50	35.50		66	87
9620	3"		20.60	.777		50	41		91	117
9622	3-1/2"		18.40	.870		78	46		124	156
9624	4"		16.70	.958		104	50.50		154.50	191
9626	6" SDR 7.4		3.70	4.324		129	228		357	485
9628	6" SDR 11		3.70	4.324		98	228		326	455
9630	8" SDR 7.4	Q-2	5.40	4.444		355	243		598	760
9632	8" SDR 11		5.40	4.444		310	243		553	715
9634	10" SDR 7.4		5	4.800		610	263		873	1,075
9636	10" SDR 11		5	4.800		530	263		793	975
9638	For reducing tee use same tee price									
9660	End cap									
9662	3/8"	1 Plum	78	.103	Ea.	1.22	6		7.22	10.45
9664	1/2"		74.60	.107		1.80	6.30		8.10	11.50
9666	3/4"		71.40	.112		2.28	6.60		8.88	12.45
9668	1"		59.50	.134		2.76	7.90		10.66	14.95
9670	1-1/4"		55.60	.144		4.36	8.45		12.81	17.55
9672	1-1/2"		49.50	.162		6	9.50		15.50	21
9674	2"	Q-1	80	.200		10.05	10.55		20.60	27
9676	2-1/2"		71.40	.224		14.55	11.85		26.40	34
9678	3"		61.70	.259		33	13.70		46.70	56.50
9680	3-1/2"		55.20	.290		39.50	15.30		54.80	66.50
9682	4"		50	.320		60.50	16.90		77.40	92
9684	6" SDR 7.4		26.50	.604		83	32		115	139
9686	6" SDR 11		26.50	.604		83	32		115	139
9688	8" SDR 7.4	Q-2	16.30	1.472		83	80.50		163.50	214
9690	8" SDR 11		16.30	1.472		72	80.50		152.50	201
9692	10" SDR 7.4		14.90	1.611		124	88		212	270
9694	10" SDR 11		14.90	1.611		106	88		194	249
9800	Accessories and tools									
9802	Pipe clamps for suspension, not including rod or beam clamp									
9804	3/8"	1 Plum	74	.108	Ea.	2.13	6.35		8.48	11.95
9805	1/2"		70	.114		2.67	6.70		9.37	13.10
9806	3/4"		68	.118		3.07	6.90		9.97	13.85
9807	1"		66	.121		3.35	7.10		10.45	14.45
9808	1-1/4"		64	.125		3.44	7.35		10.79	14.90
9809	1-1/2"		62	.129		3.76	7.55		11.31	15.60
9810	2"	Q-1	110	.145		4.66	7.70		12.36	16.75
9811	2-1/2"		104	.154		5.95	8.15		14.10	18.80
9812	3"		98	.163		6.45	8.65		15.10	20
9813	3-1/2"		92	.174		7.05	9.20		16.25	21.50

For customer support on your Plumbing Cost Data, call 877.709.6968.

231

22 11 Facility Water Distribution

22 11 13 – Facility Water Distribution Piping

22 11 13.76 Pipe Fittings, Plastic		Crew	Daily Output	Labor-Hours	Unit	Material	2015 Bare Costs Labor	Equipment	Total	Total Incl O&P
9814	4"	Q-1	86	.186	Ea.	7.90	9.85		17.75	23.50
9815	6"	↓	70	.229		9.65	12.10		21.75	29
9816	8"	Q-2	100	.240		34.50	13.15		47.65	58
9817	10"	"	94	.255	↓	39.50	14		53.50	64.50
9820	Pipe cutter									
9822	For 3/8" thru 1-1/4"				Ea.	118			118	129
9824	For 1-1/2" thru 4"				"	310			310	340
9826	Note: Pipes may be cut with standard									
9827	iron saw with blades for plastic.									
9982	For plastic hangers see Line 22 05 29.10 8000									
9986	For copper/brass fittings see Line 22 11 13.25 7000									

22 11 13.78 Pipe, High Density Polyethylene Plastic (HDPE)

		Crew	Daily Output	Labor-Hours	Unit	Material	Labor	Equipment	Total	Total Incl O&P
0010	**PIPE, HIGH DENSITY POLYETHYLENE PLASTIC (HDPE)**									
0020	Not incl. hangers, trenching, backfill, hoisting or digging equipment.									
0030	Standard length is 40', add a weld for each joint									
0040	Single wall									
0050	Straight									
0054	1" diameter DR 11				L.F.	.79			.79	.87
0058	1-1/2" diameter DR 11					1			1	1.10
0062	2" diameter DR 11					1.67			1.67	1.84
0066	3" diameter DR 11					2.01			2.01	2.21
0070	3" diameter DR 17					1.61			1.61	1.77
0074	4" diameter DR 11					3.36			3.36	3.70
0078	4" diameter DR 17					3.36			3.36	3.70
0082	6" diameter DR 11					8.35			8.35	9.20
0086	6" diameter DR 17					5.55			5.55	6.10
0090	8" diameter DR 11					13.90			13.90	15.30
0094	8" diameter DR 26					6.50			6.50	7.15
0098	10" diameter DR 11					22			22	24
0102	10" diameter DR 26					10.05			10.05	11.05
0106	12" diameter DR 11					32			32	35
0110	12" diameter DR 26					15.05			15.05	16.60
0114	16" diameter DR 11					48.50			48.50	53.50
0118	16" diameter DR 26					22			22	24
0122	18" diameter DR 11					62			62	68
0126	18" diameter DR 26					28.50			28.50	31.50
0130	20" diameter DR 11					75.50			75.50	83
0134	20" diameter DR 26					33.50			33.50	37
0138	22" diameter DR 11					92			92	101
0142	22" diameter DR 26					42			42	46
0146	24" diameter DR 11					109			109	120
0150	24" diameter DR 26					48.50			48.50	53.50
0154	28" diameter DR 17					100			100	111
0158	28" diameter DR 26					67			67	73.50
0162	30" diameter DR 21					94			94	103
0166	30" diameter DR 26					77			77	84.50
0170	36" diameter DR 26					77			77	84.50
0174	42" diameter DR 26					149			149	164
0178	48" diameter DR 26					196			196	216
0182	54" diameter DR 26				↓	246			246	271
0300	90° Elbow									
0304	1" diameter DR 11				Ea.	5.60			5.60	6.15

22 11 13 – Facility Water Distribution Piping

22 11 13.78 Pipe, High Density Polyethylene Plastic (HDPE)	Crew	Daily Output	Labor-Hours	Unit	Material	2015 Bare Costs Labor	Equipment	Total	Total Incl O&P
0308	1-1/2" diameter DR 11				Ea.	7			7	7.70
0312	2" diameter DR 11					7			7	7.70
0316	3" diameter DR 11					14			14	15.40
0320	3" diameter DR 17					14			14	15.40
0324	4" diameter DR 11					19.60			19.60	21.50
0328	4" diameter DR 17					19.60			19.60	21.50
0332	6" diameter DR 11					45			45	49.50
0336	6" diameter DR 17					45			45	49.50
0340	8" diameter DR 11					111			111	122
0344	8" diameter DR 26					98			98	108
0348	10" diameter DR 11					415			415	455
0352	10" diameter DR 26					380			380	420
0356	12" diameter DR 11					435			435	480
0360	12" diameter DR 26					395			395	435
0364	16" diameter DR 11					515			515	570
0368	16" diameter DR 26					510			510	560
0372	18" diameter DR 11					650			650	715
0376	18" diameter DR 26					610			610	670
0380	20" diameter DR 11					765			765	840
0384	20" diameter DR 26					740			740	810
0388	22" diameter DR 11					790			790	870
0392	22" diameter DR 26					765			765	840
0396	24" diameter DR 11					890			890	980
0400	24" diameter DR 26					865			865	950
0404	28" diameter DR 17					1,075			1,075	1,200
0408	28" diameter DR 26					1,025			1,025	1,125
0412	30" diameter DR 17					1,525			1,525	1,675
0416	30" diameter DR 26					1,400			1,400	1,550
0420	36" diameter DR 26					1,775			1,775	1,950
0424	42" diameter DR 26					2,300			2,300	2,525
0428	48" diameter DR 26					2,675			2,675	2,950
0432	54" diameter DR 26					6,375			6,375	7,000
0500	45° Elbow									
0512	2" diameter DR 11				Ea.	5.60			5.60	6.15
0516	3" diameter DR 11					14			14	15.40
0520	3" diameter DR 17					14			14	15.40
0524	4" diameter DR 11					19.60			19.60	21.50
0528	4" diameter DR 17					19.60			19.60	21.50
0532	6" diameter DR 11					45			45	49.50
0536	6" diameter DR 17					45			45	49.50
0540	8" diameter DR 11					111			111	122
0544	8" diameter DR 26					64.50			64.50	71
0548	10" diameter DR 11					415			415	455
0552	10" diameter DR 26					380			380	420
0556	12" diameter DR 11					435			435	480
0560	12" diameter DR 26					395			395	435
0564	16" diameter DR 11					229			229	252
0568	16" diameter DR 26					216			216	238
0572	18" diameter DR 11					243			243	268
0576	18" diameter DR 26					223			223	245
0580	20" diameter DR 11					370			370	405
0584	20" diameter DR 26					350			350	385
0588	22" diameter DR 11					510			510	560

For customer support on your Plumbing Cost Data, call 877.709.6968.

233

22 11 13.78 Pipe, High Density Polyethylene Plastic (HDPE)	Crew	Daily Output	Labor-Hours	Unit	Material	2015 Bare Costs Labor	Equipment	Total	Total Incl O&P
0592	22" diameter DR 26				Ea.	485			485	530
0596	24" diameter DR 11					625			625	685
0600	24" diameter DR 26					600			600	660
0604	28" diameter DR 17					710			710	780
0608	28" diameter DR 26					690			690	755
0612	30" diameter DR 17					880			880	965
0616	30" diameter DR 26					855			855	940
0620	36" diameter DR 26					1,075			1,075	1,200
0624	42" diameter DR 26					1,400			1,400	1,525
0628	48" diameter DR 26					1,525			1,525	1,675
0632	54" diameter DR 26					2,025			2,025	2,225
0700	Tee									
0704	1" diameter DR 11				Ea.	7.30			7.30	8.05
0708	1-1/2" diameter DR 11					10.25			10.25	11.30
0712	2" diameter DR 11					8.80			8.80	9.65
0716	3" diameter DR 11					16.10			16.10	17.70
0720	3" diameter DR 17					16.10			16.10	17.70
0724	4" diameter DR 11					23.50			23.50	26
0728	4" diameter DR 17					23.50			23.50	26
0732	6" diameter DR 11					58.50			58.50	64.50
0736	6" diameter DR 17					58.50			58.50	64.50
0740	8" diameter DR 11					145			145	159
0744	8" diameter DR 17					145			145	159
0748	10" diameter DR 11					430			430	475
0752	10" diameter DR 17					430			430	475
0756	12" diameter DR 11					575			575	635
0760	12" diameter DR 17					575			575	635
0764	16" diameter DR 11					305			305	335
0768	16" diameter DR 17					251			251	276
0772	18" diameter DR 11					430			430	470
0776	18" diameter DR 17					355			355	390
0780	20" diameter DR 11					525			525	575
0784	20" diameter DR 17					430			430	470
0788	22" diameter DR 11					670			670	740
0792	22" diameter DR 17					530			530	585
0796	24" diameter DR 11					835			835	915
0800	24" diameter DR 17					705			705	775
0804	28" diameter DR 17					1,350			1,350	1,500
0812	30" diameter DR 17					1,550			1,550	1,725
0820	36" diameter DR 17					2,575			2,575	2,825
0824	42" diameter DR 26					2,875			2,875	3,150
0828	48" diameter DR 26					3,075			3,075	3,375
1000	Flange adptr, w/back-up ring and 1/2 cost of plated bolt set									
1004	1" diameter DR 11				Ea.	28			28	30.50
1008	1-1/2" diameter DR 11					28			28	30.50
1012	2" diameter DR 11					17.55			17.55	19.35
1016	3" diameter DR 11					20.50			20.50	22.50
1020	3" diameter DR 17					20.50			20.50	22.50
1024	4" diameter DR 11					28			28	30.50
1028	4" diameter DR 17					28			28	30.50
1032	6" diameter DR 11					39.50			39.50	43.50
1036	6" diameter DR 17					39.50			39.50	43.50
1040	8" diameter DR 11					57			57	63

22 11 13 – Facility Water Distribution Piping

22 11 13.78 Pipe, High Density Polyethylene Plastic (HDPE)	Crew	Daily Output	Labor-Hours	Unit	Material	2015 Bare Costs Labor	Equipment	Total	Total Incl O&P
1044	8" diameter DR 26				Ea.	57			57	63
1048	10" diameter DR 11					91			91	100
1052	10" diameter DR 26					91			91	100
1056	12" diameter DR 11					133			133	147
1060	12" diameter DR 26					133			133	147
1064	16" diameter DR 11					286			286	315
1068	16" diameter DR 26					286			286	315
1072	18" diameter DR 11					370			370	410
1076	18" diameter DR 26					370			370	410
1080	20" diameter DR 11					515			515	565
1084	20" diameter DR 26					515			515	565
1088	22" diameter DR 11					560			560	615
1092	22" diameter DR 26					555			555	610
1096	24" diameter DR 17					605			605	665
1100	24" diameter DR 32.5					605			605	665
1104	28" diameter DR 15.5					820			820	900
1108	28" diameter DR 32.5					820			820	900
1112	30" diameter DR 11					950			950	1,050
1116	30" diameter DR 21					950			950	1,050
1120	36" diameter DR 26					1,050			1,050	1,150
1124	42" diameter DR 26					1,175			1,175	1,300
1128	48" diameter DR 26					1,450			1,450	1,600
1132	54" diameter DR 26				▼	1,750			1,750	1,925
1200	Reducer									
1208	2" x 1-1/2" diameter DR 11				Ea.	8.40			8.40	9.25
1212	3" x 2" diameter DR 11					8.40			8.40	9.25
1216	4" x 2" diameter DR 11					9.80			9.80	10.80
1220	4" x 3" diameter DR 11					12.60			12.60	13.85
1224	6" x 4" diameter DR 11					29.50			29.50	32.50
1228	8" x 6" diameter DR 11					45			45	49.50
1232	10" x 8" diameter DR 11					77			77	85
1236	12" x 8" diameter DR 11					126			126	139
1240	12" x 10" diameter DR 11					101			101	111
1244	14" x 12" diameter DR 11					112			112	123
1248	16" x 14" diameter DR 11					143			143	157
1252	18" x 16" diameter DR 11					174			174	191
1256	20" x 18" diameter DR 11					345			345	380
1260	22" x 20" diameter DR 11					425			425	465
1264	24" x 22" diameter DR 11					475			475	525
1268	26" x 24" diameter DR 11					560			560	615
1272	28" x 24" diameter DR 11					715			715	785
1276	32" x 28" diameter DR 17					925			925	1,025
1280	36" x 32" diameter DR 17				▼	1,250			1,250	1,375
4000	Welding labor per joint, not including welding machine									
4010	Pipe joint size (cost based on thickest wall for each diam.)									
4030	1" pipe size	4 Skwk	273	.117	Ea.		5.70		5.70	8.80
4040	1-1/2" pipe size		175	.183			8.90		8.90	13.75
4050	2" pipe size		128	.250			12.15		12.15	18.80
4060	3" pipe size		100	.320			15.55		15.55	24
4070	4" pipe size		77	.416			20		20	31
4080	6" pipe size	5 Skwk	63	.635			31		31	47.50
4090	8" pipe size		48	.833			40.50		40.50	62.50
4100	10" pipe size	▼	40	1	▼		48.50		48.50	75

22 11 Facility Water Distribution

22 11 13 – Facility Water Distribution Piping

22 11 13.78 Pipe, High Density Polyethylene Plastic (HDPE)	Crew	Daily Output	Labor-Hours	Unit	Material	2015 Bare Costs Labor	Equipment	Total	Total Incl O&P
4110	12" pipe size	6 Skwk	41	1.171	Ea.		57		57	88
4120	16" pipe size		34	1.412			68.50		68.50	106
4130	18" pipe size		32	1.500			73		73	113
4140	20" pipe size	8 Skwk	37	1.730			84		84	130
4150	22" pipe size		35	1.829			89		89	137
4160	24" pipe size		34	1.882			91.50		91.50	141
4170	28" pipe size		33	1.939			94.50		94.50	146
4180	30" pipe size		32	2			97.50		97.50	150
4190	36" pipe size		31	2.065			100		100	155
4200	42" pipe size		30	2.133			104		104	160
4210	48" pipe size	9 Skwk	33	2.182			106		106	164
4220	54" pipe size	"	31	2.323			113		113	175
4300	Note: Cost for set up each time welder is moved.									
4301	Add 50% of a weld cost									
4310	Welder usually remains stationary with pipe moved through it.									
4340	Weld machine, rental per day based on diam. capacity									
4350	1" thru 2" diameter				Ea.			40.50	40.50	44.50
4360	3" thru 4" diameter							46	46	50.50
4370	6" thru 8" diameter							103	103	113
4380	10" thru 12" diameter							178	178	196
4390	16" thru 18" diameter							259	259	285
4400	20" thru 24" diameter							500	500	550
4410	28" thru 32" diameter							545	545	600
4420	36" diameter							570	570	625
4430	42" thru 54" diameter							890	890	980
5000	Dual wall contained pipe									
5040	Straight									
5054	1" DR 11 x 3" DR 11				L.F.	7.10			7.10	7.80
5058	1" DR 11 x 4" DR 11					7.55			7.55	8.30
5062	1-1/2" DR 11 x 4" DR 17					7.75			7.75	8.55
5066	2" DR 11 x 4" DR 17					8.35			8.35	9.20
5070	2" DR 11 x 6" DR 17					12.50			12.50	13.75
5074	3" DR 11 x 6" DR 17					13.95			13.95	15.35
5078	3" DR 11 x 6" DR 26					12			12	13.20
5086	3" DR 17 x 8" DR 17					17.70			17.70	19.50
5090	4" DR 11 x 8" DR 17					20.50			20.50	23
5094	4" DR 17 x 8" DR 26					16.40			16.40	18.05
5098	6" DR 11 x 10" DR 17					32			32	35.50
5102	6" DR 17 x 10" DR 26					24			24	26.50
5106	6" DR 26 x 10" DR 26					23			23	25
5110	8" DR 17 x 12" DR 26					34			34	37.50
5114	8" DR 26 x 12" DR 32.5					29			29	32
5118	10" DR 17 x 14" DR 26					47.50			47.50	52.50
5122	10" DR 17 x 16" DR 26					52			52	57.50
5126	10" DR 26 x 16" DR 26					47.50			47.50	52
5130	12" DR 26 x 16" DR 26					54			54	59
5134	12" DR 17 x 18" DR 26					68.50			68.50	75.50
5138	12" DR 26 x 18" DR 26					61.50			61.50	67.50
5142	14" DR 26 x 20" DR 32.5					67.50			67.50	74
5146	16" DR 26 x 22" DR 32.5					102			102	112
5150	18" DR 26 x 24" DR 32.5					93.50			93.50	103
5154	20" DR 32.5 x 28" DR 32.5					121			121	133
5158	22" DR 32.5 x 30" DR 32.5					121			121	133

236

22 11 13 – Facility Water Distribution Piping

22 11 13.78 Pipe, High Density Polyethylene Plastic (HDPE)	Crew	Daily Output	Labor-Hours	Unit	Material	2015 Bare Costs Labor	Equipment	Total	Total Incl O&P
5162	24" DR 32.5 x 32" DR 32.5				L.F.	155			155	170
5166	36" DR 32.5 x 42" DR 32.5				↓	261			261	287
5300	Force transfer coupling									
5354	1" DR 11 x 3" DR 11				Ea.	255			255	280
5358	1" DR 11 x 4" DR 17					267			267	294
5362	1-1/2" DR 11 x 4" DR 17					280			280	310
5366	2" DR 11 x 4" DR 17					293			293	320
5370	2" DR 11 x 6" DR 17					425			425	470
5374	3" DR 11 x 6" DR 17					425			425	470
5378	3" DR 11 x 6" DR 26					425			425	470
5382	3" DR 11 x 8" DR 11					450			450	495
5386	3" DR 11 x 8" DR 17					450			450	495
5390	4" DR 11 x 8" DR 17					450			450	495
5394	4" DR 17 x 8" DR 26					395			395	435
5398	6" DR 11 x 10" DR 17					570			570	625
5402	6" DR 17 x 10" DR 26					470			470	515
5406	6" DR 26 x 10" DR 26					470			470	515
5410	8" DR 17 x 12" DR 26					625			625	685
5414	8" DR 26 x 12" DR 32.5					505			505	555
5418	10" DR 17 x 14" DR 26					820			820	900
5422	10" DR 17 x 16" DR 26					820			820	900
5426	10" DR 26 x 16" DR 26					820			820	900
5430	12" DR 26 x 16" DR 26					1,025			1,025	1,125
5434	12" DR 17 x 18" DR 26					1,050			1,050	1,150
5438	12" DR 26 x 18" DR 26					1,050			1,050	1,150
5442	14" DR 26 x 20" DR 32.5					1,100			1,100	1,225
5446	16" DR 26 x 22" DR 32.5					1,150			1,150	1,250
5450	18" DR 26 x 24" DR 32.5					1,250			1,250	1,350
5454	20" DR 32.5 x 28" DR 32.5					1,500			1,500	1,650
5458	22" DR 32.5 x 30" DR 32.5					1,675			1,675	1,825
5462	24" DR 32.5 x 32" DR 32.5					1,800			1,800	1,975
5466	36" DR 32.5 x 42" DR 32.5				↓	3,000			3,000	3,300
5600	90° Elbow									
5654	1" DR 11 x 3" DR 11				Ea.	210			210	231
5658	1" DR 11 x 4" DR 17					201			201	222
5662	1-1/2" DR 11 x 4" DR 17					220			220	242
5666	2" DR 11 x 4" DR 17					233			233	257
5670	2" DR 11 x 6" DR 17					297			297	325
5674	3" DR 11 x 6" DR 17					340			340	375
5678	3" DR 17 x 6" DR 26					271			271	298
5682	3" DR 17 x 8" DR 11					550			550	605
5686	3" DR 17 x 8" DR 17					425			425	470
5690	4" DR 11 x 8" DR 17					500			500	550
5694	4" DR 17 x 8" DR 26					400			400	440
5698	6" DR 11 x 10" DR 17					670			670	735
5702	6" DR 17 x 10" DR 26					510			510	565
5706	6" DR 26 x 10" DR 26					480			480	525
5710	8" DR 17 x 12" DR 26					760			760	835
5714	8" DR 26 x 12" DR 32.5					705			705	775
5718	10" DR 17 x 14" DR 26					1,200			1,200	1,325
5722	10" DR 17 x 16" DR 26					1,150			1,150	1,250
5726	10" DR 26 x 16" DR 26					1,050			1,050	1,175
5730	12" DR 26 x 16" DR 26					1,175			1,175	1,300

For customer support on your Plumbing Cost Data, call 877.709.6968.

237

22 11 13.78 Pipe, High Density Polyethylene Plastic (HDPE)		Crew	Daily Output	Labor-Hours	Unit	Material	2015 Bare Costs Labor	Equipment	Total	Total Incl O&P
5734	12" DR 17 x 18" DR 26				Ea.	1,400			1,400	1,550
5738	12" DR 26 x 18" DR 26					1,300			1,300	1,425
5742	14" DR 26 x 20" DR 32.5					1,700			1,700	1,875
5746	16" DR 26 x 22" DR 32.5					1,700			1,700	1,875
5750	18" DR 26 x 24" DR 32.5					2,100			2,100	2,300
5754	20" DR 32.5 x 28" DR 32.5					2,425			2,425	2,650
5758	22" DR 32.5 x 30" DR 32.5					3,300			3,300	3,625
5762	24" DR 32.5 x 32" DR 32.5					3,200			3,200	3,525
5766	36" DR 32.5 x 42" DR 32.5					5,200			5,200	5,725
5800	45° Elbow									
5804	1" DR 11 x 3" DR 11				Ea.	126			126	139
5808	1" DR 11 x 4" DR 17					125			125	138
5812	1-1/2" DR 11 x 4" DR 17					134			134	148
5816	2" DR 11 x 4" DR 17					147			147	161
5820	2" DR 11 x 6" DR 17					184			184	203
5824	3" DR 11 x 6" DR 17					204			204	225
5828	3" DR 17 x 6" DR 26					169			169	186
5832	3" DR 17 x 8" DR 11					315			315	345
5836	3" DR 17 x 8" DR 17					250			250	275
5840	4" DR 11 x 8" DR 17					292			292	320
5844	4" DR 17 x 8" DR 26					238			238	262
5848	6" DR 11 x 10" DR 17					385			385	425
5852	6" DR 17 x 10" DR 26					300			300	335
5856	6" DR 26 x 10" DR 26					285			285	315
5860	8" DR 17 x 12" DR 26					460			460	505
5864	8" DR 26 x 12" DR 32.5					430			430	475
5868	10" DR 17 x 14" DR 26					720			720	790
5872	10" DR 17 x 16" DR 26					705			705	775
5876	10" DR 26 x 16" DR 26					660			660	725
5880	12" DR 26 x 16" DR 26					745			745	820
5884	12" DR 17 x 18" DR 26					875			875	965
5888	12" DR 26 x 18" DR 26					820			820	900
5892	14" DR 26 x 20" DR 32.5					1,050			1,050	1,150
5896	16" DR 26 x 22" DR 32.5					1,100			1,100	1,225
5900	18" DR 26 x 24" DR 32.5					1,300			1,300	1,425
5904	20" DR 32.5 x 28" DR 32.5					1,600			1,600	1,750
5908	22" DR 32.5 x 30" DR 32.5					2,025			2,025	2,250
5912	24" DR 32.5 x 32" DR 32.5					2,025			2,025	2,225
5916	36" DR 32.5 x 42" DR 32.5					3,125			3,125	3,450
6000	Access port with 4" riser									
6050	1" DR 11 x 4" DR 17				Ea.	275			275	305
6054	1-1/2" DR 11 x 4" DR 17					281			281	310
6058	2" DR 11 x 6" DR 17					340			340	375
6062	3" DR 11 x 6" DR 17					345			345	380
6066	3" DR 17 x 6" DR 26					340			340	370
6070	3" DR 17 x 8" DR 11					385			385	425
6074	3" DR 17 x 8" DR 17					370			370	405
6078	4" DR 11 x 8" DR 17					380			380	420
6082	4" DR 17 x 8" DR 26					365			365	400
6086	6" DR 11 x 10" DR 17					475			475	525
6090	6" DR 17 x 10" DR 26					440			440	485
6094	6" DR 26 x 10" DR 26					430			430	475
6098	8" DR 17 x 12" DR 26					460			460	510

22 11 13 – Facility Water Distribution Piping

22 11 13.78 Pipe, High Density Polyethylene Plastic (HDPE)	Crew	Daily Output	Labor-Hours	Unit	Material	2015 Bare Costs Labor	Equipment	Total	Total Incl O&P
6102	8" DR 26 x 12" DR 32.5				Ea.	440			440	485
6200	End termination with vent plug									
6204	1" DR 11 x 3" DR 11				Ea.	228			228	250
6208	1" DR 11 x 4" DR 17					244			244	268
6212	1-1/2" DR 11 x 4" DR 17					244			244	268
6216	2" DR 11 x 4" DR 17					265			265	291
6220	2" DR 11 x 6" DR 17					300			300	330
6224	3" DR 11 x 6" DR 17					300			300	330
6228	3" DR 17 x 6" DR 26					300			300	330
6232	3" DR 17 x 8" DR 11					415			415	460
6236	3" DR 17 x 8" DR 17					415			415	460
6240	4" DR 11 x 8" DR 17					415			415	460
6244	4" DR 17 x 8" DR 26					365			365	400
6248	6" DR 11 x 10" DR 17					495			495	545
6252	6" DR 17 x 10" DR 26					405			405	450
6256	6" DR 26 x 10" DR 26					405			405	450
6260	8" DR 17 x 12" DR 26					605			605	665
6264	8" DR 26 x 12" DR 32.5					480			480	530
6268	10" DR 17 x 14" DR 26					625			625	690
6272	10" DR 17 x 16" DR 26					625			625	690
6276	10" DR 26 x 16" DR 26					625			625	690
6280	12" DR 26 x 16" DR 26					625			625	690
6284	12" DR 17 x 18" DR 26					825			825	905
6288	12" DR 26 x 18" DR 26					825			825	905
6292	14" DR 26 x 20" DR 32.5					840			840	925
6296	16" DR 26 x 22" DR 32.5					975			975	1,075
6300	18" DR 26 x 24" DR 32.5					1,000			1,000	1,100
6304	20" DR 32.5 x 28" DR 32.5					1,275			1,275	1,425
6308	22" DR 32.5 x 30" DR 32.5					1,425			1,425	1,550
6312	24" DR 32.5 x 32" DR 32.5					1,575			1,575	1,725
6316	36" DR 32.5 x 42" DR 32.5					2,275			2,275	2,500
6600	Tee									
6604	1" DR 11 x 3" DR 11				Ea.	232			232	255
6608	1" DR 11 x 4" DR 17					288			288	315
6612	1-1/2" DR 11 x 4" DR 17					320			320	350
6616	2" DR 11 x 4" DR 17					350			350	385
6620	2" DR 11 x 6" DR 17					425			425	465
6624	3" DR 11 x 6" DR 17					490			490	540
6628	3" DR 17 x 6" DR 26					475			475	525
6632	3" DR 17 x 8" DR 11					575			575	635
6636	3" DR 17 x 8" DR 17					550			550	605
6640	4" DR 11 x 8" DR 17					595			595	655
6644	4" DR 17 x 8" DR 26					575			575	630
6648	6" DR 11 x 10" DR 17					745			745	820
6652	6" DR 17 x 10" DR 26					705			705	780
6656	6" DR 26 x 10" DR 26					700			700	770
6660	8" DR 17 x 12" DR 26					700			700	770
6664	8" DR 26 x 12" DR 32.5					915			915	1,000
6668	10" DR 17 x 14" DR 26					1,050			1,050	1,150
6672	10" DR 17 x 16" DR 26					1,100			1,100	1,225
6676	10" DR 26 x 16" DR 26					1,100			1,100	1,225
6680	12" DR 26 x 16" DR 26					1,200			1,200	1,325
6684	12" DR 17 x 18" DR 26					1,200			1,200	1,325

For customer support on your Plumbing Cost Data, call 877.709.6968.

239

22 11 13 – Facility Water Distribution Piping

22 11 13.78 Pipe, High Density Polyethylene Plastic (HDPE)	Crew	Daily Output	Labor-Hours	Unit	Material	2015 Bare Costs Labor	Equipment	Total	Total Incl O&P
6688	12" DR 26 x 18" DR 26				Ea.	1,325			1,325	1,450
6692	14" DR 26 x 20" DR 32.5					1,750			1,750	1,925
6696	16" DR 26 x 22" DR 32.5					2,150			2,150	2,350
6700	18" DR 26 x 24" DR 32.5					2,325			2,325	2,575
6704	20" DR 32.5 x 28" DR 32.5					2,950			2,950	3,250
6708	22" DR 32.5 x 30" DR 32.5					3,500			3,500	3,850
6712	24" DR 32.5 x 32" DR 32.5					4,200			4,200	4,600
6716	36" DR 32.5 x 42" DR 32.5					8,525			8,525	9,400
6800	Wye									
6816	2" DR 11 x 4" DR 17				Ea.	440			440	485
6820	2" DR 11 x 6" DR 17					500			500	550
6824	3" DR 11 x 6" DR 17					505			505	555
6828	3" DR 17 x 6" DR 26					490			490	540
6832	3" DR 17 x 8" DR 11					615			615	675
6836	3" DR 17 x 8" DR 17					575			575	630
6840	4" DR 11 x 8" DR 17					625			625	690
6844	4" DR 17 x 8" DR 26					595			595	655
6848	6" DR 11 x 10" DR 17					775			775	855
6852	6" DR 17 x 10" DR 26					720			720	795
6856	6" DR 26 x 10" DR 26					715			715	790
6860	8" DR 17 x 12" DR 26					1,025			1,025	1,125
6864	8" DR 26 x 12" DR 32.5					995			995	1,100
6868	10" DR 17 x 14" DR 26					1,250			1,250	1,350
6872	10" DR 17 x 16" DR 26					1,325			1,325	1,450
6876	10" DR 26 x 16" DR 26					1,300			1,300	1,425
6880	12" DR 26 x 16" DR 26					1,450			1,450	1,600
6884	12" DR 17 x 18" DR 26					1,625			1,625	1,800
6888	12" DR 26 x 18" DR 26					1,650			1,650	1,825
6892	14" DR 26 x 20" DR 32.5					2,025			2,025	2,225
6896	16" DR 26 x 22" DR 32.5					2,900			2,900	3,200
6900	18" DR 26 x 24" DR 32.5					3,425			3,425	3,750
6904	20" DR 32.5 x 28" DR 32.5					4,300			4,300	4,725
6908	22" DR 32.5 x 30" DR 32.5					4,875			4,875	5,375
6912	24" DR 32.5 x 32" DR 32.5					5,550			5,550	6,125
9000	Welding labor per joint, not including welding machine									
9010	Pipe joint size, outer pipe (cost based on the thickest walls)									
9020	Straight pipe									
9050	3" pipe size	4 Skwk	96	.333	Ea.		16.20		16.20	25
9060	4" pipe size	"	77	.416			20		20	31
9070	6" pipe size	5 Skwk	60	.667			32.50		32.50	50
9080	8" pipe size	"	40	1			48.50		48.50	75
9090	10" pipe size	6 Skwk	41	1.171			57		57	88
9100	12" pipe size		39	1.231			60		60	92.50
9110	14" pipe size		38	1.263			61.50		61.50	95
9120	16" pipe size		35	1.371			66.50		66.50	103
9130	18" pipe size	8 Skwk	45	1.422			69		69	107
9140	20" pipe size		42	1.524			74		74	115
9150	22" pipe size		40	1.600			78		78	120
9160	24" pipe size		38	1.684			82		82	127
9170	28" pipe size		37	1.730			84		84	130
9180	30" pipe size		36	1.778			86.50		86.50	134
9190	32" pipe size		35	1.829			89		89	137
9200	42" pipe size		32	2			97.50		97.50	150

For customer support on your Plumbing Cost Data, call 877.709.6968.

22 11 Facility Water Distribution

22 11 13 – Facility Water Distribution Piping

22 11 13.78 Pipe, High Density Polyethylene Plastic (HDPE)	Crew	Daily Output	Labor-Hours	Unit	Material	2015 Bare Costs Labor	Equipment	Total	Total Incl O&P
9300	Note: Cost for set up each time welder is moved.									
9301	Add 100% of weld labor cost									
9310	For handling between fitting welds add 50% of weld labor cost									
9320	Welder usually remains stationary with pipe moved through it									
9360	Weld machine, rental per day based on diam. capacity									
9380	3" thru 4" diameter				Ea.			63.50	63.50	70
9390	6" thru 8" diameter							207	207	228
9400	10" thru 12" diameter							310	310	340
9410	14" thru 18" diameter							420	420	460
9420	20" thru 24" diameter							750	750	825
9430	28" thru 32" diameter							880	880	970
9440	42" thru 58" diameter							910	910	1,000

22 11 19 – Domestic Water Piping Specialties

22 11 19.10 Flexible Connectors

		Crew	Daily Output	Labor-Hours	Unit	Material	Labor	Equipment	Total	Total Incl O&P
0010	**FLEXIBLE CONNECTORS**, Corrugated, 7/8" O.D., 1/2" I.D.									
0050	Gas, seamless brass, steel fittings									
0200	12" long	1 Plum	36	.222	Ea.	17.40	13.05		30.45	39
0220	18" long		36	.222		21.50	13.05		34.55	43.50
0240	24" long		34	.235		25.50	13.80		39.30	49
0260	30" long		34	.235		27.50	13.80		41.30	51.50
0280	36" long		32	.250		30.50	14.70		45.20	55.50
0320	48" long		30	.267		38.50	15.65		54.15	66
0340	60" long		30	.267		46	15.65		61.65	74
0360	72" long		30	.267		53	15.65		68.65	82
2000	Water, copper tubing, dielectric separators									
2100	12" long	1 Plum	36	.222	Ea.	17.35	13.05		30.40	39
2220	15" long		36	.222		19.30	13.05		32.35	40.50
2240	18" long		36	.222		21	13.05		34.05	42.50
2260	24" long		34	.235		26	13.80		39.80	49.50

22 11 19.14 Flexible Metal Hose

		Crew	Daily Output	Labor-Hours	Unit	Material	Labor	Equipment	Total	Total Incl O&P
0010	**FLEXIBLE METAL HOSE**, Connectors, standard lengths									
0100	Bronze braided, bronze ends									
0120	3/8" diameter x 12"	1 Stpi	26	.308	Ea.	21	18.40		39.40	51
0140	1/2" diameter x 12"		24	.333		22	19.90		41.90	54
0160	3/4" diameter x 12"		20	.400		32	24		56	71
0180	1" diameter x 18"		19	.421		38	25		63	80
0200	1-1/2" diameter x 18"		13	.615		52	37		89	113
0220	2" diameter x 18"		11	.727		68	43.50		111.50	141
1000	Carbon steel ends									
1020	1/4" diameter x 12"	1 Stpi	28	.286	Ea.	15.50	17.05		32.55	43
1040	3/8" diameter x 12"		26	.308		15.90	18.40		34.30	45.50
1060	1/2" diameter x 12"		24	.333		17.55	19.90		37.45	49.50
1080	1/2" diameter x 24"		24	.333		47.50	19.90		67.40	82
1120	3/4" diameter x 12"		20	.400		28.50	24		52.50	67.50
1140	3/4" diameter x 24"		20	.400		58.50	24		82.50	101
1160	3/4" diameter x 36"		20	.400		69.50	24		93.50	112
1180	1" diameter x 18"		19	.421		35.50	25		60.50	77
1200	1" diameter x 30"		19	.421		67.50	25		92.50	112
1220	1" diameter x 36"		19	.421		87.50	25		112.50	135
1240	1-1/4" diameter x 18"		15	.533		49.50	32		81.50	102
1260	1-1/4" diameter x 36"		15	.533		109	32		141	168
1280	1-1/2" diameter x 18"		13	.615		69.50	37		106.50	132

For customer support on your Plumbing Cost Data, call 877.709.6968.

241

22 11 19.14 Flexible Metal Hose	Crew	Daily Output	Labor-Hours	Unit	Material	2015 Bare Costs Labor	Equipment	Total	Total Incl O&P
1300	1-1/2" diameter x 36"	1 Stpi	13	.615	Ea.	118	37		155	186
1320	2" diameter x 24"		11	.727		96.50	43.50		140	172
1340	2" diameter x 36"		11	.727		146	43.50		189.50	227
1360	2-1/2" diameter x 24"		9	.889		221	53		274	325
1380	2-1/2" diameter x 36"		9	.889		560	53		613	695
1400	3" diameter x 24"		7	1.143		305	68.50		373.50	440
1420	3" diameter x 36"		7	1.143		670	68.50		738.50	840
2000	Carbon steel braid, carbon steel solid ends									
2100	1/2" diameter x 12"	1 Stpi	24	.333	Ea.	42.50	19.90		62.40	77
2120	3/4" diameter x 12"		20	.400		64.50	24		88.50	107
2140	1" diameter x 12"		19	.421		92	25		117	139
2160	1-1/4" diameter x 12"		15	.533		28	32		60	79
2180	1-1/2" diameter x 12"		13	.615		30	37		67	88.50
3000	Stainless steel braid, welded on carbon steel ends									
3100	1/2" diameter x 12"	1 Stpi	24	.333	Ea.	59	19.90		78.90	95
3120	3/4" diameter x 12"		20	.400		71	24		95	115
3140	3/4" diameter x 24"		20	.400		82	24		106	126
3160	3/4" diameter x 36"		20	.400		94	24		118	140
3180	1" diameter x 12"		19	.421		84.50	25		109.50	131
3200	1" diameter x 24"		19	.421		102	25		127	150
3220	1" diameter x 36"		19	.421		117	25		142	167
3240	1-1/4" diameter x 12"		15	.533		128	32		160	188
3260	1-1/4" diameter x 24"		15	.533		139	32		171	201
3280	1-1/4" diameter x 36"		15	.533		158	32		190	222
3300	1-1/2" diameter x 12"		13	.615		44	37		81	104
3320	1-1/2" diameter x 24"		13	.615		152	37		189	223
3340	1-1/2" diameter x 36"		13	.615		178	37		215	252
3400	Metal stainless steel braid, over corrugated stainless steel, flanged ends									
3410	150 PSI									
3420	1/2" diameter x 12"	1 Stpi	24	.333	Ea.	120	19.90		139.90	162
3430	1" diameter x 12"		20	.400		180	24		204	234
3440	1-1/2" diameter x 12"		15	.533		204	32		236	272
3450	2-1/2" diameter x 9"		12	.667		108	40		148	179
3460	3" diameter x 9"		9	.889		93	53		146	182
3470	4" diameter x 9"		7	1.143		113	68.50		181.50	227
3480	4" diameter x 30"		5	1.600		560	95.50		655.50	760
3490	4" diameter x 36"		4.80	1.667		580	99.50		679.50	790
3500	6" diameter x 11"		5	1.600		188	95.50		283.50	350
3510	6" diameter x 36"		3.80	2.105		685	126		811	940
3520	8" diameter x 12"		4	2		410	120		530	630
3530	10" diameter x 13"		3	2.667		625	159		784	930
3540	12" diameter x 14"	Q-5	4	4		950	215		1,165	1,375
6000	Molded rubber with helical wire reinforcement									
6010	150 PSI									
6020	1-1/2" diameter x 12"	1 Stpi	15	.533	Ea.	88	32		120	145
6030	2" diameter x 12"		12	.667		240	40		280	325
6040	3" diameter x 12"		8	1		251	60		311	365
6050	4" diameter x 12"		6	1.333		320	79.50		399.50	470
6060	6" diameter x 18"		4	2		475	120		595	705
6070	8" diameter x 24"		3	2.667		665	159		824	975
6080	10" diameter x 24"		2	4		810	239		1,049	1,250
6090	12" diameter x 24"	Q-5	3	5.333		925	287		1,212	1,450
7000	Molded teflon with stainless steel flanges									

22 11 Facility Water Distribution

22 11 19 – Domestic Water Piping Specialties

22 11 19.14 Flexible Metal Hose

		Crew	Daily Output	Labor-Hours	Unit	Material	2015 Bare Costs Labor	Equipment	Total	Total Incl O&P
7010	150 PSI									
7020	2-1/2" diameter x 3-3/16"	Q-1	7.80	2.051	Ea.	2,325	108		2,433	2,725
7030	3" diameter x 3-5/8"		6.50	2.462		1,625	130		1,755	1,975
7040	4" diameter x 3-5/8"		5	3.200		2,100	169		2,269	2,575
7050	6" diameter x 4"		4.30	3.721		2,950	197		3,147	3,550
7060	8" diameter x 6"		3.80	4.211		4,700	222		4,922	5,500

22 11 19.18 Mixing Valve

		Crew	Daily Output	Labor-Hours	Unit	Material	2015 Bare Costs Labor	Equipment	Total	Total Incl O&P
0010	**MIXING VALVE**, Automatic, water tempering.									
0040	1/2" size	1 Stpi	19	.421	Ea.	555	25		580	650
0050	3/4" size		18	.444		555	26.50		581.50	650
0100	1" size		16	.500		845	30		875	975
0120	1-1/4" size		13	.615		1,150	37		1,187	1,325
0140	1-1/2" size		10	.800		1,400	48		1,448	1,600
0160	2" size		8	1		1,750	60		1,810	2,025
0170	2-1/2" size		6	1.333		1,750	79.50		1,829.50	2,050
0180	3" size		4	2		4,125	120		4,245	4,725
0190	4" size		3	2.667		4,125	159		4,284	4,800

22 11 19.22 Pressure Reducing Valve

		Crew	Daily Output	Labor-Hours	Unit	Material	2015 Bare Costs Labor	Equipment	Total	Total Incl O&P
0010	**PRESSURE REDUCING VALVE**, Steam, pilot operated.									
0100	Threaded, iron body									
0200	1-1/2" size	1 Stpi	8	1	Ea.	1,825	60		1,885	2,100
0220	2" size	"	5	1.600	"	2,100	95.50		2,195.50	2,450
1000	Flanged, iron body, 125 lb. flanges									
1020	2" size	1 Stpi	8	1	Ea.	2,100	60		2,160	2,425
1040	2-1/2" size	"	4	2		2,525	120		2,645	2,975
1060	3" size	Q-5	4.50	3.556		3,075	191		3,266	3,675
1080	4" size	"	3	5.333		4,425	287		4,712	5,300
1500	For 250 lb. flanges, add					5%				

22 11 19.26 Pressure Regulators

		Crew	Daily Output	Labor-Hours	Unit	Material	2015 Bare Costs Labor	Equipment	Total	Total Incl O&P
0010	**PRESSURE REGULATORS**									
0100	Gas appliance regulators									
0106	Main burner and pilot applications									
0108	Rubber seat poppet type									
0109	1/8" pipe size	1 Stpi	24	.333	Ea.	16	19.90		35.90	47.50
0110	1/4" pipe size		24	.333		18.40	19.90		38.30	50
0112	3/8" pipe size		24	.333		19.50	19.90		39.40	51.50
0113	1/2" pipe size		24	.333		21.50	19.90		41.40	54
0114	3/4" pipe size		20	.400		25.50	24		49.50	64
0122	Lever action type									
0123	3/8" pipe size	1 Stpi	24	.333	Ea.	23.50	19.90		43.40	56
0124	1/2" pipe size		24	.333		23.50	19.90		43.40	56
0125	3/4" pipe size		20	.400		46.50	24		70.50	87
0126	1" pipe size		19	.421		46.50	25		71.50	89
0132	Double diaphragm type									
0133	3/8" pipe size	1 Stpi	24	.333	Ea.	35.50	19.90		55.40	69.50
0134	1/2" pipe size		24	.333		49	19.90		68.90	84
0135	3/4" pipe size		20	.400		82	24		106	127
0136	1" pipe size		19	.421		100	25		125	148
0137	1-1/4" pipe size		15	.533		350	32		382	435
0138	1-1/2" pipe size		13	.615		640	37		677	760
0139	2" pipe size		11	.727		640	43.50		683.50	770
0140	2-1/2" pipe size	Q-5	15	1.067		1,250	57.50		1,307.50	1,450

For customer support on your Plumbing Cost Data, call 877.709.6968.

243

22 11 Facility Water Distribution

22 11 19 – Domestic Water Piping Specialties

22 11 19.26 Pressure Regulators

		Crew	Daily Output	Labor-Hours	Unit	Material	2015 Bare Costs Labor	Equipment	Total	Total Incl O&P
0141	3" pipe size	Q-5	13	1.231	Ea.	1,250	66		1,316	1,475
0142	4" pipe size (flanged)	↓	8	2	↓	2,225	108		2,333	2,575
0160	Main burner only									
0162	Straight-thru-flow design									
0163	1/2" pipe size	1 Stpi	24	.333	Ea.	40	19.90		59.90	74
0164	3/4" pipe size		20	.400		54.50	24		78.50	96
0165	1" pipe size		19	.421		77.50	25		102.50	124
0166	1-1/4" pipe size	↓	15	.533	↓	77.50	32		109.50	134
0200	Oil, light, hot water, ordinary steam, threaded									
0220	Bronze body, 1/4" size	1 Stpi	24	.333	Ea.	154	19.90		173.90	199
0230	3/8" size		24	.333		166	19.90		185.90	213
0240	1/2" size		24	.333		206	19.90		225.90	257
0250	3/4" size		20	.400		246	24		270	305
0260	1" size		19	.421		375	25		400	450
0270	1-1/4" size		15	.533		505	32		537	605
0320	Iron body, 1/4" size		24	.333		120	19.90		139.90	162
0330	3/8" size		24	.333		141	19.90		160.90	185
0340	1/2" size		24	.333		150	19.90		169.90	195
0350	3/4" size		20	.400		182	24		206	236
0360	1" size		19	.421		237	25		262	299
0370	1-1/4" size	↓	15	.533	↓	330	32		362	410
9002	For water pressure regulators, see Section 22 05 23.20									

22 11 19.30 Pressure and Temperature Safety Plug

		Crew	Daily Output	Labor-Hours	Unit	Material	2015 Bare Costs Labor	Equipment	Total	Total Incl O&P
0010	**PRESSURE & TEMPERATURE SAFETY PLUG**									
1000	3/4" external thread, 3/8" diam. element									
1020	Carbon steel									
1050	7-1/2" insertion	1 Stpi	32	.250	Ea.	45	14.95		59.95	72
1120	304 stainless steel									
1150	7-1/2" insertion	1 Stpi	32	.250	Ea.	50.50	14.95		65.45	78
1220	316 Stainless steel									
1250	7-1/2" insertion	1 Stpi	32	.250	Ea.	69.50	14.95		84.45	99

22 11 19.32 Pressure and Temperature Measurement Plug

		Crew	Daily Output	Labor-Hours	Unit	Material	2015 Bare Costs Labor	Equipment	Total	Total Incl O&P
0010	**PRESSURE & TEMPERATURE MEASUREMENT PLUG**									
0020	A permanent access port for insertion of									
0030	a pressure or temperature measuring probe									
0100	Plug, brass									
0110	1/4" MNPT, 1-1/2" long	1 Stpi	32	.250	Ea.	5.40	14.95		20.35	28.50
0120	3" long		31	.258		10.45	15.40		25.85	35
0140	1/2" MNPT, 1-1/2" long		30	.267		8.65	15.95		24.60	33.50
0150	3" long	↓	29	.276	↓	12.90	16.50		29.40	39
0200	Pressure gauge probe adapter									
0210	1/8" diameter, 1-1/2" probe				Ea.	26			26	28.50
0220	3" probe				"	53			53	58
0300	Temperature gauge, 5" stem									
0310	Analog				Ea.	69.50			69.50	76.50
0330	Digital				"	75.50			75.50	83
0400	Pressure gauge, compound									
0410	1/4" MNPT				Ea.	30.50			30.50	33.50
0500	Pressure and temperature test kit									
0510	Contains 2 thermometers and 2 pressure gauges									
0520	Kit				Ea.	355			355	390

For customer support on your Plumbing Cost Data, call 877.709.6968.

22 11 19.34 Sleeves and Escutcheons	Crew	Daily Output	Labor-Hours	Unit	Material	2015 Bare Costs Labor	Equipment	Total	Total Incl O&P
0010 **SLEEVES & ESCUTCHEONS**									
0100 Pipe sleeve									
0110 Steel, w/water stop, 12" long, with link seal									
0120 2" diam. for 1/2" carrier pipe	1 Plum	8.40	.952	Ea.	59.50	56		115.50	150
0130 2-1/2" diam. for 3/4" carrier pipe		8	1		67	58.50		125.50	163
0140 2-1/2" diam. for 1" carrier pipe		8	1		63.50	58.50		122	159
0150 3" diam. for 1-1/4" carrier pipe		7.20	1.111		81.50	65		146.50	188
0160 3-1/2" diam. for 1-1/2" carrier pipe		6.80	1.176		82	69		151	194
0170 4" diam. for 2" carrier pipe		6	1.333		88.50	78.50		167	216
0180 4" diam. for 2-1/2" carrier pipe		6	1.333		89.50	78.50		168	217
0190 5" diam. for 3" carrier pipe		5.40	1.481		102	87		189	243
0200 6" diam. for 4" carrier pipe		4.80	1.667		122	98		220	283
0210 10" diam. for 6" carrier pipe	Q-1	8	2		244	106		350	430
0220 12" diam. for 8" carrier pipe		7.20	2.222		325	117		442	535
0230 14" diam. for 10" carrier pipe		6.40	2.500		345	132		477	575
0240 16" diam. for 12" carrier pipe		5.80	2.759		380	146		526	640
0250 18" diam. for 14" carrier pipe		5.20	3.077		705	163		868	1,025
0260 24" diam. for 18" carrier pipe		4	4		1,050	211		1,261	1,475
0270 24" diam. for 20" carrier pipe		4	4		885	211		1,096	1,300
0280 30" diam. for 24" carrier pipe		3.20	5		1,300	264		1,564	1,825
0500 Wall sleeve									
0510 Ductile iron with rubber gasket seal									
0520 3"	1 Plum	8.40	.952	Ea.	700	56		756	855
0530 4"		7.20	1.111		750	65		815	925
0540 6"		6	1.333		920	78.50		998.50	1,150
0550 8"		4	2		1,125	117		1,242	1,425
0560 10"		3	2.667		1,375	157		1,532	1,750
0570 12"		2.40	3.333		1,625	196		1,821	2,100
5000 Escutcheon									
5100 Split ring, pipe									
5110 Chrome plated									
5120 1/2"	1 Plum	160	.050	Ea.	1.09	2.94		4.03	5.65
5130 3/4"		160	.050		1.21	2.94		4.15	5.75
5140 1"		135	.059		1.30	3.48		4.78	6.70
5150 1-1/2"		115	.070		1.83	4.08		5.91	8.15
5160 2"		100	.080		2.03	4.70		6.73	9.35
5170 4"		80	.100		4.50	5.85		10.35	13.80
5180 6"		68	.118		7.50	6.90		14.40	18.70
5400 Shallow flange type									
5410 Chrome plated steel									
5420 1/2" CTS	1 Plum	180	.044	Ea.	.32	2.61		2.93	4.29
5430 3/4" CTS		180	.044		.34	2.61		2.95	4.31
5440 1/2" IPS		180	.044		.36	2.61		2.97	4.34
5450 3/4" IPS		180	.044		.40	2.61		3.01	4.38
5460 1" IPS		175	.046		.46	2.68		3.14	4.56
5470 1-1/2" IPS		170	.047		.76	2.76		3.52	5
5480 2" IPS		160	.050		.91	2.94		3.85	5.45

245

22 11 19.38 Water Supply Meters	Crew	Daily Output	Labor-Hours	Unit	Material	2015 Bare Costs Labor	Equipment	Total	Total Incl O&P
0010 **WATER SUPPLY METERS**									
1000 Detector, serves dual systems such as fire and domestic or									
1020 process water, wide range cap., UL and FM approved									
1100 3" mainline x 2" by-pass, 400 GPM	Q-1	3.60	4.444	Ea.	7,450	235		7,685	8,525
1140 4" mainline x 2" by-pass, 700 GPM	"	2.50	6.400		7,450	340		7,790	8,675
1180 6" mainline x 3" by-pass, 1600 GPM	Q-2	2.60	9.231		11,400	505		11,905	13,300
1220 8" mainline x 4" by-pass, 2800 GPM		2.10	11.429		16,900	625		17,525	19,400
1260 10" mainline x 6" by-pass, 4400 GPM		2	12		24,100	655		24,755	27,500
1300 10" x 12" mainlines x 6" by-pass, 5400 GPM	↓	1.70	14.118	↓	32,700	775		33,475	37,200
2000 Domestic/commercial, bronze									
2020 Threaded									
2060 5/8" diameter, to 20 GPM	1 Plum	16	.500	Ea.	50	29.50		79.50	99.50
2080 3/4" diameter, to 30 GPM		14	.571		91	33.50		124.50	151
2100 1" diameter, to 50 GPM	↓	12	.667	↓	138	39		177	211
2300 Threaded/flanged									
2340 1-1/2" diameter, to 100 GPM	1 Plum	8	1	Ea.	340	58.50		398.50	460
2360 2" diameter, to 160 GPM	"	6	1.333	"	460	78.50		538.50	625
2600 Flanged, compound									
2640 3" diameter, 320 GPM	Q-1	3	5.333	Ea.	3,125	282		3,407	3,850
2660 4" diameter, to 500 GPM		1.50	10.667		5,000	565		5,565	6,350
2680 6" diameter, to 1,000 GPM		1	16		7,975	845		8,820	10,000
2700 8" diameter, to 1,800 GPM	↓	.80	20	↓	12,500	1,050		13,550	15,300
7000 Turbine									
7260 Flanged									
7300 2" diameter, to 160 GPM	1 Plum	7	1.143	Ea.	610	67		677	770
7320 3" diameter, to 450 GPM	Q-1	3.60	4.444		1,275	235		1,510	1,750
7340 4" diameter, to 650 GPM	"	2.50	6.400		2,100	340		2,440	2,825
7360 6" diameter, to 1800 GPM	Q-2	2.60	9.231		3,775	505		4,280	4,925
7380 8" diameter, to 2500 GPM		2.10	11.429		5,975	625		6,600	7,525
7400 10" diameter, to 5500 GPM	↓	1.70	14.118	↓	8,075	775		8,850	10,100

22 11 19.42 Backflow Preventers

	Crew	Daily Output	Labor-Hours	Unit	Material	2015 Bare Costs Labor	Equipment	Total	Total Incl O&P
0010 **BACKFLOW PREVENTERS**, Includes valves									
0020 and four test cocks, corrosion resistant, automatic operation									
1000 Double check principle									
1010 Threaded, with ball valves									
1020 3/4" pipe size	1 Plum	16	.500	Ea.	231	29.50		260.50	299
1030 1" pipe size		14	.571		264	33.50		297.50	340
1040 1-1/2" pipe size		10	.800		560	47		607	690
1050 2" pipe size	↓	7	1.143	↓	655	67		722	820
1080 Threaded, with gate valves									
1100 3/4" pipe size	1 Plum	16	.500	Ea.	1,075	29.50		1,104.50	1,250
1120 1" pipe size		14	.571		1,100	33.50		1,133.50	1,250
1140 1-1/2" pipe size		10	.800		1,400	47		1,447	1,625
1160 2" pipe size	↓	7	1.143	↓	1,725	67		1,792	2,000
1200 Flanged, valves are gate									
1210 3" pipe size	Q-1	4.50	3.556	Ea.	2,800	188		2,988	3,350
1220 4" pipe size	"	3	5.333		2,900	282		3,182	3,625
1230 6" pipe size	Q-2	3	8		4,275	440		4,715	5,350
1240 8" pipe size		2	12		7,775	655		8,430	9,550
1250 10" pipe size	↓	1	24	↓	11,500	1,325		12,825	14,700
1300 Flanged, valves are OS&Y									
1370 1" pipe size	1 Plum	5	1.600	Ea.	1,200	94		1,294	1,450

22 11 19 – Domestic Water Piping Specialties

22 11 19.42 Backflow Preventers		Crew	Daily Output	Labor-Hours	Unit	Material	2015 Bare Costs Labor	Equipment	Total	Total Incl O&P
1374	1-1/2" pipe size	1 Plum	5	1.600	Ea.	1,525	94		1,619	1,825
1378	2" pipe size	↓	4.80	1.667		1,825	98		1,923	2,175
1380	3" pipe size	Q-1	4.50	3.556		2,875	188		3,063	3,425
1400	4" pipe size	"	3	5.333		3,475	282		3,757	4,250
1420	6" pipe size	Q-2	3	8		5,450	440		5,890	6,650
1430	8" pipe size	"	2	12	↓	10,800	655		11,455	12,800
4000	Reduced pressure principle									
4100	Threaded, bronze, valves are ball									
4120	3/4" pipe size	1 Plum	16	.500	Ea.	445	29.50		474.50	535
4140	1" pipe size		14	.571		480	33.50		513.50	575
4150	1-1/4" pipe size		12	.667		845	39		884	990
4160	1-1/2" pipe size		10	.800		960	47		1,007	1,125
4180	2" pipe size	↓	7	1.143	↓	1,075	67		1,142	1,300
5000	Flanged, bronze, valves are OS&Y									
5060	2-1/2" pipe size	Q-1	5	3.200	Ea.	4,025	169		4,194	4,675
5080	3" pipe size	↓	4.50	3.556		4,650	188		4,838	5,375
5100	4" pipe size	↓	3	5.333		5,500	282		5,782	6,475
5120	6" pipe size	Q-2	3	8		8,750	440		9,190	10,300
5200	Flanged, iron, valves are gate									
5210	2-1/2" pipe size	Q-1	5	3.200	Ea.	2,650	169		2,819	3,175
5220	3" pipe size		4.50	3.556		2,775	188		2,963	3,325
5230	4" pipe size	↓	3	5.333		3,750	282		4,032	4,550
5240	6" pipe size	Q-2	3	8		5,275	440		5,715	6,475
5250	8" pipe size		2	12		9,475	655		10,130	11,400
5260	10" pipe size	↓	1	24	↓	13,400	1,325		14,725	16,700
5600	Flanged, iron, valves are OS&Y									
5660	2-1/2" pipe size	Q-1	5	3.200	Ea.	3,025	169		3,194	3,575
5680	3" pipe size		4.50	3.556		3,175	188		3,363	3,775
5700	4" pipe size	↓	3	5.333		3,975	282		4,257	4,800
5720	6" pipe size	Q-2	3	8		5,775	440		6,215	7,000
5740	8" pipe size		2	12		10,100	655		10,755	12,100
5760	10" pipe size	↓	1	24		13,600	1,325		14,925	16,900

22 11 19.50 Vacuum Breakers

		Crew	Daily Output	Labor-Hours	Unit	Material	2015 Bare Costs Labor	Equipment	Total	Total Incl O&P
0010	**VACUUM BREAKERS** R221113-40									
0013	See also backflow preventers Section 22 11 19.42									
1000	Anti-siphon continuous pressure type									
1010	Max. 150 PSI - 210°F									
1020	Bronze body									
1030	1/2" size	1 Stpi	24	.333	Ea.	187	19.90		206.90	236
1040	3/4" size		20	.400		187	24		211	242
1050	1" size		19	.421		194	25		219	251
1060	1-1/4" size		15	.533		380	32		412	470
1070	1-1/2" size		13	.615		470	37		507	570
1080	2" size	↓	11	.727	↓	485	43.50		528.50	595
1200	Max. 125 PSI with atmospheric vent									
1210	Brass, in-line construction									
1220	1/4" size	1 Stpi	24	.333	Ea.	117	19.90		136.90	159
1230	3/8" size	"	24	.333	↓	117	19.90		136.90	159
1260	For polished chrome finish, add					13%				
2000	Anti-siphon, non-continuous pressure type									
2010	Hot or cold water 125 PSI - 210°F									
2020	Bronze body									

For customer support on your Plumbing Cost Data, call 877.709.6968.

247

22 11 19.50 Vacuum Breakers		Crew	Daily Output	Labor-Hours	Unit	Material	2015 Bare Costs Labor	Equipment	Total	Total Incl O&P
2030	1/4" size	1 Stpi	24	.333	Ea.	64	19.90		83.90	101
2040	3/8" size		24	.333		64	19.90		83.90	101
2050	1/2" size		24	.333		72.50	19.90		92.40	110
2060	3/4" size		20	.400		86.50	24		110.50	131
2070	1" size		19	.421		134	25		159	185
2080	1-1/4" size		15	.533		235	32		267	305
2090	1-1/2" size		13	.615		276	37		313	360
2100	2" size		11	.727		430	43.50		473.50	535
2110	2-1/2" size		8	1		1,225	60		1,285	1,450
2120	3" size	↓	6	1.333	↓	1,625	79.50		1,704.50	1,925
2150	For polished chrome finish, add					50%				
3000	Air gap fitting									
3020	1/2" NPT size	1 Plum	19	.421	Ea.	50	24.50		74.50	92.50
3030	1" NPT size	"	15	.533		57.50	31.50		89	111
3040	2" NPT size	Q-1	21	.762		113	40.50		153.50	185
3050	3" NPT size		14	1.143		223	60.50		283.50	335
3060	4" NPT size	↓	10	1.600	↓	223	84.50		307.50	375

22 11 19.54 Water Hammer Arresters/Shock Absorbers

		Crew	Daily Output	Labor-Hours	Unit	Material	Labor	Equipment	Total	Total Incl O&P
0010	**WATER HAMMER ARRESTERS/SHOCK ABSORBERS**									
0490	Copper									
0500	3/4" male I.P.S. For 1 to 11 fixtures	1 Plum	12	.667	Ea.	28	39		67	90
0600	1" male I.P.S. For 12 to 32 fixtures		8	1		45.50	58.50		104	139
0700	1-1/4" male I.P.S. For 33 to 60 fixtures		8	1		47	58.50		105.50	140
0800	1-1/2" male I.P.S. For 61 to 113 fixtures		8	1		67.50	58.50		126	163
0900	2" male I.P.S. For 114 to 154 fixtures		8	1		98.50	58.50		157	197
1000	2-1/2" male I.P.S. For 155 to 330 fixtures	↓	4	2	↓	305	117		422	510
4000	Bellows type									
4010	3/4" FNPT, to 11 fixture units	1 Plum	10	.800	Ea.	253	47		300	350
4020	1" FNPT, to 32 fixture units		8.80	.909		510	53.50		563.50	640
4030	1" FNPT, to 60 fixture units		8.80	.909		770	53.50		823.50	925
4040	1" FNPT, to 113 fixture units		8.80	.909		1,925	53.50		1,978.50	2,200
4050	1" FNPT, to 154 fixture units		6.60	1.212		2,175	71		2,246	2,500
4060	1-1/2" FNPT, to 300 fixture units	↓	5	1.600	↓	2,675	94		2,769	3,100

22 11 19.64 Hydrants

		Crew	Daily Output	Labor-Hours	Unit	Material	Labor	Equipment	Total	Total Incl O&P
0010	**HYDRANTS**									
0050	Wall type, moderate climate, bronze, encased									
0200	3/4" IPS connection	1 Plum	16	.500	Ea.	745	29.50		774.50	860
0300	1" IPS connection		14	.571		850	33.50		883.50	985
0500	Anti-siphon type, 3/4" connection	↓	16	.500	↓	640	29.50		669.50	750
1000	Non-freeze, bronze, exposed									
1100	3/4" IPS connection, 4" to 9" thick wall	1 Plum	14	.571	Ea.	500	33.50		533.50	600
1120	10" to 14" thick wall		12	.667		545	39		584	660
1140	15" to 19" thick wall		12	.667		605	39		644	725
1160	20" to 24" thick wall	↓	10	.800	↓	640	47		687	775
1200	For 1" IPS connection, add					15%	10%			
1240	For 3/4" adapter type vacuum breaker, add				Ea.	63			63	69.50
1280	For anti-siphon type, add				"	132			132	145
2000	Non-freeze bronze, encased, anti-siphon type									
2100	3/4" IPS connection, 5" to 9" thick wall	1 Plum	14	.571	Ea.	1,275	33.50		1,308.50	1,450
2120	10" to 14" thick wall		12	.667		1,300	39		1,339	1,500
2140	15" to 19" thick wall		12	.667		1,375	39		1,414	1,550
2160	20" to 24" thick wall	↓	10	.800	↓	1,425	47		1,472	1,650

22 11 19 – Domestic Water Piping Specialties

22 11 19.64 Hydrants	Crew	Daily Output	Labor-Hours	Unit	Material	2015 Bare Costs Labor	Equipment	Total	Total Incl O&P
2200 For 1" IPS connection, add					10%	10%			
3000 Ground box type, bronze frame, 3/4" IPS connection									
3080 Non-freeze, all bronze, polished face, set flush									
3100 2 feet depth of bury	1 Plum	8	1	Ea.	950	58.50		1,008.50	1,150
3120 3 feet depth of bury		8	1		1,025	58.50		1,083.50	1,225
3140 4 feet depth of bury		8	1		1,075	58.50		1,133.50	1,300
3160 5 feet depth of bury		7	1.143		1,175	67		1,242	1,375
3180 6 feet depth of bury		7	1.143		1,250	67		1,317	1,475
3200 7 feet depth of bury		6	1.333		1,300	78.50		1,378.50	1,575
3220 8 feet depth of bury		5	1.600		1,375	94		1,469	1,675
3240 9 feet depth of bury		4	2		1,450	117		1,567	1,775
3260 10 feet depth of bury		4	2		1,525	117		1,642	1,850
3400 For 1" IPS connection, add					15%	10%			
3450 For 1-1/4" IPS connection, add					325%	14%			
3500 For 1-1/2" connection, add					370%	18%			
3550 For 2" connection, add					445%	24%			
3600 For tapped drain port in box, add					86			86	94.50
4000 Non-freeze, CI body, bronze frame & scoriated cover									
4010 with hose storage									
4100 2 feet depth of bury	1 Plum	7	1.143	Ea.	1,750	67		1,817	2,025
4120 3 feet depth of bury		7	1.143		1,825	67		1,892	2,100
4140 4 feet depth of bury		7	1.143		1,875	67		1,942	2,175
4160 5 feet depth of bury		6.50	1.231		1,925	72.50		1,997.50	2,200
4180 6 feet depth of bury		6	1.333		1,950	78.50		2,028.50	2,275
4200 7 feet depth of bury		5.50	1.455		2,025	85.50		2,110.50	2,350
4220 8 feet depth of bury		5	1.600		2,100	94		2,194	2,450
4240 9 feet depth of bury		4.50	1.778		2,150	104		2,254	2,525
4260 10 feet depth of bury		4	2		2,225	117		2,342	2,625
4280 For 1" IPS connection, add					390			390	425
4300 For tapped drain port in box, add					86			86	94.50
5000 Moderate climate, all bronze, polished face									
5020 and scoriated cover, set flush									
5100 3/4" IPS connection	1 Plum	16	.500	Ea.	655	29.50		684.50	770
5120 1" IPS connection	"	14	.571		810	33.50		843.50	940
5200 For tapped drain port in box, add					86			86	94.50
6000 Ground post type, all non-freeze, all bronze, aluminum casing									
6010 guard, exposed head, 3/4" IPS connection									
6100 2 feet depth of bury	1 Plum	8	1	Ea.	925	58.50		983.50	1,125
6120 3 feet depth of bury		8	1		1,000	58.50		1,058.50	1,200
6140 4 feet depth of bury		8	1		1,075	58.50		1,133.50	1,275
6160 5 feet depth of bury		7	1.143		1,150	67		1,217	1,375
6180 6 feet depth of bury		7	1.143		1,225	67		1,292	1,450
6200 7 feet depth of bury		6	1.333		1,325	78.50		1,403.50	1,575
6220 8 feet depth of bury		5	1.600		1,400	94		1,494	1,700
6240 9 feet depth of bury		4	2		1,500	117		1,617	1,825
6260 10 feet depth of bury		4	2		1,575	117		1,692	1,900
6300 For 1" IPS connection, add					40%	10%			
6350 For 1-1/4" IPS connection, add					140%	14%			
6400 For 1-1/2" IPS connection, add					225%	18%			
6450 For 2" IPS connection, add					315%	24%			

For customer support on your Plumbing Cost Data, call 877.709.6968.

249

22 11 23.10 General Utility Pumps	Crew	Daily Output	Labor-Hours	Unit	Material	2015 Bare Costs Labor	Equipment	Total	Total Incl O&P
0010 **GENERAL UTILITY PUMPS**									
2000 Single stage									
3000 Double suction,									
3140 50 HP, 5"D. x 6"S.	Q-2	.33	72.727	Ea.	9,175	3,975		13,150	16,100
3180 60 HP, 6"D. x 8"S.	Q-3	.30	106		13,400	5,950		19,350	23,700
3190 75 HP, to 2500 GPM		.28	114		20,100	6,375		26,475	31,800
3220 100 HP, to 3000 GPM		.26	123		25,500	6,875		32,375	38,500
3240 150 HP, to 4000 GPM		.24	133		36,000	7,450		43,450	51,000
4000 Centrifugal, end suction, mounted on base									
4010 Horizontal mounted, with drip proof motor, rated @ 100' head									
4020 Vertical split case, single stage									
4040 100 GPM, 5 HP, 1-1/2" discharge	Q-1	1.70	9.412	Ea.	3,675	495		4,170	4,775
4050 200 GPM, 10 HP, 2" discharge		1.30	12.308		4,000	650		4,650	5,375
4060 250 GPM, 10 HP, 3" discharge		1.28	12.500		4,450	660		5,110	5,900
4070 300 GPM, 15 HP, 2" discharge	Q-2	1.56	15.385		5,225	845		6,070	7,025
4080 500 GPM, 20 HP, 4" discharge		1.44	16.667		6,025	915		6,940	8,000
4090 750 GPM, 30 HP, 4" discharge		1.20	20		6,350	1,100		7,450	8,625
4100 1050 GPM, 40 HP, 5" discharge		1	24		9,175	1,325		10,500	12,100
4110 1500 GPM, 60 HP, 6" discharge		.60	40		13,400	2,200		15,600	18,000
4120 2000 GPM, 75 HP, 6" discharge		.50	48		14,600	2,625		17,225	20,000
4130 3000 GPM, 100 HP, 8" discharge		.40	60		18,900	3,275		22,175	25,800
4200 Horizontal split case, single stage									
4210 100 GPM, 7.5 HP, 1-1/2" discharge	Q-1	1.70	9.412	Ea.	3,725	495		4,220	4,850
4220 250 GPM, 15 HP, 2-1/2" discharge	"	1.30	12.308		7,900	650		8,550	9,650
4230 500 GPM, 20 HP, 4" discharge	Q-2	1.60	15		9,625	820		10,445	11,900
4240 750 GPM, 25 HP, 5" discharge		1.54	15.584		10,100	855		10,955	12,400
4250 1000 GPM, 40 HP, 5" discharge		1.20	20		14,000	1,100		15,100	17,000
4260 1500 GPM, 50 HP, 6" discharge	Q-3	1.42	22.535		16,400	1,250		17,650	20,000
4270 2000 GPM, 75 HP, 8" discharge		1.14	28.070		22,300	1,575		23,875	27,000
4280 3000 GPM, 100 HP, 10" discharge		.96	33.333		25,000	1,875		26,875	30,300
4290 3500 GPM, 150 HP, 10" discharge		.86	37.209		32,900	2,075		34,975	39,400
4300 4000 GPM, 200 HP, 10" discharge		.66	48.485		39,300	2,700		42,000	47,300
4330 Horizontal split case, two stage, 500' head									
4340 100 GPM, 40 HP, 1-1/2" discharge	Q-2	1.70	14.118	Ea.	17,000	775		17,775	19,900
4350 200 GPM, 50 HP, 1-1/2" discharge	"	1.44	16.667		17,800	915		18,715	21,000
4360 300 GPM, 75 HP, 2" discharge	Q-3	1.57	20.382		24,500	1,150		25,650	28,600
4370 400 GPM, 100 HP, 3" discharge		1.14	28.070		28,900	1,575		30,475	34,200
4380 800 GPM, 200 HP, 4" discharge		.86	37.209		38,300	2,075		40,375	45,300
5000 Centrifugal, in-line									
5006 Vertical mount, iron body, 125 lb. flgd, 1800 RPM TEFC mtr									
5010 Single stage									
5012 .5 HP, 1-1/2" suction & discharge	Q-1	3.20	5	Ea.	1,350	264		1,614	1,900
5014 .75 HP, 2" suction & discharge		2.80	5.714		2,600	300		2,900	3,300
5015 1 HP, 3" suction & discharge		2.50	6.400		2,900	340		3,240	3,700
5016 1.5 HP, 3" suction & discharge		2.30	6.957		2,950	370		3,320	3,800
5018 2 HP, 4" suction & discharge		2.10	7.619		3,825	405		4,230	4,800
5020 3 HP, 5" suction & discharge		1.90	8.421		4,475	445		4,920	5,600
5030 5 HP, 6" suction & discharge		1.60	10		5,100	530		5,630	6,400
5040 7.5 HP, 6" suction & discharge		1.30	12.308		5,850	650		6,500	7,425
5050 10 HP, 8" suction & discharge	Q-2	1.80	13.333		7,800	730		8,530	9,675
5060 15 HP, 8" suction & discharge		1.70	14.118		9,125	775		9,900	11,200
5064 20 HP, 8" suction & discharge		1.60	15		10,600	820		11,420	12,900

22 11 23 – Domestic Water Pumps

22 11 23.10 General Utility Pumps		Crew	Daily Output	Labor-Hours	Unit	Material	2015 Bare Costs Labor	Equipment	Total	Total Incl O&P
5070	30 HP, 8" suction & discharge	Q-2	1.50	16	Ea.	13,500	875		14,375	16,100
5080	40 HP, 8" suction & discharge		1.40	17.143		16,200	940		17,140	19,200
5090	50 HP, 8" suction & discharge		1.30	18.462		16,400	1,000		17,400	19,500
5094	60 HP, 8" suction & discharge	↓	.90	26.667		17,200	1,450		18,650	21,200
5100	75 HP, 8" suction & discharge	Q-3	.60	53.333		19,700	2,975		22,675	26,200
5110	100 HP, 8" suction & discharge	"	.50	64	↓	22,500	3,575		26,075	30,100

22 11 23.11 Miscellaneous Pumps

	22 11 23.11 Miscellaneous Pumps	Crew	Daily Output	Labor-Hours	Unit	Material	Labor	Equipment	Total	Total Incl O&P
0010	**MISCELLANEOUS PUMPS**									
0020	Water pump, portable, gasoline powered									
0100	170 GPH, 2" discharge	Q-1	11	1.455	Ea.	670	77		747	850
0110	343 GPH, 3" discharge		10.50	1.524		755	80.50		835.50	950
0120	608 GPH, 4" discharge	↓	10	1.600	↓	1,075	84.50		1,159.50	1,325
0500	Pump, propylene body, housing and impeller									
0510	22 GPM, 1/3 HP, 40' HD	1 Plum	5	1.600	Ea.	860	94		954	1,075
0520	33 GPM, 1/2 HP, 40' HD	Q-1	5	3.200		790	169		959	1,125
0530	53 GPM, 3/4 HP, 40' HD	"	4	4	↓	1,275	211		1,486	1,725
0600	Rotary pump, CI									
0614	282 GPH, 3/4 HP, 1" discharge	Q-1	4.50	3.556	Ea.	1,225	188		1,413	1,625
0618	277 GPH, 1 HP, 1" discharge	↓	4	4		1,225	211		1,436	1,675
0624	1100 GPH, 1.5 HP, 1-1/4" discharge		3.60	4.444		3,225	235		3,460	3,900
0628	1900 GPH, 2 HP, 1-1/4" discharge	↓	3.20	5	↓	4,550	264		4,814	5,425
1000	Turbine pump, CI									
1010	50 GPM, 2 HP, 3" discharge	Q-1	.80	20	Ea.	3,675	1,050		4,725	5,625
1020	100 GPM, 3 HP, 4" discharge	Q-2	.96	25		6,050	1,375		7,425	8,725
1030	250 GPM, 15 HP, 6" discharge	"	.94	25.532		7,500	1,400		8,900	10,400
1040	500 GPM, 25 HP, 6" discharge	Q-3	1.22	26.230		7,550	1,475		9,025	10,600
1050	1000 GPM, 50 HP, 8" discharge		1.14	28.070		10,400	1,575		11,975	13,900
1060	2000 GPM, 100 HP, 10" discharge		1	32		12,400	1,800		14,200	16,300
1070	3000 GPM, 150 HP, 10" discharge		.80	40		19,900	2,225		22,125	25,300
1080	4000 GPM, 200 HP, 12" discharge		.70	45.714		21,700	2,550		24,250	27,800
1090	6000 GPM, 300 HP, 14" discharge		.60	53.333		33,200	2,975		36,175	41,000
1100	10,000 GPM, 300 HP, 18" discharge	↓	.58	55.172	↓	38,300	3,075		41,375	46,900
2000	Centrifugal stainless steel pumps									
2100	100 GPM, 100' TDH, 3 HP	Q-1	1.80	8.889	Ea.	10,900	470		11,370	12,700
2130	250 GPM, 100' TDH, 10 HP	"	1.28	12.500		15,300	660		15,960	17,800
2160	500 GPM, 100' TDH, 20 HP	Q-2	1.44	16.667	↓	17,500	915		18,415	20,600
2200	Vertical turbine stainless steel pumps									
2220	100 GPM, 100' TDH, 7.5 HP	Q-2	.95	25.263	Ea.	61,000	1,375		62,375	69,500
2240	250 GPM, 100' TDH, 15 HP		.94	25.532		68,000	1,400		69,400	76,500
2260	500 GPM, 100' TDH, 20 HP	↓	.93	25.806		74,500	1,425		75,925	84,000
2280	750 GPM, 100' TDH, 30 HP	Q-3	1.19	26.891		81,000	1,500		82,500	91,500
2300	1000 GPM, 100' TDH, 40 HP	"	1.16	27.586		87,500	1,550		89,050	98,500
2320	100 GPM, 200' TDH, 15 HP	Q-2	.94	25.532		94,000	1,400		95,400	105,500
2340	250 GPM, 200' TDH, 25 HP	Q-3	1.22	26.230		100,500	1,475		101,975	112,500
2360	500 GPM, 200' TDH, 40 HP		1.16	27.586		107,000	1,550		108,550	120,500
2380	750 GPM, 200' TDH, 60 HP		1.13	28.319		108,000	1,575		109,575	121,000
2400	1000 GPM, 200' TDH, 75 HP	↓	1.12	28.571	↓	111,500	1,600		113,100	125,000

22 11 23.13 Domestic-Water Packaged Booster Pumps

	22 11 23.13 Domestic-Water Packaged Booster Pumps	Crew	Daily Output	Labor-Hours	Unit	Material	Labor	Equipment	Total	Total Incl O&P
0010	**DOMESTIC-WATER PACKAGED BOOSTER PUMPS**									
0200	Pump system, with diaphragm tank, control, press. switch									
0300	1 HP pump	Q-1	1.30	12.308	Ea.	6,500	650		7,150	8,125
0400	1-1/2 HP pump	↓	1.25	12.800	↓	6,550	675		7,225	8,250

22 11 Facility Water Distribution

22 11 23 – Domestic Water Pumps

22 11 23.13 Domestic-Water Packaged Booster Pumps		Crew	Daily Output	Labor-Hours	Unit	Material	2015 Bare Costs Labor	Equipment	Total	Total Incl O&P
0420	2 HP pump	Q-1	1.20	13.333	Ea.	6,725	705		7,430	8,475
0440	3 HP pump	↓	1.10	14.545		6,825	770		7,595	8,650
0460	5 HP pump	Q-2	1.50	16		7,550	875		8,425	9,625
0480	7-1/2 HP pump		1.42	16.901		8,400	925		9,325	10,700
0500	10 HP pump	↓	1.34	17.910	↓	8,800	980		9,780	11,200
2000	Pump system, variable speed, base, controls, starter									
2010	Duplex, 100' head									
2020	400 GPM, 7-1/2HP, 4" discharge	Q-2	.70	34.286	Ea.	38,900	1,875		40,775	45,500
2025	Triplex, 100' head									
2030	1000 GPM, 15HP, 6" discharge	Q-2	.50	48	Ea.	56,500	2,625		59,125	66,500
2040	1700 GPM, 30HP, 6" discharge	"	.30	80	"	68,500	4,375		72,875	81,500

22 12 Facility Potable-Water Storage Tanks

22 12 21 – Facility Underground Potable-Water Storage Tanks

22 12 21.13 Fiberglass, Undrgrnd Pot.-Water Storage Tanks

		Crew	Daily Output	Labor-Hours	Unit	Material	2015 Bare Costs Labor	Equipment	Total	Total Incl O&P
0010	**FIBERGLASS, UNDERGROUND POTABLE-WATER STORAGE TANKS**									
0020	Excludes excavation, backfill, & piping									
0030	Single wall									
2000	600 gallon capacity	B-21B	3.75	10.667	Ea.	3,300	435	174	3,909	4,475
2010	1,000 gallon capacity		3.50	11.429		4,250	465	187	4,902	5,600
2020	2,000 gallon capacity		3.25	12.308		6,200	500	201	6,901	7,800
2030	4,000 gallon capacity		3	13.333		8,575	545	218	9,338	10,500
2040	6,000 gallon capacity		2.65	15.094		9,625	615	247	10,487	11,800
2050	8,000 gallon capacity		2.30	17.391		11,500	710	284	12,494	14,100
2060	10,000 gallon capacity		2	20		13,200	815	325	14,340	16,200
2070	12,000 gallon capacity		1.50	26.667		18,300	1,100	435	19,835	22,300
2080	15,000 gallon capacity		1	40		21,200	1,625	655	23,480	26,500
2090	20,000 gallon capacity		.75	53.333		27,300	2,175	870	30,345	34,300
2100	25,000 gallon capacity		.50	80		40,800	3,275	1,300	45,375	51,500
2110	30,000 gallon capacity		.35	114		49,400	4,675	1,875	55,950	63,500
2120	40,000 gallon capacity	↓	.30	133	↓	71,500	5,450	2,175	79,125	89,500

22 12 23 – Facility Indoor Potable-Water Storage Tanks

22 12 23.13 Facility Steel, Indoor Pot.-Water Storage Tanks

		Crew	Daily Output	Labor-Hours	Unit	Material	2015 Bare Costs Labor	Equipment	Total	Total Incl O&P
0010	**FACILITY STEEL, INDOOR POT.-WATER STORAGE TANKS**									
2000	Galvanized steel, 15 gal., 14" diam., 26" LOA	1 Plum	12	.667	Ea.	1,425	39		1,464	1,625
2060	30 gal., 14" diam. x 49" LOA		11	.727		1,700	42.50		1,742.50	1,925
2080	80 gal., 20" diam. x 64" LOA		9	.889		2,625	52		2,677	2,975
2100	135 gal., 24" diam. x 75" LOA		6	1.333		3,825	78.50		3,903.50	4,325
2120	240 gal., 30" diam. x 86" LOA		4	2		7,475	117		7,592	8,400
2140	300 gal., 36" diam. x 76" LOA	↓	3	2.667		10,600	157		10,757	11,900
2160	400 gal., 36" diam. x 100" LOA	Q-1	4	4		13,100	211		13,311	14,700
2180	500 gal., 36" diam., x 126" LOA	"	3	5.333		16,200	282		16,482	18,200
3000	Glass lined, P.E., 80 gal., 20" diam. x 60" LOA	1 Plum	9	.889		2,800	52		2,852	3,150
3060	140 gal., 24" diam. x 72" LOA		6	1.333		3,825	78.50		3,903.50	4,350
3080	200 gal., 30" diam. x 71" LOA		4	2		5,225	117		5,342	5,900
3100	350 gal., 36" diam. x 86" LOA	↓	3	2.667		6,300	157		6,457	7,150
3120	450 gal., 42" diam. x 78" LOA	Q-1	4	4		8,575	211		8,786	9,750
3140	600 gal., 48" diam. x 81" LOA		3	5.333		10,500	282		10,782	11,900
3160	750 gal., 48" diam. x 105" LOA		3	5.333		11,700	282		11,982	13,300
3180	900 gal., 54" diam. x 95" LOA	↓	2.50	6.400	↓	16,400	340		16,740	18,500

For customer support on your Plumbing Cost Data, call 877.709.6968.

22 12 Facility Potable-Water Storage Tanks

22 12 23 – Facility Indoor Potable-Water Storage Tanks

22 12 23.13 Facility Steel, Indoor Pot.-Water Storage Tanks

	22 12 23.13 Facility Steel, Indoor Pot.-Water Storage Tanks	Crew	Daily Output	Labor-Hours	Unit	Material	2015 Bare Costs Labor	Equipment	Total	Total Incl O&P
3200	1250 gal., 54" diam. x 129" LOA	Q-1	2	8	Ea.	21,000	425		21,425	23,700
3220	1800 gal., 54" diam. x 181" LOA		1.50	10.667		23,500	565		24,065	26,800
3240	2000 gal., 60" diam. x 165" LOA	↓	1	16		27,900	845		28,745	32,000
3260	3500 gal., 72" diam. x 201" LOA	Q-2	1.50	16	↓	34,100	875		34,975	38,900

22 13 Facility Sanitary Sewerage

22 13 16 – Sanitary Waste and Vent Piping

22 13 16.20 Pipe, Cast Iron

			Crew	Daily Output	Labor-Hours	Unit	Material	2015 Bare Costs Labor	Equipment	Total	Total Incl O&P
0010	**PIPE, CAST IRON**, Soil, on clevis hanger assemblies, 5' O.C.	R221113-70									
0020	Single hub, service wt., lead & oakum joints 10' O.C.										
2120	2" diameter	R221316-10	Q-1	63	.254	L.F.	11.35	13.40		24.75	33
2140	3" diameter			60	.267		15	14.10		29.10	38
2160	4" diameter	R221316-20	↓	55	.291		18.55	15.35		33.90	43.50
2180	5" diameter		Q-2	76	.316		25.50	17.30		42.80	54
2200	6" diameter		"	73	.329		31	18		49	61.50
2220	8" diameter		Q-3	59	.542		47	30.50		77.50	98
2240	10" diameter			54	.593		74.50	33		107.50	132
2260	12" diameter			48	.667		105	37.50		142.50	173
2261	15" diameter		↓	40	.800		156	44.50		200.50	240
2320	For service weight, double hub, add						10%				
2340	For extra heavy, single hub, add						48%	4%			
2360	For extra heavy, double hub, add					↓	71%	4%			
2400	Lead for caulking, (1#/diam. in.)		Q-1	160	.100	Lb.	1.04	5.30		6.34	9.15
2420	Oakum for caulking, (1/8#/diam. in.)		"	40	.400	"	3.60	21		24.60	36
2960	To delete hangers, subtract										
2970	2" diam. to 4" diam.						16%	19%			
2980	5" diam. to 8" diam.						14%	14%			
2990	10" diam. to 15" diam.						13%	19%			
3000	Single hub, service wt., push-on gasket joints 10' O.C.										
3010	2" diameter		Q-1	66	.242	L.F.	12.30	12.80		25.10	33
3020	3" diameter			63	.254		16.20	13.40		29.60	38.50
3030	4" diameter		↓	57	.281		20	14.85		34.85	44.50
3040	5" diameter		Q-2	79	.304		28	16.65		44.65	55.50
3050	6" diameter		"	75	.320		33.50	17.55		51.05	63.50
3060	8" diameter		Q-3	62	.516		52.50	29		81.50	102
3070	10" diameter			56	.571		84	32		116	141
3080	12" diameter			49	.653		118	36.50		154.50	184
3082	15" diameter		↓	40	.800	↓	171	44.50		215.50	256
3100	For service weight, double hub, add						65%				
3110	For extra heavy, single hub, add						48%	4%			
3120	For extra heavy, double hub, add						29%	4%			
3130	To delete hangers, subtract										
3140	2" diam. to 4" diam.						12%	21%			
3150	5" diam. to 8" diam.						10%	16%			
3160	10" diam. to 15" diam.						9%	21%			
4000	No hub, couplings 10' O.C.										
4100	1-1/2" diameter		Q-1	71	.225	L.F.	11.05	11.90		22.95	30
4120	2" diameter			67	.239		11.25	12.60		23.85	31.50
4140	3" diameter			64	.250		15.20	13.20		28.40	36.50
4160	4" diameter		↓	58	.276		19.10	14.55		33.65	43
4180	5" diameter		Q-2	83	.289		25.50	15.85		41.35	52

22 13 16.20 Pipe, Cast Iron		Crew	Daily Output	Labor-Hours	Unit	Material	2015 Bare Costs Labor	2015 Bare Costs Equipment	Total	Total Incl O&P
4200	6" diameter	Q-2	79	.304	L.F.	32	16.65		48.65	60.50
4220	8" diameter	Q-3	69	.464		54.50	26		80.50	99
4240	10" diameter		61	.525		88.50	29.50		118	142
4244	12" diameter		58	.552		110	31		141	168
4248	15" diameter		52	.615		163	34.50		197.50	231
4280	To delete hangers, subtract									
4290	1-1/2" diam. to 6" diam.					22%	47%			
4300	8" diam. to 10" diam.					21%	44%			
4310	12" diam. to 15" diam.					19%	40%			

22 13 16.30 Pipe Fittings, Cast Iron		Crew	Daily Output	Labor-Hours	Unit	Material	2015 Bare Costs Labor	2015 Bare Costs Equipment	Total	Total Incl O&P
0010	**PIPE FITTINGS, CAST IRON**, Soil									
0040	Hub and spigot, service weight, lead & oakum joints									
0080	1/4 bend, 2"	Q-1	16	1	Ea.	19.90	53		72.90	102
0120	3"		14	1.143		26.50	60.50		87	120
0140	4"		13	1.231		41.50	65		106.50	144
0160	5"	Q-2	18	1.333		58	73		131	174
0180	6"	"	17	1.412		72	77.50		149.50	197
0200	8"	Q-3	11	2.909		217	163		380	485
0220	10"		10	3.200		320	179		499	620
0224	12"		9	3.556		430	199		629	775
0226	15"		7	4.571		1,475	256		1,731	2,000
0242	Short sweep, CI, 90°, 2"	Q-1	16	1		19.45	53		72.45	102
0243	3"		14	1.143		34.50	60.50		95	129
0244	4"		13	1.231		53	65		118	157
0245	6"	Q-2	17	1.412		108	77.50		185.50	235
0246	8"	Q-3	11	2.909		234	163		397	505
0247	10"		10	3.200		480	179		659	795
0248	12"		9	3.556		1,125	199		1,324	1,525
0251	Long sweep elbow									
0252	2"	Q-1	16	1	Ea.	30	53		83	113
0253	3"		14	1.143		43	60.50		103.50	138
0254	4"		13	1.231		63	65		128	168
0255	6"	Q-2	17	1.412		127	77.50		204.50	257
0256	8"	Q-3	11	2.909		287	163		450	560
0257	10"		10	3.200		485	179		664	805
0258	12"		9	3.556		785	199		984	1,175
0259	15"		7	4.571		2,075	256		2,331	2,650
0266	Closet bend, 3" diameter with flange 10" x 16"	Q-1	14	1.143		127	60.50		187.50	231
0268	16" x 16"		12	1.333		141	70.50		211.50	262
0270	Closet bend, 4" diameter, 2-1/2" x 4" ring, 6" x 16"		13	1.231		118	65		183	228
0280	8" x 16"		13	1.231		106	65		171	215
0290	10" x 12"		12	1.333		102	70.50		172.50	218
0300	10" x 18"		11	1.455		136	77		213	266
0310	12" x 16"		11	1.455		119	77		196	247
0330	16" x 16"		10	1.600		148	84.50		232.50	291
0340	1/8 bend, 2"		16	1		14.15	53		67.15	95.50
0350	3"		14	1.143		22	60.50		82.50	116
0360	4"		13	1.231		33.50	65		98.50	135
0380	5"	Q-2	18	1.333		45.50	73		118.50	160
0400	6"	"	17	1.412		54.50	77.50		132	177
0420	8"	Q-3	11	2.909		164	163		327	425
0440	10"		10	3.200		236	179		415	530

22 13 16.30 Pipe Fittings, Cast Iron		Crew	Daily Output	Labor-Hours	Unit	Material	2015 Bare Costs Labor	Equipment	Total	Total Incl O&P
0460	12"	Q-3	9	3.556	Ea.	445	199		644	790
0461	15"	↓	7	4.571		1,050	256		1,306	1,525
0500	Sanitary tee, 2"	Q-1	10	1.600		27.50	84.50		112	159
0540	3"	↓	9	1.778		45	94		139	192
0620	4"	↓	8	2		55	106		161	221
0700	5"	Q-2	12	2		109	110		219	285
0800	6"	"	11	2.182		126	120		246	320
0880	8"	Q-3	7	4.571		330	256		586	745
0881	10"		7	4.571		600	256		856	1,050
0882	12"		6	5.333		1,100	298		1,398	1,675
0883	15"	↓	4	8	↓	2,175	445		2,620	3,050
0900	Sanitary tee, tapped									
0901	2" x 2"	Q-1	9	1.778	Ea.	37	94		131	183
0902	3" x 2"		8	2		41.50	106		147.50	206
0903	4" x 2"	↓	7	2.286	↓	68	121		189	257
0910	Sanitary cross, tapped [double tapped sanitary tee]									
0911	2" x 2"	Q-1	7	2.286	Ea.	60.50	121		181.50	249
0912	3" x 2"		6	2.667		73	141		214	294
0913	4" x 2"	↓	5	3.200	↓	71.50	169		240.50	335
0940	Sanitary Tee, reducing									
0942	3" x 2"	Q-1	10	1.600	Ea.	38.50	84.50		123	171
0943	4" x 3"		9	1.778		50	94		144	197
0944	4" x 2"	↓	9	1.778		47	94		141	194
0945	5" x 3"	Q-2	12.50	1.920		130	105		235	300
0946	6" x 4"		10	2.400		140	131		271	355
0947	6" x 3"		11	2.182		100	120		220	291
0948	6" x 2"	↓	12.50	1.920		97	105		202	266
0949	8" x 6"	Q-3	9	3.556		277	199		476	605
0950	8" x 5"		9	3.556		277	199		476	605
0951	8" x 4"		10	3.200		210	179		389	500
0954	10" x 6"		8	4		475	224		699	860
0958	12" x 8"		7	4.571		860	256		1,116	1,325
0962	15" x 10"	↓	5	6.400		1,850	360		2,210	2,575
1000	Tee, 2"	Q-1	10	1.600		40	84.50		124.50	172
1060	3"		9	1.778		59.50	94		153.50	208
1120	4"	↓	8	2		76.50	106		182.50	244
1200	5"	Q-2	12	2		162	110		272	345
1300	6"	"	11	2.182		160	120		280	355
1380	8"	Q-3	7	4.571	↓	293	256		549	710
1400	Combination Y and 1/8 bend									
1420	2"	Q-1	10	1.600	Ea.	34.50	84.50		119	166
1460	3"		9	1.778		52.50	94		146.50	200
1520	4"	↓	8	2		72.50	106		178.50	240
1540	5"	Q-2	12	2		138	110		248	315
1560	6"	↓	11	2.182		174	120		294	375
1580	8"	↓	7	3.429		430	188		618	760
1582	12"	Q-3	6	5.333	↓	880	298		1,178	1,425
1584	Combination Y & 1/8 bend, reducing									
1586	3" x 2"	Q-1	10	1.600	Ea.	40	84.50		124.50	172
1587	4" x 2"		9.50	1.684		54	89		143	193
1588	4" x 3"	↓	9	1.778		63	94		157	212
1589	6" x 2"	Q-2	12.50	1.920		113	105		218	283
1590	6" x 3"	↓	12	2	↓	120	110		230	297

For customer support on your Plumbing Cost Data, call 877.709.6968.

255

22 13 16.30 Pipe Fittings, Cast Iron		Crew	Daily Output	Labor-Hours	Unit	Material	2015 Bare Costs Labor	Equipment	Total	Total Incl O&P
1591	6" x 4"	Q-2	11	2.182	Ea.	126	120		246	320
1592	8" x 2"	Q-3	11	2.909		237	163		400	505
1593	8" x 4"		10	3.200		209	179		388	500
1594	8" x 6"	↓	9	3.556		287	199		486	615
1600	Double Y, 2"	Q-1	8	2		61.50	106		167.50	228
1610	3"	↓	7	2.286		76.50	121		197.50	266
1620	4"	↓	6.50	2.462		100	130		230	305
1630	5"	Q-2	9	2.667		183	146		329	425
1640	6"	"	8	3		262	164		426	535
1650	8"	Q-3	5.50	5.818		635	325		960	1,175
1660	10"		5	6.400		1,425	360		1,785	2,100
1670	12"	↓	4.50	7.111	↓	1,625	395		2,020	2,400
1676	Combination double Y & 1/8 bend									
1678	2"	Q-1	8	2	Ea.	71	106		177	238
1680	3"		7	2.286		91	121		212	282
1682	4"	↓	6.50	2.462		150	130		280	360
1684	6"	Q-2	8	3	↓	515	164		679	815
1690	Combination double Y & 1/8 bend, CI, reducing									
1692	3" x 2"	Q-1	8	2	Ea.	79	106		185	247
1694	4" x 2"		7	2.286		92.50	121		213.50	284
1696	4" x 3"	↓	7	2.286		105	121		226	297
1698	6" x 4"	Q-2	9	2.667	↓	340	146		486	590
1700	Double Y, CI, reducing									
1702	3" x 2"	Q-1	8	2	Ea.	65.50	106		171.50	232
1703	4" x 2"		7	2.286		79	121		200	269
1704	4" x 3"	↓	7	2.286		83	121		204	274
1706	6" x 3"	Q-2	9.50	2.526		201	138		339	430
1707	6" x 4"	"	9	2.667		186	146		332	425
1708	8" x 4"	Q-3	8.50	3.765		440	210		650	805
1709	8" x 6"		8	4		450	224		674	835
1711	10" x 6"		8	4		960	224		1,184	1,400
1712	10" x 8"		6.50	4.923		1,375	275		1,650	1,925
1713	12" x 6"		7.50	4.267		1,725	238		1,963	2,225
1714	12" x 8"	↓	6	5.333		1,675	298		1,973	2,300
1740	Reducer, 3" x 2"	Q-1	15	1.067		19.45	56.50		75.95	107
1750	4" x 2"		14.50	1.103		22	58.50		80.50	113
1760	4" x 3"		14	1.143		25.50	60.50		86	119
1770	5" x 2"		14	1.143		53.50	60.50		114	150
1780	5" x 3"		13.50	1.185		56.50	62.50		119	157
1790	5" x 4"		13	1.231		32.50	65		97.50	134
1800	6" x 2"		13.50	1.185		51	62.50		113.50	151
1810	6" x 3"		13	1.231		51.50	65		116.50	155
1830	6" x 4"		12.50	1.280		51	67.50		118.50	158
1840	6" x 5"	↓	11	1.455		55	77		132	177
1880	8" x 3"	Q-2	13.50	1.778		99	97.50		196.50	256
1900	8" x 4"		13	1.846		85	101		186	247
1920	8" x 5"		12	2		89.50	110		199.50	264
1940	8" x 6"		12	2		87.50	110		197.50	261
1942	10" x 4"		12.50	1.920		128	105		233	300
1943	10" x 6"		11.50	2.087		143	114		257	330
1944	10" x 8"		9.50	2.526		144	138		282	365
1945	12" x 4"		11.50	2.087		209	114		323	400
1946	12" x 6"	↓	11	2.182		224	120		344	425

For customer support on your Plumbing Cost Data, call 877.709.6968.

22 13 Facility Sanitary Sewerage

22 13 16 – Sanitary Waste and Vent Piping

22 13 16.30 Pipe Fittings, Cast Iron

		Crew	Daily Output	Labor-Hours	Unit	Material	2015 Bare Costs Labor	Equipment	Total	Total Incl O&P
1947	12" x 8"	Q-2	9	2.667	Ea.	230	146		376	475
1948	12" x 10"	↓	8.50	2.824		234	155		389	490
1949	15" x 6"	Q-3	12	2.667		435	149		584	705
1950	15" x 8"		10	3.200		475	179		654	790
1951	15" x 10"		9.50	3.368		490	188		678	825
1952	15" x 12"	↓	9	3.556		495	199		694	845
1960	Increaser, 2" x 3"	Q-1	15	1.067		46.50	56.50		103	136
1980	2" x 4"		14	1.143		46.50	60.50		107	142
2000	2" x 5"		13	1.231		57	65		122	161
2020	3" x 4"		13	1.231		51	65		116	154
2040	3" x 5"		13	1.231		57	65		122	161
2060	3" x 6"		12	1.333		68.50	70.50		139	182
2070	4" x 5"		13	1.231		60.50	65		125.50	165
2080	4" x 6"	↓	12	1.333		69	70.50		139.50	182
2090	4" x 8"	Q-2	13	1.846		143	101		244	310
2100	5" x 6"	Q-1	11	1.455		103	77		180	229
2110	5" x 8"	Q-2	12	2		166	110		276	345
2120	6" x 8"		12	2		166	110		276	345
2130	6" x 10"		8	3		299	164		463	580
2140	8" x 10"		6.50	3.692		305	202		507	640
2150	10" x 12"	↓	5.50	4.364		540	239		779	955
2500	Y, 2"	Q-1	10	1.600		25.50	84.50		110	156
2510	3"		9	1.778		47	94		141	194
2520	4"	↓	8	2		62.50	106		168.50	229
2530	5"	Q-2	12	2		111	110		221	287
2540	6"	"	11	2.182		144	120		264	340
2550	8"	Q-3	7	4.571		350	256		606	770
2560	10"		6	5.333		570	298		868	1,075
2570	12"		5	6.400		1,300	360		1,660	2,000
2580	15"	↓	4	8	↓	2,875	445		3,320	3,825
2581	Y, reducing									
2582	3" x 2"	Q-1	10	1.600	Ea.	36.50	84.50		121	168
2584	4" x 2"		9.50	1.684		49	89		138	188
2586	4" x 3"	↓	9	1.778		54	94		148	201
2588	6" x 2"	Q-2	12.50	1.920		96	105		201	265
2590	6" x 3"		12.25	1.959		98	107		205	270
2592	6" x 4"	↓	12	2		97.50	110		207.50	272
2594	8" x 2"	Q-3	11	2.909		216	163		379	485
2596	8" x 3"		10.50	3.048		215	170		385	495
2598	8" x 4"		10	3.200		186	179		365	475
2600	8" x 6"		9	3.556		230	199		429	555
2602	10" x 3"		10	3.200		325	179		504	625
2604	10" x 4"		9.50	3.368		315	188		503	635
2606	10" x 6"		8	4		345	224		569	720
2608	10" x 8"		8	4		490	224		714	875
2610	12" x 4"		9	3.556		535	199		734	890
2612	12" x 6"		8.50	3.765		555	210		765	930
2614	12" x 8"		8	4		665	224		889	1,075
2616	12" x 10"		7.50	4.267		1,125	238		1,363	1,600
2618	15" x 4"		7	4.571		1,725	256		1,981	2,275
2620	15" x 6"		6.50	4.923		1,800	275		2,075	2,400
2622	15" x 8"		6.50	4.923		1,850	275		2,125	2,450
2624	15" x 10"	↓	5	6.400		1,850	360		2,210	2,575

22 13 16.30 Pipe Fittings, Cast Iron		Crew	Daily Output	Labor-Hours	Unit	Material	2015 Bare Costs Labor	Equipment	Total	Total Incl O&P
2626	15" x 12"	Q-3	4.50	7.111	Ea.	2,025	395		2,420	2,825
3000	For extra heavy, add					44%	4%			
3600	Hub and spigot, service weight gasket joint									
3605	Note: gaskets and joint labor have									
3606	been included with all listed fittings.									
3610	1/4 bend, 2"	Q-1	20	.800	Ea.	29.50	42.50		72	96
3620	3"		17	.941		38.50	49.50		88	118
3630	4"		15	1.067		56.50	56.50		113	148
3640	5"	Q-2	21	1.143		82	62.50		144.50	185
3650	6"	"	19	1.263		97	69		166	212
3660	8"	Q-3	12	2.667		272	149		421	525
3670	10"		11	2.909		415	163		578	700
3680	12"		10	3.200		550	179		729	880
3690	15"		8	4		1,625	224		1,849	2,150
3692	Short sweep, CI, 90°, 2"	Q-1	20	.800		29	42.50		71.50	96
3693	3"		17	.941		47	49.50		96.50	127
3694	4"		15	1.067		68.50	56.50		125	161
3695	6"	Q-2	19	1.263		132	69		201	251
3696	8"	Q-3	12	2.667		289	149		438	540
3697	10"		11	2.909		575	163		738	880
3698	12"		10	3.200		1,250	179		1,429	1,625
3700	Closet bend, 3" diameter with ring 10" x 16"	Q-1	17	.941		139	49.50		188.50	228
3710	16" x 16"		15	1.067		154	56.50		210.50	254
3730	Closet bend, 4" diameter, 1" x 4" ring, 6" x 16"		15	1.067		133	56.50		189.50	232
3740	8" x 16"		15	1.067		121	56.50		177.50	218
3750	10" x 12"		14	1.143		117	60.50		177.50	220
3760	10" x 18"		13	1.231		152	65		217	265
3770	12" x 16"		13	1.231		134	65		199	246
3780	16" x 16"		12	1.333		163	70.50		233.50	286
3786	Long sweep elbow									
3787	2"	Q-1	20	.800	Ea.	39.50	42.50		82	108
3788	3"		17	.941		55	49.50		104.50	136
3789	4"		15	1.067		78.50	56.50		135	171
3790	6"	Q-2	19	1.263		152	69		221	272
3791	8"	Q-3	12	2.667		340	149		489	600
3792	10"		11	2.909		580	163		743	885
3793	12"		10	3.200		910	179		1,089	1,275
3794	15"		8	4		2,225	224		2,449	2,800
3800	1/8 bend, 2"	Q-1	20	.800		23.50	42.50		66	90
3810	3"		17	.941		34.50	49.50		84	113
3820	4"		15	1.067		48.50	56.50		105	139
3830	5"	Q-2	21	1.143		69.50	62.50		132	171
3840	6"	"	19	1.263		79.50	69		148.50	193
3850	8"	Q-3	12	2.667		219	149		368	465
3860	10"		11	2.909		330	163		493	610
3870	12"		10	3.200		570	179		749	895
3880	15"		8	4		1,200	224		1,424	1,675
3882	Sanitary tee, tapped									
3884	2" x 2"	Q-1	11	1.455	Ea.	56	77		133	178
3886	3" x 2"		10	1.600		63.50	84.50		148	198
3888	4" x 2"		9	1.778		92.50	94		186.50	244
3890	Sanitary cross, tapped [double tapped sanitary tee]									
3892	2" x 2"	Q-1	9	1.778	Ea.	79	94		173	229

22 13 16.30 Pipe Fittings, Cast Iron		Crew	Daily Output	Labor-Hours	Unit	Material	2015 Bare Costs Labor	Equipment	Total	Total Incl O&P
3894	3" x 2"	Q-1	8	2	Ea.	95	106		201	264
3896	4" x 2"		7	2.286		96	121		217	288
3900	Sanitary Tee, 2"		12	1.333		46.50	70.50		117	157
3910	3"		10	1.600		69	84.50		153.50	204
3920	4"		9	1.778		85.50	94		179.50	236
3930	5"	Q-2	13	1.846		157	101		258	325
3940	6"	"	11	2.182		176	120		296	375
3950	8"	Q-3	8.50	3.765		440	210		650	800
3952	10"		8	4		795	224		1,019	1,200
3954	12"		7	4.571		1,350	256		1,606	1,850
3956	15"		6	5.333		2,450	298		2,748	3,150
3960	Sanitary Tee, reducing									
3961	3" x 2"	Q-1	10.50	1.524	Ea.	60	80.50		140.50	188
3962	4" x 2"		10	1.600		71.50	84.50		156	207
3963	4" x 3"		9.50	1.684		77.50	89		166.50	219
3964	5" x 3"	Q-2	13.50	1.778		167	97.50		264.50	330
3965	6" x 2"		13	1.846		131	101		232	297
3966	6" x 3"		12.50	1.920		137	105		242	310
3967	6" x 4"		12	2		180	110		290	365
3968	8" x 4"	Q-3	10.50	3.048		280	170		450	565
3969	8" x 5"		10	3.200		355	179		534	660
3970	8" x 6"		9.50	3.368		355	188		543	675
3971	10" x 6"		9	3.556		595	199		794	955
3972	12" x 8"		8.50	3.765		1,025	210		1,235	1,475
3973	15" x 10"		6	5.333		2,075	298		2,373	2,750
3980	Tee, 2"	Q-1	12	1.333		59	70.50		129.50	171
3990	3"		10	1.600		84	84.50		168.50	221
4000	4"		9	1.778		107	94		201	260
4010	5"	Q-2	13	1.846		210	101		311	385
4020	6"	"	11	2.182		210	120		330	410
4030	8"	Q-3	8	4		405	224		629	785
4060	Combination Y and 1/8 bend									
4070	2"	Q-1	12	1.333	Ea.	53.50	70.50		124	165
4080	3"		10	1.600		77	84.50		161.50	213
4090	4"		9	1.778		103	94		197	256
4100	5"	Q-2	13	1.846		186	101		287	355
4110	6"	"	11	2.182		224	120		344	425
4120	8"	Q-3	8	4		540	224		764	935
4121	12"	"	7	4.571		1,125	256		1,381	1,625
4130	Combination Y & 1/8 bend, reducing									
4132	3" x 2"	Q-1	10.50	1.524	Ea.	61.50	80.50		142	190
4134	4" x 2"		10	1.600		78.50	84.50		163	215
4136	4" x 3"		9.50	1.684		90.50	89		179.50	234
4138	6" x 2"	Q-2	13	1.846		147	101		248	315
4140	6" x 3"		12.50	1.920		157	105		262	330
4142	6" x 4"		12	2		166	110		276	345
4144	8" x 2"	Q-3	11	2.909		315	163		478	595
4146	8" x 4"		10.50	3.048		279	170		449	560
4148	8" x 6"		9.50	3.368		365	188		553	690
4160	Double Y, 2"	Q-1	10	1.600		90	84.50		174.50	227
4170	3"		8	2		113	106		219	284
4180	4"		7	2.286		146	121		267	345
4190	5"	Q-2	10	2.400		255	131		386	480

For customer support on your Plumbing Cost Data, call 877.709.6968.

259

22 13 16.30 Pipe Fittings, Cast Iron		Crew	Daily Output	Labor-Hours	Unit	Material	2015 Bare Costs Labor	Equipment	Total	Total Incl O&P
4200	6"	Q-2	9	2.667	Ea.	335	146		481	590
4210	8"	Q-3	6	5.333		800	298		1,098	1,325
4220	10"	↓	5	6.400		1,700	360		2,060	2,425
4230	12"	↓	4.50	7.111	↓	2,000	395		2,395	2,800
4234	Combination double Y & 1/8 bend									
4235	2"	Q-1	10	1.600	Ea.	99.50	84.50		184	237
4236	3"		8	2		128	106		234	300
4237	4"	↓	7	2.286		195	121		316	395
4238	6"	Q-2	9	2.667	↓	590	146		736	870
4242	Combination double Y & 1/8 bend, CI, reducing									
4243	3" x 2"	Q-1	9	1.778	Ea.	110	94		204	263
4244	4" x 2"		8	2		126	106		232	299
4245	4" x 3"	↓	7.50	2.133		145	113		258	330
4246	6" x 4"	Q-2	10	2.400	↓	395	131		526	635
4248	Double Y, CI, reducing									
4249	3" x 2"	Q-1	9	1.778	Ea.	96.50	94		190.50	248
4250	4" x 2"		8	2		113	106		219	284
4251	4" x 3"	↓	7.50	2.133		111	113		224	292
4252	6" x 3"	Q-2	10.50	2.286		250	125		375	465
4253	6" x 4"	"	10	2.400		242	131		373	465
4254	8" x 4"	Q-3	9.50	3.368		525	188		713	860
4255	8" x 6"		9	3.556		555	199		754	910
4256	10" x 6"		8.50	3.765		1,100	210		1,310	1,550
4257	10" x 8"		7.50	4.267		1,575	238		1,813	2,075
4258	12" x 6"		8	4		1,875	224		2,099	2,425
4259	12" x 8"	↓	7	4.571		1,925	256		2,181	2,475
4260	Reducer, 3" x 2"	Q-1	17	.941		41	49.50		90.50	120
4270	4" x 2"		16.50	.970		47	51		98	129
4280	4" x 3"		16	1		53	53		106	138
4290	5" x 2"		16	1		87	53		140	176
4300	5" x 3"		15.50	1.032		93	54.50		147.50	185
4310	5" x 4"		15	1.067		72	56.50		128.50	164
4320	6" x 2"		15.50	1.032		85	54.50		139.50	176
4330	6" x 3"		15	1.067		88.50	56.50		145	183
4336	6" x 4"		14	1.143		91	60.50		151.50	191
4340	6" x 5"	↓	13	1.231		104	65		169	212
4360	8" x 3"	Q-2	15	1.600		166	87.50		253.50	315
4370	8" x 4"		15	1.600		155	87.50		242.50	305
4380	8" x 5"		14	1.714		169	94		263	325
4390	8" x 6"		14	1.714		167	94		261	325
4394	10" x 4"		13.50	1.778		239	97.50		336.50	410
4395	10" x 6"		13	1.846		264	101		365	445
4396	10" x 8"		12.50	1.920		294	105		399	485
4397	12" x 4"		12	2		345	110		455	545
4398	12" x 6"		11.50	2.087		370	114		484	585
4399	12" x 8"		11	2.182		405	120		525	625
4400	12" x 10"	↓	10.50	2.286		450	125		575	685
4401	15" x 6"	Q-3	13	2.462		605	138		743	880
4402	15" x 8"		12	2.667		675	149		824	965
4403	15" x 10"		11	2.909		735	163		898	1,050
4404	15" x 12"	↓	10	3.200		765	179		944	1,100
4430	Increaser, 2" x 3"	Q-1	17	.941		58.50	49.50		108	140
4440	2" x 4"	↓	16	1	↓	62	53		115	148

22 13 16.30 Pipe Fittings, Cast Iron		Crew	Daily Output	Labor-Hours	Unit	Material	2015 Bare Costs Labor	Equipment	Total	Total Incl O&P
4450	2" x 5"	Q-1	15	1.067	Ea.	81	56.50		137.50	175
4460	3" x 4"		15	1.067		66.50	56.50		123	158
4470	3" x 5"		15	1.067		81	56.50		137.50	175
4480	3" x 6"		14	1.143		93.50	60.50		154	194
4490	4" x 5"		15	1.067		84.50	56.50		141	178
4500	4" x 6"		14	1.143		94	60.50		154.50	194
4510	4" x 8"	Q-2	15	1.600		198	87.50		285.50	350
4520	5" x 6"	Q-1	13	1.231		127	65		192	238
4530	5" x 8"	Q-2	14	1.714		221	94		315	385
4540	6" x 8"		14	1.714		221	94		315	385
4550	6" x 10"		10	2.400		395	131		526	635
4560	8" x 10"		8.50	2.824		400	155		555	675
4570	10" x 12"		7.50	3.200		665	175		840	995
4600	Y, 2"	Q-1	12	1.333		44	70.50		114.50	155
4610	3"		10	1.600		71	84.50		155.50	207
4620	4"		9	1.778		93.50	94		187.50	245
4630	5"	Q-2	13	1.846		159	101		260	330
4640	6"	"	11	2.182		194	120		314	395
4650	8"	Q-3	8	4		460	224		684	850
4660	10"		7	4.571		760	256		1,016	1,225
4670	12"		6	5.333		1,550	298		1,848	2,150
4672	15"		5	6.400		3,150	360		3,510	4,025
4680	Y, reducing									
4681	3" x 2"	Q-1	10.50	1.524	Ea.	58	80.50		138.50	186
4682	4" x 2"		10	1.600		73.50	84.50		158	209
4683	4" x 3"		9.50	1.684		81.50	89		170.50	224
4684	6" x 2"	Q-2	13	1.846		130	101		231	296
4685	6" x 3"		12.50	1.920		135	105		240	310
4686	6" x 4"		12	2		137	110		247	315
4687	8" x 2"	Q-3	11	2.909		281	163		444	555
4688	8" x 3"		10.75	2.977		282	166		448	560
4689	8" x 4"		10.50	3.048		256	170		426	540
4690	8" x 6"		9.50	3.368		310	188		498	625
4691	10" x 3"		10	3.200		430	179		609	745
4692	10" x 4"		9.50	3.368		430	188		618	755
4693	10" x 6"		9	3.556		465	199		664	815
4694	10" x 8"		8.80	3.636		640	203		843	1,000
4695	12" x 4"		9.50	3.368		675	188		863	1,025
4696	12" x 6"		9	3.556		700	199		899	1,075
4697	12" x 8"		8.50	3.765		840	210		1,050	1,250
4698	12" x 10"		8	4		1,350	224		1,574	1,850
4699	15" x 4"		8	4		1,900	224		2,124	2,425
4700	15" x 6"		7.50	4.267		1,975	238		2,213	2,525
4701	15" x 8"		7	4.571		2,050	256		2,306	2,625
4702	15" x 10"		6	5.333		2,100	298		2,398	2,750
4703	15" x 12"		5	6.400		2,300	360		2,660	3,075
4900	For extra heavy, add					44%	4%			
4940	Gasket and making push-on joint									
4950	2"	Q-1	40	.400	Ea.	9.40	21		30.40	42.50
4960	3"		35	.457		12.20	24		36.20	50
4970	4"		32	.500		15.30	26.50		41.80	57
4980	5"	Q-2	43	.558		24	30.50		54.50	72.50
4990	6"	"	40	.600		25	33		58	77

For customer support on your Plumbing Cost Data, call 877.709.6968.

261

22 13 16.30 Pipe Fittings, Cast Iron		Crew	Daily Output	Labor-Hours	Unit	Material	2015 Bare Costs Labor	Equipment	Total	Total Incl O&P
5000	8"	Q-3	32	1	Ea.	55	56		111	145
5010	10"		29	1.103		95.50	61.50		157	198
5020	12"		25	1.280		122	71.50		193.50	242
5022	15"		21	1.524		146	85		231	290
5030	Note: gaskets and joint labor have									
5040	Been included with all listed fittings.									
5990	No hub									
6000	Cplg. & labor required at joints not incl. in fitting									
6010	price. Add 1 coupling per joint for installed price									
6020	1/4 Bend, 1-1/2"				Ea.	10.20			10.20	11.25
6060	2"					11.15			11.15	12.25
6080	3"					16.25			16.25	17.85
6120	4"					23			23	25.50
6140	5"					55.50			55.50	61
6160	6"					55.50			55.50	61
6180	8"					156			156	172
6181	10"					335			335	370
6182	12"					975			975	1,075
6183	15"					1,275			1,275	1,400
6184	1/4 Bend, long sweep, 1-1/2"					25.50			25.50	28
6186	2"					24.50			24.50	26.50
6188	3"					29.50			29.50	32.50
6189	4"					47			47	51.50
6190	5"					90.50			90.50	99.50
6191	6"					103			103	114
6192	8"					282			282	310
6193	10"					570			570	625
6200	1/8 Bend, 1-1/2"					8.60			8.60	9.45
6210	2"					9.60			9.60	10.55
6212	3"					12.80			12.80	14.10
6214	4"					16.80			16.80	18.50
6216	5"					35			35	38.50
6218	6"					37.50			37.50	41
6220	8"					108			108	118
6222	10"					204			204	225
6364	Closet flange									
6366	4"				Ea.	15.55			15.55	17.10
6370	Closet bend, no hub									
6376	4" x 16"				Ea.	82.50			82.50	91
6380	Sanitary Tee, tapped, 1-1/2"					20.50			20.50	22.50
6382	2" x 1-1/2"					17.95			17.95	19.70
6384	2"					19.25			19.25	21
6386	3" x 2"					28.50			28.50	31.50
6388	3"					49.50			49.50	54.50
6390	4" x 1-1/2"					25.50			25.50	28
6392	4" x 2"					29			29	32
6393	4"					29			29	32
6394	6" x 1-1/2"					66			66	73
6396	6" x 2"					67.50			67.50	74
6459	Sanitary Tee, 1-1/2"					14.35			14.35	15.80
6460	2"					15.35			15.35	16.85
6470	3"					18.90			18.90	21
6472	4"					36			36	39.50

262

For customer support on your Plumbing Cost Data, call 877.709.6968.

22 13 16.30 Pipe Fittings, Cast Iron		Crew	Daily Output	Labor-Hours	Unit	Material	2015 Bare Costs Labor	Equipment	Total	Total Incl O&P
6474	5"				Ea.	83.50			83.50	92
6476	6"					85.50			85.50	94
6478	8"					345			345	380
6480	10"					425			425	465
6724	Sanitary Tee, reducing									
6725	3" x 2"				Ea.	16.75			16.75	18.40
6726	4" x 3"					27.50			27.50	30.50
6727	5" x 4"					64.50			64.50	71
6728	6" x 3"					63			63	69
6729	8" x 4"					182			182	201
6730	Y, 1-1/2"					14.50			14.50	15.95
6740	2"					14.20			14.20	15.65
6750	3"					20.50			20.50	22.50
6760	4"					33			33	36.50
6762	5"					78.50			78.50	86
6764	6"					88			88	97
6768	8"					207			207	228
6769	10"					460			460	505
6770	12"					905			905	995
6771	15"					2,025			2,025	2,225
6791	Y, reducing, 3" x 2"					15.35			15.35	16.85
6792	4" x 2"					22			22	24.50
6793	5" x 2"					48.50			48.50	53.50
6794	6" x 2"					54			54	59.50
6795	6" x 4"					70			70	77
6796	8" x 4"					121			121	133
6797	8" x 6"					149			149	163
6798	10" x 6"					335			335	370
6799	10" x 8"					400			400	440
6800	Double Y, 2"					22.50			22.50	25
6920	3"					41.50			41.50	45.50
7000	4"					84.50			84.50	92.50
7100	6"					149			149	164
7120	8"					425			425	470
7200	Combination Y and 1/8 Bend									
7220	1-1/2"				Ea.	15.45			15.45	17
7260	2"					16.15			16.15	17.80
7320	3"					25.50			25.50	28
7400	4"					49			49	54
7480	5"					101			101	111
7500	6"					135			135	148
7520	8"					315			315	345
7800	Reducer, 3" x 2"					7.85			7.85	8.60
7820	4" x 2"					12.10			12.10	13.30
7840	4" x 3"					12.10			12.10	13.30
7842	6" x 3"					32.50			32.50	36
7844	6" x 4"					29.50			29.50	32.50
7846	6" x 5"					33.50			33.50	36.50
7848	8" x 2"					51.50			51.50	57
7850	8" x 3"					47.50			47.50	52.50
7852	8" x 4"					50.50			50.50	55.50
7854	8" x 5"					57			57	62.50
7856	8" x 6"					56			56	61.50

For customer support on your Plumbing Cost Data, call 877.709.6968.

263

22 13 16 – Sanitary Waste and Vent Piping

22 13 16.30 Pipe Fittings, Cast Iron		Crew	Daily Output	Labor-Hours	Unit	Material	2015 Bare Costs Labor	Equipment	Total	Total Incl O&P
7858	10" x 4"				Ea.	99			99	109
7860	10" x 6"					104			104	115
7862	10" x 8"					122			122	134
7864	12" x 4"					204			204	225
7866	12" x 6"					219			219	241
7868	12" x 8"					225			225	248
7870	12" x 10"					225			225	248
7872	15" x 4"					425			425	470
7874	15" x 6"					400			400	440
7876	15" x 8"					465			465	510
7878	15" x 10"					480			480	530
7880	15" x 12"					485			485	535
8000	Coupling, standard (by CISPI Mfrs.)									
8020	1-1/2"	Q-1	48	.333	Ea.	13.95	17.60		31.55	42
8040	2"		44	.364		15.25	19.20		34.45	46
8080	3"		38	.421		17.10	22		39.10	52.50
8120	4"		33	.485		20	25.50		45.50	60.50
8160	5"	Q-2	44	.545		41	30		71	90
8180	6"	"	40	.600		46.50	33		79.50	101
8200	8"	Q-3	33	.970		97	54		151	189
8220	10"	"	26	1.231		124	69		193	240
8300	Coupling, cast iron clamp & neoprene gasket (by MG)									
8310	1-1/2"	Q-1	48	.333	Ea.	6.40	17.60		24	33.50
8320	2"		44	.364		8.35	19.20		27.55	38
8330	3"		38	.421		12.65	22		34.65	47.50
8340	4"		33	.485		14.60	25.50		40.10	54.50
8350	5"	Q-2	44	.545		20.50	30		50.50	67.50
8360	6"	"	40	.600		32.50	33		65.50	85.50
8380	8"	Q-3	33	.970		116	54		170	210
8400	10"	"	26	1.231		189	69		258	310
8410	Reducing, no hub									
8416	2" x 1-1/2"	Q-1	44	.364	Ea.	8.10	19.20		27.30	38
8600	Coupling, Stainless steel, heavy duty									
8620	1-1/2"	Q-1	48	.333	Ea.	10	17.60		27.60	37.50
8630	2"		44	.364		10.40	19.20		29.60	40.50
8640	2" x 1-1/2"		44	.364		12.20	19.20		31.40	42.50
8650	3"		38	.421		11.30	22		33.30	46
8660	4"		33	.485		12.75	25.50		38.25	52.50
8670	4" x 3"		33	.485		18.40	25.50		43.90	58.50
8680	5"	Q-2	44	.545		24.50	30		54.50	72
8690	6"	"	40	.600		30.50	33		63.50	83.50
8700	8"	Q-3	33	.970		52	54		106	139
8710	10"		26	1.231		65.50	69		134.50	176
8712	12"		22	1.455		102	81.50		183.50	235
8715	15"		18	1.778		120	99.50		219.50	282

22 13 16.40 Pipe Fittings, Cast Iron for Drainage

		Crew	Daily Output	Labor-Hours	Unit	Material	2015 Bare Costs Labor	Equipment	Total	Total Incl O&P
0010	**PIPE FITTINGS, CAST IRON FOR DRAINAGE**, Special									
0020	Cast iron, drainage, threaded, black									
0030	90° Elbow, straight									
0031	1-1/2" pipe size	Q-1	20	.800	Ea.	37	42.50		79.50	105
0032	2" pipe size		18	.889		55.50	47		102.50	132
0033	3" pipe size		10	1.600		209	84.50		293.50	360

22 13 16 – Sanitary Waste and Vent Piping

22 13 16.40 Pipe Fittings, Cast Iron for Drainage	Crew	Daily Output	Labor-Hours	Unit	Material	2015 Bare Costs Labor	Equipment	Total	Total Incl O&P
0034	4" pipe size	Q-1	6	2.667	Ea.	330	141		471	580
0040	90° Long turn elbow, straight									
0041	1-1/2" pipe size	Q-1	20	.800	Ea.	51.50	42.50		94	121
0042	2" pipe size		18	.889		77	47		124	156
0043	3" pipe size		10	1.600		281	84.50		365.50	440
0044	4" pipe size		6	2.667		480	141		621	740
0050	90° Street elbow, straight									
0051	1-1/2" pipe size	Q-1	20	.800	Ea.	52.50	42.50		95	122
0052	2" pipe size	"	18	.889	"	69.50	47		116.50	148
0060	45° Elbow									
0061	1-1/2" pipe size	Q-1	20	.800	Ea.	36	42.50		78.50	104
0062	2" pipe size		18	.889		52	47		99	129
0063	3" pipe size		10	1.600		204	84.50		288.50	355
0064	4" pipe size		6	2.667		320	141		461	565
0070	45° Street elbow									
0071	1-1/2" pipe size	Q-1	20	.800	Ea.	49	42.50		91.50	118
0072	2" pipe size	"	18	.889	"	79.50	47		126.50	159
0092	Tees, straight									
0093	1-1/2" pipe size	Q-1	13	1.231	Ea.	60.50	65		125.50	165
0094	2" pipe size	"	11	1.455	"	100	77		177	226
0100	TY's, straight									
0101	1-1/2" pipe size	Q-1	13	1.231	Ea.	59.50	65		124.50	164
0102	2" pipe size		11	1.455		98.50	77		175.50	224
0103	3" pipe size		6	2.667		365	141		506	615
0104	4" pipe size		4	4		535	211		746	910
0120	45° Y branch, straight									
0121	1-1/2" pipe size	Q-1	13	1.231	Ea.	73	65		138	179
0122	2" pipe size		11	1.455		85.50	77		162.50	210
0123	3" pipe size		6	2.667		470	141		611	730
0124	4" pipe size		4	4		700	211		911	1,100
0147	Double Y branch, straight									
0148	1-1/2" pipe size	Q-1	10	1.600	Ea.	222	84.50		306.50	370
0149	2" pipe size	"	7	2.286	"	206	121		327	410
0160	P trap									
0161	1-1/2" pipe size	Q-1	15	1.067	Ea.	109	56.50		165.50	205
0162	2" pipe size		13	1.231		174	65		239	289
0163	3" pipe size		7	2.286		565	121		686	805
0164	4" pipe size		5	3.200		1,275	169		1,444	1,650
0180	Tucker connection									
0181	1-1/2" pipe size	Q-1	24	.667	Ea.	106	35		141	169
0182	2" pipe size	"	21	.762	"	130	40.50		170.50	204
0205	Cast iron, drainage, threaded, galvanized									
0206	90° Elbow, straight									
0207	1-1/2" pipe size	Q-1	20	.800	Ea.	55	42.50		97.50	124
0208	2" pipe size		18	.889		79	47		126	158
0209	3" pipe size		10	1.600		315	84.50		399.50	475
0210	4" pipe size		6	2.667		380	141		521	630
0216	90° Long turn elbow, straight									
0217	1-1/2" pipe size	Q-1	20	.800	Ea.	62.50	42.50		105	133
0218	2" pipe size		18	.889		107	47		154	189
0219	3" pipe size		10	1.600		350	84.50		434.50	515
0220	4" pipe size		6	2.667		605	141		746	880
0226	90° Street elbow, straight									

For customer support on your Plumbing Cost Data, call 877.709.6968.

265

22 13 16.40 Pipe Fittings, Cast Iron for Drainage		Crew	Daily Output	Labor-Hours	Unit	Material	2015 Bare Costs Labor	Equipment	Total	Total Incl O&P
0227	1-1/2" pipe size	Q-1	20	.800	Ea.	69.50	42.50		112	141
0228	2" pipe size	"	18	.889	"	109	47		156	191
0236	45° Elbow									
0237	1-1/2" pipe size	Q-1	20	.800	Ea.	54.50	42.50		97	124
0238	2" pipe size		18	.889		77.50	47		124.50	156
0239	3" pipe size		10	1.600		270	84.50		354.50	425
0240	4" pipe size	↓	6	2.667	↓	440	141		581	700
0246	45° Street elbow									
0247	1-1/2" pipe size	Q-1	20	.800	Ea.	72	42.50		114.50	143
0248	2" pipe size	"	18	.889	"	108	47		155	190
0268	Tees, straight									
0269	1-1/2" pipe size	Q-1	13	1.231	Ea.	82.50	65		147.50	189
0270	2" pipe size	"	11	1.455	"	138	77		215	268
0276	TY's, straight									
0277	1-1/2" pipe size	Q-1	13	1.231	Ea.	130	65		195	241
0278	2" pipe size		11	1.455		136	77		213	266
0279	3" pipe size		6	2.667		550	141		691	820
0280	4" pipe size	↓	4	4	↓	655	211		866	1,050
0296	45° Y branch, straight									
0297	1-1/2" pipe size	Q-1	13	1.231	Ea.	92.50	65		157.50	200
0298	2" pipe size		11	1.455		185	77		262	320
0299	3" pipe size		6	2.667		655	141		796	940
0300	4" pipe size	↓	4	4	↓	1,000	211		1,211	1,425
0323	Double Y branch, straight									
0324	1-1/2" pipe size	Q-1	10	1.600	Ea.	184	84.50		268.50	330
0325	2" pipe size	"	7	2.286	"	315	121		436	525
0336	P Trap									
0337	1-1/2" pipe size	Q-1	15	1.067	Ea.	149	56.50		205.50	249
0338	2" pipe size		13	1.231		262	65		327	385
0339	3" pipe size		7	2.286		830	121		951	1,100
0340	4" pipe size	↓	5	3.200	↓	1,825	169		1,994	2,250
0356	Tucker connection									
0357	1-1/2" pipe size	Q-1	24	.667	Ea.	240	35		275	315
0358	2" pipe size	"	21	.762	"	257	40.50		297.50	345
1000	Drip pan elbow, (safety valve discharge elbow)									
1010	Cast iron, threaded inlet									
1014	2-1/2"	Q-1	8	2	Ea.	1,300	106		1,406	1,575
1015	3"		6.40	2.500		1,400	132		1,532	1,750
1017	4"	↓	4.80	3.333	↓	1,900	176		2,076	2,350
1018	Cast iron, flanged inlet									
1019	6"	Q-2	3.60	6.667	Ea.	1,950	365		2,315	2,700
1020	8"	"	2.60	9.231	"	2,050	505		2,555	3,025

22 13 16.50 Shower Drains

		Crew	Daily Output	Labor-Hours	Unit	Material	2015 Bare Costs Labor	Equipment	Total	Total Incl O&P
0010	**SHOWER DRAINS**									
2780	Shower, with strainer, uniform diam. trap, bronze top									
2800	2" and 3" pipe size	Q-1	8	2	Ea.	480	106		586	690
2820	4" pipe size	"	7	2.286		485	121		606	710
2840	For galvanized body, add				↓	189			189	208
2860	With strainer, backwater valve, drum trap									
2880	1-1/2", 2" & 3" pipe size	Q-1	8	2	Ea.	385	106		491	585
2890	4" pipe size	"	7	2.286		530	121		651	765
2900	For galvanized body, add				↓	193			193	213

22 13 16 – Sanitary Waste and Vent Piping

22 13 16.60 Traps		Crew	Daily Output	Labor-Hours	Unit	Material	2015 Bare Costs Labor	Equipment	Total	Total Incl O&P
0010	**TRAPS**									
0030	Cast iron, service weight									
0050	Running P trap, without vent									
1100	2"	Q-1	16	1	Ea.	143	53		196	237
1140	3"		14	1.143		143	60.50		203.50	248
1150	4"		13	1.231		143	65		208	255
1160	6"	Q-2	17	1.412		635	77.50		712.50	810
1180	Running trap, single hub, with vent									
2080	3" pipe size, 3" vent	Q-1	14	1.143	Ea.	118	60.50		178.50	220
2120	4" pipe size, 4" vent	"	13	1.231		154	65		219	267
2140	5" pipe size, 4" vent	Q-2	11	2.182		245	120		365	450
2160	6" pipe size, 4" vent		10	2.400		660	131		791	925
2180	6" pipe size, 6" vent		8	3		715	164		879	1,025
2200	8" pipe size, 4" vent	Q-3	10	3.200		3,050	179		3,229	3,625
2220	8" pipe size, 6" vent	"	8	4		2,350	224		2,574	2,950
2300	For double hub, vent, add					10%	20%			
2800	S trap,									
2850	4" pipe size	Q-1	13	1.231	Ea.	68.50	65		133.50	174
3000	P trap, B&S, 2" pipe size		16	1		34	53		87	118
3040	3" pipe size		14	1.143		50.50	60.50		111	147
3060	4" pipe size		13	1.231		73	65		138	178
3080	5" pipe size	Q-2	18	1.333		162	73		235	288
3100	6" pipe size	"	17	1.412		225	77.50		302.50	365
3120	8" pipe size	Q-3	11	2.909		680	163		843	990
3130	10" pipe size	"	10	3.200		1,275	179		1,454	1,675
3150	P trap, no hub, 1-1/2" pipe size	Q-1	17	.941		18.40	49.50		67.90	95
3160	2" pipe size		16	1		17.35	53		70.35	99
3170	3" pipe size		14	1.143		38	60.50		98.50	133
3180	4" pipe size		13	1.231		67.50	65		132.50	172
3190	6" pipe size	Q-2	17	1.412		164	77.50		241.50	297
3350	Deep seal trap, B&S									
3400	1-1/4" pipe size	Q-1	14	1.143	Ea.	53	60.50		113.50	149
3410	1-1/2" pipe size		14	1.143		53	60.50		113.50	149
3420	2" pipe size		14	1.143		49	60.50		109.50	145
3440	3" pipe size		12	1.333		62.50	70.50		133	175
3460	4" pipe size		11	1.455		99	77		176	225
3500	For trap primer connection, add					125			125	137
3540	For trap with floor cleanout, add					70%	5%			
3580	For trap with adjustable cleanout, add	Q-1	10	1.600	Ea.	284	84.50		368.50	440
4700	Copper, drainage, drum trap									
4800	3" x 5" solid, 1-1/2" pipe size	1 Plum	16	.500	Ea.	106	29.50		135.50	162
4840	3" x 6" swivel, 1-1/2" pipe size	"	16	.500	"	160	29.50		189.50	221
5100	P trap, standard pattern									
5200	1-1/4" pipe size	1 Plum	18	.444	Ea.	78.50	26		104.50	126
5240	1-1/2" pipe size		17	.471		72	27.50		99.50	121
5260	2" pipe size		15	.533		111	31.50		142.50	170
5280	3" pipe size		11	.727		281	42.50		323.50	375
5340	With cleanout, swivel joint and slip joint									
5360	1-1/4" pipe size	1 Plum	18	.444	Ea.	99.50	26		125.50	149
5400	1-1/2" pipe size		17	.471		106	27.50		133.50	159
5420	2" pipe size		15	.533		177	31.50		208.50	242
5750	Chromed brass, tubular, P trap, without cleanout, 20 Ga.									

For customer support on your Plumbing Cost Data, call 877.709.6968.

267

22 13 16.60 Traps		Crew	Daily Output	Labor-Hours	Unit	Material	2015 Bare Costs Labor	Equipment	Total	Total Incl O&P
5800	1-1/4" pipe size	1 Plum	18	.444	Ea.	19.25	26		45.25	60.50
5840	1-1/2" pipe size	"	17	.471	"	22	27.50		49.50	65.50
5900	With cleanout, 20 Ga.									
5940	1-1/4" pipe size	1 Plum	18	.444	Ea.	31	26		57	73.50
6000	1-1/2" pipe size	"	17	.471	"	33.50	27.50		61	78.50
6350	S trap, without cleanout, 20 Ga.									
6400	1-1/4" pipe size	1 Plum	18	.444	Ea.	50	26		76	94.50
6440	1-1/2" pipe size	"	17	.471	"	57.50	27.50		85	105
6550	With cleanout, 20 Ga.									
6600	1-1/4" pipe size	1 Plum	18	.444	Ea.	61.50	26		87.50	107
6640	1-1/2" pipe size	"	17	.471		64	27.50		91.50	112
6660	Corrosion resistant, glass, P trap, 1-1/2" pipe size	Q-1	17	.941		72	49.50		121.50	154
6670	2" pipe size		16	1		94	53		147	183
6680	3" pipe size		14	1.143		193	60.50		253.50	305
6690	4" pipe size		13	1.231		287	65		352	415
6700	6" pipe size	Q-2	17	1.412		1,100	77.50		1,177.50	1,350
6710	ABS DWV P trap, solvent weld joint									
6720	1-1/2" pipe size	1 Plum	18	.444	Ea.	6.30	26		32.30	46.50
6722	2" pipe size		17	.471		8.30	27.50		35.80	50.50
6724	3" pipe size		15	.533		32.50	31.50		64	83.50
6726	4" pipe size		14	.571		65	33.50		98.50	122
6732	PVC DWV P trap, solvent weld joint									
6733	1-1/2" pipe size	1 Plum	18	.444	Ea.	4.95	26		30.95	45
6734	2" pipe size		17	.471		6.70	27.50		34.20	49
6735	3" pipe size		15	.533		22.50	31.50		54	72.50
6736	4" pipe size		14	.571		51.50	33.50		85	108
6760	PP DWV, dilution trap, 1-1/2" pipe size		16	.500		226	29.50		255.50	294
6770	P trap, 1-1/2" pipe size		17	.471		56.50	27.50		84	104
6780	2" pipe size		16	.500		83	29.50		112.50	136
6790	3" pipe size		14	.571		153	33.50		186.50	219
6800	4" pipe size		13	.615		257	36		293	340
6830	S trap, 1-1/2" pipe size		16	.500		46.50	29.50		76	95.50
6840	2" pipe size		15	.533		69.50	31.50		101	124
6850	Universal trap, 1-1/2" pipe size		14	.571		92	33.50		125.50	152
6860	PVC DWV hub x hub, basin trap, 1-1/4" pipe size		18	.444		40	26		66	83.50
6870	Sink P trap, 1-1/2" pipe size		18	.444		9.65	26		35.65	50
6880	Tubular S trap, 1-1/2" pipe size		17	.471		23.50	27.50		51	67
6890	PVC sch. 40 DWV, drum trap									
6900	1-1/2" pipe size	1 Plum	16	.500	Ea.	23.50	29.50		53	70.50
6910	P trap, 1-1/2" pipe size		18	.444		5.85	26		31.85	46
6920	2" pipe size		17	.471		7.90	27.50		35.40	50
6930	3" pipe size		15	.533		27	31.50		58.50	77
6940	4" pipe size		14	.571		61	33.50		94.50	118
6950	P trap w/clean out, 1-1/2" pipe size		18	.444		9.65	26		35.65	50
6960	2" pipe size		17	.471		16.40	27.50		43.90	59.50
6970	P trap adjustable, 1-1/2" pipe size		17	.471		7.75	27.50		35.25	50
6980	P trap adj. w/union & cleanout, 1-1/2" pipe size		16	.500		30.50	29.50		60	78
7000	Trap primer, flow through type, 1/2" diameter		24	.333		42.50	19.55		62.05	76
7100	With sediment strainer		22	.364		46.50	21.50		68	83
7450	Trap primer distribution unit									
7500	2 openings	1 Plum	18	.444	Ea.	28	26		54	70.50
7540	3 openings		17	.471		30	27.50		57.50	74.50
7560	4 openings		16	.500		35.50	29.50		65	83.50

22 13 16 – Sanitary Waste and Vent Piping

22 13 16.60 Traps

		Crew	Daily Output	Labor-Hours	Unit	Material	2015 Bare Costs Labor	Equipment	Total	Total Incl O&P
7850	Trap primer manifold									
7900	2 outlet	1 Plum	18	.444	Ea.	58.50	26		84.50	104
7940	4 outlet		16	.500		94.50	29.50		124	149
7960	6 outlet		15	.533		131	31.50		162.50	192
7980	8 outlet		13	.615		166	36		202	238

22 13 16.80 Vent Flashing and Caps

		Crew	Daily Output	Labor-Hours	Unit	Material	2015 Bare Costs Labor	Equipment	Total	Total Incl O&P
0010	**VENT FLASHING AND CAPS**									
0120	Vent caps									
0140	Cast iron									
0160	1-1/4" - 1-1/2" pipe	1 Plum	23	.348	Ea.	34	20.50		54.50	68.50
0170	2" - 2-1/8" pipe		22	.364		39.50	21.50		61	75.50
0180	2-1/2" - 3-5/8" pipe		21	.381		45	22.50		67.50	83.50
0190	4" - 4-1/8" pipe		19	.421		55	24.50		79.50	98
0200	5" - 6" pipe		17	.471		83.50	27.50		111	134
0300	PVC									
0320	1-1/4" - 1-1/2" pipe	1 Plum	24	.333	Ea.	9.20	19.55		28.75	39.50
0330	2" - 2-1/8" pipe	"	23	.348	"	9.30	20.50		29.80	41
0900	Vent flashing									
1000	Aluminum with lead ring									
1020	1-1/4" pipe	1 Plum	20	.400	Ea.	8	23.50		31.50	44.50
1030	1-1/2" pipe		20	.400		8.45	23.50		31.95	45
1040	2" pipe		18	.444		8.60	26		34.60	49
1050	3" pipe		17	.471		9.55	27.50		37.05	52
1060	4" pipe		16	.500		11.50	29.50		41	57
1350	Copper with neoprene ring									
1400	1-1/4" pipe	1 Plum	20	.400	Ea.	61.50	23.50		85	103
1430	1-1/2" pipe		20	.400		61.50	23.50		85	103
1440	2" pipe		18	.444		61.50	26		87.50	107
1450	3" pipe		17	.471		74.50	27.50		102	124
1460	4" pipe		16	.500		74.50	29.50		104	127
2000	Galvanized with neoprene ring									
2020	1-1/4" pipe	1 Plum	20	.400	Ea.	16.95	23.50		40.45	54
2030	1-1/2" pipe		20	.400		16.95	23.50		40.45	54
2040	2" pipe		18	.444		17.55	26		43.55	59
2050	3" pipe		17	.471		18.55	27.50		46.05	62
2060	4" pipe		16	.500		21.50	29.50		51	68
2980	Neoprene, one piece									
3000	1-1/4" pipe	1 Plum	24	.333	Ea.	5.05	19.55		24.60	35
3030	1-1/2" pipe		24	.333		5.10	19.55		24.65	35
3040	2" pipe		23	.348		5.10	20.50		25.60	36.50
3050	3" pipe		21	.381		5.85	22.50		28.35	40.50
3060	4" pipe		20	.400		8.65	23.50		32.15	45
4000	Lead, 4#, 8" skirt, vent through roof									
4100	2" pipe	1 Plum	18	.444	Ea.	41.50	26		67.50	85.50
4110	3" pipe		17	.471		47.50	27.50		75	93.50
4120	4" pipe		16	.500		57	29.50		86.50	107
4130	6" pipe		14	.571		81	33.50		114.50	140

For customer support on your Plumbing Cost Data, call 877.709.6968.

269

22 13 19.13 Sanitary Drains	Crew	Daily Output	Labor-Hours	Unit	Material	2015 Bare Costs Labor	Equipment	Total	Total Incl O&P
0010 **SANITARY DRAINS**									
0400 Deck, auto park, C.I., 13" top									
0440 3", 4", 5", and 6" pipe size	Q-1	8	2	Ea.	1,450	106		1,556	1,725
0480 For galvanized body, add				"	780			780	855
0800 Promenade, heelproof grate, C.I., 14" top									
0840 2", 3", and 4" pipe size	Q-1	10	1.600	Ea.	550	84.50		634.50	735
0860 5" and 6" pipe size		9	1.778		685	94		779	895
0880 8" pipe size		8	2		810	106		916	1,050
0940 For galvanized body, add					400			400	440
0960 With polished bronze top, 2"-3"-4" diam.					930			930	1,025
1200 Promenade, heelproof grate, C.I., lateral, 14" top									
1240 2", 3" and 4" pipe size	Q-1	10	1.600	Ea.	720	84.50		804.50	920
1260 5" and 6" pipe size		9	1.778		855	94		949	1,075
1280 8" pipe size		8	2		980	106		1,086	1,225
1340 For galvanized body, add					410			410	450
1360 For polished bronze top, add					375			375	415
1500 Promenade, slotted grate, C.I., 11" top									
1540 2", 3", 4", 5", and 6" pipe size	Q-1	12	1.333	Ea.	495	70.50		565.50	650
1600 For galvanized body, add					230			230	253
1640 With polished bronze top					790			790	870
2000 Floor, medium duty, C.I., deep flange, 7" diam. top									
2040 2" and 3" pipe size	Q-1	12	1.333	Ea.	208	70.50		278.50	335
2080 For galvanized body, add					96.50			96.50	106
2120 With polished bronze top					315			315	345
2160 Heavy duty, C.I., 12" dia anti-tilt grate									
2180 2", 3", 4", 5" and 6" pipe size	Q-1	10	1.600	Ea.	585	84.50		669.50	770
2220 For galvanized body, add					296			296	325
2240 With polished bronze top					875			875	965
2300 X-Heavy duty, C.I., 15" antitilt grate									
2320 4", 5", 6", and 8" pipe size	Q-1	8	2	Ea.	1,250	106		1,356	1,525
2360 For galvanized body, add					480			480	530
2380 With polished bronze top					1,750			1,750	1,925
2400 Heavy duty, with sediment bucket, C.I., 12" diam. loose grate									
2420 2", 3", 4", 5", and 6" pipe size	Q-1	9	1.778	Ea.	690	94		784	900
2440 For galvanized body, add					440			440	485
2460 With polished bronze top					975			975	1,075
2500 Heavy duty, cleanout & trap w/bucket, C.I., 15" top									
2540 2", 3", and 4" pipe size	Q-1	6	2.667	Ea.	6,575	141		6,716	7,450
2560 For galvanized body, add					1,675			1,675	1,850
2580 With polished bronze top					7,300			7,300	8,025
2600 Medium duty, with perforated SS basket, C.I., body,									
2610 18" top for refuse container washing area									
2620 2" thru 6" pipe size	Q-1	4	4	Ea.	4,475	211		4,686	5,250
2630 Acid resistant									
2638 PVC									
2640 2", 3" and 4" pipe size	Q-1	16	1	Ea.	320	53		373	430
2644 Cast iron, epoxy coated									
2646 2", 3" and 4" pipe size	Q-1	14	1.143	Ea.	730	60.50		790.50	895
2650 PVC or ABS thermoplastic									
2660 3" and 4" pipe size	Q-1	16	1	Ea.	350	53		403	465
2680 Extra heavy duty, oil intercepting, gas seal cone,									
2690 with cleanout, loose grate, C.I., body 16" top									

22 13 Facility Sanitary Sewerage

22 13 19 - Sanitary Waste Piping Specialties

22 13 19.13 Sanitary Drains

		Crew	Daily Output	Labor-Hours	Unit	Material	2015 Bare Costs Labor	2015 Bare Costs Equipment	Total	Total Incl O&P
2700	3" and 4" diameter outlet, 4" slab depth	Q-1	4	4	Ea.	7,100	211		7,311	8,150
2720	4" diameter outlet, 8" slab depth		3	5.333		7,100	282		7,382	8,250
2740	4" diam. outlet, 10"-12" slab depth, 16"top		2	8		8,000	425		8,425	9,450
2910	Prison cell, vandal-proof, 1-1/2", and 2" diam. pipe		12	1.333		445	70.50		515.50	595
2920	3" pipe size		10	1.600		500	84.50		584.50	680
2930	Trap drain, light duty, backwater valve C.I. top									
2950	8" diameter top, 2" pipe size	Q-1	12	1.333	Ea.	390	70.50		460.50	530
2960	10" diameter top, 3" pipe size		10	1.600		540	84.50		624.50	720
2970	12" diameter top, 4" pipe size		8	2		745	106		851	980

22 13 19.14 Floor Receptors

		Crew	Daily Output	Labor-Hours	Unit	Material	2015 Bare Costs Labor	2015 Bare Costs Equipment	Total	Total Incl O&P
0010	**FLOOR RECEPTORS,** For connection to 2", 3" & 4" diameter pipe									
0200	12-1/2" square top, 25 sq. in. open area	Q-1	10	1.600	Ea.	970	84.50		1,054.50	1,200
0300	For grate with 4" diam. x 3-3/4" high funnel, add					174			174	191
0400	For grate with 6" diameter x 6" high funnel, add					223			223	246
0500	For full hinged grate with open center, add					81			81	89.50
0600	For aluminum bucket, add					146			146	160
0700	For acid-resisting bucket, add					274			274	300
0900	For stainless steel mesh bucket liner, add					216			216	238
1000	For bronze antisplash dome strainer, add					104			104	114
1100	For partial solid cover, add					52.50			52.50	58
1200	For trap primer connection, add					88			88	97
2000	12-5/8" diameter top, 40 sq. in. open area	Q-1	10	1.600		770	84.50		854.50	975
2100	For options, add same prices as square top									
3000	8" x 4" rectangular top, 7.5 sq. in. open area	Q-1	14	1.143	Ea.	745	60.50		805.50	910
3100	For trap primer connections, add					88			88	97
4000	24" x 16" rectangular top, 70 sq. in. open area	Q-1	4	4		4,350	211		4,561	5,125
4100	For trap primer connection, add					174			174	191

22 13 19.15 Sink Waste Treatment

		Crew	Daily Output	Labor-Hours	Unit	Material	2015 Bare Costs Labor	2015 Bare Costs Equipment	Total	Total Incl O&P
0010	**SINK WASTE TREATMENT,** System for commercial kitchens									
0100	includes clock timer, & fittings									
0200	System less chemical, wall mounted cabinet	1 Plum	16	.500	Ea.	550	29.50		579.50	650
2000	Chemical, 1 gallon, add					68.50			68.50	75.50
2100	6 gallons, add					310			310	340
2200	15 gallons, add					840			840	925
2300	30 gallons, add					1,575			1,575	1,725
2400	55 gallons, add					2,675			2,675	2,950

22 13 19.39 Floor Drain Trap Seal

		Crew	Daily Output	Labor-Hours	Unit	Material	2015 Bare Costs Labor	2015 Bare Costs Equipment	Total	Total Incl O&P
0010	**FLOOR DRAIN TRAP SEAL**									
0100	Inline									
0110	2"	1 Plum	20	.400	Ea.	45	23.50		68.50	85
0120	3"		16	.500		49	29.50		78.50	98.50
0130	3.5"		14	.571		52	33.50		85.50	108
0140	4"		10	.800		56	47		103	133

22 13 23 - Sanitary Waste Interceptors

22 13 23.10 Interceptors

		Crew	Daily Output	Labor-Hours	Unit	Material	2015 Bare Costs Labor	2015 Bare Costs Equipment	Total	Total Incl O&P
0010	**INTERCEPTORS**									
0150	Grease, fabricated steel, 4 GPM, 8 lb. fat capacity	1 Plum	4	2	Ea.	1,150	117		1,267	1,450
0200	7 GPM, 14 lb. fat capacity		4	2		1,600	117		1,717	1,950
1000	10 GPM, 20 lb. fat capacity		4	2		1,875	117		1,992	2,250
1040	15 GPM, 30 lb. fat capacity		4	2		2,800	117		2,917	3,250
1060	20 GPM, 40 lb. fat capacity		3	2.667		3,425	157		3,582	3,975

For customer support on your Plumbing Cost Data, call 877.709.6968.

271

22 13 23.10 Interceptors		Crew	Daily Output	Labor-Hours	Unit	Material	2015 Bare Costs Labor	Equipment	Total	Total Incl O&P
1080	25 GPM, 50 lb. fat capacity	Q-1	3.50	4.571	Ea.	3,825	242		4,067	4,600
1100	35 GPM, 70 lb. fat capacity		3	5.333		4,750	282		5,032	5,650
1120	50 GPM, 100 lb. fat capacity		2	8		6,300	425		6,725	7,575
1140	75 GPM, 150 lb. fat capacity		2	8		14,300	425		14,725	16,300
1160	100 GPM, 200 lb. fat capacity		2	8		15,100	425		15,525	17,200
1180	150 GPM, 300 lb. fat capacity		2	8		15,800	425		16,225	18,000
1200	200 GPM, 400 lb. fat capacity		1.50	10.667		23,000	565		23,565	26,200
1220	250 GPM, 500 lb. fat capacity		1.30	12.308		26,700	650		27,350	30,400
1240	300 GPM, 600 lb. fat capacity		1	16		31,500	845		32,345	36,000
1260	400 GPM, 800 lb. fat capacity	Q-2	1.20	20		38,600	1,100		39,700	44,100
1280	500 GPM, 1000 lb. fat capacity	"	1	24		45,700	1,325		47,025	52,000
1580	For seepage pan, add					7%				
3000	Hair, cast iron, 1-1/4" and 1-1/2" pipe connection	1 Plum	8	1	Ea.	435	58.50		493.50	565
3100	For chrome-plated cast iron, add					266			266	293
3200	For polished bronze, add					266			266	293
3400	Lint interceptor, fabricated steel									
3410	Size based on 10 GPM per machine									
3420	30 GPM, 2" pipe size	Q-1	3	5.333	Ea.	5,075	282		5,357	6,025
3430	70 GPM, 3" pipe size		2.50	6.400		5,925	340		6,265	7,000
3440	100 GPM, 4" pipe size		2	8		6,975	425		7,400	8,300
3450	200 GPM, 4" pipe size		1.50	10.667		8,425	565		8,990	10,100
3460	300 GPM, 6" pipe size		1	16		9,775	845		10,620	12,100
3470	400 GPM, 6" pipe size	Q-2	1.20	20		11,200	1,100		12,300	14,000
3480	500 GPM, 6" pipe size	"	1	24		12,500	1,325		13,825	15,800
4000	Oil, fabricated steel, 10 GPM, 2" pipe size	1 Plum	4	2		2,575	117		2,692	3,000
4100	15 GPM, 2" or 3" pipe size		4	2		3,525	117		3,642	4,075
4120	20 GPM, 2" or 3" pipe size		3	2.667		4,625	157		4,782	5,300
4140	25 GPM, 2" or 3" pipe size	Q-1	3.50	4.571		4,650	242		4,892	5,475
4160	35 GPM, 2", 3", or 4" pipe size		3	5.333		5,650	282		5,932	6,625
4180	50 GPM, 2", 3", or 4" pipe size		2	8		7,600	425		8,025	9,000
4200	75 GPM, 3" pipe size		2	8		13,200	425		13,625	15,100
4220	100 GPM, 3" pipe size		2	8		14,300	425		14,725	16,300
4240	150 GPM, 4" pipe size		2	8		17,700	425		18,125	20,000
4260	200 GPM, 4" pipe size		1.50	10.667		24,900	565		25,465	28,300
4280	250 GPM, 5" pipe size		1.30	12.308		28,900	650		29,550	32,800
4300	300 GPM, 5" pipe size		1	16		32,700	845		33,545	37,300
4320	400 GPM, 6" pipe size	Q-2	1.20	20		42,100	1,100		43,200	48,000
4340	500 GPM, 6" pipe size	"	1	24		53,000	1,325		54,325	60,500
5000	Sand interceptor, fabricated steel									
5020	20 GPM, 4" pipe size	Q-1	3	5.333	Ea.	6,350	282		6,632	7,425
5030	50 GPM, 4" pipe size		2.50	6.400		10,800	340		11,140	12,400
5040	150 GPM, 4" pipe size		2	8		32,600	425		33,025	36,500
5050	250 GPM, 6" pipe size		1.30	12.308		33,900	650		34,550	38,300
5060	500 GPM, 6" pipe size	Q-2	1	24		43,500	1,325		44,825	49,800
6000	Solids, precious metals recovery, C.I., 1-1/4" to 2" pipe	1 Plum	4	2		655	117		772	895
6100	Dental Lab., large, C.I., 1-1/2" to 2" pipe	"	3	2.667		2,275	157		2,432	2,750

22 13 26 – Sanitary Waste Separators

22 13 26.10 Separators

		Crew	Daily Output	Labor-Hours	Unit	Material	2015 Bare Costs Labor	Equipment	Total	Total Incl O&P
0010	**SEPARATORS**, Entrainment eliminator, steel body, 150 PSIG									
0100	1/4" size	1 Stpi	24	.333	Ea.	259	19.90		278.90	315
0120	1/2" size		24	.333		268	19.90		287.90	325
0140	3/4" size		20	.400		280	24		304	345

For customer support on your Plumbing Cost Data, call 877.709.6968.

22 13 26 – Sanitary Waste Separators

22 13 26.10 Separators	Crew	Daily Output	Labor-Hours	Unit	Material	2015 Bare Costs Labor	Equipment	Total	Total Incl O&P
0160	1" size	1 Stpi	19	.421	Ea.	289	25		314	360
0180	1-1/4" size		15	.533		305	32		337	390
0200	1-1/2" size		13	.615		340	37		377	425
0220	2" size		11	.727		375	43.50		418.50	480

22 13 29 – Sanitary Sewerage Pumps

22 13 29.13 Wet-Pit-Mounted, Vertical Sewerage Pumps

		Crew	Daily Output	Labor-Hours	Unit	Material	Labor	Equipment	Total	Total Incl O&P
0010	**WET-PIT-MOUNTED, VERTICAL SEWERAGE PUMPS**									
0020	Controls incl. alarm/disconnect panel w/wire. Excavation not included									
0260	Simplex, 9 GPM at 60 PSIG, 91 gal. tank				Ea.	3,325			3,325	3,675
0300	Unit with manway, 26" I.D., 18" high					3,700			3,700	4,075
0340	26" I.D., 36" high					3,750			3,750	4,125
0380	43" I.D., 4' high					3,975			3,975	4,375
0600	Simplex, 9 GPM at 60 PSIG, 150 gal. tank, indoor					3,525			3,525	3,900
0700	Unit with manway, 26" I.D., 36" high					4,225			4,225	4,650
0740	26" I.D., 4' high					4,450			4,450	4,900
2000	Duplex, 18 GPM at 60 PSIG, 150 gal. tank, indoor					7,050			7,050	7,750
2060	Unit with manway, 43" I.D., 4' high					8,100			8,100	8,925
2400	For core only					1,875			1,875	2,050
3000	Indoor residential type installation									
3020	Simplex, 9 GPM at 60 PSIG, 91 gal. HDPE tank				Ea.	3,350			3,350	3,675

22 13 29.14 Sewage Ejector Pumps

		Crew	Daily Output	Labor-Hours	Unit	Material	Labor	Equipment	Total	Total Incl O&P
0010	**SEWAGE EJECTOR PUMPS**, With operating and level controls									
0100	Simplex system incl. tank, cover, pump 15' head									
0500	37 gal. PE tank, 12 GPM, 1/2 HP, 2" discharge	Q-1	3.20	5	Ea.	480	264		744	925
0510	3" discharge		3.10	5.161		520	273		793	980
0530	87 GPM, .7 HP, 2" discharge		3.20	5		735	264		999	1,200
0540	3" discharge		3.10	5.161		795	273		1,068	1,275
0600	45 gal. coated stl. tank, 12 GPM, 1/2 HP, 2" discharge		3	5.333		855	282		1,137	1,375
0610	3" discharge		2.90	5.517		890	291		1,181	1,425
0630	87 GPM, .7 HP, 2" discharge		3	5.333		1,100	282		1,382	1,625
0640	3" discharge		2.90	5.517		1,150	291		1,441	1,725
0660	134 GPM, 1 HP, 2" discharge		2.80	5.714		1,175	300		1,475	1,750
0680	3" discharge		2.70	5.926		1,250	315		1,565	1,850
0700	70 gal. PE tank, 12 GPM, 1/2 HP, 2" discharge		2.60	6.154		920	325		1,245	1,525
0710	3" discharge		2.40	6.667		980	350		1,330	1,600
0730	87 GPM, 0.7 HP, 2" discharge		2.50	6.400		1,200	340		1,540	1,800
0740	3" discharge		2.30	6.957		1,275	370		1,645	1,950
0760	134 GPM, 1 HP, 2" discharge		2.20	7.273		1,300	385		1,685	2,000
0770	3" discharge		2	8		1,375	425		1,800	2,175
0800	75 gal. coated stl. tank, 12 GPM, 1/2 HP, 2" discharge		2.40	6.667		1,025	350		1,375	1,650
0810	3" discharge		2.20	7.273		1,075	385		1,460	1,750
0830	87 GPM, .7 HP, 2" discharge		2.30	6.957		1,300	370		1,670	1,975
0840	3" discharge		2.10	7.619		1,350	405		1,755	2,100
0860	134 GPM, 1 HP, 2" discharge		2	8		1,375	425		1,800	2,175
0880	3" discharge		1.80	8.889		1,450	470		1,920	2,300
1040	Duplex system incl. tank, covers, pumps									
1060	110 gal. fiberglass tank, 24 GPM, 1/2 HP, 2" discharge	Q-1	1.60	10	Ea.	1,850	530		2,380	2,825
1080	3" discharge		1.40	11.429		1,950	605		2,555	3,025
1100	174 GPM, .7 HP, 2" discharge		1.50	10.667		2,400	565		2,965	3,475
1120	3" discharge		1.30	12.308		2,475	650		3,125	3,700
1140	268 GPM, 1 HP, 2" discharge		1.20	13.333		2,600	705		3,305	3,925
1160	3" discharge		1	16		2,700	845		3,545	4,225

For customer support on your Plumbing Cost Data, call 877.709.6968.

273

22 13 Facility Sanitary Sewerage

22 13 29 – Sanitary Sewerage Pumps

22 13 29.14 Sewage Ejector Pumps

	22 13 29.14 Sewage Ejector Pumps	Crew	Daily Output	Labor-Hours	Unit	Material	2015 Bare Costs Labor	Equipment	Total	Total Incl O&P
1260	135 gal. coated stl. tank, 24 GPM, 1/2 HP, 2" discharge	Q-2	1.70	14.118	Ea.	1,900	775		2,675	3,275
2000	3" discharge		1.60	15		2,025	820		2,845	3,475
2640	174 GPM, .7 HP, 2" discharge		1.60	15		2,500	820		3,320	4,000
2660	3" discharge		1.50	16		2,625	875		3,500	4,225
2700	268 GPM, 1 HP, 2" discharge		1.30	18.462		2,700	1,000		3,700	4,500
3040	3" discharge		1.10	21.818		2,875	1,200		4,075	4,975
3060	275 gal. coated stl. tank, 24 GPM, 1/2 HP, 2" discharge		1.50	16		2,375	875		3,250	3,950
3080	3" discharge		1.40	17.143		2,425	940		3,365	4,075
3100	174 GPM, .7 HP, 2" discharge		1.40	17.143		3,075	940		4,015	4,800
3120	3" discharge		1.30	18.462		3,250	1,000		4,250	5,100
3140	268 GPM, 1 HP, 2" discharge		1.10	21.818		3,375	1,200		4,575	5,500
3160	3" discharge	↓	.90	26.667	↓	3,550	1,450		5,000	6,100
3260	Pump system accessories, add									
3300	Alarm horn and lights, 115 V mercury switch	Q-1	8	2	Ea.	97	106		203	266
3340	Switch, mag. contactor, alarm bell, light, 3 level control		5	3.200		475	169		644	780
3380	Alternator, mercury switch activated	↓	4	4	↓	855	211		1,066	1,250

22 14 Facility Storm Drainage

22 14 23 – Storm Drainage Piping Specialties

22 14 23.33 Backwater Valves

		Crew	Daily Output	Labor-Hours	Unit	Material	2015 Bare Costs Labor	Equipment	Total	Total Incl O&P
0010	**BACKWATER VALVES**, C.I. Body									
6980	Bronze gate and automatic flapper valves									
7000	3" and 4" pipe size	Q-1	13	1.231	Ea.	2,050	65		2,115	2,350
7100	5" and 6" pipe size	"	13	1.231	"	3,125	65		3,190	3,550
7240	Bronze flapper valve, bolted cover									
7260	2" pipe size	Q-1	16	1	Ea.	595	53		648	735
7280	3" pipe size		14.50	1.103		935	58.50		993.50	1,125
7300	4" pipe size	↓	13	1.231		1,150	65		1,215	1,375
7320	5" pipe size	Q-2	18	1.333		1,650	73		1,723	1,925
7340	6" pipe size	"	17	1.412		1,650	77.50		1,727.50	1,950
7360	8" pipe size	Q-3	10	3.200		2,250	179		2,429	2,750
7380	10" pipe size	"	9	3.556	↓	3,700	199		3,899	4,350
7500	For threaded cover, same cost									
7540	Revolving disk type, same cost as flapper type									

22 14 26 – Facility Storm Drains

22 14 26.13 Roof Drains

		Crew	Daily Output	Labor-Hours	Unit	Material	2015 Bare Costs Labor	Equipment	Total	Total Incl O&P
0010	**ROOF DRAINS**									
0140	Cornice, C.I., 45° or 90° outlet									
0200	3" and 4" pipe size	Q-1	12	1.333	Ea.	330	70.50		400.50	470
0260	For galvanized body, add					75.50			75.50	83
0280	For polished bronze dome, add				↓	85.50			85.50	94
3860	Roof, flat metal deck, C.I. body, 12" C.I. dome									
3880	2" pipe size	Q-1	15	1.067	Ea.	287	56.50		343.50	400
3890	3" pipe size		14	1.143		395	60.50		455.50	525
3900	4" pipe size		13	1.231		395	65		460	535
3910	5" pipe size		12	1.333		535	70.50		605.50	695
3920	6" pipe size		10	1.600		680	84.50		764.50	880
4280	Integral expansion joint, C.I. body, 12" C.I. dome									
4300	2" pipe size	Q-1	8	2	Ea.	590	106		696	810
4320	3" pipe size	↓	7	2.286	↓	610	121		731	850

274

22 14 26 – Facility Storm Drains

22 14 26.13 Roof Drains	Crew	Daily Output	Labor-Hours	Unit	Material	2015 Bare Costs Labor	Equipment	Total	Total Incl O&P
4340	4" pipe size	Q-1	6	2.667	Ea.	655	141		796	935
4360	5" pipe size		4	4		820	211		1,031	1,225
4380	6" pipe size		3	5.333		875	282		1,157	1,400
4400	8" pipe size		3	5.333		1,375	282		1,657	1,925
4440	For galvanized body, add					345			345	380
4620	Main, all aluminum, 12" low profile dome									
4640	2", 3" and 4" pipe size	Q-1	14	1.143	Ea.	435	60.50		495.50	570
4660	5" and 6" pipe size		13	1.231		575	65		640	735
4680	8" pipe size		10	1.600		715	84.50		799.50	920
4690	Main, CI body, 12" poly. dome, 2", 3", & 4" pipe		8	2		325	106		431	515
4710	5" and 6" pipe size		6	2.667		465	141		606	725
4720	8" pipe size		4	4		605	211		816	985
4730	For underdeck clamp, add		22	.727		224	38.50		262.50	305
4740	For vandalproof dome, add					56.50			56.50	62.50
4750	For galvanized body, add					440			440	485
4760	Main, ABS body and dome, 2" pipe size	Q-1	14	1.143		131	60.50		191.50	235
4780	3" pipe size		14	1.143		131	60.50		191.50	235
4800	4" pipe size		14	1.143		131	60.50		191.50	235
4820	For underdeck clamp, add		24	.667		27	35		62	83
4900	Terrace planting area, with perforated overflow, C.I.									
4920	2", 3" and 4" pipe size	Q-1	8	2	Ea.	590	106		696	810

22 14 26.16 Facility Area Drains

		Crew	Daily Output	Labor-Hours	Unit	Material	Labor	Equipment	Total	Total Incl O&P
0010	**FACILITY AREA DRAINS**									
4980	Scupper floor, oblique strainer, C.I.									
5000	6" x 7" top, 2", 3" and 4" pipe size	Q-1	16	1	Ea.	283	53		336	390
5100	8" x 12" top, 5" and 6" pipe size	"	14	1.143		550	60.50		610.50	695
5160	For galvanized body, add					40%				
5200	For polished bronze strainer, add					85%				

22 14 26.19 Facility Trench Drains

		Crew	Daily Output	Labor-Hours	Unit	Material	Labor	Equipment	Total	Total Incl O&P
0010	**FACILITY TRENCH DRAINS**									
5980	Trench, floor, heavy duty, modular, C.I., 12" x 12" top									
6000	2", 3", 4", 5", & 6" pipe size	Q-1	8	2	Ea.	895	106		1,001	1,150
6100	For unit with polished bronze top		8	2		1,325	106		1,431	1,625
6200	For 12" extension section, C.I. top		8	2		895	106		1,001	1,150
6240	For 12" extension section, polished bronze top		8	2		1,400	106		1,506	1,700
6600	Trench, floor, for cement concrete encasement									
6610	Not including trenching or concrete									
6640	Polyester polymer concrete									
6650	4" internal width, with grate									
6660	Light duty steel grate	Q-1	120	.133	L.F.	35	7.05		42.05	49
6670	Medium duty steel grate		115	.139		40.50	7.35		47.85	55.50
6680	Heavy duty iron grate		110	.145		62	7.70		69.70	79.50
6700	12" internal width, with grate									
6770	Heavy duty galvanized grate	Q-1	80	.200	L.F.	169	10.55		179.55	201
6800	Fiberglass									
6810	8" internal width, with grate									
6820	Medium duty galvanized grate	Q-1	115	.139	L.F.	111	7.35		118.35	133
6830	Heavy duty iron grate	"	110	.145	"	106	7.70		113.70	128

For customer support on your Plumbing Cost Data, call 877.709.6968.

22 14 29.13 Wet-Pit-Mounted, Vertical Sump Pumps	Crew	Daily Output	Labor-Hours	Unit	Material	2015 Bare Costs Labor	Equipment	Total	Total Incl O&P
0010 **WET-PIT-MOUNTED, VERTICAL SUMP PUMPS**									
0400 Molded PVC base, 21 GPM at 15' head, 1/3 HP	1 Plum	5	1.600	Ea.	135	94		229	291
0800 Iron base, 21 GPM at 15' head, 1/3 HP		5	1.600		164	94		258	320
1200 Solid brass, 21 GPM at 15' head, 1/3 HP	↓	5	1.600	↓	289	94		383	460
2000 Sump pump, single stage									
2010 25 GPM, 1 HP, 1-1/2" discharge	Q-1	1.80	8.889	Ea.	3,825	470		4,295	4,925
2020 75 GPM, 1-1/2 HP, 2" discharge		1.50	10.667		4,050	565		4,615	5,300
2030 100 GPM, 2 HP, 2-1/2" discharge		1.30	12.308		4,125	650		4,775	5,525
2040 150 GPM, 3 HP, 3" discharge		1.10	14.545		4,125	770		4,895	5,700
2050 200 GPM, 3 HP, 3" discharge	↓	1	16		4,375	845		5,220	6,100
2060 300 GPM, 10 HP, 4" discharge	Q-2	1.20	20		4,725	1,100		5,825	6,850
2070 500 GPM, 15 HP, 5" discharge		1.10	21.818		5,375	1,200		6,575	7,700
2080 800 GPM, 20 HP, 6" discharge		1	24		6,350	1,325		7,675	8,950
2090 1000 GPM, 30 HP, 6" discharge		.85	28.235		6,975	1,550		8,525	10,000
2100 1600 GPM, 50 HP, 8" discharge	↓	.72	33.333		10,900	1,825		12,725	14,800
2110 2000 GPM, 60 HP, 8" discharge	Q-3	.85	37.647		11,100	2,100		13,200	15,400
2202 For general purpose float switch, copper coated float, add	Q-1	5	3.200	↓	108	169		277	375

22 14 29.16 Submersible Sump Pumps

	Crew	Daily Output	Labor-Hours	Unit	Material	Labor	Equipment	Total	Total Incl O&P
0010 **SUBMERSIBLE SUMP PUMPS**									
1000 Elevator sump pumps, automatic									
1010 Complete systems, pump, oil detector, controls and alarm									
1020 1-1/2"discharge, does not include the sump pit/tank.									
1040 1/3 HP, 115 V	1 Plum	4.40	1.818	Ea.	1,875	107		1,982	2,200
1050 1/2 HP, 115 V		4	2		1,925	117		2,042	2,300
1060 1/2 HP, 230 V		4	2		1,975	117		2,092	2,350
1070 3/4 HP, 115 V		3.60	2.222		2,025	130		2,155	2,425
1080 3/4 HP, 230 V	↓	3.60	2.222	↓	2,075	130		2,205	2,475
1100 Sump pump only									
1110 1/3 HP, 115 V	1 Plum	6.40	1.250	Ea.	170	73.50		243.50	298
1120 1/2 HP, 115 V		5.80	1.379		243	81		324	390
1130 1/2 HP, 230 V		5.80	1.379		287	81		368	435
1140 3/4 HP, 115 V		5.40	1.481		325	87		412	485
1150 3/4 HP, 230 V	↓	5.40	1.481	↓	370	87		457	540
1200 Oil detector, control and alarm only									
1210 115 V	1 Plum	8	1	Ea.	1,700	58.50		1,758.50	1,975
1220 230 V	"	8	1	"	1,700	58.50		1,758.50	1,975
7000 Sump pump, automatic									
7100 Plastic, 1-1/4" discharge, 1/4 HP	1 Plum	6.40	1.250	Ea.	138	73.50		211.50	263
7140 1/3 HP		6	1.333		200	78.50		278.50	340
7160 1/2 HP		5.40	1.481		246	87		333	400
7180 1-1/2" discharge, 1/2 HP		5.20	1.538		281	90.50		371.50	445
7500 Cast iron, 1-1/4" discharge, 1/4 HP		6	1.333		194	78.50		272.50	330
7540 1/3 HP		6	1.333		229	78.50		307.50	370
7560 1/2 HP	↓	5	1.600	↓	277	94		371	445

22 14 53 – Rainwater Storage Tanks

22 14 53.13 Fiberglass, Rainwater Storage Tank

	Crew	Daily Output	Labor-Hours	Unit	Material	Labor	Equipment	Total	Total Incl O&P
0010 **FIBERGLASS, RAINWATER STORAGE TANK**									
2000 600 gallon	B-21B	3.75	10.667	Ea.	3,300	435	174	3,909	4,475
2010 1,000 gallon		3.50	11.429		4,250	465	187	4,902	5,600
2020 2,000 gallon		3.25	12.308		6,200	500	201	6,901	7,800
2030 4,000 gallon	↓	3	13.333		8,575	545	218	9,338	10,500

For customer support on your Plumbing Cost Data, call 877.709.6968.

22 14 Facility Storm Drainage

22 14 53 – Rainwater Storage Tanks

22 14 53.13 Fiberglass, Rainwater Storage Tank		Crew	Daily Output	Labor-Hours	Unit	Material	2015 Bare Costs Labor	Equipment	Total	Total Incl O&P
2040	6,000 gallon	B-21B	2.65	15.094	Ea.	9,625	615	247	10,487	11,800
2050	8,000 gallon		2.30	17.391		11,500	710	284	12,494	14,100
2060	10,000 gallon		2	20		13,200	815	325	14,340	16,200
2070	12,000 gallon		1.50	26.667		18,300	1,100	435	19,835	22,300
2080	15,000 gallon		1	40		21,200	1,625	655	23,480	26,500
2090	20,000 gallon		.75	53.333		27,300	2,175	870	30,345	34,300
2100	25,000 gallon		.50	80		40,800	3,275	1,300	45,375	51,500
2110	30,000 gallon		.35	114		49,400	4,675	1,875	55,950	63,500
2120	40,000 gallon		.30	133		71,500	5,450	2,175	79,125	89,500

22 15 General Service Compressed-Air Systems

22 15 13 – General Service Compressed-Air Piping

22 15 13.10 Compressor Accessories

		Crew	Daily Output	Labor-Hours	Unit	Material	2015 Bare Costs Labor	Equipment	Total	Total Incl O&P
0010	**COMPRESSOR ACCESSORIES**									
3460	Air filter, regulator, lubricator combination									
3470	Flush mount									
3480	Adjustable range 0-140 psi									
3500	1/8" NPT, 34 SCFM	1 Stpi	17	.471	Ea.	108	28		136	162
3510	1/4" NPT, 61 SCFM		17	.471		179	28		207	240
3520	3/8" NPT, 85 SCFM		16	.500		184	30		214	248
3530	1/2" NPT, 150 SCFM		15	.533		179	32		211	245
3540	3/4" NPT, 171 SCFM		14	.571		237	34		271	315
3550	1" NPT, 150 SCFM		13	.615		370	37		407	460
4000	Couplers, air line, sleeve type									
4010	Female, connection size NPT									
4020	1/4"	1 Stpi	38	.211	Ea.	9.95	12.60		22.55	30
4030	3/8"		36	.222		8.55	13.30		21.85	29.50
4040	1/2"		35	.229		21.50	13.65		35.15	44.50
4050	3/4"		34	.235		21.50	14.05		35.55	45
4100	Male									
4110	1/4"	1 Stpi	38	.211	Ea.	9	12.60		21.60	29
4120	3/8"		36	.222		10.60	13.30		23.90	31.50
4130	1/2"		35	.229		21	13.65		34.65	44
4140	3/4"		34	.235		21.50	14.05		35.55	45
4150	Coupler, combined male and female halves									
4160	1/2"	1 Stpi	17	.471	Ea.	43	28		71	89.50
4170	3/4"	"	15	.533	"	43.50	32		75.50	95.50

22 15 19 – General Service Packaged Air Compressors and Receivers

22 15 19.10 Air Compressors

		Crew	Daily Output	Labor-Hours	Unit	Material	2015 Bare Costs Labor	Equipment	Total	Total Incl O&P
0010	**AIR COMPRESSORS**									
5250	Air, reciprocating air cooled, splash lubricated, tank mounted									
5300	Single stage, 1 phase, 140 psi									
5303	1/2 HP, 30 gal. tank	1 Stpi	3	2.667	Ea.	1,950	159		2,109	2,400
5305	3/4 HP, 30 gal. tank		2.60	3.077		1,950	184		2,134	2,425
5307	1 HP, 30 gal. tank		2.20	3.636		2,450	217		2,667	3,025
5309	2 HP, 30 gal. tank	Q-5	4	4		2,625	215		2,840	3,225
5310	3 HP, 30 gal. tank		3.60	4.444		3,350	239		3,589	4,050
5314	3 HP, 60 gal. tank		3.50	4.571		3,550	246		3,796	4,275
5320	5 HP, 60 gal. tank		3.20	5		3,725	269		3,994	4,500
5330	5 HP, 80 gal. tank		3	5.333		4,100	287		4,387	4,925

For customer support on your Plumbing Cost Data, call 877.709.6968.

277

22 15 19 – General Service Packaged Air Compressors and Receivers

22 15 19.10 Air Compressors		Crew	Daily Output	Labor-Hours	Unit	Material	2015 Bare Costs Labor	Equipment	Total	Total Incl O&P
5340	7.5 HP, 80 gal. tank	Q-5	2.60	6.154	Ea.	5,300	330		5,630	6,325
5600	2 stage pkg., 3 phase									
5650	6 CFM at 125 psi 1-1/2 HP, 60 gal. tank	Q-5	3	5.333	Ea.	3,375	287		3,662	4,125
5670	10.9 CFM at 125 psi, 3 HP, 80 gal. tank		1.50	10.667		3,800	575		4,375	5,050
5680	38.7 CFM at 125 psi, 10 HP, 120 gal. tank		.60	26.667		6,825	1,425		8,250	9,675
5690	105 CFM at 125 psi, 25 HP, 250 gal. tank	Q-6	.60	40		14,100	2,225		16,325	18,900
5800	With single stage pump									
5850	8.3 CFM at 125 psi, 2 HP, 80 gal. tank	Q-6	3.50	6.857	Ea.	3,775	380		4,155	4,725
5860	38.7 CFM at 125 psi, 10 HP, 120 gal. tank	"	.90	26.667	"	6,625	1,475		8,100	9,550
6000	Reciprocating, 2 stage, tank mtd, 3 Ph., Cap. rated @175 PSIG									
6050	Pressure lubricated, hvy. duty, 9.7 CFM, 3 HP, 120 gal. tank	Q-5	1.30	12.308	Ea.	5,775	660		6,435	7,350
6054	5 CFM, 1-1/2 HP, 80 gal. tank		2.80	5.714		3,575	305		3,880	4,425
6056	6.4 CFM, 2 HP, 80 gal. tank		2	8		3,775	430		4,205	4,800
6058	8.1 CFM, 3 HP, 80 gal. tank		1.70	9.412		3,800	505		4,305	4,950
6059	14.8 CFM, 5 HP, 80 gal. tank		1	16		4,100	860		4,960	5,800
6060	16.5 CFM, 5 HP, 120 gal. tank		1	16		5,900	860		6,760	7,775
6063	13 CFM, 6 HP, 80 gal. tank		.90	17.778		5,875	955		6,830	7,925
6066	19.8 CFM, 7.5 HP, 80 gal. tank		.80	20		5,875	1,075		6,950	8,100
6070	25.8 CFM, 7-1/2 HP, 120 gal. tank		.80	20		7,975	1,075		9,050	10,400
6078	34.8 CFM, 10 HP, 80 gal. tank		.70	22.857		8,050	1,225		9,275	10,700
6080	34.8 CFM, 10 HP, 120 gal. tank		.60	26.667		8,500	1,425		9,925	11,600
6090	53.7 CFM, 15 HP, 120 gal. tank	Q-6	.80	30		10,300	1,675		11,975	13,800
6100	76.7 CFM, 20 HP, 120 gal. tank		.70	34.286		13,500	1,900		15,400	17,700
6104	76.7 CFM, 20 HP, 240 gal. tank		.68	35.294		14,800	1,975		16,775	19,300
6110	90.1 CFM, 25 HP, 120 gal. tank		.63	38.095		14,100	2,125		16,225	18,700
6120	101 CFM, 30 HP, 120 gal. tank		.57	42.105		15,200	2,350		17,550	20,300
6130	101 CFM, 30 HP, 250 gal. tank		.52	46.154		16,400	2,575		18,975	22,000
6200	Oil-less, 13.6 CFM, 5 HP, 120 gal. tank	Q-5	.88	18.182		16,900	980		17,880	20,100
6210	13.6 CFM, 5 HP, 250 gal. tank		.80	20		18,200	1,075		19,275	21,600
6220	18.2 CFM, 7.5 HP, 120 gal. tank		.73	21.918		16,900	1,175		18,075	20,400
6230	18.2 CFM, 7.5 HP, 250 gal. tank		.67	23.881		18,200	1,275		19,475	22,000
6250	30.5 CFM, 10 HP, 120 gal. tank		.57	28.070		19,900	1,500		21,400	24,200
6260	30.5 CFM, 10 HP, 250 gal. tank		.53	30.189		21,200	1,625		22,825	25,800
6270	41.3 CFM, 15 HP, 120 gal. tank	Q-6	.70	34.286		21,700	1,900		23,600	26,800
6280	41.3 CFM, 15 HP, 250 gal. tank	"	.67	35.821		22,900	2,000		24,900	28,200

22 31 Domestic Water Softeners

22 31 13 – Residential Domestic Water Softeners

22 31 13.10 Residential Water Softeners

0010	RESIDENTIAL WATER SOFTENERS									
7350	Water softener, automatic, to 30 grains per gallon	2 Plum	5	3.200	Ea.	405	188		593	730
7400	To 100 grains per gallon	"	4	4	"	660	235		895	1,075

22 31 16 – Commercial Domestic Water Softeners

22 31 16.10 Water Softeners

0010	WATER SOFTENERS									
5800	Softener systems, automatic, intermediate sizes									
5820	available, may be used in multiples.									
6000	Hardness capacity between regenerations and flow									
6060	40,000 grains, 14 GPM	Q-1	4	4	Ea.	1,850	211		2,061	2,350
6070	50,000 grains, 17 GPM		3.60	4.444		2,875	235		3,110	3,500

22 31 Domestic Water Softeners

22 31 16 – Commercial Domestic Water Softeners

22 31 16.10 Water Softeners		Crew	Daily Output	Labor-Hours	Unit	Material	2015 Bare Costs Labor	Equipment	Total	Total Incl O&P
6080	90,000 grains, 25 GPM	Q-1	2.80	5.714	Ea.	6,075	300		6,375	7,125
6100	150,000 grains, 37 GPM cont., 51 GPM peak		1.20	13.333		6,075	705		6,780	7,750
6200	300,000 grains, 81 GPM cont., 113 GPM peak		1	16		9,850	845		10,695	12,100
6300	750,000 grains, 160 GPM cont., 230 GPM peak		.80	20		12,800	1,050		13,850	15,700
6400	900,000 grains, 185 GPM cont., 270 GPM peak		.70	22.857		20,700	1,200		21,900	24,500
8000	Water treatment, salts, 50 lb. bag									
8020	Salt, water softener, bag, pelletized				Lb.	.29			.29	.32
8030	Salt, water softener, bag, crystal rock salt				"	.25			.25	.28

22 32 Domestic Water Filtration Equipment

22 32 19 – Domestic-Water Off-Floor Cartridge Filters

22 32 19.10 Water Filters

		Crew	Daily Output	Labor-Hours	Unit	Material	2015 Bare Costs Labor	Equipment	Total	Total Incl O&P
0010	**WATER FILTERS**, Purification and treatment.									
1000	Cartridge style, dirt and rust type	1 Plum	12	.667	Ea.	200	39		239	279
1200	Replacement cartridge		32	.250		19.60	14.70		34.30	43.50
1600	Taste and odor type		12	.667		246	39		285	330
1700	Replacement cartridge		32	.250		40.50	14.70		55.20	66.50
3000	Central unit, dirt/rust/odor/taste/scale		4	2		545	117		662	770
3100	Replacement cartridge, standard		20	.400		264	23.50		287.50	325
3600	Replacement cartridge, heavy duty		20	.400		209	23.50		232.50	265
8000	Commercial, fully automatic or push button automatic									
8200	Iron removal, 660 GPH, 1" pipe size	Q-1	1.50	10.667	Ea.	3,075	565		3,640	4,225
8240	1500 GPH, 1-1/4" pipe size		1	16		5,175	845		6,020	6,975
8280	2340 GPH, 1-1/2" pipe size		.80	20		5,675	1,050		6,725	7,850
8320	3420 GPH, 2" pipe size		.60	26.667		10,500	1,400		11,900	13,600
8360	4620 GPH, 2-1/2" pipe size		.50	32		16,600	1,700		18,300	20,900
8500	Neutralizer for acid water, 780 GPH, 1" pipe size		1.50	10.667		2,975	565		3,540	4,100
8540	1140 GPH, 1-1/4" pipe size		1	16		3,350	845		4,195	4,950
8580	1740 GPH, 1-1/2" pipe size		.80	20		4,850	1,050		5,900	6,925
8620	2520 GPH, 2" pipe size		.60	26.667		6,300	1,400		7,700	9,050
8660	3480 GPH, 2-1/2" pipe size		.50	32		10,500	1,700		12,200	14,100
8800	Sediment removal, 780 GPH, 1" pipe size		1.50	10.667		2,825	565		3,390	3,975
8840	1140 GPH, 1-1/4" pipe size		1	16		3,400	845		4,245	5,000
8880	1740 GPH, 1-1/2" pipe size		.80	20		4,500	1,050		5,550	6,550
8920	2520 GPH, 2" pipe size		.60	26.667		6,500	1,400		7,900	9,275
8960	3480 GPH, 2-1/2" pipe size		.50	32		10,200	1,700		11,900	13,800
9200	Taste and odor removal, 660 GPH, 1" pipe size		1.50	10.667		3,950	565		4,515	5,200
9240	1500 GPH, 1-1/4" pipe size		1	16		6,800	845		7,645	8,750
9280	2340 GPH, 1-1/2" pipe size		.80	20		7,725	1,050		8,775	10,100
9320	3420 GPH, 2" pipe size		.60	26.667		11,900	1,400		13,300	15,200
9360	4620 GPH, 2-1/2" pipe size		.50	32		18,300	1,700		20,000	22,800

For customer support on your Plumbing Cost Data, call 877.709.6968.

279

22 33 13 – Instantaneous Electric Domestic Water Heaters

22 33 13.10 Hot Water Dispensers		Crew	Daily Output	Labor-Hours	Unit	Material	2015 Bare Costs Labor	Equipment	Total	Total Incl O&P
0010	**HOT WATER DISPENSERS**									
0160	Commercial, 100 cup, 11.3 amp	1 Plum	14	.571	Ea.	510	33.50		543.50	615
3180	Household, 60 cup	"	14	.571	"	269	33.50		302.50	345

22 33 13.20 Instantaneous Electric Point-Of-Use Water Heaters

			Crew	Daily Output	Labor-Hours	Unit	Material	Labor	Equipment	Total	Total Incl O&P
0010	**INSTANTANEOUS ELECTRIC POINT-OF-USE WATER HEATERS**										
8965	Point of use, electric, glass lined										
8969	Energy saver										
8970	2.5 gal. single element	G	1 Plum	2.80	2.857	Ea.	245	168		413	525
8971	4 gal. single element	G		2.80	2.857		253	168		421	530
8974	6 gal. single element	G		2.50	3.200		261	188		449	570
8975	10 gal. single element	G		2.50	3.200		330	188		518	645
8976	15 gal. single element	G		2.40	3.333		370	196		566	700
8977	20 gal. single element	G		2.40	3.333		410	196		606	745
8978	30 gal. single element	G		2.30	3.478		480	204		684	840
8979	40 gal. single element	G		2.20	3.636		800	213		1,013	1,200
8988	Commercial (ASHRAE energy std. 90)										
8989	6 gallon	G	1 Plum	2.50	3.200	Ea.	665	188		853	1,025
8990	10 gallon	G		2.50	3.200		710	188		898	1,075
8991	15 gallon	G		2.40	3.333		750	196		946	1,125
8992	20 gallon	G		2.40	3.333		785	196		981	1,150
8993	30 gallon	G		2.30	3.478		1,700	204		1,904	2,175
8995	Under the sink, copper, w/bracket										
8996	2.5 gallon	G	1 Plum	4	2	Ea.	485	117		602	710

22 33 30 – Residential, Electric Domestic Water Heaters

22 33 30.13 Residential, Small-Capacity Elec. Water Heaters

			Crew	Daily Output	Labor-Hours	Unit	Material	Labor	Equipment	Total	Total Incl O&P
0010	**RESIDENTIAL, SMALL-CAPACITY ELECTRIC DOMESTIC WATER HEATERS**										
1000	Residential, electric, glass lined tank, 5 yr., 10 gal., single element	D2020-210	1 Plum	2.30	3.478	Ea.	330	204		534	670
1040	20 gallon, single element			2.20	3.636		410	213		623	770
1060	30 gallon, double element			2.20	3.636		475	213		688	845
1080	40 gallon, double element			2	4		800	235		1,035	1,225
1100	52 gallon, double element			2	4		895	235		1,130	1,350
1120	66 gallon, double element			1.80	4.444		1,200	261		1,461	1,725
1140	80 gallon, double element			1.60	5		1,350	294		1,644	1,925
1180	120 gallon, double element			1.40	5.714		1,900	335		2,235	2,575

22 33 33 – Light-Commercial Electric Domestic Water Heaters

22 33 33.10 Commercial Electric Water Heaters

			Crew	Daily Output	Labor-Hours	Unit	Material	Labor	Equipment	Total	Total Incl O&P
0010	**COMMERCIAL ELECTRIC WATER HEATERS**										
4000	Commercial, 100° rise. NOTE: for each size tank, a range of										
4010	heaters between the ones shown are available										
4020	Electric										
4100	5 gal., 3 kW, 12 GPH, 208 volt		1 Plum	2	4	Ea.	2,850	235		3,085	3,500
4120	10 gal., 6 kW, 25 GPH, 208 volt			2	4		3,175	235		3,410	3,825
4130	30 gal., 24 kW, 98 GPH, 208 volt			1.92	4.167		5,200	245		5,445	6,100
4136	40 gal., 36 kW, 148 GPH, 208 volt			1.88	4.255		6,225	250		6,475	7,225
4140	50 gal., 9 kW, 37 GPH, 208 volt			1.80	4.444		4,350	261		4,611	5,175
4160	50 gal., 36 kW, 148 GPH, 208 volt			1.80	4.444		6,650	261		6,911	7,700
4180	80 gal., 12 kW, 49 GPH, 208 volt			1.50	5.333		5,350	315		5,665	6,375
4200	80 gal., 36 kW, 148 GPH, 208 volt			1.50	5.333		7,400	315		7,715	8,625
4220	100 gal., 36 kW, 148 GPH, 208 volt			1.20	6.667		7,725	390		8,115	9,100
4240	120 gal., 36 kW, 148 GPH, 208 volt			1.20	6.667		8,050	390		8,440	9,450
4260	150 gal., 15 kW , 61 GPH, 480 volt			1	8		18,900	470		19,370	21,500

22 33 Electric Domestic Water Heaters

22 33 33 – Light-Commercial Electric Domestic Water Heaters

22 33 33.10 Commercial Electric Water Heaters

		Crew	Daily Output	Labor-Hours	Unit	Material	2015 Bare Costs Labor	Equipment	Total	Total Incl O&P
4280	150 gal., 120 kW, 490 GPH, 480 volt	1 Plum	1	8	Ea.	26,600	470		27,070	29,900
4300	200 gal., 15 kW, 61 GPH, 480 volt	Q-1	1.70	9.412		20,500	495		20,995	23,300
4320	200 gal., 120 kW , 490 GPH, 480 volt		1.70	9.412		28,000	495		28,495	31,600
4340	250 gal., 15 kW, 61 GPH, 480 volt		1.50	10.667		21,100	565		21,665	24,100
4360	250 gal., 150 kW, 615 GPH, 480 volt		1.50	10.667		30,800	565		31,365	34,800
4380	300 gal., 30 kW, 123 GPH, 480 volt		1.30	12.308		23,300	650		23,950	26,600
4400	300 gal., 180 kW, 738 GPH, 480 volt		1.30	12.308		41,700	650		42,350	46,900
4420	350 gal., 30 kW, 123 GPH, 480 volt		1.10	14.545		24,600	770		25,370	28,200
4440	350 gal., 180 kW, 738 GPH, 480 volt		1.10	14.545		34,700	770		35,470	39,300
4460	400 gal., 30 kW, 123 GPH, 480 volt		1	16		27,800	845		28,645	31,900
4480	400 gal., 210 kW, 860 GPH, 480 volt		1	16		40,000	845		40,845	45,300
4500	500 gal., 30 kW, 123 GPH, 480 volt		.80	20		32,500	1,050		33,550	37,300
4520	500 gal., 240 kW, 984 GPH, 480 volt		.80	20		47,800	1,050		48,850	54,000
4540	600 gal., 30 kW, 123 GPH, 480 volt	Q-2	1.20	20		24,700	1,100		25,800	28,900
4560	600 gal., 300 kW, 1230 GPH, 480 volt		1.20	20		37,600	1,100		38,700	43,000
4580	700 gal., 30 kW, 123 GPH, 480 volt		1	24		26,000	1,325		27,325	30,600
4600	700 gal., 300 kW, 1230 GPH, 480 volt		1	24		39,100	1,325		40,425	45,000
4620	800 gal., 60 kW, 245 GPH, 480 volt		.90	26.667		34,800	1,450		36,250	40,500
4640	800 gal., 300 kW, 1230 GPH, 480 volt		.90	26.667		40,000	1,450		41,450	46,200
4660	1000 gal., 60 kW, 245 GPH, 480 volt		.70	34.286		30,800	1,875		32,675	36,700
4680	1000 gal., 480 kW, 1970 GPH, 480 volt		.70	34.286		51,000	1,875		52,875	59,500
4700	1200 gal., 60 kW, 245 GPH, 480 volt		.60	40		52,500	2,200		54,700	61,000
4720	1200 gal., 480 kW, 1970 GPH, 480 volt		.60	40		79,500	2,200		81,700	91,000
4740	1500 gal., 60 kW, 245 GPH, 480 volt		.50	48		69,000	2,625		71,625	80,000
4760	1500 gal., 480 kW, 1970 GPH, 480 volt		.50	48		95,500	2,625		98,125	109,500
5400	Modulating step control for under 90 kW, 2-5 steps	1 Elec	5.30	1.509		810	82.50		892.50	1,025
5440	1 through 5 steps beyond standard		3.20	2.500		221	137		358	450
5460	6 through 10 steps beyond standard		2.70	2.963		455	162		617	745
5480	11 through 18 steps beyond standard		1.60	5		680	274		954	1,150

22 34 Fuel-Fired Domestic Water Heaters

22 34 13 – Instantaneous, Tankless, Gas Domestic Water Heaters

22 34 13.10 Instantaneous, Tankless, Gas Water Heaters

			Crew	Daily Output	Labor-Hours	Unit	Material	Labor	Equipment	Total	Total Incl O&P
0010	**INSTANTANEOUS, TANKLESS, GAS WATER HEATERS**										
9410	Natural gas/propane, 3.2 GPM	G	1 Plum	2	4	Ea.	370	235		605	760
9420	6.4 GPM	G		1.90	4.211		625	247		872	1,050
9430	8.4 GPM	G		1.80	4.444		730	261		991	1,200
9440	9.5 GPM	G		1.60	5		930	294		1,224	1,475

22 34 30 – Residential Gas Domestic Water Heaters

22 34 30.13 Residential, Atmos, Gas Domestic Wtr Heaters

		Crew	Daily Output	Labor-Hours	Unit	Material	Labor	Equipment	Total	Total Incl O&P
0010	**RESIDENTIAL, ATMOSPHERIC, GAS DOMESTIC WATER HEATERS**									
2000	Gas fired, foam lined tank, 10 yr., vent not incl.									
2040	30 gallon	1 Plum	2	4	Ea.	895	235		1,130	1,350
2060	40 gallon		1.90	4.211		895	247		1,142	1,350
2080	50 gallon		1.80	4.444		935	261		1,196	1,425
2090	60 gallon		1.70	4.706		1,325	276		1,601	1,875
2100	75 gallon		1.50	5.333		1,350	315		1,665	1,975
2120	100 gallon		1.30	6.154		1,600	360		1,960	2,325
2900	Water heater, safety-drain pan, 26" round		20	.400		37	23.50		60.50	76.50

281

22 34 36 – Commercial Gas Domestic Water Heaters

22 34 36.13 Commercial, Atmos., Gas Domestic Water Htrs.	Crew	Daily Output	Labor-Hours	Unit	Material	2015 Bare Costs Labor	Equipment	Total	Total Incl O&P
0010 **COMMERCIAL, ATMOSPHERIC, GAS DOMESTIC WATER HEATERS**									
6000 Gas fired, flush jacket, std. controls, vent not incl.									
6040 75 MBH input, 73 GPH	1 Plum	1.40	5.714	Ea.	3,500	335		3,835	4,350
6060 98 MBH input, 95 GPH		1.40	5.714		5,300	335		5,635	6,325
6080 120 MBH input, 110 GPH		1.20	6.667		5,500	390		5,890	6,625
6100 120 MBH input, 115 GPH		1.10	7.273		6,625	425		7,050	7,950
6120 140 MBH input, 130 GPH		1	8		7,225	470		7,695	8,650
6140 155 MBH input, 150 GPH		.80	10		8,200	585		8,785	9,900
6160 180 MBH input, 170 GPH		.70	11.429		8,675	670		9,345	10,600
6180 200 MBH input, 192 GPH		.60	13.333		8,925	785		9,710	11,000
6200 250 MBH input, 245 GPH		.50	16		9,325	940		10,265	11,700
6220 260 MBH input, 250 GPH	Q-1	.80	20		9,900	1,050		10,950	12,500
6240 360 MBH input, 360 GPH		.80	20		11,700	1,050		12,750	14,500
6260 500 MBH input, 480 GPH		.70	22.857		16,400	1,200		17,600	19,900
6280 725 MBH input, 690 GPH		.60	26.667		19,300	1,400		20,700	23,300
6900 For low water cutoff, add	1 Plum	8	1		350	58.50		408.50	475
6960 For bronze body hot water circulator, add	"	4	2		1,925	117		2,042	2,300

22 34 46 – Oil-Fired Domestic Water Heaters

22 34 46.10 Residential Oil-Fired Water Heaters

	Crew	Daily Output	Labor-Hours	Unit	Material	Labor	Equipment	Total	Total Incl O&P
0010 **RESIDENTIAL OIL-FIRED WATER HEATERS**									
3000 Oil fired, glass lined tank, 5 yr., vent not included, 30 gallon D2020-260	1 Plum	2	4	Ea.	1,175	235		1,410	1,625
3040 50 gallon		1.80	4.444		1,375	261		1,636	1,925
3060 70 gallon		1.50	5.333		1,975	315		2,290	2,650

22 34 46.20 Commercial Oil-Fired Water Heaters

	Crew	Daily Output	Labor-Hours	Unit	Material	Labor	Equipment	Total	Total Incl O&P
0010 **COMMERCIAL OIL-FIRED WATER HEATERS**									
8000 Oil fired, glass lined, UL listed, std. controls, vent not incl.									
8060 140 gal., 140 MBH input, 134 GPH	Q-1	2.13	7.512	Ea.	19,900	395		20,295	22,500
8080 140 gal., 199 MBH input, 191 GPH		2	8		20,600	425		21,025	23,300
8100 140 gal., 255 MBH input, 247 GPH		1.60	10		21,200	530		21,730	24,100
8120 140 gal., 270 MBH input, 259 GPH		1.20	13.333		26,200	705		26,905	29,900
8140 140 gal., 400 MBH input, 384 GPH		1	16		27,100	845		27,945	31,100
8160 140 gal., 540 MBH input, 519 GPH		.96	16.667		28,100	880		28,980	32,200
8180 140 gal., 720 MBH input, 691 GPH		.92	17.391		28,600	920		29,520	32,900
8200 221 gal., 300 MBH input, 288 GPH		.88	18.182		37,800	960		38,760	43,100
8220 221 gal., 600 MBH input, 576 GPH		.86	18.605		42,200	985		43,185	47,900
8240 221 gal., 800 MBH input, 768 GPH		.82	19.512		42,500	1,025		43,525	48,300
8260 201 gal., 1000 MBH input, 960 GPH	Q-2	1.26	19.048		43,500	1,050		44,550	49,400
8280 201 gal., 1250 MBH input, 1200 GPH		1.22	19.672		43,900	1,075		44,975	49,900
8300 201 gal., 1500 MBH input, 1441 GPH		1.16	20.690		47,800	1,125		48,925	54,000
8320 411 gal., 600 MBH input, 576 GPH		1.12	21.429		48,200	1,175		49,375	55,000
8340 411 gal., 800 MBH input, 768 GPH		1.08	22.222		48,300	1,225		49,525	55,000
8360 411 gal., 1000 MBH input, 960 GPH		1.04	23.077		51,000	1,275		52,275	58,000
8380 411 gal., 1250 MBH input, 1200 GPH		.98	24.490		52,000	1,350		53,350	59,500
8400 397 gal., 1500 MBH input, 1441 GPH		.92	26.087		55,500	1,425		56,925	63,000
8420 397 gal., 1750 MBH input, 1681 GPH		.86	27.907		57,000	1,525		58,525	65,500
8430 397 gal., 2000 MBH input, 1921 GPH		.82	29.268		62,000	1,600		63,600	70,500
8440 375 gal., 2250 MBH input, 2161 GPH		.76	31.579		63,500	1,725		65,225	72,500
8450 375 gal., 2500 MBH input, 2401 GPH		.82	29.268		66,000	1,600		67,600	75,000
8500 Oil fired, polymer lined									
8510 400 MBH, 125 gallon	Q-2	1	24	Ea.	26,300	1,325		27,625	30,900
8520 400 MBH, 600 gallon		.80	30		49,000	1,650		50,650	56,500

22 34 Fuel-Fired Domestic Water Heaters

22 34 46 – Oil-Fired Domestic Water Heaters

22 34 46.20 Commercial Oil-Fired Water Heaters	Crew	Daily Output	Labor-Hours	Unit	Material	2015 Bare Costs Labor	Equipment	Total	Total Incl O&P
8530	800 MBH, 400 gallon	Q-2	.67	35.821	Ea.	44,700	1,950		46,650	52,000
8540	800 MBH, 600 gallon		.60	40		58,000	2,200		60,200	67,500
8550	1000 MBH, 600 gallon		.50	48		60,000	2,625		62,625	70,000
8560	1200 MBH, 900 gallon		.40	60		69,000	3,275		72,275	81,000
8900	For low water cutoff, add	1 Plum	8	1		350	58.50		408.50	475
8960	For bronze body hot water circulator, add	"	4	2	▼	710	117		827	955

22 41 Residential Plumbing Fixtures

22 41 06 – Plumbing Fixtures General

22 41 06.10 Plumbing Fixture Notes

0010	**PLUMBING FIXTURE NOTES**, Incl. trim fittings unless otherwise noted R224000-30								
0080	For rough-in, supply, waste, and vent, see add for each type								
0122	For electric water coolers, see Section 22 47 16.10								
0160	For color, unless otherwise noted, add				Ea.	20%			

22 41 13 – Residential Water Closets, Urinals, and Bidets

22 41 13.13 Water Closets

		Crew	Daily Output	Labor-Hours	Unit	Material	Labor	Equipment	Total	Total Incl O&P
0010	**WATER CLOSETS** D2010–110									
0022	For seats, see Section 22 41 13.44									
0032	For automatic flush, see Line 22 42 39.10 0972									
0150	Tank type, vitreous china, incl. seat, supply pipe w/stop, 1.6 gpf or noted									
0200	Wall hung R224000-30									
0400	Two piece, close coupled	Q-1	5.30	3.019	Ea.	630	159		789	935
0960	For rough-in, supply, waste, vent and carrier	"	2.73	5.861	"	1,025	310		1,335	1,600
0999	Floor mounted									
1020	One piece, low profile	Q-1	5.30	3.019	Ea.	1,125	159		1,284	1,475
1050	One piece		5.30	3.019		1,100	159		1,259	1,450
1100	Two piece, close coupled		5.30	3.019		237	159		396	500
1102	Economy		5.30	3.019		132	159		291	385
1110	Two piece, close coupled, dual flush		5.30	3.019		310	159		469	585
1140	Two piece, close coupled, 1.28 gpf, ADA [G]	▼	5.30	3.019	▼	310	159		469	585
1960	For color, add					30%				
1961	For designer colors and trim, add					55%				
1980	For rough-in, supply, waste and vent	Q-1	3.05	5.246	Ea.	330	277		607	785

22 41 13.19 Bidets

		Crew	Daily Output	Labor-Hours	Unit	Material	Labor	Equipment	Total	Total Incl O&P
0010	**BIDETS**									
0180	Vitreous china, with trim on fixture	Q-1	5	3.200	Ea.	595	169		764	910
0200	With trim for wall mounting	"	5	3.200		710	169		879	1,025
9590	For color add					40%				
9591	For designer colors and trim add					50%				
9600	For rough-in, supply, waste and vent, add	Q-1	1.78	8.989	▼	325	475		800	1,075

22 41 13.44 Toilet Seats

		Crew	Daily Output	Labor-Hours	Unit	Material	Labor	Equipment	Total	Total Incl O&P
0010	**TOILET SEATS**									
0100	Molded composition, white									
0150	Industrial, w/o cover, open front, regular bowl	1 Plum	24	.333	Ea.	21	19.55		40.55	52.50
0200	With self-sustaining hinge		24	.333		22.50	19.55		42.05	54
0220	With self-sustaining check hinge		24	.333		22.50	19.55		42.05	54
0240	Extra heavy, with check hinge	▼	24	.333	▼	27.50	19.55		47.05	60
0260	Elongated bowl, same price									
0300	Junior size, w/o cover, open front	1 Plum	24	.333	Ea.	40.50	19.55		60.05	74
0320	Regular primary bowl, open front		24	.333		40.50	19.55		60.05	74

283

22 41 13.44 Toilet Seats		Crew	Daily Output	Labor-Hours	Unit	Material	2015 Bare Costs Labor	Equipment	Total	Total Incl O&P
0340	Regular baby bowl, open front, check hinge	1 Plum	24	.333	Ea.	37	19.55		56.55	70
0380	Open back & front, w/o cover, reg. or elongated bowl	↓	24	.333	↓	29	19.55		48.55	61.50
0400	Residential									
0420	Regular bowl, w/cover, closed front	1 Plum	24	.333	Ea.	29	19.55		48.55	61
0440	Open front	"	24	.333	"	26	19.55		45.55	58.50
0460	Elongated bowl, add					25%				
0500	Self-raising hinge, w/o cover, open front									
0520	Regular bowl	1 Plum	24	.333	Ea.	101	19.55		120.55	141
0540	Elongated bowl	"	24	.333	"	23	19.55		42.55	55
0700	Molded wood, white, with cover									
0720	Closed front, regular bowl, square back	1 Plum	24	.333	Ea.	10.90	19.55		30.45	41.50
0740	Extended back		24	.333		13.80	19.55		33.35	44.50
0780	Elongated bowl, square back		24	.333		14.10	19.55		33.65	45
0800	Open front	↓	24	.333	↓	15.05	19.55		34.60	46
0850	Decorator styles									
0890	Vinyl top, patterned	1 Plum	24	.333	Ea.	22	19.55		41.55	53.50
0900	Vinyl padded, plain colors, regular bowl		24	.333		22	19.55		41.55	53.50
0930	Elongated bowl	↓	24	.333	↓	24	19.55		43.55	56
1000	Solid plastic, white									
1030	Industrial, w/o cover, open front, regular bowl	1 Plum	24	.333	Ea.	26	19.55		45.55	58.50
1080	Extra heavy, concealed check hinge		24	.333		18.75	19.55		38.30	50
1100	Self-sustaining hinge		24	.333		23	19.55		42.55	55
1150	Elongated bowl		24	.333		29	19.55		48.55	61.50
1170	Concealed check		24	.333		17.25	19.55		36.80	48.50
1190	Self-sustaining hinge, concealed check		24	.333		51.50	19.55		71.05	86
1220	Residential, with cover, closed front, regular bowl		24	.333		45	19.55		64.55	79
1240	Elongated bowl		24	.333		55	19.55		74.55	89.50
1260	Open front, regular bowl		24	.333		40	19.55		59.55	73.50
1280	Elongated bowl	↓	24	.333	↓	48.50	19.55		68.05	83

22 41 16 – Residential Lavatories and Sinks

22 41 16.13 Lavatories

		Crew	Daily Output	Labor-Hours	Unit	Material	2015 Bare Costs Labor	Equipment	Total	Total Incl O&P
0010	**LAVATORIES**, With trim, white unless noted otherwise D2010-310									
0500	Vanity top, porcelain enamel on cast iron									
0600	20" x 18" R224000-30	Q-1	6.40	2.500	Ea.	335	132		467	570
0640	33" x 19" oval		6.40	2.500		475	132		607	725
0680	20" x 17" oval		6.40	2.500		172	132		304	390
0720	19" round		6.40	2.500		435	132		567	680
0760	20" x 12" triangular bowl	↓	6.40	2.500	↓	273	132		405	500
0860	For color, add					25%				
0861	For designer colors and trim, add					70%				
1000	Cultured marble, 19" x 17", single bowl	Q-1	6.40	2.500	Ea.	175	132		307	390
1040	25" x 19", single bowl		6.40	2.500		206	132		338	425
1080	31" x 19", single bowl		6.40	2.500		216	132		348	435
1120	25" x 22", single bowl		6.40	2.500		215	132		347	435
1160	37" x 22", single bowl		6.40	2.500		245	132		377	470
1200	49" x 22", single bowl	↓	6.40	2.500	↓	292	132		424	520
1580	For color, same price									
1900	Stainless steel, self-rimming, 25" x 22", single bowl, ledge	Q-1	6.40	2.500	Ea.	365	132		497	605
1960	17" x 22", single bowl		6.40	2.500		355	132		487	590
2040	18-3/4" round		6.40	2.500		825	132		957	1,100
2600	Steel, enameled, 20" x 17", single bowl		5.80	2.759		161	146		307	395
2660	19" round	↓	5.80	2.759	↓	171	146		317	410

For customer support on your Plumbing Cost Data, call 877.709.6968.

22 41 16 – Residential Lavatories and Sinks

22 41 16.13 Lavatories		Crew	Daily Output	Labor-Hours	Unit	Material	2015 Bare Costs Labor	Equipment	Total	Total Incl O&P
2720	18" round	Q-1	5.80	2.759	Ea.	138	146		284	370
2860	For color, add					10%				
2861	For designer colors and trim, add					20%				
2900	Vitreous china, 20" x 16", single bowl	Q-1	5.40	2.963	Ea.	260	157		417	520
2960	20" x 17", single bowl		5.40	2.963		176	157		333	430
3020	19" round, single bowl		5.40	2.963		174	157		331	430
3080	19" x 16", single bowl		5.40	2.963		267	157		424	530
3140	17" x 14", single bowl		5.40	2.963		214	157		371	470
3200	22" x 13", single bowl		5.40	2.963		267	157		424	530
3560	For color, add					50%				
3561	For designer colors and trim, add					100%				
3580	Rough-in, supply, waste and vent for all above lavatories	Q-1	2.30	6.957	Ea.	231	370		601	810
4000	Wall hung									
4040	Porcelain enamel on cast iron, 16" x 14", single bowl	Q-1	8	2	Ea.	520	106		626	730
4060	18" x 15" single bowl		8	2		405	106		511	605
4120	19" x 17", single bowl		8	2		425	106		531	630
4180	20" x 18", single bowl		8	2		277	106		383	465
4240	22" x 19", single bowl		8	2		700	106		806	930
4580	For color, add					30%				
4581	For designer colors and trim, add					75%				
6000	Vitreous china, 18" x 15", single bowl with backsplash	Q-1	7	2.286	Ea.	231	121		352	435
6060	19" x 17", single bowl		7	2.286		207	121		328	410
6120	20" x 18", single bowl		7	2.286		284	121		405	490
6210	27" x 20", wheelchair type		7	2.286		815	121		936	1,075
6500	For color, add					30%				
6501	For designer colors and trim, add					50%				
6960	Rough-in, supply, waste and vent for above lavatories	Q-1	1.66	9.639	Ea.	455	510		965	1,275
7000	Pedestal type									
7600	Vitreous china, 27" x 21", white	Q-1	6.60	2.424	Ea.	700	128		828	965
7610	27" x 21", colored		6.60	2.424		880	128		1,008	1,150
7620	27" x 21", premium color		6.60	2.424		995	128		1,123	1,300
7660	26" x 20", white		6.60	2.424		795	128		923	1,075
7670	26" x 20", colored		6.60	2.424		1,000	128		1,128	1,300
7680	26" x 20", premium color		6.60	2.424		1,125	128		1,253	1,450
7700	24" x 20", white		6.60	2.424		405	128		533	640
7710	24" x 20", colored		6.60	2.424		475	128		603	720
7720	24" x 20", premium color		6.60	2.424		495	128		623	740
7760	21" x 18", white		6.60	2.424		291	128		419	515
7770	21" x 18", colored		6.60	2.424		291	128		419	515
7990	Rough-in, supply, waste and vent for pedestal lavatories		1.66	9.639		455	510		965	1,275

22 41 16.16 Sinks

		Crew	Daily Output	Labor-Hours	Unit	Material	2015 Bare Costs Labor	Equipment	Total	Total Incl O&P
0010	**SINKS**, With faucets and drain D2010-410									
2000	Kitchen, counter top style, P.E. on C.I., 24" x 21" single bowl	Q-1	5.60	2.857	Ea.	285	151		436	545
2100	31" x 22" single bowl		5.60	2.857		615	151		766	910
2200	32" x 21" double bowl		4.80	3.333		355	176		531	655
2310	For color, add					20%				
2311	For designer colors and trim, add					50%				
3000	Stainless steel, self rimming, 19" x 18" single bowl	Q-1	5.60	2.857	Ea.	590	151		741	880
3100	25" x 22" single bowl		5.60	2.857		660	151		811	955
3200	33" x 22" double bowl		4.80	3.333		960	176		1,136	1,325
3300	43" x 22" double bowl		4.80	3.333		1,125	176		1,301	1,500
3400	22" x 43" triple bowl		4.40	3.636		1,350	192		1,542	1,775

For customer support on your Plumbing Cost Data, call 877.709.6968.

285

22 41 16.16 Sinks

		Crew	Daily Output	Labor-Hours	Unit	Material	2015 Bare Costs Labor	2015 Bare Costs Equipment	Total	Total Incl O&P
3500	Corner double bowl each 14" x 16"	Q-1	4.80	3.333	Ea.	830	176		1,006	1,175
4000	Steel, enameled, with ledge, 24" x 21" single bowl		5.60	2.857		510	151		661	790
4100	32" x 21" double bowl		4.80	3.333		495	176		671	810
4960	For color sinks except stainless steel, add					10%				
4961	For designer colors and trim add					20%				
4980	For rough-in, supply, waste and vent, counter top sinks	Q-1	2.14	7.477		260	395		655	880
5000	Kitchen, raised deck, P.E. on C.I.									
5100	32" x 21", dual level, double bowl	Q-1	2.60	6.154	Ea.	420	325		745	950
5200	42" x 21", double bowl & disposer well	"	2.20	7.273		1,175	385		1,560	1,875
5700	For color, add					20%				
5701	For designer colors and trim add					50%				
5790	For rough-in, supply, waste & vent, sinks	Q-1	1.85	8.649		260	455		715	975

22 41 19 – Residential Bathtubs

22 41 19.10 Baths

		Crew	Daily Output	Labor-Hours	Unit	Material	2015 Bare Costs Labor	2015 Bare Costs Equipment	Total	Total Incl O&P
0010	**BATHS** D2010-510									
0100	Tubs, recessed porcelain enamel on cast iron, with trim									
0180	48" x 42"	Q-1	4	4	Ea.	2,625	211		2,836	3,225
0220	72" x 36"	"	3	5.333	"	2,725	282		3,007	3,425
0300	Mat bottom									
0340	4'-6" long	Q-1	5	3.200	Ea.	1,350	169		1,519	1,750
0380	5' long		4.40	3.636		1,125	192		1,317	1,550
0420	5'-6" long		4	4		1,775	211		1,986	2,275
0480	Above floor drain, 5' long		4	4		785	211		996	1,175
0560	Corner 48" x 44"		4.40	3.636		2,625	192		2,817	3,200
0750	For color, add					30%				
0760	For designer colors & trim, add					60%				
2000	Enameled formed steel, 4'-6" long	Q-1	5.80	2.759	Ea.	495	146		641	765
2300	Above floor drain, 5' long	"	5.50	2.909	"	535	154		689	820
2350	For color, add					10%				
4000	Soaking, acrylic, w/pop-up drain 66" x 36" x 20" deep	Q-1	5.50	2.909	Ea.	2,300	154		2,454	2,750
4100	60" x 42" x 20" deep		5	3.200		1,175	169		1,344	1,550
4200	72" x 42" x 23" deep		4.80	3.333		1,750	176		1,926	2,200
4310	For color, add					5%				
4311	For designer colors & trim, add					20%				
4600	Module tub & showerwall surround, molded fiberglass									
4610	5' long x 34" wide x 76" high	Q-1	4	4	Ea.	840	211		1,051	1,250
4620	For color add					10%				
4621	For designer colors and trim add					25%				
4750	Handicap with 1-1/2" OD grab bar, antiskid bottom									
4760	60" x 32-3/4" x 72" high	Q-1	4	4	Ea.	840	211		1,051	1,250
4770	60" x 30" x 71" high with molded seat		3.50	4.571		875	242		1,117	1,325
9600	Rough-in, supply, waste and vent, for all above tubs, add		2.07	7.729		345	410		755	995

22 41 23 – Residential Showers

22 41 23.20 Showers

		Crew	Daily Output	Labor-Hours	Unit	Material	2015 Bare Costs Labor	2015 Bare Costs Equipment	Total	Total Incl O&P
0010	**SHOWERS** D2010-710									
1500	Stall, with drain only. Add for valve and door/curtain									
1510	Baked enamel, molded stone receptor, 30" square	Q-1	5.20	3.077	Ea.	1,150	163		1,313	1,525
1520	32" square		5	3.200		1,175	169		1,344	1,525
1530	36" square		4.80	3.333		2,925	176		3,101	3,475
1540	Terrazzo receptor, 32" square		5	3.200		1,350	169		1,519	1,725
1560	36" square		4.80	3.333		1,475	176		1,651	1,900
1580	36" corner angle		4.80	3.333		1,725	176		1,901	2,175

For customer support on your Plumbing Cost Data, call 877.709.6968.

22 41 Residential Plumbing Fixtures

22 41 23 – Residential Showers

22 41 23.20 Showers

	Crew	Daily Output	Labor-Hours	Unit	Material	2015 Bare Costs Labor	2015 Bare Costs Equipment	Total	Total Incl O&P
1600	For color, add					10%				
1601	For designer colors and trim, add					15%				
1604	For thermostatic valve add				Ea.	520			520	570
3000	Fiberglass, one piece, with 3 walls, 32" x 32" square	Q-1	5.50	2.909		490	154		644	770
3100	36" x 36" square	"	5.50	2.909	↓	505	154		659	785
3200	Handicap, 1-1/2" O.D. grab bars, nonskid floor									
3210	48" x 34-1/2" x 72" corner seat	Q-1	5	3.200	Ea.	720	169		889	1,050
3220	60" x 34-1/2" x 72" corner seat		4	4		750	211		961	1,150
3230	48" x 34-1/2" x 72" fold up seat		5	3.200		2,225	169		2,394	2,700
3250	64" x 65-3/4" x 81-1/2" fold. seat, whlchr.		3.80	4.211		2,325	222		2,547	2,875
3260	For thermostatic valve add					520			520	570
4000	Polypropylene, stall only, w/molded-stone floor, 30" x 30"	Q-1	2	8		635	425		1,060	1,350
4100	32" x 32"	"	2	8		650	425		1,075	1,350
4110	For thermostatic valve add					520			520	570
4200	Rough-in, supply, waste and vent for above showers	Q-1	2.05	7.805	↓	345	410		755	1,000

22 41 23.40 Shower System Components

	Crew	Daily Output	Labor-Hours	Unit	Material	2015 Bare Costs Labor	2015 Bare Costs Equipment	Total	Total Incl O&P
0010	**SHOWER SYSTEM COMPONENTS**									
4500	Receptor only									
4510	For tile, 36" x 36"	1 Plum	4	2	Ea.	365	117		482	580
4520	Fiberglass receptor only, 32" x 32"		8	1		114	58.50		172.50	214
4530	34" x 34"		7.80	1.026		128	60		188	231
4540	36" x 36"	↓	7.60	1.053	↓	133	62		195	240
4600	Rectangular									
4620	32" x 48"	1 Plum	7.40	1.081	Ea.	158	63.50		221.50	270
4630	34" x 54"		7.20	1.111		190	65		255	310
4640	34" x 60"		7	1.143		201	67		268	320
5000	Built-in, head, arm, 2.5 GPM valve		4	2		80	117		197	265
5200	Head, arm, by-pass, integral stops, handles		3.60	2.222		255	130		385	480
5500	Head, water economizer, 1.6 GPM [G]		24	.333		45	19.55		64.55	79
5800	Mixing valve, built-in		6	1.333		153	78.50		231.50	287
5900	Exposed	↓	6	1.333	↓	660	78.50		738.50	850

22 41 36 – Residential Laundry Trays

22 41 36.10 Laundry Sinks

	Crew	Daily Output	Labor-Hours	Unit	Material	2015 Bare Costs Labor	2015 Bare Costs Equipment	Total	Total Incl O&P
0010	**LAUNDRY SINKS**, With trim D2010-420									
0020	Porcelain enamel on cast iron, black iron frame									
0050	24" x 21", single compartment	Q-1	6	2.667	Ea.	580	141		721	850
0100	26" x 21", single compartment		6	2.667		610	141		751	885
0200	48" x 20", double compartment	↓	5	3.200	↓	830	169		999	1,175
2000	Molded stone, on wall hanger or legs									
2020	22" x 23", single compartment	Q-1	6	2.667	Ea.	167	141		308	395
2100	45" x 21", double compartment	"	5	3.200	"	330	169		499	620
3000	Plastic, on wall hanger or legs									
3020	18" x 23", single compartment	Q-1	6.50	2.462	Ea.	135	130		265	345
3100	20" x 24", single compartment		6.50	2.462		154	130		284	365
3200	36" x 23", double compartment		5.50	2.909		185	154		339	435
3300	40" x 24", double compartment		5.50	2.909		278	154		432	535
5000	Stainless steel, counter top, 22" x 17" single compartment		6	2.667		64	141		205	284
5200	33" x 22", double compartment		5	3.200		79	169		248	340
9600	Rough-in, supply, waste and vent, for all laundry sinks	↓	2.14	7.477		260	395		655	880

287

22 41 39.10 Faucets and Fittings	Crew	Daily Output	Labor-Hours	Unit	Material	2015 Bare Costs Labor	Equipment	Total	Total Incl O&P
0010 **FAUCETS AND FITTINGS**									
0150 Bath, faucets, diverter spout combination, sweat	1 Plum	8	1	Ea.	86.50	58.50		145	184
0200 For integral stops, IPS unions, add					109			109	120
0300 Three valve combinations, spout, head, arm, flange, sweat	1 Plum	6	1.333	↓	88	78.50		166.50	215
0400 For integral stops, IPS unions, add				Pr.	64.50			64.50	71
0420 Bath, press-bal mix valve w/diverter, spout, shower head, arm/flange	1 Plum	8	1	Ea.	168	58.50		226.50	274
0500 Drain, central lift, 1-1/2" IPS male		20	.400		71	23.50		94.50	114
0600 Trip lever, 1-1/2" IPS male		20	.400		45	23.50		68.50	85
0700 Pop up, 1-1/2" IPS male		18	.444		53	26		79	98
0800 Chain and stopper, 1-1/2" IPS male	↓	24	.333	↓	32.50	19.55		52.05	65
0810 Bidet									
0812 Fitting, over the rim, swivel spray/pop-up drain	1 Plum	8	1	Ea.	206	58.50		264.50	315
1000 Kitchen sink faucets, top mount, cast spout		10	.800		61.50	47		108.50	139
1100 For spray, add		24	.333		16.15	19.55		35.70	47.50
1110 For basket strainer w/tail piece, add		24	.333		14.85	19.55		34.40	46
1200 Wall type, swing tube spout	↓	10	.800		74.50	47		121.50	153
1240 For soap dish, add					3.60			3.60	3.96
1250 For basket strainer w/tail piece, add				↓	51			51	56
1300 Single control lever handle									
1310 With pull out spray									
1320 Polished chrome	1 Plum	10	.800	Ea.	196	47		243	286
2000 Laundry faucets, shelf type, IPS or copper unions		12	.667		49.50	39		88.50	114
2100 Lavatory faucet, centerset, without drain		10	.800		44.50	47		91.50	120
2120 With pop-up drain	↓	6.66	1.201		62.50	70.50		133	175
2130 For acrylic handles, add					5.15			5.15	5.65
2150 Concealed, 12" centers	1 Plum	10	.800		101	47		148	182
2160 With pop-up drain	"	6.66	1.201	↓	118	70.50		188.50	236
2210 Porcelain cross handles and pop-up drain									
2220 Polished chrome	1 Plum	6.66	1.201	Ea.	189	70.50		259.50	315
2230 Polished brass	"	6.66	1.201	"	298	70.50		368.50	435
2260 Single lever handle and pop-up drain									
2280 Satin nickel	1 Plum	6.66	1.201	Ea.	277	70.50		347.50	410
2290 Polished chrome		6.66	1.201		198	70.50		268.50	325
2600 Shelfback, 4" to 6" centers, 17 Ga. tailpiece		10	.800		78	47		125	157
2650 With pop-up drain		6.66	1.201		95	70.50		165.50	210
2700 Shampoo faucet with supply tube		24	.333		48.50	19.55		68.05	83
2800 Self-closing, center set		10	.800		131	47		178	215
2810 Automatic sensor and operator, with faucet head ♿ G		6.15	1.301		450	76.50		526.50	610
4000 Shower by-pass valve with union		18	.444		68.50	26		94.50	115
4100 Shower arm with flange and head		22	.364		19.25	21.50		40.75	53
4140 Shower, hand held, pin mount, massage action, chrome		22	.364		76	21.50		97.50	116
4142 Polished brass		22	.364		142	21.50		163.50	188
4144 Shower, hand held, wall mtd, adj. spray, 2 wall mounts, chrome		20	.400		102	23.50		125.50	148
4146 Polished brass		20	.400		192	23.50		215.50	248
4148 Shower, hand held head, bar mounted 24", adj. spray, chrome		20	.400		164	23.50		187.50	216
4150 Polished brass		20	.400		340	23.50		363.50	405
4200 Shower thermostatic mixing valve, concealed, with shower head trim kit	↓	8	1	↓	345	58.50		403.50	470
4220 Shower pressure balancing mixing valve,									
4230 With shower head, arm, flange and diverter tub spout									
4240 Chrome	1 Plum	6.14	1.303	Ea.	360	76.50		436.50	510
4250 Satin nickel		6.14	1.303		555	76.50		631.50	725
4260 Polished graphite		6.14	1.303	↓	555	76.50		631.50	725

22 41 39 – Residential Faucets, Supplies and Trim

22 41 39.10 Faucets and Fittings

		Crew	Daily Output	Labor-Hours	Unit	Material	2015 Bare Costs Labor	Equipment	Total	Total Incl O&P
5000	Sillcock, compact, brass, IPS or copper to hose	1 Plum	24	.333	Ea.	9.70	19.55		29.25	40
6000	Stop and waste valves, bronze									
6100	Angle, solder end 1/2"	1 Plum	24	.333	Ea.	12.05	19.55		31.60	43
6110	3/4"		20	.400		14.15	23.50		37.65	51
6300	Straightway, solder end 3/8"		24	.333		12.35	19.55		31.90	43
6310	1/2"		24	.333		12.35	19.55		31.90	43
6320	3/4"		20	.400		13.30	23.50		36.80	50
6410	Straightway, threaded 1/2"		24	.333		15.95	19.55		35.50	47
6420	3/4"		20	.400		18.30	23.50		41.80	55.50
6430	1"		19	.421		13.40	24.50		37.90	52.50
7800	Water closet, wax gasket		96	.083		1.38	4.89		6.27	8.90
7820	Gasket toilet tank to bowl		32	.250		2.38	14.70		17.08	24.50
7830	Replacement diaphragm washer assy for ballcock valve		12	.667		2.38	39		41.38	61.50
7850	Dual flush valve		12	.667		20	39		59	81
8000	Water supply stops, polished chrome plate									
8200	Angle, 3/8"	1 Plum	24	.333	Ea.	12.90	19.55		32.45	43.50
8300	1/2"		22	.364		12.90	21.50		34.40	46
8400	Straight, 3/8"		26	.308		13.55	18.05		31.60	42.50
8500	1/2"		24	.333		13.55	19.55		33.10	44.50
8600	Water closet, angle, w/flex riser, 3/8"		24	.333		44	19.55		63.55	78
9100	Miscellaneous									
9720	Teflon tape, 1/2" x 520" roll				Ea.	1.26			1.26	1.39

22 41 39.70 Washer/Dryer Accessories

		Crew	Daily Output	Labor-Hours	Unit	Material	2015 Bare Costs Labor	Equipment	Total	Total Incl O&P
0010	**WASHER/DRYER ACCESSORIES**									
1020	Valves ball type single lever									
1030	1/2" diam., IPS	1 Plum	21	.381	Ea.	54.50	22.50		77	94
1040	1/2" diam., solder	"	21	.381	"	54.50	22.50		77	94
1050	Recessed box, 16 ga., two hose valves and drain									
1060	1/2" size, 1-1/2" drain	1 Plum	18	.444	Ea.	116	26		142	168
1070	1/2" size, 2" drain	"	17	.471	"	108	27.50		135.50	161
1080	With grounding electric receptacle									
1090	1/2" size, 1-1/2" drain	1 Plum	18	.444	Ea.	127	26		153	180
1100	1/2" size, 2" drain	"	17	.471	"	138	27.50		165.50	193
1110	With grounding and dryer receptacle									
1120	1/2" size, 1-1/2" drain	1 Plum	18	.444	Ea.	157	26		183	213
1130	1/2" size, 2" drain	"	17	.471	"	159	27.50		186.50	217
1140	Recessed box 16 ga., ball valves with single lever and drain									
1150	1/2" size, 1-1/2" drain	1 Plum	19	.421	Ea.	219	24.50		243.50	279
1160	1/2" size, 2" drain	"	18	.444	"	197	26		223	257
1170	With grounding electric receptacle									
1180	1/2" size, 1-1/2" drain	1 Plum	19	.421	Ea.	190	24.50		214.50	247
1190	1/2" size, 2" drain	"	18	.444	"	211	26		237	272
1200	With grounding and dryer receptacles									
1210	1/2" size, 1-1/2" drain	1 Plum	19	.421	Ea.	208	24.50		232.50	266
1220	1/2" size, 2" drain	"	18	.444	"	230	26		256	294
1300	Recessed box, 20 ga., two hose valves and drain (economy type)									
1310	1/2" size, 1-1/2" drain	1 Plum	19	.421	Ea.	90.50	24.50		115	137
1320	1/2" size, 2" drain		18	.444		86	26		112	134
1330	Box with drain only		24	.333		52.50	19.55		72.05	87
1340	1/2" size, 1-1/2" ABS/PVC drain		19	.421		100	24.50		124.50	148
1350	1/2" size, 2" ABS/PVC drain		18	.444		108	26		134	158
1352	Box with drain and 15 A receptacle		24	.333		90.50	19.55		110.05	129

22 41 Residential Plumbing Fixtures

22 41 39 – Residential Faucets, Supplies and Trim

22 41 39.70 Washer/Dryer Accessories		Crew	Daily Output	Labor-Hours	Unit	Material	2015 Bare Costs Labor	Equipment	Total	Total Incl O&P
1360	1/2" size, 2" drain ABS/PVC, 15 A receptacle	1 Plum	24	.333	Ea.	127	19.55		146.55	170
1400	Wall mounted									
1410	1/2" size, 1-1/2" plastic drain	1 Plum	19	.421	Ea.	24.50	24.50		49	64.50
1420	1/2" size, 2" plastic drain	"	18	.444	"	21	26		47	62.50
1500	Dryer vent kit									
1510	8' flex duct, clamps and outside hood	1 Plum	20	.400	Ea.	15.30	23.50		38.80	52.50
1980	Rough-in, supply, waste, and vent for washer boxes	"	3.46	2.310	"	244	136		380	475

22 42 Commercial Plumbing Fixtures

22 42 13 – Commercial Water Closets, Urinals, and Bidets

22 42 13.13 Water Closets

		Crew	Daily Output	Labor-Hours	Unit	Material	2015 Bare Costs Labor	Equipment	Total	Total Incl O&P
0010	**WATER CLOSETS**									
3000	Bowl only, with flush valve, seat, 1.6 gpf unless noted									
3100	Wall hung	Q-1	5.80	2.759	Ea.	945	146		1,091	1,275
3200	For rough-in, supply, waste and vent, single WC		2.56	6.250		1,075	330		1,405	1,675
3300	Floor mounted		5.80	2.759		315	146		461	565
3350	With wall outlet		5.80	2.759		545	146		691	820
3360	With floor outlet, 1.28 gpf [G]		5.80	2.759		555	146		701	835
3362	With floor outlet, 1.28 gpf, ADA [G]		5.80	2.759		580	146		726	855
3370	For rough-in, supply, waste and vent, single WC	↓	2.84	5.634	↓	370	298		668	855
3390	Floor mounted children's size, 10-3/4" high									
3392	With automatic flush sensor, 1.6 gpf	Q-1	6.20	2.581	Ea.	615	136		751	885
3396	With automatic flush sensor, 1.28 gpf		6.20	2.581		620	136		756	885
3400	For rough-in, supply, waste and vent, single WC	↓	2.84	5.634	↓	370	298		668	855
3500	Gang side by side carrier system, rough-in, supply, waste & vent									
3510	For single hook-up	Q-1	1.97	8.122	Ea.	1,250	430		1,680	2,025
3520	For each additional hook-up, add	"	2.14	7.477	"	1,175	395		1,570	1,875
3550	Gang back to back carrier system, rough-in, supply, waste & vent									
3560	For pair hook-up	Q-1	1.76	9.091	Pr.	1,975	480		2,455	2,900
3570	For each additional pair hook-up, add	"	1.81	8.840	"	1,875	465		2,340	2,775

22 42 13.16 Urinals

		Crew	Daily Output	Labor-Hours	Unit	Material	2015 Bare Costs Labor	Equipment	Total	Total Incl O&P
0010	**URINALS**	D2010-210								
0102	For automatic flush see Line 22 42 39.10 0972									
3000	Wall hung, vitreous china, with self-closing valve	R224000-30								
3100	Siphon jet type	Q-1	3	5.333	Ea.	282	282		564	735
3120	Blowout type		3	5.333		465	282		747	935
3140	Water saving .5 gpf [G]		3	5.333		550	282		832	1,025
3300	Rough-in, supply, waste & vent		2.83	5.654		595	299		894	1,100
5000	Stall type, vitreous china, includes valve		2.50	6.400		740	340		1,080	1,325
6980	Rough-in, supply, waste and vent	↓	1.99	8.040	↓	365	425		790	1,050
8000	Waterless (no flush) urinal									
8010	Wall hung									
8014	Fiberglass reinforced polyester									
8020	Standard unit [G]	Q-1	21.30	.751	Ea.	385	39.50		424.50	480
8030	ADA compliant unit [G]	"	21.30	.751		400	39.50		439.50	500
8070	For solid color, add [G]					48			48	53
8080	For 2" brass flange, (new const.), add [G]	Q-1	96	.167	↓	19.20	8.80		28	34.50
8200	Vitreous china									
8220	ADA compliant unit, 14" [G]	Q-1	21.30	.751	Ea.	198	39.50		237.50	278
8240	ADA compliant unit, 18" [G]		21.30	.751		320	39.50		359.50	410
8250	ADA compliant unit, 15.5"		21.30	.751		272	39.50		311.50	360

For customer support on your Plumbing Cost Data, call 877.709.6968.

22 42 Commercial Plumbing Fixtures

22 42 13 – Commercial Water Closets, Urinals, and Bidets

22 42 13.16 Urinals

		Crew	Daily Output	Labor-Hours	Unit	Material	2015 Bare Costs Labor	Equipment	Total	Total Incl O&P
8270	For solid color, add G				Ea.	48			48	53
8290	Rough-in, supply, waste & vent G	Q-1	2.92	5.479	↓	560	289		849	1,050
8400	Trap liquid									
8410	1 quart G				Ea.	15.95			15.95	17.55
8420	1 gallon G				"	58			58	64

22 42 16 – Commercial Lavatories and Sinks

22 42 16.13 Lavatories

0010	**LAVATORIES**, With trim, white unless noted otherwise									
0020	Commercial lavatories same as residential. See Section 22 41 16									

22 42 16.16 Commercial Sinks

		Crew	Daily Output	Labor-Hours	Unit	Material	2015 Bare Costs Labor	Equipment	Total	Total Incl O&P
0010	**COMMERCIAL SINKS**									
5900	Scullery sink, stainless steel									
5910	1 bowl and drain board, 43" x 22" O.D.	Q-1	5.40	2.963	Ea.	4,050	157		4,207	4,675
5920	2 bowls and drain board, 49" x 22" O.D.		4.60	3.478		6,675	184		6,859	7,600
5930	3 bowls and drain board, 43" x 22" O.D.		4.20	3.810		6,825	201		7,026	7,825
5940	1 bowl and drain board, 50" x 28" O.D., with legs	↓	5.40	2.963	↓	3,025	157		3,182	3,550

22 42 16.30 Classroom Sinks

		Crew	Daily Output	Labor-Hours	Unit	Material	2015 Bare Costs Labor	Equipment	Total	Total Incl O&P
0010	**CLASSROOM SINKS**									
6020	Countertop, stainless steel									
6024	with faucet, bubbler and strainer, ADA compliant									
6036	25" x 17" single bowl	Q-1	5.20	3.077	Ea.	870	163		1,033	1,200
6040	28" x 22" single bowl		5.20	3.077		1,200	163		1,363	1,550
6044	31" x 19" single bowl		5.20	3.077		1,350	163		1,513	1,725
6070	37" x 17" double bowl		4.40	3.636		1,700	192		1,892	2,175
6100	For rough-in, supply, waste and vent, counter top classroom sinks	↓	2.14	7.477	↓	260	395		655	880

22 42 16.34 Laboratory Countertops and Sinks

		Crew	Daily Output	Labor-Hours	Unit	Material	2015 Bare Costs Labor	Equipment	Total	Total Incl O&P
0010	**LABORATORY COUNTERTOPS AND SINKS**									
0050	Laboratory sinks, corrosion resistant									
1000	Stainless steel sink, bench mounted, with									
1020	plug & waste fitting with 1-1/2" straight threads									
1030	Single bowl, 2 drainboards, backnut & strainer									
1050	18-1/2" x 15-1/2" x 12-1/2" sink, 54" x 24" O.D.	Q-1	3	5.333	Ea.	1,700	282		1,982	2,300
1100	Single bowl, single drainboard, backnut & strainer									
1130	18-1/2" x 15-1/2" x 12-1/2" sink, 47" x 24" O.D.	Q-1	3	5.333	Ea.	1,200	282		1,482	1,725
1146	Double bowl, single drainboard, backnut & strainer									
1150	18-1/2" x 15-1/2" x 12-1/2" sink, 70" x 24" O.D.	Q-1	3	5.333	Ea.	1,875	282		2,157	2,500
1280	Polypropylene									
1290	Flanged 1-1/4" wide, rectangular with strainer									
1300	plug & waste fitting, 1-1/2" straight threads									
1320	12" x 12" x 8" sink, 14-1/2" x 14-1/2" O.D.	Q-1	4	4	Ea.	234	211		445	575
1340	16" x 16" x 8" sink, 18-1/2" x 18-1/2" O.D.		4	4		335	211		546	690
1360	21" x 18" x 10" sink, 23-1/2" x 20-1/2" O.D.		4	4		355	211		566	710
1490	For rough-in, supply, waste & vent, add	↓	2.02	7.921	↓	208	420		628	860
1600	Polypropylene									
1620	Cup sink, oval, integral strainers									
1640	6" x 3" I.D., 7" x 4" O.D.	Q-1	6	2.667	Ea.	117	141		258	340
1660	9" x 3" I.D., 10" x 4-1/2" O.D.	"	6	2.667		139	141		280	365
1740	1-1/2" diam. x 11" long					34.50			34.50	38
1980	For rough-in, supply, waste & vent, add	Q-1	1.70	9.412	↓	211	495		706	980

For customer support on your Plumbing Cost Data, call 877.709.6968.

291

22 42 16 – Commercial Lavatories and Sinks

22 42 16.40 Service Sinks

22 42 16.40 Service Sinks	Crew	Daily Output	Labor-Hours	Unit	Material	2015 Bare Costs Labor	Equipment	Total	Total Incl O&P
0010 **SERVICE SINKS**									
6650 Service, floor, corner, P.E. on C.I., 28" x 28"	Q-1	4.40	3.636	Ea.	1,000	192		1,192	1,400
6750 Vinyl coated rim guard, add					67.50			67.50	74
6760 Mop sink, molded stone, 24" x 36"	1 Plum	3.33	2.402		273	141		414	515
6770 Mop sink, molded stone, 24" x 36", w/rim 3 sides	"	3.33	2.402		278	141		419	520
6790 For rough-in, supply, waste & vent, floor service sinks	Q-1	1.64	9.756		705	515		1,220	1,550
7000 Service, wall, P.E. on C.I., roll rim, 22" x 18"		4	4		770	211		981	1,175
7100 24" x 20"	↓	4	4		850	211		1,061	1,250
7600 For stainless steel rim guard, two sides only, add					88.50			88.50	97.50
7800 For stainless steel rim guard, front only, add					56			56	62
8600 Vitreous china, 22" x 20"	Q-1	4	4		610	211		821	995
8960 For stainless steel rim guard, front or one side, add					56			56	62
8980 For rough-in, supply, waste & vent, wall service sinks	Q-1	1.30	12.308	↓	1,100	650		1,750	2,175

22 42 23 – Commercial Showers

22 42 23.30 Group Showers

22 42 23.30 Group Showers	Crew	Daily Output	Labor-Hours	Unit	Material	2015 Bare Costs Labor	Equipment	Total	Total Incl O&P
0010 **GROUP SHOWERS**									
6000 Group, w/pressure balancing valve, rough-in and rigging not included									
6800 Column, 6 heads, no receptors, less partitions	Q-1	3	5.333	Ea.	9,200	282		9,482	10,500
6900 With stainless steel partitions		1	16		11,900	845		12,745	14,400
7600 5 heads, no receptors, less partitions		3	5.333		6,350	282		6,632	7,400
7620 4 heads (1 handicap) no receptors, less partitions		3	5.333		5,650	282		5,932	6,625
7700 With stainless steel partitions		1	16		5,650	845		6,495	7,475
8000 Wall, 2 heads, no receptors, less partitions		4	4		2,725	211		2,936	3,300
8100 With stainless steel partitions	↓	2	8	↓	5,975	425		6,400	7,200

22 42 33 – Wash Fountains

22 42 33.20 Commercial Wash Fountains

22 42 33.20 Commercial Wash Fountains	Crew	Daily Output	Labor-Hours	Unit	Material	2015 Bare Costs Labor	Equipment	Total	Total Incl O&P
0010 **COMMERCIAL WASH FOUNTAINS** D2010-610									
1900 Group, foot control									
2000 Precast terrazzo, circular, 36" diam., 5 or 6 persons	Q-2	3	8	Ea.	7,275	440		7,715	8,650
2100 54" diameter for 8 or 10 persons		2.50	9.600		9,050	525		9,575	10,800
2400 Semi-circular, 36" diam. for 3 persons		3	8		6,375	440		6,815	7,675
2500 54" diam. for 4 or 5 persons		2.50	9.600		8,575	525		9,100	10,200
2700 Quarter circle (corner), 54" for 3 persons		3.50	6.857		7,850	375		8,225	9,225
3000 Stainless steel, circular, 36" diameter		3.50	6.857		6,550	375		6,925	7,775
3100 54" diameter		2.80	8.571		8,050	470		8,520	9,550
3400 Semi-circular, 36" diameter		3.50	6.857		5,025	375		5,400	6,125
3500 54" diameter		2.80	8.571		7,000	470		7,470	8,425
5000 Thermoplastic, pre-assembled, circular, 36" diameter		6	4		4,450	219		4,669	5,225
5100 54" diameter		4	6		5,175	330		5,505	6,200
5400 Semi-circular, 36" diameter		6	4		4,100	219		4,319	4,825
5600 54" diameter	↓	4	6	↓	5,000	330		5,330	6,000
5610 Group, infrared control, barrier free									
5614 Precast terrazzo									
5620 Semi-circular 36" diam. for 3 persons	Q-2	3	8	Ea.	7,250	440		7,690	8,625
5630 46" diam. for 4 persons		2.80	8.571		7,825	470		8,295	9,300
5640 Circular, 54" diam. for 8 persons, button control	↓	2.50	9.600		9,475	525		10,000	11,200
5700 Rough-in, supply, waste and vent for above wash fountains	Q-1	1.82	8.791		370	465		835	1,100
6200 Duo for small washrooms, stainless steel		2	8		2,725	425		3,150	3,625
6400 Bowl with backsplash		2	8		2,250	425		2,675	3,125
6500 Rough-in, supply, waste & vent for duo fountains	↓	2.02	7.921	↓	209	420		629	860

22 42 39 – Commercial Faucets, Supplies, and Trim

22 42 39.10 Faucets and Fittings	Crew	Daily Output	Labor-Hours	Unit	Material	2015 Bare Costs Labor	Equipment	Total	Total Incl O&P
0010 **FAUCETS AND FITTINGS**									
0840 Flush valves, with vacuum breaker									
0850 Water closet									
0860 Exposed, rear spud	1 Plum	8	1	Ea.	148	58.50		206.50	252
0870 Top spud		8	1		149	58.50		207.50	252
0880 Concealed, rear spud		8	1		197	58.50		255.50	305
0890 Top spud		8	1		159	58.50		217.50	264
0900 Wall hung		8	1		177	58.50		235.50	284
0910 Dual flush flushometer		12	.667		197	39		236	275
0912 Flushometer retrofit kit		18	.444		16.30	26		42.30	57.50
0920 Urinal									
0930 Exposed, stall	1 Plum	8	1	Ea.	149	58.50		207.50	253
0940 Wall, (washout)		8	1		149	58.50		207.50	252
0950 Pedestal, top spud		8	1		143	58.50		201.50	246
0960 Concealed, stall		8	1		156	58.50		214.50	261
0970 Wall (washout)		8	1		168	58.50		226.50	274
0971 Automatic flush sensor and operator for									
0972 urinals or water closets, standard [G]	1 Plum	8	1	Ea.	450	58.50		508.50	585
0980 High efficiency water saving									
0984 Water closets, 1.28 gpf [G]	1 Plum	8	1	Ea.	415	58.50		473.50	550
0988 Urinals, .5 gpf [G]	"	8	1	"	415	58.50		473.50	550
2790 Faucets for lavatories									
2800 Self-closing, center set	1 Plum	10	.800	Ea.	131	47		178	215
2810 Automatic sensor and operator, with faucet head		6.15	1.301		450	76.50		526.50	610
3000 Service sink faucet, cast spout, pail hook, hose end		14	.571		80	33.50		113.50	139

22 42 39.30 Carriers and Supports

22 42 39.30 Carriers and Supports	Crew	Daily Output	Labor-Hours	Unit	Material	2015 Bare Costs Labor	Equipment	Total	Total Incl O&P
0010 **CARRIERS AND SUPPORTS**, For plumbing fixtures									
0500 Drinking fountain, wall mounted									
0600 Plate type with studs, top back plate	1 Plum	7	1.143	Ea.	95	67		162	206
0700 Top front and back plate		7	1.143		116	67		183	229
0800 Top & bottom, front & back plates, w/bearing jacks		7	1.143		193	67		260	315
3000 Lavatory, concealed arm									
3050 Floor mounted, single									
3100 High back fixture	1 Plum	6	1.333	Ea.	525	78.50		603.50	700
3200 Flat slab fixture		6	1.333		455	78.50		533.50	620
3220 Paraplegic		6	1.333		590	78.50		668.50	770
3250 Floor mounted, back to back									
3300 High back fixtures	1 Plum	5	1.600	Ea.	750	94		844	965
3400 Flat slab fixtures		5	1.600		925	94		1,019	1,175
3430 Paraplegic		5	1.600		860	94		954	1,075
3500 Wall mounted, in stud or masonry									
3600 High back fixture	1 Plum	6	1.333	Ea.	320	78.50		398.50	470
3700 Flat slab fixture	"	6	1.333	"	270	78.50		348.50	415
4000 Exposed arm type, floor mounted									
4100 Single high back or flat slab fixture	1 Plum	6	1.333	Ea.	750	78.50		828.50	945
4200 Back to back, high back or flat slab fixtures		5	1.600		1,175	94		1,269	1,450
4300 Wall mounted, High back or flat slab lavatory		6	1.333		520	78.50		598.50	695
4600 Sink, floor mounted									
4650 Exposed arm system									
4700 Single heavy fixture	1 Plum	5	1.600	Ea.	880	94		974	1,100
4750 Single heavy sink with slab		5	1.600		1,075	94		1,169	1,350
4800 Back to back, standard fixtures		5	1.600		650	94		744	850

For customer support on your Plumbing Cost Data, call 877.709.6968.

293

22 42 39.30 Carriers and Supports		Crew	Daily Output	Labor-Hours	Unit	Material	2015 Bare Costs Labor	Equipment	Total	Total Incl O&P
4850	Back to back, heavy fixtures	1 Plum	5	1.600	Ea.	920	94		1,014	1,150
4900	Back to back, heavy sink with slab	↓	5	1.600	↓	920	94		1,014	1,150
4950	Exposed offset arm system									
5000	Single heavy deep fixture	1 Plum	5	1.600	Ea.	840	94		934	1,075
5100	Plate type system									
5200	With bearing jacks, single fixture	1 Plum	5	1.600	Ea.	885	94		979	1,125
5300	With exposed arms, single heavy fixture		5	1.600		1,225	94		1,319	1,500
5400	Wall mounted, exposed arms, single heavy fixture		5	1.600		450	94		544	635
6000	Urinal, floor mounted, 2" or 3" coupling, blowout type		6	1.333		560	78.50		638.50	735
6100	With fixture or hanger bolts, blowout or washout		6	1.333		390	78.50		468.50	550
6200	With bearing plate		6	1.333		445	78.50		523.50	610
6300	Wall mounted, plate type system	↓	6	1.333	↓	345	78.50		423.50	500
6980	Water closet, siphon jet									
7000	Horizontal, adjustable, caulk									
7040	Single, 4" pipe size	1 Plum	5.33	1.501	Ea.	890	88		978	1,100
7050	4" pipe size, paraplegic		5.33	1.501		890	88		978	1,100
7060	5" pipe size		5.33	1.501		965	88		1,053	1,175
7100	Double, 4" pipe size		5	1.600		1,575	94		1,669	1,875
7110	4" pipe size, paraplegic		5	1.600		1,575	94		1,669	1,875
7120	5" pipe size	↓	5	1.600	↓	1,700	94		1,794	2,025
7160	Horizontal, adjustable, extended, caulk									
7180	Single, 4" pipe size	1 Plum	5.33	1.501	Ea.	1,000	88		1,088	1,225
7200	5" pipe size		5.33	1.501		1,275	88		1,363	1,525
7240	Double, 4" pipe size		5	1.600		1,750	94		1,844	2,075
7260	5" pipe size	↓	5	1.600	↓	2,125	94		2,219	2,500
7400	Vertical, adjustable, caulk or thread									
7440	Single, 4" pipe size	1 Plum	5.33	1.501	Ea.	890	88		978	1,100
7460	5" pipe size		5.33	1.501		1,125	88		1,213	1,350
7480	6" pipe size		5	1.600		1,325	94		1,419	1,600
7520	Double, 4" pipe size		5	1.600		1,525	94		1,619	1,850
7540	5" pipe size		5	1.600		1,775	94		1,869	2,100
7560	6" pipe size	↓	4	2	↓	1,950	117		2,067	2,325
7600	Vertical, adjustable, extended, caulk									
7620	Single, 4" pipe size	1 Plum	5.33	1.501	Ea.	1,025	88		1,113	1,250
7640	5" pipe size		5.33	1.501		1,275	88		1,363	1,525
7680	6" pipe size		5	1.600		1,450	94		1,544	1,750
7720	Double, 4" pipe size		5	1.600		1,675	94		1,769	2,000
7740	5" pipe size		5	1.600		1,925	94		2,019	2,275
7760	6" pipe size	↓	4	2	↓	2,100	117		2,217	2,500
7780	Water closet, blow out									
7800	Vertical offset, caulk or thread									
7820	Single, 4" pipe size	1 Plum	5.33	1.501	Ea.	760	88		848	970
7840	Double, 4" pipe size	"	5	1.600	"	1,300	94		1,394	1,575
7880	Vertical offset, extended, caulk									
7900	Single, 4" pipe size	1 Plum	5.33	1.501	Ea.	950	88		1,038	1,175
7920	Double, 4" pipe size	"	5	1.600	"	1,500	94		1,594	1,800
7960	Vertical, for floor mounted back-outlet									
7980	Single, 4" thread, 2" vent	1 Plum	5.33	1.501	Ea.	670	88		758	870
8000	Double, 4" thread, 2" vent	"	6	1.333	"	1,975	78.50		2,053.50	2,275
8040	Vertical, for floor mounted back-outlet, extended									
8060	Single, 4" caulk, 2" vent	1 Plum	6	1.333	Ea.	670	78.50		748.50	855
8080	Double, 4" caulk, 2" vent	"	6	1.333	"	1,975	78.50		2,053.50	2,275
8200	Water closet, residential									

22 42 Commercial Plumbing Fixtures

22 42 39 – Commercial Faucets, Supplies, and Trim

22 42 39.30 Carriers and Supports		Crew	Daily Output	Labor-Hours	Unit	Material	2015 Bare Costs Labor	Equipment	Total	Total Incl O&P
8220	Vertical centerline, floor mount									
8240	Single, 3" caulk, 2" or 3" vent	1 Plum	6	1.333	Ea.	610	78.50		688.50	790
8260	4" caulk, 2" or 4" vent		6	1.333		785	78.50		863.50	980
8280	3" copper sweat, 3" vent		6	1.333		545	78.50		623.50	720
8300	4" copper sweat, 4" vent		6	1.333		660	78.50		738.50	845
8400	Vertical offset, floor mount									
8420	Single, 3" or 4" caulk, vent	1 Plum	4	2	Ea.	760	117		877	1,000
8440	3" or 4" copper sweat, vent		5	1.600		760	94		854	975
8460	Double, 3" or 4" caulk, vent		4	2		1,300	117		1,417	1,600
8480	3" or 4" copper sweat, vent		5	1.600		1,300	94		1,394	1,575
9000	Water cooler (electric), floor mounted									
9100	Plate type with bearing plate, single	1 Plum	6	1.333	Ea.	385	78.50		463.50	545
9140	Plate type with bearing plate, back to back	"	4	2	"	535	117		652	765

22 43 Healthcare Plumbing Fixtures

22 43 13 – Healthcare Water Closets

22 43 13.40 Water Closets

		Crew	Daily Output	Labor-Hours	Unit	Material	2015 Bare Costs Labor	Equipment	Total	Total Incl O&P
0010	**WATER CLOSETS**									
1000	Bowl only, 1 piece, w/seat and flush valve, ADA compliant 18" high									
1030	Floor mounted									
1150	With wall outlet	Q-1	5.30	3.019	Ea.	385	159		544	665
1180	For rough-in, supply, waste and vent		2.84	5.634		370	298		668	855
1200	With floor outlet		5.30	3.019		310	159		469	580
1800	For rough-in, supply, waste and vent		3.05	5.246		330	277		607	785
3100	Wall hung		5.80	2.759		945	146		1,091	1,275
3150	Hospital type, slotted rim for bed pan									
3156	Elongated bowl, top spud	Q-1	5.80	2.759	Ea.	600	146		746	880
3160	Elongated bowl, rear spud		5.80	2.759		360	146		506	615
3200	For rough-in, supply, waste and vent, single WC		2.56	6.250		1,075	330		1,405	1,675
3300	Floor mounted									
3320	Bariatric, (1,200 lb. capacity), elongated bowl, ADA	Q-1	4.60	3.478	Ea.	2,775	184		2,959	3,325
3360	Hospital type, slotted rim for bed pan									
3370	Elongated bowl, top spud	Q-1	5	3.200	Ea.	320	169		489	605
3380	Elongated bowl, rear spud		5	3.200		410	169		579	705
3500	For rough-in, supply, waste and vent		3.05	5.246		330	277		607	785

22 43 16 – Healthcare Sinks

22 43 16.10 Sinks

		Crew	Daily Output	Labor-Hours	Unit	Material	2015 Bare Costs Labor	Equipment	Total	Total Incl O&P
0010	**SINKS**									
0020	Vitreous china									
6702	Hospital type, without trim (see Section 22 41 39.10)									
6710	20" x 18", contoured splash shield	Q-1	8	2	Ea.	93.50	106		199.50	263
6730	28" x 20", surgeon, side decks		8	2		575	106		681	795
6740	28" x 22", surgeon scrub-up, deep bowl		8	2		825	106		931	1,075
6750	20" x 27", patient, wheelchair		7	2.286		605	121		726	845
6760	30" x 22", all purpose		7	2.286		685	121		806	935
6770	30" x 22", plaster work		7	2.286		685	121		806	935
6820	20" x 24" clinic service, liquid/solid waste		6	2.667		965	141		1,106	1,275

For customer support on your Plumbing Cost Data, call 877.709.6968.

295

22 43 Healthcare Plumbing Fixtures

22 43 19 – Healthcare Bathtubs

22 43 19.10 Bathtubs

	Crew	Daily Output	Labor-Hours	Unit	Material	2015 Bare Costs Labor	Equipment	Total	Total Incl O&P
0010 **BATHTUBS**									
5002 Hospital type, with trim, see Section 22 41 39.10									
5050 Bathing pool, porcelain enamel on cast iron, grab bars									
5060 pop-up drain, 72" x 36"	Q-1	3	5.333	Ea.	3,750	282		4,032	4,550
5100 Perineal (sitz), vitreous china		3	5.333		1,300	282		1,582	1,850
5120 For pedestal, vitreous china, add		8	2		251	106		357	435
5300 Whirlpool, porcelain enamel on cast iron, 72" x 36"		1	16		3,575	845		4,420	5,200
5310 For color add					5%				
5311 For designer colors and trim add					15%				

22 43 23 – Healthcare Showers

22 43 23.10 Showers

	Crew	Daily Output	Labor-Hours	Unit	Material	2015 Bare Costs Labor	Equipment	Total	Total Incl O&P
0010 **SHOWERS**									
5950 Module, handicap, SS panel, fixed & hand held head, control									
5960 valves, grab bar, curtain & rod, folding seat	1 Plum	4	2	Ea.	2,875	117		2,992	3,325

22 43 39 – Healthcare Faucets

22 43 39.10 Faucets and Fittings

	Crew	Daily Output	Labor-Hours	Unit	Material	2015 Bare Costs Labor	Equipment	Total	Total Incl O&P
0010 **FAUCETS AND FITTINGS**									
2850 Medical, bedpan cleanser, with pedal valve,	1 Plum	12	.667	Ea.	760	39		799	895
2860 With screwdriver stop valve		12	.667		395	39		434	490
2870 With self-closing spray valve		12	.667		242	39		281	325
2900 Faucet, gooseneck spout, wrist handles, grid drain		10	.800		194	47		241	284
2940 Mixing valve, knee action, screwdriver stops		4	2		425	117		542	640

22 45 Emergency Plumbing Fixtures

22 45 13 – Emergency Showers

22 45 13.10 Emergency Showers

	Crew	Daily Output	Labor-Hours	Unit	Material	2015 Bare Costs Labor	Equipment	Total	Total Incl O&P
0010 **EMERGENCY SHOWERS**, Rough-in not included									
5000 Shower, single head, drench, ball valve, pull, freestanding	Q-1	4	4	Ea.	380	211		591	735
5200 Horizontal or vertical supply		4	4		555	211		766	930
6000 Multi-nozzle, eye/face wash combination		4	4		660	211		871	1,050
6400 Multi-nozzle, 12 spray, shower only		4	4		2,000	211		2,211	2,525
6600 For freeze-proof, add		6	2.667		465	141		606	730
8000 Walk-thru decontamination with eye-face wash		2	8		3,375	425		3,800	4,350
8200 For freeze proof, add		4	4		565	211		776	945

22 45 16 – Eyewash Equipment

22 45 16.10 Eyewash Safety Equipment

	Crew	Daily Output	Labor-Hours	Unit	Material	2015 Bare Costs Labor	Equipment	Total	Total Incl O&P
0010 **EYEWASH SAFETY EQUIPMENT**, Rough-in not included									
1000 Eye wash fountain									
1400 Plastic bowl, pedestal mounted	Q-1	4	4	Ea.	282	211		493	630
1600 Unmounted		4	4		247	211		458	590
1800 Wall mounted		4	4		455	211		666	820
2000 Stainless steel, pedestal mounted		4	4		350	211		561	705
2200 Unmounted		4	4		272	211		483	620
2400 Wall mounted		4	4		288	211		499	635

22 45 Emergency Plumbing Fixtures

22 45 19 – Self-Contained Eyewash Equipment

22 45 19.10 Self-Contained Eyewash Safety Equipment	Crew	Daily Output	Labor-Hours	Unit	Material	2015 Bare Costs Labor	Equipment	Total	Total Incl O&P
0010 **SELF-CONTAINED EYEWASH SAFETY EQUIPMENT**									
3000 Eye wash, portable, self-contained				Ea.	1,025			1,025	1,125

22 45 26 – Eye/Face Wash Equipment

22 45 26.10 Eye/Face Wash Safety Equipment

	Crew	Daily Output	Labor-Hours	Unit	Material	2015 Bare Costs Labor	Equipment	Total	Total Incl O&P
0010 **EYE/FACE WASH SAFETY EQUIPMENT**, Rough-in not included									
4000 Eye and face wash, combination fountain									
4200 Stainless steel, pedestal mounted	Q-1	4	4	Ea.	1,075	211		1,286	1,500
4400 Unmounted		4	4		272	211		483	620
4600 Wall mounted		4	4		246	211		457	590

22 46 Security Plumbing Fixtures

22 46 13 – Security Water Closets and Urinals

22 46 13.10 Security Water Closets and Urinals

	Crew	Daily Output	Labor-Hours	Unit	Material	2015 Bare Costs Labor	Equipment	Total	Total Incl O&P
0010 **SECURITY WATER CLOSETS AND URINALS**, Stainless steel									
2000 Urinal, back supply and flush									
2200 Wall hung	Q-1	4	4	Ea.	1,225	211		1,436	1,675
2240 Stall		2.50	6.400		2,100	340		2,440	2,800
2300 For urinal rough-in, supply, waste and vent		1.49	10.738		320	565		885	1,200
3000 Water closet, integral seat, back supply and flush									
3300 Wall hung, wall outlet	Q-1	5.80	2.759	Ea.	985	146		1,131	1,300
3400 Floor mount, wall outlet		5.80	2.759		1,275	146		1,421	1,625
3440 Floor mount, floor outlet		5.80	2.759		1,300	146		1,446	1,650
3480 For recessed tissue holder, add					88			88	96.50
3500 For water closet rough-in, supply, waste and vent	Q-1	1.19	13.445		345	710		1,055	1,450
5000 Water closet and lavatory units, push button filler valves,									
5010 soap & paper holders, seat									
5300 Wall hung	Q-1	5	3.200	Ea.	2,100	169		2,269	2,575
5400 Floor mount		5	3.200		2,000	169		2,169	2,450
6300 For unit rough-in, supply, waste and vent		1	16		390	845		1,235	1,700

22 46 16 – Security Lavatories and Sinks

22 46 16.13 Security Lavatories

	Crew	Daily Output	Labor-Hours	Unit	Material	2015 Bare Costs Labor	Equipment	Total	Total Incl O&P
0010 **SECURITY LAVATORIES**, Stainless steel									
1000 Lavatory, wall hung, push button filler valve									
1100 Rectangular bowl	Q-1	8	2	Ea.	1,075	106		1,181	1,350
1200 Oval bowl		8	2		1,125	106		1,231	1,375
1240 Oval bowl, corner mount		8	2		1,225	106		1,331	1,500
1300 For lavatory rough-in, supply, waste and vent		1.50	10.667		282	565		847	1,150

22 46 63 – Security Service Sink

22 46 63.10 Security Service Sink

	Crew	Daily Output	Labor-Hours	Unit	Material	2015 Bare Costs Labor	Equipment	Total	Total Incl O&P
0010 **SECURITY SERVICE SINK**, Stainless steel									
1700 Service sink, with soap dish									
1740 24" x 19" size	Q-1	3	5.333	Ea.	2,175	282		2,457	2,825
1790 For sink rough-in, supply, waste and vent	"	.89	17.978	"	620	950		1,570	2,100

22 46 Security Plumbing Fixtures

22 46 73 – Security Shower

22 46 73.10 Security Shower	Crew	Daily Output	Labor-Hours	Unit	Material	2015 Bare Costs Labor	Equipment	Total	Total Incl O&P
0010 **SECURITY SHOWER**, Stainless steel									
1800 Shower cabinet, unitized									
1840 36" x 36" x 88"	Q-1	2.20	7.273	Ea.	4,950	385		5,335	6,025
1900 Shower package for built-in									
1940 Hot & cold valves, recessed soap dish	Q-1	6	2.667	Ea.	540	141		681	805

22 47 Drinking Fountains and Water Coolers

22 47 13 – Drinking Fountains

22 47 13.10 Drinking Water Fountains

22 47 13.10 Drinking Water Fountains		Crew	Daily Output	Labor-Hours	Unit	Material	2015 Bare Costs Labor	Equipment	Total	Total Incl O&P
0010 **DRINKING WATER FOUNTAINS**, For connection to cold water supply										
0802 For remote water chiller, see Section 22 47 23.10										
1000 Wall mounted, non-recessed	R224000-30									
1200 Aluminum,										
1280 Dual bubbler type		1 Plum	3.20	2.500	Ea.	2,450	147		2,597	2,925
1400 Bronze, with no back			4	2		1,000	117		1,117	1,275
1600 Cast iron, enameled, low back, single bubbler	D2010-810		4	2		945	117		1,062	1,225
1640 Dual bubbler type			3.20	2.500		1,525	147		1,672	1,900
1680 Triple bubbler type			3.20	2.500		1,975	147		2,122	2,400
1800 Cast aluminum, enameled, for correctional institutions			4	2		1,675	117		1,792	2,000
2000 Fiberglass, 12" back, single bubbler unit			4	2		1,925	117		2,042	2,300
2040 Dual bubbler			3.20	2.500		2,225	147		2,372	2,650
2080 Triple bubbler			3.20	2.500		2,475	147		2,622	2,950
2200 Polymarble, no back, single bubbler			4	2		705	117		822	950
2240 Dual bubbler			3.20	2.500		1,800	147		1,947	2,200
2280 Triple bubbler			3.20	2.500		2,125	147		2,272	2,575
2400 Precast stone, no back			4	2		915	117		1,032	1,175
2700 Stainless steel, single bubbler, no back			4	2		1,050	117		1,167	1,325
2740 With back			4	2		565	117		682	795
2780 Dual handle & wheelchair projection type			4	2		745	117		862	995
2820 Dual level for handicapped type			3.20	2.500		1,575	147		1,722	1,950
2840 Vandal resistant type			4	2		650	117		767	890
3300 Vitreous china										
3340 7" back		1 Plum	4	2	Ea.	620	117		737	855
3940 For vandal-resistant bottom plate, add						77			77	84.50
3960 For freeze-proof valve system, add		1 Plum	2	4		705	235		940	1,125
3980 For rough-in, supply and waste, add		"	2.21	3.620		174	212		386	510
4000 Wall mounted, semi-recessed										
4200 Poly-marble, single bubbler		1 Plum	4	2	Ea.	955	117		1,072	1,225
4600 Stainless steel, satin finish, single bubbler			4	2		1,150	117		1,267	1,450
4900 Vitreous china, single bubbler			4	2		895	117		1,012	1,150
5980 For rough-in, supply and waste, add			1.83	4.372		174	257		431	580
6000 Wall mounted, fully recessed										
6400 Poly-marble, single bubbler		1 Plum	4	2	Ea.	1,600	117		1,717	1,925
6440 For water glass filler, add						86.50			86.50	95
6800 Stainless steel, single bubbler		1 Plum	4	2		1,550	117		1,667	1,875
6900 Fountain and cuspidor combination			2	4		3,000	235		3,235	3,650
7560 For freeze-proof valve system, add			2	4		795	235		1,030	1,225
7580 For rough-in, supply and waste, add			1.83	4.372		174	257		431	580
7600 Floor mounted, pedestal type										
7700 Aluminum, architectural style, C.I. base		1 Plum	2	4	Ea.	2,250	235		2,485	2,825

22 47 Drinking Fountains and Water Coolers

22 47 13 – Drinking Fountains

22 47 13.10 Drinking Water Fountains		Crew	Daily Output	Labor-Hours	Unit	Material	2015 Bare Costs Labor	Equipment	Total	Total Incl O&P
7780	Wheelchair handicap unit ♿	1 Plum	2	4	Ea.	1,625	235		1,860	2,125
8000	Bronze, architectural style		2	4		1,925	235		2,160	2,475
8040	Enameled steel cylindrical column style		2	4		2,200	235		2,435	2,775
8200	Precast stone/concrete, cylindrical column		1	8		1,375	470		1,845	2,200
8240	Wheelchair handicap unit ♿		1	8		2,725	470		3,195	3,700
8400	Stainless steel, architectural style		2	4		1,750	235		1,985	2,275
8600	Enameled iron, heavy duty service, 2 bubblers		2	4		2,575	235		2,810	3,200
8660	4 bubblers		2	4		3,950	235		4,185	4,675
8880	For freeze-proof valve system, add		2	4		705	235		940	1,125
8900	For rough-in, supply and waste, add		1.83	4.372		174	257		431	580
9100	Deck mounted									
9500	Stainless steel, circular receptor	1 Plum	4	2	Ea.	435	117		552	655
9540	14" x 9" receptor		4	2		360	117		477	570
9580	25" x 17" deep receptor, with water glass filler		3	2.667		305	157		462	570
9760	White enameled steel, 14" x 9" receptor		4	2		375	117		492	590
9860	White enameled cast iron, 24" x 16" receptor		3	2.667		460	157		617	740
9980	For rough-in, supply and waste, add		1.83	4.372		174	257		431	580

22 47 16 – Pressure Water Coolers

22 47 16.10 Electric Water Coolers

		Crew	Daily Output	Labor-Hours	Unit	Material	2015 Bare Costs Labor	Equipment	Total	Total Incl O&P
0010	**ELECTRIC WATER COOLERS** D2010-820									
0100	Wall mounted, non-recessed									
0140	4 GPH	Q-1	4	4	Ea.	680	211		891	1,075
0160	8 GPH, barrier free, sensor operated ♿		4	4		1,025	211		1,236	1,475
0180	8.2 GPH		4	4		750	211		961	1,150
0220	14.3 GPH		4	4		875	211		1,086	1,275
0600	8 GPH hot and cold water		4	4		1,050	211		1,261	1,500
0640	For stainless steel cabinet, add					90			90	99.50
1000	Dual height, 8.2 GPH	Q-1	3.80	4.211		2,075	222		2,297	2,600
1040	14.3 GPH	"	3.80	4.211		975	222		1,197	1,400
1240	For stainless steel cabinet, add					171			171	188
2600	Wheelchair type, 8 GPH ♿	Q-1	4	4		915	211		1,126	1,325
3000	Simulated recessed, 8 GPH		4	4		665	211		876	1,050
3040	11.5 GPH		4	4		1,025	211		1,236	1,450
3200	For glass filler, add					99			99	109
3240	For stainless steel cabinet, add					73.50			73.50	81
3300	Semi-recessed, 8.1 GPH	Q-1	4	4		750	211		961	1,150
3320	12 GPH	"	4	4		865	211		1,076	1,275
3340	For glass filler, add					129			129	142
3360	For stainless steel cabinet, add					145			145	160
3400	Full recessed, stainless steel, 8 GPH	Q-1	3.50	4.571		2,025	242		2,267	2,600
3420	11.5 GPH	"	3.50	4.571		1,650	242		1,892	2,175
3460	For glass filler, add					146			146	161
3600	For mounting can only					243			243	268
4600	Floor mounted, flush-to-wall									
4640	4 GPH	1 Plum	3	2.667	Ea.	740	157		897	1,050
4680	8.2 GPH		3	2.667		780	157		937	1,100
4720	14.3 GPH		3	2.667		890	157		1,047	1,225
4960	14 GPH hot and cold water		3	2.667		1,100	157		1,257	1,450
4980	For stainless steel cabinet, add					134			134	148
5000	Dual height, 8.2 GPH	1 Plum	2	4		1,175	235		1,410	1,625
5040	14.3 GPH	"	2	4		1,200	235		1,435	1,675
5120	For stainless steel cabinet, add					196			196	215

22 47 Drinking Fountains and Water Coolers

22 47 16 – Pressure Water Coolers

22 47 16.10 Electric Water Coolers		Crew	Daily Output	Labor-Hours	Unit	Material	2015 Bare Costs Labor	2015 Bare Costs Equipment	Total	Total Incl O&P
5600	Explosion Proof, 16 GPH	1 Plum	3	2.667	Ea.	2,075	157		2,232	2,525
6000	Refrigerator Compartment Type, 4.5 GPH		3	2.667		1,500	157		1,657	1,875
6600	Bottle Supply Type, 1.0 GPH		4	2		395	117		512	610
6640	Hot and cold, 1.0 GPH		4	2		535	117		652	765
9800	For supply, waste & vent, all coolers		2.21	3.620		174	212		386	510

22 47 23 – Remote Water Coolers

22 47 23.10 Remote Water Coolers

		Crew	Daily Output	Labor-Hours	Unit	Material	2015 Bare Costs Labor	2015 Bare Costs Equipment	Total	Total Incl O&P
0010	**REMOTE WATER COOLERS**, 80°F inlet									
0100	Air cooled, 50°F outlet, 115 V, 4.1 GPH	1 Plum	6	1.333	Ea.	525	78.50		603.50	700
0200	5.7 GPH		5.50	1.455		835	85.50		920.50	1,050
0300	8.0 GPH		5	1.600		625	94		719	830
0400	10.0 GPH		4.50	1.778		920	104		1,024	1,150
0500	13.4 GPH		4	2		1,425	117		1,542	1,750
0700	29 GPH	Q-1	5	3.200		1,675	169		1,844	2,100
1000	230V, 32 GPH	"	5	3.200		1,750	169		1,919	2,175

22 51 Swimming Pool Plumbing Systems

22 51 19 – Swimming Pool Water Treatment Equipment

22 51 19.50 Swimming Pool Filtration Equipment

		Crew	Daily Output	Labor-Hours	Unit	Material	2015 Bare Costs Labor	2015 Bare Costs Equipment	Total	Total Incl O&P
0010	**SWIMMING POOL FILTRATION EQUIPMENT**									
0900	Filter system, sand or diatomite type, incl. pump, 6,000 gal./hr.	2 Plum	1.80	8.889	Total	1,975	520		2,495	2,975
1020	Add for chlorination system, 800 S.F. pool		3	5.333	Ea.	181	315		496	675
1040	5,000 S.F. pool		3	5.333	"	1,850	315		2,165	2,500

22 52 Fountain Plumbing Systems

22 52 16 – Fountain Pumps

22 52 16.10 Fountain Water Pumps

		Crew	Daily Output	Labor-Hours	Unit	Material	2015 Bare Costs Labor	2015 Bare Costs Equipment	Total	Total Incl O&P
0010	**FOUNTAIN WATER PUMPS**									
0100	Pump w/controls									
0200	Single phase, 100' cord, 1/2 H.P. pump	2 Skwk	4.40	3.636	Ea.	1,275	177		1,452	1,675
0300	3/4 H.P. pump		4.30	3.721		2,175	181		2,356	2,675
0400	1 H.P. pump		4.20	3.810		2,350	185		2,535	2,875
0500	1-1/2 H.P. pump		4.10	3.902		2,800	190		2,990	3,375
0600	2 H.P. pump		4	4		3,775	195		3,970	4,475
0700	Three phase, 200' cord, 5 H.P. pump		3.90	4.103		5,175	200		5,375	6,000
0800	7-1/2 H.P. pump		3.80	4.211		9,000	205		9,205	10,200
0900	10 H.P. pump		3.70	4.324		13,300	210		13,510	15,000
1000	15 H.P. pump		3.60	4.444		16,800	216		17,016	18,800
2000	DESIGN NOTE: Use two horsepower per surface acre.									

22 52 33 – Fountain Ancillary

22 52 33.10 Fountain Miscellaneous

		Crew	Daily Output	Labor-Hours	Unit	Material	2015 Bare Costs Labor	2015 Bare Costs Equipment	Total	Total Incl O&P
0010	**FOUNTAIN MISCELLANEOUS**									
1300	Lights w/mounting kits, 200 watt	2 Skwk	18	.889	Ea.	1,050	43		1,093	1,225
1400	300 watt		18	.889		1,275	43		1,318	1,475
1500	500 watt		18	.889		1,425	43		1,468	1,650
1600	Color blender		12	1.333		555	65		620	710

22 62 19 – Vacuum Equipment for Laboratory and Healthcare Facilities

22 62 19.70 Healthcare Vacuum Equipment	Crew	Daily Output	Labor-Hours	Unit	Material	2015 Bare Costs Labor	Equipment	Total	Total Incl O&P
0010 **HEALTHCARE VACUUM EQUIPMENT**									
0300 Dental oral									
0310 Duplex									
0330 165 SCFM with 77 gal. separator	Q-2	1.30	18.462	Ea.	52,000	1,000		53,000	58,500
1100 Vacuum system									
1110 Vacuum outlet alarm panel	1 Plum	3.20	2.500	Ea.	1,100	147		1,247	1,450
2000 Medical, with receiver									
2100 Rotary vane type, lubricated, with controls									
2110 Simplex									
2120 1.5 HP, 80 gal tank	Q-1	8	2	Ea.	6,525	106		6,631	7,325
2130 2 HP, 80 gal tank		7	2.286		6,750	121		6,871	7,600
2140 3 HP, 80 gal tank		6.60	2.424		7,125	128		7,253	8,025
2150 5 HP, 80 gal tank		6	2.667		7,775	141		7,916	8,775
2200 Duplex									
2210 1 HP, 80 gal tank	Q-1	8.40	1.905	Ea.	10,400	101		10,501	11,700
2220 1.5 HP, 80 gal tank		7.80	2.051		10,700	108		10,808	12,000
2230 2 HP, 80 gal tank		6.80	2.353		11,600	124		11,724	12,900
2240 3 HP, 120 gal tank		6	2.667		12,600	141		12,741	14,000
2250 5 HP, 120 gal tank		5.40	2.963		14,300	157		14,457	15,900
2260 7.5 HP, 200 gal tank	Q-2	7	3.429		20,400	188		20,588	22,800
2270 10 HP, 200 gal tank		6.40	3.750		23,600	205		23,805	26,300
2280 15 HP, 200 gal tank		5.80	4.138		42,200	227		42,427	46,800
2290 20 HP, 200 gal tank		5	4.800		48,100	263		48,363	53,500
2300 25 HP, 200 gal tank		4	6		56,000	330		56,330	62,000
2400 Triplex									
2410 7.5 HP, 200 gal tank	Q-2	6.40	3.750	Ea.	32,400	205		32,605	36,000
2420 10 HP, 200 gal tank		5.70	4.211		37,800	231		38,031	41,900
2430 15 HP, 200 gal tank		4.90	4.898		64,000	268		64,268	70,500
2440 20 HP, 200 gal tank		4	6		73,000	330		73,330	80,500
2450 25 HP, 200 gal tank		3.70	6.486		84,000	355		84,355	93,000
2500 Quadruplex									
2510 7.5 HP, 200 gal tank	Q-2	5.30	4.528	Ea.	40,400	248		40,648	44,800
2520 10 HP, 200 gal tank		4.70	5.106		46,600	280		46,880	51,500
2530 15 HP, 200 gal tank		4	6		84,000	330		84,330	93,000
2540 20 HP, 200 gal tank		3.60	6.667		96,000	365		96,365	106,000
2550 25 HP, 200 gal tank		3.20	7.500		111,500	410		111,910	123,000
4000 Liquid ring type , water sealed, with controls									
4200 Duplex									
4210 1.5 HP, 120 gal tank	Q-1	6	2.667	Ea.	20,600	141		20,741	22,900
4220 3 HP, 120 gal tank	"	5.50	2.909		21,500	154		21,654	23,900
4230 4 HP, 120 gal tank	Q-2	6.80	3.529		22,900	193		23,093	25,500
4240 5 HP, 120 gal tank		6.50	3.692		24,200	202		24,402	26,900
4250 7.5 HP, 200 gal tank		6.20	3.871		29,400	212		29,612	32,700
4260 10 HP, 200 gal tank		5.80	4.138		36,700	227		36,927	40,700
4270 15 HP, 200 gal tank		5.10	4.706		44,500	258		44,758	49,400
4280 20 HP, 200 gal tank		4.60	5.217		57,500	286		57,786	64,000
4290 30 HP, 200 gal tank		4	6		69,000	330		69,330	76,000

For customer support on your Plumbing Cost Data, call 877.709.6968.

301

22 63 13.70 Healthcare Gas Piping	Crew	Daily Output	Labor-Hours	Unit	Material	2015 Bare Costs Labor	2015 Bare Costs Equipment	Total	Total Incl O&P
0010 **HEALTHCARE GAS PIPING**									
0030 Air compressor intake filter									
0034 Rooftop									
0036 Filter/silencer									
0040 1"	1 Stpi	10	.800	Ea.	210	48		258	305
0044 1-1/4"		9.60	.833		216	50		266	315
0048 1-1/2"		9.20	.870		232	52		284	335
0052 2"		8.80	.909		232	54.50		286.50	335
0056 2-1/2"		8.40	.952		310	57		367	425
0060 3"		8	1		370	60		430	500
0064 4"		7.60	1.053		385	63		448	520
0068 5"		7.40	1.081		540	64.50		604.50	695
0076 Inline									
0080 Filter/silencer									
0084 2"	1 Stpi	8.20	.976	Ea.	700	58.50		758.50	860
0088 2-1/2"		8	1		720	60		780	885
0090 3"		7.60	1.053		795	63		858	970
0092 4"		7.20	1.111		930	66.50		996.50	1,125
0094 5"		6.80	1.176		1,000	70.50		1,070.50	1,200
0096 6"		6.40	1.250		1,125	74.50		1,199.50	1,375
1000 Nitrogen or oxygen system									
1010 Cylinder manifold									
1020 5 cylinder	1 Plum	.80	10	Ea.	5,950	585		6,535	7,425
1026 10 cylinder	"	.40	20	"	7,175	1,175		8,350	9,650
1900 Vaporizers									
1910 LOX vaporizers									
1920 Nominal capacity									
1930 1410 SCFM	Q-1	2	8	Ea.	2,000	425		2,425	2,850
1940 5650 SCFM		1.40	11.429		4,000	605		4,605	5,300
1950 12703 SCFM		.80	20		6,000	1,050		7,050	8,200
1980 Removal of LOX vaporizers									
1982 Nominal capacity									
1986 1410 SCFM	Q-1	4	4	Ea.		211		211	320
1990 5650 SCFM		2.80	5.714			300		300	455
1994 12703 SCFM		1.60	10			530		530	800
3000 Outlets and valves									
3010 Recessed, wall mounted									
3012 Single outlet	1 Plum	3.20	2.500	Ea.	71.50	147		218.50	300
3100 Ceiling outlet									
3190 Zone valve with box									
3192 Cleaned for oxygen service, not including gauges									
3194 1/2" valve size	1 Plum	4.60	1.739	Ea.	214	102		316	390
3196 3/4" valve size		4.30	1.860		236	109		345	425
3198 1" valve size		4	2		261	117		378	465
3202 1-1/4" valve size		3.80	2.105		285	124		409	500
3206 1-1/2" valve size		3.60	2.222		320	130		450	545
3210 2" valve size		3.20	2.500		420	147		567	680
3214 2-1/2" valve size		3.10	2.581		970	151		1,121	1,300
3218 3" valve size		3	2.667		1,350	157		1,507	1,700
3224 Gauges for zone valve box									
3226 0-100 PSI (O2, air, N2, CO2)	1 Plum	16	.500	Ea.	18.75	29.50		48.25	65
3228 Vacuum, WAGD		16	.500		18.75	29.50		48.25	65
3230 0-300 PSI (Nitrogen)		16	.500		18.75	29.50		48.25	65

22 63 Gas Systems for Laboratory and Healthcare Facilities

22 63 13 – Gas Piping for Laboratory and Healthcare Facilities

22 63 13.70 Healthcare Gas Piping	Crew	Daily Output	Labor-Hours	Unit	Material	2015 Bare Costs Labor	Equipment	Total	Total Incl O&P
4000 Alarm panel, medical gases and vacuum									
4010 Alarm panel	1 Plum	3.20	2.500	Ea.	1,100	147		1,247	1,450
4030 Master alarm panel									
4034 Can also monitor area alarms									
4038 and communicate with PC-based alarm monitor.									
4040 10 Signal	1 Elec	3.20	2.500	Ea.	980	137		1,117	1,275
4044 20 Signal		3	2.667		1,125	146		1,271	1,475
4048 30 Signal		2.80	2.857		1,500	156		1,656	1,875
4052 40 Signal		2.60	3.077		1,675	168		1,843	2,075
4056 50 Signal		2.40	3.333		1,825	182		2,007	2,300
4060 60 Signal		2.20	3.636		2,150	199		2,349	2,650
4100 Area alarm panel									
4104 Does not include specific gas transducers.									
4108 3 Module alarm panel									
4112 P-P-P	1 Elec	2.80	2.857	Ea.	1,300	156		1,456	1,650
4116 P-P-V		2.80	2.857		875	156		1,031	1,200
4120 D-D-D		2.80	2.857		960	156		1,116	1,275
4130 6 Module alarm panel									
4132 4-P, 3-V	1 Elec	2.20	3.636	Ea.	1,400	199		1,599	1,850
4136 3-P, 2-V, B		2.20	3.636		1,250	199		1,449	1,675
4140 5-D, B		2.20	3.636		1,350	199		1,549	1,775
4144 6-D		2.20	3.636		1,775	199		1,974	2,275
4148 3-P, 3-V		2.20	3.636		1,650	199		1,849	2,125
4170 Note: P = pressure, V = vacuum, B = blank, D = dual display									
4180 Alarm transducers, gas specific									
4182 Oxygen	1 Elec	24	.333	Ea.	148	18.25		166.25	191
4184 Vacuum		24	.333		148	18.25		166.25	191
4186 Nitrous oxide		24	.333		148	18.25		166.25	191
4188 Medical air		24	.333		148	18.25		166.25	191
4190 Carbon dioxide		24	.333		148	18.25		166.25	191
4192 Nitrogen		24	.333		148	18.25		166.25	191
4194 WAGD		24	.333		158	18.25		176.25	201
4300 Ball valves cleaned for oxygen service									
4310 with copper extensions and gauge port									
4320 1/4" diam.	1 Plum	24	.333	Ea.	52.50	19.55		72.05	87.50
4330 1/2" diam.		22	.364		58.50	21.50		80	96.50
4334 3/4" diam.		20	.400		82.50	23.50		106	127
4338 1" diam.		19	.421		105	24.50		129.50	154
4342 1-1/4" diam.		15	.533		127	31.50		158.50	187
4346 1-1/2" diam.		13	.615		176	36		212	249
4350 2" diam.		11	.727		254	42.50		296.50	345
4354 2-1/2" diam.	Q-1	15	1.067		700	56.50		756.50	850
4358 3" diam.		13	1.231		1,050	65		1,115	1,250
4362 4" diam.		10	1.600		1,975	84.50		2,059.50	2,300
5000 Manifold									
5010 Automatic switchover type									
5020 Note: Both a control panel and header assembly are required									
5030 Control panel									
5040 Oxygen	Q-1	4	4	Ea.	4,000	211		4,211	4,725
5060 Header assembly									
5066 Oxygen									
5070 2 x 2	Q-1	6	2.667	Ea.	655	141		796	935
5074 3 x 3		5.50	2.909		770	154		924	1,075

For customer support on your Plumbing Cost Data, call 877.709.6968.

303

22 63 13.70 Healthcare Gas Piping		Crew	Daily Output	Labor-Hours	Unit	Material	2015 Bare Costs Labor	Equipment	Total	Total Incl O&P
5078	4 x 4	Q-1	5	3.200	Ea.	1,000	169		1,169	1,350
5082	5 x 5		4.50	3.556		1,125	188		1,313	1,500
5086	6 x 6		4	4		1,400	211		1,611	1,875
5090	7 x 7	↓	3.50	4.571	↓	1,525	242		1,767	2,050
7000	Medical air compressors									
7020	Oil-less with inlet filter, duplexed dryers and aftercoolers									
7030	Duplex systems, horizontal tank, 208/230/460/575 V, 3 Ph.									
7040	1 HP, 80 gal. tank, 3.8 ACFM @50PSIG, 2.7 ACFM @100PSIG	Q-5	4	4	Ea.	35,800	215		36,015	39,600
7050	1.5 HP, 80 gal. tank, 6.8 ACFM @50PSIG, 4.5 ACFM @100PSIG		3.80	4.211		35,800	226		36,026	39,600
7060	2 HP, 80 gal. tank, 8.2 ACFM @50PSIG, 6.3 ACFM @100PSIG		3.60	4.444		36,000	239		36,239	40,000
7070	3 HP, 120 gal. tank, 11.2 ACFM @50PSIG, 9.4 ACFM @100PSIG		3.20	5		39,600	269		39,869	43,900
7080	5 HP, 120 gal. tank, 17.4 ACFM @50PSIG, 15.6 ACFM @100PSIG		2.80	5.714		47,700	305		48,005	53,000
7090	7.5 HP, 250 gal. tank, 31.6 ACFM @50PSIG, 25.9 ACFM @100PSIG		2.40	6.667		62,000	360		62,360	68,500
7100	10 HP, 250 gal. tank, 43 ACFM @50PSIG, 35.2 ACFM @100PSIG		2	8		64,500	430		64,930	71,500
7110	15 HP, 250 gal. tank, 69 ACFM @50PSIG, 56.5 ACFM @100PSIG	↓	1.80	8.889	↓	78,000	480		78,480	86,000
7200	Aftercooler, air-cooled									
7210	Steel manifold, copper tube, aluminum fins									
7220	35 SCFM @ 100PSIG	Q-5	4	4	Ea.	895	215		1,110	1,300
7300	Air dryer system									
7310	Refrigerated type									
7320	Flow @ 125 PSI									
7330	20 SCFM	Q-5	6	2.667	Ea.	1,475	143		1,618	1,850
7332	25 SCFM		5.80	2.759		1,550	148		1,698	1,925
7334	35 SCFM		5.40	2.963		1,875	159		2,034	2,325
7336	50 SCFM		5	3.200		2,350	172		2,522	2,825
7338	75 SCFM		4.60	3.478		2,950	187		3,137	3,500
7340	100 SCFM		4.30	3.721		3,450	200		3,650	4,100
7342	125 SCFM	↓	4	4	↓	4,450	215		4,665	5,225
7360	Desiccant type									
7362	Flow with energy saving controls									
7366	40 SCFM	Q-5	3.60	4.444	Ea.	6,025	239		6,264	6,975
7368	60 SCFM		3.40	4.706		6,550	253		6,803	7,600
7370	90 SCFM		3.20	5		7,600	269		7,869	8,775
7372	115 SCFM		3	5.333		8,175	287		8,462	9,425
7374	165 SCFM	↓	2.90	5.517		8,800	297		9,097	10,100
7378	260 SCFM	Q-6	3.60	6.667		10,400	370		10,770	12,000
7382	370 SCFM		3.40	7.059		12,100	395		12,495	13,900
7386	590 SCFM		3.20	7.500		14,900	420		15,320	17,000
7390	1130 SCFM	↓	3	8	↓	23,100	445		23,545	26,200
7460	Dew point monitor									
7466	LCD readout, high dew point alarm, probe included									
7470	Monitor	1 Stpi	2	4	Ea.	2,675	239		2,914	3,300

For customer support on your Plumbing Cost Data, call 877.709.6968.

22 66 Chemical-Waste Systems for Lab. and Healthcare Facilities

22 66 53 – Laboratory Chemical-Waste and Vent Piping

22 66 53.30 Glass Pipe		Crew	Daily Output	Labor-Hours	Unit	Material	2015 Bare Costs Labor	Equipment	Total	Total Incl O&P
0010	**GLASS PIPE**, Borosilicate, couplings & clevis hanger assemblies, 10' O.C. R221113-70									
0020	Drainage									
1100	1-1/2" diameter	Q-1	52	.308	L.F.	11.40	16.25		27.65	37
1120	2" diameter		44	.364		14.65	19.20		33.85	45
1140	3" diameter		39	.410		19.50	21.50		41	54
1160	4" diameter		30	.533		34	28		62	80
1180	6" diameter		26	.615		58.50	32.50		91	114
1870	To delete coupling & hanger, subtract									
1880	1-1/2" diam. to 2" diam.					19%	22%			
1890	3" diam. to 6" diam.					20%	17%			
2000	Process supply (pressure), beaded joints									
2040	1/2" diameter	1 Plum	36	.222	L.F.	6	13.05		19.05	26.50
2060	3/4" diameter		31	.258		6.65	15.15		21.80	30.50
2080	1" diameter		27	.296		16.55	17.40		33.95	44.50
2100	1-1/2" diameter	Q-1	47	.340		14.70	18		32.70	43
2120	2" diameter		39	.410		19.70	21.50		41.20	54
2140	3" diameter		34	.471		26	25		51	66
2160	4" diameter		25	.640		39	34		73	94
2180	6" diameter		21	.762		107	40.50		147.50	179
2860	To delete coupling & hanger, subtract									
2870	1/2" diam. to 1" diam.					25%	33%			
2880	1-1/2" diam. to 3" diam.					22%	21%			
2890	4" diam. to 6" diam.					23%	15%			
3800	Conical joint, transparent									
3980	6" diameter	Q-1	21	.762	L.F.	144	40.50		184.50	219
4500	To delete couplings & hangers, subtract									
4530	6" diam.					22%	26%			

22 66 53.40 Pipe Fittings, Glass										
0010	**PIPE FITTINGS, GLASS**									
0020	Drainage, beaded ends									
0040	Coupling & labor required at joints not incl. in fitting									
0050	price. Add 1 per joint for installed price									
0070	90° Bend or sweep, 1-1/2"				Ea.	31.50			31.50	35
0090	2"					40			40	44
0100	3"					66			66	72.50
0110	4"					105			105	116
0120	6" (sweep only)					330			330	360
0200	45° Bend or sweep same as 90°									
0350	Tee, single sanitary, 1-1/2"				Ea.	51			51	56.50
0370	2"					51			51	56.50
0380	3"					76.50			76.50	84
0390	4"					137			137	150
0400	6"					365			365	405
0410	Tee, straight, 1-1/2"					63.50			63.50	69.50
0430	2"					63.50			63.50	69.50
0440	3"					91.50			91.50	101
0450	4"					107			107	117
0460	6"					395			395	435
0500	Coupling, stainless steel, TFE seal ring									
0520	1-1/2"	Q-1	32	.500	Ea.	23.50	26.50		50	65.50
0530	2"		30	.533		29.50	28		57.50	74.50
0540	3"		25	.640		39.50	34		73.50	94.50

For customer support on your Plumbing Cost Data, call 877.709.6968.

305

22 66 53 – Laboratory Chemical-Waste and Vent Piping

22 66 53.40 Pipe Fittings, Glass		Crew	Daily Output	Labor-Hours	Unit	Material	2015 Bare Costs Labor	Equipment	Total	Total Incl O&P
0550	4"	Q-1	23	.696	Ea.	67.50	37		104.50	130
0560	6"		20	.800		152	42.50		194.50	231
0600	Coupling, stainless steel, bead to plain end									
0610	1-1/2"	Q-1	36	.444	Ea.	30	23.50		53.50	68.50
0620	2"		34	.471		37	25		62	78.50
0630	3"		29	.552		63	29		92	114
0640	4"		27	.593		94	31.50		125.50	152
0650	6"		24	.667		320	35		355	405
2350	Coupling, Viton liner, for temperatures to 400°F									
2370	1/2"	Q-1	40	.400	Ea.	46.50	21		67.50	83.50
2380	3/4"		37	.432		53	23		76	93
2390	1"		35	.457		63	24		87	106
2400	1-1/2"		32	.500		23	26.50		49.50	65.50
2410	2"		30	.533		29.50	28		57.50	75
2420	3"		25	.640		39.50	34		73.50	94.50
2430	4"		23	.696		67.50	37		104.50	130
2440	6"		20	.800		152	42.50		194.50	231
2550	For beaded joint armored fittings, add					200%				
2600	Conical ends. Flange set, gasket & labor not incl. in fitting									
2620	price. Add 1 per joint for installed price.									
2650	90° Sweep elbow, 1"				Ea.	105			105	116
2670	1-1/2"					211			211	232
2680	2"					219			219	241
2690	3"					340			340	375
2700	4"					605			605	665
2710	6"					1,050			1,050	1,175
2750	Cross (straight), add					55%				
2850	Tee, add					20%				

22 66 53.60 Corrosion Resistant Pipe

		Crew	Daily Output	Labor-Hours	Unit	Material	2015 Bare Costs Labor	Equipment	Total	Total Incl O&P
0010	**CORROSION RESISTANT PIPE**, No couplings or hangers R221113-70									
0020	Iron alloy, drain, mechanical joint									
1000	1-1/2" diameter	Q-1	70	.229	L.F.	46	12.10		58.10	69
1100	2" diameter		66	.242		47	12.80		59.80	71
1120	3" diameter		60	.267		60.50	14.10		74.60	88
1140	4" diameter		52	.308		77.50	16.25		93.75	110
1980	Iron alloy, drain, B&S joint									
2000	2" diameter	Q-1	54	.296	L.F.	55.50	15.65		71.15	85
2100	3" diameter		52	.308		71	16.25		87.25	103
2120	4" diameter		48	.333		98	17.60		115.60	135
2140	6" diameter	Q-2	59	.407		158	22.50		180.50	207
2160	8" diameter	"	54	.444		297	24.50		321.50	360
2980	Plastic, epoxy, fiberglass filament wound, B&S joint									
3000	2" diameter	Q-1	62	.258	L.F.	12	13.65		25.65	33.50
3100	3" diameter		51	.314		14	16.55		30.55	40.50
3120	4" diameter		45	.356		20	18.80		38.80	50.50
3140	6" diameter		32	.500		28	26.50		54.50	71
3160	8" diameter	Q-2	38	.632		44	34.50		78.50	101
3180	10" diameter		32	.750		60	41		101	128
3200	12" diameter		28	.857		72	47		119	151
3980	Polyester, fiberglass filament wound, B&S joint									
4000	2" diameter	Q-1	62	.258	L.F.	13.05	13.65		26.70	35
4100	3" diameter		51	.314		17	16.55		33.55	43.50

22 66 53.60 Corrosion Resistant Pipe

		Crew	Daily Output	Labor-Hours	Unit	Material	2015 Bare Costs Labor	Equipment	Total	Total Incl O&P
4120	4" diameter	Q-1	45	.356	L.F.	25	18.80		43.80	56
4140	6" diameter		32	.500		36	26.50		62.50	79.50
4160	8" diameter	Q-2	38	.632		85.50	34.50		120	147
4180	10" diameter		32	.750		105	41		146	177
4200	12" diameter		28	.857		125	47		172	209
4980	Polypropylene, acid resistant, fire retardant, schedule 40									
5000	1-1/2" diameter	Q-1	68	.235	L.F.	7.90	12.45		20.35	27.50
5100	2" diameter		62	.258		12.45	13.65		26.10	34
5120	3" diameter		51	.314		22	16.55		38.55	49.50
5140	4" diameter		45	.356		28	18.80		46.80	59.50
5160	6" diameter		32	.500		56.50	26.50		83	102
5980	Proxylene, fire retardant, Schedule 40									
6000	1-1/2" diameter	Q-1	68	.235	L.F.	12.40	12.45		24.85	32.50
6100	2" diameter		62	.258		17	13.65		30.65	39
6120	3" diameter		51	.314		30.50	16.55		47.05	59
6140	4" diameter		45	.356		43.50	18.80		62.30	76
6160	6" diameter		32	.500		73.50	26.50		100	121
6820	For Schedule 80, add					35%	2%			

22 66 53.70 Pipe Fittings, Corrosion Resistant

		Crew	Daily Output	Labor-Hours	Unit	Material	2015 Bare Costs Labor	Equipment	Total	Total Incl O&P
0010	**PIPE FITTINGS, CORROSION RESISTANT**									
0030	Iron alloy									
0050	Mechanical joint									
0060	1/4 Bend, 1-1/2"	Q-1	12	1.333	Ea.	77	70.50		147.50	191
0080	2"		10	1.600		126	84.50		210.50	267
0090	3"		9	1.778		151	94		245	310
0100	4"		8	2		174	106		280	350
0110	1/8 Bend, 1-1/2"		12	1.333		49	70.50		119.50	160
0130	2"		10	1.600		84	84.50		168.50	221
0140	3"		9	1.778		112	94		206	265
0150	4"		8	2		150	106		256	325
0160	Tee and Y, sanitary, straight									
0170	1-1/2"	Q-1	8	2	Ea.	84	106		190	253
0180	2"		7	2.286		112	121		233	305
0190	3"		6	2.667		174	141		315	405
0200	4"		5	3.200		320	169		489	605
0360	Coupling, 1-1/2"		14	1.143		47.50	60.50		108	144
0380	2"		12	1.333		54	70.50		124.50	166
0390	3"		11	1.455		56.50	77		133.50	179
0400	4"		10	1.600		64.50	84.50		149	199
0500	Bell & Spigot									
0510	1/4 and 1/16 bend, 2"	Q-1	16	1	Ea.	99.50	53		152.50	189
0520	3"		14	1.143		232	60.50		292.50	345
0530	4"		13	1.231		238	65		303	360
0540	6"	Q-2	17	1.412		470	77.50		547.50	635
0550	8"	"	12	2		1,950	110		2,060	2,300
0620	1/8 bend, 2"	Q-1	16	1		112	53		165	203
0640	3"		14	1.143		207	60.50		267.50	320
0650	4"		13	1.231		238	65		303	360
0660	6"	Q-2	17	1.412		395	77.50		472.50	550
0680	8"	"	12	2		1,550	110		1,660	1,875
0700	Tee, sanitary, 2"	Q-1	10	1.600		213	84.50		297.50	360
0710	3"		9	1.778		690	94		784	900

For customer support on your Plumbing Cost Data, call 877.709.6968.

22 66 53.70 Pipe Fittings, Corrosion Resistant		Crew	Daily Output	Labor-Hours	Unit	Material	2015 Bare Costs Labor	Equipment	Total	Total Incl O&P
0720	4"	Q-1	8	2	Ea.	575	106		681	790
0730	6"	Q-2	11	2.182		715	120		835	970
0740	8"	"	8	3		1,950	164		2,114	2,400
1800	Y, sanitary, 2"	Q-1	10	1.600		224	84.50		308.50	375
1820	3"	↓	9	1.778		400	94		494	580
1830	4"	↓	8	2		360	106		466	555
1840	6"	Q-2	11	2.182		1,200	120		1,320	1,500
1850	8"	"	8	3	↓	3,300	164		3,464	3,875
3000	Epoxy, filament wound									
3030	Quick-lock joint									
3040	90° Elbow, 2"	Q-1	28	.571	Ea.	97.50	30		127.50	153
3060	3"		16	1		112	53		165	203
3070	4"		13	1.231		153	65		218	266
3080	6"	↓	8	2		223	106		329	405
3090	8"	Q-2	9	2.667		410	146		556	670
3100	10"		7	3.429		520	188		708	855
3110	12"	↓	6	4		740	219		959	1,150
3120	45° Elbow, 2"	Q-1	28	.571		75	30		105	128
3130	3"		16	1		106	53		159	197
3140	4"		13	1.231		109	65		174	218
3150	6"	↓	8	2		223	106		329	405
3160	8"	Q-2	9	2.667		410	146		556	670
3170	10"		7	3.429		515	188		703	855
3180	12"	↓	6	4		740	219		959	1,150
3190	Tee, 2"	Q-1	19	.842		233	44.50		277.50	325
3200	3"		11	1.455		280	77		357	425
3210	4"		9	1.778		335	94		429	510
3220	6"	↓	5	3.200		575	169		744	885
3230	8"	Q-2	6	4		645	219		864	1,025
3240	10"		5	4.800		900	263		1,163	1,375
3250	12"	↓	4	6	↓	1,275	330		1,605	1,900
4000	Polypropylene, acid resistant									
4020	Non-pressure, electrofusion joints									
4050	1/4 bend, 1-1/2"	1 Plum	16	.500	Ea.	13.95	29.50		43.45	60
4060	2"	Q-1	28	.571		27	30		57	75.50
4080	3"		17	.941		29.50	49.50		79	108
4090	4"		14	1.143		47.50	60.50		108	144
4110	6"	↓	8	2	↓	114	106		220	285
4150	1/4 Bend, long sweep									
4170	1-1/2"	1 Plum	16	.500	Ea.	15.10	29.50		44.60	61
4180	2"	Q-1	28	.571		27	30		57	75.50
4200	3"		17	.941		34	49.50		83.50	112
4210	4"	↓	14	1.143		49	60.50		109.50	145
4250	1/8 bend, 1-1/2"	1 Plum	16	.500		15.90	29.50		45.40	62
4260	2"	Q-1	28	.571		16.90	30		46.90	64
4280	3"		17	.941		31	49.50		80.50	109
4290	4"		14	1.143		34.50	60.50		95	129
4310	6"	↓	8	2	↓	95.50	106		201.50	265
4400	Tee, sanitary									
4420	1-1/2"	1 Plum	10	.800	Ea.	17.60	47		64.60	90.50
4430	2"	Q-1	17	.941		20.50	49.50		70	97.50
4450	3"		11	1.455		41	77		118	161
4460	4"	↓	9	1.778	↓	61.50	94		155.50	210

For customer support on your Plumbing Cost Data, call 877.709.6968.

22 66 53 – Laboratory Chemical-Waste and Vent Piping

22 66 53.70 Pipe Fittings, Corrosion Resistant		Crew	Daily Output	Labor-Hours	Unit	Material	2015 Bare Costs Labor	Equipment	Total	Total Incl O&P
4480	6"	Q-1	5	3.200	Ea.	415	169		584	710
4490	Tee, sanitary reducing, 2" x 2" x 1-1/2"		17	.941		20.50	49.50		70	97.50
4491	3" x 3" x 1-1/2"		11	1.455		36.50	77		113.50	157
4492	3" x 3" x 2"		11	1.455		41	77		118	161
4493	4" x 4" x 2"		10	1.600		55	84.50		139.50	189
4494	4" x 4" x 3"		9	1.778		64	94		158	212
4496	6" x 6" x 4"		5	3.200		202	169		371	475
4650	Wye 45°, 1-1/2"	1 Plum	10	.800		19.10	47		66.10	92
4652	2"	Q-1	17	.941		26.50	49.50		76	104
4653	3"		11	1.455		44	77		121	165
4654	4"		9	1.778		63.50	94		157.50	212
4656	6"		5	3.200		165	169		334	435
4660	Wye, reducing									
4662	2" x 2" x 1-1/2"	Q-1	17	.941	Ea.	25	49.50		74.50	103
4666	3" x 3" x 2"		11	1.455		50	77		127	171
4668	4" x 4" x 2"		10	1.600		58.50	84.50		143	193
4669	4" x 4" x 3"		9	1.778		62.50	94		156.50	211
4671	6" x 6" x 2"		6	2.667		100	141		241	325
4673	6" x 6" x 3"		5.50	2.909		110	154		264	355
4675	6" x 6" x 4"		5	3.200		114	169		283	380
4678	Combination Y & 1/8 bend									
4681	1-1/2"	1 Plum	10	.800	Ea.	23	47		70	96.50
4683	2"	Q-1	17	.941		29.50	49.50		79	108
4684	3"		11	1.455		50	77		127	171
4685	4"		9	1.778		68	94		162	217
4689	Combination Y & 1/8 bend, reducing									
4692	2" x 2" x 1-1/2"	Q-1	17	.941	Ea.	28	49.50		77.50	106
4694	3" x 3" x 1-1/2"		12	1.333		41.50	70.50		112	152
4695	3" x 3" x 2"		11	1.455		44	77		121	165
4697	4" x 4" x 2"		10	1.600		62.50	84.50		147	197
4699	4" x 4" x 3"		9	1.778		64.50	94		158.50	213
4710	Hub adapter									
4712	1-1/2"	1 Plum	16	.500	Ea.	32.50	29.50		62	80
4713	2"	Q-1	28	.571		37	30		67	86.50
4714	3"		17	.941		46.50	49.50		96	126
4715	4"		14	1.143		60	60.50		120.50	157
4719	Mechanical joint adapter									
4721	1-1/2"	1 Plum	16	.500	Ea.	19.20	29.50		48.70	65.50
4722	2"	Q-1	28	.571		20.50	30		50.50	68
4723	3"		17	.941		29.50	49.50		79	108
4724	4"		14	1.143		44	60.50		104.50	140
4728	Couplings									
4731	1-1/2"	1 Plum	16	.500	Ea.	11	29.50		40.50	56.50
4732	2"	Q-1	28	.571		13.95	30		43.95	61
4733	3"		17	.941		17.60	49.50		67.10	94.50
4734	4"		14	1.143		25	60.50		85.50	119
4736	6"		8	2		39.50	106		145.50	204

22 66 83.13 Chemical-Waste Dilution Tanks	Crew	Daily Output	Labor-Hours	Unit	Material	2015 Bare Costs Labor	Equipment	Total	Total Incl O&P
0010 **CHEMICAL-WASTE DILUTION TANKS**									
7000 Tanks, covers included									
7800 Polypropylene									
7810 Continuous service to 200°F									
7830 2 gallon, 8" x 8" x 8"	Q-1	20	.800	Ea.	124	42.50		166.50	200
7850 7 gallon, 12" x 12" x 12"		20	.800		189	42.50		231.50	272
7870 16 gallon, 18" x 12" x 18"		17	.941		243	49.50		292.50	340
8010 33 gallon, 24" x 18" x 18"		12	1.333		315	70.50		385.50	450
8070 44 gallon, 24" x 18" x 24"		10	1.600		380	84.50		464.50	550
8080 89 gallon, 36" x 24" x 24"	▼	8	2	▼	525	106		631	735
8150 Polyethylene, heavy duty walls									
8160 Continuous service to 180°F									
8180 5 gallon, 12" X 6" X 18"	Q-1	20	.800	Ea.	48.50	42.50		91	118
8210 15 gallon, 14" I.D. x 27" deep		17	.941		73.50	49.50		123	156
8230 55 gallon, 22" I.D. x 36" deep		10	1.600		154	84.50		238.50	298
8250 100 gallon 28" I.D. x 42" deep		8	2		375	106		481	570
8270 200 gallon 36" I.D. x 48" deep		6	2.667		605	141		746	880
8290 360 gallon 48" I.D. x 48" deep	▼	5	3.200	▼	700	169		869	1,025

Estimating Tips

The labor adjustment factors listed in Subdivision 22 01 02.20 also apply to Division 23.

23 10 00 Facility Fuel Systems

- The prices in this subdivision for above- and below-ground storage tanks do not include foundations or hold-down slabs, unless noted. The estimator should refer to Divisions 3 and 31 for foundation system pricing. In addition to the foundations, required tank accessories, such as tank gauges, leak detection devices, and additional manholes and piping, must be added to the tank prices.

23 50 00 Central Heating Equipment

- When estimating the cost of an HVAC system, check to see who is responsible for providing and installing the temperature control system. It is possible to overlook controls, assuming that they would be included in the electrical estimate.
- When looking up a boiler, be careful on specified capacity. Some manufacturers rate

their products on output while others use input.

- Include HVAC insulation for pipe, boiler, and duct (wrap and liner).
- Be careful when looking up mechanical items to get the correct pressure rating and connection type (thread, weld, flange).

23 70 00 Central HVAC Equipment

- Combination heating and cooling units are sized by the air conditioning requirements. (See Reference No. R236000-20 for preliminary sizing guide.)
- A ton of air conditioning is nominally 400 CFM.
- Rectangular duct is taken off by the linear foot for each size, but its cost is usually estimated by the pound. Remember that SMACNA standards now base duct on internal pressure.
- Prefabricated duct is estimated and purchased like pipe: straight sections and fittings.
- Note that cranes or other lifting equipment are not included on any lines in

Division 23. For example, if a crane is required to lift a heavy piece of pipe into place high above a gym floor, or to put a rooftop unit on the roof of a four-story building, etc., it must be added. Due to the potential for extreme variation—from nothing additional required to a major crane or helicopter— we feel that including a nominal amount for "lifting contingency" would be useless and detract from the accuracy of the estimate. When using equipment rental cost data from RSMeans, do not forget to include the cost of the operator(s).

Reference Numbers

Reference numbers are shown in shaded boxes at the beginning of some major classifications. These numbers refer to related items in the Reference Section. The reference information may be an estimating procedure, an alternate pricing method, or technical information.

Note: Not all subdivisions listed here necessarily appear in this publication. ■

*Note: **Trade Service**, in part, has been used as a reference source for some of the material prices used in Division 23.*

23 05 02 – HVAC General

23 05 02.10 Air Conditioning, General	Crew	Daily Output	Labor-Hours	Unit	Material	2015 Bare Costs Labor	Equipment	Total	Total Incl O&P
0010 **AIR CONDITIONING, GENERAL** Prices are for standard efficiencies (SEER 13)									
0020 for upgrade to SEER 14 add					10%				

23 05 05 – Selective Demolition for HVAC

23 05 05.10 HVAC Demolition	Crew	Daily Output	Labor-Hours	Unit	Material	2015 Bare Costs Labor	Equipment	Total	Total Incl O&P
0010 **HVAC DEMOLITION** R220105-10									
0100 Air conditioner, split unit, 3 ton	Q-5	2	8	Ea.		430		430	650
0150 Package unit, 3 ton R024119-10	Q-6	3	8			445		445	675
0190 Rooftop, self contained, up to 5 ton	1 Plum	1.20	6.667	↓		390		390	590
0250 Air curtain	Q-9	20	.800	L.F.		40.50		40.50	61.50
0254 Air filters, up thru 16,000 CFM		20	.800	Ea.		40.50		40.50	61.50
0256 20,000 thru 60,000 CFM	↓	16	1	↓		50.50		50.50	77
0297 Boiler blowdown	Q-5	8	2			108		108	162
0298 Boilers									
0300 Electric, up thru 148 kW	Q-19	2	12	Ea.		650		650	975
0310 150 thru 518 kW	"	1	24			1,300		1,300	1,950
0320 550 thru 2000 kW	Q-21	.40	80			4,450		4,450	6,700
0330 2070 kW and up	"	.30	106			5,925		5,925	8,925
0340 Gas and/or oil, up thru 150 MBH	Q-7	2.20	14.545			825		825	1,250
0350 160 thru 2000 MBH		.80	40			2,275		2,275	3,425
0360 2100 thru 4500 MBH		.50	64			3,650		3,650	5,500
0370 4600 thru 7000 MBH		.30	106			6,075		6,075	9,175
0380 7100 thru 12,000 MBH		.16	200			11,400		11,400	17,200
0390 12,200 thru 25,000 MBH	↓	.12	266			15,200		15,200	22,900
0400 Central station air handler unit, up thru 15 ton	Q-5	1.60	10			540		540	810
0410 17.5 thru 30 ton	"	.80	20	↓		1,075		1,075	1,625
0430 Computer room unit									
0434 Air cooled split, up thru 10 ton	Q-5	.67	23.881	Ea.		1,275		1,275	1,950
0436 12 thru 23 ton		.53	30.189			1,625		1,625	2,450
0440 Chilled water, up thru 10 ton		1.30	12.308			660		660	1,000
0444 12 thru 23 ton		1	16			860		860	1,300
0450 Glycol system, up thru 10 ton		.53	30.189			1,625		1,625	2,450
0454 12 thru 23 ton		.40	40			2,150		2,150	3,250
0460 Water cooled, not including condenser, up thru 10 ton		.80	20			1,075		1,075	1,625
0464 12 thru 23 ton		.60	26.667			1,425		1,425	2,175
0600 Condenser, up thru 50 ton	↓	1	16			860		860	1,300
0610 51 thru 100 ton	Q-6	.90	26.667			1,475		1,475	2,250
0620 101 thru 1000 ton	"	.70	34.286			1,900		1,900	2,875
0660 Condensing unit, up thru 10 ton	Q-5	1.25	12.800			690		690	1,050
0670 11 thru 50 ton	"	.40	40			2,150		2,150	3,250
0680 60 thru 100 ton	Q-6	.30	80			4,450		4,450	6,725
0700 Cooling tower, up thru 400 ton		.80	30			1,675		1,675	2,525
0710 450 thru 600 ton		.53	45.283			2,525		2,525	3,825
0720 700 thru 1300 ton	↓	.40	60			3,350		3,350	5,050
0780 Dehumidifier, up thru 155 lb./hr.	Q-1	8	2			106		106	160
0790 240 lb./hr. and up	"	2	8	↓		425		425	640
1560 Ductwork									
1570 Metal, steel, sst, fabricated	Q-9	1000	.016	~Lb.		.81		.81	1.23
1580 Aluminum, fabricated		485	.033	"		1.66		1.66	2.54
1590 Spiral, prefabricated		400	.040	L.F.		2.01		2.01	3.08
1600 Fiberglass, prefabricated		400	.040			2.01		2.01	3.08
1610 Flex, prefabricated		500	.032			1.61		1.61	2.46
1620 Glass fiber reinforced plastic, prefabricated	↓	280	.057	↓		2.88		2.88	4.40

23 05 Common Work Results for HVAC

23 05 05 – Selective Demolition for HVAC

23 05 05.10 HVAC Demolition	Crew	Daily Output	Labor-Hours	Unit	Material	2015 Bare Costs Labor	Equipment	Total	Total Incl O&P
1630 Diffusers, registers or grills, up thru 20" max dimension	1 Shee	50	.160	Ea.		8.95		8.95	13.70
1640 21 thru 36" max dimension		36	.222			12.45		12.45	19
1650 Above 36" max dimension		30	.267			14.90		14.90	23
1700 Evaporator, up thru 12,000 BTUH	Q-5	5.30	3.019			162		162	245
1710 12,500 thru 30,000 BTUH	"	2.70	5.926			320		320	480
1720 31,000 BTUH and up	Q-6	1.50	16			890		890	1,350
1730 Evaporative cooler, up thru 5 H.P.	Q-9	2.70	5.926			298		298	455
1740 10 thru 30 H.P.	"	.67	23.881			1,200		1,200	1,850
1750 Exhaust systems									
1760 Exhaust components	1 Shee	8	1	System		56		56	85.50
1770 Weld fume hoods	"	20	.400	Ea.		22.50		22.50	34
2120 Fans, up thru 1 H.P. or 2000 CFM	Q-9	8	2			101		101	154
2124 1-1/2 thru 10 H.P. or 20,000 CFM		5.30	3.019			152		152	232
2128 15 thru 30 H.P. or above 20,000 CFM		4	4			201		201	310
2150 Fan coil air conditioner, chilled water, up thru 7.5 ton	Q-5	14	1.143			61.50		61.50	93
2154 Direct expansion, up thru 10 ton		8	2			108		108	162
2158 11 thru 30 ton		2	8			430		430	650
2170 Flue shutter damper	Q-9	8	2			101		101	154
2200 Furnace, electric	Q-20	2	10			510		510	780
2300 Gas or oil, under 120 MBH	Q-9	4	4			201		201	310
2340 Over 120 MBH	"	3	5.333			269		269	410
2730 Heating and ventilating unit	Q-5	2.70	5.926			320		320	480
2740 Heater, electric, wall, baseboard and quartz	1 Elec	10	.800			44		44	65.50
2750 Heater, electric, unit, cabinet, fan and convector	"	8	1			54.50		54.50	82
2760 Heat exchanger, shell and tube type	Q-5	1.60	10			540		540	810
2770 Plate type	Q-6	.60	40			2,225		2,225	3,375
2810 Heat pump									
2820 Air source, split, up thru 10 ton	Q-5	.90	17.778	Ea.		955		955	1,450
2830 15 thru 25 ton	Q-6	.80	30			1,675		1,675	2,525
2850 Single package, up thru 12 ton	Q-5	1	16			860		860	1,300
2860 Water source, up thru 15 ton	"	.90	17.778			955		955	1,450
2870 20 thru 50 ton	Q-6	.80	30			1,675		1,675	2,525
2910 Heat recovery package, up thru 20,000 CFM	Q-5	2	8			430		430	650
2920 25,000 CFM and up		1.20	13.333			715		715	1,075
2930 Heat transfer package, up thru 130 GPM		.80	20			1,075		1,075	1,625
2934 255 thru 800 GPM		.42	38.095			2,050		2,050	3,100
2940 Humidifier		10.60	1.509			81		81	123
2961 Hydronic unit heaters, up thru 200 MBH		14	1.143			61.50		61.50	93
2962 Above 200 MBH		8	2			108		108	162
2964 Valance units		32	.500			27		27	40.50
2966 Radiant floor heating									
2967 System valves, controls, manifolds	Q-5	16	1	Ea.		54		54	81
2968 Per room distribution		8	2			108		108	162
2970 Hydronic heating, baseboard radiation		16	1			54		54	81
2976 Convectors and free standing radiators		18	.889			48		48	72
2980 Induced draft fan, up thru 1 H.P.	Q-9	4.60	3.478			175		175	268
2984 1-1/2 H.P. thru 7-1/2 H.P.	"	2.20	7.273			365		365	560
2988 Infrared unit	Q-5	16	1			54		54	81
2992 Louvers	1 Shee	46	.174	S.F.		9.75		9.75	14.85
3000 Mechanical equipment, light items. Unit is weight, not cooling.	Q-5	.90	17.778	Ton		955		955	1,450
3600 Heavy items		1.10	14.545	"		780		780	1,175
3720 Make-up air unit, up thru 6000 CFM		3	5.333	Ea.		287		287	435
3730 6500 thru 30,000 CFM		1.60	10			540		540	810

23 05 05 – Selective Demolition for HVAC

23 05 05.10 HVAC Demolition	Crew	Daily Output	Labor-Hours	Unit	Material	2015 Bare Costs Labor	Equipment	Total	Total Incl O&P
3740	35,000 thru 75,000 CFM	Q-6	1	24	Ea.		1,350		1,350	2,025
3800	Mixing boxes, constant and VAV	Q-9	18	.889			45		45	68.50
4000	Packaged terminal air conditioner, up thru 18,000 BTUH	Q-5	8	2			108		108	162
4010	24,000 thru 48,000 BTUH	"	2.80	5.714			305		305	465
5000	Refrigerant compressor, reciprocating or scroll									
5010	Up thru 5 ton	1 Stpi	6	1.333	Ea.		79.50		79.50	120
5020	5.08 thru 10 ton	Q-5	6	2.667			143		143	217
5030	15 thru 50 ton	"	3	5.333			287		287	435
5040	60 thru 130 ton	Q-6	2.80	8.571			480		480	720
5090	Remove refrigerant from system	1 Stpi	40	.200	Lb.		11.95		11.95	18.05
5100	Rooftop air conditioner, up thru 10 ton	Q-5	1.40	11.429	Ea.		615		615	930
5110	12 thru 40 ton	Q-6	1	24			1,350		1,350	2,025
5120	50 thru 140 ton		.50	48			2,675		2,675	4,050
5130	150 thru 300 ton		.30	80			4,450		4,450	6,725
6000	Self contained single package air conditioner, up thru 10 ton	Q-5	1.60	10			540		540	810
6010	15 thru 60 ton	Q-6	1.20	20			1,125		1,125	1,675
6100	Space heaters, up thru 200 MBH	Q-5	10	1.600			86		86	130
6110	Over 200 MBH		5	3.200			172		172	260
6200	Split ductless, both sections		8	2			108		108	162
6300	Steam condensate meter	1 Stpi	11	.727			43.50		43.50	65.50
6600	Thru-the-wall air conditioner	L-2	8	2			82.50		82.50	127
7000	Vent chimney, prefabricated, up thru 12" diameter	Q-9	94	.170	V.L.F.		8.55		8.55	13.10
7010	14" thru 36" diameter		40	.400			20		20	31
7020	38" thru 48" diameter		32	.500			25		25	38.50
7030	54" thru 60" diameter	Q-10	14	1.714			89.50		89.50	137
7400	Ventilators, up thru 14" neck diameter	Q-9	58	.276	Ea.		13.90		13.90	21
7410	16" thru 50" neck diameter		40	.400			20		20	31
7450	Relief vent, up thru 24" x 96"		22	.727			36.50		36.50	56
7460	48" x 60" thru 96" x 144"		10	1.600			80.50		80.50	123
8000	Water chiller up thru 10 ton	Q-5	2.50	6.400			345		345	520
8010	15 thru 100 ton	Q-6	.48	50			2,800		2,800	4,200
8020	110 thru 500 ton	Q-7	.29	110			6,275		6,275	9,475
8030	600 thru 1000 ton		.23	139			7,925		7,925	12,000
8040	1100 ton and up		.20	160			9,100		9,100	13,700
8400	Window air conditioner	1 Carp	16	.500			23.50		23.50	36

23 05 23 – General-Duty Valves for HVAC Piping

23 05 23.20 Valves, Bronze/Brass	Crew	Daily Output	Labor-Hours	Unit	Material	2015 Bare Costs Labor	Equipment	Total	Total Incl O&P
0010	**VALVES, BRONZE/BRASS**									
0020	Brass									
1300	Ball combination valves, shut-off and union									
1310	Solder, with strainer, drain and PT ports									
1320	1/2"	1 Stpi	17	.471	Ea.	75	28		103	125
1330	3/4"		16	.500		78	30		108	131
1340	1"		14	.571		123	34		157	187
1350	1-1/4"		12	.667		142	40		182	216
1360	1-1/2"		10	.800		221	48		269	315
1370	2"		8	1		260	60		320	375
1410	Threaded, with strainer, drain and PT ports									
1420	1/2"	1 Stpi	20	.400	Ea.	75	24		99	119
1430	3/4"		18	.444		78	26.50		104.50	126
1440	1"		17	.471		123	28		151	178
1450	1-1/4"		12	.667		142	40		182	216

23 05 23 – General-Duty Valves for HVAC Piping

23 05 23.20 Valves, Bronze/Brass		Crew	Daily Output	Labor-Hours	Unit	Material	2015 Bare Costs Labor	Equipment	Total	Total Incl O&P
1460	1-1/2"	1 Stpi	11	.727	Ea.	221	43.50		264.50	310
1470	2"	↓	9	.889	↓	260	53		313	365

23 05 23.30 Valves, Iron Body

	23 05 23.30 Valves, Iron Body	Crew	Daily Output	Labor-Hours	Unit	Material	2015 Bare Costs Labor	Equipment	Total	Total Incl O&P
0010	**VALVES, IRON BODY** R220523-90									
0022	For grooved joint, see Section 22 11 13.48									
0100	Angle, 125 lb.									
0110	Flanged									
0116	2"	1 Plum	5	1.600	Ea.	1,200	94		1,294	1,475
0118	4"	Q-1	3	5.333		2,000	282		2,282	2,625
0120	6"	Q-2	3	8		4,000	440		4,440	5,025
0122	8"	"	2.50	9.600	↓	5,550	525		6,075	6,900
1020	Butterfly, wafer type, gear actuator, 200 lb.									
1030	2"	1 Plum	14	.571	Ea.	96	33.50		129.50	156
1040	2-1/2"	Q-1	9	1.778		97.50	94		191.50	249
1050	3"		8	2		101	106		207	271
1060	4"	↓	5	3.200		113	169		282	380
1070	5"	Q-2	5	4.800		126	263		389	535
1080	6"		5	4.800		143	263		406	555
1090	8"		4.50	5.333		178	292		470	635
1100	10"		4	6		239	330		569	760
1110	12"	↓	3	8	↓	360	440		800	1,050
1200	Wafer type, lever actuator, 200 lb.									
1220	2"	1 Plum	14	.571	Ea.	156	33.50		189.50	222
1230	2-1/2"	Q-1	9	1.778		161	94		255	320
1240	3"		8	2		190	106		296	370
1250	4"	↓	5	3.200		207	169		376	485
1260	5"	Q-2	5	4.800		320	263		583	750
1270	6"		5	4.800		350	263		613	780
1280	8"		4.50	5.333		540	292		832	1,025
1290	10"		4	6		755	330		1,085	1,325
1300	12"	↓	3	8	↓	1,250	440		1,690	2,025
1650	Gate, 125 lb., N.R.S.									
2150	Flanged									
2200	2"	1 Plum	5	1.600	Ea.	715	94		809	930
2240	2-1/2"	Q-1	5	3.200		735	169		904	1,075
2260	3"		4.50	3.556		825	188		1,013	1,200
2280	4"	↓	3	5.333		1,175	282		1,457	1,725
2290	5"	Q-2	3.40	7.059		2,025	385		2,410	2,800
2300	6"		3	8		2,025	440		2,465	2,875
2320	8"		2.50	9.600		3,450	525		3,975	4,600
2340	10"		2.20	10.909		6,075	600		6,675	7,575
2360	12"	↓	1.70	14.118	↓	8,325	775		9,100	10,400
2420	For 250 lb. flanged, add					200%	10%			
3550	OS&Y, 125 lb., flanged									
3600	2"	1 Plum	5	1.600	Ea.	475	94		569	665
3640	2-1/2"	Q-1	5	3.200		480	169		649	785
3660	3"		4.50	3.556		530	188		718	865
3670	3-1/2"		3	5.333		680	282		962	1,175
3680	4"	↓	3	5.333		770	282		1,052	1,275
3690	5"	Q-2	3.40	7.059		1,250	385		1,635	1,950
3700	6"		3	8		1,250	440		1,690	2,025
3720	8"		2.50	9.600	↓	2,175	525		2,700	3,200

For customer support on your Plumbing Cost Data, call 877.709.6968.

315

23 05 23 – General-Duty Valves for HVAC Piping

23 05 23.30 Valves, Iron Body		Crew	Daily Output	Labor-Hours	Unit	Material	2015 Bare Costs Labor	Equipment	Total	Total Incl O&P
3740	10"	Q-2	2.20	10.909	Ea.	4,050	600		4,650	5,350
3760	12"	↓	1.70	14.118	↓	5,500	775		6,275	7,250
3900	For 175 lb., flanged, add					200%	10%			
4350	Globe, OS&Y									
4540	Class 125, flanged									
4550	2"	1 Plum	5	1.600	Ea.	935	94		1,029	1,175
4560	2-1/2"	Q-1	5	3.200		940	169		1,109	1,275
4570	3"	↓	4.50	3.556		1,150	188		1,338	1,525
4580	4"		3	5.333		1,650	282		1,932	2,225
4590	5"	Q-2	3.40	7.059		2,975	385		3,360	3,850
4600	6"		3	8		2,975	440		3,415	3,925
4610	8"		2.50	9.600		5,825	525		6,350	7,225
4612	10"		2.20	10.909		9,125	600		9,725	10,900
4614	12"	↓	1.70	14.118	↓	9,900	775		10,675	12,100
5040	Class 250, flanged									
5050	2"	1 Plum	4.50	1.778	Ea.	1,500	104		1,604	1,800
5060	2-1/2"	Q-1	4.50	3.556		1,950	188		2,138	2,425
5070	3"		4	4		2,000	211		2,211	2,550
5080	4"	↓	2.70	5.926		2,950	315		3,265	3,725
5090	5"	Q-2	3	8		5,475	440		5,915	6,675
5100	6"		2.70	8.889		5,275	485		5,760	6,550
5110	8"		2.20	10.909		8,975	600		9,575	10,800
5120	10"		2	12		13,400	655		14,055	15,800
5130	12"	↓	1.60	15		19,600	820		20,420	22,800
5240	Valve sprocket rim w/chain, for 2" valve	1 Stpi	30	.267		127	15.95		142.95	164
5250	2-1/2" valve		27	.296		127	17.70		144.70	167
5260	3-1/2" valve		25	.320		127	19.10		146.10	169
5270	6" valve		20	.400		127	24		151	176
5280	8" valve		18	.444		156	26.50		182.50	211
5290	12" valve		16	.500		156	30		186	216
5300	16" valve		12	.667		215	40		255	296
5310	20" valve		10	.800		215	48		263	310
5320	36" valve	↓	8	1	↓	380	60		440	510
5450	Swing check, 125 lb., threaded									
5500	2"	1 Plum	11	.727	Ea.	430	42.50		472.50	540
5540	2-1/2"	Q-1	15	1.067		555	56.50		611.50	695
5550	3"		13	1.231		590	65		655	750
5560	4"	↓	10	1.600	↓	955	84.50		1,039.50	1,175
5950	Flanged									
5994	1"	1 Plum	7	1.143	Ea.	210	67		277	335
5998	1-1/2"		6	1.333		299	78.50		377.50	450
6000	2"	↓	5	1.600		395	94		489	575
6040	2-1/2"	Q-1	5	3.200		380	169		549	675
6050	3"		4.50	3.556		410	188		598	735
6060	4"	↓	3	5.333		605	282		887	1,100
6070	6"	Q-2	3	8		1,025	440		1,465	1,800
6080	8"		2.50	9.600		1,950	525		2,475	2,950
6090	10"		2.20	10.909		3,325	600		3,925	4,550
6100	12"		1.70	14.118		5,175	775		5,950	6,875
6110	18"		1.30	18.462		23,700	1,000		24,700	27,600
6114	24"	↓	.75	32		52,000	1,750		53,750	60,000
6160	For 250 lb. flanged, add					200%	20%			
6600	Silent check, bronze trim									

23 05 23 – General-Duty Valves for HVAC Piping

23 05 23.30 Valves, Iron Body

		Crew	Daily Output	Labor-Hours	Unit	Material	2015 Bare Costs Labor	2015 Bare Costs Equipment	Total	Total Incl O&P
6610	Compact wafer type, for 125 or 150 lb. flanges									
6630	1-1/2"	1 Plum	11	.727	Ea.	136	42.50		178.50	214
6640	2"	"	9	.889		165	52		217	261
6650	2-1/2"	Q-1	9	1.778		180	94		274	340
6660	3"		8	2		192	106		298	370
6670	4"		5	3.200		245	169		414	525
6680	5"	Q-2	6	4		325	219		544	685
6690	6"		6	4		590	219		809	975
6700	8"		4.50	5.333		730	292		1,022	1,250
6710	10"		4	6		1,300	330		1,630	1,925
6720	12"		3	8		2,425	440		2,865	3,300
6740	For 250 or 300 lb. flanges, thru 6" no change									
6741	For 8" and 10", add				Ea.	11%	10%			
6750	Twin disc									
6752	2"	1 Plum	9	.889	Ea.	300	52		352	410
6754	4"	Q-1	5	3.200		490	169		659	795
6756	6"	Q-2	5	4.800		730	263		993	1,200
6758	8"		4.50	5.333		1,125	292		1,417	1,675
6760	10"		4	6		1,800	330		2,130	2,475
6762	12"		3	8		2,325	440		2,765	3,200
6764	18"		1.50	16		10,000	875		10,875	12,300
6766	24"		.75	32		14,100	1,750		15,850	18,200
6800	Full flange type, 150 lb.									
6900	Globe type, 125 lb.									
6911	2-1/2"	Q-1	9	1.778	Ea.	405	94		499	585
6912	3"		8	2		475	106		581	685
6913	4"		5	3.200		640	169		809	960
6914	5"	Q-2	6	4		890	219		1,109	1,300
6915	6"		5	4.800		1,125	263		1,388	1,625
6916	8"		4.50	5.333		2,050	292		2,342	2,700
6917	10"		4	6		2,575	330		2,905	3,325
6918	12"		3	8		4,275	440		4,715	5,350
6940	For 250 lb., add					40%	10%			
6980	Screwed end type, 125 lb.									
6981	1"	1 Plum	19	.421	Ea.	61.50	24.50		86	105
6982	1-1/4"		15	.533		80	31.50		111.50	136
6983	1-1/2"		13	.615		99	36		135	164
6984	2"		11	.727		134	42.50		176.50	212

23 05 23.70 Valves, Semi-Steel

		Crew	Daily Output	Labor-Hours	Unit	Material	2015 Bare Costs Labor	2015 Bare Costs Equipment	Total	Total Incl O&P
0010	**VALVES, SEMI-STEEL** R220523-90									
1020	Lubricated plug valve, threaded, 200 psi									
1030	1/2"	1 Plum	18	.444	Ea.	560	26		586	655
1040	3/4"		16	.500		560	29.50		589.50	660
1050	1"		14	.571		375	33.50		408.50	460
1060	1-1/4"		12	.667		415	39		454	515
1070	1-1/2"		11	.727		440	42.50		482.50	550
1080	2"		8	1		535	58.50		593.50	680
1090	2-1/2"	Q-1	5	3.200		590	169		759	900
1100	3"	"	4.50	3.556		845	188		1,033	1,225
6990	Flanged, 200 psi									
7000	2"	1 Plum	8	1	Ea.	217	58.50		275.50	325
7010	2-1/2"	Q-1	5	3.200		310	169		479	595

317

23 05 23.70 Valves, Semi-Steel		Crew	Daily Output	Labor-Hours	Unit	Material	2015 Bare Costs Labor	Equipment	Total	Total Incl O&P
7020	3"	Q-1	4.50	3.556	Ea.	375	188		563	700
7030	4"		3	5.333		470	282		752	945
7036	5"		2.50	6.400		840	340		1,180	1,425
7040	6"	Q-2	3	8		1,100	440		1,540	1,850
7050	8"		2.50	9.600		1,575	525		2,100	2,525
7060	10"		2.20	10.909		2,450	600		3,050	3,575
7070	12"		1.70	14.118		4,700	775		5,475	6,350

23 05 23.90 Valves, Stainless Steel		Crew	Daily Output	Labor-Hours	Unit	Material	2015 Bare Costs Labor	Equipment	Total	Total Incl O&P
0010	**VALVES, STAINLESS STEEL** R220523-90									
1700	Check, 200 lb., threaded									
1710	1/4"	1 Plum	24	.333	Ea.	143	19.55		162.55	188
1720	1/2"		22	.364		143	21.50		164.50	190
1730	3/4"		20	.400		148	23.50		171.50	198
1750	1"		19	.421		192	24.50		216.50	249
1760	1-1/2"		13	.615		360	36		396	450
1770	2"		11	.727		610	42.50		652.50	735
1800	150 lb., flanged									
1810	2-1/2"	Q-1	5	3.200	Ea.	1,875	169		2,044	2,300
1820	3"		4.50	3.556		1,875	188		2,063	2,325
1830	4"		3	5.333		2,775	282		3,057	3,500
1840	6"	Q-2	3	8		4,925	440		5,365	6,075
1850	8"	"	2.50	9.600		10,100	525		10,625	11,900
2100	Gate, OS&Y, 150 lb., flanged									
2120	1/2"	1 Plum	18	.444	Ea.	595	26		621	695
2140	3/4"		16	.500		575	29.50		604.50	680
2150	1"		14	.571		725	33.50		758.50	850
2160	1-1/2"		11	.727		1,400	42.50		1,442.50	1,625
2170	2"		8	1		1,650	58.50		1,708.50	1,925
2180	2-1/2"	Q-1	5	3.200		2,025	169		2,194	2,475
2190	3"		4.50	3.556		2,025	188		2,213	2,500
2200	4"		3	5.333		2,975	282		3,257	3,700
2205	5"		2.80	5.714		4,800	300		5,100	5,725
2210	6"	Q-2	3	8		5,575	440		6,015	6,775
2220	8"		2.50	9.600		9,675	525		10,200	11,400
2230	10"		2.30	10.435		16,800	570		17,370	19,400
2240	12"		1.90	12.632		22,500	690		23,190	25,900
2260	For 300 lb., flanged, add					120%	15%			
2600	600 lb., flanged									
2620	1/2"	1 Plum	16	.500	Ea.	213	29.50		242.50	279
2640	3/4"		14	.571		230	33.50		263.50	305
2650	1"		12	.667		277	39		316	365
2660	1-1/2"		10	.800		445	47		492	555
2670	2"		7	1.143		610	67		677	770
2680	2-1/2"	Q-1	4	4		7,375	211		7,586	8,450
2690	3"	"	3.60	4.444		7,375	235		7,610	8,475
3100	Globe, OS&Y, 150 lb., flanged									
3120	1/2"	1 Plum	18	.444	Ea.	570	26		596	665
3140	3/4"		16	.500		620	29.50		649.50	725
3150	1"		14	.571		810	33.50		843.50	940
3160	1-1/2"		11	.727		1,275	42.50		1,317.50	1,475
3170	2"		8	1		1,650	58.50		1,708.50	1,925
3180	2-1/2"	Q-1	5	3.200		3,425	169		3,594	4,000

23 05 Common Work Results for HVAC

23 05 23 – General-Duty Valves for HVAC Piping

23 05 23.90 Valves, Stainless Steel

		Crew	Daily Output	Labor-Hours	Unit	Material	2015 Bare Costs Labor	Equipment	Total	Total Incl O&P
3190	3"	Q-1	4.50	3.556	Ea.	3,425	188		3,613	4,025
3200	4"	↓	3	5.333		5,450	282		5,732	6,425
3210	6"	Q-2	3	8	↓	9,200	440		9,640	10,800

23 05 23.94 Hospital Type Valves

		Crew	Daily Output	Labor-Hours	Unit	Material	Labor	Equipment	Total	Total Incl O&P
0010	**HOSPITAL TYPE VALVES**									
0300	Chiller valves									
0330	Manual operation balancing valve									
0340	2-1/2" line size	Q-1	9	1.778	Ea.	495	94		589	680
0350	3" line size		8	2		545	106		651	760
0360	4" line size	↓	5	3.200		730	169		899	1,050
0370	5" line size	Q-2	5	4.800		920	263		1,183	1,425
0380	6" line size		5	4.800		1,150	263		1,413	1,650
0390	8" line size		4.50	5.333		2,325	292		2,617	3,000
0400	10" line size		4	6		3,725	330		4,055	4,575
0410	12" line size	↓	3	8	↓	5,175	440		5,615	6,325
0600	Automatic flow limiting valve									
0620	2-1/2" line size	Q-1	8	2	Ea.	500	106		606	710
0630	3" line size	↓	7	2.286		790	121		911	1,050
0640	4" line size	↓	6	2.667		1,125	141		1,266	1,475
0645	5" line size	Q-2	5	4.800		1,525	263		1,788	2,075
0650	6" line size		5	4.800		1,875	263		2,138	2,450
0660	8" line size		4.50	5.333		2,825	292		3,117	3,575
0670	10" line size		4	6		5,100	330		5,430	6,100
0680	12" line size	↓	3	8		7,575	440		8,015	8,975

23 05 93 – Testing, Adjusting, and Balancing for HVAC

23 05 93.50 Piping, Testing

		Crew	Daily Output	Labor-Hours	Unit	Material	Labor	Equipment	Total	Total Incl O&P
0010	**PIPING, TESTING**									
0100	Nondestructive testing									
0110	Nondestructive hydraulic pressure test, isolate & 1 hr. hold									
0120	1" - 4" pipe									
0140	0 – 250 L.F.	1 Stpi	1.33	6.015	Ea.		360		360	545
0160	250 – 500 L.F.	"	.80	10			600		600	900
0180	500 – 1000 L.F.	Q-5	1.14	14.035			755		755	1,150
0200	1000 – 2000 L.F.	"	.80	20	↓		1,075		1,075	1,625
0300	6" - 10" pipe									
0320	0 – 250 L.F.	Q-5	1	16	Ea.		860		860	1,300
0340	250 – 500 L.F.		.73	21.918			1,175		1,175	1,775
0360	500 – 1000 L.F.		.53	30.189			1,625		1,625	2,450
0380	1000 – 2000 L.F.	↓	.38	42.105	↓		2,275		2,275	3,425
1000	Pneumatic pressure test, includes soaping joints									
1120	1" - 4" pipe									
1140	0 – 250 L.F.	Q-5	2.67	5.993	Ea.	10.25	320		330.25	495
1160	250 – 500 L.F.		1.33	12.030		20.50	645		665.50	1,000
1180	500 – 1000 L.F.		.80	20		31	1,075		1,106	1,650
1200	1000 – 2000 L.F.	↓	.50	32	↓	41	1,725		1,766	2,650
1300	6" - 10" pipe									
1320	0 – 250 L.F.	Q-5	1.33	12.030	Ea.	10.25	645		655.25	985
1340	250 – 500 L.F.		.67	23.881		20.50	1,275		1,295.50	1,975
1360	500 – 1000 L.F.		.40	40		41	2,150		2,191	3,300
1380	1000 – 2000 L.F.	↓	.25	64	↓	51.50	3,450		3,501.50	5,250
2000	X-Ray of welds									
2110	2" diam.	1 Stpi	8	1	Ea.	13.50	60		73.50	105

For customer support on your Plumbing Cost Data, call 877.709.6968.

319

23 05 Common Work Results for HVAC

23 05 93 – Testing, Adjusting, and Balancing for HVAC

23 05 93.50 Piping, Testing

	23 05 93.50 Piping, Testing	Crew	Daily Output	Labor-Hours	Unit	Material	2015 Bare Costs Labor	Equipment	Total	Total Incl O&P
2120	3" diam.	1 Stpi	8	1	Ea.	13.50	60		73.50	105
2130	4" diam.		8	1		20.50	60		80.50	113
2140	6" diam.		8	1		20.50	60		80.50	113
2150	8" diam.		6.60	1.212		20.50	72.50		93	132
2160	10" diam.		6	1.333		27	79.50		106.50	150
3000	Liquid penetration of welds									
3110	2" diam.	1 Stpi	14	.571	Ea.	4.50	34		38.50	56.50
3120	3" diam.		13.60	.588		4.50	35		39.50	58
3130	4" diam.		13.40	.597		4.50	35.50		40	59
3140	6" diam.		13.20	.606		4.50	36		40.50	59.50
3150	8" diam.		13	.615		6.75	37		43.75	63
3160	10" diam.		12.80	.625		6.75	37.50		44.25	64

23 07 HVAC Insulation

23 07 13 – Duct Insulation

23 07 13.10 Duct Thermal Insulation

	23 07 13.10 Duct Thermal Insulation		Crew	Daily Output	Labor-Hours	Unit	Material	2015 Bare Costs Labor	Equipment	Total	Total Incl O&P
0010	**DUCT THERMAL INSULATION**										
0100	Rule of thumb, as a percentage of total mechanical costs					Job				10%	10%
0110	Insulation req'd. is based on the surface size/area to be covered										
3730	Sheet insulation										
3760	Polyethylene foam, closed cell, UV resistant										
3770	Standard temperature (-90°F to +212°F)										
3771	1/4" thick	G	Q-14	450	.036	S.F.	2.01	1.68		3.69	4.82
3772	3/8" thick	G		440	.036		2.48	1.71		4.19	5.40
3773	1/2" thick	G		420	.038		3.05	1.80		4.85	6.15
3774	3/4" thick	G		400	.040		4.36	1.89		6.25	7.75
3775	1" thick	G		380	.042		5.90	1.98		7.88	9.55
3776	1-1/2" thick	G		360	.044		9.30	2.09		11.39	13.50
3777	2" thick	G		340	.047		12.30	2.22		14.52	17
3778	2-1/2" thick	G		320	.050		15.80	2.36		18.16	21
3779	Adhesive (see line 7878)										
3780	Foam, rubber										
3782	1" thick	G	1 Stpi	50	.160	S.F.	2.89	9.55		12.44	17.65
7000	Board insulation										
7020	Mineral wool, 1200° F										
7022	6 lb. density, plain										
7024	1" thick	G	Q-14	370	.043	S.F.	.33	2.04		2.37	3.53
7026	1-1/2" thick	G		350	.046		.50	2.15		2.65	3.90
7028	2" thick	G		330	.048		.67	2.28		2.95	4.30
7030	3" thick	G		300	.053		.83	2.51		3.34	4.82
7032	4" thick	G		280	.057		1.33	2.69		4.02	5.65
7038	8 lb. density, plain										
7040	1" thick	G	Q-14	360	.044	S.F.	.40	2.09		2.49	3.70
7042	1-1/2" thick	G		340	.047		.61	2.22		2.83	4.12
7044	2" thick	G		320	.050		.81	2.36		3.17	4.56
7046	3" thick	G		290	.055		1.21	2.60		3.81	5.40
7048	4" thick	G		270	.059		1.61	2.79		4.40	6.10
7060	10 lb. density, plain										
7062	1" thick	G	Q-14	350	.046	S.F.	.55	2.15		2.70	3.96
7064	1-1/2" thick	G		330	.048		.83	2.28		3.11	4.47
7066	2" thick	G		310	.052		1.11	2.43		3.54	5

23 07 HVAC Insulation

23 07 13 – Duct Insulation

23 07 13.10 Duct Thermal Insulation		Crew	Daily Output	Labor-Hours	Unit	Material	2015 Bare Costs Labor	Equipment	Total	Total Incl O&P
7068	3" thick	G Q-14	280	.057	S.F.	1.66	2.69		4.35	6
7070	4" thick	G ↓	260	.062	↓	2.21	2.90		5.11	6.95
7878	Contact cement, quart can				Ea.	10.50			10.50	11.55

23 07 16 – HVAC Equipment Insulation

23 07 16.10 HVAC Equipment Thermal Insulation

		Crew	Daily Output	Labor-Hours	Unit	Material	2015 Bare Costs Labor	Equipment	Total	Total Incl O&P
0010	**HVAC EQUIPMENT THERMAL INSULATION**									
0100	Rule of thumb, as a percentage of total mechanical costs				Job				10%	10%
0110	Insulation req'd. is based on the surface size/area to be covered									
1000	Boiler, 1-1/2" calcium silicate only	G Q-14	110	.145	S.F.	4.64	6.85		11.49	15.75
1020	Plus 2" fiberglass	G "	80	.200	"	5.65	9.45		15.10	21
2000	Breeching, 2" calcium silicate									
2020	Rectangular	G Q-14	42	.381	S.F.	9.10	17.95		27.05	38
2040	Round	G "	38.70	.413	"	9.50	19.50		29	41
2300	Calcium silicate block, + 200°F to + 1200°F									
2310	On irregular surfaces, valves and fittings									
2340	1" thick	G Q-14	30	.533	S.F.	4.12	25		29.12	43.50
2360	1-1/2" thick	G ↓	25	.640		4.64	30		34.64	52
2380	2" thick	G	22	.727		6.10	34.50		40.60	60
2400	3" thick	G ↓	18	.889	↓	9.35	42		51.35	75.50
2410	On plane surfaces									
2420	1" thick	G Q-14	126	.127	S.F.	4.12	6		10.12	13.85
2430	1-1/2" thick	G ↓	120	.133		4.64	6.30		10.94	14.90
2440	2" thick	G	100	.160		6.10	7.55		13.65	18.45
2450	3" thick	G ↓	70	.229	↓	9.35	10.75		20.10	27

23 09 Instrumentation and Control for HVAC

23 09 13 – Instrumentation and Control Devices for HVAC

23 09 13.60 Water Level Controls

		Crew	Daily Output	Labor-Hours	Unit	Material	2015 Bare Costs Labor	Equipment	Total	Total Incl O&P
0010	**WATER LEVEL CONTROLS**									
3000	Low water cut-off for hot water boiler, 50 psi maximum									
3100	1" top & bottom equalizing pipes, manual reset	1 Stpi	14	.571	Ea.	350	34		384	435
3200	1" top & bottom equalizing pipes		14	.571		350	34		384	435
3300	2-1/2" side connection for nipple-to-boiler	↓	14	.571	↓	315	34		349	395

23 09 53 – Pneumatic and Electric Control System for HVAC

23 09 53.10 Control Components

		Crew	Daily Output	Labor-Hours	Unit	Material	2015 Bare Costs Labor	Equipment	Total	Total Incl O&P
0010	**CONTROL COMPONENTS** R230500-10									
0680	Controller for VAV box, includes actuator	1 Stpi	7.30	1.096	Ea.	278	65.50		343.50	405
2000	Gauges, pressure or vacuum									
2100	2" diameter dial	1 Stpi	32	.250	Ea.	8.20	14.95		23.15	31.50
2200	2-1/2" diameter dial		32	.250		8.40	14.95		23.35	32
2300	3-1/2" diameter dial		32	.250		10.40	14.95		25.35	34
2400	4-1/2" diameter dial	↓	32	.250	↓	10.70	14.95		25.65	34.50
2700	Flanged iron case, black ring									
2800	3-1/2" diameter dial	1 Stpi	32	.250	Ea.	102	14.95		116.95	135
2900	4-1/2" diameter dial		32	.250		105	14.95		119.95	139
3000	6" diameter dial	↓	32	.250	↓	167	14.95		181.95	206
3010	Steel case, 0 – 300 psi									
3012	2" diam. dial	1 Stpi	16	.500	Ea.	7.60	30		37.60	53.50
3014	4" diam. dial	"	16	.500	"	21	30		51	68
3020	Aluminum case, 0 – 300 psi									

321

23 09 53 – Pneumatic and Electric Control System for HVAC

23 09 53.10 Control Components	Crew	Daily Output	Labor-Hours	Unit	Material	2015 Bare Costs Labor	Equipment	Total	Total Incl O&P
3022	3-1/2" diam. dial	1 Stpi	16	.500	Ea.	10.15	30		40.15	56
3024	4-1/2" diam. dial		16	.500		112	30		142	169
3026	6" diam. dial		16	.500		184	30		214	247
3028	8-1/2" diam. dial		16	.500		315	30		345	390
3030	Brass case, 0 – 300 psi									
3032	2" diam. dial	1 Stpi	16	.500	Ea.	44.50	30		74.50	94
3034	4-1/2" diam. dial	"	16	.500	"	110	30		140	166
3040	Steel case, high pressure, 0 -10,000 psi									
3042	4-1/2" diam. dial	1 Stpi	16	.500	Ea.	291	30		321	365
3044	6-1/2" diam. dial		16	.500		305	30		335	380
3046	8-1/2" diam. dial		16	.500		390	30		420	475
3080	Pressure gauge, differential, magnehelic									
3084	0 – 2" W.C., with air filter kit	1 Stpi	6	1.333	Ea.	105	79.50		184.50	236
3300	For compound pressure-vacuum, add					18%				
4000	Thermometers									
4100	Dial type, 3-1/2" diameter, vapor type, union connection	1 Stpi	32	.250	Ea.	335	14.95		349.95	395
4120	Liquid type, union connection		32	.250		650	14.95		664.95	740
4500	Stem type, 6-1/2" case, 2" stem, 1/2" NPT		32	.250		68	14.95		82.95	97
4520	4" stem, 1/2" NPT		32	.250		83	14.95		97.95	114
4600	9" case, 3-1/2" stem, 3/4" NPT		28	.286		115	17.05		132.05	152
4620	6" stem, 3/4" NPT		28	.286		133	17.05		150.05	172
4640	8" stem, 3/4" NPT		28	.286		217	17.05		234.05	265
4660	12" stem, 1" NPT		26	.308		184	18.40		202.40	230
4670	Bi-metal, dial type, steel case brass stem									
4672	2" dial, 4" - 9" stem	1 Stpi	16	.500	Ea.	45	30		75	94.50
4673	2-1/2" dial, 4" - 9" stem		16	.500		45	30		75	94.50
4674	3-1/2" dial, 4" - 9" stem		16	.500		49.50	30		79.50	99.50
4680	Mercury filled, industrial, union connection type									
4682	Angle stem, 7" scale	1 Stpi	16	.500	Ea.	57	30		87	108
4683	9" scale		16	.500		153	30		183	213
4684	12" scale		16	.500		197	30		227	261
4686	Straight stem, 7" scale		16	.500		150	30		180	210
4687	9" scale		16	.500		167	30		197	229
4688	12" scale		16	.500		206	30		236	272
4690	Mercury filled, industrial, separable socket type, with well									
4692	Angle stem, with socket, 7" scale	1 Stpi	16	.500	Ea.	131	30		161	189
4693	9" scale		16	.500		126	30		156	184
4694	12" scale		16	.500		231	30		261	299
4696	Straight stem, with socket, 7" scale		16	.500		91.50	30		121.50	145
4697	9" scale		16	.500		112	30		142	168
4698	12" scale		16	.500		174	30		204	236
6000	Valves, motorized zone									
6100	Sweat connections, 1/2" C x C	1 Stpi	20	.400	Ea.	168	24		192	220
6110	3/4" C x C		20	.400		168	24		192	221
6120	1" C x C		19	.421		219	25		244	279
6140	1/2" C x C, with end switch, 2 wire		20	.400		166	24		190	218
6150	3/4" C x C, with end switch, 2 wire		20	.400		177	24		201	231
6160	1" C x C, with end switch, 2 wire		19	.421		202	25		227	260

23 11 Facility Fuel Piping

23 11 13 – Facility Fuel-Oil Piping

23 11 13.10 Fuel Oil Specialties

		Crew	Daily Output	Labor-Hours	Unit	Material	2015 Bare Costs Labor	2015 Bare Costs Equipment	Total	Total Incl O&P
0010	**FUEL OIL SPECIALTIES**									
0020	Foot valve, single poppet, metal to metal construction									
0040	Bevel seat, 1/2" diameter	1 Stpi	20	.400	Ea.	61.50	24		85.50	104
0060	3/4" diameter		18	.444		71	26.50		97.50	119
1000	Oil filters, 3/8" IPT, 20 gal. per hour		20	.400		34.50	24		58.50	74
2000	Remote tank gauging system, self contained									
3000	Valve, ball check, globe type, 3/8" diameter	1 Stpi	24	.333	Ea.	12.80	19.90		32.70	44
3500	Fusible, 3/8" diameter		24	.333		13.10	19.90		33	44.50
3600	1/2" diameter		24	.333		31.50	19.90		51.40	65
3610	3/4" diameter		20	.400		70	24		94	113
3620	1" diameter		19	.421		205	25		230	264
4000	Nonfusible, 3/8" diameter		24	.333		22.50	19.90		42.40	55
4500	Shutoff, gate type, lever handle, spring-fusible kit									
4520	1/4" diameter	1 Stpi	14	.571	Ea.	34.50	34		68.50	89.50
4540	3/8" diameter		12	.667		33.50	40		73.50	97
4560	1/2" diameter		10	.800		49.50	48		97.50	127
4570	3/4" diameter		8	1		86	60		146	185
5000	Vent alarm, whistling signal					26			26	28.50
5500	Vent protector/breather, 1-1/4" diameter	1 Stpi	32	.250		13.60	14.95		28.55	37.50

23 11 23 – Facility Natural-Gas Piping

23 11 23.10 Gas Meters

		Crew	Daily Output	Labor-Hours	Unit	Material	2015 Bare Costs Labor	2015 Bare Costs Equipment	Total	Total Incl O&P
0010	**GAS METERS**									
4000	Residential									
4010	Gas meter, residential, 3/4" pipe size	1 Plum	14	.571	Ea.	228	33.50		261.50	300
4020	Gas meter, residential, 1" pipe size		12	.667		183	39		222	260
4030	Gas meter, residential, 1-1/4" pipe size		10	.800		183	47		230	272

23 11 23.20 Gas Piping, Flexible (Csst)

		Crew	Daily Output	Labor-Hours	Unit	Material	2015 Bare Costs Labor	2015 Bare Costs Equipment	Total	Total Incl O&P
0010	**GAS PIPING, FLEXIBLE (CSST)**									
0100	Tubing with lightning protection									
0110	3/8"	1 Stpi	65	.123	L.F.	2.47	7.35		9.82	13.80
0120	1/2"		62	.129		2.77	7.70		10.47	14.70
0130	3/4"		60	.133		3.60	7.95		11.55	16
0140	1"		55	.145		5.15	8.70		13.85	18.75
0150	1-1/4"		50	.160		6.45	9.55		16	21.50
0160	1-1/2"		45	.178		11.30	10.60		21.90	28.50
0170	2"		40	.200		16.25	11.95		28.20	36
0200	Tubing for underground/underslab burial									
0210	3/8"	1 Stpi	65	.123	L.F.	3.43	7.35		10.78	14.85
0220	1/2"		62	.129		3.91	7.70		11.61	15.95
0230	3/4"		60	.133		4.96	7.95		12.91	17.50
0240	1"		55	.145		6.70	8.70		15.40	20.50
0250	1-1/4"		50	.160		8.95	9.55		18.50	24.50
0260	1-1/2"		45	.178		16.80	10.60		27.40	34.50
0270	2"		40	.200		20	11.95		31.95	40
3000	Fittings									
3010	Straight									
3100	Tube to NPT									
3110	3/8"	1 Stpi	29	.276	Ea.	9.90	16.50		26.40	36
3120	1/2"		27	.296		10.65	17.70		28.35	38
3130	3/4"		25	.320		14.50	19.10		33.60	45
3140	1"		23	.348		22.50	21		43.50	56
3150	1-1/4"		20	.400		50	24		74	91

23 11 23.20 Gas Piping, Flexible (Csst)	Crew	Daily Output	Labor-Hours	Unit	Material	2015 Bare Costs Labor	2015 Bare Costs Equipment	Total	Total Incl O&P
3160	1-1/2"	1 Stpi	17	.471	Ea.	102	28		130	156
3170	2"	↓	15	.533	↓	174	32		206	240
3200	Coupling									
3210	3/8"	1 Stpi	29	.276	Ea.	17.20	16.50		33.70	44
3220	1/2"		27	.296		19.60	17.70		37.30	48
3230	3/4"		25	.320		27	19.10		46.10	58.50
3240	1"		23	.348		41.50	21		62.50	77.50
3250	1-1/4"		20	.400		92.50	24		116.50	138
3260	1-1/2"		17	.471		194	28		222	256
3270	2"	↓	15	.533	↓	330	32		362	415
3300	Flange fitting									
3310	3/8"	1 Stpi	25	.320	Ea.	15.15	19.10		34.25	45.50
3320	1/2"		22	.364		14.45	21.50		35.95	49
3330	3/4"		19	.421		18.75	25		43.75	58.50
3340	1"		16	.500		26	30		56	73.50
3350	1-1/4"	↓	12	.667	↓	60	40		100	126
3400	90° Flange valve									
3410	3/8"	1 Stpi	25	.320	Ea.	29	19.10		48.10	61
3420	1/2"		22	.364		29.50	21.50		51	65.50
3430	3/4"	↓	19	.421	↓	36	25		61	77.50
4000	Tee									
4120	1/2"	1 Stpi	20.50	.390	Ea.	31	23.50		54.50	69
4130	3/4"		19	.421		41	25		66	83
4140	1"	↓	17.50	.457	↓	74	27.50		101.50	123
5000	Reducing									
5110	Tube to NPT									
5120	3/4" to 1/2" NPT	1 Stpi	26	.308	Ea.	15.60	18.40		34	45
5130	1" to 3/4" NPT	"	24	.333	"	24	19.90		43.90	56.50
5200	Reducing tee									
5210	1/2" x 3/8" x 3/8"	1 Stpi	21	.381	Ea.	40.50	23		63.50	79.50
5220	3/4" x 1/2" x 1/2"		20	.400		38.50	24		62.50	78
5230	1" x 3/4" x 1/2"		18	.444		63.50	26.50		90	110
5240	1-1/4" x 1-1/4" x 1"		15.60	.513		169	30.50		199.50	233
5250	1-1/2" x 1-1/2" x 1-1/4"		13.30	.602		258	36		294	340
5260	2" x 2" x 1-1/2"	↓	11.80	.678	↓	390	40.50		430.50	485
5300	Manifold with four ports and mounting bracket									
5302	Labor to mount manifold does not include making pipe									
5304	connections which are included in fitting labor.									
5310	3/4" x 1/2" x 1/2"(4)	1 Stpi	76	.105	Ea.	33.50	6.30		39.80	46
5330	1-1/4" x 1" x 3/4"(4)		72	.111		48	6.65		54.65	63
5350	2" x 1-1/2" x 1"(4)	↓	68	.118	↓	60	7.05		67.05	76.50
5600	Protective striker plate									
5610	Quarter plate, 3" x 2"	1 Stpi	88	.091	Ea.	.77	5.45		6.22	9.05
5620	Half plate, 3" x 7"		82	.098		1.50	5.85		7.35	10.45
5630	Full plate, 3" x 12"	↓	78	.103	↓	3.11	6.15		9.26	12.65

23 12 Facility Fuel Pumps

23 12 13 – Facility Fuel-Oil Pumps

23 12 13.10 Pump and Motor Sets

	Crew	Daily Output	Labor-Hours	Unit	Material	2015 Bare Costs Labor	Equipment	Total	Total Incl O&P
0010 **PUMP AND MOTOR SETS**									
1810 Light fuel and diesel oils									
1820 20 GPH 1/3 HP	Q-5	6	2.667	Ea.	1,050	143		1,193	1,375
1850 145 GPH, 1/2 HP	"	4	4	"	1,100	215		1,315	1,550

23 13 Facility Fuel-Storage Tanks

23 13 13 – Facility Underground Fuel-Oil, Storage Tanks

23 13 13.09 Single-Wall Steel Fuel-Oil Tanks

	Crew	Daily Output	Labor-Hours	Unit	Material	2015 Bare Costs Labor	Equipment	Total	Total Incl O&P
0010 **SINGLE-WALL STEEL FUEL-OIL TANKS**									
5000 Tanks, steel ugnd., sti-p3, not incl. hold-down bars									
5500 Excavation, pad, pumps and piping not included									
5510 Single wall, 500 gallon capacity, 7 ga. shell	Q-5	2.70	5.926	Ea.	2,050	320		2,370	2,725
5520 1,000 gallon capacity, 7 ga. shell	"	2.50	6.400		3,825	345		4,170	4,725
5530 2,000 gallon capacity, 1/4" thick shell	Q-7	4.60	6.957		6,200	395		6,595	7,425
5535 2,500 gallon capacity, 7 ga. shell	Q-5	3	5.333		6,850	287		7,137	7,950
5540 5,000 gallon capacity, 1/4" thick shell	Q-7	3.20	10		11,500	570		12,070	13,600
5560 10,000 gallon capacity, 1/4" thick shell		2	16		12,300	910		13,210	15,000
5580 15,000 gallon capacity, 5/16" thick shell		1.70	18.824		19,000	1,075		20,075	22,500
5600 20,000 gallon capacity, 5/16" thick shell		1.50	21.333		26,700	1,225		27,925	31,100
5610 25,000 gallon capacity, 3/8" thick shell		1.30	24.615		37,900	1,400		39,300	43,800
5620 30,000 gallon capacity, 3/8" thick shell		1.10	29.091		38,400	1,650		40,050	44,700
5630 40,000 gallon capacity, 3/8" thick shell		.90	35.556		42,000	2,025		44,025	49,300
5640 50,000 gallon capacity, 3/8" thick shell		.80	40		46,600	2,275		48,875	55,000

23 13 13.13 Dbl-Wall Steel, Undrgrnd Fuel-Oil, Stor. Tanks

	Crew	Daily Output	Labor-Hours	Unit	Material	2015 Bare Costs Labor	Equipment	Total	Total Incl O&P
0010 **DOUBLE-WALL STEEL, UNDERGROUND FUEL-OIL, STORAGE TANKS**									
6200 Steel, underground, 360°, double wall, U.L. listed,									
6210 with sti-P3 corrosion protection,									
6220 (dielectric coating, cathodic protection, electrical									
6230 isolation) 30 year warranty,									
6240 not incl. manholes or hold-downs.									
6250 500 gallon capacity	Q-5	2.40	6.667	Ea.	5,575	360		5,935	6,700
6260 1,000 gallon capactiy	"	2.25	7.111		6,900	380		7,280	8,150
6270 2,000 gallon capacity	Q-7	4.16	7.692		9,050	440		9,490	10,600
6280 3,000 gallon capacity		3.90	8.205		10,900	465		11,365	12,700
6290 4,000 gallon capacity		3.64	8.791		13,800	500		14,300	15,900
6300 5,000 gallon capacity		2.91	10.997		19,900	625		20,525	22,800
6310 6,000 gallon capacity		2.42	13.223		25,600	750		26,350	29,200
6320 8,000 gallon capacity		2.08	15.385		24,600	875		25,475	28,300
6330 10,000 gallon capacity		1.82	17.582		27,700	1,000		28,700	32,000
6340 12,000 gallon capacity		1.70	18.824		30,600	1,075		31,675	35,300
6350 15,000 gallon capacity		1.33	24.060		31,900	1,375		33,275	37,200
6360 20,000 gallon capacity		1.33	24.060		32,700	1,375		34,075	38,000
6370 25,000 gallon capacity		1.16	27.586		72,500	1,575		74,075	82,500
6380 30,000 gallon capacity		1.03	31.068		87,500	1,775		89,275	98,500
6390 40,000 gallon capacity		.80	40		112,500	2,275		114,775	127,000
6395 50,000 gallon capacity		.73	43.836		136,500	2,500		139,000	154,500
6400 For hold-downs 500-2000 gal., add		16	2	Set	183	114		297	375
6410 For hold-downs 3000-6000 gal., add		12	2.667		370	152		522	640
6420 For hold-downs 8000-12,000 gal., add		11	2.909		445	166		611	740
6430 For hold-downs 15,000 gal., add		9	3.556		640	202		842	1,000
6440 For hold-downs 20,000 gal., add		8	4		735	228		963	1,150

23 13 13 – Facility Underground Fuel-Oil, Storage Tanks

23 13 13.13 Dbl-Wall Steel, Undrgrnd Fuel-Oil, Stor. Tanks	Crew	Daily Output	Labor-Hours	Unit	Material	2015 Bare Costs Labor	Equipment	Total	Total Incl O&P
6450	For hold-downs 20,000 gal. plus, add	Q-7	6	5.333	Set	960	305		1,265	1,500
6500	For manways, add				Ea.	1,800			1,800	2,000
6600	In place with hold-downs									
6652	550 gallon capacity	Q-5	1.84	8.696	Ea.	5,775	470		6,245	7,050

23 13 13.23 Glass-Fiber-Reinfcd-Plastic, Fuel-Oil, Storage

		Crew	Daily Output	Labor-Hours	Unit	Material	Labor	Equipment	Total	Total Incl O&P
0010	**GLASS-FIBER-REINFCD-PLASTIC, UNDERGRND. FUEL-OIL, STORAGE**									
0210	Fiberglass, underground, single wall, U.L. listed, not including									
0220	manway or hold-down strap									
0225	550 gallon capacity	Q-5	2.67	5.993	Ea.	3,200	320		3,520	3,975
0230	1,000 gallon capacity	"	2.46	6.504		4,125	350		4,475	5,050
0240	2,000 gallon capacity	Q-7	4.57	7.002		5,950	400		6,350	7,150
0245	3,000 gallon capacity		3.90	8.205		7,125	465		7,590	8,550
0250	4,000 gallon capacity		3.55	9.014		8,275	515		8,790	9,875
0255	5,000 gallon capacity		3.20	10		9,200	570		9,770	11,000
0260	6,000 gallon capacity		2.67	11.985		9,400	680		10,080	11,300
0270	8,000 gallon capacity		2.29	13.974		11,200	795		11,995	13,600
0280	10,000 gallon capacity		2	16		12,800	910		13,710	15,500
0282	12,000 gallon capacity		1.88	17.021		14,500	970		15,470	17,500
0284	15,000 gallon capacity		1.68	19.048		20,500	1,075		21,575	24,100
0290	20,000 gallon capacity		1.45	22.069		26,300	1,250		27,550	30,900
0300	25,000 gallon capacity		1.28	25		54,000	1,425		55,425	61,500
0320	30,000 gallon capacity		1.14	28.070		77,500	1,600		79,100	87,500
0340	40,000 gallon capacity		.89	35.955		86,000	2,050		88,050	97,500
0360	48,000 gallon capacity		.81	39.506		128,000	2,250		130,250	144,000
0500	For manway, fittings and hold-downs, add					20%	15%			
0600	For manways, add					2,250			2,250	2,475
1000	For helical heating coil, add	Q-5	2.50	6.400		4,775	345		5,120	5,775
1020	Fiberglass, underground, double wall, U.L. listed									
1030	includes manways, not incl. hold-down straps									
1040	600 gallon capacity	Q-5	2.42	6.612	Ea.	7,200	355		7,555	8,450
1050	1,000 gallon capacity	"	2.25	7.111		9,825	380		10,205	11,400
1060	2,500 gallon capacity	Q-7	4.16	7.692		14,800	440		15,240	17,000
1070	3,000 gallon capacity		3.90	8.205		16,000	465		16,465	18,300
1080	4,000 gallon capacity		3.64	8.791		16,200	500		16,700	18,700
1090	6,000 gallon capacity		2.42	13.223		21,100	750		21,850	24,400
1100	8,000 gallon capacity		2.08	15.385		23,500	875		24,375	27,200
1110	10,000 gallon capacity		1.82	17.582		27,300	1,000		28,300	31,500
1120	12,000 gallon capacity		1.70	18.824		34,100	1,075		35,175	39,100
1122	15,000 gallon capacity		1.52	21.053		45,300	1,200		46,500	51,500
1124	20,000 gallon capacity		1.33	24.060		56,000	1,375		57,375	63,500
1126	25,000 gallon capacity		1.16	27.586		76,500	1,575		78,075	87,000
1128	30,000 gallon capacity		1.03	31.068		92,000	1,775		93,775	104,000
1140	For hold-down straps, add					2%	10%			
1150	For hold-downs 500-4000 gal., add	Q-7	16	2	Set	445	114		559	660
1160	For hold-downs 5000-15000 gal., add		8	4		895	228		1,123	1,325
1170	For hold-downs 20,000 gal., add		5.33	6.004		1,350	340		1,690	2,000
1180	For hold-downs 25,000 gal., add		4	8		1,800	455		2,255	2,650
1190	For hold-downs 30,000 gal., add		2.60	12.308		2,675	700		3,375	4,000
2210	Fiberglass, underground, single wall, U.L. listed, including									
2220	hold-down straps, no manways									
2225	550 gallon capacity	Q-5	2	8	Ea.	3,625	430		4,055	4,650
2230	1,000 gallon capacity	"	1.88	8.511		4,575	460		5,035	5,725

23 13 Facility Fuel-Storage Tanks

23 13 13 – Facility Underground Fuel-Oil, Storage Tanks

23 13 13.23 Glass-Fiber-Reinfcd-Plastic, Fuel-Oil, Storage

		Crew	Daily Output	Labor-Hours	Unit	Material	2015 Bare Costs Labor	Equipment	Total	Total Incl O&P
2240	2,000 gallon capacity	Q-7	3.55	9.014	Ea.	6,400	515		6,915	7,825
2250	4,000 gallon capacity		2.90	11.034		8,725	630		9,355	10,600
2260	6,000 gallon capacity		2	16		10,300	910		11,210	12,700
2270	8,000 gallon capacity		1.78	17.978		12,100	1,025		13,125	14,900
2280	10,000 gallon capacity		1.60	20		13,700	1,150		14,850	16,800
2282	12,000 gallon capacity		1.52	21.053		15,400	1,200		16,600	18,800
2284	15,000 gallon capacity		1.39	23.022		21,400	1,300		22,700	25,500
2290	20,000 gallon capacity		1.14	28.070		27,700	1,600		29,300	32,800
2300	25,000 gallon capacity		.96	33.333		56,000	1,900		57,900	64,500
2320	30,000 gallon capacity		.80	40		80,000	2,275		82,275	91,500
3020	Fiberglass, underground, double wall, U.L. listed									
3030	includes manways and hold-down straps									
3040	600 gallon capacity	Q-5	1.86	8.602	Ea.	7,650	465		8,115	9,100
3050	1,000 gallon capacity	"	1.70	9.412		10,300	505		10,805	12,100
3060	2,500 gallon capacity	Q-7	3.29	9.726		15,200	555		15,755	17,500
3070	3,000 gallon capacity		3.13	10.224		16,500	580		17,080	19,000
3080	4,000 gallon capacity		2.93	10.922		16,700	620		17,320	19,300
3090	6,000 gallon capacity		1.86	17.204		22,000	980		22,980	25,700
3100	8,000 gallon capacity		1.65	19.394		24,400	1,100		25,500	28,600
3110	10,000 gallon capacity		1.48	21.622		28,200	1,225		29,425	32,900
3120	12,000 gallon capacity		1.40	22.857		34,900	1,300		36,200	40,400
3122	15,000 gallon capacity		1.28	25		46,200	1,425		47,625	53,000
3124	20,000 gallon capacity		1.06	30.189		57,000	1,725		58,725	65,500
3126	25,000 gallon capacity		.90	35.556		78,500	2,025		80,525	89,500
3128	30,000 gallon capacity		.74	43.243		95,000	2,450		97,450	107,500

23 13 23 – Facility Aboveground Fuel-Oil, Storage Tanks

23 13 23.16 Horizontal, Stl, Abvgrd Fuel-Oil, Storage Tanks

		Crew	Daily Output	Labor-Hours	Unit	Material	2015 Bare Costs Labor	Equipment	Total	Total Incl O&P
0010	**HORIZONTAL, STEEL, ABOVEGROUND FUEL-OIL, STORAGE TANKS**									
3000	Steel, storage, above ground, including cradles, coating,									
3020	fittings, not including foundation, pumps or piping									
3040	Single wall, 275 gallon	Q-5	5	3.200	Ea.	490	172		662	800
3060	550 gallon	"	2.70	5.926		3,750	320		4,070	4,600
3080	1,000 gallon	Q-7	5	6.400		4,025	365		4,390	4,975
3100	1,500 gallon		4.75	6.737		8,775	385		9,160	10,300
3120	2,000 gallon		4.60	6.957		10,600	395		10,995	12,300
3320	Double wall, 500 gallon capacity	Q-5	2.40	6.667		2,625	360		2,985	3,425
3330	2000 gallon capacity	Q-7	4.15	7.711		9,975	440		10,415	11,700
3340	4000 gallon capacity		3.60	8.889		17,800	505		18,305	20,400
3350	6000 gallon capacity		2.40	13.333		21,000	760		21,760	24,300
3360	8000 gallon capacity		2	16		27,000	910		27,910	31,100
3370	10000 gallon capacity		1.80	17.778		30,200	1,000		31,200	34,700
3380	15000 gallon capacity		1.50	21.333		45,900	1,225		47,125	52,500
3390	20000 gallon capacity		1.30	24.615		52,500	1,400		53,900	59,500
3400	25000 gallon capacity		1.15	27.826		63,500	1,575		65,075	72,500
3410	30000 gallon capacity		1	32		69,500	1,825		71,325	79,500

23 13 23.26 Horizontal, Conc., Abvgrd Fuel-Oil, Stor. Tanks

		Crew	Daily Output	Labor-Hours	Unit	Material	2015 Bare Costs Labor	Equipment	Total	Total Incl O&P
0010	**HORIZONTAL, CONCRETE, ABOVEGROUND FUEL-OIL, STORAGE TANKS**									
0050	Concrete, storage, above ground, including pad & pump									
0100	500 gallon	F-3	2	20	Ea.	10,000	960	325	11,285	12,800
0200	1,000 gallon	"	2	20		14,000	960	325	15,285	17,200
0300	2,000 gallon	F-4	2	24		18,000	1,150	565	19,715	22,200

327

23 21 20 – Hydronic HVAC Piping Specialties

23 21 20.10 Air Control		Crew	Daily Output	Labor-Hours	Unit	Material	2015 Bare Costs Labor	Equipment	Total	Total Incl O&P
0010	**AIR CONTROL**									
0030	Air separator, with strainer									
0040	2" diameter	Q-5	6	2.667	Ea.	1,175	143		1,318	1,525
0080	2-1/2" diameter		5	3.200		1,325	172		1,497	1,700
0100	3" diameter	▼	4	4	▼	2,050	215		2,265	2,575
1000	Micro-bubble separator for total air removal, closed loop system									
1010	Requires bladder type tank in system.									
1020	Water (hot or chilled) or glycol system									
1030	Threaded									
1040	3/4" diameter	1 Stpi	20	.400	Ea.	90	24		114	135
1050	1" diameter		19	.421		101	25		126	150
1060	1-1/4" diameter		16	.500		140	30		170	199
1070	1-1/2" diameter		13	.615		182	37		219	257
1080	2" diameter	▼	11	.727		885	43.50		928.50	1,050
1090	2-1/2" diameter	Q-5	15	1.067		985	57.50		1,042.50	1,150
1100	3" diameter		13	1.231		1,400	66		1,466	1,650
1110	4" diameter	▼	10	1.600	▼	1,525	86		1,611	1,800

23 21 20.14 Air Purging Scoop		Crew	Daily Output	Labor-Hours	Unit	Material	2015 Bare Costs Labor	Equipment	Total	Total Incl O&P
0010	**AIR PURGING SCOOP**, With tappings.									
0020	For air vent and expansion tank connection									
0100	1" pipe size, threaded	1 Stpi	19	.421	Ea.	24.50	25		49.50	65
0110	1-1/4" pipe size, threaded		15	.533		24.50	32		56.50	75
0120	1-1/2" pipe size, threaded		13	.615		53.50	37		90.50	114
0130	2" pipe size, threaded	▼	11	.727	▼	61	43.50		104.50	133

23 21 20.18 Automatic Air Vent		Crew	Daily Output	Labor-Hours	Unit	Material	2015 Bare Costs Labor	Equipment	Total	Total Incl O&P
0010	**AUTOMATIC AIR VENT**									
0020	Cast iron body, stainless steel internals, float type									
0180	1/2" NPT inlet, 250 psi	1 Stpi	10	.800	Ea.	350	48		398	455
0220	3/4" NPT inlet, 250 psi	"	10	.800	"	350	48		398	455
0600	Forged steel body, stainless steel internals, float type									
0640	1/2" NPT inlet, 750 psi	1 Stpi	12	.667	Ea.	1,150	40		1,190	1,325
0680	3/4" NPT inlet, 750 psi	"	12	.667	"	1,150	40		1,190	1,325
1100	Formed steel body, noncorrosive									
1110	1/8" NPT inlet 150 psi	1 Stpi	32	.250	Ea.	12.45	14.95		27.40	36
1120	1/4" NPT inlet 150 psi		32	.250		42	14.95		56.95	69
1130	3/4" NPT inlet 150 psi	▼	32	.250	▼	42	14.95		56.95	69
1300	Chrome plated brass, automatic/manual, for radiators									
1310	1/8" NPT inlet, nickel plated brass	1 Stpi	32	.250	Ea.	7.25	14.95		22.20	30.50

23 21 20.26 Circuit Setter		Crew	Daily Output	Labor-Hours	Unit	Material	2015 Bare Costs Labor	Equipment	Total	Total Incl O&P
0010	**CIRCUIT SETTER**, Balance valve									
0018	Threaded									
0019	1/2" pipe size	1 Stpi	22	.364	Ea.	74	21.50		95.50	114
0020	3/4" pipe size	"	20	.400	"	79	24		103	123

23 21 20.30 Cocks, Drains and Specialties		Crew	Daily Output	Labor-Hours	Unit	Material	2015 Bare Costs Labor	Equipment	Total	Total Incl O&P
0010	**COCKS, DRAINS AND SPECIALTIES**									
1000	Boiler drain									
1010	Pipe thread to hose									
1020	Bronze									
1030	1/2" size	1 Stpi	36	.222	Ea.	9.40	13.30		22.70	30.50
1040	3/4" size	"	34	.235	"	9.85	14.05		23.90	32
1100	Solder to hose									
1110	Bronze									

23 21 20 – Hydronic HVAC Piping Specialties

23 21 20.30 Cocks, Drains and Specialties

		Crew	Daily Output	Labor-Hours	Unit	Material	2015 Bare Costs Labor	Equipment	Total	Total Incl O&P
1120	1/2" size	1 Stpi	46	.174	Ea.	11.90	10.40		22.30	29
1130	3/4" size	"	44	.182	"	12.95	10.85		23.80	30.50
1600	With built-in vacuum breaker									
1610	1/2" I.P. or solder	1 Stpi	36	.222	Ea.	40	13.30		53.30	64
1630	With tamper proof vacuum breaker									
1640	1/2" I.P. or solder	1 Stpi	36	.222	Ea.	45.50	13.30		58.80	70
1650	3/4" I.P. or solder	"	34	.235	"	46	14.05		60.05	71.50
3000	Cocks									
3010	Air, lever or tee handle									
3020	Bronze, single thread									
3030	1/8" size	1 Stpi	52	.154	Ea.	10.85	9.20	.	20.05	26
3040	1/4" size		46	.174		11.45	10.40		21.85	28.50
3050	3/8" size		40	.200		11.60	11.95		23.55	31
3060	1/2" size		36	.222		13.70	13.30		27	35
3100	Bronze, double thread									
3110	1/8" size	1 Stpi	26	.308	Ea.	13.70	18.40		32.10	43
3120	1/4" size		22	.364		14.20	21.50		35.70	48.50
3130	3/8" size		18	.444		14.80	26.50		41.30	56.50
3140	1/2" size		15	.533		18.15	32		50.15	68
4500	Gauge cock, brass									
4510	1/4" FPT	1 Stpi	24	.333	Ea.	11.15	19.90		31.05	42.50
4512	1/4" MPT	"	24	.333	"	13.75	19.90		33.65	45
4600	Pigtail, steam syphon									
4604	1/4"	1 Stpi	24	.333	Ea.	19.70	19.90		39.60	51.50
4650	Snubber valve									
4654	1/4"	1 Stpi	22	.364	Ea.	11.75	21.50		33.25	46
4660	Nipple, black steel									
4664	1/4" x 3"	1 Stpi	37	.216	Ea.	2.83	12.90		15.73	22.50

23 21 20.34 Dielectric Unions

		Crew	Daily Output	Labor-Hours	Unit	Material	2015 Bare Costs Labor	Equipment	Total	Total Incl O&P
0010	**DIELECTRIC UNIONS**, Standard gaskets for water and air									
0020	250 psi maximum pressure									
0280	Female IPT to sweat, straight									
0300	1/2" pipe size	1 Plum	24	.333	Ea.	9	19.55		28.55	39.50
0340	3/4" pipe size		20	.400		6.20	23.50		29.70	42.50
0360	1" pipe size		19	.421		8.45	24.50		32.95	47
0380	1-1/4" pipe size		15	.533		13.10	31.50		44.60	62
0400	1-1/2" pipe size		13	.615		19.70	36		55.70	76
0420	2" pipe size		11	.727		26.50	42.50		69	93.50
0580	Female IPT to brass pipe thread, straight									
0600	1/2" pipe size	1 Plum	24	.333	Ea.	12.10	19.55		31.65	43
0640	3/4" pipe size		20	.400		13.30	23.50		36.80	50
0660	1" pipe size		19	.421		24.50	24.50		49	64.50
0680	1-1/4" pipe size		15	.533		30	31.50		61.50	80.50
0700	1-1/2" pipe size		13	.615		44	36		80	103
0720	2" pipe size		11	.727		86	42.50		128.50	159
0780	Female IPT to female IPT, straight									
0800	1/2" pipe size	1 Plum	24	.333	Ea.	13.85	19.55		33.40	44.50
0840	3/4" pipe size		20	.400		15.60	23.50		39.10	52.50
0860	1" pipe size		19	.421		21	24.50		45.50	60.50
0880	1-1/4" pipe size		15	.533		28	31.50		59.50	78.50
0900	1-1/2" pipe size		13	.615		43	36		79	102
0920	2" pipe size		11	.727		63	42.50		105.50	134

For customer support on your Plumbing Cost Data, call 877.709.6968.

329

23 21 20.34 Dielectric Unions

		Crew	Daily Output	Labor-Hours	Unit	Material	2015 Bare Costs Labor	Equipment	Total	Total Incl O&P
2000	175 psi maximum pressure									
2180	Female IPT to sweat									
2240	2" pipe size	1 Plum	9	.889	Ea.	142	52		194	235
2260	2-1/2" pipe size	Q-1	15	1.067		155	56.50		211.50	256
2280	3" pipe size		14	1.143		213	60.50		273.50	325
2300	4" pipe size	▼	11	1.455	▼	565	77		642	735
2480	Female IPT to brass pipe									
2500	1-1/2" pipe size	1 Plum	11	.727	Ea.	176	42.50		218.50	259
2540	2" pipe size	"	9	.889		207	52		259	305
2560	2-1/2" pipe size	Q-1	15	1.067		305	56.50		361.50	420
2580	3" pipe size		14	1.143		355	60.50		415.50	480
2600	4" pipe size	▼	11	1.455	▼	635	77		712	815

23 21 20.42 Expansion Joints

		Crew	Daily Output	Labor-Hours	Unit	Material	2015 Bare Costs Labor	Equipment	Total	Total Incl O&P
0010	**EXPANSION JOINTS**									
0100	Bellows type, neoprene cover, flanged spool									
0140	6" face to face, 1-1/4" diameter	1 Stpi	11	.727	Ea.	255	43.50		298.50	345
0160	1-1/2" diameter	"	10.60	.755		255	45		300	350
0180	2" diameter	Q-5	13.30	1.203		258	64.50		322.50	380
0190	2-1/2" diameter		12.40	1.290		267	69.50		336.50	400
0200	3" diameter		11.40	1.404		299	75.50		374.50	445
0210	4" diameter		8.40	1.905		325	102		427	510
0220	5" diameter		7.60	2.105		395	113		508	605
0230	6" diameter		6.80	2.353		405	127		532	640
0240	8" diameter		5.40	2.963		475	159		634	760
0250	10" diameter		5	3.200		650	172		822	975
0260	12" diameter		4.60	3.478		745	187		932	1,100
0480	10" face to face, 2" diameter		13	1.231		370	66		436	510
0500	2-1/2" diameter		12	1.333		390	71.50		461.50	540
0520	3" diameter		11	1.455		400	78		478	560
0540	4" diameter		8	2		455	108		563	660
0560	5" diameter		7	2.286		540	123		663	780
0580	6" diameter		6	2.667		560	143		703	830
0600	8" diameter		5	3.200		670	172		842	995
0620	10" diameter		4.60	3.478		735	187		922	1,100
0640	12" diameter		4	4		910	215		1,125	1,325
0660	14" diameter		3.80	4.211		1,125	226		1,351	1,600
0680	16" diameter		2.90	5.517		1,300	297		1,597	1,875
0700	18" diameter		2.50	6.400		1,475	345		1,820	2,150
0720	20" diameter		2.10	7.619		1,550	410		1,960	2,325
0740	24" diameter		1.80	8.889		1,800	480		2,280	2,700
0760	26" diameter		1.40	11.429		2,025	615		2,640	3,150
0780	30" diameter		1.20	13.333		2,250	715		2,965	3,550
0800	36" diameter	▼	1	16	▼	2,800	860		3,660	4,375

23 21 20.46 Expansion Tanks

		Crew	Daily Output	Labor-Hours	Unit	Material	2015 Bare Costs Labor	Equipment	Total	Total Incl O&P
0010	**EXPANSION TANKS**									
1502	Plastic, corrosion resistant, see Section 22 66 83.13									
1505	Aboveground fuel-oil, storage tanks, see Section 23 13 23									
1507	Underground fuel-oil storage tanks, see Section 23 13 13									
1512	Tank leak detection systems, see Section 28 33 33.50									
2000	Steel, liquid expansion, ASME, painted, 15 gallon capacity	Q-5	17	.941	Ea.	640	50.50		690.50	775
2020	24 gallon capacity		14	1.143		715	61.50		776.50	880
2040	30 gallon capacity	▼	12	1.333	▼	715	71.50		786.50	895

23 21 Hydronic Piping and Pumps

23 21 20 – Hydronic HVAC Piping Specialties

23 21 20.46 Expansion Tanks		Crew	Daily Output	Labor-Hours	Unit	Material	2015 Bare Costs Labor	Equipment	Total	Total Incl O&P
2060	40 gallon capacity	Q-5	10	1.600	Ea.	835	86		921	1,050
2360	Galvanized									
2370	15 gallon capacity	Q-5	17	.941	Ea.	1,125	50.50		1,175.50	1,300
2380	24 gallon capacity		14	1.143		1,150	61.50		1,211.50	1,375
2390	30 gallon capacity		12	1.333		1,350	71.50		1,421.50	1,575
3000	Steel ASME expansion, rubber diaphragm, 19 gal. cap. accept.		12	1.333		2,425	71.50		2,496.50	2,775
3020	31 gallon capacity		8	2		2,700	108		2,808	3,125
3040	61 gallon capacity		6	2.667		3,800	143		3,943	4,400

23 21 20.50 Float Valves

23 21 20.50 Float Valves		Crew	Daily Output	Labor-Hours	Unit	Material	Labor	Equipment	Total	Total Incl O&P
0010	**FLOAT VALVES**									
0020	With ball and bracket									
0030	Single seat, threaded									
0040	Brass body									
0050	1/2"	1 Stpi	11	.727	Ea.	100	43.50		143.50	176
0060	3/4"		9	.889		112	53		165	203
0070	1"		7	1.143		144	68.50		212.50	261
0080	1-1/2"		4.50	1.778		211	106		317	390
0090	2"		3.60	2.222		215	133		348	435
0300	For condensate receivers, CI, in-line mount									
0320	1" inlet	1 Stpi	7	1.143	Ea.	144	68.50		212.50	261
0360	For condensate receiver, CI, external float, flanged tank mount									
0370	3/4" inlet	1 Stpi	5	1.600	Ea.	112	95.50		207.50	267

23 21 20.54 Flow Check Control

23 21 20.54 Flow Check Control		Crew	Daily Output	Labor-Hours	Unit	Material	Labor	Equipment	Total	Total Incl O&P
0010	**FLOW CHECK CONTROL**									
0100	Bronze body, soldered									
0110	3/4" size	1 Stpi	20	.400	Ea.	64	24		88	107
0120	1" size	"	19	.421	"	76.50	25		101.50	122
0200	Cast iron body, threaded									
0210	3/4" size	1 Stpi	20	.400	Ea.	49.50	24		73.50	90
0220	1" size		19	.421		55.50	25		80.50	99
0230	1-1/4" size		15	.533		67.50	32		99.50	122
0240	1-1/2" size		13	.615		103	37		140	169
0250	2" size		11	.727		148	43.50		191.50	229

23 21 20.58 Hydronic Heating Control Valves

23 21 20.58 Hydronic Heating Control Valves		Crew	Daily Output	Labor-Hours	Unit	Material	Labor	Equipment	Total	Total Incl O&P
0010	**HYDRONIC HEATING CONTROL VALVES**									
0050	Hot water, nonelectric, thermostatic									
0100	Radiator supply, 1/2" diameter	1 Stpi	24	.333	Ea.	68.50	19.90		88.40	105
0120	3/4" diameter	"	20	.400	"	71.50	24		95.50	115
1000	Manual, radiator supply									
1010	1/2" pipe size, angle union	1 Stpi	24	.333	Ea.	52.50	19.90		72.40	88
1020	3/4" pipe size, angle union	"	20	.400	"	66	24		90	109
1100	Radiator, balancing, straight, sweat connections									
1110	1/2" pipe size	1 Stpi	24	.333	Ea.	20	19.90		39.90	52
1120	3/4" pipe size	"	20	.400	"	28	24		52	67
1200	Steam, radiator, supply									
1210	1/2" pipe size, angle union	1 Stpi	24	.333	Ea.	49	19.90		68.90	84
1220	3/4" pipe size, angle union	"	20	.400	"	55.50	24		79.50	97.50
8000	System balancing and shut-off									
8020	Butterfly, quarter turn, calibrated, threaded or solder									
8040	Bronze, -30°F to +350°F, pressure to 175 psi									
8060	1/2" size	1 Stpi	22	.364	Ea.	18.65	21.50		40.15	53.50
8070	3/4" size	"	20	.400	"	29.50	24		53.50	68.50

23 21 20.66 Monoflow Tee Fitting		Crew	Daily Output	Labor-Hours	Unit	Material	2015 Bare Costs Labor	2015 Bare Costs Equipment	Total	Total Incl O&P
0010	**MONOFLOW TEE FITTING**									
1100	For one pipe hydronic, supply and return									
1110	Copper, soldered									
1120	3/4" x 1/2" size	1 Stpi	13	.615	Ea.	20.50	37		57.50	78

23 21 20.74 Strainers, Basket Type

		Crew	Daily Output	Labor-Hours	Unit	Material	2015 Bare Costs Labor	2015 Bare Costs Equipment	Total	Total Incl O&P
0010	**STRAINERS, BASKET TYPE**, Perforated stainless steel basket									
0100	Brass or monel available									
2000	Simplex style									
2300	Bronze body									
2320	Screwed, 3/8" pipe size	1 Stpi	22	.364	Ea.	172	21.50		193.50	222
2340	1/2" pipe size		20	.400		295	24		319	360
2360	3/4" pipe size		17	.471		400	28		428	485
2380	1" pipe size		15	.533		400	32		432	490
2400	1-1/4" pipe size		13	.615		575	37		612	690
2420	1-1/2" pipe size		12	.667		580	40		620	695
2440	2" pipe size		10	.800		855	48		903	1,000
2460	2-1/2" pipe size	Q-5	15	1.067		770	57.50		827.50	935
2480	3" pipe size	"	14	1.143		1,125	61.50		1,186.50	1,325
2600	Flanged, 2" pipe size	1 Stpi	6	1.333		820	79.50		899.50	1,025
2620	2-1/2" pipe size	Q-5	4.50	3.556		1,275	191		1,466	1,725
2640	3" pipe size		3.50	4.571		1,450	246		1,696	1,975
2660	4" pipe size		3	5.333		2,450	287		2,737	3,125
2680	5" pipe size	Q-6	3.40	7.059		3,750	395		4,145	4,725
2700	6" pipe size		3	8		4,775	445		5,220	5,925
2710	8" pipe size		2.50	9.600		7,750	535		8,285	9,325
3600	Iron body									
3700	Screwed, 3/8" pipe size	1 Stpi	22	.364	Ea.	98.50	21.50		120	142
3720	1/2" pipe size		20	.400		102	24		126	148
3740	3/4" pipe size		17	.471		130	28		158	186
3760	1" pipe size		15	.533		133	32		165	194
3780	1-1/4" pipe size		13	.615		173	37		210	246
3800	1-1/2" pipe size		12	.667		191	40		231	270
3820	2" pipe size		10	.800		228	48		276	325
3840	2-1/2" pipe size	Q-5	15	1.067		305	57.50		362.50	420
3860	3" pipe size	"	14	1.143		370	61.50		431.50	500
4000	Flanged, 2" pipe size	1 Stpi	6	1.333		355	79.50		434.50	510
4020	2-1/2" pipe size	Q-5	4.50	3.556		475	191		666	815
4040	3" pipe size		3.50	4.571		500	246		746	920
4060	4" pipe size		3	5.333		760	287		1,047	1,275
4080	5" pipe size	Q-6	3.40	7.059		1,150	395		1,545	1,875
4100	6" pipe size		3	8		1,475	445		1,920	2,300
4120	8" pipe size		2.50	9.600		2,675	535		3,210	3,750
4140	10" pipe size		2.20	10.909		5,875	610		6,485	7,400
7000	Stainless steel body									
7200	Screwed, 1" pipe size	1 Stpi	15	.533	Ea.	283	32		315	360
7210	1-1/4" pipe size		13	.615		440	37		477	540
7220	1-1/2" pipe size		12	.667		440	40		480	545
7240	2" pipe size		10	.800		655	48		703	790
7260	2-1/2" pipe size	Q-5	15	1.067		925	57.50		982.50	1,100
7280	3" pipe size	"	14	1.143		1,275	61.50		1,336.50	1,500
7400	Flanged, 2" pipe size	1 Stpi	6	1.333		1,125	79.50		1,204.50	1,350
7420	2-1/2" pipe size	Q-5	4.50	3.556		2,025	191		2,216	2,550

23 21 Hydronic Piping and Pumps

23 21 20 – Hydronic HVAC Piping Specialties

23 21 20.74 Strainers, Basket Type		Crew	Daily Output	Labor-Hours	Unit	Material	2015 Bare Costs Labor	Equipment	Total	Total Incl O&P
7440	3" pipe size	Q-5	3.50	4.571	Ea.	2,075	246		2,321	2,675
7460	4" pipe size	↓	3	5.333		3,275	287		3,562	4,025
7480	6" pipe size	Q-6	3	8		5,700	445		6,145	6,950
7500	8" pipe size	"	2.50	9.600	↓	8,300	535		8,835	9,925
8100	Duplex style									
8200	Bronze body									
8240	Screwed, 3/4" pipe size	1 Stpi	16	.500	Ea.	1,175	30		1,205	1,325
8260	1" pipe size		14	.571		1,175	34		1,209	1,325
8280	1-1/4" pipe size		12	.667		2,825	40		2,865	3,150
8300	1-1/2" pipe size		11	.727		2,825	43.50		2,868.50	3,175
8320	2" pipe size	↓	9	.889		3,700	53		3,753	4,125
8340	2-1/2" pipe size	Q-5	14	1.143		4,775	61.50		4,836.50	5,350
8420	Flanged, 2" pipe size	1 Stpi	6	1.333		3,975	79.50		4,054.50	4,500
8440	2-1/2" pipe size	Q-5	4.50	3.556		5,475	191		5,666	6,325
8460	3" pipe size		3.50	4.571		5,975	246		6,221	6,950
8480	4" pipe size	↓	3	5.333		8,575	287		8,862	9,850
8500	5" pipe size	Q-6	3.40	7.059		18,600	395		18,995	21,000
8520	6" pipe size	"	3	8	↓	19,100	445		19,545	21,700
8700	Iron body									
8740	Screwed, 3/4" pipe size	1 Stpi	16	.500	Ea.	1,050	30		1,080	1,200
8760	1" pipe size		14	.571		1,050	34		1,084	1,200
8780	1-1/4" pipe size		12	.667		1,175	40		1,215	1,325
8800	1-1/2" pipe size		11	.727		1,175	43.50		1,218.50	1,350
8820	2" pipe size	↓	9	.889		1,975	53		2,028	2,250
8840	2-1/2" pipe size	Q-5	14	1.143		2,175	61.50		2,236.50	2,500
9000	Flanged, 2" pipe size	1 Stpi	6	1.333		2,100	79.50		2,179.50	2,450
9020	2-1/2" pipe size	Q-5	4.50	3.556		2,250	191		2,441	2,775
9040	3" pipe size		3.50	4.571		2,450	246		2,696	3,075
9060	4" pipe size	↓	3	5.333		4,150	287		4,437	4,975
9080	5" pipe size	Q-6	3.40	7.059		9,025	395		9,420	10,500
9100	6" pipe size		3	8		9,025	445		9,470	10,600
9120	8" pipe size		2.50	9.600		16,700	535		17,235	19,200
9140	10" pipe size		2.20	10.909		23,900	610		24,510	27,100
9160	12" pipe size		1.70	14.118		26,200	785		26,985	30,100
9170	14" pipe size		1.40	17.143		31,000	955		31,955	35,600
9180	16" pipe size	↓	1	24	↓	38,200	1,350		39,550	44,000
9700	Stainless steel body									
9740	Screwed, 1" pipe size	1 Stpi	14	.571	Ea.	1,900	34		1,934	2,125
9760	1-1/2" pipe size		11	.727		2,850	43.50		2,893.50	3,225
9780	2" pipe size		9	.889		4,000	53		4,053	4,475
9860	Flanged, 2" pipe size		6	1.333		4,350	79.50		4,429.50	4,925
9880	2-1/2" pipe size	Q-5	4.50	3.556		7,300	191		7,491	8,325
9900	3" pipe size		3.50	4.571		7,950	246		8,196	9,125
9920	4" pipe size	↓	3	5.333		10,400	287		10,687	11,800
9940	6" pipe size	Q-6	3	8		15,800	445		16,245	18,100
9960	8" pipe size	"	2.50	9.600	↓	44,200	535		44,735	49,400

23 21 20.76 Strainers, Y Type, Bronze Body

0010	**STRAINERS, Y TYPE, BRONZE BODY**									
0050	Screwed, 125 lb., 1/4" pipe size	1 Stpi	24	.333	Ea.	23	19.90		42.90	55.50
0070	3/8" pipe size		24	.333		27.50	19.90		47.40	60.50
0100	1/2" pipe size		20	.400		27.50	24		51.50	66.50
0120	3/4" pipe size	↓	19	.421		34	25		59	75.50

333

For customer support on your Plumbing Cost Data, call 877.709.6968.

23 21 20.76 Strainers, Y Type, Bronze Body

		Crew	Daily Output	Labor-Hours	Unit	Material	2015 Bare Costs Labor	2015 Bare Costs Equipment	Total	Total Incl O&P
0140	1" pipe size	1 Stpi	17	.471	Ea.	50	28		78	97.50
0150	1-1/4" pipe size		15	.533		81	32		113	137
0160	1-1/2" pipe size		14	.571		108	34		142	171
0180	2" pipe size		13	.615		144	37		181	214
0182	3" pipe size		12	.667		845	40		885	990
0200	300 lb., 2-1/2" pipe size	Q-5	17	.941		510	50.50		560.50	635
0220	3" pipe size		16	1		1,000	54		1,054	1,175
0240	4" pipe size		15	1.067		2,300	57.50		2,357.50	2,600
0500	For 300 lb. rating 1/4" thru 2", add					15%				
1000	Flanged, 150 lb., 1-1/2" pipe size	1 Stpi	11	.727	Ea.	520	43.50		563.50	640
1020	2" pipe size	"	8	1		705	60		765	865
1030	2-1/2" pipe size	Q-5	5	3.200		950	172		1,122	1,300
1040	3" pipe size		4.50	3.556		1,175	191		1,366	1,575
1060	4" pipe size		3	5.333		1,775	287		2,062	2,375
1080	5" pipe size	Q-6	3.40	7.059		1,775	395		2,170	2,550
1100	6" pipe size		3	8		3,400	445		3,845	4,400
1106	8" pipe size		2.60	9.231		3,725	515		4,240	4,875
1500	For 300 lb. rating, add					40%				

23 21 20.78 Strainers, Y Type, Iron Body

		Crew	Daily Output	Labor-Hours	Unit	Material	2015 Bare Costs Labor	2015 Bare Costs Equipment	Total	Total Incl O&P
0010	**STRAINERS, Y TYPE, IRON BODY**									
0050	Screwed, 250 lb., 1/4" pipe size	1 Stpi	20	.400	Ea.	11.05	24		35.05	48
0070	3/8" pipe size		20	.400		11.05	24		35.05	48
0100	1/2" pipe size		20	.400		11.05	24		35.05	48
0120	3/4" pipe size		18	.444		12.95	26.50		39.45	54.50
0140	1" pipe size		16	.500		18.25	30		48.25	65
0150	1-1/4" pipe size		15	.533		24	32		56	74.50
0160	1-1/2" pipe size		12	.667		30	40		70	92.50
0180	2" pipe size		8	1		44.50	60		104.50	139
0200	2-1/2" pipe size	Q-5	12	1.333		268	71.50		339.50	400
0220	3" pipe size		11	1.455		289	78		367	440
0240	4" pipe size		5	3.200		490	172		662	800
0500	For galvanized body, add					50%				
1000	Flanged, 125 lb., 1-1/2" pipe size	1 Stpi	11	.727	Ea.	107	43.50		150.50	183
1020	2" pipe size	"	8	1		113	60		173	214
1030	2-1/2" pipe size	Q-5	5	3.200		115	172		287	385
1040	3" pipe size		4.50	3.556		186	191		377	495
1060	4" pipe size		3	5.333		305	287		592	770
1080	5" pipe size	Q-6	3.40	7.059		385	395		780	1,025
1100	6" pipe size		3	8		615	445		1,060	1,350
1120	8" pipe size		2.50	9.600		830	535		1,365	1,725
1140	10" pipe size		2	12		1,875	670		2,545	3,050
1160	12" pipe size		1.70	14.118		2,275	785		3,060	3,700
1170	14" pipe size		1.30	18.462		4,175	1,025		5,200	6,150
1180	16" pipe size		1	24		5,925	1,350		7,275	8,525
1500	For 250 lb. rating, add					20%				
2000	For galvanized body, add					50%				
2500	For steel body, add					40%				

23 21 Hydronic Piping and Pumps

23 21 20 – Hydronic HVAC Piping Specialties

23 21 20.80 Suction Diffusers		Crew	Daily Output	Labor-Hours	Unit	Material	2015 Bare Costs Labor	Equipment	Total	Total Incl O&P
0010	**SUCTION DIFFUSERS**									
0100	Cast iron body with integral straightening vanes, strainer									
1000	Flanged									
1010	2" inlet, 1-1/2" pump side	1 Stpi	6	1.333	Ea.	325	79.50		404.50	475
1020	2" pump side	"	5	1.600		335	95.50		430.50	515
1030	3" inlet, 2" pump side	Q-5	6.50	2.462		405	132		537	645

23 21 20.84 Thermoflo Indicator										
0010	**THERMOFLO INDICATOR**, For balancing									
1000	Sweat connections, 1-1/4" pipe size	1 Stpi	12	.667	Ea.	650	40		690	775
1020	1-1/2" pipe size	"	10	.800	"	660	48		708	800

23 21 20.88 Venturi Flow										
0010	**VENTURI FLOW**, Measuring device									
0050	1/2" diameter	1 Stpi	24	.333	Ea.	281	19.90		300.90	340
0100	3/4" diameter		20	.400		257	24		281	320
0120	1" diameter		19	.421		276	25		301	345
0140	1-1/4" diameter		15	.533		340	32		372	425
0160	1-1/2" diameter		13	.615		355	37		392	445
0180	2" diameter		11	.727		365	43.50		408.50	470
0200	2-1/2" diameter	Q-5	16	1		500	54		554	630
0220	3" diameter		14	1.143		515	61.50		576.50	665
0240	4" diameter		11	1.455		775	78		853	970
0260	5" diameter	Q-6	4	6		1,025	335		1,360	1,625
0280	6" diameter		3.50	6.857		1,125	380		1,505	1,825
0300	8" diameter		3	8		1,450	445		1,895	2,275
0320	10" diameter		2	12		3,425	670		4,095	4,775
0330	12" diameter		1.80	13.333		4,950	745		5,695	6,575
0340	14" diameter		1.60	15		5,600	835		6,435	7,450
0350	16" diameter		1.40	17.143		6,250	955		7,205	8,325
0500	For meter, add					2,125			2,125	2,350

23 21 23 – Hydronic Pumps

23 21 23.13 In-Line Centrifugal Hydronic Pumps										
0010	**IN-LINE CENTRIFUGAL HYDRONIC PUMPS**									
0600	Bronze, sweat connections, 1/40 HP, in line									
0640	3/4" size	Q-1	16	1	Ea.	218	53		271	320
1000	Flange connection, 3/4" to 1-1/2" size									
1040	1/12 HP	Q-1	6	2.667	Ea.	565	141		706	835
1060	1/8 HP		6	2.667		970	141		1,111	1,300
1100	1/3 HP		6	2.667		1,075	141		1,216	1,425
1140	2" size, 1/6 HP		5	3.200		1,400	169		1,569	1,775
1180	2-1/2" size, 1/4 HP		5	3.200		1,775	169		1,944	2,200
1220	3" size, 1/4 HP		4	4		1,850	211		2,061	2,375
1260	1/3 HP		4	4		2,275	211		2,486	2,825
1300	1/2 HP		4	4		2,325	211		2,536	2,875
1340	3/4 HP		4	4		2,550	211		2,761	3,125
1380	1 HP		4	4		4,100	211		4,311	4,825
2000	Cast iron, flange connection									
2040	3/4" to 1-1/2" size, in line, 1/12 HP	Q-1	6	2.667	Ea.	365	141		506	615
2060	1/8 HP		6	2.667		610	141		751	885
2100	1/3 HP		6	2.667		680	141		821	960
2140	2" size, 1/6 HP		5	3.200		745	169		914	1,075
2180	2-1/2" size, 1/4 HP		5	3.200		960	169		1,129	1,300

For customer support on your Plumbing Cost Data, call 877.709.6968.

335

23 21 23 – Hydronic Pumps

23 21 23.13 In-Line Centrifugal Hydronic Pumps		Crew	Daily Output	Labor-Hours	Unit	Material	2015 Bare Costs Labor	Equipment	Total	Total Incl O&P
2220	3" size, 1/4 HP	Q-1	4	4	Ea.	975	211		1,186	1,400
2260	1/3 HP		4	4		1,325	211		1,536	1,775
2300	1/2 HP		4	4		1,375	211		1,586	1,825
2340	3/4 HP		4	4		1,575	211		1,786	2,075
2380	1 HP		4	4		2,275	211		2,486	2,825
2600	For nonferrous impeller, add					3%				
3000	High head, bronze impeller									
3030	1-1/2" size 1/2 HP	Q-1	5	3.200	Ea.	1,150	169		1,319	1,525
3040	1-1/2" size 3/4 HP		5	3.200		1,250	169		1,419	1,625
3050	2" size 1 HP		4	4		1,525	211		1,736	2,000
3090	2" size 1-1/2 HP		4	4		1,900	211		2,111	2,400
4000	Close coupled, end suction, bronze impeller									
4040	1-1/2" size, 1-1/2 HP, to 40 GPM	Q-1	3	5.333	Ea.	2,250	282		2,532	2,900
4090	2" size, 2 HP, to 50 GPM		3	5.333		2,675	282		2,957	3,350
4100	2" size, 3 HP, to 90 GPM		2.30	6.957		2,775	370		3,145	3,600
4190	2-1/2" size, 3 HP, to 150 GPM		2	8		3,000	425		3,425	3,950
4300	3" size, 5 HP, to 225 GPM		1.80	8.889		3,425	470		3,895	4,475
4410	3" size, 10 HP, to 350 GPM		1.60	10		5,025	530		5,555	6,325
4420	4" size, 7-1/2 HP, to 350 GPM		1.60	10		5,075	530		5,605	6,375
4520	4" size, 10 HP, to 600 GPM	Q-2	1.70	14.118		5,125	775		5,900	6,800
4530	5" size, 15 HP, to 1000 GPM		1.70	14.118		5,125	775		5,900	6,825
4610	5" size, 20 HP, to 1350 GPM		1.50	16		5,450	875		6,325	7,325
4620	5" size, 25 HP, to 1550 GPM		1.50	16		7,450	875		8,325	9,525
5000	Base mounted, bronze impeller, coupling guard									
5040	1-1/2" size, 1-1/2 HP, to 40 GPM	Q-1	2.30	6.957	Ea.	5,950	370		6,320	7,100
5090	2" size, 2 HP, to 50 GPM		2.30	6.957		6,675	370		7,045	7,900
5100	2" size, 3 HP, to 90 GPM		2	8		7,325	425		7,750	8,725
5190	2-1/2" size, 3 HP, to 150 GPM		1.80	8.889		7,900	470		8,370	9,375
5300	3" size, 5 HP, to 225 GPM		1.60	10		9,050	530		9,580	10,800
5410	4" size, 5 HP, to 350 GPM		1.50	10.667		9,850	565		10,415	11,700
5420	4" size, 7-1/2 HP, to 350 GPM		1.50	10.667		11,100	565		11,665	13,100
5520	5" size, 10 HP, to 600 GPM	Q-2	1.60	15		14,400	820		15,220	17,200
5530	5" size, 15 HP, to 1000 GPM		1.60	15		15,900	820		16,720	18,800
5610	6" size, 20 HP, to 1350 GPM		1.40	17.143		17,600	940		18,540	20,700
5620	6" size, 25 HP, to 1550 GPM		1.40	17.143		19,700	940		20,640	23,100
5800	The above pump capacities are based on 1800 RPM,									
5810	at a 60 foot head. Increasing the RPM									
5820	or decreasing the head will increase the GPM.									

23 21 29 – Automatic Condensate Pump Units

23 21 29.10 Condensate Removal Pump System

0010	**CONDENSATE REMOVAL PUMP SYSTEM**									
0020	Pump with 1 gal. ABS tank									
0100	115 V									
0120	1/50 HP, 200 GPH	G	1 Stpi	12	.667	Ea.	197	40		237	277
0140	1/18 HP, 270 GPH	G		10	.800		210	48		258	305
0160	1/5 HP, 450 GPH	G		8	1		470	60		530	605
0200	230 V									
0260	1/5 HP, 450 GPH	G	1 Stpi	8	1	Ea.	520	60		580	660

23 23 Refrigerant Piping

23 23 23 − Refrigerants

23 23 23.10 Anti-Freeze	Crew	Daily Output	Labor-Hours	Unit	Material	2015 Bare Costs Labor	Equipment	Total	Total Incl O&P
0010 **ANTI-FREEZE**, Inhibited									
0900 Ethylene glycol concentrated									
1000 55 gallon drums, small quantities				Gal.	11.95			11.95	13.10
1200 Large quantities					9.40			9.40	10.35
2000 Propylene glycol, for solar heat, small quantities					15.55			15.55	17.10
2100 Large quantities					10.35			10.35	11.35

23 34 HVAC Fans

23 34 14 − Blower HVAC Fans

23 34 14.10 Blower Type HVAC Fans	Crew	Daily Output	Labor-Hours	Unit	Material	2015 Bare Costs Labor	Equipment	Total	Total Incl O&P
0010 **BLOWER TYPE HVAC FANS**									
2500 Ceiling fan, right angle, extra quiet, 0.10" S.P.									
2520 95 CFM	Q-20	20	1	Ea.	300	51		351	410
2540 210 CFM		19	1.053		355	54		409	470
2560 385 CFM	↓	18	1.111		450	57		507	580
2640 For wall or roof cap, add	1 Shee	16	.500		300	28		328	375
2660 For straight thru fan, add					10%				
2680 For speed control switch, add	1 Elec	16	.500	↓	164	27.50		191.50	222

23 34 23 − HVAC Power Ventilators

23 34 23.10 HVAC Power Circulators and Ventilators	Crew	Daily Output	Labor-Hours	Unit	Material	2015 Bare Costs Labor	Equipment	Total	Total Incl O&P
0010 **HVAC POWER CIRCULATORS AND VENTILATORS**									
6650 Residential, bath exhaust, grille, back draft damper									
6660 50 CFM	Q-20	24	.833	Ea.	63.50	42.50		106	135
6670 110 CFM		22	.909		98	46.50		144.50	179
6680 Light combination, squirrel cage, 100 watt, 70 CFM	↓	24	.833	↓	112	42.50		154.50	188
6700 Light/heater combination, ceiling mounted									
6710 70 CFM, 1450 watt	Q-20	24	.833	Ea.	162	42.50		204.50	243
6800 Heater combination, recessed, 70 CFM		24	.833		67.50	42.50		110	139
6820 With 2 infrared bulbs	↓	23	.870	↓	105	44.50		149.50	184
6940 Residential roof jacks and wall caps									
6944 Wall cap with back draft damper									
6946 3" & 4" diam. round duct	1 Shee	11	.727	Ea.	26	40.50		66.50	90.50
6948 6" diam. round duct	"	11	.727	"	64	40.50		104.50	133
6958 Roof jack with bird screen and back draft damper									
6960 3" & 4" diam. round duct	1 Shee	11	.727	Ea.	26	40.50		66.50	90.50
6962 3-1/4" x 10" rectangular duct	"	10	.800	"	48.50	45		93.50	122
6980 Transition									
6982 3-1/4" x 10" to 6" diam. round	1 Shee	20	.400	Ea.	32	22.50		54.50	69.50

For customer support on your Plumbing Cost Data, call 877.709.6968.

337

23 35 16 – Engine Exhaust Systems

23 35 16.10 Engine Exhaust Removal Systems	Crew	Daily Output	Labor-Hours	Unit	Material	2015 Bare Costs Labor	Equipment	Total	Total Incl O&P
0010 ENGINE EXHAUST REMOVAL SYSTEMS D3090-320									
0500 Engine exhaust, garage, in-floor system									
0510 Single tube outlet assemblies									
0520 For transite pipe ducting, self-storing tube									
0530 3" tubing adapter plate	1 Shee	16	.500	Ea.	240	28		268	305
0540 4" tubing adapter plate		16	.500		245	28		273	310
0550 5" tubing adapter plate	↓	16	.500	↓	245	28		273	310
0600 For vitrified tile ducting									
0610 3" tubing adapter plate, self-storing tube	1 Shee	16	.500	Ea.	241	28		269	310
0620 4" tubing adapter plate, self-storing tube		16	.500		245	28		273	310
0660 5" tubing adapter plate, self-storing tube	↓	16	.500	↓	245	28		273	310
0800 Two tube outlet assemblies									
0810 For transite pipe ducting, self-storing tube									
0820 3" tubing, dual exhaust adapter plate	1 Shee	16	.500	Ea.	240	28		268	305
0850 For vitrified tile ducting									
0860 3" tubing, dual exhaust, self-storing tube	1 Shee	16	.500	Ea.	298	28		326	375
0870 3" tubing, double outlet, non-storing tubes	"	16	.500	"	298	28		326	375
0900 Accessories for metal tubing, (overhead systems also)									
0910 Adapters, for metal tubing end									
0920 3" tail pipe type				Ea.	49.50			49.50	54.50
0930 4" tail pipe type					53			53	58.50
0940 5" tail pipe type					54			54	59.50
0990 5" diesel stack type					315			315	350
1000 6" diesel stack type				↓	330			330	360
1100 Bullnose (guide) required for in-floor assemblies									
1110 3" tubing size				Ea.	28			28	30.50
1120 4" tubing size					29.50			29.50	32.50
1130 5" tubing size				↓	31.50			31.50	34.50
1150 Plain rings, for tubing end									
1160 3" tubing size				Ea.	24.50			24.50	26.50
1170 4" tubing size				"	41.50			41.50	45.50
1200 Tubing, galvanized, flexible, (for overhead systems also)									
1210 3" ID				L.F.	11.60			11.60	12.75
1220 4" ID					14.20			14.20	15.65
1230 5" ID					16.75			16.75	18.40
1240 6" ID				↓	19.35			19.35	21.50
1250 Stainless steel, flexible, (for overhead system, also)									
1260 3" ID				L.F.	25.50			25.50	28
1270 4" ID					35			35	38.50
1280 5" ID					39.50			39.50	43.50
1290 6" ID				↓	46			46	50.50
1500 Engine exhaust, garage, overhead components, for neoprene tubing									
1510 Alternate metal tubing & accessories see above									
1550 Adapters, for neoprene tubing end									
1560 3" tail pipe, adjustable, neoprene				Ea.	56			56	61.50
1570 3" tail pipe, heavy wall neoprene					61			61	67.50
1580 4" tail pipe, heavy wall neoprene					100			100	110
1590 5" tail pipe, heavy wall neoprene				↓	105			105	116
1650 Connectors, tubing									
1660 3" interior, aluminum				Ea.	24.50			24.50	26.50
1670 4" interior, aluminum					41.50			41.50	45.50
1710 5" interior, neoprene					61			61	67.50
1750 3" spiralock, neoprene				↓	24.50			24.50	26.50

For customer support on your Plumbing Cost Data, call 877.709.6968.

23 35 Special Exhaust Systems

23 35 16 – Engine Exhaust Systems

23 35 16.10 Engine Exhaust Removal Systems

		Crew	Daily Output	Labor-Hours	Unit	Material	2015 Bare Costs Labor	Equipment	Total	Total Incl O&P
1760	4" spiralock, neoprene				Ea.	41.50			41.50	45.50
1780	Y for 3" ID tubing, neoprene, dual exhaust					195			195	215
1790	Y for 4" ID tubing, aluminum, dual exhaust				↓	195			195	215
1850	Elbows, aluminum, splice into tubing for strap									
1860	3" neoprene tubing size				Ea.	48.50			48.50	53.50
1870	4" neoprene tubing size					40.50			40.50	44.50
1900	Flange assemblies, connect tubing to overhead duct				↓	73			73	80
2000	Hardware and accessories									
2020	Cable, galvanized, 1/8" diameter				L.F.	.51			.51	.56
2040	Cleat, tie down cable or rope				Ea.	5.40			5.40	5.90
2060	Pulley					7.40			7.40	8.15
2080	Pulley hook, universal				↓	5.40			5.40	5.90
2100	Rope, nylon, 1/4" diameter				L.F.	.41			.41	.45
2120	Winch, 1" diameter				Ea.	114			114	126
2150	Lifting strap, mounts on neoprene									
2160	3" tubing size				Ea.	27			27	29.50
2170	4" tubing size					27			27	29.50
2180	5" tubing size					27			27	29.50
2190	6" tubing size				↓	27			27	29.50
2200	Tubing, neoprene, 11' lengths									
2210	3" ID				L.F.	10.90			10.90	11.95
2220	4" ID					18.15			18.15	20
2230	5" ID				↓	27			27	29.50
2500	Engine exhaust, thru-door outlet									
2510	3" tube size	1 Carp	16	.500	Ea.	59.50	23.50		83	102
2530	4" tube size	"	16	.500	"	74	23.50		97.50	117
3000	Tubing, exhaust, flex hose, with									
3010	coupler, damper and tail pipe adapter									
3020	Neoprene									
3040	3" x 20'	1 Shee	6	1.333	Ea.	330	74.50		404.50	475
3050	4" x 15'		5.40	1.481		440	83		523	610
3060	4" x 20'		5	1.600		530	89.50		619.50	720
3070	5" x 15'	↓	4.40	1.818	↓	600	102		702	815
3100	Galvanized									
3110	3" x 20'	1 Shee	6	1.333	Ea.	330	74.50		404.50	480
3120	4" x 17'		5.60	1.429		360	80		440	520
3130	4" x 20'		5	1.600		405	89.50		494.50	580
3140	5" x 17'	↓	4.60	1.739		430	97.50		527.50	625

23 35 43 – Welding Fume Elimination Systems

23 35 43.10 Welding Fume Elimination System Components

		Crew	Daily Output	Labor-Hours	Unit	Material	2015 Bare Costs Labor	Equipment	Total	Total Incl O&P
0010	**WELDING FUME ELIMINATION SYSTEM COMPONENTS**									
7500	Welding fume elimination accessories for garage exhaust systems									
7600	Cut off (blast gate)									
7610	3" tubing size, 3" x 6" opening	1 Shee	24	.333	Ea.	26	18.65		44.65	57
7620	4" tubing size, 4" x 8" opening		24	.333		26	18.65		44.65	57
7630	5" tubing size, 5" x 10" opening		24	.333		31.50	18.65		50.15	63
7640	6" tubing size		24	.333		32.50	18.65		51.15	64
7650	8" tubing size	↓	24	.333	↓	45	18.65		63.65	78
7700	Hoods, magnetic, with handle & screen									
7710	3" tubing size, 3" x 6" opening	1 Shee	24	.333	Ea.	107	18.65		125.65	147
7720	4" tubing size, 4" x 8" opening		24	.333		97	18.65		115.65	136
7730	5" tubing size, 5" x 10" opening	↓	24	.333		97	18.65		115.65	136

339

23 38 Ventilation Hoods

23 38 13 – Commercial-Kitchen Hoods

23 38 13.10 Hood and Ventilation Equipment	Crew	Daily Output	Labor-Hours	Unit	Material	2015 Bare Costs Labor	Equipment	Total	Total Incl O&P
0010 **HOOD AND VENTILATION EQUIPMENT**									
2970 Exhaust hood, sst, gutter on all sides, 4' x 4' x 2'	1 Carp	1.80	4.444	Ea.	4,725	209		4,934	5,525
2980 4' x 4' x 7'	"	1.60	5	"	7,525	235		7,760	8,625
7800 Vent hood, wall canopy with fire protection	L-3A	9	1.333	L.F.	425	68.50		493.50	570
7810 Without fire protection		10	1.200		460	61.50		521.50	600
7820 Island canopy with fire protection		7	1.714		790	88		878	1,000
7830 Without fire protection		8	1.500		800	77		877	1,000
7840 Back shelf with fire protection		11	1.091		400	56		456	525
7850 Without fire protection, black		12	1		400	51.50		451.50	520
7852 Without fire protection, stainless steel		12	1		500	51.50		551.50	630
7860 Range hood & CO$_2$ system, 30"	1 Carp	2.50	3.200	Ea.	3,975	150		4,125	4,600
7950 Hood fire protection system, electric stove	Q-1	3	5.333		1,875	282		2,157	2,500
7952 Hood fire protection system, gas stove	"	3	5.333		2,200	282		2,482	2,850

23 51 Breechings, Chimneys, and Stacks

23 51 13 – Draft Control Devices

23 51 13.16 Vent Dampers

	Crew	Daily Output	Labor-Hours	Unit	Material	2015 Bare Costs Labor	Equipment	Total	Total Incl O&P
0010 **VENT DAMPERS**									
5000 Vent damper, bi-metal, gas, 3" diameter	Q-9	24	.667	Ea.	44	33.50		77.50	100
5010 4" diameter	"	24	.667	"	44	33.50		77.50	100

23 51 13.19 Barometric Dampers

	Crew	Daily Output	Labor-Hours	Unit	Material	2015 Bare Costs Labor	Equipment	Total	Total Incl O&P
0010 **BAROMETRIC DAMPERS**									
1000 Barometric, gas fired system only, 6" size for 5" and 6" pipes	1 Shee	20	.400	Ea.	105	22.50		127.50	149
1020 7" size, for 6" and 7" pipes		19	.421		112	23.50		135.50	159
1040 8" size, for 7" and 8" pipes		18	.444		144	25		169	197
1060 9" size, for 8" and 9" pipes		16	.500		162	28		190	221
2000 All fuel, oil, oil/gas, coal									
2020 10" for 9" and 10" pipes	1 Shee	15	.533	Ea.	242	30		272	310
2040 12" for 11" and 12" pipes		15	.533		315	30		345	390
2060 14" for 13" and 14" pipes		14	.571		410	32		442	500
2080 16" for 15" and 16" pipes		13	.615		575	34.50		609.50	685
2100 18" for 17" and 18" pipes		12	.667		765	37.50		802.50	895
2120 20" for 19" and 21" pipes		10	.800		915	45		960	1,075
2140 24" for 22" and 25" pipes	Q-9	12	1.333		1,125	67		1,192	1,325
2160 28" for 26" and 30" pipes		10	1.600		1,375	80.50		1,455.50	1,650
2180 32" for 31" and 34" pipes		8	2		1,775	101		1,876	2,100
3260 For thermal switch for above, add	1 Shee	24	.333		100	18.65		118.65	139

23 51 23 – Gas Vents

23 51 23.10 Gas Chimney Vents

	Crew	Daily Output	Labor-Hours	Unit	Material	2015 Bare Costs Labor	Equipment	Total	Total Incl O&P
0010 **GAS CHIMNEY VENTS**, Prefab metal, U.L. listed									
0020 Gas, double wall, galvanized steel									
0080 3" diameter	Q-9	72	.222	V.L.F.	5.55	11.20		16.75	23
0100 4" diameter		68	.235		7.20	11.85		19.05	26
0120 5" diameter		64	.250		7.80	12.60		20.40	28
0140 6" diameter		60	.267		9.45	13.45		22.90	31
0160 7" diameter		56	.286		15.10	14.40		29.50	38.50
0180 8" diameter		52	.308		17.40	15.50		32.90	42.50
0200 10" diameter		48	.333		36	16.80		52.80	65.50
0220 12" diameter		44	.364		43	18.30		61.30	75.50
0240 14" diameter		42	.381		72	19.20		91.20	109

23 51 23 – Gas Vents

23 51 23.10 Gas Chimney Vents	Crew	Daily Output	Labor-Hours	Unit	Material	2015 Bare Costs Labor	Equipment	Total	Total Incl O&P
0260	16" diameter	Q-9	40	.400	V.L.F.	103	20		123	145
0280	18" diameter	↓	38	.421		128	21		149	173
0300	20" diameter	Q-10	36	.667		150	35		185	218
0320	22" diameter		34	.706		187	37		224	262
0340	24" diameter	↓	32	.750	↓	236	39		275	320
0600	For 4", 5" and 6" oval, add					50%				
0650	Gas, double wall, galvanized steel, fittings									
0660	Elbow 45°, 3" diameter	Q-9	36	.444	Ea.	13.50	22.50		36	49
0670	4" diameter		34	.471		16.35	23.50		39.85	54
0680	5" diameter		32	.500		19.10	25		44.10	59.50
0690	6" diameter		30	.533		23.50	27		50.50	67
0700	7" diameter		28	.571		37.50	29		66.50	85.50
0710	8" diameter		26	.615		50	31		81	103
0720	10" diameter		24	.667		107	33.50		140.50	170
0730	12" diameter		22	.727		114	36.50		150.50	181
0740	14" diameter		21	.762		179	38.50		217.50	256
0750	16" diameter		20	.800		232	40.50		272.50	315
0760	18" diameter	↓	19	.842		305	42.50		347.50	400
0770	20" diameter	Q-10	18	1.333		340	69.50		409.50	480
0780	22" diameter		17	1.412		545	73.50		618.50	715
0790	24" diameter	↓	16	1.500	↓	695	78.50		773.50	885
0916	Adjustable length									
0918	3" diameter, to 12"	Q-9	36	.444	Ea.	12.45	22.50		34.95	47.50
0920	4" diameter, to 12"		34	.471		14.50	23.50		38	52
0924	6" diameter, to 12"		30	.533		18.70	27		45.70	61.50
0928	8" diameter, to 12"		26	.615		35	31		66	86
0930	10" diameter, to 18"		24	.667		107	33.50		140.50	170
0932	12" diameter, to 18"		22	.727		110	36.50		146.50	177
0936	16" diameter, to 18"		20	.800		222	40.50		262.50	305
0938	18" diameter, to 18"	↓	19	.842		254	42.50		296.50	345
0944	24" diameter, to 18"	Q-10	16	1.500		485	78.50		563.50	655
0950	Elbow 90°, adjustable, 3" diameter	Q-9	36	.444		23	22.50		45.50	59.50
0960	4" diameter		34	.471		27	23.50		50.50	65.50
0970	5" diameter		32	.500		33.50	25		58.50	75.50
0980	6" diameter		30	.533		39	27		66	84
0990	7" diameter		28	.571		67	29		96	118
1010	8" diameter		26	.615		68	31		99	123
1020	Wall thimble, 4 to 7" adjustable, 3" diameter		36	.444		14.45	22.50		36.95	50
1022	4" diameter		34	.471		16.25	23.50		39.75	54
1024	5" diameter		32	.500		17.10	25		42.10	57.50
1026	6" diameter		30	.533		17.55	27		44.55	60.50
1028	7" diameter		28	.571		36.50	29		65.50	84
1030	8" diameter		26	.615		51.50	31		82.50	105
1040	Roof flashing, 3" diameter		36	.444		8.50	22.50		31	43.50
1050	4" diameter		34	.471		9.90	23.50		33.40	47
1060	5" diameter		32	.500		29.50	25		54.50	71
1070	6" diameter		30	.533		23.50	27		50.50	67
1080	7" diameter		28	.571		31	29		60	78.50
1090	8" diameter		26	.615		33.50	31		64.50	84
1100	10" diameter		24	.667		45	33.50		78.50	101
1110	12" diameter		22	.727		63	36.50		99.50	125
1120	14" diameter		20	.800		143	40.50		183.50	219
1130	16" diameter	↓	18	.889	↓	201	45		246	290

For customer support on your Plumbing Cost Data, call 877.709.6968.

341

23 51 Breechings, Chimneys, and Stacks

23 51 23 – Gas Vents

23 51 23.10 Gas Chimney Vents		Crew	Daily Output	Labor-Hours	Unit	Material	2015 Bare Costs Labor	Equipment	Total	Total Incl O&P
1140	18" diameter	Q-9	16	1	Ea.	229	50.50		279.50	330
1150	20" diameter	Q-10	18	1.333		305	69.50		374.50	440
1160	22" diameter		14	1.714		375	89.50		464.50	550
1170	24" diameter	↓	12	2		440	104		544	640
1200	Tee, 3" diameter	Q-9	27	.593		35	30		65	84.50
1210	4" diameter		26	.615		38.50	31		69.50	90
1220	5" diameter		25	.640		40	32		72	93.50
1230	6" diameter		24	.667		45.50	33.50		79	102
1240	7" diameter		23	.696		65.50	35		100.50	126
1250	8" diameter		22	.727		71.50	36.50		108	135
1260	10" diameter		21	.762		191	38.50		229.50	269
1270	12" diameter		20	.800		196	40.50		236.50	277
1280	14" diameter		18	.889		350	45		395	455
1290	16" diameter		16	1		515	50.50		565.50	645
1300	18" diameter	↓	14	1.143		625	57.50		682.50	775
1310	20" diameter	Q-10	17	1.412		855	73.50		928.50	1,050
1320	22" diameter		13	1.846		1,100	96.50		1,196.50	1,350
1330	24" diameter	↓	12	2		1,250	104		1,354	1,525
1460	Tee cap, 3" diameter	Q-9	45	.356		2.36	17.90		20.26	30
1470	4" diameter		42	.381		2.54	19.20		21.74	32.50
1480	5" diameter		40	.400		3.19	20		23.19	34.50
1490	6" diameter		37	.432		4.73	22		26.73	38.50
1500	7" diameter		35	.457		7.95	23		30.95	43.50
1510	8" diameter		34	.471		8.70	23.50		32.20	45.50
1520	10" diameter		32	.500		70	25		95	116
1530	12" diameter		30	.533		70	27		97	118
1540	14" diameter		28	.571		73	29		102	124
1550	16" diameter		25	.640		74.50	32		106.50	132
1560	18" diameter	↓	24	.667		88	33.50		121.50	149
1570	20" diameter	Q-10	27	.889		99.50	46.50		146	180
1580	22" diameter		22	1.091		300	57		357	415
1590	24" diameter	↓	21	1.143		340	59.50		399.50	465
1750	Top, 3" diameter	Q-9	46	.348		13.05	17.50		30.55	41.50
1760	4" diameter		44	.364		13.75	18.30		32.05	43
1770	5" diameter		42	.381		17.50	19.20		36.70	49
1780	6" diameter		40	.400		23.50	20		43.50	57
1790	7" diameter		38	.421		37.50	21		58.50	74
1800	8" diameter		36	.444		46.50	22.50		69	85
1810	10" diameter		34	.471		105	23.50		128.50	151
1820	12" diameter		32	.500		154	25		179	208
1830	14" diameter		30	.533		205	27		232	266
1840	16" diameter		28	.571		263	29		292	335
1850	18" diameter	↓	26	.615		360	31		391	450
1860	20" diameter	Q-10	28	.857		550	45		595	675
1870	22" diameter		22	1.091		840	57		897	1,000
1880	24" diameter	↓	20	1.200	↓	1,050	62.50		1,112.50	1,250
1900	Gas, double wall, galvanized steel, oval									
1904	4" x 1'	Q-9	68	.235	V.L.F.	16.65	11.85		28.50	36.50
1906	5"		64	.250		39.50	12.60		52.10	63
1908	5"/6" x 1'	↓	60	.267	↓	36.50	13.45		49.95	61
1910	Oval fittings									
1912	Adjustable length									
1914	4" diameter to 12" lg.	Q-9	34	.471	Ea.	16.10	23.50		39.60	53.50

342

For customer support on your Plumbing Cost Data, call 877.709.6968.

23 51 23 – Gas Vents

23 51 23.10 Gas Chimney Vents		Crew	Daily Output	Labor-Hours	Unit	Material	2015 Bare Costs		Total	Total Incl O&P
							Labor	Equipment		
1916	5" diameter to 12" lg.	Q-9	32	.500	Ea.	44.50	25		69.50	87.50
1918	5"/6" diameter to 12"	↓	30	.533	↓	40.50	27		67.50	85.50
1920	Elbow 45°									
1922	4"	Q-9	34	.471	Ea.	29.50	23.50		53	68.50
1924	5"		32	.500		56	25		81	100
1926	5"/6"	↓	30	.533	↓	53	27		80	99
1930	Elbow 45°, flat									
1932	4"	Q-9	34	.471	Ea.	29.50	23.50		53	68.50
1934	5"		32	.500		55.50	25		80.50	100
1936	5"/6"	↓	30	.533	↓	53	27		80	99
1940	Top									
1942	4"	Q-9	44	.364	Ea.	35	18.30		53.30	66.50
1944	5"		42	.381		24.50	19.20		43.70	56.50
1946	5"/6"	↓	40	.400	↓	59.50	20		79.50	96
1950	Adjustable flashing									
1952	4"	Q-9	34	.471	Ea.	9.15	23.50		32.65	46
1954	5"		32	.500		29	25		54	70.50
1956	5"/6"	↓	30	.533	↓	30.50	27		57.50	74.50
1960	Tee									
1962	4"	Q-9	26	.615	Ea.	43.50	31		74.50	95.50
1964	5"		25	.640		91.50	32		123.50	151
1966	5"/6"	↓	24	.667	↓	84.50	33.50		118	145
1970	Tee with short snout									
1972	4"	Q-9	26	.615	Ea.	43.50	31		74.50	95.50

23 51 26 – All-Fuel Vent Chimneys

23 51 26.10 All-Fuel Vent Chimneys, Press. Tight, Dbl. Wall		Crew	Daily Output	Labor-Hours	Unit	Material	2015 Bare Costs		Total	Total Incl O&P
							Labor	Equipment		
0010	**ALL-FUEL VENT CHIMNEYS, PRESSURE TIGHT, DOUBLE WALL**									
3200	All fuel, pressure tight, double wall, 1" insulation, U.L. listed, 1400°F.									
3210	304 stainless steel liner, aluminized steel outer jacket									
3220	6" diameter	Q-9	60	.267	L.F.	54	13.45		67.45	80
3221	8" diameter		52	.308		61	15.50		76.50	90.50
3222	10" diameter		48	.333		68	16.80		84.80	101
3223	12" diameter		44	.364		78	18.30		96.30	114
3224	14" diameter		42	.381		87.50	19.20		106.70	126
3225	16" diameter		40	.400		99	20		119	140
3226	18" diameter	↓	38	.421		112	21		133	157
3227	20" diameter	Q-10	36	.667		127	35		162	193
3228	24" diameter	"	32	.750		163	39		202	239
3260	For 316 stainless steel liner add				↓	30%				
3280	All fuel, pressure tight, double wall fittings									
3284	304 stainless steel inner, aluminized steel jacket									
3288	Adjustable 20"/29" section									
3292	6" diameter	Q-9	30	.533	Ea.	193	27		220	253
3293	8" diameter		26	.615		201	31		232	269
3294	10" diameter		24	.667		227	33.50		260.50	300
3295	12" diameter		22	.727		257	36.50		293.50	340
3296	14" diameter		21	.762		290	38.50		328.50	380
3297	16" diameter		20	.800		325	40.50		365.50	420
3298	18" diameter	↓	19	.842		370	42.50		412.50	470
3299	20" diameter	Q-10	18	1.333		415	69.50		484.50	565
3300	24" diameter	"	16	1.500		535	78.50		613.50	710
3350	Elbow 90° fixed									

23 51 26.10 All-Fuel Vent Chimneys, Press. Tight, Dbl. Wall		Crew	Daily Output	Labor-Hours	Unit	Material	2015 Bare Costs Labor	Equipment	Total	Total Incl O&P
3354	6" diameter	Q-9	30	.533	Ea.	405	27		432	485
3355	8" diameter		26	.615		455	31		486	550
3356	10" diameter		24	.667		515	33.50		548.50	615
3357	12" diameter		22	.727		585	36.50		621.50	700
3358	14" diameter		21	.762		660	38.50		698.50	785
3359	16" diameter		20	.800		745	40.50		785.50	880
3360	18" diameter	Q-9	19	.842		840	42.50		882.50	990
3361	20" diameter	Q-10	18	1.333		955	69.50		1,024.50	1,150
3362	24" diameter	"	16	1.500		1,225	78.50		1,303.50	1,475
3380	For 316 stainless steel liner, add					30%				
3400	Elbow 45°									
3404	6" diameter	Q-9	30	.533	Ea.	201	27		228	262
3405	8" diameter		26	.615		227	31		258	297
3406	10" diameter		24	.667		258	33.50		291.50	335
3407	12" diameter		22	.727		294	36.50		330.50	380
3408	14" diameter		21	.762		330	38.50		368.50	425
3409	16" diameter		20	.800		370	40.50		410.50	470
3410	18" diameter		19	.842		420	42.50		462.50	530
3411	20" diameter	Q-10	18	1.333		480	69.50		549.50	630
3412	24" diameter	"	16	1.500		610	78.50		688.50	795
3430	For 316 stainless steel liner, add					30%				
3450	Tee 90°									
3454	6" diameter	Q-9	24	.667	Ea.	237	33.50		270.50	315
3455	8" diameter		22	.727		259	36.50		295.50	340
3456	10" diameter		21	.762		292	38.50		330.50	380
3457	12" diameter		20	.800		340	40.50		380.50	430
3458	14" diameter		18	.889		385	45		430	495
3459	16" diameter		16	1		425	50.50		475.50	540
3460	18" diameter		14	1.143		500	57.50		557.50	640
3461	20" diameter	Q-10	17	1.412		580	73.50		653.50	750
3462	24" diameter	"	12	2		720	104		824	955
3480	For Tee Cap, add					35%	20%			
3500	For 316 stainless steel liner, add					30%				
3520	Plate support, galvanized									
3524	6" diameter	Q-9	26	.615	Ea.	118	31		149	178
3525	8" diameter		22	.727		138	36.50		174.50	208
3526	10" diameter		20	.800		150	40.50		190.50	227
3527	12" diameter		18	.889		158	45		203	243
3528	14" diameter		17	.941		187	47.50		234.50	279
3529	16" diameter		16	1		197	50.50		247.50	294
3530	18" diameter		15	1.067		208	53.50		261.50	310
3531	20" diameter	Q-10	16	1.500		218	78.50		296.50	360
3532	24" diameter	"	14	1.714		227	89.50		316.50	385
3570	Bellows, lined									
3574	6" diameter	Q-9	30	.533	Ea.	1,300	27		1,327	1,475
3575	8" diameter		26	.615		1,350	31		1,381	1,550
3576	10" diameter		24	.667		1,375	33.50		1,408.50	1,575
3577	12" diameter		22	.727		1,425	36.50		1,461.50	1,600
3578	14" diameter		21	.762		1,450	38.50		1,488.50	1,650
3579	16" diameter		20	.800		1,525	40.50		1,565.50	1,725
3580	18" diameter		19	.842		1,575	42.50		1,617.50	1,800
3581	20" diameter	Q-10	18	1.333		1,625	69.50		1,694.50	1,875
3590	For all 316 stainless steel construction, add					55%				

23 51 26.10 All-Fuel Vent Chimneys, Press. Tight, Dbl. Wall	Crew	Daily Output	Labor-Hours	Unit	Material	2015 Bare Costs Labor	2015 Bare Costs Equipment	Total	Total Incl O&P
3600	Ventilated roof thimble, 304 stainless steel								
3620	6" diameter	Q-9	26	.615	Ea.	247	31		278	320
3624	8" diameter		22	.727		257	36.50		293.50	340
3625	10" diameter		20	.800		264	40.50		304.50	355
3626	12" diameter		18	.889		272	45		317	370
3627	14" diameter		17	.941		282	47.50		329.50	385
3628	16" diameter		16	1		305	50.50		355.50	410
3629	18" diameter		15	1.067		330	53.50		383.50	440
3630	20" diameter	Q-10	16	1.500		345	78.50		423.50	500
3631	24" diameter	"	14	1.714		380	89.50		469.50	550
3650	For 316 stainless steel, add					30%				
3670	Exit cone, 316 stainless steel only									
3674	6" diameter	Q-9	46	.348	Ea.	187	17.50		204.50	233
3675	8" diameter		42	.381		193	19.20		212.20	242
3676	10" diameter		40	.400		204	20		224	255
3677	12" diameter		38	.421		217	21		238	272
3678	14" diameter		37	.432		232	22		254	289
3679	16" diameter		36	.444		291	22.50		313.50	355
3680	18" diameter		35	.457		315	23		338	380
3681	20" diameter	Q-10	28	.857		375	45		420	485
3682	24" diameter	"	26	.923		495	48		543	620
3720	Roof guide, 304 stainless steel									
3724	6" diameter	Q-9	25	.640	Ea.	85.50	32		117.50	144
3725	8" diameter		21	.762		99.50	38.50		138	169
3726	10" diameter		19	.842		109	42.50		151.50	185
3727	12" diameter		17	.941		113	47.50		160.50	197
3728	14" diameter		16	1		132	50.50		182.50	222
3729	16" diameter		15	1.067		139	53.50		192.50	235
3730	18" diameter		14	1.143		148	57.50		205.50	250
3731	20" diameter	Q-10	15	1.600		155	83.50		238.50	299
3732	24" diameter	"	13	1.846		162	96.50		258.50	325
3750	For 316 stainless steel, add					30%				
3770	Rain cap with bird screen									
3774	6" diameter	Q-9	46	.348	Ea.	290	17.50		307.50	345
3775	8" diameter		42	.381		335	19.20		354.20	400
3776	10" diameter		40	.400		390	20		410	460
3777	12" diameter		38	.421		450	21		471	530
3778	14" diameter		37	.432		510	22		532	600
3779	16" diameter		36	.444		580	22.50		602.50	675
3780	18" diameter		35	.457		675	23		698	775
3781	20" diameter	Q-10	28	.857		765	45		810	910
3782	24" diameter	"	26	.923		920	48		968	1,075

23 51 26.30 All-Fuel Vent Chimneys, Double Wall, St. Stl.

		Crew	Daily Output	Labor-Hours	Unit	Material	Labor	Equipment	Total	Total Incl O&P
0010	**ALL-FUEL VENT CHIMNEYS, DOUBLE WALL, STAINLESS STEEL**									
7780	All fuel, pressure tight, double wall, 4" insulation, U.L. listed, 1400°F.									
7790	304 stainless steel liner, aluminized steel outer jacket									
7800	6" diameter	Q-9	60	.267	V.L.F.	64	13.45		77.45	90.50
7804	8" diameter		52	.308		73	15.50		88.50	104
7806	10" diameter		48	.333		81.50	16.80		98.30	116
7808	12" diameter		44	.364		93.50	18.30		111.80	131
7810	14" diameter		42	.381		105	19.20		124.20	145
7880	For 316 stainless steel liner add				L.F.	30%				

For customer support on your Plumbing Cost Data, call 877.709.6968.

345

23 51 Breechings, Chimneys, and Stacks

23 51 26 – All-Fuel Vent Chimneys

23 51 26.30 All-Fuel Vent Chimneys, Double Wall, St. Stl.

		Crew	Daily Output	Labor-Hours	Unit	Material	2015 Bare Costs Labor	Equipment	Total	Total Incl O&P
8000	All fuel, double wall, stainless steel fittings									
8010	Roof support 6" diameter	Q-9	30	.533	Ea.	113	27		140	165
8030	8" diameter		26	.615		132	31		163	193
8040	10" diameter		24	.667		139	33.50		172.50	205
8050	12" diameter		22	.727		148	36.50		184.50	218
8060	14" diameter		21	.762		155	38.50		193.50	230
8100	Elbow 45°, 6" diameter		30	.533		241	27		268	305
8140	8" diameter		26	.615		271	31		302	345
8160	10" diameter		24	.667		310	33.50		343.50	390
8180	12" diameter		22	.727		350	36.50		386.50	440
8200	14" diameter		21	.762		395	38.50		433.50	495
8300	Insulated tee, 6" diameter		30	.533		283	27		310	350
8360	8" diameter		26	.615		310	31		341	390
8380	10" diameter		24	.667		350	33.50		383.50	435
8400	12" diameter		22	.727		395	36.50		431.50	490
8420	14" diameter		20	.800		460	40.50		500.50	570
8500	Boot tee, 6" diameter		28	.571		540	29		569	640
8520	8" diameter		24	.667		590	33.50		623.50	695
8530	10" diameter		22	.727		685	36.50		721.50	810
8540	12" diameter		20	.800		810	40.50		850.50	950
8550	14" diameter		18	.889		925	45		970	1,100
8600	Rain cap with bird screen, 6" diameter		30	.533		290	27		317	360
8640	8" diameter		26	.615		335	31		366	420
8660	10" diameter		24	.667		390	33.50		423.50	480
8680	12" diameter		22	.727		450	36.50		486.50	550
8700	14" diameter		21	.762		510	38.50		548.50	625
8800	Flat roof flashing, 6" diameter		30	.533		106	27		133	158
8840	8" diameter		26	.615		116	31		147	175
8860	10" diameter		24	.667		124	33.50		157.50	188
8880	12" diameter		22	.727		136	36.50		172.50	206
8900	14" diameter		21	.762		139	38.50		177.50	212

23 52 Heating Boilers

23 52 13 – Electric Boilers

23 52 13.10 Electric Boilers, ASME

		Crew	Daily Output	Labor-Hours	Unit	Material	2015 Bare Costs Labor	Equipment	Total	Total Incl O&P
0010	**ELECTRIC BOILERS, ASME**, Standard controls and trim D3020–102									
1000	Steam, 6 KW, 20.5 MBH	Q-19	1.20	20	Ea.	3,950	1,075		5,025	5,975
1040	9 KW, 30.7 MBH		1.20	20		4,025	1,075		5,100	6,050
1060	18 KW, 61.4 MBH		1.20	20		4,150	1,075		5,225	6,200
1080	24 KW, 81.8 MBH		1.10	21.818		4,800	1,175		5,975	7,075
1120	36 KW, 123 MBH		1.10	21.818		5,325	1,175		6,500	7,625
2000	Hot water, 7.5 KW, 25.6 MBH		1.30	18.462		4,975	1,000		5,975	6,975
2020	15 KW, 51.2 MBH		1.30	18.462		5,000	1,000		6,000	7,000
2040	30 KW, 102 MBH		1.20	20		5,325	1,075		6,400	7,475
2060	45 KW, 164 MBH		1.20	20		5,425	1,075		6,500	7,575
2070	60 KW, 205 MBH		1.20	20		5,525	1,075		6,600	7,700
2080	75 KW, 256 MBH		1.10	21.818		5,900	1,175		7,075	8,275

23 52 Heating Boilers

23 52 16 – Condensing Boilers

23 52 16.24 Condensing Boilers

		Crew	Daily Output	Labor-Hours	Unit	Material	2015 Bare Costs Labor	2015 Bare Costs Equipment	Total	Total Incl O&P
0010	**CONDENSING BOILERS**, Cast iron, high efficiency									
0020	Packaged with standard controls, circulator and trim									
0030	Intermittent (spark) pilot, natural or LP gas									
0040	Hot water, DOE MBH output, (AFUE)									
0100	42 MBH, (84.0%) [G]	Q-5	1.80	8.889	Ea.	1,575	480		2,055	2,475
0120	57 MBH, (84.3%) [G]		1.60	10		1,775	540		2,315	2,750
0140	85 MBH, (84.0%) [G]		1.40	11.429		1,950	615		2,565	3,050
0160	112 MBH, (83.7%) [G]		1.20	13.333		2,175	715		2,890	3,475
0180	140 MBH, (83.3%) [G]	Q-6	1.60	15		2,450	835		3,285	3,975
0200	167 MBH, (83.0%) [G]		1.40	17.143		2,775	955		3,730	4,500
0220	194 MBH, (82.7%) [G]		1.20	20		3,025	1,125		4,150	5,000

23 52 19 – Pulse Combustion Boilers

23 52 19.20 Pulse Type Combustion Boilers

		Crew	Daily Output	Labor-Hours	Unit	Material	2015 Bare Costs Labor	2015 Bare Costs Equipment	Total	Total Incl O&P
0010	**PULSE TYPE COMBUSTION BOILERS**, High efficiency									
7990	Special feature gas fired boilers									
8000	Pulse combustion, standard controls/trim									
8010	Hot water, DOE MBH output, (AFUE %)									
8030	71 MBH, 95.2%	Q-5	1.60	10	Ea.	3,450	540		3,990	4,600
8050	94 MBH, 95.3% [G]		1.40	11.429		3,825	615		4,440	5,125
8080	139 MBH, 95.6% [G]		1.20	13.333		4,300	715		5,015	5,800
8090	207 MBH, 95.4%		1.16	13.793		4,850	740		5,590	6,475
8120	270 MBH, 96.4%		1.12	14.286		6,725	770		7,495	8,550
8130	365 MBH, 91.7%		1.07	14.953		7,600	805		8,405	9,575

23 52 23 – Cast-Iron Boilers

23 52 23.20 Gas-Fired Boilers

		Crew	Daily Output	Labor-Hours	Unit	Material	2015 Bare Costs Labor	2015 Bare Costs Equipment	Total	Total Incl O&P
0010	**GAS-FIRED BOILERS**, Natural or propane, standard controls, packaged									
1000	Cast iron, with insulated jacket									
2000	Steam, gross output, 81 MBH	Q-7	1.40	22.857	Ea.	2,450	1,300		3,750	4,675
2020	102 MBH		1.30	24.615		2,675	1,400		4,075	5,075
2040	122 MBH		1	32		2,925	1,825		4,750	5,975
2060	163 MBH		.90	35.556		3,275	2,025		5,300	6,650
2080	203 MBH		.90	35.556		3,825	2,025		5,850	7,250
2100	240 MBH		.85	37.647		3,925	2,150		6,075	7,525
2120	280 MBH		.80	40		4,700	2,275		6,975	8,600
2140	320 MBH		.70	45.714		5,000	2,600		7,600	9,425
3000	Hot water, gross output, 80 MBH		1.46	21.918		1,975	1,250		3,225	4,050
3020	100 MBH		1.35	23.704		2,550	1,350		3,900	4,850
3040	122 MBH		1.10	29.091		2,675	1,650		4,325	5,425
3060	163 MBH		1	32		3,150	1,825		4,975	6,225
3080	203 MBH		1	32		3,625	1,825		5,450	6,750
3100	240 MBH		.95	33.684		3,700	1,925		5,625	6,950
3120	280 MBH		.90	35.556		4,325	2,025		6,350	7,800
3140	320 MBH		.80	40		4,675	2,275		6,950	8,575
7000	For tankless water heater, add					10%				
7050	For additional zone valves up to 312 MBH add					194			194	214

23 52 23.40 Oil-Fired Boilers

		Crew	Daily Output	Labor-Hours	Unit	Material	2015 Bare Costs Labor	2015 Bare Costs Equipment	Total	Total Incl O&P
0010	**OIL-FIRED BOILERS**, Standard controls, flame retention burner, packaged									
1000	Cast iron, with insulated flush jacket									
2000	Steam, gross output, 109 MBH	Q-7	1.20	26.667	Ea.	2,325	1,525		3,850	4,850
2020	144 MBH		1.10	29.091		2,625	1,650		4,275	5,375
2040	173 MBH		1	32		2,925	1,825		4,750	5,975

347

23 52 Heating Boilers

23 52 23 - Cast-Iron Boilers

23 52 23.40 Oil-Fired Boilers		Crew	Daily Output	Labor-Hours	Unit	Material	2015 Bare Costs Labor	Equipment	Total	Total Incl O&P
2060	207 MBH	Q-7	.90	35.556	Ea.	3,175	2,025		5,200	6,550
3000	Hot water, same price as steam									
4000	For tankless coil in smaller sizes, add				Ea.	15%				

23 52 26 - Steel Boilers

23 52 26.40 Oil-Fired Boilers

		Crew	Daily Output	Labor-Hours	Unit	Material	Labor	Equipment	Total	Total Incl O&P
0010	**OIL-FIRED BOILERS**, Standard controls, flame retention burner									
5000	Steel, with insulated flush jacket									
7000	Hot water, gross output, 103 MBH	Q-6	1.60	15	Ea.	1,775	835		2,610	3,225
7020	122 MBH		1.45	16.506		1,875	920		2,795	3,475
7040	137 MBH		1.36	17.595		2,000	980		2,980	3,675
7060	168 MBH		1.30	18.405		2,100	1,025		3,125	3,875
7080	225 MBH		1.22	19.704		2,575	1,100		3,675	4,475
7340	For tankless coil in steam or hot water, add					7%				

23 52 28 - Swimming Pool Boilers

23 52 28.10 Swimming Pool Heaters

		Crew	Daily Output	Labor-Hours	Unit	Material	Labor	Equipment	Total	Total Incl O&P
0010	**SWIMMING POOL HEATERS**, Not including wiring, external									
0020	piping, base or pad,									
0160	Gas fired, input, 155 MBH	Q-6	1.50	16	Ea.	2,000	890		2,890	3,550
0200	199 MBH		1	24		2,125	1,350		3,475	4,375
0220	250 MBH		.70	34.286		2,325	1,900		4,225	5,425
0240	300 MBH		.60	40		2,425	2,225		4,650	6,050
0260	399 MBH		.50	48		2,725	2,675		5,400	7,050
0280	500 MBH		.40	60		8,900	3,350		12,250	14,800
0300	650 MBH		.35	68.571		9,450	3,825		13,275	16,200
0320	750 MBH		.33	72.727		10,300	4,050		14,350	17,500
0360	990 MBH		.22	109		13,900	6,075		19,975	24,400
0370	1,260 MBH		.21	114		16,300	6,375		22,675	27,500
0380	1,440 MBH		.19	126		17,400	7,050		24,450	29,700
0400	1,800 MBH		.14	171		19,300	9,550		28,850	35,700
0410	2,070 MBH		.13	184		22,800	10,300		33,100	40,600
2000	Electric, 12 KW, 4,800 gallon pool	Q-19	3	8		2,075	435		2,510	2,925
2020	15 KW, 7,200 gallon pool		2.80	8.571		2,100	465		2,565	3,025
2040	24 KW, 9,600 gallon pool		2.40	10		2,425	540		2,965	3,500
2060	30 KW, 12,000 gallon pool		2	12		2,475	650		3,125	3,700
2080	36 KW, 14,400 gallon pool		1.60	15		2,850	810		3,660	4,350
2100	57 KW, 24,000 gallon pool		1.20	20		3,575	1,075		4,650	5,550
9000	To select pool heater: 12 BTUH x S.F. pool area									
9010	X temperature differential = required output									
9050	For electric, KW = gallons x 2.5 divided by 1000									
9100	For family home type pool, double the									
9110	Rated gallon capacity = 1/2°F rise per hour									

23 52 88 - Burners

23 52 88.10 Replacement Type Burners

		Crew	Daily Output	Labor-Hours	Unit	Material	Labor	Equipment	Total	Total Incl O&P
0010	**REPLACEMENT TYPE BURNERS**									
0990	Residential, conversion, gas fired, LP or natural									
1000	Gun type, atmospheric input 50 to 225 MBH	Q-1	2.50	6.400	Ea.	720	340		1,060	1,300
1020	100 to 400 MBH	"	2	8	"	1,200	425		1,625	1,975
3000	Flame retention oil fired assembly, input									
3020	.50 to 2.25 GPH	Q-1	2.40	6.667	Ea.	284	350		634	840
3040	2.0 to 5.0 GPH	"	2	8		315	425		740	985
4600	Gas safety, shut off valve, 3/4" threaded	1 Stpi	20	.400		178	24		202	232

23 52 Heating Boilers

23 52 88 – Burners

23 52 88.10 Replacement Type Burners		Crew	Daily Output	Labor-Hours	Unit	Material	2015 Bare Costs Labor	Equipment	Total	Total Incl O&P
4610	1" threaded	1 Stpi	19	.421	Ea.	172	25		197	228
4620	1-1/4" threaded		15	.533		194	32		226	261
4630	1-1/2" threaded		13	.615		210	37		247	287
4640	2" threaded	↓	11	.727		235	43.50		278.50	325
4650	2-1/2" threaded	Q-1	15	1.067		269	56.50		325.50	380
4660	3" threaded	↓	13	1.231		370	65		435	505
4670	4" flanged	↓	3	5.333		2,625	282		2,907	3,325
4680	6" flanged	Q-2	3	8	↓	5,875	440		6,315	7,125

23 54 Furnaces

23 54 16 – Fuel-Fired Furnaces

23 54 16.14 Condensing Furnaces

		Crew	Daily Output	Labor-Hours	Unit	Material	2015 Bare Costs Labor	Equipment	Total	Total Incl O&P
0010	**CONDENSING FURNACES**, High efficiency									
0020	Oil fired, packaged, complete									
0030	Upflow									
0040	Output @ 95% A.F.U.E.									
0100	49 MBH @ 1000 CFM	Q-9	3.70	4.324	Ea.	4,450	218		4,668	5,225
0110	73.5 MBH @ 2000 CFM	↓	3.60	4.444		4,650	224		4,874	5,475
0120	96 MBH @ 2000 CFM		3.40	4.706		4,650	237		4,887	5,475
0130	115.6 MBH @ 2000 CFM		3.40	4.706		4,650	237		4,887	5,475
0140	147 MBH @ 2000 CFM		3.30	4.848		10,400	244		10,644	11,900
0150	192 MBH @ 4000 CFM		2.60	6.154		10,800	310		11,110	12,400
0170	231.5 MBH @ 4000 CFM	↓	2.30	6.957		10,800	350		11,150	12,400
0260	For variable speed motor, add				↓	565			565	625
0270	Note: Also available in horizontal, counterflow and lowboy configurations.									

23 56 Solar Energy Heating Equipment

23 56 16 – Packaged Solar Heating Equipment

23 56 16.40 Solar Heating Systems

			Crew	Daily Output	Labor-Hours	Unit	Material	2015 Bare Costs Labor	Equipment	Total	Total Incl O&P
0010	**SOLAR HEATING SYSTEMS**	D2020-265									
0020	System/Package prices, not including connecting										
0030	pipe, insulation, or special heating/plumbing fixtures	D2020-270									
0152	For solar ultraviolet pipe insulation see Section 22 07 19.10										
0500	Hot water, standard package, low temperature	D2020-275									
0540	1 collector, circulator, fittings, 65 gal. tank	G	Q-1	.50	32	Ea.	3,675	1,700		5,375	6,600
0580	2 collectors, circulator, fittings, 120 gal. tank	D2020-280		.40	40		5,050	2,125		7,175	8,750
0620	3 collectors, circulator, fittings, 120 gal. tank	G	↓	.34	47.059		6,875	2,475		9,350	11,300
0700	Medium temperature package	D2020-285									
0720	1 collector, circulator, fittings, 80 gal. tank	G	Q-1	.50	32	Ea.	5,025	1,700		6,725	8,075
0740	2 collectors, circulator, fittings, 120 gal. tank	D2020-290		.40	40		6,450	2,125		8,575	10,300
0780	3 collectors, circulator, fittings, 120 gal. tank	G	↓	.30	53.333		7,325	2,825		10,150	12,300
0980	For each additional 120 gal. tank, add	D2020-295				↓	1,750			1,750	1,925

23 56 19 – Solar Heating Components

23 56 19.50 Solar Heating Ancillary

			Crew	Daily Output	Labor-Hours	Unit	Material	2015 Bare Costs Labor	Equipment	Total	Total Incl O&P
0010	**SOLAR HEATING ANCILLARY**										
2300	Circulators, air	D3010-650									
2310	Blowers										
2330	100-300 S.F. system, 1/10 HP	D3010-660	Q-9	16	1	Ea.	249	50.50		299.50	350
2340	300-500 S.F. system, 1/5 HP	G	↓	15	1.067	↓	330	53.50		383.50	440

23 56 19.50 Solar Heating Ancillary		Crew	Daily Output	Labor-Hours	Unit	Material	2015 Bare Costs Labor	Equipment	Total	Total Incl O&P
2350	Two speed, 100-300 S.F., 1/10 HP	D3010-675	Q-9	14	1.143	Ea.	142	57.50		199.50	245
2400	Reversible fan, 20" diameter, 2 speed	G		18	.889		113	45		158	193
2550	Booster fan 6" diameter, 120 CFM	G		16	1		36.50	50.50		87	118
2570	6" diameter, 225 CFM	G		16	1		45	50.50		95.50	127
2580	8" diameter, 150 CFM	G		16	1		41	50.50		91.50	123
2590	8" diameter, 310 CFM	G		14	1.143		63.50	57.50		121	158
2600	8" diameter, 425 CFM	G		14	1.143		71	57.50		128.50	166
2650	Rheostat	G		32	.500		15.45	25		40.45	55.50
2660	Shutter/damper	G		12	1.333		56	67		123	165
2670	Shutter motor	G		16	1		139	50.50		189.50	230
2800	Circulators, liquid, 1/25 HP, 5.3 GPM	G	Q-1	14	1.143		114	60.50		174.50	217
2820	1/20 HP, 17 GPM	G		12	1.333		161	70.50		231.50	283
2850	1/20 HP, 17 GPM, stainless steel	G		12	1.333		254	70.50		324.50	385
2870	1/12 HP, 30 GPM	G		10	1.600		345	84.50		429.50	510
3000	Collector panels, air with aluminum absorber plate										
3010	Wall or roof mount										
3040	Flat black, plastic glazing										
3080	4' x 8'	G	Q-9	6	2.667	Ea.	660	134		794	935
3100	4' x 10'	G		5	3.200	"	815	161		976	1,150
3200	Flush roof mount, 10' to 16' x 22" wide	G		96	.167	L.F.	132	8.40		140.40	158
3210	Manifold, by L.F. width of collectors	G		160	.100	"	144	5.05		149.05	167
3300	Collector panels, liquid with copper absorber plate										
3320	Black chrome, tempered glass glazing										
3330	Alum. frame, 4' x 8', 5/32" single glazing	G	Q-1	9.50	1.684	Ea.	995	89		1,084	1,225
3390	Alum. frame, 4' x 10', 5/32" single glazing	G		6	2.667		1,150	141		1,291	1,475
3450	Flat black, alum. frame, 3.5' x 7.5'	G		9	1.778		880	94		974	1,100
3500	4' x 8'	G		5.50	2.909		1,050	154		1,204	1,375
3520	4' x 10'	G		10	1.600		1,250	84.50		1,334.50	1,500
3540	4' x 12.5'	G		5	3.200		1,250	169		1,419	1,625
3550	Liquid with fin tube absorber plate										
3560	Alum. frame 4' x 8' tempered glass	G	Q-1	10	1.600	Ea.	580	84.50		664.50	770
3580	Liquid with vacuum tubes, 4' x 6'-10"	G		9	1.778		895	94		989	1,125
3600	Liquid, full wetted, plastic, alum. frame, 4' x 10'	G		5	3.200		320	169		489	610
3650	Collector panel mounting, flat roof or ground rack	G		7	2.286		244	121		365	450
3670	Roof clamps	G		70	.229	Set	2.80	12.10		14.90	21.50
3700	Roof strap, teflon	G	1 Plum	205	.039	L.F.	23.50	2.29		25.79	29.50
3900	Differential controller with two sensors										
3930	Thermostat, hard wired	G	1 Plum	8	1	Ea.	101	58.50		159.50	200
3950	Line cord and receptacle	G		12	.667		130	39		169	202
4050	Pool valve system	G		2.50	3.200		161	188		349	460
4070	With 12 VAC actuator	G		2	4		325	235		560	710
4080	Pool pump system, 2" pipe size	G		6	1.333		195	78.50		273.50	330
4100	Five station with digital read-out	G		3	2.667		263	157		420	525
4150	Sensors										
4200	Brass plug, 1/2" MPT	G	1 Plum	32	.250	Ea.	16	14.70		30.70	39.50
4210	Brass plug, reversed	G		32	.250		26.50	14.70		41.20	51
4220	Freeze prevention	G		32	.250		22.50	14.70		37.20	47
4240	Screw attached	G		32	.250		9.85	14.70		24.55	33
4250	Brass, immersion	G		32	.250		13.55	14.70		28.25	37
4300	Heat exchanger										
4315	includes coil, blower, circulator										
4316	and controller for DHW and space hot air										
4330	Fluid to air coil, up flow, 45 MBH	G	Q-1	4	4	Ea.	315	211		526	670

23 56 19.50 Solar Heating Ancillary		Crew	Daily Output	Labor-Hours	Unit	Material	2015 Bare Costs Labor	Equipment	Total	Total Incl O&P
4380	70 MBH	G	Q-1	3.50	4.571	Ea.	355	242		597	760
4400	80 MBH	G	↓	3	5.333	↓	475	282		757	950
4580	Fluid to fluid package includes two circulating pumps										
4590	expansion tank, check valve, relief valve										
4600	controller, high temperature cutoff and sensors	G	Q-1	2.50	6.400	Ea.	800	340		1,140	1,400
4650	Heat transfer fluid										
4700	Propylene glycol, inhibited anti-freeze	G	1 Plum	28	.286	Gal.	15.55	16.75		32.30	42.50
4800	Solar storage tanks, knocked down										
4810	Air, galvanized steel clad, double wall, 4" fiberglass insulation										
5120	45 Mil reinforced polypropylene lining,										
5140	4' high, 4' x 4' = 64 C.F./450 gallons	G	Q-9	2	8	Ea.	3,650	405		4,055	4,625
5150	4' x 8' = 128 C.F./900 gallons	G		1.50	10.667		5,450	535		5,985	6,825
5160	4' x 12' = 190 C.F./1300 gallons	G		1.30	12.308		7,275	620		7,895	8,950
5170	8' x 8' = 250 C.F./1700 gallons	G	↓	1	16		7,275	805		8,080	9,225
5190	6'-3" high, 7' x 7' = 306 C.F./2000 gallons	G	Q-10	1.20	20		14,200	1,050		15,250	17,300
5200	7' x 10'-6" = 459 C.F./3000 gallons	G		.80	30		17,800	1,575		19,375	22,000
5210	7' x 14' = 613 C.F./4000 gallons	G		.60	40		21,400	2,100		23,500	26,700
5220	10'-6" x 10'-6" = 689 C.F./4500 gallons	G		.50	48		21,400	2,500		23,900	27,300
5230	10'-6" x 14' = 919 C.F./6000 gallons	G	↓	.40	60		24,900	3,125		28,025	32,200
5240	14' x 14' = 1225 C.F./8000 gallons	G	Q-11	.40	80		28,500	4,250		32,750	37,800
5250	14' x 17'-6" = 1531 C.F./10,000 gallons	G		.30	106		32,100	5,675		37,775	44,000
5260	17'-6" x 17'-6" = 1914 C.F./12,500 gallons	G		.25	128		35,600	6,825		42,425	49,600
5270	17'-6" x 21' = 2297 C.F./15,000 gallons	G		.20	160		39,100	8,525		47,625	56,000
5280	21' x 21' = 2756 C.F./18,000 gallons	G	↓	.18	177	↓	42,700	9,475		52,175	61,500
5290	30 Mil reinforced Hypalon lining, add						.02%				
7000	Solar control valves and vents										
7050	Air purger, 1" pipe size	G	1 Plum	12	.667	Ea.	46.50	39		85.50	111
7070	Air eliminator, automatic 3/4" size	G		32	.250		30	14.70		44.70	55
7090	Air vent, automatic, 1/8" fitting	G		32	.250		17.25	14.70		31.95	41
7100	Manual, 1/8" NPT	G		32	.250		2.44	14.70		17.14	24.50
7120	Backflow preventer, 1/2" pipe size	G		16	.500		75.50	29.50		105	128
7130	3/4" pipe size	G		16	.500		77	29.50		106.50	129
7150	Balancing valve, 3/4" pipe size	G		20	.400		63	23.50		86.50	105
7180	Draindown valve, 1/2" copper tube	G		9	.889		212	52		264	310
7200	Flow control valve, 1/2" pipe size	G		22	.364		120	21.50		141.50	164
7220	Expansion tank, up to 5 gal.	G		32	.250		68	14.70		82.70	97
7250	Hydronic controller (aquastat)	G		8	1		228	58.50		286.50	340
7400	Pressure gauge, 2" dial	G		32	.250		24	14.70		38.70	48.50
7450	Relief valve, temp. and pressure 3/4" pipe size	G	↓	30	.267	↓	21	15.65		36.65	46.50
7500	Solenoid valve, normally closed										
7520	Brass, 3/4" NPT, 24V	G	1 Plum	9	.889	Ea.	123	52		175	214
7530	1" NPT, 24V	G		9	.889		193	52		245	291
7750	Vacuum relief valve, 3/4" pipe size	G	↓	32	.250	↓	29.50	14.70		44.20	54.50
7800	Thermometers										
7820	Digital temperature monitoring, 4 locations	G	1 Plum	2.50	3.200	Ea.	137	188		325	435
7900	Upright, 1/2" NPT	G		8	1		39.50	58.50		98	132
7970	Remote probe, 2" dial	G		8	1		33	58.50		91.50	125
7990	Stem, 2" dial, 9" stem	G	↓	16	.500	↓	21.50	29.50		51	68.50
8250	Water storage tank with heat exchanger and electric element										
8270	66 gal. with 2" x 2 lb. density insulation	G	1 Plum	1.60	5	Ea.	1,600	294		1,894	2,225
8300	80 gal. with 2" x 2 lb. density insulation	G		1.60	5		1,600	294		1,894	2,225
8380	120 gal. with 2" x 2 lb. density insulation	G		1.40	5.714		1,825	335		2,160	2,525
8400	120 gal. with 2" x 2 lb. density insul., 40 S.F. heat coil	G	↓	1.40	5.714	↓	2,325	335		2,660	3,050

23 56 Solar Energy Heating Equipment

23 56 19 – Solar Heating Components

23 56 19.50 Solar Heating Ancillary		Crew	Daily Output	Labor-Hours	Unit	Material	2015 Bare Costs Labor	Equipment	Total	Total Incl O&P
8500	Water storage module, plastic									
8600	Tubular, 12" diameter, 4' high	G	1 Carp	48	.167	Ea.	114	7.85		121.85	137
8610	12" diameter, 8' high	G		40	.200		175	9.40		184.40	207
8620	18" diameter, 5' high	G		38	.211		193	9.90		202.90	227
8630	18" diameter, 10' high	G		32	.250		255	11.75		266.75	299
8640	58" diameter, 5' high	G	2 Carp	32	.500		495	23.50		518.50	580
8650	Cap, 12" diameter	G					22			22	24
8660	18" diameter	G					28			28	31

23 57 Heat Exchangers for HVAC

23 57 16 – Steam-to-Water Heat Exchangers

23 57 16.10 Shell/Tube Type Steam-to-Water Heat Exch.

		Crew	Daily Output	Labor-Hours	Unit	Material	2015 Bare Costs Labor	Equipment	Total	Total Incl O&P
0010	**SHELL AND TUBE TYPE STEAM-TO-WATER HEAT EXCHANGERS**									
0016	Shell & tube type, 2 or 4 pass, 3/4" O.D. copper tubes,									
0020	C.I. heads, C.I. tube sheet, steel shell									
0100	Hot water 40°F to 180°F, by steam at 10 PSI									
0120	8 GPM	Q-5	6	2.667	Ea.	2,150	143		2,293	2,600
0140	10 GPM		5	3.200		3,250	172		3,422	3,825
0160	40 GPM		4	4		5,025	215		5,240	5,850
0500	For bronze head and tube sheet, add					50%				

23 57 19 – Liquid-to-Liquid Heat Exchangers

23 57 19.13 Plate-Type, Liquid-to-Liquid Heat Exchangers

		Crew	Daily Output	Labor-Hours	Unit	Material	2015 Bare Costs Labor	Equipment	Total	Total Incl O&P
0010	**PLATE-TYPE, LIQUID-TO-LIQUID HEAT EXCHANGERS**									
3000	Plate type,									
3100	400 GPM	Q-6	.80	30	Ea.	36,700	1,675		38,375	42,900
3120	800 GPM	"	.50	48		63,500	2,675		66,175	73,500
3140	1200 GPM	Q-7	.34	94.118		94,000	5,350		99,350	111,500
3160	1800 GPM	"	.24	133		125,000	7,575		132,575	149,000

23 57 19.16 Shell-Type, Liquid-to-Liquid Heat Exchangers

		Crew	Daily Output	Labor-Hours	Unit	Material	2015 Bare Costs Labor	Equipment	Total	Total Incl O&P
0010	**SHELL-TYPE, LIQUID-TO-LIQUID HEAT EXCHANGERS**									
1000	Hot water 40°F to 140°F, by water at 200°F									
1020	7 GPM	Q-5	6	2.667	Ea.	2,650	143		2,793	3,125
1040	16 GPM		5	3.200		3,750	172		3,922	4,375
1060	34 GPM		4	4		5,675	215		5,890	6,550
1080	55 GPM		3	5.333		8,225	287		8,512	9,475
1100	74 GPM		1.50	10.667		10,300	575		10,875	12,200
1120	86 GPM		1.40	11.429		13,700	615		14,315	16,000
1140	112 GPM	Q-6	2	12		17,100	670		17,770	19,800
1160	126 GPM		1.80	13.333		21,400	745		22,145	24,600
1180	152 GPM		1	24		27,300	1,350		28,650	32,000

23 72 Air-to-Air Energy Recovery Equipment

23 72 16 – Heat-Pipe Air-To-Air Energy-Recovery Equipment

23 72 16.10 Heat Pipes	Crew	Daily Output	Labor-Hours	Unit	Material	2015 Bare Costs Labor	Equipment	Total	Total Incl O&P
0010 **HEAT PIPES**									
8000 Heat pipe type, glycol, 50% efficient									
8010 100 MBH, 1700 CFM	1 Stpi	.80	10	Ea.	4,300	600		4,900	5,650
8020 160 MBH, 2700 CFM		.60	13.333		5,775	795		6,570	7,550
8030 620 MBH, 4000 CFM	↓	.40	20	↓	8,725	1,200		9,925	11,400

23 81 Decentralized Unitary HVAC Equipment

23 81 13 – Packaged Terminal Air-Conditioners

23 81 13.10 Packaged Cabinet Type Air-Conditioners

	Crew	Daily Output	Labor-Hours	Unit	Material	2015 Bare Costs Labor	Equipment	Total	Total Incl O&P
0010 **PACKAGED CABINET TYPE AIR-CONDITIONERS**, Cabinet, wall sleeve,									
0100 louver, electric heat, thermostat, manual changeover, 208 V									
0200 6,000 BTUH cooling, 8800 BTU heat	Q-5	6	2.667	Ea.	775	143		918	1,075
0220 9,000 BTUH cooling, 13,900 BTU heat		5	3.200		1,200	172		1,372	1,575
0240 12,000 BTUH cooling, 13,900 BTU heat		4	4		1,350	215		1,565	1,825
0260 15,000 BTUH cooling, 13,900 BTU heat	↓	3	5.333	↓	1,450	287		1,737	2,025

23 81 19 – Self-Contained Air-Conditioners

23 81 19.10 Window Unit Air Conditioners

	Crew	Daily Output	Labor-Hours	Unit	Material	2015 Bare Costs Labor	Equipment	Total	Total Incl O&P
0010 **WINDOW UNIT AIR CONDITIONERS**									
4000 Portable/window, 15 amp 125 V grounded receptacle required									
4060 5000 BTUH	1 Carp	8	1	Ea.	295	47		342	400
4340 6000 BTUH		8	1		340	47		387	450
4480 8000 BTUH		6	1.333		460	62.50		522.50	605
4500 10,000 BTUH	↓	6	1.333		590	62.50		652.50	745
4520 12,000 BTUH	L-2	8	2	↓	725	82.50		807.50	920
4600 Window/thru-the-wall, 15 amp 230 V grounded receptacle required									
4780 18,000 BTUH	L-2	6	2.667	Ea.	970	110		1,080	1,250
4940 25,000 BTUH		4	4		1,175	165		1,340	1,550
4960 29,000 BTUH	↓	4	4		1,325	165		1,490	1,725

23 81 43 – Air-Source Unitary Heat Pumps

23 81 43.10 Air-Source Heat Pumps

	Crew	Daily Output	Labor-Hours	Unit	Material	2015 Bare Costs Labor	Equipment	Total	Total Incl O&P
0010 **AIR-SOURCE HEAT PUMPS**, Not including interconnecting tubing									
1000 Air to air, split system, not including curbs, pads, fan coil and ductwork									
1010 For curbs/pads see Section 23 91 00									
1012 Outside condensing unit only, for fan coil see Section 23 82 19.10									
1015 1.5 ton cooling, 7 MBH heat @ 0°F	Q-5	2.40	6.667	Ea.	2,175	360		2,535	2,950
1020 2 ton cooling, 8.5 MBH heat @ 0°F		2	8		2,400	430		2,830	3,300
1030 2.5 ton cooling, 10 MBH heat @ 0°F		1.60	10		2,525	540		3,065	3,575
1040 3 ton cooling, 13 MBH heat @ 0°F		1.20	13.333		2,775	715		3,490	4,125
1050 3.5 ton cooling, 18 MBH heat @ 0°F		1	16		3,000	860		3,860	4,600
1054 4 ton cooling, 24 MBH heat @ 0°F		.80	20		3,250	1,075		4,325	5,200
1060 5 ton cooling, 27 MBH heat @ 0°F	↓	.50	32		3,600	1,725		5,325	6,550
1500 Single package, not including curbs, pads, or plenums									
1502 1/2 ton cooling, supplementary heat included	Q-5	8	2	Ea.	2,450	108		2,558	2,850
1504 3/4 ton cooling, supplementary heat supplementary heat included		6	2.667		2,575	143		2,718	3,050
1506 1 ton cooling, supplementary heat included		4	4		2,975	215		3,190	3,600
1510 1.5 ton cooling, 5 MBH heat @ 0°F		1.55	10.323		3,050	555		3,605	4,200
1520 2 ton cooling, 6.5 MBH heat @ 0°F		1.50	10.667		3,100	575		3,675	4,300
1540 2.5 ton cooling, 8 MBH heat @ 0°F		1.40	11.429		3,175	615		3,790	4,425
1560 3 ton cooling, 10 MBH heat @ 0°F		1.20	13.333		3,575	715		4,290	5,000
1570 3.5 ton cooling, 11 MBH heat @ 0°F	↓	1	16		3,900	860		4,760	5,575

For customer support on your Plumbing Cost Data, call 877.709.6968.

353

23 81 Decentralized Unitary HVAC Equipment

23 81 43 – Air-Source Unitary Heat Pumps

23 81 43.10 Air-Source Heat Pumps

		Crew	Daily Output	Labor-Hours	Unit	Material	2015 Bare Costs Labor	2015 Bare Costs Equipment	Total	Total Incl O&P
1580	4 ton cooling, 13 MBH heat @ 0°F	Q-5	.96	16.667	Ea.	4,200	895		5,095	5,975
1620	5 ton cooling, 27 MBH heat @ 0°F		.65	24.615		4,800	1,325		6,125	7,275
1640	7.5 ton cooling, 35 MBH heat @ 0°F	↓	.40	40	↓	7,325	2,150		9,475	11,300
6000	Air to water, single package, excluding. storage tank and ductwork									
6010	Includes circulating water pump, air duct connections, digital temperature									
6020	controller with remote tank temp. probe and sensor for storage tank.									
6040	Water heating - air cooling capacity									
6110	35.5 MBH heat water, 2.3 Ton cool air	Q-5	1.60	10	Ea.	11,400	540		11,940	13,400
6120	58 MBH heat water, 3.8 Ton cool air		1.10	14.545		13,000	780		13,780	15,500
6130	76 MBH heat water, 4.9 Ton cool air		.87	18.391		15,500	990		16,490	18,600
6140	98 MBH heat water, 6.5 Ton cool air		.62	25.806		20,100	1,400		21,500	24,200
6150	113 MBH heat water, 7.4 Ton cool air		.59	27.119		22,000	1,450		23,450	26,400
6160	142 MBH heat water, 9.2 Ton cool air		.52	30.769		26,700	1,650		28,350	31,900
6170	171 MBH heat water, 11.1 Ton cool air	↓	.49	32.653	↓	31,400	1,750		33,150	37,200

23 81 46 – Water-Source Unitary Heat Pumps

23 81 46.10 Water Source Heat Pumps

		Crew	Daily Output	Labor-Hours	Unit	Material	2015 Bare Costs Labor	2015 Bare Costs Equipment	Total	Total Incl O&P
0010	**WATER SOURCE HEAT PUMPS**, Not incl. connecting tubing or water source									
2000	Water source to air, single package									
2100	1 ton cooling, 13 MBH heat @ 75°F	Q-5	2	8	Ea.	2,075	430		2,505	2,950
2120	1.5 ton cooling, 17 MBH heat @ 75°F		1.80	8.889		2,400	480		2,880	3,350
2140	2 ton cooling, 19 MBH heat @ 75°F		1.70	9.412		2,575	505		3,080	3,600
2160	2.5 ton cooling, 25 MBH heat @ 75°F		1.60	10		2,800	540		3,340	3,900
2180	3 ton cooling, 27 MBH heat @ 75°F		1.40	11.429		2,875	615		3,490	4,075
2190	3.5 ton cooling, 29 MBH heat @ 75°F		1.30	12.308		3,100	660		3,760	4,425
2200	4 ton cooling, 31 MBH heat @ 75°F		1.20	13.333		3,475	715		4,190	4,900
2220	5 ton cooling, 29 MBH heat @ 75°F	↓	.90	17.778	↓	3,825	955		4,780	5,650
3960	For supplementary heat coil, add					10%				

23 82 Convection Heating and Cooling Units

23 82 19 – Fan Coil Units

23 82 19.10 Fan Coil Air Conditioning

		Crew	Daily Output	Labor-Hours	Unit	Material	2015 Bare Costs Labor	2015 Bare Costs Equipment	Total	Total Incl O&P
0010	**FAN COIL AIR CONDITIONING**									
0030	Fan coil AC, cabinet mounted, filters and controls									
0940	Direct expansion, for use w/air cooled condensing unit, 1.5 ton cooling	Q-5	5	3.200	Ea.	680	172		852	1,000
0950	2 ton cooling		4.80	3.333		730	179		909	1,075
0960	2.5 ton cooling		4.40	3.636		775	196		971	1,150
0970	3 ton cooling		3.80	4.211		960	226		1,186	1,400
0980	3.5 ton cooling		3.60	4.444		970	239		1,209	1,425
0990	4 ton cooling		3.40	4.706		1,100	253		1,353	1,575
1000	5 ton cooling	↓	3	5.333		1,300	287		1,587	1,875
1500	For hot water coil, add					40%	10%			

23 82 27 – Infrared Units

23 82 27.10 Infrared Type Heating Units

		Crew	Daily Output	Labor-Hours	Unit	Material	2015 Bare Costs Labor	2015 Bare Costs Equipment	Total	Total Incl O&P
0010	**INFRARED TYPE HEATING UNITS**									
0020	Gas fired, unvented, electric ignition, 100% shutoff.									
0030	Piping and wiring not included									
0100	Input, 30 MBH	Q-5	6	2.667	Ea.	945	143		1,088	1,250
0120	45 MBH		5	3.200		950	172		1,122	1,300
0140	50 MBH		4.50	3.556		950	191		1,141	1,350
0160	60 MBH		4	4	↓	950	215		1,165	1,375

23 82 27 – Infrared Units

23 82 27.10 Infrared Type Heating Units

	Crew	Daily Output	Labor-Hours	Unit	Material	2015 Bare Costs Labor	2015 Bare Costs Equipment	Total	Total Incl O&P
0180	75 MBH	Q-5	3	5.333	Ea.	950	287		1,237	1,475
0200	90 MBH		2.50	6.400		950	345		1,295	1,575
0220	105 MBH		2	8		950	430		1,380	1,700
0240	120 MBH		2	8		1,575	430		2,005	2,375
1000	Gas fired, vented, electric ignition, tubular									
1020	Piping and wiring not included, 20' to 80' lengths									
1030	Single stage, input, 60 MBH	Q-6	4.50	5.333	Ea.	1,425	297		1,722	2,025
1040	80 MBH		3.90	6.154		1,425	345		1,770	2,100
1050	100 MBH		3.40	7.059		1,425	395		1,820	2,175
1060	125 MBH		2.90	8.276		1,425	460		1,885	2,275
1070	150 MBH		2.70	8.889		1,425	495		1,920	2,325
1080	170 MBH		2.50	9.600		1,425	535		1,960	2,375
1090	200 MBH		2.20	10.909		1,625	610		2,235	2,725
1100	Note: Final pricing may vary due to									
1110	tube length and configuration package selected									
1130	Two stage, input, 60 MBH high, 45 MBH low	Q-6	4.50	5.333	Ea.	1,750	297		2,047	2,375
1140	80 MBH high, 60 MBH low		3.90	6.154		1,750	345		2,095	2,450
1150	100 MBH high, 65 MBH low		3.40	7.059		1,750	395		2,145	2,525
1160	125 MBH high, 95 MBH low		2.90	8.276		1,750	460		2,210	2,625
1170	150 MBH high, 100 MBH low		2.70	8.889		1,750	495		2,245	2,675
1180	170 MBH high, 125 MBH low		2.50	9.600		1,950	535		2,485	2,950
1190	200 MBH high, 150 MBH low		2.20	10.909		1,950	610		2,560	3,075
1220	Note: Final pricing may vary due to									
1230	tube length and configuration package selected									
2000	Electric, single or three phase									
2050	6 kW, 20,478 BTU	1 Elec	2.30	3.478	Ea.	705	190		895	1,050
2100	13.5 KW, 40,956 BTU		2.20	3.636		870	199		1,069	1,250
2150	24 KW, 81,912 BTU		2	4		1,750	219		1,969	2,250
3000	Oil fired, pump, controls, fusible valve, oil supply tank									
3050	91,000 BTU	Q-5	2.50	6.400	Ea.	6,425	345		6,770	7,600
3080	105,000 BTU		2.25	7.111		6,800	380		7,180	8,075
3110	119,000 BTU		2	8		7,450	430		7,880	8,850

23 82 29 – Radiators

23 82 29.10 Hydronic Heating

	Crew	Daily Output	Labor-Hours	Unit	Material	2015 Bare Costs Labor	2015 Bare Costs Equipment	Total	Total Incl O&P
0010	**HYDRONIC HEATING**, Terminal units, not incl. main supply pipe									
1000	Radiation									
1100	Panel, baseboard, C.I., including supports, no covers	Q-5	46	.348	L.F.	38	18.70		56.70	69.50
3000	Radiators, cast iron									
3100	Free standing or wall hung, 6 tube, 25" high	Q-5	96	.167	Section	44	8.95		52.95	62
3150	4 tube 25" high		96	.167		34	8.95		42.95	51
3200	4 tube, 19" high		96	.167		31.50	8.95		40.45	48
3250	Adj. brackets, 2 per wall radiator up to 30 sections	1 Stpi	32	.250	Ea.	53	14.95		67.95	81
3500	Recessed, 20" high x 5" deep, without grille	Q-5	60	.267	Section	37	14.35		51.35	62
3600	For inlet grille, add				"	2.58			2.58	2.84
9500	To convert SFR to BTU rating: Hot water, 150 x SFR									
9510	Forced hot water, 180 x SFR; steam, 240 x SFR									

23 82 33 – Convectors

23 82 33.10 Convector Units

	Crew	Daily Output	Labor-Hours	Unit	Material	2015 Bare Costs Labor	2015 Bare Costs Equipment	Total	Total Incl O&P
0010	**CONVECTOR UNITS**, Terminal units, not incl. main supply pipe									
2204	Convector, multifin, 2 pipe w/cabinet									
2210	17" H x 24" L	Q-5	10	1.600	Ea.	99	86		185	239
2214	17" H x 36" L		8.60	1.860		148	100		248	315

For customer support on your Plumbing Cost Data, call 877.709.6968.

355

23 82 Convection Heating and Cooling Units

23 82 33 – Convectors

23 82 33.10 Convector Units	Crew	Daily Output	Labor-Hours	Unit	Material	2015 Bare Costs Labor	Equipment	Total	Total Incl O&P
2218	17" H x 48" L	Q-5	7.40	2.162	Ea.	198	116		314	395
2222	21" H x 24" L		9	1.778		101	95.50		196.50	255
2228	21" H x 48" L		6.80	2.353		201	127		328	410
2240	For knob operated damper, add					140%				
2241	For metal trim strips, add	Q-5	64	.250	Ea.	12.90	13.45		26.35	34.50
2243	For snap-on inlet grille, add					10%	10%			
2245	For hinged access door, add	Q-5	64	.250	Ea.	35.50	13.45		48.95	59.50
2246	For air chamber, auto-venting, add	"	58	.276	"	7.75	14.85		22.60	31

23 82 36 – Finned-Tube Radiation Heaters

23 82 36.10 Finned Tube Radiation

		Crew	Daily Output	Labor-Hours	Unit	Material	Labor	Equipment	Total	Total Incl O&P
0010	**FINNED TUBE RADIATION**, Terminal units, not incl. main supply pipe									
1310	Baseboard, pkgd, 1/2" copper tube, alum. fin, 7" high	Q-5	60	.267	L.F.	11.20	14.35		25.55	34
1320	3/4" copper tube, alum. fin, 7" high	"	58	.276	"	7.35	14.85		22.20	30.50
1381	Rough in baseboard panel & fin tube, supply & balance valves	1 Stpi	1.06	7.547	Ea.	239	450		689	945
1500	Note: fin tube may also require corners, caps, etc.									

23 83 Radiant Heating Units

23 83 16 – Radiant-Heating Hydronic Piping

23 83 16.10 Radiant Floor Heating

		Crew	Daily Output	Labor-Hours	Unit	Material	Labor	Equipment	Total	Total Incl O&P
0010	**RADIANT FLOOR HEATING**									
0100	Tubing, PEX (cross-linked polyethylene)									
0110	Oxygen barrier type for systems with ferrous materials									
0120	1/2"	Q-5	800	.020	L.F.	.96	1.08		2.04	2.68
0130	3/4"		535	.030		1.35	1.61		2.96	3.92
0140	1"		400	.040		2.11	2.15		4.26	5.55
0200	Non barrier type for ferrous free systems									
0210	1/2"	Q-5	800	.020	L.F.	.61	1.08		1.69	2.29
0220	3/4"		535	.030		1.11	1.61		2.72	3.65
0230	1"		400	.040		1.90	2.15		4.05	5.35
1000	Manifolds									
1110	Brass									
1120	With supply and return valves, flow meter, thermometer,									
1122	auto air vent and drain/fill valve.									
1130	1", 2 circuit	Q-5	14	1.143	Ea.	257	61.50		318.50	375
1140	1", 3 circuit		13.50	1.185		294	63.50		357.50	420
1150	1", 4 circuit		13	1.231		320	66		386	450
1154	1", 5 circuit		12.50	1.280		365	69		434	510
1158	1", 6 circuit		12	1.333		415	71.50		486.50	565
1162	1", 7 circuit		11.50	1.391		455	75		530	615
1166	1", 8 circuit		11	1.455		500	78		578	670
1172	1", 9 circuit		10.50	1.524		540	82		622	720
1174	1", 10 circuit		10	1.600		580	86		666	770
1178	1", 11 circuit		9.50	1.684		605	90.50		695.50	800
1182	1", 12 circuit		9	1.778		670	95.50		765.50	880
1610	Copper manifold header, (cut to size)									
1620	1" header, 12 – 1/2" sweat outlets	Q-5	3.33	4.805	Ea.	91	258		349	490
1630	1-1/4" header, 12 – 1/2" sweat outlets		3.20	5		106	269		375	520
1640	1-1/4" header, 12 – 3/4" sweat outlets		3	5.333		114	287		401	560
1650	1-1/2" header, 12 – 3/4" sweat outlets		3.10	5.161		137	278		415	570
1660	2" header, 12 – 3/4" sweat outlets		2.90	5.517		201	297		498	670

23 83 16 – Radiant-Heating Hydronic Piping

23 83 16.10 Radiant Floor Heating	Crew	Daily Output	Labor-Hours	Unit	Material	2015 Bare Costs		Total	Total Incl O&P
						Labor	Equipment		
3000	Valves								
3110	Thermostatic zone valve actuator with end switch	Q-5	40	.400	Ea.	38.50	21.50		60	75
3114	Thermostatic zone valve actuator	"	36	.444	"	81	24		105	125
3120	Motorized straight zone valve with operator complete								
3130	3/4"	Q-5	35	.457	Ea.	130	24.50		154.50	180
3140	1"		32	.500		141	27		168	196
3150	1-1/4"		29.60	.541		179	29		208	241
3500	4 Way mixing valve, manual, brass								
3530	1"	Q-5	13.30	1.203	Ea.	180	64.50		244.50	296
3540	1-1/4"		11.40	1.404		195	75.50		270.50	330
3550	1-1/2"		11	1.455		249	78		327	390
3560	2"		10.60	1.509		350	81		431	515
3800	Mixing valve motor, 4 way for valves, 1" and 1-1/4"		34	.471		310	25.50		335.50	380
3810	Mixing valve motor, 4 way for valves, 1-1/2" and 2"		30	.533		355	28.50		383.50	435
5000	Radiant floor heating, zone control panel								
5120	4 Zone actuator valve control, expandable	Q-5	20	.800	Ea.	163	43		206	244
5130	6 Zone actuator valve control, expandable		18	.889		226	48		274	320
6070	Thermal track, straight panel for long continuous runs, 5.333 S.F.		40	.400		26.50	21.50		48	62
6080	Thermal track, utility panel, for direction reverse at run end, 5.333 S.F.		40	.400		26.50	21.50		48	62
6090	Combination panel, for direction reverse plus straight run, 5.333 S.F.		40	.400		26.50	21.50		48	62
7000	PEX tubing fittings								
7100	Compression type								
7116	Coupling								
7120	1/2" x 1/2"	1 Stpi	27	.296	Ea.	6.70	17.70		24.40	34
7124	3/4" x 3/4"	"	23	.348	"	10.60	21		31.60	43
7130	Adapter								
7132	1/2" x female sweat 1/2"	1 Stpi	27	.296	Ea.	4.36	17.70		22.06	31.50
7134	1/2" x female sweat 3/4"		26	.308		4.88	18.40		23.28	33.50
7136	5/8" x female sweat 3/4"		24	.333		7	19.90		26.90	37.50
7140	Elbow								
7142	1/2" x female sweat 1/2"	1 Stpi	27	.296	Ea.	6.65	17.70		24.35	34
7144	1/2" x female sweat 3/4"		26	.308		7.80	18.40		26.20	36.50
7146	5/8" x female sweat 3/4"		24	.333		8.75	19.90		28.65	39.50
7200	Insert type								
7206	PEX x male NPT								
7210	1/2" x 1/2"	1 Stpi	29	.276	Ea.	2.59	16.50		19.09	28
7220	3/4" x 3/4"		27	.296		3.81	17.70		21.51	30.50
7230	1" x 1"		26	.308		6.45	18.40		24.85	35
7300	PEX coupling								
7310	1/2" x 1/2"	1 Stpi	30	.267	Ea.	1.76	15.95		17.71	26
7320	3/4" x 3/4"		29	.276		2.15	16.50		18.65	27.50
7330	1" x 1"		28	.286		5.85	17.05		22.90	32.50
7400	PEX stainless crimp ring								
7410	1/2" x 1/2"	1 Stpi	86	.093	Ea.	.37	5.55		5.92	8.80
7420	3/4" x 3/4"		84	.095		.51	5.70		6.21	9.15
7430	1" x 1"		82	.098		.73	5.85		6.58	9.60

23 83 33 – Electric Radiant Heaters

23 83 33.10 Electric Heating

		Crew	Daily Output	Labor-Hours	Unit	Material	Labor	Equipment	Total	Total Incl O&P
0010	**ELECTRIC HEATING**, not incl. conduit or feed wiring									
1100	Rule of thumb: Baseboard units, including control	1 Elec	4.40	1.818	kW	104	99.50		203.50	263
1300	Baseboard heaters, 2' long, 350 watt		8	1	Ea.	28.50	54.50		83	114
1400	3' long, 750 watt		8	1		33	54.50		87.50	119

357

23 83 33 – Electric Radiant Heaters

23 83 33.10 Electric Heating	Crew	Daily Output	Labor-Hours	Unit	Material	2015 Bare Costs Labor	Equipment	Total	Total Incl O&P
1600	4' long, 1000 watt	1 Elec	6.70	1.194	Ea.	39	65.50		104.50	141
1800	5' long, 935 watt		5.70	1.404		48	77		125	168
2000	6' long, 1500 watt		5	1.600		54.50	87.50		142	191
2200	7' long, 1310 watt		4.40	1.818		63	99.50		162.50	219
2400	8' long, 2000 watt		4	2		66	109		175	237
2600	9' long, 1680 watt		3.60	2.222		102	122		224	294
2800	10' long, 1875 watt		3.30	2.424		190	133		323	410
2950	Wall heaters with fan, 120 to 277 volt									
3160	Recessed, residential, 750 watt	1 Elec	6	1.333	Ea.	134	73		207	256
3170	1000 watt		6	1.333		134	73		207	256
3180	1250 watt		5	1.600		134	87.50		221.50	278
3190	1500 watt		4	2		134	109		243	310
3600	Thermostats, integral		16	.500		30	27.50		57.50	74
3800	Line voltage, 1 pole		8	1		19.20	54.50		73.70	103
3810	2 pole		8	1		29.50	54.50		84	115
4000	Heat trace system, 400 degree									
4020	115 V, 2.5 watts per L.F.	1 Elec	530	.015	L.F.	7.90	.83		8.73	9.90
4030	5 watts per L.F.		530	.015		7.90	.83		8.73	9.90
4050	10 watts per L.F.		530	.015		7.90	.83		8.73	9.90
4060	208 V, 5 watts per L.F.		530	.015		7.90	.83		8.73	9.90
4080	480 V, 8 watts per L.F.		530	.015		7.90	.83		8.73	9.90
4200	Heater raceway									
4260	Heat transfer cement									
4280	1 gallon				Ea.	60			60	66
4300	5 gallon				"	234			234	257
4320	Snap band, clamp									
4340	3/4" pipe size	1 Elec	470	.017	Ea.		.93		.93	1.39
4360	1" pipe size		444	.018			.99		.99	1.48
4380	1-1/4" pipe size		400	.020			1.09		1.09	1.64
4400	1-1/2" pipe size		355	.023			1.23		1.23	1.85
4420	2" pipe size		320	.025			1.37		1.37	2.05
4440	3" pipe size		160	.050			2.74		2.74	4.10
4460	4" pipe size		100	.080			4.38		4.38	6.55
4480	Thermostat NEMA 3R, 22 amp, 0-150 Deg, 10' cap.		8	1		241	54.50		295.50	345
4500	Thermostat NEMA 4X, 25 amp, 40 Deg, 5-1/2' cap.		7	1.143		241	62.50		303.50	360
4520	Thermostat NEMA 4X, 22 amp, 25-325 Deg, 10' cap.		7	1.143		650	62.50		712.50	810
4540	Thermostat NEMA 4X, 22 amp, 15-140 Deg,		6	1.333		545	73		618	710
4580	Thermostat NEMA 4,7,9, 22 amp, 25-325 Deg, 10' cap.		3.60	2.222		800	122		922	1,050
4600	Thermostat NEMA 4,7,9, 22 amp, 15-140 Deg,		3	2.667		795	146		941	1,100
4720	Fiberglass application tape, 36 yard roll		11	.727		77.50	40		117.50	145
5000	Radiant heating ceiling panels, 2' x 4', 500 watt		16	.500		305	27.50		332.50	375
5050	750 watt		16	.500		310	27.50		337.50	385
5200	For recessed plaster frame, add		32	.250		111	13.70		124.70	143
5300	Infrared quartz heaters, 120 volts, 1000 watts		6.70	1.194		284	65.50		349.50	410
5350	1500 watt		5	1.600		284	87.50		371.50	440
5400	240 volts, 1500 watt		5	1.600		284	87.50		371.50	440
5450	2000 watt		4	2		284	109		393	475
5500	3000 watt		3	2.667		325	146		471	575
5550	4000 watt		2.60	3.077		325	168		493	605
5570	Modulating control		.80	10		146	545		691	980
5600	Unit heaters, heavy duty, with fan & mounting bracket									
5650	Single phase, 208-240-277 volt, 3 kW	1 Elec	6	1.333	Ea.	445	73		518	600
5750	5 kW		5.50	1.455		460	79.50		539.50	625

23 83 33 – Electric Radiant Heaters

23 83 33.10 Electric Heating		Crew	Daily Output	Labor-Hours	Unit	Material	2015 Bare Costs Labor	Equipment	Total	Total Incl O&P
5800	7 kW	1 Elec	5	1.600	Ea.	710	87.50		797.50	910
5850	10 kW		4	2		810	109		919	1,050
5950	15 kW		3.80	2.105		1,300	115		1,415	1,600
6000	480 volt, 3 kW		6	1.333		425	73		498	575
6020	4 kW		5.80	1.379		445	75.50		520.50	605
6040	5 kW		5.50	1.455		460	79.50		539.50	625
6060	7 kW		5	1.600		710	87.50		797.50	910
6080	10 kW		4	2		780	109		889	1,025
6100	13 kW		3.80	2.105		1,300	115		1,415	1,600
6120	15 kW		3.70	2.162		1,300	118		1,418	1,600
6140	20 kW		3.50	2.286		1,700	125		1,825	2,025
6300	3 phase, 208-240 volt, 5 kW		5.50	1.455		435	79.50		514.50	600
6320	7 kW		5	1.600		670	87.50		757.50	865
6340	10 kW		4	2		720	109		829	960
6360	15 kW		3.70	2.162		1,200	118		1,318	1,500
6380	20 kW		3.50	2.286		1,725	125		1,850	2,075
6400	25 kW		3.30	2.424		2,025	133		2,158	2,425
6500	480 volt, 5 kW		5.50	1.455		605	79.50		684.50	790
6520	7 kW		5	1.600		765	87.50		852.50	970
6540	10 kW		4	2		810	109		919	1,050
6560	13 kW		3.80	2.105		1,300	115		1,415	1,600
6580	15 kW		3.70	2.162		1,700	118		1,818	2,025
6600	20 kW		3.50	2.286		1,700	125		1,825	2,025
6800	Vertical discharge heaters, with fan									
6820	Single phase, 208-240-277 volt, 10 kW	1 Elec	4	2	Ea.	750	109		859	990
6840	15 kW		3.70	2.162		1,250	118		1,368	1,550
6900	3 phase, 208-240 volt, 10 kW		4	2		720	109		829	960
6920	15 kW		3.70	2.162		1,200	118		1,318	1,500
6940	20 kW		3.50	2.286		1,725	125		1,850	2,075
7100	480 volt, 10 kW		4	2		810	109		919	1,050
7120	15 kW		3.70	2.162		1,300	118		1,418	1,600
7140	20 kW		3.50	2.286		1,700	125		1,825	2,025
7160	25 kW		3.30	2.424		2,025	133		2,158	2,425
7900	Cabinet convector heaters, 240 volt, three phase,									
7920	3' long, 2000 watt	1 Elec	5.30	1.509	Ea.	1,900	82.50		1,982.50	2,225
7940	3000 watt		5.30	1.509		2,000	82.50		2,082.50	2,300
7960	4000 watt		5.30	1.509		2,050	82.50		2,132.50	2,375
7980	6000 watt		4.60	1.739		2,125	95		2,220	2,475
8000	8000 watt		4.60	1.739		2,200	95		2,295	2,575
8020	4' long, 4000 watt		4.60	1.739		2,075	95		2,170	2,425
8040	6000 watt		4	2		2,150	109		2,259	2,550
8060	8000 watt		4	2		2,225	109		2,334	2,625
8080	10,000 watt		4	2		2,250	109		2,359	2,650
8100	Available also in 208 or 277 volt									
8200	Cabinet unit heaters, 120 to 277 volt, 1 pole,									
8220	wall mounted, 2 kW	1 Elec	4.60	1.739	Ea.	1,875	95		1,970	2,200
8230	3 kW		4.60	1.739		1,950	95		2,045	2,300
8240	4 kW		4.40	1.818		2,000	99.50		2,099.50	2,350
8250	5 kW		4.40	1.818		2,075	99.50		2,174.50	2,450
8260	6 kW		4.20	1.905		2,125	104		2,229	2,475
8270	8 kW		4	2		2,150	109		2,259	2,550
8280	10 kW		3.80	2.105		2,225	115		2,340	2,600
8290	12 kW		3.50	2.286		2,250	125		2,375	2,650

23 83 33.10 Electric Heating	Crew	Daily Output	Labor-Hours	Unit	Material	2015 Bare Costs Labor	Equipment	Total	Total Incl O&P
8300	13.5 kW	1 Elec	2.90	2.759	Ea.	2,700	151		2,851	3,200
8310	16 kW		2.70	2.963		2,725	162		2,887	3,250
8320	20 kW		2.30	3.478		3,475	190		3,665	4,100
8330	24 kW		1.90	4.211		3,525	230		3,755	4,225
8350	Recessed, 2 kW		4.40	1.818		1,875	99.50		1,974.50	2,200
8370	3 kW		4.40	1.818		1,950	99.50		2,049.50	2,300
8380	4 kW		4.20	1.905		2,000	104		2,104	2,350
8390	5 kW		4.20	1.905		2,075	104		2,179	2,425
8400	6 kW		4	2		2,100	109		2,209	2,500
8410	8 kW		3.80	2.105		2,200	115		2,315	2,600
8420	10 kW		3.50	2.286		2,675	125		2,800	3,125
8430	12 kW		2.90	2.759		2,675	151		2,826	3,175
8440	13.5 kW		2.70	2.963		2,700	162		2,862	3,225
8450	16 kW		2.30	3.478		2,875	190		3,065	3,450
8460	20 kW		1.90	4.211		3,275	230		3,505	3,950
8470	24 kW		1.60	5		3,325	274		3,599	4,050
8490	Ceiling mounted, 2 kW		3.20	2.500		1,875	137		2,012	2,250
8510	3 kW		3.20	2.500		1,950	137		2,087	2,350
8520	4 kW		3	2.667		2,000	146		2,146	2,425
8530	5 kW		3	2.667		2,075	146		2,221	2,500
8540	6 kW		2.80	2.857		2,075	156		2,231	2,525
8550	8 kW		2.40	3.333		2,150	182		2,332	2,650
8560	10 kW		2.20	3.636		2,200	199		2,399	2,725
8570	12 kW		2	4		2,250	219		2,469	2,800
8580	13.5 kW		1.50	5.333		2,275	292		2,567	2,950
8590	16 kW		1.30	6.154		2,350	335		2,685	3,075
8600	20 kW		.90	8.889		3,275	485		3,760	4,325
8610	24 kW		.60	13.333		3,325	730		4,055	4,750
8630	208 to 480 V, 3 pole									
8650	Wall mounted, 2 kW	1 Elec	4.60	1.739	Ea.	1,975	95		2,070	2,325
8670	3 kW		4.60	1.739		2,075	95		2,170	2,425
8680	4 kW		4.40	1.818		2,125	99.50		2,224.50	2,475
8690	5 kW		4.40	1.818		2,175	99.50		2,274.50	2,550
8700	6 kW		4.20	1.905		2,200	104		2,304	2,575
8710	8 kW		4	2		2,275	109		2,384	2,675
8720	10 kW		3.80	2.105		2,325	115		2,440	2,725
8730	12 kW		3.50	2.286		2,375	125		2,500	2,775
8740	13.5 kW		2.90	2.759		2,400	151		2,551	2,875
8750	16 kW		2.70	2.963		2,450	162		2,612	2,950
8760	20 kW		2.30	3.478		3,400	190		3,590	4,000
8770	24 kW		1.90	4.211		3,600	230		3,830	4,300
8790	Recessed, 2 kW		4.40	1.818		1,975	99.50		2,074.50	2,325
8810	3 kW		4.40	1.818		2,000	99.50		2,099.50	2,350
8820	4 kW		4.20	1.905		2,425	104		2,529	2,800
8830	5 kW		4.20	1.905		2,100	104		2,204	2,475
8840	6 kW		4	2		2,125	109		2,234	2,500
8850	8 kW		3.80	2.105		2,200	115		2,315	2,600
8860	10 kW		3.50	2.286		2,350	125		2,475	2,750
8870	12 kW		2.90	2.759		2,450	151		2,601	2,925
8880	13.5 kW		2.70	2.963		2,500	162		2,662	3,000
8890	16 kW		2.30	3.478		3,625	190		3,815	4,250
8900	20 kW		1.90	4.211		3,675	230		3,905	4,400
8920	24 kW		1.60	5		3,725	274		3,999	4,500

For customer support on your Plumbing Cost Data, call 877.709.6968.

23 83 Radiant Heating Units

23 83 33 – Electric Radiant Heaters

23 83 33.10 Electric Heating		Crew	Daily Output	Labor-Hours	Unit	Material	2015 Bare Costs Labor	Equipment	Total	Total Incl O&P
8940	Ceiling mount, 2 kW	1 Elec	3.20	2.500	Ea.	2,075	137		2,212	2,475
8950	3 kW		3.20	2.500		2,200	137		2,337	2,625
8960	4 kW		3	2.667		2,200	146		2,346	2,650
8970	5 kW		3	2.667		2,275	146		2,421	2,725
8980	6 kW		2.80	2.857		2,300	156		2,456	2,750
8990	8 kW		2.40	3.333		2,375	182		2,557	2,875
9000	10 kW		2.20	3.636		2,425	199		2,624	2,950
9020	13.5 kW		1.50	5.333		2,500	292		2,792	3,175
9030	16 kW		1.30	6.154		3,625	335		3,960	4,475
9040	20 kW		.90	8.889		3,675	485		4,160	4,775
9060	24 kW		.60	13.333		4,075	730		4,805	5,600

23 91 Prefabricated Equipment Supports

23 91 10 – Prefabricated Curbs, Pads and Stands

23 91 10.10 Prefabricated Pads and Stands		Crew	Daily Output	Labor-Hours	Unit	Material	2015 Bare Costs Labor	Equipment	Total	Total Incl O&P
0010	**PREFABRICATED PADS AND STANDS**									
6000	Pad, fiberglass reinforced concrete with polystyrene foam core									
6050	Condenser, 2" thick, 20" x 38"	1 Shee	8	1	Ea.	27.50	56		83.50	116
6220	30" x 36"	"	8	1		43.50	56		99.50	134
6340	36" x 54"	Q-9	6	2.667		81.50	134		215.50	295

Division Notes

	CREW	DAILY OUTPUT	LABOR-HOURS	UNIT	BARE COSTS				TOTAL INCL O&P
					MAT.	LABOR	EQUIP.	TOTAL	

Estimating Tips

26 05 00 Common Work Results for Electrical

- Conduit should be taken off in three main categories—power distribution, branch power, and branch lighting—so the estimator can concentrate on systems and components, therefore making it easier to ensure all items have been accounted for.

- For cost modifications for elevated conduit installation, add the percentages to labor according to the height of installation, and only to the quantities exceeding the different height levels, not to the total conduit quantities.

- Remember that aluminum wiring of equal ampacity is larger in diameter than copper and may require larger conduit.

- If more than three wires at a time are being pulled, deduct percentages from the labor hours of that grouping of wires.

- When taking off grounding systems, identify separately the type and size of wire, and list each unique type of ground connection.

- The estimator should take the weights of materials into consideration when completing a takeoff. Topics to consider include: How will the materials be supported? What methods of support are available? How high will the support structure have to reach? Will the final support structure be able to withstand the total burden? Is the support material included or separate from the fixture, equipment, and material specified?

- Do not overlook the costs for equipment used in the installation. If scaffolding or highlifts are available in the field, contractors may use them in lieu of the proposed ladders and rolling staging.

26 20 00 Low-Voltage Electrical Transmission

- Supports and concrete pads may be shown on drawings for the larger equipment, or the support system may be only a piece of plywood for the back of a panelboard. In either case, it must be included in the costs.

26 40 00 Electrical and Cathodic Protection

- When taking off cathodic protections systems, identify the type and size of cable, and list each unique type of anode connection.

26 50 00 Lighting

- Fixtures should be taken off room by room, using the fixture schedule, specifications, and the ceiling plan. For large concentrations of lighting fixtures in the same area, deduct the percentages from labor hours.

Reference Numbers

Reference numbers are shown in shaded boxes at the beginning of some major classifications. These numbers refer to related items in the Reference Section. The reference information may be an estimating procedure, an alternate pricing method, or technical information.

Note: Not all subdivisions listed here necessarily appear in this publication. ∎

*Note: **Trade Service**, in part, has been used as a reference source for some of the material prices used in Division 26.*

26 05 33.95 Cutting and Drilling	Crew	Daily Output	Labor-Hours	Unit	Material	2015 Bare Costs Labor	Equipment	Total	Total Incl O&P
0010 **CUTTING AND DRILLING**									
0100 Hole drilling to 10' high, concrete wall									
0110 8" thick, 1/2" pipe size	R-31	12	.667	Ea.	.24	36.50	4.59	41.33	60
0120 3/4" pipe size		12	.667		.24	36.50	4.59	41.33	60
0130 1" pipe size		9.50	.842		.38	46	5.80	52.18	76
0140 1-1/4" pipe size		9.50	.842		.38	46	5.80	52.18	76
0150 1-1/2" pipe size		9.50	.842		.38	46	5.80	52.18	76
0160 2" pipe size		4.40	1.818		.51	99.50	12.55	112.56	163
0170 2-1/2" pipe size		4.40	1.818		.51	99.50	12.55	112.56	163
0180 3" pipe size		4.40	1.818		.51	99.50	12.55	112.56	163
0190 3-1/2" pipe size		3.30	2.424		.66	133	16.70	150.36	218
0200 4" pipe size		3.30	2.424		.66	133	16.70	150.36	218
0500 12" thick, 1/2" pipe size		9.40	.851		.35	46.50	5.85	52.70	76.50
0520 3/4" pipe size		9.40	.851		.35	46.50	5.85	52.70	76.50
0540 1" pipe size		7.30	1.096		.56	60	7.55	68.11	99
0560 1-1/4" pipe size		7.30	1.096		.56	60	7.55	68.11	99
0570 1-1/2" pipe size		7.30	1.096		.56	60	7.55	68.11	99
0580 2" pipe size		3.60	2.222		.76	122	15.30	138.06	200
0590 2-1/2" pipe size		3.60	2.222		.76	122	15.30	138.06	200
0600 3" pipe size		3.60	2.222		.76	122	15.30	138.06	200
0610 3-1/2" pipe size		2.80	2.857		.99	156	19.70	176.69	257
0630 4" pipe size		2.50	3.200		.99	175	22	197.99	288
0650 16" thick, 1/2" pipe size		7.60	1.053		.47	57.50	7.25	65.22	95
0670 3/4" pipe size		7	1.143		.47	62.50	7.85	70.82	103
0690 1" pipe size		6	1.333		.75	73	9.20	82.95	120
0710 1-1/4" pipe size		5.50	1.455		.75	79.50	10	90.25	131
0730 1-1/2" pipe size		5.50	1.455		.75	79.50	10	90.25	131
0750 2" pipe size		3	2.667		1.02	146	18.35	165.37	240
0770 2-1/2" pipe size		2.70	2.963		1.02	162	20.50	183.52	267
0790 3" pipe size		2.50	3.200		1.02	175	22	198.02	288
0810 3-1/2" pipe size		2.30	3.478		1.32	190	24	215.32	315
0830 4" pipe size		2	4		1.32	219	27.50	247.82	360
0850 20" thick, 1/2" pipe size		6.40	1.250		.59	68.50	8.60	77.69	112
0870 3/4" pipe size		6	1.333		.59	73	9.20	82.79	120
0890 1" pipe size		5	1.600		.94	87.50	11	99.44	144
0910 1-1/4" pipe size		4.80	1.667		.94	91	11.50	103.44	151
0930 1-1/2" pipe size		4.60	1.739		.94	95	12	107.94	157
0950 2" pipe size		2.70	2.963		1.27	162	20.50	183.77	267
0970 2-1/2" pipe size		2.40	3.333		1.27	182	23	206.27	300
0990 3" pipe size		2.20	3.636		1.27	199	25	225.27	325
1010 3-1/2" pipe size		2	4		1.64	219	27.50	248.14	360
1030 4" pipe size		1.70	4.706		1.64	257	32.50	291.14	420
1050 24" thick, 1/2" pipe size		5.50	1.455		.71	79.50	10	90.21	131
1070 3/4" pipe size		5.10	1.569		.71	86	10.80	97.51	142
1090 1" pipe size		4.30	1.860		1.13	102	12.80	115.93	167
1110 1-1/4" pipe size		4	2		1.13	109	13.80	123.93	180
1130 1-1/2" pipe size		4	2		1.13	109	13.80	123.93	180
1150 2" pipe size		2.40	3.333		1.52	182	23	206.52	300
1170 2-1/2" pipe size		2.20	3.636		1.52	199	25	225.52	325
1190 3" pipe size		2	4		1.52	219	27.50	248.02	360
1210 3-1/2" pipe size		1.80	4.444		1.97	243	30.50	275.47	400
1230 4" pipe size		1.50	5.333		1.97	292	37	330.97	480
1500 Brick wall, 8" thick, 1/2" pipe size		18	.444		.24	24.50	3.06	27.80	40

26 05 33 – Raceway and Boxes for Electrical Systems

26 05 33.95 Cutting and Drilling	Crew	Daily Output	Labor-Hours	Unit	Material	2015 Bare Costs Labor	Equipment	Total	Total Incl O&P
1520	3/4" pipe size	R-31	18	.444	Ea.	.24	24.50	3.06	27.80	40
1540	1" pipe size		13.30	.602		.38	33	4.14	37.52	54.50
1560	1-1/4" pipe size		13.30	.602		.38	33	4.14	37.52	54.50
1580	1-1/2" pipe size		13.30	.602		.38	33	4.14	37.52	54.50
1600	2" pipe size		5.70	1.404		.51	77	9.65	87.16	126
1620	2-1/2" pipe size		5.70	1.404		.51	77	9.65	87.16	126
1640	3" pipe size		5.70	1.404		.51	77	9.65	87.16	126
1660	3-1/2" pipe size		4.40	1.818		.66	99.50	12.55	112.71	164
1680	4" pipe size		4	2		.66	109	13.80	123.46	180
1700	12" thick, 1/2" pipe size		14.50	.552		.35	30	3.80	34.15	49.50
1720	3/4" pipe size		14.50	.552		.35	30	3.80	34.15	49.50
1740	1" pipe size		11	.727		.56	40	5	45.56	65.50
1760	1-1/4" pipe size		11	.727		.56	40	5	45.56	65.50
1780	1-1/2" pipe size		11	.727		.56	40	5	45.56	65.50
1800	2" pipe size		5	1.600		.76	87.50	11	99.26	144
1820	2-1/2" pipe size		5	1.600		.76	87.50	11	99.26	144
1840	3" pipe size		5	1.600		.76	87.50	11	99.26	144
1860	3-1/2" pipe size		3.80	2.105		.99	115	14.50	130.49	190
1880	4" pipe size		3.30	2.424		.99	133	16.70	150.69	218
1900	16" thick, 1/2" pipe size		12.30	.650		.47	35.50	4.48	40.45	59
1920	3/4" pipe size		12.30	.650		.47	35.50	4.48	40.45	59
1940	1" pipe size		9.30	.860		.75	47	5.95	53.70	78
1960	1-1/4" pipe size		9.30	.860		.75	47	5.95	53.70	78
1980	1-1/2" pipe size		9.30	.860		.75	47	5.95	53.70	78
2000	2" pipe size		4.40	1.818		1.02	99.50	12.55	113.07	164
2010	2-1/2" pipe size		4.40	1.818		1.02	99.50	12.55	113.07	164
2030	3" pipe size		4.40	1.818		1.02	99.50	12.55	113.07	164
2050	3-1/2" pipe size		3.30	2.424		1.32	133	16.70	151.02	219
2070	4" pipe size		3	2.667		1.32	146	18.35	165.67	240
2090	20" thick, 1/2" pipe size		10.70	.748		.59	41	5.15	46.74	68
2110	3/4" pipe size		10.70	.748		.59	41	5.15	46.74	68
2130	1" pipe size		8	1		.94	54.50	6.90	62.34	90.50
2150	1-1/4" pipe size		8	1		.94	54.50	6.90	62.34	90.50
2170	1-1/2" pipe size		8	1		.94	54.50	6.90	62.34	90.50
2190	2" pipe size		4	2		1.27	109	13.80	124.07	181
2210	2-1/2" pipe size		4	2		1.27	109	13.80	124.07	181
2230	3" pipe size		4	2		1.27	109	13.80	124.07	181
2250	3-1/2" pipe size		3	2.667		1.64	146	18.35	165.99	241
2270	4" pipe size		2.70	2.963		1.64	162	20.50	184.14	267
2290	24" thick, 1/2" pipe size		9.40	.851		.71	46.50	5.85	53.06	76.50
2310	3/4" pipe size		9.40	.851		.71	46.50	5.85	53.06	76.50
2330	1" pipe size		7.10	1.127		1.13	61.50	7.75	70.38	102
2350	1-1/4" pipe size		7.10	1.127		1.13	61.50	7.75	70.38	102
2370	1-1/2" pipe size		7.10	1.127		1.13	61.50	7.75	70.38	102
2390	2" pipe size		3.60	2.222		1.52	122	15.30	138.82	201
2410	2-1/2" pipe size		3.60	2.222		1.52	122	15.30	138.82	201
2430	3" pipe size		3.60	2.222		1.52	122	15.30	138.82	201
2450	3-1/2" pipe size		2.80	2.857		1.97	156	19.70	177.67	258
2470	4" pipe size		2.50	3.200		1.97	175	22	198.97	289
3000	Knockouts to 8' high, metal boxes & enclosures									
3020	With hole saw, 1/2" pipe size	1 Elec	53	.151	Ea.		8.25		8.25	12.35
3040	3/4" pipe size		47	.170			9.30		9.30	13.95
3050	1" pipe size		40	.200			10.95		10.95	16.40

For customer support on your Plumbing Cost Data, call 877.709.6968.

365

26 05 33 – Raceway and Boxes for Electrical Systems

26 05 33.95 Cutting and Drilling		Crew	Daily Output	Labor-Hours	Unit	Material	2015 Bare Costs Labor	Equipment	Total	Total Incl O&P
3060	1-1/4" pipe size	1 Elec	36	.222	Ea.		12.15		12.15	18.20
3070	1-1/2" pipe size		32	.250			13.70		13.70	20.50
3080	2" pipe size		27	.296			16.20		16.20	24.50
3090	2-1/2" pipe size		20	.400			22		22	33
4010	3" pipe size		16	.500			27.50		27.50	41
4030	3-1/2" pipe size		13	.615			33.50		33.50	50.50
4050	4" pipe size		11	.727			40		40	59.50

26 05 80 – Wiring Connections

26 05 80.10 Motor Connections

		Crew	Daily Output	Labor-Hours	Unit	Material	2015 Bare Costs Labor	Equipment	Total	Total Incl O&P
0010	**MOTOR CONNECTIONS**									
0020	Flexible conduit and fittings, 115 volt, 1 phase, up to 1 HP motor	1 Elec	8	1	Ea.	6.35	54.50		60.85	89
0050	2 HP motor		6.50	1.231		12.55	67.50		80.05	115
0100	3 HP motor		5.50	1.455		11.05	79.50		90.55	131
0110	230 volt, 3 phase, 3 HP motor		6.78	1.180		7.65	64.50		72.15	105
0112	5 HP motor		5.47	1.463		6.50	80		86.50	127
0114	7-1/2 HP motor		4.61	1.735		9.55	95		104.55	153
0120	10 HP motor		4.20	1.905		21	104		125	180
0150	15 HP motor		3.30	2.424		21	133		154	223
0200	25 HP motor		2.70	2.963		29	162		191	275
0400	50 HP motor		2.20	3.636		59	199		258	365
0600	100 HP motor		1.50	5.333		140	292		432	590

26 05 90 – Residential Applications

26 05 90.10 Residential Wiring

		Crew	Daily Output	Labor-Hours	Unit	Material	2015 Bare Costs Labor	Equipment	Total	Total Incl O&P
0010	**RESIDENTIAL WIRING**									
0020	20' avg. runs and #14/2 wiring incl. unless otherwise noted									
1000	Service & panel, includes 24' SE-AL cable, service eye, meter,									
1010	Socket, panel board, main bkr., ground rod, 15 or 20 amp									
1020	1-pole circuit breakers, and misc. hardware									
1100	100 amp, with 10 branch breakers	1 Elec	1.19	6.723	Ea.	560	370		930	1,175
1110	With PVC conduit and wire		.92	8.696		605	475		1,080	1,375
1120	With RGS conduit and wire		.73	10.959		765	600		1,365	1,750
1150	150 amp, with 14 branch breakers		1.03	7.767		860	425		1,285	1,575
1170	With PVC conduit and wire		.82	9.756		955	535		1,490	1,850
1180	With RGS conduit and wire		.67	11.940		1,275	655		1,930	2,375
1200	200 amp, with 18 branch breakers	2 Elec	1.80	8.889		1,150	485		1,635	2,000
1220	With PVC conduit and wire		1.46	10.959		1,250	600		1,850	2,275
1230	With RGS conduit and wire		1.24	12.903		1,650	705		2,355	2,850
1800	Lightning surge suppressor	1 Elec	32	.250		50.50	13.70		64.20	76.50
2000	Switch devices									
2100	Single pole, 15 amp, Ivory, with a 1-gang box, cover plate,									
2110	Type NM (Romex) cable	1 Elec	17.10	.468	Ea.	12.35	25.50		37.85	52
2120	Type MC (BX) cable		14.30	.559		21.50	30.50		52	69.50
2130	EMT & wire		5.71	1.401		30	76.50		106.50	148
2150	3-way, #14/3, type NM cable		14.55	.550		14.85	30		44.85	61.50
2170	Type MC cable		12.31	.650		27	35.50		62.50	83
2180	EMT & wire		5	1.600		32	87.50		119.50	166
2200	4-way, #14/3, type NM cable		14.55	.550		21	30		51	68
2220	Type MC cable		12.31	.650		33	35.50		68.50	90
2230	EMT & wire		5	1.600		38.50	87.50		126	173
2250	S.P., 20 amp, #12/2, type NM cable		13.33	.600		22	33		55	73
2270	Type MC cable		11.43	.700		28.50	38.50		67	89
2280	EMT & wire		4.85	1.649		41	90		131	180

26 05 90 – Residential Applications

26 05 90.10 Residential Wiring	Crew	Daily Output	Labor-Hours	Unit	Material	2015 Bare Costs Labor	Equipment	Total	Total Incl O&P
2290	S.P. rotary dimmer, 600W, no wiring	1 Elec	17	.471	Ea.	29.50	25.50		55	71
2300	S.P. rotary dimmer, 600W, type NM cable		14.55	.550		34.50	30		64.50	83
2320	Type MC cable		12.31	.650		43.50	35.50		79	102
2330	EMT & wire		5	1.600		53.50	87.50		141	190
2350	3-way rotary dimmer, type NM cable		13.33	.600		28.50	33		61.50	80.50
2370	Type MC cable		11.43	.700		37.50	38.50		76	98.50
2380	EMT & wire		4.85	1.649		47.50	90		137.50	188
2400	Interval timer wall switch, 20 amp, 1-30 min., #12/2									
2410	Type NM cable	1 Elec	14.55	.550	Ea.	56	30		86	107
2420	Type MC cable		12.31	.650		60	35.50		95.50	120
2430	EMT & wire		5	1.600		75	87.50		162.50	214
2500	Decorator style									
2510	S.P., 15 amp, type NM cable	1 Elec	17.10	.468	Ea.	16.25	25.50		41.75	56.50
2520	Type MC cable		14.30	.559		25	30.50		55.50	73.50
2530	EMT & wire		5.71	1.401		34	76.50		110.50	152
2550	3-way, #14/3, type NM cable		14.55	.550		18.75	30		48.75	65.50
2570	Type MC cable		12.31	.650		30.50	35.50		66	87
2580	EMT & wire		5	1.600		36	87.50		123.50	171
2600	4-way, #14/3, type NM cable		14.55	.550		25	30		55	72.50
2620	Type MC cable		12.31	.650		37	35.50		72.50	94
2630	EMT & wire		5	1.600		42	87.50		129.50	178
2650	S.P., 20 amp, #12/2, type NM cable		13.33	.600		26	33		59	77.50
2670	Type MC cable		11.43	.700		32.50	38.50		71	93
2680	EMT & wire		4.85	1.649		45	90		135	185
2700	S.P., slide dimmer, type NM cable		17.10	.468		28.50	25.50		54	70
2720	Type MC cable		14.30	.559		37.50	30.50		68	87.50
2730	EMT & wire		5.71	1.401		47.50	76.50		124	168
2750	S.P., touch dimmer, type NM cable		17.10	.468		32	25.50		57.50	74
2770	Type MC cable		14.30	.559		41	30.50		71.50	91.50
2780	EMT & wire		5.71	1.401		51.50	76.50		128	172
2800	3-way touch dimmer, type NM cable		13.33	.600		52	33		85	106
2820	Type MC cable		11.43	.700		60.50	38.50		99	125
2830	EMT & wire		4.85	1.649		71	90		161	213
3000	Combination devices									
3100	S.P. switch/15 amp recpt., Ivory, 1-gang box, plate									
3110	Type NM cable	1 Elec	11.43	.700	Ea.	24.50	38.50		63	84.50
3120	Type MC cable		10	.800		33.50	44		77.50	102
3130	EMT & wire		4.40	1.818		43.50	99.50		143	197
3150	S.P. switch/pilot light, type NM cable		11.43	.700		24.50	38.50		63	84.50
3170	Type MC cable		10	.800		33.50	44		77.50	103
3180	EMT & wire		4.43	1.806		43.50	99		142.50	196
3190	2-S.P. switches, 2-#14/2, no wiring		14	.571		6.90	31.50		38.40	54.50
3200	2-S.P. switches, 2-#14/2, type NM cables		10	.800		27.50	44		71.50	95.50
3220	Type MC cable		8.89	.900		41	49		90	119
3230	EMT & wire		4.10	1.951		46.50	107		153.50	211
3250	3-way switch/15 amp recpt., #14/3, type NM cable		10	.800		32	44		76	101
3270	Type MC cable		8.89	.900		44	49		93	122
3280	EMT & wire		4.10	1.951		49	107		156	214
3300	2-3 way switches, 2-#14/3, type NM cables		8.89	.900		41.50	49		90.50	120
3320	Type MC cable		8	1		60.50	54.50		115	149
3330	EMT & wire		4	2		56.50	109		165.50	226
3350	S.P. switch/20 amp recpt., #12/2, type NM cable		10	.800		32	44		76	101
3370	Type MC cable		8.89	.900		36	49		85	114

For customer support on your Plumbing Cost Data, call 877.709.6968.

367

26 05 Common Work Results for Electrical

26 05 90 – Residential Applications

26 05 90.10 Residential Wiring	Crew	Daily Output	Labor-Hours	Unit	Material	2015 Bare Costs Labor	Equipment	Total	Total Incl O&P
3380	EMT & wire	1 Elec	4.10	1.951	Ea.	51	107		158	216
3400	Decorator style									
3410	S.P. switch/15 amp recpt., type NM cable	1 Elec	11.43	.700	Ea.	28.50	38.50		67	88.50
3420	Type MC cable		10	.800		37	44		81	107
3430	EMT & wire		4.40	1.818		47.50	99.50		147	201
3450	S.P. switch/pilot light, type NM cable		11.43	.700		28.50	38.50		67	88.50
3470	Type MC cable		10	.800		37.50	44		81.50	107
3480	EMT & wire		4.40	1.818		47.50	99.50		147	202
3500	2-S.P. switches, 2-#14/2, type NM cables		10	.800		31.50	44		75.50	100
3520	Type MC cable		8.89	.900		44.50	49		93.50	123
3530	EMT & wire		4.10	1.951		50.50	107		157.50	216
3550	3-way/15 amp recpt., #14/3, type NM cable		10	.800		36	44		80	105
3570	Type MC cable		8.89	.900		47.50	49		96.50	127
3580	EMT & wire		4.10	1.951		53	107		160	218
3650	2-3 way switches, 2-#14/3, type NM cables		8.89	.900		45.50	49		94.50	124
3670	Type MC cable		8	1		64.50	54.50		119	153
3680	EMT & wire		4	2		60.50	109		169.50	231
3700	S.P. switch/20 amp recpt., #12/2, type NM cable		10	.800		35.50	44		79.50	105
3720	Type MC cable		8.89	.900		39.50	49		88.50	118
3730	EMT & wire		4.10	1.951		55	107		162	221
4000	Receptacle devices									
4010	Duplex outlet, 15 amp recpt., Ivory, 1-gang box, plate									
4015	Type NM cable	1 Elec	14.55	.550	Ea.	10.80	30		40.80	57
4020	Type MC cable		12.31	.650		19.75	35.50		55.25	75
4030	EMT & wire		5.33	1.501		28.50	82		110.50	155
4050	With #12/2, type NM cable		12.31	.650		13.30	35.50		48.80	68
4070	Type MC cable		10.67	.750		19.85	41		60.85	83.50
4080	EMT & wire		4.71	1.699		32.50	93		125.50	175
4100	20 amp recpt., #12/2, type NM cable		12.31	.650		19.95	35.50		55.45	75.50
4120	Type MC cable		10.67	.750		26.50	41		67.50	90.50
4130	EMT & wire		4.71	1.699		39	93		132	182
4140	For GFI see Section 26 05 90.10 line 4300 below									
4150	Decorator style, 15 amp recpt., type NM cable	1 Elec	14.55	.550	Ea.	14.70	30		44.70	61
4170	Type MC cable		12.31	.650		23.50	35.50		59	79.50
4180	EMT & wire		5.33	1.501		32.50	82		114.50	159
4200	With #12/2, type NM cable		12.31	.650		17.20	35.50		52.70	72.50
4220	Type MC cable		10.67	.750		23.50	41		64.50	87.50
4230	EMT & wire		4.71	1.699		36.50	93		129.50	179
4250	20 amp recpt. #12/2, type NM cable		12.31	.650		24	35.50		59.50	79.50
4270	Type MC cable		10.67	.750		30.50	41		71.50	95
4280	EMT & wire		4.71	1.699		43	93		136	187
4300	GFI, 15 amp recpt., type NM cable		12.31	.650		41.50	35.50		77	99
4320	Type MC cable		10.67	.750		50.50	41		91.50	117
4330	EMT & wire		4.71	1.699		59	93		152	204
4350	GFI with #12/2, type NM cable		10.67	.750		44	41		85	110
4370	Type MC cable		9.20	.870		50.50	47.50		98	127
4380	EMT & wire		4.21	1.900		63	104		167	226
4400	20 amp recpt., #12/2 type NM cable		10.67	.750		52	41		93	119
4420	Type MC cable		9.20	.870		58.50	47.50		106	136
4430	EMT & wire		4.21	1.900		71.50	104		175.50	235
4500	Weather-proof cover for above receptacles, add		32	.250		3.46	13.70		17.16	24.50
4550	Air conditioner outlet, 20 amp-240 volt recpt.									
4560	30' of #12/2, 2 pole circuit breaker									

368

For customer support on your Plumbing Cost Data, call 877.709.6968.

26 05 Common Work Results for Electrical

26 05 90 – Residential Applications

26 05 90.10 Residential Wiring	Crew	Daily Output	Labor-Hours	Unit	Material	2015 Bare Costs Labor	Equipment	Total	Total Incl O&P
4570	Type NM cable	1 Elec	10	.800	Ea.	63	44		107	135
4580	Type MC cable		9	.889		70.50	48.50		119	151
4590	EMT & wire		4	2		82.50	109		191.50	255
4600	Decorator style, type NM cable		10	.800		68	44		112	141
4620	Type MC cable		9	.889		75.50	48.50		124	156
4630	EMT & wire	↓	4	2	↓	87	109		196	260
4650	Dryer outlet, 30 amp-240 volt recpt., 20' of #10/3									
4660	2 pole circuit breaker									
4670	Type NM cable	1 Elec	6.41	1.248	Ea.	61.50	68.50		130	170
4680	Type MC cable	↓	5.71	1.401		64.50	76.50		141	186
4690	EMT & wire	↓	3.48	2.299	↓	76	126		202	272
4700	Range outlet, 50 amp-240 volt recpt., 30' of #8/3									
4710	Type NM cable	1 Elec	4.21	1.900	Ea.	90.50	104		194.50	256
4720	Type MC cable		4	2		124	109		233	300
4730	EMT & wire		2.96	2.703		108	148		256	340
4750	Central vacuum outlet, Type NM cable		6.40	1.250		58.50	68.50		127	167
4770	Type MC cable		5.71	1.401		69.50	76.50		146	192
4780	EMT & wire	↓	3.48	2.299	↓	83.50	126		209.50	280
4800	30 amp-110 volt locking recpt., #10/2 circ. bkr.									
4810	Type NM cable	1 Elec	6.20	1.290	Ea.	67	70.50		137.50	180
4830	EMT & wire	"	3.20	2.500	"	94	137		231	310
4900	Low voltage outlets									
4910	Telephone recpt., 20' of 4/C phone wire	1 Elec	26	.308	Ea.	9.30	16.85		26.15	35.50
4920	TV recpt., 20' of RG59U coax wire, F type connector	"	16	.500	"	18.75	27.50		46.25	61.50
4950	Door bell chime, transformer, 2 buttons, 60' of bellwire									
4970	Economy model	1 Elec	11.50	.696	Ea.	51.50	38		89.50	114
4980	Custom model		11.50	.696		99	38		137	166
4990	Luxury model, 3 buttons	↓	9.50	.842	↓	281	46		327	380
6000	Lighting outlets									
6050	Wire only (for fixture), type NM cable	1 Elec	32	.250	Ea.	6.90	13.70		20.60	28
6070	Type MC cable		24	.333		12.30	18.25		30.55	41
6080	EMT & wire		10	.800		20	44		64	87.50
6100	Box (4"), and wire (for fixture), type NM cable		25	.320		14.45	17.50		31.95	42
6120	Type MC cable		20	.400		19.90	22		41.90	55
6130	EMT & wire	↓	11	.727	↓	27.50	40		67.50	90
6200	Fixtures (use with lines 6050 or 6100 above)									
6210	Canopy style, economy grade	1 Elec	40	.200	Ea.	32	10.95		42.95	51.50
6220	Custom grade		40	.200		53.50	10.95		64.45	75.50
6250	Dining room chandelier, economy grade		19	.421		79.50	23		102.50	122
6260	Custom grade		19	.421		315	23		338	380
6270	Luxury grade		15	.533		715	29		744	830
6310	Kitchen fixture (fluorescent), economy grade		30	.267		71.50	14.60		86.10	101
6320	Custom grade		25	.320		219	17.50		236.50	267
6350	Outdoor, wall mounted, economy grade		30	.267		30	14.60		44.60	55
6360	Custom grade		30	.267		120	14.60		134.60	154
6370	Luxury grade		25	.320		248	17.50		265.50	299
6410	Outdoor PAR floodlights, 1 lamp, 150 watt		20	.400		32.50	22		54.50	68.50
6420	2 lamp, 150 watt each		20	.400		53.50	22		75.50	92
6430	For infrared security sensor, add		32	.250		132	13.70		145.70	166
6450	Outdoor, quartz-halogen, 300 watt flood		20	.400		39	22		61	75.50
6600	Recessed downlight, round, pre-wired, 50 or 75 watt trim		30	.267		80	14.60		94.60	110
6610	With shower light trim		30	.267		89	14.60		103.60	120
6620	With wall washer trim	↓	28	.286		99.50	15.65		115.15	133

For customer support on your Plumbing Cost Data, call 877.709.6968.

369

26 05 90.10 Residential Wiring	Crew	Daily Output	Labor-Hours	Unit	Material	2015 Bare Costs Labor	Equipment	Total	Total Incl O&P
6630 With eye-ball trim	1 Elec	28	.286	Ea.	99.50	15.65		115.15	133
6700 Porcelain lamp holder		40	.200		2.96	10.95		13.91	19.65
6710 With pull switch		40	.200		6.55	10.95		17.50	23.50
6750 Fluorescent strip, 2-20 watt tube, wrap around diffuser, 24"		24	.333		49.50	18.25		67.75	82
6770 2-34 watt tubes, 48"		20	.400		103	22		125	146
6800 Bathroom heat lamp, 1-250 watt		28	.286		44	15.65		59.65	71.50
6810 2-250 watt lamps		28	.286		70	15.65		85.65	101
6820 For timer switch, see Section 26 05 90.10 line 2400									
6900 Outdoor post lamp, incl. post, fixture, 35' of #14/2									
6910 Type NMC cable	1 Elec	3.50	2.286	Ea.	258	125		383	470
6920 Photo-eye, add		27	.296		32	16.20		48.20	60
6950 Clock dial time switch, 24 hr., w/enclosure, type NM cable		11.43	.700		73.50	38.50		112	139
6970 Type MC cable		11	.727		82.50	40		122.50	150
6980 EMT & wire		4.85	1.649		91	90		181	235
7000 Alarm systems									
7050 Smoke detectors, box, #14/3, type NM cable	1 Elec	14.55	.550	Ea.	34.50	30		64.50	83
7070 Type MC cable		12.31	.650		44	35.50		79.50	102
7080 EMT & wire		5	1.600		49.50	87.50		137	186
7090 For relay output to security system, add					12			12	13.20
8000 Residential equipment									
8050 Disposal hook-up, incl. switch, outlet box, 3' of flex									
8060 20 amp-1 pole circ. bkr., and 25' of #12/2									
8070 Type NM cable	1 Elec	10	.800	Ea.	32.50	44		76.50	101
8080 Type MC cable		8	1		39.50	54.50		94	126
8090 EMT & wire		5	1.600		54.50	87.50		142	191
8100 Trash compactor or dishwasher hook-up, incl. outlet box,									
8110 3' of flex, 15 amp-1 pole circ. bkr., and 25' of #14/2									
8120 Type NM cable	1 Elec	10	.800	Ea.	24	44		68	91.50
8130 Type MC cable		8	1		34	54.50		88.50	120
8140 EMT & wire		5	1.600		46	87.50		133.50	182
8150 Hot water sink dispensor hook-up, use line 8100									
8200 Vent/exhaust fan hook-up, type NM cable	1 Elec	32	.250	Ea.	6.90	13.70		20.60	28
8220 Type MC cable		24	.333		12.30	18.25		30.55	41
8230 EMT & wire		10	.800		20	44		64	87.50
8250 Bathroom vent fan, 50 CFM (use with above hook-up)									
8260 Economy model	1 Elec	15	.533	Ea.	23	29		52	69
8270 Low noise model		15	.533		41	29		70	88.50
8280 Custom model		12	.667		127	36.50		163.50	195
8300 Bathroom or kitchen vent fan, 110 CFM									
8310 Economy model	1 Elec	15	.533	Ea.	67.50	29		96.50	118
8320 Low noise model	"	15	.533	"	92	29		121	145
8350 Paddle fan, variable speed (w/o lights)									
8360 Economy model (AC motor)	1 Elec	10	.800	Ea.	109	44		153	186
8362 With light kit		10	.800		150	44		194	231
8370 Custom model (AC motor)		10	.800		227	44		271	315
8372 With light kit		10	.800		268	44		312	360
8380 Luxury model (DC motor)		8	1		330	54.50		384.50	445
8382 With light kit		8	1		375	54.50		429.50	490
8390 Remote speed switch for above, add		12	.667		37.50	36.50		74	95.50
8500 Whole house exhaust fan, ceiling mount, 36", variable speed									
8510 Remote switch, incl. shutters, 20 amp-1 pole circ. bkr.									
8520 30' of #12/2, type NM cable	1 Elec	4	2	Ea.	1,300	109		1,409	1,600
8530 Type MC cable		3.50	2.286		1,300	125		1,425	1,600

26 05 Common Work Results for Electrical

26 05 90 – Residential Applications

26 05 90.10 Residential Wiring

		Crew	Daily Output	Labor-Hours	Unit	Material	2015 Bare Costs Labor	Equipment	Total	Total Incl O&P
8540	EMT & wire	1 Elec	3	2.667	Ea.	1,325	146		1,471	1,675
8600	Whirlpool tub hook-up, incl. timer switch, outlet box									
8610	3' of flex, 20 amp-1 pole GFI circ. bkr.									
8620	30' of #12/2, type NM cable	1 Elec	5	1.600	Ea.	142	87.50		229.50	287
8630	Type MC cable		4.20	1.905		145	104		249	315
8640	EMT & wire		3.40	2.353		158	129		287	365
8650	Hot water heater hook-up, incl. 1-2 pole circ. bkr., box;									
8660	3' of flex, 20' of #10/2, type NM cable	1 Elec	5	1.600	Ea.	33	87.50		120.50	168
8670	Type MC cable		4.20	1.905		43	104		147	204
8680	EMT & wire		3.40	2.353		48.50	129		177.50	246
9000	Heating/air conditioning									
9050	Furnace/boiler hook-up, incl. firestat, local on-off switch									
9060	Emergency switch, and 40' of type NM cable	1 Elec	4	2	Ea.	52.50	109		161.50	222
9070	Type MC cable		3.50	2.286		65.50	125		190.50	260
9080	EMT & wire		1.50	5.333		83	292		375	525
9100	Air conditioner hook-up, incl. local 60 amp disc. switch									
9110	3' sealtite, 40 amp, 2 pole circuit breaker									
9130	40' of #8/2, type NM cable	1 Elec	3.50	2.286	Ea.	163	125		288	365
9140	Type MC cable		3	2.667		216	146		362	455
9150	EMT & wire		1.30	6.154		204	335		539	730
9200	Heat pump hook-up, 1-40 & 1-100 amp 2 pole circ. bkr.									
9210	Local disconnect switch, 3' sealtite									
9220	40' of #8/2 & 30' of #3/2									
9230	Type NM cable	1 Elec	1.30	6.154	Ea.	520	335		855	1,075
9240	Type MC cable		1.08	7.407		535	405		940	1,200
9250	EMT & wire		.94	8.511		585	465		1,050	1,350
9500	Thermostat hook-up, using low voltage wire									
9520	Heating only, 25' of #18-3	1 Elec	24	.333	Ea.	8.80	18.25		27.05	37
9530	Heating/cooling, 25' of #18-4	"	20	.400	"	11.45	22		33.45	45.50

26 24 Switchboards and Panelboards

26 24 19 – Motor-Control Centers

26 24 19.40 Motor Starters and Controls

		Crew	Daily Output	Labor-Hours	Unit	Material	2015 Bare Costs Labor	Equipment	Total	Total Incl O&P
0010	**MOTOR STARTERS AND CONTROLS**									
0050	Magnetic, FVNR, with enclosure and heaters, 480 volt									
0080	2 HP, size 00	1 Elec	3.50	2.286	Ea.	203	125		328	410
0100	5 HP, size 0		2.30	3.478		272	190		462	585
0200	10 HP, size 1		1.60	5		275	274		549	715
0300	25 HP, size 2	2 Elec	2.20	7.273		520	400		920	1,175
0400	50 HP, size 3		1.80	8.889		845	485		1,330	1,650
0500	100 HP, size 4		1.20	13.333		1,875	730		2,605	3,150
0600	200 HP, size 5		.90	17.778		4,375	970		5,345	6,275
0610	400 HP, size 6		.80	20		19,000	1,100		20,100	22,600
0620	NEMA 7, 5 HP, size 0	1 Elec	1.60	5		1,400	274		1,674	1,950
0630	10 HP, size 1	"	1.10	7.273		1,475	400		1,875	2,225
0640	25 HP, size 2	2 Elec	1.80	8.889		2,375	485		2,860	3,350
0650	50 HP, size 3		1.20	13.333		3,575	730		4,305	5,025
0660	100 HP, size 4		.90	17.778		5,775	970		6,745	7,800
0670	200 HP, size 5		.50	32		13,800	1,750		15,550	17,800
0700	Combination, with motor circuit protectors, 5 HP, size 0	1 Elec	1.80	4.444		880	243		1,123	1,325
0800	10 HP, size 1	"	1.30	6.154		915	335		1,250	1,500

26 24 19 – Motor-Control Centers

26 24 19.40 Motor Starters and Controls	Crew	Daily Output	Labor-Hours	Unit	Material	2015 Bare Costs Labor	Equipment	Total	Total Incl O&P
0900	25 HP, size 2	2 Elec	2	8	Ea.	1,275	440		1,715	2,050
1000	50 HP, size 3		1.32	12.121		1,850	665		2,515	3,025
1200	100 HP, size 4	↓	.80	20		4,000	1,100		5,100	6,050
1220	NEMA 7, 5 HP, size 0	1 Elec	1.30	6.154		2,825	335		3,160	3,600
1230	10 HP, size 1	"	1	8		2,900	440		3,340	3,825
1240	25 HP, size 2	2 Elec	1.32	12.121		3,850	665		4,515	5,250
1250	50 HP, size 3		.80	20		6,375	1,100		7,475	8,675
1260	100 HP, size 4		.60	26.667		9,925	1,450		11,375	13,100
1270	200 HP, size 5	↓	.40	40		21,600	2,200		23,800	27,100
1400	Combination, with fused switch, 5 HP, size 0	1 Elec	1.80	4.444		610	243		853	1,025
1600	10 HP, size 1	"	1.30	6.154		650	335		985	1,225
1800	25 HP, size 2	2 Elec	2	8		1,050	440		1,490	1,800
2000	50 HP, size 3		1.32	12.121		1,775	665		2,440	2,975
2200	100 HP, size 4	↓	.80	20	↓	3,125	1,100		4,225	5,075
3500	Magnetic FVNR with NEMA 12, enclosure & heaters, 480 volt									
3600	5 HP, size 0	1 Elec	2.20	3.636	Ea.	231	199		430	550
3700	10 HP, size 1	"	1.50	5.333		350	292		642	820
3800	25 HP, size 2	2 Elec	2	8		650	440		1,090	1,375
3900	50 HP, size 3		1.60	10		1,000	545		1,545	1,925
4000	100 HP, size 4		1	16		2,400	875		3,275	3,925
4100	200 HP, size 5	↓	.80	20		5,750	1,100		6,850	7,975
4200	Combination, with motor circuit protectors, 5 HP, size 0	1 Elec	1.70	4.706		765	257		1,022	1,225
4300	10 HP, size 1	"	1.20	6.667		795	365		1,160	1,425
4400	25 HP, size 2	2 Elec	1.80	8.889		1,200	485		1,685	2,025
4500	50 HP, size 3		1.20	13.333		1,950	730		2,680	3,225
4600	100 HP, size 4	↓	.74	21.622		4,375	1,175		5,550	6,600
4700	Combination, with fused switch, 5 HP, size 0	1 Elec	1.70	4.706		740	257		997	1,200
4800	10 HP, size 1	"	1.20	6.667		770	365		1,135	1,400
4900	25 HP, size 2	2 Elec	1.80	8.889		1,175	485		1,660	2,025
5000	50 HP, size 3		1.20	13.333		1,875	730		2,605	3,175
5100	100 HP, size 4	↓	.74	21.622	↓	3,825	1,175		5,000	5,975
5200	Factory installed controls, adders to size 0 thru 5									
5300	Start-stop push button	1 Elec	32	.250	Ea.	48.50	13.70		62.20	74
5400	Hand-off-auto-selector switch		32	.250		48.50	13.70		62.20	74
5500	Pilot light		32	.250		91	13.70		104.70	121
5600	Start-stop-pilot		32	.250		139	13.70		152.70	174
5700	Auxiliary contact, NO or NC		32	.250		66.50	13.70		80.20	93.50
5800	NO-NC	↓	32	.250	↓	133	13.70		146.70	167

26 29 Low-Voltage Controllers

26 29 13 – Enclosed Controllers

26 29 13.20 Control Stations

		Crew	Daily Output	Labor-Hours	Unit	Material	2015 Bare Costs Labor	Equipment	Total	Total Incl O&P
0010	CONTROL STATIONS									
0050	NEMA 1, heavy duty, stop/start	1 Elec	8	1	Ea.	135	54.50		189.50	231
0100	Stop/start, pilot light		6.20	1.290		184	70.50		254.50	310
0200	Hand/off/automatic		6.20	1.290		100	70.50		170.50	216
0400	Stop/start/reverse		5.30	1.509		182	82.50		264.50	325
0500	NEMA 7, heavy duty, stop/start		6	1.333		425	73		498	575
0600	Stop/start, pilot light	↓	4	2	↓	515	109		624	735

Estimating Tips

- When estimating material costs for electronic safety and security systems, it is always prudent to obtain manufacturers' quotations for equipment prices and special installation requirements that affect the total cost.

- Fire alarm systems consist of control panels, annunciator panels, battery with rack, charger, and fire alarm actuating and indicating devices. Some fire alarm systems include speakers, telephone lines, door closer controls, and other components. Be careful not to overlook the costs related to installation for these items. Also be aware of costs for integrated automation instrumentation and terminal devices, control equipment, control wiring, and programming.

- Security equipment includes items such as CCTV, access control, and other detection and identification systems to perform alert and alarm functions. Be sure to consider the costs related to installation for this security equipment, such as for integrated automation instrumentation and terminal devices, control equipment, control wiring, and programming.

Reference Numbers

Reference numbers are shown in shaded boxes at the beginning of some major classifications. These numbers refer to related items in the Reference Section. The reference information may be an estimating procedure, an alternate pricing method, or technical information.

Note: Not all subdivisions listed here necessarily appear in this publication. ■

28 16 Intrusion Detection

28 16 16 – Intrusion Detection Systems Infrastructure

28 16 16.50 Intrusion Detection	Crew	Daily Output	Labor-Hours	Unit	Material	2015 Bare Costs Labor	Equipment	Total	Total Incl O&P
0010 **INTRUSION DETECTION**, not including wires & conduits									
0100 Burglar alarm, battery operated, mechanical trigger	1 Elec	4	2	Ea.	278	109		387	470
0200 Electrical trigger		4	2		330	109		439	530
0400 For outside key control, add		8	1		84	54.50		138.50	175
0600 For remote signaling circuitry, add		8	1		125	54.50		179.50	219
0800 Card reader, flush type, standard		2.70	2.963		930	162		1,092	1,275
1000 Multi-code		2.70	2.963		1,200	162		1,362	1,575

28 31 Fire Detection and Alarm

28 31 23 – Fire Detection and Alarm Annunciation Panels and Fire Stations

28 31 23.50 Alarm Panels and Devices

	Crew	Daily Output	Labor-Hours	Unit	Material	2015 Bare Costs Labor	Equipment	Total	Total Incl O&P
0010 **ALARM PANELS AND DEVICES**, not including wires & conduits									
3594 Fire, alarm control panel									
3600 4 zone	2 Elec	2	8	Ea.	400	440		840	1,100
3800 8 zone		1	16		780	875		1,655	2,150
4000 12 zone		.67	23.988		2,400	1,300		3,700	4,600
4020 Alarm device	1 Elec	8	1		238	54.50		292.50	345
4050 Actuating device		8	1		335	54.50		389.50	450
4200 Battery and rack		4	2		410	109		519	620
4400 Automatic charger		8	1		585	54.50		639.50	720
4600 Signal bell		8	1		78.50	54.50		133	169
4800 Trouble buzzer or manual station		8	1		83.50	54.50		138	174
5600 Strobe and horn		5.30	1.509		152	82.50		234.50	291
5800 Fire alarm horn		6.70	1.194		61	65.50		126.50	165
6000 Door holder, electro-magnetic		4	2		103	109		212	277
6200 Combination holder and closer		3.20	2.500		123	137		260	340
6600 Drill switch		8	1		370	54.50		424.50	490
6800 Master box		2.70	2.963		6,400	162		6,562	7,300
7000 Break glass station		8	1		55.50	54.50		110	143
7800 Remote annunciator, 8 zone lamp		1.80	4.444		209	243		452	595
8000 12 zone lamp	2 Elec	2.60	6.154		335	335		670	875
8200 16 zone lamp	"	2.20	7.273		420	400		820	1,050

28 31 43 – Fire Detection Sensors

28 31 43.50 Fire and Heat Detectors

	Crew	Daily Output	Labor-Hours	Unit	Material	2015 Bare Costs Labor	Equipment	Total	Total Incl O&P
0010 **FIRE & HEAT DETECTORS**									
5000 Detector, rate of rise	1 Elec	8	1	Ea.	51	54.50		105.50	138
5100 Fixed temperature	"	8	1	"	51	54.50		105.50	138

28 31 46 – Smoke Detection Sensors

28 31 46.50 Smoke Detectors

	Crew	Daily Output	Labor-Hours	Unit	Material	2015 Bare Costs Labor	Equipment	Total	Total Incl O&P
0010 **SMOKE DETECTORS**									
5200 Smoke detector, ceiling type	1 Elec	6.20	1.290	Ea.	110	70.50		180.50	227
5400 Duct type	"	3.20	2.500	"	325	137		462	565

28 33 Gas Detection and Alarm

28 33 33 – Gas Detection Sensors

28 33 33.50 Tank Leak Detection Systems	Crew	Daily Output	Labor-Hours	Unit	Material	2015 Bare Costs Labor	Equipment	Total	Total Incl O&P
0010 **TANK LEAK DETECTION SYSTEMS** Liquid and vapor									
0100 For hydrocarbons and hazardous liquids/vapors									
0120 Controller, data acquisition, incl. printer, modem, RS232 port									
0140 24 channel, for use with all probes				Ea.	4,175			4,175	4,575
0160 9 channel, for external monitoring				"	915			915	1,000
0200 Probes									
0210 Well monitoring									
0220 Liquid phase detection				Ea.	475			475	525
0230 Hydrocarbon vapor, fixed position					480			480	530
0240 Hydrocarbon vapor, float mounted					480			480	530
0250 Both liquid and vapor hydrocarbon					480			480	530
0300 Secondary containment, liquid phase									
0310 Pipe trench/manway sump				Ea.	820			820	905
0320 Double wall pipe and manual sump					825			825	905
0330 Double wall fiberglass annular space					325			325	355
0340 Double wall steel tank annular space					335			335	370
0500 Accessories									
0510 Modem, non-dedicated phone line				Ea.	305			305	335
0600 Monitoring, internal									
0610 Automatic tank gauge, incl. overfill				Ea.	1,150			1,150	1,275
0620 Product line				"	1,150			1,150	1,275
0700 Monitoring, special									
0710 Cathodic protection				Ea.	725			725	795
0720 Annular space chemical monitor				"	985			985	1,075

For customer support on your Plumbing Cost Data, call 877.709.6968.

375

Division Notes

	CREW	DAILY OUTPUT	LABOR-HOURS	UNIT	BARE COSTS				TOTAL INCL O&P
					MAT.	LABOR	EQUIP.	TOTAL	

Estimating Tips

31 05 00 Common Work Results for Earthwork

- Estimating the actual cost of performing earthwork requires careful consideration of the variables involved. This includes items such as type of soil, whether water will be encountered, dewatering, whether banks need bracing, disposal of excavated earth, and length of haul to fill or spoil sites, etc. If the project has large quantities of cut or fill, consider raising or lowering the site to reduce costs, while paying close attention to the effect on site drainage and utilities.

- If the project has large quantities of fill, creating a borrow pit on the site can significantly lower the costs.

- It is very important to consider what time of year the project is scheduled for completion. Bad weather can create large cost overruns from dewatering, site repair, and lost productivity from cold weather.

Reference Numbers

Reference numbers are shown in shaded boxes at the beginning of some major classifications. These numbers refer to related items in the Reference Section. The reference information may be an estimating procedure, an alternate pricing method, or technical information.

Note: Not all subdivisions listed here necessarily appear in this publication. ■

Division 31 – Earthwork

31 23 16.13 Excavating, Trench		Crew	Daily Output	Labor-Hours	Unit	Material	2015 Bare Costs Labor	Equipment	Total	Total Incl O&P
0010	**EXCAVATING, TRENCH**	G1030–805								
0011	Or continuous footing									
0020	Common earth with no sheeting or dewatering included									
0050	1' to 4' deep, 3/8 C.Y. excavator	B-11C	150	.107	B.C.Y.		4.70	2.43	7.13	9.85
0060	1/2 C.Y. excavator	B-11M	200	.080			3.53	1.96	5.49	7.55
0090	4' to 6' deep, 1/2 C.Y. excavator	"	200	.080			3.53	1.96	5.49	7.55
0100	5/8 C.Y. excavator	B-12Q	250	.064			2.86	2.36	5.22	6.95
0300	1/2 C.Y. excavator, truck mounted	B-12J	200	.080			3.57	4.41	7.98	10.30
0500	6' to 10' deep, 3/4 C.Y. excavator	B-12F	225	.071			3.18	2.91	6.09	8.05
0600	1 C.Y. excavator, truck mounted	B-12K	400	.040			1.79	2.51	4.30	5.50
0900	10' to 14' deep, 3/4 C.Y. excavator	B-12F	200	.080			3.57	3.27	6.84	9.05
1000	1-1/2 C.Y. excavator	B-12B	540	.030			1.32	1.91	3.23	4.12
1300	14' to 20' deep, 1 C.Y. excavator	B-12A	320	.050			2.23	2.54	4.77	6.20
1340	20' to 24' deep, 1 C.Y. excavator	"	288	.056			2.48	2.82	5.30	6.90
1352	4' 6' deep, 1/2 C.Y. excavator w/trench box	B-13H	188	.085			3.80	5.10	8.90	11.45
1354	5/8 C.Y. excavator	"	235	.068			3.04	4.10	7.14	9.15
1362	6' to 10' deep, 3/4 C.Y. excavator w/trench box	B-13G	212	.075			3.37	3.47	6.84	8.95
1374	10' to 14' deep, 3/4 C.Y. excavator w/trench box	"	188	.085			3.80	3.91	7.71	10.10
1376	1-1/2 C.Y. excavator	B-13E	508	.032			1.41	2.19	3.60	4.56
1381	14' to 20' deep, 1 C.Y. excavator w/trench box	B-13D	301	.053			2.37	2.97	5.34	6.90
1386	20' to 24' deep, 1 C.Y. excavator w/trench box	"	271	.059			2.64	3.30	5.94	7.65
1400	By hand with pick and shovel 2' to 6' deep, light soil	1 Clab	8	1			37.50		37.50	58
1500	Heavy soil	"	4	2			75		75	116
1700	For tamping backfilled trenches, air tamp, add	A-1G	100	.080	E.C.Y.		3.01	.56	3.57	5.25
1900	Vibrating plate, add	B-18	180	.133	"		5.10	.26	5.36	8.15
2100	Trim sides and bottom for concrete pours, common earth		1500	.016	S.F.		.61	.03	.64	.97
2300	Hardpan		600	.040	"		1.53	.08	1.61	2.45
5020	Loam & Sandy clay with no sheeting or dewatering included									
5050	1' to 4' deep, 3/8 C.Y. tractor loader/backhoe	B-11C	162	.099	B.C.Y.		4.36	2.25	6.61	9.10
5060	1/2 C.Y. excavator	B-11M	216	.074			3.27	1.81	5.08	7
5080	4' to 6' deep, 1/2 C.Y. excavator	"	216	.074			3.27	1.81	5.08	7
5090	5/8 C.Y. excavator	B-12Q	276	.058			2.59	2.14	4.73	6.30
5130	1/2 C.Y. excavator, truck mounted	B-12J	216	.074			3.31	4.08	7.39	9.55
5140	6' to 10' deep, 3/4 C.Y. excavator	B-12F	243	.066			2.94	2.69	5.63	7.45
5160	1 C.Y. excavator, truck mounted	B-12K	432	.037			1.65	2.32	3.97	5.05
5190	10' to 14' deep, 3/4 C.Y. excavator	B-12F	216	.074			3.31	3.03	6.34	8.40
5210	1-1/2 C.Y. excavator	B-12B	583	.027			1.23	1.77	3	3.81
5250	14' to 20' deep, 1 C.Y. excavator	B-12A	346	.046			2.06	2.35	4.41	5.75
5300	20' to 24' deep, 1 C.Y. excavator	"	311	.051			2.30	2.61	4.91	6.40
5352	4' to 6' deep, 1/2 C.Y. excavator w/trench box	B-13H	205	.078			3.48	4.70	8.18	10.45
5354	5/8 C.Y. excavator	"	257	.062			2.78	3.75	6.53	8.35
5362	6' to 10' deep, 3/4 C.Y. excavator w/trench box	B-13G	231	.069			3.09	3.19	6.28	8.20
5370	10' to 14' deep, 3/4 C.Y. excavator w/trench box	"	205	.078			3.48	3.59	7.07	9.25
5374	1-1/2 C.Y. excavator	B-13E	554	.029			1.29	2.01	3.30	4.18
5382	14' to 20' deep, 1 C.Y. excavator w/trench box	B-13D	329	.049			2.17	2.72	4.89	6.30
5392	20' to 24' deep, 1 C.Y. excavator w/trench box	"	295	.054			2.42	3.03	5.45	7.05
6020	Sand & gravel with no sheeting or dewatering included									
6050	1' to 4' deep, 3/8 C.Y. excavator	B-11C	165	.097	B.C.Y.		4.28	2.21	6.49	9
6060	1/2 C.Y. excavator	B-11M	220	.073			3.21	1.78	4.99	6.85
6080	4' to 6' deep, 1/2 C.Y. excavator	"	220	.073			3.21	1.78	4.99	6.85
6090	5/8 C.Y. excavator	B-12Q	275	.058			2.60	2.14	4.74	6.35
6130	1/2 C.Y. excavator, truck mounted	B-12J	220	.073			3.25	4.01	7.26	9.35
6140	6' to 10' deep, 3/4 C.Y. excavator	B-12F	248	.065			2.88	2.64	5.52	7.30

31 23 Excavation and Fill

31 23 16 – Excavation

31 23 16.13 Excavating, Trench		Crew	Daily Output	Labor-Hours	Unit	Material	2015 Bare Costs			Total Incl O&P
							Labor	Equipment	Total	
6160	1 C.Y. excavator, truck mounted	B-12K	440	.036	B.C.Y.		1.62	2.28	3.90	4.98
6190	10' to 14' deep, 3/4 C.Y. excavator	B-12F	220	.073			3.25	2.98	6.23	8.25
6210	1-1/2 C.Y. excavator	B-12B	594	.027			1.20	1.73	2.93	3.75
6250	14' to 20' deep, 1 C.Y. excavator	B-12A	352	.045			2.03	2.31	4.34	5.65
6300	20' to 24' deep, 1 C.Y. excavator	"	317	.050			2.25	2.56	4.81	6.25
6352	4' to 6' deep, 1/2 C.Y. excavator w/trench box	B-13H	209	.077			3.42	4.61	8.03	10.25
6354	5/8 C.Y. excavator	"	261	.061			2.74	3.69	6.43	8.25
6362	6' to 10' deep, 3/4 C.Y. excavator w/trench box	B-13G	236	.068			3.03	3.12	6.15	8.05
6370	10' to 14' deep, 3/4 C.Y. excavator w/trench box	"	209	.077			3.42	3.52	6.94	9.05
6374	1-1/2 C.Y. excavator	B-13E	564	.028			1.27	1.97	3.24	4.10
6382	14' to 20' deep, 1 C.Y. excavator w/trench box	B-13D	334	.048			2.14	2.68	4.82	6.20
6392	20' to 24' deep, 1 C.Y. excavator w/trench box	"	301	.053			2.37	2.97	5.34	6.90
7020	Dense hard clay with no sheeting or dewatering included									
7050	1' to 4' deep, 3/8 C.Y. excavator	B-11C	132	.121	B.C.Y.		5.35	2.76	8.11	11.20
7060	1/2 C.Y. excavator	B-11M	176	.091			4.01	2.23	6.24	8.55
7080	4' to 6' deep, 1/2 C.Y. excavator	"	176	.091			4.01	2.23	6.24	8.55
7090	5/8 C.Y. excavator	B-12Q	220	.073			3.25	2.68	5.93	7.90
7130	1/2 C.Y. excavator, truck mounted	B-12J	176	.091			4.06	5	9.06	11.70
7140	6' to 10' deep, 3/4 C.Y. excavator	B-12F	198	.081			3.61	3.31	6.92	9.15
7160	1 C.Y. excavator, truck mounted	B-12K	352	.045			2.03	2.85	4.88	6.25
7190	10' to 14' deep, 3/4 C.Y. excavator	B-12F	176	.091			4.06	3.72	7.78	10.30
7210	1-1/2 C.Y. excavator	B-12B	475	.034			1.50	2.17	3.67	4.68
7250	14' to 20' deep, 1 C.Y. excavator	B-12A	282	.057			2.53	2.88	5.41	7.05
7300	20' to 24' deep, 1 C.Y. excavator	"	254	.063			2.81	3.20	6.01	7.80

31 23 16.14 Excavating, Utility Trench		Crew	Daily Output	Labor-Hours	Unit	Material	Labor	Equipment	Total	Total Incl O&P
0010	**EXCAVATING, UTILITY TRENCH** G1030–805									
0011	Common earth									
0050	Trenching with chain trencher, 12 H.P., operator walking									
0100	4" wide trench, 12" deep	B-53	800	.010	L.F.		.49	.09	.58	.83
0150	18" deep		750	.011			.52	.09	.61	.89
0200	24" deep		700	.011			.56	.10	.66	.95
0300	6" wide trench, 12" deep		650	.012			.60	.10	.70	1.03
0350	18" deep		600	.013			.65	.11	.76	1.10
0400	24" deep		550	.015			.71	.12	.83	1.21
0450	36" deep		450	.018			.86	.15	1.01	1.48
0600	8" wide trench, 12" deep		475	.017			.82	.14	.96	1.40
0650	18" deep		400	.020			.97	.17	1.14	1.67
0700	24" deep		350	.023			1.11	.19	1.30	1.90
0750	36" deep		300	.027			1.30	.23	1.53	2.22
1000	Backfill by hand including compaction, add									
1050	4" wide trench, 12" deep	A-1G	800	.010	L.F.		.38	.07	.45	.66
1100	18" deep		530	.015			.57	.11	.68	.99
1150	24" deep		400	.020			.75	.14	.89	1.32
1300	6" wide trench, 12" deep		540	.015			.56	.10	.66	.97
1350	18" deep		405	.020			.74	.14	.88	1.29
1400	24" deep		270	.030			1.11	.21	1.32	1.94
1450	36" deep		180	.044			1.67	.31	1.98	2.91
1600	8" wide trench, 12" deep		400	.020			.75	.14	.89	1.32
1650	18" deep		265	.030			1.14	.21	1.35	1.98
1700	24" deep		200	.040			1.50	.28	1.78	2.62
1750	36" deep		135	.059			2.23	.42	2.65	3.89
2000	Chain trencher, 40 H.P. operator riding									

31 23 Excavation and Fill

31 23 16 – Excavation

31 23 16.14 Excavating, Utility Trench

		Crew	Daily Output	Labor-Hours	Unit	Material	2015 Bare Costs Labor	Equipment	Total	Total Incl O&P
2050	6" wide trench and backfill, 12" deep	B-54	1200	.007	L.F.		.32	.28	.60	.80
2100	18" deep		1000	.008			.39	.34	.73	.96
2150	24" deep		975	.008			.40	.34	.74	.99
2200	36" deep		900	.009			.43	.37	.80	1.07
2250	48" deep		750	.011			.52	.45	.97	1.28
2300	60" deep		650	.012			.60	.52	1.12	1.48
2400	8" wide trench and backfill, 12" deep		1000	.008			.39	.34	.73	.96
2450	18" deep		950	.008			.41	.35	.76	1.01
2500	24" deep		900	.009			.43	.37	.80	1.07
2550	36" deep		800	.010			.49	.42	.91	1.20
2600	48" deep		650	.012			.60	.52	1.12	1.48
2700	12" wide trench and backfill, 12" deep		975	.008			.40	.34	.74	.99
2750	18" deep		860	.009			.45	.39	.84	1.12
2800	24" deep		800	.010			.49	.42	.91	1.20
2850	36" deep		725	.011			.54	.46	1	1.32
3000	16" wide trench and backfill, 12" deep		835	.010			.47	.40	.87	1.15
3050	18" deep		750	.011			.52	.45	.97	1.28
3100	24" deep	▼	700	.011	▼		.56	.48	1.04	1.37
3200	Compaction with vibratory plate, add								35%	35%
5100	Hand excavate and trim for pipe bells after trench excavation									
5200	8" pipe	1 Clab	155	.052	L.F.		1.94		1.94	2.99
5300	18" pipe	"	130	.062	"		2.31		2.31	3.56

31 23 19 – Dewatering

31 23 19.20 Dewatering Systems

		Crew	Daily Output	Labor-Hours	Unit	Material	2015 Bare Costs Labor	Equipment	Total	Total Incl O&P
0010	**DEWATERING SYSTEMS**									
0020	Excavate drainage trench, 2' wide, 2' deep	B-11C	90	.178	C.Y.		7.85	4.05	11.90	16.40
0100	2' wide, 3' deep, with backhoe loader	"	135	.119			5.25	2.70	7.95	10.95
0200	Excavate sump pits by hand, light soil	1 Clab	7.10	1.127			42.50		42.50	65
0300	Heavy soil	"	3.50	2.286	▼		86		86	132
0500	Pumping 8 hr., attended 2 hrs. per day, including 20 L.F.									
0550	of suction hose & 100 L.F. discharge hose									
0600	2" diaphragm pump used for 8 hours	B-10H	4	3	Day		139	18.70	157.70	232
0650	4" diaphragm pump used for 8 hours	B-10I	4	3			139	30.50	169.50	245
0800	8 hrs. attended, 2" diaphragm pump	B-10H	1	12			555	75	630	925
0900	3" centrifugal pump	B-10J	1	12			555	84	639	940
1000	4" diaphragm pump	B-10I	1	12			555	122	677	980
1100	6" centrifugal pump	B-10K	1	12	▼		555	370	925	1,250
1300	CMP, incl. excavation 3' deep, 12" diameter	B-6	115	.209	L.F.	11.05	8.60	3.17	22.82	29
1400	18" diameter		100	.240	"	16.70	9.90	3.64	30.24	37.50
1600	Sump hole construction, incl. excavation and gravel, pit		1250	.019	C.F.	1.09	.79	.29	2.17	2.73
1700	With 12" gravel collar, 12" pipe, corrugated, 16 ga.		70	.343	L.F.	21.50	14.15	5.20	40.85	50.50
1800	15" pipe, corrugated, 16 ga.		55	.436		28	18	6.60	52.60	65.50
1900	18" pipe, corrugated, 16 ga.		50	.480		32	19.80	7.30	59.10	74
2000	24" pipe, corrugated, 14 ga.		40	.600	▼	38.50	25	9.10	72.60	90.50
2200	Wood lining, up to 4' x 4', add	▼	300	.080	SFCA	16.25	3.30	1.21	20.76	24.50
9950	See Section 31 23 19.40 for wellpoints									
9960	See Section 31 23 19.30 for deep well systems									

31 23 19.30 Wells

		Crew	Daily Output	Labor-Hours	Unit	Material	2015 Bare Costs Labor	Equipment	Total	Total Incl O&P
0010	**WELLS**									
0011	For dewatering 10' to 20' deep, 2' diameter									
0020	with steel casing, minimum	B-6	165	.145	V.L.F.	38	6	2.21	46.21	53
0050	Average	▼	98	.245	▼	43	10.10	3.72	56.82	66.50

31 23 Excavation and Fill

31 23 19 – Dewatering

31 23 19.30 Wells

		Crew	Daily Output	Labor-Hours	Unit	Material	2015 Bare Costs Labor	2015 Bare Costs Equipment	Total	Total Incl O&P
0100	Maximum	B-6	49	.490	V.L.F.	47.50	20	7.45	74.95	91.50
0300	For dewatering pumps see 01 54 33 in Reference Section									
0500	For domestic water wells, see Section 33 21 13.10									

31 23 19.40 Wellpoints

		Crew	Daily Output	Labor-Hours	Unit	Material	Labor	Equipment	Total	Total Incl O&P
0010	**WELLPOINTS** R312319-90									
0011	For equipment rental, see 01 54 33 in Reference Section									
0100	Installation and removal of single stage system									
0110	Labor only, .75 labor-hours per L.F.	1 Clab	10.70	.748	LF Hdr		28		28	43.50
0200	2.0 labor-hours per L.F.	"	4	2	"		75		75	116
0400	Pump operation, 4 @ 6 hr. shifts									
0410	Per 24 hour day	4 Eqlt	1.27	25.197	Day		1,225		1,225	1,850
0500	Per 168 hour week, 160 hr. straight, 8 hr. double time		.18	177	Week		8,650		8,650	13,100
0550	Per 4.3 week month		.04	800	Month		38,900		38,900	59,000
0600	Complete installation, operation, equipment rental, fuel &									
0610	removal of system with 2" wellpoints 5' O.C.									
0700	100' long header, 6" diameter, first month	4 Eqlt	3.23	9.907	LF Hdr	159	480		639	905
0800	Thereafter, per month		4.13	7.748		127	375		502	710
1000	200' long header, 8" diameter, first month		6	5.333		145	259		404	555
1100	Thereafter, per month		8.39	3.814		71.50	185		256.50	360
1300	500' long header, 8" diameter, first month		10.63	3.010		55.50	146		201.50	283
1400	Thereafter, per month		20.91	1.530		39.50	74.50		114	157
1600	1,000' long header, 10" diameter, first month		11.62	2.754		47.50	134		181.50	256
1700	Thereafter, per month		41.81	.765		24	37		61	82.50
1900	Note: above figures include pumping 168 hrs. per week									
1910	and include the pump operator and one stand-by pump.									

31 23 23 – Fill

31 23 23.15 Borrow, Loading And/Or Spreading

		Crew	Daily Output	Labor-Hours	Unit	Material	Labor	Equipment	Total	Total Incl O&P
0010	**BORROW, LOADING AND/OR SPREADING**									
0020	Material only, bank run gravel				Ton	15.90			15.90	17.50
0500	Haul 2 mi. spread, 200 HP dozer, bank run gravel	B-15	1100	.025			1.09	2.52	3.61	4.42
1000	Hand spread, bank run gravel	A-5	33	.545			20.50	1.86	22.36	33.50
1800	Delivery charge, minimum 20 tons, 1 hr. round trip, add	B-34B	130	.062			2.46	5.30	7.76	9.60
1820	1-1/2 hr. round trip, add		93	.086			3.45	7.45	10.90	13.35
1840	2 hr. round trip, add		65	.123			4.93	10.65	15.58	19.15

31 23 23.16 Fill By Borrow and Utility Bedding

		Crew	Daily Output	Labor-Hours	Unit	Material	Labor	Equipment	Total	Total Incl O&P
0010	**FILL BY BORROW AND UTILITY BEDDING**									
0049	Utility bedding, for pipe & conduit, not incl. compaction G1030-805									
0050	Crushed or screened bank run gravel	B-6	150	.160	L.C.Y.	25.50	6.60	2.43	34.53	41
0100	Crushed stone 3/4" to 1/2"		150	.160		23.50	6.60	2.43	32.53	39
0200	Sand, dead or bank		150	.160		17.85	6.60	2.43	26.88	32.50
0500	Compacting bedding in trench	A-1D	90	.089	E.C.Y.		3.34	.40	3.74	5.60
0600	If material source exceeds 2 miles, add for extra mileage.									
0610	See Section 31 23 23.20 for hauling mileage add.									

31 23 23.17 General Fill

		Crew	Daily Output	Labor-Hours	Unit	Material	Labor	Equipment	Total	Total Incl O&P
0010	**GENERAL FILL**									
0011	Spread dumped material, no compaction									
0020	By dozer, no compaction	B-10B	1000	.012	L.C.Y.		.56	1.39	1.95	2.38
0100	By hand	1 Clab	12	.667	"		25		25	38.50
0500	Gravel fill, compacted, under floor slabs, 4" deep	B-37	10000	.005	S.F.	.42	.19	.02	.63	.77
0600	6" deep		8600	.006		.63	.22	.02	.87	1.05
0700	9" deep		7200	.007		1.05	.27	.02	1.34	1.58

For customer support on your Plumbing Cost Data, call 877.709.6968.

381

31 23 Excavation and Fill

31 23 23 – Fill

31 23 23.17 General Fill	Crew	Daily Output	Labor-Hours	Unit	Material	2015 Bare Costs Labor	Equipment	Total	Total Incl O&P
0800	12" deep	B-37	6000	.008	S.F.	1.47	.32	.03	1.82	2.14
1000	Alternate pricing method, 4" deep		120	.400	E.C.Y.	31.50	15.90	1.31	48.71	60.50
1100	6" deep		160	.300		31.50	11.95	.98	44.43	54
1200	9" deep		200	.240		31.50	9.55	.79	41.84	50
1300	12" deep		220	.218		31.50	8.70	.72	40.92	48.50
1400	Granular fill				L.C.Y.	21			21	23

31 23 23.20 Hauling

		Crew	Daily Output	Labor-Hours	Unit	Material	2015 Bare Costs Labor	Equipment	Total	Total Incl O&P
0010	**HAULING**									
0011	Excavated or borrow, loose cubic yards									
0012	no loading equipment, including hauling,waiting, loading/dumping									
0013	time per cycle (wait, load, travel, unload or dump & return)									
0014	8 C.Y. truck, 15 MPH ave, cycle 0.5 miles, 10 min. wait/Ld./Uld.	B-34A	320	.025	L.C.Y.		1	1.29	2.29	2.92
0016	cycle 1 mile		272	.029			1.18	1.51	2.69	3.44
0018	cycle 2 miles		208	.038			1.54	1.98	3.52	4.50
0020	cycle 4 miles		144	.056			2.23	2.86	5.09	6.50
0022	cycle 6 miles		112	.071			2.86	3.67	6.53	8.35
0024	cycle 8 miles		88	.091			3.64	4.67	8.31	10.65
0026	20 MPH ave,cycle 0.5 mile		336	.024			.95	1.22	2.17	2.79
0028	cycle 1 mile		296	.027			1.08	1.39	2.47	3.17
0030	cycle 2 miles		240	.033			1.33	1.71	3.04	3.90
0032	cycle 4 miles		176	.045			1.82	2.34	4.16	5.30
0034	cycle 6 miles		136	.059			2.36	3.02	5.38	6.90
0036	cycle 8 miles		112	.071			2.86	3.67	6.53	8.35
0044	25 MPH ave, cycle 4 miles		192	.042			1.67	2.14	3.81	4.88
0046	cycle 6 miles		160	.050			2	2.57	4.57	5.85
0048	cycle 8 miles		128	.063			2.50	3.21	5.71	7.30
0050	30 MPH ave, cycle 4 miles		216	.037			1.48	1.90	3.38	4.33
0052	cycle 6 miles		176	.045			1.82	2.34	4.16	5.30
0054	cycle 8 miles		144	.056			2.23	2.86	5.09	6.50
0114	15 MPH ave, cycle 0.5 mile, 15 min. wait/Ld./Uld.		224	.036			1.43	1.84	3.27	4.18
0116	cycle 1 mile		200	.040			1.60	2.06	3.66	4.68
0118	cycle 2 miles		168	.048			1.91	2.45	4.36	5.55
0120	cycle 4 miles		120	.067			2.67	3.43	6.10	7.80
0122	cycle 6 miles		96	.083			3.34	4.28	7.62	9.75
0124	cycle 8 miles		80	.100			4.01	5.15	9.16	11.70
0126	20 MPH ave, cycle 0.5 mile		232	.034			1.38	1.77	3.15	4.04
0128	cycle 1 mile		208	.038			1.54	1.98	3.52	4.50
0130	cycle 2 miles		184	.043			1.74	2.23	3.97	5.10
0132	cycle 4 miles		144	.056			2.23	2.86	5.09	6.50
0134	cycle 6 miles		112	.071			2.86	3.67	6.53	8.35
0136	cycle 8 miles		96	.083			3.34	4.28	7.62	9.75
0144	25 MPH ave, cycle 4 miles		152	.053			2.11	2.71	4.82	6.15
0146	cycle 6 miles		128	.063			2.50	3.21	5.71	7.30
0148	cycle 8 miles		112	.071			2.86	3.67	6.53	8.35
0150	30 MPH ave, cycle 4 miles		168	.048			1.91	2.45	4.36	5.55
0152	cycle 6 miles		144	.056			2.23	2.86	5.09	6.50
0154	cycle 8 miles		120	.067			2.67	3.43	6.10	7.80
0214	15 MPH ave, cycle 0.5 mile, 20 min wait/Ld./Uld.		176	.045			1.82	2.34	4.16	5.30
0216	cycle 1 mile		160	.050			2	2.57	4.57	5.85
0218	cycle 2 miles		136	.059			2.36	3.02	5.38	6.90
0220	cycle 4 miles		104	.077			3.08	3.95	7.03	9
0222	cycle 6 miles		88	.091			3.64	4.67	8.31	10.65

31 23 Excavation and Fill

31 23 23 – Fill

31 23 23.20 Hauling		Crew	Daily Output	Labor-Hours	Unit	Material	2015 Bare Costs Labor	Equipment	Total	Total Incl O&P
0224	cycle 8 miles	B-34A	72	.111	L.C.Y.		4.45	5.70	10.15	13.05
0226	20 MPH ave, cycle 0.5 mile		176	.045			1.82	2.34	4.16	5.30
0228	cycle 1 mile		168	.048			1.91	2.45	4.36	5.55
0230	cycle 2 miles		144	.056			2.23	2.86	5.09	6.50
0232	cycle 4 miles		120	.067			2.67	3.43	6.10	7.80
0234	cycle 6 miles		96	.083			3.34	4.28	7.62	9.75
0236	cycle 8 miles		88	.091			3.64	4.67	8.31	10.65
0244	25 MPH ave, cycle 4 miles		128	.063			2.50	3.21	5.71	7.30
0246	cycle 6 miles		112	.071			2.86	3.67	6.53	8.35
0248	cycle 8 miles		96	.083			3.34	4.28	7.62	9.75
0250	30 MPH ave, cycle 4 miles		136	.059			2.36	3.02	5.38	6.90
0252	cycle 6 miles		120	.067			2.67	3.43	6.10	7.80
0254	cycle 8 miles		104	.077			3.08	3.95	7.03	9
0314	15 MPH ave, cycle 0.5 mile, 25 min wait/Ld./Uld.		144	.056			2.23	2.86	5.09	6.50
0316	cycle 1 mile		128	.063			2.50	3.21	5.71	7.30
0318	cycle 2 miles		112	.071			2.86	3.67	6.53	8.35
0320	cycle 4 miles		96	.083			3.34	4.28	7.62	9.75
0322	cycle 6 miles		80	.100			4.01	5.15	9.16	11.70
0324	cycle 8 miles		64	.125			5	6.45	11.45	14.60
0326	20 MPH ave, cycle 0.5 mile		144	.056			2.23	2.86	5.09	6.50
0328	cycle 1 mile		136	.059			2.36	3.02	5.38	6.90
0330	cycle 2 miles		120	.067			2.67	3.43	6.10	7.80
0332	cycle 4 miles		104	.077			3.08	3.95	7.03	9
0334	cycle 6 miles		88	.091			3.64	4.67	8.31	10.65
0336	cycle 8 miles		80	.100			4.01	5.15	9.16	11.70
0344	25 MPH ave, cycle 4 miles		112	.071			2.86	3.67	6.53	8.35
0346	cycle 6 miles		96	.083			3.34	4.28	7.62	9.75
0348	cycle 8 miles		88	.091			3.64	4.67	8.31	10.65
0350	30 MPH ave, cycle 4 miles		112	.071			2.86	3.67	6.53	8.35
0352	cycle 6 miles		104	.077			3.08	3.95	7.03	9
0354	cycle 8 miles		96	.083			3.34	4.28	7.62	9.75
0414	15 MPH ave, cycle 0.5 mile, 30 min wait/Ld./Uld.		120	.067			2.67	3.43	6.10	7.80
0416	cycle 1 mile		112	.071			2.86	3.67	6.53	8.35
0418	cycle 2 miles		96	.083			3.34	4.28	7.62	9.75
0420	cycle 4 miles		80	.100			4.01	5.15	9.16	11.70
0422	cycle 6 miles		72	.111			4.45	5.70	10.15	13.05
0424	cycle 8 miles		64	.125			5	6.45	11.45	14.60
0426	20 MPH ave, cycle 0.5 mile		120	.067			2.67	3.43	6.10	7.80
0428	cycle 1 mile		112	.071			2.86	3.67	6.53	8.35
0430	cycle 2 miles		104	.077			3.08	3.95	7.03	9
0432	cycle 4 miles		88	.091			3.64	4.67	8.31	10.65
0434	cycle 6 miles		80	.100			4.01	5.15	9.16	11.70
0436	cycle 8 miles		72	.111			4.45	5.70	10.15	13.05
0444	25 MPH ave, cycle 4 miles		96	.083			3.34	4.28	7.62	9.75
0446	cycle 6 miles		88	.091			3.64	4.67	8.31	10.65
0448	cycle 8 miles		80	.100			4.01	5.15	9.16	11.70
0450	30 MPH ave, cycle 4 miles		96	.083			3.34	4.28	7.62	9.75
0452	cycle 6 miles		88	.091			3.64	4.67	8.31	10.65
0454	cycle 8 miles		80	.100			4.01	5.15	9.16	11.70
0514	15 MPH ave, cycle 0.5 mile, 35 min wait/Ld./Uld.		104	.077			3.08	3.95	7.03	9
0516	cycle 1 mile		96	.083			3.34	4.28	7.62	9.75
0518	cycle 2 miles		88	.091			3.64	4.67	8.31	10.65
0520	cycle 4 miles		72	.111			4.45	5.70	10.15	13.05

For customer support on your Plumbing Cost Data, call 877.709.6968.

383

31 23 23.20 Hauling		Crew	Daily Output	Labor-Hours	Unit	Material	2015 Bare Costs Labor	2015 Bare Costs Equipment	Total	Total Incl O&P
0522	cycle 6 miles	B-34A	64	.125	L.C.Y.		5	6.45	11.45	14.60
0524	cycle 8 miles		56	.143			5.70	7.35	13.05	16.75
0526	20 MPH ave, cycle 0.5 mile		104	.077			3.08	3.95	7.03	9
0528	cycle 1 mile		96	.083			3.34	4.28	7.62	9.75
0530	cycle 2 miles		96	.083			3.34	4.28	7.62	9.75
0532	cycle 4 miles		80	.100			4.01	5.15	9.16	11.70
0534	cycle 6 miles		72	.111			4.45	5.70	10.15	13.05
0536	cycle 8 miles		64	.125			5	6.45	11.45	14.60
0544	25 MPH ave, cycle 4 miles		88	.091			3.64	4.67	8.31	10.65
0546	cycle 6 miles		80	.100			4.01	5.15	9.16	11.70
0548	cycle 8 miles		72	.111			4.45	5.70	10.15	13.05
0550	30 MPH ave, cycle 4 miles		88	.091			3.64	4.67	8.31	10.65
0552	cycle 6 miles		80	.100			4.01	5.15	9.16	11.70
0554	cycle 8 miles		72	.111			4.45	5.70	10.15	13.05
1014	12 C.Y. truck, cycle 0.5 mile, 15 MPH ave, 15 min. wait/Ld./Uld.	B-34B	336	.024			.95	2.06	3.01	3.70
1016	cycle 1 mile		300	.027			1.07	2.30	3.37	4.14
1018	cycle 2 miles		252	.032			1.27	2.74	4.01	4.94
1020	cycle 4 miles		180	.044			1.78	3.84	5.62	6.90
1022	cycle 6 miles		144	.056			2.23	4.80	7.03	8.65
1024	cycle 8 miles		120	.067			2.67	5.75	8.42	10.40
1025	cycle 10 miles		96	.083			3.34	7.20	10.54	12.95
1026	20 MPH ave, cycle 0.5 mile		348	.023			.92	1.99	2.91	3.57
1028	cycle 1 mile		312	.026			1.03	2.21	3.24	3.99
1030	cycle 2 miles		276	.029			1.16	2.50	3.66	4.51
1032	cycle 4 miles		216	.037			1.48	3.20	4.68	5.75
1034	cycle 6 miles		168	.048			1.91	4.11	6.02	7.40
1036	cycle 8 miles		144	.056			2.23	4.80	7.03	8.65
1038	cycle 10 miles		120	.067			2.67	5.75	8.42	10.40
1040	25 MPH ave, cycle 4 miles		228	.035			1.41	3.03	4.44	5.45
1042	cycle 6 miles		192	.042			1.67	3.60	5.27	6.50
1044	cycle 8 miles		168	.048			1.91	4.11	6.02	7.40
1046	cycle 10 miles		144	.056			2.23	4.80	7.03	8.65
1050	30 MPH ave, cycle 4 miles		252	.032			1.27	2.74	4.01	4.94
1052	cycle 6 miles		216	.037			1.48	3.20	4.68	5.75
1054	cycle 8 miles		180	.044			1.78	3.84	5.62	6.90
1056	cycle 10 miles		156	.051			2.05	4.43	6.48	8
1060	35 MPH ave, cycle 4 miles		264	.030			1.21	2.62	3.83	4.71
1062	cycle 6 miles		228	.035			1.41	3.03	4.44	5.45
1064	cycle 8 miles		204	.039			1.57	3.39	4.96	6.10
1066	cycle 10 miles		180	.044			1.78	3.84	5.62	6.90
1068	cycle 20 miles		120	.067			2.67	5.75	8.42	10.40
1069	cycle 30 miles		84	.095			3.81	8.25	12.06	14.80
1070	cycle 40 miles		72	.111			4.45	9.60	14.05	17.30
1072	40 MPH ave, cycle 6 miles		240	.033			1.33	2.88	4.21	5.20
1074	cycle 8 miles		216	.037			1.48	3.20	4.68	5.75
1076	cycle 10 miles		192	.042			1.67	3.60	5.27	6.50
1078	cycle 20 miles		120	.067			2.67	5.75	8.42	10.40
1080	cycle 30 miles		96	.083			3.34	7.20	10.54	12.95
1082	cycle 40 miles		72	.111			4.45	9.60	14.05	17.30
1084	cycle 50 miles		60	.133			5.35	11.50	16.85	20.50
1094	45 MPH ave, cycle 8 miles		216	.037			1.48	3.20	4.68	5.75
1096	cycle 10 miles		204	.039			1.57	3.39	4.96	6.10
1098	cycle 20 miles		132	.061			2.43	5.25	7.68	9.40

31 23 23 – Fill

31 23 23.20 Hauling		Crew	Daily Output	Labor-Hours	Unit	Material	2015 Bare Costs Labor	2015 Bare Costs Equipment	Total	Total Incl O&P
1100	cycle 30 miles	B-34B	108	.074	L.C.Y.		2.97	6.40	9.37	11.55
1102	cycle 40 miles		84	.095			3.81	8.25	12.06	14.80
1104	cycle 50 miles		72	.111			4.45	9.60	14.05	17.30
1106	50 MPH ave, cycle 10 miles		216	.037			1.48	3.20	4.68	5.75
1108	cycle 20 miles		144	.056			2.23	4.80	7.03	8.65
1110	cycle 30 miles		108	.074			2.97	6.40	9.37	11.55
1112	cycle 40 miles		84	.095			3.81	8.25	12.06	14.80
1114	cycle 50 miles		72	.111			4.45	9.60	14.05	17.30
1214	15 MPH ave, cycle 0.5 mile, 20 min. wait/Ld./Uld.		264	.030			1.21	2.62	3.83	4.71
1216	cycle 1 mile		240	.033			1.33	2.88	4.21	5.20
1218	cycle 2 miles		204	.039			1.57	3.39	4.96	6.10
1220	cycle 4 miles		156	.051			2.05	4.43	6.48	8
1222	cycle 6 miles		132	.061			2.43	5.25	7.68	9.40
1224	cycle 8 miles		108	.074			2.97	6.40	9.37	11.55
1225	cycle 10 miles		96	.083			3.34	7.20	10.54	12.95
1226	20 MPH ave, cycle 0.5 mile		264	.030			1.21	2.62	3.83	4.71
1228	cycle 1 mile		252	.032			1.27	2.74	4.01	4.94
1230	cycle 2 miles		216	.037			1.48	3.20	4.68	5.75
1232	cycle 4 miles		180	.044			1.78	3.84	5.62	6.90
1234	cycle 6 miles		144	.056			2.23	4.80	7.03	8.65
1236	cycle 8 miles		132	.061			2.43	5.25	7.68	9.40
1238	cycle 10 miles		108	.074			2.97	6.40	9.37	11.55
1240	25 MPH ave, cycle 4 miles		192	.042			1.67	3.60	5.27	6.50
1242	cycle 6 miles		168	.048			1.91	4.11	6.02	7.40
1244	cycle 8 miles		144	.056			2.23	4.80	7.03	8.65
1246	cycle 10 miles		132	.061			2.43	5.25	7.68	9.40
1250	30 MPH ave, cycle 4 miles		204	.039			1.57	3.39	4.96	6.10
1252	cycle 6 miles		180	.044			1.78	3.84	5.62	6.90
1254	cycle 8 miles		156	.051			2.05	4.43	6.48	8
1256	cycle 10 miles		144	.056			2.23	4.80	7.03	8.65
1260	35 MPH ave, cycle 4 miles		216	.037			1.48	3.20	4.68	5.75
1262	cycle 6 miles		192	.042			1.67	3.60	5.27	6.50
1264	cycle 8 miles		168	.048			1.91	4.11	6.02	7.40
1266	cycle 10 miles		156	.051			2.05	4.43	6.48	8
1268	cycle 20 miles		108	.074			2.97	6.40	9.37	11.55
1269	cycle 30 miles		72	.111			4.45	9.60	14.05	17.30
1270	cycle 40 miles		60	.133			5.35	11.50	16.85	20.50
1272	40 MPH ave, cycle 6 miles		192	.042			1.67	3.60	5.27	6.50
1274	cycle 8 miles		180	.044			1.78	3.84	5.62	6.90
1276	cycle 10 miles		156	.051			2.05	4.43	6.48	8
1278	cycle 20 miles		108	.074			2.97	6.40	9.37	11.55
1280	cycle 30 miles		84	.095			3.81	8.25	12.06	14.80
1282	cycle 40 miles		72	.111			4.45	9.60	14.05	17.30
1284	cycle 50 miles		60	.133			5.35	11.50	16.85	20.50
1294	45 MPH ave, cycle 8 miles		180	.044			1.78	3.84	5.62	6.90
1296	cycle 10 miles		168	.048			1.91	4.11	6.02	7.40
1298	cycle 20 miles		120	.067			2.67	5.75	8.42	10.40
1300	cycle 30 miles		96	.083			3.34	7.20	10.54	12.95
1302	cycle 40 miles		72	.111			4.45	9.60	14.05	17.30
1304	cycle 50 miles		60	.133			5.35	11.50	16.85	20.50
1306	50 MPH ave, cycle 10 miles		180	.044			1.78	3.84	5.62	6.90
1308	cycle 20 miles		132	.061			2.43	5.25	7.68	9.40
1310	cycle 30 miles		96	.083			3.34	7.20	10.54	12.95

31 23 23.20 Hauling		Crew	Daily Output	Labor-Hours	Unit	Material	2015 Bare Costs Labor	Equipment	Total	Total Incl O&P
1312	cycle 40 miles	B-34B	84	.095	L.C.Y.		3.81	8.25	12.06	14.80
1314	cycle 50 miles		72	.111			4.45	9.60	14.05	17.30
1414	15 MPH ave, cycle 0.5 mile, 25 min. wait/Ld./Uld.		204	.039			1.57	3.39	4.96	6.10
1416	cycle 1 mile		192	.042			1.67	3.60	5.27	6.50
1418	cycle 2 miles		168	.048			1.91	4.11	6.02	7.40
1420	cycle 4 miles		132	.061			2.43	5.25	7.68	9.40
1422	cycle 6 miles		120	.067			2.67	5.75	8.42	10.40
1424	cycle 8 miles		96	.083			3.34	7.20	10.54	12.95
1425	cycle 10 miles		84	.095			3.81	8.25	12.06	14.80
1426	20 MPH ave, cycle 0.5 mile		216	.037			1.48	3.20	4.68	5.75
1428	cycle 1 mile		204	.039			1.57	3.39	4.96	6.10
1430	cycle 2 miles		180	.044			1.78	3.84	5.62	6.90
1432	cycle 4 miles		156	.051			2.05	4.43	6.48	8
1434	cycle 6 miles		132	.061			2.43	5.25	7.68	9.40
1436	cycle 8 miles		120	.067			2.67	5.75	8.42	10.40
1438	cycle 10 miles		96	.083			3.34	7.20	10.54	12.95
1440	25 MPH ave, cycle 4 miles		168	.048			1.91	4.11	6.02	7.40
1442	cycle 6 miles		144	.056			2.23	4.80	7.03	8.65
1444	cycle 8 miles		132	.061			2.43	5.25	7.68	9.40
1446	cycle 10 miles		108	.074			2.97	6.40	9.37	11.55
1450	30 MPH ave, cycle 4 miles		168	.048			1.91	4.11	6.02	7.40
1452	cycle 6 miles		156	.051			2.05	4.43	6.48	8
1454	cycle 8 miles		132	.061			2.43	5.25	7.68	9.40
1456	cycle 10 miles		120	.067			2.67	5.75	8.42	10.40
1460	35 MPH ave, cycle 4 miles		180	.044			1.78	3.84	5.62	6.90
1462	cycle 6 miles		156	.051			2.05	4.43	6.48	8
1464	cycle 8 miles		144	.056			2.23	4.80	7.03	8.65
1466	cycle 10 miles		132	.061			2.43	5.25	7.68	9.40
1468	cycle 20 miles		96	.083			3.34	7.20	10.54	12.95
1469	cycle 30 miles		72	.111			4.45	9.60	14.05	17.30
1470	cycle 40 miles		60	.133			5.35	11.50	16.85	20.50
1472	40 MPH ave, cycle 6 miles		168	.048			1.91	4.11	6.02	7.40
1474	cycle 8 miles		156	.051			2.05	4.43	6.48	8
1476	cycle 10 miles		144	.056			2.23	4.80	7.03	8.65
1478	cycle 20 miles		96	.083			3.34	7.20	10.54	12.95
1480	cycle 30 miles		84	.095			3.81	8.25	12.06	14.80
1482	cycle 40 miles		60	.133			5.35	11.50	16.85	20.50
1484	cycle 50 miles		60	.133			5.35	11.50	16.85	20.50
1494	45 MPH ave, cycle 8 miles		156	.051			2.05	4.43	6.48	8
1496	cycle 10 miles		144	.056			2.23	4.80	7.03	8.65
1498	cycle 20 miles		108	.074			2.97	6.40	9.37	11.55
1500	cycle 30 miles		84	.095			3.81	8.25	12.06	14.80
1502	cycle 40 miles		72	.111			4.45	9.60	14.05	17.30
1504	cycle 50 miles		60	.133			5.35	11.50	16.85	20.50
1506	50 MPH ave, cycle 10 miles		156	.051			2.05	4.43	6.48	8
1508	cycle 20 miles		120	.067			2.67	5.75	8.42	10.40
1510	cycle 30 miles		96	.083			3.34	7.20	10.54	12.95
1512	cycle 40 miles		72	.111			4.45	9.60	14.05	17.30
1514	cycle 50 miles		60	.133			5.35	11.50	16.85	20.50
1614	15 MPH, cycle 0.5 mile, 30 min. wait/Ld./Uld.		180	.044			1.78	3.84	5.62	6.90
1616	cycle 1 mile		168	.048			1.91	4.11	6.02	7.40
1618	cycle 2 miles		144	.056			2.23	4.80	7.03	8.65
1620	cycle 4 miles		120	.067			2.67	5.75	8.42	10.40

31 23 23.20 **Hauling**	Crew	Daily Output	Labor-Hours	Unit	Material	2015 Bare Costs Labor	Equipment	Total	Total Incl O&P
1622	cycle 6 miles	B-34B	108	.074	L.C.Y.		2.97	6.40	9.37	11.55
1624	cycle 8 miles		84	.095			3.81	8.25	12.06	14.80
1625	cycle 10 miles		84	.095			3.81	8.25	12.06	14.80
1626	20 MPH ave, cycle 0.5 mile		180	.044			1.78	3.84	5.62	6.90
1628	cycle 1 mile		168	.048			1.91	4.11	6.02	7.40
1630	cycle 2 miles		156	.051			2.05	4.43	6.48	8
1632	cycle 4 miles		132	.061			2.43	5.25	7.68	9.40
1634	cycle 6 miles		120	.067			2.67	5.75	8.42	10.40
1636	cycle 8 miles		108	.074			2.97	6.40	9.37	11.55
1638	cycle 10 miles		96	.083			3.34	7.20	10.54	12.95
1640	25 MPH ave, cycle 4 miles		144	.056			2.23	4.80	7.03	8.65
1642	cycle 6 miles		132	.061			2.43	5.25	7.68	9.40
1644	cycle 8 miles		108	.074			2.97	6.40	9.37	11.55
1646	cycle 10 miles		108	.074			2.97	6.40	9.37	11.55
1650	30 MPH ave, cycle 4 miles		144	.056			2.23	4.80	7.03	8.65
1652	cycle 6 miles		132	.061			2.43	5.25	7.68	9.40
1654	cycle 8 miles		120	.067			2.67	5.75	8.42	10.40
1656	cycle 10 miles		108	.074			2.97	6.40	9.37	11.55
1660	35 MPH ave, cycle 4 miles		156	.051			2.05	4.43	6.48	8
1662	cycle 6 miles		144	.056			2.23	4.80	7.03	8.65
1664	cycle 8 miles		132	.061			2.43	5.25	7.68	9.40
1666	cycle 10 miles		120	.067			2.67	5.75	8.42	10.40
1668	cycle 20 miles		84	.095			3.81	8.25	12.06	14.80
1669	cycle 30 miles		72	.111			4.45	9.60	14.05	17.30
1670	cycle 40 miles		60	.133			5.35	11.50	16.85	20.50
1672	40 MPH, cycle 6 miles		144	.056			2.23	4.80	7.03	8.65
1674	cycle 8 miles		132	.061			2.43	5.25	7.68	9.40
1676	cycle 10 miles		120	.067			2.67	5.75	8.42	10.40
1678	cycle 20 miles		96	.083			3.34	7.20	10.54	12.95
1680	cycle 30 miles		72	.111			4.45	9.60	14.05	17.30
1682	cycle 40 miles		60	.133			5.35	11.50	16.85	20.50
1684	cycle 50 miles		48	.167			6.70	14.40	21.10	26
1694	45 MPH ave, cycle 8 miles		144	.056			2.23	4.80	7.03	8.65
1696	cycle 10 miles		132	.061			2.43	5.25	7.68	9.40
1698	cycle 20 miles		96	.083			3.34	7.20	10.54	12.95
1700	cycle 30 miles		84	.095			3.81	8.25	12.06	14.80
1702	cycle 40 miles		60	.133			5.35	11.50	16.85	20.50
1704	cycle 50 miles		60	.133			5.35	11.50	16.85	20.50
1706	50 MPH ave, cycle 10 miles		132	.061			2.43	5.25	7.68	9.40
1708	cycle 20 miles		108	.074			2.97	6.40	9.37	11.55
1710	cycle 30 miles		84	.095			3.81	8.25	12.06	14.80
1712	cycle 40 miles		72	.111			4.45	9.60	14.05	17.30
1714	cycle 50 miles		60	.133			5.35	11.50	16.85	20.50
2000	Hauling, 8 C.Y. truck, small project cost per hour	B-34A	8	1	Hr.		40	51.50	91.50	117
2100	12 C.Y. Truck	B-34B	8	1			40	86.50	126.50	156
2150	16.5 C.Y. Truck	B-34C	8	1			40	91	131	161
2175	18 C.Y. 8 wheel Truck	B-34I	8	1			40	109	149	180
2200	20 C.Y. Truck	B-34D	8	1			40	92.50	132.50	163
2300	Grading at dump, or embankment if required, by dozer	B-10B	1000	.012	L.C.Y.		.56	1.39	1.95	2.38
2310	Spotter at fill or cut, if required	1 Clab	8	1	Hr.		37.50		37.50	58
9014	18 C.Y. truck, 8 wheels,15 min. wait/Ld./Uld.,15 MPH, cycle 0.5 mi.	B-34I	504	.016	L.C.Y.		.64	1.72	2.36	2.86
9016	cycle 1 mile		450	.018			.71	1.93	2.64	3.20
9018	cycle 2 miles		378	.021			.85	2.30	3.15	3.81

31 23 23 – Fill

31 23 23.20 Hauling		Crew	Daily Output	Labor-Hours	Unit	Material	2015 Bare Costs Labor	Equipment	Total	Total Incl O&P
9020	cycle 4 miles	B-34I	270	.030	L.C.Y.		1.19	3.22	4.41	5.35
9022	cycle 6 miles		216	.037			1.48	4.02	5.50	6.65
9024	cycle 8 miles		180	.044			1.78	4.83	6.61	8
9025	cycle 10 miles		144	.056			2.23	6.05	8.28	10
9026	20 MPH ave, cycle 0.5 mile		522	.015			.61	1.67	2.28	2.76
9028	cycle 1 mile		468	.017			.68	1.86	2.54	3.07
9030	cycle 2 miles		414	.019			.77	2.10	2.87	3.48
9032	cycle 4 miles		324	.025			.99	2.68	3.67	4.45
9034	cycle 6 miles		252	.032			1.27	3.45	4.72	5.70
9036	cycle 8 miles		216	.037			1.48	4.02	5.50	6.65
9038	cycle 10 miles		180	.044			1.78	4.83	6.61	8
9040	25 MPH ave, cycle 4 miles		342	.023			.94	2.54	3.48	4.21
9042	cycle 6 miles		288	.028			1.11	3.02	4.13	5
9044	cycle 8 miles		252	.032			1.27	3.45	4.72	5.70
9046	cycle 10 miles		216	.037			1.48	4.02	5.50	6.65
9050	30 MPH ave, cycle 4 miles		378	.021			.85	2.30	3.15	3.81
9052	cycle 6 miles		324	.025			.99	2.68	3.67	4.45
9054	cycle 8 miles		270	.030			1.19	3.22	4.41	5.35
9056	cycle 10 miles		234	.034			1.37	3.71	5.08	6.15
9060	35 MPH ave, cycle 4 miles		396	.020			.81	2.19	3	3.63
9062	cycle 6 miles		342	.023			.94	2.54	3.48	4.21
9064	cycle 8 miles		288	.028			1.11	3.02	4.13	5
9066	cycle 10 miles		270	.030			1.19	3.22	4.41	5.35
9068	cycle 20 miles		162	.049			1.98	5.35	7.33	8.90
9070	cycle 30 miles		126	.063			2.54	6.90	9.44	11.45
9072	cycle 40 miles		90	.089			3.56	9.65	13.21	16
9074	40 MPH ave, cycle 6 miles		360	.022			.89	2.41	3.30	4.01
9076	cycle 8 miles		324	.025			.99	2.68	3.67	4.45
9078	cycle 10 miles		288	.028			1.11	3.02	4.13	5
9080	cycle 20 miles		180	.044			1.78	4.83	6.61	8
9082	cycle 30 miles		144	.056			2.23	6.05	8.28	10
9084	cycle 40 miles		108	.074			2.97	8.05	11.02	13.35
9086	cycle 50 miles		90	.089			3.56	9.65	13.21	16
9094	45 MPH ave, cycle 8 miles		324	.025			.99	2.68	3.67	4.45
9096	cycle 10 miles		306	.026			1.05	2.84	3.89	4.70
9098	cycle 20 miles		198	.040			1.62	4.39	6.01	7.30
9100	cycle 30 miles		144	.056			2.23	6.05	8.28	10
9102	cycle 40 miles		126	.063			2.54	6.90	9.44	11.45
9104	cycle 50 miles		108	.074			2.97	8.05	11.02	13.35
9106	50 MPH ave, cycle 10 miles		324	.025			.99	2.68	3.67	4.45
9108	cycle 20 miles		216	.037			1.48	4.02	5.50	6.65
9110	cycle 30 miles		162	.049			1.98	5.35	7.33	8.90
9112	cycle 40 miles		126	.063			2.54	6.90	9.44	11.45
9114	cycle 50 miles		108	.074			2.97	8.05	11.02	13.35
9214	20 min. wait/Ld./Uld.,15 MPH, cycle 0.5 mi.		396	.020			.81	2.19	3	3.63
9216	cycle 1 mile		360	.022			.89	2.41	3.30	4.01
9218	cycle 2 miles		306	.026			1.05	2.84	3.89	4.70
9220	cycle 4 miles		234	.034			1.37	3.71	5.08	6.15
9222	cycle 6 miles		198	.040			1.62	4.39	6.01	7.30
9224	cycle 8 miles		162	.049			1.98	5.35	7.33	8.90
9225	cycle 10 miles		144	.056			2.23	6.05	8.28	10
9226	20 MPH ave, cycle 0.5 mile		396	.020			.81	2.19	3	3.63
9228	cycle 1 mile		378	.021			.85	2.30	3.15	3.81

31 23 23 – Fill

31 23 23.20 Hauling		Crew	Daily Output	Labor-Hours	Unit	Material	2015 Bare Costs		Total	Total Incl O&P
							Labor	Equipment		
9230	cycle 2 miles	B-34I	324	.025	L.C.Y.		.99	2.68	3.67	4.45
9232	cycle 4 miles		270	.030			1.19	3.22	4.41	5.35
9234	cycle 6 miles		216	.037			1.48	4.02	5.50	6.65
9236	cycle 8 miles		198	.040			1.62	4.39	6.01	7.30
9238	cycle 10 miles		162	.049			1.98	5.35	7.33	8.90
9240	25 MPH ave, cycle 4 miles		288	.028			1.11	3.02	4.13	5
9242	cycle 6 miles		252	.032			1.27	3.45	4.72	5.70
9244	cycle 8 miles		216	.037			1.48	4.02	5.50	6.65
9246	cycle 10 miles		198	.040			1.62	4.39	6.01	7.30
9250	30 MPH ave, cycle 4 miles		306	.026			1.05	2.84	3.89	4.70
9252	cycle 6 miles		270	.030			1.19	3.22	4.41	5.35
9254	cycle 8 miles		234	.034			1.37	3.71	5.08	6.15
9256	cycle 10 miles		216	.037			1.48	4.02	5.50	6.65
9260	35 MPH ave, cycle 4 miles		324	.025			.99	2.68	3.67	4.45
9262	cycle 6 miles		288	.028			1.11	3.02	4.13	5
9264	cycle 8 miles		252	.032			1.27	3.45	4.72	5.70
9266	cycle 10 miles		234	.034			1.37	3.71	5.08	6.15
9268	cycle 20 miles		162	.049			1.98	5.35	7.33	8.90
9270	cycle 30 miles		108	.074			2.97	8.05	11.02	13.35
9272	cycle 40 miles		90	.089			3.56	9.65	13.21	16
9274	40 MPH ave, cycle 6 miles		288	.028			1.11	3.02	4.13	5
9276	cycle 8 miles		270	.030			1.19	3.22	4.41	5.35
9278	cycle 10 miles		234	.034			1.37	3.71	5.08	6.15
9280	cycle 20 miles		162	.049			1.98	5.35	7.33	8.90
9282	cycle 30 miles		126	.063			2.54	6.90	9.44	11.45
9284	cycle 40 miles		108	.074			2.97	8.05	11.02	13.35
9286	cycle 50 miles		90	.089			3.56	9.65	13.21	16
9294	45 MPH ave, cycle 8 miles		270	.030			1.19	3.22	4.41	5.35
9296	cycle 10 miles		252	.032			1.27	3.45	4.72	5.70
9298	cycle 20 miles		180	.044			1.78	4.83	6.61	8
9300	cycle 30 miles		144	.056			2.23	6.05	8.28	10
9302	cycle 40 miles		108	.074			2.97	8.05	11.02	13.35
9304	cycle 50 miles		90	.089			3.56	9.65	13.21	16
9306	50 MPH ave, cycle 10 miles		270	.030			1.19	3.22	4.41	5.35
9308	cycle 20 miles		198	.040			1.62	4.39	6.01	7.30
9310	cycle 30 miles		144	.056			2.23	6.05	8.28	10
9312	cycle 40 miles		126	.063			2.54	6.90	9.44	11.45
9314	cycle 50 miles		108	.074			2.97	8.05	11.02	13.35
9414	25 min. wait/Ld./Uld.,15 MPH, cycle 0.5 mi.		306	.026			1.05	2.84	3.89	4.70
9416	cycle 1 mile		288	.028			1.11	3.02	4.13	5
9418	cycle 2 miles		252	.032			1.27	3.45	4.72	5.70
9420	cycle 4 miles		198	.040			1.62	4.39	6.01	7.30
9422	cycle 6 miles		180	.044			1.78	4.83	6.61	8
9424	cycle 8 miles		144	.056			2.23	6.05	8.28	10
9425	cycle 10 miles		126	.063			2.54	6.90	9.44	11.45
9426	20 MPH ave, cycle 0.5 mile		324	.025			.99	2.68	3.67	4.45
9428	cycle 1 mile		306	.026			1.05	2.84	3.89	4.70
9430	cycle 2 miles		270	.030			1.19	3.22	4.41	5.35
9432	cycle 4 miles		234	.034			1.37	3.71	5.08	6.15
9434	cycle 6 miles		198	.040			1.62	4.39	6.01	7.30
9436	cycle 8 miles		180	.044			1.78	4.83	6.61	8
9438	cycle 10 miles		144	.056			2.23	6.05	8.28	10
9440	25 MPH ave, cycle 4 miles		252	.032			1.27	3.45	4.72	5.70

31 23 23.20 Hauling		Crew	Daily Output	Labor-Hours	Unit	Material	2015 Bare Costs Labor	2015 Bare Costs Equipment	Total	Total Incl O&P
9442	cycle 6 miles	B-34I	216	.037	L.C.Y.		1.48	4.02	5.50	6.65
9444	cycle 8 miles		198	.040			1.62	4.39	6.01	7.30
9446	cycle 10 miles		180	.044			1.78	4.83	6.61	8
9450	30 MPH ave, cycle 4 miles		252	.032			1.27	3.45	4.72	5.70
9452	cycle 6 miles		234	.034			1.37	3.71	5.08	6.15
9454	cycle 8 miles		198	.040			1.62	4.39	6.01	7.30
9456	cycle 10 miles		180	.044			1.78	4.83	6.61	8
9460	35 MPH ave, cycle 4 miles		270	.030			1.19	3.22	4.41	5.35
9462	cycle 6 miles		234	.034			1.37	3.71	5.08	6.15
9464	cycle 8 miles		216	.037			1.48	4.02	5.50	6.65
9466	cycle 10 miles		198	.040			1.62	4.39	6.01	7.30
9468	cycle 20 miles		144	.056			2.23	6.05	8.28	10
9470	cycle 30 miles		108	.074			2.97	8.05	11.02	13.35
9472	cycle 40 miles		90	.089			3.56	9.65	13.21	16
9474	40 MPH ave, cycle 6 miles		252	.032			1.27	3.45	4.72	5.70
9476	cycle 8 miles		234	.034			1.37	3.71	5.08	6.15
9478	cycle 10 miles		216	.037			1.48	4.02	5.50	6.65
9480	cycle 20 miles		144	.056			2.23	6.05	8.28	10
9482	cycle 30 miles		126	.063			2.54	6.90	9.44	11.45
9484	cycle 40 miles		90	.089			3.56	9.65	13.21	16
9486	cycle 50 miles		90	.089			3.56	9.65	13.21	16
9494	45 MPH ave, cycle 8 miles		234	.034			1.37	3.71	5.08	6.15
9496	cycle 10 miles		216	.037			1.48	4.02	5.50	6.65
9498	cycle 20 miles		162	.049			1.98	5.35	7.33	8.90
9500	cycle 30 miles		126	.063			2.54	6.90	9.44	11.45
9502	cycle 40 miles		108	.074			2.97	8.05	11.02	13.35
9504	cycle 50 miles		90	.089			3.56	9.65	13.21	16
9506	50 MPH ave, cycle 10 miles		234	.034			1.37	3.71	5.08	6.15
9508	cycle 20 miles		180	.044			1.78	4.83	6.61	8
9510	cycle 30 miles		144	.056			2.23	6.05	8.28	10
9512	cycle 40 miles		108	.074			2.97	8.05	11.02	13.35
9514	cycle 50 miles		90	.089			3.56	9.65	13.21	16
9614	30 min. wait/Ld./Uld.,15 MPH, cycle 0.5 mi.		270	.030			1.19	3.22	4.41	5.35
9616	cycle 1 mile		252	.032			1.27	3.45	4.72	5.70
9618	cycle 2 miles		216	.037			1.48	4.02	5.50	6.65
9620	cycle 4 miles		180	.044			1.78	4.83	6.61	8
9622	cycle 6 miles		162	.049			1.98	5.35	7.33	8.90
9624	cycle 8 miles		126	.063			2.54	6.90	9.44	11.45
9625	cycle 10 miles		126	.063			2.54	6.90	9.44	11.45
9626	20 MPH ave, cycle 0.5 mile		270	.030			1.19	3.22	4.41	5.35
9628	cycle 1 mile		252	.032			1.27	3.45	4.72	5.70
9630	cycle 2 miles		234	.034			1.37	3.71	5.08	6.15
9632	cycle 4 miles		198	.040			1.62	4.39	6.01	7.30
9634	cycle 6 miles		180	.044			1.78	4.83	6.61	8
9636	cycle 8 miles		162	.049			1.98	5.35	7.33	8.90
9638	cycle 10 miles		144	.056			2.23	6.05	8.28	10
9640	25 MPH ave, cycle 4 miles		216	.037			1.48	4.02	5.50	6.65
9642	cycle 6 miles		198	.040			1.62	4.39	6.01	7.30
9644	cycle 8 miles		180	.044			1.78	4.83	6.61	8
9646	cycle 10 miles		162	.049			1.98	5.35	7.33	8.90
9650	30 MPH ave, cycle 4 miles		216	.037			1.48	4.02	5.50	6.65
9652	cycle 6 miles		198	.040			1.62	4.39	6.01	7.30
9654	cycle 8 miles		180	.044			1.78	4.83	6.61	8

31 23 23.20 Hauling		Crew	Daily Output	Labor-Hours	Unit	Material	2015 Bare Costs Labor	Equipment	Total	Total Incl O&P
9656	cycle 10 miles	B-34I	162	.049	L.C.Y.		1.98	5.35	7.33	8.90
9660	35 MPH ave, cycle 4 miles		234	.034			1.37	3.71	5.08	6.15
9662	cycle 6 miles		216	.037			1.48	4.02	5.50	6.65
9664	cycle 8 miles		198	.040			1.62	4.39	6.01	7.30
9666	cycle 10 miles		180	.044			1.78	4.83	6.61	8
9668	cycle 20 miles		126	.063			2.54	6.90	9.44	11.45
9670	cycle 30 miles		108	.074			2.97	8.05	11.02	13.35
9672	cycle 40 miles		90	.089			3.56	9.65	13.21	16
9674	40 MPH ave, cycle 6 miles		216	.037			1.48	4.02	5.50	6.65
9676	cycle 8 miles		198	.040			1.62	4.39	6.01	7.30
9678	cycle 10 miles		180	.044			1.78	4.83	6.61	8
9680	cycle 20 miles		144	.056			2.23	6.05	8.28	10
9682	cycle 30 miles		108	.074			2.97	8.05	11.02	13.35
9684	cycle 40 miles		90	.089			3.56	9.65	13.21	16
9686	cycle 50 miles		72	.111			4.45	12.05	16.50	20
9694	45 MPH ave, cycle 8 miles		216	.037			1.48	4.02	5.50	6.65
9696	cycle 10 miles		198	.040			1.62	4.39	6.01	7.30
9698	cycle 20 miles		144	.056			2.23	6.05	8.28	10
9700	cycle 30 miles		126	.063			2.54	6.90	9.44	11.45
9702	cycle 40 miles		108	.074			2.97	8.05	11.02	13.35
9704	cycle 50 miles		90	.089			3.56	9.65	13.21	16
9706	50 MPH ave, cycle 10 miles		198	.040			1.62	4.39	6.01	7.30
9708	cycle 20 miles		162	.049			1.98	5.35	7.33	8.90
9710	cycle 30 miles		126	.063			2.54	6.90	9.44	11.45
9712	cycle 40 miles		108	.074			2.97	8.05	11.02	13.35
9714	cycle 50 miles		90	.089			3.56	9.65	13.21	16

For customer support on your Plumbing Cost Data, call 877.709.6968.

391

Division Notes

		CREW	DAILY OUTPUT	LABOR-HOURS	UNIT	BARE COSTS				TOTAL INCL O&P
						MAT.	LABOR	EQUIP.	TOTAL	

Estimating Tips

32 01 00 Operations and Maintenance of Exterior Improvements

- Recycling of asphalt pavement is becoming very popular and is an alternative to removal and replacement. It can be a good value engineering proposal if removed pavement can be recycled, either at the project site or at another site that is reasonably close to the project site. Sections on repair of flexible and rigid pavement are included.

32 10 00 Bases, Ballasts, and Paving

- When estimating paving, keep in mind the project schedule. Also note that prices for asphalt and concrete are generally higher in the cold seasons. Lines for pavement markings, including tactile warning systems and fence lines, are included.

32 90 00 Planting

- The timing of planting and guarantee specifications often dictate the costs for establishing tree and shrub growth and a stand of grass or ground cover. Establish the work performance schedule to coincide with the local planting season. Maintenance and growth guarantees can add from 20%–100% to the total landscaping cost and can be contractually cumbersome. The cost to replace trees and shrubs can be as high as 5% of the total cost, depending on the planting zone, soil conditions, and time of year.

Reference Numbers

Reference numbers are shown in shaded boxes at the beginning of some major classifications. These numbers refer to related items in the Reference Section. The reference information may be an estimating procedure, an alternate pricing method, or technical information.

Note: Not all subdivisions listed here necessarily appear in this publication. ■

32 12 16.13 Plant-Mix Asphalt Paving	Crew	Daily Output	Labor-Hours	Unit	Material	2015 Bare Costs Labor	Equipment	Total	Total Incl O&P
0010 **PLANT-MIX ASPHALT PAVING**									
0020 And large paved areas with no hauling included									
0025 See Section 31 23 23.20 for hauling costs									
0080 Binder course, 1-1/2" thick	B-25	7725	.011	S.Y.	5.70	.47	.35	6.52	7.40
0120 2" thick		6345	.014		7.60	.57	.43	8.60	9.70
0130 2-1/2" thick		5620	.016		9.50	.65	.48	10.63	11.95
0160 3" thick		4905	.018		11.40	.74	.55	12.69	14.30
0170 3-1/2" thick		4520	.019		13.30	.80	.60	14.70	16.55
0200 4" thick		4140	.021		15.25	.88	.66	16.79	18.80
0300 Wearing course, 1" thick	B-25B	10575	.009		3.78	.38	.28	4.44	5.05
0340 1-1/2" thick		7725	.012		6.35	.52	.38	7.25	8.20
0380 2" thick		6345	.015		8.50	.64	.46	9.60	10.90
0420 2-1/2" thick		5480	.018		10.50	.74	.54	11.78	13.25
0460 3" thick		4900	.020		12.55	.82	.60	13.97	15.70
0470 3-1/2" thick		4520	.021		14.70	.89	.65	16.24	18.30
0480 4" thick		4140	.023		16.80	.98	.71	18.49	21
0500 Open graded friction course	B-25C	5000	.010		2.13	.41	.47	3.01	3.48
0800 Alternate method of figuring paving costs									
0810 Binder course, 1-1/2" thick	B-25	630	.140	Ton	70	5.75	4.31	80.06	90.50
0811 2" thick		690	.128		70	5.25	3.93	79.18	89.50
0812 3" thick		800	.110		70	4.55	3.39	77.94	87.50
0813 4" thick		900	.098		70	4.04	3.01	77.05	86.50
0850 Wearing course, 1" thick	B-25B	575	.167		69	7.05	5.15	81.20	92.50
0851 1-1/2" thick		630	.152		69	6.40	4.68	80.08	91
0852 2" thick		690	.139		69	5.85	4.28	79.13	89.50
0853 2-1/2" thick		765	.125		69	5.30	3.86	78.16	88.50
0854 3" thick		800	.120		69	5.05	3.69	77.74	88
1000 Pavement replacement over trench, 2" thick	B-37	90	.533	S.Y.	7.85	21	1.75	30.60	43
1050 4" thick		70	.686		15.55	27.50	2.25	45.30	61.50
1080 6" thick		55	.873		25	34.50	2.86	62.36	83.50

32 12 16.14 Asphaltic Concrete Paving

	Crew	Daily Output	Labor-Hours	Unit	Material	2015 Bare Costs Labor	Equipment	Total	Total Incl O&P
0011 **ASPHALTIC CONCRETE PAVING**, parking lots & driveways									
0015 No asphalt hauling included									
0018 Use 6.05 C.Y. per inch per M.S.F. for hauling									
0020 6" stone base, 2" binder course, 1" topping	B-25C	9000	.005	S.F.	1.86	.23	.26	2.35	2.68
0025 2" binder course, 2" topping		9000	.005		2.28	.23	.26	2.77	3.14
0030 3" binder course, 2" topping		9000	.005		2.71	.23	.26	3.20	3.61
0035 4" binder course, 2" topping		9000	.005		3.13	.23	.26	3.62	4.07
0040 1.5" binder course, 1" topping		9000	.005		1.65	.23	.26	2.14	2.45
0042 3" binder course, 1" topping		9000	.005		2.29	.23	.26	2.78	3.15
0045 3" binder course, 3" topping		9000	.005		3.13	.23	.26	3.62	4.07
0050 4" binder course, 3" topping		9000	.005		3.55	.23	.26	4.04	4.53
0055 4" binder course, 4" topping		9000	.005		3.96	.23	.26	4.45	4.99
0300 Binder course, 1-1/2" thick		35000	.001		.64	.06	.07	.77	.86
0400 2" thick		25000	.002		.82	.08	.09	.99	1.13
0500 3" thick		15000	.003		1.27	.14	.16	1.57	1.78
0600 4" thick		10800	.004		1.67	.19	.22	2.08	2.36
0800 Sand finish course, 3/4" thick		41000	.001		.32	.05	.06	.43	.49
0900 1" thick		34000	.001		.40	.06	.07	.53	.61
1000 Fill pot holes, hot mix, 2" thick	B-16	4200	.008		.85	.30	.16	1.31	1.57
1100 4" thick		3500	.009		1.25	.35	.20	1.80	2.13
1120 6" thick		3100	.010		1.67	.40	.22	2.29	2.70

32 12 Flexible Paving

32 12 16 – Asphalt Paving

32 12 16.14 Asphaltic Concrete Paving		Crew	Daily Output	Labor-Hours	Unit	Material	2015 Bare Costs Labor	Equipment	Total	Total Incl O&P
1140	Cold patch, 2" thick	B-51	3000	.016	S.F.	.99	.61	.08	1.68	2.12
1160	4" thick		2700	.018		1.89	.68	.09	2.66	3.22
1180	6" thick		1900	.025		2.94	.96	.13	4.03	4.86

32 84 Planting Irrigation

32 84 13 – Drip Irrigation

32 84 13.10 Subsurface Drip Irrigation

			Crew	Daily Output	Labor-Hours	Unit	Material	2015 Bare Costs Labor	Equipment	Total	Total Incl O&P
0010	**SUBSURFACE DRIP IRRIGATION**										
0011	Looped grid, pressure compensating										
0100	Preinserted PE emitter, line, hand bury, irregular area, small	G	3 Skwk	1200	.020	L.F.	.29	.97		1.26	1.82
0150	Medium	G		1800	.013		.29	.65		.94	1.32
0200	Large	G		2520	.010		.29	.46		.75	1.04
0250	Rectangular area, small	G		2040	.012		.29	.57		.86	1.20
0300	Medium	G		2640	.009		.29	.44		.73	1
0350	Large	G		3600	.007		.29	.32		.61	.82
0400	Install in trench, irregular area, small	G		4050	.006		.29	.29		.58	.77
0450	Medium	G		7488	.003		.29	.16		.45	.56
0500	Large	G		16560	.001		.29	.07		.36	.43
0550	Rectangular area, small	G		8100	.003		.29	.14		.43	.54
0600	Medium			21960	.001		.29	.05		.34	.40
0650	Large			33264	.001		.29	.04		.33	.37
0700	Trenching and backfill	G	B-53	500	.016			.78	.14	.92	1.33
0750	For vinyl tubing, 1/4", add to above						25%	10%			
0800	Vinyl tubing, 1/4", material only	G					.05			.05	.06
0850	Supply tubing, 1/2", material only, 100' coil	G					.12			.12	.13
0900	500' coil	G					.11			.11	.12
0950	Compression fittings	G	1 Skwk	90	.089	Ea.	.72	4.32		5.04	7.50
1000	Barbed fittings, 1/4"	G		360	.022		.16	1.08		1.24	1.85
1100	Flush risers	G		60	.133		3.40	6.50		9.90	13.75
1150	Flush ends, figure eight	G		180	.044		.52	2.16		2.68	3.91
1200	Ball valve, 4-1/2"	G		20	.400		9.25	19.45		28.70	40
1250	4-3/4"	G		20	.400		7.25	19.45		26.70	38
1300	Auto flush, spring loaded	G		90	.089		2.20	4.32		6.52	9.10
1350	Volumetric	G		90	.089		7.50	4.32		11.82	14.95
1400	Air relief valve, inline with compensation tee, 1/2"	G		45	.178		16	8.65		24.65	31
1450	1"	G		30	.267		17	12.95		29.95	38.50
1500	Round box for flush ends, 6"	G		30	.267		7.75	12.95		20.70	28.50
1550	Fertilizer injector, non-proportional	G		4	2		12	97.50		109.50	163
1600	Screen filter, 3/4" screen	G		12	.667		10.35	32.50		42.85	61.50
1650	1" disk	G		8	1		26	48.50		74.50	104
1700	1-1/2" disk	G		4	2		70	97.50		167.50	227
1750	2" disk	G		3	2.667		121	130		251	335
1800	Typical installation 18" O.C., small, minimum					S.F.				1.53	2.35
1850	Maximum									1.78	2.75
1900	Large, minimum									1.48	2.28
2000	Maximum									1.71	2.64
2100	For non-pressure compensating systems, deduct						10%	10%			

32 84 23.10 Sprinkler Irrigation System	Crew	Daily Output	Labor-Hours	Unit	Material	2015 Bare Costs Labor	Equipment	Total	Total Incl O&P
0010 **SPRINKLER IRRIGATION SYSTEM** G2050–710									
0011 For lawns									
0100 Golf course with fully automatic system	C-17	.05	1600	9 holes	100,000	78,500		178,500	231,000
0200 24' diam. head at 15' O.C. incl. piping, auto oper., minimum	B-20	70	.343	Head	27	14.40		41.40	51.50
0300 Maximum		40	.600		45	25		70	88.50
0600 Sprinkler irrigation sys, golf course, auto sys, 60' dia HD		23	1.043		150	44		194	233
0800 Residential system, custom, 1" supply		2000	.012	S.F.	.26	.50		.76	1.07
0900 1-1/2" supply		1800	.013	"	.49	.56		1.05	1.40
0990 For renovation work, add to above						50%			
1020 Pop up spray head w/risers, hi-pop, full circle pattern, 4"	2 Skwk	76	.211	Ea.	5.15	10.25		15.40	21.50
1030 1/2 circle pattern, 4"		76	.211		5.15	10.25		15.40	21.50
1040 6", full circle pattern		76	.211		9.50	10.25		19.75	26.50
1050 1/2 circle pattern, 6"		76	.211		9.50	10.25		19.75	26.50
1060 12", full circle pattern		76	.211		11.15	10.25		21.40	28
1070 1/2 circle pattern, 12"		76	.211		13.75	10.25		24	31
1080 Pop up bubbler head w/risers, hi-pop bubbler head, 4"		76	.211		4.40	10.25		14.65	20.50
1090 6"		76	.211		9.20	10.25		19.45	26
1100 12"		76	.211		11	10.25		21.25	28
1110 Impact full/part circle sprinklers, 28'-54' 25-60 PSI		37	.432		17.65	21		38.65	52
1120 Spaced 37'-49' @ 25-50 PSI		37	.432		22	21		43	57
1130 Spaced 43'-61' @ 30-60 PSI		37	.432		61.50	21		82.50	100
1140 Spaced 54'-78' @ 40-80 PSI		37	.432		106	21		127	150
1145 Impact rotor pop-up full/part commercial circle sprinklers									
1150 Spaced 42'-65' 35-80 PSI	2 Skwk	25	.640	Ea.	15.30	31		46.30	65
1160 Spaced 48'-76' 45-85 PSI	"	25	.640	"	16.85	31		47.85	66.50
1165 Impact rotor pop-up part. circle comm., 53'-75', 55-100 PSI, w/accessories									
1170 Plastic case, metal cover	2 Skwk	25	.640	Ea.	74	31		105	129
1180 Rubber cover		25	.640		56.50	31		87.50	111
1190 Iron case, metal cover		22	.727		122	35.50		157.50	189
1200 Rubber cover		22	.727		129	35.50		164.50	197
1250 Plastic case, 2 nozzle, metal cover		25	.640		89.50	31		120.50	146
1260 Rubber cover		25	.640		91.50	31		122.50	148
1270 Iron case, 2 nozzle, metal cover		22	.727		130	35.50		165.50	197
1280 Rubber cover		22	.727		130	35.50		165.50	197
1282 Impact rotor pop-up full circle commercial, 39'-99', 30-100 PSI									
1284 Plastic case, metal cover	2 Skwk	25	.640	Ea.	83	31		114	139
1286 Rubber cover		25	.640		94.50	31		125.50	152
1288 Iron case, metal cover		22	.727		122	35.50		157.50	189
1290 Rubber cover		22	.727		126	35.50		161.50	194
1292 Plastic case, 2 nozzle, metal cover		22	.727		90	35.50		125.50	154
1294 Rubber cover		22	.727		90	35.50		125.50	154
1296 Iron case, 2 nozzle, metal cover		20	.800		119	39		158	190
1298 Rubber cover		20	.800		123	39		162	195
1305 Electric remote control valve, plastic, 3/4"		18	.889		23.50	43		66.50	93
1310 1"		18	.889		23.50	43		66.50	93
1320 1-1/2"		18	.889		91	43		134	167
1330 2"		18	.889		109	43		152	187
1335 Quick coupling valves, brass, locking cover									
1340 Inlet coupling valve, 3/4"	2 Skwk	18.75	.853	Ea.	21.50	41.50		63	87.50
1350 1"		18.75	.853		30	41.50		71.50	97
1360 Controller valve boxes, 6" round boxes		18.75	.853		7.10	41.50		48.60	72
1370 10" round boxes		14.25	1.123		11.40	54.50		65.90	97

For customer support on your Plumbing Cost Data, call 877.709.6968.

32 84 Planting Irrigation

32 84 23 – Underground Sprinklers

32 84 23.10 Sprinkler Irrigation System	Crew	Daily Output	Labor-Hours	Unit	Material	2015 Bare Costs Labor	Equipment	Total	Total Incl O&P
1380	12" square box	2 Skwk	9.75	1.641	Ea.	16.05	80		96.05	141
1388	Electromech. control, 14 day 3-60 min., auto start to 23/day									
1390	4 station	2 Skwk	1.04	15.385	Ea.	75	750		825	1,225
1400	7 station		.64	25		140	1,225		1,365	2,025
1410	12 station		.40	40		170	1,950		2,120	3,175
1420	Dual programs, 18 station		.24	66.667		200	3,250		3,450	5,225
1430	23 station		.16	100		220	4,875		5,095	7,775
1435	Backflow preventer, bronze, 0-175 PSI, w/valves, test cocks									
1440	3/4"	2 Skwk	6	2.667	Ea.	82	130		212	290
1450	1"		6	2.667		94	130		224	305
1460	1-1/2"		6	2.667		224	130		354	445
1470	2"		6	2.667		276	130		406	505
1475	Pressure vacuum breaker, brass, 15-150 PSI									
1480	3/4"	2 Skwk	6	2.667	Ea.	25	130		155	228
1490	1"		6	2.667		30	130		160	233
1500	1-1/2"		6	2.667		70	130		200	277
1510	2"		6	2.667		120	130		250	330
2000	Pop-up spray, head & nozzle, low/medium volume, plastic	1 Skwk	30	.267		3.09	12.95		16.04	23.50
2200	Brass, economy		30	.267		10.95	12.95		23.90	32
2400	Heavy duty		30	.267		18.75	12.95		31.70	40.50
3000	Riser mounted spray, head, low/medium volume, plastic		30	.267		3.09	12.95		16.04	23.50
3200	Brass		30	.267		10.20	12.95		23.15	31.50
4000	Pop-up impact sprinkler, body & case, plastic low/medium volume		30	.267		21.50	12.95		34.45	43.50
4200	Brass, high/medium volume		30	.267		47	12.95		59.95	71.50
4400	High volume		30	.267		125	12.95		137.95	158
5000	Quick coupling valve and key, brass, thread type		25	.320		55	15.55		70.55	84.50
5200	Lug type		25	.320		58	15.55		73.55	88
5400	Pop-up gear drive sprinkler, nozzle & case plastic, medium volume		30	.267		24.50	12.95		37.45	47
6000	Riser mounted gear drive sprinkler, nozzle and case									
6200	Plastic, medium volume	1 Skwk	30	.267	Ea.	14.05	12.95		27	35.50
7000	Riser mounted impact sprinkler, body, part or full circle									
7200	Full circle plastic, low/medium volume	1 Skwk	25	.320	Ea.	14	15.55		29.55	39.50
7400	Brass, low/medium volume		25	.320		37	15.55		52.55	64.50
7600	Medium volume		25	.320		47.50	15.55		63.05	76
7800	Female thread		25	.320		87.50	15.55		103.05	120
8000	Riser mounted impact sprinkler, body, full circle only									
8200	Plastic, low/medium volume	1 Skwk	25	.320	Ea.	8.25	15.55		23.80	33
8400	Brass, low/medium volume		25	.320		35.50	15.55		51.05	63
8600	High volume		25	.320		72.50	15.55		88.05	104
8800	Very high volume		25	.320		143	15.55		158.55	181

For customer support on your Plumbing Cost Data, call 877.709.6968.

397

Division Notes

		CREW	DAILY OUTPUT	LABOR-HOURS	UNIT	BARE COSTS				TOTAL INCL O&P
						MAT.	LABOR	EQUIP.	TOTAL	

Estimating Tips

33 10 00 Water Utilities
33 30 00 Sanitary Sewerage Utilities
33 40 00 Storm Drainage Utilities

- Never assume that the water, sewer, and drainage lines will go in at the early stages of the project. Consider the site access needs before dividing the site in half with open trenches, loose pipe, and machinery obstructions. Always inspect the site to establish that the site drawings are complete. Check off all existing utilities on your drawings as you locate them. Be especially careful with underground utilities because appurtenances are sometimes buried during regrading or repaving operations. If you find any discrepancies, mark up the site plan for further research. Differing site conditions can be very costly if discovered later in the project.

- See also Section 33 01 00 for restoration of pipe where removal/replacement may be undesirable. Use of new types of piping materials can reduce the overall project cost. Owners/design engineers should consider the installing contractor as a valuable source of current information on utility products and local conditions that could lead to significant cost savings.

Reference Numbers

Reference numbers are shown in shaded boxes at the beginning of some major classifications. These numbers refer to related items in the Reference Section. The reference information may be an estimating procedure, an alternate pricing method, or technical information.

Note: Not all subdivisions listed here necessarily appear in this publication. ■

*Note: **Trade Service**, in part, has been used as a reference source for some of the material prices used in Division 33.*

33 01 10.10 Corrosion Resistance	Crew	Daily Output	Labor-Hours	Unit	Material	2015 Bare Costs Labor	Equipment	Total	Total Incl O&P
0010 **CORROSION RESISTANCE**									
0012 Wrap & coat, add to pipe, 4" diameter				L.F.	2.21			2.21	2.43
0020 5" diameter					2.66			2.66	2.93
0040 6" diameter					3.28			3.28	3.61
0060 8" diameter					4.04			4.04	4.44
0080 10" diameter					4.91			4.91	5.40
0100 12" diameter					6.20			6.20	6.80
0120 14" diameter					7.30			7.30	8
0140 16" diameter					9.75			9.75	10.70
0160 18" diameter					10			10	11
0180 20" diameter					10.50			10.50	11.55
0200 24" diameter					12.95			12.95	14.25
0220 Small diameter pipe, 1" diameter, add					1.02			1.02	1.12
0240 2" diameter					1.38			1.38	1.52
0260 2-1/2" diameter					1.62			1.62	1.78
0280 3" diameter					2.01			2.01	2.21
0300 Fittings, field covered, add				S.F.	10.40			10.40	11.45
0500 Coating, bituminous, per diameter inch, 1 coat, add				L.F.	.63			.63	.69
0540 3 coat					1.92			1.92	2.11
0560 Coal tar epoxy, per diameter inch, 1 coat, add					.22			.22	.24
0600 3 coat					.68			.68	.75
1000 Polyethylene H.D. extruded, .025" thk., 1/2" diameter add					.10			.10	.11
1020 3/4" diameter					.15			.15	.17
1040 1" diameter					.20			.20	.22
1060 1-1/4" diameter					.25			.25	.28
1080 1-1/2" diameter					.30			.30	.33
1100 .030" thk., 2" diameter					.40			.40	.44
1120 2-1/2" diameter					.50			.50	.55
1140 .035" thk., 3" diameter					.60			.60	.66
1160 3-1/2" diameter					.70			.70	.77
1180 4" diameter					.80			.80	.88
1200 .040" thk, 5" diameter					1			1	1.10
1220 6" diameter					1.20			1.20	1.32
1240 8" diameter					1.60			1.60	1.76
1260 10" diameter					2			2	2.20
1280 12" diameter					2.40			2.40	2.64
1300 .060" thk., 14" diameter					2.80			2.80	3.08
1320 16" diameter					3.20			3.20	3.52
1340 18" diameter					3.60			3.60	3.96
1360 20" diameter					4			4	4.40
1380 Fittings, field wrapped, add				S.F.	4.79			4.79	5.25

33 01 10.20 Pipe Repair

	Crew	Daily Output	Labor-Hours	Unit	Material	2015 Bare Costs Labor	Equipment	Total	Total Incl O&P
0010 **PIPE REPAIR**									
0020 Not including excavation or backfill									
0100 Clamp, stainless steel, lightweight, for steel pipe									
0110 3" long, 1/2" diameter pipe	1 Plum	34	.235	Ea.	12.80	13.80		26.60	35
0120 3/4" diameter pipe		32	.250		13.05	14.70		27.75	36.50
0130 1" diameter pipe		30	.267		14.25	15.65		29.90	39
0140 1-1/4" diameter pipe		28	.286		14.45	16.75		31.20	41.50
0150 1-1/2" diameter pipe		26	.308		15.35	18.05		33.40	44.50
0160 2" diameter pipe		24	.333		16.50	19.55		36.05	47.50
0170 2-1/2" diameter pipe		23	.348		17.50	20.50		38	50.50

33 01 10 – Operation and Maintenance of Water Utilities

33 01 10.20 Pipe Repair		Crew	Daily Output	Labor-Hours	Unit	Material	2015 Bare Costs Labor	Equipment	Total	Total Incl O&P
0180	3" diameter pipe	1 Plum	22	.364	Ea.	21	21.50		42.50	55
0190	3-1/2" diameter pipe	↓	21	.381		25	22.50		47.50	61.50
0200	4" diameter pipe	B-20	44	.545		25	23		48	63
0210	5" diameter pipe		42	.571		29.50	24		53.50	69.50
0220	6" diameter pipe		38	.632		34	26.50		60.50	78.50
0230	8" diameter pipe		30	.800		40.50	33.50		74	96
0240	10" diameter pipe		28	.857		115	36		151	183
0250	12" diameter pipe		24	1		120	42		162	197
0260	14" diameter pipe		22	1.091		125	46		171	209
0270	16" diameter pipe		20	1.200		135	50.50		185.50	227
0280	18" diameter pipe		18	1.333		145	56		201	246
0290	20" diameter pipe		16	1.500		150	63		213	262
0300	24" diameter pipe	↓	14	1.714		170	72		242	298
0360	For 6" long, add					100%	40%			
0370	For 9" long, add					200%	100%			
0380	For 12" long, add					300%	150%			
0390	For 18" long, add				↓	500%	200%			
0400	Pipe Freezing for live repairs of systems 3/8 inch to 6 inch									
0410	Note: Pipe Freezing can also be used to install a valve into a live system									
0420	Pipe Freezing each side 3/8 inch	2 Skwk	8	2	Ea.	540	97.50		637.50	745
0425	Pipe Freezing each side 3/8 inch, second location same kit		8	2		22.50	97.50		120	175
0430	Pipe Freezing each side 3/4 inch		8	2		495	97.50		592.50	695
0435	Pipe Freezing each side 3/4 inch, second location same kit		8	2		22.50	97.50		120	175
0440	Pipe Freezing each side 1-1/2 inch		6	2.667		495	130		625	745
0445	Pipe Freezing each side 1-1/2 inch , second location same kit		6	2.667		22.50	130		152.50	225
0450	Pipe Freezing each side 2 inch		6	2.667		890	130		1,020	1,175
0455	Pipe Freezing each side 2 inch, second location same kit		6	2.667		22.50	130		152.50	225
0460	Pipe Freezing each side 2-1/2 inch-3 inch		6	2.667		925	130		1,055	1,225
0465	Pipe Freeze each side 2-1/2 -3 inch, second location same kit		6	2.667		45.50	130		175.50	250
0470	Pipe Freezing each side 4 inch		4	4		1,475	195		1,670	1,925
0475	Pipe Freezing each side 4 inch, second location same kit		4	4		72	195		267	380
0480	Pipe Freezing each side 5-6 inch		4	4		4,000	195		4,195	4,700
0485	Pipe Freezing each side 5-6 inch, second location same kit	↓	4	4		216	195		411	540
0490	Pipe Freezing extra 20 lb. CO_2 cylinders (3/8" to 2" - 1 ea, 3" -2 ea)					222			222	244
0500	Pipe Freezing extra 50 lb. CO_2 cylinders (4" - 2 ea, 5"-6" -6 ea)				↓	440			440	485
1000	Clamp, stainless steel, with threaded service tap									
1040	Full seal for iron, steel, PVC pipe									
1100	6" long, 2" diameter pipe	1 Plum	17	.471	Ea.	87	27.50		114.50	137
1110	2-1/2" diameter pipe		16	.500		90	29.50		119.50	144
1120	3" diameter pipe		15.60	.513		104	30		134	161
1130	3-1/2" diameter pipe		15	.533		110	31.50		141.50	169
1140	4" diameter pipe	B-20	32	.750		117	31.50		148.50	177
1150	6" diameter pipe		28	.857		145	36		181	216
1160	8" diameter pipe		21	1.143		171	48		219	262
1170	10" diameter pipe		20	1.200		224	50.50		274.50	325
1180	12" diameter pipe	↓	17	1.412		261	59		320	380
1200	8" long, 2" diameter pipe	1 Plum	11.72	.683		178	40		218	257
1210	2-1/2" diameter pipe		11	.727		183	42.50		225.50	266
1220	3" diameter pipe		10.75	.744		189	43.50		232.50	274
1230	3-1/2" diameter pipe		10.34	.774		199	45.50		244.50	288
1240	4" diameter pipe	B-20	22	1.091		206	46		252	297
1250	6" diameter pipe		19.31	1.243		234	52		286	340
1260	8" diameter pipe	↓	14.48	1.657		264	69.50		333.50	400

For customer support on your Plumbing Cost Data, call 877.709.6968.

401

33 01 10.20 Pipe Repair		Crew	Daily Output	Labor-Hours	Unit	Material	2015 Bare Costs Labor	Equipment	Total	Total Incl O&P
1270	10" diameter pipe	B-20	13.80	1.739	Ea.	330	73		403	475
1280	12" diameter pipe	↓	11.72	2.048		370	86		456	535
1300	12" long, 2" diameter pipe	1 Plum	9.44	.847		257	50		307	360
1310	2-1/2" diameter pipe		8.89	.900		262	53		315	370
1320	3" diameter pipe		8.67	.923		269	54		323	380
1330	3-1/2" diameter pipe	↓	8.33	.960		280	56.50		336.50	395
1340	4" diameter pipe	B-20	17.78	1.350		450	56.50		506.50	585
1350	6" diameter pipe		15.56	1.542		340	64.50		404.50	470
1360	8" diameter pipe		11.67	2.057		390	86.50		476.50	560
1370	10" diameter pipe		11.11	2.160		485	90.50		575.50	675
1380	12" diameter pipe	↓	9.44	2.542		555	107		662	775
1400	20" long, 2" diameter pipe	1 Plum	8.10	.988		299	58		357	420
1410	2-1/2" diameter pipe		7.62	1.050		325	61.50		386.50	450
1420	3" diameter pipe		7.43	1.077		365	63		428	495
1430	3-1/2" diameter pipe	↓	7.14	1.120		380	66		446	520
1440	4" diameter pipe	B-20	15.24	1.575		450	66		516	595
1450	6" diameter pipe		13.33	1.800		525	75.50		600.50	690
1460	8" diameter pipe		10	2.400		590	101		691	805
1470	10" diameter pipe		9.52	2.521		715	106		821	950
1480	12" diameter pipe	↓	8.10	2.963	↓	835	124		959	1,100
1600	Clamp, stainless steel, single section									
1640	Full seal for iron, steel, PVC pipe									
1700	6" long, 2" diameter pipe	1 Plum	17	.471	Ea.	87	27.50		114.50	137
1710	2-1/2" diameter pipe		16	.500		90	29.50		119.50	144
1720	3" diameter pipe		15.60	.513		104	30		134	161
1730	3-1/2" diameter pipe	↓	15	.533		110	31.50		141.50	169
1740	4" diameter pipe	B-20	32	.750		117	31.50		148.50	177
1750	6" diameter pipe		27	.889		145	37.50		182.50	218
1760	8" diameter pipe		21	1.143		171	48		219	262
1770	10" diameter pipe		20	1.200		224	50.50		274.50	325
1780	12" diameter pipe	↓	17	1.412		261	59		320	380
1800	8" long, 2" diameter pipe	1 Plum	11.72	.683		118	40		158	191
1805	2-1/2" diameter pipe		11.03	.725		123	42.50		165.50	200
1810	3" diameter pipe		10.76	.743		129	43.50		172.50	208
1815	3-1/2" diameter pipe	↓	10.34	.774		139	45.50		184.50	222
1820	4" diameter pipe	B-20	22.07	1.087		146	45.50		191.50	231
1825	6" diameter pipe		19.31	1.243		174	52		226	273
1830	8" diameter pipe		14.48	1.657		204	69.50		273.50	330
1835	10" diameter pipe		13.79	1.740		272	73		345	410
1840	12" diameter pipe	↓	11.72	2.048		310	86		396	470
1850	12" long, 2" diameter pipe	1 Plum	9.44	.847		197	50		247	291
1855	2-1/2" diameter pipe		8.89	.900		201	53		254	300
1860	3" diameter pipe		8.67	.923		209	54		263	310
1865	3-1/2" diameter pipe	↓	8.33	.960		220	56.50		276.50	325
1870	4" diameter pipe	B-20	17.78	1.350		231	56.50		287.50	340
1875	6" diameter pipe		15.56	1.542		278	64.50		342.50	405
1880	8" diameter pipe		11.67	2.057		330	86.50		416.50	495
1885	10" diameter pipe		11.11	2.160		425	90.50		515.50	610
1890	12" diameter pipe	↓	9.44	2.542		495	107		602	710
1900	20" long, 2" diameter pipe	1 Plum	8.10	.988		239	58		297	350
1905	2-1/2" diameter pipe		7.62	1.050		264	61.50		325.50	385
1910	3" diameter pipe		7.43	1.077		305	63		368	430
1915	3-1/2" diameter pipe	↓	7.14	1.120	↓	320	66		386	450

402

For customer support on your Plumbing Cost Data, call 877.709.6968.

33 01 Operation and Maintenance of Utilities

33 01 10 – Operation and Maintenance of Water Utilities

33 01 10.20 Pipe Repair

33 01 10.20 Pipe Repair		Crew	Daily Output	Labor-Hours	Unit	Material	2015 Bare Costs Labor	Equipment	Total	Total Incl O&P
1920	4" diameter pipe	B-20	15.24	1.575	Ea.	390	66		456	530
1925	6" diameter pipe		13.33	1.800		465	75.50		540.50	625
1930	8" diameter pipe		10	2.400		530	101		631	740
1935	10" diameter pipe		9.52	2.521		655	106		761	885
1940	12" diameter pipe		8.10	2.963		775	124		899	1,050
2000	Clamp, stainless steel, two section									
2040	Full seal, for iron, steel, PVC pipe									
2100	6" long, 4" diameter pipe	B-20	24	1	Ea.	230	42		272	320
2110	6" diameter pipe		20	1.200		264	50.50		314.50	370
2120	8" diameter pipe		13	1.846		300	77.50		377.50	450
2130	10" diameter pipe		12	2		305	84		389	465
2140	12" diameter pipe		10	2.400		390	101		491	585
2200	9" long, 4" diameter pipe		16	1.500		300	63		363	425
2210	6" diameter pipe		13	1.846		340	77.50		417.50	495
2220	8" diameter pipe		9	2.667		375	112		487	585
2230	10" diameter pipe		8	3		495	126		621	740
2240	12" diameter pipe		7	3.429		565	144		709	845
2250	14" diameter pipe		6.40	3.750		760	157		917	1,075
2260	16" diameter pipe		6	4		705	168		873	1,050
2270	18" diameter pipe		5	4.800		800	201		1,001	1,200
2280	20" diameter pipe		4.60	5.217		885	219		1,104	1,300
2290	24" diameter pipe		4	6		1,175	252		1,427	1,700
2320	For 12" long, add to 9"					15%	25%			
2330	For 18" long, add to 9"					70%	55%			
8000	For internal cleaning and inspection, see Section 33 01 30.16									
8100	For pipe testing, see Section 23 05 93.50									

33 01 30 – Operation and Maintenance of Sewer Utilities

33 01 30.16 TV Inspection of Sewer Pipelines

		Crew	Daily Output	Labor-Hours	Unit	Material	2015 Bare Costs Labor	Equipment	Total	Total Incl O&P
0010	**TV INSPECTION OF SEWER PIPELINES**									
0100	Pipe internal cleaning & inspection, cleaning, pressure pipe systems									
0120	Pig method, lengths 1000' to 10,000'									
0140	4" diameter thru 24" diameter, minimum				L.F.				3.60	4.14
0160	Maximum				"				18	21
6000	Sewage/sanitary systems									
6100	Power rodder with header & cutters									
6110	Mobilization charge, minimum				Total				695	800
6120	Mobilization charge, maximum				"				9,125	10,600
6140	Cleaning 4"-12" diameter				L.F.				3.39	3.90
6190	14"-24" diameter								3.98	4.59
6240	30" diameter								5.80	6.65
6250	36" diameter								6.75	7.75
6260	48" diameter								7.70	8.85
6270	60" diameter								8.70	9.95
6280	72" diameter								9.65	11.10
9000	Inspection, television camera with video									
9060	up to 500 linear feet				Total				715	820

33 01 30.71 Rehabilitation of Sewer Utilities

		Crew	Daily Output	Labor-Hours	Unit	Material	2015 Bare Costs Labor	Equipment	Total	Total Incl O&P
0010	**REHABILITATION OF SEWER UTILITIES**									
0011	300' runs, replace with HDPE pipe									
0020	Not including excavation, backfill, shoring, or dewatering									
0100	6" to 15" diameter, minimum				L.F.				104	114
0200	Maximum								214	236

For customer support on your Plumbing Cost Data, call 877.709.6968.

403

33 01 30 – Operation and Maintenance of Sewer Utilities

33 01 30.71 Rehabilitation of Sewer Utilities	Crew	Daily Output	Labor-Hours	Unit	Material	2015 Bare Costs Labor	Equipment	Total	Total Incl O&P
0300	18" to 36" diameter, minimum				L.F.				208	229
0400	Maximum				↓				435	510
0500	Mobilize and demobilize, minimum				Job				3,000	3,300
0600	Maximum				"				32,100	35,400

33 01 30.72 Relining Sewers

		Crew	Daily Output	Labor-Hours	Unit	Material	Labor	Equipment	Total	Total Incl O&P
0010	**RELINING SEWERS**									
0011	With cement incl. bypass & cleaning									
0020	Less than 10,000 L.F., urban, 6" to 10"	C-17E	130	.615	L.F.	9.55	30	.74	40.29	58
0050	10" to 12"		125	.640		11.70	31.50	.77	43.97	62.50
0070	12" to 16"		115	.696		12.05	34	.84	46.89	66.50
0100	16" to 20"		95	.842		14.15	41.50	1.02	56.67	80.50
0200	24" to 36"		90	.889		15.25	43.50	1.08	59.83	85.50
0300	48" to 72"		80	1		24.50	49	1.21	74.71	104

33 01 30.74 HDPE Pipe Lining

		Crew	Daily Output	Labor-Hours	Unit	Material	Labor	Equipment	Total	Total Incl O&P
0010	**HDPE PIPE LINING**, excludes cleaning and video inspection									
0020	Pipe relined with one pipe size smaller than original (4" for 6")									
0100	6" diameter, original size	B-6B	600	.080	L.F.	2.47	3.06	1.50	7.03	9.10
0150	8" diameter, original size		600	.080		5.85	3.06	1.50	10.41	12.80
0200	10" diameter, original size		600	.080		8.55	3.06	1.50	13.11	15.75
0250	12" diameter, original size		400	.120		14.65	4.59	2.26	21.50	25.50
0300	14" diameter, original size		400	.120		17.75	4.59	2.26	24.60	29
0350	16" diameter, original size	B-6C	300	.160		27.50	6.10	5.35	38.95	45.50
0400	18" diameter, original size	"	300	.160		32.50	6.10	5.35	43.95	51
1000	Pipe HDPE lining, make service line taps	B-6	4	6	Ea.	97	248	91	436	585

33 05 Common Work Results for Utilities

33 05 16 – Utility Structures

33 05 16.13 Precast Concrete Utility Boxes

		Crew	Daily Output	Labor-Hours	Unit	Material	Labor	Equipment	Total	Total Incl O&P
0010	**PRECAST CONCRETE UTILITY BOXES**, 6" thick									
0050	5' x 10' x 6' high, I.D.	B-13	2	28	Ea.	3,750	1,150	370	5,270	6,275
0100	6' x 10' x 6' high, I.D.		2	28		3,900	1,150	370	5,420	6,425
0150	5' x 12' x 6' high, I.D.		2	28		4,125	1,150	370	5,645	6,675
0200	6' x 12' x 6' high, I.D.		1.80	31.111		4,600	1,275	410	6,285	7,475
0250	6' x 13' x 6' high, I.D.		1.50	37.333		6,050	1,525	490	8,065	9,550
0300	8' x 14' x 7' high, I.D.		1	56		6,525	2,300	735	9,560	11,500
0350	Hand hole, precast concrete, 1-1/2" thick									
0400	1'-0" x 2'-0" x 1'-9", I.D., light duty	B-1	4	6	Ea.	400	230		630	795
0450	4'-6" x 3'-2" x 2'-0", O.D., heavy duty	B-6	3	8	"	1,475	330	121	1,926	2,250

33 05 23 – Trenchless Utility Installation

33 05 23.19 Microtunneling

		Crew	Daily Output	Labor-Hours	Unit	Material	Labor	Equipment	Total	Total Incl O&P
0010	**MICROTUNNELING**									
0011	Not including excavation, backfill, shoring,									
0020	or dewatering, average 50'/day, slurry method									
0100	24" to 48" outside diameter, minimum				L.F.				875	965
0110	Adverse conditions, add				%				50%	50%
1000	Rent microtunneling machine, average monthly lease				Month				97,500	107,000
1010	Operating technician				Day				630	705
1100	Mobilization and demobilization, minimum				Job				41,200	45,900
1110	Maximum				"				445,500	490,500

33 05 Common Work Results for Utilities

33 05 23 – Trenchless Utility Installation

33 05 23.20 Horizontal Boring

		Crew	Daily Output	Labor-Hours	Unit	Material	2015 Bare Costs Labor	Equipment	Total	Total Incl O&P
0010	**HORIZONTAL BORING**									
0011	Casing only, 100' minimum,									
0020	not incl. jacking pits or dewatering									
0100	Roadwork, 1/2" thick wall, 24" diameter casing	B-42	20	3.200	L.F.	121	136	68	325	420
0200	36" diameter		16	4		223	170	85	478	605
0300	48" diameter		15	4.267		310	181	90.50	581.50	725
0500	Railroad work, 24" diameter		15	4.267		121	181	90.50	392.50	515
0600	36" diameter		14	4.571		223	194	97	514	655
0700	48" diameter	▼	12	5.333		310	226	113	649	820
0900	For ledge, add								20%	20%
1000	Small diameter boring, 3", sandy soil	B-82	900	.018		22	.77	.09	22.86	25.50
1040	Rocky soil	"	500	.032	▼	22	1.38	.17	23.55	26.50
1100	Prepare jacking pits, incl. mobilization & demobilization, minimum				Ea.				3,225	3,700
1101	Maximum				"				22,000	25,500

33 05 26 – Utility Identification

33 05 26.05 Utility Connection

		Crew	Daily Output	Labor-Hours	Unit	Material	2015 Bare Costs Labor	Equipment	Total	Total Incl O&P
0010	**UTILITY CONNECTION**									
0020	Water, sanitary, stormwater, gas, single connection	B-14	1	48	Ea.	3,225	1,900	365	5,490	6,875
0030	Telecommunication	"	1	48	"	395	1,900	365	2,660	3,750

33 05 26.10 Utility Accessories

		Crew	Daily Output	Labor-Hours	Unit	Material	2015 Bare Costs Labor	Equipment	Total	Total Incl O&P
0010	**UTILITY ACCESSORIES**									
0400	Underground tape, detectable, reinforced, alum. foil core, 2"	1 Clab	150	.053	C.L.F.	6.50	2.01		8.51	10.25
0500	6"	"	140	.057	"	27.50	2.15		29.65	33.50

33 11 Water Utility Distribution Piping

33 11 13 – Public Water Utility Distribution Piping

33 11 13.15 Water Supply, Ductile Iron Pipe

		Crew	Daily Output	Labor-Hours	Unit	Material	2015 Bare Costs Labor	Equipment	Total	Total Incl O&P
0010	**WATER SUPPLY, DUCTILE IRON PIPE** R331113-80									
0020	Not including excavation or backfill									
2000	Pipe, class 50 water piping, 18' lengths									
2020	Mechanical joint, 4" diameter	B-21A	200	.200	L.F.	30.50	9.40	2.37	42.27	50.50
2040	6" diameter		160	.250		32	11.75	2.96	46.71	56
2060	8" diameter		133.33	.300		44.50	14.05	3.55	62.10	74.50
2080	10" diameter		114.29	.350		58.50	16.40	4.14	79.04	93.50
2100	12" diameter		105.26	.380		79	17.85	4.50	101.35	119
2120	14" diameter		100	.400		93	18.75	4.74	116.49	136
2140	16" diameter		72.73	.550		94.50	26	6.50	127	151
2160	18" diameter		68.97	.580		126	27	6.85	159.85	188
2170	20" diameter		57.14	.700		127	33	8.30	168.30	199
2180	24" diameter		47.06	.850		141	40	10.05	191.05	227
3000	Push-on joint, 4" diameter		400	.100		21	4.69	1.18	26.87	31.50
3020	6" diameter		333.33	.120		17	5.65	1.42	24.07	29
3040	8" diameter		200	.200		23	9.40	2.37	34.77	42.50
3060	10" diameter		181.82	.220		36.50	10.30	2.60	49.40	59
3080	12" diameter		160	.250		38.50	11.75	2.96	53.21	63.50
3100	14" diameter		133.33	.300		42.50	14.05	3.55	60.10	72.50
3120	16" diameter		114.29	.350		51.50	16.40	4.14	72.04	86
3140	18" diameter		100	.400		57	18.75	4.74	80.49	96
3160	20" diameter		88.89	.450		59.50	21	5.35	85.85	103
3180	24" diameter	▼	76.92	.520	▼	59	24.50	6.15	89.65	109

33 11 13 – Public Water Utility Distribution Piping

33 11 13.15 Water Supply, Ductile Iron Pipe	Crew	Daily Output	Labor-Hours	Unit	Material	2015 Bare Costs Labor	Equipment	Total	Total Incl O&P
8000	Piping, fittings, mechanical joint, AWWA C110									
8006	90° bend, 4" diameter	B-20A	16	2	Ea.	155	91.50		246.50	310
8020	6" diameter		12.80	2.500		229	114		343	425
8040	8" diameter		10.67	2.999		450	137		587	705
8060	10" diameter	B-21A	11.43	3.500		620	164	41.50	825.50	975
8080	12" diameter		10.53	3.799		880	178	45	1,103	1,300
8100	14" diameter		10	4		1,200	188	47.50	1,435.50	1,650
8120	16" diameter		7.27	5.502		1,525	258	65	1,848	2,150
8140	18" diameter		6.90	5.797		2,125	272	68.50	2,465.50	2,850
8160	20" diameter		5.71	7.005		2,650	330	83	3,063	3,525
8180	24" diameter		4.70	8.511		4,200	400	101	4,701	5,325
8200	Wye or tee, 4" diameter	B-20A	10.67	2.999		355	137		492	600
8220	6" diameter		8.53	3.751		535	171		706	850
8240	8" diameter		7.11	4.501		850	206		1,056	1,250
8260	10" diameter	B-21A	7.62	5.249		1,225	246	62	1,533	1,800
8280	12" diameter		7.02	5.698		1,625	267	67.50	1,959.50	2,250
8300	14" diameter		6.67	5.997		2,600	281	71	2,952	3,375
8320	16" diameter		4.85	8.247		2,900	385	97.50	3,382.50	3,875
8340	18" diameter		4.60	8.696		3,900	410	103	4,413	5,025
8360	20" diameter		3.81	10.499		5,475	490	124	6,089	6,900
8380	24" diameter		3.14	12.739		9,275	600	151	10,026	11,300
8398	45° bends, 4" diameter	B-20A	16	2		255	91.50		346.50	420
8400	6" diameter		12.80	2.500		199	114		313	395
8405	8" diameter		10.67	2.999		291	137		428	530
8410	12" diameter	B-21A	10.53	3.799		730	178	45	953	1,125
8420	16" diameter		7.27	5.502		1,275	258	65	1,598	1,875
8430	20" diameter		5.71	7.005		2,075	330	83	2,488	2,875
8440	24" diameter		4.70	8.511		2,900	400	101	3,401	3,900
8450	Decreaser, 6" x 4" diameter	B-20A	14.22	2.250		207	103		310	385
8460	8" x 6" diameter	"	11.64	2.749		310	126		436	530
8470	10" x 6" diameter	B-21A	13.33	3.001		390	141	35.50	566.50	685
8480	12" x 6" diameter		12.70	3.150		550	148	37.50	735.50	870
8490	16" x 6" diameter		10	4		890	188	47.50	1,125.50	1,325
8500	20" x 6" diameter		8.42	4.751		1,625	223	56.50	1,904.50	2,175
8552	For water utility valves see Section 33 12 16									
8700	Joint restraint, ductile iron mechanical joints									
8710	4" diameter	B-20A	32	1	Ea.	29	45.50		74.50	102
8720	6" diameter		25.60	1.250		31.50	57		88.50	122
8730	8" diameter		21.33	1.500		30	68.50		98.50	137
8740	10" diameter		18.28	1.751		38.50	80		118.50	165
8750	12" diameter		16.84	1.900		50	87		137	187
8760	14" diameter		16	2		53	91.50		144.50	198
8770	16" diameter		11.64	2.749		58	126		184	255
8780	18" diameter		11.03	2.901		73	133		206	282
8785	20" diameter		9.14	3.501		78	160		238	330
8790	24" diameter		7.53	4.250		104	194		298	410
9600	Steel sleeve with tap, 4" diameter	B-20	3	8		435	335		770	995
9620	6" diameter		2	12		475	505		980	1,300
9630	8" diameter		2	12		515	505		1,020	1,350

33 11 Water Utility Distribution Piping

33 11 13 – Public Water Utility Distribution Piping

33 11 13.20 Water Supply, Polyethylene Pipe, C901

		Crew	Daily Output	Labor-Hours	Unit	Material	2015 Bare Costs Labor	Equipment	Total	Total Incl O&P
0010	**WATER SUPPLY, POLYETHYLENE PIPE, C901**									
0020	Not including excavation or backfill									
1000	Piping, 160 PSI, 3/4" diameter	Q-1A	525	.019	L.F.	.46	1.13		1.59	2.21
1120	1" diameter		485	.021		.60	1.22		1.82	2.50
1140	1-1/2" diameter		450	.022		1.35	1.31		2.66	3.47
1160	2" diameter	↓	365	.027	↓	2.20	1.62		3.82	4.87
2000	Fittings, insert type, nylon, 160 & 250 psi, cold water									
2220	Clamp ring, stainless steel, 3/4" diameter	Q-1A	345	.029	Ea.	1.27	1.71		2.98	3.99
2240	1" diameter		321	.031		1.47	1.84		3.31	4.40
2260	1-1/2" diameter		285	.035		2.64	2.07		4.71	6.05
2280	2" diameter		255	.039		3.47	2.32		5.79	7.30
2300	Coupling, 3/4" diameter		66	.152		1.40	8.95		10.35	15.05
2320	1" diameter		57	.175		1.63	10.35		11.98	17.45
2340	1-1/2" diameter		51	.196		3.30	11.60		14.90	21
2360	2" diameter		48	.208		5.15	12.30		17.45	24.50
2400	Elbow, 90°, 3/4" diameter		66	.152		2.27	8.95		11.22	16
2420	1" diameter		57	.175		2.74	10.35		13.09	18.65
2440	1-1/2" diameter		51	.196		4.51	11.60		16.11	22.50
2460	2" diameter	↓	48	.208		6.35	12.30		18.65	25.50

33 11 13.25 Water Supply, Polyvinyl Chloride Pipe

		Crew	Daily Output	Labor-Hours	Unit	Material	2015 Bare Costs Labor	Equipment	Total	Total Incl O&P
0010	**WATER SUPPLY, POLYVINYL CHLORIDE PIPE** R331113-80									
0020	Not including excavation or backfill, unless specified									
2100	PVC pipe, Class 150, 1-1/2" diameter	Q-1A	750	.013	L.F.	.45	.79		1.24	1.69
2120	2" diameter		686	.015		.68	.86		1.54	2.05
2140	2-1/2" diameter	↓	500	.020		1.30	1.18		2.48	3.22
2160	3" diameter	B-20	430	.056	↓	1.40	2.34		3.74	5.15
3010	AWWA C905, PR 100, DR 25									
3030	14" diameter	B-20A	213	.150	L.F.	13.65	6.85		20.50	25.50
3040	16" diameter		200	.160		19.25	7.30		26.55	32
3050	18" diameter		160	.200		24.50	9.15		33.65	40.50
3060	20" diameter		133	.241		30	11		41	50
3070	24" diameter		107	.299		43	13.65		56.65	68.50
3080	30" diameter		80	.400		81	18.30		99.30	117
3090	36" diameter		80	.400		126	18.30		144.30	166
3100	42" diameter		60	.533		168	24.50		192.50	222
3200	48" diameter	↓	60	.533		220	24.50		244.50	279
3960	Pressure pipe, class 200, ASTM 2241, SDR 21, 3/4" diameter	Q-1A	1000	.010		.23	.59		.82	1.14
3980	1" diameter		900	.011		.33	.66		.99	1.35
4000	1-1/2" diameter		750	.013		.72	.79		1.51	1.98
4010	2" diameter		686	.015		.88	.86		1.74	2.27
4020	2-1/2" diameter	↓	500	.020		1.37	1.18		2.55	3.30
4030	3" diameter	B-20A	430	.074		1.61	3.40		5.01	6.95
4040	4" diameter		375	.085		2.48	3.90		6.38	8.70
4050	6" diameter		316	.101		4.85	4.63		9.48	12.40
4060	8" diameter	↓	260	.123		8.40	5.65		14.05	17.80
4090	Including trenching to 3' deep, 3/4" diameter	Q-1C	300	.080		.23	4.17	11.35	15.75	19
4100	1" diameter		280	.086		.33	4.46	12.15	16.94	20.50
4110	1-1/2" diameter		260	.092		.72	4.81	13.10	18.63	22.50
4120	2" diameter		220	.109		.88	5.70	15.45	22.03	26.50
4130	2-1/2" diameter		200	.120		1.37	6.25	17	24.62	29.50
4140	3" diameter		175	.137		1.61	7.15	19.45	28.21	34
4150	4" diameter	↓	150	.160		2.48	8.35	22.50	33.33	40.50

407

33 11 13 – Public Water Utility Distribution Piping

33 11 13.25 Water Supply, Polyvinyl Chloride Pipe	Crew	Daily Output	Labor-Hours	Unit	Material	2015 Bare Costs Labor	Equipment	Total	Total Incl O&P
4160	6" diameter	Q-1C	125	.192	L.F.	4.85	10	27	41.85	50.50
4165	Fittings									
4170	Elbow, 90°, 3/4"	Q-1A	114	.088	Ea.	1.58	5.20		6.78	9.60
4180	1"		100	.100		4.58	5.90		10.48	14
4190	1-1/2"		80	.125		16.25	7.40		23.65	29
4200	2"		72	.139		26	8.20		34.20	41
4210	3"	B-20A	46	.696		31	32		63	82.50
4220	4"		36	.889		38.50	40.50		79	105
4230	6"		24	1.333		70.50	61		131.50	171
4240	8"		14	2.286		145	104		249	320
4250	Elbow, 45°, 3/4"	Q-1A	114	.088		2.65	5.20		7.85	10.75
4260	1"		100	.100		6.35	5.90		12.25	15.95
4270	1-1/2"		80	.125		18.75	7.40		26.15	31.50
4280	2"		72	.139		19.95	8.20		28.15	34.50
4290	2-1/2"		54	.185		21	10.95		31.95	39.50
4300	3"	B-20A	46	.696		25	32		57	76
4310	4"		36	.889		37	40.50		77.50	103
4320	6"		24	1.333		75.50	61		136.50	176
4330	8"		14	2.286		136	104		240	310
4340	Tee, 3/4"	Q-1A	76	.132		3.40	7.80		11.20	15.50
4350	1"		66	.152		16.80	8.95		25.75	32
4360	1-1/2"		54	.185		24.50	10.95		35.45	43.50
4370	2"		48	.208		29.50	12.30		41.80	51
4380	2-1/2"		36	.278		32	16.40		48.40	60.50
4390	3"	B-20A	30	1.067		33.50	49		82.50	111
4400	4"		24	1.333		53.50	61		114.50	152
4410	6"		14.80	2.162		117	99		216	279
4420	8"		9	3.556		210	163		373	480
4430	Coupling, 3/4"	Q-1A	114	.088		1.50	5.20		6.70	9.50
4440	1"		100	.100		6.65	5.90		12.55	16.25
4450	1-1/2"		80	.125		10.75	7.40		18.15	23
4460	2"		72	.139		11.65	8.20		19.85	25.50
4470	2-1/2"		54	.185		12.85	10.95		23.80	30.50
4480	3"	B-20A	46	.696		16.35	32		48.35	66.50
4490	4"		36	.889		25	40.50		65.50	89.50
4500	6"		24	1.333		38	61		99	135
4510	8"		14	2.286		87	104		191	255
4520	Pressure pipe Class 150, SDR 18, AWWA C900, 4" diameter		380	.084	L.F.	2.73	3.85		6.58	8.85
4530	6" diameter		316	.101		5.35	4.63		9.98	12.95
4540	8" diameter		264	.121		8.80	5.55		14.35	18.10
4550	10" diameter		220	.145		14.05	6.65		20.70	25.50
4560	12" diameter		186	.172		19.90	7.85		27.75	34
8000	Fittings with rubber gasket									
8003	Class 150, DR 18									
8006	90° Bend , 4" diameter	B-20	100	.240	Ea.	42.50	10.05		52.55	62
8020	6" diameter		90	.267		75	11.20		86.20	100
8040	8" diameter		80	.300		145	12.60		157.60	178
8060	10" diameter		50	.480		330	20		350	395
8080	12" diameter		30	.800		425	33.50		458.50	515
8100	Tee, 4" diameter		90	.267		58.50	11.20		69.70	82
8120	6" diameter		80	.300		144	12.60		156.60	177
8140	8" diameter		70	.343		185	14.40		199.40	225
8160	10" diameter		40	.600		590	25		615	690

33 11 Water Utility Distribution Piping

33 11 13 – Public Water Utility Distribution Piping

33 11 13.25 Water Supply, Polyvinyl Chloride Pipe		Crew	Daily Output	Labor-Hours	Unit	Material	2015 Bare Costs Labor	Equipment	Total	Total Incl O&P
8180	12" diameter	B-20	20	1.200	Ea.	765	50.50		815.50	920
8200	45° Bend, 4" diameter		100	.240		42	10.05		52.05	62
8220	6" diameter		90	.267		73	11.20		84.20	97.50
8240	8" diameter		50	.480		139	20		159	183
8260	10" diameter		50	.480		278	20		298	335
8280	12" diameter		30	.800		360	33.50		393.50	445
8300	Reducing tee 6" x 4"		100	.240		104	10.05		114.05	131
8320	8" x 6"		90	.267		166	11.20		177.20	199
8330	10" x 6"		90	.267		196	11.20		207.20	232
8340	10" x 8"		90	.267		216	11.20		227.20	255
8350	12" x 6"		90	.267		245	11.20		256.20	286
8360	12" x 8"		90	.267		265	11.20		276.20	310
8400	Tapped service tee (threaded type) 6" x 6" x 3/4"		100	.240		95.50	10.05		105.55	121
8430	6" x 6" x 1"		90	.267		95.50	11.20		106.70	122
8440	6" x 6" x 1-1/2"		90	.267		95.50	11.20		106.70	122
8450	6" x 6" x 2"		90	.267		95.50	11.20		106.70	122
8460	8" x 8" x 3/4"		90	.267		140	11.20		151.20	171
8470	8" x 8" x 1"		90	.267		140	11.20		151.20	171
8480	8" x 8" x 1-1/2"		90	.267		140	11.20		151.20	171
8490	8" x 8" x 2"		90	.267		140	11.20		151.20	171
8500	Repair coupling 4"		100	.240		26	10.05		36.05	44
8520	6" diameter		90	.267		40.50	11.20		51.70	62
8540	8" diameter		50	.480		96.50	20		116.50	137
8560	10" diameter		50	.480		203	20		223	254
8580	12" diameter		50	.480		296	20		316	355
8600	Plug end 4"		100	.240		23	10.05		33.05	41
8620	6" diameter		90	.267		41	11.20		52.20	62.50
8640	8" diameter		50	.480		69	20		89	107
8660	10" diameter		50	.480		97	20		117	137
8680	12" diameter		50	.480		119	20		139	162
8700	PVC pipe, joint restraint									
8710	4" diameter	B-20A	32	1	Ea.	45.50	45.50		91	120
8720	6" diameter		25.60	1.250		56	57		113	149
8730	8" diameter		21.33	1.500		74.50	68.50		143	186
8740	10" diameter		18.28	1.751		130	80		210	265
8750	12" diameter		16.84	1.900		145	87		232	291
8760	14" diameter		16	2		204	91.50		295.50	365
8770	16" diameter		11.64	2.749		265	126		391	485
8780	18" diameter		11.03	2.901		340	133		473	575
8785	20" diameter		9.14	3.501		420	160		580	705
8790	24" diameter		7.53	4.250		490	194		684	835

33 11 13.35 Water Supply, HDPE

		Crew	Daily Output	Labor-Hours	Unit	Material	2015 Bare Costs Labor	Equipment	Total	Total Incl O&P
0010	**WATER SUPPLY, HDPE**									
0011	Butt fusion joints, SDR 21 40' lengths not including excavation or backfill									
0100	4" diameter	B-22A	400	.100	L.F.	2.47	4.30	1.64	8.41	11.15
0200	6" diameter		380	.105		5.85	4.53	1.73	12.11	15.30
0300	8" diameter		320	.125		8.55	5.40	2.06	16.01	19.90
0400	10" diameter		300	.133		14.65	5.75	2.19	22.59	27.50
0500	12" diameter		260	.154		17.75	6.60	2.53	26.88	32.50
0600	14" diameter	B-22B	220	.182		27.50	7.80	5	40.30	47.50
0700	16" diameter		180	.222		32.50	9.55	6.15	48.20	57
0800	18" diameter		140	.286		38	12.30	7.90	58.20	69

For customer support on your Plumbing Cost Data, call 877.709.6968.

409

33 11 13 – Public Water Utility Distribution Piping

33 11 13.35 Water Supply, HDPE

		Crew	Daily Output	Labor-Hours	Unit	Material	2015 Bare Costs Labor	Equipment	Total	Total Incl O&P
0900	24" diameter	B-22B	100	.400	L.F.	59	17.20	11.05	87.25	104
1000	Fittings									
1100	Elbows, 90 degrees									
1200	4" diameter	B-22A	32	1.250	Ea.	18.45	54	20.50	92.95	126
1300	6" diameter		28	1.429		59	61.50	23.50	144	186
1400	8" diameter		24	1.667		121	71.50	27.50	220	274
1500	10" diameter		18	2.222		385	95.50	36.50	517	610
1600	12" diameter		12	3.333		400	143	55	598	720
1700	14" diameter	B-22B	9	4.444		490	191	123	804	965
1800	16" diameter		6	6.667		570	287	184	1,041	1,275
1900	18" diameter		4	10		660	430	276	1,366	1,700
2000	24" diameter		3	13.333		985	575	370	1,930	2,350
2100	Tees									
2200	4" diameter	B-22A	30	1.333	Ea.	23.50	57.50	22	103	138
2300	6" diameter		26	1.538		58	66	25.50	149.50	194
2400	8" diameter		22	1.818		146	78	30	254	315
2500	10" diameter		15	2.667		193	115	44	352	435
2600	12" diameter		10	4		545	172	66	783	935
2700	14" diameter	B-22B	8	5		640	215	138	993	1,175
2800	16" diameter		6	6.667		755	287	184	1,226	1,475
2900	18" diameter		4	10		985	430	276	1,691	2,050
3000	24" diameter		2	20		1,500	860	555	2,915	3,575
4100	Caps									
4110	4" diameter	B-22A	34	1.176	Ea.	13.60	50.50	19.35	83.45	114
4120	6" diameter		30	1.333		35.50	57.50	22	115	151
4130	8" diameter		26	1.538		52	66	25.50	143.50	187
4150	10" diameter		20	2		111	86	33	230	290
4160	12" diameter		14	2.857		124	123	47	294	380

33 11 13.45 Water Supply, Copper Pipe

		Crew	Daily Output	Labor-Hours	Unit	Material	2015 Bare Costs Labor	Equipment	Total	Total Incl O&P
0010	**WATER SUPPLY, COPPER PIPE**									
0020	Not including excavation or backfill									
2000	Tubing, type K, 20' joints, 3/4" diameter	Q-1	400	.040	L.F.	6.30	2.11		8.41	10.15
2200	1" diameter		320	.050		8.40	2.64		11.04	13.25
3000	1-1/2" diameter		265	.060		13.40	3.19		16.59	19.50
3020	2" diameter		230	.070		20.50	3.68		24.18	28
3040	2-1/2" diameter		146	.110		30.50	5.80		36.30	42.50
3060	3" diameter		134	.119		42	6.30		48.30	56
4012	4" diameter		95	.168		69.50	8.90		78.40	90
4016	6" diameter	Q-2	80	.300		150	16.45		166.45	190
5000	Tubing, type L									
5108	2" diameter	Q-1	230	.070	L.F.	16.05	3.68		19.73	23
6010	3" diameter		134	.119		31.50	6.30		37.80	44
6012	4" diameter		95	.168		55.50	8.90		64.40	74.50
6016	6" diameter	Q-2	80	.300		132	16.45		148.45	170
7165	Fittings, brass, corporation stops, no lead, 3/4" diameter	1 Plum	19	.421	Ea.	82	24.50		106.50	128
7166	1" diameter		16	.500		108	29.50		137.50	164
7167	1-1/2" diameter		13	.615		240	36		276	320
7168	2" diameter		11	.727		350	42.50		392.50	450
7170	Curb stops, no lead, 3/4" diameter		19	.421		112	24.50		136.50	161
7171	1" diameter		16	.500		156	29.50		185.50	217
7172	1-1/2" diameter		13	.615		296	36		332	380
7173	2" diameter		11	.727		315	42.50		357.50	410

33 12 13 – Water Service Connections

33 12 13.15 Tapping, Crosses and Sleeves	Crew	Daily Output	Labor-Hours	Unit	Material	2015 Bare Costs Labor	Equipment	Total	Total Incl O&P
0010 **TAPPING, CROSSES AND SLEEVES**									
4000 Drill and tap pressurized main (labor only)									
4100 6" main, 1" to 2" service	Q-1	3	5.333	Ea.		282		282	425
4150 8" main, 1" to 2" service	"	2.75	5.818	"		305		305	465
4500 Tap and insert gate valve									
4600 8" main, 4" branch	B-21	3.20	8.750	Ea.		380	43	423	635
4650 6" branch		2.70	10.370			450	51	501	745
4700 10" Main, 4" branch		2.70	10.370			450	51	501	745
4750 6" branch		2.35	11.915			515	58.50	573.50	860
4800 12" main, 6" branch		2.35	11.915			515	58.50	573.50	860
4850 8" branch		2.35	11.915			515	58.50	573.50	860
7020 Crosses, 4" x 4"		37	.757		1,375	33	3.72	1,411.72	1,550
7030 6" x 4"		25	1.120		1,150	48.50	5.50	1,204	1,350
7040 6" x 6"		25	1.120		1,175	48.50	5.50	1,229	1,375
7060 8" x 6"		21	1.333		1,500	58	6.55	1,564.55	1,750
7080 8" x 8"		21	1.333		1,800	58	6.55	1,864.55	2,075
7100 10" x 6"		21	1.333		2,600	58	6.55	2,664.55	2,950
7120 10" x 10"		21	1.333		2,650	58	6.55	2,714.55	3,025
7140 12" x 6"		18	1.556		2,600	67.50	7.65	2,675.15	2,950
7160 12" x 12"		18	1.556		3,575	67.50	7.65	3,650.15	4,050
7180 14" x 6"		16	1.750		6,475	76	8.60	6,559.60	7,250
7200 14" x 14"		16	1.750		6,900	76	8.60	6,984.60	7,700
7220 16" x 6"		14	2		6,900	86.50	9.85	6,996.35	7,750
7240 16" x 10"		14	2		7,000	86.50	9.85	7,096.35	7,850
7260 16" x 16"		14	2		7,425	86.50	9.85	7,521.35	8,325
7280 18" x 6"		10	2.800		9,875	121	13.80	10,009.80	11,100
7300 18" x 12"		10	2.800		10,100	121	13.80	10,234.80	11,300
7320 18" x 18"		10	2.800		9,650	121	13.80	9,784.80	10,800
7340 20" x 6"		8	3.500		8,475	152	17.20	8,644.20	9,575
7360 20" x 12"		8	3.500		8,800	152	17.20	8,969.20	9,925
7380 20" x 20"		8	3.500		13,400	152	17.20	13,569.20	15,000
7400 24" x 6"		6	4.667		12,400	202	23	12,625	13,900
7420 24" x 12"		6	4.667		11,600	202	23	11,825	13,100
7440 24" x 18"		6	4.667		17,400	202	23	17,625	19,500
7460 24" x 24"		6	4.667		18,000	202	23	18,225	20,100
7600 Cut-in sleeves with rubber gaskets, 4"		18	1.556		430	67.50	7.65	505.15	585
7620 6"		12	2.333		505	101	11.50	617.50	725
7640 8"		10	2.800		650	121	13.80	784.80	915
7660 10"		10	2.800		960	121	13.80	1,094.80	1,250
7680 12"		9	3.111		1,225	135	15.30	1,375.30	1,575
7800 Cut-in valves with rubber gaskets, 4"		18	1.556		505	67.50	7.65	580.15	665
7820 6"		12	2.333		675	101	11.50	787.50	910
7840 8"		10	2.800		1,050	121	13.80	1,184.80	1,375
7860 10"		10	2.800		1,150	121	13.80	1,284.80	1,450
7880 12"		9	3.111		1,300	135	15.30	1,450.30	1,650
7900 Tapping Valve 4 inch, MJ, ductile iron		18	1.556		555	67.50	7.65	630.15	720
7920 6 inch, MJ, ductile iron		12	2.333		780	101	11.50	892.50	1,025
8000 Sleeves with rubber gaskets, 4" x 4"		37	.757		845	33	3.72	881.72	985
8010 6" x 4"		25	1.120		900	48.50	5.50	954	1,075
8020 6" x 6"		25	1.120		1,125	48.50	5.50	1,179	1,325
8030 8" x 4"		21	1.333		960	58	6.55	1,024.55	1,150
8040 8" x 6"		21	1.333		1,075	58	6.55	1,139.55	1,275
8060 8" x 8"		21	1.333		1,250	58	6.55	1,314.55	1,475

For customer support on your Plumbing Cost Data, call 877.709.6968.

411

33 12 Water Utility Distribution Equipment

33 12 13 – Water Service Connections

33 12 13.15 Tapping, Crosses and Sleeves		Crew	Daily Output	Labor-Hours	Unit	Material	2015 Bare Costs Labor	Equipment	Total	Total Incl O&P
8070	10" x 4"	B-21	21	1.333	Ea.	1,075	58	6.55	1,139.55	1,300
8080	10" x 6"		21	1.333		1,200	58	6.55	1,264.55	1,425
8090	10" x 8"		21	1.333		1,325	58	6.55	1,389.55	1,550
8100	10" x 10"		21	1.333		1,950	58	6.55	2,014.55	2,250
8110	12" x 4"		18	1.556		1,150	67.50	7.65	1,225.15	1,375
8120	12" x 6"		18	1.556		1,175	67.50	7.65	1,250.15	1,400
8130	12" x 8"		18	1.556		1,475	67.50	7.65	1,550.15	1,725
8135	12" x 10"		18	1.556		2,325	67.50	7.65	2,400.15	2,650
8140	12" x 12"		18	1.556		2,350	67.50	7.65	2,425.15	2,700
8160	14" x 6"		16	1.750		1,900	76	8.60	1,984.60	2,225
8180	14" x 14"		16	1.750		2,500	76	8.60	2,584.60	2,875
8200	16" x 6"		14	2		2,100	86.50	9.85	2,196.35	2,450
8220	16" x 10"		14	2		3,525	86.50	9.85	3,621.35	4,025
8240	16" x 16"		14	2		8,650	86.50	9.85	8,746.35	9,675
8260	18" x 6"		10	2.800		2,250	121	13.80	2,384.80	2,675
8280	18" x 12"		10	2.800		3,625	121	13.80	3,759.80	4,175
8300	18" x 18"		10	2.800		11,600	121	13.80	11,734.80	13,000
8320	20" x 6"		8	3.500		2,550	152	17.20	2,719.20	3,075
8340	20" x 12"		8	3.500		4,400	152	17.20	4,569.20	5,100
8360	20" x 20"		8	3.500		11,000	152	17.20	11,169.20	12,400
8380	24" x 6"		6	4.667		2,700	202	23	2,925	3,300
8400	24" x 12"		6	4.667		4,400	202	23	4,625	5,175
8420	24" x 18"		6	4.667		11,700	202	23	11,925	13,200
8440	24" x 24"	▼	6	4.667		14,800	202	23	15,025	16,600
8800	Hydrant valve box, 6' long	B-20	20	1.200		280	50.50		330.50	390
8820	8' long		18	1.333		355	56		411	475
8830	Valve box w/lid 4' deep		14	1.714		107	72		179	228
8840	Valve box and large base w/lid	▼	14	1.714	▼	325	72		397	470

33 12 16 – Water Utility Distribution Valves

33 12 16.10 Valves		Crew	Daily Output	Labor-Hours	Unit	Material	2015 Bare Costs Labor	Equipment	Total	Total Incl O&P
0010	**VALVES**, water distribution									
0011	See Sections 22 05 23.20 and 22 05 23.60									
3000	Butterfly valves with boxes, cast iron, mech. jt.									
3100	4" diameter	B-6	6	4	Ea.	575	165	60.50	800.50	955
3140	6" diameter		6	4		750	165	60.50	975.50	1,150
3180	8" diameter		6	4		940	165	60.50	1,165.50	1,350
3300	10" diameter		6	4		1,275	165	60.50	1,500.50	1,725
3340	12" diameter		6	4		1,700	165	60.50	1,925.50	2,200
3400	14" diameter		4	6		3,400	248	91	3,739	4,200
3440	16" diameter		4	6		5,025	248	91	5,364	6,000
3460	18" diameter		4	6		6,225	248	91	6,564	7,325
3480	20" diameter		4	6		8,200	248	91	8,539	9,475
3500	24" diameter	▼	4	6	▼	13,700	248	91	14,039	15,600
3600	With lever operator									
3610	4" diameter	B-6	6	4	Ea.	470	165	60.50	695.50	835
3614	6" diameter		6	4		645	165	60.50	870.50	1,025
3616	8" diameter		6	4		835	165	60.50	1,060.50	1,250
3618	10" diameter		6	4		1,175	165	60.50	1,400.50	1,600
3620	12" diameter		6	4		1,600	165	60.50	1,825.50	2,100
3622	14" diameter		4	6		3,075	248	91	3,414	3,850
3624	16" diameter		4	6		4,700	248	91	5,039	5,625
3626	18" diameter		4	6		5,900	248	91	6,239	6,950

33 12 16 – Water Utility Distribution Valves

33 12 16.10 Valves		Crew	Daily Output	Labor-Hours	Unit	Material	2015 Bare Costs Labor	Equipment	Total	Total Incl O&P
3628	20" diameter	B-6	4	6	Ea.	7,875	248	91	8,214	9,125
3630	24" diameter	↓	4	6	↓	13,400	248	91	13,739	15,200
3700	Check valves, flanged									
3710	4" diameter	B-6	6	4	Ea.	605	165	60.50	830.50	985
3714	6" diameter		6	4		1,025	165	60.50	1,250.50	1,475
3716	8" diameter		6	4		1,950	165	60.50	2,175.50	2,475
3718	10" diameter		6	4		3,325	165	60.50	3,550.50	3,975
3720	12" diameter		6	4		5,175	165	60.50	5,400.50	6,025
3726	18" diameter		4	6		23,700	248	91	24,039	26,600
3730	24" diameter	↓	4	6	↓	52,000	248	91	52,339	58,000
3800	Gate valves, C.I., 250 PSI, mechanical joint, w/boxes									
3810	4" diameter	B-6	6	4	Ea.	880	165	60.50	1,105.50	1,275
3814	6" diameter		6	4		1,350	165	60.50	1,575.50	1,800
3816	8" diameter		6	4		2,300	165	60.50	2,525.50	2,850
3818	10" diameter		6	4		4,150	165	60.50	4,375.50	4,875
3820	12" diameter		6	4		5,625	165	60.50	5,850.50	6,500
3822	14" diameter		4	6		12,300	248	91	12,639	14,100
3824	16" diameter		4	6		17,300	248	91	17,639	19,500
3826	18" diameter		4	6		21,100	248	91	21,439	23,700
3828	20" diameter		4	6		23,200	248	91	23,539	26,000
3830	24" diameter		4	6		25,200	248	91	25,539	28,200
3831	30" diameter		4	6		39,200	248	91	39,539	43,600
3832	36" diameter	↓	4	6		57,000	248	91	57,339	63,000
3880	Sleeve, for tapping mains, 8" x 4", add					960			960	1,050
3884	10" x 6", add					1,200			1,200	1,325
3888	12" x 6", add					1,175			1,175	1,300
3892	12" x 8", add				▼	1,475			1,475	1,625

33 12 16.20 Valves		Crew	Daily Output	Labor-Hours	Unit	Material	2015 Bare Costs Labor	Equipment	Total	Total Incl O&P
0010	**VALVES**									
0011	Special trim or use									
9000	Valves, gate valve, N.R.S. PIV with post, 4" diameter	B-6	6	4	Ea.	1,875	165	60.50	2,100.50	2,375
9020	6" diameter		6	4		2,350	165	60.50	2,575.50	2,900
9040	8" diameter		6	4		3,275	165	60.50	3,500.50	3,950
9060	10" diameter		6	4		5,150	165	60.50	5,375.50	5,975
9080	12" diameter		6	4		6,600	165	60.50	6,825.50	7,600
9100	14" diameter		6	4		13,100	165	60.50	13,325.50	14,700
9120	OS&Y, 4" diameter		6	4		730	165	60.50	955.50	1,125
9140	6" diameter		6	4		990	165	60.50	1,215.50	1,425
9160	8" diameter		6	4		1,550	165	60.50	1,775.50	2,025
9180	10" diameter		6	4		2,475	165	60.50	2,700.50	3,025
9200	12" diameter		6	4		3,725	165	60.50	3,950.50	4,425
9220	14" diameter		4	6		5,075	248	91	5,414	6,075
9400	Check valves, rubber disc, 2-1/2" diameter		6	4		710	165	60.50	935.50	1,100
9420	3" diameter		6	4		725	165	60.50	950.50	1,125
9440	4" diameter		6	4		605	165	60.50	830.50	985
9480	6" diameter		6	4		1,025	165	60.50	1,250.50	1,475
9500	8" diameter		6	4		1,950	165	60.50	2,175.50	2,475
9520	10" diameter		6	4		3,325	165	60.50	3,550.50	3,975
9540	12" diameter		6	4		5,175	165	60.50	5,400.50	6,025
9542	14" diameter		4	6		5,250	248	91	5,589	6,250
9700	Detector check valves, reducing, 4" diameter		6	4		1,350	165	60.50	1,575.50	1,800
9720	6" diameter	↓	6	4	↓	2,075	165	60.50	2,300.50	2,625

For customer support on your Plumbing Cost Data, call 877.709.6968.

413

33 12 Water Utility Distribution Equipment

33 12 16 – Water Utility Distribution Valves

33 12 16.20 Valves		Crew	Daily Output	Labor-Hours	Unit	Material	2015 Bare Costs		Total	Total Incl O&P
							Labor	Equipment		
9740	8" diameter	B-6	6	4	Ea.	3,050	165	60.50	3,275.50	3,675
9760	10" diameter		6	4		5,400	165	60.50	5,625.50	6,275
9800	Galvanized, 4" diameter		6	4		1,575	165	60.50	1,800.50	2,075
9820	6" diameter		6	4		2,125	165	60.50	2,350.50	2,675
9840	8" diameter		6	4		3,350	165	60.50	3,575.50	4,000
9860	10" diameter		6	4		5,125	165	60.50	5,350.50	5,975

33 12 19 – Water Utility Distribution Fire Hydrants

33 12 19.10 Fire Hydrants

33 12 19.10 Fire Hydrants		Crew	Daily Output	Labor-Hours	Unit	Material	2015 Bare Costs		Total	Total Incl O&P
							Labor	Equipment		
0010	**FIRE HYDRANTS** G3010-410									
0020	Mechanical joints unless otherwise noted									
1000	Fire hydrants, two way; excavation and backfill not incl.									
1100	4-1/2" valve size, depth 2'-0"	B-21	10	2.800	Ea.	1,650	121	13.80	1,784.80	2,025
1200	4'-6"		9	3.111		1,975	135	15.30	2,125.30	2,400
1260	6'-0"		7	4		2,100	173	19.70	2,292.70	2,600
1340	8'-0"		6	4.667		2,225	202	23	2,450	2,775
1420	10'-0"		5	5.600		2,375	243	27.50	2,645.50	3,025
2000	5-1/4" valve size, depth 2'-0"		10	2.800		1,775	121	13.80	1,909.80	2,150
2080	4'-0"		9	3.111		1,900	135	15.30	2,050.30	2,325
2160	6'-0"		7	4		2,050	173	19.70	2,242.70	2,550
2240	8'-0"		6	4.667		2,175	202	23	2,400	2,725
2320	10'-0"		5	5.600		2,500	243	27.50	2,770.50	3,150
2350	For threeway valves, add					7%				
2400	Lower barrel extensions with stems, 1'-0"	B-20	14	1.714		320	72		392	460
2440	2'-0"		13	1.846		390	77.50		467.50	550
2480	3'-0"		12	2		770	84		854	975
2520	4'-0"		10	2.400		740	101		841	970
5000	Indicator post									
5020	Adjustable, valve size 4" to 14", 4' bury	B-21	10	2.800	Ea.	815	121	13.80	949.80	1,100
5060	8' bury		7	4		1,100	173	19.70	1,292.70	1,500
5080	10' bury		6	4.667		1,150	202	23	1,375	1,600
5100	12' bury		5	5.600		1,400	243	27.50	1,670.50	1,950
5120	14' bury		4	7		1,525	305	34.50	1,864.50	2,175
5500	Non-adjustable, valve size 4" to 14", 3' bury		10	2.800		815	121	13.80	949.80	1,100
5520	3'-6" bury		10	2.800		815	121	13.80	949.80	1,100
5540	4' bury		9	3.111		815	135	15.30	965.30	1,125

33 16 Water Utility Storage Tanks

33 16 13 – Aboveground Water Utility Storage Tanks

33 16 13.16 Prestressed Conc. Water Storage Tanks

33 16 13.16 Prestressed Conc. Water Storage Tanks		Crew	Daily Output	Labor-Hours	Unit	Material	2015 Bare Costs		Total	Total Incl O&P
							Labor	Equipment		
0010	**PRESTRESSED CONC. WATER STORAGE TANKS**									
0020	Not including fdn., pipe or pumps, 250,000 gallons				Ea.				299,000	329,500

33 21 Water Supply Wells

33 21 13 – Public Water Supply Wells

33 21 13.10 Wells and Accessories	Crew	Daily Output	Labor-Hours	Unit	Material	2015 Bare Costs Labor	Equipment	Total	Total Incl O&P
0010 **WELLS & ACCESSORIES**									
0011 Domestic									
0100 Drilled, 4" to 6" diameter	B-23	120	.333	L.F.		12.65	23.50	36.15	45.50
0200 8" diameter	"	95.20	.420	"		15.95	30	45.95	57.50
0400 Gravel pack well, 40' deep, incl. gravel & casing, complete									
0500 24" diameter casing x 18" diameter screen	B-23	.13	307	Total	38,800	11,700	21,900	72,400	85,000
0600 36" diameter casing x 18" diameter screen		.12	333	"	40,000	12,700	23,700	76,400	89,500
0800 Observation wells, 1-1/4" riser pipe	↓	163	.245	V.L.F.	20.50	9.35	17.45	47.30	56
0900 For flush Buffalo roadway box, add	1 Skwk	16.60	.482	Ea.	51	23.50		74.50	92
1200 Test well, 2-1/2" diameter, up to 50' deep (15 to 50 GPM)	B-23	1.51	26.490	"	805	1,000	1,875	3,680	4,525
1300 Over 50' deep, add	"	121.80	.328	L.F.	21.50	12.50	23.50	57.50	68
1500 Pumps, installed in wells to 100' deep, 4" submersible									
1510 1/2 H.P.	Q-1	3.22	4.969	Ea.	350	263		613	780
1520 3/4 H.P.		2.66	6.015		400	320		720	920
1600 1 H.P.	↓	2.29	6.987		425	370		795	1,025
1700 1-1/2 H.P.	Q-22	1.60	10		800	530	410	1,740	2,125
1800 2 H.P.		1.33	12.030		830	635	490	1,955	2,425
1900 3 H.P.		1.14	14.035		1,150	740	575	2,465	3,025
2000 5 H.P.		1.14	14.035		1,600	740	575	2,915	3,500
3000 Pump, 6" submersible, 25' to 150' deep, 25 H.P., 249 to 297 GPM		.89	17.978		3,350	950	735	5,035	5,900
3100 25' to 500' deep, 30 H.P., 100 to 300 GPM	↓	.73	21.918	↓	4,100	1,150	895	6,145	7,225
8000 Steel well casing	B-23A	3020	.008	Lb.	1.21	.34	.89	2.44	2.83
8110 Well screen assembly, stainless steel, 2" diameter		273	.088	L.F.	81	3.75	9.85	94.60	106
8120 3" diameter		253	.095		142	4.04	10.60	156.64	174
8130 4" diameter		200	.120		177	5.10	13.45	195.55	217
8140 5" diameter		168	.143		187	6.10	16	209.10	232
8150 6" diameter		126	.190		212	8.10	21.50	241.60	269
8160 8" diameter		98.50	.244		288	10.40	27.50	325.90	360
8170 10" diameter		73	.329		360	14	37	411	455
8180 12" diameter		62.50	.384		410	16.35	43	469.35	525
8190 14" diameter		54.30	.442		460	18.85	49.50	528.35	590
8200 16" diameter		48.30	.497		510	21	55.50	586.50	655
8210 18" diameter		39.20	.612		635	26	68.50	729.50	815
8220 20" diameter		31.20	.769		710	33	86	829	930
8230 24" diameter		23.80	1.008		855	43	113	1,011	1,125
8240 26" diameter		21	1.143		930	48.50	128	1,106.50	1,250
8244 Well casing or drop pipe, PVC, 0.5" dia		550	.044		1.55	1.86	4.88	8.29	9.90
8245 0.75" diameter		550	.044		1.13	1.86	4.88	7.87	9.45
8246 1" diameter		550	.044		1.17	1.86	4.88	7.91	9.50
8247 1.25" diameter		520	.046		1.35	1.97	5.15	8.47	10.20
8248 1.5" diameter		490	.049		1.58	2.09	5.50	9.17	11
8249 1.75" diameter		380	.063		1.48	2.69	7.05	11.22	13.55
8250 2" diameter		280	.086		1.65	3.65	9.60	14.90	17.95
8252 3" diameter		260	.092		3.39	3.93	10.35	17.67	21
8254 4" diameter		205	.117		4.90	4.99	13.10	22.99	27.50
8255 5" diameter		170	.141		3.20	6	15.80	25	30
8256 6" diameter		130	.185		8.50	7.85	20.50	36.85	44
8258 8" diameter		100	.240		12.95	10.20	27	50.15	59.50
8260 10" diameter		73	.329		9	14	37	60	72
8262 12" diameter		62	.387		12	16.50	43.50	72	85.50
8300 Slotted PVC, 1-1/4" diameter		521	.046		2.40	1.96	5.15	9.51	11.30
8310 1-1/2" diameter		488	.049		2.85	2.10	5.50	10.45	12.40
8320 2" diameter	↓	273	.088	↓	3.67	3.75	9.85	17.27	20.50

33 21 Water Supply Wells

33 21 13 – Public Water Supply Wells

33 21 13.10 Wells and Accessories

		Crew	Daily Output	Labor-Hours	Unit	Material	2015 Bare Costs Labor	Equipment	Total	Total Incl O&P
8330	3" diameter	B-23A	253	.095	L.F.	5.85	4.04	10.60	20.49	24.50
8340	4" diameter		200	.120		6.75	5.10	13.45	25.30	30
8350	5" diameter		168	.143		14.05	6.10	16	36.15	42.50
8360	6" diameter		126	.190		16.65	8.10	21.50	46.25	54.50
8370	8" diameter		98.50	.244		26	10.40	27.50	63.90	74.50
8400	Artificial gravel pack, 2" screen, 6" casing	B-23B	174	.138		3.60	5.90	17.45	26.95	32
8405	8" casing		111	.216		4.63	9.20	27.50	41.33	49
8410	10" casing		74.50	.322		6.60	13.70	40.50	60.80	73.50
8415	12" casing		60	.400		9.25	17.05	50.50	76.80	91.50
8420	14" casing		50.20	.478		10.75	20.50	60.50	91.75	109
8425	16" casing		40.70	.590		13.65	25	74.50	113.15	136
8430	18" casing		36	.667		16.40	28.50	84.50	129.40	154
8435	20" casing		29.50	.814		19.25	34.50	103	156.75	187
8440	24" casing		25.70	.934		22	40	118	180	215
8445	26" casing		24.60	.976		24.50	41.50	123	189	227
8450	30" casing		20	1.200		28	51	152	231	276
8455	36" casing		16.40	1.463		31.50	62.50	185	279	335
8500	Develop well		8	3	Hr.	262	128	380	770	900
8550	Pump test well		8	3		82.50	128	380	590.50	700
8560	Standby well	B-23A	8	3		78	128	335	541	650
8570	Standby, drill rig		8	3			128	335	463	565
8580	Surface seal well, concrete filled		1	24	Ea.	850	1,025	2,675	4,550	5,450
8590	Well test pump, install & remove	B-23	1	40			1,525	2,850	4,375	5,475
8600	Well sterilization, chlorine	2 Clab	1	16		122	600		722	1,050
8610	Well water pressure switch	1 Clab	12	.667		50.50	25		75.50	94
8630	Well, water pressure switch with manual reset	"	12	.667		35.50	25		60.50	77.50
9950	See Section 31 23 19.40 for wellpoints									
9960	See Section 31 23 19.30 for drainage wells									

33 21 13.20 Water Supply Wells, Pumps

		Crew	Daily Output	Labor-Hours	Unit	Material	2015 Bare Costs Labor	Equipment	Total	Total Incl O&P
0010	**WATER SUPPLY WELLS, PUMPS**									
0011	With pressure control									
1000	Deep well, jet, 42 gal. galvanized tank									
1040	3/4 HP	1 Plum	.80	10	Ea.	1,100	585		1,685	2,100
3000	Shallow well, jet, 30 gal. galvanized tank									
3040	1/2 HP	1 Plum	2	4	Ea.	895	235		1,130	1,350

33 31 Sanitary Utility Sewerage Piping

33 31 13 – Public Sanitary Utility Sewerage Piping

33 31 13.10 Sewage Collection, Valves

		Crew	Daily Output	Labor-Hours	Unit	Material	2015 Bare Costs Labor	Equipment	Total	Total Incl O&P
0010	**SEWAGE COLLECTION, VALVES**									
1000	Backwater sewer line valve									
1010	Offset type, bronze swing check assy and bronze cover									
1020	B&S connections									
1040	2" size	2 Skwk	14	1.143	Ea.	555	55.50		610.50	695
1050	3" size		12.50	1.280		870	62.50		932.50	1,050
1060	4" size		11	1.455		1,075	71		1,146	1,275
1070	6" size	3 Skwk	15	1.600		1,550	78		1,628	1,825
1080	8" size	"	7.50	3.200		2,150	156		2,306	2,625
1110	Backwater drainage control									
1120	Offset type, w/bronze manually operated shear gate									
1130	B&S connections									

33 31 Sanitary Utility Sewerage Piping

33 31 13 – Public Sanitary Utility Sewerage Piping

33 31 13.10 Sewage Collection, Valves

		Crew	Daily Output	Labor-Hours	Unit	Material	2015 Bare Costs Labor	Equipment	Total	Total Incl O&P
1160	4" size	2 Skwk	11	1.455	Ea.	2,300	71		2,371	2,625
1170	6" size	3 Skwk	15	1.600	"	3,000	78		3,078	3,425

33 31 13.15 Sewage Collection, Concrete Pipe

0010	**SEWAGE COLLECTION, CONCRETE PIPE**									
0020	See Section 33 41 13.60 for sewage/drainage collection, concrete pipe									

33 31 13.25 Sewage Collection, Polyvinyl Chloride Pipe

		Crew	Daily Output	Labor-Hours	Unit	Material	Labor	Equipment	Total	Total Incl O&P
0010	**SEWAGE COLLECTION, POLYVINYL CHLORIDE PIPE**									
0020	Not including excavation or backfill									
2000	20' lengths, SDR 35, B&S, 4" diameter	B-20	375	.064	L.F.	1.45	2.68		4.13	5.75
2040	6" diameter	"	350	.069		3.28	2.88		6.16	8.05
2120	10" diameter	B-21	330	.085		11.40	3.68	.42	15.50	18.60
2160	12" diameter		320	.088		12.75	3.79	.43	16.97	20.50
2200	15" diameter		240	.117		12.95	5.05	.57	18.57	22.50
4000	Piping, DWV PVC, no exc./bkfill., 10' L, Sch 40, 4" diameter	B-20	375	.064		3.84	2.68		6.52	8.35
4010	6" diameter		350	.069		7.65	2.88		10.53	12.85
4020	8" diameter		335	.072		12	3.01		15.01	17.85

33 36 Utility Septic Tanks

33 36 13 – Utility Septic Tank and Effluent Wet Wells

33 36 13.13 Concrete Utility Septic Tank

		Crew	Daily Output	Labor-Hours	Unit	Material	Labor	Equipment	Total	Total Incl O&P
0010	**CONCRETE UTILITY SEPTIC TANK** G3020-300									
0011	Not including excavation or piping									
0015	Septic tanks, precast, 1,000 gallon	B-21	8	3.500	Ea.	1,000	152	17.20	1,169.20	1,350
0060	1,500 gallon		7	4		1,325	173	19.70	1,517.70	1,775
0100	2,000 gallon		5	5.600		1,925	243	27.50	2,195.50	2,500
0200	5,000 gallon	B-13	3.50	16		9,500	655	210	10,365	11,700
0300	15,000 gallon, 4 piece	B-13B	1.70	32.941		20,000	1,350	665	22,015	24,800
0400	25,000 gallon, 4 piece		1.10	50.909		38,800	2,075	1,025	41,900	47,000
0500	40,000 gallon, 4 piece		.80	70		49,700	2,875	1,400	53,975	60,500
0520	50,000 gallon, 5 piece	B-13C	.60	93.333		57,000	3,825	2,775	63,600	72,000
0640	75,000 gallon, cast in place	C-14C	.25	448		69,500	20,200	125	89,825	107,500
0660	100,000 gallon	"	.15	746		86,000	33,700	209	119,909	146,500
1150	Leaching field chambers, 13' x 3'-7" x 1'-4", standard	B-13	16	3.500		485	143	46	674	800
1200	Heavy duty, 8' x 4' x 1'-6"		14	4		284	164	52.50	500.50	620
1300	13' x 3'-9" x 1'-6"		12	4.667		1,125	191	61.50	1,377.50	1,575
1350	20' x 4' x 1'-6"		5	11.200		1,175	460	147	1,782	2,150
1400	Leaching pit, precast concrete, 3' diameter, 3' deep	B-21	8	3.500		710	152	17.20	879.20	1,025
1500	6' diameter, 3' section		4.70	5.957		885	258	29.50	1,172.50	1,400
2000	Velocity reducing pit, precast conc., 6' diameter, 3' deep		4.70	5.957		1,600	258	29.50	1,887.50	2,175

33 36 13.19 Polyethylene Utility Septic Tank

		Crew	Daily Output	Labor-Hours	Unit	Material	Labor	Equipment	Total	Total Incl O&P
0010	**POLYETHYLENE UTILITY SEPTIC TANK**									
0015	High density polyethylene, 1,000 gallon	B-21	8	3.500	Ea.	1,550	152	17.20	1,719.20	1,950
0020	1,250 gallon		8	3.500		1,200	152	17.20	1,369.20	1,575
0025	1,500 gallon		7	4		1,475	173	19.70	1,667.70	1,925

33 36 19 – Utility Septic Tank Effluent Filter

33 36 19.13 Utility Septic Tank Effluent Tube Filter

		Crew	Daily Output	Labor-Hours	Unit	Material	Labor	Equipment	Total	Total Incl O&P
0010	**UTILITY SEPTIC TANK EFFLUENT TUBE FILTER**									
3000	Effluent filter, 4" diameter	1 Skwk	8	1	Ea.	42	48.50		90.50	121
3020	6" diameter		7	1.143		249	55.50		304.50	360
3030	8" diameter		7	1.143		225	55.50		280.50	335

For customer support on your Plumbing Cost Data, call 877.709.6968.

417

33 36 Utility Septic Tanks

33 36 19 – Utility Septic Tank Effluent Filter

33 36 19.13 Utility Septic Tank Effluent Tube Filter		Crew	Daily Output	Labor-Hours	Unit	Material	2015 Bare Costs Labor	Equipment	Total	Total Incl O&P
3040	8" diameter, very fine	1 Skwk	7	1.143	Ea.	460	55.50		515.50	590
3050	10" diameter, very fine		6	1.333		235	65		300	360
3060	10" diameter		6	1.333		280	65		345	410
3080	12" diameter		6	1.333		670	65		735	840
3090	15" diameter		5	1.600		1,150	78		1,228	1,375

33 36 33 – Utility Septic Tank Drainage Field

33 36 33.13 Utility Septic Tank Tile Drainage Field

		Crew	Daily Output	Labor-Hours	Unit	Material	Labor	Equipment	Total	Total Incl O&P
0010	**UTILITY SEPTIC TANK TILE DRAINAGE FIELD**									
0015	Distribution box, concrete, 5 outlets	2 Clab	20	.800	Ea.	80.50	30		110.50	135
0020	7 outlets		16	1		80.50	37.50		118	147
0025	9 outlets		8	2		490	75		565	655
0115	Distribution boxes, HDPE, 5 outlets		20	.800		64.50	30		94.50	118
0117	6 outlets		15	1.067		65	40		105	133
0118	7 outlets		15	1.067		65	40		105	133
0120	8 outlets		10	1.600		68	60		128	168
0240	Distribution boxes, Outlet Flow Leveler	1 Clab	50	.160		2.25	6		8.25	11.75
0300	Precast concrete, galley, 4' x 4' x 4'	B-21	16	1.750		250	76	8.60	334.60	400
0350	HDPE infiltration chamber 12" H X 15" W	2 Clab	300	.053	L.F.	7.30	2.01		9.31	11.15
0351	12" H X 15" W End Cap	1 Clab	32	.250	Ea.	19.75	9.40		29.15	36
0355	chamber 12" H X 22" W	2 Clab	300	.053	L.F.	6.70	2.01		8.71	10.45
0356	12" H X 22" W End Cap	1 Clab	32	.250	Ea.	16.85	9.40		26.25	33
0360	chamber 13" H X 34" W	2 Clab	300	.053	L.F.	15	2.01		17.01	19.60
0361	13" H X 34" W End Cap	1 Clab	32	.250	Ea.	53.50	9.40		62.90	73
0365	chamber 16" H X 34" W	2 Clab	300	.053	L.F.	10.70	2.01		12.71	14.90
0366	16" H X 34" W End Cap	1 Clab	32	.250	Ea.	8	9.40		17.40	23.50
0370	chamber 8" H X 16" W	2 Clab	300	.053	L.F.	9.65	2.01		11.66	13.70
0371	8" H X 16" W End Cap	1 Clab	32	.250	Ea.	10.80	9.40		20.20	26.50

33 36 50 – Drainage Field Systems

33 36 50.10 Drainage Field Excavation and Fill

		Crew	Daily Output	Labor-Hours	Unit	Material	Labor	Equipment	Total	Total Incl O&P
0010	**DRAINAGE FIELD EXCAVATION AND FILL**									
2200	Septic tank & drainage field excavation with 3/4 c.y backhoe	B-12F	145	.110	C.Y.		4.93	4.52	9.45	12.45
2400	4' trench for disposal field, 3/4 C.Y. backhoe	"	335	.048	L.F.		2.13	1.95	4.08	5.40
2600	Gravel fill, run of bank	B-6	150	.160	C.Y.	20	6.60	2.43	29.03	35
2800	Crushed stone, 3/4"	"	150	.160	"	36.50	6.60	2.43	45.53	53

33 41 Storm Utility Drainage Piping

33 41 13 – Public Storm Utility Drainage Piping

33 41 13.40 Piping, Storm Drainage, Corrugated Metal

		Crew	Daily Output	Labor-Hours	Unit	Material	Labor	Equipment	Total	Total Incl O&P
0010	**PIPING, STORM DRAINAGE, CORRUGATED METAL**									
0020	Not including excavation or backfill									
2000	Corrugated metal pipe, galvanized									
2020	Bituminous coated with paved invert, 20' lengths									
2040	8" diameter, 16 ga.	B-14	330	.145	L.F.	8.65	5.80	1.10	15.55	19.60
2060	10" diameter, 16 ga.		260	.185		9	7.35	1.40	17.75	22.50
2080	12" diameter, 16 ga.		210	.229		11.05	9.10	1.73	21.88	28
2100	15" diameter, 16 ga.		200	.240		15.15	9.55	1.82	26.52	33.50
2120	18" diameter, 16 ga.		190	.253		16.70	10.05	1.92	28.67	36
2140	24" diameter, 14 ga.		160	.300		21.50	11.95	2.28	35.73	44.50
2160	30" diameter, 14 ga.	B-13	120	.467		27.50	19.15	6.15	52.80	67
2180	36" diameter, 12 ga.		120	.467		35.50	19.15	6.15	60.80	75.50

33 41 13 – Public Storm Utility Drainage Piping

33 41 13.40 Piping, Storm Drainage, Corrugated Metal		Crew	Daily Output	Labor-Hours	Unit	Material	2015 Bare Costs Labor	Equipment	Total	Total Incl O&P
2200	48" diameter, 12 ga.	B-13	100	.560	L.F.	52.50	23	7.35	82.85	101
2220	60" diameter, 10 ga.	B-13B	75	.747		80	30.50	15.05	125.55	151
2240	72" diameter, 8 ga.	"	45	1.244		95.50	51	25	171.50	211
2250	End sections, 8" diameter, 16 ga.	B-14	20	2.400	Ea.	43.50	95.50	18.20	157.20	214
2255	10" diameter, 16 ga.		20	2.400		60	95.50	18.20	173.70	232
2260	12" diameter, 16 ga.		18	2.667		120	106	20	246	320
2265	15" diameter, 16 ga.		18	2.667		205	106	20	331	410
2270	18" diameter, 16 ga.		16	3		240	119	23	382	470
2275	24" diameter, 16 ga.	B-13	16	3.500		280	143	46	469	580
2280	30" diameter, 16 ga.		14	4		490	164	52.50	706.50	845
2285	36" diameter, 14 ga.		14	4		730	164	52.50	946.50	1,100
2290	48" diameter, 14 ga.		10	5.600		1,850	230	73.50	2,153.50	2,450
2292	60" diameter, 14 ga.		6	9.333		1,950	385	123	2,458	2,875
2294	72" diameter, 14 ga.	B-13B	5	11.200		3,125	460	226	3,811	4,400
2300	Bends or elbows, 8" diameter	B-14	28	1.714		117	68	13	198	248
2320	10" diameter		25	1.920		146	76.50	14.55	237.05	293
2340	12" diameter, 16 ga.		23	2.087		169	83	15.85	267.85	330
2342	18" diameter, 16 ga.		20	2.400		244	95.50	18.20	357.70	435
2344	24" diameter, 14 ga.		16	3		335	119	23	477	580
2346	30" diameter, 14 ga.		15	3.200		430	127	24.50	581.50	690
2348	36" diameter, 14 ga.	B-13	15	3.733		580	153	49	782	925
2350	48" diameter, 12 ga.	"	12	4.667		805	191	61.50	1,057.50	1,250
2352	60" diameter, 10 ga.	B-13B	10	5.600		1,075	230	113	1,418	1,650
2354	72" diameter, 10 ga.	"	6	9.333		1,325	385	188	1,898	2,275
2360	Wyes or tees, 8" diameter	B-14	25	1.920		164	76.50	14.55	255.05	315
2380	10" diameter		21	2.286		205	91	17.35	313.35	385
2400	12" diameter, 16 ga.		19	2.526		244	100	19.15	363.15	445
2410	18" diameter, 16 ga.		16	3		325	119	23	467	565
2412	24" diameter, 14 ga.		16	3		485	119	23	627	745
2414	30" diameter, 14 ga.	B-13	12	4.667		645	191	61.50	897.50	1,075
2416	36" diameter, 14 ga.		11	5.091		795	209	67	1,071	1,275
2418	48" diameter, 12 ga.		10	5.600		1,175	230	73.50	1,478.50	1,700
2420	60" diameter, 10 ga.	B-13B	8	7		1,700	287	141	2,128	2,475
2422	72" diameter, 10 ga.	"	5	11.200		2,050	460	226	2,736	3,200
2500	Galvanized, uncoated, 20' lengths									
2520	8" diameter, 16 ga.	B-14	355	.135	L.F.	7.80	5.40	1.03	14.23	18
2540	10" diameter, 16 ga.		280	.171		8.95	6.80	1.30	17.05	21.50
2560	12" diameter, 16 ga.		220	.218		9.95	8.70	1.66	20.31	26
2580	15" diameter, 16 ga.		220	.218		12.45	8.70	1.66	22.81	29
2600	18" diameter, 16 ga.		205	.234		15.05	9.30	1.78	26.13	33
2620	24" diameter, 14 ga.		175	.274		18.95	10.90	2.08	31.93	40
2640	30" diameter, 14 ga.	B-13	130	.431		25	17.65	5.65	48.30	61
2660	36" diameter, 12 ga.		130	.431		32	17.65	5.65	55.30	68.50
2680	48" diameter, 12 ga.		110	.509		47.50	21	6.70	75.20	91.50
2690	60" diameter, 10 ga.	B-13B	78	.718		72	29.50	14.45	115.95	140
2695	72" diameter, 10 ga.	"	60	.933		86	38.50	18.80	143.30	174
2711	Bends or elbows, 12" diameter, 16 ga.	B-14	30	1.600	Ea.	145	63.50	12.15	220.65	271
2712	15" diameter, 16 ga.		25.04	1.917		180	76	14.55	270.55	330
2714	18" diameter, 16 ga.		20	2.400		202	95.50	18.20	315.70	390
2716	24" diameter, 14 ga.		16	3		293	119	23	435	530
2718	30" diameter, 14 ga.		15	3.200		340	127	24.50	491.50	595
2720	36" diameter, 14 ga.	B-13	15	3.733		515	153	49	717	855
2722	48" diameter, 12 ga.		12	4.667		680	191	61.50	932.50	1,100

For customer support on your Plumbing Cost Data, call 877.709.6968.

419

33 41 13.40 Piping, Storm Drainage, Corrugated Metal

		Crew	Daily Output	Labor-Hours	Unit	Material	2015 Bare Costs Labor	Equipment	Total	Total Incl O&P
2724	60" diameter, 10 ga.	B-13	10	5.600	Ea.	1,050	230	73.50	1,353.50	1,575
2726	72" diameter, 10 ga.	↓	6	9.333		1,350	385	123	1,858	2,200
2728	Wyes or tees, 12" diameter, 16 ga.	B-14	22.48	2.135		193	85	16.20	294.20	360
2730	18" diameter, 16 ga.		15	3.200		282	127	24.50	433.50	530
2732	24" diameter, 14 ga.		15	3.200		450	127	24.50	601.50	715
2734	30" diameter, 14 ga.	↓	14	3.429		580	136	26	742	875
2736	36" diameter, 14 ga.	B-13	14	4		755	164	52.50	971.50	1,150
2738	48" diameter, 12 ga.		12	4.667		1,075	191	61.50	1,327.50	1,550
2740	60" diameter, 10 ga.		10	5.600		1,575	230	73.50	1,878.50	2,150
2742	72" diameter, 10 ga.	↓	6	9.333		1,850	385	123	2,358	2,775
2780	End sections, 8" diameter	B-14	35	1.371		73	54.50	10.40	137.90	175
2785	10" diameter		35	1.371		77	54.50	10.40	141.90	179
2790	12" diameter		35	1.371		114	54.50	10.40	178.90	220
2800	18" diameter	↓	30	1.600		115	63.50	12.15	190.65	237
2810	24" diameter	B-13	25	2.240		215	92	29.50	336.50	410
2820	30" diameter		25	2.240		330	92	29.50	451.50	535
2825	36" diameter		20	2.800		480	115	37	632	745
2830	48" diameter	↓	10	5.600		950	230	73.50	1,253.50	1,475
2835	60" diameter	B-13B	5	11.200		1,650	460	226	2,336	2,750
2840	72" diameter	"	4	14		1,975	575	282	2,832	3,350
2850	Couplings, 12" diameter					10.55			10.55	11.60
2855	18" diameter					15.10			15.10	16.60
2860	24" diameter					19.95			19.95	22
2865	30" diameter					25			25	27.50
2870	36" diameter					32			32	35
2875	48" diameter					47.50			47.50	52
2880	60" diameter					60			60	66
2885	72" diameter	↓			↓	72.50			72.50	79.50

33 41 13.60 Sewage/Drainage Collection, Concrete Pipe

		Crew	Daily Output	Labor-Hours	Unit	Material	2015 Bare Costs Labor	Equipment	Total	Total Incl O&P
0010	**SEWAGE/DRAINAGE COLLECTION, CONCRETE PIPE**									
0020	Not including excavation or backfill									
1000	Non-reinforced pipe, extra strength, B&S or T&G joints									
1010	6" diameter	B-14	265.04	.181	L.F.	6	7.20	1.37	14.57	19.15
1020	8" diameter		224	.214		6.60	8.50	1.63	16.73	22
1030	10" diameter		216	.222		7.30	8.85	1.69	17.84	23.50
1040	12" diameter		200	.240		8.20	9.55	1.82	19.57	25.50
1050	15" diameter		180	.267		12	10.60	2.02	24.62	31.50
1060	18" diameter		144	.333		15	13.25	2.53	30.78	40
1070	21" diameter		112	.429		17	17.05	3.25	37.30	48.50
1080	24" diameter	↓	100	.480	↓	22	19.10	3.64	44.74	57.50
2000	Reinforced culvert, class 3, no gaskets									
2010	12" diameter	B-14	150	.320	L.F.	11	12.75	2.43	26.18	34.50
2020	15" diameter		150	.320		14	12.75	2.43	29.18	37.50
2030	18" diameter		132	.364		18	14.45	2.76	35.21	45
2035	21" diameter		120	.400		22	15.90	3.04	40.94	52
2040	24" diameter	↓	100	.480		26	19.10	3.64	48.74	62
2045	27" diameter	B-13	92	.609		37	25	8	70	87.50
2050	30" diameter		88	.636		42	26	8.35	76.35	95
2060	36" diameter	↓	72	.778		56	32	10.25	98.25	122
2070	42" diameter	B-13B	72	.778		75	32	15.65	122.65	149
2080	48" diameter		64	.875		89	36	17.60	142.60	172
2090	60" diameter	↓	48	1.167	↓	136	48	23.50	207.50	249

33 41 Storm Utility Drainage Piping

33 41 13 – Public Storm Utility Drainage Piping

33 41 13.60 Sewage/Drainage Collection, Concrete Pipe		Crew	Daily Output	Labor-Hours	Unit	Material	2015 Bare Costs Labor	Equipment	Total	Total Incl O&P
2100	72" diameter	B-13B	40	1.400	L.F.	206	57.50	28	291.50	345
2120	84" diameter		32	1.750		275	71.50	35.50	382	455
2140	96" diameter	↓	24	2.333		330	95.50	47	472.50	565
2200	With gaskets, class 3, 12" diameter	B-21	168	.167		12.10	7.20	.82	20.12	25.50
2220	15" diameter		160	.175		15.40	7.60	.86	23.86	29.50
2230	18" diameter		152	.184		19.80	8	.91	28.71	35.50
2240	24" diameter	↓	136	.206		33	8.90	1.01	42.91	51.50
2260	30" diameter	B-13	88	.636		50	26	8.35	84.35	104
2270	36" diameter	"	72	.778		65.50	32	10.25	107.75	132
2290	48" diameter	B-13B	64	.875		102	36	17.60	155.60	186
2310	72" diameter	"	40	1.400	↓	225	57.50	28	310.50	365
2330	Flared ends, 12" diameter	B-21	31	.903	Ea.	226	39	4.44	269.44	315
2340	15" diameter		25	1.120		267	48.50	5.50	321	375
2400	18" diameter	↓	20	1.400		305	60.50	6.90	372.40	440
2420	24" diameter		14	2		370	86.50	9.85	466.35	555
2440	36" diameter	B-13	10	5.600	↓	790	230	73.50	1,093.50	1,300
3080	Radius pipe, add to pipe prices, 12" to 60" diameter				L.F.	50%				
3090	Over 60" diameter, add				"	20%				
3500	Reinforced elliptical, 8' lengths, C507 class 3									
3520	14" x 23" inside, round equivalent 18" diameter	B-21	82	.341	L.F.	41	14.80	1.68	57.48	70
3530	24" x 38" inside, round equivalent 30" diameter	B-13	58	.966		62	39.50	12.70	114.20	142
3540	29" x 45" inside, round equivalent 36" diameter		52	1.077		78	44	14.15	136.15	169
3550	38" x 60" inside, round equivalent 48" diameter		38	1.474		134	60.50	19.40	213.90	261
3560	48" x 76" inside, round equivalent 60" diameter		26	2.154		186	88.50	28.50	303	370
3570	58" x 91" inside, round equivalent 72" diameter	↓	22	2.545	↓	272	104	33.50	409.50	495
3780	Concrete slotted pipe, class 4 mortar joint									
3800	12" diameter	B-21	168	.167	L.F.	28	7.20	.82	36.02	43
3840	18" diameter	"	152	.184	"	32	8	.91	40.91	48.50
3900	Concrete slotted pipe, Class 4 O-ring joint									
3940	12" diameter	B-21	168	.167	L.F.	28	7.20	.82	36.02	43
3960	18" diameter	"	152	.184	"	32	8	.91	40.91	48.50

33 42 Culverts

33 42 16 – Concrete Culverts

33 42 16.15 Oval Arch Culverts

		Crew	Daily Output	Labor-Hours	Unit	Material	2015 Bare Costs Labor	Equipment	Total	Total Incl O&P
0010	**OVAL ARCH CULVERTS**									
3000	Corrugated galvanized or aluminum, coated & paved									
3020	17" x 13", 16 ga., 15" equivalent	B-14	200	.240	L.F.	13.20	9.55	1.82	24.57	31
3040	21" x 15", 16 ga., 18" equivalent		150	.320		16.05	12.75	2.43	31.23	40
3060	28" x 20", 14 ga., 24" equivalent		125	.384		24	15.25	2.91	42.16	53
3080	35" x 24", 14 ga., 30" equivalent	↓	100	.480		29.50	19.10	3.64	52.24	66
3100	42" x 29", 12 ga., 36" equivalent	B-13	100	.560		35.50	23	7.35	65.85	82
3120	49" x 33", 12 ga., 42" equivalent		90	.622		41	25.50	8.20	74.70	93.50
3140	57" x 38", 12 ga., 48" equivalent	↓	75	.747	↓	57	30.50	9.80	97.30	120
3160	Steel, plain oval arch culverts, plain									
3180	17" x 13", 16 ga., 15" equivalent	B-14	225	.213	L.F.	11.95	8.50	1.62	22.07	28
3200	21" x 15", 16 ga., 18" equivalent		175	.274		14.45	10.90	2.08	27.43	35
3220	28" x 20", 14 ga., 24" equivalent	↓	150	.320		21.50	12.75	2.43	36.68	45.50
3240	35" x 24", 14 ga., 30" equivalent	B-13	108	.519		26.50	21.50	6.80	54.80	69
3260	42" x 29", 12 ga., 36" equivalent		108	.519		32	21.50	6.80	60.30	75.50
3280	49" x 33", 12 ga., 42" equivalent	↓	92	.609		37	25	8	70	88

421

33 42 Culverts

33 42 16 – Concrete Culverts

33 42 16.15 Oval Arch Culverts	Crew	Daily Output	Labor-Hours	Unit	Material	2015 Bare Costs Labor	Equipment	Total	Total Incl O&P
3300	57" x 38", 12 ga., 48" equivalent	B-13	75	.747	L.F.	51.50	30.50	9.80	91.80	114
3320	End sections, 17" x 13"		22	2.545	Ea.	144	104	33.50	281.50	355
3340	42" x 29"	↓	17	3.294	"	395	135	43.50	573.50	690
3360	Multi-plate arch, steel	B-20	1690	.014	Lb.	1.25	.60		1.85	2.30

33 44 Storm Utility Water Drains

33 44 13 – Utility Area Drains

33 44 13.13 Catchbasins

		Crew	Daily Output	Labor-Hours	Unit	Material	2015 Bare Costs Labor	Equipment	Total	Total Incl O&P
0010	**CATCHBASINS**									
0011	Not including footing & excavation									
1600	Frames & grates, C.I., 24" square, 500 lb.	B-6	7.80	3.077	Ea.	340	127	46.50	513.50	620
1700	26" D shape, 600 lb.		7	3.429		505	142	52	699	830
1800	Light traffic, 18" diameter, 100 lb.		10	2.400		123	99	36.50	258.50	325
1900	24" diameter, 300 lb.		8.70	2.759		196	114	42	352	435
2000	36" diameter, 900 lb.		5.80	4.138		570	171	63	804	955
2100	Heavy traffic, 24" diameter, 400 lb.		7.80	3.077		244	127	46.50	417.50	515
2200	36" diameter, 1150 lb.		3	8		795	330	121	1,246	1,525
2300	Mass. State standard, 26" diameter, 475 lb.		7	3.429		266	142	52	460	565
2400	30" diameter, 620 lb.		7	3.429		345	142	52	539	655
2500	Watertight, 24" diameter, 350 lb.		7.80	3.077		320	127	46.50	493.50	595
2600	26" diameter, 500 lb.		7	3.429		425	142	52	619	745
2700	32" diameter, 575 lb.	↓	6	4	↓	850	165	60.50	1,075.50	1,250
2800	3 piece cover & frame, 10" deep,									
2900	1200 lb., for heavy equipment	B-6	3	8	Ea.	1,050	330	121	1,501	1,825
3000	Raised for paving 1-1/4" to 2" high									
3100	4 piece expansion ring									
3200	20" to 26" diameter	1 Clab	3	2.667	Ea.	158	100		258	330
3300	30" to 36" diameter	"	3	2.667	"	217	100		317	395
3320	Frames and covers, existing, raised for paving, 2", including									
3340	row of brick, concrete collar, up to 12" wide frame	B-6	18	1.333	Ea.	45	55	20	120	156
3360	20" to 26" wide frame		11	2.182		67.50	90	33	190.50	249
3380	30" to 36" wide frame	↓	9	2.667		83.50	110	40.50	234	305
3400	Inverts, single channel brick	D-1	3	5.333		97	225		322	450
3500	Concrete		5	3.200		104	135		239	320
3600	Triple channel, brick		2	8		148	335		483	680
3700	Concrete	↓	3	5.333	↓	139	225		364	500

33 46 Subdrainage

33 46 16 – Subdrainage Piping

33 46 16.25 Piping, Subdrainage, Corrugated Metal

		Crew	Daily Output	Labor-Hours	Unit	Material	2015 Bare Costs Labor	Equipment	Total	Total Incl O&P
0010	**PIPING, SUBDRAINAGE, CORRUGATED METAL** G1030–805									
0021	Not including excavation and backfill									
2010	Aluminum, perforated									
2020	6" diameter, 18 ga.	B-20	380	.063	L.F.	6.50	2.65		9.15	11.25
2200	8" diameter, 16 ga.	"	370	.065		8.55	2.72		11.27	13.60
2220	10" diameter, 16 ga.	B-21	360	.078		10.70	3.37	.38	14.45	17.35
2240	12" diameter, 16 ga.		285	.098		11.95	4.26	.48	16.69	20
2260	18" diameter, 16 ga.	↓	205	.137	↓	17.95	5.90	.67	24.52	29.50
3000	Uncoated galvanized, perforated									

33 46 Subdrainage

33 46 16 – Subdrainage Piping

33 46 16.25 Piping, Subdrainage, Corrugated Metal	Crew	Daily Output	Labor-Hours	Unit	Material	2015 Bare Costs Labor	Equipment	Total	Total Incl O&P
3020	6" diameter, 18 ga.	B-20	380	.063	L.F.	6.05	2.65		8.70	10.75
3200	8" diameter, 16 ga.	"	370	.065		8.30	2.72		11.02	13.35
3220	10" diameter, 16 ga.	B-21	360	.078		8.80	3.37	.38	12.55	15.30
3240	12" diameter, 16 ga.		285	.098		9.80	4.26	.48	14.54	17.85
3260	18" diameter, 16 ga.		205	.137		15	5.90	.67	21.57	26.50
4000	Steel, perforated, asphalt coated									
4020	6" diameter 18 ga.	B-20	380	.063	L.F.	6.50	2.65		9.15	11.25
4030	8" diameter 18 ga.	"	370	.065		8.55	2.72		11.27	13.60
4040	10" diameter 16 ga.	B-21	360	.078		10.05	3.37	.38	13.80	16.65
4050	12" diameter 16 ga.		285	.098		11.05	4.26	.48	15.79	19.25
4060	18" diameter 16 ga.		205	.137		17.10	5.90	.67	23.67	28.50

33 46 16.30 Piping, Subdrainage, Plastic

		Crew	Daily Output	Labor-Hours	Unit	Material	2015 Bare Costs Labor	Equipment	Total	Total Incl O&P
0010	**PIPING, SUBDRAINAGE, PLASTIC**									
0020	Not including excavation and backfill									
2100	Perforated PVC, 4" diameter	B-14	314	.153	L.F.	1.45	6.10	1.16	8.71	12.25
2110	6" diameter		300	.160		3.28	6.35	1.21	10.84	14.70
2120	8" diameter		290	.166		6.30	6.60	1.26	14.16	18.40
2130	10" diameter		280	.171		9.10	6.80	1.30	17.20	22
2140	12" diameter		270	.178		12.75	7.05	1.35	21.15	26.50

33 49 Storm Drainage Structures

33 49 13 – Storm Drainage Manholes, Frames, and Covers

33 49 13.10 Storm Drainage Manholes, Frames and Covers	Crew	Daily Output	Labor-Hours	Unit	Material	2015 Bare Costs Labor	Equipment	Total	Total Incl O&P
0010	**STORM DRAINAGE MANHOLES, FRAMES & COVERS**									
0020	Excludes footing, excavation, backfill (See line items for frame & cover)									
0050	Brick, 4' inside diameter, 4' deep	D-1	1	16	Ea.	520	675		1,195	1,600
0100	6' deep		.70	22.857		740	960		1,700	2,300
0150	8' deep		.50	32		955	1,350		2,305	3,100
0200	For depths over 8', add		4	4	V.L.F.	83.50	168		251.50	350
0400	Concrete blocks (radial), 4' I.D., 4' deep		1.50	10.667	Ea.	415	450		865	1,150
0500	6' deep		1	16		565	675		1,240	1,650
0600	8' deep		.70	22.857		710	960		1,670	2,250
0700	For depths over 8', add		5.50	2.909	V.L.F.	77	122		199	272
0800	Concrete, cast in place, 4' x 4', 8" thick, 4' deep	C-14H	2	24	Ea.	505	1,100	15.60	1,620.60	2,275
0900	6' deep		1.50	32		730	1,475	21	2,226	3,100
1000	8' deep		1	48		1,050	2,225	31	3,306	4,575
1100	For depths over 8', add		8	6	V.L.F.	119	278	3.90	400.90	560
1110	Precast, 4' I.D., 4' deep	B-22	4.10	7.317	Ea.	725	320	50.50	1,095.50	1,350
1120	6' deep		3	10		925	440	69	1,434	1,775
1130	8' deep		2	15		1,075	660	103	1,838	2,300
1140	For depths over 8', add		16	1.875	V.L.F.	127	82.50	12.90	222.40	280
1150	5' I.D., 4' deep	B-6	3	8	Ea.	1,650	330	121	2,101	2,450
1160	6' deep		2	12		1,925	495	182	2,602	3,075
1170	8' deep		1.50	16		2,400	660	243	3,303	3,925
1180	For depths over 8', add		12	2	V.L.F.	280	82.50	30.50	393	470
1190	6' I.D., 4' deep		2	12	Ea.	2,150	495	182	2,827	3,325
1200	6' deep		1.50	16		2,600	660	243	3,503	4,125
1210	8' deep		1	24		3,200	990	365	4,555	5,425
1220	For depths over 8', add		8	3	V.L.F.	380	124	45.50	549.50	655
1250	Slab tops, precast, 8" thick									
1300	4' diameter manhole	B-6	8	3	Ea.	252	124	45.50	421.50	515

423

33 49 Storm Drainage Structures

33 49 13 – Storm Drainage Manholes, Frames, and Covers

33 49 13.10 Storm Drainage Manholes, Frames and Covers		Crew	Daily Output	Labor-Hours	Unit	Material	2015 Bare Costs Labor	Equipment	Total	Total Incl O&P
1400	5' diameter manhole	B-6	7.50	3.200	Ea.	410	132	48.50	590.50	710
1500	6' diameter manhole	↓	7	3.429		635	142	52	829	975
3800	Steps, heavyweight cast iron, 7" x 9"	1 Bric	40	.200		19.25	9.25		28.50	35
3900	8" x 9"		40	.200		23	9.25		32.25	39.50
3928	12" x 10-1/2"		40	.200		27	9.25		36.25	43.50
4000	Standard sizes, galvanized steel		40	.200		22	9.25		31.25	38
4100	Aluminum		40	.200		24	9.25		33.25	40.50
4150	Polyethylene	↓	40	.200	↓	26	9.25		35.25	42.50

33 51 Natural-Gas Distribution

33 51 13 – Natural-Gas Piping

33 51 13.10 Piping, Gas Service and Distribution, P.E.

		Crew	Daily Output	Labor-Hours	Unit	Material	2015 Bare Costs Labor	Equipment	Total	Total Incl O&P
0010	**PIPING, GAS SERVICE AND DISTRIBUTION, POLYETHYLENE**									
0020	Not including excavation or backfill									
1000	60 psi coils, compression coupling @ 100', 1/2" diameter, SDR 11	B-20A	608	.053	L.F.	.45	2.41		2.86	4.16
1010	1" diameter, SDR 11		544	.059		1.05	2.69		3.74	5.25
1040	1-1/4" diameter, SDR 11		544	.059		1.62	2.69		4.31	5.85
1100	2" diameter, SDR 11		488	.066		2.82	3		5.82	7.65
1160	3" diameter, SDR 11	↓	408	.078		6.45	3.59		10.04	12.55
1500	60 PSI 40' joints with coupling, 3" diameter, SDR 11	B-21A	408	.098		10.05	4.60	1.16	15.81	19.35
1540	4" diameter, SDR 11		352	.114		14.25	5.35	1.35	20.95	25
1600	6" diameter, SDR 11		328	.122		33.50	5.70	1.44	40.64	47.50
1640	8" diameter, SDR 11	↓	272	.147	↓	51.50	6.90	1.74	60.14	69

33 52 Liquid Fuel Distribution

33 52 16 – Gasoline Distribution

33 52 16.13 Gasoline Piping

		Crew	Daily Output	Labor-Hours	Unit	Material	2015 Bare Costs Labor	Equipment	Total	Total Incl O&P
0010	**GASOLINE PIPING**									
0020	Primary containment pipe, fiberglass-reinforced									
0030	Plastic pipe 15' & 30' lengths									
0040	2" diameter	Q-6	425	.056	L.F.	5.95	3.15		9.10	11.25
0050	3" diameter		400	.060		10.35	3.35		13.70	16.45
0060	4" diameter	↓	375	.064	↓	13.65	3.57		17.22	20.50
0100	Fittings									
0110	Elbows, 90° & 45°, bell-ends, 2"	Q-6	24	1	Ea.	43.50	56		99.50	132
0120	3" diameter		22	1.091		55	61		116	153
0130	4" diameter		20	1.200		70	67		137	178
0200	Tees, bell ends, 2"		21	1.143		60.50	63.50		124	163
0210	3" diameter		18	1.333		64	74.50		138.50	183
0230	Flanges bell ends, 2"		24	1		33.50	56		89.50	121
0240	3" diameter		22	1.091		39	61		100	135
0250	4" diameter		20	1.200		45	67		112	151
0260	Sleeve couplings, 2"		21	1.143		12.40	63.50		75.90	110
0270	3" diameter		18	1.333		17.80	74.50		92.30	132
0280	4" diameter		15	1.600		23	89		112	160
0290	Threaded adapters 2"		21	1.143		17.95	63.50		81.45	116
0300	3" diameter		18	1.333		34	74.50		108.50	150
0310	4" diameter		15	1.600		38	89		127	177
0320	Reducers, 2"	↓	27	.889	↓	27.50	49.50		77	105

33 52 Liquid Fuel Distribution

33 52 16 – Gasoline Distribution

33 52 16.13 Gasoline Piping

33 52 16.13 Gasoline Piping		Crew	Daily Output	Labor-Hours	Unit	Material	2015 Bare Costs Labor	Equipment	Total	Total Incl O&P
0330	3" diameter	Q-6	22	1.091	Ea.	27.50	61		88.50	122
0340	4" diameter	↓	20	1.200	↓	37	67		104	142
1010	Gas station product line for secondary containment (double wall)									
1100	Fiberglass reinforced plastic pipe 25' lengths									
1120	Pipe, plain end, 3" diameter	Q-6	375	.064	L.F.	26.50	3.57		30.07	34.50
1130	4" diameter		350	.069		32	3.82		35.82	41.50
1140	5" diameter		325	.074		35.50	4.12		39.62	45
1150	6" diameter	↓	300	.080	↓	39	4.46		43.46	50
1200	Fittings									
1230	Elbows, 90° & 45°, 3" diameter	Q-6	18	1.333	Ea.	136	74.50		210.50	262
1240	4" diameter		16	1.500		167	83.50		250.50	310
1250	5" diameter		14	1.714		184	95.50		279.50	345
1260	6" diameter		12	2		204	112		316	390
1270	Tees, 3" diameter		15	1.600		166	89		255	315
1280	4" diameter		12	2		202	112		314	390
1290	5" diameter		9	2.667		315	149		464	575
1300	6" diameter		6	4		375	223		598	750
1310	Couplings, 3" diameter		18	1.333		55	74.50		129.50	173
1320	4" diameter		16	1.500		119	83.50		202.50	257
1330	5" diameter		14	1.714		214	95.50		309.50	380
1340	6" diameter		12	2		315	112		427	520
1350	Cross-over nipples, 3" diameter		18	1.333		10.60	74.50		85.10	124
1360	4" diameter		16	1.500		12.70	83.50		96.20	140
1370	5" diameter		14	1.714		15.90	95.50		111.40	162
1380	6" diameter		12	2		19.05	112		131.05	189
1400	Telescoping, reducers, concentric 4" x 3"		18	1.333		48	74.50		122.50	165
1410	5" x 4"		17	1.412		95.50	78.50		174	224
1420	6" x 5"	↓	16	1.500	↓	234	83.50		317.50	385

33 61 Hydronic Energy Distribution

33 61 13 – Underground Hydronic Energy Distribution

33 61 13.20 Pipe Conduit, Prefabricated/Preinsulated

	33 61 13.20 Pipe Conduit, Prefabricated/Preinsulated	Crew	Daily Output	Labor-Hours	Unit	Material	2015 Bare Costs Labor	Equipment	Total	Total Incl O&P
0010	**PIPE CONDUIT, PREFABRICATED/PREINSULATED** R221113-70									
0020	Does not include trenching, fittings or crane.									
0300	For cathodic protection, add 12 to 14%									
0310	of total built-up price (casing plus service pipe)									
0580	Polyurethane insulated system, 250°F. max. temp.									
0620	Black steel service pipe, standard wt., 1/2" insulation									
0660	3/4" diam. pipe size	Q-17	54	.296	L.F.	57	15.95	1.07	74.02	88
0670	1" diam. pipe size		50	.320		63	17.20	1.16	81.36	96.50
0680	1-1/4" diam. pipe size		47	.340		70	18.30	1.23	89.53	106
0690	1-1/2" diam. pipe size		45	.356		76	19.10	1.28	96.38	114
0700	2" diam. pipe size		42	.381		79	20.50	1.38	100.88	119
0710	2-1/2" diam. pipe size		34	.471		80	25.50	1.70	107.20	128
0720	3" diam. pipe size		28	.571		93	30.50	2.06	125.56	151
0730	4" diam. pipe size		22	.727		117	39	2.63	158.63	191
0740	5" diam. pipe size	↓	18	.889		149	48	3.21	200.21	240
0750	6" diam. pipe size	Q-18	23	1.043		173	58	2.50	233.50	282
0760	8" diam. pipe size		19	1.263		254	70.50	3.03	327.53	390
0770	10" diam. pipe size		16	1.500		320	83.50	3.60	407.10	485
0780	12" diam. pipe size		13	1.846		400	103	4.43	507.43	600

For customer support on your Plumbing Cost Data, call 877.709.6968.

425

33 61 Hydronic Energy Distribution

33 61 13 – Underground Hydronic Energy Distribution

33 61 13.20 Pipe Conduit, Prefabricated/Preinsulated	Crew	Daily Output	Labor-Hours	Unit	Material	2015 Bare Costs Labor	Equipment	Total	Total Incl O&P
0790	14" diam. pipe size	Q-18	11	2.182	L.F.	445	122	5.25	572.25	680
0800	16" diam. pipe size		10	2.400		510	134	5.75	649.75	775
0810	18" diam. pipe size		8	3		590	167	7.20	764.20	910
0820	20" diam. pipe size		7	3.429		655	191	8.25	854.25	1,025
0830	24" diam. pipe size		6	4		805	223	9.60	1,037.60	1,225
0900	For 1" thick insulation, add					10%				
0940	For 1-1/2" thick insulation, add					13%				
0980	For 2" thick insulation, add					20%				
1500	Gland seal for system, 3/4" diam. pipe size	Q-17	32	.500	Ea.	775	27	1.81	803.81	890
1510	1" diam. pipe size		32	.500		775	27	1.81	803.81	890
1540	1-1/4" diam. pipe size		30	.533		825	28.50	1.93	855.43	955
1550	1-1/2" diam. pipe size		30	.533		825	28.50	1.93	855.43	955
1560	2" diam. pipe size		28	.571		980	30.50	2.06	1,012.56	1,125
1570	2-1/2" diam. pipe size		26	.615		1,050	33	2.22	1,085.22	1,225
1580	3" diam. pipe size		24	.667		1,125	36	2.41	1,163.41	1,300
1590	4" diam. pipe size		22	.727		1,350	39	2.63	1,391.63	1,525
1600	5" diam. pipe size		19	.842		1,650	45.50	3.04	1,698.54	1,900
1610	6" diam. pipe size	Q-18	26	.923		1,750	51.50	2.22	1,803.72	2,000
1620	8" diam. pipe size		25	.960		2,050	53.50	2.30	2,105.80	2,325
1630	10" diam. pipe size		23	1.043		2,375	58	2.50	2,435.50	2,725
1640	12" diam. pipe size		21	1.143		2,625	63.50	2.74	2,691.24	3,000
1650	14" diam. pipe size		19	1.263		2,950	70.50	3.03	3,023.53	3,350
1660	16" diam. pipe size		18	1.333		3,500	74.50	3.20	3,577.70	3,975
1670	18" diam. pipe size		16	1.500		3,700	83.50	3.60	3,787.10	4,200
1680	20" diam. pipe size		14	1.714		4,200	95.50	4.11	4,299.61	4,775
1690	24" diam. pipe size		12	2		4,675	112	4.80	4,791.80	5,325
2000	Elbow, 45° for system									
2020	3/4" diam. pipe size	Q-17	14	1.143	Ea.	485	61.50	4.13	550.63	635
2040	1" diam. pipe size		13	1.231		500	66	4.44	570.44	655
2050	1-1/4" diam. pipe size		11	1.455		570	78	5.25	653.25	750
2060	1-1/2" diam. pipe size		9	1.778		595	95.50	6.40	696.90	805
2070	2" diam. pipe size		6	2.667		620	143	9.65	772.65	915
2080	2-1/2" diam. pipe size		4	4		670	215	14.45	899.45	1,075
2090	3" diam. pipe size		3.50	4.571		775	246	16.50	1,037.50	1,250
2100	4" diam. pipe size		3	5.333		905	287	19.25	1,211.25	1,450
2110	5" diam. pipe size		2.80	5.714		1,175	305	20.50	1,500.50	1,775
2120	6" diam. pipe size	Q-18	4	6		1,325	335	14.40	1,674.40	1,975
2130	8" diam. pipe size		3	8		1,925	445	19.20	2,389.20	2,800
2140	10" diam. pipe size		2.40	10		2,350	560	24	2,934	3,450
2150	12" diam. pipe size		2	12		3,075	670	29	3,774	4,425
2160	14" diam. pipe size		1.80	13.333		3,825	745	32	4,602	5,375
2170	16" diam. pipe size		1.60	15		4,550	835	36	5,421	6,350
2180	18" diam. pipe size		1.30	18.462		5,725	1,025	44.50	6,794.50	7,900
2190	20" diam. pipe size		1	24		7,200	1,350	57.50	8,607.50	10,000
2200	24" diam. pipe size		.70	34.286		9,050	1,900	82.50	11,032.50	12,900
2260	For elbow, 90°, add					25%				
2300	For tee, straight, add					85%	30%			
2340	For tee, reducing, add					170%	30%			
2380	For weldolet, straight, add					50%				
2400	Polyurethane insulation, 1"									
2410	FRP carrier and casing									
2420	4"	Q-5	18	.889	L.F.	54.50	48		102.50	132
2422	6"	Q-6	23	1.043		94.50	58		152.50	192

426

For customer support on your Plumbing Cost Data, call 877.709.6968.

33 61 Hydronic Energy Distribution

33 61 13 – Underground Hydronic Energy Distribution

33 61 13.20 Pipe Conduit, Prefabricated/Preinsulated	Crew	Daily Output	Labor-Hours	Unit	Material	2015 Bare Costs Labor	Equipment	Total	Total Incl O&P
2424	8"	Q-6	19	1.263	L.F.	154	70.50		224.50	275
2426	10"		16	1.500		207	83.50		290.50	355
2428	12"		13	1.846		270	103		373	450
2430	FRP carrier and PVC casing									
2440	4"	Q-5	18	.889	L.F.	21.50	48		69.50	95.50
2444	8"	Q-6	19	1.263		45	70.50		115.50	155
2446	10"		16	1.500		64.50	83.50		148	197
2448	12"		13	1.846		82.50	103		185.50	246
2450	PVC carrier and casing									
2460	4"	Q-1	36	.444	L.F.	10.45	23.50		33.95	47
2462	6"	"	29	.552		14.55	29		43.55	60
2464	8"	Q-2	36	.667		20.50	36.50		57	77.50
2466	10"		32	.750		28	41		69	92.50
2468	12"		31	.774		32.50	42.50		75	99.50

33 63 Steam Energy Distribution

33 63 13 – Underground Steam and Condensate Distribution Piping

33 63 13.10 Calcium Silicate Insulated System

		Crew	Daily Output	Labor-Hours	Unit	Material	2015 Bare Costs Labor	Equipment	Total	Total Incl O&P
0010	**CALCIUM SILICATE INSULATED SYSTEM**									
0011	High temp. (1200 degrees F)									
2840	Steel casing with protective exterior coating									
2850	6-5/8" diameter	Q-18	52	.462	L.F.	106	25.50	1.11	132.61	157
2860	8-5/8" diameter		50	.480		115	27	1.15	143.15	169
2870	10-3/4" diameter		47	.511		135	28.50	1.23	164.73	193
2880	12-3/4" diameter		44	.545		147	30.50	1.31	178.81	209
2890	14" diameter		41	.585		166	32.50	1.40	199.90	234
2900	16" diameter		39	.615		178	34.50	1.48	213.98	250
2910	18" diameter		36	.667		196	37	1.60	234.60	274
2920	20" diameter		34	.706		221	39.50	1.69	262.19	305
2930	22" diameter		32	.750		305	42	1.80	348.80	400
2940	24" diameter		29	.828		350	46	1.99	397.99	455
2950	26" diameter		26	.923		400	51.50	2.22	453.72	520
2960	28" diameter		23	1.043		495	58	2.50	555.50	635
2970	30" diameter		21	1.143		530	63.50	2.74	596.24	685
2980	32" diameter		19	1.263		595	70.50	3.03	668.53	765
2990	34" diameter		18	1.333		600	74.50	3.20	677.70	775
3000	36" diameter		16	1.500		645	83.50	3.60	732.10	840
3040	For multi-pipe casings, add					10%				
3060	For oversize casings, add					2%				
3400	Steel casing gland seal, single pipe									
3420	6-5/8" diameter	Q-18	25	.960	Ea.	1,475	53.50	2.30	1,530.80	1,700
3440	8-5/8" diameter		23	1.043		1,725	58	2.50	1,785.50	2,000
3450	10-3/4" diameter		21	1.143		1,925	63.50	2.74	1,991.24	2,225
3460	12-3/4" diameter		19	1.263		2,275	70.50	3.03	2,348.53	2,600
3470	14" diameter		17	1.412		2,525	78.50	3.39	2,606.89	2,900
3480	16" diameter		16	1.500		2,950	83.50	3.60	3,037.10	3,375
3490	18" diameter		15	1.600		3,275	89	3.84	3,367.84	3,750
3500	20" diameter		13	1.846		3,625	103	4.43	3,732.43	4,150
3510	22" diameter		12	2		4,075	112	4.80	4,191.80	4,650
3520	24" diameter		11	2.182		4,575	122	5.25	4,702.25	5,225
3530	26" diameter		10	2.400		5,225	134	5.75	5,364.75	5,950

427

33 63 13 — Underground Steam and Condensate Distribution Piping

33 63 13.10 Calcium Silicate Insulated System	Crew	Daily Output	Labor-Hours	Unit	Material	2015 Bare Costs Labor	Equipment	Total	Total Incl O&P
3540	28" diameter	Q-18	9.50	2.526	Ea.	5,900	141	6.05	6,047.05	6,725
3550	30" diameter		9	2.667		6,000	149	6.40	6,155.40	6,825
3560	32" diameter		8.50	2.824		6,750	157	6.80	6,913.80	7,675
3570	34" diameter		8	3		7,350	167	7.20	7,524.20	8,325
3580	36" diameter		7	3.429		7,825	191	8.25	8,024.25	8,900
3620	For multi-pipe casings, add					5%				
4000	Steel casing anchors, single pipe									
4020	6-5/8" diameter	Q-18	8	3	Ea.	1,325	167	7.20	1,499.20	1,700
4040	8-5/8" diameter		7.50	3.200		1,375	178	7.70	1,560.70	1,800
4050	10-3/4" diameter		7	3.429		1,800	191	8.25	1,999.25	2,275
4060	12-3/4" diameter		6.50	3.692		1,925	206	8.85	2,139.85	2,450
4070	14" diameter		6	4		2,275	223	9.60	2,507.60	2,850
4080	16" diameter		5.50	4.364		2,625	243	10.45	2,878.45	3,275
4090	18" diameter		5	4.800		2,975	268	11.50	3,254.50	3,700
4100	20" diameter		4.50	5.333		3,275	297	12.80	3,584.80	4,075
4110	22" diameter		4	6		3,650	335	14.40	3,999.40	4,550
4120	24" diameter		3.50	6.857		3,975	380	16.45	4,371.45	4,975
4130	26" diameter		3	8		4,500	445	19.20	4,964.20	5,650
4140	28" diameter		2.50	9.600		4,950	535	23	5,508	6,275
4150	30" diameter		2	12		5,250	670	29	5,949	6,800
4160	32" diameter		1.50	16		6,275	890	38.50	7,203.50	8,300
4170	34" diameter		1	24		7,050	1,350	57.50	8,457.50	9,850
4180	36" diameter		1	24		7,675	1,350	57.50	9,082.50	10,500
4220	For multi-pipe, add					5%	20%			
4800	Steel casing elbow									
4820	6-5/8" diameter	Q-18	15	1.600	Ea.	1,800	89	3.84	1,892.84	2,125
4830	8-5/8" diameter		15	1.600		1,925	89	3.84	2,017.84	2,275
4850	10-3/4" diameter		14	1.714		2,325	95.50	4.11	2,424.61	2,700
4860	12-3/4" diameter		13	1.846		2,750	103	4.43	2,857.43	3,175
4870	14" diameter		12	2		2,950	112	4.80	3,066.80	3,425
4880	16" diameter		11	2.182		3,225	122	5.25	3,352.25	3,750
4890	18" diameter		10	2.400		3,675	134	5.75	3,814.75	4,250
4900	20" diameter		9	2.667		3,925	149	6.40	4,080.40	4,550
4910	22" diameter		8	3		4,225	167	7.20	4,399.20	4,900
4920	24" diameter		7	3.429		4,650	191	8.25	4,849.25	5,425
4930	26" diameter		6	4		5,100	223	9.60	5,332.60	5,950
4940	28" diameter		5	4.800		5,500	268	11.50	5,779.50	6,475
4950	30" diameter		4	6		5,525	335	14.40	5,874.40	6,600
4960	32" diameter		3	8		6,275	445	19.20	6,739.20	7,600
4970	34" diameter		2	12		6,900	670	29	7,599	8,625
4980	36" diameter		2	12		7,350	670	29	8,049	9,100
5500	Black steel service pipe, std. wt., 1" thick insulation									
5510	3/4" diameter pipe size	Q-17	54	.296	L.F.	62	15.95	1.07	79.02	93
5540	1" diameter pipe size		50	.320		64	17.20	1.16	82.36	98
5550	1-1/4" diameter pipe size		47	.340		70	18.30	1.23	89.53	106
5560	1-1/2" diameter pipe size		45	.356		74	19.10	1.28	94.38	112
5570	2" diameter pipe size		42	.381		81	20.50	1.38	102.88	122
5580	2-1/2" diameter pipe size		34	.471		85	25.50	1.70	112.20	133
5590	3" diameter pipe size		28	.571		93	30.50	2.06	125.56	151
5600	4" diameter pipe size		22	.727		98	39	2.63	139.63	170
5610	5" diameter pipe size		18	.889		107	48	3.21	158.21	194
5620	6" diameter pipe size	Q-18	23	1.043		115	58	2.50	175.50	218
6000	Black steel service pipe, std. wt., 1-1/2" thick insul.									

33 63 Steam Energy Distribution

33 63 13 – Underground Steam and Condensate Distribution Piping

33 63 13.10 Calcium Silicate Insulated System		Crew	Daily Output	Labor-Hours	Unit	Material	2015 Bare Costs			Total	Total Incl O&P
							Labor	Equipment		Total	
6010	3/4" diameter pipe size	Q-17	54	.296	L.F.	62	15.95	1.07		79.02	93
6040	1" diameter pipe size		50	.320		67	17.20	1.16		85.36	101
6050	1-1/4" diameter pipe size		47	.340		73	18.30	1.23		92.53	109
6060	1-1/2" diameter pipe size		45	.356		79	19.10	1.28		99.38	117
6070	2" diameter pipe size		42	.381		85	20.50	1.38		106.88	126
6080	2-1/2" diameter pipe size		34	.471		89	25.50	1.70		116.20	138
6090	3" diameter pipe size		28	.571		99	30.50	2.06		131.56	158
6100	4" diameter pipe size		22	.727		107	39	2.63		148.63	180
6110	5" diameter pipe size		18	.889		115	48	3.21		166.21	203
6120	6" diameter pipe size	Q-18	23	1.043		130	58	2.50		190.50	234
6130	8" diameter pipe size		19	1.263		180	70.50	3.03		253.53	305
6140	10" diameter pipe size		16	1.500		240	83.50	3.60		327.10	395
6150	12" diameter pipe size		13	1.846		300	103	4.43		407.43	490
6190	For 2" thick insulation, add					15%					
6220	For 2-1/2" thick insulation, add					25%					
6260	For 3" thick insulation, add					30%					
6800	Black steel service pipe, ex. hvy. wt., 1" thick insul.										
6820	3/4" diameter pipe size	Q-17	50	.320	L.F.	64	17.20	1.16		82.36	98
6840	1" diameter pipe size		47	.340		68	18.30	1.23		87.53	104
6850	1-1/4" diameter pipe size		44	.364		76	19.55	1.31		96.86	114
6860	1-1/2" diameter pipe size		42	.381		78	20.50	1.38		99.88	119
6870	2" diameter pipe size		40	.400		85	21.50	1.44		107.94	128
6880	2-1/2" diameter pipe size		31	.516		89	28	1.86		118.86	142
6890	3" diameter pipe size		27	.593		97	32	2.14		131.14	157
6900	4" diameter pipe size		21	.762		102	41	2.75		145.75	177
6910	5" diameter pipe size		17	.941		110	50.50	3.40		163.90	201
6920	6" diameter pipe size	Q-18	22	1.091		120	61	2.62		183.62	227
7400	Black steel service pipe, ex. hvy. wt., 1-1/2" thick insul.										
7420	3/4" diameter pipe size	Q-17	50	.320	L.F.	55	17.20	1.16		73.36	88
7440	1" diameter pipe size		47	.340		60	18.30	1.23		79.53	95
7450	1-1/4" diameter pipe size		44	.364		67	19.55	1.31		87.86	104
7460	1-1/2" diameter pipe size		42	.381		72	20.50	1.38		93.88	112
7470	2" diameter pipe size		40	.400		77	21.50	1.44		99.94	119
7480	2-1/2" diameter pipe size		31	.516		82	28	1.86		111.86	134
7490	3" diameter pipe size		27	.593		91	32	2.14		125.14	150
7500	4" diameter pipe size		21	.762		110	41	2.75		153.75	186
7510	5" diameter pipe size		17	.941		152	50.50	3.40		205.90	247
7520	6" diameter pipe size	Q-18	22	1.091		160	61	2.62		223.62	271
7530	8" diameter pipe size		18	1.333		215	74.50	3.20		292.70	355
7540	10" diameter pipe size		15	1.600		275	89	3.84		367.84	445
7550	12" diameter pipe size		13	1.846		330	103	4.43		437.43	525
7590	For 2" thick insulation, add					13%					
7640	For 2-1/2" thick insulation, add					18%					
7680	For 3" thick insulation, add					24%					

33 63 13.20 Combined Steam Pipe With Condensate Return

		Crew	Daily Output	Labor-Hours	Unit	Material	Labor	Equipment		Total	Total Incl O&P
0010	**COMBINED STEAM PIPE WITH CONDENSATE RETURN**										
9010	8" and 4" in 24" case	Q-18	100	.240	L.F.	395	13.40	.58		408.98	450
9020	6" and 3" in 20" case		104	.231		365	12.85	.55		378.40	425
9030	3" and 1-1/2" in 16" case		110	.218		275	12.15	.52		287.67	325
9040	2" and 1-1/4" in 12-3/4" case		114	.211		235	11.75	.51		247.26	277
9050	1-1/2" and 1-1/4" in 10-3/4" case		116	.207		235	11.55	.50		247.05	277
9100	Steam pipe only (no return)										

33 63 Steam Energy Distribution

33 63 13 – Underground Steam and Condensate Distribution Piping

33 63 13.20 Combined Steam Pipe With Condensate Return	Crew	Daily Output	Labor-Hours	Unit	Material	2015 Bare Costs Labor	Equipment	Total	Total Incl O&P
9110	6" in 18" case	Q-18	108	.222	L.F.	300	12.40	.53	312.93	350
9120	4" in 14" case		112	.214		235	11.95	.51	247.46	278
9130	3" in 14" case		114	.211		220	11.75	.51	232.26	260
9140	2-1/2" in 12-3/4" case		118	.203		205	11.35	.49	216.84	244
9150	2" in 12-3/4" case		122	.197		195	10.95	.47	206.42	232

For customer support on your Plumbing Cost Data, call 877.709.6968.

Estimating Tips

Products such as conveyors, material handling cranes and hoists, as well as other items specified in this division, require trained installers. The general contractor may not have any choice as to who will perform the installation or when it will be performed. Long lead times are often required for these products, making early decisions in purchasing and scheduling necessary. The installation of this type of equipment may require the embedment of mounting hardware during construction of floors, structural walls, or interior walls/partitions. Electrical connections will require coordination with the electrical contractor.

Reference Numbers

Reference numbers are shown in shaded boxes at the beginning of some major classifications. These numbers refer to related items in the Reference Section. The reference information may be an estimating procedure, an alternate pricing method, or technical information.

Note: Not all subdivisions listed here necessarily appear in this publication. ■

41 22 23.10 Material Handling	Crew	Daily Output	Labor-Hours	Unit	Material	2015 Bare Costs Labor	Equipment	Total	Total Incl O&P
0010 **MATERIAL HANDLING**, cranes, hoists and lifts									
1500 Cranes, portable hydraulic, floor type, 2,000 lb. capacity				Ea.	3,500			3,500	3,875
1600 4,000 lb. capacity					4,025			4,025	4,425
1800 Movable gantry type, 12' to 15' range, 2,000 lb. capacity					3,950			3,950	4,325
1900 6,000 lb. capacity					5,975			5,975	6,575
2100 Hoists, electric overhead, chain, hook hung, 15' lift, 1 ton cap.					2,500			2,500	2,750
2200 3 ton capacity					3,425			3,425	3,775
2500 5 ton capacity					6,925			6,925	7,625
2600 For hand-pushed trolley, add					15%				
2700 For geared trolley, add					30%				
2800 For motor trolley, add					75%				
3000 For lifts over 15', 1 ton, add				L.F.	24.50			24.50	27
3100 5 ton, add				"	56.50			56.50	62
3300 Lifts, scissor type, portable, electric, 36" high, 2,000 lb.				Ea.	3,475			3,475	3,825
3400 48" high, 4,000 lb.				"	4,175			4,175	4,600

Estimating Tips

This section involves equipment and construction costs for air noise and odor pollution control systems. These systems may be interrelated and care must be taken that the complete systems are estimated. For example, air pollution equipment may include dust and air-entrained particles that have to be collected. The vacuum systems could be noisy, requiring silencers to reduce noise pollution, and the collected solids have to be disposed of to prevent solid pollution.

Reference Numbers

Reference numbers are shown in shaded boxes at the beginning of some major classifications. These numbers refer to related items in the Reference Section. The reference information may be an estimating procedure, an alternate pricing method, or technical information.

Note: Not all subdivisions listed here necessarily appear in this publication. ■

44 11 16.10 Dust Collection Systems	Crew	Daily Output	Labor-Hours	Unit	Material	2015 Bare Costs Labor	Equipment	Total	Total Incl O&P
0010 **DUST COLLECTION SYSTEMS** Commercial/Industrial									
0120 Central vacuum units									
0130 Includes stand, filters and motorized shaker									
0200 500 CFM, 10" inlet, 2 HP	Q-20	2.40	8.333	Ea.	4,175	425		4,600	5,250
0220 1000 CFM, 10" inlet, 3 HP		2.20	9.091		4,400	465		4,865	5,525
0240 1500 CFM, 10" inlet, 5 HP		2	10		4,625	510		5,135	5,875
0260 3000 CFM, 13" inlet, 10 HP		1.50	13.333		12,900	685		13,585	15,300
0280 5000 CFM, 16" inlet, 2 @ 10 HP		1	20		13,700	1,025		14,725	16,700
1000 Vacuum tubing, galvanized									
1100 2-1/8" OD, 16 ga.	Q-9	440	.036	L.F.	2.95	1.83		4.78	6.05
1110 2-1/2" OD, 16 ga.		420	.038		3.24	1.92		5.16	6.50
1120 3" OD, 16 ga.		400	.040		4.13	2.01		6.14	7.60
1130 3-1/2" OD, 16 ga.		380	.042		5.85	2.12		7.97	9.70
1140 4" OD, 16 ga.		360	.044		6	2.24		8.24	10
1150 5" OD, 14 ga.		320	.050		10.80	2.52		13.32	15.70
1160 6" OD, 14 ga.		280	.057		12.10	2.88		14.98	17.70
1170 8" OD, 14 ga.		200	.080		18.30	4.03		22.33	26
1180 10" OD, 12 ga.		160	.100		35.50	5.05		40.55	46.50
1190 12" OD, 12 ga.		120	.133		47	6.70		53.70	62
1200 14" OD, 12 ga.		80	.200		53.50	10.05		63.55	74
1940 Hose, flexible wire reinforced rubber									
1956 3" diam.	Q-9	400	.040	L.F.	9.15	2.01		11.16	13.15
1960 4" diam.		360	.044		10.60	2.24		12.84	15.05
1970 5" diam.		320	.050		13.75	2.52		16.27	19
1980 6" diam.		280	.057		14.40	2.88		17.28	20
2000 90° Elbow, slip fit									
2110 2-1/8" diam.	Q-9	70	.229	Ea.	10.85	11.50		22.35	29.50
2120 2-1/2" diam.		65	.246		15.20	12.40		27.60	35.50
2130 3" diam.		60	.267		20.50	13.45		33.95	43
2140 3-1/2" diam.		55	.291		25.50	14.65		40.15	50.50
2150 4" diam.		50	.320		31.50	16.10		47.60	59
2160 5" diam.		45	.356		60	17.90		77.90	93.50
2170 6" diam.		40	.400		82.50	20		102.50	122
2180 8" diam.		30	.533		157	27		184	213
2400 45° Elbow, slip fit									
2410 2-1/8" diam.	Q-9	70	.229	Ea.	9.50	11.50		21	28
2420 2-1/2" diam.		65	.246		14	12.40		26.40	34.50
2430 3" diam.		60	.267		16.90	13.45		30.35	39
2440 3-1/2" diam.		55	.291		21.50	14.65		36.15	46
2450 4" diam.		50	.320		27.50	16.10		43.60	55
2460 5" diam.		45	.356		46.50	17.90		64.40	78.50
2470 6" diam.		40	.400		63	20		83	100
2480 8" diam.		35	.457		123	23		146	171
2800 90° TY, slip fit thru 6" diam.									
2810 2-1/8" diam.	Q-9	42	.381	Ea.	17.75	19.20		36.95	49
2820 2-1/2" diam.		39	.410		26.50	20.50		47	60.50
2830 3" diam.		36	.444		37	22.50		59.50	74.50
2840 3-1/2" diam.		33	.485		39	24.50		63.50	80
2850 4" diam.		30	.533		41.50	27		68.50	87
2860 5" diam.		27	.593		70.50	30		100.50	123
2870 6" diam.		24	.667		101	33.50		134.50	163
2880 8" diam., butt end					155			155	170
2890 10" diam., butt end					475			475	525

44 11 16.10 Dust Collection Systems	Crew	Daily Output	Labor-Hours	Unit	Material	2015 Bare Costs Labor	Equipment	Total	Total Incl O&P
2900	12" diam., butt end				Ea.	475			475	525
2910	14" diam., butt end					855			855	940
2920	6" x 4" diam., butt end					106			106	116
2930	8" x 4" diam., butt end					200			200	220
2940	10" x 4" diam., butt end					263			263	289
2950	12" x 4" diam., butt end					300			300	330
3100	90° Elbow, butt end segmented									
3110	8" diam., butt end, segmented				Ea.	157			157	172
3120	10" diam., butt end, segmented					390			390	430
3130	12" diam., butt end, segmented					495			495	545
3140	14" diam., butt end, segmented					620			620	680
3200	45° Elbow, butt end segmented									
3210	8" diam., butt end, segmented				Ea.	123			123	136
3220	10" diam., butt end, segmented					282			282	310
3230	12" diam., butt end, segmented					288			288	315
3240	14" diam., butt end, segmented					585			585	645
3400	All butt end fittings require one coupling per joint.									
3410	Labor for fitting included with couplings.									
3460	Compression coupling, galvanized, neoprene gasket									
3470	2-1/8" diam.	Q-9	44	.364	Ea.	10.85	18.30		29.15	40
3480	2-1/2" diam.		44	.364		10.85	18.30		29.15	40
3490	3" diam.		38	.421		15.60	21		36.60	49.50
3500	3-1/2" diam.		35	.457		17.45	23		40.45	54
3510	4" diam.		33	.485		18.95	24.50		43.45	58.50
3520	5" diam.		29	.552		21.50	28		49.50	66.50
3530	6" diam.		26	.615		25.50	31		56.50	75.50
3540	8" diam.		22	.727		46	36.50		82.50	107
3550	10" diam.		20	.800		67.50	40.50		108	136
3560	12" diam.		18	.889		84	45		129	161
3570	14" diam.		16	1		119	50.50		169.50	208
3800	Air gate valves, galvanized									
3810	2-1/8" diam.	Q-9	30	.533	Ea.	111	27		138	163
3820	2-1/2" diam.		28	.571		119	29		148	175
3830	3" diam.		26	.615		127	31		158	188
3840	4" diam.		23	.696		147	35		182	216
3850	6" diam.		18	.889		205	45		250	294

For customer support on your Plumbing Cost Data, call 877.709.6968.

435

Division Notes

		CREW	DAILY OUTPUT	LABOR-HOURS	UNIT	BARE COSTS				TOTAL INCL O&P
						MAT.	LABOR	EQUIP.	TOTAL	

Estimating Tips

This division contains information about water and wastewater equipment and systems, which was formerly located in Division 44. The main areas of focus are total wastewater treatment plants and components of wastewater treatment plants. In addition, there are assemblies such as sewage treatment lagoons that can be found in some publications under G30 Site Mechanical Utilities. Also included in this section are oil/water separators for wastewater treatment.

Reference Numbers

Reference numbers are shown in shaded boxes at the beginning of some major classifications. These numbers refer to related items in the Reference Section. The reference information may be an estimating procedure, an alternate pricing method, or technical information.

Note: Not all subdivisions listed here necessarily appear in this publication. ■

46 07 Packaged Water and Wastewater Treatment Equipment

46 07 53 – Packaged Wastewater Treatment Equipment

46 07 53.10 Biological Pkg. Wastewater Treatment Plants	Crew	Daily Output	Labor-Hours	Unit	Material	2015 Bare Costs Labor	Equipment	Total	Total Incl O&P
0010 **BIOLOGICAL PACKAGED WASTEWATER TREATMENT PLANTS**									
0011 Not including fencing or external piping									
0020 Steel packaged, blown air aeration plants									
0100 1,000 GPD				Gal.				55	60.50
0200 5,000 GPD								22	24
0300 15,000 GPD								22	24
0400 30,000 GPD								15.40	16.95
0500 50,000 GPD								11	12.10
0600 100,000 GPD								9.90	10.90
0700 200,000 GPD								8.80	9.70
0800 500,000 GPD								7.70	8.45
1000 Concrete, extended aeration, primary and secondary treatment									
1010 10,000 GPD				Gal.				22	24
1100 30,000 GPD								15.40	16.95
1200 50,000 GPD								11	12.10
1400 100,000 GPD								9.90	10.90
1500 500,000 GPD								7.70	8.45
1700 Municipal wastewater treatment facility									
1720 1.0 MGD				Gal.				11	12.10
1740 1.5 MGD								10.60	11.65
1760 2.0 MGD								10	11
1780 3.0 MGD								7.80	8.60
1800 5.0 MGD								5.80	6.70
2000 Holding tank system, not incl. excavation or backfill									
2010 Recirculating chemical water closet	2 Plum	4	4	Ea.	530	235		765	940
2100 For voltage converter, add	"	16	1		300	58.50		358.50	420
2200 For high level alarm, add	1 Plum	7.80	1.026		117	60		177	219

46 07 53.20 Wastewater Treatment System

	Crew	Daily Output	Labor-Hours	Unit	Material	Labor	Equipment	Total	Total Incl O&P
0010 **WASTEWATER TREATMENT SYSTEM**									
0020 Fiberglass, 1,000 gallon	B-21	1.29	21.705	Ea.	4,250	940	107	5,297	6,250
0100 1,500 gallon	"	1.03	27.184	"	8,475	1,175	134	9,784	11,300

46 23 Grit Removal And Handling Equipment

46 23 23 – Vortex Grit Removal Equipment

46 23 23.10 Rainwater Filters

	Crew	Daily Output	Labor-Hours	Unit	Material	Labor	Equipment	Total	Total Incl O&P
0010 **RAINWATER FILTERS**									
0100 42 gal./min	B-21	3.50	8	Ea.	45,100	345	39.50	45,484.50	50,000
0200 65 gal./min		3.50	8		45,200	345	39.50	45,584.50	50,500
0300 208 gal./min		3.50	8		45,200	345	39.50	45,584.50	50,500

46 25 Oil and Grease Separation and Removal Equipment

46 25 13 – Coalescing Oil-Water Separators

46 25 13.20 Oil/Water Separators	Crew	Daily Output	Labor-Hours	Unit	Material	2015 Bare Costs Labor	Equipment	Total	Total Incl O&P
0010 **OIL/WATER SEPARATORS**									
0020 Underground, tank only									
0030 Excludes excavation, backfill, & piping									
0100 200 GPM	B-21	3.50	8	Ea.	44,900	345	39.50	45,284.50	50,000
0110 400 GPM		3.25	8.615		62,500	375	42.50	62,917.50	69,500
0120 600 GPM		2.75	10.182		70,000	440	50	70,490	77,500
0130 800 GPM		2.50	11.200		79,000	485	55	79,540	88,000
0140 1000 GPM		2	14		88,500	605	69	89,174	98,500
0150 1200 GPM		1.50	18.667		101,000	810	92	101,902	112,500
0160 1500 GPM		1	28		119,500	1,225	138	120,863	133,000

Division Notes

	CREW	DAILY OUTPUT	LABOR-HOURS	UNIT	BARE COSTS				TOTAL INCL O&P
					MAT.	LABOR	EQUIP.	TOTAL	

Assemblies Section

Table of Contents

How RSMeans Assemblies Data Works

Assemblies estimating provides a fast and reasonably accurate way to develop construction costs. An assembly is the grouping of individual work items, with appropriate quantities, to provide a cost for a major construction component in a convenient unit of measure.

An assemblies estimate is often used during early stages of design development to compare the cost impact of various design alternatives on total building cost.

Assemblies estimates are also used as an efficient tool to verify construction estimates.

Assemblies estimates do not require a completed design or detailed drawings. Instead, they are based on the general size of the structure and other known parameters of the project. The degree of accuracy of an assemblies estimate is generally within +/- 15%.

D40 Fire Protection

D4020   Standpipes

System Components			COST PER FLOOR		
	QUANTITY	UNIT	MAT.	INST.	TOTAL
SYSTEM D4020 310 0560					
WET STANDPIPE RISER, CLASS I, STEEL, BLACK, SCH. 40 PIPE, 10' HEIGHT					
4" DIAMETER PIPE, ONE FLOOR					
Pipe, steel, black, schedule 40, threaded, 4" diam.	20.000	L.F.	590	710	1,300
Pipe, Tee, malleable iron, black, 150 lb. threaded, 4" pipe size	2.000	Ea.	630	640	1,270
Pipe, 90° elbow, malleable iron, black, 150 lb threaded 4" pipe size	1.000	Ea.	200	213	413
Pipe, nipple, steel, black, schedule 40, 2-1/2" pipe size x 3" long	2.000	Ea.	24.50	160	184.50
Fire valve, gate, 300 lb., brass w/handwheel, 2-1/2" pipe size	1.000	Ea.	208	97	305
Fire valve, pressure restricting, adj, rgh brs, 2-1/2" pipe size	1.000	Ea.	352	194	546
Valve, swing check, w/ball drip, CI w/brs. ftngs., 4" pipe size	1.000	Ea.	385	410	795
Standpipe conn wall dble. flush brs. w/plugs & chains 2-1/2"x2-1/2"x4"	1.000	Ea.	570	245	815
Valve, swing check, bronze, 125 lb, regrinding disc, 2-1/2" pipe size	1.000	Ea.	880	85	965
Roof manifold, fire, w/valves & caps, horiz/vert brs 2-1/2"x2-1/2"x4"	1.000	Ea.	185	255	440
Fire, hydrolator, vent & drain, 2-1/2" pipe size	1.000	Ea.	92.50	56.50	149
Valve, gate, iron body 125 lb., OS&Y, threaded, 4" pipe size	1.000	Ea.	850	425	1,275
TOTAL			4,967	3,490.50	8,457.50

D4020 310	Wet Standpipe Risers, Class I		COST PER FLOOR		
			MAT.	INST.	TOTAL
0550	Wet standpipe risers, Class I, steel, black sch. 40, 10' height				
0560	4" diameter pipe, one floor		4,975	3,500	8,475
0580	Additional floors		1,100	1,075	2,175
0600	6" diameter pipe, one floor		8,550	6,125	14,675
0620	Additional floors	R211226 -10	2,075	1,725	3,800
0640	8" diameter pipe, one floor		13,200	7,425	20,625
0660	Additional floors	R211226 -20	3,100	2,075	5,175
0680					

RSMeans assemblies are identified by a **unique 12-character identifier**. The assemblies are numbered using UNIFORMAT II, ASTM Standard E1557. The first 6 characters represent this system to Level 4. The last 6 characters represent further breakdown by RSMeans in order to arrange items in understandable groups of similar tasks. Line numbers are consistent across all RSMeans publications, so a line number in any RSMeans assemblies data set will always refer to the same work.

RSMeans assemblies **Narrative descriptions** are shown in a hierarchical structure to make them readable. In order to read a complete description, read up through the indents to the top of the section. Include everything that is above and to the left that is not contradicted by information below.

Information is available in the Reference Section to assist the estimator with estimating procedures, alternate pricing methods, and additional technical information.

The **Reference Box** indicates the exact location of this information in the Reference Section at the back of the book. The "R" stands for "reference," and the remaining characters are the RSMeans line numbers.

For supplemental customizable square foot estimating forms, visit: **http://www.reedconstructiondata.com/rsmeans/assemblies**

Most assemblies consist of three major elements: a graphic, the system components, and the cost data itself. The **Graphic** is a visual representation showing the typical appearance of the assembly in question, frequently accompanied by additional explanatory technical information describing the class of items. The **System Components** is a listing of the individual tasks that make up the assembly, including the quantity and unit of measure for each item, along with the cost of material and installation. The **Assemblies Data** below lists prices for other similar systems with dimensional and/or size variations.

All RSMeans assemblies costs represent the cost for the installing contractor. An allowance for profit has been added to all material, labor, and equipment rental costs. A markup for labor burdens, including workers' compensation, fixed overhead, and business overhead, is included with installation costs.

The data published in RSMeans print books represents a "national average" cost. This data should be modified to the project location using the **City Cost Indexes** or **Location Factors** tables found in the Reference Section in the back of the book.

All RSMeans assemblies data includes a typical **Unit of Measure** used for estimating that item. For instance, while the unit of measure for the graphic shown here is Each, some A/C systems are estimated by the square foot (S.F.). The estimator needs to take special care that the unit in the data matches the unit in the takeoff. Abbreviations can be found in the Reference Section.

System components are listed separately to detail what is included in the development of the total system price.

D40 Fire Protection

D4020 Standpipes

System Components	QUANTITY	UNIT	COST PER FLOOR		
			MAT.	INST.	TOTAL
SYSTEM D4020 310 0560					
WET STANDPIPE RISER, CLASS I, STEEL, BLACK, SCH. 40 PIPE, 10' HEIGHT					
4" DIAMETER PIPE, ONE FLOOR					
Pipe, steel, black, schedule 40, threaded, 4" diam.	20.000	L.F.	590	710	1,300
Pipe, Tee, malleable iron, black, 150 lb. threaded, 4" pipe size	2.000	Ea.	630	640	1,270
Pipe, 90° elbow, malleable iron, black, 150 lb threaded 4" pipe size	1.000	Ea.	200	213	413
Pipe, nipple, steel, black, schedule 40, 2-1/2" pipe size x 3" long	2.000	Ea.	24.50	160	184.50
Fire valve, gate, 300 lb., brass w/handwheel, 2-1/2" pipe size	1.000	Ea.	208	97	305
Fire valve, pressure restricting, adj, rgh brs, 2-1/2" pipe size	1.000	Ea.	352	194	546
Valve, swing check, w/ball drip, CI w/brs. ftngs., 4" pipe size	1.000	Ea.	385	410	795
Standpipe conn wall dble. flush brs. w/plugs & chains 2-1/2"x2-1/2"x4"	1.000	Ea.	570	245	815
Valve, swing check, bronze, 125 lb, regrinding disc, 2-1/2" pipe size	1.000	Ea.	880	85	965
Roof manifold, fire, w/valves & caps, horiz/vert brs 2-1/2"x2-1/2"x4"	1.000	Ea.	185	255	440
Fire, hydrolator, vent & drain, 2-1/2" pipe size	1.000	Ea.	92.50	56.50	149
Valve, gate, iron body 125 lb., OS&Y, threaded, 4" pipe size	1.000	Ea.	850	425	1,275
TOTAL			4,967	3,490.50	8,457.50

D4020 310	Wet Standpipe Risers, Class I		COST PER FLOOR		
			MAT.	INST.	TOTAL
0550	Wet standpipe risers, Class I, steel, black sch. 40, 10' height				
0560	4" diameter pipe, one floor		4,975	3,500	8,475
0580	Additional floors		1,100	1,075	2,175
0600	6" diameter pipe, one floor		8,550	6,125	14,675
0620	Additional floors	R211226 -10	2,075	1,725	3,800
0640	8" diameter pipe, one floor		13,200	7,425	20,625
0660	Additional floors	R211226 -20	3,100	2,075	5,175
0680					

How RSMeans Assemblies Data Works (Continued)

Sample Estimate

This sample demonstrates the elements of an estimate, including a tally of the RSMeans data lines. Published assemblies costs include all markups for labor burden and profit for the installing contractor. This estimate adds a summary of the markups applied by a general contractor on the installing contractors' work. These figures represent the total cost to the owner. The RSMeans location factor is added at the bottom of the estimate to adjust the cost of the work to a specific location

Work performed: The body of the estimate shows the RSMeans data selected, including line numbers, a brief description of each item, its takeoff quantity and unit, and the total installed cost, including the installing contractor's overhead and profit.

Location Factor: RSMeans published data is based on national average costs. If necessary, adjust the total cost of the project using a location factor from the "Location Factor" table or the "City Cost Indexes" table, found in the Reference Section. Use location factors if the work is general, covering the work of multiple trades. If the work is by a single trade (e.g. masonry) use the more specific data found in the City Cost Indexes.

To adjust costs by location factors, multiply the base cost by the factor and divide by 100.

Contingency: A factor for contingency may be added to any estimate to represent the cost of unknowns that may occur between the time that the estimate is performed and the time the project is constructed. The amount of the allowance will depend on the stage of design at which the estimate is done, and the contractor's assessment of the risk invloved.

Project Name:	Interior Fit-out, ABC Office
Location:	**Anywhere, USA**
Assembly Number	**Description**
C1010 124 1200	Wood partition, 2 x 4 @ 16" OC w/5/8" FR gypsum board
C1020 114 1800	Metal door & frame, flush hollow core, 3'-0" x 7'-0"
C3010 230 0080	Painting, brushwork, primer & 2 coats
C3020 410 0140	Carpet, tufted, nylon, roll goods, 12' wide, 26 oz
C3030 210 6000	Acoustic ceilings, 24" x 48" tile, tee grid suspension
D5020 125 0560	Receptacles incl plate, box, conduit, wire, 20 A duplex
D5020 125 0720	Light switch incl plate, box, conduit, wire, 20 A single pole
D5020 210 0560	Fluorescent fixtures, recess mounted, 20 per 1000 SF
Assembly Subtotal	
	Sales Tax @
	General Requirements @
Subtotal A	
	GC Overhead @
Subtotal B	
	GC Profit @
Subtotal C	
Adjusted by Location Factor	
	Architects Fee @
	Contingency @
Project Total Cost	

This estimate is based on an interactive spreadsheet. A copy of this spreadsheet is located on the RSMeans website at **http://www.reedconstructiondata.com/rsmeans/extras/546011**. You are free to download it and adjust it to your methodology.

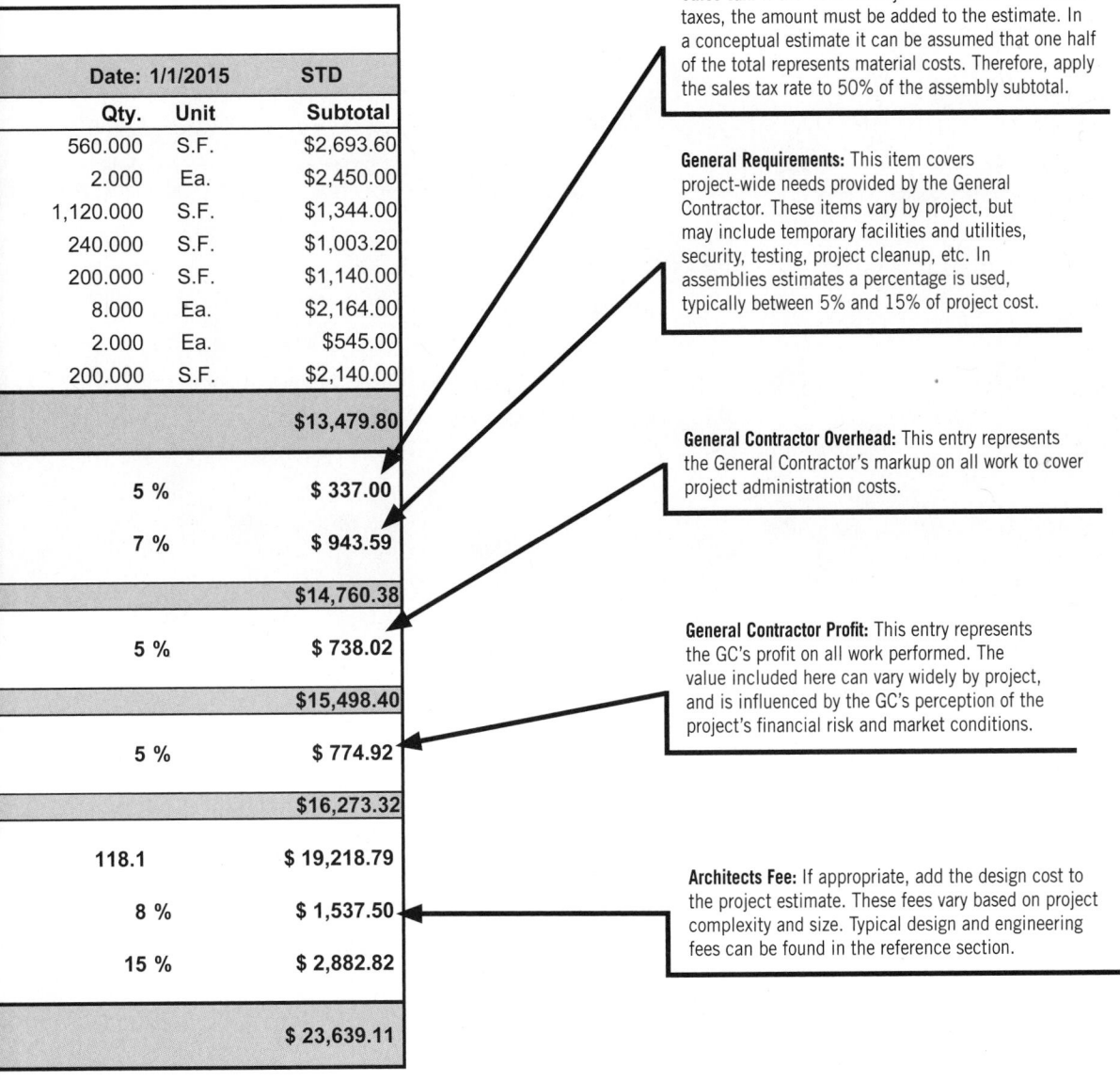

Date: 1/1/2015		STD
Qty.	Unit	Subtotal
560.000	S.F.	$2,693.60
2.000	Ea.	$2,450.00
1,120.000	S.F.	$1,344.00
240.000	S.F.	$1,003.20
200.000	S.F.	$1,140.00
8.000	Ea.	$2,164.00
2.000	Ea.	$545.00
200.000	S.F.	$2,140.00
		$13,479.80
5 %		$ 337.00
7 %		$ 943.59
		$14,760.38
5 %		$ 738.02
		$15,498.40
5 %		$ 774.92
		$16,273.32
118.1		$ 19,218.79
8 %		$ 1,537.50
15 %		$ 2,882.82
		$ 23,639.11

Sales Tax: If the work is subject to state or local sales taxes, the amount must be added to the estimate. In a conceptual estimate it can be assumed that one half of the total represents material costs. Therefore, apply the sales tax rate to 50% of the assembly subtotal.

General Requirements: This item covers project-wide needs provided by the General Contractor. These items vary by project, but may include temporary facilities and utilities, security, testing, project cleanup, etc. In assemblies estimates a percentage is used, typically between 5% and 15% of project cost.

General Contractor Overhead: This entry represents the General Contractor's markup on all work to cover project administration costs.

General Contractor Profit: This entry represents the GC's profit on all work performed. The value included here can vary widely by project, and is influenced by the GC's perception of the project's financial risk and market conditions.

Architects Fee: If appropriate, add the design cost to the project estimate. These fees vary based on project complexity and size. Typical design and engineering fees can be found in the reference section.

D2010 Plumbing Fixtures

Systems are complete with trim seat and rough-in (supply, waste and vent) for connection to supply branches and waste mains.

One Piece Wall Hung   **Supply**   **Waste/Vent**   **Floor Mount**

System Components	QUANTITY	UNIT	COST EACH		
			MAT.	INST.	TOTAL
SYSTEM D2010 110 1880					
WATER CLOSET, VITREOUS CHINA					
TANK TYPE, WALL HUNG, TWO PIECE					
Water closet, tank type vit china wall hung 2 pc. w/seat supply & stop	1.000	Ea.	695	241	936
Pipe Steel galvanized, schedule 40, threaded, 2" diam.	4.000	L.F.	45.40	79.80	125.20
Pipe, CI soil, no hub, cplg 10' OC, hanger 5' OC, 4" diam.	2.000	L.F.	42	44	86
Pipe, coupling, standard coupling, CI soil, no hub, 4" diam.	2.000	Ea.	44	77	121
Copper tubing type L solder joint, hangar 10' O.C., 1/2" diam.	6.000	L.F.	24.42	52.50	76.92
Wrought copper 90° elbow for solder joints 1/2" diam.	2.000	Ea.	2.42	71	73.42
Wrought copper Tee for solder joints 1/2" diam.	1.000	Ea.	2.06	54.50	56.56
Supports/carrier, water closet, siphon jet, horiz, single, 4" waste	1.000	Ea.	975	133	1,108
TOTAL			1,830.30	752.80	2,583.10

D2010 110	Water Closet Systems		COST EACH		
			MAT.	INST.	TOTAL
1800	Water closet, vitreous china				
1840	Tank type, wall hung				
1880	Close coupled two piece	R224000 -30	1,825	755	2,580
1920	Floor mount, one piece		1,600	800	2,400
1960	One piece low profile		1,625	800	2,425
2000	Two piece close coupled		670	800	1,470
2040	Bowl only with flush valve				
2080	Wall hung		2,300	855	3,155
2120	Floor mount		795	815	1,610
2160	Floor mount, ADA compliant with 18" high bowl		790	835	1,625

D2010 Plumbing Fixtures

Systems are complete with trim, seat, flush valve and rough-in (supply, waste and vent) for connection to supply branches and waste mains.

Side by Side

Back to Back

Supply   Waste/Vent   Supply   Waste/Vent

System Components	QUANTITY	UNIT	COST EACH		
			MAT.	INST.	TOTAL
SYSTEM D2010 120 1760					
WATER CLOSETS, BATTERY MOUNT, WALL HUNG, SIDE BY SIDE, FIRST CLOSET					
Water closet, bowl only w/flush valve, seat, wall hung	1.000	Ea.	1,050	220	1,270
Pipe, CI soil, no hub, cplg 10' OC, hanger 5' OC, 4" diam.	3.000	L.F.	63	66	129
Coupling, standard, CI, soil, no hub, 4" diam.	2.000	Ea.	44	77	121
Copper tubing, type L, solder joints, hangers 10' OC, 1" diam.	6.000	L.F.	51.60	62.70	114.30
Copper tubing, type DWV, solder joints, hangers 10'OC, 2" diam.	6.000	L.F.	96.60	96.60	193.20
Wrought copper 90° elbow for solder joints 1" diam.	1.000	Ea.	6.65	44.50	51.15
Wrought copper Tee for solder joints 1" diam.	1.000	Ea.	15.30	71	86.30
Support/carrier, siphon jet, horiz, adjustable single, 4" pipe	1.000	Ea.	975	133	1,108
Valve, gate, bronze, 125 lb, NRS, soldered 1" diam.	1.000	Ea.	84	37.50	121.50
Wrought copper, DWV, 90° elbow, 2" diam.	1.000	Ea.	31	71	102
TOTAL			2,417.15	879.30	3,296.45

D2010 120	Water Closets, Group		COST EACH		
			MAT.	INST.	TOTAL
1760	Water closets, battery mount, wall hung, side by side, first closet	R224000 -30	2,425	880	3,305
1800	Each additional water closet, add		2,300	830	3,130
3000	Back to back, first pair of closets		4,250	1,175	5,425
3100	Each additional pair of closets, back to back		4,150	1,150	5,300
9000	Back to back, first pair of closets, auto sensor flush valve, 1.28 gpf		4,725	1,250	5,975
9100	Ea additional pair of cls, back to back, auto sensor flush valve, 1.28 gpf		4,575	1,175	5,750

449

For customer support on your Plumbing Cost Data, call 877.709.6968.

Systems are complete with trim, flush valve and rough-in (supply, waste and vent) for connection to supply branches and waste mains.

Stall Type

Supply   **Waste/Vent**

Wall Hung

System Components	QUANTITY	UNIT	COST EACH		
			MAT.	INST.	TOTAL
SYSTEM D2010 210 2000					
URINAL, VITREOUS CHINA, WALL HUNG					
Urinal, wall hung, vitreous china, incl. hanger	1.000	Ea.	310	425	735
Pipe, steel, galvanized, schedule 40, threaded, 1-1/2" diam.	5.000	L.F.	44.25	79.75	124
Copper tubing type DWV, solder joint, hangers 10' OC, 2" diam.	3.000	L.F.	48.30	48.30	96.60
Combination Y & 1/8 bend for CI soil pipe, no hub, 3" diam.	1.000	Ea.	21		21
Pipe, CI, no hub, cplg. 10' OC, hanger 5' OC, 3" diam.	4.000	L.F.	66.80	79.80	146.60
Pipe coupling standard, CI soil, no hub, 3" diam.	3.000	Ea.	37.60	67	104.60
Copper tubing type L, solder joint, hanger 10' OC 3/4" diam.	5.000	L.F.	28.50	46.75	75.25
Wrought copper 90° elbow for solder joints 3/4" diam.	1.000	Ea.	2.71	37.50	40.21
Wrought copper Tee for solder joints, 3/4" diam.	1.000	Ea.	4.96	59	63.96
TOTAL			564.12	843.10	1,407.22

D2010 210	Urinal Systems		COST EACH		
			MAT.	INST.	TOTAL
2000	Urinal, vitreous china, wall hung	R224000 -30	565	845	1,410
2040	Stall type		1,200	1,000	2,200

450

For customer support on your Plumbing Cost Data, call 877.709.6968.

D2010 Plumbing Fixtures

Systems are complete with trim, flush valve and rough-in (supply, waste and vent) for connection to supply branches and waste mains.

Side by Side   Back to Back

Waste/Vent   Supply

Supply

Waste/Vent

System Components	QUANTITY	UNIT	COST EACH		
			MAT.	INST.	TOTAL
SYSTEM D2010 220 1760					
URINALS, BATTERY MOUNT, WALL HUNG, SIDE BY SIDE, FIRST URINAL					
Urinal, wall hung, vitreous china, with hanger & trim	1.000	Ea.	310	425	735
No hub cast iron soil pipe, 3″ diameter	4.000	L.F.	66.80	79.80	146.60
No hub cast iron sanitary tee, 3″ diameter	1.000	Ea.	21		21
No hub coupling, 3″ diameter	2.000	Ea.	37.60	67	104.60
Copper tubing, type L, 3/4″ diameter	5.000	L.F.	28.50	46.75	75.25
Copper tubing, type DWV, 2″ diameter	2.000	L.F.	32.20	32.20	64.40
Copper 90° elbow, 3/4″ diameter	2.000	Ea.	5.42	75	80.42
Copper 90° elbow, type DWV, 2″ diameter	2.000	Ea.	31	71	102
Galvanized steel pipe, 1-1/2″ diameter	5.000	L.F.	44.25	79.75	124
Cast iron drainage elbow, 90°, 1-1/2″ diameter	1.000	Ea.	4.96	59	63.96
TOTAL			581.73	935.50	1,517.23

D2010 220	Urinal Systems, Battery Mount		COST EACH		
			MAT.	INST.	TOTAL
1760	Urinals, battery mount, side by side, first urinal		580	935	1,515
1800	Each additional urinal, add		550	910	1,460
2000	Back to back, first pair of urinals	R224000 -30	1,150	1,525	2,675
2100	Each additional pair of urinals, back to back		860	1,200	2,060

451

D2010 Plumbing Fixtures

Systems are complete with trim and rough-in (supply, waste and vent) to connect to supply branches and waste mains.

Vanity Top

Supply   **Waste/Vent**

Wall Hung

System Components	QUANTITY	UNIT	COST EACH		
			MAT.	INST.	TOTAL
SYSTEM D2010 310 1560					
LAVATORY W/TRIM, VANITY TOP, P.E. ON C.I., 20″ X 18″					
Lavatory w/trim, PE on CI, white, vanity top, 20″ x 18″ oval	1.000	Ea.	370	199	569
Pipe, steel, galvanized, schedule 40, threaded, 1-1/4″ diam.	4.000	L.F.	30.80	57.40	88.20
Copper tubing type DWV, solder joint, hanger 10′ OC 1-1/4″ diam.	4.000	L.F.	40.20	47.20	87.40
Wrought copper DWV, Tee, sanitary, 1-1/4″ diam.	1.000	Ea.	23	79	102
P trap w/cleanout, 20 ga., 1-1/4″ diam.	1.000	Ea.	86	39.50	125.50
Copper tubing type L, solder joint, hanger 10′ OC 1/2″ diam.	10.000	L.F.	40.70	87.50	128.20
Wrought copper 90° elbow for solder joints 1/2″ diam.	2.000	Ea.	2.42	71	73.42
Wrought copper Tee for solder joints, 1/2″ diam.	2.000	Ea.	4.12	109	113.12
Stop, chrome, angle supply, 1/2″ diam.	2.000	Ea.	28.40	64	92.40
TOTAL			625.64	753.60	1,379.24

D2010 310	Lavatory Systems	COST EACH		
		MAT.	INST.	TOTAL
1560	Lavatory w/trim, vanity top, PE on CI, 20″ x 18″, Vanity top by others.	625	755	1,380
1600	19″ x 16″ oval	445	755	1,200
1640	18″ round	735	755	1,490
1680	Cultured marble, 19″ x 17″	450	755	1,205
1720	25″ x 19″	485	755	1,240
1760	Stainless, self-rimming, 25″ x 22″	660	755	1,415
1800	17″ x 22″	645	755	1,400
1840	Steel enameled, 20″ x 17″	435	775	1,210
1880	19″ round	445	775	1,220
1920	Vitreous china, 20″ x 16″	540	790	1,330
1960	19″ x 16″	550	790	1,340
2000	22″ x 13″	550	790	1,340
2040	Wall hung, PE on CI, 18″ x 15″	945	835	1,780
2080	19″ x 17″	970	835	1,805
2120	20″ x 18″	805	835	1,640
2160	Vitreous china, 18″ x 15″	755	855	1,610
2200	19″ x 17″	730	855	1,585
2240	24″ x 20″	810	855	1,665
2300	20″ x 27″, handicap	1,475	920	2,395

(Note: row 1800/1840 contains reference box: R224000 -30)

D2010 Plumbing Fixtures

Systems are complete with trim, flush valve and rough-in (supply, waste and vent) for connection to supply branches and waste mains.

Side by Side

Back to Back

Waste/Vent

Supply
(Two supply systems required)

Supply
(Two supply systems required)

Waste/Vent

System Components	QUANTITY	UNIT	COST EACH		
			MAT.	INST.	TOTAL
SYSTEM D2010 320 1760					
LAVATORIES, BATTERY MOUNT, WALL HUNG, SIDE BY SIDE, FIRST LAVATORY					
Lavatory w/trim wall hung PE on CI 20" x 18"	1.000	Ea.	305	160	465
Stop, chrome, angle supply, 3/8" diameter	2.000	Ea.	28.40	59	87.40
Concealed arm support	1.000	Ea.	580	118	698
P trap w/cleanout, 20 ga. C.P., 1-1/4" diameter	1.000	Ea.	34	39.50	73.50
Copper tubing, type L, 1/2" diameter	10.000	L.F.	40.70	87.50	128.20
Copper tubing, type DWV, 1-1/4" diameter	4.000	L.F.	40.20	47.20	87.40
Copper 90° elbow, 1/2" diameter	2.000	Ea.	2.42	71	73.42
Copper tee, 1/2" diameter	2.000	Ea.	4.12	109	113.12
DWV copper sanitary tee, 1-1/4" diameter	2.000	Ea.	46	158	204
Galvanized steel pipe, 1-1/4" diameter	4.000	L.F.	30.80	57.40	88.20
Black cast iron 90° elbow, 1-1/4" diameter	1.000	Ea.	11.85	58	69.85
TOTAL			1,123.49	964.60	2,088.09

D2010 320	Lavatory Systems, Battery Mount	COST EACH		
		MAT.	INST.	TOTAL
1760	Lavatories, battery mount, side by side, first lavatory	1,125	965	2,090
1800	Each additional lavatory, add	1,025	670	1,695
2000	Back to back, first pair of lavatories	1,775	1,525	3,300
2100	Each additional pair of lavatories, back to back	1,725	1,275	3,000

453

D2010 Plumbing Fixtures

Systems are complete with trim and rough-in (supply, waste and vent) to connect to supply branches and waste mains.

**Countertop
Single Bowl**

Supply

Waste/Vent

**Countertop
Double Bowl**

System Components	QUANTITY	UNIT	COST EACH		
			MAT.	INST.	TOTAL
SYSTEM D2010 410 1720					
KITCHEN SINK W/TRIM, COUNTERTOP, P.E. ON C.I., 24" X 21", SINGLE BOWL					
Kitchen sink, counter top, PE on CI, 1 bowl, 24" x 21" OD	1.000	Ea.	315	228	543
Pipe, steel, galvanized, schedule 40, threaded, 1-1/4" diam.	4.000	L.F.	30.80	57.40	88.20
Copper tubing, type DWV, solder, hangers 10' OC 1-1/2" diam.	6.000	L.F.	73.80	78.90	152.70
Wrought copper, DWV, Tee, sanitary, 1-1/2" diam.	1.000	Ea.	28.50	88.50	117
P trap, standard, copper, 1-1/2" diam.	1.000	Ea.	79	41.50	120.50
Copper tubing, type L, solder joints, hangers 10' OC 1/2" diam.	10.000	L.F.	40.70	87.50	128.20
Wrought copper 90° elbow for solder joints 1/2" diam.	2.000	Ea.	2.42	71	73.42
Wrought copper Tee for solder joints, 1/2" diam.	2.000	Ea.	4.12	109	113.12
Stop, angle supply, chrome, 1/2" CTS	2.000	Ea.	28.40	64	92.40
TOTAL			602.74	825.80	1,428.54

D2010 410	Kitchen Sink Systems	COST EACH		
		MAT.	INST.	TOTAL
1720	Kitchen sink w/trim, countertop, PE on CI, 24"x21", single bowl	605	825	1,430
1760	30" x 21" single bowl	970	825	1,795
1800	32" x 21" double bowl	700	890	1,590
1880	Stainless steel, 19" x 18" single bowl	940	825	1,765
1920	25" x 22" single bowl	1,025	825	1,850
1960	33" x 22" double bowl	1,350	890	2,240
2000	43" x 22" double bowl	1,550	905	2,455
2040	44" x 22" triple bowl	1,800	940	2,740
2080	44" x 24" corner double bowl	1,250	905	2,155
2120	Steel, enameled, 24" x 21" single bowl	850	825	1,675
2160	32" x 21" double bowl	855	890	1,745
2240	Raised deck, PE on CI, 32" x 21", dual level, double bowl	785	1,125	1,910
2280	42" x 21" dual level, triple bowl	1,625	1,225	2,850

D2010 Plumbing Fixtures

Systems are complete with trim and rough-in (supply, waste and vent) to connect to supply branches and waste mains.

Single Compartment Sink **Supply** **Waste/Vent** **Double Compartment Sink**

System Components	QUANTITY	UNIT	COST EACH		
			MAT.	INST.	TOTAL
SYSTEM D2010 420 1760					
LAUNDRY SINK W/TRIM, PE ON CI, BLACK IRON FRAME					
24″ X 20″ OD, SINGLE COMPARTMENT					
Laundry sink PE on CI w/trim & frame, 24″ x 21″ OD, 1 compartment	1.000	Ea.	635	213	848
Pipe, steel, galvanized, schedule 40, threaded, 1-1/4″ diam	4.000	L.F.	30.80	57.40	88.20
Copper tubing, type DWV, solder joint, hanger 10′ OC 1-1/2″diam	6.000	L.F.	73.80	78.90	152.70
Wrought copper, DWV, Tee, sanitary, 1-1/2″ diam	1.000	Ea.	28.50	88.50	117
P trap, standard, copper, 1-1/2″ diam	1.000	Ea.	79	41.50	120.50
Copper tubing type L, solder joints, hangers 10′ OC, 1/2″ diam	10.000	L.F.	40.70	87.50	128.20
Wrought copper 90° elbow for solder joints 1/2″ diam	2.000	Ea.	2.42	71	73.42
Wrought copper Tee for solder joints, 1/2″ diam	2.000	Ea.	4.12	109	113.12
Stop, angle supply, 1/2″ diam	2.000	Ea.	28.40	64	92.40
TOTAL			922.74	810.80	1,733.54

D2010 420	Laundry Sink Systems		COST EACH		
			MAT.	INST.	TOTAL
1740	Laundry sink w/trim, PE on CI, black iron frame				
1760	24″ x 20″, single compartment		925	810	1,735
1800	24″ x 23″ single compartment	R224000 -30	960	810	1,770
1840	48″ x 21″ double compartment		1,225	880	2,105
1920	Molded stone, on wall, 22″ x 21″ single compartment		470	810	1,280
1960	45″x 21″ double compartment		675	880	1,555
2040	Plastic, on wall or legs, 18″ x 23″ single compartment		435	795	1,230
2080	20″ x 24″ single compartment		460	795	1,255
2120	36″ x 23″ double compartment		515	855	1,370
2160	40″ x 24″ double compartment		615	855	1,470

D2010 Plumbing Fixtures

Corrosion resistant laboratory sink systems are complete with trim and rough-in (supply, waste and vent) to connect to supply branches and waste mains.

Laboratory Sink   **Supply**   **Waste/Vent**   **Polypropylene Cup Sink**

System Components	QUANTITY	UNIT	COST EACH MAT.	COST EACH INST.	COST EACH TOTAL
SYSTEM D2010 430 1600					
LABORATORY SINK W/TRIM, STAINLESS STEEL, SINGLE BOWL					
DOUBLE DRAINBOARD, 54″ X 24″ O.D.					
Sink w/trim, stainless steel, 1 bowl, 2 drainboards 54″ x 24″ OD	1.000	Ea.	1,875	425	2,300
Pipe, polypropylene, schedule 40, acid resistant 1-1/2″ diam.	10.000	L.F.	87	187.50	274.50
Tee, sanitary, polypropylene, acid resistant, 1-1/2″ diam.	1.000	Ea.	19.40	71	90.40
P trap, polypropylene, acid resistant, 1-1/2″ diam.	1.000	Ea.	62	41.50	103.50
Copper tubing type L, solder joint, hanger 10′ O.C. 1/2″ diam.	10.000	L.F.	40.70	87.50	128.20
Wrought copper 90° elbow for solder joints 1/2″ diam.	2.000	Ea.	2.42	71	73.42
Wrought copper Tee for solder joints, 1/2″ diam.	2.000	Ea.	4.12	109	113.12
Stop, angle supply, chrome, 1/2″ diam.	2.000	Ea.	28.40	64	92.40
TOTAL			2,119.04	1,056.50	3,175.54

D2010 430	Laboratory Sink Systems	COST EACH MAT.	COST EACH INST.	COST EACH TOTAL
1580	Laboratory sink w/trim, stainless steel, single bowl,			
1590	Stainless steel, single bowl,			
1600	Double drainboard, 54″ x 24″ O.D.	2,125	1,050	3,175
1640	Single drainboard, 47″ x 24″O.D.	1,550	1,050	2,600
1670	Stainless steel, double bowl,			
1680	70″ x 24″ O.D.	2,325	1,050	3,375
1750	Polyethylene, single bowl,			
1760	Flanged, 14-1/2″ x 14-1/2″ O.D.	500	950	1,450
1800	18-1/2″ x 18-1/2″ O.D.	615	950	1,565
1840	23-1/2″ x 20-1/2″ O.D.	635	950	1,585
1920	Polypropylene, cup sink, oval, 7″ x 4″ O.D.	370	840	1,210
1960	10″ x 4-1/2″ O.D.	395	840	1,235

D2010 Plumbing Fixtures

Corrosion resistant laboratory sink systems are complete with trim and rough–in (supply, waste and vent) to connect to supply branches and waste mains.

Wall Hung   **Supply**   **Waste/Vent**   **Corner, Floor**

System Components	QUANTITY	UNIT	COST EACH		
			MAT.	INST.	TOTAL
SYSTEM D2010 440 4260					
SERVICE SINK, PE ON CI, CORNER FLOOR, 28"X28", W/RIM GUARD & TRIM					
Service sink, corner floor, PE on CI, 28" x 28", w/rim guard & trim	1.000	Ea.	1,100	290	1,390
Copper tubing type DWV, solder joint, hanger 10'OC 3" diam.	6.000	L.F.	174	132	306
Copper tubing type DWV, solder joint, hanger 10'OC 2" diam	4.000	L.F.	64.40	64.40	128.80
Wrought copper DWV, Tee, sanitary, 3" diam.	1.000	Ea.	151	182	333
P trap with cleanout & slip joint, copper 3" diam	1.000	Ea.	310	64.50	374.50
Copper tubing, type L, solder joints, hangers 10' OC, 1/2" diam	10.000	L.F.	40.70	87.50	128.20
Wrought copper 90° elbow for solder joints 1/2" diam	2.000	Ea.	2.42	71	73.42
Wrought copper Tee for solder joints, 1/2" diam	2.000	Ea.	4.12	109	113.12
Stop, angle supply, chrome, 1/2" diam	2.000	Ea.	28.40	64	92.40
TOTAL			1,875.04	1,064.40	2,939.44

D2010 440	Service Sink Systems		COST EACH		
			MAT.	INST.	TOTAL
4260	Service sink w/trim, PE on CI, corner floor, 28" x 28", w/rim guard	R224000 -30	1,875	1,075	2,950
4300	Wall hung w/rim guard, 22" x 18"		2,125	1,225	3,350
4340	24" x 20"		2,200	1,225	3,425
4380	Vitreous china, wall hung 22" x 20"		1,950	1,225	3,175

457

For customer support on your Plumbing Cost Data, call 877.709.6968.

D2010 Plumbing Fixtures

Systems are complete with trim and rough-in (supply, waste and vent) to connect to supply branches and waste mains.

Recessed Bathtub   **Supply**   **Waste/Vent**   **Corner Bathtub**

System Components	QUANTITY	UNIT	COST EACH MAT.	COST EACH INST.	COST EACH TOTAL
SYSTEM D2010 510 2000					
BATHTUB, RECESSED, PORCELAIN ENAMEL ON CAST IRON,, 48" x 42"					
Bath tub, porcelain enamel on cast iron, w/fittings, 48" x 42"	1.000	Ea.	2,900	320	3,220
Pipe, steel, galvanized, schedule 40, threaded, 1-1/4" diam.	4.000	L.F.	30.80	57.40	88.20
Pipe, CI no hub soil w/couplings 10' OC, hangers 5' OC, 4" diam.	3.000	L.F.	63	66	129
Combination Y and 1/8 bend for C.I. soil pipe, no hub, 4" pipe size	1.000	Ea.	54		54
Drum trap, 3" x 5", copper, 1-1/2" diam.	1.000	Ea.	117	44.50	161.50
Copper tubing type L, solder joints, hangers 10' OC 1/2" diam.	10.000	L.F.	40.70	87.50	128.20
Wrought copper 90° elbow, solder joints, 1/2" diam.	2.000	Ea.	2.42	71	73.42
Wrought copper Tee, solder joints, 1/2" diam.	2.000	Ea.	4.12	109	113.12
Stop, angle supply, 1/2" diameter	2.000	Ea.	28.40	64	92.40
Copper tubing type DWV, solder joints, hanger 10' OC 1-1/2" diam.	3.000	L.F.	36.90	39.45	76.35
Pipe coupling, standard, C.I. soil no hub, 4" pipe size	2.000	Ea.	44	77	121
TOTAL			3,321.34	935.85	4,257.19

D2010 510	Bathtub Systems		COST EACH MAT.	COST EACH INST.	COST EACH TOTAL
2000	Bathtub, recessed, P.E. on Cl., 48" x 42"	R224000 -30	3,325	935	4,260
2040	72" x 36"		3,425	1,050	4,475
2080	Mat bottom, 5' long		1,675	905	2,580
2120	5'-6" long		2,375	935	3,310
2160	Corner, 48" x 42"		3,325	905	4,230
2200	Formed steel, enameled, 4'-6" long		965	835	1,800

D2010 Plumbing Fixtures

Circular Fountain

Supply

Waste/Vent

Systems are complete with trim, flush valve and rough-in (supply, waste and vent) for connection to supply branches and waste mains.

Semi-Circular Fountain

System Components	QUANTITY	UNIT	COST EACH		
			MAT.	INST.	TOTAL
SYSTEM D2010 610 1760					
GROUP WASH FOUNTAIN, PRECAST TERRAZZO					
CIRCULAR, 36" DIAMETER					
Wash fountain, group, precast terrazzo, foot control 36" diam.	1.000	Ea.	8,000	660	8,660
Copper tubing type DWV, solder joint, hanger 10' OC, 2" diam.	10.000	L.F.	161	161	322
P trap, standard, copper, 2" diam.	1.000	Ea.	122	47.50	169.50
Wrought copper, Tee, sanitary, 2" diam.	1.000	Ea.	33	101	134
Copper tubing type L, solder joint, hanger 10' OC 1/2" diam.	20.000	L.F.	81.40	175	256.40
Wrought copper 90° elbow for solder joints 1/2" diam.	3.000	Ea.	3.63	106.50	110.13
Wrought copper Tee for solder joints, 1/2" diam.	2.000	Ea.	4.12	109	113.12
TOTAL			8,405.15	1,360	9,765.15

D2010 610	Group Wash Fountain Systems		COST EACH		
			MAT.	INST.	TOTAL
1740	Group wash fountain, precast terrazzo				
1760	Circular, 36" diameter		8,400	1,350	9,750
1800	54" diameter	R224000 -30	10,400	1,500	11,900
1840	Semi-circular, 36" diameter		7,425	1,350	8,775
1880	54" diameter		9,825	1,500	11,325
1960	Stainless steel, circular, 36" diameter		7,600	1,275	8,875
2000	54" diameter		9,250	1,400	10,650
2040	Semi-circular, 36" diameter		5,950	1,275	7,225
2080	54" diameter		8,125	1,400	9,525
2160	Thermoplastic, circular, 36" diameter		5,300	1,025	6,325
2200	54" diameter		6,100	1,200	7,300
2240	Semi-circular, 36" diameter		4,900	1,025	5,925
2280	54" diameter		5,900	1,200	7,100

459

For customer support on your Plumbing Cost Data, call 877.709.6968.

D20 Plumbing

D2010 Plumbing Fixtures

Three Wall

Systems are complete with trim and rough-in (supply, waste and vent) for connection to supply branches and waste mains.

Supply   **Waste/Vent**

Corner Angle

System Components	QUANTITY	UNIT	COST EACH		
			MAT.	INST.	TOTAL
SYSTEM D2010 710 1560					
SHOWER, STALL, BAKED ENAMEL, MOLDED STONE RECEPTOR, 30" SQUARE					
Shower stall, enameled steel, molded stone receptor, 30" square	1.000	Ea.	1,275	245	1,520
Copper tubing type DWV, solder joints, hangers 10' OC, 2" diam.	6.000	L.F.	73.80	78.90	152.70
Wrought copper DWV, Tee, sanitary, 2" diam.	1.000	Ea.	28.50	88.50	117
Trap, standard, copper, 2" diam.	1.000	Ea.	79	41.50	120.50
Copper tubing type L, solder joint, hanger 10' OC 1/2" diam.	16.000	L.F.	65.12	140	205.12
Wrought copper 90° elbow for solder joints 1/2" diam.	3.000	Ea.	3.63	106.50	110.13
Wrought copper Tee for solder joints, 1/2" diam.	2.000	Ea.	4.12	109	113.12
Stop and waste, straightway, bronze, solder joint 1/2" diam.	2.000	Ea.	27.10	59	86.10
TOTAL			1,556.27	868.40	2,424.67

D2010 710	Shower Systems		COST EACH		
			MAT.	INST.	TOTAL
1560	Shower, stall, baked enamel, molded stone receptor, 30" square		1,550	870	2,420
1600	32" square		1,550	880	2,430
1640	Terrazzo receptor, 32" square	R224000	1,750	880	2,630
1680	36" square	-30	1,900	890	2,790
1720	36" corner angle		2,175	890	3,065
1800	Fiberglass one piece, three walls, 32" square		820	855	1,675
1840	36" square		835	855	1,690
1880	Polypropylene, molded stone receptor, 30" square		980	1,275	2,255
1920	32" square		995	1,275	2,270
1960	Built-in head, arm, bypass, stops and handles		112	325	437
2050	Shower, stainless steel panels, handicap				
2100	w/fixed and handheld head, control valves, grab bar, and seat		5,400	3,925	9,325
2500	Shower, group with six heads, thermostatic mix valves & balancing valve		12,400	955	13,355
2520	Five heads		8,875	870	9,745

D2010 Plumbing Fixtures

Systems are complete with trim and rough-in (supply, waste and vent) to connect to supply branches and waste mains.

Wall Mounted, No Back

Supply   **Waste/Vent**

Wall Mounted, Low Back

System Components	QUANTITY	UNIT	COST EACH		
			MAT.	INST.	TOTAL
SYSTEM D2010 810 1800					
DRINKING FOUNTAIN, ONE BUBBLER, WALL MOUNTED					
NON RECESSED, BRONZE, NO BACK					
Drinking fountain, wall mount, bronze, 1 bubbler	1.000	Ea.	1,100	177	1,277
Copper tubing, type L, solder joint, hanger 10' OC 3/8" diam.	5.000	L.F.	19.25	42.25	61.50
Stop, supply, straight, chrome, 3/8" diam.	1.000	Ea.	13.55	29.50	43.05
Wrought copper 90° elbow for solder joints 3/8" diam.	1.000	Ea.	3.42	32	35.42
Wrought copper Tee for solder joints, 3/8" diam.	1.000	Ea.	5.80	50.50	56.30
Copper tubing, type DWV, solder joint, hanger 10' OC 1-1/4" diam.	4.000	L.F.	40.20	47.20	87.40
P trap, standard, copper drainage, 1-1/4" diam.	1.000	Ea.	86	39.50	125.50
Wrought copper, DWV, Tee, sanitary, 1-1/4" diam.	1.000	Ea.	23	79	102
TOTAL			1,291.22	496.95	1,788.17

D2010 810	Drinking Fountain Systems		COST EACH		
			MAT.	INST.	TOTAL
1740	Drinking fountain, one bubbler, wall mounted				
1760	Non recessed				
1800	Bronze, no back	R224000 -30	1,300	495	1,795
1840	Cast iron, enameled, low back		1,250	495	1,745
1880	Fiberglass, 12" back		2,325	495	2,820
1920	Stainless steel, no back		1,350	495	1,845
1960	Semi-recessed, poly marble		1,250	495	1,745
2040	Stainless steel		1,475	495	1,970
2080	Vitreous china		1,175	495	1,670
2120	Full recessed, poly marble		1,950	495	2,445
2200	Stainless steel		1,900	495	2,395
2240	Floor mounted, pedestal type, aluminum		2,675	675	3,350
2320	Bronze		2,325	675	3,000
2360	Stainless steel		2,125	675	2,800

D2010 Plumbing Fixtures

Systems are complete with trim and rough-in (supply, waste and vent) for connection to supply branches and waste mains.

Wall Hung   **Supply**   **Waste/Vent**   **Floor Mounted**

System Components	QUANTITY	UNIT	COST EACH		
			MAT.	INST.	TOTAL
SYSTEM D2010 820 1840					
WATER COOLER, ELECTRIC, SELF CONTAINED, WALL HUNG, 8.2 G.P.H.					
Water cooler, wall mounted, 8.2 GPH	1.000	Ea.	825	320	1,145
Copper tubing type DWV, solder joint, hanger 10' OC 1-1/4" diam.	4.000	L.F.	40.20	47.20	87.40
Wrought copper DWV, Tee, sanitary 1-1/4" diam.	1.000	Ea.	23	79	102
P trap, copper drainage, 1-1/4" diam.	1.000	Ea.	86	39.50	125.50
Copper tubing type L, solder joint, hanger 10' OC 3/8" diam.	5.000	L.F.	19.25	42.25	61.50
Wrought copper 90° elbow for solder joints 3/8" diam.	1.000	Ea.	3.42	32	35.42
Wrought copper Tee for solder joints, 3/8" diam.	1.000	Ea.	5.80	50.50	56.30
Stop and waste, straightway, bronze, solder, 3/8" diam.	1.000	Ea.	13.55	29.50	43.05
TOTAL			1,016.22	639.95	1,656.17

D2010 820	Water Cooler Systems		COST EACH		
			MAT.	INST.	TOTAL
1840	Water cooler, electric, wall hung, 8.2 G.P.H.	R224000 -30	1,025	640	1,665
1880	Dual height, 14.3 G.P.H.		1,275	655	1,930
1920	Wheelchair type, 7.5 G.P.H.		1,200	640	1,840
1960	Semi recessed, 8.1 G.P.H.		1,025	640	1,665
2000	Full recessed, 8 G.P.H.		2,425	685	3,110
2040	Floor mounted, 14.3 G.P.H.		1,175	555	1,730
2080	Dual height, 14.3 G.P.H.		1,525	675	2,200
2120	Refrigerated compartment type, 1.5 G.P.H.		1,850	555	2,405

462

For customer support on your Plumbing Cost Data, call 877.709.6968.

D2010 Plumbing Fixtures

Two Fixture Bathroom Systems consisting of a lavatory, water closet, and rough-in service piping.

• Prices for plumbing and fixtures only.

*Common wall is with an adjacent bathroom.

System Components	QUANTITY	UNIT	COST EACH MAT.	COST EACH INST.	COST EACH TOTAL
SYSTEM D2010 920 1180					
BATHROOM, LAVATORY & WATER CLOSET, 2 WALL PLUMBING, STAND ALONE					
Water closet, two piece, close coupled	1.000	Ea.	261	241	502
Water closet, rough-in waste & vent	1.000	Set	365	420	785
Lavatory w/ftngs., wall hung, white, PE on CI, 20" x 18"	1.000	Ea.	305	160	465
Lavatory, rough-in waste & vent	1.000	Set	500	770	1,270
Copper tubing type L, solder joint, hanger 10' OC 1/2" diam.	10.000	L.F.	40.70	87.50	128.20
Pipe, steel, galvanized, schedule 40, threaded, 2" diam.	12.000	L.F.	136.20	239.40	375.60
Pipe, CI soil, no hub, coupling 10' OC, hanger 5' OC, 4" diam.	7.000	L.F.	143.50	161	304.50
TOTAL			1,751.40	2,078.90	3,830.30

D2010 920	Two Fixture Bathroom, Two Wall Plumbing		COST EACH MAT.	COST EACH INST.	COST EACH TOTAL
1180	Bathroom, lavatory & water closet, 2 wall plumbing, stand alone		1,750	2,075	3,825
1200	Share common plumbing wall*		1,600	1,800	3,400

D2010 922	Two Fixture Bathroom, One Wall Plumbing		COST EACH MAT.	COST EACH INST.	COST EACH TOTAL
2220	Bathroom, lavatory & water closet, one wall plumbing, stand alone	R224000 -30	1,650	1,875	3,525
2240	Share common plumbing wall*		1,425	1,600	3,025
2260					
2280					

D2010 Plumbing Fixtures

Three Fixture Bathroom Systems consisting of a lavatory, water closet, bathtub or shower and rough-in service piping.

• Prices for plumbing and fixtures only.

*Common wall is with an adjacent bathroom.

System Components	QUANTITY	UNIT	COST EACH		
			MAT.	INST.	TOTAL
SYSTEM D2010 924 1170					
BATHROOM, LAVATORY, WATER CLOSET & BATHTUB					
ONE WALL PLUMBING, STAND ALONE					
Wtr closet, rough-in, supply, waste and vent	1.000	Set	365	420	785
Wtr closet, 2 pc close cpld vit china flr mntd w/seat supply & stop	1.000	Ea.	261	241	502
Lavatory w/ftngs, wall hung, white, PE on CI, 20" x 18"	1.000	Ea.	305	160	465
Lavatory, rough-in waste & vent	1.000	Set	500	770	1,270
Bathtub, white PE on CI, w/ftgs, mat bottom, recessed, 5' long	1.000	Ea.	1,250	290	1,540
Baths, rough-in waste and vent	1.000	Set	342	553.50	895.50
TOTAL			3,023	2,434.50	5,457.50

D2010 924	Three Fixture Bathroom, One Wall Plumbing		COST EACH		
			MAT.	INST.	TOTAL
1150	Bathroom, three fixture, one wall plumbing	R224000 -30			
1160	Lavatory, water closet & bathtub				
1170	Stand alone		3,025	2,425	5,450
1180	Share common plumbing wall *		2,575	1,750	4,325

D2010 926	Three Fixture Bathroom, Two Wall Plumbing	COST EACH		
		MAT.	INST.	TOTAL
2130	Bathroom, three fixture, two wall plumbing			
2140	Lavatory, water closet & bathtub			
2160	Stand alone	3,050	2,475	5,525
2180	Long plumbing wall common *	2,725	1,975	4,700
3610	Lavatory, bathtub & water closet			
3620	Stand alone	3,250	2,800	6,050
3640	Long plumbing wall common *	3,075	2,525	5,600
4660	Water closet, corner bathtub & lavatory			
4680	Stand alone	4,700	2,500	7,200
4700	Long plumbing wall common *	4,325	1,875	6,200
6100	Water closet, stall shower & lavatory			
6120	Stand alone	3,325	2,800	6,125
6140	Long plumbing wall common *	3,175	2,600	5,775
7060	Lavatory, corner stall shower & water closet			
7080	Stand alone	3,700	2,475	6,175
7100	Short plumbing wall common *	3,200	1,675	4,875

D20 Plumbing

D2010 Plumbing Fixtures

Four Fixture Bathroom Systems consisting of a lavatory, water closet, bathtub, shower and rough-in service piping.

• Prices for plumbing and fixtures only.

*Common wall is with an adjacent bathroom.

System Components	QUANTITY	UNIT	COST EACH		
			MAT.	INST.	TOTAL
SYSTEM D2010 928 1160					
BATHROOM, BATHTUB, WATER CLOSET, STALL SHOWER & LAVATORY					
TWO WALL PLUMBING, STAND ALONE					
Wtr closet, 2 pc close cpld vit china flr mntd w/seat supply & stop	1.000	Ea.	261	241	502
Water closet, rough-in waste & vent	1.000	Set	365	420	785
Lavatory w/ftngs, wall hung, white PE on CI, 20" x 18"	1.000	Ea.	305	160	465
Lavatory, rough-in waste & vent	1.000	Set	50	77	127
Bathtub, white PE on CI, w/ftgs, mat bottom, recessed, 5' long	1.000	Ea.	1,250	290	1,540
Baths, rough-in waste and vent	1.000	Set	380	615	995
Shower stall, bkd enam, molded stone receptor, door & trim 32" sq.	1.000	Ea.	1,275	255	1,530
Shower stall, rough-in supply, waste & vent	1.000	Set	375	625	1,000
TOTAL			4,261	2,683	6,944

D2010 928	Four Fixture Bathroom, Two Wall Plumbing		COST EACH		
			MAT.	INST.	TOTAL
1140	Bathroom, four fixture, two wall plumbing	R224000 -30			
1150	Bathtub, water closet, stall shower & lavatory				
1160	Stand alone		4,250	2,675	6,925
1180	Long plumbing wall common *		3,875	2,075	5,950
2260	Bathtub, lavatory, corner stall shower & water closet				
2280	Stand alone		4,875	2,700	7,575
2320	Long plumbing wall common *		4,500	2,075	6,575
3620	Bathtub, stall shower, lavatory & water closet				
3640	Stand alone		4,700	3,375	8,075
3660	Long plumbing wall (opp. door) common *		4,325	2,750	7,075

D2010 930	Four Fixture Bathroom, Three Wall Plumbing	COST EACH		
		MAT.	INST.	TOTAL
4680	Bathroom, four fixture, three wall plumbing			
4700	Bathtub, stall shower, lavatory & water closet			
4720	Stand alone	5,525	3,725	9,250
4760	Long plumbing wall (opposite door) common *	5,350	3,425	8,775

For customer support on your Plumbing Cost Data, call 877.709.6968.

465

D2010 Plumbing Fixtures

Five Fixture Bathroom Systems consisting of two lavatories, a water closet, bathtub, shower and rough-in service piping.

• Prices for plumbing and fixtures only.

*Common wall is with an adjacent bathroom.

System Components	QUANTITY	UNIT	COST EACH		
			MAT.	INST.	TOTAL
SYSTEM D2010 932 1360					
BATHROOM, BATHTUB, WATER CLOSET, STALL SHOWER & TWO LAVATORIES					
TWO WALL PLUMBING, STAND ALONE					
Wtr closet, 2 pc close cpld vit china flr mntd incl seat,supply & stop	1.000	Ea.	261	241	502
Water closet, rough-in waste & vent	1.000	Set	365	420	785
Lavatory w/ftngs, wall hung, white PE on CI, 20" x 18"	2.000	Ea.	610	320	930
Lavatory, rough-in waste & vent	2.000	Set	1,000	1,540	2,540
Bathtub, white PE on CI, w/ftgs, mat bottom, recessed, 5' long	1.000	Ea.	1,250	290	1,540
Baths, rough-in waste and vent	1.000	Set	380	615	995
Shower stall, bkd enam molded stone receptor, door & ftng, 32" sq.	1.000	Ea.	1,275	255	1,530
Shower stall, rough-in supply, waste & vent	1.000	Set	375	625	1,000
TOTAL			5,516	4,306	9,822

D2010 932	Five Fixture Bathroom, Two Wall Plumbing	COST EACH		
		MAT.	INST.	TOTAL
1320	Bathroom, five fixture, two wall plumbing			
1340	Bathtub, water closet, stall shower & two lavatories	R224000 -30		
1360	Stand alone	5,525	4,300	9,825
1400	One short plumbing wall common *	5,125	3,700	8,825
1500	Bathtub, two lavatories, corner stall shower & water closet			
1520	Stand alone	6,150	4,325	10,475
1540	Long plumbing wall common*	5,525	3,350	8,875

D2010 934	Five Fixture Bathroom, Three Wall Plumbing	COST EACH		
		MAT.	INST.	TOTAL
2360	Bathroom, five fixture, three wall plumbing			
2380	Water closet, bathtub, two lavatories & stall shower			
2400	Stand alone	6,150	4,325	10,475
2440	One short plumbing wall common *	5,750	3,700	9,450

D2010 936	Five Fixture Bathroom, One Wall Plumbing	COST EACH		
		MAT.	INST.	TOTAL
4080	Bathroom, five fixture, one wall plumbing			
4100	Bathtub, two lavatories, corner stall shower & water closet			
4120	Stand alone	5,850	3,850	9,700
4160	Share common wall *	4,900	2,500	7,400

D20 Plumbing

D2010 Plumbing Fixtures

Example of Plumbing Cost Calculations: The bathroom system includes the individual fixtures such as bathtub, lavatory, shower and water closet. These fixtures are listed below as separate items merely as a checklist.

D2010 951	Plumbing Systems 20 Unit, 2 Story Apartment Building							
	FIXTURE	SYSTEM	LINE	QUANTITY	UNIT	COST EACH		
						MAT.	INST.	TOTAL
0440	Bathroom	D2010 926	3640	20	Ea.	61,500	50,500	112,000
0480	Bathtub							
0520	Booster pump[1]	not req'd.						
0560	Drinking fountain							
0600	Garbage disposal[1]	not incl.						
0660								
0680	Grease interceptor							
0720	Water heater	D2020 250	2140	1	Ea.	13,100	3,450	16,550
0760	Kitchen sink	D2010 410	1960	20	Ea.	27,200	17,800	45,000
0800	Laundry sink	D2010 420	1840	4	Ea.	4,900	3,525	8,425
0840	Lavatory							
0900								
0920	Roof drain, 1 floor	D2040 210	4200	2	Ea.	2,000	1,825	3,825
0960	Roof drain, add'l floor	D2040 210	4240	20	L.F.	410	460	870
1000	Service sink	D2010 440	4300	1	Ea.	2,125	1,225	3,350
1040	Sewage ejector[1]	not req'd.						
1080	Shower							
1100								
1160	Sump pump							
1200	Urinal							
1240	Water closet							
1320								
1360	**SUB TOTAL**					111,500	79,000	190,500
1481	Water controls	R221113-40		10%[2]		11,100	7,900	19,000
1521	Pipe & fittings[3]	R221113-40		30%[2]		33,400	23,700	57,100
1560	Other							
1601	Quality/complexity	R221113-40		15%[2]		16,700	11,900	28,600
1680								
1720	**TOTAL**					173,000	122,500	295,500
1741								

[1]**Note:** Cost for items such as booster pumps, backflow preventers, sewage ejectors, water meters, etc., may be obtained from the unit price section in the front of this book.

Water controls, pipe and fittings, and the Quality/Complexity factors come from Table R221113-40.

[2]Percentage of subtotal.

[3]Long, easily discernable runs of pipe would be more accurately priced from Unit Price Section 22 11 13. If this is done, reduce the miscellaneous percentage in proportion.

D2020 Domestic Water Distribution

Installation includes piping and fittings within 10' of heater. Electric water heaters do not require venting.

1 Kilowatt hour will raise:			
Gallons of Water	Degrees F	Gallons of Water	Degrees F
4.1	100°	6.8	60°
4.5	90°	8.2	50°
5.1	80°	10.0	40°
5.9	70°		

System Components			COST EACH		
	QUANTITY	UNIT	MAT.	INST.	TOTAL
SYSTEM D2020 210 1780					
ELECTRIC WATER HEATER, RESIDENTIAL, 100°F RISE					
10 GALLON TANK, 7 GPH					
Water heater, residential electric, glass lined tank, 10 gal.	1.000	Ea.	360	310	670
Copper tubing, type L, solder joint, hanger 10' OC 1/2" diam.	30.000	L.F.	122.10	262.50	384.60
Wrought copper 90° elbow for solder joints 1/2" diam.	4.000	Ea.	4.84	142	146.84
Wrought copper Tee for solder joints, 1/2" diam.	2.000	Ea.	4.12	109	113.12
Union, wrought copper, 1/2" diam.	2.000	Ea.	30.90	75	105.90
Valve, gate, bronze, 125 lb, NRS, soldered 1/2" diam.	2.000	Ea.	119	59	178
Relief valve, bronze, press & temp, self-close, 3/4" IPS	1.000	Ea.	278	25.50	303.50
Wrought copper adapter, CTS to MPT 3/4" IPS	1.000	Ea.	4.22	41.50	45.72
Copper tubing, type L, solder joints, 3/4" diam.	1.000	L.F.	5.70	9.35	15.05
Wrought copper 90° elbow for solder joints 3/4" diam.	1.000	Ea.	2.71	37.50	40.21
TOTAL			931.59	1,071.35	2,002.94

D2020 210	Electric Water Heaters - Residential Systems		COST EACH		
			MAT.	INST.	TOTAL
1760	Electric water heater, residential, 100°F rise				
1780	10 gallon tank, 7 GPH	R224000 -10	930	1,075	2,005
1820	20 gallon tank, 7 GPH		1,125	1,150	2,275
1860	30 gallon tank, 7 GPH	R224000 -20	1,400	1,200	2,600
1900	40 gallon tank, 8 GPH		1,925	1,325	3,250
1940	52 gallon tank, 10 GPH		2,050	1,325	3,375
1980	66 gallon tank, 13 GPH		3,200	1,525	4,725
2020	80 gallon tank, 16 GPH		3,375	1,600	4,975
2060	120 gallon tank, 23 GPH		4,675	1,850	6,525

D2020 Domestic Water Distribution

Installation includes piping and fittings within 10′ of heater. Gas heaters require vent piping (not included with these units).

System Components	QUANTITY	UNIT	COST EACH		
			MAT.	INST.	TOTAL
SYSTEM D2020 220 2260					
GAS FIRED WATER HEATER, RESIDENTIAL, 100°F RISE					
30 GALLON TANK, 32 GPH					
Water heater, residential, gas, glass lined tank, 30 gallon	1.000	Ea.	985	355	1,340
Copper tubing, type L, solder joint, hanger 10′ OC, 3/4″ diam	33.000	L.F.	188.10	308.55	496.65
Wrought copper 90° elbow for solder joints, 3/4″ diam	5.000	Ea.	13.55	187.50	201.05
Wrought copper Tee for solder joints, 3/4″ diam	2.000	Ea.	9.92	118	127.92
Wrought copper union for soldered joints, 3/4″ diam.	2.000	Ea.	38.70	79	117.70
Valve bronze, 125 lb., NRS, soldered 3/4″ diam	2.000	Ea.	140	71	211
Relief valve, press & temp, bronze, self-close, 3/4″ diam	1.000	Ea.	278	25.50	303.50
Wrought copper, adapter, CTS to MPT 3/4″ IPS	1.000	Ea.	4.22	41.50	45.72
Pipe steel black, schedule 40, threaded, 1/2″ diam	10.000	L.F.	36.60	112.50	149.10
Pipe, 90° elbow, malleable iron black, 150 lb., threaded, 1/2″ diam	2.000	Ea.	7.26	95	102.26
Pipe, union with brass seat, malleable iron black, 1/2″ diam	1.000	Ea.	16.10	50.50	66.60
Valve, gas stop w/o check, brass, 1/2″ IPS	1.000	Ea.	14.50	29.50	44
TOTAL			1,731.95	1,473.55	3,205.50

D2020 220	Gas Fired Water Heaters - Residential Systems	COST EACH		
		MAT.	INST.	TOTAL
2200	Gas fired water heater, residential, 100°F rise			
2260	30 gallon tank, 32 GPH R224000-10	1,725	1,475	3,200
2300	40 gallon tank, 32 GPH	1,925	1,650	3,575
2340	50 gallon tank, 63 GPH R224000-20	1,975	1,650	3,625
2380	75 gallon tank, 63 GPH	2,725	1,850	4,575
2420	100 gallon tank, 63 GPH	3,000	1,950	4,950

D2020 Domestic Water Distribution

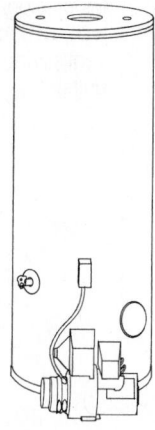

Installation includes piping and fittings within 10' of heater. Oil fired heaters require vent piping (not included in these prices).

System Components	QUANTITY	UNIT	COST EACH		
			MAT.	INST.	TOTAL
SYSTEM D2020 230 2220					
OIL FIRED WATER HEATER, RESIDENTIAL, 100°F RISE					
30 GALLON TANK, 103 GPH					
Water heater, residential, oil glass lined tank, 30 Gal	1.000	Ea.	1,275	355	1,630
Copper tubing, type L, solder joint, hanger 10' O.C. 3/4" diam.	33.000	L.F.	188.10	308.55	496.65
Wrought copper 90° elbow for solder joints 3/4" diam.	5.000	Ea.	13.55	187.50	201.05
Wrought copper Tee for solder joints, 3/4" diam.	2.000	Ea.	9.92	118	127.92
Wrought copper union for soldered joints, 3/4" diam.	2.000	Ea.	38.70	79	117.70
Valve, gate, bronze, 125 lb, NRS, soldered 3/4" diam.	2.000	Ea.	140	71	211
Relief valve, bronze, press & temp, self-close, 3/4" IPS	1.000	Ea.	278	25.50	303.50
Wrought copper adapter, CTS to MPT, 3/4" IPS	1.000	Ea.	4.22	41.50	45.72
Copper tubing, type L, solder joint, hanger 10' OC 3/8" diam.	10.000	L.F.	38.50	84.50	123
Wrought copper 90° elbow for solder joints 3/8" diam.	2.000	Ea.	6.84	64	70.84
Valve, globe, fusible, 3/8" diam.	1.000	Ea.	14.40	30	44.40
TOTAL			2,007.23	1,364.55	3,371.78

D2020 230	Oil Fired Water Heaters - Residential Systems		COST EACH		
			MAT.	INST.	TOTAL
2200	Oil fired water heater, residential, 100°F rise				
2220	30 gallon tank, 103 GPH		2,000	1,375	3,375
2260	50 gallon tank, 145 GPH	R224000 -20	2,450	1,525	3,975
2300	70 gallon tank, 164 GPH		3,350	1,725	5,075
2340	85 gallon tank, 181 GPH		10,900	1,775	12,675

Systems below include piping and fittings within 10′ of heater. Electric water heaters do not require venting.

System Components	QUANTITY	UNIT	COST EACH		
			MAT.	INST.	TOTAL
SYSTEM D2020 240 1820					
ELECTRIC WATER HEATER, COMMERCIAL, 100°F RISE					
50 GALLON TANK, 9 KW, 37 GPH					
Water heater, commercial, electric, 50 Gal, 9 KW, 37 GPH	1.000	Ea.	4,775	395	5,170
Copper tubing, type L, solder joint, hanger 10′ OC, 3/4″ diam	34.000	L.F.	193.80	317.90	511.70
Wrought copper 90° elbow for solder joints 3/4″ diam	5.000	Ea.	13.55	187.50	201.05
Wrought copper Tee for solder joints, 3/4″ diam	2.000	Ea.	9.92	118	127.92
Wrought copper union for soldered joints, 3/4″ diam.	2.000	Ea.	38.70	79	117.70
Valve, gate, bronze, 125 lb, NRS, soldered 3/4″ diam	2.000	Ea.	140	71	211
Relief valve, bronze, press & temp, self-close, 3/4″ IPS	1.000	Ea.	278	25.50	303.50
Wrought copper adapter, copper tubing to male, 3/4″ IPS	1.000	Ea.	4.22	41.50	45.72
TOTAL			5,453.19	1,235.40	6,688.59

D2020 240	Electric Water Heaters - Commercial Systems		COST EACH		
			MAT.	INST.	TOTAL
1800	Electric water heater, commercial, 100°F rise				
1820	50 gallon tank, 9 KW 37 GPH		5,450	1,225	6,675
1860	80 gal, 12 KW 49 GPH	R224000 -10	7,025	1,525	8,550
1900	36 KW 147 GPH		9,475	1,650	11,125
1940	120 gal, 36 KW 147 GPH	R224000 -20	10,200	1,775	11,975
1980	150 gal, 120 KW 490 GPH		30,500	1,900	32,400
2020	200 gal, 120 KW 490 GPH		32,100	1,950	34,050
2060	250 gal, 150 KW 615 GPH		35,900	2,275	38,175
2100	300 gal, 180 KW 738 GPH		47,900	2,400	50,300
2140	350 gal, 30 KW 123 GPH		29,000	2,575	31,575
2180	180 KW 738 GPH		40,100	2,575	42,675
2220	500 gal, 30 KW 123 GPH		37,700	3,050	40,750
2260	240 KW 984 GPH		54,500	3,050	57,550
2300	700 gal, 30 KW 123 GPH		31,200	3,475	34,675
2340	300 KW 1230 GPH		45,600	3,475	49,075
2380	1000 gal, 60 KW 245 GPH		38,400	4,825	43,225
2420	480 KW 1970 GPH		61,000	4,850	65,850
2460	1500 gal, 60 KW 245 GPH		80,500	6,000	86,500
2500	480 KW 1970 GPH		110,000	6,000	116,000

471

D2020 Domestic Water Distribution

Units may be installed in multiples for increased capacity.

Included below is the heater with self-energizing gas controls, safety pilots, insulated jacket, hi-limit aquastat and pressure relief valve.

Installation includes piping and fittings within 10′ of heater. Gas heaters require vent piping (not included in these prices).

System Components	QUANTITY	UNIT	COST EACH		
			MAT.	INST.	TOTAL
SYSTEM D2020 250 1780					
GAS FIRED WATER HEATER, COMMERCIAL, 100°F RISE					
75.5 MBH INPUT, 63 GPH					
Water heater, commercial, gas, 75.5 MBH, 63 GPH	1.000	Ea.	3,850	505	4,355
Copper tubing, type L, solder joint, hanger 10′ OC, 1-1/4″ diam	30.000	L.F.	340.50	367.50	708
Wrought copper 90° elbow for solder joints 1-1/4″ diam	4.000	Ea.	39.60	190	229.60
Wrought copper tee for solder joints, 1-1/4″ diam	2.000	Ea.	41	158	199
Wrought copper union for soldered joints, 1-1/4″ diam	2.000	Ea.	109	101	210
Valve, gate, bronze, 125 lb, NRS, soldered 1-1/4″ diam	2.000	Ea.	278	95	373
Relief valve, bronze, press & temp, self-close, 3/4″ IPS	1.000	Ea.	278	25.50	303.50
Copper tubing, type L, solder joints, 3/4″ diam	8.000	L.F.	45.60	74.80	120.40
Wrought copper 90° elbow for solder joints 3/4″ diam	1.000	Ea.	2.71	37.50	40.21
Wrought copper, adapter, CTS to MPT, 3/4″ IPS	1.000	Ea.	4.22	41.50	45.72
Pipe steel black, schedule 40, threaded, 3/4″ diam	10.000	L.F.	43	116.50	159.50
Pipe, 90° elbow, malleable iron black, 150 lb threaded, 3/4″ diam	2.000	Ea.	8.78	101	109.78
Pipe, union with brass seat, malleable iron black, 3/4″ diam	1.000	Ea.	18.50	54.50	73
Valve, gas stop w/o check, brass, 3/4″ IPS	1.000	Ea.	17.05	32	49.05
TOTAL			5,075.96	1,899.80	6,975.76

D2020 250	Gas Fired Water Heaters - Commercial Systems		COST EACH		
			MAT.	INST.	TOTAL
1760	Gas fired water heater, commercial, 100°F rise				
1780	75.5 MBH input, 63 GPH		5,075	1,900	6,975
1820	95 MBH input, 86 GPH		7,050	1,900	8,950
1860	100 MBH input, 91 GPH	R224000 -10	7,250	1,975	9,225
1900	115 MBH input, 110 GPH		8,525	2,050	10,575
1980	155 MBH input, 150 GPH		10,300	2,300	12,600
2020	175 MBH input, 168 GPH	R224000 -20	11,000	2,475	13,475
2060	200 MBH input, 192 GPH		11,500	2,775	14,275
2100	240 MBH input, 230 GPH		12,000	3,025	15,025
2140	300 MBH input, 278 GPH		13,100	3,450	16,550
2180	390 MBH input, 374 GPH		15,700	3,500	19,200
2220	500 MBH input, 480 GPH		21,000	3,750	24,750
2260	600 MBH input, 576 GPH		24,100	4,050	28,150

472

For customer support on your Plumbing Cost Data, call 877.709.6968.

D2020 Domestic Water Distribution

Units may be installed in multiples for increased capacity.

Included below is the heater, wired-in flame retention burners, cadmium cell primary controls, hi-limit controls, ASME pressure relief valves, draft controls, and insulated jacket.

Oil fired water heater systems include piping and fittings within 10' of heater. Oil fired heaters require vent piping (not included in these systems).

System Components	QUANTITY	UNIT	COST EACH		
			MAT.	INST.	TOTAL
SYSTEM D2020 260 1820					
OIL FIRED WATER HEATER, COMMERCIAL, 100°F RISE					
140 GAL., 140 MBH INPUT, 134 GPH					
Water heater, commercial, oil, 140 gal., 140 MBH input, 134 GPH	1.000	Ea.	21,900	600	22,500
Copper tubing, type L, solder joint, hanger 10' OC, 3/4" diam.	34.000	L.F.	193.80	317.90	511.70
Wrought copper 90° elbow for solder joints 3/4" diam.	5.000	Ea.	13.55	187.50	201.05
Wrought copper Tee for solder joints, 3/4" diam.	2.000	Ea.	9.92	118	127.92
Wrought copper union for soldered joints, 3/4" diam.	2.000	Ea.	38.70	79	117.70
Valve, bronze, 125 lb, NRS, soldered 3/4" diam.	2.000	Ea.	140	71	211
Relief valve, bronze, press & temp, self-close, 3/4" IPS	1.000	Ea.	278	25.50	303.50
Wrought copper adapter, copper tubing to male, 3/4" IPS	1.000	Ea.	4.22	41.50	45.72
Copper tubing, type L, solder joint, hanger 10' OC, 3/8" diam.	10.000	L.F.	38.50	84.50	123
Wrought copper 90° elbow for solder joints 3/8" diam.	2.000	Ea.	6.84	64	70.84
Valve, globe, fusible, 3/8" IPS	1.000	Ea.	14.40	30	44.40
TOTAL			22,637.93	1,618.90	24,256.83

D2020 260	Oil Fired Water Heaters - Commercial Systems		COST EACH		
			MAT.	INST.	TOTAL
1800	Oil fired water heater, commercial, 100°F rise				
1820	140 gal., 140 MBH input, 134 GPH		22,600	1,625	24,225
1900	140 gal., 255 MBH input, 247 GPH	R224000 -10	24,500	2,025	26,525
1940	140 gal., 270 MBH input, 259 GPH		30,000	2,325	32,325
1980	140 gal., 400 MBH input, 384 GPH	R224000 -20	31,400	2,675	34,075
2060	140 gal., 720 MBH input, 691 GPH		33,100	2,775	35,875
2100	221 gal., 300 MBH input, 288 GPH		43,700	3,050	46,750
2140	221 gal., 600 MBH input, 576 GPH		48,500	3,075	51,575
2180	221 gal., 800 MBH input, 768 GPH		49,400	3,200	52,600
2220	201 gal., 1000 MBH input, 960 GPH		50,500	3,225	53,725
2260	201 gal., 1250 MBH input, 1200 GPH		51,500	3,325	54,825
2300	201 gal., 1500 MBH input, 1441 GPH		55,500	3,400	58,900
2340	411 gal., 600 MBH input, 576 GPH		56,000	3,475	59,475
2380	411 gal., 800 MBH input, 768 GPH		56,500	3,575	60,075
2420	411 gal., 1000 MBH input, 960 GPH		61,000	4,100	65,100
2460	411 gal., 1250 MBH input, 1200 GPH		62,500	4,225	66,725
2500	397 gal., 1500 MBH input, 1441 GPH		66,000	4,350	70,350

473

For customer support on your Plumbing Cost Data, call 877.709.6968.

In this closed-loop indirect collection system, fluid with a low freezing temperature, such as propylene glycol, transports heat from the collectors to water storage. The transfer fluid is contained in a closed-loop consisting of collectors, supply and return piping, and a remote heat exchanger. The heat exchanger transfers heat energy from the

fluid in the collector loop to potable water circulated in a storage loop. A typical two-or-three panel system contains 5 to 6 gallons of heat transfer fluid.

When the collectors become approximately 20°F warmer than the storage temperature, a controller activates the circulator on the collector and storage

loops. The circulators will move the fluid and potable water through the heat exchanger until heat collection no longer occurs. At that point, the system shuts down. Since the heat transfer medium is a fluid with a very low freezing temperature, there is no need for it to be drained from the system between periods of collection.

D2020 Domestic Water Distribution

System Components	QUANTITY	UNIT	COST EACH		
			MAT.	INST.	TOTAL
SYSTEM D2020 265 2760					
SOLAR, CLOSED LOOP, ADD-ON HOT WATER SYS., EXTERNAL HEAT EXCHANGER					
3/4″ TUBING, TWO 3′X7′ BLACK CHROME COLLECTORS					
A,B,G,L,K,M Heat exchanger fluid-fluid pkg incl 2 circulators, expansion tank,					
Check valve, relief valve, controller, hi temp cutoff, & 2 sensors	1.000	Ea.	880	510	1,390
C Thermometer, 2″ dial	3.000	Ea.	72	133.50	205.50
D, T Fill & drain valve, brass, 3/4″ connection	1.000	Ea.	10.70	29.50	40.20
E Air vent, manual, 1/8″ fitting	2.000	Ea.	5.36	44	49.36
F Air purger	1.000	Ea.	51.50	59	110.50
H Strainer, Y type, bronze body, 3/4″ IPS	1.000	Ea.	37.50	38	75.50
I Valve, gate, bronze, NRS, soldered 3/4″ diam	6.000	Ea.	420	213	633
J Neoprene vent flashing	2.000	Ea.	19	71	90
N-1, N Relief valve temp & press, 150 psi 210°F self-closing 3/4″ IPS	1.000	Ea.	23	23.50	46.50
O Pipe covering, urethane, ultraviolet cover, 1″ wall 3/4″ diam	20.000	L.F.	62.20	128	190.20
P Pipe covering, fiberglass, all service jacket, 1″ wall, 3/4″ diam	50.000	L.F.	49.50	255	304.50
Q Collector panel solar energy blk chrome on copper, 1/8″ temp glass 3′x7′	2.000	Ea.	2,200	268	2,468
Roof clamps for solar energy collector panels	2.000	Set	6.16	36.50	42.66
R Valve, swing check, bronze, regrinding disc, 3/4″ diam	2.000	Ea.	220	71	291
S Pressure gauge, 60 psi, 2″ dial	1.000	Ea.	26.50	22	48.50
U Valve, water tempering, bronze, sweat connections, 3/4″ diam	1.000	Ea.	139	35.50	174.50
W-2, V Tank water storage w/heating element, drain, relief valve, existing	1.000	Ea.			
Copper tubing type L, solder joint, hanger 10′ OC 3/4″ diam	20.000	L.F.	114	187	301
Copper tubing, type M, solder joint, hanger 10′ OC 3/4″ diam	70.000	L.F.	317.80	637	954.80
Sensor wire, #22-2 conductor multistranded	.500	C.L.F.	7.83	32.75	40.58
Solar energy heat transfer fluid, propylene glycol anti-freeze	6.000	Gal.	102.60	153	255.60
Wrought copper fittings & solder, 3/4″ diam	76.000	Ea.	205.96	2,850	3,055.96
TOTAL			4,970.61	5,797.25	10,767.86

D2020 265	Solar, Closed Loop, Add-On Hot Water Systems		COST EACH		
			MAT.	INST.	TOTAL
2550	Solar, closed loop, add-on hot water system, external heat exchanger				
2570	3/8″ tubing, 3 ea. 4′ x 4′-4″ vacuum tube collectors		5,525	5,400	10,925
2580	1/2″ tubing, 4 ea. 4 x 4′-4″ vacuum tube collectors	R235616 -60	6,325	5,850	12,175
2600	2 ea. 3′x7′ black chrome collectors		4,575	5,525	10,100
2620	3 ea. 3′x7′ black chrome collectors		5,675	5,675	11,350
2640	2 ea. 3′x7′ flat black collectors		4,300	5,525	9,825
2660	3 ea. 3′x7′ flat black collectors		5,275	5,700	10,975
2700	3/4″ tubing, 3 ea. 3′x7′ black chrome collectors		6,075	5,950	12,025
2720	3 ea. 3′x7′ flat black absorber plate collectors		5,675	5,975	11,650
2740	2 ea. 4′x9′ flat black w/plastic glazing collectors		5,075	6,000	11,075
2760	2 ea. 3′x7′ black chrome collectors		4,975	5,800	10,775
2780	1″ tubing, 4 ea 2′x9′ plastic absorber & glazing collectors		6,100	6,750	12,850
2800	4 ea. 3′x7′ black chrome absorber collectors		7,925	6,775	14,700
2820	4 ea. 3′x7′ flat black absorber collectors		7,400	6,800	14,200

For customer support on your Plumbing Cost Data, call 877.709.6968.

475

In the drainback indirect-collection system, the heat transfer fluid is distilled water contained in a loop consisting of collectors, supply and return piping, and an unpressurized holding tank. A large heat exchanger containing incoming potable water is immersed in the holding tank. When a controller activates solar collection, the distilled water is pumped through the collectors and heated and pumped back down to the holding tank. When the temperature differential between the water in the collectors and water in storage is such that collection no longer occurs, the pump turns off and gravity causes the distilled water in the collector loop to drain back to the holding tank. All the loop piping is pitched so that the water can drain out of the collectors and piping and not freeze there. As hot water is needed in the home, incoming water first flows through the holding tank with the immersed heat exchanger and is warmed and then flows through a conventional heater for any supplemental heating that is necessary.

D2020 Domestic Water Distribution

System Components	QUANTITY	UNIT	COST EACH		
			MAT.	INST.	TOTAL
SYSTEM D2020 270 2760					
SOLAR, DRAINBACK, ADD ON, HOT WATER, IMMERSED HEAT EXCHANGER					
3/4″ TUBING, THREE EA 3′X7′ BLACK CHROME COLLECTOR					
A, B Differential controller 2 sensors, thermostat, solar energy system	1.000	Ea.	143	59	202
C Thermometer 2″ dial	3.000	Ea.	72	133.50	205.50
D, T Fill & drain valve, brass, 3/4″ connection	1.000	Ea.	10.70	29.50	40.20
E-1 Automatic air vent 1/8″ fitting	1.000	Ea.	18.95	22	40.95
H Strainer, Y type, bronze body, 3/4″ IPS	1.000	Ea.	37.50	38	75.50
I Valve, gate, bronze, NRS, soldered 3/4″ diam	2.000	Ea.	140	71	211
J Neoprene vent flashing	2.000	Ea.	19	71	90
L Circulator, solar heated liquid, 1/20 HP	1.000	Ea.	177	106	283
N Relief valve temp. & press. 150 psi 210°F self-closing 3/4″ IPS	1.000	Ea.	23	23.50	46.50
O Pipe covering, urethane, ultraviolet cover, 1″ wall, 3/4″ diam	20.000	L.F.	62.20	128	190.20
P Pipe covering, fiberglass, all service jacket, 1″ wall, 3/4″ diam	50.000	L.F.	49.50	255	304.50
Q Collector panel solar energy blk chrome on copper, 1/8″ temp glas 3′x7′	3.000	Ea.	3,300	402	3,702
Roof clamps for solar energy collector panels	3.000	Set	9.24	54.75	63.99
R Valve, swing check, bronze, regrinding disc, 3/4″ diam	1.000	Ea.	110	35.50	145.50
U Valve, water tempering, bronze sweat connections, 3/4″ diam	1.000	Ea.	139	35.50	174.50
V Tank, water storage w/heating element, drain, relief valve, existing	1.000	Ea.			
W Tank, water storage immersed heat exchr elec 2″x1/2# insul 120 gal	1.000	Ea.	2,025	505	2,530
X Valve, globe, bronze, rising stem, 3/4″ diam, soldered	3.000	Ea.	327	106.50	433.50
Y Flow control valve	1.000	Ea.	132	32	164
Z Valve, ball, bronze, solder 3/4″ diam, solar loop flow control	1.000	Ea.	25.50	35.50	61
Copper tubing, type L, solder joint, hanger 10′ OC 3/4″ diam	20.000	L.F.	114	187	301
Copper tubing, type M, solder joint, hanger 10′ OC 3/4″ diam	70.000	L.F.	317.80	637	954.80
Sensor wire, #22-2 conductor, multistranded	.500	C.L.F.	7.83	32.75	40.58
Wrought copper fittings & solder, 3/4″ diam	76.000	Ea.	205.96	2,850	3,055.96
TOTAL			**7,466.18**	**5,850**	**13,316.18**

D2020 270	Solar, Drainback, Hot Water Systems		COST EACH		
			MAT.	INST.	TOTAL
2550	Solar, drainback, hot water, immersed heat exchanger				
2560	3/8″ tubing, 3 ea. 4′ x 4′-4″ vacuum tube collectors		6,700	5,250	11,950
2580	1/2″ tubing, 4 ea 4′ x 4′-4″ vacuum tube collectors, 80 gal tank	R235616	7,500	5,700	13,200
2600	120 gal tank	-60	7,750	5,775	13,525
2640	2 ea. 3′x7′ blk chrome collectors, 80 gal tank		5,750	5,375	11,125
2660	3 ea. 3′x7′ blk chrome collectors, 120 gal tank		7,100	5,575	12,675
2700	2 ea. 3′x7′ flat blk collectors, 120 gal tank		5,725	5,450	11,175
2720	3 ea. 3′x7′ flat blk collectors, 120 gal tank		6,700	5,600	12,300
2760	3/4″ tubing, 3 ea 3′x7′ black chrome collectors, 120 gal tank		7,475	5,850	13,325
2780	3 ea. 3′x7′ flat black absorber collectors, 120 gal tank		7,050	5,875	12,925
2800	2 ea. 4′x9′ flat blk w/plastic glazing collectors 120 gal tank		6,475	5,900	12,375
2840	1″ tubing, 4 ea. 2′x9′ plastic absorber & glazing collectors, 120 gal tank		7,575	6,650	14,225
2860	4 ea. 3′x7′ black chrome absorber collectors, 120 gal tank		9,425	6,675	16,100
2880	4 ea. 3′x7′ flat black absorber collectors, 120 gal tank		8,875	6,700	15,575

477

For customer support on your Plumbing Cost Data, call 877.709.6968.

In the draindown direct-collection system, incoming domestic water is heated in the collectors. When the controller activates solar collection, domestic water is first heated as it flows through the collectors and is then pumped to storage. When conditions are no longer suitable for heat collection, the pump shuts off and the water in the loop drains down and out of the system by means of solenoid valves and properly pitched piping.

478

For customer support on your Plumbing Cost Data, call 877.709.6968.

D2020 Domestic Water Distribution

System Components	QUANTITY	UNIT	COST EACH		
			MAT.	INST.	TOTAL
SYSTEM D2020 275 2760					
SOLAR, DRAINDOWN, HOT WATER, DIRECT COLLECTION					
3/4" TUBING, THREE 3'X7' BLACK CHROME COLLECTORS					
A, B Differential controller, 2 sensors, thermostat, solar energy system	1.000	Ea.	143	59	202
A-1 Solenoid valve, solar heating loop, brass, 3/4" diam, 24 volts	3.000	Ea.	405	237	642
B-1 Solar energy sensor, freeze prevention	1.000	Ea.	25	22	47
C Thermometer, 2" dial	3.000	Ea.	72	133.50	205.50
E-1 Vacuum relief valve, 3/4" diam	1.000	Ea.	32.50	22	54.50
F-1 Air vent, automatic, 1/8" fitting	1.000	Ea.	18.95	22	40.95
H Strainer, Y type, bronze body, 3/4" IPS	1.000	Ea.	37.50	38	75.50
I Valve, gate, bronze, NRS, soldered, 3/4" diam	2.000	Ea.	140	71	211
J Vent flashing neoprene	2.000	Ea.	19	71	90
K Circulator, solar heated liquid, 1/25 HP	1.000	Ea.	126	91	217
N Relief valve temp & press 150 psi 210°F self-closing 3/4" IPS	1.000	Ea.	23	23.50	46.50
O Pipe covering, urethane, ultraviolet cover, 1" wall, 3/4" diam	20.000	L.F.	62.20	128	190.20
P Pipe covering, fiberglass, all service jacket, 1" wall, 3/4" diam	50.000	L.F.	49.50	255	304.50
Roof clamps for solar energy collector panels	3.000	Set	9.24	54.75	63.99
Q Collector panel solar energy blk chrome on copper, 1/8" temp glass 3'x7'	3.000	Ea.	3,300	402	3,702
R Valve, swing check, bronze, regrinding disc, 3/4" diam, soldered	2.000	Ea.	220	71	291
T Drain valve, brass, 3/4" connection	2.000	Ea.	21.40	59	80.40
U Valve, water tempering, bronze, sweat connections, 3/4" diam	1.000	Ea.	139	35.50	174.50
W-2, W Tank, water storage elec elem 2"x1/2# insul 120 gal	1.000	Ea.	2,025	505	2,530
X Valve, globe, bronze, rising stem, 3/4" diam, soldered	1.000	Ea.	109	35.50	144.50
Copper tubing, type L, solder joints, hangers 10' OC 3/4" diam	20.000	L.F.	114	187	301
Copper tubing, type M, solder joints, hangers 10' OC 3/4" diam	70.000	L.F.	317.80	637	954.80
Sensor wire, #22-2 conductor, multistranded	.500	C.L.F.	7.83	32.75	40.58
Wrought copper fittings & solder, 3/4" diam	76.000	Ea.	205.96	2,850	3,055.96
TOTAL			7,622.88	6,042.50	13,665.38

D2020 275	Solar, Draindown, Hot Water Systems		COST EACH		
			MAT.	INST.	TOTAL
2550	Solar, draindown, hot water				
2560	3/8" tubing, 3 ea. 4' x 4'-4" vacuum tube collectors, 80 gal tank		6,675	5,750	12,425
2580	1/2" tubing, 4 ea. 4' x 4'-4" vacuum tube collectors, 80 gal tank	R235616	7,650	5,900	13,550
2600	120 gal tank	-60	7,900	5,975	13,875
2640	2 ea. 3'x7' black chrome collectors, 80 gal tank		5,900	5,575	11,475
2660	3 ea. 3'x7' black chrome collectors, 120 gal tank		7,250	5,775	13,025
2700	2 ea. 3'x7' flat black collectors, 120 gal tank		5,875	5,650	11,525
2720	3 ea. 3'x7' flat black collectors, 120 gal tank		6,775	5,775	12,550
2760	3/4" tubing, 3 ea. 3'x7' black chrome collectors, 120 gal tank		7,625	6,050	13,675
2780	3 ea. 3'x7' flat collectors, 120 gal tank		7,225	6,075	13,300
2800	2 ea. 4'x9' flat black & plastic glazing collectors, 120 gal tank		6,625	6,075	12,700
2840	1" tubing, 4 ea. 2'x9' plastic absorber & glazing collectors, 120 gal tank		11,300	8,000	19,300
2860	4 ea. 3'x7' black chrome absorber collectors, 120 gal tank		10,800	7,575	18,375
2880	4 ea. 3'x7' flat black absorber collectors, 120 gal tank		9,175	6,875	16,050

In the recirculation system, (a direct-collection system), incoming domestic water is heated in the collectors. When the controller activates solar collection, domestic water is heated as it flows through the collectors and then it flows back to storage. When conditions are not suitable for heat collection, the pump shuts off and the flow of the water stops. In this type of system, water remains in the collector loop at all times. A "frost sensor" at the collector activates the circulation of warm water from storage through the collectors when protection from freezing is required.

480

For customer support on your Plumbing Cost Data, call 877.709.6968.

D2020 Domestic Water Distribution

System Components	QUANTITY	UNIT	COST EACH		
			MAT.	INST.	TOTAL
SYSTEM D2020 280 2820					
SOLAR, RECIRCULATION, HOT WATER					
3/4" TUBING, TWO 3'X7' BLACK CHROME COLLECTORS					
A, B Differential controller 2 sensors, thermostat, for solar energy system	1.000	Ea.	143	59	202
A-1 Solenoid valve, solar heating loop, brass, 3/4" IPS, 24 volts	2.000	Ea.	270	158	428
B-1 Solar energy sensor freeze prevention	1.000	Ea.	25	22	47
C Thermometer, 2" dial	3.000	Ea.	72	133.50	205.50
D Drain valve, brass, 3/4" connection	1.000	Ea.	10.70	29.50	40.20
F-1 Air vent, automatic, 1/8" fitting	1.000	Ea.	18.95	22	40.95
H Strainer, Y type, bronze body, 3/4" IPS	1.000	Ea.	37.50	38	75.50
I Valve, gate, bronze, 125 lb, soldered 3/4" diam	3.000	Ea.	210	106.50	316.50
J Vent flashing, neoprene	2.000	Ea.	19	71	90
L Circulator, solar heated liquid, 1/20 HP	1.000	Ea.	177	106	283
N Relief valve, temp & press 150 psi 210° F self-closing 3/4" IPS	2.000	Ea.	46	47	93
O Pipe covering, urethane, ultraviolet cover, 1" wall, 3/4" diam	20.000	L.F.	62.20	128	190.20
P Pipe covering, fiberglass, all service jacket, 1" wall 3/4" diam	50.000	L.F.	49.50	255	304.50
Q Collector panel solar energy blk chrome on copper, 1/8" temp glass 3'x7'	2.000	Ea.	2,200	268	2,468
Roof clamps for solar energy collector panels	2.000	Set	6.16	36.50	42.66
R Valve, swing check, bronze, 125 lb, regrinding disc, soldered 3/4" diam	2.000	Ea.	220	71	291
U Valve, water tempering, bronze, sweat connections, 3/4" diam	1.000	Ea.	139	35.50	174.50
V Tank, water storage, w/heating element, drain, relief valve, existing	1.000	Ea.			
X Valve, globe, bronze, 125 lb, 3/4" diam	2.000	Ea.	218	71	289
Copper tubing, type L, solder joints, hangers 10' OC 3/4" diam	20.000	L.F.	114	187	301
Copper tubing, type M, solder joints, hangers 10' OC 3/4" diam	70.000	L.F.	317.80	637	954.80
Wrought copper fittings & solder, 3/4" diam	76.000	Ea.	205.96	2,850	3,055.96
Sensor wire, #22-2 conductor, multistranded	.500	C.L.F.	7.83	32.75	40.58
TOTAL			4,569.60	5,364.25	9,933.85

D2020 280	Solar Recirculation, Domestic Hot Water Systems		COST EACH		
			MAT.	INST.	TOTAL
2550	Solar recirculation, hot water				
2560	3/8" tubing, 3 ea. 4' x 4'-4" vacuum tube collectors		5,125	4,975	10,100
2580	1/2" tubing, 4 ea. 4' x 4'-4" vacuum tube collectors		5,925	5,425	11,350
2640	2 ea. 3'x7' black chrome collectors	R235616 -60	4,200	5,100	9,300
2660	3 ea. 3'x7' black chrome collectors		5,300	5,250	10,550
2700	2 ea. 3'x7' flat black collectors		3,925	5,100	9,025
2720	3 ea. 3'x7' flat black collectors		4,900	5,275	10,175
2760	3/4" tubing, 3 ea. 3'x7' black chrome collectors		5,675	5,525	11,200
2780	3 ea. 3'x7' flat black absorber plate collectors		5,275	5,550	10,825
2800	2 ea. 4'x9' flat black w/plastic glazing collectors		4,675	5,550	10,225
2820	2 ea. 3'x7' black chrome collectors		4,575	5,375	9,950
2840	1" tubing, 4 ea. 2'x9' black plastic absorber & glazing collectors		5,925	6,300	12,225
2860	4 ea. 3'x7' black chrome absorber collectors		7,775	6,325	14,100
2880	4 ea. 3'x7' flat black absorber collectors		7,225	6,350	13,575

481

The thermosyphon domestic hot water system, a direct collection system, operates under city water pressure and does not require pumps for system operation. An insulated water storage tank is located above the collectors. As the sun heats the collectors, warm water in them rises by means of natural convection; the colder water in the storage tank flows into the collectors by means of gravity. As long as the sun is shining the water continues to flow through the collectors and to become warmer.

To prevent freezing, the system must be drained or the collectors covered with an insulated lid when the temperature drops below 32°F.

482

For customer support on your Plumbing Cost Data, call 877.709.6968.

D2020 Domestic Water Distribution

System Components	QUANTITY	UNIT	COST EACH		
			MAT.	INST.	TOTAL
SYSTEM D2020 285 0960					
SOLAR, THERMOSYPHON, WATER HEATER					
3/4" TUBING, TWO 3'X7' BLACK CHROME COLLECTORS					
D-1 Framing lumber, fir, 2" x 6" x 8', tank cradle	.008	M.B.F.	5.88	7.40	13.28
F-1 Framing lumber, fir, 2" x 4" x 24', sleepers	.016	M.B.F.	11.20	22.40	33.60
I Valve, gate, bronze, 125 lb, soldered 1/2" diam	2.000	Ea.	119	59	178
J Vent flashing, neoprene	4.000	Ea.	38	142	180
O Pipe covering, urethane, ultraviolet cover, 1" wall, 1/2"diam	40.000	L.F.	95.20	250	345.20
P Pipe covering fiberglass all service jacket 1" wall 1/2" diam	160.000	L.F.	145.60	782.40	928
Q Collector panel solar, blk chrome on copper, 3/16" temp glass 3'-6" x 7.5'	2.000	Ea.	1,930	284	2,214
Y Flow control valve, globe, bronze, 125#, soldered, 1/2" diam	1.000	Ea.	105	29.50	134.50
U Valve, water tempering, bronze,sweat connections, 1/2" diam	1.000	Ea.	108	29.50	137.50
W Tank, water storage, solar energy system, 80 Gal, 2" x 1/2 lb insul	1.000	Ea.	1,775	445	2,220
Copper tubing type L, solder joints, hangers 10' OC 1/2" diam	150.000	L.F.	610.50	1,312.50	1,923
Copper tubing type M, solder joints, hangers 10' OC 1/2" diam	50.000	L.F.	168	422.50	590.50
Sensor wire, #22-2 gauge multistranded	.500	C.L.F.	7.83	32.75	40.58
Wrought copper fittings & solder, 1/2" diam	75.000	Ea.	90.75	2,662.50	2,753.25
TOTAL			5,209.96	6,481.45	11,691.41

D2020 285	Thermosyphon, Hot Water		COST EACH		
			MAT.	INST.	TOTAL
0960 0970	Solar, thermosyphon, hot water, two collector system	R235616 -60	5,200	6,475	11,675

483

For customer support on your Plumbing Cost Data, call 877.709.6968.

ATTIC FLOOR

CWS

HWS

This domestic hot water pre-heat system includes heat exchanger with a circulating pump, blower, air-to-water, coil and controls, mounted in the upper collector manifold. Heat from the hot air coming out of the collectors is transferred through the heat exchanger. For each degree of DHW preheating gained, one degree less heating is needed from the fuel fired water heater. The system is simple, inexpensive to operate and can provide a substantial portion of DHW requirements for modest additional cost.

484

For customer support on your Plumbing Cost Data, call 877.709.6968.

D2020 Domestic Water Distribution

System Components	QUANTITY	UNIT	COST EACH		
			MAT.	INST.	TOTAL
SYSTEM D2020 290 2560					
SOLAR, HOT WATER, AIR TO WATER HEAT EXCHANGE					
THREE COLLECTORS, OPTICAL BLACK ON ALUM., 7.5' x 3.5',80 GAL TANK					
A, B Differential controller, 2 sensors, thermostat, solar energy system	1.000	Ea.	143	59	202
C Thermometer, 2" dial	2.000	Ea.	48	89	137
C-1 Heat exchanger, air to fluid, up flow 70 MBH	1.000	Ea.	395	365	760
E Air vent, manual, for solar energy system 1/8" fitting	1.000	Ea.	2.68	22	24.68
F Air purger	1.000	Ea.	51.50	59	110.50
G Expansion tank	1.000	Ea.	75	22	97
H Strainer, Y type, bronze body, 1/2" IPS	1.000	Ea.	30.50	36	66.50
I Valve, gate, bronze, 125 lb, NRS, soldered 1/2" diam	5.000	Ea.	297.50	147.50	445
N Relief valve, temp & pressure solar 150 psi 210°F self-closing	1.000	Ea.	23	23.50	46.50
P Pipe covering, fiberglass, all service jacket, 1" wall, 1/2" diam	60.000	L.F.	54.60	293.40	348
Q Collector panel solar energy, air, black on alum plate, 7.5' x 3.5'	3.000	Ea.	2,895	426	3,321
B-1, R-1 Shutter damper for solar heater circulator	1.000	Ea.	61.50	103	164.50
B-1, R-1 Shutter motor for solar heater circulator	1.000	Ea.	153	77	230
R Backflow preventer, 1/2" pipe size	2.000	Ea.	166	89	255
T Drain valve, brass, 3/4" connection	1.000	Ea.	10.70	29.50	40.20
U Valve, water tempering, bronze, sweat connections, 1/2" diam	1.000	Ea.	108	29.50	137.50
W Tank, water storage, solar energy system, 80 Gal, 2" x 2 lb insul	1.000	Ea.	1,775	445	2,220
V Tank, water storage, w/heating element, drain, relief valve, existing	1.000	System			
X Valve, globe, bronze, 125 lb, rising stem, 1/2" diam	1.000	Ea.	105	29.50	134.50
Copper tubing type M, solder joints, hangers 10' OC, 1/2" diam	50.000	L.F.	168	422.50	590.50
Copper tubing type L, solder joints, hangers 10' OC 1/2" diam	10.000	L.F.	40.70	87.50	128.20
Wrought copper fittings & solder, 1/2" diam	10.000	Ea.	12.10	355	367.10
Sensor wire, #22-2 conductor multistranded	.500	C.L.F.	7.83	32.75	40.58
Q-1, Q-2 Ductwork, fiberglass, aluminized jacket, 1-1/2" thick, 8" diam	32.000	S.F.	180.80	219.20	400
Q-3 Manifold for flush mount solar energy collector panels, air	6.000	L.F.	954	46.20	1,000.20
TOTAL			7,758.41	3,508.05	11,266.46

D2020 290	Solar Hot Water, Air To Water Heat Exchange		COST EACH		
			MAT.	INST.	TOTAL
2550	Solar hot water, air to water heat exchange				
2560	Three collectors, optical black on aluminum, 7.5' x 3.5', 80 Gal tank		7,750	3,500	11,250
2580	Four collectors, optical black on aluminum, 7.5' x 3.5', 80 Gal tank	R235616 -60	9,100	3,725	12,825
2600	Four collectors, optical black on aluminum, 7.5' x 3.5', 120 Gal tank		9,350	3,775	13,125

485

For customer support on your Plumbing Cost Data, call 877.709.6968.

In this closed-loop indirect collection system, fluid with a low freezing temperature, such as propylene glycol, transports heat from the collectors to water storage. The transfer fluid is contained in a closed-loop consisting of collectors, supply and return piping, and a heat exchanger immersed in the storage tank. A typical two-or-three panel system contains 5 to 6 gallons of heat transfer fluid.

When the collectors become approximately 20°F warmer than the storage temperature, a controller activates the circulator. The circulator moves the fluid continuously through the collectors until the temperature difference between the collectors and storage is such that heat collection no longer occurs; at that point, the circulator shuts off. Since the heat transfer fluid has a very low freezing temperature, there is no need for it to be drained from the collectors between periods of collection.

D20 Plumbing

D2020 Domestic Water Distribution

System Components	QUANTITY	UNIT	COST EACH		
			MAT.	INST.	TOTAL
SYSTEM D2020 295 2760					
SOLAR, CLOSED LOOP, HOT WATER SYSTEM, IMMERSED HEAT EXCHANGER					
3/4" TUBING, THREE 3' X 7' BLACK CHROME COLLECTORS					
A, B Differential controller, 2 sensors, thermostat, solar energy system	1.000	Ea.	143	59	202
C Thermometer 2" dial	3.000	Ea.	72	133.50	205.50
D, T Fill & drain valves, brass, 3/4" connection	3.000	Ea.	32.10	88.50	120.60
E Air vent, manual, 1/8" fitting	1.000	Ea.	2.68	22	24.68
F Air purger	1.000	Ea.	51.50	59	110.50
G Expansion tank	1.000	Ea.	75	22	97
I Valve, gate, bronze, NRS, soldered 3/4" diam	3.000	Ea.	210	106.50	316.50
J Neoprene vent flashing	2.000	Ea.	19	71	90
K Circulator, solar heated liquid, 1/25 HP	1.000	Ea.	126	91	217
N-1, N Relief valve, temp & press 150 psi 210°F self-closing 3/4" IPS	2.000	Ea.	46	47	93
O Pipe covering, urethane ultraviolet cover, 1" wall, 3/4" diam	20.000	L.F.	62.20	128	190.20
P Pipe covering, fiberglass, all service jacket, 1" wall, 3/4" diam	50.000	L.F.	49.50	255	304.50
Roof clamps for solar energy collector panel	3.000	Set	9.24	54.75	63.99
Q Collector panel solar blk chrome on copper, 1/8" temp glass, 3'x7'	3.000	Ea.	3,300	402	3,702
R-1 Valve, swing check, bronze, regrinding disc, 3/4" diam, soldered	1.000	Ea.	110	35.50	145.50
S Pressure gauge, 60 psi, 2-1/2" dial	1.000	Ea.	26.50	22	48.50
U Valve, water tempering, bronze, sweat connections, 3/4" diam	1.000	Ea.	139	35.50	174.50
W-2, W Tank, water storage immersed heat exchr elec elem 2"x2# insul 120 Gal	1.000	Ea.	2,025	505	2,530
X Valve, globe, bronze, rising stem, 3/4" diam, soldered	1.000	Ea.	109	35.50	144.50
Copper tubing type L, solder joint, hanger 10' OC 3/4" diam	20.000	L.F.	114	187	301
Copper tubing, type M, solder joint, hanger 10' OC 3/4" diam	70.000	L.F.	317.80	637	954.80
Sensor wire, #22-2 conductor multistranded	.500	C.L.F.	7.83	32.75	40.58
Solar energy heat transfer fluid, propylene glycol, anti-freeze	6.000	Gal.	102.60	153	255.60
Wrought copper fittings & solder, 3/4" diam	76.000	Ea.	205.96	2,850	3,055.96
TOTAL			7,355.91	6,032.50	13,388.41

D2020 295	Solar, Closed Loop, Hot Water Systems		COST EACH		
			MAT.	INST.	TOTAL
2550	Solar, closed loop, hot water system, immersed heat exchanger				
2560	3/8" tubing, 3 ea. 4' x 4'-4" vacuum tube collectors, 80 gal. tank		6,600	5,450	12,050
2580	1/2" tubing, 4 ea. 4' x 4'-4" vacuum tube collectors, 80 gal. tank		7,400	5,900	13,300
2600	120 gal. tank		7,650	5,950	13,600
2640	2 ea. 3'x7' black chrome collectors, 80 gal. tank	R235616 -60	5,725	5,625	11,350
2660	120 gal. tank		5,900	5,625	11,525
2700	2 ea. 3'x7' flat black collectors, 120 gal. tank		5,650	5,625	11,275
2720	3 ea. 3'x7' flat black collectors, 120 gal. tank		6,600	5,800	12,400
2760	3/4" tubing, 3 ea. 3'x7' black chrome collectors, 120 gal. tank		7,350	6,025	13,375
2780	3 ea. 3'x7' flat black collectors, 120 gal. tank		6,950	6,050	13,000
2800	2 ea. 4'x9' flat black w/plastic glazing collectors 120 gal. tank		6,350	6,075	12,425
2840	1" tubing, 4 ea. 2'x9' plastic absorber & glazing collectors 120 gal. tank		7,275	6,800	14,075
2860	4 ea. 3'x7' black chrome collectors, 120 gal. tank		9,100	6,825	15,925
2880	4 ea. 3'x7' flat black absorber collectors, 120 gal. tank		8,575	6,850	15,425

For customer support on your Plumbing Cost Data, call 877.709.6968.

487

D2040 Rain Water Drainage

Design Assumptions: Vertical conductor size is based on a maximum rate of rainfall of 4″ per hour. To convert roof area to other rates multiply "Max. S.F. Roof Area" shown by four and divide the result by desired local rate. The answer is the local roof area that may be handled by the indicated pipe diameter.

Basic cost is for roof drain, 10′ of vertical leader and 10′ of horizontal, plus connection to the main.

Pipe Dia.	Max. S.F. Roof Area	Gallons per Min.
2″	544	23
3″	1610	67
4″	3460	144
5″	6280	261
6″	10,200	424
8″	22,000	913

System Components			COST EACH		
	QUANTITY	UNIT	MAT.	INST.	TOTAL
SYSTEM D2040 210 1880					
ROOF DRAIN, DWV PVC PIPE, 2″ DIAM., 10′ HIGH					
Drain, roof, main, ABS, dome type 2″ pipe size	1.000	Ea.	144	91	235
Clamp, roof drain, underdeck	1.000	Ea.	30	53	83
Pipe, Tee, PVC DWV, schedule 40, 2″ pipe size	1.000	Ea.	4.18	64	68.18
Pipe, PVC, DWV, schedule 40, 2″ diam.	20.000	L.F.	124	430	554
Pipe, elbow, PVC schedule 40, 2″ diam.	2.000	Ea.	6.36	70	76.36
TOTAL			308.54	708	1,016.54

D2040 210	Roof Drain Systems	COST EACH		
		MAT.	INST.	TOTAL
1880	Roof drain, DWV PVC, 2″ diam., piping, 10′ high	310	710	1,020
1920	For each additional foot add	6.20	21.50	27.70
1960	3″ diam., 10′ high	385	825	1,210
2000	For each additional foot add	8.85	24	32.85
2040	4″ diam., 10′ high	455	930	1,385
2080	For each additional foot add	11	26.50	37.50
2120	5″ diam., 10′ high	1,400	1,075	2,475
2160	For each additional foot add	29	29.50	58.50
2200	6″ diam., 10′ high	1,375	1,200	2,575
2240	For each additional foot add	18.05	32.50	50.55
2280	8″ diam., 10′ high	3,000	2,025	5,025
2320	For each additional foot add	40	41.50	81.50
3940	C.I., soil, single hub, service wt., 2″ diam. piping, 10′ high	635	785	1,420
3980	For each additional foot add	12.50	20.50	33
4120	3″ diam., 10′ high	875	845	1,720
4160	For each additional foot add	16.50	21.50	38
4200	4″ diam., 10′ high	1,000	915	1,915
4240	For each additional foot add	20.50	23	43.50
4280	5″ diam., 10′ high	1,400	1,000	2,400
4320	For each additional foot add	28	26	54
4360	6″ diam., 10′ high	1,750	1,075	2,825
4400	For each additional foot add	34.50	27	61.50
4440	8″ diam., 10′ high	3,400	2,225	5,625
4480	For each additional foot add	52	46	98
6040	Steel galv. sch 40 threaded, 2″ diam. piping, 10′ high	715	755	1,470
6080	For each additional foot add	11.35	19.95	31.30
6120	3″ diam., 10′ high	1,400	1,075	2,475
6160	For each additional foot add	23	29.50	52.50

D2040 Rain Water Drainage

D2040 210	Roof Drain Systems	COST EACH		
		MAT.	INST.	TOTAL
6200	4" diam., 10' high	1,975	1,400	3,375
6240	For each additional foot add	33	35.50	68.50
6280	5" diam., 10' high	2,500	1,150	3,650
6320	For each additional foot add	65.50	34.50	100
6360	6" diam, 10' high	2,925	1,500	4,425
6400	For each additional foot add	71	47.50	118.50
6440	8" diam., 10' high	4,675	2,175	6,850
6480	For each additional foot add	83.50	53.50	137

489

For customer support on your Plumbing Cost Data, call 877.709.6968.

D3010 Energy Supply

Basis for Heat Loss Estimate, Apartment Type Structures:

1. Masonry walls and flat roof are insulated. U factor is assumed at .08.
2. Window glass area taken as BOCA minimum, 1/10th of floor area. Double insulating glass with 1/4" air space, U = .65.
3. Infiltration = 0.3 C.F. per hour per S.F. of net wall.
4. Concrete floor loss is 2 BTUH per S.F.
5. Temperature difference taken as 70°F.
6. Ventilating or makeup air has not been included and must be added if desired. Air shafts are not used.

System Components	QUANTITY	UNIT	COST EACH		
			MAT.	INST.	TOTAL
SYSTEM D3010 510 1760					
HEATING SYSTEM, FIN TUBE RADIATION, FORCED HOT WATER					
1,000 S.F. AREA, 10,000 C.F. VOLUME					
Boiler, oil fired, CI, burner, ctrls/insul/breech/pipe/ftng/valves, 109 MBH	1.000	Ea.	4,462.50	4,025	8,487.50
Circulating pump, CI flange connection, 1/12 HP	1.000	Ea.	400	213	613
Expansion tank, painted steel, ASME 18 Gal capacity	1.000	Ea.	785	93	878
Storage tank, steel, above ground, 275 Gal capacity w/supports	1.000	Ea.	540	260	800
Copper tubing type L, solder joint, hanger 10' OC, 3/4" diam	100.000	L.F.	570	935	1,505
Radiation, 3/4" copper tube w/alum fin baseboard pkg, 7" high	30.000	L.F.	243	675	918
Pipe covering, calcium silicate w/cover, 1' wall, 3/4' diam	100.000	L.F.	448	690	1,138
TOTAL			7,448.50	6,891	14,339.50
COST PER S.F.			7.45	6.89	14.34

D3010 510	Apartment Building Heating - Fin Tube Radiation	COST PER S.F.		
		MAT.	INST.	TOTAL
1740	Heating systems, fin tube radiation, forced hot water			
1760	1,000 S.F. area, 10,000 C.F. volume	7.45	6.90	14.35
1800	10,000 S.F. area, 100,000 C.F. volume	3.13	4.13	7.26
1840	20,000 S.F. area, 200,000 C.F. volume	3.45	4.64	8.09
1880	30,000 S.F. area, 300,000 C.F. volume	3.34	4.49	7.83

Many styles of active solar energy systems exist. Those shown on the following page represent the majority of systems now being installed in different regions of the country.

The five active domestic hot water (DHW) systems which follow are typically specified as two or three panel systems with additional variations being the type of glazing and size of the storage tanks. Various combinations have been costed for the user's evaluation and comparison. The basic specifications from which the following systems were developed satisfy the construction detail requirements specified in the HUD Intermediate Minimum Property Standards (IMPS). If these standards are not complied with, the renewable energy system's costs could be significantly lower than shown.

To develop the system's specifications and costs it was necessary to make a number of assumptions about the systems. Certain systems are more appropriate to one climatic region than another or the systems may require modifications to be usable in particular locations. Specific instances in which the systems are not appropriate throughout the country as specified, or in which modification will be needed include the following:

- The freeze protection mechanisms provided in the DHW systems vary greatly. In harsh climates, where freeze is a major concern, a closed-loop indirect collection system may be more appropriate than a direct collection system.
- The thermosyphon water heater system described cannot be used when temperatures drop below 32°F.
- In warm climates it may be necessary to modify the systems installed to prevent overheating.

For each renewable resource (solar) system a schematic diagram and descriptive summary of the system is provided along with a list of all the components priced as part of the system. The costs were developed based on these specifications.

Considerations affecting costs which may increase or decrease beyond the estimates presented here include the following:

- Special structural qualities (allowance for earthquake, future expansion, high winds, and unusual spans or shapes);
- Isolated building site or rough terrain that would affect the transportation of personnel, material, or equipment;
- Unusual climatic conditions during the construction process;
- Substitution of other materials or system components for those used in the system specifications.

491

In this closed-loop indirect collection system, fluid with a low freezing temperature, propylene glycol, transports heat from the collectors to water storage. The transfer fluid is contained in a closed loop consisting of collectors, supply and return piping, and a heat exchanger immersed in the storage tank.

When the collectors become approximately 20°F warmer than the storage temperature, the controller activates the circulator. The circulator moves the fluid continuously until the temperature difference between fluid in the collectors and storage is such that the collection will no longer occur and then the circulator turns off. Since the heat transfer fluid has a very low freezing temperature, there is no need for it to be drained from the collectors between periods of collection.

D3010 Energy Supply

System Components	QUANTITY	UNIT	COST EACH MAT.	COST EACH INST.	COST EACH TOTAL
SYSTEM D3010 650 2750					
SOLAR, CLOSED LOOP, SPACE/HOT WATER					
1" TUBING, TEN 3'X7' BLK CHROME ON COPPER ABSORBER COLLECTORS					
A, B Differential controller 2 sensors, thermostat, solar energy system	2.000	Ea.	286	118	404
C Thermometer, 2" dial	10.000	Ea.	240	445	685
C-1 Heat exchanger, solar energy system, fluid to air, up flow, 80 MBH	1.000	Ea.	525	425	950
D, T Fill & drain valves, brass, 3/4" connection	5.000	Ea.	53.50	147.50	201
D-1 Fan center	1.000	Ea.	130	146	276
E Air vent, manual, 1/8" fitting	1.000	Ea.	2.68	22	24.68
E-2 Thermostat, 2 stage for sensing room temperature	1.000	Ea.	205	90	295
F Air purger	2.000	Ea.	103	118	221
F-1 Controller, liquid temperature, solar energy system	1.000	Ea.	118	142	260
G Expansion tank	2.000	Ea.	150	44	194
I Valve, gate, bronze, 125 lb, soldered, 1" diam	5.000	Ea.	420	187.50	607.50
J Vent flashing, neoprene	2.000	Ea.	19	71	90
K Circulator, solar heated liquid, 1/25 HP	2.000	Ea.	252	182	434
L Circulator, solar heated liquid, 1/20 HP	1.000	Ea.	177	106	283
N Relief valve, temp & pressure 150 psi 210°F self-closing	4.000	Ea.	92	94	186
N-1 Relief valve, pressure poppet, bronze, 30 psi, 3/4" IPS	3.000	Ea.	241.50	76.50	318
O Pipe covering, urethane, ultraviolet cover, 1" wall, 1" diam	50.000	L.F.	155.50	320	475.50
P Pipe covering, fiberglass, all service jacket, 1" wall, 1" diam	60.000	L.F.	59.40	306	365.40
Q Collector panel solar energy blk chrome on copper 1/8" temp glass 3'x7'	10.000	Ea.	11,000	1,340	12,340
Roof clamps for solar energy collector panel	10.000	Set	30.80	182.50	213.30
R Valve, swing check, bronze, 125 lb, regrinding disc, 3/4" & 1" diam	4.000	Ea.	632	150	782
S Pressure gage, 0-60 psi, for solar energy system	2.000	Ea.	53	44	97
U Valve, water tempering, bronze, sweat connections, 3/4" diam	1.000	Ea.	139	35.50	174.50
W-2, W-1, W Tank, water storage immersed heat xchr elec elem 2"x1/2# ins 120 gal	4.000	Ea.	8,100	2,020	10,120
X Valve, globe, bronze, 125 lb, rising stem, 1" diam	3.000	Ea.	507	112.50	619.50
Y Valve, flow control	1.000	Ea.	132	32	164
Copper tubing, type M, solder joint, hanger 10' OC 1" diam	110.000	L.F.	797.50	1,116.50	1,914
Copper tubing, type L, solder joint, hanger 10' OC 3/4" diam	20.000	L.F.	114	187	301
Wrought copper fittings & solder, 3/4" & 1" diam	121.000	Ea.	804.65	5,384.50	6,189.15
Sensor, wire, #22-2 conductor, multistranded	.700	C.L.F.	10.96	45.85	56.81
Ductwork, galvanized steel, for heat exchanger	8.000	Lb.	5.76	65.20	70.96
Solar energy heat transfer fluid propylene glycol, anti-freeze	25.000	Gal.	427.50	637.50	1,065
TOTAL			25,983.75	14,393.55	40,377.30

D3010 650	Solar, Closed Loop, Space/Hot Water Systems		COST EACH MAT.	COST EACH INST.	COST EACH TOTAL
2540	Solar, closed loop, space/hot water				
2550	1/2" tubing, 12 ea. 4'x4'4" vacuum tube collectors		25,100	13,400	38,500
2600	3/4" tubing, 12 ea. 4'x4'4" vacuum tube collectors	R235616 -60	26,300	13,800	40,100
2650	10 ea. 3' x 7' black chrome absorber collectors		25,400	13,400	38,800
2700	10 ea. 3' x 7' flat black absorber collectors		24,100	13,500	37,600
2750	1" tubing, 10 ea. 3' x 7' black chrome absorber collectors		26,000	14,400	40,400
2800	10 ea. 3' x 7' flat black absorber collectors		24,800	14,500	39,300
2850	6 ea. 4' x 9' flat black w/plastic glazing collectors		22,000	14,400	36,400
2900	12 ea. 2' x 9' plastic absorber and glazing collectors		22,800	14,600	37,400

This draindown pool system uses a differential thermostat similar to those used in solar domestic hot water and space heating applications. To heat the pool, the pool water passes through the conventional pump-filter loop and then flows through the collectors. When collection is not possible, or when the pool temperature is reached, all water drains from the solar loop back to the pool through the existing piping. The modes are controlled by solenoid valves or other automatic valves in conjunction with a vacuum breaker relief valve, which facilitates draindown.

D30 HVAC

D3010 Energy Supply

System Components	QUANTITY	UNIT	COST EACH		
			MAT.	INST.	TOTAL
SYSTEM D3010 660 2640					
SOLAR SWIMMING POOL HEATER, ROOF MOUNTED COLLECTORS					
TEN 4' X 10' FULLY WETTED UNGLAZED PLASTIC ABSORBERS					
A Differential thermostat/controller, 110V, adj pool pump system	1.000	Ea.	355	355	710
A-1 Solenoid valve, PVC, normally 1 open 1 closed (included)	2.000	Ea.			
B Sensor, thermistor type (included)	2.000	Ea.			
E-1 Valve, vacuum relief	1.000	Ea.	32.50	22	54.50
Q Collector panel, solar energy, plastic, liquid full wetted, 4' x 10'	10.000	Ea.	3,550	2,550	6,100
R Valve, ball check, PVC, socket, 1-1/2" diam	1.000	Ea.	118	35.50	153.50
Z Valve, ball, PVC, socket, 1-1/2" diam	3.000	Ea.	276	106.50	382.50
Pipe, PVC, sch 40, 1-1/2" diam	80.000	L.F.	604	1,576	2,180
Pipe fittings, PVC sch 40, socket joint, 1-1/2" diam	10.000	Ea.	20.40	355	375.40
Sensor wire, #22-2 conductor, multistranded	.500	C.L.F.	7.83	32.75	40.58
Roof clamps for solar energy collector panels	10.000	Set	30.80	182.50	213.30
Roof strap, teflon for solar energy collector panels	26.000	L.F.	676	89.96	765.96
TOTAL			5,670.53	5,305.21	10,975.74

D3010 660	Solar Swimming Pool Heater Systems		COST EACH		
			MAT.	INST.	TOTAL
2530	Solar swimming pool heater systems, roof mounted collectors				
2540	10 ea. 3'x7' black chrome absorber, 1/8" temp. glass		13,100	4,100	17,200
2560	10 ea. 4'x8' black chrome absorber, 3/16" temp. glass	R235616 -60	14,600	4,875	19,475
2580	10 ea. 3'8"x6' flat black absorber, 3/16" temp. glass		11,800	4,175	15,975
2600	10 ea. 4'x9' flat black absorber, plastic glazing		13,600	5,075	18,675
2620	10 ea. 2'x9' rubber absorber, plastic glazing		15,000	5,500	20,500
2640	10 ea. 4'x10' fully wetted unglazed plastic absorber		5,675	5,300	10,975
2660	Ground mounted collectors				
2680	10 ea. 3'x7' black chrome absorber, 1/8" temp. glass		13,100	4,575	17,675
2700	10 ea. 4'x8' black chrome absorber, 3/16" temp. glass		14,600	5,350	19,950
2720	10 ea. 3'8"x6' flat blk absorber, 3/16" temp. glass		11,800	4,650	16,450
2740	10 ea. 4'x9' flat blk absorber, plastic glazing		13,600	5,550	19,150
2760	10 ea. 2'x9' rubber absorber, plastic glazing		14,900	5,775	20,675
2780	10 ea. 4'x10' fully wetted unglazed plastic absorber		5,700	5,775	11,475

495

For customer support on your Plumbing Cost Data, call 877.709.6968.

The complete Solar Air Heating System provides maximum savings of conventional fuel with both space heating and year-round domestic hot water heating. It allows for the home air conditioning to operate simultaneously and independently from the solar domestic water heating in summer. The system's modes of operation are:

Mode 1: The building is heated directly from the collectors with air circulated by the Solar Air Mover.

Mode 2: When heat is not needed in the building, the dampers change within the air mover to circulate the air from the collectors to the rock storage bin.

Mode 3: When heat is not available from the collector array and is available in rock storage, the air mover draws heated air from rock storage and directs it into the building. When heat is not available from the collectors or the rock storage bin, the auxiliary heating unit will provide heat for the building. The size of the collector array is typically 25% the size of the main floor area.

D3010 Energy Supply

System Components	QUANTITY	UNIT	COST EACH		
			MAT.	INST.	TOTAL
SYSTEM D3010 675 1210					
SOLAR, SPACE/HOT WATER, AIR TO WATER HEAT EXCHANGE					
A, B — Differential controller 2 sensors thermos., solar energy sys liquid loop	1.000	Ea.	143	59	202
A-1, B — Differential controller 2 sensors 6 station solar energy sys air loop	1.000	Ea.	289	236	525
B-1 — Solar energy sensor, freeze prevention	1.000	Ea.	25	22	47
C — Thermometer for solar energy system, 2" dial	2.000	Ea.	48	89	137
C-1 — Heat exchanger, solar energy system, air to fluid, up flow, 70 MBH	1.000	Ea.	395	365	760
D — Drain valve, brass, 3/4" connection	2.000	Ea.	21.40	59	80.40
E — Air vent, manual, for solar energy system 1/8" fitting	1.000	Ea.	2.68	22	24.68
E-1 — Thermostat, 2 stage for sensing room temperature	1.000	Ea.	205	90	295
F — Air purger	1.000	Ea.	51.50	59	110.50
G — Expansion tank, for solar energy system	1.000	Ea.	75	22	97
I — Valve, gate, bronze, 125 lb, NRS, soldered 3/4" diam	2.000	Ea.	140	71	211
K — Circulator, solar heated liquid, 1/25 HP	1.000	Ea.	126	91	217
N — Relief valve temp & press 150 psi 210°F self-closing, 3/4" IPS	1.000	Ea.	23	23.50	46.50
N-1 — Relief valve, pressure, poppet, bronze, 30 psi, 3/4" IPS	1.000	Ea.	80.50	25.50	106
P — Pipe covering, fiberglass, all service jacket, 1" wall, 3/4" diam	60.000	L.F.	59.40	306	365.40
Q-3 — Manifold for flush mount solar energy collector panels	20.000	L.F.	3,180	154	3,334
Q — Collector panel solar energy, air, black on alum. plate, 7.5' x 3.5'	10.000	Ea.	9,650	1,420	11,070
R — Valve, swing check, bronze, 125 lb, regrinding disc, 3/4" diam	2.000	Ea.	220	71	291
S — Pressure gage, 2" dial, for solar energy system	1.000	Ea.	26.50	22	48.50
U — Valve, water tempering, bronze, sweat connections, 3/4" diam	1.000	Ea.	139	35.50	174.50
W-2, W — Tank, water storage, solar, elec element 2"x1/2# insul, 80 Gal	1.000	Ea.	1,775	445	2,220
X — Valve, globe, bronze, 125 lb, soldered, 3/4" diam	1.000	Ea.	109	35.50	144.50
Copper tubing type L, solder joints, hangers 10' OC 3/4" diam	10.000	L.F.	57	93.50	150.50
Copper tubing type M, solder joints, hangers 10' OC 3/4" diam	60.000	L.F.	272.40	546	818.40
Wrought copper fittings & solder, 3/4" diam	26.000	Ea.	70.46	975	1,045.46
Sensor wire, #22-2 conductor multistranded	1.200	C.L.F.	18.78	78.60	97.38
Q-1 — Duct work, rigid fiberglass, rectangular	400.000	S.F.	348	2,180	2,528
Duct work, spiral preformed, steel, PVC coated both sides, 12" x 10"	8.000	Ea.	208	412	620
R-2 — Shutter/damper for solar heater circulator	9.000	Ea.	553.50	927	1,480.50
K-1 — Shutter motor for solar heater circulator blower	9.000	Ea.	1,377	693	2,070
K-1 — Fan, solar energy heated air circulator, space & DHW system	1.000	Ea.	1,700	2,450	4,150
Z-1 — Tank, solar energy air storage, 6'-3"H 7'x7' = 306 CF/2000 Gal	1.000	Ea.	15,700	1,600	17,300
Z-2 — Crushed stone 1-1/2"	11.000	C.Y.	286	89.32	375.32
C-2 — Thermometer, remote probe, 2" dial	1.000	Ea.	36.50	88.50	125
R-1 — Solenoid valve	1.000	Ea.	135	79	214
V, R-4, R-3 — Furnace, supply diffusers, return grilles, existing	1.000	Ea.	15,700	1,600	17,300
TOTAL			53,246.62	15,534.92	68,781.54

D3010 675	Air To Water Heat Exchange		COST EACH		
			MAT.	INST.	TOTAL
1210 1220	Solar, air to water heat exchange, for space/hot water heating	R235616 -60	53,000	15,600	68,600

D3020 Heat Generating Systems

Small Electric Boiler
System Considerations:
1. Terminal units are fin tube baseboard radiation rated at 720 BTU/hr with 200° water temperature or 820 BTU/hr steam.
2. Primary use being for residential or smaller supplementary areas, the floor levels are based on 7-1/2' ceiling heights.
3. All distribution piping is copper for boilers through 205 MBH. All piping for larger systems is steel pipe.

Boiler   **Baseboard Radiation**

System Components	QUANTITY	UNIT	COST EACH		
			MAT.	INST.	TOTAL
SYSTEM D3020 102 1120					
SMALL HEATING SYSTEM, HYDRONIC, ELECTRIC BOILER					
1,480 S.F., 61 MBH, STEAM, 1 FLOOR					
Boiler, electric steam, std cntrls, trim, ftngs and valves, 18 KW, 61.4 MBH	1.000	Ea.	5,032.50	1,787.50	6,820
Copper tubing type L, solder joint, hanger 10'OC, 1-1/4" diam	160.000	L.F.	1,816	1,960	3,776
Radiation, 3/4" copper tube w/alum fin baseboard pkg 7" high	60.000	L.F.	486	1,350	1,836
Rough in baseboard panel or fin tube with valves & traps	10.000	Set	2,630	6,800	9,430
Pipe covering, calcium silicate w/cover, 1" wall 1-1/4" diam	160.000	L.F.	716.80	1,136	1,852.80
Low water cut-off, quick hookup, in gage glass tappings	1.000	Ea.	281	45	326
TOTAL			10,962.30	13,078.50	24,040.80
COST PER S.F.			7.41	8.84	16.25

D3020 102	Small Heating Systems, Hydronic, Electric Boilers	COST PER S.F.		
		MAT.	INST.	TOTAL
1100	Small heating systems, hydronic, electric boilers			
1120	Steam, 1 floor, 1480 S.F., 61 M.B.H.	7.43	8.85	16.28
1160	3,000 S.F., 123 M.B.H.	5.65	7.75	13.40
1200	5,000 S.F., 205 M.B.H.	4.93	7.15	12.08
1240	2 floors, 12,400 S.F., 512 M.B.H.	3.92	7.10	11.02
1280	3 floors, 24,800 S.F., 1023 M.B.H.	4.41	7.05	11.46
1360	Hot water, 1 floor, 1,000 S.F., 41 M.B.H.	12.55	4.91	17.46
1400	2,500 S.F., 103 M.B.H.	8.80	8.85	17.65
1440	2 floors, 4,850 S.F., 205 M.B.H.	8.50	10.60	19.10
1480	3 floors, 9,700 S.F., 410 M.B.H.	8.80	10.95	19.75

498

For customer support on your Plumbing Cost Data, call 877.709.6968.

D3090 Other HVAC Systems/Equip

Cast Iron Garage Exhaust System

Dual Exhaust System

System Components		QUANTITY	UNIT	COST EACH		
				MAT.	INST.	TOTAL
SYSTEM D3090 320 1040						
GARAGE, EXHAUST, SINGLE 3″ EXHAUST OUTLET, CARS & LIGHT TRUCKS						
A	Outlet top assy, for engine exhaust system with adapters and ftngs, 3″ diam	1.000	Ea.	265	43	308
F	Bullnose (guide) for engine exhaust system, 3″ diam	1.000	Ea.	30.50		30.50
G	Galvanized flexible tubing for engine exhaust system, 3″ diam	8.000	L.F.	102		102
H	Adapter for metal tubing end of engine exhaust system, 3″ tail pipe	1.000	Ea.	54.50		54.50
J	Pipe, sewer, cast iron, push-on joint, 8″ diam.	18.000	L.F.	1,044	783	1,827
	Excavating utility trench, chain trencher, 8″ wide, 24″ deep	18.000	L.F.		34.20	34.20
	Backfill utility trench by hand, incl. compaction, 8″ wide 24″ deep	18.000	L.F.		47.16	47.16
	Stand for blower, concrete over polystyrene core, 6″ high	1.000	Ea.	6.05	21.50	27.55
	AC&V duct spiral reducer 10″x8″	1.000	Ea.	22.50	38.50	61
	AC&V duct spiral reducer 12″x10″	1.000	Ea.	26	51.50	77.50
	AC&V duct spiral preformed 45° elbow, 8″ diam	1.000	Ea.	8.90	44	52.90
	AC&V utility fan, belt drive, 3 phase, 2000 CFM, 1 HP	2.000	Ea.	2,700	680	3,380
	Safety switch, heavy duty fused, 240V, 3 pole, 30 amp	1.000	Ea.	130	205	335
	TOTAL			4,389.45	1,947.86	6,337.31

D3090 320	Garage Exhaust Systems	COST PER BAY		
		MAT.	INST.	TOTAL
1040	Garage, single 3″ exhaust outlet, cars & light trucks, one bay	4,400	1,950	6,350
1060	Additional bays up to seven bays	1,025	525	1,550
1500	4″ outlet, trucks, one bay	4,425	1,950	6,375
1520	Additional bays up to six bays	1,075	525	1,600
1600	5″ outlet, diesel trucks, one bay	4,750	1,950	6,700
1650	Additional single bays up to six	1,500	620	2,120
1700	Two adjoining bays	4,750	1,950	6,700
2000	Dual exhaust, 3″ outlets, pair of adjoining bays	5,500	2,450	7,950
2100	Additional pairs of adjoining bays	1,675	620	2,295

499

For customer support on your Plumbing Cost Data, call 877.709.6968.

Dry Pipe System: A system employing automatic sprinklers attached to a piping system containing air under pressure, the release of which from the opening of sprinklers permits the water pressure to open a valve known as a "dry pipe valve". The water then flows into the piping system and out the opened sprinklers.

All areas are assumed to be open.

System Components	QUANTITY	UNIT	COST EACH		
			MAT.	INST.	TOTAL
SYSTEM D4010 310 0580					
DRY PIPE SPRINKLER, STEEL, BLACK, SCH. 40 PIPE					
LIGHT HAZARD, ONE FLOOR, 2000 S.F.					
Valve, gate, iron body 125 lb., OS&Y, flanged, 4″ pipe size	1.000	Ea.	637.50	318.75	956.25
Valve, swing check, bronze, 125 lb, regrinding disc, 2-1/2″ pipe size	1.000	Ea.	660	63.75	723.75
Valve, angle, bronze, 150 lb., rising stem, threaded, 2″ pipe size	1.000	Ea.	622.50	48.38	670.88
*Alarm valve, 2-1/2″ pipe size	1.000	Ea.	1,350	307.50	1,657.50
Alarm, water motor, complete with gong	1.000	Ea.	341.25	127.50	468.75
Fire alarm horn, electric	1.000	Ea.	50.25	73.50	123.75
Valve swing check w/balldrip CI with brass trim, 4″ pipe size	1.000	Ea.	288.75	307.50	596.25
Pipe, steel, black, schedule 40, 4″ diam.	10.000	L.F.	180	271.65	451.65
Dry pipe valve, trim & gauges, 4″ pipe size	1.000	Ea.	2,362.50	918.75	3,281.25
Pipe, steel, black, schedule 40, threaded, cplg & hngr 10'OC 2-1/2″ diam.	20.000	L.F.	231	382.50	613.50
Pipe, steel, black, schedule 40, threaded, cplg & hngr 10'OC 2″ diam.	12.500	L.F.	92.81	187.03	279.84
Pipe, steel, black, schedule 40, threaded, cplg & hngr 10'OC 1-1/4″ diam.	37.500	L.F.	191.25	403.59	594.84
Pipe, steel, black, schedule 40, threaded, cplg & hngr 10'OC 1″ diam.	112.000	L.F.	466.20	1,125.60	1,591.80
Pipe Tee, malleable iron black, 150 lb. threaded, 4″ pipe size	2.000	Ea.	472.50	480	952.50
Pipe Tee, malleable iron black, 150 lb. threaded, 2-1/2″ pipe size	2.000	Ea.	132.75	213	345.75
Pipe Tee, malleable iron black, 150 lb. threaded, 2″ pipe size	1.000	Ea.	30.75	87	117.75
Pipe Tee, malleable iron black, 150 lb. threaded, 1-1/4″ pipe size	5.000	Ea.	72.56	341.25	413.81
Pipe Tee, malleable iron black, 150 lb. threaded, 1″ pipe size	4.000	Ea.	35.85	265.50	301.35
Pipe 90° elbow malleable iron black, 150 lb. threaded, 1″ pipe size	6.000	Ea.	34.20	245.25	279.45
Sprinkler head dry 1/2″ orifice 1″ NPT, 3″ to 4-3/4″ length	12.000	Ea.	1,632	582	2,214
Air compressor, 200 Gal sprinkler system capacity, 1/3 HP	1.000	Ea.	630	393.75	1,023.75
*Standpipe connection, wall, flush, brs. w/plug & chain 2-1/2″x2-1/2″	1.000	Ea.	131.25	183.75	315
Valve gate bronze, 300 psi, NRS, class 150, threaded, 1″ pipe size	1.000	Ea.	84	28.13	112.13
TOTAL			10,729.87	7,355.63	18,085.50
COST PER S.F.			5.36	3.68	9.04

*Not included in systems under 2000 S.F.

D4010 310	Dry Pipe Sprinkler Systems		COST PER S.F.		
			MAT.	INST.	TOTAL
0520	Dry pipe sprinkler systems, steel, black, sch. 40 pipe				
0530	Light hazard, one floor, 500 S.F.		9.75	6.25	16
0560	1000 S.F.		5.55	3.63	9.18
0580	2000 S.F.	R211313 -10	5.35	3.69	9.04
0600	5000 S.F.		2.85	2.50	5.35
0620	10,000 S.F.	R211313 -20	2.02	2.05	4.07

D4010 Sprinklers

D4010 310	Dry Pipe Sprinkler Systems	COST PER S.F.		
		MAT.	INST.	TOTAL
0640	50,000 S.F.	1.55	1.85	3.40
0660	Each additional floor, 500 S.F.	2.13	3.07	5.20
0680	1000 S.F.	1.86	2.50	4.36
0700	2000 S.F.	1.85	2.32	4.17
0720	5000 S.F.	1.51	2	3.51
0740	10,000 S.F.	1.39	1.82	3.21
0760	50,000 S.F.	1.26	1.64	2.90
1000	Ordinary hazard, one floor, 500 S.F.	9.95	6.30	16.25
1020	1000 S.F.	5.70	3.69	9.39
1040	2000 S.F.	5.50	3.82	9.32
1060	5000 S.F.	3.28	2.68	5.96
1080	10,000 S.F.	2.63	2.73	5.36
1100	50,000 S.F.	2.42	2.59	5.01
1140	Each additional floor, 500 S.F.	2.34	3.14	5.48
1160	1000 S.F.	2.17	2.83	5
1180	2000 S.F.	2.19	2.56	4.75
1200	5000 S.F.	2.02	2.22	4.24
1220	10,000 S.F.	1.84	2.16	4
1240	50,000 S.F.	1.79	1.88	3.67
1500	Extra hazard, one floor, 500 S.F.	13.85	7.85	21.70
1520	1000 S.F.	8.50	5.65	14.15
1540	2000 S.F.	6.15	4.93	11.08
1560	5000 S.F.	3.79	3.76	7.55
1580	10,000 S.F.	4	3.56	7.56
1600	50,000 S.F.	4.34	3.46	7.80
1660	Each additional floor, 500 S.F.	3.31	3.90	7.21
1680	1000 S.F.	3.24	3.69	6.93
1700	2000 S.F.	3	3.69	6.69
1720	5000 S.F.	2.55	3.24	5.79
1740	10,000 S.F.	3.04	2.93	5.97
1760	50,000 S.F.	3.10	2.82	5.92
2020	Grooved steel, black, sch. 40 pipe, light hazard, one floor, 2000 S.F.	5.15	3.15	8.30
2060	10,000 S.F.	2.01	1.77	3.78
2100	Each additional floor, 2000 S.F.	1.83	1.89	3.72
2150	10,000 S.F.	1.38	1.54	2.92
2200	Ordinary hazard, one floor, 2000 S.F.	5.45	3.34	8.79
2250	10,000 S.F.	2.48	2.30	4.78
2300	Each additional floor, 2000 S.F.	2.10	2.08	4.18
2350	10,000 S.F.	1.86	2.07	3.93
2400	Extra hazard, one floor, 2000 S.F.	6.15	4.22	10.37
2450	10,000 S.F.	3.50	2.96	6.46
2500	Each additional floor, 2000 S.F.	3.01	3.05	6.06
2550	10,000 S.F.	2.71	2.58	5.29
3050	Grooved steel, black, sch. 10 pipe, light hazard, one floor, 2000 S.F.	5.10	3.13	8.23
3100	10,000 S.F.	1.96	1.74	3.70
3150	Each additional floor, 2000 S.F.	1.75	1.87	3.62
3200	10,000 S.F.	1.33	1.51	2.84
3250	Ordinary hazard, one floor, 2000 S.F.	5.35	3.31	8.66
3300	10,000 S.F.	2.38	2.25	4.63
3350	Each additional floor, 2000 S.F.	2.02	2.05	4.07
3400	10,000 S.F.	1.76	2.02	3.78
3450	Extra hazard, one floor, 2000 S.F.	6.10	4.19	10.29
3500	10,000 S.F.	3.31	2.90	6.21
3550	Each additional floor, 2000 S.F.	2.95	3.02	5.97
3600	10,000 S.F.	2.63	2.53	5.16
4050	Copper tubing, type M, light hazard, one floor, 2000 S.F.	5.45	3.12	8.57
4100	10,000 S.F.	2.38	1.78	4.16
4150	Each additional floor, 2000 S.F.	2.14	1.90	4.04

501

For customer support on your Plumbing Cost Data, call 877.709.6968.

D4010 Sprinklers

D4010 310	Dry Pipe Sprinkler Systems	COST PER S.F.		
		MAT.	INST.	TOTAL
4200	10,000 S.F.	1.76	1.56	3.32
4250	Ordinary hazard, one floor, 2000 S.F.	5.85	3.47	9.32
4300	10,000 S.F.	2.94	2.11	5.05
4350	Each additional floor, 2000 S.F.	2.79	2.18	4.97
4400	10,000 S.F.	2.26	1.84	4.10
4450	Extra hazard, one floor, 2000 S.F.	6.70	4.25	10.95
4500	10,000 S.F.	5.10	3.22	8.32
4550	Each additional floor, 2000 S.F.	3.57	3.08	6.65
4600	10,000 S.F.	3.76	2.81	6.57
5050	Copper tubing, type M, T-drill system, light hazard, one floor			
5060	2000 S.F.	5.45	2.90	8.35
5100	10,000 S.F.	2.26	1.48	3.74
5150	Each additional floor, 2000 S.F.	2.13	1.68	3.81
5200	10,000 S.F.	1.64	1.26	2.90
5250	Ordinary hazard, one floor, 2000 S.F.	5.65	2.96	8.61
5300	10,000 S.F.	2.84	1.89	4.73
5350	Each additional floor, 2000 S.F.	2.30	1.70	4
5400	10,000 S.F.	2.12	1.58	3.70
5450	Extra hazard, one floor, 2000 S.F.	6.20	3.51	9.71
5500	10,000 S.F.	4.32	2.37	6.69
5550	Each additional floor, 2000 S.F.	3.06	2.34	5.40
5600	10,000 S.F.	2.99	1.96	4.95

Pre-Action System: A system employing automatic sprinklers attached to a piping system containing air that may or may not be under pressure, with a supplemental heat responsive system of generally more sensitive characteristics than the automatic sprinklers themselves, installed in the same areas as the sprinklers. Actuation of the heat responsive system, as from a fire, opens a valve which permits water to flow into the sprinkler piping system and to be discharged from those sprinklers which were opened by heat from the fire.

All areas are assumed to be open.

System Components	QUANTITY	UNIT	COST EACH MAT.	COST EACH INST.	COST EACH TOTAL
SYSTEM D4010 350 0580					
PREACTION SPRINKLER SYSTEM, STEEL BLACK SCH. 40 PIPE					
LIGHT HAZARD, 1 FLOOR, 2000 S.F.					
Valve, gate, iron body 125 lb., OS&Y, flanged, 4″ pipe size	1.000	Ea.	637.50	318.75	956.25
*Valve, swing check w/ball drip CI with brass trim 4″ pipe size	1.000	Ea.	288.75	307.50	596.25
Valve, swing check, bronze, 125 lb, regrinding disc, 2-1/2″ pipe size	1.000	Ea.	660	63.75	723.75
Valve, angle, bronze, 150 lb., rising stem, threaded, 2″ pipe size	1.000	Ea.	622.50	48.38	670.88
*Alarm valve, 2-1/2″ pipe size	1.000	Ea.	1,350	307.50	1,657.50
Alarm, water motor, complete with gong	1.000	Ea.	341.25	127.50	468.75
Fire alarm horn, electric	1.000	Ea.	50.25	73.50	123.75
Thermostatic release for release line	2.000	Ea.	1,110	51	1,161
Pipe, steel, black, schedule 40, 4″ diam.	10.000	L.F.	180	271.65	451.65
Dry pipe valve, trim & gauges, 4″ pipe size	1.000	Ea.	2,362.50	918.75	3,281.25
Pipe, steel, black, schedule 40, threaded, cplg. & hngr. 10'OC 2-1/2″ diam.	20.000	L.F.	231	382.50	613.50
Pipe steel black, schedule 40, threaded, cplg. & hngr. 10'OC 2″ diam.	12.500	L.F.	92.81	187.03	279.84
Pipe, steel, black, schedule 40, threaded, cplg. & hngr. 10'OC 1-1/4″ diam.	37.500	L.F.	191.25	403.59	594.84
Pipe, steel, black, schedule 40, threaded, cplg. & hngr. 10'OC 1″ diam.	112.000	L.F.	466.20	1,125.60	1,591.80
Pipe, Tee, malleable iron, black, 150 lb. threaded, 4″ diam.	2.000	Ea.	472.50	480	952.50
Pipe, Tee, malleable iron, black, 150 lb. threaded, 2-1/2″ pipe size	2.000	Ea.	132.75	213	345.75
Pipe, Tee, malleable iron, black, 150 lb. threaded, 2″ pipe size	1.000	Ea.	30.75	87	117.75
Pipe, Tee, malleable iron, black, 150 lb. threaded, 1-1/4″ pipe size	5.000	Ea.	72.56	341.25	413.81
Pipe, Tee, malleable iron, black, 150 lb. threaded, 1″ pipe size	4.000	Ea.	35.85	265.50	301.35
Pipe, 90° elbow, malleable iron, blk., 150 lb. threaded, 1″ pipe size	6.000	Ea.	34.20	245.25	279.45
Sprinkler head, std. spray, brass 135°-286°F 1/2″ NPT, 3/8″ orifice	12.000	Ea.	203.40	510	713.40
Air compressor auto complete 200 Gal sprinkler sys. cap., 1/3 HP	1.000	Ea.	630	393.75	1,023.75
*Standpipe conn.,wall, flush, brass w/plug & chain 2-1/2″ x 2-1/2″	1.000	Ea.	131.25	183.75	315
Valve, gate, bronze, 300 psi, NRS, class 150, threaded, 1″ pipe size	1.000	Ea.	84	28.13	112.13
TOTAL			10,411.27	7,334.63	17,745.90
COST PER S.F.			5.21	3.67	8.88

*Not included in systems under 2000 S.F.

D4010 350	Preaction Sprinkler Systems		COST PER S.F. MAT.	COST PER S.F. INST.	COST PER S.F. TOTAL
0520	Preaction sprinkler systems, steel, black, sch. 40 pipe				
0530	Light hazard, one floor, 500 S.F.		9.60	4.97	14.57
0560	1000 S.F.	R211313 -10	5.60	3.72	9.32
0580	2000 S.F.		5.20	3.68	8.88

D4010 Sprinklers

D4010 350	Preaction Sprinkler Systems		COST PER S.F.		
			MAT.	INST.	TOTAL
0600	5000 S.F.	R211313 -20	2.72	2.49	5.21
0620	10,000 S.F.		1.91	2.05	3.96
0640	50,000 S.F.		1.45	1.85	3.30
0660	Each additional floor, 500 S.F.		2.42	2.73	5.15
0680	1000 S.F.		1.89	2.50	4.39
0700	2000 S.F.		1.87	2.32	4.19
0720	5000 S.F.		1.38	1.99	3.37
0740	10,000 S.F.		1.28	1.82	3.10
0760	50,000 S.F.		1.26	1.70	2.96
1000	Ordinary hazard, one floor, 500 S.F.		9.90	5.40	15.30
1020	1000 S.F.		5.55	3.69	9.24
1040	2000 S.F.		5.65	3.82	9.47
1060	5000 S.F.		3	2.66	5.66
1080	10,000 S.F.		2.30	2.72	5.02
1100	50,000 S.F.		2.10	2.57	4.67
1140	Each additional floor, 500 S.F.		2.73	3.15	5.88
1160	1000 S.F.		1.88	2.56	4.44
1180	2000 S.F.		1.75	2.53	4.28
1200	5000 S.F.		1.86	2.39	4.25
1220	10,000 S.F.		1.68	2.49	4.17
1240	50,000 S.F.		1.63	2.23	3.86
1500	Extra hazard, one floor, 500 S.F.		13.40	6.90	20.30
1520	1000 S.F.		7.75	5.20	12.95
1540	2000 S.F.		5.60	4.90	10.50
1560	5000 S.F.		3.52	4.06	7.58
1580	10,000 S.F.		3.51	3.95	7.46
1600	50,000 S.F.		3.76	3.84	7.60
1660	Each additional floor, 500 S.F.		3.35	3.89	7.24
1680	1000 S.F.		2.73	3.66	6.39
1700	2000 S.F.		2.48	3.66	6.14
1720	5000 S.F.		2.05	3.26	5.31
1740	10,000 S.F.		2.39	2.97	5.36
1760	50,000 S.F.		2.33	2.78	5.11
2020	Grooved steel, black, sch. 40 pipe, light hazard, one floor, 2000 S.F.		5.20	3.15	8.35
2060	10,000 S.F.		1.90	1.77	3.67
2100	Each additional floor of 2000 S.F.		1.85	1.89	3.74
2150	10,000 S.F.		1.27	1.54	2.81
2200	Ordinary hazard, one floor, 2000 S.F.		5.25	3.33	8.58
2250	10,000 S.F.		2.15	2.29	4.44
2300	Each additional floor, 2000 S.F.		1.93	2.07	4
2350	10,000 S.F.		1.53	2.06	3.59
2400	Extra hazard, one floor, 2000 S.F.		5.60	4.19	9.79
2450	10,000 S.F.		2.85	2.94	5.79
2500	Each additional floor, 2000 S.F.		2.49	3.02	5.51
2550	10,000 S.F.		2.02	2.55	4.57
3050	Grooved steel, black, sch. 10 pipe light hazard, one floor, 2000 S.F.		5.10	3.13	8.23
3100	10,000 S.F.		1.85	1.74	3.59
3150	Each additional floor, 2000 S.F.		1.77	1.87	3.64
3200	10,000 S.F.		1.22	1.51	2.73
3250	Ordinary hazard, one floor, 2000 S.F.		5.10	3.10	8.20
3300	10,000 S.F.		1.77	2.23	4
3350	Each additional floor, 2000 S.F.		1.85	2.04	3.89
3400	10,000 S.F.		1.43	2.01	3.44
3450	Extra hazard, one floor, 2000 S.F.		5.55	4.16	9.71
3500	10,000 S.F.		2.62	2.87	5.49
3550	Each additional floor, 2000 S.F.		2.43	2.99	5.42
3600	10,000 S.F.		1.94	2.50	4.44
4050	Copper tubing, type M, light hazard, one floor, 2000 S.F.		5.45	3.12	8.57

D4010 Sprinklers

D4010 350	Preaction Sprinkler Systems	COST PER S.F.		
		MAT.	INST.	TOTAL
4100	10,000 S.F.	2.27	1.78	4.05
4150	Each additional floor, 2000 S.F.	2.17	1.90	4.07
4200	10,000 S.F.	1.37	1.55	2.92
4250	Ordinary hazard, one floor, 2000 S.F.	5.65	3.46	9.11
4300	10,000 S.F.	2.61	2.10	4.71
4350	Each additional floor, 2000 S.F.	2.13	1.93	4.06
4400	10,000 S.F.	1.76	1.69	3.45
4450	Extra hazard, one floor, 2000 S.F.	6.20	4.22	10.42
4500	10,000 S.F.	4.36	3.18	7.54
4550	Each additional floor, 2000 S.F.	3.05	3.05	6.10
4600	10,000 S.F.	3.07	2.78	5.85
5050	Copper tubing, type M, T-drill system, light hazard, one floor			
5060	2000 S.F.	5.45	2.90	8.35
5100	10,000 S.F.	2.15	1.48	3.63
5150	Each additional floor, 2000 S.F.	2.15	1.68	3.83
5200	10,000 S.F.	1.53	1.26	2.79
5250	Ordinary hazard, one floor, 2000 S.F.	5.45	2.95	8.40
5300	10,000 S.F.	2.51	1.88	4.39
5350	Each additional floor, 2000 S.F.	2.14	1.70	3.84
5400	10,000 S.F.	1.89	1.65	3.54
5450	Extra hazard, one floor, 2000 S.F.	5.65	3.48	9.13
5500	10,000 S.F.	3.59	2.33	5.92
5550	Each additional floor, 2000 S.F.	2.54	2.31	4.85
5600	10,000 S.F.	2.30	1.93	4.23

D4010 Sprinklers

Deluge System: A system employing open sprinklers attached to a piping system connected to a water supply through a valve which is opened by the operation of a heat responsive system installed in the same areas as the sprinklers. When this valve opens, water flows into the piping system and discharges from all sprinklers attached thereto.

All areas are assumed to be open.

System Components	QUANTITY	UNIT	COST EACH		
			MAT.	INST.	TOTAL
SYSTEM D4010 370 0580					
DELUGE SPRINKLER SYSTEM, STEEL BLACK SCH. 40 PIPE					
LIGHT HAZARD, 1 FLOOR, 2000 S.F.					
Valve, gate, iron body 125 lb., OS&Y, flanged, 4" pipe size	1.000	Ea.	637.50	318.75	956.25
Valve, swing check w/ball drip, CI w/brass ftngs., 4" pipe size	1.000	Ea.	288.75	307.50	596.25
Valve, swing check, bronze, 125 lb, regrinding disc, 2-1/2" pipe size	1.000	Ea.	660	63.75	723.75
Valve, angle, bronze, 150 lb., rising stem, threaded, 2" pipe size	1.000	Ea.	622.50	48.38	670.88
*Alarm valve, 2-1/2" pipe size	1.000	Ea.	1,350	307.50	1,657.50
Alarm, water motor, complete with gong	1.000	Ea.	341.25	127.50	468.75
Fire alarm horn, electric	1.000	Ea.	50.25	73.50	123.75
Thermostatic release for release line	2.000	Ea.	1,110	51	1,161
Pipe, steel, black, schedule 40, 4" diam.	10.000	L.F.	180	271.65	451.65
Deluge valve trim, pressure relief, emergency release, gauge, 4" pipe size	1.000	Ea.	3,862.50	918.75	4,781.25
Deluge system, monitoring panel w/deluge valve & trim	1.000	Ea.	8,925	28.50	8,953.50
Pipe, steel, black, schedule 40, threaded, cplg & hngr 10' OC 2-1/2" diam.	20.000	L.F.	231	382.50	613.50
Pipe, steel, black, schedule 40, threaded, cplg & hngr 10' OC 2" diam.	12.500	L.F.	92.81	187.03	279.84
Pipe, steel, black, schedule 40, threaded, cplg & hngr 10' OC 1-1/4" diam.	37.500	L.F.	191.25	403.59	594.84
Pipe, steel, black, schedule 40, threaded, cplg & hngr 10' OC 1" diam.	112.000	L.F.	466.20	1,125.60	1,591.80
Pipe, Tee, malleable iron, black, 150 lb. threaded, 4" pipe size	2.000	Ea.	472.50	480	952.50
Pipe, Tee, malleable iron, black, 150 lb. threaded, 2-1/2" pipe size	2.000	Ea.	132.75	213	345.75
Pipe, Tee, malleable iron, black, 150 lb. threaded, 2" pipe size	1.000	Ea.	30.75	87	117.75
Pipe, Tee, malleable iron, black, 150 lb. threaded, 1-1/4" pipe size	5.000	Ea.	72.56	341.25	413.81
Pipe, Tee, malleable iron, black, 150 lb. threaded, 1" pipe size	4.000	Ea.	35.85	265.50	301.35
Pipe, 90° elbow, malleable iron, black, 150 lb. threaded 1" pipe size	6.000	Ea.	34.20	245.25	279.45
Sprinkler head, std spray, brass 135°-286°F 1/2" NPT, 3/8" orifice	9.720	Ea.	203.40	510	713.40
Air compressor, auto, complete, 200 Gal sprinkler sys. cap., 1/3 HP	1.000	Ea.	630	393.75	1,023.75
*Standpipe connection, wall, flush w/plug & chain 2-1/2" x 2-1/2"	1.000	Ea.	131.25	183.75	315
Valve, gate, bronze, 300 psi, NRS, class 150, threaded, 1" pipe size	1.000	Ea.	84	28.13	112.13
TOTAL			20,836.27	7,363.13	28,199.40
COST PER S.F.			10.42	3.68	14.10

*Not included in systems under 2000 S.F.

D4010 370	Deluge Sprinkler Systems		COST PER S.F.		
			MAT.	INST.	TOTAL
0520	Deluge sprinkler systems, steel, black, sch. 40 pipe				
0530	Light hazard, one floor, 500 S.F.		29	5.05	34.05
0560	1000 S.F.		15.35	3.58	18.93
0580	2000 S.F.	R211313 -10	10.40	3.69	14.09

D40 Fire Protection

D4010 Sprinklers

D4010 370	Deluge Sprinkler Systems		MAT.	INST.	TOTAL
0600	5000 S.F.	R211313 -20	4.80	2.50	7.30
0620	10,000 S.F.		2.95	2.05	5
0640	50,000 S.F.		1.66	1.85	3.51
0660	Each additional floor, 500 S.F.		2.42	2.73	5.15
0680	1000 S.F.		1.89	2.50	4.39
0700	2000 S.F.		1.87	2.32	4.19
0720	5000 S.F.		1.38	1.99	3.37
0740	10,000 S.F.		1.28	1.82	3.10
0760	50,000 S.F.		1.26	1.70	2.96
1000	Ordinary hazard, one floor, 500 S.F.		30.50	5.75	36.25
1020	1000 S.F.		15.35	3.72	19.07
1040	2000 S.F.		10.85	3.83	14.68
1060	5000 S.F.		5.10	2.67	7.77
1080	10,000 S.F.		3.34	2.72	6.06
1100	50,000 S.F.		2.36	2.61	4.97
1140	Each additional floor, 500 S.F.		2.73	3.15	5.88
1160	1000 S.F.		1.88	2.56	4.44
1180	2000 S.F.		1.75	2.53	4.28
1200	5000 S.F.		1.74	2.20	3.94
1220	10,000 S.F.		1.64	2.20	3.84
1240	50,000 S.F.		1.54	2.05	3.59
1500	Extra hazard, one floor, 500 S.F.		33	7	40
1520	1000 S.F.		18	5.40	23.40
1540	2000 S.F.		10.85	4.91	15.76
1560	5000 S.F.		5.35	3.74	9.09
1580	10,000 S.F.		4.41	3.61	8.02
1600	50,000 S.F.		4.46	3.53	7.99
1660	Each additional floor, 500 S.F.		3.35	3.89	7.24
1680	1000 S.F.		2.73	3.66	6.39
1700	2000 S.F.		2.48	3.66	6.14
1720	5000 S.F.		2.05	3.26	5.31
1740	10,000 S.F.		2.44	3.08	5.52
1760	50,000 S.F.		2.45	2.97	5.42
2000	Grooved steel, black, sch. 40 pipe, light hazard, one floor				
2020	2000 S.F.		10.40	3.16	13.56
2060	10,000 S.F.		2.96	1.79	4.75
2100	Each additional floor, 2,000 S.F.		1.85	1.89	3.74
2150	10,000 S.F.		1.27	1.54	2.81
2200	Ordinary hazard, one floor, 2000 S.F.		5.25	3.33	8.58
2250	10,000 S.F.		3.19	2.29	5.48
2300	Each additional floor, 2000 S.F.		1.93	2.07	4
2350	10,000 S.F.		1.53	2.06	3.59
2400	Extra hazard, one floor, 2000 S.F.		10.85	4.20	15.05
2450	10,000 S.F.		3.92	2.95	6.87
2500	Each additional floor, 2000 S.F.		2.49	3.02	5.51
2550	10,000 S.F.		2.02	2.55	4.57
3000	Grooved steel, black, sch. 10 pipe, light hazard, one floor				
3050	2000 S.F.		9.70	3.01	12.71
3100	10,000 S.F.		2.89	1.74	4.63
3150	Each additional floor, 2000 S.F.		1.77	1.87	3.64
3200	10,000 S.F.		1.22	1.51	2.73
3250	Ordinary hazard, one floor, 2000 S.F.		10.40	3.31	13.71
3300	10,000 S.F.		2.81	2.23	5.04
3350	Each additional floor, 2000 S.F.		1.85	2.04	3.89
3400	10,000 S.F.		1.43	2.01	3.44
3450	Extra hazard, one floor, 2000 S.F.		10.75	4.17	14.92
3500	10,000 S.F.		3.66	2.87	6.53
3550	Each additional floor, 2000 S.F.		2.43	2.99	5.42

D4010 Sprinklers

D4010 370	Deluge Sprinkler Systems	COST PER S.F.		
		MAT.	INST.	TOTAL
3600	10,000 S.F.	1.94	2.50	4.44
4000	Copper tubing, type M, light hazard, one floor			
4050	2000 S.F.	10.65	3.13	13.78
4100	10,000 S.F.	3.31	1.78	5.09
4150	Each additional floor, 2000 S.F.	2.15	1.90	4.05
4200	10,000 S.F.	1.37	1.55	2.92
4250	Ordinary hazard, one floor, 2000 S.F.	10.85	3.47	14.32
4300	10,000 S.F.	3.65	2.10	5.75
4350	Each additional floor, 2000 S.F.	2.13	1.93	4.06
4400	10,000 S.F.	1.76	1.69	3.45
4450	Extra hazard, one floor, 2000 S.F.	11.40	4.23	15.63
4500	10,000 S.F.	5.45	3.20	8.65
4550	Each additional floor, 2000 S.F.	3.05	3.05	6.10
4600	10,000 S.F.	3.07	2.78	5.85
5000	Copper tubing, type M, T-drill system, light hazard, one floor			
5050	2000 S.F.	10.65	2.91	13.56
5100	10,000 S.F.	3.19	1.48	4.67
5150	Each additional floor, 2000 S.F.	2.17	1.70	3.87
5200	10,000 S.F.	1.53	1.26	2.79
5250	Ordinary hazard, one floor, 2000 S.F.	10.65	2.96	13.61
5300	10,000 S.F.	3.55	1.88	5.43
5350	Each additional floor, 2000 S.F.	2.13	1.69	3.82
5400	10,000 S.F.	1.89	1.65	3.54
5450	Extra hazard, one floor, 2000 S.F.	10.90	3.49	14.39
5500	10,000 S.F.	4.63	2.33	6.96
5550	Each additional floor, 2000 S.F.	2.54	2.31	4.85
5600	10,000 S.F.	2.30	1.93	4.23

D4010 Sprinklers

Firecycle is a fixed fire protection sprinkler system utilizing water as its extinguishing agent. It is a time delayed, recycling, preaction type which automatically shuts the water off when heat is reduced below the detector operating temperature and turns the water back on when that temperature is exceeded.

The system senses a fire condition through a closed circuit electrical detector system which controls water flow to the fire automatically. Batteries supply up to 90 hour emergency power supply for system operation. The piping system is dry (until water is required) and is monitored with pressurized air. Should

any leak in the system piping occur, an alarm will sound, but water will not enter the system until heat is sensed by a Firecycle detector.

All areas are assumed to be open.

System Components	QUANTITY	UNIT	COST EACH		
			MAT.	INST.	TOTAL
SYSTEM D4010 390 0580					
FIRECYCLE SPRINKLER SYSTEM, STEEL BLACK SCH. 40 PIPE					
LIGHT HAZARD, ONE FLOOR, 2000 S.F.					
Valve, gate, iron body 125 lb., OS&Y, flanged, 4″ pipe size	1.000	Ea.	637.50	318.75	956.25
Valve, angle, bronze, 150 lb., rising stem, threaded, 2″ pipe size	1.000	Ea.	622.50	48.38	670.88
Valve, swing check, bronze, 125 lb, regrinding disc, 2-1/2″ pipe size	1.000	Ea.	660	63.75	723.75
*Alarm valve, 2-1/2″ pipe size	1.000	Ea.	1,350	307.50	1,657.50
Alarm, water motor, complete with gong	1.000	Ea.	341.25	127.50	468.75
Pipe, steel, black, schedule 40, 4″ diam.	10.000	L.F.	180	271.65	451.65
Fire alarm, horn, electric	1.000	Ea.	50.25	73.50	123.75
Pipe, steel, black, schedule 40, threaded, cplg & hngr 10′ OC 2-1/2″ diam.	20.000	L.F.	231	382.50	613.50
Pipe, steel, black, schedule 40, threaded, cplg & hngr 10′ OC 2″ diam.	12.500	L.F.	92.81	187.03	279.84
Pipe, steel, black, schedule 40, threaded, cplg & hngr 10′ OC 1-1/4″ diam.	37.500	L.F.	191.25	403.59	594.84
Pipe, steel, black, schedule 40, threaded, cplg & hngr 10′ OC 1″ diam.	112.000	L.F.	466.20	1,125.60	1,591.80
Pipe, Tee, malleable iron, black, 150 lb. threaded, 4″ pipe size	2.000	Ea.	472.50	480	952.50
Pipe, Tee, malleable iron, black, 150 lb. threaded, 2-1/2″ pipe size	2.000	Ea.	132.75	213	345.75
Pipe, Tee, malleable iron, black, 150 lb. threaded, 2″ pipe size	1.000	Ea.	30.75	87	117.75
Pipe, Tee, malleable iron, black, 150 lb. threaded, 1-1/4″ pipe size	5.000	Ea.	72.56	341.25	413.81
Pipe, Tee, malleable iron, black, 150 lb. threaded, 1″ pipe size	4.000	Ea.	35.85	265.50	301.35
Pipe, 90° elbow, malleable iron, black, 150 lb. threaded, 1″ pipe size	6.000	Ea.	34.20	245.25	279.45
Sprinkler head std spray, brass 135°-286°F 1/2″ NPT, 3/8″ orifice	12.000	Ea.	203.40	510	713.40
Firecycle controls, incls panel, battery, solenoid valves, press switches	1.000	Ea.	17,250	1,950	19,200
Detector, firecycle system	2.000	Ea.	1,177.50	63.75	1,241.25
Firecycle pkg, swing check & flow control valves w/trim 4″ pipe size	1.000	Ea.	4,668.75	918.75	5,587.50
Air compressor, auto, complete, 200 Gal sprinkler sys. cap., 1/3 HP	1.000	Ea.	630	393.75	1,023.75
*Standpipe connection, wall, flush, brass w/plug & chain 2-1/2″x2-1/2″	1.000	Ea.	131.25	183.75	315
Valve, gate, bronze 300 psi, NRS, class 150, threaded, 1″ diam.	1.000	Ea.	84	28.13	112.13
TOTAL			29,746.27	8,989.88	38,736.15
COST PER S.F.			14.87	4.49	19.36

*Not included in systems under 2000 S.F.

D4010 390	Firecycle Sprinkler Systems	COST PER S.F.		
		MAT.	INST.	TOTAL
0520	Firecycle sprinkler systems, steel black sch. 40 pipe			
0530	Light hazard, one floor, 500 S.F.	49	9.75	58.75

D4010 Sprinklers

D4010 390	Firecycle Sprinkler Systems	COST PER S.F.		
		MAT.	INST.	TOTAL
0560	1000 S.F.	25	6.10	31.10
0580	2000 S.F. _R211313 -10_	14.85	4.50	19.35
0600	5000 S.F.	6.60	2.83	9.43
0620	10,000 S.F. _R211313 -20_	3.92	2.23	6.15
0640	50,000 S.F.	1.88	1.88	3.76
0660	Each additional floor of 500 S.F.	2.49	2.74	5.23
0680	1000 S.F.	1.92	2.50	4.42
0700	2000 S.F.	1.61	2.31	3.92
0720	5000 S.F.	1.41	2	3.41
0740	10,000 S.F.	1.36	1.83	3.19
0760	50,000 S.F.	1.30	1.69	2.99
1000	Ordinary hazard, one floor, 500 S.F.	49	10.15	59.15
1020	1000 S.F.	25	6.05	31.05
1040	2000 S.F.	15	4.63	19.63
1060	5000 S.F.	6.90	3	9.90
1080	10,000 S.F.	4.31	2.90	7.21
1100	50,000 S.F.	2.79	2.91	5.70
1140	Each additional floor, 500 S.F.	2.80	3.16	5.96
1160	1000 S.F.	1.91	2.56	4.47
1180	2000 S.F.	1.96	2.33	4.29
1200	5000 S.F.	1.77	2.21	3.98
1220	10,000 S.F.	1.59	2.16	3.75
1240	50,000 S.F.	1.55	1.94	3.49
1500	Extra hazard, one floor, 500 S.F.	52.50	11.65	64.15
1520	1000 S.F.	27	7.55	34.55
1540	2000 S.F.	15.30	5.70	21
1560	5000 S.F.	7.15	4.07	11.22
1580	10,000 S.F.	5.45	4.10	9.55
1600	50,000 S.F.	4.59	4.53	9.12
1660	Each additional floor, 500 S.F.	3.42	3.90	7.32
1680	1000 S.F.	2.76	3.66	6.42
1700	2000 S.F.	2.52	3.66	6.18
1720	5000 S.F.	2.08	3.27	5.35
1740	10,000 S.F.	2.47	2.98	5.45
1760	50,000 S.F.	2.46	2.86	5.32
2020	Grooved steel, black, sch. 40 pipe, light hazard, one floor			
2030	2000 S.F.	14.85	3.97	18.82
2060	10,000 S.F.	4.23	2.69	6.92
2100	Each additional floor, 2000 S.F.	1.89	1.89	3.78
2150	10,000 S.F.	1.35	1.55	2.90
2200	Ordinary hazard, one floor, 2000 S.F.	14.95	4.15	19.10
2250	10,000 S.F.	4.46	2.60	7.06
2300	Each additional floor, 2000 S.F.	1.97	2.07	4.04
2350	10,000 S.F.	1.61	2.07	3.68
2400	Extra hazard, one floor, 2000 S.F.	15.30	5	20.30
2450	10,000 S.F.	4.80	3.09	7.89
2500	Each additional floor, 2000 S.F.	2.53	3.02	5.55
2550	10,000 S.F.	2.10	2.56	4.66
3050	Grooved steel, black, sch. 10 pipe light hazard, one floor,			
3060	2000 S.F.	14.75	3.95	18.70
3100	10,000 S.F.	3.86	1.92	5.78
3150	Each additional floor, 2000 S.F.	1.81	1.87	3.68
3200	10,000 S.F.	1.30	1.52	2.82
3250	Ordinary hazard, one floor, 2000 S.F.	14.85	4.12	18.97
3300	10,000 S.F.	4.06	2.42	6.48
3350	Each additional floor, 2000 S.F.	1.89	2.04	3.93
3400	10,000 S.F.	1.51	2.02	3.53
3450	Extra hazard, one floor, 2000 S.F.	15.25	4.98	20.23

D40 Fire Protection

D4010 Sprinklers

D4010 390	Firecycle Sprinkler Systems	COST PER S.F.		
		MAT.	INST.	TOTAL
3500	10,000 S.F.	4.61	3.03	7.64
3550	Each additional floor, 2000 S.F.	2.47	2.99	5.46
3600	10,000 S.F.	2.02	2.51	4.53
4060	Copper tubing, type M, light hazard, one floor, 2000 S.F.	15.15	3.94	19.09
4100	10,000 S.F.	4.28	1.96	6.24
4150	Each additional floor, 2000 S.F.	2.20	1.90	4.10
4200	10,000 S.F.	1.73	1.57	3.30
4250	Ordinary hazard, one floor, 2000 S.F.	15.35	4.28	19.63
4300	10,000 S.F.	4.62	2.28	6.90
4350	Each additional floor, 2000 S.F.	2.17	1.93	4.10
4400	10,000 S.F.	1.81	1.68	3.49
4450	Extra hazard, one floor, 2000 S.F.	15.85	5.05	20.90
4500	10,000 S.F.	6.40	3.39	9.79
4550	Each additional floor, 2000 S.F.	3.09	3.05	6.14
4600	10,000 S.F.	3.15	2.79	5.94
5060	Copper tubing, type M, T-drill system, light hazard, one floor 2000 S.F.	15.10	3.72	18.82
5100	10,000 S.F.	4.16	1.66	5.82
5150	Each additional floor, 2000 S.F.	2.39	1.77	4.16
5200	10,000 S.F.	1.61	1.27	2.88
5250	Ordinary hazard, one floor, 2000 S.F.	15.15	3.77	18.92
5300	10,000 S.F.	4.52	2.06	6.58
5350	Each additional floor, 2000 S.F.	2.17	1.69	3.86
5400	10,000 S.F.	1.97	1.66	3.63
5450	Extra hazard, one floor, 2000 S.F.	15.35	4.30	19.65
5500	10,000 S.F.	5.60	2.49	8.09
5550	Each additional floor, 2000 S.F.	2.58	2.31	4.89
5600	10,000 S.F.	2.38	1.94	4.32

Wet Pipe System. A system employing automatic sprinklers attached to a piping system containing water and connected to a water supply so that water discharges immediately from sprinklers opened by heat from a fire.

All areas are assumed to be open.

System Components	QUANTITY	UNIT	COST EACH		
			MAT.	INST.	TOTAL
SYSTEM D4010 410 0580					
WET PIPE SPRINKLER, STEEL, BLACK, SCH. 40 PIPE					
LIGHT HAZARD, ONE FLOOR, 2000 S.F.					
Valve, gate, iron body, 125 lb., OS&Y, flanged, 4" diam.	1.000	Ea.	637.50	318.75	956.25
Valve, swing check, bronze, 125 lb, regrinding disc, 2-1/2" pipe size	1.000	Ea.	660	63.75	723.75
Valve, angle, bronze, 150 lb., rising stem, threaded, 2" diam.	1.000	Ea.	622.50	48.38	670.88
*Alarm valve, 2-1/2" pipe size	1.000	Ea.	1,350	307.50	1,657.50
Alarm, water motor, complete with gong	1.000	Ea.	341.25	127.50	468.75
Valve, swing check, w/balldrip CI with brass trim 4" pipe size	1.000	Ea.	288.75	307.50	596.25
Pipe, steel, black, schedule 40, 4" diam.	10.000	L.F.	180	271.65	451.65
*Flow control valve, trim & gauges, 4" pipe size	1.000	Set	4,762.50	693.75	5,456.25
Fire alarm horn, electric	1.000	Ea.	50.25	73.50	123.75
Pipe, steel, black, schedule 40, threaded, cplg & hngr 10' OC, 2-1/2" diam.	20.000	L.F.	231	382.50	613.50
Pipe, steel, black, schedule 40, threaded, cplg & hngr 10' OC, 2" diam.	12.500	L.F.	92.81	187.03	279.84
Pipe, steel, black, schedule 40, threaded, cplg & hngr 10' OC, 1-1/4" diam.	37.500	L.F.	191.25	403.59	594.84
Pipe, steel, black, schedule 40, threaded cplg & hngr 10' OC, 1" diam.	112.000	L.F.	466.20	1,125.60	1,591.80
Pipe Tee, malleable iron black, 150 lb. threaded, 4" pipe size	2.000	Ea.	472.50	480	952.50
Pipe Tee, malleable iron black, 150 lb. threaded, 2-1/2" pipe size	2.000	Ea.	132.75	213	345.75
Pipe Tee, malleable iron black, 150 lb. threaded, 2" pipe size	1.000	Ea.	30.75	87	117.75
Pipe Tee, malleable iron black, 150 lb. threaded, 1-1/4" pipe size	5.000	Ea.	72.56	341.25	413.81
Pipe Tee, malleable iron black, 150 lb. threaded, 1" pipe size	4.000	Ea.	35.85	265.50	301.35
Pipe 90° elbow, malleable iron black, 150 lb. threaded, 1" pipe size	6.000	Ea.	34.20	245.25	279.45
Sprinkler head, standard spray, brass 135°-286°F 1/2" NPT, 3/8" orifice	12.000	Ea.	203.40	510	713.40
Valve, gate, bronze, NRS, class 150, threaded, 1" pipe size	1.000	Ea.	84	28.13	112.13
*Standpipe connection, wall, single, flush w/plug & chain 2-1/2"x2-1/2"	1.000	Ea.	131.25	183.75	315
TOTAL			11,071.27	6,664.88	17,736.15
COST PER S.F.			5.54	3.33	8.87

*Not included in systems under 2000 S.F.

D4010 410	Wet Pipe Sprinkler Systems		COST PER S.F.		
			MAT.	INST.	TOTAL
0520	Wet pipe sprinkler systems, steel, black, sch. 40 pipe				
0530	Light hazard, one floor, 500 S.F.		2.95	3.22	6.17
0560	1000 S.F.		6.15	3.30	9.45
0580	2000 S.F.	R211313 -10	5.55	3.34	8.89
0600	5000 S.F.		2.63	2.35	4.98
0620	10,000 S.F.	R211313 -20	1.70	1.97	3.67

D40 Fire Protection

D4010 Sprinklers

D4010 410	Wet Pipe Sprinkler Systems	COST PER S.F.		
		MAT.	INST.	TOTAL
0640	50,000 S.F.	1.13	1.81	2.94
0660	Each additional floor, 500 S.F.	1.33	2.74	4.07
0680	1000 S.F.	1.35	2.54	3.89
0700	2000 S.F.	1.32	2.29	3.61
0720	5000 S.F.	.94	1.97	2.91
0740	10,000 S.F.	.89	1.80	2.69
0760	50,000 S.F.	.74	1.43	2.17
1000	Ordinary hazard, one floor, 500 S.F.	3.23	3.45	6.68
1020	1000 S.F.	6.10	3.27	9.37
1040	2000 S.F.	5.70	3.47	9.17
1060	5000 S.F.	2.91	2.52	5.43
1080	10,000 S.F.	2.09	2.64	4.73
1100	50,000 S.F.	1.75	2.50	4.25
1140	Each additional floor, 500 S.F.	1.62	3.10	4.72
1160	1000 S.F.	1.32	2.53	3.85
1180	2000 S.F.	1.47	2.52	3.99
1200	5000 S.F.	1.45	2.41	3.86
1220	10,000 S.F.	1.29	2.47	3.76
1240	50,000 S.F.	1.27	2.21	3.48
1500	Extra hazard, one floor, 500 S.F.	11.80	5.30	17.10
1520	1000 S.F.	7.35	4.60	11.95
1540	2000 S.F.	6	4.70	10.70
1560	5000 S.F.	3.63	4.15	7.78
1580	10,000 S.F.	3.23	3.91	7.14
1600	50,000 S.F.	3.57	3.78	7.35
1660	Each additional floor, 500 S.F.	2.24	3.84	6.08
1680	1000 S.F.	2.17	3.63	5.80
1700	2000 S.F.	1.93	3.63	5.56
1720	5000 S.F.	1.61	3.24	4.85
1740	10,000 S.F.	2	2.95	4.95
1760	50,000 S.F.	1.99	2.81	4.80
2020	Grooved steel, black sch. 40 pipe, light hazard, one floor, 2000 S.F.	5.50	2.81	8.31
2060	10,000 S.F.	2.16	1.76	3.92
2100	Each additional floor, 2000 S.F.	1.30	1.86	3.16
2150	10,000 S.F.	.88	1.52	2.40
2200	Ordinary hazard, one floor, 2000 S.F.	5.60	2.99	8.59
2250	10,000 S.F.	1.94	2.21	4.15
2300	Each additional floor, 2000 S.F.	1.38	2.04	3.42
2350	10,000 S.F.	1.14	2.04	3.18
2400	Extra hazard, one floor, 2000 S.F.	5.95	3.85	9.80
2450	10,000 S.F.	2.60	2.85	5.45
2500	Each additional floor, 2000 S.F.	1.94	2.99	4.93
2550	10,000 S.F.	1.63	2.53	4.16
3050	Grooved steel, black sch. 10 pipe, light hazard, one floor, 2000 S.F.	5.45	2.79	8.24
3100	10,000 S.F.	1.64	1.66	3.30
3150	Each additional floor, 2000 S.F.	1.22	1.84	3.06
3200	10,000 S.F.	.83	1.49	2.32
3250	Ordinary hazard, one floor, 2000 S.F.	5.50	2.96	8.46
3300	10,000 S.F.	1.84	2.16	4
3350	Each additional floor, 2000 S.F.	1.30	2.01	3.31
3400	10,000 S.F.	1.04	1.99	3.03
3450	Extra hazard, one floor, 2000 S.F.	5.90	3.82	9.72
3500	10,000 S.F.	2.41	2.79	5.20
3550	Each additional floor, 2000 S.F.	1.88	2.96	4.84
3600	10,000 S.F.	1.55	2.48	4.03
4050	Copper tubing, type M, light hazard, one floor, 2000 S.F.	5.80	2.78	8.58
4100	10,000 S.F.	2.06	1.70	3.76
4150	Each additional floor, 2000 S.F.	1.61	1.87	3.48

R211313 -40

D40 Fire Protection

D4010 Sprinklers

D4010 410	Wet Pipe Sprinkler Systems	COST PER S.F.		
		MAT.	INST.	TOTAL
4200	10,000 S.F.	1.26	1.54	2.80
4250	Ordinary hazard, one floor, 2000 S.F.	6	3.12	9.12
4300	10,000 S.F.	2.40	2.02	4.42
4350	Each additional floor, 2000 S.F.	1.81	2.07	3.88
4400	10,000 S.F.	1.54	1.81	3.35
4450	Extra hazard, one floor, 2000 S.F.	6.50	3.88	10.38
4500	10,000 S.F.	4.15	3.10	7.25
4550	Each additional floor, 2000 S.F.	2.50	3.02	5.52
4600	10,000 S.F.	2.68	2.76	5.44
5050	Copper tubing, type M, T-drill system, light hazard, one floor			
5060	2000 S.F.	5.80	2.56	8.36
5100	10,000 S.F.	1.94	1.40	3.34
5150	Each additional floor, 2000 S.F.	1.60	1.65	3.25
5200	10,000 S.F.	1.14	1.24	2.38
5250	Ordinary hazard, one floor, 2000 S.F.	5.80	2.61	8.41
5300	10,000 S.F.	2.30	1.80	4.10
5350	Each additional floor, 2000 S.F.	1.58	1.66	3.24
5400	10,000 S.F.	1.50	1.63	3.13
5450	Extra hazard, one floor, 2000 S.F.	6	3.14	9.14
5500	10,000 S.F.	3.38	2.25	5.63
5550	Each additional floor, 2000 S.F.	2.09	2.33	4.42
5600	10,000 S.F.	1.91	1.91	3.82

D4010 Sprinklers

System Components	QUANTITY	UNIT	COST PER EACH		
			MAT.	INST.	TOTAL
SYSTEM D4010 412 1000					
WET SPRINKLER SYSTEM, SCHEDULE 10 BLACK STEEL GROOVED PIPE					
SEPARATION ASSEMBLY, 2 INCH DIAMETER					
Hanger support for 2 inch pipe	2.000	Ea.	47	78	125
Pipe roller support yoke	2.000	Ea.	23.30	18.20	41.50
Spool pieces 2 inch	3.000	Ea.	23.55	49.50	73.05
2 inch grooved elbows	6.000	Ea.	210	171	381
Coupling grooved joint 2 inch pipe	10.000	Ea.	345	142	487
Roll grooved joint labor only	8.000	Ea.		88	88
TOTAL			648.85	546.70	1,195.55

D4010 412		Wet Sprinkler Seismic Components	COST PER EACH		
			MAT.	INST.	TOTAL
1000		Wet sprinkler sys, Sch. 10, blk steel grooved, separation assembly , 2 inch	650	545	1,195
1100		2-1/2 inch	715	605	1,320
1200		3 inch	960	695	1,655
1300		4 inch	1,200	865	2,065
1400		6 inch	2,450	1,250	3,700
1500		8 inch	4,375	1,450	5,825
2000		Wet sprinkler sys, Sch. 10, blk steel grooved, flexible coupling,2-1/2 inch	500	300	800
2100		3 inch	520	355	875

515

For customer support on your Plumbing Cost Data, call 877.709.6968.

D4010 Sprinklers

D4010 412	Wet Sprinkler Seismic Components	COST PER EACH		
		MAT.	INST.	TOTAL
2200	4 inch	700	470	1,170
2300	6 inch	1,000	665	1,665
2400	8 inch	1,425	830	2,255
2990	Allowance for seismic movement by providing annular space in concrete wall			
2995	If additional cores are performed for the same size and at the same time reduce			
2996	price of additional cores by 50 %. If different size cores reduce by 30 %.			
3000	Wet pipe sprinkler systems, 2 inch black steel, annular space 6 inch wall	18.90	630	648.90
3010	8 inch wall	19.15	635	654.15
3020	10 inch wall	19.35	645	664.35
3030	12 inch wall	19.60	650	669.60
3100	Wet pipe sprinkler systems, 2-1/2/3 inch black steel, annular space 6" wall	18.90	645	663.90
3110	8 inch wall	19.15	655	674.15
3120	10 inch wall	19.35	660	679.35
3130	12 inch wall	19.60	670	689.60
3200	Wet pipe sprinkler systems, 4 inch black steel, annular space 6" wall	18.90	720	738.90
3210	8 inch wall	19.15	725	744.15
3220	10 inch wall	19.35	730	749.35
3230	12 inch wall	19.60	740	759.60
3300	Wet pipe sprinkler systems, 6 inch black steel, annular space 6" wall	18.90	745	763.90
3310	8 inch wall	19.15	755	774.15
3320	10 inch wall	19.35	765	784.35
3330	12 inch wall	19.60	770	789.60
3400	Wet pipe sprinkler systems, 8 inch black steel, annular space 6" wall	18.90	780	798.90
3410	8 inch wall	19.15	785	804.15
3420	10 inch wall	19.35	790	809.35
3430	12 inch wall	19.60	795	814.60
3996	Lateral, longitudinal and 4-way seismic strut braces			
4000	Wet pipe sprinkler systems, 2 inch pipe double lateral strut brace	148	296	444
4010	2 -1/2 inch	152	296	448
4020	3 inch	155	297	452
4030	4 inch	179	300	479
4040	6 inch	211	300	511
4050	8 inch	234	300	534
4100	Wet pipe sprinkler systems, 2 inch pipe longitudinal strut brace	148	296	444
4110	2 -1/2 inch	152	296	448
4120	3 inch	155	297	452
4130	4 inch	179	300	479
4140	6 inch	211	300	511
4150	8 inch	234	300	534
4200	Wet pipe sprinkler systems, 2 inch pipe single lateral strut brace	103	195	298
4210	2 -1/2 inch	107	196	303
4220	3 inch	110	196	306
4230	4 inch	131	199	330
4240	6 inch	163	200	363
4250	8 inch	186	200	386
4300	Wet pipe sprinkler systems, 2 inch pipe longitudinal wire brace	86	204	290
4310	2 -1/2 inch	90	205	295
4320	3 inch	93.50	205	298.50
4330	4 inch	118	212	330
4340	6 inch	149	210	359
4350	8 inch	172	210	382
4400	Wet pipe sprinkler systems, 2 inch pipe 4-way wire brace	115	315	430
4410	2 -1/2 inch	119	315	434
4420	3 inch	122	315	437
4430	4 inch	153	325	478
4440	6 inch	183	320	503
4450	8 inch	206	320	526
4500	Wet pipe sprinkler systems, 2 inch pipe 4-way strut brace	239	500	739

D4010 Sprinklers

D4010 412	Wet Sprinkler Seismic Components	COST PER EACH		
		MAT.	INST.	TOTAL
4510	2 -1/2 inch	243	500	743
4520	3 inch	246	500	746
4530	4 inch	274	505	779
4540	6 inch	305	505	810
4550	8 inch	330	505	835

Wet Pipe System. A system employing automatic sprinklers attached to a piping system containing water and connected to a water supply so that water discharges immediately from sprinklers opened by heat from a fire.

All areas are assumed to be open.

System Components	QUANTITY	UNIT	COST PER S.F.		
			MAT.	INST.	TOTAL
SYSTEM D4010 413 0580					
SEISMIC WET PIPE SPRINKLER, STEEL, BLACK, SCH. 40 PIPE					
LIGHT HAZARD, ONE FLOOR, 3′ FLEXIBLE FEED, 2000 S.F.					
Valve, gate, iron body, 125 lb., OS&Y, flanged, 4″ diam.	1.000	Ea.	.24	.12	.36
Valve, swing check, bronze, 125 lb, regrinding disc, 2-1/2″ pipe size	1.000	Ea.	.25	.02	.27
Valve, angle, bronze, 150 lb., rising stem, threaded, 2″ diam.	1.000	Ea.	.23	.02	.25
*Alarm valve, 2-1/2″ pipe size	1.000	Ea.	.51	.12	.63
Alarm, water motor, complete with gong	1.000	Ea.	.13	.05	.18
Valve, swing check, w/balldrip CI with brass trim 4″ pipe size	1.000	Ea.	.11	.12	.23
Pipe, steel, black, schedule 40, 4″ diam.	10.000	L.F.	.07	.10	.17
*Flow control valve, trim & gauges, 4″ pipe size	1.000	Set	1.79	.26	2.05
Fire alarm horn, electric	1.000	Ea.	.02	.03	.05
Pipe, steel, black, schedule 40, threaded, cplg & hngr 10′ OC, 2-1/2″ diam.	20.000	L.F.	.09	.14	.23
Pipe, steel, black, schedule 40, threaded, cplg & hngr 10′ OC, 2″ diam.	12.500	L.F.	.03	.07	.10
Pipe, steel, black, schedule 40, threaded, cplg & hngr 10′ OC, 1-1/4″ diam.	37.500	L.F.	.07	.15	.22
Pipe Tee, malleable iron black, 150 lb. threaded, 4″ pipe size	2.000	Ea.	.18	.18	.36
Pipe Tee, malleable iron black, 150 lb. threaded, 2-1/2″ pipe size	2.000	Ea.	.05	.08	.13
Pipe Tee, malleable iron black, 150 lb. threaded, 2″ pipe size	1.000	Ea.	.01	.03	.04
Pipe Tee, malleable iron black, 150 lb. threaded, 1-1/4″ pipe size	5.000	Ea.	.03	.13	.16
Pipe Tee, malleable iron black, 150 lb. threaded, 1″ pipe size	4.000	Ea.	.01	.10	.11
Pipe 90° elbow, malleable iron black, 150 lb. threaded, 1″ pipe size	6.000	Ea.	.01	.09	.10
Sprinkler head, standard spray, brass 135°-286°F 1/2″ NPT, 3/8″ orifice	12.000	Ea.	.08	.19	.27
Valve, gate, bronze, NRS, class 150, threaded, 1″ pipe size	1.000	Ea.	.03	.01	.04
*Standpipe connection, wall, single, flush w/plug & chain 2-1/2″x2-1/2″	1.000	Ea.	.05	.07	.12
Flexible connector 36 inches long	12.000	Ea.	.35	.10	.45
TOTAL			4.34	2.18	6.52

*Not included in systems under 2000 S.F.

D4010 413	Seismic Wet Pipe Flexible Feed Sprinkler Systems	COST PER S.F.		
		MAT.	INST.	TOTAL
0520	Seismic wet pipe 3′ flexible feed sprinkler systems, steel, black, sch. 40 pipe			
0525	Note: See D4010-412 for strut braces, flexible joints and separation assemblies			
0530	Light hazard, one floor, 3′ flexible feed, 500 S.F.	3	2.60	5.60
0560	1000 S.F.	6.30	2.90	9.20
0580	2000 S.F.	4.34	2.18	6.52
0600	5000 S.F.	2.83	2.06	4.89
0620	10,000 S.F.	1.93	1.83	3.76
0640	50,000 S.F.	1.29	1.57	2.86

For customer support on your Plumbing Cost Data, call 877.709.6968.

D4010 Sprinklers

D4010 413	Seismic Wet Pipe Flexible Feed Sprinkler Systems	COST PER S.F.		
		MAT.	INST.	TOTAL
0660	Each additional floor, 500 S.F.	1.49	2.15	3.64
0680	1000 S.F.	1.61	2.18	3.79
0700	2000 S.F.	1.44	1.83	3.27
0720	5000 S.F.	1.14	1.68	2.82
0740	10,000 S.F.	1.12	1.66	2.78
0760	50,000 S.F.	.97	1.34	2.31
1000	Ordinary hazard, one floor, 500 S.F.	3.39	2.86	6.25
1020	1000 S.F.	6.35	2.84	9.19
1040	2000 S.F.	6	3.20	9.20
1060	5000 S.F.	3.24	2.33	5.57
1080	10,000 S.F.	2.45	2.52	4.97
1100	50,000 S.F.	2.04	2.25	4.29
1140	Each additional floor, 500 S.F.	1.78	2.51	4.29
1160	1000 S.F.	1.55	2.10	3.65
1180	2000 S.F.	1.76	2.25	4.01
1200	5000 S.F.	1.78	2.22	4
1220	10,000 S.F.	1.65	2.35	4
1240	50,000 S.F.	1.56	1.96	3.52
1500	Extra hazard, one floor, 500 S.F.	12.05	4.39	16.44
1520	1000 S.F.	7.55	3.70	11.25
1540	2000 S.F.	6.40	4.28	10.68
1560	5000 S.F.	4.07	3.88	7.95
1580	10,000 S.F.	3.75	3.70	7.45
1600	50,000 S.F.	4.11	3.54	7.65
1660	Each additional floor, 500 S.F.	2.45	2.92	5.37
1680	1000 S.F.	2.39	2.73	5.12
1700	2000 S.F.	2.36	3.21	5.57
1720	5000 S.F.	2.05	2.97	5.02
1740	10,000 S.F.	2.52	2.74	5.26
1760	50,000 S.F.	2.53	2.57	5.10
2020	Grooved steel, black sch. 40 pipe, light hazard, one floor, 2000 S.F.	5.65	2.44	8.09
2060	10,000 S.F.	2.15	1.60	3.75
2100	Each additional floor, 2000 S.F.	1.41	1.49	2.90
2150	10,000 S.F.	.87	1.36	2.23
2200	Ordinary hazard, one floor, 2000 S.F.	5.90	2.78	8.68
2250	10,000 S.F.	2.30	2.13	4.43
2300	Each additional floor, 2000 S.F.	1.67	1.83	3.50
2350	10,000 S.F.	1.50	1.96	3.46
2400	Extra hazard, one floor, 2000 S.F.	6.35	3.52	9.87
2450	10,000 S.F.	3.10	2.70	5.80
2500	Each additional floor, 2000 S.F.	2.35	2.66	5.01
2550	10,000 S.F.	2.13	2.38	4.51
3050	Grooved steel black sch. 10 pipe, light hazard, one floor, 2000 S.F.	5.55	2.42	7.97
3100	10,000 S.F.	1.87	1.56	3.43
3150	Each additional floor, 2000 S.F.	1.33	1.47	2.80
3200	10,000 S.F.	.82	1.33	2.15
3250	Ordinary hazard, one floor, 2000 S.F.	5.80	2.75	8.55
3300	10,000 S.F.	2.20	2.08	4.28
3350	Each additional floor, 2000 S.F.	1.59	1.80	3.39
3400	10,000 S.F.	1.40	1.91	3.31
3450	Extra hazard, one floor, 2000 S.F.	6.30	3.49	9.79
3500	10,000 S.F.	2.91	2.64	5.55
3550	Each additional floor, 2000 S.F.	2.29	2.63	4.92
3600	10,000 S.F.	2.05	2.33	4.38

519

For customer support on your Plumbing Cost Data, call 877.709.6968.

Wet Pipe System. A system employing automatic sprinklers attached to a piping system containing water and connected to a water supply so that water discharges immediately from sprinklers opened by heat from a fire.

All areas are assumed to be open.

System Components	QUANTITY	UNIT	COST PER S.F.		
			MAT.	INST.	TOTAL
SYSTEM D4010 414 0580					
SEISMIC WET PIPE SPRINKLER, STEEL, BLACK, SCH. 40 PIPE					
LIGHT HAZARD, ONE FLOOR, 4' FLEXIBLE FEED, 2000 S.F.					
Valve, gate, iron body, 125 lb., OS&Y, flanged, 4″ diam.	1.000	Ea.	.24	.12	.36
Valve, swing check, bronze, 125 lb, regrinding disc, 2-1/2″ pipe size	1.000	Ea.	.25	.02	.27
Valve, angle, bronze, 150 lb., rising stem, threaded, 2″ diam.	1.000	Ea.	.23	.02	.25
*Alarm valve, 2-1/2″ pipe size	1.000	Ea.	.51	.12	.63
Alarm, water motor, complete with gong	1.000	Ea.	.13	.05	.18
Valve, swing check, w/balldrip CI with brass trim 4″ pipe size	1.000	Ea.	.11	.12	.23
Pipe, steel, black, schedule 40, 4″ diam.	10.000	L.F.	.07	.10	.17
*Flow control valve, trim & gauges, 4″ pipe size	1.000	Set	1.79	.26	2.05
Fire alarm horn, electric	1.000	Ea.	.02	.03	.05
Pipe, steel, black, schedule 40, threaded, cplg & hngr 10' OC, 2-1/2″ diam.	20.000	L.F.	.09	.14	.23
Pipe, steel, black, schedule 40, threaded, cplg & hngr 10' OC, 2″ diam.	12.500	L.F.	.03	.07	.10
Pipe, steel, black, schedule 40, threaded, cplg & hngr 10' OC, 1-1/4″ diam.	37.500	L.F.	.07	.15	.22
Pipe Tee, malleable iron black, 150 lb. threaded, 4″ pipe size	2.000	Ea.	.18	.18	.36
Pipe Tee, malleable iron black, 150 lb. threaded, 2-1/2″ pipe size	2.000	Ea.	.05	.08	.13
Pipe Tee, malleable iron black, 150 lb. threaded, 2″ pipe size	1.000	Ea.	.01	.03	.04
Pipe Tee, malleable iron black, 150 lb. threaded, 1-1/4″ pipe size	5.000	Ea.	.03	.13	.16
Pipe Tee, malleable iron black, 150 lb. threaded, 1″ pipe size	4.000	Ea.	.01	.10	.11
Pipe 90° elbow, malleable iron black, 150 lb. threaded, 1″ pipe size	6.000	Ea.	.01	.09	.10
Sprinkler head, standard spray, brass 135°-286°F 1/2″ NPT, 3/8″ orifice	12.000	Ea.	.08	.19	.27
Valve, gate, bronze, NRS, class 150, threaded, 1″ pipe size	1.000	Ea.	.03	.01	.04
*Standpipe connection, wall, single, flush w/plug & chain 2-1/2″x2-1/2″	1.000	Ea.	.05	.07	.12
Flexible connector 48 inches long	12.000	Ea.	.40	.14	.54
TOTAL			4.39	2.22	6.61

*Not included in systems under 2000 S.F.

D4010 414	Seismic Wet Pipe Flexible Feed Sprinkler Systems	COST PER S.F.		
		MAT.	INST.	TOTAL
0520	Seismic wet pipe 4' flexible feed sprinkler systems, steel, black, sch. 40 pipe			
0525	Note: See D4010-412 for strut braces, flexible joints and separation assemblies			
0530	Light hazard, one floor, 4' flexible feed, 500 S.F.	3.05	2.64	5.69
0560	1000 S.F.	6.35	2.94	9.29
0580	2000 S.F.	4.39	2.22	6.61
0600	5000 S.F.	2.89	2.10	4.99
0620	10,000 S.F.	1.97	1.87	3.84
0640	50,000 S.F.	1.34	1.60	2.94

D4010 Sprinklers

D4010 414	Seismic Wet Pipe Flexible Feed Sprinkler Systems	COST PER S.F.		
		MAT.	INST.	TOTAL
0660	Each additional floor, 500 S.F.	1.56	2.21	3.77
0680	1000 S.F.	1.67	2.23	3.90
0700	2000 S.F.	1.49	1.87	3.36
0720	5000 S.F.	1.20	1.72	2.92
0740	10,000 S.F.	1.16	1.70	2.86
0760	50,000 S.F.	1.02	1.37	2.39
1000	Ordinary hazard, one floor, 500 S.F.	3.46	2.92	6.38
1020	1000 S.F.	6.40	2.89	9.29
1040	2000 S.F.	6.05	3.26	9.31
1060	5000 S.F.	3.31	2.39	5.70
1080	10,000 S.F.	2.51	2.57	5.08
1100	50,000 S.F.	2.11	2.30	4.41
1140	Each additional floor, 500 S.F.	1.85	2.57	4.42
1160	1000 S.F.	1.61	2.15	3.76
1180	2000 S.F.	1.83	2.31	4.14
1200	5000 S.F.	1.85	2.28	4.13
1220	10,000 S.F.	1.71	2.40	4.11
1240	50,000 S.F.	1.63	2.01	3.64
1500	Extra hazard, one floor, 500 S.F.	12.15	4.47	16.62
1520	1000 S.F.	7.65	3.78	11.43
1540	2000 S.F.	6.50	4.36	10.86
1560	5000 S.F.	4.16	3.95	8.11
1580	10,000 S.F.	3.86	3.78	7.64
1600	50,000 S.F.	4.22	3.62	7.84
1660	Each additional floor, 500 S.F.	2.55	3	5.55
1680	1000 S.F.	2.49	2.81	5.30
1700	2000 S.F.	2.46	3.29	5.75
1720	5000 S.F.	2.14	3.04	5.18
1740	10,000 S.F.	2.63	2.82	5.45
1760	50,000 S.F.	2.64	2.65	5.29
2020	Grooved steel, black sch. 40 pipe, light hazard, one floor, 2000 S.F.	5.70	2.48	8.18
2060	10,000 S.F.	2.17	1.61	3.78
2100	Each additional floor, 2000 S.F.	1.46	1.53	2.99
2150	10,000 S.F.	.89	1.37	2.26
2200	Ordinary hazard, one floor, 2000 S.F.	5.95	2.84	8.79
2250	10,000 S.F.	2.36	2.18	4.54
2300	Each additional floor, 2000 S.F.	1.74	1.89	3.63
2350	10,000 S.F.	1.56	2.01	3.57
2400	Extra hazard, one floor, 2000 S.F.	6.45	3.60	10.05
2450	10,000 S.F.	3.21	2.78	5.99
2500	Each additional floor, 2000 S.F.	2.45	2.74	5.19
2550	10,000 S.F.	2.24	2.46	4.70
3050	Grooved steel black sch. 10 pipe, light hazard, one floor, 2000 S.F.	5.60	2.46	8.06
3100	10,000 S.F.	1.91	1.60	3.51
3150	Each additional floor, 2000 S.F.	1.38	1.51	2.89
3200	10,000 S.F.	.84	1.34	2.18
3250	Ordinary hazard, one floor, 2000 S.F.	5.90	2.81	8.71
3300	10,000 S.F.	2.26	2.13	4.39
3350	Each additional floor, 2000 S.F.	1.66	1.86	3.52
3400	10,000 S.F.	1.46	1.96	3.42
3450	Extra hazard, one floor, 2000 S.F.	6.40	3.57	9.97
3500	10,000 S.F.	3.02	2.72	5.74
3550	Each additional floor, 2000 S.F.	2.39	2.71	5.10
3600	10,000 S.F.	2.16	2.41	4.57

D40 Fire Protection

D4020 Standpipes

Roof

Roof connections with hose gate valves (for combustible roof)

Hose connections on each floor (size based on class of service)

Check Valve

Siamese inlet connections (for fire department use)

System Components	QUANTITY	UNIT	COST PER FLOOR		
			MAT.	INST.	TOTAL
SYSTEM D4020 310 0560					
WET STANDPIPE RISER, CLASS I, STEEL, BLACK, SCH. 40 PIPE, 10' HEIGHT					
4" DIAMETER PIPE, ONE FLOOR					
Pipe, steel, black, schedule 40, threaded, 4" diam.	20.000	L.F.	590	710	1,300
Pipe, Tee, malleable iron, black, 150 lb. threaded, 4" pipe size	2.000	Ea.	630	640	1,270
Pipe, 90° elbow, malleable iron, black, 150 lb threaded 4" pipe size	1.000	Ea.	200	213	413
Pipe, nipple, steel, black, schedule 40, 2-1/2" pipe size x 3" long	2.000	Ea.	24.50	160	184.50
Fire valve, gate, 300 lb., brass w/handwheel, 2-1/2" pipe size	1.000	Ea.	208	97	305
Fire valve, pressure restricting, adj, rgh brs, 2-1/2" pipe size	1.000	Ea.	352	194	546
Valve, swing check, w/ball drip, CI w/brs. ftngs., 4" pipe size	1.000	Ea.	385	410	795
Standpipe conn wall dble. flush brs w/plugs & chains 2-1/2"x2-1/2"x4"	1.000	Ea.	570	245	815
Valve, swing check, bronze, 125 lb, regrinding disc, 2-1/2" pipe size	1.000	Ea.	880	85	965
Roof manifold, fire, w/valves & caps, horiz/vert brs 2-1/2"x2-1/2"x4"	1.000	Ea.	185	255	440
Fire, hydrolator, vent & drain, 2-1/2" pipe size	1.000	Ea.	92.50	56.50	149
Valve, gate, iron body 125 lb., OS&Y, threaded, 4" pipe size	1.000	Ea.	850	425	1,275
TOTAL			4,967	3,490.50	8,457.50

D4020 310	Wet Standpipe Risers, Class I		COST PER FLOOR		
			MAT.	INST.	TOTAL
0550	Wet standpipe risers, Class I, steel, black sch. 40, 10' height				
0560	4" diameter pipe, one floor		4,975	3,500	8,475
0580	Additional floors		1,100	1,075	2,175
0600	6" diameter pipe, one floor		8,550	6,125	14,675
0620	Additional floors		2,075	1,725	3,800
0640	8" diameter pipe, one floor	R211226 -10	13,200	7,425	20,625
0660	Additional floors		3,100	2,075	5,175
0680		R211226 -20			

D4020 310	Wet Standpipe Risers, Class II		COST PER FLOOR		
			MAT.	INST.	TOTAL
1030	Wet standpipe risers, Class II, steel, black sch. 40, 10' height				
1040	2" diameter pipe, one floor		1,950	1,250	3,200
1060	Additional floors		570	485	1,055
1080	2-1/2" diameter pipe, one floor		2,875	1,850	4,725
1100	Additional floors		660	565	1,225
1120					

D4020 Standpipes

D4020 310	Wet Standpipe Risers, Class III	COST PER FLOOR		
		MAT.	INST.	TOTAL
1530	Wet standpipe risers, Class III, steel, black sch. 40, 10' height			
1540	4" diameter pipe, one floor	5,100	3,500	8,600
1560	Additional floors	950	910	1,860
1580	6" diameter pipe, one floor	8,675	6,125	14,800
1600	Additional floors	2,125	1,725	3,850
1620	8" diameter pipe, one floor	13,300	7,425	20,725
1640	Additional floors	3,150	2,075	5,225

D40 Fire Protection

D4020 Standpipes

Roof

Roof connections with hose gate valves (for combustible roof)

Hose connections on each floor (size based on class of service)

Check Valve

Siamese inlet connections (for fire department use)

System Components	QUANTITY	UNIT	COST PER FLOOR		
			MAT.	INST.	TOTAL
SYSTEM D4020 330 0540					
DRY STANDPIPE RISER, CLASS I, PIPE, STEEL, BLACK, SCH 40, 10′ HEIGHT					
4″ DIAMETER PIPE, ONE FLOOR					
Pipe, steel, black, schedule 40, threaded, 4″ diam.	20.000	L.F.	590	710	1,300
Pipe, Tee, malleable iron, black, 150 lb. threaded, 4″ pipe size	2.000	Ea.	630	640	1,270
Pipe, 90° elbow, malleable iron, black, 150 lb threaded 4″ pipe size	1.000	Ea.	200	213	413
Pipe, nipple, steel, black, schedule 40, 2-1/2″ pipe size x 3″ long	2.000	Ea.	24.50	160	184.50
Fire valve gate NRS 300 lb., brass w/handwheel, 2-1/2″ pipe size	1.000	Ea.	208	97	305
Fire valve, pressure restricting, adj, rgh brs, 2-1/2″ pipe size	1.000	Ea.	176	97	273
Standpipe conn wall dble. flush brs. w/plugs & chains 2-1/2″x2-1/2″x4″	1.000	Ea.	570	245	815
Valve swing check w/ball drip CI w/brs. ftngs., 4″pipe size	1.000	Ea.	385	410	795
Roof manifold, fire, w/valves & caps, horiz/vert brs 2-1/2″x2-1/2″x4″	1.000	Ea.	185	255	440
TOTAL			2,968.50	2,827	5,795.50

D4020 330	Dry Standpipe Risers, Class I		COST PER FLOOR		
			MAT.	INST.	TOTAL
0530	Dry standpipe riser, Class I, steel, black sch. 40, 10′ height				
0540	4″ diameter pipe, one floor		2,975	2,825	5,800
0560	Additional floors		1,025	1,025	2,050
0580	6″ diameter pipe, one floor		6,100	4,850	10,950
0600	Additional floors	R211226 -10	1,975	1,650	3,625
0620	8″ diameter pipe, one floor		9,700	5,900	15,600
0640	Additional floors	R211226 -20	3,000	2,025	5,025
0660					

D4020 330	Dry Standpipe Risers, Class II		COST PER FLOOR		
			MAT.	INST.	TOTAL
1030	Dry standpipe risers, Class II, steel, black sch. 40, 10′ height				
1040	2″ diameter pipe, one floor		1,575	1,300	2,875
1060	Additional floor		480	430	910
1080	2-1/2″ diameter pipe, one floor		2,250	1,525	3,775
1100	Additional floors		570	510	1,080
1120					

D4020 Standpipes

D4020 330	Dry Standpipe Risers, Class III	COST PER FLOOR		
		MAT.	INST.	TOTAL
1530	Dry standpipe risers, Class III, steel, black sch. 40, 10′ height			
1540	4″ diameter pipe, one floor	3,025	2,775	5,800
1560	Additional floors	870	930	1,800
1580	6″ diameter pipe, one floor	6,150	4,850	11,000
1600	Additional floors	2,025	1,650	3,675
1620	8″ diameter pipe, one floor	9,775	5,900	15,675
1640	Additional floor	3,075	2,025	5,100

D4020 410	Fire Hose Equipment	COST EACH		
		MAT.	INST.	TOTAL
0100	Adapters, reducing, 1 piece, FxM, hexagon, cast brass, 2-1/2" x 1-1/2"	61		61
0200	Pin lug, 1-1/2" x 1"	52.50		52.50
0250	3" x 2-1/2"	137		137
0300	For polished chrome, add 75% mat.			
0400	Cabinets, D.S. glass in door, recessed, steel box, not equipped			
0500	Single extinguisher, steel door & frame	128	153	281
0550	Stainless steel door & frame	226	153	379
0600	Valve, 2-1/2" angle, steel door & frame	149	102	251
0650	Aluminum door & frame	180	102	282
0700	Stainless steel door & frame	243	102	345
0750	Hose rack assy, 2-1/2" x 1-1/2" valve & 100' hose, steel door & frame	278	204	482
0800	Aluminum door & frame	410	204	614
0850	Stainless steel door & frame	545	204	749
0900	Hose rack assy & extinguisher,2-1/2"x1-1/2" valve & hose,steel door & frame	291	245	536
0950	Aluminum	525	245	770
1000	Stainless steel	590	245	835
1550	Compressor, air, dry pipe system, automatic, 200 gal., 3/4 H.P.	840	525	1,365
1600	520 gal., 1 H.P.	870	525	1,395
1650	Alarm, electric pressure switch (circuit closer)	97.50	26	123.50
2500	Couplings, hose, rocker lug, cast brass, 1-1/2"	58.50		58.50
2550	2-1/2"	77		77
3000	Escutcheon plate, for angle valves, polished brass, 1-1/2"	16.70		16.70
3050	2-1/2"	27		27
3500	Fire pump, electric, w/controller, fittings, relief valve			
3550	4" pump, 30 H.P., 500 G.P.M.	16,400	3,800	20,200
3600	5" pump, 40 H.P., 1000 G.P.M.	18,200	4,325	22,525
3650	5" pump, 100 H.P., 1000 G.P.M.	24,800	4,800	29,600
3700	For jockey pump system, add	2,850	610	3,460
5000	Hose, per linear foot, synthetic jacket, lined,			
5100	300 lb. test, 1-1/2" diameter	3.56	.47	4.03
5150	2-1/2" diameter	6.15	.56	6.71
5200	500 lb. test, 1-1/2" diameter	3.69	.47	4.16
5250	2-1/2" diameter	6.50	.56	7.06
5500	Nozzle, plain stream, polished brass, 1-1/2" x 10"	52.50		52.50
5550	2-1/2" x 15" x 13/16" or 1-1/2"	110		110
5600	Heavy duty combination adjustable fog and straight stream w/handle 1-1/2"	495		495
5650	2-1/2" direct connection	615		615
6000	Rack, for 1-1/2" diameter hose 100 ft. long, steel	67	61	128
6050	Brass	115	61	176
6500	Reel, steel, for 50 ft. long 1-1/2" diameter hose	144	87.50	231.50
6550	For 75 ft. long 2-1/2" diameter hose	234	87.50	321.50
7050	Siamese, w/plugs & chains, polished brass, sidewalk, 4" x 2-1/2" x 2-1/2"	695	490	1,185
7100	6" x 2-1/2" x 2-1/2"	860	610	1,470
7200	Wall type, flush, 4" x 2-1/2" x 2-1/2"	570	245	815
7250	6" x 2-1/2" x 2-1/2"	795	266	1,061
7300	Projecting, 4" x 2-1/2" x 2-1/2"	520	245	765
7350	6" x 2-1/2" x 2-1/2"	880	266	1,146
7400	For chrome plate, add 15% mat.			
8000	Valves, angle, wheel handle, 300 Lb., rough brass, 1-1/2"	104	56.50	160.50
8050	2-1/2"	192	97	289
8100	Combination pressure restricting, 1-1/2"	94	56.50	150.50
8150	2-1/2"	192	97	289
8200	Pressure restricting, adjustable, satin brass, 1-1/2"	117	56.50	173.50
8250	2-1/2"	176	97	273
8300	Hydrolator, vent and drain, rough brass, 1-1/2"	92.50	56.50	149
8350	2-1/2"	92.50	56.50	149
8400	Cabinet assy, incls. adapter, rack, hose, and nozzle	875	370	1,245

D4090 Other Fire Protection Systems

General: Automatic fire protection (suppression) systems other than water sprinklers may be desired for special environments, high risk areas, isolated locations or unusual hazards. Some typical applications would include:

Paint dip tanks
Securities vaults
Electronic data processing
Tape and data storage
Transformer rooms
Spray booths
Petroleum storage
High rack storage

Piping and wiring costs are dependent on the individual application and must be added to the component costs shown below.

All areas are assumed to be open.

D4090 910	Fire Suppression Unit Components	COST EACH		
		MAT.	INST.	TOTAL
0020	Detectors with brackets			
0040	Fixed temperature heat detector	56	82	138
0060	Rate of temperature rise detector	56	82	138
0080	Ion detector (smoke) detector	121	106	227
0200	Extinguisher agent			
0240	200 lb FM200, container	8,550	320	8,870
0280	75 lb carbon dioxide cylinder	1,400	213	1,613
0320	Dispersion nozzle			
0340	FM200 1-1/2″ dispersion nozzle	73.50	50.50	124
0380	Carbon dioxide 3″ x 5″ dispersion nozzle	73.50	39.50	113
0420	Control station			
0440	Single zone control station with batteries	1,900	655	2,555
0470	Multizone (4) control station with batteries	3,600	1,300	4,900
0490				
0500	Electric mechanical release	184	340	524
0520				
0550	Manual pull station	66.50	118	184.50
0570				
0640	Battery standby power 10″ x 10″ x 17″	455	164	619
0700				
0740	Bell signalling device	86.50	82	168.50

D4090 920	FM200 Systems	COST PER C.F.		
		MAT.	INST.	TOTAL
0820	Average FM200 system, minimum			1.94
0840	Maximum			3.85

527

G1030 Site Earthwork

Trenching Systems are shown on a cost per linear foot basis. The systems include: excavation; backfill and removal of spoil; and compaction for various depths and trench bottom widths. The backfill has been reduced to accommodate a pipe of suitable diameter and bedding.

The slope for trench sides varies from none to 1:1.

The Expanded System Listing shows Trenching Systems that range from 2' to 12' in width. Depths range from 2' to 25'.

System Components			COST PER L.F.		
	QUANTITY	UNIT	EQUIP.	LABOR	TOTAL
SYSTEM G1030 805 1310					
TRENCHING COMMON EARTH, NO SLOPE, 2' WIDE, 2' DP, 3/8 C.Y. BUCKET					
Excavation, trench, hyd. backhoe, track mtd., 3/8 C.Y. bucket	.148	B.C.Y.	.40	1.07	1.47
Backfill and load spoil, from stockpile	.153	L.C.Y.	.13	.32	.45
Compaction by vibrating plate, 6" lifts, 4 passes	.118	E.C.Y.	.03	.39	.42
Remove excess spoil, 8 C.Y. dump truck, 2 mile roundtrip	.040	L.C.Y.	.15	.16	.31
TOTAL			.71	1.94	2.65

G1030 805	Trenching Common Earth	COST PER L.F.		
		EQUIP.	LABOR	TOTAL
1310	Trenching, common earth, no slope, 2' wide, 2' deep, 3/8 C.Y. bucket	.71	1.94	2.65
1320	3' deep, 3/8 C.Y. bucket	1	2.93	3.93
1330	4' deep, 3/8 C.Y. bucket	1.29	3.90	5.19
1340	6' deep, 3/8 C.Y. bucket	1.67	5.05	6.72
1350	8' deep, 1/2 C.Y. bucket	2.21	6.70	8.91
1360	10' deep, 1 C.Y. bucket	3.46	7.95	11.41
1400	4' wide, 2' deep, 3/8 C.Y. bucket	1.60	3.83	5.43
1410	3' deep, 3/8 C.Y. bucket	2.20	5.80	8
1420	4' deep, 1/2 C.Y. bucket	2.56	6.50	9.06
1430	6' deep, 1/2 C.Y. bucket	4.12	10.45	14.57
1440	8' deep, 1/2 C.Y. bucket	6.60	13.45	20.05
1450	10' deep, 1 C.Y. bucket	7.95	16.70	24.65
1460	12' deep, 1 C.Y. bucket	10.25	21.50	31.75
1470	15' deep, 1-1/2 C.Y. bucket	9.40	19.05	28.45
1480	18' deep, 2-1/2 C.Y. bucket	13	26.50	39.50
1520	6' wide, 6' deep, 5/8 C.Y. bucket w/trench box	8.50	15.45	23.95
1530	8' deep, 3/4 C.Y. bucket	11.30	20.50	31.80
1540	10' deep, 1 C.Y. bucket	11.50	21	32.50
1550	12' deep, 1-1/2 C.Y. bucket	12.35	22.50	34.85
1560	16' deep, 2-1/2 C.Y. bucket	16.90	28.50	45.40
1570	20' deep, 3-1/2 C.Y. bucket	22	34	56
1580	24' deep, 3-1/2 C.Y. bucket	26	40.50	66.50
1640	8' wide, 12' deep, 1-1/2 C.Y. bucket w/trench box	17.30	28.50	45.80
1650	15' deep, 1-1/2 C.Y. bucket	22.50	37.50	60
1660	18' deep, 2-1/2 C.Y. bucket	24.50	38	62.50
1680	24' deep, 3-1/2 C.Y. bucket	35.50	52.50	88
1730	10' wide, 20' deep, 3-1/2 C.Y. bucket w/trench box	28.50	50	78.50
1740	24' deep, 3-1/2 C.Y. bucket	42.50	60	102.50
1780	12' wide, 20' deep, 3-1/2 C.Y. bucket w/trench box	45	63.50	108.50
1790	25' deep, bucket	55.50	81	136.50
1800	1/2 to 1 slope, 2' wide, 2' deep, 3/8 C.Y. bucket	1	2.93	3.93
1810	3' deep, 3/8 C.Y. bucket	1.67	5.15	6.82
1820	4' deep, 3/8 C.Y. bucket	2.49	7.85	10.34
1840	6' deep, 3/8 C.Y. bucket	4	12.70	16.70

G1030 Site Earthwork

G1030 805	Trenching Common Earth	COST PER L.F.		
		EQUIP.	LABOR	TOTAL
1860	8' deep, 1/2 C.Y. bucket	6.35	20.50	26.85
1880	10' deep, 1 C.Y. bucket	11.90	28	39.90
2300	4' wide, 2' deep, 3/8 C.Y. bucket	1.90	4.81	6.71
2310	3' deep, 3/8 C.Y. bucket	2.87	8	10.87
2320	4' deep, 1/2 C.Y. bucket	3.61	9.90	13.51
2340	6' deep, 1/2 C.Y. bucket	6.90	18.55	25.45
2360	8' deep, 1/2 C.Y. bucket	12.75	27	39.75
2380	10' deep, 1 C.Y. bucket	17.60	38.50	56.10
2400	12' deep, 1 C.Y. bucket	24	52	76
2430	15' deep, 1-1/2 C.Y. bucket	26.50	55.50	82
2460	18' deep, 2-1/2 C.Y. bucket	46	87	133
2840	6' wide, 6' deep, 5/8 C.Y. bucket w/trench box	12.55	23	35.55
2860	8' deep, 3/4 C.Y. bucket	18.20	34.50	52.70
2880	10' deep, 1 C.Y. bucket	18.55	35	53.55
2900	12' deep, 1-1/2 C.Y. bucket	23.50	46	69.50
2940	16' deep, 2-1/2 C.Y. bucket	38.50	67.50	106
2980	20' deep, 3-1/2 C.Y. bucket	55	91	146
3020	24' deep, 3-1/2 C.Y. bucket	77.50	125	202.50
3100	8' wide, 12' deep, 1-1/2 C.Y. bucket w/trench box	29	52.50	81.50
3120	15' deep, 1-1/2 C.Y. bucket	42	76	118
3140	18' deep, 2-1/2 C.Y. bucket	53	90.50	143.50
3180	24' deep, 3-1/2 C.Y. bucket	86.50	137	223.50
3270	10' wide, 20' deep, 3-1/2 C.Y. bucket w/trench box	55.50	105	160.50
3280	24' deep, 3-1/2 C.Y. bucket	96	149	245
3370	12' wide, 20' deep, 3-1/2 C.Y. bucket w/trench box	80.50	122	202.50
3380	25' deep, 3-1/2 C.Y. bucket	112	174	286
3500	1 to 1 slope, 2' wide, 2' deep, 3/8 C.Y. bucket	1.29	3.90	5.19
3520	3' deep, 3/8 C.Y. bucket	3.70	8.80	12.50
3540	4' deep, 3/8 C.Y. bucket	3.68	11.75	15.43
3560	6' deep, 1/2 C.Y. bucket	4	12.70	16.70
3580	8' deep, 1/2 C.Y. bucket	7.90	25.50	33.40
3600	10' deep, 1 C.Y. bucket	20.50	48.50	69
3800	4' wide, 2' deep, 3/8 C.Y. bucket	2.20	5.80	8
3820	3' deep, 3/8 C.Y. bucket	3.54	10.20	13.74
3840	4' deep, 1/2 C.Y. bucket	4.64	13.30	17.94
3860	6' deep, 1/2 C.Y. bucket	9.65	26.50	36.15
3880	8' deep, 1/2 C.Y. bucket	18.95	41	59.95
3900	10' deep, 1 C.Y. bucket	27.50	60	87.50
3920	12' deep, 1 C.Y. bucket	40	86.50	126.50
3940	15' deep, 1-1/2 C.Y. bucket	43.50	92.50	136
3960	18' deep, 2-1/2 C.Y. bucket	63.50	120	183.50
4030	6' wide, 6' deep, 5/8 C.Y. bucket w/trench box	16.50	31	47.50
4040	8' deep, 3/4 C.Y. bucket	23.50	42.50	66
4050	10' deep, 1 C.Y. bucket	27	52	79
4060	12' deep, 1-1/2 C.Y. bucket	36	70.50	106.50
4070	16' deep, 2-1/2 C.Y. bucket	60.50	107	167.50
4080	20' deep, 3-1/2 C.Y. bucket	88.50	149	237.50
4090	24' deep, 3-1/2 C.Y. bucket	129	209	338
4500	8' wide, 12' deep, 1-1/2 C.Y. bucket w/trench box	41	76	117
4550	15' deep, 1-1/2 C.Y. bucket	62	114	176
4600	18' deep, 2-1/2 C.Y. bucket	80.50	140	220.50
4650	24' deep, 3-1/2 C.Y. bucket	138	221	359
4800	10' wide, 20' deep, 3-1/2 C.Y. bucket w/trench box	82.50	160	242.50
4850	24' deep, 3-1/2 C.Y. bucket	147	233	380
4950	12' wide, 20' deep, 3-1/2 C.Y. bucket w/trench box	116	180	296
4980	25' deep, 3-1/2 C.Y. bucket	167	263	430

G10 Site Preparation

G1030 Site Earthwork

Trenching Systems are shown on a cost per linear foot basis. The systems include: excavation; backfill and removal of spoil; and compaction for various depths and trench bottom widths. The backfill has been reduced to accommodate a pipe of suitable diameter and bedding.

The slope for trench sides varies from none to 1:1.

The Expanded System Listing shows Trenching Systems that range from 2' to 12' in width. Depths range from 2' to 25'.

System Components			COST PER L.F.		
	QUANTITY	UNIT	EQUIP.	LABOR	TOTAL
SYSTEM G1030 806 1310					
TRENCHING LOAM & SANDY CLAY, NO SLOPE, 2' WIDE, 2' DP, 3/8 C.Y. BUCKET					
Excavation, trench, hyd. backhoe, track mtd., 3/8 C.Y. bucket	.148	B.C.Y.	.37	.98	1.35
Backfill and load spoil, from stockpile	.165	L.C.Y.	.14	.35	.49
Compaction by vibrating plate 18" wide, 6" lifts, 4 passes	.118	E.C.Y.	.03	.39	.42
Remove excess spoil, 8 C.Y. dump truck, 2 mile roundtrip	.042	L.C.Y.	.16	.17	.33
TOTAL			.70	1.89	2.59

G1030 806	Trenching Loam & Sandy Clay	COST PER L.F.		
		EQUIP.	LABOR	TOTAL
1310	Trenching, loam & sandy clay, no slope, 2' wide, 2' deep, 3/8 C.Y. bucket	.70	1.89	2.59
1320	3' deep, 3/8 C.Y. bucket	1.08	3.13	4.21
1330	4' deep, 3/8 C.Y. bucket	1.27	3.81	5.10
1340	6' deep, 3/8 C.Y. bucket	1.80	4.51	6.30
1350	8' deep, 1/2 C.Y. bucket	2.38	5.95	8.35
1360	10' deep, 1 C.Y. bucket	2.69	6.40	9.10
1400	4' wide, 2' deep, 3/8 C.Y. bucket	1.60	3.75	5.35
1410	3' deep, 3/8 C.Y. bucket	2.18	5.65	7.85
1420	4' deep, 1/2 C.Y. bucket	2.55	6.40	8.95
1430	6' deep, 1/2 C.Y. bucket	4.39	9.35	13.75
1440	8' deep, 1/2 C.Y. bucket	6.45	13.25	19.70
1450	10' deep, 1 C.Y. bucket	6.45	13.60	20
1460	12' deep, 1 C.Y. bucket	8.10	16.90	25
1470	15' deep, 1-1/2 C.Y. bucket	9.80	19.70	29.50
1480	18' deep, 2-1/2 C.Y. bucket	11.50	21.50	33
1520	6' wide, 6' deep, 5/8 C.Y. bucket w/trench box	8.25	15.20	23.50
1530	8' deep, 3/4 C.Y. bucket	10.90	19.95	31
1540	10' deep, 1 C.Y. bucket	10.55	20.50	31
1550	12' deep, 1-1/2 C.Y. bucket	12.05	23	35
1560	16' deep, 2-1/2 C.Y. bucket	16.50	28.50	45
1570	20' deep, 3-1/2 C.Y. bucket	20	34	54
1580	24' deep, 3-1/2 C.Y. bucket	25.50	41.50	67
1640	8' wide, 12' deep, 1-1/4 C.Y. bucket w/trench box	17.05	29	46
1650	15' deep, 1-1/2 C.Y. bucket	21	36.50	57.50
1660	18' deep, 2-1/2 C.Y. bucket	25	41.50	66.50
1680	24' deep, 3-1/2 C.Y. bucket	34.50	53.50	88
1730	10' wide, 20' deep, 3-1/2 C.Y. bucket w/trench box	34.50	53.50	88
1740	24' deep, 3-1/2 C.Y. bucket	43.50	66	110
1780	12' wide, 20' deep, 3-1/2 C.Y. bucket w/trench box	42	63.50	106
1790	25' deep, 3-1/2 C.Y. bucket	54.50	82	137
1800	1/2:1 slope, 2' wide, 2' deep, 3/8 C.Y. bucket	.98	2.86	3.84
1810	3' deep, 3/8 C.Y. bucket	1.63	5	6.65
1820	4' deep, 3/8 C.Y. bucket	2.43	7.65	10.10
1840	6' deep, 3/8 C.Y. bucket	4.31	11.30	15.60

532

G1030 Site Earthwork

G1030 806	Trenching Loam & Sandy Clay	COST PER L.F.		
		EQUIP.	LABOR	TOTAL
1860	8' deep, 1/2 C.Y. bucket	6.85	18.10	25
1880	10' deep, 1 C.Y. bucket	9.15	22.50	31.50
2300	4' wide, 2' deep, 3/8 C.Y. bucket	1.88	4.70	6.60
2310	3' deep, 3/8 C.Y. bucket	2.83	7.80	10.65
2320	4' deep, 1/2 C.Y. bucket	3.57	9.70	13.25
2340	6' deep, 1/2 C.Y. bucket	7.35	16.65	24
2360	8' deep, 1/2 C.Y. bucket	12.40	27	39.50
2380	10' deep, 1 C.Y. bucket	14.10	31	45
2400	12' deep, 1 C.Y. bucket	23.50	51.50	75
2430	15' deep, 1-1/2 C.Y. bucket	27.50	57.50	85
2460	18' deep, 2-1/2 C.Y. bucket	45.50	88	134
2840	6' wide, 6' deep, 5/8 C.Y. bucket w/trench box	11.95	23	35
2860	8' deep, 3/4 C.Y. bucket	17.50	34	51.50
2880	10' deep, 1 C.Y. bucket	18.90	38	57
2900	12' deep, 1-1/2 C.Y. bucket	23.50	46.50	70
2940	16' deep, 2-1/2 C.Y. bucket	37.50	68.50	106
2980	20' deep, 3-1/2 C.Y. bucket	53	92.50	146
3020	24' deep, 3-1/2 C.Y. bucket	75	127	202
3100	8' wide, 12' deep, 1-1/2 C.Y. bucket w/trench box	28.50	52.50	81
3120	15' deep, 1-1/2 C.Y. bucket	39	74	113
3140	18' deep, 2-1/2 C.Y. bucket	52	91.50	144
3180	24' deep, 3-1/2 C.Y. bucket	84	139	223
3270	10' wide, 20' deep, 3-1/2 C.Y. bucket w/trench box	67.50	112	180
3280	24' deep, 3-1/2 C.Y. bucket	93	151	244
3320	12' wide, 20' deep, 3-1/2 C.Y. bucket w/trench box	75	122	197
3380	25' deep, 3-1/2 C.Y. bucket w/trench box	102	164	266
3500	1:1 slope, 2' wide, 2' deep, 3/8 C.Y. bucket	1.27	3.81	5.10
3520	3' deep, 3/8 C.Y. bucket	2.28	7.15	9.45
3540	4' deep, 3/8 C.Y. bucket	3.59	11.45	15.05
3560	6' deep, 1/2 C.Y. bucket	4.31	11.30	15.60
3580	8' deep, 1/2 C.Y. bucket	11.35	30	41.50
3600	10' deep, 1 C.Y. bucket	15.65	39	54.50
3800	4' wide, 2' deep, 3/8 C.Y. bucket	2.18	5.65	7.85
3820	3' deep, 1/2 C.Y. bucket	3.48	9.95	13.45
3840	4' deep, 1/2 C.Y. bucket	4.58	13.05	17.65
3860	6' deep, 1/2 C.Y. bucket	10.30	24	34.50
3880	8' deep, 1/2 C.Y. bucket	18.40	40.50	59
3900	10' deep, 1 C.Y. bucket	22	49	71
3920	12' deep, 1 C.Y. bucket	31.50	69	101
3940	15' deep, 1-1/2 C.Y. bucket	45.50	95.50	141
3960	18' deep, 2-1/2 C.Y. bucket	62	121	183
4030	6' wide, 6' deep, 5/8 C.Y. bucket w/trench box	15.70	31	46.50
4040	8' deep, 3/4 C.Y. bucket	24	47.50	71.50
4050	10' deep, 1 C.Y. bucket	27	56	83
4060	12' deep, 1-1/2 C.Y. bucket	35	70.50	106
4070	16' deep, 2-1/2 C.Y. bucket	59	108	167
4080	20' deep, 3-1/2 C.Y. bucket	86	151	237
4090	24' deep, 3-1/2 C.Y. bucket	125	212	335
4500	8' wide, 12' deep, 1-1/4 C.Y. bucket w/trench box	40	76.50	117
4550	15' deep, 1-1/2 C.Y. bucket	56.50	111	168
4600	18' deep, 2-1/2 C.Y. bucket	78.50	142	221
4650	24' deep, 3-1/2 C.Y. bucket	134	224	360
4800	10' wide, 20' deep, 3-1/2 C.Y. bucket w/trench box	100	171	271
4850	24' deep, 3-1/2 C.Y. bucket	143	237	380
4950	12' wide, 20' deep, 3-1/2 C.Y. bucket w/trench box	108	181	289
4980	25' deep, 3-1/2 C.Y. bucket	162	267	430

533

For customer support on your Plumbing Cost Data, call 877.709.6968.

Trenching Systems are shown on a cost per linear foot basis. The systems include: excavation; backfill and removal of spoil; and compaction for various depths and trench bottom widths. The backfill has been reduced to accommodate a pipe of suitable diameter and bedding.

The slope for trench sides varies from none to 1:1.

The Expanded System Listing shows Trenching Systems that range from 2' to 12' in width. Depths range from 2' to 25'.

System Components	QUANTITY	UNIT	COST PER L.F.		
			EQUIP.	LABOR	TOTAL
SYSTEM G1030 807 1310					
TRENCHING SAND & GRAVEL, NO SLOPE, 2' WIDE, 2' DEEP, 3/8 C.Y. BUCKET					
Excavation, trench, hyd. backhoe, track mtd., 3/8 C.Y. bucket	.148	B.C.Y.	.36	.97	1.33
Backfill and load spoil, from stockpile	.140	L.C.Y.	.11	.30	.41
Compaction by vibrating plate 18" wide, 6" lifts, 4 passes	.118	E.C.Y.	.03	.39	.42
Remove excess spoil, 8 C.Y. dump truck, 2 mile roundtrip	.035	L.C.Y.	.13	.14	.27
TOTAL			.63	1.80	2.43

G1030 807	Trenching Sand & Gravel	COST PER L.F.		
		EQUIP.	LABOR	TOTAL
1310	Trenching, sand & gravel, no slope, 2' wide, 2' deep, 3/8 C.Y. bucket	.63	1.80	2.43
1320	3' deep, 3/8 C.Y. bucket	1.01	3.01	4.02
1330	4' deep, 3/8 C.Y. bucket	1.18	3.62	4.80
1340	6' deep, 3/8 C.Y. bucket	1.70	4.30	6
1350	8' deep, 1/2 C.Y. bucket	2.25	5.70	7.95
1360	10' deep, 1 C.Y. bucket	2.50	6.05	8.55
1400	4' wide, 2' deep, 3/8 C.Y. bucket	1.46	3.54	5
1410	3' deep, 3/8 C.Y. bucket	2.01	5.35	7.35
1420	4' deep, 1/2 C.Y. bucket	2.33	6.05	8.40
1430	6' deep, 1/2 C.Y. bucket	4.16	8.95	13.10
1440	8' deep, 1/2 C.Y. bucket	6.05	12.60	18.65
1450	10' deep, 1 C.Y. bucket	6.05	12.85	18.90
1460	12' deep, 1 C.Y. bucket	7.65	16	23.50
1470	15' deep, 1-1/2 C.Y. bucket	9.25	18.55	28
1480	18' deep, 2-1/2 C.Y. bucket	10.85	20.50	31.50
1520	6' wide, 6' deep, 5/8 C.Y. bucket w/trench box	7.75	14.35	22
1530	8' deep, 3/4 C.Y. bucket	10.30	18.95	29.50
1540	10' deep, 1 C.Y. bucket	9.95	19.15	29
1550	12' deep, 1-1/2 C.Y. bucket	11.40	21.50	33
1560	16' deep, 2 C.Y. bucket	15.75	27	43
1570	20' deep, 3-1/2 C.Y. bucket	19.10	32	51
1580	24' deep, 3-1/2 C.Y. bucket	24	39	63
1640	8' wide, 12' deep, 1-1/2 C.Y. bucket w/trench box	16.05	27	43
1650	15' deep, 1-1/2 C.Y. bucket	19.50	34.50	54
1660	18' deep, 2-1/2 C.Y. bucket	23.50	39	62.50
1680	24' deep, 3-1/2 C.Y. bucket	32.50	50	82.50
1730	10' wide, 20' deep, 3-1/2 C.Y. bucket w/trench box	32.50	50	82.50
1740	24' deep, 3-1/2 C.Y. bucket	41	62	103
1780	12' wide, 20' deep, 3-1/2 C.Y. bucket w/trench box	39.50	59.50	99
1790	25' deep, 3-1/2 C.Y. bucket	51	77	128
1800	1/2:1 slope, 2' wide, 2' deep, 3/8 C.Y. bucket	.91	2.72	3.63
1810	3' deep, 3/8 C.Y. bucket	1.53	4.77	6.30
1820	4' deep, 3/8 C.Y. bucket	2.27	7.30	9.55
1840	6' deep, 3/8 C.Y. bucket	4.10	10.80	14.90

G1030 Site Earthwork

G1030 807	Trenching Sand & Gravel	COST PER L.F.		
		EQUIP.	LABOR	TOTAL
1860	8' deep, 1/2 C.Y. bucket	6.55	17.30	24
1880	10' deep, 1 C.Y. bucket	8.55	21.50	30
2300	4' wide, 2' deep, 3/8 C.Y. bucket	1.74	4.45	6.20
2310	3' deep, 3/8 C.Y. bucket	2.62	7.40	10
2320	4' deep, 1/2 C.Y. bucket	3.29	9.20	12.50
2340	6' deep, 1/2 C.Y. bucket	7	15.95	23
2360	8' deep, 1/2 C.Y. bucket	11.75	25.50	37.50
2380	10' deep, 1 C.Y. bucket	13.35	29.50	43
2400	12' deep, 1 C.Y. bucket	22.50	49	71.50
2430	15' deep, 1-1/2 C.Y. bucket	26	54.50	80.50
2460	18' deep, 2-1/2 C.Y. bucket	43	83	126
2840	6' wide, 6' deep, 5/8 C.Y. bucket w/trench box	11.35	22	33.50
2860	8' deep, 3/4 C.Y. bucket	16.60	32	48.50
2880	10' deep, 1 C.Y. bucket	17.85	36	54
2900	12' deep, 1-1/2 C.Y. bucket	22.50	44	66.50
2940	16' deep, 2 C.Y. bucket	36	64.50	101
2980	20' deep, 3-1/2 C.Y. bucket	50.50	87	138
3020	24' deep, 3-1/2 C.Y. bucket	71.50	119	191
3100	8' wide, 12' deep, 1-1/4 C.Y. bucket w/trench box	27	50	77
3120	15' deep, 1-1/2 C.Y. bucket	36.50	69.50	106
3140	18' deep, 2-1/2 C.Y. bucket	49	86	135
3180	24' deep, 3-1/2 C.Y. bucket	80	130	210
3270	10' wide, 20' deep, 3-1/2 C.Y. bucket w/trench box	64	105	169
3280	24' deep, 3-1/2 C.Y. bucket	88	142	230
3370	12' wide, 20' deep, 3-1/2 C.Y. bucket w/trench box	71.50	117	189
3380	25' deep, 3-1/2 C.Y. bucket	103	164	267
3500	1:1 slope, 2' wide, 2' deep, 3/8 C.Y. bucket	2.05	4.46	6.50
3520	3' deep, 3/8 C.Y. bucket	2.14	6.85	9
3540	4' deep, 3/8 C.Y. bucket	3.36	10.95	14.30
3560	6' deep, 3/8 C.Y. bucket	4.10	10.80	14.90
3580	8' deep, 1/2 C.Y. bucket	10.80	29	40
3600	10' deep, 1 C.Y. bucket	14.65	36.50	51
3800	4' wide, 2' deep, 3/8 C.Y. bucket	2.01	5.35	7.35
3820	3' deep, 3/8 C.Y. bucket	3.24	9.50	12.75
3840	4' deep, 1/2 C.Y. bucket	4.23	12.35	16.60
3860	6' deep, 1/2 C.Y. bucket	9.85	23	33
3880	8' deep, 1/2 C.Y. bucket	17.45	38.50	56
3900	10' deep, 1 C.Y. bucket	20.50	46	66.50
3920	12' deep, 1 C.Y. bucket	30	65.50	95.50
3940	15' deep, 1-1/2 C.Y. bucket	43	90	133
3960	18' deep, 2-1/2 C.Y. bucket	59	114	173
4030	6' wide, 6' deep, 5/8 C.Y. bucket w/trench box	14.90	29.50	44.50
4040	8' deep, 3/4 C.Y. bucket	23	45.50	68.50
4050	10' deep, 1 C.Y. bucket	25.50	53	78.50
4060	12' deep, 1-1/2 C.Y. bucket	33	66.50	99.50
4070	16' deep, 2 C.Y. bucket	56.50	102	159
4080	20' deep, 3-1/2 C.Y. bucket	81.50	142	224
4090	24' deep, 3-1/2 C.Y. bucket	119	199	320
4500	8' wide, 12' deep, 1-1/2 C.Y. bucket w/trench box	38	72.50	111
4550	15' deep, 1-1/2 C.Y. bucket	53.50	105	159
4600	18' deep, 2-1/2 C.Y. bucket	74.50	134	209
4650	24' deep, 3-1/2 C.Y. bucket	127	210	335
4800	10' wide, 20' deep, 3-1/2 C.Y. bucket w/trench box	95	160	255
4850	24' deep, 3-1/2 C.Y. bucket	135	222	355
4950	12' wide, 20' deep, 3-1/2 C.Y. bucket w/trench box	102	170	272
4980	25' deep, 3-1/2 C.Y. bucket	154	251	405

The Pipe Bedding System is shown for various pipe diameters. Compacted bank sand is used for pipe bedding and to fill 12″ over the pipe. No backfill is included. Various side slopes are shown to accommodate different soil conditions. Pipe sizes vary from 6″ to 84″ diameter.

System Components			QUANTITY	UNIT	COST PER L.F.		
					MAT.	INST.	TOTAL
SYSTEM G1030 815 1440							
PIPE BEDDING, SIDE SLOPE 0 TO 1, 1′ WIDE, PIPE SIZE 6″ DIAMETER							
Borrow, bank sand, 2 mile haul, machine spread			.086	C.Y.	1.68	.70	2.38
Compaction, vibrating plate			.086	C.Y.		.22	.22
		TOTAL			1.68	.92	2.60

G1030 815	Pipe Bedding	COST PER L.F.		
		MAT.	INST.	TOTAL
1440	Pipe bedding, side slope 0 to 1, 1′ wide, pipe size 6″ diameter	1.68	.92	2.60
1460	2′ wide, pipe size 8″ diameter	3.63	1.97	5.60
1480	Pipe size 10″ diameter	3.70	2.01	5.71
1500	Pipe size 12″ diameter	3.78	2.05	5.83
1520	3′ wide, pipe size 14″ diameter	6.15	3.33	9.48
1540	Pipe size 15″ diameter	6.20	3.35	9.55
1560	Pipe size 16″ diameter	6.25	3.39	9.64
1580	Pipe size 18″ diameter	6.35	3.46	9.81
1600	4′ wide, pipe size 20″ diameter	9.05	4.92	13.97
1620	Pipe size 21″ diameter	9.15	4.96	14.11
1640	Pipe size 24″ diameter	9.35	5.05	14.40
1660	Pipe size 30″ diameter	9.50	5.15	14.65
1680	6′ wide, pipe size 32″ diameter	16.25	8.85	25.10
1700	Pipe size 36″ diameter	16.65	9	25.65
1720	7′ wide, pipe size 48″ diameter	26.50	14.35	40.85
1740	8′ wide, pipe size 60″ diameter	32	17.45	49.45
1760	10′ wide, pipe size 72″ diameter	45	24.50	69.50
1780	12′ wide, pipe size 84″ diameter	59	32	91
2140	Side slope 1/2 to 1, 1′ wide, pipe size 6″ diameter	3.13	1.69	4.82
2160	2′ wide, pipe size 8″ diameter	5.30	2.88	8.18
2180	Pipe size 10″ diameter	5.70	3.09	8.79
2200	Pipe size 12″ diameter	6.05	3.27	9.32
2220	3′ wide, pipe size 14″ diameter	8.70	4.73	13.43
2240	Pipe size 15″ diameter	8.90	4.83	13.73
2260	Pipe size 16″ diameter	9.15	4.96	14.11
2280	Pipe size 18″ diameter	9.60	5.20	14.80
2300	4′ wide, pipe size 20″ diameter	12.70	6.90	19.60
2320	Pipe size 21″ diameter	13	7.05	20.05
2340	Pipe size 24″ diameter	13.80	7.50	21.30
2360	Pipe size 30″ diameter	15.30	8.30	23.60
2380	6′ wide, pipe size 32″ diameter	22.50	12.25	34.75
2400	Pipe size 36″ diameter	24	13.05	37.05
2420	7′ wide, pipe size 48″ diameter	37.50	20.50	58
2440	8′ wide, pipe size 60″ diameter	47.50	26	73.50
2460	10′ wide, pipe size 72″ diameter	65	35.50	100.50
2480	12′ wide, pipe size 84″ diameter	85.50	46	131.50
2620	Side slope 1 to 1, 1′ wide, pipe size 6″ diameter	4.58	2.49	7.07
2640	2′ wide, pipe size 8″ diameter	7.05	3.83	10.88

G10 Site Preparation

G1030 Site Earthwork

G1030 815	Pipe Bedding	COST PER L.F.		
		MAT.	INST.	TOTAL
2660	Pipe size 10" diameter	7.65	4.15	11.80
2680	Pipe size 12" diameter	8.35	4.53	12.88
2700	3' wide, pipe size 14" diameter	11.30	6.15	17.45
2720	Pipe size 15" diameter	11.65	6.30	17.95
2740	Pipe size 16" diameter	12.05	6.55	18.60
2760	Pipe size 18" diameter	12.90	7	19.90
2780	4' wide, pipe size 20" diameter	16.35	8.85	25.20
2800	Pipe size 21" diameter	16.80	9.10	25.90
2820	Pipe size 24" diameter	18.25	9.90	28.15
2840	Pipe size 30" diameter	21	11.40	32.40
2860	6' wide, pipe size 32" diameter	29	15.65	44.65
2880	Pipe size 36" diameter	31.50	17	48.50
2900	7' wide, pipe size 48" diameter	48.50	26.50	75
2920	8' wide, pipe size 60" diameter	63	34	97
2940	10' wide, pipe size 72" diameter	85.50	46.50	132
2960	12' wide, pipe size 84" diameter	112	60.50	172.50

For customer support on your Plumbing Cost Data, call 877.709.6968.

G2040 Site Development

Chain Link Fence

Concrete Sidewalk

Pool Equipment
(Building not included in price)

12' x 30' Swimming Pool

Fence Gate

The Swimming Pool System is a complete package. Everything from excavation to deck hardware is included in system costs. Below are three basic types of pool systems: residential, motel, and municipal. Systems elements include: excavation, pool materials, installation, deck hardware, pumps and filters, sidewalk, and fencing.

The Expanded System Listing shows three basic types of pools with a variety of finishes and basic materials. These systems are either vinyl lined with metal sides; gunite shell with a cement plaster finish or tile finish; or concrete sided with vinyl lining. Pool sizes listed here vary from 12' x 30' to 60' x 82.5'. All costs are on a per unit basis.

System Components	QUANTITY	UNIT	COST EACH		
			MAT.	INST.	TOTAL
SYSTEM G2040 920 1000					
SWIMMING POOL, RESIDENTIAL, CONC. SIDES, VINYL LINED, 12' X 30'					
Swimming pools,resi in-ground,vyl lined,conc sides,W/ equip,sand bot	360.000	S.F.	9,180	5,266.80	14,446.80
4" thick reinforced concrete sidewalk, broom finish, no base	400.000	S.F.	744	1,048	1,792
Chain link fence, residential, 3' high	124.000	L.F.	288.92	416.64	705.56
Fence gate, chain link	1.000	Ea.	89.50	140.50	230
TOTAL			10,302.42	6,871.94	17,174.36

G2040 920	Swimming Pools	COST EACH		
		MAT.	INST.	TOTAL
1000	Swimming pool, residential class, concrete sides, vinyl lined, 12' x 30'	10,300	6,875	17,175
1100	16' x 32'	11,700	7,775	19,475
1200	20' x 40'	15,100	9,925	25,025
1500	Tile finish, 12' x 30'	23,800	28,800	52,600
1600	16' x 32'	24,300	30,400	54,700
1700	20' x 40'	36,400	43,000	79,400
2000	Metal sides, vinyl lined, 12' x 30'	8,850	4,525	13,375
2100	16' x 32'	10,100	5,100	15,200
2200	20' x 40'	13,000	6,450	19,450
3000	Gunite shell, cement plaster finish, 12' x 30'	18,400	12,600	31,000
3100	16' x 32'	23,300	16,300	39,600
3200	20' x 40'	32,300	16,100	48,400
4000	Motel class, concrete sides, vinyl lined, 20' x 40'	21,900	13,800	35,700
4100	28' x 60'	30,700	19,300	50,000
4500	Tile finish, 20' x 40'	42,800	49,800	92,600
4600	28' x 60'	68,500	80,500	149,000
5000	Metal sides, vinyl lined, 20' x 40'	23,000	10,200	33,200
5100	28' x 60'	30,800	13,800	44,600
6000	Gunite shell, cement plaster finish, 20' x 40'	60,500	40,300	100,800
6100	28' x 60'	90,500	60,000	150,500
7000	Municipal class, gunite shell, cement plaster finish, 42' x 75'	242,000	142,500	384,500
7100	60' x 82.5'	311,000	182,500	493,500
7500	Concrete walls, tile finish, 42' x 75'	274,000	189,000	463,000
7600	60' x 82.5'	357,500	250,000	607,500
7700	Tile finish and concrete gutter, 42' x 75'	300,500	189,000	489,500
7800	60' x 82.5'	389,500	250,000	639,500
7900	Tile finish and stainless gutter, 42' x 75'	351,500	189,000	540,500
8000	60'x 82.5'	520,000	289,500	809,500

G2050 Landscaping

There are three basic types of Site Irrigation Systems: pop-up, riser mounted and quick coupling. Sprinkler heads are spray, impact or gear driven. Each system includes: the hardware for spraying the water; the pipe and fittings needed to deliver the water; and all other accessory equipment such as valves, couplings, nipples, and nozzles. Excavation heads and backfill costs are also included in the system.

The Expanded System Listing shows a wide variety of Site Irrigation Systems.

System Components	QUANTITY	UNIT	COST PER S.F.		
			MAT.	INST.	TOTAL
SYSTEM G2050 710 1000					
SITE IRRIGATION, POP UP SPRAY, PLASTIC, 10' RADIUS, 1000 S.F., PVC PIPE					
Excavation, chain trencher	54.000	L.F.		44.82	44.82
Pipe, fittings & nipples, PVC Schedule 40, 1" diameter	5.000	Ea.	4.25	112.50	116.75
Fittings, bends or elbows, 4" diameter	5.000	Ea.	54	250	304
Couplings PVC plastic, high pressure, 1" diameter	5.000	Ea.	4.25	142.50	146.75
Valves, bronze, globe, 125 lb. rising stem, threaded, 1" diameter	1.000	Ea.	169	37.50	206.50
Head & nozzle, pop-up spray, PVC plastic	5.000	Ea.	17	100	117
Backfill by hand with compaction	54.000	L.F.		35.64	35.64
Total cost per 1,000 S.F.			248.50	722.96	971.46
Total cost per S.F.			.25	.72	.97

G2050 710	Site Irrigation	COST PER S.F.		
		MAT.	INST.	TOTAL
1000	Site irrigation, pop up spray, 10' radius, 1000 S.F., PVC pipe	.25	.73	.98
1100	Polyethylene pipe	.28	.62	.90
1200	14' radius, 8000 S.F., PVC pipe	.09	.31	.40
1300	Polyethylene pipe	.09	.29	.38
1400	18' square, 1000 S.F. PVC pipe	.24	.49	.73
1500	Polyethylene pipe	.25	.42	.67
1600	24' square, 8000 S.F., PVC pipe	.07	.20	.27
1700	Polyethylene pipe	.07	.19	.26
1800	4' x 30' strip, 200 S.F., PVC pipe	1.15	2.04	3.19
1900	Polyethylene pipe	1.17	1.89	3.06
2000	6' x 40' strip, 800 S.F., PVC pipe	.31	.67	.98
2200	Economy brass, 11' radius, 1000 S.F., PVC pipe	.31	.68	.99
2300	Polyethylene pipe	.32	.62	.94
2400	14' radius, 8000 S.F., PVC pipe	.11	.31	.42
2500	Polyethylene pipe	.11	.29	.40
2600	3' x 28' strip, 200 S.F., PVC pipe	1.28	2.04	3.32
2700	Polyethylene pipe	1.30	1.89	3.19
2800	7' x 36' strip, 1000 S.F., PVC pipe	.28	.54	.82
2900	Polyethylene pipe	.28	.49	.77
3000	Hd brass, 11' radius, 1000 S.F., PVC pipe	.35	.68	1.03
3100	Polyethylene pipe	.36	.62	.98
3200	14' radius, 8000 S.F., PVC pipe	.13	.31	.44
3300	Polyethylene pipe	.13	.29	.42
3400	Riser mounted spray, plastic,10'radius, 1000S.F., PVC pipe	.27	.68	.95
3500	Polyethylene pipe	.28	.62	.90
3600	12' radius, 5000 S.F., PVC pipe	.13	.41	.54

G2050 Landscaping

G2050 710	Site Irrigation	COST PER S.F.		
		MAT.	INST.	TOTAL
3700	Polyethylene pipe	.13	.38	.51
3800	19' square, 2000 S.F., PVC pipe	.12	.22	.34
3900	Polyethylene pipe	.12	.20	.32
4000	24' square, 8000 S.F., PVC pipe	.07	.20	.27
4100	Polyethylene pipe	.07	.19	.26
4200	5' x 32' strip, 300 S.F., PVC pipe	.76	1.38	2.14
4300	Polyethylene pipe	.77	1.28	2.05
4400	6' x 40' strip, 800 S.F., PVC pipe	.31	.67	.98
4500	Polyethylene pipe	.32	.61	.93
4600	Brass, 11' radius, 1000 S.F., PVC pipe	.31	.68	.99
4700	Polyethylene pipe	.32	.62	.94
4800	14' radius, 8000 S.F., PVC pipe	.11	.31	.42
4900	Polythylene pipe	.11	.29	.40
5000	Pop up gear drive stream type,plastic, 30'radius, 10,000 S.F.,PVC pipe	.28	.60	.88
5020	Polyethylene pipe	.12	.17	.29
5040	40,000 S.F., PVC pipe	.25	.61	.86
5060	Polyethylene pipe	.09	.18	.27
5080	Riser mounted gear drive stream type,plas.,30'rad.,10,000S.F.,PVC pipe	.27	.60	.87
5100	Polyethylene pipe	.11	.17	.28
5120	40,000 S.F., PVC pipe	.25	.61	.86
5140	Polyethylene pipe	.09	.18	.27
5200	Q.C. valve thread type w/impact head,brass,75'rad,20,000S.F.,PVC pipe	.12	.23	.35
5250	Polyethylene pipe	.05	.05	.10
5300	100,000 S.F., PVC pipe	.17	.37	.54
5350	Polyethylene pipe	.07	.08	.15
5400	Q.C. valve lug type w/impact head,brass, 75'rad, 20,000 S.F.,PVC pipe	.12	.23	.35
5450	Polyethylene pipe	.05	.05	.10
5500	100,000 S.F. PVC pipe	.17	.37	.54
5550	Polyethylene pipe	.07	.08	.15
6000	Site irrigation,pop up impact type,plastic,40'rad.,10000 S.F.,PVC pipe	.18	.38	.56
6100	Polyethylene pipe	.07	.10	.17
6200	45,000 S.F., PVC pipe	.17	.43	.60
6300	Polyethylene pipe	.06	.11	.17
6400	High medium volume brass, 40' radius, 10000 S.F., PVC pipe	.19	.38	.57
6500	Polyethylene pipe	.08	.10	.18
6600	45,000 S.F. PVC pipe	.17	.43	.60
6700	Polyethylene pipe	.06	.11	.17
6800	High volume brass, 60' radius, 25,000 S.F., PVC pipe	.14	.30	.44
6900	Polyethylene pipe	.06	.08	.14
7000	100,000 S.F., PVC pipe	.14	.30	.44
7100	Polyethylene pipe	.06	.07	.13
7200	Riser mounted part/full impact type,plas.,40'rad.,10,000 S.F.,PVC pipe	.17	.38	.55
7300	Polyethylene pipe	.07	.20	.27
7400	45,000 S.F., PVC pipe	.16	.43	.59
7500	Polyethylene	.05	.11	.16
7600	Low medium volume brass, 40' radius, 10,000 S.F., PVC pipe	.18	.38	.56
7700	Polyethylene pipe	.07	.10	.17
7800	45,000 S.F., PVC pipe	.17	.43	.60
7900	Polyethylene pipe	.06	.11	.17
8000	Medium volume brass, 50' radius, 30,000 S.F., PVC pipe	.15	.32	.47
8100	Polyethylene pipe	.06	.08	.14
8200	70,000 S.F., PVC pipe	.21	.52	.73
8300	Polyethylene pipe	.06	.12	.18
8400	Riser mounted full only impact type,plas.,40'rad.,10000 S.F., PVC pipe	.17	.38	.55
8500	Polyethylene pipe	.06	.10	.16
8600	45,000 S.F., PVC pipe	.16	.43	.59
8700	Polyethylene pipe	.05	.11	.16
8800	Low medium volume brass, 40' radius, 10,000 S.F., PVC pipe	.18	.38	.56

G2050 Landscaping

G2050 710	Site Irrigation	COST PER S.F.		
		MAT.	INST.	TOTAL
8900	Polyethylene pipe	.07	.10	.17
9000	45,000 S.F., PVC pipe	.17	.43	.60
9100	Polyethylene pipe	.06	.11	.17
9200	High volume brass, 80' radius, 75,000 S.F., PVC pipe	.10	.22	.32
9300	Polyethylene pipe	.01	.02	.03
9400	150,000 S.F., PVC pipe	.10	.20	.30
9500	Polyethylene pipe	.01	.02	.03
9600	Very high volume brass, 100' radius, 100,000 S.F., PVC pipe	.09	.17	.26
9700	Polyethylene pipe	.01	.02	.03
9800	200,000 S.F., PVC pipe	.09	.17	.26
9900	Polyethylene pipe	.01	.02	.03

541

G20 Site Improvements

G2050 Landscaping

Valve
Backflow Preventer
Pressure Regulator
Filter
Tubing Adapter
Drip Tubing

Drip irrigation systems can be for either lawns or individual plants. Shown below are drip irrigation systems for lawns. These drip type systems are most cost effective where there are water restrictions on use or the cost of water is very high. The supply uses PVC piping with the laterals polyethylene drip lines.

System Components	QUANTITY	UNIT	COST PER S.F.		
			MAT.	INST.	TOTAL
SYSTEM G2050 720 1000					
SITE IRRIGATION, DRIP SYSTEM, 800 S.F. (20' X40')					
Excavate supply and header lines 4" x 18" deep	50.000	L.F.		44.50	44.50
Excavate drip lines 4" x 4" deep	5.000	L.F.		744.80	744.80
Install supply manifold (stop valve, strainer, PRV, adapters)	1.000	Ea.	24.50	150	174.50
Install 3/4" PVC supply line and risers to 4" drip level	75.000	L.F.	24	150	174
Install 3/4" PVC fittings for supply line	33.000	Ea.	55.77	496.65	552.42
Install PVC to drip line adapters	14.000	Ea.	15.26	210.70	225.96
Install 1/4 inch drip lines with hose clamps	560.000	L.F.	123.20	280	403.20
Backfill trenches	3.180	C.Y.		133.56	133.56
Cleanup area when completed	1.000	Ea.		116	116
Total cost per 800 S.F.			242.73	2,326.21	2,568.94
Total cost per S.F.			.30	2.91	3.21

G2050 720	Site Irrigation	COST PER S.F.		
		MAT.	INST.	TOTAL
1000	Site irrigation, drip lawn watering system, 800 S.F. (20' x 40' area)	.30	2.91	3.21
1100	1000 S.F. (20' x 50' area)	.27	2.63	2.90
1200	1600 S.F. (40' x 40' area), 2 zone	.30	2.77	3.07
1300	2000 S.F. (40' x 50' area), 2 zone	.27	2.54	2.81
1400	2400 S.F. (60' x 40' area), 3 zone	.29	2.76	3.05
1500	3000 S.F. (60' x 50' area), 3 zone	.27	2.48	2.75
1600	3200 S.F. (2 x 40' x 40'), 4 zone	.28	2.72	3
1700	4000 S.F. (2 x 40' x 50'), 4 zone	.27	2.49	2.76
1800	4800 S.F. (2 x 60' x 40'), 6 zone with control	.33	2.72	3.05
1900	6000 S.F. (2 x 60' x 50'), 6 zone with control	.28	2.46	2.74
2000	6400 S.F. (4 x 40' x 40'), 8 zone with control	.32	2.75	3.07
2100	8000 S.F. (4 x 40' x 50'), 8 zone with control	.29	2.50	2.79
2200	3200 S.F. (2 x 40' x 40'), 4 zone with control	.32	2.79	3.11
2300	4000 S.F. (2 x 40' x 50'), 4 zone with control	.29	2.54	2.83
2400	2400 S.F. (60' x 40'), 3 zone with control	.34	2.85	3.19
2500	3000 S.F. (60' x 50'), 3 zone with control	.30	2.55	2.85
2600	4800 S.F. (2 x 60' x 40'), 6 zone, manual	.29	2.69	2.98
2700	6000 S.F. (2 x 60' x 50'), 6 zone, manual	.26	2.42	2.68
2800	1000 S.F. (10' x 100' area)	.17	1.63	1.80
2900	2000 S.F. (2 x 10' x 100' area)	.17	1.58	1.75

G3010 Water Supply

The Water Service Systems are for copper service taps from 1″ to 2″ diameter into pressurized mains from 6″ to 8″ diameter. Costs are given for two offsets with depths varying from 2′ to 10′. Included in system components are excavation and backfill and required curb stops with boxes.

System Components	QUANTITY	UNIT	COST EACH		
			MAT.	INST.	TOTAL
SYSTEM G3010 121 1000					
WATER SERVICE, LEAD FREE, 6″ MAIN, 1″ COPPER SERVICE, 10′ OFFSET, 2′ DEEP					
Trench excavation, 1/2 C.Y. backhoe, 1 laborer	2.220	C.Y.		16.79	16.79
Drill & tap pressurized main, 6″, 1″ to 2″ service	1.000	Ea.		425	425
Corporation stop, 1″ diameter	1.000	Ea.	119	44.50	163.50
Saddles, 3/4″ & 1″ diameter, add	1.000	Ea.	84.50		84.50
Copper tubing, type K, 1″ diameter	12.000	L.F.	111	47.88	158.88
Piping, curb stops, no lead, 1″ diameter	1.000	Ea.	172	44.50	216.50
Curb box, cast iron, 1/2″ to 1″ curb stops	1.000	Ea.	56.50	59	115.50
Backfill by hand, no compaction, heavy soil	2.890	C.Y.		7.60	7.60
Compaction, vibrating plate	2.220	C.Y.		18.07	18.07
TOTAL			543	663.34	1,206.34

G3010 121	Water Service, Lead Free	COST EACH		
		MAT.	INST.	TOTAL
1000	Water service, Lead Free, 6″ main, 1″ copper service, 10′ offset, 2′ deep	545	665	1,210
1040	4′ deep	545	705	1,250
1060	6′ deep	425	850	1,275
1080	8′ deep	545	985	1,530
1100	10′ deep	545	1,075	1,620
1200	20′ offset, 2′ deep	545	705	1,250
1240	4′ deep	545	790	1,335
1260	6′ deep	425	1,125	1,550
1280	8′ deep	545	1,350	1,895
1300	10′ deep	545	1,525	2,070
1400	1-1/2″ copper service, 10′ offset, 2′ deep	985	720	1,705
1440	4′ deep	985	765	1,750
1460	6′ deep	985	955	1,940
1480	8′ deep	985	1,050	2,035
1500	10′ deep	985	1,125	2,110
1520	20′ offset, 2′ deep	985	765	1,750
1530	4′ deep	985	850	1,835
1540	6′ deep	985	1,225	2,210
1560	8′ deep	985	1,400	2,385
1580	10′ deep	985	1,575	2,560
1600	2″ copper service, 10′ offset, 2′ deep	1,250	750	2,000
1610	4′ deep	1,225	795	2,020
1620	6′ deep	1,225	985	2,210
1630	8′ deep	1,225	1,075	2,300
1650	10′ deep	1,225	1,175	2,400
1660	20′ offset, 2′ deep	1,225	795	2,020
1670	4′ deep	1,225	880	2,105
1680	6′ deep	1,225	1,275	2,500

543

G3010 Water Supply

G3010 121	Water Service, Lead Free	COST EACH		
		MAT.	INST.	TOTAL
1690	8' deep	1,225	1,425	2,650
1695	10' deep	1,225	1,625	2,850
1700	8" main, 1" copper service, 10' offset, 2' deep	545	705	1,250
1710	4' deep	545	745	1,290
1720	6' deep	425	890	1,315
1730	8' deep	545	1,025	1,570
1740	10' deep	545	1,125	1,670
1750	20' offset, 2' deep	545	745	1,290
1760	4' deep	545	830	1,375
1770	6' deep	425	1,175	1,600
1780	8' deep	545	1,400	1,945
1790	10' deep	545	1,575	2,120
1800	1-1/2" copper service, 10' offset, 2' deep	985	760	1,745
1810	4' deep	985	805	1,790
1820	6' deep	985	995	1,980
1830	8' deep	985	1,450	2,435
1840	10' deep	985	1,175	2,160
1850	20' offset, 2' deep	985	805	1,790
1860	4' deep	1,225	835	2,060
1870	6' deep	985	1,275	2,260
1880	8' deep	985	1,450	2,435
1890	10' deep	985	1,625	2,610
1900	2" copper service, 10' offset, 2' deep	1,250	790	2,040
1910	4' deep	1,225	835	2,060
1920	6' deep	1,225	1,025	2,250
1930	8' deep	1,225	1,100	2,325
1940	10' deep	1,225	1,225	2,450
1950	20' offset, 2' deep	1,225	835	2,060
1960	4' deep	1,225	920	2,145
1970	6' deep	1,225	1,300	2,525
1980	8' deep	1,225	1,475	2,700
1990	10' deep	1,225	1,650	2,875

The Fire Hydrant Systems include: four different hydrants with three different lengths of offsets, at several depths of trenching. Excavation and backfill is included with each system as well as thrust blocks as necessary and pipe bedding. Finally spreading of excess material and fine grading complete the components.

System Components	QUANTITY	UNIT	COST PER EACH		
			MAT.	INST.	TOTAL
SYSTEM G3010 410 1500					
HYDRANT, 4-1/2″ VALVE SIZE, TWO WAY, 10′ OFFSET, 2′ DEEP					
Excavation, trench, 1 C.Y. hydraulic backhoe	3.556	C.Y.		28.63	28.63
Sleeve, 12″ x 6″	1.000	Ea.	1,300	112.40	1,412.40
Ductile iron pipe, 6″ diameter	8.000	L.F.	280	168.88	448.88
Gate valve, 6″	1.000	Ea.	2,575	320	2,895
Hydrant valve box, 6′ long	1.000	Ea.	310	77.50	387.50
4-1/2″ valve size, depth 2′-0″	1.000	Ea.	1,825	202.15	2,027.15
Thrust blocks, at valve, shoe and sleeve	1.600	C.Y.	292.80	445.18	737.98
Borrow, crushed stone, 3/8″	4.089	C.Y.	122	32.48	154.48
Backfill and compact, dozer, air tamped	4.089	C.Y.		70.74	70.74
Spread excess excavated material	.204	C.Y.		8.57	8.57
Fine grade, hand	400.000	S.F.		269.50	269.50
TOTAL			6,704.80	1,736.03	8,440.83

G3010 410	Fire Hydrants	COST PER EACH		
		MAT.	INST.	TOTAL
1000	Hydrant, 4-1/2″ valve size, two way, 0′ offset, 2′ deep	6,425	1,200	7,625
1100	4′ deep	6,750	1,225	7,975
1200	6′ deep	6,900	1,275	8,175
1300	8′ deep	7,125	1,325	8,450
1400	10′ deep	7,300	1,400	8,700
1500	10′ offset, 2′ deep	6,700	1,750	8,450
1600	4′ deep	7,025	2,050	9,075
1700	6′ deep	7,175	2,575	9,750
1800	8′ deep	7,400	3,400	10,800
1900	10′ deep	7,575	5,075	12,650
2000	20′ offset, 2′ deep	7,050	2,300	9,350
2100	4′ deep	7,375	2,775	10,150
2200	6′ deep	7,525	3,600	11,125
2300	8′ deep	7,750	4,775	12,525
2400	10′ deep	7,925	6,375	14,300
2500	Hydrant, 4-1/2″ valve size, three way, 0′ offset, 2′ deep	6,550	1,225	7,775
2600	4′ deep	6,900	1,225	8,125
2700	6′ deep	7,050	1,300	8,350
2800	8′ deep	7,300	1,350	8,650
2900	10′ deep	7,500	1,425	8,925
3000	10′ offset, 2′ deep	6,825	1,750	8,575
3100	4′ deep	7,175	2,050	9,225
3200	6′ deep	7,350	2,600	9,950
3300	8′ deep	7,575	3,400	10,975

545

G3010 410	Fire Hydrants	COST PER EACH		
		MAT.	INST.	TOTAL
3400	10' deep	7,775	5,100	12,875
3500	20' offset, 2' deep	7,175	2,325	9,500
3600	4' deep	7,525	2,800	10,325
3700	6' deep	7,700	3,625	11,325
3800	8' deep	7,925	4,800	12,725
3900	10' deep	8,125	6,400	14,525
5000	5-1/4" valve size, two way, 0' offset, 2' deep	6,550	1,200	7,750
5100	4' deep	6,700	1,225	7,925
5200	6' deep	6,850	1,275	8,125
5300	8' deep	7,075	1,325	8,400
5400	10' deep	7,425	1,400	8,825
5500	10' offset, 2' deep	6,825	1,750	8,575
5600	4' deep	6,975	2,050	9,025
5700	6' deep	7,125	2,575	9,700
5800	8' deep	7,350	3,400	10,750
5900	10' deep	7,700	5,075	12,775
6000	20' offset, 2' deep	7,175	2,300	9,475
6100	4' deep	7,325	2,775	10,100
6200	6' deep	7,475	3,600	11,075
6300	8' deep	7,700	4,775	12,475
6400	10' deep	8,050	6,375	14,425
6500	Three way, 0' offset, 2' deep	6,675	1,225	7,900
6600	4' deep	6,850	1,225	8,075
6700	6' deep	7,000	1,300	8,300
6800	8' deep	7,250	1,350	8,600
6900	10' deep	7,625	1,425	9,050
7000	10' offset, 2' deep	6,975	1,750	8,725
7100	4' deep	7,125	2,050	9,175
7200	6' deep	7,275	2,600	9,875
7300	8' deep	7,525	3,400	10,925
7400	10' deep	7,900	5,100	13,000
7500	20' offset, 2' deep	7,325	2,325	9,650
7600	4' deep	7,475	2,525	10,000
7700	6' deep	7,625	3,625	11,250

G30 Site Mechanical Utilities

G3020 Sanitary Sewer

The Septic System includes: a septic tank; leaching field; concrete distribution boxes; plus excavation and gravel backfill.

The Expanded System Listing shows systems with tanks ranging from 1000 gallons to 2000 gallons. Tanks are either concrete or fiberglass. Cost is on a complete unit basis.

System Components	QUANTITY	UNIT	COST EACH MAT.	COST EACH INST.	COST EACH TOTAL
SYSTEM G3020 302 0300					
1000 GALLON TANK, LEACHING FIELD, 600 S.F., 50′ FROM BLDG.					
Mobilization	2.000	Ea.		1,590	1,590
Septic tank, precast, 1000 gallon	1.000	Ea.	1,100	251.95	1,351.95
Effluent filter	1.000	Ea.	46	75	121
Distribution box, precast, 5 outlets	1.000	Ea.	88.50	58	146.50
Flow leveler	3.000	Ea.	7.44	27.75	35.19
Sewer pipe, PVC, SDR 35, 4″ diameter	40.000	L.F.	256	662.40	918.40
Tee	1.000	Ea.	14.80	75	89.80
Elbow	2.000	Ea.	21.60	100	121.60
Viewport cap	1.000	Ea.	8.40	25	33.40
Filter fabric	67.000	S.Y.	110.55	26.13	136.68
Detectable marking tape	1.600	C.L.F.	11.44	4.94	16.38
Excavation	160.000	C.Y.		1,995.20	1,995.20
Backfill	133.000	L.C.Y.		389.69	389.69
Spoil	55.000	L.C.Y.		357.50	357.50
Compaction	113.000	E.C.Y.		944.68	944.68
Stone fill, 3/4″ to 1-1/2″	39.000	C.Y.	1,560	498.03	2,058.03
TOTAL			3,224.73	7,081.27	10,306

G3020 302	Septic Systems	MAT.	INST.	TOTAL
0300	1000 gal. concrete septic tank, 600 S.F. leaching field, 50 feet from bldg.	3,225	7,075	10,300
0301	200 feet from building	3,475	8,750	12,225
0305	750 S.F. leaching field, 50 feet from building	3,675	7,700	11,375
0306	200 feet from building	3,925	9,400	13,325
0400	1250 gal. concrete septic tank, 600 S.F. leaching field, 50 feet from bldg.	3,275	7,350	10,625
0401	200 feet from building	3,525	9,050	12,575
0405	750 S.F. leaching field, 50 feet from building	3,725	8,000	11,725
0406	200 feet from building	3,975	9,725	13,700
0410	1000 S.F. leaching field, 50 feet from building	4,500	9,150	13,650
0411	200 feet from building	4,750	10,900	15,650
0500	1500 gal. concrete septic tank, 600 S.F. leaching field, 50 feet from bldg.	3,625	7,625	11,250
0501	200 feet from building	3,850	9,325	13,175
0505	750 S.F. leaching field, 50 feet from building	4,050	8,275	12,325
0506	200 feet from building	4,300	10,800	15,100
0510	1000 S.F. leaching field, 50 feet from building	4,825	9,425	14,250
0511	200 feet from building	5,075	11,200	16,275
0515	1100 S.F. leaching field, 50 feet from building	5,125	9,850	14,975
0516	200 feet from building	5,375	11,700	17,075
0600	2000 gal concrete tank, 1200 S.F. leaching field, 50 feet from building	6,075	10,500	16,575
0601	200 feet from building	6,325	12,100	18,425

547

For customer support on your Plumbing Cost Data, call 877.709.6968.

G3020 Sanitary Sewer

G3020 302	Septic Systems	COST EACH		
		MAT.	INST.	TOTAL
0605	1500 S.F. leaching field, 50 feet from building	6,975	11,800	18,775
0606	200 feet from building	7,225	13,500	20,725
0710	2500 gal concrete tank, 1500 S.F. leaching field, 50 feet from building	7,100	12,500	19,600
0711	200 feet from building	7,350	14,200	21,550
0715	1600 S.F. leaching field, 50 feet from building	7,450	13,000	20,450
0716	200 feet from building	7,700	14,700	22,400
0720	1800 S.F. leaching field, 50 feet from building	8,050	14,000	22,050
0721	200 feet from building	8,300	15,700	24,000
1300	1000 gal. 2 compartment tank, 600 S.F. leaching field, 50 feet from bldg.	2,875	7,075	9,950
1301	200 feet from building	3,100	8,750	11,850
1305	750 S.F. leaching field, 50 feet from building	3,300	7,700	11,000
1306	200 feet from building	3,550	9,400	12,950
1400	1250 gal. 2 compartment tank, 600 S.F. leaching field, 50 feet from bldg.	3,100	7,350	10,450
1401	200 feet from building	3,350	9,050	12,400
1405	750 S.F. leaching field, 50 feet from building	3,525	8,000	11,525
1406	200 feet from building	3,775	9,725	13,500
1410	1000 S.F. leaching field, 50 feet from building	4,325	9,150	13,475
1411	200 feet from building	4,575	10,900	15,475
1500	1500 gal. 2 compartment tank, 600 S.F. leaching field, 50 feet from bldg.	3,950	7,625	11,575
1501	200 feet from building	4,175	9,325	13,500
1505	750 S.F. leaching field, 50 feet from building	4,375	8,275	12,650
1506	200 feet from building	4,625	10,800	15,425
1510	1000 S.F. leaching field, 50 feet from building	5,150	9,425	14,575
1511	200 feet from building	5,400	11,200	16,600
1515	1100 S.F. leaching field, 50 feet from building	5,450	9,850	15,300
1516	200 feet from building	5,700	11,700	17,400
1600	2000 gal 2 compartment tank, 1200 S.F. leaching field, 50 feet from bldg.	6,200	10,500	16,700
1601	200 feet from building	6,450	12,100	18,550
1605	1500 S.F. leaching field, 50 feet from building	7,100	11,800	18,900
1606	200 feet from building	7,350	13,500	20,850
1710	2500 gal 2 compartment tank, 1500 S.F. leaching field, 50 feet from bldg.	7,800	12,500	20,300
1711	200 feet from building	8,050	14,200	22,250
1715	1600 S.F. leaching field, 50 feet from building	8,150	13,000	21,150
1716	200 feet from building	8,400	14,700	23,100
1720	1800 S.F. leaching field, 50 feet from building	8,750	14,000	22,750
1721	200 feet from building	9,000	15,700	24,700

G3030 Storm Sewer

Manhole   **Catch Basin**

The Manhole and Catch Basin System includes: excavation with a backhoe; a formed concrete footing; frame and cover; cast iron steps and compacted backfill.

The Expanded System Listing shows manholes that have a 4', 5' and 6' inside diameter riser. Depths range from 4' to 14'. Construction material shown is either concrete, concrete block, precast concrete, or brick.

System Components	QUANTITY	UNIT	COST PER EACH		
			MAT.	INST.	TOTAL
SYSTEM G3030 210 1920					
MANHOLE/CATCH BASIN, BRICK, 4' I.D. RISER, 4' DEEP					
Excavation, hydraulic backhoe, 3/8 C.Y. bucket	14.815	B.C.Y.		119.26	119.26
Trim sides and bottom of excavation	64.000	S.F.		62.08	62.08
Forms in place, manhole base, 4 uses	20.000	SFCA	16.40	106	122.40
Reinforcing in place footings, #4 to #7	.019	Ton	20.43	23.75	44.18
Concrete, 3000 psi	.925	C.Y.	103.60		103.60
Place and vibrate concrete, footing, direct chute	.925	C.Y.		49.26	49.26
Catch basin or MH, brick, 4' ID, 4' deep	1.000	Ea.	575	1,025	1,600
Catch basin or MH steps; heavy galvanized cast iron	1.000	Ea.	21	14.10	35.10
Catch basin or MH frame and cover	1.000	Ea.	268	245.50	513.50
Fill, granular	12.954	L.C.Y.	297.94		297.94
Backfill, spread with wheeled front end loader	12.954	L.C.Y.		34.07	34.07
Backfill compaction, 12" lifts, air tamp	12.954	E.C.Y.		117.88	117.88
TOTAL			1,302.37	1,796.90	3,099.27

G3030 210	Manholes & Catch Basins	COST PER EACH		
		MAT.	INST.	TOTAL
1920	Manhole/catch basin, brick, 4' I.D. riser, 4' deep	1,300	1,775	3,075
1940	6' deep	1,850	2,500	4,350
1960	8' deep	2,450	3,425	5,875
1980	10' deep	2,900	4,200	7,100
3000	12' deep	3,625	4,575	8,200
3020	14' deep	4,450	6,425	10,875
3200	Block, 4' I.D. riser, 4' deep	1,175	1,450	2,625
3220	6' deep	1,650	2,050	3,700
3240	8' deep	2,200	2,850	5,050
3260	10' deep	2,625	3,500	6,125
3280	12' deep	3,325	4,450	7,775
3300	14' deep	4,150	5,425	9,575
4620	Concrete, cast-in-place, 4' I.D. riser, 4' deep	1,275	2,475	3,750
4640	6' deep	1,825	3,325	5,150
4660	8' deep	2,550	4,800	7,350
4680	10' deep	3,075	5,950	9,025
4700	12' deep	3,875	7,375	11,250
4720	14' deep	4,800	8,825	13,625
5820	Concrete, precast, 4' I.D. riser, 4' deep	1,525	1,325	2,850
5840	6' deep	2,050	1,775	3,825

549

G3030 Storm Sewer

G3030 210	Manholes & Catch Basins	COST PER EACH		
		MAT.	INST.	TOTAL
5860	8' deep	2,575	2,475	5,050
5880	10' deep	3,125	3,050	6,175
5900	12' deep	3,925	3,775	7,700
5920	14' deep	4,875	4,775	9,650
6000	5' I.D. riser, 4' deep	2,600	1,500	4,100
6020	6' deep	3,225	2,100	5,325
6040	8' deep	4,175	2,775	6,950
6060	10' deep	5,300	3,525	8,825
6080	12' deep	6,525	4,525	11,050
6100	14' deep	7,900	5,500	13,400
6200	6' I.D. riser, 4' deep	3,325	1,975	5,300
6220	6' deep	4,150	2,600	6,750
6240	8' deep	5,275	3,675	8,950
6260	10' deep	6,675	4,650	11,325
6280	12' deep	8,200	5,850	14,050
6300	14' deep	9,850	7,100	16,950

Reference Section

All the reference information is in one section, making it easy to find what you need to know . . . and easy to use the book on a daily basis. This section is visually identified by a vertical black bar on the page edges.

In this Reference Section, we've included Equipment Rental Costs, a listing of rental and operating costs; Crew Listings, a full listing of all crews and equipment, and their costs; Historical Cost Indexes for cost comparisons over time; City Cost Indexes and Location Factors for adjusting costs to the region you are in; Reference Tables, where you will find explanations, estimating information and procedures, or technical data; Change Orders, information on pricing changes to contract documents; Square Foot Costs that allow you to make a rough estimate for the overall cost of a project; and an explanation of all the Abbreviations in the book.

Table of Contents

Estimating Tips

- This section contains the average costs to rent and operate hundreds of pieces of construction equipment. This is useful information when estimating the time and material requirements of any particular operation in order to establish a unit or total cost. Equipment costs include not only rental, but also operating costs for equipment under normal use.

Rental Costs

- Equipment rental rates are obtained from the following industry sources throughout North America: contractors, suppliers, dealers, manufacturers, and distributors.

- Rental rates vary throughout the country, with larger cities generally having lower rates. Lease plans for new equipment are available for periods in excess of six months, with a percentage of payments applying toward purchase.

- Monthly rental rates vary from 2% to 5% of the purchase price of the equipment depending on the anticipated life of the equipment and its wearing parts.

- Weekly rental rates are about 1/3 the monthly rates, and daily rental rates are about 1/3 the weekly rate.

- Rental rates can also be treated as reimbursement costs for contractor-owned equipment. Owned equipment costs include depreciation, loan payments, interest, taxes, insurance, storage, and major repairs.

Operating Costs

- The operating costs include parts and labor for routine servicing, such as repair and replacement of pumps, filters and worn lines. Normal operating expendables, such as fuel, lubricants, tires and electricity (where applicable), are also included.

- Extraordinary operating expendables with highly variable wear patterns, such as diamond bits and blades, are excluded. These costs can be found as material costs in the Unit Price section.

- The hourly operating costs listed do not include the operator's wages.

Equipment Cost/Day

- Any power equipment required by a crew is shown in the Crew Listings with a daily cost.

- The daily cost of equipment needed by a crew is based on dividing the weekly rental rate by 5 (number of working days in the week), and then adding the hourly operating cost times 8 (the number of hours in a day). This "Equipment Cost/ Day" is shown in the far right column of the Equipment Rental pages.

- If equipment is needed for only one or two days, it is best to develop your own cost by including components for daily rent and hourly operating cost. This is important when the listed Crew for a task does not contain the equipment needed, such as a crane for lifting mechanical heating/ cooling equipment up onto a roof.

- If the quantity of work is less than the crew's Daily Output shown for a Unit Price line item that includes a bare unit equipment cost, it is recommended to estimate one day's rental cost and operating cost for equipment shown in the Crew Listing for that line item.

Mobilization/ Demobilization

- The cost to move construction equipment from an equipment yard or rental company to the job site and back again is not included in equipment rental costs listed in the Reference Section, nor in the bare equipment cost of any unit price line item, nor in any equipment costs shown in the Crew listings.

- Mobilization (to the site) and demobilization (from the site) costs can be found in the Unit Price Section.

- If a piece of equipment is already at the job site, it is not appropriate to utilize mobilization/demobilization. costs again in an estimate. ■

01 54 33 | Equipment Rental

			UNIT	HOURLY OPER. COST	RENT PER DAY	RENT PER WEEK	RENT PER MONTH	EQUIPMENT COST/DAY	
10	0010	**CONCRETE EQUIPMENT RENTAL** without operators	R015433 -10						10
	0200	Bucket, concrete lightweight, 1/2 C.Y.	Ea.	.80	23.50	70	210	20.40	
	0300	1 C.Y.		.90	27.50	83	249	23.80	
	0400	1-1/2 C.Y.		1.10	36.50	110	330	30.80	
	0500	2 C.Y.		1.25	45	135	405	37	
	0580	8 C.Y.		6.10	257	770	2,300	202.80	
	0600	Cart, concrete, self-propelled, operator walking, 10 C.F.		3.25	56.50	170	510	60	
	0700	Operator riding, 18 C.F.		5.35	95	285	855	99.80	
	0800	Conveyer for concrete, portable, gas, 16" wide, 26' long		12.60	123	370	1,100	174.80	
	0900	46' long		13.00	148	445	1,325	193	
	1000	56' long		13.10	157	470	1,400	198.80	
	1100	Core drill, electric, 2-1/2 H.P., 1" to 8" bit diameter		1.77	68.50	205	615	55.15	
	1150	11 H.P., 8" to 18" cores		5.95	113	340	1,025	115.60	
	1200	Finisher, concrete floor, gas, riding trowel; 96" wide		12.75	145	435	1,300	189	
	1300	Gas, walk-behind, 3 blade, 36" trowel		2.15	20	60	180	29.20	
	1400	4 blade, 48" trowel		4.20	27.50	83	249	50.20	
	1500	Float, hand-operated (Bull float) 48" wide		.08	13.35	40	120	8.65	
	1570	Curb builder, 14 H.P., gas, single screw		14.65	248	745	2,225	266.20	
	1590	Double screw		15.35	290	870	2,600	296.80	
	1600	Floor grinder, concrete and terrazzo, electric, 22" path		2.57	158	475	1,425	115.55	
	1700	Edger, concrete, electric, 7" path		1.04	51.50	155	465	39.30	
	1750	Vacuum pick-up system for floor grinders, wet/dry		1.49	81.50	245	735	60.90	
	1800	Mixer, powered, mortar and concrete, gas, 6 C.F., 18 H.P.		8.65	118	355	1,075	140.20	
	1900	10 C.F., 25 H.P.		10.80	143	430	1,300	172.40	
	2000	16 C.F.		11.15	165	495	1,475	188.20	
	2100	Concrete, stationary, tilt drum, 2 C.Y.		6.90	232	695	2,075	194.20	
	2120	Pump, concrete, truck mounted 4" line 80' boom		25.00	865	2,600	7,800	720	
	2140	5" line, 110' boom		32.40	1,150	3,435	10,300	946.20	
	2160	Mud jack, 50 C.F. per hr.		7.30	125	375	1,125	133.40	
	2180	225 C.F. per hr.		9.50	145	435	1,300	163	
	2190	Shotcrete pump rig, 12 C.Y./hr.		14.85	220	660	1,975	250.80	
	2200	35 C.Y./hr.		17.50	235	705	2,125	281	
	2600	Saw, concrete, manual, gas, 18 H.P.		6.75	45	135	405	81	
	2650	Self-propelled, gas, 30 H.P.		13.30	102	305	915	167.40	
	2675	V-groove crack chaser, manual, gas, 6 H.P.		2.35	17.35	52	156	29.20	
	2700	Vibrators, concrete, electric, 60 cycle, 2 H.P.		.46	8.65	26	78	8.90	
	2800	3 H.P.		.60	11.65	35	105	11.80	
	2900	Gas engine, 5 H.P.		2.00	16	48	144	25.60	
	3000	8 H.P.		2.75	15.35	46	138	31.20	
	3050	Vibrating screed, gas engine, 8 H.P.		2.91	71.50	215	645	66.30	
	3120	Concrete transit mixer, 6 x 4, 250 H.P., 8 C.Y., rear discharge		61.55	570	1,715	5,150	835.40	
	3200	Front discharge		72.45	700	2,095	6,275	998.60	
	3300	6 x 6, 285 H.P., 12 C.Y., rear discharge		71.50	660	1,980	5,950	968	
	3400	Front discharge		74.00	705	2,120	6,350	1,016	
20	0010	**EARTHWORK EQUIPMENT RENTAL** without operators	R015433 -10						20
	0040	Aggregate spreader, push type 8' to 12' wide	Ea.	3.05	25.50	76	228	39.60	
	0045	Tailgate type, 8' wide		2.90	32.50	98	294	42.80	
	0055	Earth auger, truck-mounted, for fence & sign posts, utility poles		19.15	440	1,320	3,950	417.20	
	0060	For borings and monitoring wells		46.65	675	2,030	6,100	779.20	
	0070	Portable, trailer mounted		3.30	32.50	98	294	46	
	0075	Truck-mounted, for caissons, water wells		100.15	2,900	8,705	26,100	2,542	
	0080	Horizontal boring machine, 12" to 36" diameter, 45 H.P.		24.20	192	575	1,725	308.60	
	0090	12" to 48" diameter, 65 H.P.		34.35	335	1,000	3,000	474.80	
	0095	Auger, for fence posts, gas engine, hand held		.60	6	18	54	8.40	
	0100	Excavator, diesel hydraulic, crawler mounted, 1/2 C.Y. cap.		24.65	420	1,255	3,775	448.20	
	0120	5/8 C.Y. capacity		32.55	550	1,645	4,925	589.40	
	0140	3/4 C.Y. capacity		35.60	615	1,850	5,550	654.80	
	0150	1 C.Y. capacity		49.70	690	2,075	6,225	812.60	

01 54 33 | Equipment Rental

		UNIT	HOURLY OPER. COST	RENT PER DAY	RENT PER WEEK	RENT PER MONTH	EQUIPMENT COST/DAY
0200	1-1/2 C.Y. capacity	Ea.	59.80	920	2,760	8,275	1,030
0300	2 C.Y. capacity		67.35	1,050	3,180	9,550	1,175
0320	2-1/2 C.Y. capacity		102.55	1,300	3,925	11,800	1,605
0325	3-1/2 C.Y. capacity		144.45	2,150	6,420	19,300	2,440
0330	4-1/2 C.Y. capacity		173.40	2,625	7,880	23,600	2,963
0335	6 C.Y. capacity		219.80	2,900	8,680	26,000	3,494
0340	7 C.Y. capacity		222.05	3,025	9,070	27,200	3,590
0342	Excavator attachments, bucket thumbs		3.20	245	735	2,200	172.60
0345	Grapples		2.75	193	580	1,750	138
0346	Hydraulic hammer for boom mounting, 4000 ft lb.		12.40	350	1,045	3,125	308.20
0347	5000 ft lb.		14.40	425	1,280	3,850	371.20
0348	8000 ft lb.		21.30	625	1,875	5,625	545.40
0349	12,000 ft lb.		23.35	750	2,245	6,725	635.80
0350	Gradall type, truck mounted, 3 ton @ 15' radius, 5/8 C.Y.		43.65	890	2,665	8,000	882.20
0370	1 C.Y. capacity		47.85	1,025	3,095	9,275	1,002
0400	Backhoe-loader, 40 to 45 H.P., 5/8 C.Y. capacity		14.90	240	720	2,150	263.20
0450	45 H.P. to 60 H.P., 3/4 C.Y. capacity		23.40	295	885	2,650	364.20
0460	80 H.P., 1-1/4 C.Y. capacity		25.60	310	935	2,800	391.80
0470	112 H.P., 1-1/2 C.Y. capacity		40.95	645	1,930	5,800	713.60
0482	Backhoe-loader attachment, compactor, 20,000 lb.		5.80	142	425	1,275	131.40
0485	Hydraulic hammer, 750 ft lb.		3.30	96.50	290	870	84.40
0486	Hydraulic hammer, 1200 ft lb.		6.30	217	650	1,950	180.40
0500	Brush chipper, gas engine, 6" cutter head, 35 H.P.		11.10	105	315	945	151.80
0550	Diesel engine, 12" cutter head, 130 H.P.		27.55	285	855	2,575	391.40
0600	15" cutter head, 165 H.P.		33.10	335	1,005	3,025	465.80
0750	Bucket, clamshell, general purpose, 3/8 C.Y.		1.30	38.50	115	345	33.40
0800	1/2 C.Y.		1.40	45	135	405	38.20
0850	3/4 C.Y.		1.55	55	165	495	45.40
0900	1 C.Y.		1.60	58.50	175	525	47.80
0950	1-1/2 C.Y.		2.55	81.50	245	735	69.40
1000	2 C.Y.		2.70	90	270	810	75.60
1010	Bucket, dragline, medium duty, 1/2 C.Y.		.75	23.50	70	210	20
1020	3/4 C.Y.		.75	24.50	73	219	20.60
1030	1 C.Y.		.80	26	78	234	22
1040	1-1/2 C.Y.		1.20	40	120	360	33.60
1050	2 C.Y.		1.30	45	135	405	37.40
1070	3 C.Y.		1.95	61.50	185	555	52.60
1200	Compactor, manually guided 2-drum vibratory smooth roller, 7.5 H.P.		7.25	203	610	1,825	180
1250	Rammer/tamper, gas, 8"		2.75	46.50	140	420	50
1260	15"		3.05	53.50	160	480	56.40
1300	Vibratory plate, gas, 18" plate, 3000 lb. blow		2.70	24.50	73	219	36.20
1350	21" plate, 5000 lb. blow		3.40	32.50	98	294	46.80
1370	Curb builder/extruder, 14 H.P., gas, single screw		14.65	248	745	2,225	266.20
1390	Double screw		15.35	290	870	2,600	296.80
1500	Disc harrow attachment, for tractor		.44	73.50	221	665	47.70
1810	Feller buncher, shearing & accumulating trees, 100 H.P.		48.30	755	2,260	6,775	838.40
1860	Grader, self-propelled, 25,000 lb.		39.50	640	1,925	5,775	701
1910	30,000 lb.		43.15	650	1,955	5,875	736.20
1920	40,000 lb.		64.90	1,100	3,275	9,825	1,174
1930	55,000 lb.		84.15	1,700	5,130	15,400	1,699
1950	Hammer, pavement breaker, self-propelled, diesel, 1000 to 1250 lb.		29.90	350	1,055	3,175	450.20
2000	1300 to 1500 lb.		44.85	705	2,110	6,325	780.80
2050	Pile driving hammer, steam or air, 4150 ft lb. @ 225 bpm		10.40	475	1,420	4,250	367.20
2100	8750 ft lb. @ 145 bpm		12.40	660	1,975	5,925	494.20
2150	15,000 ft lb. @ 60 bpm		14.00	790	2,370	7,100	586
2200	24,450 ft lb. @ 111 bpm		15.00	875	2,630	7,900	646
2250	Leads, 60' high for pile driving hammers up to 20,000 ft lb.		3.30	80.50	242	725	74.80
2300	90' high for hammers over 20,000 ft lb.		4.95	141	424	1,275	124.40

For customer support on your Plumbing Cost Data, call 877.709.6968.

555

01 54 33 | Equipment Rental

		UNIT	HOURLY OPER. COST	RENT PER DAY	RENT PER WEEK	RENT PER MONTH	EQUIPMENT COST/DAY	
20	2350	Diesel type hammer, 22,400 ft lb.	Ea.	22.55	460	1,380	4,150	456.40	**20**
	2400	41,300 ft lb.		32.00	540	1,625	4,875	581	
	2450	141,000 ft lb.		51.65	960	2,875	8,625	988.20	
	2500	Vib. elec. hammer/extractor, 200 kW diesel generator, 34 H.P.		54.90	635	1,900	5,700	819.20	
	2550	80 H.P.		100.65	925	2,775	8,325	1,360	
	2600	150 H.P.		190.90	1,775	5,300	15,900	2,587	
	2800	Log chipper, up to 22" diameter, 600 H.P.		63.30	640	1,915	5,750	889.40	
	2850	Logger, for skidding & stacking logs, 150 H.P.		52.75	815	2,445	7,325	911	
	2860	Mulcher, diesel powered, trailer mounted		24.05	212	635	1,900	319.40	
	2900	Rake, spring tooth, with tractor		18.23	345	1,042	3,125	354.25	
	3000	Roller, vibratory, tandem, smooth drum, 20 H.P.		8.80	145	435	1,300	157.40	
	3050	35 H.P.		11.10	247	740	2,225	236.80	
	3100	Towed type vibratory compactor, smooth drum, 50 H.P.		26.00	315	940	2,825	396	
	3150	Sheepsfoot, 50 H.P.		27.35	345	1,035	3,100	425.80	
	3170	Landfill compactor, 220 H.P.		91.00	1,475	4,440	13,300	1,616	
	3200	Pneumatic tire roller, 80 H.P.		16.80	350	1,050	3,150	344.40	
	3250	120 H.P.		25.20	600	1,800	5,400	561.60	
	3300	Sheepsfoot vibratory roller, 240 H.P.		68.95	1,100	3,270	9,800	1,206	
	3320	340 H.P.		100.85	1,575	4,730	14,200	1,753	
	3350	Smooth drum vibratory roller, 75 H.P.		24.85	585	1,755	5,275	549.80	
	3400	125 H.P.		32.55	710	2,135	6,400	687.40	
	3410	Rotary mower, brush, 60", with tractor		22.60	315	950	2,850	370.80	
	3420	Rototiller, walk-behind, gas, 5 H.P.		1.96	51.50	155	465	46.70	
	3422	8 H.P.		3.08	80	240	720	72.65	
	3440	Scrapers, towed type, 7 C.Y. capacity		5.80	113	340	1,025	114.40	
	3450	10 C.Y. capacity		6.60	157	470	1,400	146.80	
	3500	15 C.Y. capacity		7.10	180	540	1,625	164.80	
	3525	Self-propelled, single engine, 14 C.Y. capacity		114.76	1,600	4,830	14,500	1,884	
	3550	Dual engine, 21 C.Y. capacity		173.80	2,175	6,490	19,500	2,688	
	3600	31 C.Y. capacity		231.70	3,075	9,195	27,600	3,693	
	3640	44 C.Y. capacity		287.80	3,950	11,885	35,700	4,679	
	3650	Elevating type, single engine, 11 C.Y. capacity		71.45	985	2,955	8,875	1,163	
	3700	22 C.Y. capacity		139.55	2,350	7,025	21,100	2,521	
	3710	Screening plant 110 H.P. w/5' x 10' screen		25.35	400	1,200	3,600	442.80	
	3720	5' x 16' screen		29.65	505	1,515	4,550	540.20	
	3850	Shovel, crawler-mounted, front-loading, 7 C.Y. capacity		258.85	3,200	9,610	28,800	3,993	
	3855	12 C.Y. capacity		386.45	4,400	13,220	39,700	5,736	
	3860	Shovel/backhoe bucket, 1/2 C.Y.		2.50	66.50	200	600	60	
	3870	3/4 C.Y.		2.55	75	225	675	65.40	
	3880	1 C.Y.		2.65	83.50	250	750	71.20	
	3890	1-1/2 C.Y.		2.75	96.50	290	870	80	
	3910	3 C.Y.		3.15	130	390	1,175	103.20	
	3950	Stump chipper, 18" deep, 30 H.P.		7.82	205	615	1,850	185.55	
	4110	Dozer, crawler, torque converter, diesel 80 H.P.		29.05	400	1,200	3,600	472.40	
	4150	105 H.P.		35.60	530	1,590	4,775	602.80	
	4200	140 H.P.		51.25	795	2,380	7,150	886	
	4260	200 H.P.		77.25	1,275	3,845	11,500	1,387	
	4310	300 H.P.		100.65	1,825	5,460	16,400	1,897	
	4360	410 H.P.		132.60	2,250	6,720	20,200	2,405	
	4370	500 H.P.		171.15	3,225	9,650	29,000	3,299	
	4380	700 H.P.		280.90	4,550	13,660	41,000	4,979	
	4400	Loader, crawler, torque conv., diesel, 1-1/2 C.Y., 80 H.P.		31.25	455	1,360	4,075	522	
	4450	1-1/2 to 1-3/4 C.Y., 95 H.P.		34.60	590	1,765	5,300	629.80	
	4510	1-3/4 to 2-1/4 C.Y., 130 H.P.		53.80	875	2,630	7,900	956.40	
	4530	2-1/2 to 3-1/4 C.Y., 190 H.P.		65.95	1,100	3,300	9,900	1,188	
	4560	3-1/2 to 5 C.Y., 275 H.P.		87.80	1,425	4,295	12,900	1,561	
	4610	Front end loader, 4WD, articulated frame, diesel, 1 to 1-1/4 C.Y., 70 H.P.		19.75	235	705	2,125	299	
	4620	1-1/2 to 1-3/4 C.Y., 95 H.P.		24.80	300	900	2,700	378.40	

01 54 33 | Equipment Rental

		UNIT	HOURLY OPER. COST	RENT PER DAY	RENT PER WEEK	RENT PER MONTH	EQUIPMENT COST/DAY	
20	4650	1-3/4 to 2 C.Y., 130 H.P.	Ea.	29.15	380	1,140	3,425	461.20	**20**
	4710	2-1/2 to 3-1/2 C.Y., 145 H.P.		33.10	425	1,275	3,825	519.80	
	4730	3 to 4-1/2 C.Y., 185 H.P.		42.40	550	1,650	4,950	669.20	
	4760	5-1/4 to 5-3/4 C.Y., 270 H.P.		67.50	905	2,710	8,125	1,082	
	4810	7 to 9 C.Y., 475 H.P.		114.05	1,700	5,125	15,400	1,937	
	4870	9 - 11 C.Y., 620 H.P.		160.55	2,775	8,290	24,900	2,942	
	4880	Skid steer loader, wheeled, 10 C.F., 30 H.P. gas		10.45	150	450	1,350	173.60	
	4890	1 C.Y., 78 H.P., diesel		20.20	262	785	2,350	318.60	
	4892	Skid-steer attachment, auger		.48	80.50	242	725	52.25	
	4893	Backhoe		.66	111	332	995	71.70	
	4894	Broom		.74	124	372	1,125	80.30	
	4895	Forks		.22	36	108	325	23.35	
	4896	Grapple		.56	92.50	278	835	60.10	
	4897	Concrete hammer		1.06	177	531	1,600	114.70	
	4898	Tree spade		1.03	172	515	1,550	111.25	
	4899	Trencher		.54	90	270	810	58.30	
	4900	Trencher, chain, boom type, gas, operator walking, 12 H.P.		5.00	46.50	140	420	68	
	4910	Operator riding, 40 H.P.		20.15	290	870	2,600	335.20	
	5000	Wheel type, diesel, 4' deep, 12" wide		85.95	845	2,535	7,600	1,195	
	5100	6' deep, 20" wide		92.75	1,925	5,740	17,200	1,890	
	5150	Chain type, diesel, 5' deep, 8" wide		38.35	560	1,675	5,025	641.80	
	5200	Diesel, 8' deep, 16" wide		155.95	3,600	10,770	32,300	3,402	
	5202	Rock trencher, wheel type, 6" wide x 18" deep		21.50	335	1,000	3,000	372	
	5206	Chain type, 18" wide x 7' deep		112.00	2,800	8,395	25,200	2,575	
	5210	Tree spade, self-propelled		14.38	267	800	2,400	275.05	
	5250	Truck, dump, 2-axle, 12 ton, 8 C.Y. payload, 220 H.P.		34.40	227	680	2,050	411.20	
	5300	Three axle dump, 16 ton, 12 C.Y. payload, 400 H.P.		61.00	340	1,015	3,050	691	
	5310	Four axle dump, 25 ton, 18 C.Y. payload, 450 H.P.		71.75	490	1,475	4,425	869	
	5350	Dump trailer only, rear dump, 16-1/2 C.Y.		5.45	138	415	1,250	126.60	
	5400	20 C.Y.		5.90	157	470	1,400	141.20	
	5450	Flatbed, single axle, 1-1/2 ton rating		25.55	68.50	205	615	245.40	
	5500	3 ton rating		30.65	96.50	290	870	303.20	
	5550	Off highway rear dump, 25 ton capacity		73.90	1,275	3,800	11,400	1,351	
	5600	35 ton capacity		82.50	1,425	4,260	12,800	1,512	
	5610	50 ton capacity		102.65	1,700	5,080	15,200	1,837	
	5620	65 ton capacity		105.65	1,700	5,090	15,300	1,863	
	5630	100 ton capacity		151.40	2,850	8,570	25,700	2,925	
	6000	Vibratory plow, 25 H.P., walking		8.45	60	180	540	103.60	
40	0010	**GENERAL EQUIPMENT RENTAL** without operators [R015433 -10]							**40**
	0150	Aerial lift, scissor type, to 15' high, 1000 lb. cap., electric	Ea.	3.05	51.50	155	465	55.40	
	0160	To 25' high, 2000 lb. capacity		3.45	66.50	200	600	67.60	
	0170	Telescoping boom to 40' high, 500 lb. capacity, diesel		13.55	320	965	2,900	301.40	
	0180	To 45' high, 500 lb. capacity		14.75	350	1,055	3,175	329	
	0190	To 60' high, 600 lb. capacity		17.20	495	1,490	4,475	435.60	
	0195	Air compressor, portable, 6.5 CFM, electric		.67	12.65	38	114	12.95	
	0196	Gasoline		.81	19	57	171	17.90	
	0200	Towed type, gas engine, 60 CFM		13.55	48.50	145	435	137.40	
	0300	160 CFM		15.80	50	150	450	156.40	
	0400	Diesel engine, rotary screw, 250 CFM		17.05	108	325	975	201.40	
	0500	365 CFM		23.10	132	395	1,175	263.80	
	0550	450 CFM		29.35	165	495	1,475	333.80	
	0600	600 CFM		51.70	228	685	2,050	550.60	
	0700	750 CFM		51.90	237	710	2,125	557.20	
	0800	For silenced models, small sizes, add to rent		3%	5%	5%	5%		
	0900	Large sizes, add to rent		5%	7%	7%	7%		
	0930	Air tools, breaker, pavement, 60 lb.	Ea.	.50	9.65	29	87	9.80	
	0940	80 lb.		.50	10.35	31	93	10.20	
	0950	Drills, hand (jackhammer) 65 lb.		.60	17.35	52	156	15.20	

For customer support on your Plumbing Cost Data, call 877.709.6968.

557

01 54 33 | Equipment Rental	UNIT	HOURLY OPER. COST	RENT PER DAY	RENT PER WEEK	RENT PER MONTH	EQUIPMENT COST/DAY
40							**40**
0960 Track or wagon, swing boom, 4" drifter	Ea.	62.50	880	2,640	7,925	1,028
0970 5" drifter		74.45	1,075	3,235	9,700	1,243
0975 Track mounted quarry drill, 6" diameter drill		128.10	1,600	4,765	14,300	1,978
0980 Dust control per drill		1.02	23.50	71	213	22.35
0990 Hammer, chipping, 12 lb.		.55	26	78	234	20
1000 Hose, air with couplings, 50' long, 3/4" diameter		.03	5	15	45	3.25
1100 1" diameter		.04	6.35	19	57	4.10
1200 1-1/2" diameter		.05	9	27	81	5.80
1300 2" diameter		.07	12	36	108	7.75
1400 2-1/2" diameter		.11	19	57	171	12.30
1410 3" diameter		.14	23	69	207	14.90
1450 Drill, steel, 7/8" x 2'		.05	8.65	26	78	5.60
1460 7/8" x 6'		.05	9	27	81	5.80
1520 Moil points		.02	3.33	10	30	2.15
1525 Pneumatic nailer w/accessories		.58	38.50	115	345	27.65
1530 Sheeting driver for 60 lb. breaker		.04	6	18	54	3.90
1540 For 90 lb. breaker		.12	8	24	72	5.75
1550 Spade, 25 lb.		.45	6.65	20	60	7.60
1560 Tamper, single, 35 lb.		.55	36.50	109	325	26.20
1570 Triple, 140 lb.		.82	54.50	164	490	39.35
1580 Wrenches, impact, air powered, up to 3/4" bolt		.40	12.65	38	114	10.80
1590 Up to 1-1/4" bolt		.50	23.50	70	210	18
1600 Barricades, barrels, reflectorized, 1 to 99 barrels		.03	4.60	13.80	41.50	3
1610 100 to 200 barrels		.02	3.53	10.60	32	2.30
1620 Barrels with flashers, 1 to 99 barrels		.03	5.25	15.80	47.50	3.40
1630 100 to 200 barrels		.03	4.20	12.60	38	2.75
1640 Barrels with steady burn type C lights		.04	7	21	63	4.50
1650 Illuminated board, trailer mounted, with generator		3.50	130	390	1,175	106
1670 Portable barricade, stock, with flashers, 1 to 6 units		.03	5.25	15.80	47.50	3.40
1680 25 to 50 units		.03	4.90	14.70	44	3.20
1685 Butt fusion machine, wheeled, 1.5 HP electric, 2" - 8" diameter pipe		2.63	167	500	1,500	121.05
1690 Tracked, 20 HP diesel, 4"-12" diameter pipe		11.14	490	1,465	4,400	382.10
1695 83 HP diesel, 8" - 24" diameter pipe		30.46	975	2,930	8,800	829.70
1700 Carts, brick, hand powered, 1000 lb. capacity		.50	83.50	251	755	54.20
1800 Gas engine, 1500 lb., 7-1/2' lift		4.17	115	345	1,025	102.35
1822 Dehumidifier, medium, 6 lb./hr., 150 CFM		1.00	62	186	560	45.20
1824 Large, 18 lb./hr., 600 CFM		2.02	126	378	1,125	91.75
1830 Distributor, asphalt, trailer mounted, 2000 gal., 38 H.P. diesel		9.95	325	980	2,950	275.60
1840 3000 gal., 38 H.P. diesel		11.45	355	1,070	3,200	305.60
1850 Drill, rotary hammer, electric		1.02	26	78	234	23.75
1860 Carbide bit, 1-1/2" diameter, add to electric rotary hammer		.02	3.61	10.84	32.50	2.35
1865 Rotary, crawler, 250 H.P.		148.50	2,050	6,185	18,600	2,425
1870 Emulsion sprayer, 65 gal., 5 H.P. gas engine		2.91	97	291	875	81.50
1880 200 gal., 5 H.P. engine		7.85	162	485	1,450	159.80
1900 Floor auto-scrubbing machine, walk-behind, 28" path		4.98	325	970	2,900	233.85
1930 Floodlight, mercury vapor, or quartz, on tripod, 1000 watt		.44	20.50	62	186	15.90
1940 2000 watt		.83	41	123	370	31.25
1950 Floodlights, trailer mounted with generator, 1 - 300 watt light		3.65	71.50	215	645	72.20
1960 2 - 1000 watt lights		4.85	96.50	290	870	96.80
2000 4 - 300 watt lights		4.55	93.50	280	840	92.40
2005 Foam spray rig, incl. box trailer, compressor, generator, proportioner		35.03	490	1,465	4,400	573.25
2020 Forklift, straight mast, 12' lift, 5000 lb., 2 wheel drive, gas		26.35	202	605	1,825	331.80
2040 21' lift, 5000 lb., 4 wheel drive, diesel		20.90	240	720	2,150	311.20
2050 For rough terrain, 42' lift, 35' reach, 9000 lb., 110 H.P.		29.55	485	1,450	4,350	526.40
2060 For plant, 4 ton capacity, 80 H.P., 2 wheel drive, gas		16.15	93.50	280	840	185.20
2080 10 ton capacity, 120 H.P., 2 wheel drive, diesel		24.10	162	485	1,450	289.80
2100 Generator, electric, gas engine, 1.5 kW to 3 kW		3.75	11.35	34	102	36.80
2200 5 kW		4.80	15.35	46	138	47.60

01 54 33 | Equipment Rental

		UNIT	HOURLY OPER. COST	RENT PER DAY	RENT PER WEEK	RENT PER MONTH	EQUIPMENT COST/DAY	
40	2300	10 kW	Ea.	9.10	36.50	110	330	94.80	40
	2400	25 kW		10.60	85	255	765	135.80	
	2500	Diesel engine, 20 kW		12.10	66.50	200	600	136.80	
	2600	50 kW		23.35	103	310	930	248.80	
	2700	100 kW		43.30	128	385	1,150	423.40	
	2800	250 kW		85.15	235	705	2,125	822.20	
	2850	Hammer, hydraulic, for mounting on boom, to 500 ft lb.		2.55	75	225	675	65.40	
	2860	1000 ft lb.		4.35	127	380	1,150	110.80	
	2900	Heaters, space, oil or electric, 50 MBH		2.01	7.65	23	69	20.70	
	3000	100 MBH		3.76	10.65	32	96	36.50	
	3100	300 MBH		11.03	38.50	115	345	111.25	
	3150	500 MBH		18.15	45	135	405	172.20	
	3200	Hose, water, suction with coupling, 20' long, 2" diameter		.02	3	9	27	1.95	
	3210	3" diameter		.03	4.33	13	39	2.85	
	3220	4" diameter		.03	5	15	45	3.25	
	3230	6" diameter		.11	17.65	53	159	11.50	
	3240	8" diameter		.27	44.50	133	400	28.75	
	3250	Discharge hose with coupling, 50' long, 2" diameter		.01	1.33	4	12	.90	
	3260	3" diameter		.01	2.33	7	21	1.50	
	3270	4" diameter		.02	3.67	11	33	2.35	
	3280	6" diameter		.06	9.35	28	84	6.10	
	3290	8" diameter		.18	30	90	270	19.45	
	3295	Insulation blower		.78	6	18	54	9.85	
	3300	Ladders, extension type, 16' to 36' long		.14	22.50	68	204	14.70	
	3400	40' to 60' long		.19	31	93	279	20.10	
	3405	Lance for cutting concrete		2.35	71.50	215	645	61.80	
	3407	Lawn mower, rotary, 22", 5 H.P.		1.80	49	147	440	43.80	
	3408	48" self propelled		2.98	75	225	675	68.85	
	3410	Level, electronic, automatic, with tripod and leveling rod		1.20	75.50	226	680	54.80	
	3430	Laser type, for pipe and sewer line and grade		.73	48.50	145	435	34.85	
	3440	Rotating beam for interior control		.78	52	156	470	37.45	
	3460	Builder's optical transit, with tripod and rod		.09	14.65	44	132	9.50	
	3500	Light towers, towable, with diesel generator, 2000 watt		4.55	93.50	280	840	92.40	
	3600	4000 watt		4.85	96.50	290	870	96.80	
	3700	Mixer, powered, plaster and mortar, 6 C.F., 7 H.P.		2.70	20	60	180	33.60	
	3800	10 C.F., 9 H.P.		2.80	31.50	94	282	41.20	
	3850	Nailer, pneumatic		.58	38.50	115	345	27.65	
	3900	Paint sprayers complete, 8 CFM		1.04	69.50	208	625	49.90	
	4000	17 CFM		1.90	127	380	1,150	91.20	
	4020	Pavers, bituminous, rubber tires, 8' wide, 50 H.P., diesel		31.15	490	1,465	4,400	542.20	
	4030	10' wide, 150 H.P.		106.10	1,800	5,370	16,100	1,923	
	4050	Crawler, 8' wide, 100 H.P., diesel		88.65	1,800	5,385	16,200	1,786	
	4060	10' wide, 150 H.P.		113.15	2,200	6,610	19,800	2,227	
	4070	Concrete paver, 12' to 24' wide, 250 H.P.		101.40	1,550	4,675	14,000	1,746	
	4080	Placer-spreader-trimmer, 24' wide, 300 H.P.		150.25	2,650	7,960	23,900	2,794	
	4100	Pump, centrifugal gas pump, 1-1/2" diam., 65 GPM		3.90	48.50	145	435	60.20	
	4200	2" diameter, 130 GPM		5.35	53.50	160	480	74.80	
	4300	3" diameter, 250 GPM		5.65	55	165	495	78.20	
	4400	6" diameter, 1500 GPM		30.45	172	515	1,550	346.60	
	4500	Submersible electric pump, 1-1/4" diameter, 55 GPM		.39	16.35	49	147	12.90	
	4600	1-1/2" diameter, 83 GPM		.43	18.65	56	168	14.65	
	4700	2" diameter, 120 GPM		1.50	23.50	70	210	26	
	4800	3" diameter, 300 GPM		2.69	41.50	125	375	46.50	
	4900	4" diameter, 560 GPM		12.34	158	475	1,425	193.70	
	5000	6" diameter, 1590 GPM		18.40	212	635	1,900	274.20	
	5100	Diaphragm pump, gas, single, 1-1/2" diameter		1.18	49.50	148	445	39.05	
	5200	2" diameter		4.25	61.50	185	555	71	
	5300	3" diameter		4.25	61.50	185	555	71	

For customer support on your Plumbing Cost Data, call 877.709.6968.

559

01 54 33 | Equipment Rental

		UNIT	HOURLY OPER. COST	RENT PER DAY	RENT PER WEEK	RENT PER MONTH	EQUIPMENT COST/DAY	
40	5400	Double, 4" diameter	Ea.	6.40	105	315	945	114.20	40
	5450	Pressure washer 5 GPM, 3000 psi		4.90	51.50	155	465	70.20	
	5460	7 GPM, 3000 psi		6.45	60	180	540	87.60	
	5500	Trash pump, self-priming, gas, 2" diameter		4.60	21	63	189	49.40	
	5600	Diesel, 4" diameter		9.15	88.50	265	795	126.20	
	5650	Diesel, 6" diameter		25.25	147	440	1,325	290	
	5655	Grout Pump		26.35	268	805	2,425	371.80	
	5700	Salamanders, L.P. gas fired, 100,000 Btu		3.96	13.65	41	123	39.90	
	5705	50,000 Btu		2.21	10.35	31	93	23.90	
	5720	Sandblaster, portable, open top, 3 C.F. capacity		.55	26.50	80	240	20.40	
	5730	6 C.F. capacity		.95	40	120	360	31.60	
	5740	Accessories for above		.13	21.50	65	195	14.05	
	5750	Sander, floor		.70	14.35	43	129	14.20	
	5760	Edger		.50	14.35	43	129	12.60	
	5800	Saw, chain, gas engine, 18" long		2.30	21.50	64	192	31.20	
	5900	Hydraulic powered, 36" long		.75	65	195	585	45	
	5950	60" long		.75	66.50	200	600	46	
	6000	Masonry, table mounted, 14" diameter, 5 H.P.		1.32	56.50	170	510	44.55	
	6050	Portable cut-off, 8 H.P.		2.50	33.50	100	300	40	
	6100	Circular, hand held, electric, 7-1/4" diameter		.23	4.67	14	42	4.65	
	6200	12" diameter		.23	8	24	72	6.65	
	6250	Wall saw, w/hydraulic power, 10 H.P.		9.70	61.50	185	555	114.60	
	6275	Shot blaster, walk-behind, 20" wide		4.75	293	880	2,650	214	
	6280	Sidewalk broom, walk-behind		2.52	78.50	235	705	67.15	
	6300	Steam cleaner, 100 gallons per hour		3.70	76.50	230	690	75.60	
	6310	200 gallons per hour		5.35	95	285	855	99.80	
	6340	Tar Kettle/Pot, 400 gallons		15.60	75	225	675	169.80	
	6350	Torch, cutting, acetylene-oxygen, 150' hose, excludes gases		.30	15	45	135	11.40	
	6360	Hourly operating cost includes tips and gas		19.00				152	
	6410	Toilet, portable chemical		.13	21	63	189	13.65	
	6420	Recycle flush type		.15	25	75	225	16.20	
	6430	Toilet, fresh water flush, garden hose,		.18	30.50	91	273	19.65	
	6440	Hoisted, non-flush, for high rise		.15	24.50	74	222	16	
	6465	Tractor, farm with attachment		21.50	297	890	2,675	350	
	6480	Trailers, platform, flush deck, 2 axle, 3 ton capacity		1.45	20	60	180	23.60	
	6500	25 ton capacity		5.45	117	350	1,050	113.60	
	6600	40 ton capacity		7.00	163	490	1,475	154	
	6700	3 axle, 50 ton capacity		7.55	180	540	1,625	168.40	
	6800	75 ton capacity		9.40	235	705	2,125	216.20	
	6810	Trailer mounted cable reel for high voltage line work		5.45	260	779	2,325	199.40	
	6820	Trailer mounted cable tensioning rig		10.85	515	1,550	4,650	396.80	
	6830	Cable pulling rig		73.05	2,900	8,680	26,000	2,320	
	6900	Water tank trailer, engine driven discharge, 5000 gallons		7.00	142	425	1,275	141	
	6925	10,000 gallons		9.50	197	590	1,775	194	
	6950	Water truck, off highway, 6000 gallons		88.15	770	2,315	6,950	1,168	
	7010	Tram car for high voltage line work, powered, 2 conductor		6.60	141	423	1,275	137.40	
	7020	Transit (builder's level) with tripod		.09	14.65	44	132	9.50	
	7030	Trench box, 3000 lb., 6' x 8'		.56	93.50	280	840	60.50	
	7040	7200 lb., 6' x 20'		.75	125	375	1,125	81	
	7050	8000 lb., 8' x 16'		1.08	180	540	1,625	116.65	
	7060	9500 lb., 8' x 20'		1.21	201	603	1,800	130.30	
	7065	11,000 lb., 8' x 24'		1.27	211	633	1,900	136.75	
	7070	12,000 lb., 10' x 20'		1.50	249	748	2,250	161.60	
	7100	Truck, pickup, 3/4 ton, 2 wheel drive		13.65	58.50	175	525	144.20	
	7200	4 wheel drive		13.95	73.50	220	660	155.60	
	7250	Crew carrier, 9 passenger		19.30	86.50	260	780	206.40	
	7290	Flat bed truck, 20,000 lb. GVW		21.70	125	375	1,125	248.60	
	7300	Tractor, 4 x 2, 220 H.P.		30.20	197	590	1,775	359.60	

01 54 33 | Equipment Rental

		UNIT	HOURLY OPER. COST	RENT PER DAY	RENT PER WEEK	RENT PER MONTH	EQUIPMENT COST/DAY	
40	7410	330 H.P.	Ea.	44.75	270	810	2,425	520	**40**
	7500	6 x 4, 380 H.P.		51.35	315	945	2,825	599.80	
	7600	450 H.P.		62.30	380	1,145	3,425	727.40	
	7610	Tractor, with A frame, boom and winch, 225 H.P.		33.00	272	815	2,450	427	
	7620	Vacuum truck, hazardous material, 2500 gallons		12.05	290	870	2,600	270.40	
	7625	5,000 gallons		14.26	405	1,220	3,650	358.10	
	7650	Vacuum, HEPA, 16 gallon, wet/dry		.85	18	54	162	17.60	
	7655	55 gallon, wet/dry		.80	27	81	243	22.60	
	7660	Water tank, portable		.16	26.50	80	240	17.30	
	7690	Sewer/catch basin vacuum, 14 C.Y., 1500 gallons		18.53	610	1,830	5,500	514.25	
	7700	Welder, electric, 200 amp		3.88	16.35	49	147	40.85	
	7800	300 amp		5.71	20	60	180	57.70	
	7900	Gas engine, 200 amp		14.65	23.50	70	210	131.20	
	8000	300 amp		16.38	24.50	74	222	145.85	
	8100	Wheelbarrow, any size		.08	13	39	117	8.45	
	8200	Wrecking ball, 4000 lb.		2.35	68.50	205	615	59.80	
50	0010	**HIGHWAY EQUIPMENT RENTAL** without operators R015433-10							**50**
	0050	Asphalt batch plant, portable drum mixer, 100 ton/hr.	Ea.	78.06	1,425	4,270	12,800	1,478	
	0060	200 ton/hr.		89.18	1,500	4,530	13,600	1,619	
	0070	300 ton/hr.		106.34	1,775	5,325	16,000	1,916	
	0100	Backhoe attachment, long stick, up to 185 H.P., 10.5' long		.35	23.50	70	210	16.80	
	0140	Up to 250 H.P., 12' long		.38	25	75	225	18.05	
	0180	Over 250 H.P., 15' long		.53	35	105	315	25.25	
	0200	Special dipper arm, up to 100 H.P., 32' long		1.08	71.50	215	645	51.65	
	0240	Over 100 H.P., 33' long		1.34	89.50	268	805	64.30	
	0280	Catch basin/sewer cleaning truck, 3 ton, 9 C.Y., 1000 gal.		42.70	385	1,160	3,475	573.60	
	0300	Concrete batch plant, portable, electric, 200 C.Y./hr.		22.87	515	1,550	4,650	492.95	
	0520	Grader/dozer attachment, ripper/scarifier, rear mounted, up to 135 H.P.		2.90	56.50	170	510	57.20	
	0540	Up to 180 H.P.		3.90	88.50	265	795	84.20	
	0580	Up to 250 H.P.		5.00	117	350	1,050	110	
	0700	Pvmt. removal bucket, for hyd. excavator, up to 90 H.P.		1.90	53.50	160	480	47.20	
	0740	Up to 200 H.P.		2.15	73.50	220	660	61.20	
	0780	Over 200 H.P.		2.25	85	255	765	69	
	0900	Aggregate spreader, self-propelled, 187 H.P.		53.00	685	2,050	6,150	834	
	1000	Chemical spreader, 3 C.Y.		3.30	43.50	130	390	52.40	
	1900	Hammermill, traveling, 250 H.P.		78.83	2,075	6,240	18,700	1,879	
	2000	Horizontal borer, 3" diameter, 13 H.P. gas driven		6.35	55	165	495	83.80	
	2150	Horizontal directional drill, 20,000 lb. thrust, 78 H.P. diesel		31.15	675	2,025	6,075	654.20	
	2160	30,000 lb. thrust, 115 H.P.		38.90	1,025	3,100	9,300	931.20	
	2170	50,000 lb. thrust, 170 H.P.		55.65	1,325	3,960	11,900	1,237	
	2190	Mud trailer for HDD, 1500 gallons, 175 H.P., gas		31.90	152	455	1,375	346.20	
	2200	Hydromulcher, diesel, 3000 gallon, for truck mounting		23.40	245	735	2,200	334.20	
	2300	Gas, 600 gallon		8.60	98.50	295	885	127.80	
	2400	Joint & crack cleaner, walk behind, 25 H.P.		3.95	50	150	450	61.60	
	2500	Filler, trailer mounted, 400 gallons, 20 H.P.		9.50	210	630	1,900	202	
	3000	Paint striper, self-propelled, 40 gallon, 22 H.P.		7.30	155	465	1,400	151.40	
	3100	120 gallon, 120 H.P.		23.00	395	1,185	3,550	421	
	3200	Post drivers, 6" I-Beam frame, for truck mounting		19.05	380	1,135	3,400	379.40	
	3400	Road sweeper, self-propelled, 8' wide, 90 H.P.		39.80	575	1,720	5,150	662.40	
	3450	Road sweeper, vacuum assisted, 4 C.Y., 220 gallons		75.75	635	1,900	5,700	986	
	4000	Road mixer, self-propelled, 130 H.P.		46.85	765	2,295	6,875	833.80	
	4100	310 H.P.		83.25	2,100	6,315	18,900	1,929	
	4220	Cold mix paver, incl. pug mill and bitumen tank, 165 H.P.		97.00	2,275	6,855	20,600	2,147	
	4240	Pavement brush, towed		3.20	93.50	280	840	81.60	
	4250	Paver, asphalt, wheel or crawler, 130 H.P., diesel		96.65	2,275	6,795	20,400	2,132	
	4300	Paver, road widener, gas 1' to 6', 67 H.P.		48.50	895	2,680	8,050	924	
	4400	Diesel, 2' to 14', 88 H.P.		62.70	1,050	3,135	9,400	1,129	
	4600	Slipform pavers, curb and gutter, 2 track, 75 H.P.		57.20	900	2,700	8,100	997.60	

01 54 33 | Equipment Rental

			UNIT	HOURLY OPER. COST	RENT PER DAY	RENT PER WEEK	RENT PER MONTH	EQUIPMENT COST/DAY	
50	4700	4 track, 165 H.P.	Ea.	41.50	740	2,220	6,650	776	50
	4800	Median barrier, 215 H.P.		60.50	1,050	3,150	9,450	1,114	
	4901	Trailer, low bed, 75 ton capacity		10.10	235	705	2,125	221.80	
	5000	Road planer, walk behind, 10" cutting width, 10 H.P.		3.70	33.50	100	300	49.60	
	5100	Self-propelled, 12" cutting width, 64 H.P.		10.40	113	340	1,025	151.20	
	5120	Traffic line remover, metal ball blaster, truck mounted, 115 H.P.		47.95	730	2,190	6,575	821.60	
	5140	Grinder, truck mounted, 115 H.P.		54.60	790	2,375	7,125	911.80	
	5160	Walk-behind, 11 H.P.		4.20	53.50	160	480	65.60	
	5200	Pavement profiler, 4' to 6' wide, 450 H.P.		254.50	3,325	9,960	29,900	4,028	
	5300	8' to 10' wide, 750 H.P.		399.20	4,350	13,055	39,200	5,805	
	5400	Roadway plate, steel, 1" x 8' x 20'		.08	13.35	40	120	8.65	
	5600	Stabilizer, self-propelled, 150 H.P.		49.30	600	1,795	5,375	753.40	
	5700	310 H.P.		99.05	1,675	5,030	15,100	1,798	
	5800	Striper, truck mounted, 120 gallon paint, 460 H.P.		64.20	485	1,455	4,375	804.60	
	5900	Thermal paint heating kettle, 115 gallons		8.20	25.50	77	231	81	
	6000	Tar kettle, 330 gallon, trailer mounted		12.25	56.50	170	510	132	
	7000	Tunnel locomotive, diesel, 8 to 12 ton		31.85	585	1,750	5,250	604.80	
	7005	Electric, 10 ton		26.00	665	1,995	5,975	607	
	7010	Muck cars, 1/2 C.Y. capacity		2.00	24.50	73	219	30.60	
	7020	1 C.Y. capacity		2.25	32.50	97	291	37.40	
	7030	2 C.Y. capacity		2.35	36.50	110	330	40.80	
	7040	Side dump, 2 C.Y. capacity		2.60	45	135	405	47.80	
	7050	3 C.Y. capacity		3.45	50	150	450	57.60	
	7060	5 C.Y. capacity		4.95	63.50	190	570	77.60	
	7100	Ventilating blower for tunnel, 7-1/2 H.P.		2.05	51.50	155	465	47.40	
	7110	10 H.P.		2.28	51.50	155	465	49.25	
	7120	20 H.P.		3.48	67.50	202	605	68.25	
	7140	40 H.P.		5.76	96.50	290	870	104.10	
	7160	60 H.P.		8.77	152	455	1,375	161.15	
	7175	75 H.P.		11.66	207	620	1,850	217.30	
	7180	200 H.P.		23.65	305	910	2,725	371.20	
	7800	Windrow loader, elevating		55.00	1,300	3,935	11,800	1,227	
60	0010	**LIFTING AND HOISTING EQUIPMENT RENTAL** without operators							60
	0120	Aerial lift truck, 2 person, to 80'	Ea.	26.30	705	2,120	6,350	634.40	
	0140	Boom work platform, 40' snorkel		12.00	275	825	2,475	261	
	0150	Crane, flatbed mounted, 3 ton capacity		16.35	188	565	1,700	243.80	
	0200	Crane, climbing, 106' jib, 6000 lb. capacity, 410 fpm		39.35	1,650	4,960	14,900	1,307	
	0300	101' jib, 10,250 lb. capacity, 270 fpm		45.95	2,100	6,280	18,800	1,624	
	0500	Tower, static, 130' high, 106' jib, 6200 lb. capacity at 400 fpm		43.20	1,900	5,730	17,200	1,492	
	0600	Crawler mounted, lattice boom, 1/2 C.Y., 15 tons at 12' radius		36.98	625	1,870	5,600	669.85	
	0700	3/4 C.Y., 20 tons at 12' radius		49.31	780	2,340	7,025	862.50	
	0800	1 C.Y., 25 tons at 12' radius		65.75	1,050	3,120	9,350	1,150	
	0900	1-1/2 C.Y., 40 tons at 12' radius		65.75	1,050	3,150	9,450	1,156	
	1000	2 C.Y., 50 tons at 12' radius		69.70	1,225	3,680	11,000	1,294	
	1100	3 C.Y., 75 tons at 12' radius		74.70	1,425	4,305	12,900	1,459	
	1200	100 ton capacity, 60' boom		84.60	1,650	4,950	14,900	1,667	
	1300	165 ton capacity, 60' boom		107.75	1,925	5,785	17,400	2,019	
	1400	200 ton capacity, 70' boom		130.55	2,400	7,210	21,600	2,486	
	1500	350 ton capacity, 80' boom		182.90	3,625	10,845	32,500	3,632	
	1600	Truck mounted, lattice boom, 6 x 4, 20 tons at 10' radius		37.11	1,075	3,260	9,775	948.90	
	1700	25 tons at 10' radius		40.09	1,175	3,550	10,700	1,031	
	1800	8 x 4, 30 tons at 10' radius		43.61	1,250	3,780	11,300	1,105	
	1900	40 tons at 12' radius		46.70	1,325	3,950	11,900	1,164	
	2000	60 tons at 15' radius		53.07	1,400	4,180	12,500	1,261	
	2050	82 tons at 15' radius		59.80	1,475	4,460	13,400	1,370	
	2100	90 tons at 15' radius		67.36	1,625	4,860	14,600	1,511	
	2200	115 tons at 15' radius		76.11	1,800	5,430	16,300	1,695	
	2300	150 tons at 18' radius		83.65	1,900	5,720	17,200	1,813	

Reference boxes: R015433 -10, R015433 -15, R312316 -45

01 54 33 | Equipment Rental

		UNIT	HOURLY OPER. COST	RENT PER DAY	RENT PER WEEK	RENT PER MONTH	EQUIPMENT COST/DAY	
60	2350	165 tons at 18' radius	Ea.	90.05	2,025	6,060	18,200	1,932	60
	2400	Truck mounted, hydraulic, 12 ton capacity		42.60	520	1,565	4,700	653.80	
	2500	25 ton capacity		44.95	630	1,885	5,650	736.60	
	2550	33 ton capacity		45.50	645	1,930	5,800	750	
	2560	40 ton capacity		58.85	750	2,250	6,750	920.80	
	2600	55 ton capacity		76.80	855	2,570	7,700	1,128	
	2700	80 ton capacity		100.45	1,375	4,105	12,300	1,625	
	2720	100 ton capacity		94.20	1,425	4,250	12,800	1,604	
	2740	120 ton capacity		100.05	1,525	4,575	13,700	1,715	
	2760	150 ton capacity		127.60	2,000	5,995	18,000	2,220	
	2800	Self-propelled, 4 x 4, with telescoping boom, 5 ton		17.55	225	675	2,025	275.40	
	2900	12-1/2 ton capacity		32.30	360	1,075	3,225	473.40	
	3000	15 ton capacity		33.00	375	1,130	3,400	490	
	3050	20 ton capacity		35.95	445	1,335	4,000	554.60	
	3100	25 ton capacity		37.55	500	1,495	4,475	599.40	
	3150	40 ton capacity		46.00	560	1,675	5,025	703	
	3200	Derricks, guy, 20 ton capacity, 60' boom, 75' mast		27.55	405	1,214	3,650	463.20	
	3300	100' boom, 115' mast		43.21	695	2,090	6,275	763.70	
	3400	Stiffleg, 20 ton capacity, 70' boom, 37' mast		30.11	525	1,580	4,750	556.90	
	3500	100' boom, 47' mast		46.29	845	2,530	7,600	876.30	
	3550	Helicopter, small, lift to 1250 lb. maximum, w/pilot		103.70	3,275	9,800	29,400	2,790	
	3600	Hoists, chain type, overhead, manual, 3/4 ton		.10	.33	1	3	1	
	3900	10 ton		.70	6	18	54	9.20	
	4000	Hoist and tower, 5000 lb. cap., portable electric, 40' high		4.95	233	699	2,100	179.40	
	4100	For each added 10' section, add		.11	18.35	55	165	11.90	
	4200	Hoist and single tubular tower, 5000 lb. electric, 100' high		6.70	325	976	2,925	248.80	
	4300	For each added 6'-6" section, add		.19	31.50	95	285	20.50	
	4400	Hoist and double tubular tower, 5000 lb., 100' high		7.19	360	1,075	3,225	272.50	
	4500	For each added 6'-6" section, add		.21	35	105	315	22.70	
	4550	Hoist and tower, mast type, 6000 lb., 100' high		7.76	370	1,115	3,350	285.10	
	4570	For each added 10' section, add		.13	21.50	65	195	14.05	
	4600	Hoist and tower, personnel, electric, 2000 lb., 100' @ 125 fpm		16.31	990	2,970	8,900	724.50	
	4700	3000 lb., 100' @ 200 fpm		18.62	1,125	3,360	10,100	820.95	
	4800	3000 lb., 150' @ 300 fpm		20.62	1,250	3,760	11,300	916.95	
	4900	4000 lb., 100' @ 300 fpm		21.38	1,275	3,840	11,500	939.05	
	5000	6000 lb., 100' @ 275 fpm		23.06	1,350	4,030	12,100	990.50	
	5100	For added heights up to 500', add	L.F.	.01	1.67	5	15	1.10	
	5200	Jacks, hydraulic, 20 ton	Ea.	.05	2	6	18	1.60	
	5500	100 ton		.40	11.65	35	105	10.20	
	6100	Jacks, hydraulic, climbing w/50' jackrods, control console, 30 ton cap.		2.01	134	402	1,200	96.50	
	6150	For each added 10' jackrod section, add		.05	3.33	10	30	2.40	
	6300	50 ton capacity		3.23	215	646	1,950	155.05	
	6350	For each added 10' jackrod section, add		.06	4	12	36	2.90	
	6500	125 ton capacity		8.45	565	1,690	5,075	405.60	
	6550	For each added 10' jackrod section, add		.58	38.50	115	345	27.65	
	6600	Cable jack, 10 ton capacity with 200' cable		1.68	112	336	1,000	80.65	
	6650	For each added 50' of cable, add		.20	13.35	40	120	9.60	
70	0010	**WELLPOINT EQUIPMENT RENTAL** without operators							70
	0020	Based on 2 months rental							
	0100	Combination jetting & wellpoint pump, 60 H.P. diesel	Ea.	18.40	330	996	3,000	346.40	
	0200	High pressure gas jet pump, 200 H.P., 300 psi	"	44.06	284	851	2,550	522.70	
	0300	Discharge pipe, 8" diameter	L.F.	.01	.54	1.62	4.86	.40	
	0350	12" diameter		.01	.79	2.38	7.15	.55	
	0400	Header pipe, flows up to 150 GPM, 4" diameter		.01	.49	1.47	4.41	.35	
	0500	400 GPM, 6" diameter		.01	.57	1.72	5.15	.40	
	0600	800 GPM, 8" diameter		.01	.79	2.38	7.15	.55	
	0700	1500 GPM, 10" diameter		.01	.84	2.51	7.55	.60	

R015433 -10

01 54 33 | Equipment Rental

			UNIT	HOURLY OPER. COST	RENT PER DAY	RENT PER WEEK	RENT PER MONTH	EQUIPMENT COST/DAY	
70	0800	2500 GPM, 12" diameter	L.F.	.02	1.58	4.74	14.20	1.10	70
	0900	4500 GPM, 16" diameter		.03	2.02	6.07	18.20	1.45	
	0950	For quick coupling aluminum and plastic pipe, add	↓	.03	2.09	6.28	18.85	1.50	
	1100	Wellpoint, 25' long, with fittings & riser pipe, 1-1/2" or 2" diameter	Ea.	.06	4.18	12.54	37.50	3	
	1200	Wellpoint pump, diesel powered, 4" suction, 20 H.P.		7.83	191	574	1,725	177.45	
	1300	6" suction, 30 H.P.		10.70	237	712	2,125	228	
	1400	8" suction, 40 H.P.		14.45	325	976	2,925	310.80	
	1500	10" suction, 75 H.P.		22.27	380	1,141	3,425	406.35	
	1600	12" suction, 100 H.P.		31.72	605	1,810	5,425	615.75	
	1700	12" suction, 175 H.P.	↓	47.41	670	2,010	6,025	781.30	
80	0010	**MARINE EQUIPMENT RENTAL** without operators							80
	0200	Barge, 400 Ton, 30' wide x 90' long [R015433-10]	Ea.	17.50	1,050	3,180	9,550	776	
	0240	800 Ton, 45' wide x 90' long		21.25	1,300	3,870	11,600	944	
	2000	Tugboat, diesel, 100 H.P.		38.20	217	650	1,950	435.60	
	2040	250 H.P.		79.25	395	1,180	3,550	870	
	2080	380 H.P.		156.80	1,175	3,535	10,600	1,961	
	3000	Small work boat, gas, 16-foot, 50 H.P.		17.15	60	180	540	173.20	
	4000	Large, diesel, 48-foot, 200 H.P.	↓	89.40	1,250	3,740	11,200	1,463	

Crews

Crew No.	Bare Costs Hr.	Daily	Incl. Subs O&P Hr.	Daily	Cost Per Labor-Hour Bare Costs	Incl. O&P
Crew A-1	Hr.	Daily	Hr.	Daily	Bare Costs	Incl. O&P
1 Building Laborer	$37.60	$300.80	$57.85	$462.80	$37.60	$57.85
1 Concrete Saw, Gas Manual		81.00		89.10	10.13	11.14
8 L.H., Daily Totals		$381.80		$551.90	$47.73	$68.99
Crew A-1A	Hr.	Daily	Hr.	Daily	Bare Costs	Incl. O&P
1 Skilled Worker	$48.65	$389.20	$75.15	$601.20	$48.65	$75.15
1 Shot Blaster, 20"		214.00		235.40	26.75	29.43
8 L.H., Daily Totals		$603.20		$836.60	$75.40	$104.58
Crew A-1B	Hr.	Daily	Hr.	Daily	Bare Costs	Incl. O&P
1 Building Laborer	$37.60	$300.80	$57.85	$462.80	$37.60	$57.85
1 Concrete Saw		167.40		184.14	20.93	23.02
8 L.H., Daily Totals		$468.20		$646.94	$58.52	$80.87
Crew A-1C	Hr.	Daily	Hr.	Daily	Bare Costs	Incl. O&P
1 Building Laborer	$37.60	$300.80	$57.85	$462.80	$37.60	$57.85
1 Chain Saw, Gas, 18"		31.20		34.32	3.90	4.29
8 L.H., Daily Totals		$332.00		$497.12	$41.50	$62.14
Crew A-1D	Hr.	Daily	Hr.	Daily	Bare Costs	Incl. O&P
1 Building Laborer	$37.60	$300.80	$57.85	$462.80	$37.60	$57.85
1 Vibrating Plate, Gas, 18"		36.20		39.82	4.53	4.98
8 L.H., Daily Totals		$337.00		$502.62	$42.13	$62.83
Crew A-1E	Hr.	Daily	Hr.	Daily	Bare Costs	Incl. O&P
1 Building Laborer	$37.60	$300.80	$57.85	$462.80	$37.60	$57.85
1 Vibrating Plate, Gas, 21"		46.80		51.48	5.85	6.43
8 L.H., Daily Totals		$347.60		$514.28	$43.45	$64.28
Crew A-1F	Hr.	Daily	Hr.	Daily	Bare Costs	Incl. O&P
1 Building Laborer	$37.60	$300.80	$57.85	$462.80	$37.60	$57.85
1 Rammer/Tamper, Gas, 8"		50.00		55.00	6.25	6.88
8 L.H., Daily Totals		$350.80		$517.80	$43.85	$64.72
Crew A-1G	Hr.	Daily	Hr.	Daily	Bare Costs	Incl. O&P
1 Building Laborer	$37.60	$300.80	$57.85	$462.80	$37.60	$57.85
1 Rammer/Tamper, Gas, 15"		56.40		62.04	7.05	7.75
8 L.H., Daily Totals		$357.20		$524.84	$44.65	$65.61
Crew A-1H	Hr.	Daily	Hr.	Daily	Bare Costs	Incl. O&P
1 Building Laborer	$37.60	$300.80	$57.85	$462.80	$37.60	$57.85
1 Exterior Steam Cleaner		75.60		83.16	9.45	10.40
8 L.H., Daily Totals		$376.40		$545.96	$47.05	$68.25
Crew A-1J	Hr.	Daily	Hr.	Daily	Bare Costs	Incl. O&P
1 Building Laborer	$37.60	$300.80	$57.85	$462.80	$37.60	$57.85
1 Cultivator, Walk-Behind, 5 H.P.		46.70		51.37	5.84	6.42
8 L.H., Daily Totals		$347.50		$514.17	$43.44	$64.27
Crew A-1K	Hr.	Daily	Hr.	Daily	Bare Costs	Incl. O&P
1 Building Laborer	$37.60	$300.80	$57.85	$462.80	$37.60	$57.85
1 Cultivator, Walk-Behind, 8 H.P.		72.65		79.92	9.08	9.99
8 L.H., Daily Totals		$373.45		$542.72	$46.68	$67.84
Crew A-1M	Hr.	Daily	Hr.	Daily	Bare Costs	Incl. O&P
1 Building Laborer	$37.60	$300.80	$57.85	$462.80	$37.60	$57.85
1 Snow Blower, Walk-Behind		67.15		73.86	8.39	9.23
8 L.H., Daily Totals		$367.95		$536.66	$45.99	$67.08

Crew No.	Bare Costs Hr.	Daily	Incl. Subs O&P Hr.	Daily	Cost Per Labor-Hour Bare Costs	Incl. O&P
Crew A-2	Hr.	Daily	Hr.	Daily	Bare Costs	Incl. O&P
2 Laborers	$37.60	$601.60	$57.85	$925.60	$38.10	$58.27
1 Truck Driver (light)	39.10	312.80	59.10	472.80		
1 Flatbed Truck, Gas, 1.5 Ton		245.40		269.94	10.23	11.25
24 L.H., Daily Totals		$1159.80		$1668.34	$48.33	$69.51
Crew A-2A	Hr.	Daily	Hr.	Daily	Bare Costs	Incl. O&P
2 Laborers	$37.60	$601.60	$57.85	$925.60	$38.10	$58.27
1 Truck Driver (light)	39.10	312.80	59.10	472.80		
1 Flatbed Truck, Gas, 1.5 Ton		245.40		269.94		
1 Concrete Saw		167.40		184.14	17.20	18.92
24 L.H., Daily Totals		$1327.20		$1852.48	$55.30	$77.19
Crew A-2B	Hr.	Daily	Hr.	Daily	Bare Costs	Incl. O&P
1 Truck Driver (light)	$39.10	$312.80	$59.10	$472.80	$39.10	$59.10
1 Flatbed Truck, Gas, 1.5 Ton		245.40		269.94	30.68	33.74
8 L.H., Daily Totals		$558.20		$742.74	$69.78	$92.84
Crew A-3A	Hr.	Daily	Hr.	Daily	Bare Costs	Incl. O&P
1 Equip. Oper. (light)	$48.60	$388.80	$73.75	$590.00	$48.60	$73.75
1 Pickup Truck, 4 x 4, 3/4 Ton		155.60		171.16	19.45	21.40
8 L.H., Daily Totals		$544.40		$761.16	$68.05	$95.14
Crew A-3B	Hr.	Daily	Hr.	Daily	Bare Costs	Incl. O&P
1 Equip. Oper. (medium)	$50.60	$404.80	$76.75	$614.00	$45.33	$68.65
1 Truck Driver (heavy)	40.05	320.40	60.55	484.40		
1 Dump Truck, 12 C.Y., 400 H.P.		691.00		760.10		
1 F.E. Loader, W.M.,2.5 C.Y.		519.80		571.78	75.67	83.24
16 L.H., Daily Totals		$1936.00		$2430.28	$121.00	$151.89
Crew A-3C	Hr.	Daily	Hr.	Daily	Bare Costs	Incl. O&P
1 Equip. Oper. (light)	$48.60	$388.80	$73.75	$590.00	$48.60	$73.75
1 Loader, Skid Steer, 78 H.P.		318.60		350.46	39.83	43.81
8 L.H., Daily Totals		$707.40		$940.46	$88.42	$117.56
Crew A-3D	Hr.	Daily	Hr.	Daily	Bare Costs	Incl. O&P
1 Truck Driver (light)	$39.10	$312.80	$59.10	$472.80	$39.10	$59.10
1 Pickup Truck, 4 x 4, 3/4 Ton		155.60		171.16		
1 Flatbed Trailer, 25 Ton		113.60		124.96	33.65	37.02
8 L.H., Daily Totals		$582.00		$768.92	$72.75	$96.11
Crew A-3E	Hr.	Daily	Hr.	Daily	Bare Costs	Incl. O&P
1 Equip. Oper. (crane)	$51.70	$413.60	$78.45	$627.60	$45.88	$69.50
1 Truck Driver (heavy)	40.05	320.40	60.55	484.40		
1 Pickup Truck, 4 x 4, 3/4 Ton		155.60		171.16	9.72	10.70
16 L.H., Daily Totals		$889.60		$1283.16	$55.60	$80.20
Crew A-3F	Hr.	Daily	Hr.	Daily	Bare Costs	Incl. O&P
1 Equip. Oper. (crane)	$51.70	$413.60	$78.45	$627.60	$45.88	$69.50
1 Truck Driver (heavy)	40.05	320.40	60.55	484.40		
1 Pickup Truck, 4 x 4, 3/4 Ton		155.60		171.16		
1 Truck Tractor, 6x4, 380 H.P.		599.80		659.78		
1 Lowbed Trailer, 75 Ton		221.80		243.98	61.08	67.18
16 L.H., Daily Totals		$1711.20		$2186.92	$106.95	$136.68

Crews

Crew No.	Bare Costs Hr.	Daily	Incl. Subs O&P Hr.	Daily	Cost Per Labor-Hour Bare Costs	Incl. O&P
Crew A-3G	Hr.	Daily	Hr.	Daily	Bare Costs	Incl. O&P
1 Equip. Oper. (crane)	$51.70	$413.60	$78.45	$627.60	$45.88	$69.50
1 Truck Driver (heavy)	40.05	320.40	60.55	484.40		
1 Pickup Truck, 4 x 4, 3/4 Ton		155.60		171.16		
1 Truck Tractor, 6x4, 450 H.P.		727.40		800.14		
1 Lowbed Trailer, 75 Ton		221.80		243.98	69.05	75.95
16 L.H., Daily Totals		$1838.80		$2327.28	$114.93	$145.46
Crew A-3H	Hr.	Daily	Hr.	Daily	Bare Costs	Incl. O&P
1 Equip. Oper. (crane)	$51.70	$413.60	$78.45	$627.60	$51.70	$78.45
1 Hyd. Crane, 12 Ton (Daily)		860.80		946.88	107.60	118.36
8 L.H., Daily Totals		$1274.40		$1574.48	$159.30	$196.81
Crew A-3I	Hr.	Daily	Hr.	Daily	Bare Costs	Incl. O&P
1 Equip. Oper. (crane)	$51.70	$413.60	$78.45	$627.60	$51.70	$78.45
1 Hyd. Crane, 25 Ton (Daily)		989.60		1088.56	123.70	136.07
8 L.H., Daily Totals		$1403.20		$1716.16	$175.40	$214.52
Crew A-3J	Hr.	Daily	Hr.	Daily	Bare Costs	Incl. O&P
1 Equip. Oper. (crane)	$51.70	$413.60	$78.45	$627.60	$51.70	$78.45
1 Hyd. Crane, 40 Ton (Daily)		1221.00		1343.10	152.63	167.89
8 L.H., Daily Totals		$1634.60		$1970.70	$204.32	$246.34
Crew A-3K	Hr.	Daily	Hr.	Daily	Bare Costs	Incl. O&P
1 Equip. Oper. (crane)	$51.70	$413.60	$78.45	$627.60	$48.45	$73.50
1 Equip. Oper. (oiler)	45.20	361.60	68.55	548.40		
1 Hyd. Crane, 55 Ton (Daily)		1469.00		1615.90		
1 P/U Truck, 3/4 Ton (Daily)		167.20		183.92	102.26	112.49
16 L.H., Daily Totals		$2411.40		$2975.82	$150.71	$185.99
Crew A-3L	Hr.	Daily	Hr.	Daily	Bare Costs	Incl. O&P
1 Equip. Oper. (crane)	$51.70	$413.60	$78.45	$627.60	$48.45	$73.50
1 Equip. Oper. (oiler)	45.20	361.60	68.55	548.40		
1 Hyd. Crane, 80 Ton (Daily)		2174.00		2391.40		
1 P/U Truck, 3/4 Ton (Daily)		167.20		183.92	146.32	160.96
16 L.H., Daily Totals		$3116.40		$3751.32	$194.78	$234.46
Crew A-3M	Hr.	Daily	Hr.	Daily	Bare Costs	Incl. O&P
1 Equip. Oper. (crane)	$51.70	$413.60	$78.45	$627.60	$48.45	$73.50
1 Equip. Oper. (oiler)	45.20	361.60	68.55	548.40		
1 Hyd. Crane, 100 Ton (Daily)		2169.00		2385.90		
1 P/U Truck, 3/4 Ton (Daily)		167.20		183.92	146.01	160.61
16 L.H., Daily Totals		$3111.40		$3745.82	$194.46	$234.11
Crew A-3N	Hr.	Daily	Hr.	Daily	Bare Costs	Incl. O&P
1 Equip. Oper. (crane)	$51.70	$413.60	$78.45	$627.60	$51.70	$78.45
1 Tower Cane (monthly)		1128.00		1240.80	141.00	155.10
8 L.H., Daily Totals		$1541.60		$1868.40	$192.70	$233.55
Crew A-3P	Hr.	Daily	Hr.	Daily	Bare Costs	Incl. O&P
1 Equip. Oper. (light)	$48.60	$388.80	$73.75	$590.00	$48.60	$73.75
1 A.T. Forklift, 42' lift		526.40		579.04	65.80	72.38
8 L.H., Daily Totals		$915.20		$1169.04	$114.40	$146.13
Crew A-3Q	Hr.	Daily	Hr.	Daily	Bare Costs	Incl. O&P
1 Equip. Oper. (light)	$48.60	$388.80	$73.75	$590.00	$48.60	$73.75
1 Pickup Truck, 4 x 4, 3/4 Ton		155.60		171.16		
1 Flatbed trailer, 3 Ton		23.60		25.96	22.40	24.64
8 L.H., Daily Totals		$568.00		$787.12	$71.00	$98.39

Crew No.	Bare Costs Hr.	Daily	Incl. Subs O&P Hr.	Daily	Cost Per Labor-Hour Bare Costs	Incl. O&P
Crew A-4	Hr.	Daily	Hr.	Daily	Bare Costs	Incl. O&P
2 Carpenters	$46.95	$751.20	$72.25	$1156.00	$44.75	$68.43
1 Painter, Ordinary	40.35	322.80	60.80	486.40		
24 L.H., Daily Totals		$1074.00		$1642.40	$44.75	$68.43
Crew A-5	Hr.	Daily	Hr.	Daily	Bare Costs	Incl. O&P
2 Laborers	$37.60	$601.60	$57.85	$925.60	$37.77	$57.99
.25 Truck Driver (light)	39.10	78.20	59.10	118.20		
.25 Flatbed Truck, Gas, 1.5 Ton		61.35		67.48	3.41	3.75
18 L.H., Daily Totals		$741.15		$1111.29	$41.17	$61.74
Crew A-6	Hr.	Daily	Hr.	Daily	Bare Costs	Incl. O&P
1 Instrument Man	$48.65	$389.20	$75.15	$601.20	$46.88	$71.92
1 Rodman/Chainman	45.10	360.80	68.70	549.60		
1 Level, Electronic		54.80		60.28	3.42	3.77
16 L.H., Daily Totals		$804.80		$1211.08	$50.30	$75.69
Crew A-7	Hr.	Daily	Hr.	Daily	Bare Costs	Incl. O&P
1 Chief of Party	$60.90	$487.20	$93.20	$745.60	$51.55	$79.02
1 Instrument Man	48.65	389.20	75.15	601.20		
1 Rodman/Chainman	45.10	360.80	68.70	549.60		
1 Level, Electronic		54.80		60.28	2.28	2.51
24 L.H., Daily Totals		$1292.00		$1956.68	$53.83	$81.53
Crew A-8	Hr.	Daily	Hr.	Daily	Bare Costs	Incl. O&P
1 Chief of Party	$60.90	$487.20	$93.20	$745.60	$49.94	$76.44
1 Instrument Man	48.65	389.20	75.15	601.20		
2 Rodmen/Chainmen	45.10	721.60	68.70	1099.20		
1 Level, Electronic		54.80		60.28	1.71	1.88
32 L.H., Daily Totals		$1652.80		$2506.28	$51.65	$78.32
Crew A-9	Hr.	Daily	Hr.	Daily	Bare Costs	Incl. O&P
1 Asbestos Foreman	$52.85	$422.80	$82.30	$658.40	$52.41	$81.60
7 Asbestos Workers	52.35	2931.60	81.50	4564.00		
64 L.H., Daily Totals		$3354.40		$5222.40	$52.41	$81.60
Crew A-10A	Hr.	Daily	Hr.	Daily	Bare Costs	Incl. O&P
1 Asbestos Foreman	$52.85	$422.80	$82.30	$658.40	$52.52	$81.77
2 Asbestos Workers	52.35	837.60	81.50	1304.00		
24 L.H., Daily Totals		$1260.40		$1962.40	$52.52	$81.77
Crew A-10B	Hr.	Daily	Hr.	Daily	Bare Costs	Incl. O&P
1 Asbestos Foreman	$52.85	$422.80	$82.30	$658.40	$52.48	$81.70
3 Asbestos Workers	52.35	1256.40	81.50	1956.00		
32 L.H., Daily Totals		$1679.20		$2614.40	$52.48	$81.70
Crew A-10C	Hr.	Daily	Hr.	Daily	Bare Costs	Incl. O&P
3 Asbestos Workers	$52.35	$1256.40	$81.50	$1956.00	$52.35	$81.50
1 Flatbed Truck, Gas, 1.5 Ton		245.40		269.94	10.23	11.25
24 L.H., Daily Totals		$1501.80		$2225.94	$62.58	$92.75
Crew A-10D	Hr.	Daily	Hr.	Daily	Bare Costs	Incl. O&P
2 Asbestos Workers	$52.35	$837.60	$81.50	$1304.00	$50.40	$77.50
1 Equip. Oper. (crane)	51.70	413.60	78.45	627.60		
1 Equip. Oper. (oiler)	45.20	361.60	68.55	548.40		
1 Hydraulic Crane, 33 Ton		750.00		825.00	23.44	25.78
32 L.H., Daily Totals		$2362.80		$3305.00	$73.84	$103.28

566

For customer support on your Plumbing Cost Data, call 877.709.6968.

Crews

Crew A-11	Hr.	Daily	Hr.	Daily	Bare Costs	Incl. O&P
1 Asbestos Foreman	$52.85	$422.80	$82.30	$658.40	$52.41	$81.60
7 Asbestos Workers	52.35	2931.60	81.50	4564.00		
2 Chip. Hammers, 12 Lb., Elec.		40.00		44.00	.63	.69
64 L.H., Daily Totals		$3394.40		$5266.40	$53.04	$82.29

Crew A-12	Hr.	Daily	Hr.	Daily	Bare Costs	Incl. O&P
1 Asbestos Foreman	$52.85	$422.80	$82.30	$658.40	$52.41	$81.60
7 Asbestos Workers	52.35	2931.60	81.50	4564.00		
1 Trk-Mtd Vac, 14 CY, 1500 Gal.		514.25		565.67		
1 Flatbed Truck, 20,000 GVW		248.60		273.46	11.92	13.11
64 L.H., Daily Totals		$4117.25		$6061.53	$64.33	$94.71

Crew A-13	Hr.	Daily	Hr.	Daily	Bare Costs	Incl. O&P
1 Equip. Oper. (light)	$48.60	$388.80	$73.75	$590.00	$48.60	$73.75
1 Trk-Mtd Vac, 14 CY, 1500 Gal.		514.25		565.67		
1 Flatbed Truck, 20,000 GVW		248.60		273.46	95.36	104.89
8 L.H., Daily Totals		$1151.65		$1429.14	$143.96	$178.64

Crew B-1	Hr.	Daily	Hr.	Daily	Bare Costs	Incl. O&P
1 Labor Foreman (outside)	$39.60	$316.80	$60.95	$487.60	$38.27	$58.88
2 Laborers	37.60	601.60	57.85	925.60		
24 L.H., Daily Totals		$918.40		$1413.20	$38.27	$58.88

Crew B-1A	Hr.	Daily	Hr.	Daily	Bare Costs	Incl. O&P
1 Labor Foreman (outside)	$39.60	$316.80	$60.95	$487.60	$38.27	$58.88
2 Laborers	37.60	601.60	57.85	925.60		
2 Cutting Torches		22.80		25.08		
2 Sets of Gases		304.00		334.40	13.62	14.98
24 L.H., Daily Totals		$1245.20		$1772.68	$51.88	$73.86

Crew B-1B	Hr.	Daily	Hr.	Daily	Bare Costs	Incl. O&P
1 Labor Foreman (outside)	$39.60	$316.80	$60.95	$487.60	$41.63	$63.77
2 Laborers	37.60	601.60	57.85	925.60		
1 Equip. Oper. (crane)	51.70	413.60	78.45	627.60		
2 Cutting Torches		22.80		25.08		
2 Sets of Gases		304.00		334.40		
1 Hyd. Crane, 12 Ton		653.80		719.18	30.64	33.71
32 L.H., Daily Totals		$2312.60		$3119.46	$72.27	$97.48

Crew B-1C	Hr.	Daily	Hr.	Daily	Bare Costs	Incl. O&P
1 Labor Foreman (outside)	$39.60	$316.80	$60.95	$487.60	$38.27	$58.88
2 Laborers	37.60	601.60	57.85	925.60		
1 Aerial Lift Truck, 60' Boom		435.60		479.16	18.15	19.97
24 L.H., Daily Totals		$1354.00		$1892.36	$56.42	$78.85

Crew B-1D	Hr.	Daily	Hr.	Daily	Bare Costs	Incl. O&P
2 Laborers	$37.60	$601.60	$57.85	$925.60	$37.60	$57.85
1 Small Work Boat, Gas, 50 H.P.		173.20		190.52		
1 Pressure Washer, 7 GPM		87.60		96.36	16.30	17.93
16 L.H., Daily Totals		$862.40		$1212.48	$53.90	$75.78

Crew B-1E	Hr.	Daily	Hr.	Daily	Bare Costs	Incl. O&P
1 Labor Foreman (outside)	$39.60	$316.80	$60.95	$487.60	$38.10	$58.63
3 Laborers	37.60	902.40	57.85	1388.40		
1 Work Boat, Diesel, 200 H.P.		1463.00		1609.30		
2 Pressure Washer, 7 GPM		175.20		192.72	51.19	56.31
32 L.H., Daily Totals		$2857.40		$3678.02	$89.29	$114.94

Crew B-1F	Hr.	Daily	Hr.	Daily	Bare Costs	Incl. O&P
2 Skilled Workers	$48.65	$778.40	$75.15	$1202.40	$44.97	$69.38
1 Laborer	37.60	300.80	57.85	462.80		
1 Small Work Boat, Gas, 50 H.P.		173.20		190.52		
1 Pressure Washer, 7 GPM		87.60		96.36	10.87	11.95
24 L.H., Daily Totals		$1340.00		$1952.08	$55.83	$81.34

Crew B-1G	Hr.	Daily	Hr.	Daily	Bare Costs	Incl. O&P
2 Laborers	$37.60	$601.60	$57.85	$925.60	$37.60	$57.85
1 Small Work Boat, Gas, 50 H.P.		173.20		190.52	10.82	11.91
16 L.H., Daily Totals		$774.80		$1116.12	$48.42	$69.76

Crew B-1H	Hr.	Daily	Hr.	Daily	Bare Costs	Incl. O&P
2 Skilled Workers	$48.65	$778.40	$75.15	$1202.40	$44.97	$69.38
1 Laborer	37.60	300.80	57.85	462.80		
1 Small Work Boat, Gas, 50 H.P.		173.20		190.52	7.22	7.94
24 L.H., Daily Totals		$1252.40		$1855.72	$52.18	$77.32

Crew B-1J	Hr.	Daily	Hr.	Daily	Bare Costs	Incl. O&P
1 Labor Foreman (inside)	$38.10	$304.80	$58.65	$469.20	$37.85	$58.25
1 Laborer	37.60	300.80	57.85	462.80		
16 L.H., Daily Totals		$605.60		$932.00	$37.85	$58.25

Crew B-1K	Hr.	Daily	Hr.	Daily	Bare Costs	Incl. O&P
1 Carpenter Foreman (inside)	$47.45	$379.60	$73.05	$584.40	$47.20	$72.65
1 Carpenter	46.95	375.60	72.25	578.00		
16 L.H., Daily Totals		$755.20		$1162.40	$47.20	$72.65

Crew B-2	Hr.	Daily	Hr.	Daily	Bare Costs	Incl. O&P
1 Labor Foreman (outside)	$39.60	$316.80	$60.95	$487.60	$38.00	$58.47
4 Laborers	37.60	1203.20	57.85	1851.20		
40 L.H., Daily Totals		$1520.00		$2338.80	$38.00	$58.47

Crew B-2A	Hr.	Daily	Hr.	Daily	Bare Costs	Incl. O&P
1 Labor Foreman (outside)	$39.60	$316.80	$60.95	$487.60	$38.27	$58.88
2 Laborers	37.60	601.60	57.85	925.60		
1 Aerial Lift Truck, 60' Boom		435.60		479.16	18.15	19.97
24 L.H., Daily Totals		$1354.00		$1892.36	$56.42	$78.85

Crew B-3	Hr.	Daily	Hr.	Daily	Bare Costs	Incl. O&P
1 Labor Foreman (outside)	$39.60	$316.80	$60.95	$487.60	$40.92	$62.42
2 Laborers	37.60	601.60	57.85	925.60		
1 Equip. Oper. (medium)	50.60	404.80	76.75	614.00		
2 Truck Drivers (heavy)	40.05	640.80	60.55	968.80		
1 Crawler Loader, 3 C.Y.		1188.00		1306.80		
2 Dump Trucks, 12 C.Y., 400 H.P.		1382.00		1520.20	53.54	58.90
48 L.H., Daily Totals		$4534.00		$5823.00	$94.46	$121.31

Crew B-3A	Hr.	Daily	Hr.	Daily	Bare Costs	Incl. O&P
4 Laborers	$37.60	$1203.20	$57.85	$1851.20	$40.20	$61.63
1 Equip. Oper. (medium)	50.60	404.80	76.75	614.00		
1 Hyd. Excavator, 1.5 C.Y.		1030.00		1133.00	25.75	28.32
40 L.H., Daily Totals		$2638.00		$3598.20	$65.95	$89.95

Crew B-3B	Hr.	Daily	Hr.	Daily	Bare Costs	Incl. O&P
2 Laborers	$37.60	$601.60	$57.85	$925.60	$41.46	$63.25
1 Equip. Oper. (medium)	50.60	404.80	76.75	614.00		
1 Truck Driver (heavy)	40.05	320.40	60.55	484.40		
1 Backhoe Loader, 80 H.P.		391.80		430.98		
1 Dump Truck, 12 C.Y., 400 H.P.		691.00		760.10	33.84	37.22
32 L.H., Daily Totals		$2409.60		$3215.08	$75.30	$100.47

567

Crew No.	Bare Costs Hr.	Daily	Incl. Subs O&P Hr.	Daily	Cost Per Labor-Hour Bare Costs	Incl. O&P
Crew B-3C	Hr.	Daily	Hr.	Daily	Bare Costs	Incl. O&P
3 Laborers	$37.60	$902.40	$57.85	$1388.40	$40.85	$62.58
1 Equip. Oper. (medium)	50.60	404.80	76.75	614.00		
1 Crawler Loader, 4 C.Y.		1561.00		1717.10	48.78	53.66
32 L.H., Daily Totals		$2868.20		$3719.50	$89.63	$116.23

Crew B-4	Hr.	Daily	Hr.	Daily	Bare Costs	Incl. O&P
1 Labor Foreman (outside)	$39.60	$316.80	$60.95	$487.60	$38.34	$58.82
4 Laborers	37.60	1203.20	57.85	1851.20		
1 Truck Driver (heavy)	40.05	320.40	60.55	484.40		
1 Truck Tractor, 220 H.P.		359.60		395.56		
1 Flatbed Trailer, 40 Ton		154.00		169.40	10.70	11.77
48 L.H., Daily Totals		$2354.00		$3388.16	$49.04	$70.59

Crew B-5	Hr.	Daily	Hr.	Daily	Bare Costs	Incl. O&P
1 Labor Foreman (outside)	$39.60	$316.80	$60.95	$487.60	$41.60	$63.69
4 Laborers	37.60	1203.20	57.85	1851.20		
2 Equip. Oper. (medium)	50.60	809.60	76.75	1228.00		
1 Air Compressor, 250 cfm		201.40		221.54		
2 Breakers, Pavement, 60 lb.		19.60		21.56		
2 -50' Air Hoses, 1.5"		11.60		12.76		
1 Crawler Loader, 3 C.Y.		1188.00		1306.80	25.37	27.90
56 L.H., Daily Totals		$3750.20		$5129.46	$66.97	$91.60

Crew B-5A	Hr.	Daily	Hr.	Daily	Bare Costs	Incl. O&P
1 Labor Foreman (outside)	$39.60	$316.80	$60.95	$487.60	$41.26	$63.03
6 Laborers	37.60	1804.80	57.85	2776.80		
2 Equip. Oper. (medium)	50.60	809.60	76.75	1228.00		
1 Equip. Oper. (light)	48.60	388.80	73.75	590.00		
2 Truck Drivers (heavy)	40.05	640.80	60.55	968.80		
1 Air Compressor, 365 cfm		263.80		290.18		
2 Breakers, Pavement, 60 lb.		19.60		21.56		
8 -50' Air Hoses, 1"		32.80		36.08		
2 Dump Trucks, 8 C.Y., 220 H.P.		822.40		904.64	11.86	13.05
96 L.H., Daily Totals		$5099.40		$7303.66	$53.12	$76.08

Crew B-5B	Hr.	Daily	Hr.	Daily	Bare Costs	Incl. O&P
1 Powderman	$48.65	$389.20	$75.15	$601.20	$45.00	$68.38
2 Equip. Oper. (medium)	50.60	809.60	76.75	1228.00		
3 Truck Drivers (heavy)	40.05	961.20	60.55	1453.20		
1 F.E. Loader, W.M.,2.5 C.Y.		519.80		571.78		
3 Dump Trucks, 12 C.Y., 400 H.P.		2073.00		2280.30		
1 Air Compressor, 365 CFM		263.80		290.18	59.51	65.46
48 L.H., Daily Totals		$5016.60		$6424.66	$104.51	$133.85

Crew B-5C	Hr.	Daily	Hr.	Daily	Bare Costs	Incl. O&P
3 Laborers	$37.60	$902.40	$57.85	$1388.40	$42.55	$64.80
1 Equip. Oper. (medium)	50.60	404.80	76.75	614.00		
2 Truck Drivers (heavy)	40.05	640.80	60.55	968.80		
1 Equip. Oper. (crane)	51.70	413.60	78.45	627.60		
1 Equip. Oper. (oiler)	45.20	361.60	68.55	548.40		
2 Dump Trucks, 12 C.Y., 400 H.P.		1382.00		1520.20		
1 Crawler Loader, 4 C.Y.		1561.00		1717.10		
1 S.P. Crane, 4x4, 25 Ton		599.40		659.34	55.35	60.88
64 L.H., Daily Totals		$6265.60		$8043.84	$97.90	$125.69

Crew B-5D	Hr.	Daily	Hr.	Daily	Bare Costs	Incl. O&P
1 Labor Foreman (outside)	$39.60	$316.80	$60.95	$487.60	$41.41	$63.30
4 Laborers	37.60	1203.20	57.85	1851.20		
2 Equip. Oper. (medium)	50.60	809.60	76.75	1228.00		
1 Truck Driver (heavy)	40.05	320.40	60.55	484.40		
1 Air Compressor, 250 cfm		201.40		221.54		
2 Breakers, Pavement, 60 lb.		19.60		21.56		
2 -50' Air Hoses, 1.5"		11.60		12.76		
1 Crawler Loader, 3 C.Y.		1188.00		1306.80		
1 Dump Truck, 12 C.Y., 400 H.P.		691.00		760.10	32.99	36.29
64 L.H., Daily Totals		$4761.60		$6373.96	$74.40	$99.59

Crew B-6	Hr.	Daily	Hr.	Daily	Bare Costs	Incl. O&P
2 Laborers	$37.60	$601.60	$57.85	$925.60	$41.27	$63.15
1 Equip. Oper. (light)	48.60	388.80	73.75	590.00		
1 Backhoe Loader, 48 H.P.		364.20		400.62	15.18	16.69
24 L.H., Daily Totals		$1354.60		$1916.22	$56.44	$79.84

Crew B-6A	Hr.	Daily	Hr.	Daily	Bare Costs	Incl. O&P
.5 Labor Foreman (outside)	$39.60	$158.40	$60.95	$243.80	$43.20	$66.03
1 Laborer	37.60	300.80	57.85	462.80		
1 Equip. Oper. (medium)	50.60	404.80	76.75	614.00		
1 Vacuum Truck, 5000 Gal.		358.10		393.91	17.91	19.70
20 L.H., Daily Totals		$1222.10		$1714.51	$61.10	$85.73

Crew B-6B	Hr.	Daily	Hr.	Daily	Bare Costs	Incl. O&P
2 Labor Foremen (outside)	$39.60	$633.60	$60.95	$975.20	$38.27	$58.88
4 Laborers	37.60	1203.20	57.85	1851.20		
1 S.P. Crane, 4x4, 5 Ton		275.40		302.94		
1 Flatbed Truck, Gas, 1.5 Ton		245.40		269.94		
1 Butt Fusion Mach., 4"-12" diam.		382.10		420.31	18.81	20.69
48 L.H., Daily Totals		$2739.70		$3819.59	$57.08	$79.57

Crew B-6C	Hr.	Daily	Hr.	Daily	Bare Costs	Incl. O&P
2 Labor Foremen (outside)	$39.60	$633.60	$60.95	$975.20	$38.27	$58.88
4 Laborers	37.60	1203.20	57.85	1851.20		
1 S.P. Crane, 4x4, 12 Ton		473.40		520.74		
1 Flatbed Truck, Gas, 3 Ton		303.20		333.52		
1 Butt Fusion Mach., 8"-24" diam.		829.70		912.67	33.46	36.81
48 L.H., Daily Totals		$3443.10		$4593.33	$71.73	$95.69

Crew B-7	Hr.	Daily	Hr.	Daily	Bare Costs	Incl. O&P
1 Labor Foreman (outside)	$39.60	$316.80	$60.95	$487.60	$40.10	$61.52
4 Laborers	37.60	1203.20	57.85	1851.20		
1 Equip. Oper. (medium)	50.60	404.80	76.75	614.00		
1 Brush Chipper, 12", 130 H.P.		391.40		430.54		
1 Crawler Loader, 3 C.Y.		1188.00		1306.80		
2 Chain Saws, Gas, 36" Long		90.00		99.00	34.78	38.26
48 L.H., Daily Totals		$3594.20		$4789.14	$74.88	$99.77

Crew B-7A	Hr.	Daily	Hr.	Daily	Bare Costs	Incl. O&P
2 Laborers	$37.60	$601.60	$57.85	$925.60	$41.27	$63.15
1 Equip. Oper. (light)	48.60	388.80	73.75	590.00		
1 Rake w/Tractor		354.25		389.68		
2 Chain Saw, Gas, 18"		62.40		68.64	17.36	19.10
24 L.H., Daily Totals		$1407.05		$1973.92	$58.63	$82.25

Crew B-7B

Crew No.	Bare Costs Hr.	Bare Costs Daily	Incl. Subs O&P Hr.	Incl. Subs O&P Daily	Cost Per Labor-Hour Bare Costs	Cost Per Labor-Hour Incl. O&P
1 Labor Foreman (outside)	$39.60	$316.80	$60.95	$487.60	$40.09	$61.38
4 Laborers	37.60	1203.20	57.85	1851.20		
1 Equip. Oper. (medium)	50.60	404.80	76.75	614.00		
1 Truck Driver (heavy)	40.05	320.40	60.55	484.40		
1 Brush Chipper, 12", 130 H.P.		391.40		430.54		
1 Crawler Loader, 3 C.Y.		1188.00		1306.80		
2 Chain Saws, Gas, 36" Long		90.00		99.00		
1 Dump Truck, 8 C.Y., 220 H.P.		411.20		452.32	37.15	40.87
56 L.H., Daily Totals		$4325.80		$5725.86	$77.25	$102.25

Crew B-7C

Crew No.	Bare Costs Hr.	Bare Costs Daily	Incl. Subs O&P Hr.	Incl. Subs O&P Daily	Cost Per Labor-Hour Bare Costs	Cost Per Labor-Hour Incl. O&P
1 Labor Foreman (outside)	$39.60	$316.80	$60.95	$487.60	$40.09	$61.38
4 Laborers	37.60	1203.20	57.85	1851.20		
1 Equip. Oper. (medium)	50.60	404.80	76.75	614.00		
1 Truck Driver (heavy)	40.05	320.40	60.55	484.40		
1 Brush Chipper, 12", 130 H.P.		391.40		430.54		
1 Crawler Loader, 3 C.Y.		1188.00		1306.80		
2 Chain Saws, Gas, 36" Long		90.00		99.00		
1 Dump Truck, 12 C.Y., 400 H.P.		691.00		760.10	42.15	46.37
56 L.H., Daily Totals		$4605.60		$6033.64	$82.24	$107.74

Crew B-8

Crew No.	Bare Costs Hr.	Bare Costs Daily	Incl. Subs O&P Hr.	Incl. Subs O&P Daily	Cost Per Labor-Hour Bare Costs	Cost Per Labor-Hour Incl. O&P
1 Labor Foreman (outside)	$39.60	$316.80	$60.95	$487.60	$42.66	$64.97
2 Laborers	37.60	601.60	57.85	925.60		
2 Equip. Oper. (medium)	50.60	809.60	76.75	1228.00		
1 Equip. Oper. (oiler)	45.20	361.60	68.55	548.40		
2 Truck Drivers (heavy)	40.05	640.80	60.55	968.80		
1 Hyd. Crane, 25 Ton		736.60		810.26		
1 Crawler Loader, 3 C.Y.		1188.00		1306.80		
2 Dump Trucks, 12 C.Y., 400 H.P.		1382.00		1520.20	51.67	56.83
64 L.H., Daily Totals		$6037.00		$7795.66	$94.33	$121.81

Crew B-9

Crew No.	Bare Costs Hr.	Bare Costs Daily	Incl. Subs O&P Hr.	Incl. Subs O&P Daily	Cost Per Labor-Hour Bare Costs	Cost Per Labor-Hour Incl. O&P
1 Labor Foreman (outside)	$39.60	$316.80	$60.95	$487.60	$38.00	$58.47
4 Laborers	37.60	1203.20	57.85	1851.20		
1 Air Compressor, 250 cfm		201.40		221.54		
2 Breakers, Pavement, 60 lb.		19.60		21.56		
2 -50' Air Hoses, 1.5"		11.60		12.76	5.82	6.40
40 L.H., Daily Totals		$1752.60		$2594.66	$43.81	$64.87

Crew B-9A

Crew No.	Bare Costs Hr.	Bare Costs Daily	Incl. Subs O&P Hr.	Incl. Subs O&P Daily	Cost Per Labor-Hour Bare Costs	Cost Per Labor-Hour Incl. O&P
2 Laborers	$37.60	$601.60	$57.85	$925.60	$38.42	$58.75
1 Truck Driver (heavy)	40.05	320.40	60.55	484.40		
1 Water Tank Trailer, 5000 Gal.		141.00		155.10		
1 Truck Tractor, 220 H.P.		359.60		395.56		
2 -50' Discharge Hoses, 3"		3.00		3.30	20.98	23.08
24 L.H., Daily Totals		$1425.60		$1963.96	$59.40	$81.83

Crew B-9B

Crew No.	Bare Costs Hr.	Bare Costs Daily	Incl. Subs O&P Hr.	Incl. Subs O&P Daily	Cost Per Labor-Hour Bare Costs	Cost Per Labor-Hour Incl. O&P
2 Laborers	$37.60	$601.60	$57.85	$925.60	$38.42	$58.75
1 Truck Driver (heavy)	40.05	320.40	60.55	484.40		
2 -50' Discharge Hoses, 3"		3.00		3.30		
1 Water Tank Trailer, 5000 Gal.		141.00		155.10		
1 Truck Tractor, 220 H.P.		359.60		395.56		
1 Pressure Washer		70.20		77.22	23.91	26.30
24 L.H., Daily Totals		$1495.80		$2041.18	$62.33	$85.05

Crew B-9D

Crew No.	Bare Costs Hr.	Bare Costs Daily	Incl. Subs O&P Hr.	Incl. Subs O&P Daily	Cost Per Labor-Hour Bare Costs	Cost Per Labor-Hour Incl. O&P
1 Labor Foreman (outside)	$39.60	$316.80	$60.95	$487.60	$38.00	$58.47
4 Common Laborers	37.60	1203.20	57.85	1851.20		
1 Air Compressor, 250 cfm		201.40		221.54		
2 -50' Air Hoses, 1.5"		11.60		12.76		
2 Air Powered Tampers		52.40		57.64	6.63	7.30
40 L.H., Daily Totals		$1785.40		$2630.74	$44.63	$65.77

Crew B-10

Crew No.	Bare Costs Hr.	Bare Costs Daily	Incl. Subs O&P Hr.	Incl. Subs O&P Daily	Cost Per Labor-Hour Bare Costs	Cost Per Labor-Hour Incl. O&P
1 Equip. Oper. (medium)	$50.60	$404.80	$76.75	$614.00	$46.27	$70.45
.5 Laborer	37.60	150.40	57.85	231.40		
12 L.H., Daily Totals		$555.20		$845.40	$46.27	$70.45

Crew B-10A

Crew No.	Bare Costs Hr.	Bare Costs Daily	Incl. Subs O&P Hr.	Incl. Subs O&P Daily	Cost Per Labor-Hour Bare Costs	Cost Per Labor-Hour Incl. O&P
1 Equip. Oper. (medium)	$50.60	$404.80	$76.75	$614.00	$46.27	$70.45
.5 Laborer	37.60	150.40	57.85	231.40		
1 Roller, 2-Drum, W.B., 7.5 H.P.		180.00		198.00	15.00	16.50
12 L.H., Daily Totals		$735.20		$1043.40	$61.27	$86.95

Crew B-10B

Crew No.	Bare Costs Hr.	Bare Costs Daily	Incl. Subs O&P Hr.	Incl. Subs O&P Daily	Cost Per Labor-Hour Bare Costs	Cost Per Labor-Hour Incl. O&P
1 Equip. Oper. (medium)	$50.60	$404.80	$76.75	$614.00	$46.27	$70.45
.5 Laborer	37.60	150.40	57.85	231.40		
1 Dozer, 200 H.P.		1387.00		1525.70	115.58	127.14
12 L.H., Daily Totals		$1942.20		$2371.10	$161.85	$197.59

Crew B-10C

Crew No.	Bare Costs Hr.	Bare Costs Daily	Incl. Subs O&P Hr.	Incl. Subs O&P Daily	Cost Per Labor-Hour Bare Costs	Cost Per Labor-Hour Incl. O&P
1 Equip. Oper. (medium)	$50.60	$404.80	$76.75	$614.00	$46.27	$70.45
.5 Laborer	37.60	150.40	57.85	231.40		
1 Dozer, 200 H.P.		1387.00		1525.70		
1 Vibratory Roller, Towed, 23 Ton		396.00		435.60	148.58	163.44
12 L.H., Daily Totals		$2338.20		$2806.70	$194.85	$233.89

Crew B-10D

Crew No.	Bare Costs Hr.	Bare Costs Daily	Incl. Subs O&P Hr.	Incl. Subs O&P Daily	Cost Per Labor-Hour Bare Costs	Cost Per Labor-Hour Incl. O&P
1 Equip. Oper. (medium)	$50.60	$404.80	$76.75	$614.00	$46.27	$70.45
.5 Laborer	37.60	150.40	57.85	231.40		
1 Dozer, 200 H.P.		1387.00		1525.70		
1 Sheepsft. Roller, Towed		425.80		468.38	151.07	166.17
12 L.H., Daily Totals		$2368.00		$2839.48	$197.33	$236.62

Crew B-10E

Crew No.	Bare Costs Hr.	Bare Costs Daily	Incl. Subs O&P Hr.	Incl. Subs O&P Daily	Cost Per Labor-Hour Bare Costs	Cost Per Labor-Hour Incl. O&P
1 Equip. Oper. (medium)	$50.60	$404.80	$76.75	$614.00	$46.27	$70.45
.5 Laborer	37.60	150.40	57.85	231.40		
1 Tandem Roller, 5 Ton		157.40		173.14	13.12	14.43
12 L.H., Daily Totals		$712.60		$1018.54	$59.38	$84.88

Crew B-10F

Crew No.	Bare Costs Hr.	Bare Costs Daily	Incl. Subs O&P Hr.	Incl. Subs O&P Daily	Cost Per Labor-Hour Bare Costs	Cost Per Labor-Hour Incl. O&P
1 Equip. Oper. (medium)	$50.60	$404.80	$76.75	$614.00	$46.27	$70.45
.5 Laborer	37.60	150.40	57.85	231.40		
1 Tandem Roller, 10 Ton		236.80		260.48	19.73	21.71
12 L.H., Daily Totals		$792.00		$1105.88	$66.00	$92.16

Crew B-10G

Crew No.	Bare Costs Hr.	Bare Costs Daily	Incl. Subs O&P Hr.	Incl. Subs O&P Daily	Cost Per Labor-Hour Bare Costs	Cost Per Labor-Hour Incl. O&P
1 Equip. Oper. (medium)	$50.60	$404.80	$76.75	$614.00	$46.27	$70.45
.5 Laborer	37.60	150.40	57.85	231.40		
1 Sheepsfoot Roller, 240 H.P.		1206.00		1326.60	100.50	110.55
12 L.H., Daily Totals		$1761.20		$2172.00	$146.77	$181.00

569

Crews

Crew B-10H	Hr.	Daily	Hr.	Daily	Bare Costs	Incl. O&P
1 Equip. Oper. (medium)	$50.60	$404.80	$76.75	$614.00	$46.27	$70.45
.5 Laborer	37.60	150.40	57.85	231.40		
1 Diaphragm Water Pump, 2"		71.00		78.10		
1 -20' Suction Hose, 2"		1.95		2.15		
2 -50' Discharge Hoses, 2"		1.80		1.98	6.23	6.85
12 L.H., Daily Totals		$629.95		$927.63	$52.50	$77.30

Crew B-10I	Hr.	Daily	Hr.	Daily	Bare Costs	Incl. O&P
1 Equip. Oper. (medium)	$50.60	$404.80	$76.75	$614.00	$46.27	$70.45
.5 Laborer	37.60	150.40	57.85	231.40		
1 Diaphragm Water Pump, 4"		114.20		125.62		
1 -20' Suction Hose, 4"		3.25		3.58		
2 -50' Discharge Hoses, 4"		4.70		5.17	10.18	11.20
12 L.H., Daily Totals		$677.35		$979.76	$56.45	$81.65

Crew B-10J	Hr.	Daily	Hr.	Daily	Bare Costs	Incl. O&P
1 Equip. Oper. (medium)	$50.60	$404.80	$76.75	$614.00	$46.27	$70.45
.5 Laborer	37.60	150.40	57.85	231.40		
1 Centrifugal Water Pump, 3"		78.20		86.02		
1 -20' Suction Hose, 3"		2.85		3.13		
2 -50' Discharge Hoses, 3"		3.00		3.30	7.00	7.70
12 L.H., Daily Totals		$639.25		$937.86	$53.27	$78.15

Crew B-10K	Hr.	Daily	Hr.	Daily	Bare Costs	Incl. O&P
1 Equip. Oper. (medium)	$50.60	$404.80	$76.75	$614.00	$46.27	$70.45
.5 Laborer	37.60	150.40	57.85	231.40		
1 Centr. Water Pump, 6"		346.60		381.26		
1 -20' Suction Hose, 6"		11.50		12.65		
2 -50' Discharge Hoses, 6"		12.20		13.42	30.86	33.94
12 L.H., Daily Totals		$925.50		$1252.73	$77.13	$104.39

Crew B-10L	Hr.	Daily	Hr.	Daily	Bare Costs	Incl. O&P
1 Equip. Oper. (medium)	$50.60	$404.80	$76.75	$614.00	$46.27	$70.45
.5 Laborer	37.60	150.40	57.85	231.40		
1 Dozer, 80 H.P.		472.40		519.64	39.37	43.30
12 L.H., Daily Totals		$1027.60		$1365.04	$85.63	$113.75

Crew B-10M	Hr.	Daily	Hr.	Daily	Bare Costs	Incl. O&P
1 Equip. Oper. (medium)	$50.60	$404.80	$76.75	$614.00	$46.27	$70.45
.5 Laborer	37.60	150.40	57.85	231.40		
1 Dozer, 300 H.P.		1897.00		2086.70	158.08	173.89
12 L.H., Daily Totals		$2452.20		$2932.10	$204.35	$244.34

Crew B-10N	Hr.	Daily	Hr.	Daily	Bare Costs	Incl. O&P
1 Equip. Oper. (medium)	$50.60	$404.80	$76.75	$614.00	$46.27	$70.45
.5 Laborer	37.60	150.40	57.85	231.40		
1 F.E. Loader, T.M., 1.5 C.Y		522.00		574.20	43.50	47.85
12 L.H., Daily Totals		$1077.20		$1419.60	$89.77	$118.30

Crew B-10O	Hr.	Daily	Hr.	Daily	Bare Costs	Incl. O&P
1 Equip. Oper. (medium)	$50.60	$404.80	$76.75	$614.00	$46.27	$70.45
.5 Laborer	37.60	150.40	57.85	231.40		
1 F.E. Loader, T.M., 2.25 C.Y.		956.40		1052.04	79.70	87.67
12 L.H., Daily Totals		$1511.60		$1897.44	$125.97	$158.12

Crew B-10P	Hr.	Daily	Hr.	Daily	Bare Costs	Incl. O&P
1 Equip. Oper. (medium)	$50.60	$404.80	$76.75	$614.00	$46.27	$70.45
.5 Laborer	37.60	150.40	57.85	231.40		
1 Crawler Loader, 3 C.Y.		1188.00		1306.80	99.00	108.90
12 L.H., Daily Totals		$1743.20		$2152.20	$145.27	$179.35

Crew B-10Q	Hr.	Daily	Hr.	Daily	Bare Costs	Incl. O&P
1 Equip. Oper. (medium)	$50.60	$404.80	$76.75	$614.00	$46.27	$70.45
.5 Laborer	37.60	150.40	57.85	231.40		
1 Crawler Loader, 4 C.Y.		1561.00		1717.10	130.08	143.09
12 L.H., Daily Totals		$2116.20		$2562.50	$176.35	$213.54

Crew B-10R	Hr.	Daily	Hr.	Daily	Bare Costs	Incl. O&P
1 Equip. Oper. (medium)	$50.60	$404.80	$76.75	$614.00	$46.27	$70.45
.5 Laborer	37.60	150.40	57.85	231.40		
1 F.E. Loader, W.M., 1 C.Y.		299.00		328.90	24.92	27.41
12 L.H., Daily Totals		$854.20		$1174.30	$71.18	$97.86

Crew B-10S	Hr.	Daily	Hr.	Daily	Bare Costs	Incl. O&P
1 Equip. Oper. (medium)	$50.60	$404.80	$76.75	$614.00	$46.27	$70.45
.5 Laborer	37.60	150.40	57.85	231.40		
1 F.E. Loader, W.M., 1.5 C.Y.		378.40		416.24	31.53	34.69
12 L.H., Daily Totals		$933.60		$1261.64	$77.80	$105.14

Crew B-10T	Hr.	Daily	Hr.	Daily	Bare Costs	Incl. O&P
1 Equip. Oper. (medium)	$50.60	$404.80	$76.75	$614.00	$46.27	$70.45
.5 Laborer	37.60	150.40	57.85	231.40		
1 F.E. Loader, W.M.,2.5 C.Y.		519.80		571.78	43.32	47.65
12 L.H., Daily Totals		$1075.00		$1417.18	$89.58	$118.10

Crew B-10U	Hr.	Daily	Hr.	Daily	Bare Costs	Incl. O&P
1 Equip. Oper. (medium)	$50.60	$404.80	$76.75	$614.00	$46.27	$70.45
.5 Laborer	37.60	150.40	57.85	231.40		
1 F.E. Loader, W.M., 5.5 C.Y.		1082.00		1190.20	90.17	99.18
12 L.H., Daily Totals		$1637.20		$2035.60	$136.43	$169.63

Crew B-10V	Hr.	Daily	Hr.	Daily	Bare Costs	Incl. O&P
1 Equip. Oper. (medium)	$50.60	$404.80	$76.75	$614.00	$46.27	$70.45
.5 Laborer	37.60	150.40	57.85	231.40		
1 Dozer, 700 H.P.		4979.00		5476.90	414.92	456.41
12 L.H., Daily Totals		$5534.20		$6322.30	$461.18	$526.86

Crew B-10W	Hr.	Daily	Hr.	Daily	Bare Costs	Incl. O&P
1 Equip. Oper. (medium)	$50.60	$404.80	$76.75	$614.00	$46.27	$70.45
.5 Laborer	37.60	150.40	57.85	231.40		
1 Dozer, 105 H.P.		602.80		663.08	50.23	55.26
12 L.H., Daily Totals		$1158.00		$1508.48	$96.50	$125.71

Crew B-10X	Hr.	Daily	Hr.	Daily	Bare Costs	Incl. O&P
1 Equip. Oper. (medium)	$50.60	$404.80	$76.75	$614.00	$46.27	$70.45
.5 Laborer	37.60	150.40	57.85	231.40		
1 Dozer, 410 H.P.		2405.00		2645.50	200.42	220.46
12 L.H., Daily Totals		$2960.20		$3490.90	$246.68	$290.91

Crew B-10Y	Hr.	Daily	Hr.	Daily	Bare Costs	Incl. O&P
1 Equip. Oper. (medium)	$50.60	$404.80	$76.75	$614.00	$46.27	$70.45
.5 Laborer	37.60	150.40	57.85	231.40		
1 Vibr. Roller, Towed, 12 Ton		549.80		604.78	45.82	50.40
12 L.H., Daily Totals		$1105.00		$1450.18	$92.08	$120.85

Crew B-11A	Hr.	Daily	Hr.	Daily	Bare Costs	Incl. O&P
1 Equipment Oper. (med.)	$50.60	$404.80	$76.75	$614.00	$44.10	$67.30
1 Laborer	37.60	300.80	57.85	462.80		
1 Dozer, 200 H.P.		1387.00		1525.70	86.69	95.36
16 L.H., Daily Totals		$2092.60		$2602.50	$130.79	$162.66

For customer support on your Plumbing Cost Data, call 877.709.6968.

Crew B-11B

Crew No.	Bare Costs Hr.	Daily	Incl. Subs O&P Hr.	Daily	Cost Per LH Bare Costs	Incl. O&P
1 Equipment Oper. (light)	$48.60	$388.80	$73.75	$590.00	$43.10	$65.80
1 Laborer	37.60	300.80	57.85	462.80		
1 Air Powered Tamper		26.20		28.82		
1 Air Compressor, 365 cfm		263.80		290.18		
2 -50' Air Hoses, 1.5"		11.60		12.76	18.85	20.73
16 L.H., Daily Totals		$991.20		$1384.56	$61.95	$86.53

Crew B-11C

Crew No.	Bare Costs Hr.	Daily	Incl. Subs O&P Hr.	Daily	Cost Per LH Bare Costs	Incl. O&P
1 Equipment Oper. (med.)	$50.60	$404.80	$76.75	$614.00	$44.10	$67.30
1 Laborer	37.60	300.80	57.85	462.80		
1 Backhoe Loader, 48 H.P.		364.20		400.62	22.76	25.04
16 L.H., Daily Totals		$1069.80		$1477.42	$66.86	$92.34

Crew B-11J

Crew No.	Bare Costs Hr.	Daily	Incl. Subs O&P Hr.	Daily	Cost Per LH Bare Costs	Incl. O&P
1 Equipment Oper. (med.)	$50.60	$404.80	$76.75	$614.00	$44.10	$67.30
1 Laborer	37.60	300.80	57.85	462.80		
1 Grader, 30,000 Lbs.		736.20		809.82		
1 Ripper, Beam & 1 Shank		84.20		92.62	51.27	56.40
16 L.H., Daily Totals		$1526.00		$1979.24	$95.38	$123.70

Crew B-11K

Crew No.	Bare Costs Hr.	Daily	Incl. Subs O&P Hr.	Daily	Cost Per LH Bare Costs	Incl. O&P
1 Equipment Oper. (med.)	$50.60	$404.80	$76.75	$614.00	$44.10	$67.30
1 Laborer	37.60	300.80	57.85	462.80		
1 Trencher, Chain Type, 8' D		3402.00		3742.20	212.63	233.89
16 L.H., Daily Totals		$4107.60		$4819.00	$256.73	$301.19

Crew B-11L

Crew No.	Bare Costs Hr.	Daily	Incl. Subs O&P Hr.	Daily	Cost Per LH Bare Costs	Incl. O&P
1 Equipment Oper. (med.)	$50.60	$404.80	$76.75	$614.00	$44.10	$67.30
1 Laborer	37.60	300.80	57.85	462.80		
1 Grader, 30,000 Lbs.		736.20		809.82	46.01	50.61
16 L.H., Daily Totals		$1441.80		$1886.62	$90.11	$117.91

Crew B-11M

Crew No.	Bare Costs Hr.	Daily	Incl. Subs O&P Hr.	Daily	Cost Per LH Bare Costs	Incl. O&P
1 Equipment Oper. (med.)	$50.60	$404.80	$76.75	$614.00	$44.10	$67.30
1 Laborer	37.60	300.80	57.85	462.80		
1 Backhoe Loader, 80 H.P.		391.80		430.98	24.49	26.94
16 L.H., Daily Totals		$1097.40		$1507.78	$68.59	$94.24

Crew B-11N

Crew No.	Bare Costs Hr.	Daily	Incl. Subs O&P Hr.	Daily	Cost Per LH Bare Costs	Incl. O&P
1 Labor Foreman (outside)	$39.60	$316.80	$60.95	$487.60	$42.34	$64.19
2 Equipment Operators (med.)	50.60	809.60	76.75	1228.00		
6 Truck Drivers (heavy)	40.05	1922.40	60.55	2906.40		
1 F.E. Loader, W.M., 5.5 C.Y.		1082.00		1190.20		
1 Dozer, 410 H.P.		2405.00		2645.50		
6 Dump Trucks, Off Hwy., 50 Ton		11022.00		12124.20	201.51	221.67
72 L.H., Daily Totals		$17557.80		$20581.90	$243.86	$285.86

Crew B-11Q

Crew No.	Bare Costs Hr.	Daily	Incl. Subs O&P Hr.	Daily	Cost Per LH Bare Costs	Incl. O&P
1 Equipment Operator (med.)	$50.60	$404.80	$76.75	$614.00	$46.27	$70.45
.5 Laborer	37.60	150.40	57.85	231.40		
1 Dozer, 140 H.P.		886.00		974.60	73.83	81.22
12 L.H., Daily Totals		$1441.20		$1820.00	$120.10	$151.67

Crew B-11R

Crew No.	Bare Costs Hr.	Daily	Incl. Subs O&P Hr.	Daily	Cost Per LH Bare Costs	Incl. O&P
1 Equipment Operator (med.)	$50.60	$404.80	$76.75	$614.00	$46.27	$70.45
.5 Laborer	37.60	150.40	57.85	231.40		
1 Dozer, 200 H.P.		1387.00		1525.70	115.58	127.14
12 L.H., Daily Totals		$1942.20		$2371.10	$161.85	$197.59

Crew B-11S

Crew No.	Bare Costs Hr.	Daily	Incl. Subs O&P Hr.	Daily	Cost Per LH Bare Costs	Incl. O&P
1 Equipment Operator (med.)	$50.60	$404.80	$76.75	$614.00	$46.27	$70.45
.5 Laborer	37.60	150.40	57.85	231.40		
1 Dozer, 300 H.P.		1897.00		2086.70		
1 Ripper, Beam & 1 Shank		84.20		92.62	165.10	181.61
12 L.H., Daily Totals		$2536.40		$3024.72	$211.37	$252.06

Crew B-11T

Crew No.	Bare Costs Hr.	Daily	Incl. Subs O&P Hr.	Daily	Cost Per LH Bare Costs	Incl. O&P
1 Equipment Operator (med.)	$50.60	$404.80	$76.75	$614.00	$46.27	$70.45
.5 Laborer	37.60	150.40	57.85	231.40		
1 Dozer, 410 H.P.		2405.00		2645.50		
1 Ripper, Beam & 2 Shanks		110.00		121.00	209.58	230.54
12 L.H., Daily Totals		$3070.20		$3611.90	$255.85	$300.99

Crew B-11U

Crew No.	Bare Costs Hr.	Daily	Incl. Subs O&P Hr.	Daily	Cost Per LH Bare Costs	Incl. O&P
1 Equipment Operator (med.)	$50.60	$404.80	$76.75	$614.00	$46.27	$70.45
.5 Laborer	37.60	150.40	57.85	231.40		
1 Dozer, 520 H.P.		3299.00		3628.90	274.92	302.41
12 L.H., Daily Totals		$3854.20		$4474.30	$321.18	$372.86

Crew B-11V

Crew No.	Bare Costs Hr.	Daily	Incl. Subs O&P Hr.	Daily	Cost Per LH Bare Costs	Incl. O&P
3 Laborers	$37.60	$902.40	$57.85	$1388.40	$37.60	$57.85
1 Roller, 2-Drum, W.B., 7.5 H.P.		180.00		198.00	7.50	8.25
24 L.H., Daily Totals		$1082.40		$1586.40	$45.10	$66.10

Crew B-11W

Crew No.	Bare Costs Hr.	Daily	Incl. Subs O&P Hr.	Daily	Cost Per LH Bare Costs	Incl. O&P
1 Equipment Operator (med.)	$50.60	$404.80	$76.75	$614.00	$40.73	$61.67
1 Common Laborer	37.60	300.80	57.85	462.80		
10 Truck Drivers (heavy)	40.05	3204.00	60.55	4844.00		
1 Dozer, 200 H.P.		1387.00		1525.70		
1 Vibratory Roller, Towed, 23 Ton		396.00		435.60		
10 Dump Trucks, 8 C.Y., 220 H.P.		4112.00		4523.20	61.41	67.55
96 L.H., Daily Totals		$9804.60		$12405.30	$102.13	$129.22

Crew B-11Y

Crew No.	Bare Costs Hr.	Daily	Incl. Subs O&P Hr.	Daily	Cost Per LH Bare Costs	Incl. O&P
1 Labor Foreman (outside)	$39.60	$316.80	$60.95	$487.60	$42.16	$64.49
5 Common Laborers	37.60	1504.00	57.85	2314.00		
3 Equipment Operators (med.)	50.60	1214.40	76.75	1842.00		
1 Dozer, 80 H.P.		472.40		519.64		
2 Roller, 2-Drum, W.B., 7.5 H.P.		360.00		396.00		
4 Vibrating Plate, Gas, 21"		187.20		205.92	14.16	15.58
72 L.H., Daily Totals		$4054.80		$5765.16	$56.32	$80.07

Crew B-12A

Crew No.	Bare Costs Hr.	Daily	Incl. Subs O&P Hr.	Daily	Cost Per LH Bare Costs	Incl. O&P
1 Equip. Oper. (crane)	$51.70	$413.60	$78.45	$627.60	$44.65	$68.15
1 Laborer	37.60	300.80	57.85	462.80		
1 Hyd. Excavator, 1 C.Y.		812.60		893.86	50.79	55.87
16 L.H., Daily Totals		$1527.00		$1984.26	$95.44	$124.02

Crew B-12B

Crew No.	Bare Costs Hr.	Daily	Incl. Subs O&P Hr.	Daily	Cost Per LH Bare Costs	Incl. O&P
1 Equip. Oper. (crane)	$51.70	$413.60	$78.45	$627.60	$44.65	$68.15
1 Laborer	37.60	300.80	57.85	462.80		
1 Hyd. Excavator, 1.5 C.Y.		1030.00		1133.00	64.38	70.81
16 L.H., Daily Totals		$1744.40		$2223.40	$109.03	$138.96

Crew B-12C

Crew No.	Bare Costs Hr.	Daily	Incl. Subs O&P Hr.	Daily	Cost Per LH Bare Costs	Incl. O&P
1 Equip. Oper. (crane)	$51.70	$413.60	$78.45	$627.60	$44.65	$68.15
1 Laborer	37.60	300.80	57.85	462.80		
1 Hyd. Excavator, 2 C.Y.		1175.00		1292.50	73.44	80.78
16 L.H., Daily Totals		$1889.40		$2382.90	$118.09	$148.93

571

For customer support on your Plumbing Cost Data, call 877.709.6968.

Crew No.	Bare Costs Hr.	Daily	Incl. Subs O&P Hr.	Daily	Cost Per Labor-Hour Bare Costs	Incl. O&P
Crew B-12D	Hr.	Daily	Hr.	Daily	Bare Costs	Incl. O&P
1 Equip. Oper. (crane)	$51.70	$413.60	$78.45	$627.60	$44.65	$68.15
1 Laborer	37.60	300.80	57.85	462.80		
1 Hyd. Excavator, 3.5 C.Y.		2440.00		2684.00	152.50	167.75
16 L.H., Daily Totals		$3154.40		$3774.40	$197.15	$235.90

Crew No.	Bare Costs Hr.	Daily	Incl. Subs O&P Hr.	Daily	Cost Per Labor-Hour Bare Costs	Incl. O&P
Crew B-12E	Hr.	Daily	Hr.	Daily	Bare Costs	Incl. O&P
1 Equip. Oper. (crane)	$51.70	$413.60	$78.45	$627.60	$44.65	$68.15
1 Laborer	37.60	300.80	57.85	462.80		
1 Hyd. Excavator, .5 C.Y.		448.20		493.02	28.01	30.81
16 L.H., Daily Totals		$1162.60		$1583.42	$72.66	$98.96

Crew No.	Bare Costs Hr.	Daily	Incl. Subs O&P Hr.	Daily	Cost Per Labor-Hour Bare Costs	Incl. O&P
Crew B-12F	Hr.	Daily	Hr.	Daily	Bare Costs	Incl. O&P
1 Equip. Oper. (crane)	$51.70	$413.60	$78.45	$627.60	$44.65	$68.15
1 Laborer	37.60	300.80	57.85	462.80		
1 Hyd. Excavator, .75 C.Y.		654.80		720.28	40.92	45.02
16 L.H., Daily Totals		$1369.20		$1810.68	$85.58	$113.17

Crew No.	Bare Costs Hr.	Daily	Incl. Subs O&P Hr.	Daily	Cost Per Labor-Hour Bare Costs	Incl. O&P
Crew B-12G	Hr.	Daily	Hr.	Daily	Bare Costs	Incl. O&P
1 Equip. Oper. (crane)	$51.70	$413.60	$78.45	$627.60	$44.65	$68.15
1 Laborer	37.60	300.80	57.85	462.80		
1 Crawler Crane, 15 Ton		669.85		736.84		
1 Clamshell Bucket, .5 C.Y.		38.20		42.02	44.25	48.68
16 L.H., Daily Totals		$1422.45		$1869.26	$88.90	$116.83

Crew No.	Bare Costs Hr.	Daily	Incl. Subs O&P Hr.	Daily	Cost Per Labor-Hour Bare Costs	Incl. O&P
Crew B-12H	Hr.	Daily	Hr.	Daily	Bare Costs	Incl. O&P
1 Equip. Oper. (crane)	$51.70	$413.60	$78.45	$627.60	$44.65	$68.15
1 Laborer	37.60	300.80	57.85	462.80		
1 Crawler Crane, 25 Ton		1150.00		1265.00		
1 Clamshell Bucket, 1 C.Y.		47.80		52.58	74.86	82.35
16 L.H., Daily Totals		$1912.20		$2407.98	$119.51	$150.50

Crew No.	Bare Costs Hr.	Daily	Incl. Subs O&P Hr.	Daily	Cost Per Labor-Hour Bare Costs	Incl. O&P
Crew B-12I	Hr.	Daily	Hr.	Daily	Bare Costs	Incl. O&P
1 Equip. Oper. (crane)	$51.70	$413.60	$78.45	$627.60	$44.65	$68.15
1 Laborer	37.60	300.80	57.85	462.80		
1 Crawler Crane, 20 Ton		862.50		948.75		
1 Dragline Bucket, .75 C.Y.		20.60		22.66	55.19	60.71
16 L.H., Daily Totals		$1597.50		$2061.81	$99.84	$128.86

Crew No.	Bare Costs Hr.	Daily	Incl. Subs O&P Hr.	Daily	Cost Per Labor-Hour Bare Costs	Incl. O&P
Crew B-12J	Hr.	Daily	Hr.	Daily	Bare Costs	Incl. O&P
1 Equip. Oper. (crane)	$51.70	$413.60	$78.45	$627.60	$44.65	$68.15
1 Laborer	37.60	300.80	57.85	462.80		
1 Gradall, 5/8 C.Y.		882.20		970.42	55.14	60.65
16 L.H., Daily Totals		$1596.60		$2060.82	$99.79	$128.80

Crew No.	Bare Costs Hr.	Daily	Incl. Subs O&P Hr.	Daily	Cost Per Labor-Hour Bare Costs	Incl. O&P
Crew B-12K	Hr.	Daily	Hr.	Daily	Bare Costs	Incl. O&P
1 Equip. Oper. (crane)	$51.70	$413.60	$78.45	$627.60	$44.65	$68.15
1 Laborer	37.60	300.80	57.85	462.80		
1 Gradall, 3 Ton, 1 C.Y.		1002.00		1102.20	62.63	68.89
16 L.H., Daily Totals		$1716.40		$2192.60	$107.28	$137.04

Crew No.	Bare Costs Hr.	Daily	Incl. Subs O&P Hr.	Daily	Cost Per Labor-Hour Bare Costs	Incl. O&P
Crew B-12L	Hr.	Daily	Hr.	Daily	Bare Costs	Incl. O&P
1 Equip. Oper. (crane)	$51.70	$413.60	$78.45	$627.60	$44.65	$68.15
1 Laborer	37.60	300.80	57.85	462.80		
1 Crawler Crane, 15 Ton		669.85		736.84		
1 F.E. Attachment, .5 C.Y.		60.00		66.00	45.62	50.18
16 L.H., Daily Totals		$1444.25		$1893.23	$90.27	$118.33

Crew No.	Bare Costs Hr.	Daily	Incl. Subs O&P Hr.	Daily	Cost Per Labor-Hour Bare Costs	Incl. O&P
Crew B-12M	Hr.	Daily	Hr.	Daily	Bare Costs	Incl. O&P
1 Equip. Oper. (crane)	$51.70	$413.60	$78.45	$627.60	$44.65	$68.15
1 Laborer	37.60	300.80	57.85	462.80		
1 Crawler Crane, 20 Ton		862.50		948.75		
1 F.E. Attachment, .75 C.Y.		65.40		71.94	57.99	63.79
16 L.H., Daily Totals		$1642.30		$2111.09	$102.64	$131.94

Crew No.	Bare Costs Hr.	Daily	Incl. Subs O&P Hr.	Daily	Cost Per Labor-Hour Bare Costs	Incl. O&P
Crew B-12N	Hr.	Daily	Hr.	Daily	Bare Costs	Incl. O&P
1 Equip. Oper. (crane)	$51.70	$413.60	$78.45	$627.60	$44.65	$68.15
1 Laborer	37.60	300.80	57.85	462.80		
1 Crawler Crane, 25 Ton		1150.00		1265.00		
1 F.E. Attachment, 1 C.Y.		71.20		78.32	76.33	83.96
16 L.H., Daily Totals		$1935.60		$2433.72	$120.97	$152.11

Crew No.	Bare Costs Hr.	Daily	Incl. Subs O&P Hr.	Daily	Cost Per Labor-Hour Bare Costs	Incl. O&P
Crew B-12O	Hr.	Daily	Hr.	Daily	Bare Costs	Incl. O&P
1 Equip. Oper. (crane)	$51.70	$413.60	$78.45	$627.60	$44.65	$68.15
1 Laborer	37.60	300.80	57.85	462.80		
1 Crawler Crane, 40 Ton		1156.00		1271.60		
1 F.E. Attachment, 1.5 C.Y.		80.00		88.00	77.25	84.97
16 L.H., Daily Totals		$1950.40		$2450.00	$121.90	$153.13

Crew No.	Bare Costs Hr.	Daily	Incl. Subs O&P Hr.	Daily	Cost Per Labor-Hour Bare Costs	Incl. O&P
Crew B-12P	Hr.	Daily	Hr.	Daily	Bare Costs	Incl. O&P
1 Equip. Oper. (crane)	$51.70	$413.60	$78.45	$627.60	$44.65	$68.15
1 Laborer	37.60	300.80	57.85	462.80		
1 Crawler Crane, 40 Ton		1156.00		1271.60		
1 Dragline Bucket, 1.5 C.Y.		33.60		36.96	74.35	81.78
16 L.H., Daily Totals		$1904.00		$2398.96	$119.00	$149.94

Crew No.	Bare Costs Hr.	Daily	Incl. Subs O&P Hr.	Daily	Cost Per Labor-Hour Bare Costs	Incl. O&P
Crew B-12Q	Hr.	Daily	Hr.	Daily	Bare Costs	Incl. O&P
1 Equip. Oper. (crane)	$51.70	$413.60	$78.45	$627.60	$44.65	$68.15
1 Laborer	37.60	300.80	57.85	462.80		
1 Hyd. Excavator, 5/8 C.Y.		589.40		648.34	36.84	40.52
16 L.H., Daily Totals		$1303.80		$1738.74	$81.49	$108.67

Crew No.	Bare Costs Hr.	Daily	Incl. Subs O&P Hr.	Daily	Cost Per Labor-Hour Bare Costs	Incl. O&P
Crew B-12S	Hr.	Daily	Hr.	Daily	Bare Costs	Incl. O&P
1 Equip. Oper. (crane)	$51.70	$413.60	$78.45	$627.60	$44.65	$68.15
1 Laborer	37.60	300.80	57.85	462.80		
1 Hyd. Excavator, 2.5 C.Y.		1605.00		1765.50	100.31	110.34
16 L.H., Daily Totals		$2319.40		$2855.90	$144.96	$178.49

Crew No.	Bare Costs Hr.	Daily	Incl. Subs O&P Hr.	Daily	Cost Per Labor-Hour Bare Costs	Incl. O&P
Crew B-12T	Hr.	Daily	Hr.	Daily	Bare Costs	Incl. O&P
1 Equip. Oper. (crane)	$51.70	$413.60	$78.45	$627.60	$44.65	$68.15
1 Laborer	37.60	300.80	57.85	462.80		
1 Crawler Crane, 75 Ton		1459.00		1604.90		
1 F.E. Attachment, 3 C.Y.		103.20		113.52	97.64	107.40
16 L.H., Daily Totals		$2276.60		$2808.82	$142.29	$175.55

Crew No.	Bare Costs Hr.	Daily	Incl. Subs O&P Hr.	Daily	Cost Per Labor-Hour Bare Costs	Incl. O&P
Crew B-12V	Hr.	Daily	Hr.	Daily	Bare Costs	Incl. O&P
1 Equip. Oper. (crane)	$51.70	$413.60	$78.45	$627.60	$44.65	$68.15
1 Laborer	37.60	300.80	57.85	462.80		
1 Crawler Crane, 75 Ton		1459.00		1604.90		
1 Dragline Bucket, 3 C.Y.		52.60		57.86	94.47	103.92
16 L.H., Daily Totals		$2226.00		$2753.16	$139.13	$172.07

Crew No.	Bare Costs Hr.	Daily	Incl. Subs O&P Hr.	Daily	Cost Per Labor-Hour Bare Costs	Incl. O&P
Crew B-12Y	Hr.	Daily	Hr.	Daily	Bare Costs	Incl. O&P
1 Equip. Oper. (crane)	$51.70	$413.60	$78.45	$627.60	$42.30	$64.72
2 Laborers	37.60	601.60	57.85	925.60		
1 Hyd. Excavator, 3.5 C.Y.		2440.00		2684.00	101.67	111.83
24 L.H., Daily Totals		$3455.20		$4237.20	$143.97	$176.55

For customer support on your Plumbing Cost Data, call 877.709.6968.

Crews

Crew B-12Z	Bare Costs Hr.	Daily	Incl. Subs O&P Hr.	Daily	Cost/LH Bare Costs	Incl. O&P
1 Equip. Oper. (crane)	$51.70	$413.60	$78.45	$627.60	$42.30	$64.72
2 Laborers	37.60	601.60	57.85	925.60		
1 Hyd. Excavator, 2.5 C.Y.		1605.00		1765.50	66.88	73.56
24 L.H., Daily Totals		$2620.20		$3318.70	$109.18	$138.28

Crew B-13	Bare Costs Hr.	Daily	Incl. Subs O&P Hr.	Daily	Cost/LH Bare Costs	Incl. O&P
1 Labor Foreman (outside)	$39.60	$316.80	$60.95	$487.60	$40.99	$62.76
4 Laborers	37.60	1203.20	57.85	1851.20		
1 Equip. Oper. (crane)	51.70	413.60	78.45	627.60		
1 Equip. Oper. (oiler)	45.20	361.60	68.55	548.40		
1 Hyd. Crane, 25 Ton		736.60		810.26	13.15	14.47
56 L.H., Daily Totals		$3031.80		$4325.06	$54.14	$77.23

Crew B-13A	Bare Costs Hr.	Daily	Incl. Subs O&P Hr.	Daily	Cost/LH Bare Costs	Incl. O&P
1 Labor Foreman (outside)	$39.60	$316.80	$60.95	$487.60	$42.30	$64.46
2 Laborers	37.60	601.60	57.85	925.60		
2 Equipment Operators (med.)	50.60	809.60	76.75	1228.00		
2 Truck Drivers (heavy)	40.05	640.80	60.55	968.80		
1 Crawler Crane, 75 Ton		1459.00		1604.90		
1 Crawler Loader, 4 C.Y.		1561.00		1717.10		
2 Dump Trucks, 8 C.Y., 220 H.P.		822.40		904.64	68.61	75.48
56 L.H., Daily Totals		$6211.20		$7836.64	$110.91	$139.94

Crew B-13B	Bare Costs Hr.	Daily	Incl. Subs O&P Hr.	Daily	Cost/LH Bare Costs	Incl. O&P
1 Labor Foreman (outside)	$39.60	$316.80	$60.95	$487.60	$40.99	$62.76
4 Laborers	37.60	1203.20	57.85	1851.20		
1 Equip. Oper. (crane)	51.70	413.60	78.45	627.60		
1 Equip. Oper. (oiler)	45.20	361.60	68.55	548.40		
1 Hyd. Crane, 55 Ton		1128.00		1240.80	20.14	22.16
56 L.H., Daily Totals		$3423.20		$4755.60	$61.13	$84.92

Crew B-13C	Bare Costs Hr.	Daily	Incl. Subs O&P Hr.	Daily	Cost/LH Bare Costs	Incl. O&P
1 Labor Foreman (outside)	$39.60	$316.80	$60.95	$487.60	$40.99	$62.76
4 Laborers	37.60	1203.20	57.85	1851.20		
1 Equip. Oper. (crane)	51.70	413.60	78.45	627.60		
1 Equip. Oper. (oiler)	45.20	361.60	68.55	548.40		
1 Crawler Crane, 100 Ton		1667.00		1833.70	29.77	32.74
56 L.H., Daily Totals		$3962.20		$5348.50	$70.75	$95.51

Crew B-13D	Bare Costs Hr.	Daily	Incl. Subs O&P Hr.	Daily	Cost/LH Bare Costs	Incl. O&P
1 Laborer	$37.60	$300.80	$57.85	$462.80	$44.65	$68.15
1 Equip. Oper. (crane)	51.70	413.60	78.45	627.60		
1 Hyd. Excavator, 1 C.Y.		812.60		893.86		
1 Trench Box		81.00		89.10	55.85	61.44
16 L.H., Daily Totals		$1608.00		$2073.36	$100.50	$129.59

Crew B-13E	Bare Costs Hr.	Daily	Incl. Subs O&P Hr.	Daily	Cost/LH Bare Costs	Incl. O&P
1 Laborer	$37.60	$300.80	$57.85	$462.80	$44.65	$68.15
1 Equip. Oper. (crane)	51.70	413.60	78.45	627.60		
1 Hyd. Excavator, 1.5 C.Y.		1030.00		1133.00		
1 Trench Box		81.00		89.10	69.44	76.38
16 L.H., Daily Totals		$1825.40		$2312.50	$114.09	$144.53

Crew B-13F	Bare Costs Hr.	Daily	Incl. Subs O&P Hr.	Daily	Cost/LH Bare Costs	Incl. O&P
1 Laborer	$37.60	$300.80	$57.85	$462.80	$44.65	$68.15
1 Equip. Oper. (crane)	51.70	413.60	78.45	627.60		
1 Hyd. Excavator, 3.5 C.Y.		2440.00		2684.00		
1 Trench Box		81.00		89.10	157.56	173.32
16 L.H., Daily Totals		$3235.40		$3863.50	$202.21	$241.47

Crew B-13G	Bare Costs Hr.	Daily	Incl. Subs O&P Hr.	Daily	Cost/LH Bare Costs	Incl. O&P
1 Laborer	$37.60	$300.80	$57.85	$462.80	$44.65	$68.15
1 Equip. Oper. (crane)	51.70	413.60	78.45	627.60		
1 Hyd. Excavator, .75 C.Y.		654.80		720.28		
1 Trench Box		81.00		89.10	45.99	50.59
16 L.H., Daily Totals		$1450.20		$1899.78	$90.64	$118.74

Crew B-13H	Bare Costs Hr.	Daily	Incl. Subs O&P Hr.	Daily	Cost/LH Bare Costs	Incl. O&P
1 Laborer	$37.60	$300.80	$57.85	$462.80	$44.65	$68.15
1 Equip. Oper. (crane)	51.70	413.60	78.45	627.60		
1 Gradall, 5/8 C.Y.		882.20		970.42		
1 Trench Box		81.00		89.10	60.20	66.22
16 L.H., Daily Totals		$1677.60		$2149.92	$104.85	$134.37

Crew B-13I	Bare Costs Hr.	Daily	Incl. Subs O&P Hr.	Daily	Cost/LH Bare Costs	Incl. O&P
1 Laborer	$37.60	$300.80	$57.85	$462.80	$44.65	$68.15
1 Equip. Oper. (crane)	51.70	413.60	78.45	627.60		
1 Gradall, 3 Ton, 1 C.Y.		1002.00		1102.20		
1 Trench Box		81.00		89.10	67.69	74.46
16 L.H., Daily Totals		$1797.40		$2281.70	$112.34	$142.61

Crew B-13J	Bare Costs Hr.	Daily	Incl. Subs O&P Hr.	Daily	Cost/LH Bare Costs	Incl. O&P
1 Laborer	$37.60	$300.80	$57.85	$462.80	$44.65	$68.15
1 Equip. Oper. (crane)	51.70	413.60	78.45	627.60		
1 Hyd. Excavator, 2.5 C.Y.		1605.00		1765.50		
1 Trench Box		81.00		89.10	105.38	115.91
16 L.H., Daily Totals		$2400.40		$2945.00	$150.03	$184.06

Crew B-13K	Bare Costs Hr.	Daily	Incl. Subs O&P Hr.	Daily	Cost/LH Bare Costs	Incl. O&P
2 Equip. Opers. (crane)	$51.70	$827.20	$78.45	$1255.20	$51.70	$78.45
1 Hyd. Excavator, .75 C.Y.		654.80		720.28		
1 Hyd. Hammer, 4000 ft-lb		308.20		339.02		
1 Hyd. Excavator, .75 C.Y.		654.80		720.28	101.11	111.22
16 L.H., Daily Totals		$2445.00		$3034.78	$152.81	$189.67

Crew B-13L	Bare Costs Hr.	Daily	Incl. Subs O&P Hr.	Daily	Cost/LH Bare Costs	Incl. O&P
2 Equip. Opers. (crane)	$51.70	$827.20	$78.45	$1255.20	$51.70	$78.45
1 Hyd. Excavator, 1.5 C.Y.		1030.00		1133.00		
1 Hyd. Hammer, 5000 ft-lb		371.20		408.32		
1 Hyd. Excavator, .75 C.Y.		654.80		720.28	128.50	141.35
16 L.H., Daily Totals		$2883.20		$3516.80	$180.20	$219.80

Crew B-13M	Bare Costs Hr.	Daily	Incl. Subs O&P Hr.	Daily	Cost/LH Bare Costs	Incl. O&P
2 Equip. Opers. (crane)	$51.70	$827.20	$78.45	$1255.20	$51.70	$78.45
1 Hyd. Excavator, 2.5 C.Y.		1605.00		1765.50		
1 Hyd. Hammer, 8000 ft-lb		545.40		599.94		
1 Hyd. Excavator, 1.5 C.Y.		1030.00		1133.00	198.78	218.65
16 L.H., Daily Totals		$4007.60		$4753.64	$250.47	$297.10

Crew B-13N	Bare Costs Hr.	Daily	Incl. Subs O&P Hr.	Daily	Cost/LH Bare Costs	Incl. O&P
2 Equip. Opers. (crane)	$51.70	$827.20	$78.45	$1255.20	$51.70	$78.45
1 Hyd. Excavator, 3.5 C.Y.		2440.00		2684.00		
1 Hyd. Hammer, 12,000 ft-lb		635.80		699.38		
1 Hyd. Excavator, 1.5 C.Y.		1030.00		1133.00	256.61	282.27
16 L.H., Daily Totals		$4933.00		$5771.58	$308.31	$360.72

573

Crew B-14

Crew B-14	Bare Costs Hr.	Daily	Incl. Subs O&P Hr.	Daily	Bare Costs	Incl. O&P
1 Labor Foreman (outside)	$39.60	$316.80	$60.95	$487.60	$39.77	$61.02
4 Laborers	37.60	1203.20	57.85	1851.20		
1 Equip. Oper. (light)	48.60	388.80	73.75	590.00		
1 Backhoe Loader, 48 H.P.		364.20		400.62	7.59	8.35
48 L.H., Daily Totals		$2273.00		$3329.42	$47.35	$69.36

Crew B-14A

Crew B-14A	Bare Costs Hr.	Daily	Incl. Subs O&P Hr.	Daily	Bare Costs	Incl. O&P
1 Equip. Oper. (crane)	$51.70	$413.60	$78.45	$627.60	$47.00	$71.58
.5 Laborer	37.60	150.40	57.85	231.40		
1 Hyd. Excavator, 4.5 C.Y.		2963.00		3259.30	246.92	271.61
12 L.H., Daily Totals		$3527.00		$4118.30	$293.92	$343.19

Crew B-14B

Crew B-14B	Bare Costs Hr.	Daily	Incl. Subs O&P Hr.	Daily	Bare Costs	Incl. O&P
1 Equip. Oper. (crane)	$51.70	$413.60	$78.45	$627.60	$47.00	$71.58
.5 Laborer	37.60	150.40	57.85	231.40		
1 Hyd. Excavator, 6 C.Y.		3494.00		3843.40	291.17	320.28
12 L.H., Daily Totals		$4058.00		$4702.40	$338.17	$391.87

Crew B-14C

Crew B-14C	Bare Costs Hr.	Daily	Incl. Subs O&P Hr.	Daily	Bare Costs	Incl. O&P
1 Equip. Oper. (crane)	$51.70	$413.60	$78.45	$627.60	$47.00	$71.58
.5 Laborer	37.60	150.40	57.85	231.40		
1 Hyd. Excavator, 7 C.Y.		3590.00		3949.00	299.17	329.08
12 L.H., Daily Totals		$4154.00		$4808.00	$346.17	$400.67

Crew B-14F

Crew B-14F	Bare Costs Hr.	Daily	Incl. Subs O&P Hr.	Daily	Bare Costs	Incl. O&P
1 Equip. Oper. (crane)	$51.70	$413.60	$78.45	$627.60	$47.00	$71.58
.5 Laborer	37.60	150.40	57.85	231.40		
1 Hyd. Shovel, 7 C.Y.		3993.00		4392.30	332.75	366.02
12 L.H., Daily Totals		$4557.00		$5251.30	$379.75	$437.61

Crew B-14G

Crew B-14G	Bare Costs Hr.	Daily	Incl. Subs O&P Hr.	Daily	Bare Costs	Incl. O&P
1 Equip. Oper. (crane)	$51.70	$413.60	$78.45	$627.60	$47.00	$71.58
.5 Laborer	37.60	150.40	57.85	231.40		
1 Hyd. Shovel, 12 C.Y.		5736.00		6309.60	478.00	525.80
12 L.H., Daily Totals		$6300.00		$7168.60	$525.00	$597.38

Crew B-14J

Crew B-14J	Bare Costs Hr.	Daily	Incl. Subs O&P Hr.	Daily	Bare Costs	Incl. O&P
1 Equip. Oper. (medium)	$50.60	$404.80	$76.75	$614.00	$46.27	$70.45
.5 Laborer	37.60	150.40	57.85	231.40		
1 F.E. Loader, 8 C.Y.		1937.00		2130.70	161.42	177.56
12 L.H., Daily Totals		$2492.20		$2976.10	$207.68	$248.01

Crew B-14K

Crew B-14K	Bare Costs Hr.	Daily	Incl. Subs O&P Hr.	Daily	Bare Costs	Incl. O&P
1 Equip. Oper. (medium)	$50.60	$404.80	$76.75	$614.00	$46.27	$70.45
.5 Laborer	37.60	150.40	57.85	231.40		
1 F.E. Loader, 10 C.Y.		2942.00		3236.20	245.17	269.68
12 L.H., Daily Totals		$3497.20		$4081.60	$291.43	$340.13

Crew B-15

Crew B-15	Bare Costs Hr.	Daily	Incl. Subs O&P Hr.	Daily	Bare Costs	Incl. O&P
1 Equipment Oper. (med.)	$50.60	$404.80	$76.75	$614.00	$42.71	$64.79
.5 Laborer	37.60	150.40	57.85	231.40		
2 Truck Drivers (heavy)	40.05	640.80	60.55	968.80		
2 Dump Trucks, 12 C.Y., 400 H.P.		1382.00		1520.20		
1 Dozer, 200 H.P.		1387.00		1525.70	98.89	108.78
28 L.H., Daily Totals		$3965.00		$4860.10	$141.61	$173.57

Crew B-16

Crew B-16	Bare Costs Hr.	Daily	Incl. Subs O&P Hr.	Daily	Bare Costs	Incl. O&P
1 Labor Foreman (outside)	$39.60	$316.80	$60.95	$487.60	$38.71	$59.30
2 Laborers	37.60	601.60	57.85	925.60		
1 Truck Driver (heavy)	40.05	320.40	60.55	484.40		
1 Dump Truck, 12 C.Y., 400 H.P.		691.00		760.10	21.59	23.75
32 L.H., Daily Totals		$1929.80		$2657.70	$60.31	$83.05

Crew B-17

Crew B-17	Bare Costs Hr.	Daily	Incl. Subs O&P Hr.	Daily	Bare Costs	Incl. O&P
2 Laborers	$37.60	$601.60	$57.85	$925.60	$40.96	$62.50
1 Equip. Oper. (light)	48.60	388.80	73.75	590.00		
1 Truck Driver (heavy)	40.05	320.40	60.55	484.40		
1 Backhoe Loader, 48 H.P.		364.20		400.62		
1 Dump Truck, 8 C.Y., 220 H.P.		411.20		452.32	24.23	26.65
32 L.H., Daily Totals		$2086.20		$2852.94	$65.19	$89.15

Crew B-17A

Crew B-17A	Bare Costs Hr.	Daily	Incl. Subs O&P Hr.	Daily	Bare Costs	Incl. O&P
2 Labor Foremen (outside)	$39.60	$633.60	$60.95	$975.20	$40.41	$62.24
6 Laborers	37.60	1804.80	57.85	2776.80		
1 Skilled Worker Foreman (out)	50.65	405.20	78.25	626.00		
1 Skilled Worker	48.65	389.20	75.15	601.20		
80 L.H., Daily Totals		$3232.80		$4979.20	$40.41	$62.24

Crew B-17B

Crew B-17B	Bare Costs Hr.	Daily	Incl. Subs O&P Hr.	Daily	Bare Costs	Incl. O&P
2 Laborers	$37.60	$601.60	$57.85	$925.60	$40.96	$62.50
1 Equip. Oper. (light)	48.60	388.80	73.75	590.00		
1 Truck Driver (heavy)	40.05	320.40	60.55	484.40		
1 Backhoe Loader, 48 H.P.		364.20		400.62		
1 Dump Truck, 12 C.Y., 400 H.P.		691.00		760.10	32.98	36.27
32 L.H., Daily Totals		$2366.00		$3160.72	$73.94	$98.77

Crew B-18

Crew B-18	Bare Costs Hr.	Daily	Incl. Subs O&P Hr.	Daily	Bare Costs	Incl. O&P
1 Labor Foreman (outside)	$39.60	$316.80	$60.95	$487.60	$38.27	$58.88
2 Laborers	37.60	601.60	57.85	925.60		
1 Vibrating Plate, Gas, 21"		46.80		51.48	1.95	2.15
24 L.H., Daily Totals		$965.20		$1464.68	$40.22	$61.03

Crew B-19

Crew B-19	Bare Costs Hr.	Daily	Incl. Subs O&P Hr.	Daily	Bare Costs	Incl. O&P
1 Pile Driver Foreman (outside)	$48.10	$384.80	$76.15	$609.20	$47.64	$74.20
4 Pile Drivers	46.10	1475.20	73.00	2336.00		
2 Equip. Oper. (crane)	51.70	827.20	78.45	1255.20		
1 Equip. Oper. (oiler)	45.20	361.60	68.55	548.40		
1 Crawler Crane, 40 Ton		1156.00		1271.60		
1 Lead, 90' High		124.40		136.84		
1 Hammer, Diesel, 22k ft-lb		456.40		502.04	27.14	29.85
64 L.H., Daily Totals		$4785.60		$6659.28	$74.78	$104.05

Crew B-19A

Crew B-19A	Bare Costs Hr.	Daily	Incl. Subs O&P Hr.	Daily	Bare Costs	Incl. O&P
1 Pile Driver Foreman (outside)	$48.10	$384.80	$76.15	$609.20	$47.64	$74.20
4 Pile Drivers	46.10	1475.20	73.00	2336.00		
2 Equip. Oper. (crane)	51.70	827.20	78.45	1255.20		
1 Equip. Oper. (oiler)	45.20	361.60	68.55	548.40		
1 Crawler Crane, 75 Ton		1459.00		1604.90		
1 Lead, 90' high		124.40		136.84		
1 Hammer, Diesel, 41k ft-lb		581.00		639.10	33.82	37.20
64 L.H., Daily Totals		$5213.20		$7129.64	$81.46	$111.40

Crew B-19B

Crew B-19B	Hr.	Daily	Hr.	Daily	Bare Costs	Incl. O&P
1 Pile Driver Foreman (outside)	$48.10	$384.80	$76.15	$609.20	$47.64	$74.20
4 Pile Drivers	46.10	1475.20	73.00	2336.00		
2 Equip. Oper. (crane)	51.70	827.20	78.45	1255.20		
1 Equip. Oper. (oiler)	45.20	361.60	68.55	548.40		
1 Crawler Crane, 40 Ton		1156.00		1271.60		
1 Lead, 90' High		124.40		136.84		
1 Hammer, Diesel, 22k ft-lb		456.40		502.04		
1 Barge, 400 Ton		776.00		853.60	39.26	43.19
64 L.H., Daily Totals		$5561.60		$7512.88	$86.90	$117.39

Crew B-19C

Crew B-19C	Hr.	Daily	Hr.	Daily	Bare Costs	Incl. O&P
1 Pile Driver Foreman (outside)	$48.10	$384.80	$76.15	$609.20	$47.64	$74.20
4 Pile Drivers	46.10	1475.20	73.00	2336.00		
2 Equip. Oper. (crane)	51.70	827.20	78.45	1255.20		
1 Equip. Oper. (oiler)	45.20	361.60	68.55	548.40		
1 Crawler Crane, 75 Ton		1459.00		1604.90		
1 Lead, 90' High		124.40		136.84		
1 Hammer, Diesel, 41k ft-lb		581.00		639.10		
1 Barge, 400 Ton		776.00		853.60	45.94	50.54
64 L.H., Daily Totals		$5989.20		$7983.24	$93.58	$124.74

Crew B-20

Crew B-20	Hr.	Daily	Hr.	Daily	Bare Costs	Incl. O&P
1 Labor Foreman (outside)	$39.60	$316.80	$60.95	$487.60	$41.95	$64.65
1 Skilled Worker	48.65	389.20	75.15	601.20		
1 Laborer	37.60	300.80	57.85	462.80		
24 L.H., Daily Totals		$1006.80		$1551.60	$41.95	$64.65

Crew B-20A

Crew B-20A	Hr.	Daily	Hr.	Daily	Bare Costs	Incl. O&P
1 Labor Foreman (outside)	$39.60	$316.80	$60.95	$487.60	$45.71	$69.59
1 Laborer	37.60	300.80	57.85	462.80		
1 Plumber	58.70	469.60	88.65	709.20		
1 Plumber Apprentice	46.95	375.60	70.90	567.20		
32 L.H., Daily Totals		$1462.80		$2226.80	$45.71	$69.59

Crew B-21

Crew B-21	Hr.	Daily	Hr.	Daily	Bare Costs	Incl. O&P
1 Labor Foreman (outside)	$39.60	$316.80	$60.95	$487.60	$43.34	$66.62
1 Skilled Worker	48.65	389.20	75.15	601.20		
1 Laborer	37.60	300.80	57.85	462.80		
.5 Equip. Oper. (crane)	51.70	206.80	78.45	313.80		
.5 S.P. Crane, 4x4, 5 Ton		137.70		151.47	4.92	5.41
28 L.H., Daily Totals		$1351.30		$2016.87	$48.26	$72.03

Crew B-21A

Crew B-21A	Hr.	Daily	Hr.	Daily	Bare Costs	Incl. O&P
1 Labor Foreman (outside)	$39.60	$316.80	$60.95	$487.60	$46.91	$71.36
1 Laborer	37.60	300.80	57.85	462.80		
1 Plumber	58.70	469.60	88.65	709.20		
1 Plumber Apprentice	46.95	375.60	70.90	567.20		
1 Equip. Oper. (crane)	51.70	413.60	78.45	627.60		
1 S.P. Crane, 4x4, 12 Ton		473.40		520.74	11.84	13.02
40 L.H., Daily Totals		$2349.80		$3375.14	$58.74	$84.38

Crew B-21B

Crew B-21B	Hr.	Daily	Hr.	Daily	Bare Costs	Incl. O&P
1 Labor Foreman (outside)	$39.60	$316.80	$60.95	$487.60	$40.82	$62.59
3 Laborers	37.60	902.40	57.85	1388.40		
1 Equip. Oper. (crane)	51.70	413.60	78.45	627.60		
1 Hyd. Crane, 12 Ton		653.80		719.18	16.34	17.98
40 L.H., Daily Totals		$2286.60		$3222.78	$57.16	$80.57

Crew B-21C

Crew B-21C	Hr.	Daily	Hr.	Daily	Bare Costs	Incl. O&P
1 Labor Foreman (outside)	$39.60	$316.80	$60.95	$487.60	$40.99	$62.76
4 Laborers	37.60	1203.20	57.85	1851.20		
1 Equip. Oper. (crane)	51.70	413.60	78.45	627.60		
1 Equip. Oper. (oiler)	45.20	361.60	68.55	548.40		
2 Cutting Torches		22.80		25.08		
2 Sets of Gases		304.00		334.40		
1 Lattice Boom Crane, 90 Ton		1511.00		1662.10	32.82	36.10
56 L.H., Daily Totals		$4133.00		$5536.38	$73.80	$98.86

Crew B-22

Crew B-22	Hr.	Daily	Hr.	Daily	Bare Costs	Incl. O&P
1 Labor Foreman (outside)	$39.60	$316.80	$60.95	$487.60	$43.90	$67.41
1 Skilled Worker	48.65	389.20	75.15	601.20		
1 Laborer	37.60	300.80	57.85	462.80		
.75 Equip. Oper. (crane)	51.70	310.20	78.45	470.70		
.75 S.P. Crane, 4x4, 5 Ton		206.55		227.21	6.88	7.57
30 L.H., Daily Totals		$1523.55		$2249.51	$50.78	$74.98

Crew B-22A

Crew B-22A	Hr.	Daily	Hr.	Daily	Bare Costs	Incl. O&P
1 Labor Foreman (outside)	$39.60	$316.80	$60.95	$487.60	$43.03	$66.05
1 Skilled Worker	48.65	389.20	75.15	601.20		
2 Laborers	37.60	601.60	57.85	925.60		
1 Equipment Operator, Crane	51.70	413.60	78.45	627.60		
1 S.P. Crane, 4x4, 5 Ton		275.40		302.94		
1 Butt Fusion Mach., 4"-12" diam.		382.10		420.31	16.44	18.08
40 L.H., Daily Totals		$2378.70		$3365.25	$59.47	$84.13

Crew B-22B

Crew B-22B	Hr.	Daily	Hr.	Daily	Bare Costs	Incl. O&P
1 Labor Foreman (outside)	$39.60	$316.80	$60.95	$487.60	$43.03	$66.05
1 Skilled Worker	48.65	389.20	75.15	601.20		
2 Laborers	37.60	601.60	57.85	925.60		
1 Equip. Oper. (crane)	51.70	413.60	78.45	627.60		
1 S.P. Crane, 4x4, 5 Ton		275.40		302.94		
1 Butt Fusion Mach., 8"-24" diam.		829.70		912.67	27.63	30.39
40 L.H., Daily Totals		$2826.30		$3857.61	$70.66	$96.44

Crew B-22C

Crew B-22C	Hr.	Daily	Hr.	Daily	Bare Costs	Incl. O&P
1 Skilled Worker	$48.65	$389.20	$75.15	$601.20	$43.13	$66.50
1 Laborer	37.60	300.80	57.85	462.80		
1 Butt Fusion Mach., 2"-8" diam.		121.05		133.16	7.57	8.32
16 L.H., Daily Totals		$811.05		$1197.16	$50.69	$74.82

Crew B-23

Crew B-23	Hr.	Daily	Hr.	Daily	Bare Costs	Incl. O&P
1 Labor Foreman (outside)	$39.60	$316.80	$60.95	$487.60	$38.00	$58.47
4 Laborers	37.60	1203.20	57.85	1851.20		
1 Drill Rig, Truck-Mounted		2542.00		2796.20		
1 Flatbed Truck, Gas, 3 Ton		303.20		333.52	71.13	78.24
40 L.H., Daily Totals		$4365.20		$5468.52	$109.13	$136.71

Crew B-23A

Crew B-23A	Hr.	Daily	Hr.	Daily	Bare Costs	Incl. O&P
1 Labor Foreman (outside)	$39.60	$316.80	$60.95	$487.60	$42.60	$65.18
1 Laborer	37.60	300.80	57.85	462.80		
1 Equip. Oper. (medium)	50.60	404.80	76.75	614.00		
1 Drill Rig, Truck-Mounted		2542.00		2796.20		
1 Pickup Truck, 3/4 Ton		144.20		158.62	111.93	123.12
24 L.H., Daily Totals		$3708.60		$4519.22	$154.53	$188.30

Crew B-23B

Crew No.	Bare Costs Hr.	Daily	Incl. Subs O&P Hr.	Daily	Cost Per Labor-Hour Bare Costs	Incl. O&P
1 Labor Foreman (outside)	$39.60	$316.80	$60.95	$487.60	$42.60	$65.18
1 Laborer	37.60	300.80	57.85	462.80		
1 Equip. Oper. (medium)	50.60	404.80	76.75	614.00		
1 Drill Rig, Truck-Mounted		2542.00		2796.20		
1 Pickup Truck, 3/4 Ton		144.20		158.62		
1 Centr. Water Pump, 6"		346.60		381.26	126.37	139.00
24 L.H., Daily Totals		$4055.20		$4900.48	$168.97	$204.19

Crew B-24

Crew No.	Bare Costs Hr.	Daily	Incl. Subs O&P Hr.	Daily	Cost Per Labor-Hour Bare Costs	Incl. O&P
1 Cement Finisher	$45.00	$360.00	$66.65	$533.20	$43.18	$65.58
1 Laborer	37.60	300.80	57.85	462.80		
1 Carpenter	46.95	375.60	72.25	578.00		
24 L.H., Daily Totals		$1036.40		$1574.00	$43.18	$65.58

Crew B-25

Crew No.	Bare Costs Hr.	Daily	Incl. Subs O&P Hr.	Daily	Cost Per Labor-Hour Bare Costs	Incl. O&P
1 Labor Foreman (outside)	$39.60	$316.80	$60.95	$487.60	$41.33	$63.29
7 Laborers	37.60	2105.60	57.85	3239.60		
3 Equip. Oper. (medium)	50.60	1214.40	76.75	1842.00		
1 Asphalt Paver, 130 H.P.		2132.00		2345.20		
1 Tandem Roller, 10 Ton		236.80		260.48		
1 Roller, Pneum. Whl., 12 Ton		344.40		378.84	30.83	33.91
88 L.H., Daily Totals		$6350.00		$8553.72	$72.16	$97.20

Crew B-25B

Crew No.	Bare Costs Hr.	Daily	Incl. Subs O&P Hr.	Daily	Cost Per Labor-Hour Bare Costs	Incl. O&P
1 Labor Foreman (outside)	$39.60	$316.80	$60.95	$487.60	$42.10	$64.41
7 Laborers	37.60	2105.60	57.85	3239.60		
4 Equip. Oper. (medium)	50.60	1619.20	76.75	2456.00		
1 Asphalt Paver, 130 H.P.		2132.00		2345.20		
2 Tandem Rollers, 10 Ton		473.60		520.96		
1 Roller, Pneum. Whl., 12 Ton		344.40		378.84	30.73	33.80
96 L.H., Daily Totals		$6991.60		$9428.20	$72.83	$98.21

Crew B-25C

Crew No.	Bare Costs Hr.	Daily	Incl. Subs O&P Hr.	Daily	Cost Per Labor-Hour Bare Costs	Incl. O&P
1 Labor Foreman (outside)	$39.60	$316.80	$60.95	$487.60	$42.27	$64.67
3 Laborers	37.60	902.40	57.85	1388.40		
2 Equip. Oper. (medium)	50.60	809.60	76.75	1228.00		
1 Asphalt Paver, 130 H.P.		2132.00		2345.20		
1 Tandem Roller, 10 Ton		236.80		260.48	49.35	54.28
48 L.H., Daily Totals		$4397.60		$5709.68	$91.62	$118.95

Crew B-25D

Crew No.	Bare Costs Hr.	Daily	Incl. Subs O&P Hr.	Daily	Cost Per Labor-Hour Bare Costs	Incl. O&P
1 Labor Foreman (outside)	$39.60	$316.80	$60.95	$487.60	$42.39	$64.83
3 Laborers	37.60	902.40	57.85	1388.40		
2.125 Equip. Oper. (medium)	50.60	860.20	76.75	1304.75		
.125 Truck Driver (heavy)	40.05	40.05	60.55	60.55		
.125 Truck Tractor, 6x4, 380 H.P.		74.97		82.47		
.125 Dist. Tanker, 3000 Gallon		38.20		42.02		
1 Asphalt Paver, 130 H.P.		2132.00		2345.20		
1 Tandem Roller, 10 Ton		236.80		260.48	49.64	54.60
50 L.H., Daily Totals		$4601.43		$5971.47	$92.03	$119.43

Crew B-25E

Crew No.	Bare Costs Hr.	Daily	Incl. Subs O&P Hr.	Daily	Cost Per Labor-Hour Bare Costs	Incl. O&P
1 Labor Foreman (outside)	$39.60	$316.80	$60.95	$487.60	$42.50	$64.97
3 Laborers	37.60	902.40	57.85	1388.40		
2.250 Equip. Oper. (medium)	50.60	910.80	76.75	1381.50		
.25 Truck Driver (heavy)	40.05	80.10	60.55	121.10		
.25 Truck Tractor, 6x4, 380 H.P.		149.95		164.94		
.25 Dist. Tanker, 3000 Gallon		76.40		84.04		
1 Asphalt Paver, 130 H.P.		2132.00		2345.20		
1 Tandem Roller, 10 Ton		236.80		260.48	49.91	54.90
52 L.H., Daily Totals		$4805.25		$6233.27	$92.41	$119.87

Crew B-26

Crew No.	Bare Costs Hr.	Daily	Incl. Subs O&P Hr.	Daily	Cost Per Labor-Hour Bare Costs	Incl. O&P
1 Labor Foreman (outside)	$39.60	$316.80	$60.95	$487.60	$42.18	$64.57
6 Laborers	37.60	1804.80	57.85	2776.80		
2 Equip. Oper. (medium)	50.60	809.60	76.75	1228.00		
1 Rodman (reinf.)	52.55	420.40	82.10	656.80		
1 Cement Finisher	45.00	360.00	66.65	533.20		
1 Grader, 30,000 Lbs.		736.20		809.82		
1 Paving Mach. & Equip.		2794.00		3073.40	40.12	44.13
88 L.H., Daily Totals		$7241.80		$9565.62	$82.29	$108.70

Crew B-26A

Crew No.	Bare Costs Hr.	Daily	Incl. Subs O&P Hr.	Daily	Cost Per Labor-Hour Bare Costs	Incl. O&P
1 Labor Foreman (outside)	$39.60	$316.80	$60.95	$487.60	$42.18	$64.57
6 Laborers	37.60	1804.80	57.85	2776.80		
2 Equip. Oper. (medium)	50.60	809.60	76.75	1228.00		
1 Rodman (reinf.)	52.55	420.40	82.10	656.80		
1 Cement Finisher	45.00	360.00	66.65	533.20		
1 Grader, 30,000 Lbs.		736.20		809.82		
1 Paving Mach. & Equip.		2794.00		3073.40		
1 Concrete Saw		167.40		184.14	42.02	46.22
88 L.H., Daily Totals		$7409.20		$9749.76	$84.20	$110.79

Crew B-26B

Crew No.	Bare Costs Hr.	Daily	Incl. Subs O&P Hr.	Daily	Cost Per Labor-Hour Bare Costs	Incl. O&P
1 Labor Foreman (outside)	$39.60	$316.80	$60.95	$487.60	$42.88	$65.59
6 Laborers	37.60	1804.80	57.85	2776.80		
3 Equip. Oper. (medium)	50.60	1214.40	76.75	1842.00		
1 Rodman (reinf.)	52.55	420.40	82.10	656.80		
1 Cement Finisher	45.00	360.00	66.65	533.20		
1 Grader, 30,000 Lbs.		736.20		809.82		
1 Paving Mach. & Equip.		2794.00		3073.40		
1 Concrete Pump, 110' Boom		946.20		1040.82	46.63	51.29
96 L.H., Daily Totals		$8592.80		$11220.44	$89.51	$116.88

Crew B-26C

Crew No.	Bare Costs Hr.	Daily	Incl. Subs O&P Hr.	Daily	Cost Per Labor-Hour Bare Costs	Incl. O&P
1 Labor Foreman (outside)	$39.60	$316.80	$60.95	$487.60	$41.34	$63.35
6 Laborers	37.60	1804.80	57.85	2776.80		
1 Equip. Oper. (medium)	50.60	404.80	76.75	614.00		
1 Rodman (reinf.)	52.55	420.40	82.10	656.80		
1 Cement Finisher	45.00	360.00	66.65	533.20		
1 Paving Mach. & Equip.		2794.00		3073.40		
1 Concrete Saw		167.40		184.14	37.02	40.72
80 L.H., Daily Totals		$6268.20		$8325.94	$78.35	$104.07

Crew B-27

Crew No.	Bare Costs Hr.	Daily	Incl. Subs O&P Hr.	Daily	Cost Per Labor-Hour Bare Costs	Incl. O&P
1 Labor Foreman (outside)	$39.60	$316.80	$60.95	$487.60	$38.10	$58.63
3 Laborers	37.60	902.40	57.85	1388.40		
1 Berm Machine		296.80		326.48	9.28	10.20
32 L.H., Daily Totals		$1516.00		$2202.48	$47.38	$68.83

Crew B-28

Crew No.	Bare Costs Hr.	Daily	Incl. Subs O&P Hr.	Daily	Cost Per Labor-Hour Bare Costs	Incl. O&P
2 Carpenters	$46.95	$751.20	$72.25	$1156.00	$43.83	$67.45
1 Laborer	37.60	300.80	57.85	462.80		
24 L.H., Daily Totals		$1052.00		$1618.80	$43.83	$67.45

Crew B-29

Crew No.	Bare Costs Hr.	Daily	Incl. Subs O&P Hr.	Daily	Cost Per Labor-Hour Bare Costs	Incl. O&P
1 Labor Foreman (outside)	$39.60	$316.80	$60.95	$487.60	$40.99	$62.76
4 Laborers	37.60	1203.20	57.85	1851.20		
1 Equip. Oper. (crane)	51.70	413.60	78.45	627.60		
1 Equip. Oper. (oiler)	45.20	361.60	68.55	548.40		
1 Gradall, 5/8 C.Y.		882.20		970.42	15.75	17.33
56 L.H., Daily Totals		$3177.40		$4485.22	$56.74	$80.09

Crew B-30

Crew No.	Hr.	Daily	Hr.	Daily	Bare Costs	Incl. O&P
1 Equip. Oper. (medium)	$50.60	$404.80	$76.75	$614.00	$43.57	$65.95
2 Truck Drivers (heavy)	40.05	640.80	60.55	968.80		
1 Hyd. Excavator, 1.5 C.Y.		1030.00		1133.00		
2 Dump Trucks, 12 C.Y., 400 H.P.		1382.00		1520.20	100.50	110.55
24 L.H., Daily Totals		$3457.60		$4236.00	$144.07	$176.50

Crew B-31

Crew No.	Hr.	Daily	Hr.	Daily	Bare Costs	Incl. O&P
1 Labor Foreman (outside)	$39.60	$316.80	$60.95	$487.60	$39.87	$61.35
3 Laborers	37.60	902.40	57.85	1388.40		
1 Carpenter	46.95	375.60	72.25	578.00		
1 Air Compressor, 250 cfm		201.40		221.54		
1 Sheeting Driver		5.75		6.33		
2 -50' Air Hoses, 1.5"		11.60		12.76	5.47	6.02
40 L.H., Daily Totals		$1813.55		$2694.63	$45.34	$67.37

Crew B-32

Crew No.	Hr.	Daily	Hr.	Daily	Bare Costs	Incl. O&P
1 Laborer	$37.60	$300.80	$57.85	$462.80	$47.35	$72.03
3 Equip. Oper. (medium)	50.60	1214.40	76.75	1842.00		
1 Grader, 30,000 Lbs.		736.20		809.82		
1 Tandem Roller, 10 Ton		236.80		260.48		
1 Dozer, 200 H.P.		1387.00		1525.70	73.75	81.13
32 L.H., Daily Totals		$3875.20		$4900.80	$121.10	$153.15

Crew B-32A

Crew No.	Hr.	Daily	Hr.	Daily	Bare Costs	Incl. O&P
1 Laborer	$37.60	$300.80	$57.85	$462.80	$46.27	$70.45
2 Equip. Oper. (medium)	50.60	809.60	76.75	1228.00		
1 Grader, 30,000 Lbs.		736.20		809.82		
1 Roller, Vibratory, 25 Ton		687.40		756.14	59.32	65.25
24 L.H., Daily Totals		$2534.00		$3256.76	$105.58	$135.70

Crew B-32B

Crew No.	Hr.	Daily	Hr.	Daily	Bare Costs	Incl. O&P
1 Laborer	$37.60	$300.80	$57.85	$462.80	$46.27	$70.45
2 Equip. Oper. (medium)	50.60	809.60	76.75	1228.00		
1 Dozer, 200 H.P.		1387.00		1525.70		
1 Roller, Vibratory, 25 Ton		687.40		756.14	86.43	95.08
24 L.H., Daily Totals		$3184.80		$3972.64	$132.70	$165.53

Crew B-32C

Crew No.	Hr.	Daily	Hr.	Daily	Bare Costs	Incl. O&P
1 Labor Foreman (outside)	$39.60	$316.80	$60.95	$487.60	$44.43	$67.82
2 Laborers	37.60	601.60	57.85	925.60		
3 Equip. Oper. (medium)	50.60	1214.40	76.75	1842.00		
1 Grader, 30,000 Lbs.		736.20		809.82		
1 Tandem Roller, 10 Ton		236.80		260.48		
1 Dozer, 200 H.P.		1387.00		1525.70	49.17	54.08
48 L.H., Daily Totals		$4492.80		$5851.20	$93.60	$121.90

Crew B-33A

Crew No.	Hr.	Daily	Hr.	Daily	Bare Costs	Incl. O&P
1 Equip. Oper. (medium)	$50.60	$404.80	$76.75	$614.00	$46.89	$71.35
.5 Laborer	37.60	150.40	57.85	231.40		
.25 Equip. Oper. (medium)	50.60	101.20	76.75	153.50		
1 Scraper, Towed, 7 C.Y.		114.40		125.84		
1.25 Dozers, 300 H.P.		2371.25		2608.38	177.55	195.30
14 L.H., Daily Totals		$3142.05		$3733.11	$224.43	$266.65

Crew B-33B

Crew No.	Hr.	Daily	Hr.	Daily	Bare Costs	Incl. O&P
1 Equip. Oper. (medium)	$50.60	$404.80	$76.75	$614.00	$46.89	$71.35
.5 Laborer	37.60	150.40	57.85	231.40		
.25 Equip. Oper. (medium)	50.60	101.20	76.75	153.50		
1 Scraper, Towed, 10 C.Y.		146.80		161.48		
1.25 Dozers, 300 H.P.		2371.25		2608.38	179.86	197.85
14 L.H., Daily Totals		$3174.45		$3768.76	$226.75	$269.20

Crew B-33C

Crew No.	Hr.	Daily	Hr.	Daily	Bare Costs	Incl. O&P
1 Equip. Oper. (medium)	$50.60	$404.80	$76.75	$614.00	$46.89	$71.35
.5 Laborer	37.60	150.40	57.85	231.40		
.25 Equip. Oper. (medium)	50.60	101.20	76.75	153.50		
1 Scraper, Towed, 15 C.Y.		164.80		181.28		
1.25 Dozers, 300 H.P.		2371.25		2608.38	181.15	199.26
14 L.H., Daily Totals		$3192.45		$3788.55	$228.03	$270.61

Crew B-33D

Crew No.	Hr.	Daily	Hr.	Daily	Bare Costs	Incl. O&P
1 Equip. Oper. (medium)	$50.60	$404.80	$76.75	$614.00	$46.89	$71.35
.5 Laborer	37.60	150.40	57.85	231.40		
.25 Equip. Oper. (medium)	50.60	101.20	76.75	153.50		
1 S.P. Scraper, 14 C.Y.		1884.00		2072.40		
.25 Dozer, 300 H.P.		474.25		521.67	168.45	185.29
14 L.H., Daily Totals		$3014.65		$3592.97	$215.33	$256.64

Crew B-33E

Crew No.	Hr.	Daily	Hr.	Daily	Bare Costs	Incl. O&P
1 Equip. Oper. (medium)	$50.60	$404.80	$76.75	$614.00	$46.89	$71.35
.5 Laborer	37.60	150.40	57.85	231.40		
.25 Equip. Oper. (medium)	50.60	101.20	76.75	153.50		
1 S.P. Scraper, 21 C.Y.		2688.00		2956.80		
.25 Dozer, 300 H.P.		474.25		521.67	225.88	248.46
14 L.H., Daily Totals		$3818.65		$4477.38	$272.76	$319.81

Crew B-33F

Crew No.	Hr.	Daily	Hr.	Daily	Bare Costs	Incl. O&P
1 Equip. Oper. (medium)	$50.60	$404.80	$76.75	$614.00	$46.89	$71.35
.5 Laborer	37.60	150.40	57.85	231.40		
.25 Equip. Oper. (medium)	50.60	101.20	76.75	153.50		
1 Elev. Scraper, 11 C.Y.		1163.00		1279.30		
.25 Dozer, 300 H.P.		474.25		521.67	116.95	128.64
14 L.H., Daily Totals		$2293.65		$2799.88	$163.83	$199.99

Crew B-33G

Crew No.	Hr.	Daily	Hr.	Daily	Bare Costs	Incl. O&P
1 Equip. Oper. (medium)	$50.60	$404.80	$76.75	$614.00	$46.89	$71.35
.5 Laborer	37.60	150.40	57.85	231.40		
.25 Equip. Oper. (medium)	50.60	101.20	76.75	153.50		
1 Elev. Scraper, 22 C.Y.		2521.00		2773.10		
.25 Dozer, 300 H.P.		474.25		521.67	213.95	235.34
14 L.H., Daily Totals		$3651.65		$4293.68	$260.83	$306.69

Crew B-33H

Crew No.	Hr.	Daily	Hr.	Daily	Bare Costs	Incl. O&P
.5 Laborer	$37.60	$150.40	$57.85	$231.40	$46.89	$71.35
1 Equipment Operator (med.)	50.60	404.80	76.75	614.00		
.25 Equipment Operator (med.)	50.60	101.20	76.75	153.50		
1 S.P. Scraper, 44 C.Y.		4679.00		5146.90		
.25 Dozer, 410 H.P.		601.25		661.38	377.16	414.88
14 L.H., Daily Totals		$5936.65		$6807.18	$424.05	$486.23

Crew B-33J

Crew No.	Hr.	Daily	Hr.	Daily	Bare Costs	Incl. O&P
1 Equipment Operator (med.)	$50.60	$404.80	$76.75	$614.00	$50.60	$76.75
1 S.P. Scraper, 14 C.Y.		1884.00		2072.40	235.50	259.05
8 L.H., Daily Totals		$2288.80		$2686.40	$286.10	$335.80

Crew B-33K

Crew No.	Hr.	Daily	Hr.	Daily	Bare Costs	Incl. O&P
1 Equipment Operator (med.)	$50.60	$404.80	$76.75	$614.00	$46.89	$71.35
.25 Equipment Operator (med.)	50.60	101.20	76.75	153.50		
.5 Laborer	37.60	150.40	57.85	231.40		
1 S.P. Scraper, 31 C.Y.		3693.00		4062.30		
.25 Dozer, 410 H.P.		601.25		661.38	306.73	337.41
14 L.H., Daily Totals		$4950.65		$5722.57	$353.62	$408.76

Crew No.	Bare Costs Hr.	Daily	Incl. Subs O&P Hr.	Daily	Cost Per Labor-Hour Bare Costs	Incl. O&P
Crew B-34A	Hr.	Daily	Hr.	Daily	Bare Costs	Incl. O&P
1 Truck Driver (heavy)	$40.05	$320.40	$60.55	$484.40	$40.05	$60.55
1 Dump Truck, 8 C.Y., 220 H.P.		411.20		452.32	51.40	56.54
8 L.H., Daily Totals		$731.60		$936.72	$91.45	$117.09
Crew B-34B	Hr.	Daily	Hr.	Daily	Bare Costs	Incl. O&P
1 Truck Driver (heavy)	$40.05	$320.40	$60.55	$484.40	$40.05	$60.55
1 Dump Truck, 12 C.Y., 400 H.P.		691.00		760.10	86.38	95.01
8 L.H., Daily Totals		$1011.40		$1244.50	$126.43	$155.56
Crew B-34C	Hr.	Daily	Hr.	Daily	Bare Costs	Incl. O&P
1 Truck Driver (heavy)	$40.05	$320.40	$60.55	$484.40	$40.05	$60.55
1 Truck Tractor, 6x4, 380 H.P.		599.80		659.78		
1 Dump Trailer, 16.5 C.Y.		126.60		139.26	90.80	99.88
8 L.H., Daily Totals		$1046.80		$1283.44	$130.85	$160.43
Crew B-34D	Hr.	Daily	Hr.	Daily	Bare Costs	Incl. O&P
1 Truck Driver (heavy)	$40.05	$320.40	$60.55	$484.40	$40.05	$60.55
1 Truck Tractor, 6x4, 380 H.P.		599.80		659.78		
1 Dump Trailer, 20 C.Y.		141.20		155.32	92.63	101.89
8 L.H., Daily Totals		$1061.40		$1299.50	$132.68	$162.44
Crew B-34E	Hr.	Daily	Hr.	Daily	Bare Costs	Incl. O&P
1 Truck Driver (heavy)	$40.05	$320.40	$60.55	$484.40	$40.05	$60.55
1 Dump Truck, Off Hwy., 25 Ton		1351.00		1486.10	168.88	185.76
8 L.H., Daily Totals		$1671.40		$1970.50	$208.93	$246.31
Crew B-34F	Hr.	Daily	Hr.	Daily	Bare Costs	Incl. O&P
1 Truck Driver (heavy)	$40.05	$320.40	$60.55	$484.40	$40.05	$60.55
1 Dump Truck, Off Hwy., 35 Ton		1512.00		1663.20	189.00	207.90
8 L.H., Daily Totals		$1832.40		$2147.60	$229.05	$268.45
Crew B-34G	Hr.	Daily	Hr.	Daily	Bare Costs	Incl. O&P
1 Truck Driver (heavy)	$40.05	$320.40	$60.55	$484.40	$40.05	$60.55
1 Dump Truck, Off Hwy., 50 Ton		1837.00		2020.70	229.63	252.59
8 L.H., Daily Totals		$2157.40		$2505.10	$269.68	$313.14
Crew B-34H	Hr.	Daily	Hr.	Daily	Bare Costs	Incl. O&P
1 Truck Driver (heavy)	$40.05	$320.40	$60.55	$484.40	$40.05	$60.55
1 Dump Truck, Off Hwy., 65 Ton		1863.00		2049.30	232.88	256.16
8 L.H., Daily Totals		$2183.40		$2533.70	$272.93	$316.71
Crew B-34I	Hr.	Daily	Hr.	Daily	Bare Costs	Incl. O&P
1 Truck Driver (heavy)	$40.05	$320.40	$60.55	$484.40	$40.05	$60.55
1 Dump Truck, 18 C.Y., 450 H.P.		869.00		955.90	108.63	119.49
8 L.H., Daily Totals		$1189.40		$1440.30	$148.68	$180.04
Crew B-34J	Hr.	Daily	Hr.	Daily	Bare Costs	Incl. O&P
1 Truck Driver (heavy)	$40.05	$320.40	$60.55	$484.40	$40.05	$60.55
1 Dump Truck, Off Hwy., 100 Ton		2925.00		3217.50	365.63	402.19
8 L.H., Daily Totals		$3245.40		$3701.90	$405.68	$462.74
Crew B-34K	Hr.	Daily	Hr.	Daily	Bare Costs	Incl. O&P
1 Truck Driver (heavy)	$40.05	$320.40	$60.55	$484.40	$40.05	$60.55
1 Truck Tractor, 6x4, 450 H.P.		727.40		800.14		
1 Lowbed Trailer, 75 Ton		221.80		243.98	118.65	130.51
8 L.H., Daily Totals		$1269.60		$1528.52	$158.70	$191.07

Crew No.	Bare Costs Hr.	Daily	Incl. Subs O&P Hr.	Daily	Cost Per Labor-Hour Bare Costs	Incl. O&P
Crew B-34L	Hr.	Daily	Hr.	Daily	Bare Costs	Incl. O&P
1 Equip. Oper. (light)	$48.60	$388.80	$73.75	$590.00	$48.60	$73.75
1 Flatbed Truck, Gas, 1.5 Ton		245.40		269.94	30.68	33.74
8 L.H., Daily Totals		$634.20		$859.94	$79.28	$107.49
Crew B-34M	Hr.	Daily	Hr.	Daily	Bare Costs	Incl. O&P
1 Equip. Oper. (light)	$48.60	$388.80	$73.75	$590.00	$48.60	$73.75
1 Flatbed Truck, Gas, 3 Ton		303.20		333.52	37.90	41.69
8 L.H., Daily Totals		$692.00		$923.52	$86.50	$115.44
Crew B-34N	Hr.	Daily	Hr.	Daily	Bare Costs	Incl. O&P
1 Truck Driver (heavy)	$40.05	$320.40	$60.55	$484.40	$45.33	$68.65
1 Equip. Oper. (medium)	50.60	404.80	76.75	614.00		
1 Truck Tractor, 6x4, 380 H.P.		599.80		659.78		
1 Flatbed Trailer, 40 Ton		154.00		169.40	47.11	51.82
16 L.H., Daily Totals		$1479.00		$1927.58	$92.44	$120.47
Crew B-34P	Hr.	Daily	Hr.	Daily	Bare Costs	Incl. O&P
1 Pipe Fitter	$59.75	$478.00	$90.20	$721.60	$49.82	$75.35
1 Truck Driver (light)	39.10	312.80	59.10	472.80		
1 Equip. Oper. (medium)	50.60	404.80	76.75	614.00		
1 Flatbed Truck, Gas, 3 Ton		303.20		333.52		
1 Backhoe Loader, 48 H.P.		364.20		400.62	27.81	30.59
24 L.H., Daily Totals		$1863.00		$2542.54	$77.63	$105.94
Crew B-34Q	Hr.	Daily	Hr.	Daily	Bare Costs	Incl. O&P
1 Pipe Fitter	$59.75	$478.00	$90.20	$721.60	$50.18	$75.92
1 Truck Driver (light)	39.10	312.80	59.10	472.80		
1 Equip. Oper. (crane)	51.70	413.60	78.45	627.60		
1 Flatbed Trailer, 25 Ton		113.60		124.96		
1 Dump Truck, 8 C.Y., 220 H.P.		411.20		452.32		
1 Hyd. Crane, 25 Ton		736.60		810.26	52.56	57.81
24 L.H., Daily Totals		$2465.80		$3209.54	$102.74	$133.73
Crew B-34R	Hr.	Daily	Hr.	Daily	Bare Costs	Incl. O&P
1 Pipe Fitter	$59.75	$478.00	$90.20	$721.60	$50.18	$75.92
1 Truck Driver (light)	39.10	312.80	59.10	472.80		
1 Equip. Oper. (crane)	51.70	413.60	78.45	627.60		
1 Flatbed Trailer, 25 Ton		113.60		124.96		
1 Dump Truck, 8 C.Y., 220 H.P.		411.20		452.32		
1 Hyd. Crane, 25 Ton		736.60		810.26		
1 Hyd. Excavator, 1 C.Y.		812.60		893.86	86.42	95.06
24 L.H., Daily Totals		$3278.40		$4103.40	$136.60	$170.97
Crew B-34S	Hr.	Daily	Hr.	Daily	Bare Costs	Incl. O&P
2 Pipe Fitters	$59.75	$956.00	$90.20	$1443.20	$52.81	$79.85
1 Truck Driver (heavy)	40.05	320.40	60.55	484.40		
1 Equip. Oper. (crane)	51.70	413.60	78.45	627.60		
1 Flatbed Trailer, 40 Ton		154.00		169.40		
1 Truck Tractor, 6x4, 380 H.P.		599.80		659.78		
1 Hyd. Crane, 80 Ton		1625.00		1787.50		
1 Hyd. Excavator, 2 C.Y.		1175.00		1292.50	111.06	122.16
32 L.H., Daily Totals		$5243.80		$6464.38	$163.87	$202.01
Crew B-34T	Hr.	Daily	Hr.	Daily	Bare Costs	Incl. O&P
2 Pipe Fitters	$59.75	$956.00	$90.20	$1443.20	$52.81	$79.85
1 Truck Driver (heavy)	40.05	320.40	60.55	484.40		
1 Equip. Oper. (crane)	51.70	413.60	78.45	627.60		
1 Flatbed Trailer, 40 Ton		154.00		169.40		
1 Truck Tractor, 6x4, 380 H.P.		599.80		659.78		
1 Hyd. Crane, 80 Ton		1625.00		1787.50	74.34	81.77
32 L.H., Daily Totals		$4068.80		$5171.88	$127.15	$161.62

Crew No.	Bare Costs		Incl. Subs O&P		Cost Per Labor-Hour	
	Hr.	Daily	Hr.	Daily	Bare Costs	Incl. O&P
Crew B-34U	Hr.	Daily	Hr.	Daily	Bare Costs	Incl. O&P
1 Truck Driver (heavy)	$40.05	$320.40	$60.55	$484.40	$44.33	$67.15
1 Equip. Oper. (light)	48.60	388.80	73.75	590.00		
1 Truck Tractor, 220 H.P.		359.60		395.56		
1 Flatbed Trailer, 25 Ton		113.60		124.96	29.57	32.53
16 L.H., Daily Totals		$1182.40		$1594.92	$73.90	$99.68
Crew B-34V	Hr.	Daily	Hr.	Daily	Bare Costs	Incl. O&P
1 Truck Driver (heavy)	$40.05	$320.40	$60.55	$484.40	$46.78	$70.92
1 Equip. Oper. (crane)	51.70	413.60	78.45	627.60		
1 Equip. Oper. (light)	48.60	388.80	73.75	590.00		
1 Truck Tractor, 6x4, 450 H.P.		727.40		800.14		
1 Equipment Trailer, 50 Ton		168.40		185.24		
1 Pickup Truck, 4 x 4, 3/4 Ton		155.60		171.16	43.81	48.19
24 L.H., Daily Totals		$2174.20		$2858.54	$90.59	$119.11
Crew B-34W	Hr.	Daily	Hr.	Daily	Bare Costs	Incl. O&P
5 Truck Drivers (heavy)	$40.05	$1602.00	$60.55	$2422.00	$43.66	$66.21
2 Equip. Opers. (crane)	51.70	827.20	78.45	1255.20		
1 Equip. Oper. (mechanic)	51.65	413.20	78.35	626.80		
1 Laborer	37.60	300.80	57.85	462.80		
4 Truck Tractor, 6x4, 380 H.P.		2399.20		2639.12		
2 Equipment Trailer, 50 Ton		336.80		370.48		
2 Flatbed Trailer, 40 Ton		308.00		338.80		
1 Pickup Truck, 4 x 4, 3/4 Ton		155.60		171.16		
1 S.P. Crane, 4x4, 20 Ton		554.60		610.06	52.14	57.36
72 L.H., Daily Totals		$6897.40		$8896.42	$95.80	$123.56
Crew B-35	Hr.	Daily	Hr.	Daily	Bare Costs	Incl. O&P
1 Labor Foreman (outside)	$39.60	$316.80	$60.95	$487.60	$46.91	$71.60
1 Skilled Worker	48.65	389.20	75.15	601.20		
1 Welder (plumber)	58.70	469.60	88.65	709.20		
1 Laborer	37.60	300.80	57.85	462.80		
1 Equip. Oper. (crane)	51.70	413.60	78.45	627.60		
1 Equip. Oper. (oiler)	45.20	361.60	68.55	548.40		
1 Welder, Electric, 300 amp		57.70		63.47		
1 Hyd. Excavator, .75 C.Y.		654.80		720.28	14.84	16.33
48 L.H., Daily Totals		$2964.10		$4220.55	$61.75	$87.93
Crew B-35A	Hr.	Daily	Hr.	Daily	Bare Costs	Incl. O&P
1 Labor Foreman (outside)	$39.60	$316.80	$60.95	$487.60	$45.58	$69.64
2 Laborers	37.60	601.60	57.85	925.60		
1 Skilled Worker	48.65	389.20	75.15	601.20		
1 Welder (plumber)	58.70	469.60	88.65	709.20		
1 Equip. Oper. (crane)	51.70	413.60	78.45	627.60		
1 Equip. Oper. (oiler)	45.20	361.60	68.55	548.40		
1 Welder, Gas Engine, 300 amp		145.85		160.44		
1 Crawler Crane, 75 Ton		1459.00		1604.90	28.66	31.52
56 L.H., Daily Totals		$4157.25		$5664.94	$74.24	$101.16
Crew B-36	Hr.	Daily	Hr.	Daily	Bare Costs	Incl. O&P
1 Labor Foreman (outside)	$39.60	$316.80	$60.95	$487.60	$43.20	$66.03
2 Laborers	37.60	601.60	57.85	925.60		
2 Equip. Oper. (medium)	50.60	809.60	76.75	1228.00		
1 Dozer, 200 H.P.		1387.00		1525.70		
1 Aggregate Spreader		39.60		43.56		
1 Tandem Roller, 10 Ton		236.80		260.48	41.59	45.74
40 L.H., Daily Totals		$3391.40		$4470.94	$84.78	$111.77

Crew No.	Bare Costs		Incl. Subs O&P		Cost Per Labor-Hour	
	Hr.	Daily	Hr.	Daily	Bare Costs	Incl. O&P
Crew B-36A	Hr.	Daily	Hr.	Daily	Bare Costs	Incl. O&P
1 Labor Foreman (outside)	$39.60	$316.80	$60.95	$487.60	$45.31	$69.09
2 Laborers	37.60	601.60	57.85	925.60		
4 Equip. Oper. (medium)	50.60	1619.20	76.75	2456.00		
1 Dozer, 200 H.P.		1387.00		1525.70		
1 Aggregate Spreader		39.60		43.56		
1 Tandem Roller, 10 Ton		236.80		260.48		
1 Roller, Pneum. Whl., 12 Ton		344.40		378.84	35.85	39.44
56 L.H., Daily Totals		$4545.40		$6077.78	$81.17	$108.53
Crew B-36B	Hr.	Daily	Hr.	Daily	Bare Costs	Incl. O&P
1 Labor Foreman (outside)	$39.60	$316.80	$60.95	$487.60	$44.66	$68.03
2 Laborers	37.60	601.60	57.85	925.60		
4 Equip. Oper. (medium)	50.60	1619.20	76.75	2456.00		
1 Truck Driver (heavy)	40.05	320.40	60.55	484.40		
1 Grader, 30,000 Lbs.		736.20		809.82		
1 F.E. Loader, Crl, 1.5 C.Y.		629.80		692.78		
1 Dozer, 300 H.P.		1897.00		2086.70		
1 Roller, Vibratory, 25 Ton		687.40		756.14		
1 Truck Tractor, 6x4, 450 H.P.		727.40		800.14		
1 Water Tank Trailer, 5000 Gal.		141.00		155.10	75.29	82.82
64 L.H., Daily Totals		$7676.80		$9654.28	$119.95	$150.85
Crew B-36C	Hr.	Daily	Hr.	Daily	Bare Costs	Incl. O&P
1 Labor Foreman (outside)	$39.60	$316.80	$60.95	$487.60	$46.29	$70.35
3 Equip. Oper. (medium)	50.60	1214.40	76.75	1842.00		
1 Truck Driver (heavy)	40.05	320.40	60.55	484.40		
1 Grader, 30,000 Lbs.		736.20		809.82		
1 Dozer, 300 H.P.		1897.00		2086.70		
1 Roller, Vibratory, 25 Ton		687.40		756.14		
1 Truck Tractor, 6x4, 450 H.P.		727.40		800.14		
1 Water Tank Trailer, 5000 Gal.		141.00		155.10	104.72	115.20
40 L.H., Daily Totals		$6040.60		$7421.90	$151.01	$185.55
Crew B-36D	Hr.	Daily	Hr.	Daily	Bare Costs	Incl. O&P
1 Labor Foreman (outside)	$39.60	$316.80	$60.95	$487.60	$47.85	$72.80
3 Equip. Oper. (medium)	50.60	1214.40	76.75	1842.00		
1 Grader, 30,000 Lbs.		736.20		809.82		
1 Dozer, 300 H.P.		1897.00		2086.70		
1 Roller, Vibratory, 25 Ton		687.40		756.14	103.77	114.15
32 L.H., Daily Totals		$4851.80		$5982.26	$151.62	$186.95
Crew B-37	Hr.	Daily	Hr.	Daily	Bare Costs	Incl. O&P
1 Labor Foreman (outside)	$39.60	$316.80	$60.95	$487.60	$39.77	$61.02
4 Laborers	37.60	1203.20	57.85	1851.20		
1 Equip. Oper. (light)	48.60	388.80	73.75	590.00		
1 Tandem Roller, 5 Ton		157.40		173.14	3.28	3.61
48 L.H., Daily Totals		$2066.20		$3101.94	$43.05	$64.62
Crew B-37A	Hr.	Daily	Hr.	Daily	Bare Costs	Incl. O&P
2 Laborers	$37.60	$601.60	$57.85	$925.60	$38.10	$58.27
1 Truck Driver (light)	39.10	312.80	59.10	472.80		
1 Flatbed Truck, Gas, 1.5 Ton		245.40		269.94		
1 Tar Kettle, T.M.		132.00		145.20	15.73	17.30
24 L.H., Daily Totals		$1291.80		$1813.54	$53.83	$75.56
Crew B-37B	Hr.	Daily	Hr.	Daily	Bare Costs	Incl. O&P
3 Laborers	$37.60	$902.40	$57.85	$1388.40	$37.98	$58.16
1 Truck Driver (light)	39.10	312.80	59.10	472.80		
1 Flatbed Truck, Gas, 1.5 Ton		245.40		269.94		
1 Tar Kettle, T.M.		132.00		145.20	11.79	12.97
32 L.H., Daily Totals		$1592.60		$2276.34	$49.77	$71.14

579

Crew No.	Bare Costs		Incl. Subs O&P		Cost Per Labor-Hour	
Crew B-37C	Hr.	Daily	Hr.	Daily	Bare Costs	Incl. O&P
2 Laborers	$37.60	$601.60	$57.85	$925.60	$38.35	$58.48
2 Truck Drivers (light)	39.10	625.60	59.10	945.60		
2 Flatbed Trucks, Gas, 1.5 Ton		490.80		539.88		
1 Tar Kettle, T.M.		132.00		145.20	19.46	21.41
32 L.H., Daily Totals		$1850.00		$2556.28	$57.81	$79.88

	Hr.	Daily	Hr.	Daily	Bare Costs	Incl. O&P
Crew B-37D						
1 Laborer	$37.60	$300.80	$57.85	$462.80	$38.35	$58.48
1 Truck Driver (light)	39.10	312.80	59.10	472.80		
1 Pickup Truck, 3/4 Ton		144.20		158.62	9.01	9.91
16 L.H., Daily Totals		$757.80		$1094.22	$47.36	$68.39

	Hr.	Daily	Hr.	Daily	Bare Costs	Incl. O&P
Crew B-37E						
3 Laborers	$37.60	$902.40	$57.85	$1388.40	$41.46	$63.18
1 Equip. Oper. (light)	48.60	388.80	73.75	590.00		
1 Equip. Oper. (medium)	50.60	404.80	76.75	614.00		
2 Truck Drivers (light)	39.10	625.60	59.10	945.60		
4 Barrels w/ Flasher		13.60		14.96		
1 Concrete Saw		167.40		184.14		
1 Rotary Hammer Drill		23.75		26.13		
1 Hammer Drill Bit		2.35		2.59		
1 Loader, Skid Steer, 30 H.P.		173.60		190.96		
1 Conc. Hammer Attach.		114.70		126.17		
1 Vibrating Plate, Gas, 18"		36.20		39.82		
2 Flatbed Trucks, Gas, 1.5 Ton		490.80		539.88	18.26	20.08
56 L.H., Daily Totals		$3344.00		$4662.64	$59.71	$83.26

	Hr.	Daily	Hr.	Daily	Bare Costs	Incl. O&P
Crew B-37F						
3 Laborers	$37.60	$902.40	$57.85	$1388.40	$37.98	$58.16
1 Truck Driver (light)	39.10	312.80	59.10	472.80		
4 Barrels w/ Flasher		13.60		14.96		
1 Concrete Mixer, 10 C.F.		172.40		189.64		
1 Air Compressor, 60 cfm		137.40		151.14		
1 -50' Air Hose, 3/4"		3.25		3.58		
1 Spade (Chipper)		7.60		8.36		
1 Flatbed Truck, Gas, 1.5 Ton		245.40		269.94	18.11	19.93
32 L.H., Daily Totals		$1794.85		$2498.82	$56.09	$78.09

	Hr.	Daily	Hr.	Daily	Bare Costs	Incl. O&P
Crew B-37G						
1 Labor Foreman (outside)	$39.60	$316.80	$60.95	$487.60	$39.77	$61.02
4 Laborers	37.60	1203.20	57.85	1851.20		
1 Equip. Oper. (light)	48.60	388.80	73.75	590.00		
1 Berm Machine		296.80		326.48		
1 Tandem Roller, 5 Ton		157.40		173.14	9.46	10.41
48 L.H., Daily Totals		$2363.00		$3428.42	$49.23	$71.43

	Hr.	Daily	Hr.	Daily	Bare Costs	Incl. O&P
Crew B-37H						
1 Labor Foreman (outside)	$39.60	$316.80	$60.95	$487.60	$39.77	$61.02
4 Laborers	37.60	1203.20	57.85	1851.20		
1 Equip. Oper. (light)	48.60	388.80	73.75	590.00		
1 Tandem Roller, 5 Ton		157.40		173.14		
1 Flatbed Trucks, Gas, 1.5 Ton		245.40		269.94		
1 Tar Kettle, T.M.		132.00		145.20	11.14	12.26
48 L.H., Daily Totals		$2443.60		$3517.08	$50.91	$73.27

Crew No.	Bare Costs		Incl. Subs O&P		Cost Per Labor-Hour	
Crew B-37I	Hr.	Daily	Hr.	Daily	Bare Costs	Incl. O&P
3 Laborers	$37.60	$902.40	$57.85	$1388.40	$41.46	$63.18
1 Equip. Oper. (light)	48.60	388.80	73.75	590.00		
1 Equip. Oper. (medium)	50.60	404.80	76.75	614.00		
2 Truck Drivers (light)	39.10	625.60	59.10	945.60		
4 Barrels w/ Flasher		13.60		14.96		
1 Concrete Saw		167.40		184.14		
1 Rotary Hammer Drill		23.75		26.13		
1 Hammer Drill Bit		2.35		2.59		
1 Air Compressor, 60 cfm		137.40		151.14		
1 -50' Air Hose, 3/4"		3.25		3.58		
1 Spade (Chipper)		7.60		8.36		
1 Loader, Skid Steer, 30 H.P.		173.60		190.96		
1 Conc. Hammer Attach.		114.70		126.17		
1 Concrete Mixer, 10 C.F.		172.40		189.64		
1 Vibrating Plate, Gas, 18"		36.20		39.82		
2 Flatbed Trucks, Gas, 1.5 Ton		490.80		539.88	23.98	26.38
56 L.H., Daily Totals		$3664.65		$5015.35	$65.44	$89.56

	Hr.	Daily	Hr.	Daily	Bare Costs	Incl. O&P
Crew B-37J						
1 Labor Foreman (outside)	$39.60	$316.80	$60.95	$487.60	$39.77	$61.02
4 Laborers	37.60	1203.20	57.85	1851.20		
1 Equip. Oper. (light)	48.60	388.80	73.75	590.00		
1 Air Compressor, 60 cfm		137.40		151.14		
1 -50' Air Hose, 3/4"		3.25		3.58		
2 Concrete Mixer, 10 C.F.		344.80		379.28		
2 Flatbed Trucks, Gas, 1.5 Ton		490.80		539.88		
1 Shot Blaster, 20"		214.00		235.40	24.80	27.28
48 L.H., Daily Totals		$3099.05		$4238.07	$64.56	$88.29

	Hr.	Daily	Hr.	Daily	Bare Costs	Incl. O&P
Crew B-37K						
1 Labor Foreman (outside)	$39.60	$316.80	$60.95	$487.60	$39.77	$61.02
4 Laborers	37.60	1203.20	57.85	1851.20		
1 Equip. Oper. (light)	48.60	388.80	73.75	590.00		
1 Air Compressor, 60 cfm		137.40		151.14		
1 -50' Air Hose, 3/4"		3.25		3.58		
2 Flatbed Trucks, Gas, 1.5 Ton		490.80		539.88		
1 Shot Blaster, 20"		214.00		235.40	17.61	19.37
48 L.H., Daily Totals		$2754.25		$3858.80	$57.38	$80.39

	Hr.	Daily	Hr.	Daily	Bare Costs	Incl. O&P
Crew B-38						
1 Labor Foreman (outside)	$39.60	$316.80	$60.95	$487.60	$42.80	$65.43
2 Laborers	37.60	601.60	57.85	925.60		
1 Equip. Oper. (light)	48.60	388.80	73.75	590.00		
1 Equip. Oper. (medium)	50.60	404.80	76.75	614.00		
1 Backhoe Loader, 48 H.P.		364.20		400.62		
1 Hyd. Hammer, (1200 lb.)		180.40		198.44		
1 F.E. Loader, W.M., 4 C.Y.		669.20		736.12		
1 Pvmt. Rem. Bucket		61.20		67.32	31.88	35.06
40 L.H., Daily Totals		$2987.00		$4019.70	$74.67	$100.49

	Hr.	Daily	Hr.	Daily	Bare Costs	Incl. O&P
Crew B-39						
1 Labor Foreman (outside)	$39.60	$316.80	$60.95	$487.60	$39.77	$61.02
4 Laborers	37.60	1203.20	57.85	1851.20		
1 Equip. Oper. (light)	48.60	388.80	73.75	590.00		
1 Air Compressor, 250 cfm		201.40		221.54		
2 Breakers, Pavement, 60 lb.		19.60		21.56		
2 -50' Air Hoses, 1.5"		11.60		12.76	4.85	5.33
48 L.H., Daily Totals		$2141.40		$3184.66	$44.61	$66.35

For customer support on your Plumbing Cost Data, call 877.709.6968.

Crew No.	Bare Costs Hr.	Daily	Incl. Subs O&P Hr.	Daily	Cost Per Labor-Hour Bare Costs	Incl. O&P
Crew B-40	Hr.	Daily	Hr.	Daily	Bare Costs	Incl. O&P
1 Pile Driver Foreman (outside)	$48.10	$384.80	$76.15	$609.20	$47.64	$74.20
4 Pile Drivers	46.10	1475.20	73.00	2336.00		
2 Equip. Oper. (crane)	51.70	827.20	78.45	1255.20		
1 Equip. Oper. (oiler)	45.20	361.60	68.55	548.40		
1 Crawler Crane, 40 Ton		1156.00		1271.60		
1 Vibratory Hammer & Gen.		2587.00		2845.70	58.48	64.33
64 L.H., Daily Totals		$6791.80		$8866.10	$106.12	$138.53
Crew B-40B	Hr.	Daily	Hr.	Daily	Bare Costs	Incl. O&P
1 Labor Foreman (outside)	$39.60	$316.80	$60.95	$487.60	$41.55	$63.58
3 Laborers	37.60	902.40	57.85	1388.40		
1 Equip. Oper. (crane)	51.70	413.60	78.45	627.60		
1 Equip. Oper. (oiler)	45.20	361.60	68.55	548.40		
1 Lattice Boom Crane, 40 Ton		1164.00		1280.40	24.25	26.68
48 L.H., Daily Totals		$3158.40		$4332.40	$65.80	$90.26
Crew B-41	Hr.	Daily	Hr.	Daily	Bare Costs	Incl. O&P
1 Labor Foreman (outside)	$39.60	$316.80	$60.95	$487.60	$38.95	$59.84
4 Laborers	37.60	1203.20	57.85	1851.20		
.25 Equip. Oper. (crane)	51.70	103.40	78.45	156.90		
.25 Equip. Oper. (oiler)	45.20	90.40	68.55	137.10		
.25 Crawler Crane, 40 Ton		289.00		317.90	6.57	7.22
44 L.H., Daily Totals		$2002.80		$2950.70	$45.52	$67.06
Crew B-42	Hr.	Daily	Hr.	Daily	Bare Costs	Incl. O&P
1 Labor Foreman (outside)	$39.60	$316.80	$60.95	$487.60	$42.44	$66.35
4 Laborers	37.60	1203.20	57.85	1851.20		
1 Equip. Oper. (crane)	51.70	413.60	78.45	627.60		
1 Equip. Oper. (oiler)	45.20	361.60	68.55	548.40		
1 Welder	52.65	421.20	91.45	731.60		
1 Hyd. Crane, 25 Ton		736.60		810.26		
1 Welder, Gas Engine, 300 amp		145.85		160.44		
1 Horz. Boring Csg. Mch.		474.80		522.28	21.21	23.33
64 L.H., Daily Totals		$4073.65		$5739.38	$63.65	$89.68
Crew B-43	Hr.	Daily	Hr.	Daily	Bare Costs	Incl. O&P
1 Labor Foreman (outside)	$39.60	$316.80	$60.95	$487.60	$41.55	$63.58
3 Laborers	37.60	902.40	57.85	1388.40		
1 Equip. Oper. (crane)	51.70	413.60	78.45	627.60		
1 Equip. Oper. (oiler)	45.20	361.60	68.55	548.40		
1 Drill Rig, Truck-Mounted		2542.00		2796.20	52.96	58.25
48 L.H., Daily Totals		$4536.40		$5848.20	$94.51	$121.84
Crew B-44	Hr.	Daily	Hr.	Daily	Bare Costs	Incl. O&P
1 Pile Driver Foreman (outside)	$48.10	$384.80	$76.15	$609.20	$46.69	$72.86
4 Pile Drivers	46.10	1475.20	73.00	2336.00		
2 Equip. Oper. (crane)	51.70	827.20	78.45	1255.20		
1 Laborer	37.60	300.80	57.85	462.80		
1 Crawler Crane, 40 Ton		1156.00		1271.60		
1 Lead, 60' High		74.80		82.28		
1 Hammer, Diesel, 15K ft.-lbs.		586.00		644.60	28.39	31.23
64 L.H., Daily Totals		$4804.80		$6661.68	$75.08	$104.09
Crew B-45	Hr.	Daily	Hr.	Daily	Bare Costs	Incl. O&P
1 Equip. Oper. (medium)	$50.60	$404.80	$76.75	$614.00	$45.33	$68.65
1 Truck Driver (heavy)	40.05	320.40	60.55	484.40		
1 Dist. Tanker, 3000 Gallon		305.60		336.16		
1 Truck Tractor, 6x4, 380 H.P.		599.80		659.78	56.59	62.25
16 L.H., Daily Totals		$1630.60		$2094.34	$101.91	$130.90

Crew No.	Bare Costs Hr.	Daily	Incl. Subs O&P Hr.	Daily	Cost Per Labor-Hour Bare Costs	Incl. O&P
Crew B-46	Hr.	Daily	Hr.	Daily	Bare Costs	Incl. O&P
1 Pile Driver Foreman (outside)	$48.10	$384.80	$76.15	$609.20	$42.18	$65.95
2 Pile Drivers	46.10	737.60	73.00	1168.00		
3 Laborers	37.60	902.40	57.85	1388.40		
1 Chain Saw, Gas, 36" Long		45.00		49.50	.94	1.03
48 L.H., Daily Totals		$2069.80		$3215.10	$43.12	$66.98
Crew B-47	Hr.	Daily	Hr.	Daily	Bare Costs	Incl. O&P
1 Blast Foreman (outside)	$39.60	$316.80	$60.95	$487.60	$41.93	$64.18
1 Driller	37.60	300.80	57.85	462.80		
1 Equip. Oper. (light)	48.60	388.80	73.75	590.00		
1 Air Track Drill, 4"		1028.00		1130.80		
1 Air Compressor, 600 cfm		550.60		605.66		
2 -50' Air Hoses, 3"		29.80		32.78	67.02	73.72
24 L.H., Daily Totals		$2614.80		$3309.64	$108.95	$137.90
Crew B-47A	Hr.	Daily	Hr.	Daily	Bare Costs	Incl. O&P
1 Drilling Foreman (outside)	$39.60	$316.80	$60.95	$487.60	$45.50	$69.32
1 Equip. Oper. (heavy)	51.70	413.60	78.45	627.60		
1 Equip. Oper. (oiler)	45.20	361.60	68.55	548.40		
1 Air Track Drill, 5"		1243.00		1367.30	51.79	56.97
24 L.H., Daily Totals		$2335.00		$3030.90	$97.29	$126.29
Crew B-47C	Hr.	Daily	Hr.	Daily	Bare Costs	Incl. O&P
1 Laborer	$37.60	$300.80	$57.85	$462.80	$43.10	$65.80
1 Equip. Oper. (light)	48.60	388.80	73.75	590.00		
1 Air Compressor, 750 cfm		557.20		612.92		
2 -50' Air Hoses, 3"		29.80		32.78		
1 Air Track Drill, 4"		1028.00		1130.80	100.94	111.03
16 L.H., Daily Totals		$2304.60		$2829.30	$144.04	$176.83
Crew B-47E	Hr.	Daily	Hr.	Daily	Bare Costs	Incl. O&P
1 Labor Foreman (outside)	$39.60	$316.80	$60.95	$487.60	$38.10	$58.63
3 Laborers	37.60	902.40	57.85	1388.40		
1 Flatbed Truck, Gas, 3 Ton		303.20		333.52	9.47	10.42
32 L.H., Daily Totals		$1522.40		$2209.52	$47.58	$69.05
Crew B-47G	Hr.	Daily	Hr.	Daily	Bare Costs	Incl. O&P
1 Labor Foreman (outside)	$39.60	$316.80	$60.95	$487.60	$40.85	$62.60
2 Laborers	37.60	601.60	57.85	925.60		
1 Equip. Oper. (light)	48.60	388.80	73.75	590.00		
1 Air Track Drill, 4"		1028.00		1130.80		
1 Air Compressor, 600 cfm		550.60		605.66		
2 -50' Air Hoses, 3"		29.80		32.78		
1 Gunite Pump Rig		371.80		408.98	61.88	68.07
32 L.H., Daily Totals		$3287.40		$4181.42	$102.73	$130.67
Crew B-47H	Hr.	Daily	Hr.	Daily	Bare Costs	Incl. O&P
1 Skilled Worker Foreman (out)	$50.65	$405.20	$78.25	$626.00	$49.15	$75.92
3 Skilled Workers	48.65	1167.60	75.15	1803.60		
1 Flatbed Truck, Gas, 3 Ton		303.20		333.52	9.47	10.42
32 L.H., Daily Totals		$1876.00		$2763.12	$58.63	$86.35

Crew No.	Bare Costs		Incl. Subs O&P		Cost Per Labor-Hour	

Crew B-48	Hr.	Daily	Hr.	Daily	Bare Costs	Incl. O&P
1 Labor Foreman (outside)	$39.60	$316.80	$60.95	$487.60	$42.56	$65.04
3 Laborers	37.60	902.40	57.85	1388.40		
1 Equip. Oper. (crane)	51.70	413.60	78.45	627.60		
1 Equip. Oper. (oiler)	45.20	361.60	68.55	548.40		
1 Equip. Oper. (light)	48.60	388.80	73.75	590.00		
1 Centr. Water Pump, 6"		346.60		381.26		
1 -20' Suction Hose, 6"		11.50		12.65		
1 -50' Discharge Hose, 6"		6.10		6.71		
1 Drill Rig, Truck-Mounted		2542.00		2796.20	51.90	57.09
56 L.H., Daily Totals		$5289.40		$6838.82	$94.45	$122.12

Crew B-49	Hr.	Daily	Hr.	Daily	Bare Costs	Incl. O&P
1 Labor Foreman (outside)	$39.60	$316.80	$60.95	$487.60	$44.27	$68.02
3 Laborers	37.60	902.40	57.85	1388.40		
2 Equip. Oper. (crane)	51.70	827.20	78.45	1255.20		
2 Equip. Oper. (oilers)	45.20	723.20	68.55	1096.80		
1 Equip. Oper. (light)	48.60	388.80	73.75	590.00		
2 Pile Drivers	46.10	737.60	73.00	1168.00		
1 Hyd. Crane, 25 Ton		736.60		810.26		
1 Centr. Water Pump, 6"		346.60		381.26		
1 -20' Suction Hose, 6"		11.50		12.65		
1 -50' Discharge Hose, 6"		6.10		6.71		
1 Drill Rig, Truck-Mounted		2542.00		2796.20	41.40	45.53
88 L.H., Daily Totals		$7538.80		$9993.08	$85.67	$113.56

Crew B-50	Hr.	Daily	Hr.	Daily	Bare Costs	Incl. O&P
2 Pile Driver Foremen (outside)	$48.10	$769.60	$76.15	$1218.40	$45.30	$70.66
6 Pile Drivers	46.10	2212.80	73.00	3504.00		
2 Equip. Oper. (crane)	51.70	827.20	78.45	1255.20		
1 Equip. Oper. (oiler)	45.20	361.60	68.55	548.40		
3 Laborers	37.60	902.40	57.85	1388.40		
1 Crawler Crane, 40 Ton		1156.00		1271.60		
1 Lead, 60' High		74.80		82.28		
1 Hammer, Diesel, 15K ft.-lbs.		586.00		644.60		
1 Air Compressor, 600 cfm		550.60		605.66		
2 -50' Air Hoses, 3"		29.80		32.78		
1 Chain Saw, Gas, 36" Long		45.00		49.50	21.81	23.99
112 L.H., Daily Totals		$7515.80		$10600.82	$67.11	$94.65

Crew B-51	Hr.	Daily	Hr.	Daily	Bare Costs	Incl. O&P
1 Labor Foreman (outside)	$39.60	$316.80	$60.95	$487.60	$38.18	$58.58
4 Laborers	37.60	1203.20	57.85	1851.20		
1 Truck Driver (light)	39.10	312.80	59.10	472.80		
1 Flatbed Truck, Gas, 1.5 Ton		245.40		269.94	5.11	5.62
48 L.H., Daily Totals		$2078.20		$3081.54	$43.30	$64.20

Crew B-52	Hr.	Daily	Hr.	Daily	Bare Costs	Incl. O&P
1 Carpenter Foreman (outside)	$48.95	$391.60	$75.35	$602.80	$43.61	$66.75
1 Carpenter	46.95	375.60	72.25	578.00		
3 Laborers	37.60	902.40	57.85	1388.40		
1 Cement Finisher	45.00	360.00	66.65	533.20		
.5 Rodman (reinf.)	52.55	210.20	82.10	328.40		
.5 Equip. Oper. (medium)	50.60	202.40	76.75	307.00		
.5 Crawler Loader, 3 C.Y.		594.00		653.40	10.61	11.67
56 L.H., Daily Totals		$3036.20		$4391.20	$54.22	$78.41

Crew B-53	Hr.	Daily	Hr.	Daily	Bare Costs	Incl. O&P
1 Equip. Oper. (light)	$48.60	$388.80	$73.75	$590.00	$48.60	$73.75
1 Trencher, Chain, 12 H.P.		68.00		74.80	8.50	9.35
8 L.H., Daily Totals		$456.80		$664.80	$57.10	$83.10

Crew B-54	Hr.	Daily	Hr.	Daily	Bare Costs	Incl. O&P
1 Equip. Oper. (light)	$48.60	$388.80	$73.75	$590.00	$48.60	$73.75
1 Trencher, Chain, 40 H.P.		335.20		368.72	41.90	46.09
8 L.H., Daily Totals		$724.00		$958.72	$90.50	$119.84

Crew B-54A	Hr.	Daily	Hr.	Daily	Bare Costs	Incl. O&P
.17 Labor Foreman (outside)	$39.60	$53.86	$60.95	$82.89	$49.00	$74.45
1 Equipment Operator (med.)	50.60	404.80	76.75	614.00		
1 Wheel Trencher, 67 H.P.		1195.00		1314.50	127.67	140.44
9.36 L.H., Daily Totals		$1653.66		$2011.39	$176.67	$214.89

Crew B-54B	Hr.	Daily	Hr.	Daily	Bare Costs	Incl. O&P
.25 Labor Foreman (outside)	$39.60	$79.20	$60.95	$121.90	$48.40	$73.59
1 Equipment Operator (med.)	50.60	404.80	76.75	614.00		
1 Wheel Trencher, 150 H.P.		1890.00		2079.00	189.00	207.90
10 L.H., Daily Totals		$2374.00		$2814.90	$237.40	$281.49

Crew B-54C	Hr.	Daily	Hr.	Daily	Bare Costs	Incl. O&P
1 Laborer	$37.60	$300.80	$57.85	$462.80	$44.10	$67.30
1 Equipment Operator (med.)	50.60	404.80	76.75	614.00		
1 Wheel Trencher, 67 H.P.		1195.00		1314.50	74.69	82.16
16 L.H., Daily Totals		$1900.60		$2391.30	$118.79	$149.46

Crew B-54D	Hr.	Daily	Hr.	Daily	Bare Costs	Incl. O&P
1 Laborer	$37.60	$300.80	$57.85	$462.80	$44.10	$67.30
1 Equipment Operator (med.)	50.60	404.80	76.75	614.00		
1 Rock Trencher, 6" Width		372.00		409.20	23.25	25.57
16 L.H., Daily Totals		$1077.60		$1486.00	$67.35	$92.88

Crew B-54E	Hr.	Daily	Hr.	Daily	Bare Costs	Incl. O&P
1 Laborer	$37.60	$300.80	$57.85	$462.80	$44.10	$67.30
1 Equipment Operator (med.)	50.60	404.80	76.75	614.00		
1 Rock Trencher, 18" Width		2575.00		2832.50	160.94	177.03
16 L.H., Daily Totals		$3280.60		$3909.30	$205.04	$244.33

Crew B-55	Hr.	Daily	Hr.	Daily	Bare Costs	Incl. O&P
2 Laborers	$37.60	$601.60	$57.85	$925.60	$38.10	$58.27
1 Truck Driver (light)	39.10	312.80	59.10	472.80		
1 Truck-Mounted Earth Auger		779.20		857.12		
1 Flatbed Truck, Gas, 3 Ton		303.20		333.52	45.10	49.61
24 L.H., Daily Totals		$1996.80		$2589.04	$83.20	$107.88

Crew B-56	Hr.	Daily	Hr.	Daily	Bare Costs	Incl. O&P
1 Laborer	$37.60	$300.80	$57.85	$462.80	$43.10	$65.80
1 Equip. Oper. (light)	48.60	388.80	73.75	590.00		
1 Air Track Drill, 4"		1028.00		1130.80		
1 Air Compressor, 600 cfm		550.60		605.66		
1 -50' Air Hose, 3"		14.90		16.39	99.59	109.55
16 L.H., Daily Totals		$2283.10		$2805.65	$142.69	$175.35

Crew No.	Bare Costs		Incl. Subs O&P		Cost Per Labor-Hour	

Crew B-57

Crew B-57	Hr.	Daily	Hr.	Daily	Bare Costs	Incl. O&P
1 Labor Foreman (outside)	$39.60	$316.80	$60.95	$487.60	$43.38	$66.23
2 Laborers	37.60	601.60	57.85	925.60		
1 Equip. Oper. (crane)	51.70	413.60	78.45	627.60		
1 Equip. Oper. (light)	48.60	388.80	73.75	590.00		
1 Equip. Oper. (oiler)	45.20	361.60	68.55	548.40		
1 Crawler Crane, 25 Ton		1150.00		1265.00		
1 Clamshell Bucket, 1 C.Y.		47.80		52.58		
1 Centr. Water Pump, 6"		346.60		381.26		
1 -20' Suction Hose, 6"		11.50		12.65		
20 -50' Discharge Hoses, 6"		122.00		134.20	34.96	38.45
48 L.H., Daily Totals		$3760.30		$5024.89	$78.34	$104.69

Crew B-58

Crew B-58	Hr.	Daily	Hr.	Daily	Bare Costs	Incl. O&P
2 Laborers	$37.60	$601.60	$57.85	$925.60	$41.27	$63.15
1 Equip. Oper. (light)	48.60	388.80	73.75	590.00		
1 Backhoe Loader, 48 H.P.		364.20		400.62		
1 Small Helicopter, w/ Pilot		2790.00		3069.00	131.43	144.57
24 L.H., Daily Totals		$4144.60		$4985.22	$172.69	$207.72

Crew B-59

Crew B-59	Hr.	Daily	Hr.	Daily	Bare Costs	Incl. O&P
1 Truck Driver (heavy)	$40.05	$320.40	$60.55	$484.40	$40.05	$60.55
1 Truck Tractor, 220 H.P.		359.60		395.56		
1 Water Tank Trailer, 5000 Gal.		141.00		155.10	62.58	68.83
8 L.H., Daily Totals		$821.00		$1035.06	$102.63	$129.38

Crew B-59A

Crew B-59A	Hr.	Daily	Hr.	Daily	Bare Costs	Incl. O&P
2 Laborers	$37.60	$601.60	$57.85	$925.60	$38.42	$58.75
1 Truck Driver (heavy)	40.05	320.40	60.55	484.40		
1 Water Tank Trailer, 5000 Gal.		141.00		155.10		
1 Truck Tractor, 220 H.P.		359.60		395.56	20.86	22.94
24 L.H., Daily Totals		$1422.60		$1960.66	$59.27	$81.69

Crew B-60

Crew B-60	Hr.	Daily	Hr.	Daily	Bare Costs	Incl. O&P
1 Labor Foreman (outside)	$39.60	$316.80	$60.95	$487.60	$44.13	$67.31
2 Laborers	37.60	601.60	57.85	925.60		
1 Equip. Oper. (crane)	51.70	413.60	78.45	627.60		
2 Equip. Oper. (light)	48.60	777.60	73.75	1180.00		
1 Equip. Oper. (oiler)	45.20	361.60	68.55	548.40		
1 Crawler Crane, 40 Ton		1156.00		1271.60		
1 Lead, 60' High		74.80		82.28		
1 Hammer, Diesel, 15K ft.-lbs.		586.00		644.60		
1 Backhoe Loader, 48 H.P.		364.20		400.62	38.95	42.84
56 L.H., Daily Totals		$4652.20		$6168.30	$83.08	$110.15

Crew B-61

Crew B-61	Hr.	Daily	Hr.	Daily	Bare Costs	Incl. O&P
1 Labor Foreman (outside)	$39.60	$316.80	$60.95	$487.60	$40.20	$61.65
3 Laborers	37.60	902.40	57.85	1388.40		
1 Equip. Oper. (light)	48.60	388.80	73.75	590.00		
1 Cement Mixer, 2 C.Y.		194.20		213.62		
1 Air Compressor, 160 cfm		156.40		172.04	8.77	9.64
40 L.H., Daily Totals		$1958.60		$2851.66	$48.97	$71.29

Crew B-62

Crew B-62	Hr.	Daily	Hr.	Daily	Bare Costs	Incl. O&P
2 Laborers	$37.60	$601.60	$57.85	$925.60	$41.27	$63.15
1 Equip. Oper. (light)	48.60	388.80	73.75	590.00		
1 Loader, Skid Steer, 30 H.P.		173.60		190.96	7.23	7.96
24 L.H., Daily Totals		$1164.00		$1706.56	$48.50	$71.11

Crew B-62A

Crew B-62A	Hr.	Daily	Hr.	Daily	Bare Costs	Incl. O&P
2 Laborers	$37.60	$601.60	$57.85	$925.60	$41.27	$63.15
1 Equip. Oper. (light)	48.60	388.80	73.75	590.00		
1 Loader, Skid Steer, 30 H.P.		173.60		190.96		
1 Trencher Attachment		58.30		64.13	9.66	10.63
24 L.H., Daily Totals		$1222.30		$1770.69	$50.93	$73.78

Crew B-63

Crew B-63	Hr.	Daily	Hr.	Daily	Bare Costs	Incl. O&P
4 Laborers	$37.60	$1203.20	$57.85	$1851.20	$39.80	$61.03
1 Equip. Oper. (light)	48.60	388.80	73.75	590.00		
1 Loader, Skid Steer, 30 H.P.		173.60		190.96	4.34	4.77
40 L.H., Daily Totals		$1765.60		$2632.16	$44.14	$65.80

Crew B-63B

Crew B-63B	Hr.	Daily	Hr.	Daily	Bare Costs	Incl. O&P
1 Labor Foreman (inside)	$38.10	$304.80	$58.65	$469.20	$40.48	$62.02
2 Laborers	37.60	601.60	57.85	925.60		
1 Equip. Oper. (light)	48.60	388.80	73.75	590.00		
1 Loader, Skid Steer, 78 H.P.		318.60		350.46	9.96	10.95
32 L.H., Daily Totals		$1613.80		$2335.26	$50.43	$72.98

Crew B-64

Crew B-64	Hr.	Daily	Hr.	Daily	Bare Costs	Incl. O&P
1 Laborer	$37.60	$300.80	$57.85	$462.80	$38.35	$58.48
1 Truck Driver (light)	39.10	312.80	59.10	472.80		
1 Power Mulcher (small)		151.80		166.98		
1 Flatbed Truck, Gas, 1.5 Ton		245.40		269.94	24.82	27.31
16 L.H., Daily Totals		$1010.80		$1372.52	$63.17	$85.78

Crew B-65

Crew B-65	Hr.	Daily	Hr.	Daily	Bare Costs	Incl. O&P
1 Laborer	$37.60	$300.80	$57.85	$462.80	$38.35	$58.48
1 Truck Driver (light)	39.10	312.80	59.10	472.80		
1 Power Mulcher (Large)		319.40		351.34		
1 Flatbed Truck, Gas, 1.5 Ton		245.40		269.94	35.30	38.83
16 L.H., Daily Totals		$1178.40		$1556.88	$73.65	$97.31

Crew B-66

Crew B-66	Hr.	Daily	Hr.	Daily	Bare Costs	Incl. O&P
1 Equip. Oper. (light)	$48.60	$388.80	$73.75	$590.00	$48.60	$73.75
1 Loader-Backhoe, 40 H.P.		263.20		289.52	32.90	36.19
8 L.H., Daily Totals		$652.00		$879.52	$81.50	$109.94

Crew B-67

Crew B-67	Hr.	Daily	Hr.	Daily	Bare Costs	Incl. O&P
1 Millwright	$49.15	$393.20	$72.35	$578.80	$48.88	$73.05
1 Equip. Oper. (light)	48.60	388.80	73.75	590.00		
1 Forklift, R/T, 4,000 Lb.		311.20		342.32	19.45	21.40
16 L.H., Daily Totals		$1093.20		$1511.12	$68.33	$94.44

Crew B-67B

Crew B-67B	Hr.	Daily	Hr.	Daily	Bare Costs	Incl. O&P
1 Millwright Foreman (inside)	$49.65	$397.20	$73.10	$584.80	$49.40	$72.72
1 Millwright	49.15	393.20	72.35	578.80		
16 L.H., Daily Totals		$790.40		$1163.60	$49.40	$72.72

Crew B-68

Crew B-68	Hr.	Daily	Hr.	Daily	Bare Costs	Incl. O&P
2 Millwrights	$49.15	$786.40	$72.35	$1157.60	$48.97	$72.82
1 Equip. Oper. (light)	48.60	388.80	73.75	590.00		
1 Forklift, R/T, 4,000 Lb.		311.20		342.32	12.97	14.26
24 L.H., Daily Totals		$1486.40		$2089.92	$61.93	$87.08

Crew B-68A

Crew B-68A	Hr.	Daily	Hr.	Daily	Bare Costs	Incl. O&P
1 Millwright Foreman (inside)	$49.65	$397.20	$73.10	$584.80	$49.32	$72.60
2 Millwrights	49.15	786.40	72.35	1157.60		
1 Forklift, 8,000 Lb.		185.20		203.72	7.72	8.49
24 L.H., Daily Totals		$1368.80		$1946.12	$57.03	$81.09

Crew No.	Bare Costs Hr.	Daily	Incl. Subs O&P Hr.	Daily	Cost Per Labor-Hour Bare Costs	Incl. O&P
Crew B-68B	**Hr.**	**Daily**	**Hr.**	**Daily**	**Bare Costs**	**Incl. O&P**
1 Millwright Foreman (inside)	$49.65	$397.20	$73.10	$584.80	$53.54	$79.86
2 Millwrights	49.15	786.40	72.35	1157.60		
2 Electricians	54.70	875.20	81.95	1311.20		
2 Plumbers	58.70	939.20	88.65	1418.40		
1 Forklift, 5,000 Lb.		331.80		364.98	5.92	6.52
56 L.H., Daily Totals		$3329.80		$4836.98	$59.46	$86.37
Crew B-68C	**Hr.**	**Daily**	**Hr.**	**Daily**	**Bare Costs**	**Incl. O&P**
1 Millwright Foreman (inside)	$49.65	$397.20	$73.10	$584.80	$53.05	$79.01
1 Millwright	49.15	393.20	72.35	578.80		
1 Electrician	54.70	437.60	81.95	655.60		
1 Plumber	58.70	469.60	88.65	709.20		
1 Forklift, 5,000 Lb.		331.80		364.98	10.37	11.41
32 L.H., Daily Totals		$2029.40		$2893.38	$63.42	$90.42
Crew B-68D	**Hr.**	**Daily**	**Hr.**	**Daily**	**Bare Costs**	**Incl. O&P**
1 Labor Foreman (inside)	$38.10	$304.80	$58.65	$469.20	$41.43	$63.42
1 Laborer	37.60	300.80	57.85	462.80		
1 Equip. Oper. (light)	48.60	388.80	73.75	590.00		
1 Forklift, 5,000 Lb.		331.80		364.98	13.82	15.21
24 L.H., Daily Totals		$1326.20		$1886.98	$55.26	$78.62
Crew B-68E	**Hr.**	**Daily**	**Hr.**	**Daily**	**Bare Costs**	**Incl. O&P**
1 Struc. Steel Foreman (inside)	$53.15	$425.20	$92.30	$738.40	$52.75	$91.62
3 Struc. Steel Workers	52.65	1263.60	91.45	2194.80		
1 Welder	52.65	421.20	91.45	731.60		
1 Forklift, 8,000 Lb.		185.20		203.72	4.63	5.09
40 L.H., Daily Totals		$2295.20		$3868.52	$57.38	$96.71
Crew B-68F	**Hr.**	**Daily**	**Hr.**	**Daily**	**Bare Costs**	**Incl. O&P**
1 Skilled Worker Foreman (out)	$50.65	$405.20	$78.25	$626.00	$49.32	$76.18
2 Skilled Workers	48.65	778.40	75.15	1202.40		
1 Forklift, 5,000 Lb.		331.80		364.98	13.82	15.21
24 L.H., Daily Totals		$1515.40		$2193.38	$63.14	$91.39
Crew B-68G	**Hr.**	**Daily**	**Hr.**	**Daily**	**Bare Costs**	**Incl. O&P**
2 Structural Steel Workers	$52.65	$842.40	$91.45	$1463.20	$52.65	$91.45
1 Forklift, 5,000 Lb.		331.80		364.98	20.74	22.81
16 L.H., Daily Totals		$1174.20		$1828.18	$73.39	$114.26
Crew B-69	**Hr.**	**Daily**	**Hr.**	**Daily**	**Bare Costs**	**Incl. O&P**
1 Labor Foreman (outside)	$39.60	$316.80	$60.95	$487.60	$41.55	$63.58
3 Laborers	37.60	902.40	57.85	1388.40		
1 Equip. Oper. (crane)	51.70	413.60	78.45	627.60		
1 Equip. Oper. (oiler)	45.20	361.60	68.55	548.40		
1 Hyd. Crane, 80 Ton		1625.00		1787.50	33.85	37.24
48 L.H., Daily Totals		$3619.40		$4839.50	$75.40	$100.82
Crew B-69A	**Hr.**	**Daily**	**Hr.**	**Daily**	**Bare Costs**	**Incl. O&P**
1 Labor Foreman (outside)	$39.60	$316.80	$60.95	$487.60	$41.33	$62.98
3 Laborers	37.60	902.40	57.85	1388.40		
1 Equip. Oper. (medium)	50.60	404.80	76.75	614.00		
1 Concrete Finisher	45.00	360.00	66.65	533.20		
1 Curb/Gutter Paver, 2-Track		997.60		1097.36	20.78	22.86
48 L.H., Daily Totals		$2981.60		$4120.56	$62.12	$85.84

Crew No.	Bare Costs Hr.	Daily	Incl. Subs O&P Hr.	Daily	Cost Per Labor-Hour Bare Costs	Incl. O&P
Crew B-69B	**Hr.**	**Daily**	**Hr.**	**Daily**	**Bare Costs**	**Incl. O&P**
1 Labor Foreman (outside)	$39.60	$316.80	$60.95	$487.60	$41.33	$62.98
3 Laborers	37.60	902.40	57.85	1388.40		
1 Equip. Oper. (medium)	50.60	404.80	76.75	614.00		
1 Cement Finisher	45.00	360.00	66.65	533.20		
1 Curb/Gutter Paver, 4-Track		776.00		853.60	16.17	17.78
48 L.H., Daily Totals		$2760.00		$3876.80	$57.50	$80.77
Crew B-70	**Hr.**	**Daily**	**Hr.**	**Daily**	**Bare Costs**	**Incl. O&P**
1 Labor Foreman (outside)	$39.60	$316.80	$60.95	$487.60	$43.46	$66.39
3 Laborers	37.60	902.40	57.85	1388.40		
3 Equip. Oper. (medium)	50.60	1214.40	76.75	1842.00		
1 Grader, 30,000 Lbs.		736.20		809.82		
1 Ripper, Beam & 1 Shank		84.20		92.62		
1 Road Sweeper, S.P., 8' wide		662.40		728.64		
1 F.E. Loader, W.M., 1.5 C.Y.		378.40		416.24	33.24	36.56
56 L.H., Daily Totals		$4294.80		$5765.32	$76.69	$102.95
Crew B-70A	**Hr.**	**Daily**	**Hr.**	**Daily**	**Bare Costs**	**Incl. O&P**
1 Laborer	$37.60	$300.80	$57.85	$462.80	$48.00	$72.97
4 Equip. Oper. (medium)	50.60	1619.20	76.75	2456.00		
1 Grader, 40,000 Lbs.		1174.00		1291.40		
1 F.E. Loader, W.M.,2.5 C.Y.		519.80		571.78		
1 Dozer, 80 H.P.		472.40		519.64		
1 Roller, Pneum. Whl., 12 Ton		344.40		378.84	62.77	69.04
40 L.H., Daily Totals		$4430.60		$5680.46	$110.77	$142.01
Crew B-71	**Hr.**	**Daily**	**Hr.**	**Daily**	**Bare Costs**	**Incl. O&P**
1 Labor Foreman (outside)	$39.60	$316.80	$60.95	$487.60	$43.46	$66.39
3 Laborers	37.60	902.40	57.85	1388.40		
3 Equip. Oper. (medium)	50.60	1214.40	76.75	1842.00		
1 Pvmt. Profiler, 750 H.P.		5805.00		6385.50		
1 Road Sweeper, S.P., 8' wide		662.40		728.64		
1 F.E. Loader, W.M., 1.5 C.Y.		378.40		416.24	122.25	134.47
56 L.H., Daily Totals		$9279.40		$11248.38	$165.70	$200.86
Crew B-72	**Hr.**	**Daily**	**Hr.**	**Daily**	**Bare Costs**	**Incl. O&P**
1 Labor Foreman (outside)	$39.60	$316.80	$60.95	$487.60	$44.35	$67.69
3 Laborers	37.60	902.40	57.85	1388.40		
4 Equip. Oper. (medium)	50.60	1619.20	76.75	2456.00		
1 Pvmt. Profiler, 750 H.P.		5805.00		6385.50		
1 Hammermill, 250 H.P.		1879.00		2066.90		
1 Windrow Loader		1227.00		1349.70		
1 Mix Paver 165 H.P.		2147.00		2361.70		
1 Roller, Pneum. Whl., 12 Ton		344.40		378.84	178.16	195.98
64 L.H., Daily Totals		$14240.80		$16874.64	$222.51	$263.67
Crew B-73	**Hr.**	**Daily**	**Hr.**	**Daily**	**Bare Costs**	**Incl. O&P**
1 Labor Foreman (outside)	$39.60	$316.80	$60.95	$487.60	$45.98	$70.05
2 Laborers	37.60	601.60	57.85	925.60		
5 Equip. Oper. (medium)	50.60	2024.00	76.75	3070.00		
1 Road Mixer, 310 H.P.		1929.00		2121.90		
1 Tandem Roller, 10 Ton		236.80		260.48		
1 Hammermill, 250 H.P.		1879.00		2066.90		
1 Grader, 30,000 Lbs.		736.20		809.82		
.5 F.E. Loader, W.M. 1.5 C.Y.		189.20		208.12		
.5 Truck Tractor, 220 H.P.		179.80		197.78		
.5 Water Tank Trailer, 5000 Gal.		70.50		77.55	81.57	89.73
64 L.H., Daily Totals		$8162.90		$10225.75	$127.55	$159.78

Crew No.	Bare Costs		Incl. Subs O&P		Cost Per Labor-Hour	

Crew B-74

Crew B-74	Hr.	Daily	Hr.	Daily	Bare Costs	Incl. O&P
1 Labor Foreman (outside)	$39.60	$316.80	$60.95	$487.60	$44.96	$68.36
1 Laborer	37.60	300.80	57.85	462.80		
4 Equip. Oper. (medium)	50.60	1619.20	76.75	2456.00		
2 Truck Drivers (heavy)	40.05	640.80	60.55	968.80		
1 Grader, 30,000 Lbs.		736.20		809.82		
1 Ripper, Beam & 1 Shank		84.20		92.62		
2 Stabilizers, 310 H.P.		3596.00		3955.60		
1 Flatbed Truck, Gas, 3 Ton		303.20		333.52		
1 Chem. Spreader, Towed		52.40		57.64		
1 Roller, Vibratory, 25 Ton		687.40		756.14		
1 Water Tank Trailer, 5000 Gal.		141.00		155.10		
1 Truck Tractor, 220 H.P.		359.60		395.56	93.13	102.44
64 L.H., Daily Totals		$8837.60		$10931.20	$138.09	$170.80

Crew B-75	Hr.	Daily	Hr.	Daily	Bare Costs	Incl. O&P
1 Labor Foreman (outside)	$39.60	$316.80	$60.95	$487.60	$45.66	$69.48
1 Laborer	37.60	300.80	57.85	462.80		
4 Equip. Oper. (medium)	50.60	1619.20	76.75	2456.00		
1 Truck Driver (heavy)	40.05	320.40	60.55	484.40		
1 Grader, 30,000 Lbs.		736.20		809.82		
1 Ripper, Beam & 1 Shank		84.20		92.62		
2 Stabilizers, 310 H.P.		3596.00		3955.60		
1 Dist. Tanker, 3000 Gallon		305.60		336.16		
1 Truck Tractor, 6x4, 380 H.P.		599.80		659.78		
1 Roller, Vibratory, 25 Ton		687.40		756.14	107.31	118.04
56 L.H., Daily Totals		$8566.40		$10500.92	$152.97	$187.52

Crew B-76	Hr.	Daily	Hr.	Daily	Bare Costs	Incl. O&P
1 Dock Builder Foreman (outside)	$48.10	$384.80	$76.15	$609.20	$47.47	$74.07
5 Dock Builders	46.10	1844.00	73.00	2920.00		
2 Equip. Oper. (crane)	51.70	827.20	78.45	1255.20		
1 Equip. Oper. (oiler)	45.20	361.60	68.55	548.40		
1 Crawler Crane, 50 Ton		1294.00		1423.40		
1 Barge, 400 Ton		776.00		853.60		
1 Hammer, Diesel, 15K ft.-lbs.		586.00		644.60		
1 Lead, 60' High		74.80		82.28		
1 Air Compressor, 600 cfm		550.60		605.66		
2 -50' Air Hoses, 3"		29.80		32.78	45.99	50.59
72 L.H., Daily Totals		$6728.80		$8975.12	$93.46	$124.65

Crew B-76A	Hr.	Daily	Hr.	Daily	Bare Costs	Incl. O&P
1 Labor Foreman (outside)	$39.60	$316.80	$60.95	$487.60	$40.56	$62.15
5 Laborers	37.60	1504.00	57.85	2314.00		
1 Equip. Oper. (crane)	51.70	413.60	78.45	627.60		
1 Equip. Oper. (oiler)	45.20	361.60	68.55	548.40		
1 Crawler Crane, 50 Ton		1294.00		1423.40		
1 Barge, 400 Ton		776.00		853.60	32.34	35.58
64 L.H., Daily Totals		$4666.00		$6254.60	$72.91	$97.73

Crew B-77	Hr.	Daily	Hr.	Daily	Bare Costs	Incl. O&P
1 Labor Foreman (outside)	$39.60	$316.80	$60.95	$487.60	$38.30	$58.72
3 Laborers	37.60	902.40	57.85	1388.40		
1 Truck Driver (light)	39.10	312.80	59.10	472.80		
1 Crack Cleaner, 25 H.P.		61.60		67.76		
1 Crack Filler, Trailer Mtd.		202.00		222.20		
1 Flatbed Truck, Gas, 3 Ton		303.20		333.52	14.17	15.59
40 L.H., Daily Totals		$2098.80		$2972.28	$52.47	$74.31

Crew B-78	Hr.	Daily	Hr.	Daily	Bare Costs	Incl. O&P
1 Labor Foreman (outside)	$39.60	$316.80	$60.95	$487.60	$38.18	$58.58
4 Laborers	37.60	1203.20	57.85	1851.20		
1 Truck Driver (light)	39.10	312.80	59.10	472.80		
1 Paint Striper, S.P., 40 Gallon		151.40		166.54		
1 Flatbed Truck, Gas, 3 Ton		303.20		333.52		
1 Pickup Truck, 3/4 Ton		144.20		158.62	12.48	13.72
48 L.H., Daily Totals		$2431.60		$3470.28	$50.66	$72.30

Crew B-78A	Hr.	Daily	Hr.	Daily	Bare Costs	Incl. O&P
1 Equip. Oper. (light)	$48.60	$388.80	$73.75	$590.00	$48.60	$73.75
1 Line Rem. (Metal Balls) 115 H.P.		821.60		903.76	102.70	112.97
8 L.H., Daily Totals		$1210.40		$1493.76	$151.30	$186.72

Crew B-78B	Hr.	Daily	Hr.	Daily	Bare Costs	Incl. O&P
2 Laborers	$37.60	$601.60	$57.85	$925.60	$38.82	$59.62
.25 Equip. Oper. (light)	48.60	97.20	73.75	147.50		
1 Pickup Truck, 3/4 Ton		144.20		158.62		
1 Line Rem.,11 H.P.,Walk Behind		65.60		72.16		
.25 Road Sweeper, S.P., 8' wide		165.60		182.16	20.86	22.94
18 L.H., Daily Totals		$1074.20		$1486.04	$59.68	$82.56

Crew B-78C	Hr.	Daily	Hr.	Daily	Bare Costs	Incl. O&P
1 Labor Foreman (outside)	$39.60	$316.80	$60.95	$487.60	$38.18	$58.58
4 Laborers	37.60	1203.20	57.85	1851.20		
1 Truck Driver (light)	39.10	312.80	59.10	472.80		
1 Paint Striper, T.M., 120 Gal.		804.60		885.06		
1 Flatbed Truck, Gas, 3 Ton		303.20		333.52		
1 Pickup Truck, 3/4 Ton		144.20		158.62	26.08	28.69
48 L.H., Daily Totals		$3084.80		$4188.80	$64.27	$87.27

Crew B-78D	Hr.	Daily	Hr.	Daily	Bare Costs	Incl. O&P
2 Labor Foremen (outside)	$39.60	$633.60	$60.95	$975.20	$38.15	$58.59
7 Laborers	37.60	2105.60	57.85	3239.60		
1 Truck Driver (light)	39.10	312.80	59.10	472.80		
1 Paint Striper, T.M., 120 Gal.		804.60		885.06		
1 Flatbed Truck, Gas, 3 Ton		303.20		333.52		
3 Pickup Trucks, 3/4 Ton		432.60		475.86		
1 Air Compressor, 60 cfm		137.40		151.14		
1 -50' Air Hose, 3/4"		3.25		3.58		
1 Breakers, Pavement, 60 lb.		9.80		10.78	21.14	23.25
80 L.H., Daily Totals		$4742.85		$6547.53	$59.29	$81.84

Crew B-78E	Hr.	Daily	Hr.	Daily	Bare Costs	Incl. O&P
2 Labor Foremen (outside)	$39.60	$633.60	$60.95	$975.20	$38.06	$58.47
9 Laborers	37.60	2707.20	57.85	4165.20		
1 Truck Driver (light)	39.10	312.80	59.10	472.80		
1 Paint Striper, T.M., 120 Gal.		804.60		885.06		
1 Flatbed Truck, Gas, 3 Ton		303.20		333.52		
4 Pickup Trucks, 3/4 Ton		576.80		634.48		
2 Air Compressor, 60 cfm		274.80		302.28		
2 -50' Air Hose, 3/4"		6.50		7.15		
2 Breakers, Pavement, 60 lb.		19.60		21.56	20.68	22.75
96 L.H., Daily Totals		$5639.10		$7797.25	$58.74	$81.22

Crews

Crew No.	Bare Costs		Incl. Subs O&P		Cost Per Labor-Hour	

Crew B-78F

	Hr.	Daily	Hr.	Daily	Bare Costs	Incl. O&P
2 Labor Foremen (outside)	$39.60	$633.60	$60.95	$975.20	$37.99	$58.38
11 Laborers	37.60	3308.80	57.85	5090.80		
1 Truck Driver (light)	39.10	312.80	59.10	472.80		
1 Paint Striper, T.M., 120 Gal.		804.60		885.06		
1 Flatbed Truck, Gas, 3 Ton		303.20		333.52		
7 Pickup Trucks, 3/4 Ton		1009.40		1110.34		
3 Air Compressor, 60 cfm		412.20		453.42		
3 -50' Air Hose, 3/4"		9.75		10.73		
3 Breakers, Pavement, 60 lb.		29.40		32.34	22.93	25.23
112 L.H., Daily Totals		$6823.75		$9364.20	$60.93	$83.61

Crew B-79

	Hr.	Daily	Hr.	Daily	Bare Costs	Incl. O&P
1 Labor Foreman (outside)	$39.60	$316.80	$60.95	$487.60	$38.30	$58.72
3 Laborers	37.60	902.40	57.85	1388.40		
1 Truck Driver (light)	39.10	312.80	59.10	472.80		
1 Paint Striper, T.M., 120 Gal.		804.60		885.06		
1 Heating Kettle, 115 Gallon		81.00		89.10		
1 Flatbed Truck, Gas, 3 Ton		303.20		333.52		
2 Pickup Trucks, 3/4 Ton		288.40		317.24	36.93	40.62
40 L.H., Daily Totals		$3009.20		$3973.72	$75.23	$99.34

Crew B-79A

	Hr.	Daily	Hr.	Daily	Bare Costs	Incl. O&P
1.5 Equip. Oper. (light)	$48.60	$583.20	$73.75	$885.00	$48.60	$73.75
.5 Line Remov. (Grinder) 115 H.P.		455.90		501.49		
1 Line Rem. (Metal Balls) 115 H.P.		821.60		903.76	106.46	117.10
12 L.H., Daily Totals		$1860.70		$2290.25	$155.06	$190.85

Crew B-79B

	Hr.	Daily	Hr.	Daily	Bare Costs	Incl. O&P
1 Laborer	$37.60	$300.80	$57.85	$462.80	$37.60	$57.85
1 Set of Gases		152.00		167.20	19.00	20.90
8 L.H., Daily Totals		$452.80		$630.00	$56.60	$78.75

Crew B-79C

	Hr.	Daily	Hr.	Daily	Bare Costs	Incl. O&P
1 Labor Foreman (outside)	$39.60	$316.80	$60.95	$487.60	$38.10	$58.47
5 Laborers	37.60	1504.00	57.85	2314.00		
1 Truck Driver (light)	39.10	312.80	59.10	472.80		
1 Paint Striper, T.M., 120 Gal.		804.60		885.06		
1 Heating Kettle, 115 Gallon		81.00		89.10		
1 Flatbed Truck, Gas, 3 Ton		303.20		333.52		
3 Pickup Trucks, 3/4 Ton		432.60		475.86		
1 Air Compressor, 60 cfm		137.40		151.14		
1 -50' Air Hose, 3/4"		3.25		3.58		
1 Breakers, Pavement, 60 lb.		9.80		10.78	31.64	34.80
56 L.H., Daily Totals		$3905.45		$5223.44	$69.74	$93.28

Crew B-79D

	Hr.	Daily	Hr.	Daily	Bare Costs	Incl. O&P
2 Labor Foremen (outside)	$39.60	$633.60	$60.95	$975.20	$38.29	$58.78
5 Laborers	37.60	1504.00	57.85	2314.00		
1 Truck Driver (light)	39.10	312.80	59.10	472.80		
1 Paint Striper, T.M., 120 Gal.		804.60		885.06		
1 Heating Kettle, 115 Gallon		81.00		89.10		
1 Flatbed Truck, Gas, 3 Ton		303.20		333.52		
4 Pickup Trucks, 3/4 Ton		576.80		634.48		
1 Air Compressor, 60 cfm		137.40		151.14		
1 -50' Air Hose, 3/4"		3.25		3.58		
1 Breakers, Pavement, 60 lb.		9.80		10.78	29.94	32.93
64 L.H., Daily Totals		$4366.45		$5869.65	$68.23	$91.71

Crew B-79E

	Hr.	Daily	Hr.	Daily	Bare Costs	Incl. O&P
2 Labor Foremen (outside)	$39.60	$633.60	$60.95	$975.20	$38.15	$58.59
7 Laborers	37.60	2105.60	57.85	3239.60		
1 Truck Driver (light)	39.10	312.80	59.10	472.80		
1 Paint Striper, T.M., 120 Gal.		804.60		885.06		
1 Heating Kettle, 115 Gallon		81.00		89.10		
1 Flatbed Truck, Gas, 3 Ton		303.20		333.52		
5 Pickup Trucks, 3/4 Ton		721.00		793.10		
2 Air Compressors, 60 cfm		274.80		302.28		
2 -50' Air Hoses, 3/4"		6.50		7.15		
2 Breakers, Pavement, 60 lb.		19.60		21.56	27.63	30.40
80 L.H., Daily Totals		$5262.70		$7119.37	$65.78	$88.99

Crew B-80

	Hr.	Daily	Hr.	Daily	Bare Costs	Incl. O&P
1 Labor Foreman (outside)	$39.60	$316.80	$60.95	$487.60	$41.23	$62.91
1 Laborer	37.60	300.80	57.85	462.80		
1 Truck Driver (light)	39.10	312.80	59.10	472.80		
1 Equip. Oper. (light)	48.60	388.80	73.75	590.00		
1 Flatbed Truck, Gas, 3 Ton		303.20		333.52		
1 Earth Auger, Truck-Mtd.		417.20		458.92	22.51	24.76
32 L.H., Daily Totals		$2039.60		$2805.64	$63.74	$87.68

Crew B-80A

	Hr.	Daily	Hr.	Daily	Bare Costs	Incl. O&P
3 Laborers	$37.60	$902.40	$57.85	$1388.40	$37.60	$57.85
1 Flatbed Truck, Gas, 3 Ton		303.20		333.52	12.63	13.90
24 L.H., Daily Totals		$1205.60		$1721.92	$50.23	$71.75

Crew B-80B

	Hr.	Daily	Hr.	Daily	Bare Costs	Incl. O&P
3 Laborers	$37.60	$902.40	$57.85	$1388.40	$40.35	$61.83
1 Equip. Oper. (light)	48.60	388.80	73.75	590.00		
1 Crane, Flatbed Mounted, 3 Ton		243.80		268.18	7.62	8.38
32 L.H., Daily Totals		$1535.00		$2246.58	$47.97	$70.21

Crew B-80C

	Hr.	Daily	Hr.	Daily	Bare Costs	Incl. O&P
2 Laborers	$37.60	$601.60	$57.85	$925.60	$38.10	$58.27
1 Truck Driver (light)	39.10	312.80	59.10	472.80		
1 Flatbed Truck, Gas, 1.5 Ton		245.40		269.94		
1 Manual Fence Post Auger, Gas		8.40		9.24	10.57	11.63
24 L.H., Daily Totals		$1168.20		$1677.58	$48.67	$69.90

Crew B-81

	Hr.	Daily	Hr.	Daily	Bare Costs	Incl. O&P
1 Laborer	$37.60	$300.80	$57.85	$462.80	$42.75	$65.05
1 Equip. Oper. (medium)	50.60	404.80	76.75	614.00		
1 Truck Driver (heavy)	40.05	320.40	60.55	484.40		
1 Hydromulcher, T.M., 3000 Gal.		334.20		367.62		
1 Truck Tractor, 220 H.P.		359.60		395.56	28.91	31.80
24 L.H., Daily Totals		$1719.80		$2324.38	$71.66	$96.85

Crew B-81A

	Hr.	Daily	Hr.	Daily	Bare Costs	Incl. O&P
1 Laborer	$37.60	$300.80	$57.85	$462.80	$38.35	$58.48
1 Truck Driver (light)	39.10	312.80	59.10	472.80		
1 Hydromulcher, T.M., 600 Gal.		127.80		140.58		
1 Flatbed Truck, Gas, 3 Ton		303.20		333.52	26.94	29.63
16 L.H., Daily Totals		$1044.60		$1409.70	$65.29	$88.11

Crew B-82

	Hr.	Daily	Hr.	Daily	Bare Costs	Incl. O&P
1 Laborer	$37.60	$300.80	$57.85	$462.80	$43.10	$65.80
1 Equip. Oper. (light)	48.60	388.80	73.75	590.00		
1 Horiz. Borer, 6 H.P.		83.80		92.18	5.24	5.76
16 L.H., Daily Totals		$773.40		$1144.98	$48.34	$71.56

Crews

Crew No.	Bare Costs Hr.	Bare Costs Daily	Incl. Subs O&P Hr.	Incl. Subs O&P Daily	Cost Per Labor-Hour Bare Costs	Cost Per Labor-Hour Incl. O&P
Crew B-82A	Hr.	Daily	Hr.	Daily	Bare Costs	Incl. O&P
2 Laborers	$37.60	$601.60	$57.85	$925.60	$43.10	$65.80
2 Equip. Opers. (light)	48.60	777.60	73.75	1180.00		
2 Dump Truck, 8 C.Y., 220 H.P.		822.40		904.64		
1 Flatbed Trailer, 25 Ton		113.60		124.96		
1 Horiz. Dir. Drill, 20k lb. Thrust		654.20		719.62		
1 Mud Trailer for HDD, 1500 Gal.		346.20		380.82		
1 Pickup Truck, 4 x 4, 3/4 Ton		155.60		171.16		
1 Flatbed trailer, 3 Ton		23.60		25.96		
1 Loader, Skid Steer, 78 H.P.		318.60		350.46	76.07	83.68
32 L.H., Daily Totals		$3813.40		$4783.22	$119.17	$149.48
Crew B-82B	Hr.	Daily	Hr.	Daily	Bare Costs	Incl. O&P
2 Laborers	$37.60	$601.60	$57.85	$925.60	$43.10	$65.80
2 Equip. Opers. (light)	48.60	777.60	73.75	1180.00		
2 Dump Truck, 8 C.Y., 220 H.P.		822.40		904.64		
1 Flatbed Trailer, 25 Ton		113.60		124.96		
1 Horiz. Dir. Drill, 30k lb. Thrust		931.20		1024.32		
1 Mud Trailer for HDD, 1500 Gal.		346.20		380.82		
1 Pickup Truck, 4 x 4, 3/4 Ton		155.60		171.16		
1 Flatbed trailer, 3 Ton		23.60		25.96		
1 Loader, Skid Steer, 78 H.P.		318.60		350.46	84.72	93.20
32 L.H., Daily Totals		$4090.40		$5087.92	$127.83	$159.00
Crew B-82C	Hr.	Daily	Hr.	Daily	Bare Costs	Incl. O&P
2 Laborers	$37.60	$601.60	$57.85	$925.60	$43.10	$65.80
2 Equip. Opers. (light)	48.60	777.60	73.75	1180.00		
2 Dump Truck, 8 C.Y., 220 H.P.		822.40		904.64		
1 Flatbed Trailer, 25 Ton		113.60		124.96		
1 Horiz. Dir. Drill, 50k lb. Thrust		1237.00		1360.70		
1 Mud Trailer for HDD, 1500 Gal.		346.20		380.82		
1 Pickup Truck, 4 x 4, 3/4 Ton		155.60		171.16		
1 Flatbed trailer, 3 Ton		23.60		25.96		
1 Loader, Skid Steer, 78 H.P.		318.60		350.46	94.28	103.71
32 L.H., Daily Totals		$4396.20		$5424.30	$137.38	$169.51
Crew B-82D	Hr.	Daily	Hr.	Daily	Bare Costs	Incl. O&P
1 Equip. Oper. (light)	$48.60	$388.80	$73.75	$590.00	$48.60	$73.75
1 Mud Trailer for HDD, 1500 Gal.		346.20		380.82	43.27	47.60
8 L.H., Daily Totals		$735.00		$970.82	$91.88	$121.35
Crew B-83	Hr.	Daily	Hr.	Daily	Bare Costs	Incl. O&P
1 Tugboat Captain	$50.60	$404.80	$76.75	$614.00	$44.10	$67.30
1 Tugboat Hand	37.60	300.80	57.85	462.80		
1 Tugboat, 250 H.P.		870.00		957.00	54.38	59.81
16 L.H., Daily Totals		$1575.60		$2033.80	$98.47	$127.11
Crew B-84	Hr.	Daily	Hr.	Daily	Bare Costs	Incl. O&P
1 Equip. Oper. (medium)	$50.60	$404.80	$76.75	$614.00	$50.60	$76.75
1 Rotary Mower/Tractor		370.80		407.88	46.35	50.98
8 L.H., Daily Totals		$775.60		$1021.88	$96.95	$127.74
Crew B-85	Hr.	Daily	Hr.	Daily	Bare Costs	Incl. O&P
3 Laborers	$37.60	$902.40	$57.85	$1388.40	$40.69	$62.17
1 Equip. Oper. (medium)	50.60	404.80	76.75	614.00		
1 Truck Driver (heavy)	40.05	320.40	60.55	484.40		
1 Aerial Lift Truck, 80'		634.40		697.84		
1 Brush Chipper, 12", 130 H.P.		391.40		430.54		
1 Pruning Saw, Rotary		6.65		7.32	25.81	28.39
40 L.H., Daily Totals		$2660.05		$3622.49	$66.50	$90.56
Crew B-86	Hr.	Daily	Hr.	Daily	Bare Costs	Incl. O&P
1 Equip. Oper. (medium)	$50.60	$404.80	$76.75	$614.00	$50.60	$76.75
1 Stump Chipper, S.P.		185.55		204.10	23.19	25.51
8 L.H., Daily Totals		$590.35		$818.11	$73.79	$102.26
Crew B-86A	Hr.	Daily	Hr.	Daily	Bare Costs	Incl. O&P
1 Equip. Oper. (medium)	$50.60	$404.80	$76.75	$614.00	$50.60	$76.75
1 Grader, 30,000 Lbs.		736.20		809.82	92.03	101.23
8 L.H., Daily Totals		$1141.00		$1423.82	$142.63	$177.98
Crew B-86B	Hr.	Daily	Hr.	Daily	Bare Costs	Incl. O&P
1 Equip. Oper. (medium)	$50.60	$404.80	$76.75	$614.00	$50.60	$76.75
1 Dozer, 200 H.P.		1387.00		1525.70	173.38	190.71
8 L.H., Daily Totals		$1791.80		$2139.70	$223.97	$267.46
Crew B-87	Hr.	Daily	Hr.	Daily	Bare Costs	Incl. O&P
1 Laborer	$37.60	$300.80	$57.85	$462.80	$48.00	$72.97
4 Equip. Oper. (medium)	50.60	1619.20	76.75	2456.00		
2 Feller Bunchers, 100 H.P.		1676.80		1844.48		
1 Log Chipper, 22" Tree		889.40		978.34		
1 Dozer, 105 H.P.		602.80		663.08		
1 Chain Saw, Gas, 36" Long		45.00		49.50	80.35	88.39
40 L.H., Daily Totals		$5134.00		$6454.20	$128.35	$161.35
Crew B-88	Hr.	Daily	Hr.	Daily	Bare Costs	Incl. O&P
1 Laborer	$37.60	$300.80	$57.85	$462.80	$48.74	$74.05
6 Equip. Oper. (medium)	50.60	2428.80	76.75	3684.00		
2 Feller Bunchers, 100 H.P.		1676.80		1844.48		
1 Log Chipper, 22" Tree		889.40		978.34		
2 Log Skidders, 50 H.P.		1822.00		2004.20		
1 Dozer, 105 H.P.		602.80		663.08		
1 Chain Saw, Gas, 36" Long		45.00		49.50	89.93	98.92
56 L.H., Daily Totals		$7765.60		$9686.40	$138.67	$172.97
Crew B-89	Hr.	Daily	Hr.	Daily	Bare Costs	Incl. O&P
1 Equip. Oper. (light)	$48.60	$388.80	$73.75	$590.00	$43.85	$66.42
1 Truck Driver (light)	39.10	312.80	59.10	472.80		
1 Flatbed Truck, Gas, 3 Ton		303.20		333.52		
1 Concrete Saw		167.40		184.14		
1 Water Tank, 65 Gal.		17.30		19.03	30.49	33.54
16 L.H., Daily Totals		$1189.50		$1599.49	$74.34	$99.97
Crew B-89A	Hr.	Daily	Hr.	Daily	Bare Costs	Incl. O&P
1 Skilled Worker	$48.65	$389.20	$75.15	$601.20	$43.13	$66.50
1 Laborer	37.60	300.80	57.85	462.80		
1 Core Drill (Large)		115.60		127.16	7.22	7.95
16 L.H., Daily Totals		$805.60		$1191.16	$50.35	$74.45
Crew B-89B	Hr.	Daily	Hr.	Daily	Bare Costs	Incl. O&P
1 Equip. Oper. (light)	$48.60	$388.80	$73.75	$590.00	$43.85	$66.42
1 Truck Driver (light)	39.10	312.80	59.10	472.80		
1 Wall Saw, Hydraulic, 10 H.P.		114.60		126.06		
1 Generator, Diesel, 100 kW		423.40		465.74		
1 Water Tank, 65 Gal.		17.30		19.03		
1 Flatbed Truck, Gas, 3 Ton		303.20		333.52	53.66	59.02
16 L.H., Daily Totals		$1560.10		$2007.15	$97.51	$125.45

Crew No.	Bare Costs		Incl. Subs O&P		Cost Per Labor-Hour	

Crew B-90

Crew B-90	Hr.	Daily	Hr.	Daily	Bare Costs	Incl. O&P
1 Labor Foreman (outside)	$39.60	$316.80	$60.95	$487.60	$41.21	$62.89
3 Laborers	37.60	902.40	57.85	1388.40		
2 Equip. Oper. (light)	48.60	777.60	73.75	1180.00		
2 Truck Drivers (heavy)	40.05	640.80	60.55	968.80		
1 Road Mixer, 310 H.P.		1929.00		2121.90		
1 Dist. Truck, 2000 Gal.		275.60		303.16	34.45	37.89
64 L.H., Daily Totals		$4842.20		$6449.86	$75.66	$100.78

Crew B-90A	Hr.	Daily	Hr.	Daily	Bare Costs	Incl. O&P
1 Labor Foreman (outside)	$39.60	$316.80	$60.95	$487.60	$45.31	$69.09
2 Laborers	37.60	601.60	57.85	925.60		
4 Equip. Oper. (medium)	50.60	1619.20	76.75	2456.00		
2 Graders, 30,000 Lbs.		1472.40		1619.64		
1 Tandem Roller, 10 Ton		236.80		260.48		
1 Roller, Pneum. Whl., 12 Ton		344.40		378.84	36.67	40.34
56 L.H., Daily Totals		$4591.20		$6128.16	$81.99	$109.43

Crew B-90B	Hr.	Daily	Hr.	Daily	Bare Costs	Incl. O&P
1 Labor Foreman (outside)	$39.60	$316.80	$60.95	$487.60	$44.43	$67.82
2 Laborers	37.60	601.60	57.85	925.60		
3 Equip. Oper. (medium)	50.60	1214.40	76.75	1842.00		
1 Roller, Pneum. Whl., 12 Ton		344.40		378.84		
1 Road Mixer, 310 H.P.		1929.00		2121.90	47.36	52.10
48 L.H., Daily Totals		$4406.20		$5755.94	$91.80	$119.92

Crew B-90C	Hr.	Daily	Hr.	Daily	Bare Costs	Incl. O&P
1 Labor Foreman (outside)	$39.60	$316.80	$60.95	$487.60	$42.00	$64.02
4 Laborers	37.60	1203.20	57.85	1851.20		
3 Equip. Oper. (medium)	50.60	1214.40	76.75	1842.00		
3 Truck Drivers (heavy)	40.05	961.20	60.55	1453.20		
3 Road Mixers, 310 H.P.		5787.00		6365.70	65.76	72.34
88 L.H., Daily Totals		$9482.60		$11999.70	$107.76	$136.36

Crew B-90D	Hr.	Daily	Hr.	Daily	Bare Costs	Incl. O&P
1 Labor Foreman (outside)	$39.60	$316.80	$60.95	$487.60	$41.32	$63.07
6 Laborers	37.60	1804.80	57.85	2776.80		
3 Equip. Oper. (medium)	50.60	1214.40	76.75	1842.00		
3 Truck Drivers (heavy)	40.05	961.20	60.55	1453.20		
3 Road Mixers, 310 H.P.		5787.00		6365.70	55.64	61.21
104 L.H., Daily Totals		$10084.20		$12925.30	$96.96	$124.28

Crew B-90E	Hr.	Daily	Hr.	Daily	Bare Costs	Incl. O&P
1 Labor Foreman (outside)	$39.60	$316.80	$60.95	$487.60	$42.43	$64.79
4 Laborers	37.60	1203.20	57.85	1851.20		
3 Equip. Oper. (medium)	50.60	1214.40	76.75	1842.00		
1 Truck Driver (heavy)	40.05	320.40	60.55	484.40		
1 Road Mixers, 310 H.P.		1929.00		2121.90	26.79	29.47
72 L.H., Daily Totals		$4983.80		$6787.10	$69.22	$94.27

Crew B-91	Hr.	Daily	Hr.	Daily	Bare Costs	Incl. O&P
1 Labor Foreman (outside)	$39.60	$316.80	$60.95	$487.60	$44.66	$68.03
2 Laborers	37.60	601.60	57.85	925.60		
4 Equip. Oper. (medium)	50.60	1619.20	76.75	2456.00		
1 Truck Driver (heavy)	40.05	320.40	60.55	484.40		
1 Dist. Tanker, 3000 Gallon		305.60		336.16		
1 Truck Tractor, 6x4, 380 H.P.		599.80		659.78		
1 Aggreg. Spreader, S.P.		834.00		917.40		
1 Roller, Pneum. Whl., 12 Ton		344.40		378.84		
1 Tandem Roller, 10 Ton		236.80		260.48	36.26	39.89
64 L.H., Daily Totals		$5178.60		$6906.26	$80.92	$107.91

Crew B-91B	Hr.	Daily	Hr.	Daily	Bare Costs	Incl. O&P
1 Laborer	$37.60	$300.80	$57.85	$462.80	$44.10	$67.30
1 Equipment Oper. (med.)	50.60	404.80	76.75	614.00		
1 Road Sweeper, Vac. Assist.		986.00		1084.60	61.63	67.79
16 L.H., Daily Totals		$1691.60		$2161.40	$105.72	$135.09

Crew B-91C	Hr.	Daily	Hr.	Daily	Bare Costs	Incl. O&P
1 Laborer	$37.60	$300.80	$57.85	$462.80	$38.35	$58.48
1 Truck Driver (light)	39.10	312.80	59.10	472.80		
1 Catch Basin Cleaning Truck		573.60		630.96	35.85	39.44
16 L.H., Daily Totals		$1187.20		$1566.56	$74.20	$97.91

Crew B-91D	Hr.	Daily	Hr.	Daily	Bare Costs	Incl. O&P
1 Labor Foreman (outside)	$39.60	$316.80	$60.95	$487.60	$43.13	$65.77
5 Laborers	37.60	1504.00	57.85	2314.00		
5 Equip. Oper. (medium)	50.60	2024.00	76.75	3070.00		
2 Truck Drivers (heavy)	40.05	640.80	60.55	968.80		
1 Aggreg. Spreader, S.P.		834.00		917.40		
2 Truck Tractor, 6x4, 380 H.P.		1199.60		1319.56		
2 Dist. Tanker, 3000 Gallon		611.20		672.32		
2 Pavement Brush, Towed		163.20		179.52		
2 Roller, Pneum. Whl., 12 Ton		688.80		757.68	33.62	36.99
104 L.H., Daily Totals		$7982.40		$10686.88	$76.75	$102.76

Crew B-92	Hr.	Daily	Hr.	Daily	Bare Costs	Incl. O&P
1 Labor Foreman (outside)	$39.60	$316.80	$60.95	$487.60	$38.10	$58.63
3 Laborers	37.60	902.40	57.85	1388.40		
1 Crack Cleaner, 25 H.P.		61.60		67.76		
1 Air Compressor, 60 cfm		137.40		151.14		
1 Tar Kettle, T.M.		132.00		145.20		
1 Flatbed Truck, Gas, 3 Ton		303.20		333.52	19.82	21.80
32 L.H., Daily Totals		$1853.40		$2573.62	$57.92	$80.43

Crew B-93	Hr.	Daily	Hr.	Daily	Bare Costs	Incl. O&P
1 Equip. Oper. (medium)	$50.60	$404.80	$76.75	$614.00	$50.60	$76.75
1 Feller Buncher, 100 H.P.		838.40		922.24	104.80	115.28
8 L.H., Daily Totals		$1243.20		$1536.24	$155.40	$192.03

Crew B-94A	Hr.	Daily	Hr.	Daily	Bare Costs	Incl. O&P
1 Laborer	$37.60	$300.80	$57.85	$462.80	$37.60	$57.85
1 Diaphragm Water Pump, 2"		71.00		78.10		
1 -20' Suction Hose, 2"		1.95		2.15		
2 -50' Discharge Hoses, 2"		1.80		1.98	9.34	10.28
8 L.H., Daily Totals		$375.55		$545.02	$46.94	$68.13

Crew B-94B	Hr.	Daily	Hr.	Daily	Bare Costs	Incl. O&P
1 Laborer	$37.60	$300.80	$57.85	$462.80	$37.60	$57.85
1 Diaphragm Water Pump, 4"		114.20		125.62		
1 -20' Suction Hose, 4"		3.25		3.58		
2 -50' Discharge Hoses, 4"		4.70		5.17	15.27	16.80
8 L.H., Daily Totals		$422.95		$597.16	$52.87	$74.65

Crew B-94C	Hr.	Daily	Hr.	Daily	Bare Costs	Incl. O&P
1 Laborer	$37.60	$300.80	$57.85	$462.80	$37.60	$57.85
1 Centrifugal Water Pump, 3"		78.20		86.02		
1 -20' Suction Hose, 3"		2.85		3.13		
2 -50' Discharge Hoses, 3"		3.00		3.30	10.51	11.56
8 L.H., Daily Totals		$384.85		$555.26	$48.11	$69.41

Crew No.	Bare Costs		Incl. Subs O&P		Cost Per Labor-Hour	

Crew B-94D

	Hr.	Daily	Hr.	Daily	Bare Costs	Incl. O&P
1 Laborer	$37.60	$300.80	$57.85	$462.80	$37.60	$57.85
1 Centr. Water Pump, 6"		346.60		381.26		
1 -20' Suction Hose, 6"		11.50		12.65		
2 -50' Discharge Hoses, 6"		12.20		13.42	46.29	50.92
8 L.H., Daily Totals		$671.10		$870.13	$83.89	$108.77

Crew C-1

	Hr.	Daily	Hr.	Daily	Bare Costs	Incl. O&P
3 Carpenters	$46.95	$1126.80	$72.25	$1734.00	$44.61	$68.65
1 Laborer	37.60	300.80	57.85	462.80		
32 L.H., Daily Totals		$1427.60		$2196.80	$44.61	$68.65

Crew C-2

	Hr.	Daily	Hr.	Daily	Bare Costs	Incl. O&P
1 Carpenter Foreman (outside)	$48.95	$391.60	$75.35	$602.80	$45.73	$70.37
4 Carpenters	46.95	1502.40	72.25	2312.00		
1 Laborer	37.60	300.80	57.85	462.80		
48 L.H., Daily Totals		$2194.80		$3377.60	$45.73	$70.37

Crew C-2A

	Hr.	Daily	Hr.	Daily	Bare Costs	Incl. O&P
1 Carpenter Foreman (outside)	$48.95	$391.60	$75.35	$602.80	$45.40	$69.43
3 Carpenters	46.95	1126.80	72.25	1734.00		
1 Cement Finisher	45.00	360.00	66.65	533.20		
1 Laborer	37.60	300.80	57.85	462.80		
48 L.H., Daily Totals		$2179.20		$3332.80	$45.40	$69.43

Crew C-3

	Hr.	Daily	Hr.	Daily	Bare Costs	Incl. O&P
1 Rodman Foreman (outside)	$54.55	$436.40	$85.20	$681.60	$48.57	$75.38
4 Rodmen (reinf.)	52.55	1681.60	82.10	2627.20		
1 Equip. Oper. (light)	48.60	388.80	73.75	590.00		
2 Laborers	37.60	601.60	57.85	925.60		
3 Stressing Equipment		30.60		33.66		
.5 Grouting Equipment		81.50		89.65	1.75	1.93
64 L.H., Daily Totals		$3220.50		$4947.71	$50.32	$77.31

Crew C-4

	Hr.	Daily	Hr.	Daily	Bare Costs	Incl. O&P
1 Rodman Foreman (outside)	$54.55	$436.40	$85.20	$681.60	$53.05	$82.88
3 Rodmen (reinf.)	52.55	1261.20	82.10	1970.40		
3 Stressing Equipment		30.60		33.66	.96	1.05
32 L.H., Daily Totals		$1728.20		$2685.66	$54.01	$83.93

Crew C-4A

	Hr.	Daily	Hr.	Daily	Bare Costs	Incl. O&P
2 Rodmen (reinf.)	$52.55	$840.80	$82.10	$1313.60	$52.55	$82.10
4 Stressing Equipment		40.80		44.88	2.55	2.81
16 L.H., Daily Totals		$881.60		$1358.48	$55.10	$84.91

Crew C-5

	Hr.	Daily	Hr.	Daily	Bare Costs	Incl. O&P
1 Rodman Foreman (outside)	$54.55	$436.40	$85.20	$681.60	$51.66	$80.09
4 Rodmen (reinf.)	52.55	1681.60	82.10	2627.20		
1 Equip. Oper. (crane)	51.70	413.60	78.45	627.60		
1 Equip. Oper. (oiler)	45.20	361.60	68.55	548.40		
1 Hyd. Crane, 25 Ton		736.60		810.26	13.15	14.47
56 L.H., Daily Totals		$3629.80		$5295.06	$64.82	$94.55

Crew C-6

	Hr.	Daily	Hr.	Daily	Bare Costs	Incl. O&P
1 Labor Foreman (outside)	$39.60	$316.80	$60.95	$487.60	$39.17	$59.83
4 Laborers	37.60	1203.20	57.85	1851.20		
1 Cement Finisher	45.00	360.00	66.65	533.20		
2 Gas Engine Vibrators		62.40		68.64	1.30	1.43
48 L.H., Daily Totals		$1942.40		$2940.64	$40.47	$61.26

Crew C-7

	Hr.	Daily	Hr.	Daily	Bare Costs	Incl. O&P
1 Labor Foreman (outside)	$39.60	$316.80	$60.95	$487.60	$40.93	$62.46
5 Laborers	37.60	1504.00	57.85	2314.00		
1 Cement Finisher	45.00	360.00	66.65	533.20		
1 Equip. Oper. (medium)	50.60	404.80	76.75	614.00		
1 Equip. Oper. (oiler)	45.20	361.60	68.55	548.40		
2 Gas Engine Vibrators		62.40		68.64		
1 Concrete Bucket, 1 C.Y.		23.80		26.18		
1 Hyd. Crane, 55 Ton		1128.00		1240.80	16.86	18.55
72 L.H., Daily Totals		$4161.40		$5832.82	$57.80	$81.01

Crew C-7A

	Hr.	Daily	Hr.	Daily	Bare Costs	Incl. O&P
1 Labor Foreman (outside)	$39.60	$316.80	$60.95	$487.60	$38.46	$58.91
5 Laborers	37.60	1504.00	57.85	2314.00		
2 Truck Drivers (heavy)	40.05	640.80	60.55	968.80		
2 Conc. Transit Mixers		2032.00		2235.20	31.75	34.92
64 L.H., Daily Totals		$4493.60		$6005.60	$70.21	$93.84

Crew C-7B

	Hr.	Daily	Hr.	Daily	Bare Costs	Incl. O&P
1 Labor Foreman (outside)	$39.60	$316.80	$60.95	$487.60	$40.56	$62.15
5 Laborers	37.60	1504.00	57.85	2314.00		
1 Equipment Operator, Crane	51.70	413.60	78.45	627.60		
1 Equipment Oiler	45.20	361.60	68.55	548.40		
1 Conc. Bucket, 2 C.Y.		37.00		40.70		
1 Lattice Boom Crane, 165 Ton		1932.00		2125.20	30.77	33.84
64 L.H., Daily Totals		$4565.00		$6143.50	$71.33	$95.99

Crew C-7C

	Hr.	Daily	Hr.	Daily	Bare Costs	Incl. O&P
1 Labor Foreman (outside)	$39.60	$316.80	$60.95	$487.60	$41.10	$62.96
5 Laborers	37.60	1504.00	57.85	2314.00		
2 Equipment Operators (med.)	50.60	809.60	76.75	1228.00		
2 F.E. Loaders, W.M., 4 C.Y.		1338.40		1472.24	20.91	23.00
64 L.H., Daily Totals		$3968.80		$5501.84	$62.01	$85.97

Crew C-7D

	Hr.	Daily	Hr.	Daily	Bare Costs	Incl. O&P
1 Labor Foreman (outside)	$39.60	$316.80	$60.95	$487.60	$39.74	$60.99
5 Laborers	37.60	1504.00	57.85	2314.00		
1 Equip. Oper. (medium)	50.60	404.80	76.75	614.00		
1 Concrete Conveyer		198.80		218.68	3.55	3.90
56 L.H., Daily Totals		$2424.40		$3634.28	$43.29	$64.90

Crew C-8

	Hr.	Daily	Hr.	Daily	Bare Costs	Incl. O&P
1 Labor Foreman (outside)	$39.60	$316.80	$60.95	$487.60	$41.86	$63.51
3 Laborers	37.60	902.40	57.85	1388.40		
2 Cement Finishers	45.00	720.00	66.65	1066.40		
1 Equip. Oper. (medium)	50.60	404.80	76.75	614.00		
1 Concrete Pump (Small)		720.00		792.00	12.86	14.14
56 L.H., Daily Totals		$3064.00		$4348.40	$54.71	$77.65

Crew C-8A

	Hr.	Daily	Hr.	Daily	Bare Costs	Incl. O&P
1 Labor Foreman (outside)	$39.60	$316.80	$60.95	$487.60	$40.40	$61.30
3 Laborers	37.60	902.40	57.85	1388.40		
2 Cement Finishers	45.00	720.00	66.65	1066.40		
48 L.H., Daily Totals		$1939.20		$2942.40	$40.40	$61.30

589

Crew No.	Bare Costs		Incl. Subs O&P		Cost Per Labor-Hour	
Crew C-8B	**Hr.**	**Daily**	**Hr.**	**Daily**	**Bare Costs**	**Incl. O&P**
1 Labor Foreman (outside)	$39.60	$316.80	$60.95	$487.60	$40.60	$62.25
3 Laborers	37.60	902.40	57.85	1388.40		
1 Equip. Oper. (medium)	50.60	404.80	76.75	614.00		
1 Vibrating Power Screed		66.30		72.93		
1 Roller, Vibratory, 25 Ton		687.40		756.14		
1 Dozer, 200 H.P.		1387.00		1525.70	53.52	58.87
40 L.H., Daily Totals		$3764.70		$4844.77	$94.12	$121.12
Crew C-8C	**Hr.**	**Daily**	**Hr.**	**Daily**	**Bare Costs**	**Incl. O&P**
1 Labor Foreman (outside)	$39.60	$316.80	$60.95	$487.60	$41.33	$62.98
3 Laborers	37.60	902.40	57.85	1388.40		
1 Cement Finisher	45.00	360.00	66.65	533.20		
1 Equip. Oper. (medium)	50.60	404.80	76.75	614.00		
1 Shotcrete Rig, 12 C.Y./hr		250.80		275.88		
1 Air Compressor, 160 cfm		156.40		172.04		
4 -50' Air Hoses, 1"		16.40		18.04		
4 -50' Air Hoses, 2"		31.00		34.10	9.47	10.42
48 L.H., Daily Totals		$2438.60		$3523.26	$50.80	$73.40
Crew C-8D	**Hr.**	**Daily**	**Hr.**	**Daily**	**Bare Costs**	**Incl. O&P**
1 Labor Foreman (outside)	$39.60	$316.80	$60.95	$487.60	$42.70	$64.80
1 Laborer	37.60	300.80	57.85	462.80		
1 Cement Finisher	45.00	360.00	66.65	533.20		
1 Equipment Oper. (light)	48.60	388.80	73.75	590.00		
1 Air Compressor, 250 cfm		201.40		221.54		
2 -50' Air Hoses, 1"		8.20		9.02	6.55	7.21
32 L.H., Daily Totals		$1576.00		$2304.16	$49.25	$72.00
Crew C-8E	**Hr.**	**Daily**	**Hr.**	**Daily**	**Bare Costs**	**Incl. O&P**
1 Labor Foreman (outside)	$39.60	$316.80	$60.95	$487.60	$41.00	$62.48
3 Laborers	37.60	902.40	57.85	1388.40		
1 Cement Finisher	45.00	360.00	66.65	533.20		
1 Equipment Oper. (light)	48.60	388.80	73.75	590.00		
1 Shotcrete Rig, 35 C.Y./hr		281.00		309.10		
1 Air Compressor, 250 cfm		201.40		221.54		
4 -50' Air Hoses, 1"		16.40		18.04		
4 -50' Air Hoses, 2"		31.00		34.10	11.04	12.14
48 L.H., Daily Totals		$2497.80		$3581.98	$52.04	$74.62
Crew C-10	**Hr.**	**Daily**	**Hr.**	**Daily**	**Bare Costs**	**Incl. O&P**
1 Laborer	$37.60	$300.80	$57.85	$462.80	$42.53	$63.72
2 Cement Finishers	45.00	720.00	66.65	1066.40		
24 L.H., Daily Totals		$1020.80		$1529.20	$42.53	$63.72
Crew C-10B	**Hr.**	**Daily**	**Hr.**	**Daily**	**Bare Costs**	**Incl. O&P**
3 Laborers	$37.60	$902.40	$57.85	$1388.40	$40.56	$61.37
2 Cement Finishers	45.00	720.00	66.65	1066.40		
1 Concrete Mixer, 10 C.F.		172.40		189.64		
2 Trowels, 48" Walk-Behind		100.40		110.44	6.82	7.50
40 L.H., Daily Totals		$1895.20		$2754.88	$47.38	$68.87
Crew C-10C	**Hr.**	**Daily**	**Hr.**	**Daily**	**Bare Costs**	**Incl. O&P**
1 Laborer	$37.60	$300.80	$57.85	$462.80	$42.53	$63.72
2 Cement Finishers	45.00	720.00	66.65	1066.40		
1 Trowel, 48" Walk-Behind		50.20		55.22	2.09	2.30
24 L.H., Daily Totals		$1071.00		$1584.42	$44.63	$66.02

Crew No.	Bare Costs		Incl. Subs O&P		Cost Per Labor-Hour	
Crew C-10D	**Hr.**	**Daily**	**Hr.**	**Daily**	**Bare Costs**	**Incl. O&P**
1 Laborer	$37.60	$300.80	$57.85	$462.80	$42.53	$63.72
2 Cement Finishers	45.00	720.00	66.65	1066.40		
1 Vibrating Power Screed		66.30		72.93		
1 Trowel, 48" Walk-Behind		50.20		55.22	4.85	5.34
24 L.H., Daily Totals		$1137.30		$1657.35	$47.39	$69.06
Crew C-10E	**Hr.**	**Daily**	**Hr.**	**Daily**	**Bare Costs**	**Incl. O&P**
1 Laborer	$37.60	$300.80	$57.85	$462.80	$42.53	$63.72
2 Cement Finishers	45.00	720.00	66.65	1066.40		
1 Vibrating Power Screed		66.30		72.93		
1 Cement Trowel, 96" Ride-On		189.00		207.90	10.64	11.70
24 L.H., Daily Totals		$1276.10		$1810.03	$53.17	$75.42
Crew C-10F	**Hr.**	**Daily**	**Hr.**	**Daily**	**Bare Costs**	**Incl. O&P**
1 Laborer	$37.60	$300.80	$57.85	$462.80	$42.53	$63.72
2 Cement Finishers	45.00	720.00	66.65	1066.40		
1 Aerial Lift Truck, 60' Boom		435.60		479.16	18.15	19.97
24 L.H., Daily Totals		$1456.40		$2008.36	$60.68	$83.68
Crew C-11	**Hr.**	**Daily**	**Hr.**	**Daily**	**Bare Costs**	**Incl. O&P**
1 Struc. Steel Foreman (outside)	$54.65	$437.20	$94.95	$759.60	$51.94	$87.85
6 Struc. Steel Workers	52.65	2527.20	91.45	4389.60		
1 Equip. Oper. (crane)	51.70	413.60	78.45	627.60		
1 Equip. Oper. (oiler)	45.20	361.60	68.55	548.40		
1 Lattice Boom Crane, 150 Ton		1813.00		1994.30	25.18	27.70
72 L.H., Daily Totals		$5552.60		$8319.50	$77.12	$115.55
Crew C-12	**Hr.**	**Daily**	**Hr.**	**Daily**	**Bare Costs**	**Incl. O&P**
1 Carpenter Foreman (outside)	$48.95	$391.60	$75.35	$602.80	$46.52	$71.40
3 Carpenters	46.95	1126.80	72.25	1734.00		
1 Laborer	37.60	300.80	57.85	462.80		
1 Equip. Oper. (crane)	51.70	413.60	78.45	627.60		
1 Hyd. Crane, 12 Ton		653.80		719.18	13.62	14.98
48 L.H., Daily Totals		$2886.60		$4146.38	$60.14	$86.38
Crew C-13	**Hr.**	**Daily**	**Hr.**	**Daily**	**Bare Costs**	**Incl. O&P**
1 Struc. Steel Worker	$52.65	$421.20	$91.45	$731.60	$50.75	$85.05
1 Welder	52.65	421.20	91.45	731.60		
1 Carpenter	46.95	375.60	72.25	578.00		
1 Welder, Gas Engine, 300 amp		145.85		160.44	6.08	6.68
24 L.H., Daily Totals		$1363.85		$2201.64	$56.83	$91.73
Crew C-14	**Hr.**	**Daily**	**Hr.**	**Daily**	**Bare Costs**	**Incl. O&P**
1 Carpenter Foreman (outside)	$48.95	$391.60	$75.35	$602.80	$46.18	$70.93
5 Carpenters	46.95	1878.00	72.25	2890.00		
4 Laborers	37.60	1203.20	57.85	1851.20		
4 Rodmen (reinf.)	52.55	1681.60	82.10	2627.20		
2 Cement Finishers	45.00	720.00	66.65	1066.40		
1 Equip. Oper. (crane)	51.70	413.60	78.45	627.60		
1 Equip. Oper. (oiler)	45.20	361.60	68.55	548.40		
1 Hyd. Crane, 80 Ton		1625.00		1787.50	11.28	12.41
144 L.H., Daily Totals		$8274.60		$12001.10	$57.46	$83.34

Crews

Crew C-14A

Crew C-14A	Hr.	Daily	Hr.	Daily	Bare Costs	Incl. O&P
1 Carpenter Foreman (outside)	$48.95	$391.60	$75.35	$602.80	$47.25	$72.75
16 Carpenters	46.95	6009.60	72.25	9248.00		
4 Rodmen (reinf.)	52.55	1681.60	82.10	2627.20		
2 Laborers	37.60	601.60	57.85	925.60		
1 Cement Finisher	45.00	360.00	66.65	533.20		
1 Equip. Oper. (medium)	50.60	404.80	76.75	614.00		
1 Gas Engine Vibrator		31.20		34.32		
1 Concrete Pump (Small)		720.00		792.00	3.76	4.13
200 L.H., Daily Totals		$10200.40		$15377.12	$51.00	$76.89

Crew C-14B

Crew C-14B	Hr.	Daily	Hr.	Daily	Bare Costs	Incl. O&P
1 Carpenter Foreman (outside)	$48.95	$391.60	$75.35	$602.80	$47.16	$72.52
16 Carpenters	46.95	6009.60	72.25	9248.00		
4 Rodmen (reinf.)	52.55	1681.60	82.10	2627.20		
2 Laborers	37.60	601.60	57.85	925.60		
2 Cement Finishers	45.00	720.00	66.65	1066.40		
1 Equip. Oper. (medium)	50.60	404.80	76.75	614.00		
1 Gas Engine Vibrator		31.20		34.32		
1 Concrete Pump (Small)		720.00		792.00	3.61	3.97
208 L.H., Daily Totals		$10560.40		$15910.32	$50.77	$76.49

Crew C-14C

Crew C-14C	Hr.	Daily	Hr.	Daily	Bare Costs	Incl. O&P
1 Carpenter Foreman (outside)	$48.95	$391.60	$75.35	$602.80	$45.08	$69.36
6 Carpenters	46.95	2253.60	72.25	3468.00		
2 Rodmen (reinf.)	52.55	840.80	82.10	1313.60		
4 Laborers	37.60	1203.20	57.85	1851.20		
1 Cement Finisher	45.00	360.00	66.65	533.20		
1 Gas Engine Vibrator		31.20		34.32	.28	.31
112 L.H., Daily Totals		$5080.40		$7803.12	$45.36	$69.67

Crew C-14D

Crew C-14D	Hr.	Daily	Hr.	Daily	Bare Costs	Incl. O&P
1 Carpenter Foreman (outside)	$48.95	$391.60	$75.35	$602.80	$46.80	$71.97
18 Carpenters	46.95	6760.80	72.25	10404.00		
2 Rodmen (reinf.)	52.55	840.80	82.10	1313.60		
2 Laborers	37.60	601.60	57.85	925.60		
1 Cement Finisher	45.00	360.00	66.65	533.20		
1 Equip. Oper. (medium)	50.60	404.80	76.75	614.00		
1 Gas Engine Vibrator		31.20		34.32		
1 Concrete Pump (Small)		720.00		792.00	3.76	4.13
200 L.H., Daily Totals		$10110.80		$15219.52	$50.55	$76.10

Crew C-14E

Crew C-14E	Hr.	Daily	Hr.	Daily	Bare Costs	Incl. O&P
1 Carpenter Foreman (outside)	$48.95	$391.60	$75.35	$602.80	$46.44	$71.68
2 Carpenters	46.95	751.20	72.25	1156.00		
4 Rodmen (reinf.)	52.55	1681.60	82.10	2627.20		
3 Laborers	37.60	902.40	57.85	1388.40		
1 Cement Finisher	45.00	360.00	66.65	533.20		
1 Gas Engine Vibrator		31.20		34.32	.35	.39
88 L.H., Daily Totals		$4118.00		$6341.92	$46.80	$72.07

Crew C-14F

Crew C-14F	Hr.	Daily	Hr.	Daily	Bare Costs	Incl. O&P
1 Labor Foreman (outside)	$39.60	$316.80	$60.95	$487.60	$42.76	$64.06
2 Laborers	37.60	601.60	57.85	925.60		
6 Cement Finishers	45.00	2160.00	66.65	3199.20		
1 Gas Engine Vibrator		31.20		34.32	.43	.48
72 L.H., Daily Totals		$3109.60		$4646.72	$43.19	$64.54

Crew C-14G

Crew C-14G	Hr.	Daily	Hr.	Daily	Bare Costs	Incl. O&P
1 Labor Foreman (outside)	$39.60	$316.80	$60.95	$487.60	$42.11	$63.32
2 Laborers	37.60	601.60	57.85	925.60		
4 Cement Finishers	45.00	1440.00	66.65	2132.80		
1 Gas Engine Vibrator		31.20		34.32	.56	.61
56 L.H., Daily Totals		$2389.60		$3580.32	$42.67	$63.93

Crew C-14H

Crew C-14H	Hr.	Daily	Hr.	Daily	Bare Costs	Incl. O&P
1 Carpenter Foreman (outside)	$48.95	$391.60	$75.35	$602.80	$46.33	$71.08
2 Carpenters	46.95	751.20	72.25	1156.00		
1 Rodman (reinf.)	52.55	420.40	82.10	656.80		
1 Laborer	37.60	300.80	57.85	462.80		
1 Cement Finisher	45.00	360.00	66.65	533.20		
1 Gas Engine Vibrator		31.20		34.32	.65	.71
48 L.H., Daily Totals		$2255.20		$3445.92	$46.98	$71.79

Crew C-14L

Crew C-14L	Hr.	Daily	Hr.	Daily	Bare Costs	Incl. O&P
1 Carpenter Foreman (outside)	$48.95	$391.60	$75.35	$602.80	$43.84	$67.24
6 Carpenters	46.95	2253.60	72.25	3468.00		
4 Laborers	37.60	1203.20	57.85	1851.20		
1 Cement Finisher	45.00	360.00	66.65	533.20		
1 Gas Engine Vibrator		31.20		34.32	.33	.36
96 L.H., Daily Totals		$4239.60		$6489.52	$44.16	$67.60

Crew C-14M

Crew C-14M	Hr.	Daily	Hr.	Daily	Bare Costs	Incl. O&P
1 Carpenter Foreman (outside)	$48.95	$391.60	$75.35	$602.80	$45.77	$70.13
2 Carpenters	46.95	751.20	72.25	1156.00		
1 Rodman (reinf.)	52.55	420.40	82.10	656.80		
2 Laborers	37.60	601.60	57.85	925.60		
1 Cement Finisher	45.00	360.00	66.65	533.20		
1 Equip. Oper. (medium)	50.60	404.80	76.75	614.00		
1 Gas Engine Vibrator		31.20		34.32		
1 Concrete Pump (Small)		720.00		792.00	11.74	12.91
64 L.H., Daily Totals		$3680.80		$5314.72	$57.51	$83.04

Crew C-15

Crew C-15	Hr.	Daily	Hr.	Daily	Bare Costs	Incl. O&P
1 Carpenter Foreman (outside)	$48.95	$391.60	$75.35	$602.80	$44.24	$67.64
2 Carpenters	46.95	751.20	72.25	1156.00		
3 Laborers	37.60	902.40	57.85	1388.40		
2 Cement Finishers	45.00	720.00	66.65	1066.40		
1 Rodman (reinf.)	52.55	420.40	82.10	656.80		
72 L.H., Daily Totals		$3185.60		$4870.40	$44.24	$67.64

Crew C-16

Crew C-16	Hr.	Daily	Hr.	Daily	Bare Costs	Incl. O&P
1 Labor Foreman (outside)	$39.60	$316.80	$60.95	$487.60	$41.86	$63.51
3 Laborers	37.60	902.40	57.85	1388.40		
2 Cement Finishers	45.00	720.00	66.65	1066.40		
1 Equip. Oper. (medium)	50.60	404.80	76.75	614.00		
1 Gunite Pump Rig		371.80		408.98		
2 -50' Air Hoses, 3/4"		6.50		7.15		
2 -50' Air Hoses, 2"		15.50		17.05	7.03	7.74
56 L.H., Daily Totals		$2737.80		$3989.58	$48.89	$71.24

Crew C-16A

Crew C-16A	Hr.	Daily	Hr.	Daily	Bare Costs	Incl. O&P
1 Laborer	$37.60	$300.80	$57.85	$462.80	$44.55	$66.97
2 Cement Finishers	45.00	720.00	66.65	1066.40		
1 Equip. Oper. (medium)	50.60	404.80	76.75	614.00		
1 Gunite Pump Rig		371.80		408.98		
2 -50' Air Hoses, 3/4"		6.50		7.15		
2 -50' Air Hoses, 2"		15.50		17.05		
1 Aerial Lift Truck, 60' Boom		435.60		479.16	25.92	28.51
32 L.H., Daily Totals		$2255.00		$3055.54	$70.47	$95.49

591

For customer support on your Plumbing Cost Data, call 877.709.6968.

Crew C-17

Crew No.	Bare Costs Hr.	Daily	Incl. Subs O&P Hr.	Daily	Cost Per Labor-Hour Bare Costs	Incl. O&P
2 Skilled Worker Foremen (out)	$50.65	$810.40	$78.25	$1252.00	$49.05	$75.77
8 Skilled Workers	48.65	3113.60	75.15	4809.60		
80 L.H., Daily Totals		$3924.00		$6061.60	$49.05	$75.77

Crew C-17A

Crew No.	Bare Costs Hr.	Daily	Incl. Subs O&P Hr.	Daily	Cost Per Labor-Hour Bare Costs	Incl. O&P
2 Skilled Worker Foremen (out)	$50.65	$810.40	$78.25	$1252.00	$49.08	$75.80
8 Skilled Workers	48.65	3113.60	75.15	4809.60		
.125 Equip. Oper. (crane)	51.70	51.70	78.45	78.45		
.125 Hyd. Crane, 80 Ton		203.13		223.44	2.51	2.76
81 L.H., Daily Totals		$4178.82		$6363.49	$51.59	$78.56

Crew C-17B

Crew No.	Bare Costs Hr.	Daily	Incl. Subs O&P Hr.	Daily	Cost Per Labor-Hour Bare Costs	Incl. O&P
2 Skilled Worker Foremen (out)	$50.65	$810.40	$78.25	$1252.00	$49.11	$75.84
8 Skilled Workers	48.65	3113.60	75.15	4809.60		
.25 Equip. Oper. (crane)	51.70	103.40	78.45	156.90		
.25 Hyd. Crane, 80 Ton		406.25		446.88		
.25 Trowel, 48" Walk-Behind		12.55		13.81	5.11	5.62
82 L.H., Daily Totals		$4446.20		$6679.18	$54.22	$81.45

Crew C-17C

Crew No.	Bare Costs Hr.	Daily	Incl. Subs O&P Hr.	Daily	Cost Per Labor-Hour Bare Costs	Incl. O&P
2 Skilled Worker Foremen (out)	$50.65	$810.40	$78.25	$1252.00	$49.15	$75.87
8 Skilled Workers	48.65	3113.60	75.15	4809.60		
.375 Equip. Oper. (crane)	51.70	155.10	78.45	235.35		
.375 Hyd. Crane, 80 Ton		609.38		670.31	7.34	8.08
83 L.H., Daily Totals		$4688.48		$6967.26	$56.49	$83.94

Crew C-17D

Crew No.	Bare Costs Hr.	Daily	Incl. Subs O&P Hr.	Daily	Cost Per Labor-Hour Bare Costs	Incl. O&P
2 Skilled Worker Foremen (out)	$50.65	$810.40	$78.25	$1252.00	$49.18	$75.90
8 Skilled Workers	48.65	3113.60	75.15	4809.60		
.5 Equip. Oper. (crane)	51.70	206.80	78.45	313.80		
.5 Hyd. Crane, 80 Ton		812.50		893.75	9.67	10.64
84 L.H., Daily Totals		$4943.30		$7269.15	$58.85	$86.54

Crew C-17E

Crew No.	Bare Costs Hr.	Daily	Incl. Subs O&P Hr.	Daily	Cost Per Labor-Hour Bare Costs	Incl. O&P
2 Skilled Worker Foremen (out)	$50.65	$810.40	$78.25	$1252.00	$49.05	$75.77
8 Skilled Workers	48.65	3113.60	75.15	4809.60		
1 Hyd. Jack with Rods		96.50		106.15	1.21	1.33
80 L.H., Daily Totals		$4020.50		$6167.75	$50.26	$77.10

Crew C-18

Crew No.	Bare Costs Hr.	Daily	Incl. Subs O&P Hr.	Daily	Cost Per Labor-Hour Bare Costs	Incl. O&P
.125 Labor Foreman (outside)	$39.60	$39.60	$60.95	$60.95	$37.82	$58.19
1 Laborer	37.60	300.80	57.85	462.80		
1 Concrete Cart, 10 C.F.		60.00		66.00	6.67	7.33
9 L.H., Daily Totals		$400.40		$589.75	$44.49	$65.53

Crew C-19

Crew No.	Bare Costs Hr.	Daily	Incl. Subs O&P Hr.	Daily	Cost Per Labor-Hour Bare Costs	Incl. O&P
.125 Labor Foreman (outside)	$39.60	$39.60	$60.95	$60.95	$37.82	$58.19
1 Laborer	37.60	300.80	57.85	462.80		
1 Concrete Cart, 18 C.F.		99.80		109.78	11.09	12.20
9 L.H., Daily Totals		$440.20		$633.53	$48.91	$70.39

Crew C-20

Crew No.	Bare Costs Hr.	Daily	Incl. Subs O&P Hr.	Daily	Cost Per Labor-Hour Bare Costs	Incl. O&P
1 Labor Foreman (outside)	$39.60	$316.80	$60.95	$487.60	$40.40	$61.70
5 Laborers	37.60	1504.00	57.85	2314.00		
1 Cement Finisher	45.00	360.00	66.65	533.20		
1 Equip. Oper. (medium)	50.60	404.80	76.75	614.00		
2 Gas Engine Vibrators		62.40		68.64		
1 Concrete Pump (Small)		720.00		792.00	12.23	13.45
64 L.H., Daily Totals		$3368.00		$4809.44	$52.63	$75.15

Crew C-21

Crew No.	Bare Costs Hr.	Daily	Incl. Subs O&P Hr.	Daily	Cost Per Labor-Hour Bare Costs	Incl. O&P
1 Labor Foreman (outside)	$39.60	$316.80	$60.95	$487.60	$40.40	$61.70
5 Laborers	37.60	1504.00	57.85	2314.00		
1 Cement Finisher	45.00	360.00	66.65	533.20		
1 Equip. Oper. (medium)	50.60	404.80	76.75	614.00		
2 Gas Engine Vibrators		62.40		68.64		
1 Concrete Conveyer		198.80		218.68	4.08	4.49
64 L.H., Daily Totals		$2846.80		$4236.12	$44.48	$66.19

Crew C-22

Crew No.	Bare Costs Hr.	Daily	Incl. Subs O&P Hr.	Daily	Cost Per Labor-Hour Bare Costs	Incl. O&P
1 Rodman Foreman (outside)	$54.55	$436.40	$85.20	$681.60	$52.74	$82.28
4 Rodmen (reinf.)	52.55	1681.60	82.10	2627.20		
.125 Equip. Oper. (crane)	51.70	51.70	78.45	78.45		
.125 Equip. Oper. (oiler)	45.20	45.20	68.55	68.55		
.125 Hyd. Crane, 25 Ton		92.08		101.28	2.19	2.41
42 L.H., Daily Totals		$2306.97		$3557.08	$54.93	$84.69

Crew C-23

Crew No.	Bare Costs Hr.	Daily	Incl. Subs O&P Hr.	Daily	Cost Per Labor-Hour Bare Costs	Incl. O&P
2 Skilled Worker Foremen (out)	$50.65	$810.40	$78.25	$1252.00	$49.01	$75.44
6 Skilled Workers	48.65	2335.20	75.15	3607.20		
1 Equip. Oper. (crane)	.51.70	413.60	78.45	627.60		
1 Equip. Oper. (oiler)	45.20	361.60	68.55	548.40		
1 Lattice Boom Crane, 90 Ton		1511.00		1662.10	18.89	20.78
80 L.H., Daily Totals		$5431.80		$7697.30	$67.90	$96.22

Crew C-23A

Crew No.	Bare Costs Hr.	Daily	Incl. Subs O&P Hr.	Daily	Cost Per Labor-Hour Bare Costs	Incl. O&P
1 Labor Foreman (outside)	$39.60	$316.80	$60.95	$487.60	$42.34	$64.73
2 Laborers	37.60	601.60	57.85	925.60		
1 Equip. Oper. (crane)	51.70	413.60	78.45	627.60		
1 Equip. Oper. (oiler)	45.20	361.60	68.55	548.40		
1 Crawler Crane, 100 Ton		1667.00		1833.70		
3 Conc. Buckets, 8 C.Y.		608.40		669.24	56.88	62.57
40 L.H., Daily Totals		$3969.00		$5092.14	$99.22	$127.30

Crew C-24

Crew No.	Bare Costs Hr.	Daily	Incl. Subs O&P Hr.	Daily	Cost Per Labor-Hour Bare Costs	Incl. O&P
2 Skilled Worker Foremen (out)	$50.65	$810.40	$78.25	$1252.00	$49.01	$75.44
6 Skilled Workers	48.65	2335.20	75.15	3607.20		
1 Equip. Oper. (crane)	51.70	413.60	78.45	627.60		
1 Equip. Oper. (oiler)	45.20	361.60	68.55	548.40		
1 Lattice Boom Crane, 150 Ton		1813.00		1994.30	22.66	24.93
80 L.H., Daily Totals		$5733.80		$8029.50	$71.67	$100.37

Crew C-25

Crew No.	Bare Costs Hr.	Daily	Incl. Subs O&P Hr.	Daily	Cost Per Labor-Hour Bare Costs	Incl. O&P
2 Rodmen (reinf.)	$52.55	$840.80	$82.10	$1313.60	$41.30	$66.70
2 Rodmen Helpers	30.05	480.80	51.30	820.80		
32 L.H., Daily Totals		$1321.60		$2134.40	$41.30	$66.70

Crew C-27

Crew No.	Bare Costs Hr.	Daily	Incl. Subs O&P Hr.	Daily	Cost Per Labor-Hour Bare Costs	Incl. O&P
2 Cement Finishers	$45.00	$720.00	$66.65	$1066.40	$45.00	$66.65
1 Concrete Saw		167.40		184.14	10.46	11.51
16 L.H., Daily Totals		$887.40		$1250.54	$55.46	$78.16

Crew C-28

Crew No.	Bare Costs Hr.	Daily	Incl. Subs O&P Hr.	Daily	Cost Per Labor-Hour Bare Costs	Incl. O&P
1 Cement Finisher	$45.00	$360.00	$66.65	$533.20	$45.00	$66.65
1 Portable Air Compressor, Gas		17.90		19.69	2.24	2.46
8 L.H., Daily Totals		$377.90		$552.89	$47.24	$69.11

Crew C-29

Crew No.	Bare Costs Hr.	Daily	Incl. Subs O&P Hr.	Daily	Cost Per Labor-Hour Bare Costs	Incl. O&P
1 Laborer	$37.60	$300.80	$57.85	$462.80	$37.60	$57.85
1 Pressure Washer		70.20		77.22	8.78	9.65
8 L.H., Daily Totals		$371.00		$540.02	$46.38	$67.50

Crew C-30

	Bare Costs Hr.	Daily	Incl. Subs O&P Hr.	Daily	Cost Per Labor-Hour Bare Costs	Incl. O&P
1 Laborer	$37.60	$300.80	$57.85	$462.80	$37.60	$57.85
1 Concrete Mixer, 10 C.F.		172.40		189.64	21.55	23.70
8 L.H., Daily Totals		$473.20		$652.44	$59.15	$81.56

Crew C-31

	Bare Costs Hr.	Daily	Incl. Subs O&P Hr.	Daily	Cost Per Labor-Hour Bare Costs	Incl. O&P
1 Cement Finisher	$45.00	$360.00	$66.65	$533.20	$45.00	$66.65
1 Grout Pump		371.80		408.98	46.48	51.12
8 L.H., Daily Totals		$731.80		$942.18	$91.47	$117.77

Crew C-32

	Bare Costs Hr.	Daily	Incl. Subs O&P Hr.	Daily	Cost Per Labor-Hour Bare Costs	Incl. O&P
1 Cement Finisher	$45.00	$360.00	$66.65	$533.20	$41.30	$62.25
1 Laborer	37.60	300.80	57.85	462.80		
1 Crack Chaser Saw, Gas, 6 H.P.		29.20		32.12		
1 Vacuum Pick-Up System		60.90		66.99	5.63	6.19
16 L.H., Daily Totals		$750.90		$1095.11	$46.93	$68.44

Crew D-1

	Bare Costs Hr.	Daily	Incl. Subs O&P Hr.	Daily	Cost Per Labor-Hour Bare Costs	Incl. O&P
1 Bricklayer	$46.15	$369.20	$70.45	$563.60	$42.10	$64.28
1 Bricklayer Helper	38.05	304.40	58.10	464.80		
16 L.H., Daily Totals		$673.60		$1028.40	$42.10	$64.28

Crew D-2

	Bare Costs Hr.	Daily	Incl. Subs O&P Hr.	Daily	Cost Per Labor-Hour Bare Costs	Incl. O&P
3 Bricklayers	$46.15	$1107.60	$70.45	$1690.80	$43.28	$66.12
2 Bricklayer Helpers	38.05	608.80	58.10	929.60		
.5 Carpenter	46.95	187.80	72.25	289.00		
44 L.H., Daily Totals		$1904.20		$2909.40	$43.28	$66.12

Crew D-3

	Bare Costs Hr.	Daily	Incl. Subs O&P Hr.	Daily	Cost Per Labor-Hour Bare Costs	Incl. O&P
3 Bricklayers	$46.15	$1107.60	$70.45	$1690.80	$43.10	$65.83
2 Bricklayer Helpers	38.05	608.80	58.10	929.60		
.25 Carpenter	46.95	93.90	72.25	144.50		
42 L.H., Daily Totals		$1810.30		$2764.90	$43.10	$65.83

Crew D-4

	Bare Costs Hr.	Daily	Incl. Subs O&P Hr.	Daily	Cost Per Labor-Hour Bare Costs	Incl. O&P
1 Bricklayer	$46.15	$369.20	$70.45	$563.60	$42.71	$65.10
2 Bricklayer Helpers	38.05	608.80	58.10	929.60		
1 Equip. Oper. (light)	48.60	388.80	73.75	590.00		
1 Grout Pump, 50 C.F./hr.		133.40		146.74	4.17	4.59
32 L.H., Daily Totals		$1500.20		$2229.94	$46.88	$69.69

Crew D-5

	Bare Costs Hr.	Daily	Incl. Subs O&P Hr.	Daily	Cost Per Labor-Hour Bare Costs	Incl. O&P
1 Bricklayer	46.15	369.20	70.45	563.60	46.15	70.45
8 L.H., Daily Totals		$369.20		$563.60	$46.15	$70.45

Crew D-6

	Bare Costs Hr.	Daily	Incl. Subs O&P Hr.	Daily	Cost Per Labor-Hour Bare Costs	Incl. O&P
3 Bricklayers	$46.15	$1107.60	$70.45	$1690.80	$42.29	$64.59
3 Bricklayer Helpers	38.05	913.20	58.10	1394.40		
.25 Carpenter	46.95	93.90	72.25	144.50		
50 L.H., Daily Totals		$2114.70		$3229.70	$42.29	$64.59

Crew D-7

	Bare Costs Hr.	Daily	Incl. Subs O&P Hr.	Daily	Cost Per Labor-Hour Bare Costs	Incl. O&P
1 Tile Layer	$42.80	$342.40	$63.30	$506.40	$38.17	$56.45
1 Tile Layer Helper	33.55	268.40	49.60	396.80		
16 L.H., Daily Totals		$610.80		$903.20	$38.17	$56.45

Crew D-8

	Bare Costs Hr.	Daily	Incl. Subs O&P Hr.	Daily	Cost Per Labor-Hour Bare Costs	Incl. O&P
3 Bricklayers	$46.15	$1107.60	$70.45	$1690.80	$42.91	$65.51
2 Bricklayer Helpers	38.05	608.80	58.10	929.60		
40 L.H., Daily Totals		$1716.40		$2620.40	$42.91	$65.51

Crew D-9

	Bare Costs Hr.	Daily	Incl. Subs O&P Hr.	Daily	Cost Per Labor-Hour Bare Costs	Incl. O&P
3 Bricklayers	$46.15	$1107.60	$70.45	$1690.80	$42.10	$64.28
3 Bricklayer Helpers	38.05	913.20	58.10	1394.40		
48 L.H., Daily Totals		$2020.80		$3085.20	$42.10	$64.28

Crew D-10

	Bare Costs Hr.	Daily	Incl. Subs O&P Hr.	Daily	Cost Per Labor-Hour Bare Costs	Incl. O&P
1 Bricklayer Foreman (outside)	$48.15	$385.20	$73.55	$588.40	$46.01	$70.14
1 Bricklayer	46.15	369.20	70.45	563.60		
1 Bricklayer Helper	38.05	304.40	58.10	464.80		
1 Equip. Oper. (crane)	51.70	413.60	78.45	627.60		
1 S.P. Crane, 4x4, 12 Ton		473.40		520.74	14.79	16.27
32 L.H., Daily Totals		$1945.80		$2765.14	$60.81	$86.41

Crew D-11

	Bare Costs Hr.	Daily	Incl. Subs O&P Hr.	Daily	Cost Per Labor-Hour Bare Costs	Incl. O&P
1 Bricklayer Foreman (outside)	$48.15	$385.20	$73.55	$588.40	$44.12	$67.37
1 Bricklayer	46.15	369.20	70.45	563.60		
1 Bricklayer Helper	38.05	304.40	58.10	464.80		
24 L.H., Daily Totals		$1058.80		$1616.80	$44.12	$67.37

Crew D-12

	Bare Costs Hr.	Daily	Incl. Subs O&P Hr.	Daily	Cost Per Labor-Hour Bare Costs	Incl. O&P
1 Bricklayer Foreman (outside)	$48.15	$385.20	$73.55	$588.40	$42.60	$65.05
1 Bricklayer	46.15	369.20	70.45	563.60		
2 Bricklayer Helpers	38.05	608.80	58.10	929.60		
32 L.H., Daily Totals		$1363.20		$2081.60	$42.60	$65.05

Crew D-13

	Bare Costs Hr.	Daily	Incl. Subs O&P Hr.	Daily	Cost Per Labor-Hour Bare Costs	Incl. O&P
1 Bricklayer Foreman (outside)	$48.15	$385.20	$73.55	$588.40	$44.84	$68.48
1 Bricklayer	46.15	369.20	70.45	563.60		
2 Bricklayer Helpers	38.05	608.80	58.10	929.60		
1 Carpenter	46.95	375.60	72.25	578.00		
1 Equip. Oper. (crane)	51.70	413.60	78.45	627.60		
1 S.P. Crane, 4x4, 12 Ton		473.40		520.74	9.86	10.85
48 L.H., Daily Totals		$2625.80		$3807.94	$54.70	$79.33

Crew E-1

	Bare Costs Hr.	Daily	Incl. Subs O&P Hr.	Daily	Cost Per Labor-Hour Bare Costs	Incl. O&P
1 Welder Foreman (outside)	$54.65	$437.20	$94.95	$759.60	$51.97	$86.72
1 Welder	52.65	421.20	91.45	731.60		
1 Equip. Oper. (light)	48.60	388.80	73.75	590.00		
1 Welder, Gas Engine, 300 amp		145.85		160.44	6.08	6.68
24 L.H., Daily Totals		$1393.05		$2241.64	$58.04	$93.40

Crew E-2

	Bare Costs Hr.	Daily	Incl. Subs O&P Hr.	Daily	Cost Per Labor-Hour Bare Costs	Incl. O&P
1 Struc. Steel Foreman (outside)	$54.65	$437.20	$94.95	$759.60	$51.74	$86.82
4 Struc. Steel Workers	52.65	1684.80	91.45	2926.40		
1 Equip. Oper. (crane)	51.70	413.60	78.45	627.60		
1 Equip. Oper. (oiler)	45.20	361.60	68.55	548.40		
1 Lattice Boom Crane, 90 Ton		1511.00		1662.10	26.98	29.68
56 L.H., Daily Totals		$4408.20		$6524.10	$78.72	$116.50

Crew E-3

	Bare Costs Hr.	Daily	Incl. Subs O&P Hr.	Daily	Cost Per Labor-Hour Bare Costs	Incl. O&P
1 Struc. Steel Foreman (outside)	$54.65	$437.20	$94.95	$759.60	$53.32	$92.62
1 Struc. Steel Worker	52.65	421.20	91.45	731.60		
1 Welder	52.65	421.20	91.45	731.60		
1 Welder, Gas Engine, 300 amp		145.85		160.44	6.08	6.68
24 L.H., Daily Totals		$1425.45		$2383.24	$59.39	$99.30

Crew No.	Bare Costs		Incl. Subs O&P		Cost Per Labor-Hour	

Crew E-3A	Hr.	Daily	Hr.	Daily	Bare Costs	Incl. O&P
1 Struc. Steel Foreman (outside)	$54.65	$437.20	$94.95	$759.60	$53.32	$92.62
1 Struc. Steel Worker	52.65	421.20	91.45	731.60		
1 Welder	52.65	421.20	91.45	731.60		
1 Welder, Gas Engine, 300 amp		145.85		160.44		
1 Aerial Lift Truck, 40' Boom		301.40		331.54	18.64	20.50
24 L.H., Daily Totals		$1726.85		$2714.78	$71.95	$113.12

Crew E-4	Hr.	Daily	Hr.	Daily	Bare Costs	Incl. O&P
1 Struc. Steel Foreman (outside)	$54.65	$437.20	$94.95	$759.60	$53.15	$92.33
3 Struc. Steel Workers	52.65	1263.60	91.45	2194.80		
1 Welder, Gas Engine, 300 amp		145.85		160.44	4.56	5.01
32 L.H., Daily Totals		$1846.65		$3114.84	$57.71	$97.34

Crew E-5	Hr.	Daily	Hr.	Daily	Bare Costs	Incl. O&P
2 Struc. Steel Foremen (outside)	$54.65	$874.40	$94.95	$1519.20	$52.21	$88.56
5 Struc. Steel Workers	52.65	2106.00	91.45	3658.00		
1 Equip. Oper. (crane)	51.70	413.60	78.45	627.60		
1 Welder	52.65	421.20	91.45	731.60		
1 Equip. Oper. (oiler)	45.20	361.60	68.55	548.40		
1 Lattice Boom Crane, 90 Ton		1511.00		1662.10		
1 Welder, Gas Engine, 300 amp		145.85		160.44	20.71	22.78
80 L.H., Daily Totals		$5833.65		$8907.33	$72.92	$111.34

Crew E-6	Hr.	Daily	Hr.	Daily	Bare Costs	Incl. O&P
3 Struc. Steel Foremen (outside)	$54.65	$1311.60	$94.95	$2278.80	$52.25	$88.76
9 Struc. Steel Workers	52.65	3790.80	91.45	6584.40		
1 Equip. Oper. (crane)	51.70	413.60	78.45	627.60		
1 Welder	52.65	421.20	91.45	731.60		
1 Equip. Oper. (oiler)	45.20	361.60	68.55	548.40		
1 Equip. Oper. (light)	48.60	388.80	73.75	590.00		
1 Lattice Boom Crane, 90 Ton		1511.00		1662.10		
1 Welder, Gas Engine, 300 amp		145.85		160.44		
1 Air Compressor, 160 cfm		156.40		172.04		
2 Impact Wrenches		36.00		39.60	14.45	15.89
128 L.H., Daily Totals		$8536.85		$13394.98	$66.69	$104.65

Crew E-7	Hr.	Daily	Hr.	Daily	Bare Costs	Incl. O&P
1 Struc. Steel Foreman (outside)	$54.65	$437.20	$94.95	$759.60	$52.21	$88.56
4 Struc. Steel Workers	52.65	1684.80	91.45	2926.40		
1 Equip. Oper. (crane)	51.70	413.60	78.45	627.60		
1 Equip. Oper. (oiler)	45.20	361.60	68.55	548.40		
1 Welder Foreman (outside)	54.65	437.20	94.95	759.60		
2 Welders	52.65	842.40	91.45	1463.20		
1 Lattice Boom Crane, 90 Ton		1511.00		1662.10		
2 Welder, Gas Engine, 300 amp		291.70		320.87	22.53	24.79
80 L.H., Daily Totals		$5979.50		$9067.77	$74.74	$113.35

Crew E-8	Hr.	Daily	Hr.	Daily	Bare Costs	Incl. O&P
1 Struc. Steel Foreman (outside)	$54.65	$437.20	$94.95	$759.60	$52.00	$87.87
4 Struc. Steel Workers	52.65	1684.80	91.45	2926.40		
1 Welder Foreman (outside)	54.65	437.20	94.95	759.60		
4 Welders	52.65	1684.80	91.45	2926.40		
1 Equip. Oper. (crane)	51.70	413.60	78.45	627.60		
1 Equip. Oper. (oiler)	45.20	361.60	68.55	548.40		
1 Equip. Oper. (light)	48.60	388.80	73.75	590.00		
1 Lattice Boom Crane, 90 Ton		1511.00		1662.10		
4 Welder, Gas Engine, 300 amp		583.40		641.74	20.14	22.15
104 L.H., Daily Totals		$7502.40		$11441.84	$72.14	$110.02

Crew E-9	Hr.	Daily	Hr.	Daily	Bare Costs	Incl. O&P
2 Struc. Steel Foremen (outside)	$54.65	$874.40	$94.95	$1519.20	$52.25	$88.76
5 Struc. Steel Workers	52.65	2106.00	91.45	3658.00		
1 Welder Foreman (outside)	54.65	437.20	94.95	759.60		
5 Welders	52.65	2106.00	91.45	3658.00		
1 Equip. Oper. (crane)	51.70	413.60	78.45	627.60		
1 Equip. Oper. (oiler)	45.20	361.60	68.55	548.40		
1 Equip. Oper. (light)	48.60	388.80	73.75	590.00		
1 Lattice Boom Crane, 90 Ton		1511.00		1662.10		
5 Welder, Gas Engine, 300 amp		729.25		802.17	17.50	19.25
128 L.H., Daily Totals		$8927.85		$13825.08	$69.75	$108.01

Crew E-10	Hr.	Daily	Hr.	Daily	Bare Costs	Incl. O&P
1 Welder Foreman (outside)	$54.65	$437.20	$94.95	$759.60	$53.65	$93.20
1 Welder	52.65	421.20	91.45	731.60		
1 Welder, Gas Engine, 300 amp		145.85		160.44		
1 Flatbed Truck, Gas, 3 Ton		303.20		333.52	28.07	30.87
16 L.H., Daily Totals		$1307.45		$1985.16	$81.72	$124.07

Crew E-11	Hr.	Daily	Hr.	Daily	Bare Costs	Incl. O&P
2 Painters, Struc. Steel	$41.45	$663.20	$72.95	$1167.20	$42.27	$69.38
1 Building Laborer	37.60	300.80	57.85	462.80		
1 Equip. Oper. (light)	48.60	388.80	73.75	590.00		
1 Air Compressor, 250 cfm		201.40		221.54		
1 Sandblaster, Portable, 3 C.F.		20.40		22.44		
1 Set Sand Blasting Accessories		14.05		15.46	7.37	8.11
32 L.H., Daily Totals		$1588.65		$2479.43	$49.65	$77.48

Crew E-11A	Hr.	Daily	Hr.	Daily	Bare Costs	Incl. O&P
2 Painters, Struc. Steel	$41.45	$663.20	$72.95	$1167.20	$42.27	$69.38
1 Building Laborer	37.60	300.80	57.85	462.80		
1 Equip. Oper. (light)	48.60	388.80	73.75	590.00		
1 Air Compressor, 250 cfm		201.40		221.54		
1 Sandblaster, Portable, 3 C.F.		20.40		22.44		
1 Set Sand Blasting Accessories		14.05		15.46		
1 Aerial Lift Truck, 60' Boom		435.60		479.16	20.98	23.08
32 L.H., Daily Totals		$2024.25		$2958.59	$63.26	$92.46

Crew E-11B	Hr.	Daily	Hr.	Daily	Bare Costs	Incl. O&P
2 Painters, Struc. Steel	$41.45	$663.20	$72.95	$1167.20	$40.17	$67.92
1 Building Laborer	37.60	300.80	57.85	462.80		
2 Paint Sprayer, 8 C.F.M.		99.80		109.78		
1 Aerial Lift Truck, 60' Boom		435.60		479.16	22.31	24.54
24 L.H., Daily Totals		$1499.40		$2218.94	$62.48	$92.46

Crew E-12	Hr.	Daily	Hr.	Daily	Bare Costs	Incl. O&P
1 Welder Foreman (outside)	$54.65	$437.20	$94.95	$759.60	$51.63	$84.35
1 Equip. Oper. (light)	48.60	388.80	73.75	590.00		
1 Welder, Gas Engine, 300 amp		145.85		160.44	9.12	10.03
16 L.H., Daily Totals		$971.85		$1510.04	$60.74	$94.38

Crew E-13	Hr.	Daily	Hr.	Daily	Bare Costs	Incl. O&P
1 Welder Foreman (outside)	$54.65	$437.20	$94.95	$759.60	$52.63	$87.88
.5 Equip. Oper. (light)	48.60	194.40	73.75	295.00		
1 Welder, Gas Engine, 300 amp		145.85		160.44	12.15	13.37
12 L.H., Daily Totals		$777.45		$1215.04	$64.79	$101.25

Crew E-14	Hr.	Daily	Hr.	Daily	Bare Costs	Incl. O&P
1 Welder Foreman (outside)	$54.65	$437.20	$94.95	$759.60	$54.65	$94.95
1 Welder, Gas Engine, 300 amp		145.85		160.44	18.23	20.05
8 L.H., Daily Totals		$583.05		$920.03	$72.88	$115.00

For customer support on your Plumbing Cost Data, call 877.709.6968.

Crew No.	Bare Costs		Incl. Subs O&P		Cost Per Labor-Hour	
Crew E-16	Hr.	Daily	Hr.	Daily	Bare Costs	Incl. O&P
1 Welder Foreman (outside)	$54.65	$437.20	$94.95	$759.60	$53.65	$93.20
1 Welder	52.65	421.20	91.45	731.60		
1 Welder, Gas Engine, 300 amp		145.85		160.44	9.12	10.03
16 L.H., Daily Totals		$1004.25		$1651.64	$62.77	$103.23

Crew No.	Hr.	Daily	Hr.	Daily	Bare Costs	Incl. O&P
Crew E-17						
1 Struc. Steel Foreman (outside)	$54.65	$437.20	$94.95	$759.60	$53.65	$93.20
1 Structural Steel Worker	52.65	421.20	91.45	731.60		
16 L.H., Daily Totals		$858.40		$1491.20	$53.65	$93.20

Crew No.	Hr.	Daily	Hr.	Daily	Bare Costs	Incl. O&P
Crew E-18						
1 Struc. Steel Foreman (outside)	$54.65	$437.20	$94.95	$759.60	$52.64	$89.21
3 Structural Steel Workers	52.65	1263.60	91.45	2194.80		
1 Equipment Operator (med.)	50.60	404.80	76.75	614.00		
1 Lattice Boom Crane, 20 Ton		948.90		1043.79	23.72	26.09
40 L.H., Daily Totals		$3054.50		$4612.19	$76.36	$115.30

Crew No.	Hr.	Daily	Hr.	Daily	Bare Costs	Incl. O&P
Crew E-19						
1 Struc. Steel Foreman (outside)	$54.65	$437.20	$94.95	$759.60	$51.97	$86.72
1 Structural Steel Worker	52.65	421.20	91.45	731.60		
1 Equip. Oper. (light)	48.60	388.80	73.75	590.00		
1 Lattice Boom Crane, 20 Ton		948.90		1043.79	39.54	43.49
24 L.H., Daily Totals		$2196.10		$3124.99	$91.50	$130.21

Crew No.	Hr.	Daily	Hr.	Daily	Bare Costs	Incl. O&P
Crew E-20						
1 Struc. Steel Foreman (outside)	$54.65	$437.20	$94.95	$759.60	$51.85	$87.40
5 Structural Steel Workers	52.65	2106.00	91.45	3658.00		
1 Equip. Oper. (crane)	51.70	413.60	78.45	627.60		
1 Equip. Oper. (oiler)	45.20	361.60	68.55	548.40		
1 Lattice Boom Crane, 40 Ton		1164.00		1280.40	18.19	20.01
64 L.H., Daily Totals		$4482.40		$6874.00	$70.04	$107.41

Crew No.	Hr.	Daily	Hr.	Daily	Bare Costs	Incl. O&P
Crew E-22						
1 Skilled Worker Foreman (out)	$50.65	$405.20	$78.25	$626.00	$49.32	$76.18
2 Skilled Workers	48.65	778.40	75.15	1202.40		
24 L.H., Daily Totals		$1183.60		$1828.40	$49.32	$76.18

Crew No.	Hr.	Daily	Hr.	Daily	Bare Costs	Incl. O&P
Crew E-24						
3 Structural Steel Workers	$52.65	$1263.60	$91.45	$2194.80	$52.14	$87.78
1 Equipment Operator (med.)	50.60	404.80	76.75	614.00		
1 Hyd. Crane, 25 Ton		736.60		810.26	23.02	25.32
32 L.H., Daily Totals		$2405.00		$3619.06	$75.16	$113.10

Crew No.	Hr.	Daily	Hr.	Daily	Bare Costs	Incl. O&P
Crew E-25						
1 Welder Foreman (outside)	$54.65	$437.20	$94.95	$759.60	$54.65	$94.95
1 Cutting Torch		11.40		12.54	1.43	1.57
8 L.H., Daily Totals		$448.60		$772.14	$56.08	$96.52

Crew No.	Hr.	Daily	Hr.	Daily	Bare Costs	Incl. O&P
Crew F-3						
4 Carpenters	$46.95	$1502.40	$72.25	$2312.00	$47.90	$73.49
1 Equip. Oper. (crane)	51.70	413.60	78.45	627.60		
1 Hyd. Crane, 12 Ton		653.80		719.18	16.34	17.98
40 L.H., Daily Totals		$2569.80		$3658.78	$64.25	$91.47

Crew No.	Hr.	Daily	Hr.	Daily	Bare Costs	Incl. O&P
Crew F-4						
4 Carpenters	$46.95	$1502.40	$72.25	$2312.00	$47.45	$72.67
1 Equip. Oper. (crane)	51.70	413.60	78.45	627.60		
1 Equip. Oper. (oiler)	45.20	361.60	68.55	548.40		
1 Hyd. Crane, 55 Ton		1128.00		1240.80	23.50	25.85
48 L.H., Daily Totals		$3405.60		$4728.80	$70.95	$98.52

Crew No.	Bare Costs		Incl. Subs O&P		Cost Per Labor-Hour	
Crew F-5	Hr.	Daily	Hr.	Daily	Bare Costs	Incl. O&P
1 Carpenter Foreman (outside)	$48.95	$391.60	$75.35	$602.80	$47.45	$73.03
3 Carpenters	46.95	1126.80	72.25	1734.00		
32 L.H., Daily Totals		$1518.40		$2336.80	$47.45	$73.03

Crew No.	Hr.	Daily	Hr.	Daily	Bare Costs	Incl. O&P
Crew F-6						
2 Carpenters	$46.95	$751.20	$72.25	$1156.00	$44.16	$67.73
2 Building Laborers	37.60	601.60	57.85	925.60		
1 Equip. Oper. (crane)	51.70	413.60	78.45	627.60		
1 Hyd. Crane, 12 Ton		653.80		719.18	16.34	17.98
40 L.H., Daily Totals		$2420.20		$3428.38	$60.51	$85.71

Crew No.	Hr.	Daily	Hr.	Daily	Bare Costs	Incl. O&P
Crew F-7						
2 Carpenters	$46.95	$751.20	$72.25	$1156.00	$42.27	$65.05
2 Building Laborers	37.60	601.60	57.85	925.60		
32 L.H., Daily Totals		$1352.80		$2081.60	$42.27	$65.05

Crew No.	Hr.	Daily	Hr.	Daily	Bare Costs	Incl. O&P
Crew G-1						
1 Roofer Foreman (outside)	$42.10	$336.80	$71.85	$574.80	$37.51	$64.04
4 Roofers Composition	40.10	1283.20	68.45	2190.40		
2 Roofer Helpers	30.05	480.80	51.30	820.80		
1 Application Equipment		193.00		212.30		
1 Tar Kettle/Pot		169.80		186.78		
1 Crew Truck		206.40		227.04	10.16	11.18
56 L.H., Daily Totals		$2670.00		$4212.12	$47.68	$75.22

Crew No.	Hr.	Daily	Hr.	Daily	Bare Costs	Incl. O&P
Crew G-2						
1 Plasterer	$42.95	$343.60	$64.40	$515.20	$39.55	$59.78
1 Plasterer Helper	38.10	304.80	57.10	456.80		
1 Building Laborer	37.60	300.80	57.85	462.80		
1 Grout Pump, 50 C.F./hr.		133.40		146.74	5.56	6.11
24 L.H., Daily Totals		$1082.60		$1581.54	$45.11	$65.90

Crew No.	Hr.	Daily	Hr.	Daily	Bare Costs	Incl. O&P
Crew G-2A						
1 Roofer Composition	$40.10	$320.80	$68.45	$547.60	$35.92	$59.20
1 Roofer Helper	30.05	240.40	51.30	410.40		
1 Building Laborer	37.60	300.80	57.85	462.80		
1 Foam Spray Rig, Trailer-Mtd.		573.25		630.58		
1 Pickup Truck, 3/4 Ton		144.20		158.62	29.89	32.88
24 L.H., Daily Totals		$1579.45		$2209.99	$65.81	$92.08

Crew No.	Hr.	Daily	Hr.	Daily	Bare Costs	Incl. O&P
Crew G-3						
2 Sheet Metal Workers	$55.95	$895.20	$85.50	$1368.00	$46.77	$71.67
2 Building Laborers	37.60	601.60	57.85	925.60		
32 L.H., Daily Totals		$1496.80		$2293.60	$46.77	$71.67

Crew No.	Hr.	Daily	Hr.	Daily	Bare Costs	Incl. O&P
Crew G-4						
1 Labor Foreman (outside)	$39.60	$316.80	$60.95	$487.60	$38.27	$58.88
2 Building Laborers	37.60	601.60	57.85	925.60		
1 Flatbed Truck, Gas, 1.5 Ton		245.40		269.94		
1 Air Compressor, 160 cfm		156.40		172.04	16.74	18.42
24 L.H., Daily Totals		$1320.20		$1855.18	$55.01	$77.30

Crew No.	Hr.	Daily	Hr.	Daily	Bare Costs	Incl. O&P
Crew G-5						
1 Roofer Foreman (outside)	$42.10	$336.80	$71.85	$574.80	$36.48	$62.27
2 Roofers Composition	40.10	641.60	68.45	1095.20		
2 Roofer Helpers	30.05	480.80	51.30	820.80		
1 Application Equipment		193.00		212.30	4.83	5.31
40 L.H., Daily Totals		$1652.20		$2703.10	$41.31	$67.58

Crews

Crew No.	Bare Costs Hr.	Daily	Incl. Subs O&P Hr.	Daily	Cost Per Labor-Hour Bare Costs	Incl. O&P
Crew G-6A	Hr.	Daily	Hr.	Daily	Bare Costs	Incl. O&P
2 Roofers Composition	$40.10	$641.60	$68.45	$1095.20	$40.10	$68.45
1 Small Compressor, Electric		12.95		14.24		
2 Pneumatic Nailers		55.30		60.83	4.27	4.69
16 L.H., Daily Totals		$709.85		$1170.28	$44.37	$73.14
Crew G-7	Hr.	Daily	Hr.	Daily	Bare Costs	Incl. O&P
1 Carpenter	$46.95	$375.60	$72.25	$578.00	$46.95	$72.25
1 Small Compressor, Electric		12.95		14.24		
1 Pneumatic Nailer		27.65		30.41	5.08	5.58
8 L.H., Daily Totals		$416.20		$622.66	$52.02	$77.83
Crew H-1	Hr.	Daily	Hr.	Daily	Bare Costs	Incl. O&P
2 Glaziers	$45.10	$721.60	$68.70	$1099.20	$48.88	$80.08
2 Struc. Steel Workers	52.65	842.40	91.45	1463.20		
32 L.H., Daily Totals		$1564.00		$2562.40	$48.88	$80.08
Crew H-2	Hr.	Daily	Hr.	Daily	Bare Costs	Incl. O&P
2 Glaziers	$45.10	$721.60	$68.70	$1099.20	$42.60	$65.08
1 Building Laborer	37.60	300.80	57.85	462.80		
24 L.H., Daily Totals		$1022.40		$1562.00	$42.60	$65.08
Crew H-3	Hr.	Daily	Hr.	Daily	Bare Costs	Incl. O&P
1 Glazier	$45.10	$360.80	$68.70	$549.60	$40.27	$61.83
1 Helper	35.45	283.60	54.95	439.60		
16 L.H., Daily Totals		$644.40		$989.20	$40.27	$61.83
Crew H-4	Hr.	Daily	Hr.	Daily	Bare Costs	Incl. O&P
1 Carpenter	$46.95	$375.60	$72.25	$578.00	$43.90	$67.27
1 Carpenter Helper	35.45	283.60	54.95	439.60		
.5 Electrician	54.70	218.80	81.95	327.80		
20 L.H., Daily Totals		$878.00		$1345.40	$43.90	$67.27
Crew J-1	Hr.	Daily	Hr.	Daily	Bare Costs	Incl. O&P
3 Plasterers	$42.95	$1030.80	$64.40	$1545.60	$41.01	$61.48
2 Plasterer Helpers	38.10	609.60	57.10	913.60		
1 Mixing Machine, 6 C.F.		140.20		154.22	3.50	3.86
40 L.H., Daily Totals		$1780.60		$2613.42	$44.52	$65.34
Crew J-2	Hr.	Daily	Hr.	Daily	Bare Costs	Incl. O&P
3 Plasterers	$42.95	$1030.80	$64.40	$1545.60	$41.34	$61.80
2 Plasterer Helpers	38.10	609.60	57.10	913.60		
1 Lather	43.00	344.00	63.40	507.20		
1 Mixing Machine, 6 C.F.		140.20		154.22	2.92	3.21
48 L.H., Daily Totals		$2124.60		$3120.62	$44.26	$65.01
Crew J-3	Hr.	Daily	Hr.	Daily	Bare Costs	Incl. O&P
1 Terrazzo Worker	$42.95	$343.60	$63.50	$508.00	$39.23	$58.00
1 Terrazzo Helper	35.50	284.00	52.50	420.00		
1 Floor Grinder, 22" Path		115.55		127.11		
1 Terrazzo Mixer		188.20		207.02	18.98	20.88
16 L.H., Daily Totals		$931.35		$1262.13	$58.21	$78.88
Crew J-4	Hr.	Daily	Hr.	Daily	Bare Costs	Incl. O&P
2 Cement Finishers	$45.00	$720.00	$66.65	$1066.40	$42.53	$63.72
1 Laborer	37.60	300.80	57.85	462.80		
1 Floor Grinder, 22" Path		115.55		127.11		
1 Floor Edger, 7" Path		39.30		43.23		
1 Vacuum Pick-Up System		60.90		66.99	8.99	9.89
24 L.H., Daily Totals		$1236.55		$1766.53	$51.52	$73.61

Crew No.	Bare Costs Hr.	Daily	Incl. Subs O&P Hr.	Daily	Cost Per Labor-Hour Bare Costs	Incl. O&P
Crew J-4A	Hr.	Daily	Hr.	Daily	Bare Costs	Incl. O&P
2 Cement Finishers	$45.00	$720.00	$66.65	$1066.40	$41.30	$62.25
2 Laborers	37.60	601.60	57.85	925.60		
1 Floor Grinder, 22" Path		115.55		127.11		
1 Floor Edger, 7" Path		39.30		43.23		
1 Vacuum Pick-Up System		60.90		66.99		
1 Floor Auto Scrubber		233.85		257.24	14.05	15.46
32 L.H., Daily Totals		$1771.20		$2486.56	$55.35	$77.70
Crew J-4B	Hr.	Daily	Hr.	Daily	Bare Costs	Incl. O&P
1 Laborer	$37.60	$300.80	$57.85	$462.80	$37.60	$57.85
1 Floor Auto Scrubber		233.85		257.24	29.23	32.15
8 L.H., Daily Totals		$534.65		$720.03	$66.83	$90.00
Crew J-6	Hr.	Daily	Hr.	Daily	Bare Costs	Incl. O&P
2 Painters	$40.35	$645.60	$60.80	$972.80	$41.73	$63.30
1 Building Laborer	37.60	300.80	57.85	462.80		
1 Equip. Oper. (light)	48.60	388.80	73.75	590.00		
1 Air Compressor, 250 cfm		201.40		221.54		
1 Sandblaster, Portable, 3 C.F.		20.40		22.44		
1 Set Sand Blasting Accessories		14.05		15.46	7.37	8.11
32 L.H., Daily Totals		$1571.05		$2285.03	$49.10	$71.41
Crew J-7	Hr.	Daily	Hr.	Daily	Bare Costs	Incl. O&P
2 Painters	$40.35	$645.60	$60.80	$972.80	$40.35	$60.80
1 Floor Belt Sander		14.20		15.62		
1 Floor Sanding Edger		12.60		13.86	1.68	1.84
16 L.H., Daily Totals		$672.40		$1002.28	$42.02	$62.64
Crew K-1	Hr.	Daily	Hr.	Daily	Bare Costs	Incl. O&P
1 Carpenter	$46.95	$375.60	$72.25	$578.00	$43.02	$65.67
1 Truck Driver (light)	39.10	312.80	59.10	472.80		
1 Flatbed Truck, Gas, 3 Ton		303.20		333.52	18.95	20.84
16 L.H., Daily Totals		$991.60		$1384.32	$61.98	$86.52
Crew K-2	Hr.	Daily	Hr.	Daily	Bare Costs	Incl. O&P
1 Struc. Steel Foreman (outside)	$54.65	$437.20	$94.95	$759.60	$48.80	$81.83
1 Struc. Steel Worker	52.65	421.20	91.45	731.60		
1 Truck Driver (light)	39.10	312.80	59.10	472.80		
1 Flatbed Truck, Gas, 3 Ton		303.20		333.52	12.63	13.90
24 L.H., Daily Totals		$1474.40		$2297.52	$61.43	$95.73
Crew L-1	Hr.	Daily	Hr.	Daily	Bare Costs	Incl. O&P
1 Electrician	$54.70	$437.60	$81.95	$655.60	$56.70	$85.30
1 Plumber	58.70	469.60	88.65	709.20		
16 L.H., Daily Totals		$907.20		$1364.80	$56.70	$85.30
Crew L-2	Hr.	Daily	Hr.	Daily	Bare Costs	Incl. O&P
1 Carpenter	$46.95	$375.60	$72.25	$578.00	$41.20	$63.60
1 Carpenter Helper	35.45	283.60	54.95	439.60		
16 L.H., Daily Totals		$659.20		$1017.60	$41.20	$63.60
Crew L-3	Hr.	Daily	Hr.	Daily	Bare Costs	Incl. O&P
1 Carpenter	$46.95	$375.60	$72.25	$578.00	$51.14	$77.99
.5 Electrician	54.70	218.80	81.95	327.80		
.5 Sheet Metal Worker	55.95	223.80	85.50	342.00		
16 L.H., Daily Totals		$818.20		$1247.80	$51.14	$77.99

For customer support on your Plumbing Cost Data, call 877.709.6968.

Crew No.	Bare Costs		Incl. Subs O&P		Cost Per Labor-Hour	
Crew L-3A	Hr.	Daily	Hr.	Daily	Bare Costs	Incl. O&P
1 Carpenter Foreman (outside)	$48.95	$391.60	$75.35	$602.80	$51.28	$78.73
.5 Sheet Metal Worker	55.95	223.80	85.50	342.00		
12 L.H., Daily Totals		$615.40		$944.80	$51.28	$78.73
Crew L-4	Hr.	Daily	Hr.	Daily	Bare Costs	Incl. O&P
2 Skilled Workers	$48.65	$778.40	$75.15	$1202.40	$44.25	$68.42
1 Helper	35.45	283.60	54.95	439.60		
24 L.H., Daily Totals		$1062.00		$1642.00	$44.25	$68.42
Crew L-5	Hr.	Daily	Hr.	Daily	Bare Costs	Incl. O&P
1 Struc. Steel Foreman (outside)	$54.65	$437.20	$94.95	$759.60	$52.80	$90.09
5 Struc. Steel Workers	52.65	2106.00	91.45	3658.00		
1 Equip. Oper. (crane)	51.70	413.60	78.45	627.60		
1 Hyd. Crane, 25 Ton		736.60		810.26	13.15	14.47
56 L.H., Daily Totals		$3693.40		$5855.46	$65.95	$104.56
Crew L-5A	Hr.	Daily	Hr.	Daily	Bare Costs	Incl. O&P
1 Struc. Steel Foreman (outside)	$54.65	$437.20	$94.95	$759.60	$52.91	$89.08
2 Structural Steel Workers	52.65	842.40	91.45	1463.20		
1 Equip. Oper. (crane)	51.70	413.60	78.45	627.60		
1 S.P. Crane, 4x4, 25 Ton		599.40		659.34	18.73	20.60
32 L.H., Daily Totals		$2292.60		$3509.74	$71.64	$109.68
Crew L-5B	Hr.	Daily	Hr.	Daily	Bare Costs	Incl. O&P
1 Struc. Steel Foreman (outside)	$54.65	$437.20	$94.95	$759.60	$53.97	$85.46
2 Structural Steel Workers	52.65	842.40	91.45	1463.20		
2 Electricians	54.70	875.20	81.95	1311.20		
2 Steamfitters/Pipefitters	59.75	956.00	90.20	1443.20		
1 Equip. Oper. (crane)	51.70	413.60	78.45	627.60		
1 Equip. Oper. (oiler)	45.20	361.60	68.55	548.40		
1 Hyd. Crane, 80 Ton		1625.00		1787.50	22.57	24.83
72 L.H., Daily Totals		$5511.00		$7940.70	$76.54	$110.29
Crew L-6	Hr.	Daily	Hr.	Daily	Bare Costs	Incl. O&P
1 Plumber	$58.70	$469.60	$88.65	$709.20	$57.37	$86.42
.5 Electrician	54.70	218.80	81.95	327.80		
12 L.H., Daily Totals		$688.40		$1037.00	$57.37	$86.42
Crew L-7	Hr.	Daily	Hr.	Daily	Bare Costs	Incl. O&P
2 Carpenters	$46.95	$751.20	$72.25	$1156.00	$45.39	$69.52
1 Building Laborer	37.60	300.80	57.85	462.80		
.5 Electrician	54.70	218.80	81.95	327.80		
28 L.H., Daily Totals		$1270.80		$1946.60	$45.39	$69.52
Crew L-8	Hr.	Daily	Hr.	Daily	Bare Costs	Incl. O&P
2 Carpenters	$46.95	$751.20	$72.25	$1156.00	$49.30	$75.53
.5 Plumber	58.70	234.80	88.65	354.60		
20 L.H., Daily Totals		$986.00		$1510.60	$49.30	$75.53
Crew L-9	Hr.	Daily	Hr.	Daily	Bare Costs	Incl. O&P
1 Labor Foreman (inside)	$38.10	$304.80	$58.65	$469.20	$42.96	$68.17
2 Building Laborers	37.60	601.60	57.85	925.60		
1 Struc. Steel Worker	52.65	421.20	91.45	731.60		
.5 Electrician	54.70	218.80	81.95	327.80		
36 L.H., Daily Totals		$1546.40		$2454.20	$42.96	$68.17
Crew L-10	Hr.	Daily	Hr.	Daily	Bare Costs	Incl. O&P
1 Struc. Steel Foreman (outside)	$54.65	$437.20	$94.95	$759.60	$53.00	$88.28
1 Structural Steel Worker	52.65	421.20	91.45	731.60		
1 Equip. Oper. (crane)	51.70	413.60	78.45	627.60		
1 Hyd. Crane, 12 Ton		653.80		719.18	27.24	29.97
24 L.H., Daily Totals		$1925.80		$2837.98	$80.24	$118.25
Crew L-11	Hr.	Daily	Hr.	Daily	Bare Costs	Incl. O&P
2 Wreckers	$37.60	$601.60	$62.10	$993.60	$43.88	$69.10
1 Equip. Oper. (crane)	51.70	413.60	78.45	627.60		
1 Equip. Oper. (light)	48.60	388.80	73.75	590.00		
1 Hyd. Excavator, 2.5 C.Y.		1605.00		1765.50		
1 Loader, Skid Steer, 78 H.P.		318.60		350.46	60.11	66.12
32 L.H., Daily Totals		$3327.60		$4327.16	$103.99	$135.22
Crew M-1	Hr.	Daily	Hr.	Daily	Bare Costs	Incl. O&P
3 Elevator Constructors	$76.50	$1836.00	$114.20	$2740.80	$72.67	$108.49
1 Elevator Apprentice	61.20	489.60	91.35	730.80		
5 Hand Tools		46.00		50.60	1.44	1.58
32 L.H., Daily Totals		$2371.60		$3522.20	$74.11	$110.07
Crew M-3	Hr.	Daily	Hr.	Daily	Bare Costs	Incl. O&P
1 Electrician Foreman (outside)	$56.70	$453.60	$84.95	$679.60	$57.56	$86.48
1 Common Laborer	37.60	300.80	57.85	462.80		
.25 Equipment Operator (med.)	50.60	101.20	76.75	153.50		
1 Elevator Constructor	76.50	612.00	114.20	913.60		
1 Elevator Apprentice	61.20	489.60	91.35	730.80		
.25 S.P. Crane, 4x4, 20 Ton		138.65		152.51	4.08	4.49
34 L.H., Daily Totals		$2095.85		$3092.82	$61.64	$90.97
Crew M-4	Hr.	Daily	Hr.	Daily	Bare Costs	Incl. O&P
1 Electrician Foreman (outside)	$56.70	$453.60	$84.95	$679.60	$56.94	$85.58
1 Common Laborer	37.60	300.80	57.85	462.80		
.25 Equipment Operator, Crane	51.70	103.40	78.45	156.90		
.25 Equip. Oper. (oiler)	45.20	90.40	68.55	137.10		
1 Elevator Constructor	76.50	612.00	114.20	913.60		
1 Elevator Apprentice	61.20	489.60	91.35	730.80		
.25 S.P. Crane, 4x4, 40 Ton		175.75		193.32	4.88	5.37
36 L.H., Daily Totals		$2225.55		$3274.13	$61.82	$90.95
Crew Q-1	Hr.	Daily	Hr.	Daily	Bare Costs	Incl. O&P
1 Plumber	$58.70	$469.60	$88.65	$709.20	$52.83	$79.78
1 Plumber Apprentice	46.95	375.60	70.90	567.20		
16 L.H., Daily Totals		$845.20		$1276.40	$52.83	$79.78
Crew Q-1A	Hr.	Daily	Hr.	Daily	Bare Costs	Incl. O&P
.25 Plumber Foreman (outside)	$60.70	$121.40	$91.65	$183.30	$59.10	$89.25
1 Plumber	58.70	469.60	88.65	709.20		
10 L.H., Daily Totals		$591.00		$892.50	$59.10	$89.25
Crew Q-1C	Hr.	Daily	Hr.	Daily	Bare Costs	Incl. O&P
1 Plumber	$58.70	$469.60	$88.65	$709.20	$52.08	$78.77
1 Plumber Apprentice	46.95	375.60	70.90	567.20		
1 Equip. Oper. (medium)	50.60	404.80	76.75	614.00		
1 Trencher, Chain Type, 8' D		3402.00		3742.20	141.75	155.93
24 L.H., Daily Totals		$4652.00		$5632.60	$193.83	$234.69
Crew Q-2	Hr.	Daily	Hr.	Daily	Bare Costs	Incl. O&P
2 Plumbers	$58.70	$939.20	$88.65	$1418.40	$54.78	$82.73
1 Plumber Apprentice	46.95	375.60	70.90	567.20		
24 L.H., Daily Totals		$1314.80		$1985.60	$54.78	$82.73

Crews

Crew No.	Bare Costs		Incl. Subs O&P		Cost Per Labor-Hour	
	Hr.	Daily	Hr.	Daily	Bare Costs	Incl. O&P
Crew Q-3	Hr.	Daily	Hr.	Daily	Bare Costs	Incl. O&P
1 Plumber Foreman (inside)	$59.20	$473.60	$89.40	$715.20	$55.89	$84.40
2 Plumbers	58.70	939.20	88.65	1418.40		
1 Plumber Apprentice	46.95	375.60	70.90	567.20		
32 L.H., Daily Totals		$1788.40		$2700.80	$55.89	$84.40
Crew Q-4	Hr.	Daily	Hr.	Daily	Bare Costs	Incl. O&P
1 Plumber Foreman (inside)	$59.20	$473.60	$89.40	$715.20	$55.89	$84.40
1 Plumber	58.70	469.60	88.65	709.20		
1 Welder (plumber)	58.70	469.60	88.65	709.20		
1 Plumber Apprentice	46.95	375.60	70.90	567.20		
1 Welder, Electric, 300 amp		57.70		63.47	1.80	1.98
32 L.H., Daily Totals		$1846.10		$2764.27	$57.69	$86.38
Crew Q-5	Hr.	Daily	Hr.	Daily	Bare Costs	Incl. O&P
1 Steamfitter	$59.75	$478.00	$90.20	$721.60	$53.77	$81.20
1 Steamfitter Apprentice	47.80	382.40	72.20	577.60		
16 L.H., Daily Totals		$860.40		$1299.20	$53.77	$81.20
Crew Q-6	Hr.	Daily	Hr.	Daily	Bare Costs	Incl. O&P
2 Steamfitters	$59.75	$956.00	$90.20	$1443.20	$55.77	$84.20
1 Steamfitter Apprentice	47.80	382.40	72.20	577.60		
24 L.H., Daily Totals		$1338.40		$2020.80	$55.77	$84.20
Crew Q-7	Hr.	Daily	Hr.	Daily	Bare Costs	Incl. O&P
1 Steamfitter Foreman (inside)	$60.25	$482.00	$91.00	$728.00	$56.89	$85.90
2 Steamfitters	59.75	956.00	90.20	1443.20		
1 Steamfitter Apprentice	47.80	382.40	72.20	577.60		
32 L.H., Daily Totals		$1820.40		$2748.80	$56.89	$85.90
Crew Q-8	Hr.	Daily	Hr.	Daily	Bare Costs	Incl. O&P
1 Steamfitter Foreman (inside)	$60.25	$482.00	$91.00	$728.00	$56.89	$85.90
1 Steamfitter	59.75	478.00	90.20	721.60		
1 Welder (steamfitter)	59.75	478.00	90.20	721.60		
1 Steamfitter Apprentice	47.80	382.40	72.20	577.60		
1 Welder, Electric, 300 amp		57.70		63.47	1.80	1.98
32 L.H., Daily Totals		$1878.10		$2812.27	$58.69	$87.88
Crew Q-9	Hr.	Daily	Hr.	Daily	Bare Costs	Incl. O&P
1 Sheet Metal Worker	$55.95	$447.60	$85.50	$684.00	$50.35	$76.95
1 Sheet Metal Apprentice	44.75	358.00	68.40	547.20		
16 L.H., Daily Totals		$805.60		$1231.20	$50.35	$76.95
Crew Q-10	Hr.	Daily	Hr.	Daily	Bare Costs	Incl. O&P
2 Sheet Metal Workers	$55.95	$895.20	$85.50	$1368.00	$52.22	$79.80
1 Sheet Metal Apprentice	44.75	358.00	68.40	547.20		
24 L.H., Daily Totals		$1253.20		$1915.20	$52.22	$79.80
Crew Q-11	Hr.	Daily	Hr.	Daily	Bare Costs	Incl. O&P
1 Sheet Metal Foreman (inside)	$56.45	$451.60	$86.25	$690.00	$53.27	$81.41
2 Sheet Metal Workers	55.95	895.20	85.50	1368.00		
1 Sheet Metal Apprentice	44.75	358.00	68.40	547.20		
32 L.H., Daily Totals		$1704.80		$2605.20	$53.27	$81.41
Crew Q-12	Hr.	Daily	Hr.	Daily	Bare Costs	Incl. O&P
1 Sprinkler Installer	$56.15	$449.20	$84.95	$679.60	$50.52	$76.45
1 Sprinkler Apprentice	44.90	359.20	67.95	543.60		
16 L.H., Daily Totals		$808.40		$1223.20	$50.52	$76.45

Crew No.	Bare Costs		Incl. Subs O&P		Cost Per Labor-Hour	
	Hr.	Daily	Hr.	Daily	Bare Costs	Incl. O&P
Crew Q-13	Hr.	Daily	Hr.	Daily	Bare Costs	Incl. O&P
1 Sprinkler Foreman (inside)	$56.65	$453.20	$85.70	$685.60	$53.46	$80.89
2 Sprinkler Installers	56.15	898.40	84.95	1359.20		
1 Sprinkler Apprentice	44.90	359.20	67.95	543.60		
32 L.H., Daily Totals		$1710.80		$2588.40	$53.46	$80.89
Crew Q-14	Hr.	Daily	Hr.	Daily	Bare Costs	Incl. O&P
1 Asbestos Worker	$52.35	$418.80	$81.50	$652.00	$47.13	$73.38
1 Asbestos Apprentice	41.90	335.20	65.25	522.00		
16 L.H., Daily Totals		$754.00		$1174.00	$47.13	$73.38
Crew Q-15	Hr.	Daily	Hr.	Daily	Bare Costs	Incl. O&P
1 Plumber	$58.70	$469.60	$88.65	$709.20	$52.83	$79.78
1 Plumber Apprentice	46.95	375.60	70.90	567.20		
1 Welder, Electric, 300 amp		57.70		63.47	3.61	3.97
16 L.H., Daily Totals		$902.90		$1339.87	$56.43	$83.74
Crew Q-16	Hr.	Daily	Hr.	Daily	Bare Costs	Incl. O&P
2 Plumbers	$58.70	$939.20	$88.65	$1418.40	$54.78	$82.73
1 Plumber Apprentice	46.95	375.60	70.90	567.20		
1 Welder, Electric, 300 amp		57.70		63.47	2.40	2.64
24 L.H., Daily Totals		$1372.50		$2049.07	$57.19	$85.38
Crew Q-17	Hr.	Daily	Hr.	Daily	Bare Costs	Incl. O&P
1 Steamfitter	$59.75	$478.00	$90.20	$721.60	$53.77	$81.20
1 Steamfitter Apprentice	47.80	382.40	72.20	577.60		
1 Welder, Electric, 300 amp		57.70		63.47	3.61	3.97
16 L.H., Daily Totals		$918.10		$1362.67	$57.38	$85.17
Crew Q-17A	Hr.	Daily	Hr.	Daily	Bare Costs	Incl. O&P
1 Steamfitter	$59.75	$478.00	$90.20	$721.60	$53.08	$80.28
1 Steamfitter Apprentice	47.80	382.40	72.20	577.60		
1 Equip. Oper. (crane)	51.70	413.60	78.45	627.60		
1 Hyd. Crane, 12 Ton		653.80		719.18		
1 Welder, Electric, 300 amp		57.70		63.47	29.65	32.61
24 L.H., Daily Totals		$1985.50		$2709.45	$82.73	$112.89
Crew Q-18	Hr.	Daily	Hr.	Daily	Bare Costs	Incl. O&P
2 Steamfitters	$59.75	$956.00	$90.20	$1443.20	$55.77	$84.20
1 Steamfitter Apprentice	47.80	382.40	72.20	577.60		
1 Welder, Electric, 300 amp		57.70		63.47	2.40	2.64
24 L.H., Daily Totals		$1396.10		$2084.27	$58.17	$86.84
Crew Q-19	Hr.	Daily	Hr.	Daily	Bare Costs	Incl. O&P
1 Steamfitter	$59.75	$478.00	$90.20	$721.60	$54.08	$81.45
1 Steamfitter Apprentice	47.80	382.40	72.20	577.60		
1 Electrician	54.70	437.60	81.95	655.60		
24 L.H., Daily Totals		$1298.00		$1954.80	$54.08	$81.45
Crew Q-20	Hr.	Daily	Hr.	Daily	Bare Costs	Incl. O&P
1 Sheet Metal Worker	$55.95	$447.60	$85.50	$684.00	$51.22	$77.95
1 Sheet Metal Apprentice	44.75	358.00	68.40	547.20		
.5 Electrician	54.70	218.80	81.95	327.80		
20 L.H., Daily Totals		$1024.40		$1559.00	$51.22	$77.95
Crew Q-21	Hr.	Daily	Hr.	Daily	Bare Costs	Incl. O&P
2 Steamfitters	$59.75	$956.00	$90.20	$1443.20	$55.50	$83.64
1 Steamfitter Apprentice	47.80	382.40	72.20	577.60		
1 Electrician	54.70	437.60	81.95	655.60		
32 L.H., Daily Totals		$1776.00		$2676.40	$55.50	$83.64

For customer support on your Plumbing Cost Data, call 877.709.6968.

Crew No.	Bare Costs		Incl. Subs O&P		Cost Per Labor-Hour	
	Hr.	Daily	Hr.	Daily	Bare Costs	Incl. O&P
Crew Q-22						
1 Plumber	$58.70	$469.60	$88.65	$709.20	$52.83	$79.78
1 Plumber Apprentice	46.95	375.60	70.90	567.20		
1 Hyd. Crane, 12 Ton		653.80		719.18	40.86	44.95
16 L.H., Daily Totals		$1499.00		$1995.58	$93.69	$124.72
Crew Q-22A	Hr.	Daily	Hr.	Daily	Bare Costs	Incl. O&P
1 Plumber	$58.70	$469.60	$88.65	$709.20	$48.74	$73.96
1 Plumber Apprentice	46.95	375.60	70.90	567.20		
1 Laborer	37.60	300.80	57.85	462.80		
1 Equip. Oper. (crane)	51.70	413.60	78.45	627.60		
1 Hyd. Crane, 12 Ton		653.80		719.18	20.43	22.47
32 L.H., Daily Totals		$2213.40		$3085.98	$69.17	$96.44
Crew Q-23	Hr.	Daily	Hr.	Daily	Bare Costs	Incl. O&P
1 Plumber Foreman (outside)	$60.70	$485.60	$91.65	$733.20	$56.67	$85.68
1 Plumber	58.70	469.60	88.65	709.20		
1 Equip. Oper. (medium)	50.60	404.80	76.75	614.00		
1 Lattice Boom Crane, 20 Ton		948.90		1043.79	39.54	43.49
24 L.H., Daily Totals		$2308.90		$3100.19	$96.20	$129.17
Crew R-1	Hr.	Daily	Hr.	Daily	Bare Costs	Incl. O&P
1 Electrician Foreman	$55.20	$441.60	$82.70	$661.60	$48.37	$73.08
3 Electricians	54.70	1312.80	81.95	1966.80		
2 Helpers	35.45	567.20	54.95	879.20		
48 L.H., Daily Totals		$2321.60		$3507.60	$48.37	$73.08
Crew R-1A	Hr.	Daily	Hr.	Daily	Bare Costs	Incl. O&P
1 Electrician	$54.70	$437.60	$81.95	$655.60	$45.08	$68.45
1 Helper	35.45	283.60	54.95	439.60		
16 L.H., Daily Totals		$721.20		$1095.20	$45.08	$68.45
Crew R-2	Hr.	Daily	Hr.	Daily	Bare Costs	Incl. O&P
1 Electrician Foreman	$55.20	$441.60	$82.70	$661.60	$48.84	$73.84
3 Electricians	54.70	1312.80	81.95	1966.80		
2 Helpers	35.45	567.20	54.95	879.20		
1 Equip. Oper. (crane)	51.70	413.60	78.45	627.60		
1 S.P. Crane, 4x4, 5 Ton		275.40		302.94	4.92	5.41
56 L.H., Daily Totals		$3010.60		$4438.14	$53.76	$79.25
Crew R-3	Hr.	Daily	Hr.	Daily	Bare Costs	Incl. O&P
1 Electrician Foreman	$55.20	$441.60	$82.70	$661.60	$54.30	$81.55
1 Electrician	54.70	437.60	81.95	655.60		
.5 Equip. Oper. (crane)	51.70	206.80	78.45	313.80		
.5 S.P. Crane, 4x4, 5 Ton		137.70		151.47	6.88	7.57
20 L.H., Daily Totals		$1223.70		$1782.47	$61.19	$89.12
Crew R-4	Hr.	Daily	Hr.	Daily	Bare Costs	Incl. O&P
1 Struc. Steel Foreman (outside)	$54.65	$437.20	$94.95	$759.60	$53.46	$90.25
3 Struc. Steel Workers	52.65	1263.60	91.45	2194.80		
1 Electrician	54.70	437.60	81.95	655.60		
1 Welder, Gas Engine, 300 amp		145.85		160.44	3.65	4.01
40 L.H., Daily Totals		$2284.25		$3770.43	$57.11	$94.26

Crew No.	Bare Costs		Incl. Subs O&P		Cost Per Labor-Hour	
	Hr.	Daily	Hr.	Daily	Bare Costs	Incl. O&P
Crew R-5						
1 Electrician Foreman	$55.20	$441.60	$82.70	$661.60	$47.75	$72.20
4 Electrician Linemen	54.70	1750.40	81.95	2622.40		
2 Electrician Operators	54.70	875.20	81.95	1311.20		
4 Electrician Groundmen	35.45	1134.40	54.95	1758.40		
1 Crew Truck		206.40		227.04		
1 Flatbed Truck, 20,000 GVW		248.60		273.46		
1 Pickup Truck, 3/4 Ton		144.20		158.62		
.2 Hyd. Crane, 55 Ton		225.60		248.16		
.2 Hyd. Crane, 12 Ton		130.76		143.84		
.2 Earth Auger, Truck-Mtd.		83.44		91.78		
1 Tractor w/Winch		427.00		469.70	16.66	18.32
88 L.H., Daily Totals		$5667.60		$7966.20	$64.40	$90.53
Crew R-6	Hr.	Daily	Hr.	Daily	Bare Costs	Incl. O&P
1 Electrician Foreman	$55.20	$441.60	$82.70	$661.60	$47.75	$72.20
4 Electrician Linemen	54.70	1750.40	81.95	2622.40		
2 Electrician Operators	54.70	875.20	81.95	1311.20		
4 Electrician Groundmen	35.45	1134.40	54.95	1758.40		
1 Crew Truck		206.40		227.04		
1 Flatbed Truck, 20,000 GVW		248.60		273.46		
1 Pickup Truck, 3/4 Ton		144.20		158.62		
.2 Hyd. Crane, 55 Ton		225.60		248.16		
.2 Hyd. Crane, 12 Ton		130.76		143.84		
.2 Earth Auger, Truck-Mtd.		83.44		91.78		
1 Tractor w/Winch		427.00		469.70		
3 Cable Trailers		598.20		658.02		
.5 Tensioning Rig		198.40		218.24		
.5 Cable Pulling Rig		1160.00		1276.00	38.89	42.78
88 L.H., Daily Totals		$7624.20		$10118.46	$86.64	$114.98
Crew R-7	Hr.	Daily	Hr.	Daily	Bare Costs	Incl. O&P
1 Electrician Foreman	$55.20	$441.60	$82.70	$661.60	$38.74	$59.58
5 Electrician Groundmen	35.45	1418.00	54.95	2198.00		
1 Crew Truck		206.40		227.04	4.30	4.73
48 L.H., Daily Totals		$2066.00		$3086.64	$43.04	$64.31
Crew R-8	Hr.	Daily	Hr.	Daily	Bare Costs	Incl. O&P
1 Electrician Foreman	$55.20	$441.60	$82.70	$661.60	$48.37	$73.08
3 Electrician Linemen	54.70	1312.80	81.95	1966.80		
2 Electrician Groundmen	35.45	567.20	54.95	879.20		
1 Pickup Truck, 3/4 Ton		144.20		158.62		
1 Crew Truck		206.40		227.04	7.30	8.03
48 L.H., Daily Totals		$2672.20		$3893.26	$55.67	$81.11
Crew R-9	Hr.	Daily	Hr.	Daily	Bare Costs	Incl. O&P
1 Electrician Foreman	$55.20	$441.60	$82.70	$661.60	$45.14	$68.54
1 Electrician Lineman	54.70	437.60	81.95	655.60		
2 Electrician Operators	54.70	875.20	81.95	1311.20		
4 Electrician Groundmen	35.45	1134.40	54.95	1758.40		
1 Pickup Truck, 3/4 Ton		144.20		158.62		
1 Crew Truck		206.40		227.04	5.48	6.03
64 L.H., Daily Totals		$3239.40		$4772.46	$50.62	$74.57
Crew R-10	Hr.	Daily	Hr.	Daily	Bare Costs	Incl. O&P
1 Electrician Foreman	$55.20	$441.60	$82.70	$661.60	$51.58	$77.58
4 Electrician Linemen	54.70	1750.40	81.95	2622.40		
1 Electrician Groundman	35.45	283.60	54.95	439.60		
1 Crew Truck		206.40		227.04		
3 Tram Cars		412.20		453.42	12.89	14.18
48 L.H., Daily Totals		$3094.20		$4404.06	$64.46	$91.75

Crew No.	Bare Costs		Incl. Subs O&P		Cost Per Labor-Hour	
Crew R-11	Hr.	Daily	Hr.	Daily	Bare Costs	Incl. O&P
1 Electrician Foreman	$55.20	$441.60	$82.70	$661.60	$51.90	$78.11
4 Electricians	54.70	1750.40	81.95	2622.40		
1 Equip. Oper. (crane)	51.70	413.60	78.45	627.60		
1 Common Laborer	37.60	300.80	57.85	462.80		
1 Crew Truck		206.40		227.04		
1 Hyd. Crane, 12 Ton		653.80		719.18	15.36	16.90
56 L.H., Daily Totals		$3766.60		$5320.62	$67.26	$95.01
Crew R-12	Hr.	Daily	Hr.	Daily	Bare Costs	Incl. O&P
1 Carpenter Foreman (inside)	$47.45	$379.60	$73.05	$584.40	$44.45	$69.24
4 Carpenters	46.95	1502.40	72.25	2312.00		
4 Common Laborers	37.60	1203.20	57.85	1851.20		
1 Equip. Oper. (medium)	50.60	404.80	76.75	614.00		
1 Steel Worker	52.65	421.20	91.45	731.60		
1 Dozer, 200 H.P.		1387.00		1525.70		
1 Pickup Truck, 3/4 Ton		144.20		158.62	17.40	19.14
88 L.H., Daily Totals		$5442.40		$7777.52	$61.85	$88.38
Crew R-13	Hr.	Daily	Hr.	Daily	Bare Costs	Incl. O&P
1 Electrician Foreman	$55.20	$441.60	$82.70	$661.60	$52.84	$79.37
3 Electricians	54.70	1312.80	81.95	1966.80		
.25 Equip. Oper. (crane)	51.70	103.40	78.45	156.90		
1 Equipment Oiler	45.20	361.60	68.55	548.40		
.25 Hydraulic Crane, 33 Ton		187.50		206.25	4.46	4.91
42 L.H., Daily Totals		$2406.90		$3539.95	$57.31	$84.28
Crew R-15	Hr.	Daily	Hr.	Daily	Bare Costs	Incl. O&P
1 Electrician Foreman	$55.20	$441.60	$82.70	$661.60	$53.77	$80.71
4 Electricians	54.70	1750.40	81.95	2622.40		
1 Equipment Oper. (light)	48.60	388.80	73.75	590.00		
1 Aerial Lift Truck, 40' Boom		301.40		331.54	6.28	6.91
48 L.H., Daily Totals		$2882.20		$4205.54	$60.05	$87.62
Crew R-18	Hr.	Daily	Hr.	Daily	Bare Costs	Incl. O&P
.25 Electrician Foreman	$55.20	$110.40	$82.70	$165.40	$42.89	$65.39
1 Electrician	54.70	437.60	81.95	655.60		
2 Helpers	35.45	567.20	54.95	879.20		
26 L.H., Daily Totals		$1115.20		$1700.20	$42.89	$65.39
Crew R-19	Hr.	Daily	Hr.	Daily	Bare Costs	Incl. O&P
.5 Electrician Foreman	$55.20	$220.80	$82.70	$330.80	$54.80	$82.10
2 Electricians	54.70	875.20	81.95	1311.20		
20 L.H., Daily Totals		$1096.00		$1642.00	$54.80	$82.10
Crew R-21	Hr.	Daily	Hr.	Daily	Bare Costs	Incl. O&P
1 Electrician Foreman	$55.20	$441.60	$82.70	$661.60	$54.72	$82.01
3 Electricians	54.70	1312.80	81.95	1966.80		
.1 Equip. Oper. (medium)	50.60	40.48	76.75	61.40		
.1 S.P. Crane, 4x4, 25 Ton		59.94		65.93	1.83	2.01
32.8 L.H., Daily Totals		$1854.82		$2755.73	$56.55	$84.02
Crew R-22	Hr.	Daily	Hr.	Daily	Bare Costs	Incl. O&P
.66 Electrician Foreman	$55.20	$291.46	$82.70	$436.66	$46.51	$70.47
2 Electricians	54.70	875.20	81.95	1311.20		
2 Helpers	35.45	567.20	54.95	879.20		
37.28 L.H., Daily Totals		$1733.86		$2627.06	$46.51	$70.47

Crew No.	Bare Costs		Incl. Subs O&P		Cost Per Labor-Hour	
Crew R-30	Hr.	Daily	Hr.	Daily	Bare Costs	Incl. O&P
.25 Electrician Foreman (outside)	$56.70	$113.40	$84.95	$169.90	$44.33	$67.35
1 Electrician	54.70	437.60	81.95	655.60		
2 Laborers, (Semi-Skilled)	37.60	601.60	57.85	925.60		
26 L.H., Daily Totals		$1152.60		$1751.10	$44.33	$67.35
Crew R-31	Hr.	Daily	Hr.	Daily	Bare Costs	Incl. O&P
1 Electrician	$54.70	$437.60	$81.95	$655.60	$54.70	$81.95
1 Core Drill, Electric, 2.5 H.P.		55.15		60.66	6.89	7.58
8 L.H., Daily Totals		$492.75		$716.26	$61.59	$89.53
Crew W-41E	Hr.	Daily	Hr.	Daily	Bare Costs	Incl. O&P
.5 Plumber Foreman (outside)	$60.70	$242.80	$91.65	$366.60	$50.66	$76.93
1 Plumber	58.70	469.60	88.65	709.20		
1 Laborer	37.60	300.80	57.85	462.80		
20 L.H., Daily Totals		$1013.20		$1538.60	$50.66	$76.93

Historical Cost Indexes

The table below lists both the RSMeans® historical cost index based on Jan. 1, 1993 = 100 as well as the computed value of an index based on Jan. 1, 2015 costs. Since the Jan. 1, 2015 figure is estimated, space is left to write in the actual index figures as they become available through either the quarterly *RSMeans Construction Cost Indexes* or as printed in the *Engineering News-Record*. To compute the actual index based on Jan. 1, 2015 = 100, divide the historical cost index for a particular year by the actual Jan. 1, 2015 construction cost index. Space has been left to advance the index figures as the year progresses.

Year	Historical Cost Index Jan. 1, 1993 = 100		Current Index Based on Jan. 1, 2015 = 100		Year	Historical Cost Index Jan. 1, 1993 = 100	Current Index Based on Jan. 1, 2015 = 100		Year	Historical Cost Index Jan. 1, 1993 = 100	Current Index Based on Jan. 1, 2015 = 100	
	Est.	Actual	Est.	Actual		Actual	Est.	Actual		Actual	Est.	Actual
Oct 2015*					July 2000	120.9	58.5		July 1982	76.1	36.8	
July 2015*					1999	117.6	56.9		1981	70.0	33.9	
April 2015*					1998	115.1	55.7		1980	62.9	30.4	
Jan 2015*	206.7		100.0	100.0	1997	112.8	54.6		1979	57.8	28.0	
July 2014		204.9	99.1		1996	110.2	53.3		1978	53.5	25.9	
2013		201.2	97.3		1995	107.6	52.1		1977	49.5	23.9	
2012		194.6	94.1		1994	104.4	50.5		1976	46.9	22.7	
2011		191.2	92.5		1993	101.7	49.2		1975	44.8	21.7	
2010		183.5	88.8		1992	99.4	48.1		1974	41.4	20.0	
2009		180.1	87.1		1991	96.8	46.8		1973	37.7	18.2	
2008		180.4	87.3		1990	94.3	45.6		1972	34.8	16.8	
2007		169.4	82.0		1989	92.1	44.6		1971	32.1	15.5	
2006		162.0	78.4		1988	89.9	43.5		1970	28.7	13.9	
2005		151.6	73.3		1987	87.7	42.4		1969	26.9	13.0	
2004		143.7	69.5		1986	84.2	40.7		1968	24.9	12.0	
2003		132.0	63.9		1985	82.6	40.0		1967	23.5	11.4	
2002		128.7	62.3		1984	82.0	39.7		1966	22.7	11.0	
2001		125.1	60.5		1983	80.2	38.8		1965	21.7	10.5	

Adjustments to Costs

The "Historical Cost Index" can be used to convert national average building costs at a particular time to the approximate building costs for some other time.

Example:

Estimate and compare construction costs for different years in the same city.

To estimate the national average construction cost of a building in 1970, knowing that it cost $900,000 in 2015:

INDEX in 1970 = 28.7

INDEX in 2015 = 206.7

Note: The city cost indexes for Canada can be used to convert U.S. national averages to local costs in Canadian dollars.

Example:

To estimate and compare the cost of a building in Toronto, ON in 2015 with the known cost of $600,000 (US$) in New York, NY in 2015:

INDEX Toronto = 110.9

INDEX New York = 131.8

$$\frac{\text{INDEX Toronto}}{\text{INDEX New York}} \times \text{Cost New York} = \text{Cost Toronto}$$

$$\frac{110.9}{131.8} \times \$600,000 = .841 \times \$600,000 = \$504,600$$

The construction cost of the building in Toronto is $504,600 (CN$).

Time Adjustment Using the Historical Cost Indexes:

$$\frac{\text{Index for Year A}}{\text{Index for Year B}} \times \text{Cost in Year B} = \text{Cost in Year A}$$

$$\frac{\text{INDEX 1970}}{\text{INDEX 2015}} \times \text{Cost 2015} = \text{Cost 1970}$$

$$\frac{28.7}{206.7} \times \$900,000 = .139 \times \$900,000 = \$125,100$$

The construction cost of the building in 1970 is $125,100.

*Historical Cost Index updates and other resources are provided on the following website.
http://info.thegordiangroup.com/RSMeans.html

601

How to Use the City Cost Indexes

What you should know before you begin

RSMeans City Cost Indexes (CCI) are an extremely useful tool to use when you want to compare costs from city to city and region to region.

This publication contains average construction cost indexes for 731 U.S. and Canadian cities covering over 930 three-digit zip code locations, as listed directly under each city.

Keep in mind that a City Cost Index number is a percentage ratio of a specific city's cost to the national average cost of the same item at a stated time period.

In other words, these index figures represent relative construction factors (or, if you prefer, multipliers) for Material and Installation costs, as well as the weighted average for Total In Place costs for each CSI MasterFormat division. Installation costs include both labor and equipment rental costs. When estimating equipment rental rates only, for a specific location, use 01 54 33 EQUIPMENT RENTAL COSTS in the Reference Section at the back of the book.

The 30 City Average Index is the average of 30 major U.S. cities and serves as a National Average.

Index figures for both material and installation are based on the 30 major city average of 100 and represent the cost relationship as of July 1, 2014. The index for each division is computed from representative material and labor quantities for that division. The weighted average for each city is a weighted total of the components listed above it, but does not include relative productivity between trades or cities.

As changes occur in local material prices, labor rates, and equipment rental rates (including fuel costs), the impact of these changes should be accurately measured by the change in the City Cost Index for each particular city (as compared to the 30 City Average).

Therefore, if you know (or have estimated) building costs in one city today, you can easily convert those costs to expected building costs in another city.

In addition, by using the Historical Cost Index, you can easily convert National Average building costs at a particular time to the approximate building costs for some other time. The City Cost Indexes can then be applied to calculate the costs for a particular city.

Quick Calculations

Location Adjustment Using the City Cost Indexes:

$$\frac{\text{Index for City A}}{\text{Index for City B}} \times \text{Cost in City B} = \text{Cost in City A}$$

Time Adjustment for the National Average Using the Historical Cost Index:

$$\frac{\text{Index for Year A}}{\text{Index for Year B}} \times \text{Cost in Year B} = \text{Cost in Year A}$$

Adjustment from the National Average:

$$\frac{\text{Index for City A}}{100} \times \text{National Average Cost} = \text{Cost in City A}$$

Since each of the other RSMeans publications contains many different items, any *one* item multiplied by the particular city index may give incorrect results. However, the larger the number of items compiled, the closer the results should be to actual costs for that particular city.

The City Cost Indexes for Canadian cities are calculated using Canadian material and equipment prices and labor rates, in Canadian dollars. Therefore, indexes for Canadian cities can be used to convert U.S. National Average prices to local costs in Canadian dollars.

How to use this section

1. Compare costs from city to city.

In using the RSMeans Indexes, remember that an index number is not a fixed number but a ratio: It's a percentage ratio of a building component's cost at any stated time to the National Average cost of that same component at the same time period. Put in the form of an equation:

$$\frac{\text{Specific City Cost}}{\text{National Average Cost}} \times 100 = \text{City Index Number}$$

Therefore, when making cost comparisons between cities, do not subtract one city's index number from the index number of another city and read the result as a percentage difference. Instead, divide one city's index number by that of the other city. The resulting number may then be used as a multiplier to calculate cost differences from city to city.

The formula used to find cost differences between cities for the purpose of comparison is as follows:

$$\frac{\text{City A Index}}{\text{City B Index}} \times \text{City B Cost (Known)} = \text{City A Cost (Unknown)}$$

In addition, you can use RSMeans CCI to calculate and compare costs division by division between cities using the same basic formula. (Just be sure that you're comparing similar divisions.)

2. Compare a specific city's construction costs with the National Average.

When you're studying construction location feasibility, it's advisable to compare a prospective project's cost index with an index of the National Average cost.

For example, divide the weighted average index of construction costs of a specific city by that of the 30 City Average, which = 100.

$$\frac{\text{City Index}}{100} = \text{\% of National Average}$$

As a result, you get a ratio that indicates the relative cost of construction in that city in comparison with the National Average.

3. Convert U.S. National Average to actual costs in Canadian City.

$$\frac{\text{Index for Canadian City}}{100} \times \text{National Average Cost} = \text{Cost in Canadian City in \$ CAN}$$

4. Adjust construction cost data based on a National Average.

When you use a source of construction cost data which is based on a National Average (such as RSMeans cost data publications), it is necessary to adjust those costs to a specific location.

$$\frac{\text{City Index}}{100} \times \frac{\text{``Book'' Cost Based on}}{\text{National Average Costs}} = \frac{\text{City Cost}}{\text{(Unknown)}}$$

5. When applying the City Cost Indexes to demolition projects, use the appropriate division installation index. For example, for removal of existing doors and windows, use Division 8 (Openings) index.

What you might like to know about how we developed the Indexes

The information presented in the CCI is organized according to the Construction Specifications Institute (CSI) MasterFormat 2014 classification system.

To create a reliable index, RSMeans researched the building type most often constructed in the United States and Canada. Because it was concluded that no one type of building completely represented the building construction industry, nine different types of buildings were combined to create a composite model.

The exact material, labor, and equipment quantities are based on detailed analyses of these nine building types, and then each quantity is weighted in proportion to expected usage. These various material items, labor hours, and equipment rental rates are thus combined to form a composite building representing as closely as possible the actual usage of materials, labor, and equipment used in the North American building construction industry.

The following structures were chosen to make up that composite model:

1. Factory, 1 story
2. Office, 2–4 story
3. Store, Retail
4. Town Hall, 2–3 story
5. High School, 2–3 story
6. Hospital, 4–8 story
7. Garage, Parking
8. Apartment, 1–3 story
9. Hotel/Motel, 2–3 story

For the purposes of ensuring the timeliness of the data, the components of the index for the composite model have been streamlined. They currently consist of:

- specific quantities of 66 commonly used construction materials;
- specific labor-hours for 21 building construction trades; and
- specific days of equipment rental for 6 types of construction equipment (normally used to install the 66 material items by the 21 trades.) Fuel costs and routine maintenance costs are included in the equipment cost.

A sophisticated computer program handles the updating of all costs for each city on a quarterly basis. Material and equipment price quotations are gathered quarterly from cities in the United States and Canada. These prices and the latest negotiated labor wage rates for 21 different building trades are used to compile the quarterly update of the City Cost Index.

The 30 major U.S. cities used to calculate the National Average are:

Atlanta, GA	Memphis, TN
Baltimore, MD	Milwaukee, WI
Boston, MA	Minneapolis, MN
Buffalo, NY	Nashville, TN
Chicago, IL	New Orleans, LA
Cincinnati, OH	New York, NY
Cleveland, OH	Philadelphia, PA
Columbus, OH	Phoenix, AZ
Dallas, TX	Pittsburgh, PA
Denver, CO	St. Louis, MO
Detroit, MI	San Antonio, TX
Houston, TX	San Diego, CA
Indianapolis, IN	San Francisco, CA
Kansas City, MO	Seattle, WA
Los Angeles, CA	Washington, DC

What the CCI does not indicate

The weighted average for each city is a total of the divisional components weighted to reflect typical usage, but it does not include the productivity variations between trades or cities.

In addition, the CCI does not take into consideration factors such as the following:

- managerial efficiency
- competitive conditions
- automation
- restrictive union practices
- unique local requirements
- regional variations due to specific building codes

		UNITED STATES			ALABAMA												
		30 CITY AVERAGE			ANNISTON 362			BIRMINGHAM 350 - 352			BUTLER 369			DECATUR 356			DOTHAN 363		
DIVISION		MAT.	INST.	TOTAL	MAT.	INST.	TOTAL	MAT.	INST.	TOTAL	MAT.	INST.	TOTAL	MAT.	INST.	TOTAL	MAT.	INST.	TOTAL
015433	CONTRACTOR EQUIPMENT		100.0	100.0		101.0	101.0		101.1	101.1		98.2	98.2		101.0	101.0		98.2	98.2
0241, 31 - 34	SITE & INFRASTRUCTURE, DEMOLITION	100.0	100.0	100.0	90.2	93.1	92.3	97.4	93.8	94.8	103.5	87.1	91.8	88.8	91.5	90.7	101.2	87.1	91.1
0310	Concrete Forming & Accessories	100.0	100.0	100.0	90.1	47.1	53.0	93.1	75.9	78.2	86.4	42.6	48.6	94.6	45.4	52.1	95.5	42.0	49.4
0320	Concrete Reinforcing	100.0	100.0	100.0	87.8	86.3	87.1	94.7	86.9	90.7	92.8	45.9	68.9	88.6	77.0	82.7	92.8	45.4	68.7
0330	Cast-in-Place Concrete	100.0	100.0	100.0	101.8	49.7	80.4	110.0	74.6	95.4	99.3	54.8	81.0	103.4	64.1	87.2	99.3	46.3	77.5
03	CONCRETE	100.0	100.0	100.0	102.1	57.0	79.9	102.2	78.4	90.5	103.0	49.3	76.6	98.3	59.3	79.1	102.3	46.0	74.7
04	MASONRY	100.0	100.0	100.0	100.1	66.8	79.3	98.8	76.0	84.6	105.9	48.6	70.2	97.3	51.3	68.6	107.1	39.7	65.1
05	METALS	100.0	100.0	100.0	104.0	91.5	100.1	104.0	94.1	101.0	102.8	76.6	94.8	102.6	86.1	100.0	102.9	75.3	94.4
06	WOOD, PLASTICS & COMPOSITES	100.0	100.0	100.0	90.5	42.9	63.7	97.4	76.1	85.4	85.0	42.4	61.1	98.9	42.9	67.6	97.4	42.4	66.6
07	THERMAL & MOISTURE PROTECTION	100.0	100.0	100.0	97.9	52.9	79.5	99.7	81.5	92.2	98.0	56.7	81.0	96.7	58.3	81.0	98.0	49.1	77.9
08	OPENINGS	100.0	100.0	100.0	98.6	50.1	87.3	103.2	76.0	96.8	98.6	44.2	85.9	106.5	50.4	93.5	98.6	44.2	86.0
0920	Plaster & Gypsum Board	100.0	100.0	100.0	90.8	41.4	57.4	94.8	75.8	81.9	87.9	41.0	56.2	95.8	41.6	59.2	98.6	41.0	59.7
0950, 0980	Ceilings & Acoustic Treatment	100.0	100.0	100.0	80.4	41.4	54.7	89.8	75.8	80.6	80.4	41.0	54.5	87.2	41.6	57.2	80.4	41.0	54.5
0960	Flooring	100.0	100.0	100.0	91.5	40.2	76.8	102.5	76.7	95.1	95.8	56.0	84.4	99.7	50.6	85.6	100.6	27.5	79.7
0970, 0990	Wall Finishes & Painting/Coating	100.0	100.0	100.0	103.7	31.5	60.1	103.9	66.5	81.3	103.7	53.3	73.3	99.6	66.8	79.8	103.7	53.3	73.3
09	FINISHES	100.0	100.0	100.0	87.0	43.0	62.8	95.4	74.8	84.1	89.9	45.7	65.5	91.9	47.7	67.6	92.5	40.1	63.6
COVERS	DIVS. 10 - 14, 25, 28, 41, 43, 44, 46	100.0	100.0	100.0	100.0	68.3	93.6	100.0	88.4	97.7	100.0	41.4	88.2	100.0	43.2	88.5	100.0	41.5	88.2
21, 22, 23	FIRE SUPPRESSION, PLUMBING & HVAC	100.0	100.0	100.0	100.0	55.4	82.0	100.0	70.4	88.1	97.3	33.3	71.5	100.0	40.8	76.1	97.3	33.2	71.5
26, 27, 3370	ELECTRICAL, COMMUNICATIONS & UTIL.	100.0	100.0	100.0	93.1	57.3	74.2	99.0	61.6	79.3	95.0	40.2	66.0	94.2	64.9	78.7	93.8	57.4	74.6
MF2014	WEIGHTED AVERAGE	100.0	100.0	100.0	98.6	61.8	82.6	100.6	76.7	90.2	98.9	49.4	77.3	99.8	57.9	81.5	99.0	49.4	77.4

		ALABAMA															
		EVERGREEN 364			GADSDEN 359			HUNTSVILLE 357 - 358			JASPER 355			MOBILE 365 - 366			MONTGOMERY 360 - 361		
DIVISION		MAT.	INST.	TOTAL	MAT.	INST.	TOTAL	MAT.	INST.	TOTAL	MAT.	INST.	TOTAL	MAT.	INST.	TOTAL	MAT.	INST.	TOTAL
015433	CONTRACTOR EQUIPMENT		98.2	98.2		101.0	101.0		101.0	101.0		101.0	101.0		98.2	98.2		98.2	98.2
0241, 31 - 34	SITE & INFRASTRUCTURE, DEMOLITION	104.0	87.4	92.2	94.9	92.7	93.3	88.6	92.7	91.5	94.6	92.4	93.1	96.4	88.1	90.5	96.5	88.1	90.5
0310	Concrete Forming & Accessories	83.0	44.2	49.5	86.9	42.5	48.6	94.6	69.9	73.3	92.1	34.6	42.5	94.5	54.0	59.6	94.1	44.1	51.0
0320	Concrete Reinforcing	92.9	45.9	69.0	94.0	86.8	90.3	88.6	81.6	85.1	88.6	85.6	87.1	90.6	83.6	87.0	95.8	86.1	90.9
0330	Cast-in-Place Concrete	99.3	57.3	82.1	103.4	66.7	88.3	100.7	69.5	87.9	114.5	44.1	85.6	104.1	66.4	88.6	105.2	56.5	85.2
03	CONCRETE	103.3	50.9	77.5	102.9	60.8	82.2	97.0	72.9	85.2	106.6	49.4	78.5	98.9	65.3	82.4	100.2	57.9	79.4
04	MASONRY	105.9	53.2	73.0	95.6	62.4	74.9	98.8	67.7	79.4	92.9	41.2	60.7	103.7	54.2	72.9	100.0	49.7	68.6
05	METALS	102.9	76.5	94.7	104.0	93.2	100.7	106.2	90.6	101.4	104.0	90.2	99.7	105.0	91.6	100.9	104.0	91.9	100.3
06	WOOD, PLASTICS & COMPOSITES	81.3	42.4	59.5	89.3	38.7	61.0	98.9	72.6	84.2	95.9	31.6	59.9	95.9	52.8	71.8	95.3	42.4	65.6
07	THERMAL & MOISTURE PROTECTION	97.9	56.9	81.1	96.8	71.7	86.5	96.6	77.6	88.8	96.8	45.4	75.7	97.6	70.1	86.3	97.9	66.6	85.0
08	OPENINGS	98.6	44.2	85.9	102.8	49.0	90.3	106.5	67.4	97.4	102.8	48.0	90.0	102.2	58.8	92.1	103.3	53.1	91.7
0920	Plaster & Gypsum Board	87.2	41.0	56.0	88.2	37.3	53.8	95.8	72.2	79.9	92.1	30.0	50.1	95.2	51.8	65.9	94.6	41.0	58.4
0950, 0980	Ceilings & Acoustic Treatment	80.4	41.0	54.5	83.8	37.3	53.2	89.0	72.2	78.0	83.8	30.0	48.4	85.6	51.8	63.4	88.0	41.0	57.1
0960	Flooring	93.9	56.0	83.1	95.5	76.7	90.1	99.7	76.7	93.1	97.8	40.2	81.3	100.5	58.5	88.5	97.3	58.5	86.2
0970, 0990	Wall Finishes & Painting/Coating	103.7	53.3	73.3	99.6	60.7	76.1	99.6	65.9	79.3	99.6	38.6	62.8	107.3	54.3	75.3	103.7	53.3	73.1
09	FINISHES	89.2	46.8	65.9	89.4	48.7	67.0	92.3	70.7	80.4	90.5	34.4	59.6	92.7	53.5	71.1	92.8	46.1	67.0
COVERS	DIVS. 10 - 14, 25, 28, 41, 43, 44, 46	100.0	42.9	88.5	100.0	79.8	95.9	100.0	85.3	97.0	100.0	39.8	87.8	100.0	83.1	96.6	100.0	79.6	95.9
21, 22, 23	FIRE SUPPRESSION, PLUMBING & HVAC	97.3	35.6	72.4	102.0	36.3	75.5	100.0	61.6	84.5	102.0	61.4	85.6	99.9	60.7	84.1	100.0	34.2	73.4
26, 27, 3370	ELECTRICAL, COMMUNICATIONS & UTIL.	92.5	40.2	64.8	94.3	61.6	77.0	95.3	64.9	79.3	93.8	61.3	76.6	95.7	59.1	76.3	96.3	60.9	77.6
MF2014	WEIGHTED AVERAGE	98.6	50.7	77.7	99.9	60.0	82.5	99.8	72.1	87.8	100.2	57.5	81.6	99.9	65.4	84.9	99.9	57.1	81.2

		ALABAMA									ALASKA								
		PHENIX CITY 368			SELMA 367			TUSCALOOSA 354			ANCHORAGE 995 - 996			FAIRBANKS 997			JUNEAU 998		
DIVISION		MAT.	INST.	TOTAL	MAT.	INST.	TOTAL	MAT.	INST.	TOTAL	MAT.	INST.	TOTAL	MAT.	INST.	TOTAL	MAT.	INST.	TOTAL
015433	CONTRACTOR EQUIPMENT		98.2	98.2		98.2	98.2		101.0	101.0		114.7	114.7		114.7	114.7		114.7	114.7
0241, 31 - 34	SITE & INFRASTRUCTURE, DEMOLITION	107.8	88.2	93.9	101.0	88.1	91.8	89.1	92.9	91.8	127.2	129.6	128.9	119.2	129.6	126.6	135.6	129.6	131.3
0310	Concrete Forming & Accessories	90.1	39.1	46.1	87.6	42.6	48.8	94.5	47.9	54.3	123.7	119.9	120.3	131.6	119.9	121.5	130.1	119.8	121.2
0320	Concrete Reinforcing	92.8	61.4	76.8	92.8	85.2	89.0	88.6	86.8	87.7	149.7	110.7	129.9	150.9	110.7	130.5	135.7	110.7	123.0
0330	Cast-in-Place Concrete	99.3	57.0	81.9	99.3	46.5	77.6	104.9	68.4	89.9	132.0	117.8	126.2	128.3	118.2	124.1	133.2	117.8	126.9
03	CONCRETE	106.2	51.3	79.2	101.7	53.7	78.1	99.0	63.8	81.7	138.6	116.7	127.9	123.8	116.9	120.4	136.0	116.7	126.5
04	MASONRY	105.9	41.1	65.5	109.6	38.7	65.4	97.6	65.5	77.6	184.6	125.3	147.6	189.5	125.3	149.5	176.6	125.3	144.6
05	METALS	102.8	81.9	96.4	102.8	89.3	98.7	105.3	93.2	101.6	113.6	103.4	110.5	117.6	103.5	113.3	116.3	103.4	112.3
06	WOOD, PLASTICS & COMPOSITES	90.2	35.4	59.5	87.0	42.4	62.0	98.9	44.4	68.4	129.2	118.7	123.4	136.1	118.7	126.4	129.4	118.7	123.4
07	THERMAL & MOISTURE PROTECTION	98.3	63.8	84.1	97.8	52.1	79.1	96.7	73.7	87.3	159.5	118.0	142.5	166.5	119.1	147.1	168.0	118.0	147.5
08	OPENINGS	98.6	43.5	85.8	98.6	53.1	88.0	106.5	58.8	95.4	132.7	115.6	128.7	128.9	115.7	125.8	129.8	115.6	126.5
0920	Plaster & Gypsum Board	92.2	33.9	52.8	90.1	41.0	56.9	95.8	43.1	60.2	140.9	119.1	126.2	164.8	119.1	133.9	148.1	119.1	128.5
0950, 0980	Ceilings & Acoustic Treatment	80.4	33.9	49.8	80.4	41.0	54.5	89.0	43.1	58.9	119.3	119.1	119.3	119.2	119.1	119.2	125.9	119.1	121.4
0960	Flooring	97.6	58.5	86.4	96.2	28.5	76.8	99.7	76.7	93.1	134.6	133.0	134.1	127.2	133.0	128.9	131.0	133.0	131.6
0970, 0990	Wall Finishes & Painting/Coating	103.7	53.3	73.3	103.7	53.3	73.3	99.6	45.9	67.2	136.8	116.3	124.4	133.5	121.2	126.1	132.1	116.3	122.6
09	FINISHES	91.4	42.2	64.3	90.0	40.2	62.6	92.3	51.2	69.7	134.3	122.2	127.7	132.4	122.8	127.1	134.4	122.2	127.7
COVERS	DIVS. 10 - 14, 25, 28, 41, 43, 44, 46	100.0	79.3	95.8	100.0	41.4	88.2	100.0	81.4	96.2	100.0	112.8	102.6	100.0	112.8	102.6	100.0	112.8	102.6
21, 22, 23	FIRE SUPPRESSION, PLUMBING & HVAC	97.3	34.5	72.0	97.3	33.6	71.6	100.0	34.0	73.4	100.3	105.0	102.2	100.2	108.0	103.4	100.3	105.0	102.2
26, 27, 3370	ELECTRICAL, COMMUNICATIONS & UTIL.	94.4	69.5	81.3	93.6	40.2	65.4	94.8	61.6	77.3	117.7	117.8	117.7	130.0	117.8	123.5	119.9	117.8	118.8
MF2014	WEIGHTED AVERAGE	99.5	54.6	79.9	98.7	49.9	77.4	99.8	61.2	83.0	121.1	115.6	118.7	121.0	116.4	119.0	121.3	115.6	118.8

For customer support on your Plumbing Cost Data, call 877.709.6968.

City Cost Indexes

		ALASKA			ARIZONA													
		KETCHIKAN			CHAMBERS			FLAGSTAFF			GLOBE			KINGMAN			MESA/TEMPE		
DIVISION		999			865			860			855			864			852		
		MAT.	INST.	TOTAL	MAT.	INST.	TOTAL	MAT.	INST.	TOTAL	MAT.	INST.	TOTAL	MAT.	INST.	TOTAL	MAT.	INST.	TOTAL
015433	CONTRACTOR EQUIPMENT		114.7	114.7		93.0	93.0		93.0	93.0		92.2	92.2		93.0	93.0		92.2	92.2
0241, 31 - 34	SITE & INFRASTRUCTURE, DEMOLITION	173.8	129.6	142.4	69.6	96.1	88.4	87.3	96.3	93.7	97.3	95.6	96.1	69.6	96.3	88.6	89.0	95.8	93.8
0310	Concrete Forming & Accessories	122.9	119.8	120.2	98.8	58.1	63.7	104.4	65.1	70.5	99.0	58.1	63.7	97.0	65.1	69.5	102.2	69.3	73.9
0320	Concrete Reinforcing	117.4	110.7	114.0	97.1	85.2	91.1	97.0	85.3	91.0	108.0	85.2	96.4	97.2	85.3	91.1	108.7	85.3	96.8
0330	Cast-in-Place Concrete	259.2	117.8	201.2	90.7	73.2	83.5	90.8	73.4	83.7	94.3	72.0	85.1	90.4	73.4	83.4	95.0	72.3	85.7
03	CONCRETE	208.7	116.7	163.5	95.0	68.6	82.0	114.1	71.8	93.3	110.0	68.2	89.5	94.6	71.8	83.4	101.8	73.3	87.8
04	MASONRY	198.4	125.3	152.8	92.5	62.3	73.7	92.6	62.4	73.8	109.9	62.2	80.1	92.5	62.4	73.7	110.1	62.2	80.3
05	METALS	117.8	103.4	113.4	96.2	75.6	89.8	96.7	76.3	90.4	93.3	76.0	88.0	96.8	76.3	90.5	93.6	76.8	88.4
06	WOOD, PLASTICS & COMPOSITES	126.3	118.7	122.1	101.0	55.0	75.2	107.3	64.2	83.1	97.6	55.0	73.8	96.1	64.2	78.2	101.1	69.8	83.6
07	THERMAL & MOISTURE PROTECTION	169.3	118.0	148.3	94.6	65.0	82.5	96.2	68.1	84.7	101.9	63.0	85.9	94.6	65.6	82.7	101.2	65.1	86.4
08	OPENINGS	130.8	115.6	127.3	108.1	65.3	98.2	108.3	70.3	99.4	100.0	65.3	91.9	108.3	70.3	99.5	100.1	73.3	93.8
0920	Plaster & Gypsum Board	152.7	119.1	130.0	90.2	53.7	65.5	93.6	63.2	73.0	96.6	53.7	67.6	82.9	63.2	69.6	98.7	68.9	78.6
0950, 0980	Ceilings & Acoustic Treatment	112.7	119.1	116.9	99.7	53.7	69.5	100.5	63.2	76.0	88.9	53.7	65.8	100.5	63.2	76.0	88.9	68.9	75.8
0960	Flooring	127.2	133.0	128.9	93.3	39.5	77.9	95.4	39.7	79.5	103.5	39.5	85.2	92.1	53.7	81.1	104.7	49.5	88.9
0970, 0990	Wall Finishes & Painting/Coating	133.5	116.3	123.1	98.3	55.1	72.2	98.3	55.1	72.2	103.4	55.1	74.3	98.3	55.1	72.2	103.4	55.1	74.3
09	FINISHES	133.4	122.2	127.2	93.7	52.9	71.2	96.6	58.4	75.6	97.1	53.0	72.8	92.5	60.7	75.0	96.8	63.6	78.5
COVERS	DIVS. 10 - 14, 25, 28, 41, 43, 44, 46	100.0	112.8	102.6	100.0	82.3	96.4	100.0	83.3	96.6	100.0	82.4	96.4	100.0	83.3	96.6	100.0	84.0	96.8
21, 22, 23	FIRE SUPPRESSION, PLUMBING & HVAC	98.4	105.0	101.1	97.0	78.9	89.7	100.2	79.0	91.6	95.2	78.9	88.7	97.0	79.0	89.7	100.0	79.0	91.5
26, 27, 3370	ELECTRICAL, COMMUNICATIONS & UTIL.	129.9	117.8	123.5	104.5	70.9	86.7	103.4	61.3	81.1	97.5	61.2	78.3	104.5	61.3	81.7	94.2	61.3	76.8
MF2014	WEIGHTED AVERAGE	132.0	115.6	124.8	97.6	71.5	86.2	101.3	71.8	88.4	98.8	70.0	86.3	97.6	72.0	86.4	98.6	72.8	87.4

		ARIZONA												ARKANSAS				
		PHOENIX			PRESCOTT			SHOW LOW			TUCSON			BATESVILLE			CAMDEN		
DIVISION		850,853			863			859			856 - 857			725			717		
		MAT.	INST.	TOTAL	MAT.	INST.	TOTAL	MAT.	INST.	TOTAL	MAT.	INST.	TOTAL	MAT.	INST.	TOTAL	MAT.	INST.	TOTAL
015433	CONTRACTOR EQUIPMENT		92.7	92.7		93.0	93.0		92.2	92.2		92.2	92.2		89.1	89.1		89.1	89.1
0241, 31 - 34	SITE & INFRASTRUCTURE, DEMOLITION	89.4	96.0	94.1	76.0	96.0	90.2	99.2	95.6	96.6	85.2	95.8	92.7	77.1	84.9	82.6	78.3	84.4	82.7
0310	Concrete Forming & Accessories	103.1	68.3	73.1	100.4	52.5	59.1	106.3	69.1	74.2	102.5	68.1	72.9	85.9	44.9	50.5	83.7	30.7	38.0
0320	Concrete Reinforcing	107.0	85.4	96.0	97.0	85.2	91.0	108.7	85.2	96.8	89.9	85.3	87.5	87.4	67.5	77.3	93.6	67.4	80.3
0330	Cast-in-Place Concrete	95.1	72.4	85.8	90.7	73.1	83.5	94.3	72.1	85.2	97.8	72.3	87.3	77.4	45.3	64.2	82.6	38.3	64.4
03	CONCRETE	101.4	73.0	87.4	100.1	66.1	83.4	112.4	73.1	93.1	100.0	72.8	86.6	83.5	50.1	67.1	87.5	41.3	64.8
04	MASONRY	97.5	64.1	76.7	92.6	62.3	73.7	109.9	62.2	80.1	95.5	62.2	74.8	99.8	40.7	63.0	116.4	31.4	63.5
05	METALS	95.1	77.6	89.7	96.7	75.3	90.1	93.1	76.1	87.8	94.3	76.7	88.9	99.5	66.9	89.4	103.9	66.5	92.4
06	WOOD, PLASTICS & COMPOSITES	102.1	68.3	83.2	102.4	47.4	71.6	105.4	69.8	85.6	101.4	68.3	82.8	89.9	45.5	65.0	90.3	29.7	56.4
07	THERMAL & MOISTURE PROTECTION	100.9	67.0	87.0	95.1	64.2	82.4	102.1	65.0	86.9	102.4	64.2	86.7	98.2	43.9	76.0	94.2	35.7	70.2
08	OPENINGS	102.0	72.5	95.2	108.3	61.1	97.3	99.2	73.3	93.2	96.3	72.5	90.7	97.7	45.8	85.6	102.1	40.6	87.8
0920	Plaster & Gypsum Board	101.0	67.4	78.2	90.3	45.9	60.3	100.9	68.9	79.3	104.0	67.4	79.2	83.3	44.3	56.9	85.0	28.0	46.5
0950, 0980	Ceilings & Acoustic Treatment	96.5	67.4	77.3	98.8	45.9	64.0	88.9	68.9	75.8	89.8	67.4	75.0	83.2	44.3	57.6	83.3	28.0	47.0
0960	Flooring	104.9	51.7	89.7	94.1	39.5	78.5	106.2	45.0	88.7	95.2	44.2	80.6	96.7	59.1	85.9	99.6	39.6	82.4
0970, 0990	Wall Finishes & Painting/Coating	103.4	61.7	78.3	98.3	55.1	72.2	103.4	55.1	74.3	104.5	55.1	74.7	107.4	41.2	67.4	104.4	50.4	71.8
09	FINISHES	98.9	63.8	79.6	94.2	48.5	69.0	98.7	62.7	78.8	94.9	61.6	76.5	86.6	46.7	64.6	88.5	33.5	58.2
COVERS	DIVS. 10 - 14, 25, 28, 41, 43, 44, 46	100.0	83.9	96.7	100.0	81.4	96.3	100.0	84.0	96.8	100.0	83.9	96.7	100.0	39.5	87.8	100.0	35.6	87.0
21, 22, 23	FIRE SUPPRESSION, PLUMBING & HVAC	99.9	79.0	91.5	100.2	78.9	91.6	95.2	79.0	88.7	100.0	79.0	91.5	95.3	51.6	77.7	95.3	53.1	78.3
26, 27, 3370	ELECTRICAL, COMMUNICATIONS & UTIL.	101.0	66.7	82.9	103.0	61.2	80.9	94.6	61.2	77.0	96.5	61.3	77.9	96.5	63.5	79.0	96.8	62.1	78.5
MF2014	WEIGHTED AVERAGE	99.2	73.8	88.1	99.2	68.9	86.0	98.9	72.5	87.4	97.4	72.4	86.5	94.6	54.6	77.1	97.1	50.1	76.6

		ARKANSAS																
		FAYETTEVILLE			FORT SMITH			HARRISON			HOT SPRINGS			JONESBORO			LITTLE ROCK		
DIVISION		727			729			726			719			724			720 - 722		
		MAT.	INST.	TOTAL	MAT.	INST.	TOTAL	MAT.	INST.	TOTAL	MAT.	INST.	TOTAL	MAT.	INST.	TOTAL	MAT.	INST.	TOTAL
015433	CONTRACTOR EQUIPMENT		89.1	89.1		89.1	89.1		89.1	89.1		89.1	89.1		108.3	108.3		89.1	89.1
0241, 31 - 34	SITE & INFRASTRUCTURE, DEMOLITION	76.4	86.3	83.5	81.8	86.7	85.3	82.1	84.9	84.1	81.4	85.8	84.5	103.2	101.1	101.7	93.1	86.8	88.6
0310	Concrete Forming & Accessories	81.2	38.5	44.4	101.3	57.3	63.4	90.6	44.8	51.1	81.1	35.7	41.9	89.6	48.2	53.9	96.4	60.0	65.0
0320	Concrete Reinforcing	87.4	71.3	79.2	88.4	72.0	80.1	87.0	63.7	75.1	91.8	67.5	79.4	84.5	70.1	77.2	93.4	68.3	80.6
0330	Cast-in-Place Concrete	77.4	49.4	65.9	88.5	70.3	81.0	85.8	43.4	68.4	84.5	38.8	65.7	84.3	55.0	72.2	87.3	70.3	80.3
03	CONCRETE	83.2	49.3	66.5	91.3	65.0	78.4	90.8	48.7	70.1	91.2	43.7	67.9	88.1	56.0	72.3	92.8	65.5	79.4
04	MASONRY	90.5	41.4	59.9	96.9	50.9	68.3	100.1	39.0	62.0	87.6	33.5	53.9	91.9	41.9	60.7	96.8	50.9	68.2
05	METALS	99.5	67.8	89.7	101.8	70.6	92.2	100.6	66.5	90.1	103.8	66.9	92.5	95.9	80.0	91.0	102.0	69.4	92.0
06	WOOD, PLASTICS & COMPOSITES	85.8	35.5	57.6	107.6	59.0	80.4	96.0	45.5	67.7	87.5	35.7	58.5	93.9	48.3	68.4	99.9	62.5	78.9
07	THERMAL & MOISTURE PROTECTION	99.0	45.3	77.0	99.4	57.5	82.2	98.5	42.6	75.6	94.4	37.9	71.2	103.3	49.1	81.1	95.7	57.8	80.2
08	OPENINGS	97.7	44.7	85.4	98.5	56.2	88.7	98.5	46.5	86.4	102.1	42.3	88.2	100.6	54.0	89.7	99.5	58.4	89.9
0920	Plaster & Gypsum Board	81.8	34.0	49.5	89.4	58.3	68.4	88.3	44.3	58.6	84.0	34.2	50.4	97.7	46.9	63.4	94.6	61.8	72.5
0950, 0980	Ceilings & Acoustic Treatment	83.2	34.0	50.9	84.8	58.3	67.4	84.8	44.3	58.2	83.3	34.2	51.0	86.2	46.9	60.3	88.0	61.8	70.8
0960	Flooring	94.0	59.1	84.0	102.9	60.8	90.9	98.8	59.1	87.5	98.5	59.1	87.2	68.9	53.4	64.5	101.0	60.8	89.5
0970, 0990	Wall Finishes & Painting/Coating	107.4	29.4	60.3	107.4	53.8	75.1	107.4	41.2	67.4	104.4	55.2	74.7	94.9	48.4	66.8	108.8	55.2	76.5
09	FINISHES	85.7	40.0	60.5	89.9	57.6	72.1	88.6	46.8	65.6	88.4	41.0	62.3	83.7	48.5	64.3	92.7	59.8	74.6
COVERS	DIVS. 10 - 14, 25, 28, 41, 43, 44, 46	100.0	48.7	89.6	100.0	78.8	95.7	100.0	49.1	89.7	100.0	36.5	87.2	100.0	45.2	88.9	100.0	79.2	95.8
21, 22, 23	FIRE SUPPRESSION, PLUMBING & HVAC	95.3	50.2	77.1	100.1	50.6	80.1	95.3	49.4	76.8	95.3	48.7	76.5	100.3	52.5	81.0	99.9	56.2	82.3
26, 27, 3370	ELECTRICAL, COMMUNICATIONS & UTIL.	90.1	51.2	69.5	93.7	64.5	78.3	95.0	37.1	64.4	99.0	67.8	82.5	100.2	63.5	80.8	100.7	70.2	84.6
MF2014	WEIGHTED AVERAGE	93.4	52.0	75.4	97.2	61.6	81.7	95.9	50.4	76.0	96.4	51.8	76.9	96.7	59.1	80.3	98.5	64.0	83.4

ARKANSAS / CALIFORNIA

DIVISION		PINE BLUFF 716 MAT.	INST.	TOTAL	RUSSELLVILLE 728 MAT.	INST.	TOTAL	TEXARKANA 718 MAT.	INST.	TOTAL	WEST MEMPHIS 723 MAT.	INST.	TOTAL	ALHAMBRA 917-918 MAT.	INST.	TOTAL	ANAHEIM 928 MAT.	INST.	TOTAL
015433	CONTRACTOR EQUIPMENT		89.1	89.1		89.1	89.1		89.9	89.9		108.3	108.3		100.1	100.1		100.8	100.8
0241, 31 - 34	SITE & INFRASTRUCTURE, DEMOLITION	83.7	86.8	85.9	78.3	84.9	83.0	94.9	87.1	89.3	110.5	101.1	103.8	97.9	110.9	107.1	99.4	108.3	105.7
0310	Concrete Forming & Accessories	80.7	60.0	62.8	86.6	54.7	59.1	87.3	40.1	46.6	95.5	48.5	54.9	117.1	116.4	116.5	105.4	124.2	121.6
0320	Concrete Reinforcing	93.5	68.2	80.7	88.0	67.3	77.5	93.1	67.5	80.1	84.5	70.1	77.2	106.6	113.3	110.0	93.8	113.2	103.7
0330	Cast-in-Place Concrete	84.5	70.3	78.7	81.1	45.3	66.4	92.1	43.3	72.0	88.4	55.1	74.7	95.1	120.8	105.6	92.9	123.3	105.4
03	CONCRETE	92.1	65.5	79.0	86.7	54.4	70.8	89.8	47.2	68.9	95.5	56.2	76.2	101.1	116.5	108.7	101.8	120.8	111.2
04	MASONRY	125.0	50.9	78.8	96.1	37.3	59.5	104.0	33.1	59.2	79.4	41.9	56.0	122.7	121.1	121.7	80.0	118.8	104.2
05	METALS	104.6	69.4	93.8	99.5	66.5	89.3	96.8	66.9	87.6	95.0	80.3	90.5	85.3	100.9	90.1	102.7	101.1	102.2
06	WOOD, PLASTICS & COMPOSITES	87.0	62.5	73.3	91.3	59.0	73.2	95.5	42.0	65.5	100.1	48.3	71.1	100.0	112.1	106.7	101.9	122.6	113.5
07	THERMAL & MOISTURE PROTECTION	94.5	57.8	79.5	99.2	43.9	76.5	95.2	44.5	74.4	103.7	49.1	81.3	95.1	118.1	104.5	99.4	122.0	108.7
08	OPENINGS	102.1	58.4	91.9	97.7	55.7	87.9	107.3	46.5	93.2	100.6	54.0	89.7	91.8	114.8	97.2	104.2	120.6	108.0
0920	Plaster & Gypsum Board	83.6	61.8	68.9	83.3	58.3	66.4	86.6	40.7	55.6	100.2	46.9	64.2	97.4	112.5	107.6	107.1	123.2	118.0
0950, 0980	Ceilings & Acoustic Treatment	83.3	61.8	69.2	83.2	58.3	66.8	86.6	40.7	56.4	84.3	46.9	59.7	100.2	112.5	108.3	105.0	123.2	117.0
0960	Flooring	98.2	60.8	87.5	96.2	59.1	85.6	100.4	51.0	86.3	71.0	53.4	66.0	95.4	110.2	99.6	101.9	112.0	104.8
0970, 0990	Wall Finishes & Painting/Coating	104.4	55.2	74.7	107.4	35.0	63.7	104.4	30.3	59.7	94.9	50.4	68.0	100.5	111.5	107.1	98.9	106.8	103.7
09	FINISHES	88.3	59.8	72.6	86.7	54.1	68.7	90.7	40.4	62.9	85.0	48.8	65.0	99.8	114.0	107.6	100.5	120.1	111.3
COVERS	DIVS. 10 - 14, 25, 28, 41, 43, 44, 46	100.0	79.2	95.8	100.0	41.0	88.1	100.0	32.6	86.4	100.0	45.2	88.9	100.0	110.5	102.1	100.0	112.0	102.4
21, 22, 23	FIRE SUPPRESSION, PLUMBING & HVAC	100.1	52.9	81.1	95.3	51.5	77.6	100.1	53.5	81.3	95.6	65.8	83.6	95.2	115.6	103.4	100.0	118.6	107.5
26, 27, 3370	ELECTRICAL, COMMUNICATIONS & UTIL.	97.0	70.2	82.8	93.7	44.7	67.8	98.9	36.0	65.7	101.9	65.7	82.8	118.4	120.5	119.5	92.2	105.7	99.3
MF2014	WEIGHTED AVERAGE	99.4	63.3	83.6	94.6	53.7	76.8	98.2	49.2	76.8	96.1	62.4	81.4	98.3	114.8	105.5	99.4	114.9	106.2

CALIFORNIA

DIVISION		BAKERSFIELD 932-933 MAT.	INST.	TOTAL	BERKELEY 947 MAT.	INST.	TOTAL	EUREKA 955 MAT.	INST.	TOTAL	FRESNO 936-938 MAT.	INST.	TOTAL	INGLEWOOD 903-905 MAT.	INST.	TOTAL	LONG BEACH 906-908 MAT.	INST.	TOTAL
015433	CONTRACTOR EQUIPMENT		98.8	98.8		100.3	100.3		98.5	98.5		98.8	98.8		96.5	96.5		96.5	96.5
0241, 31 - 34	SITE & INFRASTRUCTURE, DEMOLITION	100.4	106.0	104.4	118.4	107.6	110.7	110.8	103.9	105.9	103.1	105.4	104.7	89.3	103.5	99.4	95.9	103.5	101.3
0310	Concrete Forming & Accessories	104.0	124.3	121.5	115.9	149.9	145.2	114.8	133.6	131.0	104.3	134.2	130.1	111.8	115.4	114.9	106.5	115.4	114.1
0320	Concrete Reinforcing	103.4	113.2	108.4	94.5	114.4	104.6	102.5	114.1	108.4	80.6	113.7	97.5	102.7	113.3	108.1	101.9	113.3	107.7
0330	Cast-in-Place Concrete	94.0	122.6	105.7	123.4	124.8	124.0	100.8	118.0	107.9	99.4	118.9	107.4	84.5	121.4	99.7	96.2	121.4	106.5
03	CONCRETE	97.9	120.6	109.1	107.3	132.7	119.8	112.8	123.2	117.9	98.6	123.8	111.0	90.9	116.3	103.4	99.8	116.3	107.9
04	MASONRY	100.2	118.3	111.5	113.4	134.3	126.5	104.0	132.6	121.8	103.6	122.0	115.1	76.3	121.2	104.3	85.3	121.2	107.7
05	METALS	105.3	100.9	104.0	106.9	100.5	104.9	102.6	98.9	101.4	105.6	100.5	104.1	94.4	101.5	96.6	94.3	101.5	96.5
06	WOOD, PLASTICS & COMPOSITES	96.0	122.7	111.0	111.4	154.9	135.8	116.6	138.5	128.9	108.1	138.5	125.1	103.8	110.7	107.6	96.9	110.7	104.6
07	THERMAL & MOISTURE PROTECTION	98.1	116.1	105.5	105.2	137.2	118.3	103.4	119.8	110.1	88.7	116.0	99.9	98.3	118.0	106.4	98.5	118.0	106.5
08	OPENINGS	97.8	116.1	102.0	93.0	140.6	104.1	103.5	113.5	105.8	98.1	124.7	104.3	88.8	114.1	94.7	88.8	114.1	94.7
0920	Plaster & Gypsum Board	101.9	123.2	116.3	111.6	156.0	141.6	113.4	139.5	131.0	100.6	139.5	126.9	103.8	111.0	108.7	99.9	111.0	107.4
0950, 0980	Ceilings & Acoustic Treatment	96.3	123.2	114.0	106.2	156.0	139.0	110.1	139.5	129.4	92.8	139.5	123.5	100.3	111.0	107.4	100.3	111.0	107.4
0960	Flooring	106.8	110.2	107.7	108.2	127.6	113.8	106.0	116.5	109.0	113.3	133.5	119.1	104.4	110.2	106.0	101.8	110.2	104.2
0970, 0990	Wall Finishes & Painting/Coating	111.4	102.0	105.7	102.5	143.9	127.5	100.5	46.2	67.7	129.0	107.2	115.8	100.9	111.5	107.3	100.9	111.5	107.3
09	FINISHES	100.0	120.5	111.3	105.4	147.2	128.4	106.1	123.6	115.8	102.1	133.2	119.2	102.7	113.2	108.5	101.8	113.2	108.1
COVERS	DIVS. 10 - 14, 25, 28, 41, 43, 44, 46	100.0	111.9	102.4	100.0	126.4	105.3	100.0	122.2	104.5	100.0	122.2	104.5	100.0	110.4	102.1	100.0	110.4	102.1
21, 22, 23	FIRE SUPPRESSION, PLUMBING & HVAC	100.1	117.1	107.0	95.3	147.8	116.5	95.2	110.6	101.4	100.2	111.3	104.7	94.8	115.6	103.2	94.8	115.6	103.2
26, 27, 3370	ELECTRICAL, COMMUNICATIONS & UTIL.	103.6	102.5	103.1	109.4	145.0	128.2	99.1	116.7	108.4	93.8	100.8	97.5	103.8	120.5	112.6	103.5	120.5	112.5
MF2014	WEIGHTED AVERAGE	100.7	113.6	106.3	102.7	135.1	116.8	102.1	116.4	108.4	100.1	115.3	106.7	94.8	114.1	103.2	96.2	114.1	104.0

CALIFORNIA

DIVISION		LOS ANGELES 900-902 MAT.	INST.	TOTAL	MARYSVILLE 959 MAT.	INST.	TOTAL	MODESTO 953 MAT.	INST.	TOTAL	MOJAVE 935 MAT.	INST.	TOTAL	OAKLAND 946 MAT.	INST.	TOTAL	OXNARD 930 MAT.	INST.	TOTAL
015433	CONTRACTOR EQUIPMENT		100.1	100.1		98.5	98.5		98.5	98.5		98.8	98.8		100.3	100.3		97.7	97.7
0241, 31 - 34	SITE & INFRASTRUCTURE, DEMOLITION	96.6	107.1	104.1	107.2	105.0	105.7	102.2	105.1	104.3	96.4	106.0	103.2	124.7	107.6	112.5	103.3	104.0	103.8
0310	Concrete Forming & Accessories	108.6	124.3	122.2	104.6	134.4	130.3	100.7	134.3	129.7	116.1	112.1	112.6	104.9	149.9	143.7	107.5	124.3	122.0
0320	Concrete Reinforcing	103.6	113.4	108.6	102.5	113.8	108.2	106.2	113.8	110.0	97.3	113.1	105.3	96.7	114.4	105.7	95.5	113.1	104.4
0330	Cast-in-Place Concrete	91.2	122.0	103.9	112.6	119.2	115.3	100.8	119.2	108.4	86.0	122.4	101.0	117.0	124.8	120.2	100.1	122.7	109.4
03	CONCRETE	95.9	120.6	108.0	113.9	124.0	118.9	104.7	124.0	114.2	91.0	115.1	102.8	106.8	132.7	119.5	99.2	120.6	109.7
04	MASONRY	90.5	122.6	110.5	104.9	123.1	116.2	102.9	123.1	115.5	101.9	118.3	112.1	120.6	134.3	129.2	104.8	116.5	112.1
05	METALS	101.1	102.9	101.7	102.0	101.8	102.0	99.1	101.7	99.9	102.8	100.3	102.0	101.4	100.4	101.1	100.7	100.7	100.7
06	WOOD, PLASTICS & COMPOSITES	103.0	122.6	113.9	103.0	138.5	122.9	98.5	138.5	120.9	107.4	106.8	107.1	98.8	154.9	130.2	102.1	122.7	113.6
07	THERMAL & MOISTURE PROTECTION	97.6	121.2	107.3	102.9	121.5	110.5	102.5	118.7	109.1	94.4	110.9	101.2	103.1	137.2	117.1	97.4	119.8	106.6
08	OPENINGS	95.4	120.7	101.3	102.8	125.3	108.0	101.6	125.4	107.1	93.2	107.5	96.5	93.1	140.6	104.1	95.7	120.6	101.5
0920	Plaster & Gypsum Board	103.6	123.2	116.9	105.0	139.5	128.3	107.5	139.5	129.1	110.8	106.8	108.1	105.7	156.0	139.7	104.7	123.2	117.2
0950, 0980	Ceilings & Acoustic Treatment	110.5	123.2	118.8	109.2	139.5	129.1	105.0	139.5	127.7	94.2	106.8	102.5	108.9	156.0	139.9	96.6	123.2	114.1
0960	Flooring	102.0	112.0	104.9	102.0	117.6	106.4	102.4	112.8	105.3	112.6	110.2	111.9	103.4	127.6	110.3	104.8	112.0	106.9
0970, 0990	Wall Finishes & Painting/Coating	100.0	111.5	106.9	100.5	115.8	109.7	100.5	117.0	110.5	111.1	102.0	105.6	102.5	143.9	127.5	111.1	101.1	105.1
09	FINISHES	104.2	120.6	113.3	103.1	131.5	118.8	102.2	130.9	118.0	101.8	109.9	106.2	104.3	147.2	127.9	99.4	119.6	110.5
COVERS	DIVS. 10 - 14, 25, 28, 41, 43, 44, 46	100.0	111.9	102.4	100.0	122.2	104.5	100.0	122.2	104.5	100.0	107.8	101.6	100.0	126.3	105.3	100.0	112.2	102.5
21, 22, 23	FIRE SUPPRESSION, PLUMBING & HVAC	99.9	118.6	107.5	95.2	109.0	100.8	100.0	111.3	104.6	95.3	115.1	103.3	100.1	147.8	119.4	100.1	118.6	107.6
26, 27, 3370	ELECTRICAL, COMMUNICATIONS & UTIL.	102.0	121.3	112.2	95.7	107.4	101.9	98.2	104.0	101.3	92.0	99.9	96.2	108.5	139.3	124.8	97.7	107.3	102.8
MF2014	WEIGHTED AVERAGE	99.2	117.5	107.2	101.3	115.9	107.7	100.8	115.7	107.3	96.8	109.8	102.5	103.1	134.3	116.7	99.6	114.4	106.0

CALIFORNIA

	DIVISION	PALM SPRINGS 922			PALO ALTO 943			PASADENA 910-912			REDDING 960			RICHMOND 948			RIVERSIDE 925		
		MAT.	INST.	TOTAL	MAT.	INST.	TOTAL	MAT.	INST.	TOTAL	MAT.	INST.	TOTAL	MAT.	INST.	TOTAL	MAT.	INST.	TOTAL
015433	CONTRACTOR EQUIPMENT		99.6	99.6		100.3	100.3		100.1	100.1		98.5	98.5		100.3	100.3		99.6	99.6
0241, 31-34	SITE & INFRASTRUCTURE, DEMOLITION	90.9	106.3	101.8	114.4	107.6	109.6	94.9	110.9	106.3	125.1	105.0	110.8	123.8	107.6	112.3	97.9	106.3	103.9
0310	Concrete Forming & Accessories	101.9	115.3	113.4	103.0	146.3	140.3	106.2	115.3	114.1	107.8	142.7	137.9	118.7	149.7	145.5	105.8	124.2	121.7
0320	Concrete Reinforcing	107.8	113.1	110.5	94.5	114.4	104.6	107.5	113.3	110.5	118.7	113.8	116.2	94.5	114.3	104.6	104.7	113.1	109.0
0330	Cast-in-Place Concrete	88.6	123.2	102.8	104.4	124.8	112.8	90.2	120.8	102.8	118.0	119.2	118.5	120.0	124.8	122.0	96.3	123.3	107.4
03	CONCRETE	96.6	116.8	106.5	96.5	131.1	113.5	96.6	116.0	106.2	122.7	127.7	125.2	108.8	132.6	120.5	102.6	120.8	111.6
04	MASONRY	77.6	118.5	103.1	97.4	134.3	120.4	106.5	121.1	115.6	130.0	123.1	125.7	113.2	132.8	125.4	78.6	118.1	103.2
05	METALS	103.3	100.8	102.5	98.9	100.3	99.3	85.3	100.8	90.1	101.6	101.8	101.7	98.9	100.2	99.3	102.8	101.0	102.3
06	WOOD, PLASTICS & COMPOSITES	96.6	110.8	104.5	96.1	150.2	126.4	86.5	110.6	100.0	110.6	149.8	132.6	115.3	154.9	137.5	101.9	122.6	113.5
07	THERMAL & MOISTURE PROTECTION	99.1	118.7	107.2	102.7	137.5	117.0	94.9	117.4	104.1	121.6	122.1	121.8	103.3	135.9	116.7	99.6	121.1	108.4
08	OPENINGS	100.1	114.1	103.4	93.1	136.9	103.3	91.8	114.0	97.0	114.1	131.5	118.1	93.1	139.4	103.9	102.8	120.6	107.0
0920	Plaster & Gypsum Board	102.0	111.0	108.1	103.9	151.1	135.9	90.2	110.4	104.3	106.2	151.1	136.6	113.6	156.0	142.3	106.3	123.2	117.7
0950, 0980	Ceilings & Acoustic Treatment	101.7	111.0	107.8	107.1	151.1	136.0	100.2	111.0	107.3	131.5	151.1	144.4	107.1	156.0	139.3	109.2	123.2	118.4
0960	Flooring	104.6	107.2	105.3	102.4	127.6	109.6	90.8	110.2	96.3	98.0	117.6	103.6	109.9	127.6	115.0	106.0	112.0	107.7
0970, 0990	Wall Finishes & Painting/Coating	97.2	111.3	105.7	102.5	143.9	127.5	100.5	111.5	107.1	115.9	115.8	115.8	102.5	143.9	127.5	97.2	106.8	103.0
09	FINISHES	99.1	112.8	106.6	102.7	144.4	125.6	97.3	113.2	106.0	109.3	138.2	125.3	106.9	147.2	129.1	102.1	120.1	112.0
COVERS	DIVS. 10-14, 25, 28, 41, 43, 44, 46	100.0	110.7	102.2	100.0	125.9	105.2	100.0	110.3	102.1	100.0	123.4	104.7	100.0	126.3	105.3	100.0	112.0	102.4
21, 22, 23	FIRE SUPPRESSION, PLUMBING & HVAC	95.2	115.5	103.4	95.3	145.5	115.6	95.2	115.6	103.4	100.2	109.0	103.7	95.3	147.8	116.5	100.0	118.6	107.5
26, 27, 3370	ELECTRICAL, COMMUNICATIONS & UTIL.	95.3	105.4	100.7	108.4	150.2	130.5	115.1	120.5	118.0	99.2	107.4	103.5	109.0	132.7	121.5	91.9	105.4	99.1
MF2014	WEIGHTED AVERAGE	97.2	112.0	103.6	98.9	134.5	114.4	96.4	114.6	104.3	107.8	117.7	112.1	101.8	133.1	115.4	99.4	114.6	106.0

CALIFORNIA

	DIVISION	SACRAMENTO 942,956-958			SALINAS 939			SAN BERNARDINO 923-924			SAN DIEGO 919-921			SAN FRANCISCO 940-941			SAN JOSE 951		
		MAT.	INST.	TOTAL	MAT.	INST.	TOTAL	MAT.	INST.	TOTAL	MAT.	INST.	TOTAL	MAT.	INST.	TOTAL	MAT.	INST.	TOTAL
015433	CONTRACTOR EQUIPMENT		99.9	99.9		98.8	98.8		99.6	99.6		100.1	100.1		110.7	110.7		99.3	99.3
0241, 31-34	SITE & INFRASTRUCTURE, DEMOLITION	100.6	113.3	109.6	116.3	105.7	108.8	77.7	106.3	98.0	101.7	104.0	103.4	127.0	113.6	117.5	133.5	100.4	109.9
0310	Concrete Forming & Accessories	103.3	137.0	132.4	111.0	137.4	133.7	109.6	115.2	114.4	105.6	113.3	112.2	104.6	150.9	144.6	107.0	149.8	143.9
0320	Concrete Reinforcing	89.4	113.8	101.8	96.0	114.2	105.2	104.7	113.0	108.9	104.2	113.1	108.7	110.1	115.0	112.6	93.5	114.5	104.1
0330	Cast-in-Place Concrete	99.0	120.3	107.7	98.9	119.6	107.4	66.6	123.2	89.8	98.9	107.8	102.6	120.0	126.4	122.6	117.1	124.2	120.0
03	CONCRETE	97.0	125.2	110.9	108.5	125.6	116.9	76.8	116.8	96.4	101.8	110.7	106.2	110.3	134.4	122.1	110.8	132.9	121.7
04	MASONRY	98.4	123.1	113.8	101.6	128.1	118.1	84.9	115.9	104.2	97.4	115.6	108.7	121.1	141.5	133.8	131.4	134.4	133.3
05	METALS	97.0	96.1	96.7	105.5	102.9	104.7	102.8	100.5	102.1	101.1	101.4	101.2	107.6	110.8	108.6	97.5	107.5	100.6
06	WOOD, PLASTICS & COMPOSITES	92.9	141.7	120.2	107.3	141.5	126.4	105.8	110.8	108.6	100.6	110.2	106.0	98.8	155.1	130.3	111.3	154.7	135.6
07	THERMAL & MOISTURE PROTECTION	110.8	121.3	115.1	95.0	126.1	107.7	98.3	117.7	106.3	99.3	107.1	102.5	104.9	141.9	120.1	98.9	139.3	115.4
08	OPENINGS	106.3	127.2	111.1	96.9	133.3	105.4	100.2	114.1	103.4	99.1	111.8	102.1	97.3	140.7	107.4	92.8	140.5	103.9
0920	Plaster & Gypsum Board	101.0	142.6	129.1	105.8	142.6	130.6	108.2	111.0	110.1	97.8	110.3	106.3	108.3	156.0	140.6	102.7	156.0	138.7
0950, 0980	Ceilings & Acoustic Treatment	107.1	142.6	130.4	94.2	142.6	126.0	105.0	111.0	109.0	107.7	110.3	109.4	117.1	156.0	142.7	103.3	156.0	138.0
0960	Flooring	102.6	117.6	106.9	107.1	121.1	111.1	107.8	110.2	108.5	98.5	112.0	102.4	103.4	127.6	110.3	95.4	127.6	104.6
0970, 0990	Wall Finishes & Painting/Coating	100.2	116.7	110.1	112.3	143.9	131.4	97.2	104.0	101.3	97.1	111.5	105.8	102.5	153.2	133.1	100.8	143.9	126.8
09	FINISHES	101.4	133.6	119.1	101.4	137.0	121.0	100.5	112.5	107.1	102.9	112.9	108.4	106.5	148.3	129.6	100.7	147.0	126.2
COVERS	DIVS. 10-14, 25, 28, 41, 43, 44, 46	100.0	123.0	104.6	100.0	122.6	104.6	100.0	108.1	101.6	100.0	109.7	102.0	100.0	126.8	105.4	100.0	125.9	105.2
21, 22, 23	FIRE SUPPRESSION, PLUMBING & HVAC	100.0	120.8	108.4	95.3	117.7	104.4	95.2	115.6	103.4	99.9	116.9	106.8	100.1	175.0	130.3	100.0	147.3	119.1
26, 27, 3370	ELECTRICAL, COMMUNICATIONS & UTIL.	103.5	108.7	106.2	93.2	120.9	107.8	95.3	103.4	99.6	101.2	98.5	99.8	108.6	157.8	134.6	101.1	155.9	130.1
MF2014	WEIGHTED AVERAGE	100.5	119.3	108.7	100.2	121.6	109.5	95.1	111.3	102.1	100.5	109.6	104.5	105.2	145.4	122.7	102.5	136.6	117.4

CALIFORNIA

	DIVISION	SAN LUIS OBISPO 934			SAN MATEO 944			SAN RAFAEL 949			SANTA ANA 926-927			SANTA BARBARA 931			SANTA CRUZ 950		
		MAT.	INST.	TOTAL	MAT.	INST.	TOTAL	MAT.	INST.	TOTAL	MAT.	INST.	TOTAL	MAT.	INST.	TOTAL	MAT.	INST.	TOTAL
015433	CONTRACTOR EQUIPMENT		98.8	98.8		100.3	100.3		100.3	100.3		99.6	99.6		98.8	98.8		99.3	99.3
0241, 31-34	SITE & INFRASTRUCTURE, DEMOLITION	108.4	106.0	106.7	121.3	107.6	111.6	113.2	113.4	113.4	89.4	106.3	101.4	103.3	106.0	105.2	133.1	100.0	109.6
0310	Concrete Forming & Accessories	117.9	115.4	115.8	109.0	149.8	144.2	112.9	149.9	144.9	109.9	115.3	114.5	108.3	124.2	122.1	107.0	137.6	133.4
0320	Concrete Reinforcing	97.3	113.1	105.4	94.5	114.6	104.7	95.1	114.6	105.1	108.3	113.1	110.8	95.5	113.2	104.5	115.7	114.2	114.9
0330	Cast-in-Place Concrete	106.3	122.5	112.9	116.2	124.8	119.7	135.2	124.1	130.7	85.0	123.2	100.7	99.8	122.6	109.1	116.3	121.4	118.4
03	CONCRETE	106.9	116.6	111.7	105.4	132.7	118.8	126.0	132.4	129.1	94.2	116.8	105.3	99.0	120.6	109.6	113.9	126.5	120.1
04	MASONRY	103.5	118.3	112.7	112.9	137.3	128.1	93.2	137.4	120.7	74.6	118.8	102.1	102.2	119.1	112.7	135.3	128.2	130.9
05	METALS	103.5	100.6	102.6	98.7	100.6	99.3	100.1	98.9	99.7	102.9	100.8	102.2	101.2	100.9	101.1	104.5	106.3	105.0
06	WOOD, PLASTICS & COMPOSITES	109.9	154.9	132.5	104.0	154.9	132.5	101.8	154.7	131.4	107.8	110.8	109.5	102.1	122.7	113.6	111.3	141.6	128.3
07	THERMAL & MOISTURE PROTECTION	95.1	116.9	104.0	103.1	138.8	117.7	107.2	138.9	120.2	99.4	118.3	107.2	94.6	118.9	104.5	98.8	129.2	111.3
08	OPENINGS	95.1	109.7	98.5	93.1	139.4	103.8	103.8	139.3	112.1	99.5	114.1	102.9	96.6	120.6	102.2	94.0	133.4	103.2
0920	Plaster & Gypsum Board	112.2	111.0	111.4	108.9	156.0	140.8	111.5	156.0	141.6	109.6	111.0	110.6	104.7	123.2	117.2	110.1	142.6	132.0
0950, 0980	Ceilings & Acoustic Treatment	94.2	111.0	105.3	107.1	156.0	139.3	115.4	156.0	142.1	105.0	111.0	109.0	96.6	123.2	114.1	106.7	142.6	130.3
0960	Flooring	113.5	110.2	112.6	105.1	127.6	111.5	115.0	121.1	116.8	108.3	110.2	108.8	106.2	108.6	106.9	99.6	121.1	105.7
0970, 0990	Wall Finishes & Painting/Coating	111.1	101.7	105.5	102.5	143.9	127.5	98.9	142.6	125.3	97.2	106.8	103.0	111.1	101.1	105.1	101.0	143.9	126.9
09	FINISHES	103.2	112.3	108.2	104.7	147.2	128.1	107.6	145.8	128.6	102.0	112.8	107.9	100.0	119.0	110.5	103.7	137.1	122.1
COVERS	DIVS. 10-14, 25, 28, 41, 43, 44, 46	100.0	110.7	102.2	100.0	126.4	105.3	100.0	125.7	105.2	100.0	110.7	102.2	100.0	112.2	102.5	100.0	122.9	104.6
21, 22, 23	FIRE SUPPRESSION, PLUMBING & HVAC	95.3	115.6	103.5	95.3	142.2	114.2	95.3	169.5	125.2	95.2	114.3	102.9	100.1	118.6	107.6	100.0	117.8	107.2
26, 27, 3370	ELECTRICAL, COMMUNICATIONS & UTIL.	92.0	106.4	99.6	108.4	148.0	129.3	105.4	120.0	113.1	95.4	105.7	100.8	91.0	110.5	101.3	100.3	120.9	111.2
MF2014	WEIGHTED AVERAGE	99.4	112.0	104.9	100.9	134.6	115.5	103.5	136.7	118.0	96.9	111.8	103.4	98.9	115.1	106.0	104.5	121.8	112.0

For customer support on your Plumbing Cost Data, call 877.709.6968.

607

City Cost Indexes

CALIFORNIA / COLORADO

DIVISION		SANTA ROSA 954			STOCKTON 952			SUSANVILLE 961			VALLEJO 945			VAN NUYS 913 - 916			ALAMOSA 811		
		MAT.	INST.	TOTAL	MAT.	INST.	TOTAL	MAT.	INST.	TOTAL	MAT.	INST.	TOTAL	MAT.	INST.	TOTAL	MAT.	INST.	TOTAL
015433	CONTRACTOR EQUIPMENT		99.0	99.0		98.5	98.5		98.5	98.5		100.3	100.3		100.1	100.1		93.8	93.8
0241, 31 - 34	SITE & INFRASTRUCTURE, DEMOLITION	103.0	105.1	104.5	102.0	105.1	104.2	131.8	105.0	112.8	100.5	113.2	109.5	111.4	110.9	111.0	134.7	88.0	101.5
0310	Concrete Forming & Accessories	102.9	148.1	141.9	104.8	136.6	132.2	109.1	142.8	138.1	103.4	147.7	141.6	112.7	115.3	115.0	105.0	67.5	72.7
0320	Concrete Reinforcing	103.4	114.8	109.2	106.2	113.8	110.1	118.7	113.2	115.9	96.3	114.5	105.6	107.5	113.3	110.5	105.1	76.4	90.5
0330	Cast-in-Place Concrete	110.6	120.8	114.8	98.2	119.2	106.9	107.4	119.3	112.3	107.7	121.5	113.4	95.1	120.8	105.7	101.1	78.4	91.8
03	CONCRETE	113.7	130.9	122.1	103.7	125.0	114.2	125.2	127.7	126.4	102.2	130.5	116.1	110.2	116.0	113.0	113.5	73.3	93.8
04	MASONRY	103.1	137.2	124.3	102.8	123.1	115.5	127.9	112.7	118.4	71.6	133.1	109.9	122.7	121.1	121.7	126.0	71.9	92.3
05	METALS	103.2	105.0	103.8	99.3	101.8	100.0	100.6	101.1	100.7	100.1	98.1	99.5	84.4	100.8	89.5	98.2	80.2	92.7
06	WOOD, PLASTICS & COMPOSITES	98.0	154.5	129.6	104.3	141.5	125.1	112.5	149.8	133.4	90.6	154.7	126.5	95.0	110.6	103.7	98.0	67.5	80.9
07	THERMAL & MOISTURE PROTECTION	99.8	136.2	114.7	102.6	119.9	109.7	122.2	120.4	121.4	105.0	135.3	117.4	91.6	117.4	104.6	104.9	78.1	93.9
08	OPENINGS	101.0	140.4	110.2	101.6	127.1	107.5	113.9	131.5	118.0	105.6	140.5	113.7	91.6	114.0	96.8	98.3	74.1	92.7
0920	Plaster & Gypsum Board	103.9	156.0	139.2	107.5	142.6	131.2	107.1	151.1	136.9	105.8	156.0	139.8	95.3	111.0	105.9	79.0	66.3	70.4
0950, 0980	Ceilings & Acoustic Treatment	105.0	156.0	138.6	112.5	142.6	132.3	124.1	151.1	141.9	117.3	156.0	142.8	97.7	111.0	106.5	96.3	66.3	76.6
0960	Flooring	105.0	115.2	107.9	102.4	112.8	105.3	98.4	122.1	105.2	110.6	127.6	115.4	93.2	110.2	98.0	113.5	54.8	96.7
0970, 0990	Wall Finishes & Painting/Coating	97.2	142.6	124.6	100.5	117.0	110.5	115.9	115.8	115.8	99.8	142.6	125.6	100.5	111.5	107.1	114.1	39.8	69.3
09	FINISHES	101.1	143.6	124.5	103.8	132.7	119.7	108.9	139.0	125.5	104.6	145.8	127.3	99.3	113.2	107.0	102.5	61.7	80.0
COVERS	DIVS. 10 - 14, 25, 28, 41, 43, 44, 46	100.0	123.8	104.8	100.0	122.5	104.6	100.0	123.5	104.7	100.0	124.3	104.9	100.0	110.3	102.1	100.0	88.5	97.7
21, 22, 23	FIRE SUPPRESSION, PLUMBING & HVAC	95.2	167.3	124.3	100.0	111.3	104.6	95.4	109.0	100.9	100.1	127.5	111.1	95.2	115.6	103.4	95.2	72.0	85.9
26, 27, 3370	ELECTRICAL, COMMUNICATIONS & UTIL.	95.7	114.3	105.5	98.2	110.6	104.8	99.6	118.4	109.5	100.9	122.4	112.3	115.1	120.5	118.0	99.1	72.5	85.0
MF2014	WEIGHTED AVERAGE	100.8	134.7	115.6	100.9	117.2	108.0	106.8	118.1	111.7	100.1	127.2	111.9	99.1	114.6	105.9	102.2	73.7	89.8

COLORADO

DIVISION		BOULDER 803			COLORADO SPRINGS 808 - 809			DENVER 800 - 802			DURANGO 813			FORT COLLINS 805			FORT MORGAN 807		
		MAT.	INST.	TOTAL	MAT.	INST.	TOTAL	MAT.	INST.	TOTAL	MAT.	INST.	TOTAL	MAT.	INST.	TOTAL	MAT.	INST.	TOTAL
015433	CONTRACTOR EQUIPMENT		97.5	97.5		95.8	95.8		100.4	100.4		93.8	93.8		97.5	97.5		97.5	97.5
0241, 31 - 34	SITE & INFRASTRUCTURE, DEMOLITION	94.8	96.4	95.9	96.9	94.7	95.3	96.1	102.6	100.7	128.5	88.0	99.7	107.1	96.0	99.2	97.7	95.9	96.4
0310	Concrete Forming & Accessories	103.1	79.9	83.1	93.6	79.4	81.3	100.0	76.0	79.3	111.4	67.7	73.7	100.7	74.7	78.3	103.6	74.9	78.9
0320	Concrete Reinforcing	98.9	76.6	87.5	98.1	80.2	89.0	98.1	80.2	89.0	105.1	76.5	90.5	99.0	76.6	87.6	99.1	76.5	87.6
0330	Cast-in-Place Concrete	101.8	80.0	92.9	104.6	88.3	97.9	99.0	82.8	92.4	116.3	78.5	100.8	114.9	78.9	100.1	99.9	78.9	91.3
03	CONCRETE	103.1	79.6	91.5	106.3	82.9	94.8	100.7	79.5	90.3	116.1	73.4	95.1	113.5	76.7	95.4	101.5	76.8	89.4
04	MASONRY	91.0	72.1	79.2	91.3	81.9	85.4	93.3	72.2	80.1	113.6	71.9	87.6	108.5	75.7	88.1	104.9	72.2	84.5
05	METALS	96.0	83.5	92.2	99.1	85.7	95.0	101.6	85.6	96.7	98.2	80.4	92.7	97.2	80.4	92.0	95.7	80.3	91.0
06	WOOD, PLASTICS & COMPOSITES	102.2	83.0	91.4	91.7	77.3	83.6	99.8	77.2	87.2	107.6	67.5	85.1	99.6	77.2	87.0	102.2	77.2	88.2
07	THERMAL & MOISTURE PROTECTION	103.1	81.7	94.3	103.9	85.6	96.4	102.4	75.8	91.5	104.9	78.1	93.9	103.5	73.5	91.2	103.0	81.2	94.1
08	OPENINGS	98.2	82.5	94.5	102.4	80.5	97.3	103.0	80.4	97.8	105.3	74.1	98.0	98.2	79.4	93.8	98.1	79.4	93.8
0920	Plaster & Gypsum Board	105.1	82.7	89.9	87.9	76.7	80.3	100.7	76.8	84.5	92.5	66.3	74.8	99.1	76.7	84.0	105.1	76.7	85.9
0950, 0980	Ceilings & Acoustic Treatment	96.9	82.7	87.5	104.4	76.7	86.2	107.3	76.8	87.2	96.3	66.3	76.6	96.9	76.7	83.6	96.9	76.7	83.6
0960	Flooring	99.7	83.8	95.2	92.0	68.2	85.2	96.7	84.8	93.3	118.2	54.8	100.0	96.6	54.8	84.6	100.1	54.8	87.1
0970, 0990	Wall Finishes & Painting/Coating	101.7	66.6	80.5	101.4	40.5	64.7	101.7	75.7	86.0	114.1	39.8	69.3	101.7	40.2	64.6	101.7	53.6	72.6
09	FINISHES	101.7	79.4	89.4	98.8	72.8	84.5	102.4	77.3	88.6	105.0	61.7	81.1	100.5	67.4	82.3	101.8	68.9	83.7
COVERS	DIVS. 10 - 14, 25, 28, 41, 43, 44, 46	100.0	89.4	97.9	100.0	92.2	98.4	100.0	88.8	97.7	100.0	88.4	97.7	100.0	88.8	97.7	100.0	88.8	97.7
21, 22, 23	FIRE SUPPRESSION, PLUMBING & HVAC	95.3	77.3	88.1	100.2	88.2	95.4	100.0	79.9	91.9	95.2	83.1	90.3	100.1	77.2	90.9	95.3	77.2	88.0
26, 27, 3370	ELECTRICAL, COMMUNICATIONS & UTIL.	97.4	84.5	90.6	100.8	82.1	90.9	102.6	83.2	92.4	98.5	69.4	83.1	97.4	84.5	90.6	97.8	84.5	90.7
MF2014	WEIGHTED AVERAGE	97.8	81.3	90.6	100.4	84.0	93.2	100.8	81.7	92.5	102.7	75.7	90.9	103.1	79.0	91.6	98.3	79.0	89.9

COLORADO

DIVISION		GLENWOOD SPRINGS 816			GOLDEN 804			GRAND JUNCTION 815			GREELEY 806			MONTROSE 814			PUEBLO 810		
		MAT.	INST.	TOTAL	MAT.	INST.	TOTAL	MAT.	INST.	TOTAL	MAT.	INST.	TOTAL	MAT.	INST.	TOTAL	MAT.	INST.	TOTAL
015433	CONTRACTOR EQUIPMENT		96.7	96.7		97.5	97.5		96.7	96.7		97.5	97.5		95.2	95.2		93.8	93.8
0241, 31 - 34	SITE & INFRASTRUCTURE, DEMOLITION	143.2	95.1	109.0	107.3	96.2	99.4	128.4	94.7	104.5	94.0	95.4	95.0	137.1	91.2	104.5	120.6	91.0	99.6
0310	Concrete Forming & Accessories	102.0	75.0	78.7	96.1	74.8	77.7	110.2	74.4	79.3	98.6	78.4	81.2	101.4	74.8	78.4	107.3	79.4	83.3
0320	Concrete Reinforcing	104.0	76.5	90.0	99.1	76.4	87.6	104.3	76.3	90.1	98.9	75.1	86.8	103.9	76.4	89.9	100.6	80.2	90.2
0330	Cast-in-Place Concrete	101.1	77.9	91.6	100.0	78.9	91.3	111.9	77.1	97.6	96.1	59.1	80.9	101.1	77.8	91.5	100.4	87.8	95.2
03	CONCRETE	118.5	76.5	97.9	111.9	76.7	94.6	112.5	75.9	94.5	98.3	71.3	85.0	109.7	76.3	93.3	102.7	82.7	92.9
04	MASONRY	98.9	72.1	82.2	107.7	71.8	85.3	132.6	71.5	94.5	102.6	49.0	69.2	106.2	71.9	84.9	95.4	81.7	86.9
05	METALS	97.9	80.9	92.6	95.9	80.1	91.0	99.5	79.2	93.2	97.2	77.5	91.1	97.1	80.0	91.8	101.1	86.2	96.5
06	WOOD, PLASTICS & COMPOSITES	93.2	77.3	84.3	94.3	77.2	84.7	105.3	77.3	89.6	96.8	83.0	89.1	94.3	77.4	84.8	100.9	77.6	87.8
07	THERMAL & MOISTURE PROTECTION	104.8	79.2	94.3	103.9	75.4	92.2	104.0	68.5	89.5	102.9	66.4	87.9	105.0	79.2	94.4	103.5	83.6	95.3
08	OPENINGS	104.2	79.4	98.5	98.2	78.8	93.7	105.0	78.8	98.9	98.1	82.4	94.5	105.5	79.5	99.4	100.2	80.6	95.6
0920	Plaster & Gypsum Board	117.8	76.7	90.0	96.9	76.7	83.3	131.0	76.7	94.3	97.6	82.7	87.5	78.3	76.7	77.2	83.9	76.7	79.1
0950, 0980	Ceilings & Acoustic Treatment	95.5	76.7	83.2	96.9	76.7	83.6	95.5	76.7	83.2	96.9	82.7	87.5	96.3	76.7	83.4	104.7	76.7	86.3
0960	Flooring	112.8	50.4	95.0	94.7	54.8	83.3	117.5	54.8	99.6	95.7	54.8	84.0	115.8	45.4	95.7	114.6	84.8	106.1
0970, 0990	Wall Finishes & Painting/Coating	114.1	66.5	85.4	101.7	66.6	80.5	114.1	66.6	85.4	101.7	25.0	55.4	114.1	39.8	69.3	114.1	37.5	67.9
09	FINISHES	107.9	69.6	86.8	100.4	71.0	84.2	109.3	70.4	87.9	99.3	69.2	82.7	103.2	65.7	82.5	103.7	75.6	88.2
COVERS	DIVS. 10 - 14, 25, 28, 41, 43, 44, 46	100.0	88.8	97.7	100.0	88.8	97.7	100.0	88.8	97.7	100.0	89.1	97.8	100.0	89.1	97.8	100.0	90.5	98.5
21, 22, 23	FIRE SUPPRESSION, PLUMBING & HVAC	95.2	83.0	90.3	95.3	76.7	87.8	99.9	82.5	92.9	100.1	77.2	90.8	95.2	83.0	90.3	99.9	76.8	90.6
26, 27, 3370	ELECTRICAL, COMMUNICATIONS & UTIL.	95.7	69.4	81.8	97.8	84.5	90.7	98.2	53.8	74.7	97.4	84.5	90.6	98.2	56.4	76.1	99.1	73.0	85.3
MF2014	WEIGHTED AVERAGE	102.3	78.1	91.8	99.8	78.9	90.7	104.8	75.3	91.9	98.9	75.5	88.7	101.4	75.4	90.1	101.1	80.3	92.0

COLORADO / CONNECTICUT

DIVISION		SALIDA 812 MAT.	INST.	TOTAL	BRIDGEPORT 066 MAT.	INST.	TOTAL	BRISTOL 060 MAT.	INST.	TOTAL	HARTFORD 061 MAT.	INST.	TOTAL	MERIDEN 064 MAT.	INST.	TOTAL	NEW BRITAIN 060 MAT.	INST.	TOTAL
015433	CONTRACTOR EQUIPMENT		95.2	95.2		100.9	100.9		100.9	100.9		100.9	100.9		101.3	101.3		100.9	100.9
0241, 31 - 34	SITE & INFRASTRUCTURE, DEMOLITION	128.2	91.5	102.1	109.8	105.1	106.4	108.9	105.0	106.2	104.4	105.0	104.9	106.7	105.8	106.0	109.1	105.0	106.2
0310	Concrete Forming & Accessories	110.1	74.7	79.6	98.9	124.5	121.0	98.9	124.3	120.8	95.7	124.3	120.4	98.6	124.3	120.8	99.2	124.3	120.9
0320	Concrete Reinforcing	103.7	76.4	89.8	105.1	127.3	116.4	105.1	127.3	116.4	104.6	127.3	116.2	105.1	127.3	116.4	105.1	127.3	116.4
0330	Cast-in-Place Concrete	115.8	77.8	100.2	107.9	127.1	115.8	101.1	127.1	111.8	103.0	127.1	112.9	97.3	127.1	109.5	102.8	127.1	112.7
03	CONCRETE	111.2	76.3	94.1	106.6	125.7	116.0	103.4	125.6	114.3	104.0	125.6	114.6	101.6	125.6	113.4	104.2	125.6	114.7
04	MASONRY	133.9	71.9	95.3	111.8	134.0	125.6	102.9	134.0	122.3	107.7	134.0	124.1	102.5	134.0	122.1	105.2	134.0	123.1
05	METALS	96.7	79.9	91.6	100.5	125.5	108.2	100.5	125.4	108.1	105.6	125.4	111.7	97.4	125.4	106.0	96.5	125.4	105.4
06	WOOD, PLASTICS & COMPOSITES	101.9	77.4	88.2	98.8	123.4	112.6	98.8	123.4	112.6	92.3	123.4	109.7	98.8	123.4	112.6	98.8	123.4	112.6
07	THERMAL & MOISTURE PROTECTION	104.0	79.2	93.8	98.6	129.1	111.1	98.7	126.0	109.9	102.8	126.0	112.3	98.7	126.0	109.9	98.7	126.0	109.9
08	OPENINGS	98.4	79.5	94.0	101.2	132.1	108.4	101.2	132.1	108.4	100.3	132.1	107.7	103.6	132.1	110.2	101.2	132.1	108.4
0920	Plaster & Gypsum Board	78.6	76.7	77.3	101.7	123.5	116.4	101.7	123.5	116.4	97.4	123.5	115.0	103.2	123.5	116.9	101.7	123.5	116.4
0950, 0980	Ceilings & Acoustic Treatment	96.3	76.7	83.4	86.4	123.5	110.8	86.4	123.5	110.8	89.7	123.5	111.9	90.4	123.5	112.2	86.4	123.5	110.8
0960	Flooring	120.6	45.4	99.1	101.9	131.2	110.3	101.9	131.2	110.3	100.5	131.2	109.2	101.9	131.2	110.3	101.9	131.2	110.3
0970, 0990	Wall Finishes & Painting/Coating	114.1	41.7	70.4	104.9	122.9	115.8	104.9	122.9	115.8	105.9	122.9	116.2	104.9	122.9	115.8	104.9	122.9	115.8
09	FINISHES	103.6	65.9	82.8	97.2	125.0	112.6	97.3	125.0	112.6	96.3	125.0	112.1	98.4	125.0	113.1	97.3	125.0	112.6
COVERS	DIVS. 10 - 14, 25, 28, 41, 43, 44, 46	100.0	89.2	97.8	100.0	112.4	102.5	100.0	112.4	102.5	100.0	112.4	102.5	100.0	112.4	102.5	100.0	112.4	102.5
21, 22, 23	FIRE SUPPRESSION, PLUMBING & HVAC	95.2	72.0	85.8	100.0	116.4	106.6	100.0	116.4	106.6	100.0	116.4	106.6	95.2	116.4	103.8	100.0	116.4	106.6
26, 27, 3370	ELECTRICAL, COMMUNICATIONS & UTIL.	98.4	72.5	84.7	98.5	112.2	105.7	98.5	111.8	105.6	97.9	112.8	105.8	98.4	111.8	105.5	98.6	111.8	105.6
MF2014	WEIGHTED AVERAGE	101.9	75.3	90.3	101.3	120.8	109.8	100.5	120.6	109.3	101.4	120.8	109.8	98.9	120.7	108.4	100.1	120.6	109.1

CONNECTICUT

DIVISION		NEW HAVEN 065 MAT.	INST.	TOTAL	NEW LONDON 063 MAT.	INST.	TOTAL	NORWALK 068 MAT.	INST.	TOTAL	STAMFORD 069 MAT.	INST.	TOTAL	WATERBURY 067 MAT.	INST.	TOTAL	WILLIMANTIC 062 MAT.	INST.	TOTAL
015433	CONTRACTOR EQUIPMENT		101.3	101.3		101.3	101.3		100.9	100.9		100.9	100.9		100.9	100.9		100.9	100.9
0241, 31 - 34	SITE & INFRASTRUCTURE, DEMOLITION	108.9	105.8	106.7	101.1	105.8	104.4	109.5	105.1	106.4	110.2	105.1	106.6	109.5	105.0	106.3	109.5	105.0	106.3
0310	Concrete Forming & Accessories	98.6	124.3	120.8	98.6	124.3	120.8	98.9	124.9	121.3	98.9	124.9	121.3	98.9	124.3	120.8	98.9	124.1	120.6
0320	Concrete Reinforcing	105.1	127.3	116.4	82.4	127.3	105.2	105.1	127.4	116.5	105.1	127.4	116.5	105.1	127.3	116.4	105.1	127.2	116.4
0330	Cast-in-Place Concrete	104.5	127.1	114.8	89.1	127.1	104.7	106.2	128.6	115.4	107.9	128.6	116.4	107.9	127.1	115.8	100.8	125.8	111.0
03	CONCRETE	118.7	125.6	122.1	91.4	125.6	108.2	105.8	126.4	115.9	106.6	126.4	116.3	106.6	125.6	115.9	103.3	125.6	114.0
04	MASONRY	103.2	134.0	122.4	101.5	134.0	121.7	102.7	135.4	123.1	103.4	135.4	123.3	103.4	134.0	122.5	102.7	134.0	122.2
05	METALS	96.8	125.4	105.6	96.5	125.4	105.4	100.5	126.1	108.3	100.5	126.1	108.3	100.5	125.4	108.1	100.2	125.2	107.9
06	WOOD, PLASTICS & COMPOSITES	98.8	123.4	112.6	98.8	123.4	112.6	98.8	123.4	112.6	98.8	123.4	112.6	98.8	123.4	112.6	98.8	123.4	112.6
07	THERMAL & MOISTURE PROTECTION	98.8	126.1	110.0	98.7	126.0	109.9	98.8	129.6	111.4	98.7	129.6	111.4	98.7	126.1	110.0	98.9	124.6	109.5
08	OPENINGS	101.2	132.1	108.4	104.0	132.1	110.6	101.2	132.1	108.4	101.2	132.1	108.4	101.2	132.1	108.4	104.0	132.1	110.6
0920	Plaster & Gypsum Board	101.7	123.5	116.4	101.7	123.5	116.4	101.7	123.5	116.4	101.7	123.5	116.4	101.7	123.5	116.4	101.7	123.5	116.4
0950, 0980	Ceilings & Acoustic Treatment	86.4	123.5	110.2	84.5	123.5	110.2	86.4	123.5	110.8	86.4	123.5	110.8	86.4	123.5	110.8	84.5	123.5	110.2
0960	Flooring	101.9	131.2	110.3	101.9	131.2	110.3	101.9	131.2	110.3	101.9	131.2	110.3	101.9	131.2	110.3	101.9	133.2	110.9
0970, 0990	Wall Finishes & Painting/Coating	104.9	122.9	115.8	104.9	122.9	115.8	104.9	122.9	115.8	104.9	122.9	115.8	104.9	122.9	115.8	104.9	122.9	115.8
09	FINISHES	97.3	125.0	112.6	96.4	125.0	112.2	97.3	125.0	112.6	97.4	125.0	112.6	97.2	125.0	112.5	97.0	125.4	112.6
COVERS	DIVS. 10 - 14, 25, 28, 41, 43, 44, 46	100.0	112.4	102.5	100.0	112.4	102.5	100.0	112.6	102.6	100.0	112.6	102.6	100.0	112.4	102.5	100.0	112.4	102.5
21, 22, 23	FIRE SUPPRESSION, PLUMBING & HVAC	100.0	116.4	106.6	95.2	116.4	103.8	100.0	116.4	106.6	100.0	116.4	106.6	100.0	116.4	106.6	100.0	116.1	106.5
26, 27, 3370	ELECTRICAL, COMMUNICATIONS & UTIL.	98.4	111.8	105.5	95.3	111.8	104.0	98.5	167.2	134.8	98.5	167.2	134.8	98.0	112.2	105.5	98.5	109.6	104.4
MF2014	WEIGHTED AVERAGE	101.6	120.7	109.9	97.1	120.7	107.4	100.8	128.7	113.0	100.9	128.7	113.0	100.9	120.7	109.5	100.8	120.2	109.2

D.C. / DELAWARE / FLORIDA

DIVISION		WASHINGTON 200 - 205 MAT.	INST.	TOTAL	DOVER 199 MAT.	INST.	TOTAL	NEWARK 197 MAT.	INST.	TOTAL	WILMINGTON 198 MAT.	INST.	TOTAL	DAYTONA BEACH 321 MAT.	INST.	TOTAL	FORT LAUDERDALE 333 MAT.	INST.	TOTAL
015433	CONTRACTOR EQUIPMENT		105.5	105.5		118.0	118.0		118.0	118.0		118.2	118.2		98.2	98.2		91.0	91.0
0241, 31 - 34	SITE & INFRASTRUCTURE, DEMOLITION	103.8	94.5	97.2	102.3	113.3	110.1	102.6	113.3	110.2	99.5	113.6	109.5	105.9	89.2	94.0	96.4	77.1	82.7
0310	Concrete Forming & Accessories	97.4	78.3	80.9	96.3	102.2	101.4	97.0	102.2	101.5	98.1	102.2	101.6	97.6	68.1	72.1	95.7	68.4	72.1
0320	Concrete Reinforcing	105.4	93.8	99.5	94.0	102.9	98.5	91.5	102.9	97.3	93.4	102.9	98.2	91.2	76.6	83.8	88.2	72.1	80.0
0330	Cast-in-Place Concrete	115.4	88.0	104.2	104.6	104.6	104.6	90.0	104.6	96.0	99.2	104.6	101.4	90.2	70.8	82.2	94.6	77.1	87.4
03	CONCRETE	107.7	85.9	97.0	102.2	104.0	103.1	94.6	104.0	99.2	99.6	104.0	101.8	91.6	71.8	81.9	94.5	73.3	84.0
04	MASONRY	98.2	80.0	86.9	107.5	98.1	101.6	103.2	98.1	100.0	108.5	98.1	102.0	99.1	65.7	78.3	102.8	68.3	81.3
05	METALS	99.9	108.6	102.6	102.0	116.7	106.5	103.5	116.7	107.5	102.0	116.7	106.5	104.7	91.1	100.5	102.0	89.6	98.2
06	WOOD, PLASTICS & COMPOSITES	95.3	76.2	84.6	94.0	101.9	98.4	95.8	101.9	99.2	92.0	101.9	97.6	96.3	69.2	81.1	84.3	67.2	74.7
07	THERMAL & MOISTURE PROTECTION	101.7	85.8	95.2	98.8	110.0	103.4	101.7	110.0	105.1	97.9	110.0	102.9	96.2	73.6	86.9	100.3	77.2	90.8
08	OPENINGS	99.9	87.9	97.1	91.3	110.9	95.9	91.6	110.9	96.1	89.9	110.9	94.8	98.6	67.4	91.3	97.5	65.5	90.1
0920	Plaster & Gypsum Board	108.0	75.5	86.0	97.1	101.8	100.3	99.5	101.8	101.1	101.7	101.8	101.8	95.6	68.7	77.4	105.4	66.6	79.2
0950, 0980	Ceilings & Acoustic Treatment	112.2	75.5	88.1	91.3	101.8	98.2	88.7	101.8	97.3	92.4	101.8	98.6	84.9	68.7	74.2	87.9	66.6	73.9
0960	Flooring	104.7	93.6	101.5	99.8	109.7	102.7	96.8	109.7	100.5	104.3	109.7	105.9	110.5	73.6	100.0	105.1	69.3	94.9
0970, 0990	Wall Finishes & Painting/Coating	108.0	82.2	92.4	97.2	106.4	102.8	97.3	106.4	102.8	95.0	106.4	101.9	104.5	74.1	86.2	98.9	70.2	81.6
09	FINISHES	100.8	80.5	89.6	96.2	103.6	100.3	95.0	103.6	99.7	98.7	103.6	101.4	96.9	69.3	81.7	94.9	68.1	80.1
COVERS	DIVS. 10 - 14, 25, 28, 41, 43, 44, 46	100.0	98.6	99.7	100.0	88.0	97.6	100.0	88.0	97.6	100.0	88.0	97.6	100.0	84.4	96.9	100.0	87.4	97.5
21, 22, 23	FIRE SUPPRESSION, PLUMBING & HVAC	100.1	92.1	96.9	100.0	117.4	107.0	100.1	117.4	107.1	100.1	117.4	107.1	99.9	76.1	90.3	100.0	66.5	86.5
26, 27, 3370	ELECTRICAL, COMMUNICATIONS & UTIL.	101.5	105.8	103.8	96.6	111.9	104.7	98.7	111.9	105.7	99.8	111.9	106.2	95.8	54.9	74.2	96.0	72.7	83.7
MF2014	WEIGHTED AVERAGE	101.1	91.9	97.1	99.3	109.2	103.6	98.8	109.2	103.3	99.3	109.2	103.6	99.0	72.9	87.6	98.6	72.6	87.2

City Cost Indexes

FLORIDA

DIVISION		FORT MYERS 339,341			GAINESVILLE 326,344			JACKSONVILLE 320,322			LAKELAND 338			MELBOURNE 329			MIAMI 330 - 332,340		
		MAT.	INST.	TOTAL	MAT.	INST.	TOTAL	MAT.	INST.	TOTAL	MAT.	INST.	TOTAL	MAT.	INST.	TOTAL	MAT.	INST.	TOTAL
015433	CONTRACTOR EQUIPMENT		98.2	98.2		98.2	98.2		98.2	98.2		98.2	98.2		98.2	98.2		91.0	91.0
0241, 31 - 34	SITE & INFRASTRUCTURE, DEMOLITION	107.3	88.3	93.8	113.9	88.3	95.7	106.0	88.6	93.6	109.2	88.7	94.6	112.7	88.5	95.5	99.5	76.9	83.4
0310	Concrete Forming & Accessories	91.5	74.8	77.1	92.7	54.3	59.6	97.4	54.7	60.5	88.0	75.3	77.1	93.8	69.8	73.1	100.7	68.7	73.1
0320	Concrete Reinforcing	89.3	92.7	91.0	96.8	64.8	80.5	91.2	64.9	77.8	91.5	93.6	92.5	92.3	76.6	84.3	94.3	72.2	83.0
0330	Cast-in-Place Concrete	98.7	68.2	86.2	103.6	62.6	86.7	91.1	68.3	81.8	100.9	69.6	88.1	108.7	73.1	94.1	95.5	78.1	88.3
03	CONCRETE	95.1	76.9	86.2	102.6	60.7	82.0	92.1	62.8	77.7	96.8	77.8	87.5	103.0	73.3	88.4	96.1	73.7	85.1
04	MASONRY	95.8	61.9	74.7	114.0	61.2	81.1	99.1	61.2	75.5	114.1	75.8	90.2	97.5	69.8	80.2	103.1	71.9	83.7
05	METALS	104.2	96.8	102.0	103.6	85.0	97.9	103.2	85.4	97.7	104.1	97.8	102.2	113.4	91.3	106.6	102.4	88.5	98.1
06	WOOD, PLASTICS & COMPOSITES	81.1	77.4	79.0	89.9	51.8	68.5	96.3	51.8	71.4	76.5	77.4	77.0	91.5	69.2	79.0	90.6	67.2	77.5
07	THERMAL & MOISTURE PROTECTION	100.1	80.0	91.9	96.5	61.7	82.2	96.4	62.3	82.4	100.1	84.6	93.7	96.6	76.0	88.2	101.6	71.6	89.3
08	OPENINGS	98.4	74.4	92.8	96.9	52.7	86.6	98.6	55.8	88.6	98.4	75.1	93.0	97.8	70.7	91.5	99.7	65.5	91.7
0920	Plaster & Gypsum Board	101.0	77.1	84.9	91.3	50.7	63.9	95.6	50.7	65.3	97.4	77.1	83.7	91.3	68.7	76.0	103.2	66.6	78.4
0950, 0980	Ceilings & Acoustic Treatment	82.7	77.1	79.0	79.5	50.7	60.6	84.9	50.7	62.5	82.7	77.1	79.0	83.3	68.7	73.7	92.2	66.6	75.4
0960	Flooring	102.3	55.1	88.8	108.1	43.1	89.5	110.5	64.5	97.4	100.3	56.5	87.8	108.3	73.6	98.4	107.3	73.0	97.5
0970, 0990	Wall Finishes & Painting/Coating	103.7	67.6	81.9	104.5	67.6	82.3	104.5	67.6	82.3	103.7	67.6	81.9	104.5	91.9	96.9	96.0	70.2	80.5
09	FINISHES	94.4	70.7	81.3	95.3	52.5	71.7	97.0	56.7	74.8	93.5	71.0	81.1	95.9	72.3	82.9	97.1	69.2	81.7
COVERS	DIVS. 10 - 14, 25, 28, 41, 43, 44, 46	100.0	72.6	94.5	100.0	82.8	96.5	100.0	81.0	96.2	100.0	72.6	94.5	100.0	85.8	97.1	100.0	88.0	97.6
21, 22, 23	FIRE SUPPRESSION, PLUMBING & HVAC	97.4	64.3	84.1	98.8	64.1	84.8	99.9	64.2	85.5	97.4	80.8	90.7	99.9	78.2	91.2	100.0	66.4	86.4
26, 27, 3370	ELECTRICAL, COMMUNICATIONS & UTIL.	98.1	62.8	79.5	96.1	71.8	83.3	95.5	62.2	77.9	96.3	61.6	77.9	97.0	67.8	81.6	100.1	74.9	86.8
MF2014	WEIGHTED AVERAGE	98.6	72.6	87.3	100.3	66.7	85.7	98.8	66.4	84.7	99.4	77.8	90.0	101.7	76.3	90.6	99.8	73.1	88.2

FLORIDA

DIVISION		ORLANDO 327 - 328,347			PANAMA CITY 324			PENSACOLA 325			SARASOTA 342			ST. PETERSBURG 337			TALLAHASSEE 323		
		MAT.	INST.	TOTAL	MAT.	INST.	TOTAL	MAT.	INST.	TOTAL	MAT.	INST.	TOTAL	MAT.	INST.	TOTAL	MAT.	INST.	TOTAL
015433	CONTRACTOR EQUIPMENT		98.2	98.2		98.2	98.2		98.2	98.2		98.2	98.2		98.2	98.2		98.2	98.2
0241, 31 - 34	SITE & INFRASTRUCTURE, DEMOLITION	107.6	88.4	93.9	117.6	87.4	96.1	117.6	87.9	96.5	114.0	88.4	95.8	110.9	88.1	94.7	105.1	87.7	92.7
0310	Concrete Forming & Accessories	101.4	72.0	76.1	96.7	43.9	51.1	94.6	51.8	57.7	96.0	75.0	77.9	94.9	51.0	57.1	99.9	44.1	51.8
0320	Concrete Reinforcing	96.6	73.9	85.0	95.3	72.5	83.7	97.7	73.0	85.1	92.5	93.5	93.0	91.5	86.1	88.7	98.3	64.7	81.2
0330	Cast-in-Place Concrete	112.1	70.6	95.1	95.7	56.7	79.7	118.2	65.9	96.8	106.6	69.5	91.4	102.0	64.5	86.6	97.2	56.4	80.5
03	CONCRETE	103.4	73.0	88.4	100.8	55.4	78.5	110.7	62.2	86.9	100.2	77.5	89.1	98.5	63.8	81.4	97.3	54.0	76.0
04	MASONRY	100.6	65.7	78.8	103.9	47.3	68.6	124.7	54.5	80.9	99.8	75.8	84.8	157.9	49.0	90.0	103.6	53.8	72.6
05	METALS	102.4	89.6	98.5	104.4	86.5	98.9	105.6	87.9	100.2	105.2	97.5	102.8	105.0	92.9	101.3	102.2	84.6	96.8
06	WOOD, PLASTICS & COMPOSITES	95.8	75.2	84.2	95.1	41.8	65.2	92.7	51.5	69.7	95.7	77.4	85.4	85.5	50.1	65.7	95.0	41.3	64.9
07	THERMAL & MOISTURE PROTECTION	94.9	74.3	86.4	96.7	56.6	80.2	96.6	62.3	82.5	98.1	84.6	92.5	100.3	57.8	82.8	102.6	71.5	89.9
08	OPENINGS	101.4	69.2	93.9	96.5	46.5	84.9	96.5	56.9	87.3	99.7	74.2	93.8	98.4	60.9	89.6	100.2	47.3	87.9
0920	Plaster & Gypsum Board	99.7	74.9	82.9	94.5	40.4	57.9	97.4	50.5	65.7	97.8	77.1	83.8	103.5	49.0	66.7	108.1	40.0	62.0
0950, 0980	Ceilings & Acoustic Treatment	91.4	74.9	80.5	83.3	40.4	55.1	83.3	50.5	61.7	86.2	77.1	80.3	84.5	49.0	61.2	94.1	40.0	58.5
0960	Flooring	104.1	73.6	95.4	110.1	43.0	90.9	106.1	63.0	93.7	111.1	57.9	95.9	104.1	55.2	90.1	111.1	62.0	97.0
0970, 0990	Wall Finishes & Painting/Coating	103.4	70.2	83.4	104.5	64.9	80.6	104.5	67.6	82.3	109.5	67.6	84.2	103.7	60.8	77.8	99.5	67.6	80.3
09	FINISHES	98.2	72.4	84.0	97.5	44.8	68.5	96.4	54.8	73.5	99.9	71.2	84.1	95.9	51.7	71.6	100.7	48.5	72.0
COVERS	DIVS. 10 - 14, 25, 28, 41, 43, 44, 46	100.0	85.1	97.0	100.0	46.2	89.1	100.0	46.3	89.1	100.0	72.6	94.5	100.0	55.8	91.1	100.0	65.8	93.1
21, 22, 23	FIRE SUPPRESSION, PLUMBING & HVAC	99.9	56.6	82.4	99.9	52.3	80.7	99.9	52.5	80.8	99.9	64.8	85.7	100.0	58.4	83.2	100.0	38.9	75.4
26, 27, 3370	ELECTRICAL, COMMUNICATIONS & UTIL.	97.6	60.0	77.7	94.5	59.3	75.9	98.5	55.8	75.9	97.2	61.6	78.4	96.3	61.6	77.9	103.3	60.0	80.4
MF2014	WEIGHTED AVERAGE	100.5	70.0	87.2	100.2	57.8	81.7	102.7	61.2	84.6	100.8	74.3	89.2	102.6	63.3	85.5	100.8	56.9	81.7

FLORIDA / GEORGIA

DIVISION		TAMPA 335 - 336,346			WEST PALM BEACH 334,349			ALBANY 317,398			ATHENS 306			ATLANTA 300 - 303,399			AUGUSTA 308 - 309		
		MAT.	INST.	TOTAL	MAT.	INST.	TOTAL	MAT.	INST.	TOTAL	MAT.	INST.	TOTAL	MAT.	INST.	TOTAL	MAT.	INST.	TOTAL
015433	CONTRACTOR EQUIPMENT		98.2	98.2		91.0	91.0		91.9	91.9		94.2	94.2		94.7	94.7		94.2	94.2
0241, 31 - 34	SITE & INFRASTRUCTURE, DEMOLITION	111.4	88.6	95.2	93.2	77.1	81.8	98.8	78.7	84.5	100.9	93.4	95.6	97.6	94.8	95.7	94.5	93.2	93.6
0310	Concrete Forming & Accessories	97.7	75.6	78.6	99.2	68.1	72.4	90.4	43.6	50.0	92.9	45.9	52.4	96.7	73.5	76.7	94.2	65.5	69.4
0320	Concrete Reinforcing	88.2	93.6	91.0	90.7	71.8	81.1	90.6	80.2	85.3	95.3	77.9	86.5	94.6	81.2	87.8	95.7	71.8	83.6
0330	Cast-in-Place Concrete	99.7	69.7	87.4	90.0	74.3	83.5	92.4	54.5	76.8	107.8	55.6	86.4	107.8	70.8	92.6	101.9	49.9	80.5
03	CONCRETE	97.0	77.9	87.6	91.3	72.1	81.9	93.1	55.8	74.8	104.8	56.0	80.8	102.2	74.1	88.4	97.5	61.7	79.9
04	MASONRY	103.9	75.8	86.4	102.3	66.6	80.0	102.5	49.1	69.2	81.5	53.8	64.2	95.2	66.2	77.1	95.5	43.3	63.0
05	METALS	104.0	98.1	102.2	101.1	89.3	97.4	105.0	85.5	99.0	91.9	73.3	86.2	92.8	77.0	88.0	91.6	71.6	85.5
06	WOOD, PLASTICS & COMPOSITES	89.3	77.4	82.6	89.2	67.2	76.9	85.0	38.0	58.7	91.8	40.9	63.3	86.1	75.8	84.7	93.3	70.9	80.8
07	THERMAL & MOISTURE PROTECTION	100.5	84.6	94.0	100.1	70.2	87.8	95.9	60.3	81.3	94.1	52.8	77.2	94.0	72.0	85.0	93.7	57.0	78.7
08	OPENINGS	99.7	79.2	94.9	96.8	65.5	89.5	91.1	44.2	80.2	89.5	45.7	79.3	94.7	71.9	89.4	89.5	62.2	83.1
0920	Plaster & Gypsum Board	106.1	77.1	86.5	110.2	66.6	80.7	97.5	36.5	56.3	98.7	39.4	58.7	100.9	75.4	83.6	99.8	70.4	79.9
0950, 0980	Ceilings & Acoustic Treatment	87.9	77.1	80.8	82.7	66.6	72.1	84.0	36.5	52.8	97.2	39.4	59.2	97.2	75.4	82.9	98.1	70.4	79.9
0960	Flooring	105.1	56.5	91.2	106.8	66.4	95.3	110.9	47.8	92.9	96.4	53.9	84.2	97.7	65.2	88.4	96.6	46.6	82.3
0970, 0990	Wall Finishes & Painting/Coating	103.7	67.6	81.9	98.9	70.2	81.6	105.3	52.8	73.6	106.0	46.0	69.8	106.0	85.1	93.4	106.0	46.0	69.8
09	FINISHES	97.4	71.0	82.8	94.7	67.5	79.7	98.1	43.2	67.9	95.4	45.7	68.0	95.7	73.4	83.4	95.2	60.5	76.1
COVERS	DIVS. 10 - 14, 25, 28, 41, 43, 44, 46	100.0	85.1	97.0	100.0	80.2	96.0	100.0	80.2	96.0	100.0	77.3	95.4	100.0	83.6	97.1	100.0	78.5	95.7
21, 22, 23	FIRE SUPPRESSION, PLUMBING & HVAC	100.0	80.8	92.2	97.4	62.7	83.4	99.9	68.0	87.0	95.2	69.6	84.9	99.9	70.6	88.1	100.0	61.2	84.4
26, 27, 3370	ELECTRICAL, COMMUNICATIONS & UTIL.	96.0	61.6	77.8	97.2	72.7	84.3	96.9	58.7	76.7	99.7	69.5	83.8	99.0	71.9	84.7	100.4	61.8	80.0
MF2014	WEIGHTED AVERAGE	100.1	78.4	90.6	97.4	71.1	85.9	98.5	61.3	82.3	95.5	63.9	81.7	97.6	74.4	87.5	96.3	63.6	82.1

City Cost Indexes

GEORGIA

DIVISION		COLUMBUS 318-319 MAT.	INST.	TOTAL	DALTON 307 MAT.	INST.	TOTAL	GAINESVILLE 305 MAT.	INST.	TOTAL	MACON 310-312 MAT.	INST.	TOTAL	SAVANNAH 313-314 MAT.	INST.	TOTAL	STATESBORO 304 MAT.	INST.	TOTAL
015433	CONTRACTOR EQUIPMENT		91.9	91.9		106.1	106.1		94.2	94.2		102.3	102.3		92.8	92.8		93.4	93.4
0241, 31-34	SITE & INFRASTRUCTURE, DEMOLITION	98.7	78.7	84.5	101.1	97.8	98.8	100.8	93.3	95.5	99.8	93.5	95.3	100.6	80.1	86.1	102.0	77.5	84.6
0310	Concrete Forming & Accessories	90.3	54.3	59.3	85.7	46.7	52.1	96.4	42.9	50.2	89.9	51.9	57.1	92.1	50.0	55.8	80.3	51.4	55.4
0320	Concrete Reinforcing	90.9	80.6	85.7	94.8	73.9	84.2	95.1	77.8	86.3	92.1	80.3	86.1	98.1	71.9	84.7	94.4	41.8	67.6
0330	Cast-in-Place Concrete	92.1	53.9	76.4	104.7	50.2	82.3	113.3	52.9	88.5	90.9	65.6	80.5	100.2	55.2	81.7	107.6	59.7	87.9
03	CONCRETE	93.0	60.4	77.0	103.9	54.6	79.7	106.7	53.7	80.6	92.6	63.3	78.2	98.0	57.3	78.0	104.2	54.0	79.6
04	MASONRY	102.6	55.2	73.1	83.8	36.7	54.4	89.5	54.2	67.5	115.9	44.5	71.4	98.8	50.9	69.0	84.7	40.6	57.2
05	METALS	104.5	86.3	98.9	92.8	82.8	89.7	91.2	72.5	85.4	100.1	85.9	95.7	101.3	82.4	95.5	96.3	72.6	89.0
06	WOOD, PLASTICS & COMPOSITES	85.0	52.1	66.6	75.2	47.5	59.7	95.9	37.9	63.4	91.8	51.2	69.0	96.0	45.9	68.0	68.5	53.9	60.3
07	THERMAL & MOISTURE PROTECTION	95.9	63.5	82.6	96.1	51.5	77.8	94.1	54.8	78.0	94.4	62.2	81.2	95.0	56.6	79.2	94.7	51.1	76.8
08	OPENINGS	91.1	59.0	83.7	90.1	49.5	80.6	89.5	40.0	77.9	90.0	52.8	81.3	94.9	48.4	84.1	91.1	40.6	79.4
0920	Plaster & Gypsum Board	97.5	51.1	66.1	86.2	46.2	59.2	100.9	36.3	57.3	103.0	50.1	67.2	95.6	44.7	61.2	86.9	52.8	63.9
0950, 0980	Ceilings & Acoustic Treatment	84.0	51.1	62.4	109.3	46.2	67.8	97.2	36.3	57.2	79.2	50.1	60.1	91.5	44.7	60.8	105.7	52.8	71.0
0960	Flooring	110.9	50.4	93.6	96.8	46.9	82.5	97.6	46.6	83.0	87.9	47.8	76.5	108.6	46.4	90.8	115.6	44.9	95.4
0970, 0990	Wall Finishes & Painting/Coating	105.3	66.5	81.9	96.3	59.8	74.2	106.0	46.0	69.8	107.7	52.8	74.6	103.8	58.3	76.3	103.5	39.2	64.7
09	FINISHES	98.0	53.5	73.5	103.0	47.1	72.2	95.9	41.6	66.0	86.7	49.9	66.4	98.2	48.9	71.0	106.9	49.0	75.0
COVERS	DIVS. 10-14, 25, 28, 41, 43, 44, 46	100.0	81.8	96.3	100.0	22.8	84.4	100.0	39.4	87.8	100.0	80.2	96.0	100.0	78.0	95.6	100.0	43.5	88.6
21, 22, 23	FIRE SUPPRESSION, PLUMBING & HVAC	99.9	63.4	85.2	95.2	57.0	79.8	95.2	55.4	84.8	99.9	66.7	86.6	100.0	61.8	84.6	95.7	56.7	80.0
26, 27, 3370	ELECTRICAL, COMMUNICATIONS & UTIL.	97.1	69.6	82.5	109.7	67.7	87.5	99.7	69.5	83.8	96.1	61.2	77.7	101.7	57.4	78.3	99.9	57.4	77.4
MF2014	WEIGHTED AVERAGE	98.4	65.4	84.0	97.2	59.0	80.5	96.0	61.6	81.0	97.2	64.7	83.0	99.2	60.9	82.5	97.3	55.8	79.2

GEORGIA / HAWAII / IDAHO

DIVISION		VALDOSTA 316 MAT.	INST.	TOTAL	WAYCROSS 315 MAT.	INST.	TOTAL	HILO 967 MAT.	INST.	TOTAL	HONOLULU 968 MAT.	INST.	TOTAL	STATES & POSS., GUAM 969 MAT.	INST.	TOTAL	BOISE 836-837 MAT.	INST.	TOTAL
015433	CONTRACTOR EQUIPMENT		91.9	91.9		91.9	91.9		99.5	99.5		99.5	99.5		164.5	164.5		98.2	98.2
0241, 31-34	SITE & INFRASTRUCTURE, DEMOLITION	107.9	78.8	87.2	104.7	77.4	85.3	144.8	106.5	117.6	155.4	106.5	120.6	184.9	103.7	127.2	85.5	96.3	93.2
0310	Concrete Forming & Accessories	81.0	44.1	49.2	82.8	64.7	67.2	111.2	135.6	132.2	123.6	135.6	134.0	113.7	63.6	70.5	100.8	77.4	80.7
0320	Concrete Reinforcing	92.8	76.4	84.4	92.8	72.7	82.5	123.5	117.9	120.7	132.5	117.9	125.1	214.4	30.2	120.7	99.2	80.4	89.6
0330	Cast-in-Place Concrete	90.5	59.5	76.3	102.4	48.2	80.0	198.3	125.7	168.5	163.7	125.7	148.1	171.9	105.5	144.7	91.6	88.4	90.3
03	CONCRETE	98.1	55.8	77.3	101.2	61.4	81.6	158.1	127.7	143.2	150.8	127.7	139.5	159.9	72.3	116.9	99.1	81.9	90.7
04	MASONRY	108.9	50.5	72.5	109.7	39.3	65.8	150.5	128.3	136.7	150.6	128.3	136.7	213.9	43.5	107.7	121.4	84.2	98.2
05	METALS	104.1	84.0	97.9	103.1	77.7	95.3	108.1	107.3	107.8	120.7	107.3	116.5	139.2	76.2	119.8	102.2	80.9	95.6
06	WOOD, PLASTICS & COMPOSITES	73.4	37.6	53.4	75.0	72.4	73.5	114.5	139.2	128.3	134.4	139.2	137.1	127.1	67.4	93.7	93.1	75.9	83.4
07	THERMAL & MOISTURE PROTECTION	96.1	62.4	82.3	95.9	50.3	77.2	120.8	124.6	122.4	138.4	124.6	132.7	140.4	65.2	109.6	94.6	81.4	89.2
08	OPENINGS	87.3	42.5	76.9	87.4	57.8	80.5	111.3	132.8	116.3	121.8	132.8	124.4	116.6	54.1	102.1	99.1	71.2	92.6
0920	Plaster & Gypsum Board	90.4	36.2	53.8	90.4	72.0	78.0	97.9	140.1	130.3	151.3	140.1	143.7	219.9	55.5	108.8	92.4	75.1	80.7
0950, 0980	Ceilings & Acoustic Treatment	81.5	36.2	51.7	79.6	72.0	74.6	121.5	140.1	133.8	132.0	140.1	137.3	237.3	55.5	117.8	99.1	75.1	83.3
0960	Flooring	104.7	47.8	88.4	105.9	29.9	84.1	116.0	139.9	122.8	132.2	139.9	134.4	135.5	46.4	110.0	96.3	83.7	92.7
0970, 0990	Wall Finishes & Painting/Coating	105.3	52.2	73.2	105.3	46.0	69.5	110.4	143.8	130.6	119.4	143.8	134.2	115.9	35.9	67.6	103.2	39.5	64.7
09	FINISHES	95.6	43.6	67.0	95.2	57.4	74.4	113.7	138.6	127.4	128.4	138.6	134.0	188.8	58.9	117.2	96.0	74.8	84.3
COVERS	DIVS. 10-14, 25, 28, 41, 43, 44, 46	100.0	77.1	95.4	100.0	51.5	90.2	100.0	116.0	103.2	100.0	116.0	103.2	100.0	75.2	95.0	100.0	87.4	97.5
21, 22, 23	FIRE SUPPRESSION, PLUMBING & HVAC	99.9	70.0	87.9	96.9	57.1	80.8	100.2	107.7	103.2	100.3	107.7	103.3	102.6	37.4	76.3	100.0	71.8	88.6
26, 27, 3370	ELECTRICAL, COMMUNICATIONS & UTIL.	95.0	56.2	74.5	99.6	57.4	77.3	106.5	121.1	114.2	107.9	121.1	114.9	153.5	41.2	94.1	98.5	72.8	84.9
MF2014	WEIGHTED AVERAGE	98.5	61.4	82.3	98.3	59.4	81.4	114.8	120.2	117.2	119.5	120.2	119.8	136.2	58.0	102.1	100.1	78.6	90.7

IDAHO / ILLINOIS

DIVISION		COEUR D'ALENE 838 MAT.	INST.	TOTAL	IDAHO FALLS 834 MAT.	INST.	TOTAL	LEWISTON 835 MAT.	INST.	TOTAL	POCATELLO 832 MAT.	INST.	TOTAL	TWIN FALLS 833 MAT.	INST.	TOTAL	BLOOMINGTON 617 MAT.	INST.	TOTAL
015433	CONTRACTOR EQUIPMENT		92.8	92.8		98.2	98.2		92.8	92.8		98.2	98.2		98.2	98.2		101.6	101.6
0241, 31-34	SITE & INFRASTRUCTURE, DEMOLITION	83.1	91.7	89.2	83.1	96.1	92.4	89.9	92.5	91.7	85.8	96.3	93.3	92.1	97.2	95.8	97.4	98.4	98.1
0310	Concrete Forming & Accessories	112.6	81.1	85.5	94.7	78.1	80.4	117.9	82.2	87.1	101.0	77.1	80.4	102.0	54.7	61.2	84.3	117.2	112.7
0320	Concrete Reinforcing	106.2	96.5	101.3	101.0	80.0	90.3	106.2	96.8	101.4	99.6	80.2	89.7	101.3	79.9	90.4	94.6	110.7	102.8
0330	Cast-in-Place Concrete	99.0	87.2	94.1	87.2	74.3	81.9	102.9	86.0	95.9	94.1	88.3	91.7	96.6	63.9	83.1	102.8	114.8	107.7
03	CONCRETE	105.7	86.1	96.1	91.3	77.2	84.4	109.3	86.2	98.0	98.3	81.7	90.1	105.8	63.3	84.9	100.1	115.2	107.5
04	MASONRY	122.9	83.9	98.6	116.5	81.0	94.4	123.4	85.9	100.0	118.8	81.1	95.3	121.6	81.1	96.4	120.3	118.3	119.1
05	METALS	96.3	86.9	93.4	100.9	79.1	100.5	95.7	88.0	93.4	110.1	80.3	100.9	110.1	78.9	100.5	97.6	113.6	102.5
06	WOOD, PLASTICS & COMPOSITES	96.8	81.0	88.0	86.8	78.8	82.3	102.4	81.0	90.4	93.1	75.9	83.4	94.2	47.2	67.9	86.4	115.4	102.6
07	THERMAL & MOISTURE PROTECTION	148.3	80.8	120.6	94.3	71.7	85.0	148.5	81.4	121.0	94.7	73.7	86.1	95.4	74.7	86.9	98.4	112.8	104.3
08	OPENINGS	117.8	73.0	107.4	102.8	68.0	94.7	117.8	75.6	108.0	99.8	66.3	92.0	102.8	48.8	90.2	94.9	104.9	97.2
0920	Plaster & Gypsum Board	161.4	80.5	106.7	78.5	78.1	78.2	162.5	80.5	107.0	80.7	75.1	76.9	82.3	45.6	57.5	91.3	115.8	107.9
0950, 0980	Ceilings & Acoustic Treatment	129.0	80.5	97.1	97.2	78.1	84.6	129.0	80.5	97.1	104.7	75.1	85.2	99.7	45.6	64.1	88.0	115.8	106.3
0960	Flooring	138.3	45.9	111.9	96.4	43.1	81.2	141.2	93.8	127.7	99.6	83.7	95.1	100.7	43.1	84.3	93.8	118.7	100.9
0970, 0990	Wall Finishes & Painting/Coating	123.8	67.8	90.0	103.3	40.3	65.3	123.8	67.8	90.0	103.1	41.1	65.7	103.3	37.3	63.4	93.8	126.4	113.5
09	FINISHES	118.1	73.3	113.0	93.3	68.7	79.7	162.9	82.8	118.8	96.6	75.0	84.7	96.7	49.7	70.8	93.3	118.9	107.4
COVERS	DIVS. 10-14, 25, 28, 41, 43, 44, 46	100.0	87.4	97.5	100.0	47.7	89.4	100.0	87.6	97.5	100.0	87.4	97.5	100.0	44.2	88.7	100.0	105.2	101.0
21, 22, 23	FIRE SUPPRESSION, PLUMBING & HVAC	99.5	82.6	92.7	100.9	71.7	89.1	100.7	85.6	94.6	99.9	71.8	88.6	99.9	69.3	87.6	95.1	108.8	100.8
26, 27, 3370	ELECTRICAL, COMMUNICATIONS & UTIL.	91.0	77.6	83.9	90.8	70.4	80.0	88.9	80.7	84.5	96.2	70.4	82.6	92.3	60.1	75.3	94.2	94.6	94.4
MF2014	WEIGHTED AVERAGE	107.9	82.2	96.7	99.8	74.7	88.9	108.7	84.9	98.3	101.1	77.4	90.8	102.2	67.3	87.0	97.5	109.4	102.7

For customer support on your Plumbing Cost Data, call 877.709.6968.

City Cost Indexes

ILLINOIS

DIVISION		CARBONDALE 629 MAT.	INST.	TOTAL	CENTRALIA 628 MAT.	INST.	TOTAL	CHAMPAIGN 618-619 MAT.	INST.	TOTAL	CHICAGO 606-608 MAT.	INST.	TOTAL	DECATUR 625 MAT.	INST.	TOTAL	EAST ST. LOUIS 620-622 MAT.	INST.	TOTAL
015433	CONTRACTOR EQUIPMENT		108.0	108.0		108.0	108.0		102.4	102.4		94.3	94.3		102.4	102.4		108.0	108.0
0241, 31 - 34	SITE & INFRASTRUCTURE, DEMOLITION	98.7	99.5	99.2	99.0	100.1	99.8	106.4	99.0	101.2	107.0	95.7	99.0	93.2	99.1	97.4	101.2	99.8	100.2
0310	Concrete Forming & Accessories	90.0	109.0	106.4	91.6	112.8	109.9	90.7	115.2	111.8	96.6	156.7	148.4	91.8	114.9	111.8	87.6	114.8	111.1
0320	Concrete Reinforcing	91.2	111.6	101.6	91.2	111.8	101.7	94.6	105.1	99.9	99.0	158.1	129.1	89.3	102.7	96.1	91.1	109.9	100.6
0330	Cast-in-Place Concrete	97.2	102.5	99.4	97.7	117.3	105.7	119.2	109.5	115.2	107.5	149.4	124.7	106.1	110.5	107.9	99.3	117.0	106.6
03	CONCRETE	90.1	107.9	98.9	90.6	114.8	102.5	113.1	111.3	112.2	102.6	153.2	127.5	101.3	111.1	106.1	91.7	115.2	103.2
04	MASONRY	83.6	108.6	99.2	83.6	116.7	104.2	145.3	117.7	128.1	101.1	156.6	135.7	79.6	114.7	101.5	83.9	116.6	104.3
05	METALS	96.2	119.6	103.4	96.2	120.8	103.8	97.6	108.7	101.0	94.3	134.3	106.6	99.9	108.1	102.4	97.3	119.4	104.1
06	WOOD, PLASTICS & COMPOSITES	91.7	106.1	99.8	94.1	109.7	102.8	93.5	113.9	104.9	104.0	155.9	133.1	91.9	113.9	104.2	89.2	112.1	102.0
07	THERMAL & MOISTURE PROTECTION	97.1	101.5	98.9	97.2	111.7	103.1	99.0	113.2	104.9	98.0	145.1	117.3	103.2	109.3	105.7	97.2	110.2	102.5
08	OPENINGS	89.4	114.2	95.2	89.4	116.2	95.7	95.5	111.2	99.2	105.3	158.7	117.8	100.9	110.6	103.2	89.5	116.9	95.9
0920	Plaster & Gypsum Board	97.2	106.2	103.3	98.3	110.0	106.2	93.7	114.2	107.6	92.7	157.6	136.6	100.0	114.2	109.6	96.1	112.4	107.2
0950, 0980	Ceilings & Acoustic Treatment	91.4	106.2	101.2	91.4	110.0	103.6	88.0	114.2	105.2	99.2	157.6	137.6	98.1	114.2	108.7	91.4	112.4	105.2
0960	Flooring	122.2	120.2	121.6	123.2	115.7	121.0	96.7	120.2	103.4	96.9	148.0	111.5	109.1	116.7	111.3	121.2	115.7	119.6
0970, 0990	Wall Finishes & Painting/Coating	112.6	97.4	103.5	112.6	106.6	109.0	93.8	110.0	103.6	91.9	153.0	128.8	102.5	106.5	104.9	112.6	103.5	107.1
09	FINISHES	101.8	108.6	105.5	102.2	112.6	107.9	95.2	115.9	106.6	98.0	155.6	129.7	101.8	115.0	109.1	101.4	113.8	108.2
COVERS	DIVS. 10 - 14, 25, 28, 41, 43, 44, 46	100.0	102.6	100.5	100.0	104.1	100.8	100.0	104.4	100.9	100.0	124.5	105.0	100.0	104.3	100.9	100.0	104.4	100.9
21, 22, 23	FIRE SUPPRESSION, PLUMBING & HVAC	95.1	106.1	99.5	95.1	95.3	95.2	95.1	105.5	99.3	99.8	133.9	113.6	99.9	98.5	99.4	99.9	98.8	99.4
26, 27, 3370	ELECTRICAL, COMMUNICATIONS & UTIL.	95.4	107.7	101.9	96.8	107.7	102.6	97.4	94.9	96.1	98.3	133.9	117.1	99.5	90.4	94.7	96.4	103.5	100.1
MF2014	WEIGHTED AVERAGE	94.7	107.9	100.5	95.0	108.5	100.9	101.0	107.6	103.9	99.8	139.7	117.2	99.2	104.9	101.7	96.4	108.7	101.8

ILLINOIS

DIVISION		EFFINGHAM 624 MAT.	INST.	TOTAL	GALESBURG 614 MAT.	INST.	TOTAL	JOLIET 604 MAT.	INST.	TOTAL	KANKAKEE 609 MAT.	INST.	TOTAL	LA SALLE 613 MAT.	INST.	TOTAL	NORTH SUBURBAN 600-603 MAT.	INST.	TOTAL
015433	CONTRACTOR EQUIPMENT		102.4	102.4		101.6	101.6		92.5	92.5		92.5	92.5		101.6	101.6		92.5	92.5
0241, 31 - 34	SITE & INFRASTRUCTURE, DEMOLITION	97.6	98.9	98.5	99.9	98.3	98.8	106.9	94.9	98.4	100.5	94.6	96.3	99.2	99.2	99.2	106.1	94.9	98.1
0310	Concrete Forming & Accessories	96.3	114.0	111.6	90.6	116.6	113.1	98.4	159.6	151.2	91.8	143.2	136.2	104.6	125.3	122.4	97.8	154.0	146.3
0320	Concrete Reinforcing	92.2	99.9	96.1	94.1	110.6	102.5	99.0	150.8	125.4	99.8	147.4	124.0	94.3	144.7	119.9	99.0	156.5	128.3
0330	Cast-in-Place Concrete	105.7	108.4	106.8	105.9	106.2	106.0	107.4	146.5	123.5	100.1	132.5	113.4	105.7	122.5	112.6	107.5	141.2	121.3
03	CONCRETE	102.1	109.4	105.7	103.2	112.0	107.5	102.7	152.1	126.9	96.7	139.3	117.7	104.1	127.9	115.8	102.7	148.8	125.4
04	MASONRY	88.2	108.5	100.8	120.5	118.0	118.9	104.3	148.5	131.8	100.6	141.2	125.9	120.5	125.1	123.4	101.1	144.1	127.9
05	METALS	97.0	105.6	99.7	97.6	113.0	102.4	92.2	129.1	103.6	92.2	126.6	102.8	97.7	131.6	108.1	93.4	131.4	105.1
06	WOOD, PLASTICS & COMPOSITES	94.2	113.9	105.2	93.3	115.5	105.7	105.7	161.0	136.7	97.8	142.2	122.7	108.8	123.7	117.1	104.0	155.7	132.9
07	THERMAL & MOISTURE PROTECTION	102.7	106.6	104.3	98.5	107.6	102.2	97.8	141.6	115.8	97.0	135.7	112.9	98.7	117.8	106.5	98.2	139.3	115.1
08	OPENINGS	94.9	109.9	98.4	94.9	111.0	98.6	102.8	159.6	116.0	95.4	148.4	107.7	94.9	129.5	102.9	102.9	158.2	115.8
0920	Plaster & Gypsum Board	99.7	114.2	109.5	93.7	115.9	108.7	89.9	162.9	139.2	86.7	143.4	125.0	100.1	124.3	116.5	92.7	157.3	136.4
0950, 0980	Ceilings & Acoustic Treatment	91.4	114.2	106.4	88.0	115.9	106.3	99.2	162.9	141.0	99.2	143.4	128.3	88.0	124.3	111.9	99.2	157.3	137.4
0960	Flooring	110.1	120.2	113.0	96.6	118.7	102.9	96.5	140.3	109.1	93.7	131.9	104.6	102.8	122.6	108.5	96.9	140.3	109.3
0970, 0990	Wall Finishes & Painting/Coating	102.5	104.2	103.5	93.8	95.1	94.5	90.1	152.5	127.8	90.1	126.4	112.0	93.8	126.4	113.5	91.9	152.5	128.4
09	FINISHES	100.6	115.2	108.7	94.6	115.4	106.1	97.4	156.9	130.2	95.7	137.9	119.0	97.3	124.3	112.2	97.9	152.7	128.1
COVERS	DIVS. 10 - 14, 25, 28, 41, 43, 44, 46	100.0	70.8	94.1	100.0	105.1	101.0	100.0	124.5	105.0	100.0	120.8	104.2	100.0	102.5	100.5	100.0	122.3	104.5
21, 22, 23	FIRE SUPPRESSION, PLUMBING & HVAC	95.2	102.8	98.2	95.1	105.3	99.2	99.9	131.2	112.5	95.1	129.0	108.7	95.1	123.4	106.6	99.8	128.6	111.4
26, 27, 3370	ELECTRICAL, COMMUNICATIONS & UTIL.	97.3	107.6	102.8	95.1	86.6	90.6	97.5	137.9	118.9	92.3	137.1	116.0	92.2	137.1	116.0	97.3	128.5	113.8
MF2014	WEIGHTED AVERAGE	97.3	106.0	101.1	98.2	106.7	101.9	99.2	138.4	116.3	95.6	131.7	111.3	98.4	124.4	109.7	99.3	135.1	114.9

ILLINOIS

DIVISION		PEORIA 615-616 MAT.	INST.	TOTAL	QUINCY 623 MAT.	INST.	TOTAL	ROCK ISLAND 612 MAT.	INST.	TOTAL	ROCKFORD 610-611 MAT.	INST.	TOTAL	SOUTH SUBURBAN 605 MAT.	INST.	TOTAL	SPRINGFIELD 626-627 MAT.	INST.	TOTAL
015433	CONTRACTOR EQUIPMENT		101.6	101.6		102.4	102.4		101.6	101.6		101.6	101.6		92.5	92.5		102.4	102.4
0241, 31 - 34	SITE & INFRASTRUCTURE, DEMOLITION	100.4	98.3	98.9	96.5	98.7	98.0	98.0	97.4	97.5	99.8	99.7	99.7	106.1	94.9	98.1	99.0	99.1	99.1
0310	Concrete Forming & Accessories	93.7	117.4	114.2	94.2	111.8	109.4	92.3	104.2	102.5	98.1	131.3	126.7	97.8	154.0	146.3	92.6	115.3	112.1
0320	Concrete Reinforcing	91.7	110.8	101.4	91.8	105.3	98.7	94.1	103.7	99.0	86.7	136.9	112.2	99.0	156.5	128.3	94.2	105.1	99.7
0330	Cast-in-Place Concrete	102.7	114.1	107.3	105.9	102.9	104.7	103.6	97.3	101.0	105.1	126.4	113.8	107.5	141.2	121.3	100.8	109.2	104.2
03	CONCRETE	100.2	115.0	107.5	101.7	107.6	104.6	101.0	102.0	101.5	100.9	130.4	115.4	102.7	148.8	125.4	99.1	111.3	105.1
04	MASONRY	120.1	118.1	118.8	111.7	104.1	106.9	120.3	97.0	105.8	94.1	135.8	120.0	101.1	144.1	127.9	91.0	117.8	107.7
05	METALS	100.4	113.8	104.5	97.1	108.0	100.5	97.7	108.4	101.0	100.4	128.2	108.9	93.4	131.4	105.1	97.5	109.3	101.1
06	WOOD, PLASTICS & COMPOSITES	101.0	115.5	109.2	91.8	113.9	104.2	95.0	104.4	100.3	101.1	128.3	116.4	104.0	155.7	132.9	89.2	113.9	103.0
07	THERMAL & MOISTURE PROTECTION	99.3	112.4	104.7	102.7	103.5	103.0	98.5	98.3	98.4	101.8	129.0	113.0	98.2	139.3	115.1	104.4	112.2	107.6
08	OPENINGS	101.3	116.0	104.7	95.8	111.4	99.4	94.9	103.2	96.8	101.3	132.1	108.5	102.9	158.2	115.8	103.0	111.2	104.9
0920	Plaster & Gypsum Board	97.1	115.9	109.8	98.3	114.2	109.0	93.7	104.5	101.0	97.1	129.1	118.8	92.7	157.3	136.4	98.7	114.2	109.2
0950, 0980	Ceilings & Acoustic Treatment	93.0	115.9	108.1	91.4	114.2	106.4	88.0	104.5	98.8	93.0	129.1	116.7	99.2	157.3	137.4	102.1	114.2	110.1
0960	Flooring	99.9	118.7	105.3	109.1	108.1	108.8	97.7	105.6	100.0	99.9	122.6	106.4	96.9	140.3	109.3	113.5	107.8	111.9
0970, 0990	Wall Finishes & Painting/Coating	93.8	126.4	113.5	102.5	106.5	104.9	93.8	95.1	94.5	93.8	133.1	117.5	91.9	152.5	128.4	101.1	106.5	104.3
09	FINISHES	97.2	118.9	109.2	100.1	111.8	106.6	94.9	104.7	99.7	97.2	130.0	115.3	97.9	152.7	128.1	104.9	113.5	109.6
COVERS	DIVS. 10 - 14, 25, 28, 41, 43, 44, 46	100.0	105.2	101.0	100.0	71.2	94.2	100.0	99.0	99.8	100.0	114.3	102.9	100.0	122.3	104.5	100.0	104.5	100.9
21, 22, 23	FIRE SUPPRESSION, PLUMBING & HVAC	99.9	104.9	101.9	95.2	101.0	97.5	95.1	99.9	97.1	100.0	116.5	106.7	99.8	128.6	111.4	99.9	103.3	101.3
26, 27, 3370	ELECTRICAL, COMMUNICATIONS & UTIL.	96.1	96.7	96.4	94.7	80.8	87.3	87.3	95.5	91.6	96.4	131.1	114.7	97.3	128.5	113.8	102.4	92.0	96.9
MF2014	WEIGHTED AVERAGE	100.5	109.3	104.4	98.1	101.0	99.3	97.2	100.5	98.6	99.5	124.8	110.5	99.3	135.1	114.9	100.0	106.5	102.8

INDIANA

DIVISION		ANDERSON 460 MAT.	INST.	TOTAL	BLOOMINGTON 474 MAT.	INST.	TOTAL	COLUMBUS 472 MAT.	INST.	TOTAL	EVANSVILLE 476-477 MAT.	INST.	TOTAL	FORT WAYNE 467-468 MAT.	INST.	TOTAL	GARY 463-464 MAT.	INST.	TOTAL
015433	CONTRACTOR EQUIPMENT		97.0	97.0		86.5	86.5		86.5	86.5		116.0	116.0		97.0	97.0		97.0	97.0
0241, 31 - 34	SITE & INFRASTRUCTURE, DEMOLITION	93.9	96.1	95.5	85.8	94.4	91.9	82.2	94.3	90.8	91.0	123.9	114.4	94.8	96.0	95.7	94.5	99.6	98.1
0310	Concrete Forming & Accessories	97.8	81.5	83.7	101.0	80.7	83.5	95.0	78.6	80.8	94.4	82.7	84.3	96.1	75.2	78.1	97.9	116.3	113.8
0320	Concrete Reinforcing	95.8	82.8	89.2	86.8	80.9	83.8	87.2	80.9	84.0	95.1	78.1	86.4	95.8	75.5	85.5	95.8	110.8	103.5
0330	Cast-in-Place Concrete	109.4	80.2	97.4	103.6	78.2	93.1	103.1	71.2	90.0	99.0	87.9	94.4	116.3	82.8	102.5	114.3	113.9	114.1
03	CONCRETE	100.8	81.7	91.5	105.0	79.7	92.6	104.2	76.3	90.5	105.3	83.8	94.7	104.0	78.4	91.4	103.2	114.2	108.6
04	MASONRY	94.1	80.1	85.4	95.6	76.5	83.7	95.4	76.5	83.6	91.2	82.5	85.8	98.5	77.9	85.6	95.6	114.6	107.5
05	METALS	92.8	89.3	91.7	95.8	77.6	90.2	95.9	77.0	90.0	89.3	85.0	87.9	92.8	85.9	90.7	92.8	108.7	97.7
06	WOOD, PLASTICS & COMPOSITES	100.2	81.9	89.9	112.9	80.9	95.0	107.7	78.2	91.2	93.5	82.0	87.0	99.9	74.7	85.8	97.7	115.7	107.8
07	THERMAL & MOISTURE PROTECTION	107.7	75.9	94.7	95.1	78.3	88.2	94.7	78.1	87.9	99.4	83.7	93.0	107.5	77.7	95.3	106.3	108.6	107.2
08	OPENINGS	98.5	82.1	94.7	105.8	81.1	100.1	101.5	79.6	96.4	99.1	80.8	94.9	98.5	74.2	92.9	98.5	120.3	103.6
0920	Plaster & Gypsum Board	102.5	81.6	88.4	97.5	81.0	86.4	95.0	78.2	83.6	93.4	80.8	84.9	101.8	74.2	83.1	96.1	116.4	109.8
0950, 0980	Ceilings & Acoustic Treatment	87.9	81.6	83.7	80.9	81.0	81.0	80.9	78.2	79.1	85.5	80.8	82.4	87.9	74.2	78.9	87.9	116.4	106.6
0960	Flooring	100.6	85.0	96.2	105.5	75.0	96.8	100.8	75.0	93.4	100.3	80.7	94.7	100.6	78.8	94.4	100.6	123.4	107.1
0970, 0990	Wall Finishes & Painting/Coating	105.6	69.4	83.7	95.3	82.4	87.5	95.3	82.4	87.5	101.2	84.9	91.4	105.6	73.9	86.5	105.6	123.9	116.7
09	FINISHES	95.6	81.0	87.6	94.8	79.8	86.5	93.0	78.2	84.9	93.8	82.6	87.6	95.4	75.7	84.5	94.7	118.7	107.9
COVERS	DIVS. 10 - 14, 25, 28, 41, 43, 44, 46	100.0	91.8	98.3	100.0	90.8	98.1	100.0	90.5	98.1	100.0	95.9	99.2	100.0	91.7	98.3	100.0	106.1	101.2
21, 22, 23	FIRE SUPPRESSION, PLUMBING & HVAC	100.0	78.8	91.5	99.7	78.9	91.3	94.9	78.8	88.4	99.9	79.6	91.7	100.0	72.1	88.7	100.0	106.6	102.7
26, 27, 3370	ELECTRICAL, COMMUNICATIONS & UTIL.	87.0	88.4	87.7	98.9	88.0	93.2	98.1	87.9	92.7	95.1	88.3	91.5	87.7	79.2	83.2	98.7	107.6	103.4
MF2014	WEIGHTED AVERAGE	97.0	83.8	91.3	99.3	81.8	91.6	97.2	80.9	90.1	97.1	86.9	92.6	97.6	79.1	89.5	98.3	110.4	103.6

INDIANA

DIVISION		INDIANAPOLIS 461-462 MAT.	INST.	TOTAL	KOKOMO 469 MAT.	INST.	TOTAL	LAFAYETTE 479 MAT.	INST.	TOTAL	LAWRENCEBURG 470 MAT.	INST.	TOTAL	MUNCIE 473 MAT.	INST.	TOTAL	NEW ALBANY 471 MAT.	INST.	TOTAL
015433	CONTRACTOR EQUIPMENT		93.7	93.7		97.0	97.0		86.5	86.5		104.4	104.4		95.0	95.0		94.5	94.5
0241, 31 - 34	SITE & INFRASTRUCTURE, DEMOLITION	93.5	99.3	97.7	90.3	96.0	94.4	83.1	94.3	91.0	81.2	110.4	102.0	85.6	94.6	92.0	78.2	96.8	91.4
0310	Concrete Forming & Accessories	98.6	85.6	87.4	101.2	76.6	80.0	92.6	82.6	84.0	91.5	75.4	77.6	92.6	81.0	82.6	89.5	73.3	75.5
0320	Concrete Reinforcing	99.4	83.0	91.1	86.7	81.1	83.9	86.8	82.7	84.7	86.1	74.2	80.1	96.0	82.7	89.3	87.4	77.1	82.2
0330	Cast-in-Place Concrete	101.9	86.4	95.5	108.3	82.7	97.8	103.7	81.8	94.7	97.0	74.8	87.9	108.8	78.6	96.4	100.1	74.0	89.4
03	CONCRETE	100.9	85.1	93.1	97.5	80.1	89.0	104.5	82.1	93.5	97.4	75.5	86.6	103.5	81.0	92.4	102.8	74.6	89.0
04	MASONRY	95.8	80.3	86.2	93.7	79.1	84.6	101.2	80.3	88.2	79.9	74.7	76.7	97.7	80.1	86.7	86.9	68.3	75.3
05	METALS	93.5	80.5	89.5	89.3	88.3	89.0	94.3	78.4	89.4	91.0	83.5	88.7	97.6	89.2	95.0	92.9	81.0	89.3
06	WOOD, PLASTICS & COMPOSITES	99.7	86.2	92.2	103.4	75.0	87.5	104.7	83.1	92.6	91.9	75.2	82.6	106.3	81.5	92.4	94.1	74.3	83.0
07	THERMAL & MOISTURE PROTECTION	100.7	80.8	92.5	107.3	75.7	94.3	94.7	80.6	88.9	100.1	76.7	90.5	97.7	76.8	89.2	86.8	69.6	79.7
08	OPENINGS	105.7	84.5	100.8	93.3	77.9	89.8	99.9	82.8	95.9	101.4	75.3	95.3	99.0	81.9	95.0	98.7	76.3	93.5
0920	Plaster & Gypsum Board	97.0	85.8	89.5	107.5	74.5	85.2	92.2	83.3	86.1	71.5	75.0	73.9	93.4	81.6	85.4	91.6	73.8	79.6
0950, 0980	Ceilings & Acoustic Treatment	93.8	85.8	88.6	87.9	74.5	79.1	76.7	83.3	81.0	89.7	75.0	80.0	81.7	81.6	81.6	85.5	73.8	77.8
0960	Flooring	101.4	85.0	96.7	104.6	93.3	101.3	99.7	88.8	96.6	74.1	85.0	77.2	99.9	85.0	95.6	97.9	62.8	87.9
0970, 0990	Wall Finishes & Painting/Coating	103.6	82.4	90.8	105.6	71.7	85.2	95.3	93.1	94.0	96.2	71.6	81.4	95.3	69.4	79.7	101.2	82.7	90.0
09	FINISHES	96.1	85.4	90.2	97.3	78.8	87.1	91.4	85.0	87.9	83.7	77.2	80.1	92.4	80.6	85.9	93.1	72.7	81.8
COVERS	DIVS. 10 - 14, 25, 28, 41, 43, 44, 46	100.0	93.3	98.7	100.0	91.1	98.2	100.0	91.1	98.2	100.0	43.6	88.6	100.0	90.8	98.1	100.0	43.1	88.5
21, 22, 23	FIRE SUPPRESSION, PLUMBING & HVAC	99.9	79.4	91.6	95.2	79.0	88.7	94.9	79.4	88.6	95.8	74.1	87.1	99.7	78.7	91.2	95.2	75.2	87.1
26, 27, 3370	ELECTRICAL, COMMUNICATIONS & UTIL.	101.0	88.4	94.4	91.4	78.7	84.7	97.6	82.5	89.6	93.0	73.5	82.7	90.9	79.6	84.9	93.6	75.8	84.2
MF2014	WEIGHTED AVERAGE	99.1	84.8	92.9	94.8	81.5	89.0	96.9	82.7	90.7	94.0	77.7	86.9	97.9	82.3	91.1	95.1	75.4	86.5

INDIANA / IOWA

DIVISION		SOUTH BEND 465-466 MAT.	INST.	TOTAL	TERRE HAUTE 478 MAT.	INST.	TOTAL	WASHINGTON 475 MAT.	INST.	TOTAL	BURLINGTON 526 MAT.	INST.	TOTAL	CARROLL 514 MAT.	INST.	TOTAL	CEDAR RAPIDS 522-524 MAT.	INST.	TOTAL
015433	CONTRACTOR EQUIPMENT		105.3	105.3		116.0	116.0		116.0	116.0		99.4	99.4		99.4	99.4		96.1	96.1
0241, 31 - 34	SITE & INFRASTRUCTURE, DEMOLITION	96.7	96.2	96.3	92.9	124.1	115.1	92.4	121.3	113.0	98.3	96.5	97.0	87.5	96.5	93.9	100.0	95.1	96.5
0310	Concrete Forming & Accessories	99.5	81.4	83.9	95.4	82.2	84.0	96.2	78.8	81.2	94.8	75.5	78.2	82.6	50.4	54.8	100.5	82.3	84.8
0320	Concrete Reinforcing	97.3	79.9	88.4	95.1	82.9	88.9	87.9	48.8	68.0	91.0	82.7	86.8	91.8	78.7	85.1	91.7	84.2	87.9
0330	Cast-in-Place Concrete	106.6	79.8	95.6	95.9	85.6	91.6	104.3	84.8	96.3	111.7	55.1	88.4	111.7	59.7	90.3	111.9	79.7	98.7
03	CONCRETE	98.3	81.8	90.2	108.4	83.7	96.2	114.3	75.3	95.1	102.6	70.8	86.9	101.3	60.2	81.1	102.7	82.3	92.7
04	MASONRY	105.7	77.8	88.3	98.9	79.6	86.9	91.3	80.8	84.8	102.3	64.9	79.0	104.2	74.1	85.4	108.2	78.0	89.4
05	METALS	92.8	99.9	95.0	90.0	87.4	89.2	84.6	68.2	79.5	88.8	93.0	90.1	88.9	88.7	88.8	91.3	92.2	91.5
06	WOOD, PLASTICS & COMPOSITES	94.8	81.6	87.4	95.7	82.1	88.1	96.1	78.9	86.5	95.1	76.6	84.8	81.6	44.9	61.1	101.9	82.6	91.1
07	THERMAL & MOISTURE PROTECTION	101.9	80.6	93.2	99.5	80.8	91.8	99.6	82.2	92.5	102.9	73.9	91.0	103.2	69.4	89.3	103.9	80.9	94.5
08	OPENINGS	96.0	80.5	92.4	99.7	82.2	95.6	96.4	67.6	89.7	94.8	70.4	89.1	99.3	53.4	88.6	100.3	81.9	96.0
0920	Plaster & Gypsum Board	90.8	81.3	84.4	93.4	81.0	85.0	93.4	77.7	82.8	102.3	75.9	84.5	97.7	43.2	60.9	106.9	82.3	90.3
0950, 0980	Ceilings & Acoustic Treatment	88.9	81.3	83.9	85.5	81.0	82.6	79.6	77.7	78.3	98.7	75.9	83.7	98.7	43.2	62.3	101.2	82.3	88.8
0960	Flooring	99.3	91.1	96.9	100.3	85.0	95.9	101.2	76.4	94.1	107.9	37.8	87.8	102.2	33.2	82.5	124.0	78.3	110.9
0970, 0990	Wall Finishes & Painting/Coating	99.2	87.1	91.9	101.2	83.1	90.3	101.2	83.4	90.5	106.3	78.5	89.5	106.3	78.7	89.6	107.6	72.2	86.3
09	FINISHES	95.4	83.9	89.1	93.8	82.8	87.7	93.1	79.6	85.6	102.5	68.9	84.0	98.7	48.3	71.0	108.4	80.6	93.1
COVERS	DIVS. 10 - 14, 25, 28, 41, 43, 44, 46	100.0	92.8	98.5	100.0	93.7	98.7	100.0	95.3	99.0	100.0	86.8	97.3	100.0	65.1	92.9	100.0	91.8	98.3
21, 22, 23	FIRE SUPPRESSION, PLUMBING & HVAC	99.9	77.1	90.7	99.9	79.6	91.7	95.2	77.6	88.1	95.4	75.2	87.3	95.4	72.6	86.2	100.2	80.5	92.2
26, 27, 3370	ELECTRICAL, COMMUNICATIONS & UTIL.	99.6	86.6	92.7	93.3	86.7	89.8	93.8	84.2	88.8	100.6	67.7	83.2	101.3	81.3	90.7	98.1	80.6	88.8
MF2014	WEIGHTED AVERAGE	98.0	84.4	92.1	97.9	86.5	92.9	95.8	81.8	89.7	97.2	75.3	87.6	97.0	71.2	85.8	100.0	83.2	92.7

For customer support on your Plumbing Cost Data, call 877.709.6968.

613

City Cost Indexes

IOWA

DIVISION		COUNCIL BLUFFS 515			CRESTON 508			DAVENPORT 527 - 528			DECORAH 521			DES MOINES 500 - 503,509			DUBUQUE 520		
		MAT.	INST.	TOTAL	MAT.	INST.	TOTAL	MAT.	INST.	TOTAL	MAT.	INST.	TOTAL	MAT.	INST.	TOTAL	MAT.	INST.	TOTAL
015433	CONTRACTOR EQUIPMENT		95.5	95.5		99.4	99.4		99.4	99.4		99.4	99.4		101.1	101.1		94.9	94.9
0241, 31 - 34	SITE & INFRASTRUCTURE, DEMOLITION	103.8	91.5	95.0	93.9	95.5	95.1	98.7	98.6	98.6	96.9	95.5	95.9	102.9	99.8	100.7	97.9	92.4	94.0
0310	Concrete Forming & Accessories	82.0	71.2	72.7	78.4	65.8	67.5	100.0	91.3	92.5	92.3	45.4	51.8	95.4	81.8	83.6	83.3	74.9	76.1
0320	Concrete Reinforcing	93.6	79.0	86.1	89.3	82.4	85.8	91.7	97.0	94.4	91.0	76.5	83.6	98.2	86.7	92.3	90.4	84.1	87.2
0330	Cast-in-Place Concrete	116.4	73.6	98.8	115.2	64.1	94.2	107.8	90.1	100.5	108.6	58.0	87.8	101.4	87.5	95.7	109.6	98.6	105.1
03	CONCRETE	105.0	74.3	89.9	102.9	69.3	86.4	100.7	92.5	96.7	100.5	57.0	79.1	100.0	85.4	92.8	99.4	85.5	92.6
04	MASONRY	109.8	75.0	88.1	108.1	81.9	91.7	105.3	86.8	93.7	124.0	71.4	91.2	101.0	81.2	88.7	109.2	73.9	87.2
05	METALS	96.3	89.3	94.1	93.8	90.5	92.8	91.3	102.7	94.8	89.0	85.8	88.0	99.9	96.5	98.8	89.8	92.0	90.5
06	WOOD, PLASTICS & COMPOSITES	80.4	70.9	75.1	74.1	61.6	67.1	101.9	91.0	95.8	92.2	37.1	61.3	91.8	81.1	85.8	82.1	73.5	77.3
07	THERMAL & MOISTURE PROTECTION	103.3	67.1	88.5	104.2	77.6	93.3	103.4	87.5	96.8	103.1	53.4	82.7	98.5	79.4	90.6	103.6	73.9	91.4
08	OPENINGS	99.3	75.9	93.9	108.4	63.9	98.1	100.3	90.8	98.1	97.8	48.6	86.4	102.7	86.5	98.9	99.3	79.3	94.7
0920	Plaster & Gypsum Board	97.7	70.3	79.2	93.6	60.5	71.2	106.9	90.6	96.0	101.2	35.2	56.6	90.4	80.6	83.8	97.7	72.9	81.0
0950, 0980	Ceilings & Acoustic Treatment	98.7	70.3	80.0	90.8	60.5	70.9	101.2	90.8	94.4	98.7	35.2	57.0	94.2	80.6	85.2	98.7	72.9	81.8
0960	Flooring	100.8	80.7	95.0	97.4	33.2	79.1	110.3	97.0	106.5	107.6	47.2	90.3	102.5	89.6	98.8	114.9	77.0	104.1
0970, 0990	Wall Finishes & Painting/Coating	102.2	65.8	80.2	101.0	78.7	87.5	106.3	95.5	99.8	106.3	32.9	62.0	96.2	88.6	91.6	106.7	63.8	80.8
09	FINISHES	99.4	72.1	84.4	94.2	60.2	75.4	104.3	92.7	97.9	102.2	42.6	69.4	97.1	83.7	89.7	103.7	73.5	87.1
COVERS	DIVS. 10 - 14, 25, 28, 41, 43, 44, 46	100.0	88.6	97.7	100.0	69.5	93.8	100.0	94.5	98.9	100.0	82.9	96.5	100.0	92.1	98.4	100.0	90.2	98.0
21, 22, 23	FIRE SUPPRESSION, PLUMBING & HVAC	100.2	74.2	89.7	95.2	79.7	89.0	100.2	92.8	97.2	95.4	72.9	86.3	99.9	79.7	91.7	100.2	75.3	90.1
26, 27, 3370	ELECTRICAL, COMMUNICATIONS & UTIL.	103.6	81.5	91.9	93.8	81.3	87.2	95.6	91.0	93.2	98.1	44.3	69.7	105.4	81.3	92.7	102.1	78.2	89.4
MF2014	WEIGHTED AVERAGE	100.7	78.1	90.9	98.2	77.3	89.1	99.0	93.1	96.4	98.0	64.2	83.3	100.5	85.2	93.8	99.1	80.2	90.9

IOWA

DIVISION		FORT DODGE 505			MASON CITY 504			OTTUMWA 525			SHENANDOAH 516			SIBLEY 512			SIOUX CITY 510 - 511		
		MAT.	INST.	TOTAL	MAT.	INST.	TOTAL	MAT.	INST.	TOTAL	MAT.	INST.	TOTAL	MAT.	INST.	TOTAL	MAT.	INST.	TOTAL
015433	CONTRACTOR EQUIPMENT		99.4	99.4		99.4	99.4		94.9	94.9		95.5	95.5		99.4	99.4		99.4	99.4
0241, 31 - 34	SITE & INFRASTRUCTURE, DEMOLITION	102.2	94.4	96.6	102.3	95.4	97.4	98.1	90.5	92.7	102.2	90.5	93.9	108.1	94.2	98.2	109.9	95.0	99.3
0310	Concrete Forming & Accessories	78.9	45.1	49.7	83.1	45.5	50.6	90.3	73.7	75.9	83.6	56.5	60.2	84.0	37.9	44.2	100.5	66.4	71.1
0320	Concrete Reinforcing	89.3	67.2	78.1	89.2	81.8	85.4	91.0	86.1	88.5	93.6	68.4	80.7	93.6	65.3	79.2	91.7	77.9	84.7
0330	Cast-in-Place Concrete	108.2	44.2	81.9	108.2	57.5	87.4	112.4	53.7	88.3	112.5	59.8	90.9	110.3	44.7	83.4	111.0	55.5	88.2
03	CONCRETE	98.2	50.4	74.7	98.5	57.9	78.5	102.1	69.9	86.3	102.4	61.0	82.1	101.4	47.0	74.6	102.0	65.8	84.2
04	MASONRY	106.9	37.9	63.9	120.5	69.8	88.9	105.8	57.8	75.9	109.4	75.0	88.0	128.6	38.1	72.2	102.2	54.6	72.6
05	METALS	93.9	81.3	90.1	94.0	88.7	92.4	88.8	91.7	89.7	95.3	83.3	91.6	89.1	79.7	86.2	91.3	87.4	90.1
06	WOOD, PLASTICS & COMPOSITES	74.5	45.2	58.1	78.4	37.1	55.3	89.3	80.9	84.6	82.1	52.8	65.7	82.9	35.9	56.6	101.9	66.6	82.1
07	THERMAL & MOISTURE PROTECTION	103.5	57.1	84.5	103.0	63.5	86.8	103.7	64.6	87.7	102.6	65.2	87.2	102.9	48.3	80.5	103.4	64.3	87.4
08	OPENINGS	101.9	49.5	89.7	93.5	50.1	83.4	99.3	77.7	94.3	90.3	54.7	82.0	95.9	43.8	83.8	100.3	65.9	92.3
0920	Plaster & Gypsum Board	93.6	43.5	59.8	93.6	35.2	54.1	98.8	80.6	86.5	97.7	51.6	66.5	97.7	33.9	54.6	106.9	65.7	79.0
0950, 0980	Ceilings & Acoustic Treatment	90.8	43.5	59.7	90.8	35.2	54.2	98.7	80.6	86.8	98.7	51.6	67.7	98.7	33.9	56.1	101.2	65.7	77.9
0960	Flooring	98.9	47.2	84.1	100.9	47.2	85.6	117.9	50.2	98.5	101.5	34.3	82.3	103.1	34.0	83.3	110.3	53.7	94.1
0970, 0990	Wall Finishes & Painting/Coating	101.0	61.7	77.3	101.0	30.1	58.2	106.7	78.7	89.8	102.2	65.8	80.2	106.3	61.7	79.4	106.3	63.5	80.5
09	FINISHES	96.1	46.1	68.5	96.7	42.0	66.5	104.9	70.4	85.9	99.5	52.3	73.4	101.9	38.0	66.7	105.8	63.6	82.5
COVERS	DIVS. 10 - 14, 25, 28, 41, 43, 44, 46	100.0	81.7	96.3	100.0	86.5	97.3	100.0	82.4	96.4	100.0	66.1	93.2	100.0	80.8	96.1	100.0	88.2	97.6
21, 22, 23	FIRE SUPPRESSION, PLUMBING & HVAC	95.2	65.5	83.3	95.2	70.5	85.2	95.4	70.9	85.5	95.4	73.5	86.6	95.4	68.0	84.3	100.2	75.8	90.4
26, 27, 3370	ELECTRICAL, COMMUNICATIONS & UTIL.	100.3	41.9	69.4	99.4	55.6	76.3	100.4	72.0	85.4	98.1	81.5	89.3	98.1	41.4	68.1	98.1	73.0	84.8
MF2014	WEIGHTED AVERAGE	97.9	58.2	80.6	97.6	65.8	83.8	97.9	73.7	87.3	97.6	71.2	86.1	98.3	56.4	80.0	99.7	72.6	87.9

IOWA / KANSAS

DIVISION		IOWA SPENCER 513			WATERLOO 506 - 507			KANSAS BELLEVILLE 669			COLBY 677			DODGE CITY 678			EMPORIA 668		
		MAT.	INST.	TOTAL	MAT.	INST.	TOTAL	MAT.	INST.	TOTAL	MAT.	INST.	TOTAL	MAT.	INST.	TOTAL	MAT.	INST.	TOTAL
015433	CONTRACTOR EQUIPMENT		99.4	99.4		99.4	99.4		103.4	103.4		103.4	103.4		103.4	103.4		101.6	101.6
0241, 31 - 34	SITE & INFRASTRUCTURE, DEMOLITION	108.1	94.2	98.2	107.6	95.3	98.8	111.0	95.0	99.7	112.3	95.2	100.1	114.9	94.9	100.7	103.0	92.5	95.5
0310	Concrete Forming & Accessories	90.2	37.9	45.1	94.1	55.1	60.5	96.2	54.1	59.8	99.5	58.3	64.0	93.1	58.2	63.0	87.2	65.3	68.3
0320	Concrete Reinforcing	93.6	67.1	80.1	89.9	83.9	86.8	96.6	55.1	75.5	98.9	55.2	76.7	96.5	55.1	75.4	95.3	55.7	75.2
0330	Cast-in-Place Concrete	110.3	44.7	83.4	115.8	59.2	92.6	124.2	55.7	96.1	127.1	55.8	97.8	129.3	55.5	99.0	120.1	53.6	92.8
03	CONCRETE	101.8	47.3	75.0	104.4	63.1	84.1	119.3	56.2	88.3	119.9	58.1	89.5	121.3	57.9	90.2	111.5	60.5	86.5
04	MASONRY	128.6	38.1	72.2	107.7	73.1	86.1	99.2	58.7	73.9	108.1	59.9	78.1	118.9	58.9	81.5	105.5	68.5	82.4
05	METALS	89.0	80.5	86.4	96.4	91.1	94.7	95.6	77.3	90.0	96.0	77.7	90.4	97.4	76.7	91.1	95.3	78.9	90.3
06	WOOD, PLASTICS & COMPOSITES	89.2	35.9	59.3	91.6	48.0	67.2	97.9	52.1	72.3	103.6	57.9	78.0	95.8	57.9	74.5	88.7	66.6	76.3
07	THERMAL & MOISTURE PROTECTION	103.8	48.2	81.0	103.3	71.3	90.2	94.8	60.7	80.8	99.1	61.6	83.7	99.1	60.7	83.4	93.1	75.9	86.0
08	OPENINGS	107.5	44.4	92.9	94.4	58.2	86.0	98.6	48.4	86.9	103.2	51.5	91.2	103.1	51.5	91.1	96.3	56.3	87.0
0920	Plaster & Gypsum Board	98.8	33.9	54.9	101.9	46.5	64.4	97.2	50.6	65.7	99.7	56.5	70.5	94.0	56.5	68.7	94.4	65.5	74.9
0950, 0980	Ceilings & Acoustic Treatment	98.7	33.9	56.1	94.2	46.5	62.8	86.3	50.6	62.8	84.0	56.5	65.9	84.0	56.5	65.9	86.3	65.5	72.7
0960	Flooring	105.7	34.0	85.2	106.2	64.9	94.4	103.0	39.2	84.8	103.1	39.2	84.9	99.6	39.2	82.3	98.0	36.4	80.4
0970, 0990	Wall Finishes & Painting/Coating	106.3	61.7	79.4	101.0	77.8	87.0	100.0	38.9	63.1	104.7	38.9	65.0	104.7	38.9	65.0	100.0	38.9	63.1
09	FINISHES	102.8	38.0	67.1	100.3	57.6	76.8	96.9	49.3	70.6	96.6	52.7	72.4	94.9	52.7	71.6	94.0	57.3	73.8
COVERS	DIVS. 10 - 14, 25, 28, 41, 43, 44, 46	100.0	80.8	96.1	100.0	87.8	97.5	100.0	40.8	88.2	100.0	41.4	88.2	100.0	41.4	88.2	100.0	40.5	88.0
21, 22, 23	FIRE SUPPRESSION, PLUMBING & HVAC	95.4	68.0	84.3	100.0	79.8	91.9	95.2	71.6	85.7	95.2	69.0	84.6	100.0	69.8	87.5	95.2	73.4	86.4
26, 27, 3370	ELECTRICAL, COMMUNICATIONS & UTIL.	99.9	41.4	69.0	95.8	55.7	74.6	96.5	67.7	81.3	100.0	64.5	81.2	96.8	73.8	84.6	93.9	71.4	82.0
MF2014	WEIGHTED AVERAGE	99.9	56.6	81.0	99.5	71.8	87.4	99.6	64.9	84.4	101.1	65.0	85.4	102.7	66.0	86.7	97.9	69.2	85.4

614

City Cost Indexes

KANSAS

DIVISION		FORT SCOTT 667			HAYS 676			HUTCHINSON 675			INDEPENDENCE 673			KANSAS CITY 660 - 662			LIBERAL 679		
		MAT.	INST.	TOTAL	MAT.	INST.	TOTAL	MAT.	INST.	TOTAL	MAT.	INST.	TOTAL	MAT.	INST.	TOTAL	MAT.	INST.	TOTAL
015433	CONTRACTOR EQUIPMENT		102.5	102.5		103.4	103.4		103.4	103.4		103.4	103.4		99.9	99.9		103.4	103.4
0241, 31 - 34	SITE & INFRASTRUCTURE, DEMOLITION	99.8	93.4	95.2	117.7	95.2	101.7	96.2	95.0	95.4	116.0	95.1	101.1	94.4	93.1	93.5	117.3	95.0	101.5
0310	Concrete Forming & Accessories	104.2	76.8	80.6	97.2	58.3	63.7	87.9	58.2	62.2	107.9	66.3	72.0	100.9	94.2	95.1	93.5	58.2	63.0
0320	Concrete Reinforcing	94.6	93.3	94.0	96.5	55.2	75.5	96.5	55.1	75.4	95.9	62.7	79.0	91.8	97.2	94.6	97.8	55.1	76.1
0330	Cast-in-Place Concrete	111.4	54.5	88.0	100.3	55.8	82.0	92.9	53.1	76.6	129.9	53.4	98.5	95.4	97.4	96.3	100.3	53.1	80.9
03	CONCRETE	106.6	72.9	90.1	110.4	58.1	84.7	92.4	57.1	75.0	122.6	62.2	92.9	98.4	96.2	97.3	112.5	57.1	85.2
04	MASONRY	106.7	60.3	77.8	118.0	59.9	81.8	107.9	58.6	77.2	105.3	64.5	79.9	107.8	99.4	102.6	116.6	58.6	80.4
05	METALS	95.3	92.8	94.5	95.6	77.7	90.1	95.4	76.7	89.6	95.3	80.5	90.8	102.9	100.1	102.1	95.9	76.7	90.0
06	WOOD, PLASTICS & COMPOSITES	108.7	83.5	94.6	100.6	57.9	76.7	90.9	57.9	72.4	113.7	68.0	88.1	104.4	94.5	98.8	96.4	57.9	74.8
07	THERMAL & MOISTURE PROTECTION	94.0	80.8	88.6	99.4	61.6	83.9	98.0	60.6	82.7	99.1	80.5	91.5	93.7	98.0	95.5	99.5	60.6	83.6
08	OPENINGS	96.3	78.2	92.1	103.1	51.5	91.1	103.0	51.5	91.0	100.7	58.8	90.9	97.7	89.0	95.7	103.1	51.5	91.1
0920	Plaster & Gypsum Board	99.7	82.9	88.4	97.3	56.5	69.7	93.0	56.5	68.3	106.9	67.0	79.9	92.9	94.2	93.8	94.8	56.5	68.9
0950, 0980	Ceilings & Acoustic Treatment	86.3	82.9	84.1	84.0	56.5	65.9	84.0	56.5	65.9	84.0	67.0	72.8	86.3	94.2	91.5	84.0	56.5	65.9
0960	Flooring	113.7	39.4	92.5	102.0	39.2	84.0	96.8	39.2	80.3	107.3	39.2	87.8	91.4	98.7	93.5	99.8	39.2	82.5
0970, 0990	Wall Finishes & Painting/Coating	101.8	38.9	63.8	104.7	38.9	65.0	104.7	38.9	65.0	104.7	38.9	65.0	108.5	67.7	83.9	104.7	38.9	65.0
09	FINISHES	99.6	66.9	81.6	96.4	52.7	72.3	92.4	52.7	70.5	98.9	58.7	76.8	94.3	91.0	92.5	95.7	52.7	72.0
COVERS	DIVS. 10 - 14, 25, 28, 41, 43, 44, 46	100.0	46.2	89.1	100.0	41.4	88.2	100.0	41.4	88.2	100.0	42.5	88.4	100.0	61.9	92.3	100.0	41.4	88.2
21, 22, 23	FIRE SUPPRESSION, PLUMBING & HVAC	95.2	68.3	84.3	95.2	69.0	84.6	95.2	69.0	84.6	95.2	70.1	85.1	99.9	96.4	98.5	95.2	67.2	83.9
26, 27, 3370	ELECTRICAL, COMMUNICATIONS & UTIL.	93.2	71.4	81.6	98.9	67.7	82.4	93.9	67.7	80.0	96.1	71.4	83.0	98.7	97.2	97.9	96.8	73.8	84.6
MF2014	WEIGHTED AVERAGE	97.9	73.0	87.0	100.5	65.4	85.2	96.5	65.0	82.8	100.9	69.2	87.1	99.5	94.8	97.5	100.4	65.5	85.2

KANSAS / KENTUCKY

DIVISION		SALINA 674			TOPEKA 664 - 666			WICHITA 670 - 672			ASHLAND 411 - 412			BOWLING GREEN 421 - 422			CAMPTON 413 - 414		
		MAT.	INST.	TOTAL	MAT.	INST.	TOTAL	MAT.	INST.	TOTAL	MAT.	INST.	TOTAL	MAT.	INST.	TOTAL	MAT.	INST.	TOTAL
015433	CONTRACTOR EQUIPMENT		103.4	103.4		101.6	101.6		103.4	103.4		97.3	97.3		94.5	94.5		101.1	101.1
0241, 31 - 34	SITE & INFRASTRUCTURE, DEMOLITION	105.2	95.2	98.1	98.2	91.5	93.4	101.8	93.8	96.1	112.3	86.3	93.8	78.6	97.1	91.8	87.0	98.3	95.0
0310	Concrete Forming & Accessories	89.7	54.9	59.7	98.4	40.9	48.8	95.3	51.0	57.1	87.9	105.2	102.8	85.7	83.0	83.4	89.4	82.6	83.6
0320	Concrete Reinforcing	95.9	59.4	77.3	94.9	99.5	97.2	95.3	77.3	86.1	88.8	108.7	98.9	86.2	91.3	88.8	87.0	101.8	94.5
0330	Cast-in-Place Concrete	112.3	53.7	88.2	100.4	45.5	77.9	105.3	51.7	83.3	91.1	101.1	95.2	90.4	95.5	92.5	100.8	73.6	89.6
03	CONCRETE	107.3	56.7	82.5	99.4	55.1	77.7	101.6	57.4	79.9	98.6	104.8	101.6	96.8	88.9	92.9	100.7	83.2	92.1
04	MASONRY	134.8	59.9	88.2	102.5	55.9	73.4	106.3	50.2	71.4	97.8	103.7	101.5	99.7	80.2	87.5	96.9	69.0	79.5
05	METALS	97.3	80.6	92.1	99.4	96.9	98.6	99.4	84.2	94.7	92.1	110.3	97.7	93.6	87.3	91.7	92.9	92.0	92.6
06	WOOD, PLASTICS & COMPOSITES	92.4	52.1	69.8	97.1	37.3	63.6	95.1	51.3	70.6	76.0	105.2	92.3	88.6	83.1	85.5	87.1	89.0	88.2
07	THERMAL & MOISTURE PROTECTION	98.6	61.4	83.3	96.7	67.4	84.7	97.2	56.1	80.4	91.0	97.5	93.7	86.6	80.2	84.0	99.7	67.5	86.5
08	OPENINGS	103.1	48.5	90.4	100.8	55.4	90.2	105.3	56.2	93.9	97.9	99.6	98.3	98.7	79.3	94.2	100.2	87.6	97.3
0920	Plaster & Gypsum Board	93.0	50.6	64.3	96.8	35.3	55.2	91.8	49.8	63.4	60.9	105.4	91.0	87.4	82.9	84.3	87.4	88.1	87.9
0950, 0980	Ceilings & Acoustic Treatment	84.0	50.6	62.0	92.0	35.3	54.7	88.8	49.8	63.2	80.8	105.4	97.0	85.5	82.9	83.8	85.5	88.1	87.2
0960	Flooring	98.2	61.9	87.8	103.4	73.7	94.9	105.4	63.0	93.3	79.4	102.4	86.0	95.8	86.0	93.0	98.0	41.2	81.7
0970, 0990	Wall Finishes & Painting/Coating	104.7	38.9	65.0	101.8	52.5	72.1	103.6	48.0	70.0	103.1	101.2	102.0	101.2	72.7	84.0	101.2	65.4	79.6
09	FINISHES	93.6	53.6	71.6	98.5	45.9	69.5	97.7	52.3	72.7	80.9	104.8	94.1	91.8	82.8	86.8	92.6	74.0	82.4
COVERS	DIVS. 10 - 14, 25, 28, 41, 43, 44, 46	100.0	86.0	97.2	100.0	47.2	89.3	100.0	83.0	96.6	100.0	94.1	98.8	100.0	62.3	92.4	100.0	55.8	91.1
21, 22, 23	FIRE SUPPRESSION, PLUMBING & HVAC	100.0	69.1	87.5	100.0	69.1	87.5	99.8	64.8	85.7	95.0	92.5	94.0	99.9	82.7	93.0	95.2	79.8	89.0
26, 27, 3370	ELECTRICAL, COMMUNICATIONS & UTIL.	96.6	73.8	84.5	97.5	71.8	83.9	100.9	73.8	86.5	91.4	97.3	94.5	94.0	81.5	87.4	91.4	66.1	78.0
MF2014	WEIGHTED AVERAGE	101.5	67.6	86.7	99.5	66.0	84.9	100.7	66.0	85.6	94.6	99.3	96.6	96.2	83.9	90.8	95.8	78.6	88.3

KENTUCKY

DIVISION		CORBIN 407 - 409			COVINGTON 410			ELIZABETHTOWN 427			FRANKFORT 406			HAZARD 417 - 418			HENDERSON 424		
		MAT.	INST.	TOTAL	MAT.	INST.	TOTAL	MAT.	INST.	TOTAL	MAT.	INST.	TOTAL	MAT.	INST.	TOTAL	MAT.	INST.	TOTAL
015433	CONTRACTOR EQUIPMENT		101.1	101.1		104.4	104.4		94.5	94.5		101.1	101.1		101.1	101.1		116.0	116.0
0241, 31 - 34	SITE & INFRASTRUCTURE, DEMOLITION	87.9	98.1	95.1	82.6	112.3	103.7	73.1	97.2	90.2	88.9	99.2	96.2	84.8	99.6	95.3	81.1	124.0	111.6
0310	Concrete Forming & Accessories	84.5	70.6	72.5	85.0	82.0	82.4	80.7	80.5	80.5	94.2	71.8	74.8	86.1	83.5	83.8	92.5	81.5	83.0
0320	Concrete Reinforcing	86.0	69.8	77.7	85.7	91.1	88.5	86.6	90.8	88.8	97.7	90.8	94.2	87.4	107.6	97.7	86.3	84.8	85.6
0330	Cast-in-Place Concrete	93.6	58.6	79.2	96.5	97.6	96.9	81.8	73.0	78.2	89.4	69.1	81.1	97.0	82.5	91.0	79.9	90.6	84.3
03	CONCRETE	93.3	66.8	80.3	98.9	89.7	94.4	88.6	80.0	84.4	93.8	74.8	84.5	97.3	87.7	92.6	93.9	85.4	89.7
04	MASONRY	94.4	61.9	74.1	111.8	101.5	105.4	83.5	75.4	78.5	94.6	79.4	85.1	92.9	71.5	80.3	103.6	92.9	96.9
05	METALS	90.4	78.2	86.6	91.0	96.9	92.8	92.8	87.2	91.1	93.0	87.8	91.4	92.9	94.2	93.3	84.4	86.9	85.1
06	WOOD, PLASTICS & COMPOSITES	76.9	75.9	76.4	85.0	71.7	77.6	83.7	82.0	82.7	91.4	68.7	78.7	84.1	89.0	86.8	91.3	79.0	84.4
07	THERMAL & MOISTURE PROTECTION	99.3	63.8	84.7	100.3	92.3	97.0	86.3	76.9	82.5	101.0	75.1	90.4	99.7	69.5	87.3	99.0	91.3	95.9
08	OPENINGS	97.9	65.6	90.4	102.4	81.7	97.6	98.7	83.2	95.1	104.5	76.7	98.0	100.6	79.9	95.8	96.8	79.9	92.9
0920	Plaster & Gypsum Board	92.3	74.6	80.3	68.8	71.4	70.6	86.7	81.7	83.3	96.5	67.2	76.7	86.7	88.1	87.6	90.2	77.8	81.8
0950, 0980	Ceilings & Acoustic Treatment	80.4	74.6	76.6	88.8	71.4	77.4	85.5	81.7	83.0	89.6	67.2	74.8	85.5	88.1	87.2	79.6	77.8	78.4
0960	Flooring	97.2	41.2	81.2	71.8	88.6	76.6	93.4	86.0	91.3	106.0	54.4	91.2	96.4	42.0	80.9	99.5	86.0	95.6
0970, 0990	Wall Finishes & Painting/Coating	107.2	49.4	72.3	96.2	83.2	88.3	101.2	76.2	86.1	109.1	79.8	91.4	101.2	65.4	79.6	101.1	96.0	98.1
09	FINISHES	90.7	63.6	75.8	82.7	82.0	82.4	90.6	80.7	85.1	97.6	68.8	81.7	91.9	74.8	82.5	91.4	83.8	87.2
COVERS	DIVS. 10 - 14, 25, 28, 41, 43, 44, 46	100.0	48.4	89.6	100.0	98.7	99.7	100.0	84.2	96.8	100.0	65.5	93.0	100.0	56.6	91.2	100.0	65.4	93.0
21, 22, 23	FIRE SUPPRESSION, PLUMBING & HVAC	95.2	71.4	85.6	95.9	92.4	94.5	95.5	80.1	89.3	100.1	82.3	92.9	95.2	80.7	89.4	95.5	77.6	88.3
26, 27, 3370	ELECTRICAL, COMMUNICATIONS & UTIL.	91.4	80.1	85.4	95.2	79.3	86.8	91.2	80.2	85.4	101.3	87.1	93.8	91.4	58.4	73.9	93.4	80.3	86.5
MF2014	WEIGHTED AVERAGE	94.0	71.8	84.3	95.9	91.4	93.9	92.8	82.0	88.1	98.1	80.7	90.5	95.2	78.7	88.0	93.7	86.1	90.4

For customer support on your Plumbing Cost Data, call 877.709.6968.

615

KENTUCKY

DIVISION		LEXINGTON 403 - 405			LOUISVILLE 400 - 402			OWENSBORO 423			PADUCAH 420			PIKEVILLE 415 - 416			SOMERSET 425 - 426		
		MAT.	INST.	TOTAL	MAT.	INST.	TOTAL	MAT.	INST.	TOTAL	MAT.	INST.	TOTAL	MAT.	INST.	TOTAL	MAT.	INST.	TOTAL
015433	CONTRACTOR EQUIPMENT		101.1	101.1		94.5	94.5		116.0	116.0		116.0	116.0		97.3	97.3		101.1	101.1
0241, 31 - 34	SITE & INFRASTRUCTURE, DEMOLITION	90.1	100.6	97.5	84.8	97.3	93.7	91.0	124.5	114.8	83.7	123.4	111.9	123.2	85.3	96.3	77.9	98.8	92.8
0310	Concrete Forming & Accessories	96.4	73.0	76.2	95.8	82.0	83.9	90.9	85.1	85.9	88.9	79.9	81.1	96.8	88.9	90.0	87.0	75.9	77.4
0320	Concrete Reinforcing	94.4	92.2	93.3	96.8	92.6	94.7	86.3	91.8	89.1	86.8	86.2	86.5	89.3	108.4	99.0	86.6	90.8	88.7
0330	Cast-in-Place Concrete	95.8	92.7	94.5	96.4	73.9	87.2	93.2	91.1	92.4	85.2	86.5	85.7	100.1	96.7	98.7	79.9	98.3	87.5
03	CONCRETE	96.5	83.7	90.2	97.1	81.3	89.4	106.3	88.5	97.5	98.6	83.5	91.2	112.7	96.0	104.5	83.6	86.6	85.1
04	MASONRY	93.7	73.3	81.0	91.6	78.8	83.6	95.9	91.5	93.2	98.8	88.4	92.3	95.3	96.8	96.2	89.8	76.7	81.6
05	METALS	92.8	89.0	91.6	93.9	88.2	92.1	85.8	90.6	87.3	82.9	86.7	84.1	92.0	108.8	97.2	92.8	87.6	91.2
06	WOOD, PLASTICS & COMPOSITES	92.0	68.7	79.0	90.9	83.1	86.5	89.2	83.3	85.9	86.8	79.1	82.5	85.5	88.3	87.1	84.6	75.9	79.7
07	THERMAL & MOISTURE PROTECTION	99.5	84.1	93.2	96.3	78.2	88.9	99.4	92.6	96.7	99.1	78.8	90.8	91.7	79.5	86.7	99.0	71.2	87.6
08	OPENINGS	98.1	73.9	92.5	93.5	84.3	91.4	96.8	85.7	94.2	96.0	77.5	91.7	98.6	87.8	96.1	99.5	77.7	94.4
0920	Plaster & Gypsum Board	101.9	67.2	78.4	98.0	82.9	87.8	88.7	82.2	84.3	87.7	77.9	81.1	63.8	88.1	80.2	86.7	74.6	78.5
0950, 0980	Ceilings & Acoustic Treatment	84.5	67.2	73.1	88.9	82.9	84.9	79.6	82.2	81.3	79.6	77.9	78.5	80.8	88.1	85.6	85.5	74.6	78.3
0960	Flooring	101.9	70.3	92.9	98.6	86.0	95.0	98.9	86.0	95.2	97.8	59.4	86.8	83.4	102.4	88.8	96.7	41.2	80.8
0970, 0990	Wall Finishes & Painting/Coating	107.2	81.0	91.4	104.7	76.2	87.5	101.2	96.0	98.1	101.2	79.7	88.2	103.1	81.2	89.9	101.2	76.2	86.1
09	FINISHES	94.4	72.3	82.2	94.9	82.3	88.0	91.5	86.0	88.5	90.7	76.1	82.7	83.3	90.9	87.5	91.2	69.4	79.2
COVERS	DIVS. 10 - 14, 25, 28, 41, 43, 44, 46	100.0	86.5	97.3	100.0	84.9	96.9	100.0	102.9	100.6	100.0	61.8	92.3	100.0	56.3	91.2	100.0	58.5	91.6
21, 22, 23	FIRE SUPPRESSION, PLUMBING & HVAC	100.0	79.5	91.7	100.0	80.9	92.3	99.9	78.6	91.4	99.5	81.1	89.7	95.0	87.2	91.8	95.5	77.4	88.2
26, 27, 3370	ELECTRICAL, COMMUNICATIONS & UTIL.	94.2	81.5	87.5	99.4	81.5	89.9	93.5	81.4	87.1	95.8	79.9	87.4	94.4	73.4	83.3	91.8	80.2	85.6
MF2014	WEIGHTED AVERAGE	96.6	81.4	90.0	96.6	83.3	90.8	96.3	88.9	93.1	93.9	84.5	89.8	96.9	88.6	93.3	93.2	79.9	87.4

LOUISIANA

DIVISION		ALEXANDRIA 713 - 714			BATON ROUGE 707 - 708			HAMMOND 704			LAFAYETTE 705			LAKE CHARLES 706			MONROE 712		
		MAT.	INST.	TOTAL	MAT.	INST.	TOTAL	MAT.	INST.	TOTAL	MAT.	INST.	TOTAL	MAT.	INST.	TOTAL	MAT.	INST.	TOTAL
015433	CONTRACTOR EQUIPMENT		89.9	89.9		89.5	89.5		90.1	90.1		90.1	90.1		89.5	89.5		89.9	89.9
0241, 31 - 34	SITE & INFRASTRUCTURE, DEMOLITION	101.2	87.1	91.1	104.0	87.2	92.0	102.1	87.8	91.9	103.2	88.1	92.5	103.9	87.0	91.9	101.2	87.0	91.1
0310	Concrete Forming & Accessories	82.9	43.4	48.9	97.6	60.8	65.8	78.4	46.2	50.7	96.0	54.1	59.9	96.8	56.7	62.2	82.4	43.3	48.7
0320	Concrete Reinforcing	94.9	54.3	74.2	101.1	58.9	79.6	97.9	59.3	78.3	99.3	58.9	78.7	99.3	59.1	78.9	93.8	54.0	73.6
0330	Cast-in-Place Concrete	96.2	49.9	77.2	97.2	58.8	81.4	95.8	44.0	74.5	95.3	47.9	75.8	100.3	67.3	86.8	96.2	56.8	80.0
03	CONCRETE	95.5	48.8	72.6	98.9	60.4	80.0	95.2	48.9	72.5	96.4	53.7	75.4	98.8	61.5	80.5	95.3	51.1	73.6
04	MASONRY	121.3	53.4	79.0	94.9	54.1	69.5	95.7	51.2	68.0	95.7	52.4	68.7	95.0	58.8	72.5	115.6	47.9	73.4
05	METALS	95.4	70.4	87.7	100.6	72.1	91.8	95.6	71.2	88.1	94.8	72.1	87.8	94.8	72.6	87.9	95.4	69.9	87.6
06	WOOD, PLASTICS & COMPOSITES	90.5	41.7	63.2	99.0	63.8	79.3	84.2	47.0	63.3	105.7	55.7	77.7	103.8	57.7	78.0	89.8	42.2	63.1
07	THERMAL & MOISTURE PROTECTION	95.6	62.9	82.2	96.2	65.4	83.5	96.5	61.4	82.1	97.1	63.8	83.4	96.1	66.8	84.1	95.6	60.8	81.3
08	OPENINGS	109.1	44.7	94.1	97.9	59.1	88.9	93.0	51.7	83.4	96.7	53.2	86.6	96.7	54.9	87.0	109.1	47.6	94.8
0920	Plaster & Gypsum Board	84.2	40.4	54.6	96.7	63.0	74.0	97.5	45.7	62.5	105.7	54.7	71.2	105.7	56.8	72.6	83.8	40.9	54.8
0950, 0980	Ceilings & Acoustic Treatment	84.1	40.4	55.4	91.3	63.0	72.7	94.1	45.7	62.3	92.4	54.7	67.6	93.3	56.8	69.3	84.1	40.9	55.7
0960	Flooring	98.5	65.7	89.1	101.8	65.7	91.5	96.3	65.7	87.5	104.9	65.7	93.7	104.9	72.9	95.7	98.1	57.1	86.4
0970, 0990	Wall Finishes & Painting/Coating	104.4	64.4	80.3	94.7	46.3	65.5	99.4	48.3	68.6	99.4	57.9	74.4	99.4	50.0	69.6	104.4	48.1	70.4
09	FINISHES	89.8	48.9	67.2	94.7	60.2	75.7	94.8	49.7	70.0	98.1	56.3	75.0	98.3	58.6	76.4	89.6	45.5	65.3
COVERS	DIVS. 10 - 14, 25, 28, 41, 43, 44, 46	100.0	50.0	89.9	100.0	80.8	96.1	100.0	45.4	89.0	100.0	79.4	95.8	100.0	80.3	96.0	100.0	45.7	89.0
21, 22, 23	FIRE SUPPRESSION, PLUMBING & HVAC	100.1	55.3	82.1	100.0	59.5	83.7	95.3	42.8	74.1	100.1	61.1	84.4	100.1	62.0	84.7	100.1	54.3	81.6
26, 27, 3370	ELECTRICAL, COMMUNICATIONS & UTIL.	95.7	57.9	75.7	100.8	61.6	80.1	96.7	56.0	75.2	97.9	67.4	81.8	97.4	65.0	80.3	97.7	59.6	77.6
MF2014	WEIGHTED AVERAGE	99.4	57.2	81.0	99.1	63.7	83.7	95.8	54.5	77.8	97.9	63.0	82.7	98.1	65.0	83.7	99.3	56.5	80.6

LOUISIANA / MAINE

DIVISION		NEW ORLEANS 700 - 701			SHREVEPORT 710 - 711			THIBODAUX 703			AUGUSTA 043			BANGOR 044			BATH 045		
		MAT.	INST.	TOTAL	MAT.	INST.	TOTAL	MAT.	INST.	TOTAL	MAT.	INST.	TOTAL	MAT.	INST.	TOTAL	MAT.	INST.	TOTAL
015433	CONTRACTOR EQUIPMENT		90.4	90.4		89.9	89.9		90.1	90.1		100.9	100.9		100.9	100.9		100.9	100.9
0241, 31 - 34	SITE & INFRASTRUCTURE, DEMOLITION	103.4	91.1	94.7	105.5	87.1	92.4	104.4	88.1	92.8	91.6	101.1	98.4	93.3	101.3	99.0	91.0	101.1	98.2
0310	Concrete Forming & Accessories	95.1	65.5	69.5	98.5	44.2	51.7	90.4	64.6	68.1	95.9	94.8	94.9	93.1	96.6	96.1	88.9	94.9	94.1
0320	Concrete Reinforcing	100.2	60.1	79.8	98.3	54.4	76.0	97.9	59.1	78.2	96.1	106.7	101.5	87.7	108.1	98.1	86.8	107.9	97.5
0330	Cast-in-Place Concrete	98.0	69.4	86.3	101.1	53.6	81.6	102.9	51.6	81.8	101.5	58.8	84.0	82.2	110.6	93.9	82.2	58.9	72.7
03	CONCRETE	99.0	66.3	82.9	100.4	50.5	75.9	100.5	59.7	80.4	103.1	84.1	93.8	94.1	102.9	98.4	94.1	84.4	89.4
04	MASONRY	97.7	59.8	74.1	107.1	50.5	71.8	120.5	48.8	75.8	107.4	63.6	80.1	118.0	103.9	109.2	125.6	86.1	101.0
05	METALS	110.3	73.3	98.9	99.5	70.6	90.6	95.6	71.6	88.2	106.0	86.6	100.1	96.1	89.3	94.0	94.6	88.4	92.7
06	WOOD, PLASTICS & COMPOSITES	97.9	66.3	80.2	101.7	42.7	68.7	92.3	69.7	79.6	91.1	105.3	99.0	90.1	96.4	93.6	84.6	105.3	96.2
07	THERMAL & MOISTURE PROTECTION	94.9	68.8	84.2	95.3	59.8	80.8	96.4	61.8	82.2	100.1	65.0	85.7	97.8	84.4	92.3	97.8	71.8	87.1
08	OPENINGS	97.8	65.4	90.2	108.0	45.3	93.4	97.6	64.1	89.8	106.3	87.9	102.0	102.8	83.0	98.2	102.8	87.8	99.3
0920	Plaster & Gypsum Board	101.4	65.7	77.2	94.0	41.5	58.5	99.7	69.2	79.0	100.6	104.8	103.5	99.4	95.7	96.9	95.5	104.8	101.8
0950, 0980	Ceilings & Acoustic Treatment	98.2	65.7	76.8	88.1	41.5	57.4	94.1	69.2	77.7	97.4	104.8	102.3	86.2	95.7	92.5	84.3	104.8	97.8
0960	Flooring	105.7	65.7	94.2	101.9	61.3	90.3	102.3	44.1	85.7	103.9	55.3	90.0	97.1	114.3	102.0	95.3	55.3	83.9
0970, 0990	Wall Finishes & Painting/Coating	101.9	64.1	79.1	100.4	44.6	66.7	100.7	49.6	69.9	112.7	63.3	82.9	105.0	44.1	68.3	105.0	44.1	68.3
09	FINISHES	101.2	65.1	81.3	94.5	46.3	67.9	97.2	59.8	76.6	100.2	85.8	92.3	95.7	94.8	95.2	94.2	83.7	88.4
COVERS	DIVS. 10 - 14, 25, 28, 41, 43, 44, 46	100.0	83.1	96.6	100.0	78.1	95.6	100.0	81.4	96.2	100.0	100.5	100.1	100.0	109.8	102.0	100.0	100.5	100.1
21, 22, 23	FIRE SUPPRESSION, PLUMBING & HVAC	100.0	65.1	85.9	99.9	58.6	83.3	95.3	61.7	81.8	99.9	64.2	85.5	100.1	74.3	89.7	95.4	64.3	82.8
26, 27, 3370	ELECTRICAL, COMMUNICATIONS & UTIL.	102.1	73.3	86.9	103.5	66.6	84.0	95.3	73.3	83.7	100.0	80.1	89.9	97.2	75.8	85.9	95.2	80.1	87.2
MF2014	WEIGHTED AVERAGE	101.4	69.5	87.5	101.0	59.6	83.0	98.2	65.4	83.9	102.1	79.5	92.3	99.0	89.5	94.9	97.5	81.9	90.7

City Cost Indexes

MAINE

DIVISION		HOULTON 047 MAT.	INST.	TOTAL	KITTERY 039 MAT.	INST.	TOTAL	LEWISTON 042 MAT.	INST.	TOTAL	MACHIAS 046 MAT.	INST.	TOTAL	PORTLAND 040-041 MAT.	INST.	TOTAL	ROCKLAND 048 MAT.	INST.	TOTAL
015433	CONTRACTOR EQUIPMENT		100.9	100.9		100.9	100.9		100.9	100.9		100.9	100.9		100.9	100.9		100.9	100.9
0241, 31 - 34	SITE & INFRASTRUCTURE, DEMOLITION	93.0	102.2	99.5	84.8	101.2	96.5	90.9	101.3	98.3	92.3	102.2	99.4	90.1	101.3	98.1	88.7	102.2	98.3
0310	Concrete Forming & Accessories	96.7	101.6	101.0	89.2	95.5	94.6	98.5	96.7	96.9	93.9	101.6	100.5	98.5	96.7	96.9	94.9	101.6	100.7
0320	Concrete Reinforcing	87.7	106.5	97.2	84.1	107.9	96.2	107.8	108.1	107.9	87.7	106.5	97.2	102.2	108.1	105.2	87.7	106.5	97.2
0330	Cast-in-Place Concrete	82.3	71.7	77.9	81.2	59.8	72.4	83.9	110.7	94.9	82.2	71.1	77.7	96.7	110.7	102.4	83.9	71.1	78.7
03	CONCRETE	95.1	91.5	93.4	89.7	85.0	87.4	94.7	102.9	98.7	94.6	91.3	93.0	101.9	102.9	102.4	92.0	91.3	91.6
04	MASONRY	101.3	76.3	85.7	113.3	87.1	97.0	101.9	103.9	103.2	101.3	76.3	85.7	112.8	103.9	107.2	95.2	76.3	83.4
05	METALS	94.8	86.6	92.3	85.5	88.6	86.4	99.5	89.3	96.4	94.8	86.5	92.3	107.9	89.3	102.2	94.7	86.5	92.2
06	WOOD, PLASTICS & COMPOSITES	94.0	105.3	100.3	89.2	105.3	98.2	96.0	96.4	96.2	91.0	105.3	99.0	93.7	96.4	95.2	91.9	105.3	99.4
07	THERMAL & MOISTURE PROTECTION	97.9	72.6	87.5	99.2	71.1	87.7	97.6	84.4	92.2	97.8	71.6	87.1	100.1	84.4	93.7	97.5	71.7	86.9
08	OPENINGS	102.9	86.1	99.0	102.0	87.8	98.7	106.2	83.0	100.8	102.9	86.1	99.0	104.5	83.0	99.5	102.8	86.1	98.9
0920	Plaster & Gypsum Board	101.6	104.8	103.8	100.0	104.8	103.3	104.9	95.7	98.7	100.1	104.8	103.3	102.5	95.7	97.9	100.1	104.8	103.3
0950, 0980	Ceilings & Acoustic Treatment	84.3	104.8	97.8	94.2	104.8	101.2	95.4	95.7	95.6	84.3	104.8	97.8	99.1	95.7	96.9	84.3	104.8	97.8
0960	Flooring	98.3	51.7	85.0	99.8	57.7	87.7	100.0	114.3	104.1	97.5	51.7	84.4	98.6	114.3	103.1	97.9	51.7	84.7
0970, 0990	Wall Finishes & Painting/Coating	105.0	137.1	124.4	96.4	38.4	61.4	105.0	44.1	68.3	105.0	137.1	124.4	102.0	44.1	67.1	105.0	137.1	124.4
09	FINISHES	95.9	97.8	97.0	96.5	83.7	89.5	98.9	94.8	96.6	95.4	97.8	96.7	98.8	94.8	96.6	95.2	97.8	96.6
COVERS	DIVS. 10 - 14, 25, 28, 41, 43, 44, 46	100.0	106.4	101.3	100.0	100.8	100.2	100.0	109.8	102.0	100.0	106.4	101.3	100.0	109.8	102.0	100.0	106.4	101.3
21, 22, 23	FIRE SUPPRESSION, PLUMBING & HVAC	95.4	73.2	86.4	95.3	76.6	87.8	100.1	74.3	89.7	95.4	73.2	86.4	100.0	74.3	89.6	95.4	73.2	86.4
26, 27, 3370	ELECTRICAL, COMMUNICATIONS & UTIL.	99.2	80.1	89.1	97.0	80.1	88.1	99.3	80.1	89.1	99.2	80.1	89.1	98.9	80.1	89.0	99.1	80.1	89.1
MF2014	WEIGHTED AVERAGE	97.2	85.6	92.1	95.2	84.7	90.6	99.6	90.1	95.5	97.1	88.7	92.0	102.0	90.1	96.8	96.3	85.5	91.6

MAINE / MARYLAND

DIVISION		WATERVILLE 049 MAT.	INST.	TOTAL	ANNAPOLIS 214 MAT.	INST.	TOTAL	BALTIMORE 210-212 MAT.	INST.	TOTAL	COLLEGE PARK 207-208 MAT.	INST.	TOTAL	CUMBERLAND 215 MAT.	INST.	TOTAL	EASTON 216 MAT.	INST.	TOTAL
015433	CONTRACTOR EQUIPMENT		100.9	100.9		99.8	99.8		103.4	103.4		105.5	105.5		99.8	99.8		99.8	99.8
0241, 31 - 34	SITE & INFRASTRUCTURE, DEMOLITION	92.8	101.1	98.7	103.0	91.5	94.9	101.5	95.6	97.3	100.5	94.5	96.2	94.3	91.8	92.5	101.4	88.6	92.3
0310	Concrete Forming & Accessories	88.4	94.8	93.9	97.2	73.6	76.8	101.4	71.1	75.3	84.3	70.2	72.2	91.4	81.0	82.4	89.2	70.2	72.8
0320	Concrete Reinforcing	87.7	106.7	97.3	100.3	80.7	90.3	105.6	80.7	92.9	105.4	78.1	91.5	86.6	72.3	79.3	86.0	79.0	82.4
0330	Cast-in-Place Concrete	82.3	58.8	72.6	110.0	75.7	95.9	107.5	76.4	94.7	116.5	77.5	100.4	91.7	85.6	89.2	101.8	48.1	79.8
03	CONCRETE	95.6	84.1	90.0	104.8	76.6	90.9	104.8	75.8	90.6	107.5	75.5	91.8	90.2	81.8	86.1	98.0	65.2	81.9
04	MASONRY	112.2	63.6	81.9	103.5	71.6	83.6	100.1	71.7	82.4	111.8	69.0	85.1	99.1	83.8	89.5	113.4	42.9	69.4
05	METALS	94.8	86.6	92.3	104.0	93.4	100.8	100.6	93.9	98.5	86.8	97.4	90.0	98.5	90.5	96.1	98.8	87.2	95.2
06	WOOD, PLASTICS & COMPOSITES	84.0	105.3	95.9	92.9	75.4	83.1	99.6	72.1	84.2	80.0	70.0	74.4	84.9	80.1	82.2	82.7	77.4	79.7
07	THERMAL & MOISTURE PROTECTION	97.9	65.0	84.4	100.2	78.8	91.4	101.9	78.6	92.3	101.9	78.7	92.4	100.2	78.9	91.5	100.3	59.0	83.4
08	OPENINGS	102.9	87.9	99.4	105.8	80.3	99.9	99.0	78.5	94.2	94.3	74.8	89.7	99.7	78.0	94.7	98.0	71.4	91.8
0920	Plaster & Gypsum Board	95.5	104.8	101.8	100.5	74.9	83.2	102.3	71.3	81.4	99.0	69.0	78.8	102.3	79.8	87.1	102.3	77.0	85.2
0950, 0980	Ceilings & Acoustic Treatment	84.3	104.8	97.8	91.3	74.9	80.6	94.0	71.3	79.1	102.2	69.0	80.4	93.3	79.8	84.4	93.3	77.0	82.6
0960	Flooring	95.0	55.3	83.6	99.8	77.2	93.3	100.7	77.2	94.0	97.8	79.7	92.6	95.5	90.8	94.2	94.7	51.7	82.4
0970, 0990	Wall Finishes & Painting/Coating	105.0	63.3	79.8	92.1	79.2	84.3	97.9	79.2	86.6	108.0	77.6	89.7	96.5	73.4	82.5	96.5	77.6	85.1
09	FINISHES	94.3	85.8	89.6	94.8	74.3	83.5	100.1	72.3	84.8	95.3	71.9	82.4	96.5	81.9	88.5	96.7	69.1	81.5
COVERS	DIVS. 10 - 14, 25, 28, 41, 43, 44, 46	100.0	100.5	100.1	100.0	86.2	97.2	100.0	86.2	97.2	100.0	85.5	97.1	100.0	90.5	98.1	100.0	75.3	95.0
21, 22, 23	FIRE SUPPRESSION, PLUMBING & HVAC	95.4	64.2	82.8	100.1	80.5	92.2	100.0	80.6	92.2	95.4	83.6	90.7	95.2	73.0	86.3	95.2	68.3	84.4
26, 27, 3370	ELECTRICAL, COMMUNICATIONS & UTIL.	99.2	80.1	89.1	99.5	91.9	95.5	102.0	91.9	96.6	101.1	99.7	100.4	98.3	81.8	89.6	97.8	65.2	80.5
MF2014	WEIGHTED AVERAGE	97.5	79.5	89.7	101.5	82.0	93.0	100.8	81.9	92.6	97.2	83.4	91.2	96.8	81.8	90.2	98.3	68.6	85.3

MARYLAND / MASSACHUSETTS

DIVISION		ELKTON 219 MAT.	INST.	TOTAL	HAGERSTOWN 217 MAT.	INST.	TOTAL	SALISBURY 218 MAT.	INST.	TOTAL	SILVER SPRING 209 MAT.	INST.	TOTAL	WALDORF 206 MAT.	INST.	TOTAL	BOSTON 020-022, 024 MAT.	INST.	TOTAL
015433	CONTRACTOR EQUIPMENT		99.8	99.8		99.8	99.8		99.8	99.8		98.1	98.1		98.1	98.1		106.5	106.5
0241, 31 - 34	SITE & INFRASTRUCTURE, DEMOLITION	88.4	89.5	89.2	92.6	92.3	92.4	101.3	88.6	92.3	89.0	87.6	88.0	95.4	87.3	89.7	98.9	109.1	106.2
0310	Concrete Forming & Accessories	95.4	81.0	82.9	90.3	75.1	77.2	103.9	51.7	58.9	92.8	70.4	73.5	100.2	68.7	73.0	102.7	143.8	138.1
0320	Concrete Reinforcing	86.0	105.3	95.8	86.6	72.3	79.3	86.0	63.5	74.6	104.1	77.9	90.8	104.8	78.0	91.1	108.9	155.6	132.8
0330	Cast-in-Place Concrete	82.5	75.2	79.5	87.4	85.7	86.7	101.8	46.7	79.2	119.3	79.3	102.8	133.6	76.5	110.2	104.0	151.2	123.4
03	CONCRETE	83.3	84.3	83.8	86.8	79.2	83.0	99.0	53.7	76.8	105.9	76.1	91.2	116.5	74.4	95.8	105.7	147.5	126.2
04	MASONRY	98.3	58.9	73.8	105.3	83.8	91.9	113.1	47.3	72.1	110.9	72.3	86.8	95.4	67.2	77.9	106.6	164.2	142.5
05	METALS	98.8	101.3	99.6	98.7	90.7	96.2	98.8	81.2	93.4	91.0	93.8	91.9	91.0	93.9	91.9	100.2	132.4	110.1
06	WOOD, PLASTICS & COMPOSITES	90.0	86.8	88.3	84.0	71.9	77.2	100.6	55.1	75.1	86.5	69.2	76.8	94.0	69.2	80.1	98.6	143.3	123.6
07	THERMAL & MOISTURE PROTECTION	99.9	73.5	89.1	100.0	76.8	90.5	100.6	63.7	85.4	105.4	84.9	97.0	105.8	82.7	96.4	103.4	151.9	123.3
08	OPENINGS	98.0	81.3	94.1	98.0	73.0	92.2	98.2	61.7	89.7	85.7	74.3	83.0	86.2	74.3	83.5	101.4	146.2	111.8
0920	Plaster & Gypsum Board	105.2	86.7	92.7	102.3	71.3	81.4	111.6	54.0	72.6	105.1	69.0	80.7	108.3	69.0	81.8	107.5	144.1	132.2
0950, 0980	Ceilings & Acoustic Treatment	93.3	86.7	89.0	94.3	71.3	79.2	93.3	54.0	67.5	112.2	69.0	83.8	112.2	69.0	83.8	106.1	144.1	131.1
0960	Flooring	97.0	58.5	86.0	95.2	90.8	93.9	100.7	64.7	90.4	104.1	79.7	97.2	107.7	79.0	99.5	97.9	182.7	122.2
0970, 0990	Wall Finishes & Painting/Coating	96.5	77.6	85.1	96.5	77.6	85.1	96.5	77.6	85.1	115.8	77.6	92.8	115.8	77.6	92.8	104.4	155.9	135.4
09	FINISHES	96.9	76.9	85.9	96.4	77.5	86.0	99.8	56.3	75.8	96.9	71.4	82.8	98.6	70.7	83.2	104.2	152.7	130.9
COVERS	DIVS. 10 - 14, 25, 28, 41, 43, 44, 46	100.0	59.9	91.9	100.0	89.6	97.9	100.0	38.7	87.6	100.0	84.7	96.9	100.0	83.0	96.6	100.0	119.4	103.9
21, 22, 23	FIRE SUPPRESSION, PLUMBING & HVAC	95.2	80.5	89.3	100.0	86.2	94.4	95.2	59.9	81.0	95.4	84.8	91.2	95.4	82.3	90.1	100.1	132.4	113.1
26, 27, 3370	ELECTRICAL, COMMUNICATIONS & UTIL.	99.6	91.9	95.5	98.1	81.8	89.5	96.5	67.1	81.0	98.4	99.7	99.1	95.7	99.7	97.8	100.8	135.4	119.1
MF2014	WEIGHTED AVERAGE	95.9	81.9	89.8	97.6	83.3	91.4	98.7	62.1	82.7	96.5	83.2	90.7	97.1	81.7	90.4	101.6	139.5	118.1

For customer support on your Plumbing Cost Data, call 877.709.6968.

617

		MASSACHUSETTS																
	DIVISION	BROCKTON			BUZZARDS BAY			FALL RIVER			FITCHBURG			FRAMINGHAM			GREENFIELD		
		023			025			027			014			017			013		
		MAT.	INST.	TOTAL	MAT.	INST.	TOTAL	MAT.	INST.	TOTAL	MAT.	INST.	TOTAL	MAT.	INST.	TOTAL	MAT.	INST.	TOTAL
015433	CONTRACTOR EQUIPMENT		102.7	102.7		102.7	102.7		103.7	103.7		100.9	100.9		101.9	101.9		100.9	100.9
0241, 31 - 34	SITE & INFRASTRUCTURE, DEMOLITION	94.5	105.7	102.4	84.8	105.6	99.6	93.5	105.8	102.2	86.3	105.6	100.0	83.0	105.3	98.9	90.0	104.3	100.1
0310	Concrete Forming & Accessories	101.6	138.7	133.6	99.2	138.5	133.1	101.6	138.8	133.6	93.8	130.8	125.7	101.2	138.8	133.6	92.1	113.2	110.3
0320	Concrete Reinforcing	106.4	155.6	131.4	85.3	160.1	123.4	106.4	160.2	133.7	83.9	151.4	118.3	83.9	155.6	120.4	87.3	124.4	106.2
0330	Cast-in-Place Concrete	97.7	151.5	119.8	81.2	151.5	110.1	94.5	152.0	118.1	85.9	150.5	112.4	85.9	148.3	111.5	88.3	131.2	105.9
03	CONCRETE	101.8	145.2	123.1	86.2	145.9	115.5	100.3	146.2	122.9	85.9	140.4	112.7	88.7	144.1	116.0	89.6	120.9	104.9
04	MASONRY	101.7	160.0	138.0	94.4	160.0	135.3	102.6	159.9	138.3	101.0	159.0	137.2	107.2	160.1	140.1	105.4	136.0	124.4
05	METALS	97.4	130.0	107.5	92.3	131.7	104.4	97.4	132.1	108.1	95.5	125.5	104.7	95.5	130.0	106.2	97.8	110.4	101.7
06	WOOD, PLASTICS & COMPOSITES	99.1	138.4	121.1	96.1	138.4	119.8	99.1	138.7	121.3	93.7	128.3	113.1	100.1	138.2	121.4	91.5	110.9	102.4
07	THERMAL & MOISTURE PROTECTION	100.9	148.7	120.5	100.1	147.0	119.4	100.8	146.3	119.5	99.2	141.5	116.5	99.3	149.1	119.7	99.2	121.9	108.5
08	OPENINGS	102.2	143.5	111.9	97.9	139.2	107.5	102.2	139.3	110.9	104.6	137.0	112.2	94.9	143.4	106.2	104.7	115.0	107.1
0920	Plaster & Gypsum Board	93.7	139.0	124.3	89.5	139.0	123.0	93.7	139.0	124.3	100.3	128.6	119.4	102.8	139.0	127.3	101.0	110.7	107.5
0950, 0980	Ceilings & Acoustic Treatment	102.2	139.0	126.4	88.2	139.0	121.6	102.2	139.0	126.4	89.0	128.6	115.0	89.0	139.0	121.9	97.4	110.7	106.1
0960	Flooring	98.6	182.7	122.7	96.7	182.7	121.3	97.7	182.7	122.0	98.1	182.7	122.3	99.7	182.7	123.5	97.3	154.0	113.5
0970, 0990	Wall Finishes & Painting/Coating	99.9	155.1	133.2	99.9	155.1	133.2	99.9	155.1	133.2	99.5	155.1	133.1	100.5	155.1	133.5	99.5	115.8	109.4
09	FINISHES	98.9	149.1	126.6	94.2	149.1	124.4	98.7	149.3	126.6	94.6	143.1	121.3	95.3	148.9	124.8	96.7	121.0	110.0
COVERS	DIVS. 10 - 14, 25, 28, 41, 43, 44, 46	100.0	117.9	103.6	100.0	117.9	103.6	100.0	118.5	103.7	100.0	109.0	101.8	100.0	117.5	103.5	100.0	104.6	100.9
21, 22, 23	FIRE SUPPRESSION, PLUMBING & HVAC	100.1	111.1	104.6	95.4	111.1	101.7	100.1	111.2	104.6	96.0	113.6	103.1	96.0	126.1	108.1	96.0	102.8	98.7
26, 27, 3370	ELECTRICAL, COMMUNICATIONS & UTIL.	99.3	100.1	99.7	96.3	100.1	98.3	99.2	100.1	99.7	100.1	105.1	102.8	96.6	128.9	113.7	100.1	98.9	99.5
MF2014	WEIGHTED AVERAGE	99.9	128.1	112.2	94.4	128.1	109.1	99.7	128.3	112.2	96.5	126.5	109.6	95.7	135.1	112.9	97.7	112.3	104.1

		MASSACHUSETTS																
	DIVISION	HYANNIS			LAWRENCE			LOWELL			NEW BEDFORD			PITTSFIELD			SPRINGFIELD		
		026			019			018			027			012			010 - 011		
		MAT.	INST.	TOTAL	MAT.	INST.	TOTAL	MAT.	INST.	TOTAL	MAT.	INST.	TOTAL	MAT.	INST.	TOTAL	MAT.	INST.	TOTAL
015433	CONTRACTOR EQUIPMENT		102.7	102.7		102.7	102.7		100.9	100.9		103.7	103.7		100.9	100.9		100.9	100.9
0241, 31 - 34	SITE & INFRASTRUCTURE, DEMOLITION	90.9	105.6	101.3	95.2	105.7	102.7	94.2	105.6	102.3	92.1	105.8	101.8	95.2	104.2	101.6	94.6	104.4	101.6
0310	Concrete Forming & Accessories	93.5	138.5	132.3	102.4	139.0	134.0	99.0	139.0	133.5	101.6	138.8	133.6	99.0	112.2	110.4	99.3	113.6	111.6
0320	Concrete Reinforcing	85.3	160.1	123.4	103.8	150.4	127.5	104.6	150.4	127.9	106.4	160.2	133.7	86.6	122.3	104.8	104.6	124.4	114.7
0330	Cast-in-Place Concrete	89.1	151.5	114.7	99.4	148.9	119.7	90.3	148.9	114.4	83.3	152.0	111.5	98.5	129.7	111.3	94.0	131.7	109.5
03	CONCRETE	92.4	145.9	118.7	103.7	143.5	123.2	95.1	143.3	118.8	95.0	146.2	120.2	96.2	119.5	107.6	96.9	121.2	108.8
04	MASONRY	100.6	160.0	137.6	112.3	160.7	142.5	100.2	160.0	137.5	100.6	159.9	137.6	100.9	133.2	121.0	100.5	136.9	123.2
05	METALS	93.7	131.7	105.4	98.2	128.3	107.5	98.2	126.0	106.7	97.4	132.1	108.1	98.0	109.5	101.5	100.9	110.4	103.8
06	WOOD, PLASTICS & COMPOSITES	89.3	138.4	116.6	100.6	138.4	121.8	99.8	138.4	121.4	99.1	138.7	121.3	99.8	110.9	106.0	99.8	110.9	106.0
07	THERMAL & MOISTURE PROTECTION	100.4	147.0	119.5	99.7	148.5	119.7	99.5	148.6	119.7	100.7	146.3	119.4	99.6	120.7	108.2	99.5	122.2	108.8
08	OPENINGS	98.5	139.2	108.0	99.0	142.1	109.0	105.9	142.1	114.4	102.2	139.3	110.9	105.9	114.4	107.9	105.9	115.0	108.0
0920	Plaster & Gypsum Board	85.2	139.0	121.6	105.6	139.0	128.2	105.6	139.0	128.2	93.7	139.0	124.3	105.6	110.7	109.0	105.6	110.7	109.0
0950, 0980	Ceilings & Acoustic Treatment	93.8	139.0	123.5	99.3	139.0	125.4	99.3	139.0	125.4	102.2	139.0	126.4	99.3	110.7	106.8	99.3	110.7	106.8
0960	Flooring	94.4	182.7	119.6	100.2	182.7	123.8	100.2	182.7	123.8	97.7	182.7	122.0	100.6	154.0	115.9	99.7	154.0	115.3
0970, 0990	Wall Finishes & Painting/Coating	99.9	155.1	133.2	99.6	155.1	133.1	99.5	155.1	133.1	99.9	155.1	133.2	99.5	115.8	109.4	100.9	115.8	109.9
09	FINISHES	94.4	149.1	124.6	98.8	149.1	126.5	98.7	149.1	126.5	98.6	149.3	126.5	98.8	120.3	110.6	98.7	121.2	111.1
COVERS	DIVS. 10 - 14, 25, 28, 41, 43, 44, 46	100.0	117.9	103.6	100.0	118.1	103.6	100.0	118.1	103.6	100.0	118.5	103.7	100.0	103.7	100.7	100.0	104.9	101.0
21, 22, 23	FIRE SUPPRESSION, PLUMBING & HVAC	100.1	111.1	104.6	100.1	124.8	110.0	100.1	125.7	110.4	100.1	111.2	104.6	100.1	101.4	100.6	100.1	103.3	101.4
26, 27, 3370	ELECTRICAL, COMMUNICATIONS & UTIL.	96.8	100.1	98.5	99.1	128.9	114.9	99.6	128.9	115.1	100.0	100.1	100.1	99.6	98.9	99.2	99.7	98.9	99.2
MF2014	WEIGHTED AVERAGE	97.0	128.1	110.6	100.3	134.6	115.3	99.5	134.5	114.8	99.1	128.3	111.8	99.7	111.3	104.8	100.2	112.6	105.6

		MASSACHUSETTS			MICHIGAN														
	DIVISION	WORCESTER			ANN ARBOR			BATTLE CREEK			BAY CITY			DEARBORN			DETROIT		
		015 - 016			481			490			487			481			482		
		MAT.	INST.	TOTAL	MAT.	INST.	TOTAL	MAT.	INST.	TOTAL	MAT.	INST.	TOTAL	MAT.	INST.	TOTAL	MAT.	INST.	TOTAL
015433	CONTRACTOR EQUIPMENT		100.9	100.9		110.7	110.7		102.8	102.8		110.7	110.7		110.7	110.7		98.8	98.8
0241, 31 - 34	SITE & INFRASTRUCTURE, DEMOLITION	94.6	105.6	102.4	82.3	98.8	94.0	93.3	87.5	89.2	73.9	97.8	90.9	82.1	98.9	94.1	94.7	100.6	98.9
0310	Concrete Forming & Accessories	99.6	130.7	126.5	98.3	111.6	109.8	97.6	87.6	89.0	98.4	87.1	88.6	98.2	115.8	113.3	100.8	115.8	113.7
0320	Concrete Reinforcing	104.6	150.6	128.0	92.3	120.1	106.5	91.4	93.3	92.4	92.3	119.2	106.0	92.3	120.2	106.5	94.9	120.1	107.8
0330	Cast-in-Place Concrete	93.5	150.5	116.9	88.8	107.4	96.4	99.0	100.4	99.6	85.0	90.6	87.3	86.8	109.9	96.3	94.1	109.9	100.6
03	CONCRETE	96.6	140.2	118.0	93.5	112.0	102.6	98.1	92.5	95.4	91.7	95.2	93.4	92.6	114.7	103.5	96.6	113.6	105.0
04	MASONRY	100.0	159.0	136.8	104.4	108.4	106.9	105.2	87.2	94.0	104.0	87.0	93.4	104.3	111.6	108.9	97.9	111.6	106.5
05	METALS	101.0	125.2	108.4	94.2	118.8	101.8	98.3	87.6	95.0	94.8	115.4	101.2	94.3	119.1	101.9	94.9	102.1	97.1
06	WOOD, PLASTICS & COMPOSITES	100.3	128.3	116.0	97.5	112.8	106.0	96.5	87.2	91.3	97.5	86.5	91.3	97.5	116.8	108.3	103.0	116.8	110.7
07	THERMAL & MOISTURE PROTECTION	99.5	141.5	116.7	100.4	108.2	103.6	94.6	84.8	90.6	98.2	92.4	95.8	98.9	115.7	105.8	97.7	115.7	105.1
08	OPENINGS	105.9	136.7	113.1	99.4	110.4	101.9	96.6	82.3	93.3	99.4	93.9	98.1	99.4	112.6	102.5	99.2	112.9	102.4
0920	Plaster & Gypsum Board	105.6	128.6	121.2	103.8	112.4	109.6	93.7	82.9	86.4	103.8	85.3	91.3	103.8	116.6	112.4	100.5	116.6	111.3
0950, 0980	Ceilings & Acoustic Treatment	99.3	128.6	118.5	88.7	112.4	104.3	96.3	82.9	87.5	89.7	85.3	86.8	88.7	116.6	107.0	91.2	116.6	107.9
0960	Flooring	100.2	179.8	123.0	97.5	117.6	103.2	105.7	91.9	101.7	97.5	81.5	92.9	96.9	113.9	101.8	95.9	113.9	101.1
0970, 0990	Wall Finishes & Painting/Coating	99.5	155.1	133.1	96.2	98.7	97.7	107.9	81.9	92.2	96.2	82.0	87.6	96.2	100.8	99.0	97.3	100.8	99.4
09	FINISHES	98.7	142.5	122.9	92.8	111.6	103.1	98.7	87.6	92.3	92.6	84.8	88.3	92.6	114.3	104.5	93.9	114.3	105.1
COVERS	DIVS. 10 - 14, 25, 28, 41, 43, 44, 46	100.0	109.0	101.8	100.0	105.8	101.2	100.0	96.4	99.3	100.0	94.1	98.8	100.0	107.1	101.4	100.0	107.1	101.4
21, 22, 23	FIRE SUPPRESSION, PLUMBING & HVAC	100.1	113.6	105.5	100.0	98.8	99.5	100.0	87.4	94.9	100.0	83.6	93.4	100.0	109.2	103.7	100.0	110.2	104.1
26, 27, 3370	ELECTRICAL, COMMUNICATIONS & UTIL.	99.7	105.1	102.5	96.3	108.8	102.9	94.4	83.7	88.8	95.2	90.3	92.6	96.3	107.0	101.9	98.2	107.0	102.9
MF2014	WEIGHTED AVERAGE	100.2	126.4	111.6	97.1	107.5	101.7	98.3	87.6	93.7	96.6	91.7	94.5	97.0	111.0	103.1	97.8	109.6	102.9

MICHIGAN

DIVISION		FLINT 484-485 MAT.	INST.	TOTAL	GAYLORD 497 MAT.	INST.	TOTAL	GRAND RAPIDS 493,495 MAT.	INST.	TOTAL	IRON MOUNTAIN 498-499 MAT.	INST.	TOTAL	JACKSON 492 MAT.	INST.	TOTAL	KALAMAZOO 491 MAT.	INST.	TOTAL
015433	CONTRACTOR EQUIPMENT		110.7	110.7		104.7	104.7		102.8	102.8		93.7	93.7		104.7	104.7		102.8	102.8
0241, 31 - 34	SITE & INFRASTRUCTURE, DEMOLITION	71.7	98.0	90.4	88.5	85.0	86.0	93.2	87.6	89.2	96.1	93.7	94.4	109.4	87.3	93.7	93.6	87.5	89.3
0310	Concrete Forming & Accessories	101.3	89.9	91.5	95.5	77.3	79.8	97.1	84.6	86.3	87.8	85.0	85.4	92.5	86.7	87.5	97.6	87.3	88.7
0320	Concrete Reinforcing	92.3	119.6	106.2	85.2	118.6	102.2	97.8	93.2	95.5	85.0	99.8	92.5	82.8	119.4	101.4	91.4	93.4	92.4
0330	Cast-in-Place Concrete	89.4	92.4	90.6	98.7	86.1	93.5	104.1	97.6	101.4	116.8	85.4	103.9	98.6	92.2	96.0	101.0	98.8	100.1
03	CONCRETE	94.0	97.1	95.5	94.6	88.9	91.8	101.6	90.2	96.0	104.3	87.9	96.2	89.3	95.4	92.3	101.6	91.8	96.8
04	MASONRY	104.5	96.5	99.5	116.2	79.1	93.1	102.0	87.4	92.9	101.2	86.1	91.8	95.1	91.6	92.9	103.8	89.1	94.6
05	METALS	94.3	116.3	101.0	99.7	109.9	102.8	95.3	87.1	92.8	99.0	89.6	96.1	99.9	114.1	104.2	98.3	86.5	94.7
06	WOOD, PLASTICS & COMPOSITES	100.9	88.2	93.8	89.4	76.8	82.4	97.7	83.2	89.6	85.0	85.1	85.0	88.1	84.7	86.2	96.5	87.2	91.3
07	THERMAL & MOISTURE PROTECTION	98.2	96.3	97.4	93.2	74.5	85.5	97.4	76.6	88.9	96.7	81.4	90.4	92.5	92.6	92.5	94.6	85.1	90.7
08	OPENINGS	99.4	95.6	98.5	97.5	75.1	92.3	103.4	84.7	99.0	104.6	76.0	98.0	96.5	93.1	95.7	96.6	81.7	93.1
0920	Plaster & Gypsum Board	105.2	87.1	93.0	93.4	74.9	80.9	95.4	78.8	84.2	51.7	85.2	74.3	91.7	82.9	85.8	93.7	82.9	86.4
0950, 0980	Ceilings & Acoustic Treatment	88.7	87.1	87.6	95.4	74.9	81.9	104.3	78.8	87.5	94.4	85.2	88.3	95.4	82.9	87.2	96.3	82.9	87.5
0960	Flooring	97.5	95.8	97.0	97.8	92.4	96.2	105.5	85.7	99.8	122.4	93.8	114.2	96.5	82.7	92.6	105.7	82.7	99.1
0970, 0990	Wall Finishes & Painting/Coating	96.2	85.4	89.7	103.8	82.0	90.6	109.7	80.8	92.3	126.1	72.1	93.5	103.8	96.8	99.6	107.9	81.9	92.2
09	FINISHES	92.1	89.8	90.9	97.4	79.1	87.3	101.6	84.2	92.0	100.0	85.3	91.9	98.3	86.2	91.6	98.0	85.6	91.2
COVERS	DIVS. 10 - 14, 25, 28, 41, 43, 44, 46	100.0	95.5	99.1	100.0	88.6	97.7	100.0	96.4	99.3	100.0	87.2	97.4	100.0	100.9	100.2	100.0	96.4	99.3
21, 22, 23	FIRE SUPPRESSION, PLUMBING & HVAC	100.0	89.6	95.8	95.6	78.4	88.6	100.0	83.0	93.1	95.5	85.3	91.4	95.6	88.4	92.7	100.0	80.8	92.3
26, 27, 3370	ELECTRICAL, COMMUNICATIONS & UTIL.	96.3	96.0	96.1	92.4	79.1	85.4	99.7	79.1	88.8	98.7	84.3	91.1	96.3	108.8	102.9	94.3	81.8	87.7
MF2014	WEIGHTED AVERAGE	96.8	96.0	96.4	97.2	83.6	91.3	99.7	85.1	93.3	99.3	86.2	93.6	96.5	95.1	95.9	98.7	85.7	93.0

MICHIGAN / MINNESOTA

DIVISION		LANSING 488-489 MAT.	INST.	TOTAL	MUSKEGON 494 MAT.	INST.	TOTAL	ROYAL OAK 480,483 MAT.	INST.	TOTAL	SAGINAW 486 MAT.	INST.	TOTAL	TRAVERSE CITY 496 MAT.	INST.	TOTAL	BEMIDJI 566 MAT.	INST.	TOTAL
015433	CONTRACTOR EQUIPMENT		110.7	110.7		102.8	102.8		96.2	96.2		110.7	110.7		93.7	93.7		98.2	98.2
0241, 31 - 34	SITE & INFRASTRUCTURE, DEMOLITION	95.6	98.0	97.3	91.2	87.5	88.5	86.5	98.4	95.0	74.9	97.8	91.2	82.4	93.2	90.0	95.4	97.5	96.9
0310	Concrete Forming & Accessories	98.2	89.7	90.9	98.0	84.1	86.0	94.1	112.6	110.1	98.3	87.6	89.1	87.8	75.2	76.9	86.1	89.6	89.1
0320	Concrete Reinforcing	94.3	119.4	107.1	92.1	93.3	92.7	83.7	119.9	102.1	92.3	119.2	106.0	86.3	92.5	89.5	93.7	106.6	100.2
0330	Cast-in-Place Concrete	103.5	92.1	98.8	98.6	94.8	97.1	77.8	106.3	89.5	87.7	90.5	88.9	91.2	78.2	85.9	106.0	103.0	104.8
03	CONCRETE	100.8	96.9	98.9	96.4	89.1	92.8	80.5	110.8	95.4	93.0	95.4	94.2	85.8	79.8	82.9	97.6	98.5	98.1
04	MASONRY	106.6	94.2	98.9	102.2	88.5	93.7	98.2	112.0	106.8	106.0	87.0	94.1	99.2	82.4	88.7	102.5	103.7	103.3
05	METALS	93.9	115.6	100.6	96.0	87.2	93.3	97.3	98.0	97.5	94.3	115.2	100.7	99.0	88.2	95.7	92.1	120.0	100.7
06	WOOD, PLASTICS & COMPOSITES	96.3	88.8	92.1	93.2	83.4	87.7	93.0	113.1	104.3	93.8	87.6	90.3	85.0	75.1	79.4	71.3	85.5	79.3
07	THERMAL & MOISTURE PROTECTION	98.6	89.6	94.9	93.2	78.9	87.6	96.8	111.2	102.7	99.0	92.6	96.4	95.8	71.7	85.9	104.7	95.4	100.9
08	OPENINGS	103.2	95.1	101.3	95.8	84.5	93.2	99.3	109.4	101.7	97.4	94.5	96.7	104.6	70.6	96.7	100.2	107.2	101.8
0920	Plaster & Gypsum Board	96.9	87.7	90.7	75.7	79.0	77.9	101.3	112.7	109.0	103.8	86.4	92.1	51.7	74.9	67.3	102.2	85.3	90.8
0950, 0980	Ceilings & Acoustic Treatment	94.0	87.7	89.8	97.1	79.0	85.2	88.0	112.7	104.3	88.7	86.4	87.2	94.4	74.9	81.6	122.8	85.3	98.2
0960	Flooring	106.6	85.7	100.6	104.5	85.7	99.1	94.8	113.9	100.3	97.5	81.5	92.9	122.4	94.9	114.5	104.4	120.8	109.1
0970, 0990	Wall Finishes & Painting/Coating	108.0	80.8	91.6	106.3	80.8	90.9	97.6	96.8	97.1	96.2	82.0	87.6	126.1	44.7	76.9	102.2	95.9	98.4
09	FINISHES	98.3	87.6	92.4	94.7	83.2	88.4	91.7	111.5	102.6	92.4	85.5	88.6	99.0	74.1	85.3	106.2	95.2	100.1
COVERS	DIVS. 10 - 14, 25, 28, 41, 43, 44, 46	100.0	100.6	100.1	100.0	96.0	99.2	100.0	103.0	100.6	100.0	94.3	98.8	100.0	84.6	96.9	100.0	96.9	99.4
21, 22, 23	FIRE SUPPRESSION, PLUMBING & HVAC	99.9	88.6	95.3	99.9	82.3	92.8	95.6	107.2	100.3	100.0	83.1	93.2	95.5	77.7	88.3	95.6	83.4	90.6
26, 27, 3370	ELECTRICAL, COMMUNICATIONS & UTIL.	98.6	93.3	95.8	94.8	79.1	86.5	98.4	104.4	101.6	93.9	90.5	92.1	94.0	79.1	86.1	104.3	104.6	104.5
MF2014	WEIGHTED AVERAGE	99.3	94.7	97.3	97.2	84.8	91.8	94.9	106.9	100.1	96.4	91.8	94.4	96.3	80.1	89.3	98.2	98.2	98.2

MINNESOTA

DIVISION		BRAINERD 564 MAT.	INST.	TOTAL	DETROIT LAKES 565 MAT.	INST.	TOTAL	DULUTH 556-558 MAT.	INST.	TOTAL	MANKATO 560 MAT.	INST.	TOTAL	MINNEAPOLIS 553-555 MAT.	INST.	TOTAL	ROCHESTER 559 MAT.	INST.	TOTAL
015433	CONTRACTOR EQUIPMENT		100.8	100.8		98.2	98.2		99.1	99.1		100.8	100.8		102.6	102.6		99.1	99.1
0241, 31 - 34	SITE & INFRASTRUCTURE, DEMOLITION	96.9	102.2	100.7	93.6	97.9	96.6	100.6	99.4	99.7	93.8	102.9	100.3	97.9	104.7	102.8	97.4	99.0	98.5
0310	Concrete Forming & Accessories	87.7	91.6	91.1	83.1	91.6	90.4	98.7	109.9	108.4	95.6	99.1	98.6	100.7	128.6	124.8	99.3	106.5	105.5
0320	Concrete Reinforcing	92.5	106.7	99.7	93.7	106.6	100.2	97.4	107.1	102.3	92.4	116.2	104.5	92.2	116.6	104.6	91.7	116.4	104.3
0330	Cast-in-Place Concrete	115.2	106.7	111.7	92.9	106.1	100.2	114.8	106.0	111.2	106.2	107.3	106.6	109.6	117.4	112.8	112.3	99.4	107.0
03	CONCRETE	101.7	100.7	101.2	95.2	100.4	97.8	106.5	108.7	107.6	97.2	106.1	101.5	103.4	122.8	112.9	102.5	106.8	104.6
04	MASONRY	126.4	112.0	117.4	126.3	110.1	116.2	107.6	117.6	113.9	114.5	110.4	112.0	109.0	126.7	120.1	104.5	114.1	110.5
05	METALS	93.2	120.0	101.5	92.1	120.0	100.6	101.7	122.4	108.0	93.1	125.5	103.0	99.1	128.9	108.3	100.4	127.3	108.7
06	WOOD, PLASTICS & COMPOSITES	89.4	85.3	87.1	68.4	85.5	78.0	98.2	109.3	104.4	98.8	96.4	97.5	106.7	127.4	118.3	103.2	105.3	104.4
07	THERMAL & MOISTURE PROTECTION	103.4	107.7	105.2	104.6	108.3	106.1	101.6	114.9	107.1	103.8	99.0	101.8	102.8	127.4	112.9	104.9	100.9	103.2
08	OPENINGS	86.9	107.1	91.6	100.2	107.2	101.8	102.5	116.5	105.8	91.5	115.5	97.1	97.1	133.6	105.6	96.4	121.5	102.2
0920	Plaster & Gypsum Board	86.9	85.3	85.8	101.1	85.3	90.5	90.2	110.0	103.6	91.5	96.8	95.1	95.8	128.6	117.9	97.6	105.8	103.2
0950, 0980	Ceilings & Acoustic Treatment	59.3	85.3	76.4	122.8	85.3	98.2	96.1	110.0	105.3	59.3	96.8	84.0	98.2	128.6	118.2	93.2	105.8	101.5
0960	Flooring	103.3	120.8	108.3	103.2	120.8	108.2	101.9	122.1	107.6	105.0	120.8	109.5	97.9	120.8	104.4	102.9	120.8	108.0
0970, 0990	Wall Finishes & Painting/Coating	96.0	95.9	95.9	102.2	95.9	98.4	101.9	110.3	107.0	108.3	104.9	106.3	98.2	126.1	115.1	97.2	104.9	101.9
09	FINISHES	89.4	96.4	93.2	105.5	96.5	100.6	98.3	112.5	106.1	91.1	103.4	97.9	98.6	127.6	114.6	97.4	109.1	103.8
COVERS	DIVS. 10 - 14, 25, 28, 41, 43, 44, 46	100.0	98.3	99.7	100.0	98.6	99.7	100.0	97.6	99.5	100.0	98.6	99.7	100.0	109.7	102.0	100.0	102.8	100.6
21, 22, 23	FIRE SUPPRESSION, PLUMBING & HVAC	94.8	86.1	91.3	95.6	86.0	91.7	99.8	96.5	98.5	94.8	87.7	91.9	99.9	115.6	106.2	99.9	96.2	98.4
26, 27, 3370	ELECTRICAL, COMMUNICATIONS & UTIL.	101.8	102.4	102.1	104.0	67.5	84.7	98.4	102.4	100.6	108.6	89.2	98.3	101.2	111.4	106.6	98.6	89.2	93.6
MF2014	WEIGHTED AVERAGE	96.9	100.5	98.5	98.9	95.1	97.3	101.3	107.3	103.9	97.2	101.3	99.0	100.4	120.0	108.9	99.9	104.7	102.0

619

MINNESOTA / MISSISSIPPI

DIVISION		SAINT PAUL 550-551 MAT.	INST.	TOTAL	ST. CLOUD 563 MAT.	INST.	TOTAL	THIEF RIVER FALLS 567 MAT.	INST.	TOTAL	WILLMAR 562 MAT.	INST.	TOTAL	WINDOM 561 MAT.	INST.	TOTAL	BILOXI 395 MAT.	INST.	TOTAL
015433	CONTRACTOR EQUIPMENT		99.1	99.1		100.8	100.8		98.2	98.2		100.8	100.8		100.8	100.8		98.8	98.8
0241, 31 - 34	SITE & INFRASTRUCTURE, DEMOLITION	95.2	100.1	98.7	92.5	104.3	100.9	94.4	97.5	96.6	91.8	102.0	99.1	86.0	101.0	96.6	100.6	88.4	91.9
0310	Concrete Forming & Accessories	93.9	123.0	119.0	84.9	117.7	113.2	86.8	89.0	88.7	84.7	91.1	90.2	89.1	84.8	85.4	94.7	52.9	58.7
0320	Concrete Reinforcing	98.9	116.6	107.9	92.6	116.4	104.7	94.0	106.5	100.4	92.2	116.0	104.3	92.2	115.3	104.0	95.0	58.3	76.4
0330	Cast-in-Place Concrete	124.0	116.7	121.0	101.7	114.8	107.1	105.1	88.1	98.1	103.3	85.3	95.9	89.4	66.3	79.9	108.4	54.0	86.0
03	CONCRETE	111.2	120.1	115.5	92.9	116.9	104.7	96.3	93.1	94.7	92.9	94.9	93.9	83.6	85.5	84.5	101.6	55.9	79.2
04	MASONRY	111.7	126.7	121.1	111.1	119.9	116.6	102.4	104.5	103.7	115.4	112.0	113.3	125.8	92.1	104.8	102.7	44.3	66.3
05	METALS	101.1	128.6	109.6	93.9	126.8	104.0	92.2	119.4	100.6	93.0	124.5	102.7	92.9	122.4	102.0	94.9	81.7	90.8
06	WOOD, PLASTICS & COMPOSITES	93.4	119.9	108.2	86.7	114.9	102.5	72.3	85.5	79.7	86.4	85.5	85.9	90.7	82.8	86.3	99.1	56.1	75.0
07	THERMAL & MOISTURE PROTECTION	104.1	126.4	113.3	103.6	117.5	109.3	105.6	98.4	102.7	103.4	108.6	105.5	103.4	89.6	97.7	97.4	56.1	80.4
08	OPENINGS	99.8	129.5	106.7	91.5	126.7	99.7	100.2	107.2	101.8	89.0	91.9	89.7	92.7	90.4	92.2	99.6	57.7	89.8
0920	Plaster & Gypsum Board	88.9	120.9	110.6	86.9	115.8	106.4	101.8	85.3	90.7	86.9	85.6	86.0	86.9	82.8	84.1	101.7	55.2	70.3
0950, 0980	Ceilings & Acoustic Treatment	95.6	120.9	112.3	59.3	115.8	96.4	122.8	85.3	98.2	59.3	85.6	76.6	59.3	82.8	74.7	87.3	55.2	66.2
0960	Flooring	103.7	120.8	108.6	100.0	120.8	105.9	104.1	120.8	108.8	101.6	120.8	107.1	103.9	120.8	108.8	101.4	55.1	88.1
0970, 0990	Wall Finishes & Painting/Coating	105.3	126.1	117.8	108.3	126.1	119.1	102.2	95.9	98.4	102.2	95.9	98.4	102.2	104.9	103.8	99.5	43.4	65.7
09	FINISHES	97.9	123.1	111.8	88.9	119.2	105.6	106.0	95.2	100.0	88.9	94.2	91.8	89.1	91.3	90.3	93.3	52.4	70.8
COVERS	DIVS. 10 - 14, 25, 28, 41, 43, 44, 46	100.0	108.6	101.7	100.0	104.3	100.9	100.0	96.8	99.3	100.0	98.2	99.6	100.0	93.7	98.7	100.0	55.4	91.0
21, 22, 23	FIRE SUPPRESSION, PLUMBING & HVAC	99.9	113.8	105.5	99.6	113.8	105.3	95.6	83.1	90.5	94.8	106.8	99.6	94.8	84.1	90.5	100.0	49.1	79.5
26, 27, 3370	ELECTRICAL, COMMUNICATIONS & UTIL.	97.1	111.4	104.7	101.8	111.4	106.9	101.3	67.5	83.5	101.8	85.2	93.0	108.6	89.2	98.3	103.8	56.9	79.0
MF2014	WEIGHTED AVERAGE	101.4	117.9	108.6	96.9	116.0	105.2	97.8	92.4	95.4	95.5	101.3	98.0	95.9	92.3	94.3	99.2	58.1	81.3

MISSISSIPPI

DIVISION		CLARKSDALE 386 MAT.	INST.	TOTAL	COLUMBUS 397 MAT.	INST.	TOTAL	GREENVILLE 387 MAT.	INST.	TOTAL	GREENWOOD 389 MAT.	INST.	TOTAL	JACKSON 390 - 392 MAT.	INST.	TOTAL	LAUREL 394 MAT.	INST.	TOTAL
015433	CONTRACTOR EQUIPMENT		98.8	98.8		98.8	98.8		98.8	98.8		98.8	98.8		98.8	98.8		98.8	98.8
0241, 31 - 34	SITE & INFRASTRUCTURE, DEMOLITION	98.3	88.2	91.1	98.9	88.4	91.5	103.9	88.4	92.9	101.0	87.9	91.7	97.6	88.4	91.1	104.4	88.3	92.9
0310	Concrete Forming & Accessories	85.0	49.2	54.2	83.3	51.2	55.6	81.7	63.7	66.1	94.2	49.9	55.4	93.6	57.5	62.5	83.3	52.9	57.1
0320	Concrete Reinforcing	99.8	41.6	70.1	101.8	42.5	71.6	100.3	60.3	79.9	99.8	52.5	75.7	100.2	56.1	77.7	102.4	39.1	70.2
0330	Cast-in-Place Concrete	105.6	53.5	84.2	110.5	58.6	89.2	108.7	55.6	86.9	113.5	53.2	88.7	98.7	56.8	81.5	108.1	55.2	86.4
03	CONCRETE	98.7	50.9	75.2	103.2	53.7	78.9	104.2	61.6	83.3	105.2	52.7	79.4	97.7	58.5	78.5	106.0	52.7	79.8
04	MASONRY	100.6	45.0	66.0	131.6	51.6	81.7	149.8	49.5	87.3	101.3	44.9	66.1	109.2	49.5	72.0	127.1	48.1	77.9
05	METALS	95.4	73.4	88.6	91.9	74.2	86.5	96.4	82.5	92.1	95.4	74.8	89.0	101.8	80.8	95.4	92.0	73.3	86.3
06	WOOD, PLASTICS & COMPOSITES	82.1	51.2	64.8	83.9	52.2	66.1	78.7	68.1	72.8	95.0	51.2	70.5	96.3	59.6	75.8	85.1	54.6	68.0
07	THERMAL & MOISTURE PROTECTION	95.7	47.2	75.8	97.3	51.5	78.5	96.0	60.0	81.2	96.1	47.1	76.0	97.5	59.2	81.8	97.4	53.1	79.3
08	OPENINGS	97.0	49.4	85.9	98.9	49.5	87.4	97.0	61.4	88.7	97.0	50.7	86.2	103.7	55.6	92.5	95.6	50.3	85.0
0920	Plaster & Gypsum Board	90.2	50.2	63.1	91.3	51.2	64.2	89.8	67.6	74.8	100.9	50.2	66.6	89.7	58.8	68.8	91.3	53.7	65.8
0950, 0980	Ceilings & Acoustic Treatment	82.9	50.2	61.4	82.0	51.2	61.8	85.7	67.6	73.8	82.9	50.2	61.4	89.7	58.8	69.4	82.0	53.7	63.4
0960	Flooring	105.7	52.8	90.6	95.5	59.3	85.1	104.1	52.8	89.4	111.1	52.8	94.4	99.3	55.1	86.7	94.2	52.8	82.4
0970, 0990	Wall Finishes & Painting/Coating	104.4	46.1	69.2	99.5	53.0	71.4	104.4	65.9	81.2	104.4	46.1	69.2	96.5	65.9	78.0	99.5	60.9	76.2
09	FINISHES	94.0	49.9	69.7	89.1	53.2	69.3	94.6	62.9	77.1	97.4	49.9	71.2	93.2	58.3	74.0	89.2	54.3	70.0
COVERS	DIVS. 10 - 14, 25, 28, 41, 43, 44, 46	100.0	55.7	91.1	100.0	56.8	91.3	100.0	58.5	91.6	100.0	55.7	91.1	100.0	57.5	91.4	100.0	37.5	87.4
21, 22, 23	FIRE SUPPRESSION, PLUMBING & HVAC	97.9	41.3	75.1	97.4	37.5	73.3	100.0	47.6	78.8	97.9	39.3	74.3	100.1	57.0	82.7	97.5	45.8	76.6
26, 27, 3370	ELECTRICAL, COMMUNICATIONS & UTIL.	97.1	46.4	70.3	100.8	49.1	73.5	97.1	62.7	78.9	97.1	43.4	68.8	105.4	62.7	82.8	102.6	56.9	78.5
MF2014	WEIGHTED AVERAGE	97.3	52.6	77.8	98.8	53.9	79.2	101.0	61.7	83.9	98.4	52.2	78.3	100.7	62.2	83.9	98.9	55.8	80.1

MISSISSIPPI / MISSOURI

DIVISION		MCCOMB 396 MAT.	INST.	TOTAL	MERIDIAN 393 MAT.	INST.	TOTAL	TUPELO 388 MAT.	INST.	TOTAL	BOWLING GREEN 633 MAT.	INST.	TOTAL	CAPE GIRARDEAU 637 MAT.	INST.	TOTAL	CHILLICOTHE 646 MAT.	INST.	TOTAL
015433	CONTRACTOR EQUIPMENT		98.8	98.8		98.8	98.8		98.8	98.8		106.9	106.9		106.9	106.9		101.9	101.9
0241, 31 - 34	SITE & INFRASTRUCTURE, DEMOLITION	92.2	88.0	89.2	96.2	88.7	90.8	95.6	88.1	90.3	89.6	95.6	93.9	91.3	95.4	94.2	105.2	94.6	97.7
0310	Concrete Forming & Accessories	83.3	50.9	55.4	80.4	63.3	65.7	82.2	51.2	55.5	92.3	87.3	88.0	85.2	85.1	85.1	89.0	93.0	92.5
0320	Concrete Reinforcing	103.0	41.0	71.4	101.8	56.1	78.5	97.6	52.5	74.7	101.4	102.8	102.1	102.7	88.7	95.5	96.8	106.5	101.8
0330	Cast-in-Place Concrete	96.0	53.0	78.3	102.7	59.2	84.8	105.6	54.9	84.8	91.2	84.7	88.5	90.2	86.0	88.5	103.0	85.8	95.9
03	CONCRETE	92.9	51.3	72.5	97.2	61.9	79.8	98.2	54.2	76.6	95.8	90.5	93.2	94.9	87.3	91.2	105.6	93.6	99.7
04	MASONRY	133.0	44.4	77.7	102.4	53.5	71.9	138.4	47.9	82.0	117.6	98.7	105.8	113.9	80.0	92.8	102.9	98.8	100.3
05	METALS	92.2	71.6	85.8	93.0	81.0	89.3	95.3	75.1	89.1	94.6	114.4	100.7	95.6	107.0	99.1	95.1	108.5	99.2
06	WOOD, PLASTICS & COMPOSITES	83.9	54.0	67.2	81.2	65.1	72.2	79.3	52.2	64.1	90.8	86.0	88.1	83.5	85.5	84.6	93.6	93.3	93.4
07	THERMAL & MOISTURE PROTECTION	96.9	46.4	76.2	97.0	61.6	82.5	95.7	52.9	78.2	100.3	98.9	99.7	100.2	84.2	93.6	98.3	94.5	96.8
08	OPENINGS	99.0	51.0	87.8	98.9	64.1	90.8	97.0	50.6	86.2	97.7	97.5	97.6	97.7	82.9	94.2	95.0	95.7	95.2
0920	Plaster & Gypsum Board	91.3	53.0	65.4	91.3	64.5	73.2	89.8	51.2	63.7	95.2	85.6	88.7	94.2	85.1	88.0	102.0	92.9	95.8
0950, 0980	Ceilings & Acoustic Treatment	82.0	53.0	63.0	83.9	64.5	71.2	82.9	51.2	62.0	92.6	85.6	88.0	92.6	85.1	87.6	89.1	92.9	91.6
0960	Flooring	95.5	52.8	83.3	94.2	55.1	83.0	104.4	52.8	89.6	96.4	96.9	96.5	93.3	84.0	90.6	97.3	102.5	98.8
0970, 0990	Wall Finishes & Painting/Coating	99.5	42.3	65.0	99.5	76.1	85.4	104.4	58.7	76.8	97.1	102.8	100.5	97.1	73.2	82.7	87.9	106.3	99.0
09	FINISHES	88.5	51.0	67.8	88.8	63.7	75.0	93.5	52.6	71.0	97.8	89.7	93.4	96.7	83.0	89.2	100.2	96.1	97.9
COVERS	DIVS. 10 - 14, 25, 28, 41, 43, 44, 46	100.0	59.1	91.7	100.0	59.5	91.8	100.0	56.8	91.3	100.0	88.2	97.6	100.0	93.5	98.7	100.0	88.5	97.7
21, 22, 23	FIRE SUPPRESSION, PLUMBING & HVAC	97.4	34.6	72.1	100.0	59.1	83.5	98.1	43.2	76.0	95.2	96.2	95.6	99.9	96.8	98.7	95.3	97.0	96.0
26, 27, 3370	ELECTRICAL, COMMUNICATIONS & UTIL.	99.2	49.4	72.9	102.6	57.8	78.9	96.9	49.3	71.7	96.9	82.6	89.3	96.9	101.5	99.3	92.4	78.6	85.1
MF2014	WEIGHTED AVERAGE	97.4	51.7	77.5	97.6	64.1	83.0	98.8	54.9	79.7	97.3	94.4	96.0	98.2	92.4	95.7	97.6	94.5	96.3

DIVISION		MISSOURI																
		COLUMBIA			FLAT RIVER			HANNIBAL			HARRISONVILLE			JEFFERSON CITY			JOPLIN		
		652			636			634			647			650 - 651			648		
		MAT.	INST.	TOTAL	MAT.	INST.	TOTAL	MAT.	INST.	TOTAL	MAT.	INST.	TOTAL	MAT.	INST.	TOTAL	MAT.	INST.	TOTAL
015433	CONTRACTOR EQUIPMENT		108.0	108.0		106.9	106.9		106.9	106.9		101.9	101.9		108.0	108.0		105.7	105.7
0241, 31 - 34	SITE & INFRASTRUCTURE, DEMOLITION	102.8	97.2	98.8	92.2	95.6	94.6	87.6	95.7	93.4	96.6	95.7	95.9	102.3	97.2	98.6	105.6	99.7	101.4
0310	Concrete Forming & Accessories	86.9	81.8	82.5	98.3	84.3	86.2	90.5	79.0	80.6	86.2	98.7	97.0	98.3	81.8	84.1	101.3	75.3	78.9
0320	Concrete Reinforcing	96.3	119.0	107.9	102.7	107.7	105.3	100.9	97.7	99.3	96.4	115.3	106.0	98.5	111.9	105.3	99.9	98.5	99.2
0330	Cast-in-Place Concrete	96.1	87.8	92.7	94.2	93.8	94.0	86.3	86.0	86.2	105.5	103.9	104.8	102.0	87.0	95.9	111.5	77.9	97.7
03	CONCRETE	93.2	92.2	92.7	98.8	93.2	96.0	92.1	86.3	89.2	101.8	104.0	102.9	98.9	90.6	94.8	105.2	81.4	93.5
04	MASONRY	154.1	90.0	114.2	114.5	81.4	93.9	109.0	101.3	104.2	97.4	103.4	101.2	112.0	90.0	98.3	96.5	85.9	89.9
05	METALS	98.4	121.2	105.4	94.5	115.6	101.0	94.6	111.9	99.9	95.6	113.2	101.0	98.0	118.3	104.2	98.2	100.5	98.9
06	WOOD, PLASTICS & COMPOSITES	91.7	78.7	84.4	99.1	82.6	89.9	88.9	73.8	80.4	89.9	97.5	94.2	100.0	78.7	88.1	106.0	74.2	88.2
07	THERMAL & MOISTURE PROTECTION	96.2	87.5	92.6	100.5	93.8	97.7	100.1	97.0	98.8	97.6	103.2	99.9	101.8	87.6	96.0	97.7	78.9	90.0
08	OPENINGS	99.5	95.0	98.4	97.7	97.2	97.6	97.7	82.3	94.1	95.1	102.8	96.9	99.1	93.1	97.7	96.2	78.5	92.1
0920	Plaster & Gypsum Board	87.2	78.0	81.0	101.3	82.1	88.3	94.5	73.0	79.9	97.3	97.3	97.3	91.3	78.0	82.3	108.2	73.2	84.5
0950, 0980	Ceilings & Acoustic Treatment	93.2	78.0	83.2	92.6	82.1	85.7	92.6	73.0	79.7	89.1	97.3	94.5	97.2	78.0	84.6	90.0	73.2	78.9
0960	Flooring	97.6	98.6	97.9	99.4	84.0	95.0	95.8	96.8	96.1	92.6	102.5	95.4	104.4	98.6	102.7	123.8	74.9	109.8
0970, 0990	Wall Finishes & Painting/Coating	105.2	77.1	88.2	97.1	78.4	85.8	97.1	102.8	100.5	92.5	103.8	99.3	102.3	77.1	87.1	87.5	74.9	79.9
09	FINISHES	92.8	83.0	87.4	99.7	82.2	90.1	97.4	83.1	89.5	97.8	99.5	98.7	97.2	83.0	89.4	107.0	75.3	89.6
COVERS	DIVS. 10 - 14, 25, 28, 41, 43, 44, 46	100.0	96.2	99.2	100.0	91.8	98.3	100.0	87.8	97.5	100.0	90.6	98.1	100.0	96.2	99.2	100.0	87.9	97.6
21, 22, 23	FIRE SUPPRESSION, PLUMBING & HVAC	99.8	96.4	98.4	95.2	97.8	96.2	95.2	97.5	96.1	95.2	100.0	97.2	99.8	98.1	99.1	100.1	72.1	88.8
26, 27, 3370	ELECTRICAL, COMMUNICATIONS & UTIL.	94.8	86.5	90.4	101.1	101.5	101.3	95.7	82.6	88.8	98.8	99.6	99.2	100.4	86.5	93.0	90.4	77.5	83.6
MF2014	WEIGHTED AVERAGE	100.2	93.9	97.5	98.1	95.0	96.8	96.2	92.4	94.6	97.1	101.5	99.1	99.9	93.7	97.2	99.5	81.7	91.8

DIVISION		MISSOURI																
		KANSAS CITY			KIRKSVILLE			POPLAR BLUFF			ROLLA			SEDALIA			SIKESTON		
		640 - 641			635			639			654 - 655			653			638		
		MAT.	INST.	TOTAL	MAT.	INST.	TOTAL	MAT.	INST.	TOTAL	MAT.	INST.	TOTAL	MAT.	INST.	TOTAL	MAT.	INST.	TOTAL
015433	CONTRACTOR EQUIPMENT		103.3	103.3		99.1	99.1		101.5	101.5		108.0	108.0		99.9	99.9		101.5	101.5
0241, 31 - 34	SITE & INFRASTRUCTURE, DEMOLITION	98.9	98.2	98.4	91.8	91.4	91.5	79.1	95.4	90.7	101.5	97.1	98.3	99.0	92.8	94.6	82.4	95.4	91.7
0310	Concrete Forming & Accessories	100.1	105.4	104.6	83.2	80.6	81.0	83.3	85.0	84.7	95.0	91.9	92.3	92.4	81.4	82.9	84.3	84.9	84.9
0320	Concrete Reinforcing	94.9	119.8	107.6	101.6	98.8	100.2	104.7	94.7	99.6	96.7	94.7	95.7	95.3	106.2	100.8	104.0	98.4	101.2
0330	Cast-in-Place Concrete	103.7	106.6	104.9	94.1	83.8	89.9	72.4	86.5	78.2	98.3	96.2	97.5	102.7	83.8	94.9	77.4	86.5	81.1
03	CONCRETE	101.1	108.7	104.8	110.8	86.0	98.6	85.9	88.1	86.9	95.2	95.0	95.1	109.8	87.7	99.0	89.5	88.7	89.1
04	MASONRY	99.5	107.5	104.5	121.0	88.9	101.0	112.5	80.1	92.3	126.3	87.9	102.4	132.9	88.3	105.1	112.2	80.1	92.2
05	METALS	105.9	115.8	108.9	94.2	103.8	97.2	94.8	101.7	96.9	97.8	110.5	101.7	96.5	108.0	100.1	95.1	103.2	97.6
06	WOOD, PLASTICS & COMPOSITES	105.3	105.0	105.1	77.0	78.7	77.9	76.1	85.5	81.4	100.3	92.8	96.1	93.1	78.7	85.1	77.7	85.5	82.1
07	THERMAL & MOISTURE PROTECTION	97.8	106.8	101.5	106.8	88.2	99.2	105.2	85.9	97.3	96.4	93.4	95.1	102.5	93.1	98.7	105.3	84.5	96.8
08	OPENINGS	102.5	109.2	104.0	102.7	85.8	98.8	103.7	84.5	99.2	99.5	88.3	96.9	104.5	87.4	100.5	103.7	85.5	99.5
0920	Plaster & Gypsum Board	105.2	105.0	105.0	89.8	78.0	81.8	90.2	85.1	86.7	89.3	92.5	91.5	82.5	78.0	79.5	92.0	85.1	87.3
0950, 0980	Ceilings & Acoustic Treatment	96.7	105.0	102.1	90.9	78.0	82.4	92.6	85.1	87.6	93.2	92.5	92.7	93.2	78.0	83.2	92.6	85.1	87.6
0960	Flooring	98.0	109.0	101.2	74.4	95.8	80.5	88.3	82.0	86.5	101.2	95.8	99.6	78.8	95.8	83.7	88.7	82.0	86.8
0970, 0990	Wall Finishes & Painting/Coating	92.5	112.2	104.4	92.5	82.1	86.2	92.0	73.2	80.7	105.2	100.6	102.5	105.2	106.3	105.9	92.0	73.2	80.7
09	FINISHES	102.0	106.6	104.5	96.5	82.7	88.9	96.5	82.7	88.9	94.3	93.4	93.8	91.9	85.4	88.3	97.1	82.7	89.2
COVERS	DIVS. 10 - 14, 25, 28, 41, 43, 44, 46	100.0	100.6	100.1	100.0	87.1	97.4	100.0	93.6	98.7	100.0	89.0	97.8	100.0	90.2	98.0	100.0	94.2	98.8
21, 22, 23	FIRE SUPPRESSION, PLUMBING & HVAC	100.0	104.3	101.7	95.2	97.3	96.1	95.2	96.3	95.6	95.0	98.2	96.3	95.0	95.9	95.4	95.2	96.3	95.6
26, 27, 3370	ELECTRICAL, COMMUNICATIONS & UTIL.	100.3	100.1	100.2	95.8	82.5	88.8	96.1	101.5	99.0	93.2	86.5	89.6	94.7	128.8	112.7	95.3	101.5	98.6
MF2014	WEIGHTED AVERAGE	101.4	105.7	103.3	99.5	89.9	95.3	96.2	92.0	94.4	97.9	94.6	96.5	100.2	97.3	98.9	96.7	92.2	94.8

DIVISION		MISSOURI									MONTANA								
		SPRINGFIELD			ST. JOSEPH			ST. LOUIS			BILLINGS			BUTTE			GREAT FALLS		
		656 - 658			644 - 645			630 - 631			590 - 591			597			594		
		MAT.	INST.	TOTAL	MAT.	INST.	TOTAL	MAT.	INST.	TOTAL	MAT.	INST.	TOTAL	MAT.	INST.	TOTAL	MAT.	INST.	TOTAL
015433	CONTRACTOR EQUIPMENT		102.5	102.5		101.9	101.9		107.9	107.9		98.5	98.5		98.2	98.2		98.2	98.2
0241, 31 - 34	SITE & INFRASTRUCTURE, DEMOLITION	101.6	95.1	97.0	100.5	93.9	95.8	91.7	98.9	96.8	95.0	96.3	95.9	102.0	96.1	97.8	105.7	96.2	98.9
0310	Concrete Forming & Accessories	101.4	79.1	82.2	100.0	89.6	91.0	98.4	104.8	103.9	98.4	71.5	75.2	85.5	71.8	73.7	98.4	71.5	75.2
0320	Concrete Reinforcing	92.7	118.7	105.9	93.8	115.0	104.6	93.9	112.7	103.4	90.2	81.7	85.9	97.8	81.9	89.7	90.2	81.8	85.9
0330	Cast-in-Place Concrete	104.3	79.6	94.2	103.6	103.3	103.5	90.2	105.0	96.3	129.3	67.6	103.9	142.3	69.0	112.2	150.4	63.7	114.8
03	CONCRETE	105.8	87.4	96.8	100.9	99.7	100.3	94.4	107.1	100.7	109.8	72.9	91.7	113.8	73.5	94.0	119.5	71.5	96.0
04	MASONRY	102.9	89.8	94.8	99.3	96.1	97.3	95.4	111.9	105.7	128.0	73.7	94.2	123.5	76.6	94.3	128.0	75.4	95.2
05	METALS	102.8	108.8	104.7	101.9	112.4	105.2	100.3	120.2	106.4	105.8	90.0	100.9	100.1	90.1	97.1	103.2	90.0	99.2
06	WOOD, PLASTICS & COMPOSITES	101.8	77.1	88.0	106.0	87.6	95.7	99.0	103.0	101.2	98.6	70.5	82.8	85.7	70.8	77.4	99.9	70.5	83.4
07	THERMAL & MOISTURE PROTECTION	100.7	82.2	93.2	98.2	95.6	97.1	100.2	106.9	103.0	106.5	69.9	91.5	106.3	71.0	91.8	106.9	68.5	91.2
08	OPENINGS	107.2	84.3	101.9	100.6	98.6	100.1	97.6	110.9	100.7	97.3	68.7	90.6	95.6	70.0	89.6	98.5	68.0	91.4
0920	Plaster & Gypsum Board	90.4	76.3	80.9	109.7	87.0	94.4	102.2	103.1	102.8	98.0	69.8	78.9	98.7	70.2	79.4	107.3	69.8	82.0
0950, 0980	Ceilings & Acoustic Treatment	93.2	76.3	82.1	95.8	87.0	90.0	97.6	103.1	101.2	81.5	69.8	73.8	88.3	70.2	76.4	90.0	69.8	76.7
0960	Flooring	99.9	74.9	92.8	101.5	107.3	103.2	99.2	97.6	98.7	106.6	70.2	96.2	104.4	74.2	95.7	111.5	74.2	100.8
0970, 0990	Wall Finishes & Painting/Coating	99.0	74.9	84.4	87.9	86.7	87.2	97.1	106.0	102.5	105.2	86.6	94.0	103.4	51.2	71.9	103.4	86.6	93.3
09	FINISHES	96.7	78.0	86.4	103.2	91.7	96.9	100.8	103.1	102.1	96.6	72.6	83.4	97.2	69.7	82.0	101.1	73.3	85.8
COVERS	DIVS. 10 - 14, 25, 28, 41, 43, 44, 46	100.0	94.8	99.0	100.0	96.9	99.4	100.0	102.4	100.5	100.0	94.7	98.9	100.0	94.8	98.9	100.0	94.7	98.9
21, 22, 23	FIRE SUPPRESSION, PLUMBING & HVAC	99.8	74.6	89.6	100.1	89.0	95.6	100.0	107.4	103.0	100.1	74.7	89.9	100.2	75.6	90.3	100.2	70.7	88.3
26, 27, 3370	ELECTRICAL, COMMUNICATIONS & UTIL.	99.2	69.7	83.6	98.8	78.6	88.2	99.2	108.3	104.0	98.5	72.2	84.6	105.5	71.7	87.7	97.7	71.9	84.1
MF2014	WEIGHTED AVERAGE	101.7	83.7	93.8	100.6	93.5	97.5	98.7	107.8	102.7	102.7	77.1	91.5	102.6	77.3	91.6	104.1	76.2	91.9

For customer support on your Plumbing Cost Data, call 877.709.6968.

621

MONTANA

DIVISION		HAVRE 595 MAT.	INST.	TOTAL	HELENA 596 MAT.	INST.	TOTAL	KALISPELL 599 MAT.	INST.	TOTAL	MILES CITY 593 MAT.	INST.	TOTAL	MISSOULA 598 MAT.	INST.	TOTAL	WOLF POINT 592 MAT.	INST.	TOTAL
015433	CONTRACTOR EQUIPMENT		98.2	98.2		98.2	98.2		98.2	98.2		98.2	98.2		98.2	98.2		98.2	98.2
0241, 31 - 34	SITE & INFRASTRUCTURE, DEMOLITION	109.2	95.8	99.7	98.5	95.8	96.6	92.2	96.0	94.9	98.5	95.4	96.3	85.1	95.8	92.7	115.5	95.6	101.3
0310	Concrete Forming & Accessories	78.4	69.0	70.3	101.5	65.2	70.2	88.7	71.4	73.8	96.8	67.6	71.6	88.7	71.4	73.8	89.2	67.9	70.8
0320	Concrete Reinforcing	98.6	81.7	90.0	102.5	81.7	91.9	100.4	83.4	91.8	98.2	81.8	89.8	99.5	83.4	91.3	99.6	81.8	90.5
0330	Cast-in-Place Concrete	153.3	60.4	115.1	115.0	62.9	93.6	123.6	62.7	98.6	135.3	60.3	104.5	104.9	64.9	88.4	151.5	60.6	114.2
03	CONCRETE	122.7	69.3	96.5	106.7	68.4	87.9	102.6	71.5	87.3	110.4	68.6	89.9	90.0	72.2	81.3	126.5	68.9	98.2
04	MASONRY	124.5	69.8	90.4	120.1	75.7	92.4	122.2	77.0	94.0	129.7	65.3	89.6	147.9	76.6	103.4	131.0	66.0	90.5
05	METALS	96.3	89.6	94.2	102.1	88.8	98.0	96.1	90.6	94.4	95.4	89.9	93.7	96.6	90.7	94.8	95.5	89.2	93.6
06	WOOD, PLASTICS & COMPOSITES	77.4	70.5	73.5	102.9	62.3	80.2	89.1	70.5	78.7	96.9	70.5	82.1	89.1	70.5	78.7	88.5	70.5	78.4
07	THERMAL & MOISTURE PROTECTION	106.6	68.0	90.8	103.9	70.7	90.3	105.9	72.8	92.3	106.2	65.7	89.6	105.9	79.1	94.7	107.1	66.0	90.3
08	OPENINGS	95.6	68.0	89.2	99.8	65.2	91.7	95.6	68.5	89.3	95.1	68.0	88.8	95.6	68.3	89.2	95.1	67.4	88.7
0920	Plaster & Gypsum Board	94.8	69.8	77.9	96.0	61.4	72.6	98.7	69.8	79.2	106.7	69.8	81.8	98.7	69.8	79.2	102.1	69.8	80.3
0950, 0980	Ceilings & Acoustic Treatment	88.3	69.8	76.2	90.8	61.4	71.5	88.3	69.8	76.2	87.5	69.8	75.9	88.3	69.8	76.2	87.5	69.8	75.9
0960	Flooring	101.8	74.2	93.9	106.7	59.1	93.1	106.2	74.2	97.1	111.3	70.2	99.5	106.2	74.2	97.1	107.6	70.2	96.9
0970, 0990	Wall Finishes & Painting/Coating	103.4	48.6	70.4	99.7	47.2	68.0	103.4	47.2	69.5	103.4	47.2	69.5	103.4	62.0	78.4	103.4	48.6	70.4
09	FINISHES	96.5	67.7	80.6	98.6	61.6	78.2	97.2	69.1	81.7	100.0	65.7	81.1	96.7	70.7	82.3	99.6	66.0	81.1
COVERS	DIVS. 10 - 14, 25, 28, 41, 43, 44, 46	100.0	70.8	94.1	100.0	94.0	98.8	100.0	93.9	98.8	100.0	90.4	98.1	100.0	93.7	98.7	100.0	90.6	98.1
21, 22, 23	FIRE SUPPRESSION, PLUMBING & HVAC	95.4	67.8	84.3	100.1	70.1	88.0	95.4	69.2	84.8	95.4	69.6	85.0	100.2	69.2	87.7	95.4	70.0	85.1
26, 27, 3370	ELECTRICAL, COMMUNICATIONS & UTIL.	97.7	71.2	83.7	104.6	71.3	87.0	102.2	69.3	84.8	97.7	76.9	86.7	103.2	67.7	84.5	97.7	76.9	86.7
MF2014	WEIGHTED AVERAGE	101.2	73.1	89.0	102.4	73.8	89.9	99.0	75.3	88.7	100.1	74.0	88.7	99.9	75.5	89.3	102.3	74.2	90.0

NEBRASKA

DIVISION		ALLIANCE 693 MAT.	INST.	TOTAL	COLUMBUS 686 MAT.	INST.	TOTAL	GRAND ISLAND 688 MAT.	INST.	TOTAL	HASTINGS 689 MAT.	INST.	TOTAL	LINCOLN 683 - 685 MAT.	INST.	TOTAL	MCCOOK 690 MAT.	INST.	TOTAL
015433	CONTRACTOR EQUIPMENT		97.8	97.8		101.6	101.6		101.6	101.6		101.6	101.6		101.6	101.6		101.6	101.6
0241, 31 - 34	SITE & INFRASTRUCTURE, DEMOLITION	100.5	98.8	99.3	100.2	92.3	94.6	105.0	93.1	96.5	103.7	92.3	95.6	92.7	93.0	92.9	103.8	92.2	95.6
0310	Concrete Forming & Accessories	87.3	56.4	60.7	95.5	67.4	71.2	95.1	72.4	75.5	98.2	71.9	75.5	95.7	68.2	71.9	93.0	56.1	61.2
0320	Concrete Reinforcing	108.8	85.1	96.7	98.3	76.5	87.2	97.7	76.4	86.9	97.7	77.1	87.2	96.7	76.0	86.1	101.8	76.2	88.8
0330	Cast-in-Place Concrete	113.7	60.0	91.6	115.7	60.0	92.8	122.5	67.2	99.8	122.5	63.2	98.2	92.4	69.6	83.1	122.6	59.3	96.6
03	CONCRETE	123.9	63.5	94.2	108.8	67.3	88.4	113.7	72.1	93.3	113.9	70.6	92.6	95.6	71.0	83.5	112.9	62.0	87.9
04	MASONRY	115.6	74.1	89.7	122.7	77.8	94.7	115.4	76.1	90.9	124.9	88.0	101.9	104.8	67.2	81.4	110.7	73.9	87.8
05	METALS	101.5	78.7	94.5	96.1	84.6	92.6	97.8	86.3	94.3	98.7	85.8	94.7	99.9	85.5	95.5	96.3	83.6	92.4
06	WOOD, PLASTICS & COMPOSITES	87.0	52.2	67.5	101.2	66.7	81.9	100.4	72.2	84.6	104.1	72.2	86.3	98.4	66.7	80.7	95.7	52.2	71.4
07	THERMAL & MOISTURE PROTECTION	104.3	67.8	89.4	100.8	69.8	88.1	100.9	72.8	89.4	100.9	83.4	93.8	98.1	69.9	86.5	98.9	66.6	85.6
08	OPENINGS	94.5	58.6	86.1	95.4	66.1	88.6	95.5	69.7	89.5	95.5	69.1	89.3	107.8	63.0	97.4	95.0	56.6	86.1
0920	Plaster & Gypsum Board	82.7	50.7	61.1	90.9	65.6	73.8	90.1	71.3	77.4	91.9	71.3	78.0	96.2	65.6	75.5	93.7	50.7	64.7
0950, 0980	Ceilings & Acoustic Treatment	91.9	50.7	64.8	80.2	65.6	70.6	80.2	71.3	74.3	80.2	71.3	74.3	90.4	65.6	74.1	80.0	50.7	63.5
0960	Flooring	99.9	82.1	94.8	92.9	100.4	95.0	92.7	107.2	96.8	93.9	100.4	95.7	101.0	83.6	96.0	97.6	82.1	93.1
0970, 0990	Wall Finishes & Painting/Coating	168.5	54.8	99.9	85.5	63.7	72.4	85.5	67.6	74.7	85.5	63.7	72.4	103.8	77.9	88.1	93.8	47.7	66.0
09	FINISHES	98.4	59.5	77.0	89.9	71.9	80.0	90.0	77.1	82.9	90.6	75.2	82.1	96.9	71.7	83.0	95.4	58.6	75.2
COVERS	DIVS. 10 - 14, 25, 28, 41, 43, 44, 46	100.0	65.3	93.0	100.0	86.1	97.2	100.0	88.7	97.7	100.0	86.7	97.3	100.0	88.1	97.6	100.0	64.5	92.8
21, 22, 23	FIRE SUPPRESSION, PLUMBING & HVAC	95.3	70.5	85.3	95.2	76.0	87.4	100.0	78.0	91.1	95.2	76.0	87.5	99.9	78.0	91.0	95.1	75.9	87.4
26, 27, 3370	ELECTRICAL, COMMUNICATIONS & UTIL.	92.3	69.4	80.2	93.6	82.3	87.6	92.2	67.0	78.9	91.6	83.6	87.4	105.4	67.0	85.1	94.4	69.4	81.2
MF2014	WEIGHTED AVERAGE	101.0	70.5	87.7	98.4	77.1	89.1	100.0	77.1	90.0	99.4	79.8	90.9	100.5	74.9	89.4	98.7	71.1	86.7

NEBRASKA / NEVADA

DIVISION		NORFOLK 687 MAT.	INST.	TOTAL	NORTH PLATTE 691 MAT.	INST.	TOTAL	OMAHA 680 - 681 MAT.	INST.	TOTAL	VALENTINE 692 MAT.	INST.	TOTAL	CARSON CITY 897 MAT.	INST.	TOTAL	ELKO 898 MAT.	INST.	TOTAL
015433	CONTRACTOR EQUIPMENT		92.8	92.8		101.6	101.6		92.8	92.8		95.9	95.9		98.2	98.2		98.2	98.2
0241, 31 - 34	SITE & INFRASTRUCTURE, DEMOLITION	83.6	91.5	89.2	105.2	92.3	96.0	90.0	92.1	91.5	88.6	96.5	94.2	84.5	99.0	94.8	66.9	97.6	88.8
0310	Concrete Forming & Accessories	81.7	71.1	72.5	95.4	71.5	74.8	94.3	72.4	75.4	83.7	55.8	59.6	103.3	92.8	94.2	109.2	81.3	85.2
0320	Concrete Reinforcing	98.4	66.5	82.2	101.2	77.1	88.9	99.7	76.6	87.9	101.8	66.1	83.7	103.2	118.4	111.0	102.7	113.2	108.1
0330	Cast-in-Place Concrete	116.7	62.4	94.4	122.6	64.5	98.7	101.1	75.5	90.6	108.4	57.2	87.4	100.4	85.7	94.4	95.1	80.2	89.0
03	CONCRETE	107.3	67.5	87.8	113.0	70.8	92.3	100.1	74.5	87.5	110.2	58.9	85.0	100.9	95.2	98.1	95.4	87.2	91.4
04	MASONRY	129.4	79.1	98.0	98.6	87.9	92.0	106.8	79.0	89.4	110.9	73.9	87.8	117.3	94.6	103.2	115.0	71.9	88.1
05	METALS	99.6	73.5	91.6	95.5	85.4	92.4	99.9	78.9	93.4	107.6	73.4	97.1	104.7	104.7	97.3	96.9	99.4	97.7
06	WOOD, PLASTICS & COMPOSITES	84.0	71.7	77.1	97.8	72.2	83.5	94.1	72.0	81.7	81.7	51.7	64.9	92.1	94.4	93.4	100.6	81.8	90.1
07	THERMAL & MOISTURE PROTECTION	100.8	73.6	89.6	98.9	82.8	92.3	95.5	78.7	88.6	99.6	67.6	86.5	103.5	92.3	99.0	99.9	76.1	90.1
08	OPENINGS	97.2	66.5	90.0	94.3	69.1	88.4	100.0	71.5	93.3	96.9	55.6	87.3	99.5	107.9	101.4	100.9	85.9	97.4
0920	Plaster & Gypsum Board	90.8	71.3	77.6	93.7	71.3	78.6	100.5	71.6	81.0	95.6	50.7	65.3	100.9	94.1	96.3	105.3	81.2	89.0
0950, 0980	Ceilings & Acoustic Treatment	92.0	71.3	78.4	88.0	71.3	77.0	93.0	71.6	78.9	104.9	50.7	69.3	94.0	94.1	94.1	94.4	81.2	85.7
0960	Flooring	117.4	100.4	112.5	98.5	100.4	99.1	113.4	81.9	104.4	127.4	79.8	113.8	101.9	107.9	103.6	105.4	51.1	89.9
0970, 0990	Wall Finishes & Painting/Coating	151.7	63.7	98.6	93.8	63.7	75.6	126.2	67.2	90.6	167.2	66.1	106.1	101.2	80.8	88.9	102.6	95.9	98.5
09	FINISHES	108.8	75.2	90.3	95.9	75.2	84.5	106.0	73.3	88.0	118.2	59.7	86.0	97.9	94.3	95.9	97.6	77.2	86.3
COVERS	DIVS. 10 - 14, 25, 28, 41, 43, 44, 46	100.0	85.5	97.1	100.0	66.8	93.3	100.0	87.6	97.5	100.0	62.5	92.4	100.0	89.8	97.9	100.0	62.8	92.5
21, 22, 23	FIRE SUPPRESSION, PLUMBING & HVAC	95.0	75.8	87.2	99.9	76.0	90.2	99.9	76.8	90.6	94.8	75.7	87.1	100.0	80.1	92.0	97.7	79.6	90.4
26, 27, 3370	ELECTRICAL, COMMUNICATIONS & UTIL.	92.4	82.3	87.1	92.6	77.1	84.4	99.1	82.3	90.2	89.7	86.6	88.0	103.2	96.7	99.7	100.7	96.6	98.6
MF2014	WEIGHTED AVERAGE	100.0	76.7	89.8	99.0	78.3	90.0	100.2	78.6	90.8	101.2	72.5	88.7	99.7	93.5	97.0	98.3	84.9	92.4

NEVADA / NEW HAMPSHIRE

DIVISION		ELY 893			LAS VEGAS 889 - 891			RENO 894 - 895			CHARLESTON 036			CLAREMONT 037			CONCORD 032 - 033		
		MAT.	INST.	TOTAL	MAT.	INST.	TOTAL	MAT.	INST.	TOTAL	MAT.	INST.	TOTAL	MAT.	INST.	TOTAL	MAT.	INST.	TOTAL
015433	CONTRACTOR EQUIPMENT		98.2	98.2		98.2	98.2		98.2	98.2		100.9	100.9		100.9	100.9		100.9	100.9
0241, 31 - 34	SITE & INFRASTRUCTURE, DEMOLITION	72.2	99.1	91.3	75.4	101.0	93.6	72.2	99.0	91.3	86.9	99.6	96.0	80.9	99.6	94.2	93.7	103.9	100.9
0310	Concrete Forming & Accessories	102.3	105.7	105.2	103.5	109.6	108.8	98.7	92.7	93.5	86.8	81.1	81.9	92.6	81.1	82.7	95.5	93.6	93.8
0320	Concrete Reinforcing	101.5	111.6	106.7	93.7	119.9	107.1	96.0	119.9	108.1	84.1	93.5	88.9	84.1	93.5	88.9	95.3	94.1	94.7
0330	Cast-in-Place Concrete	102.1	105.4	103.4	98.9	107.8	102.5	108.0	85.7	98.8	97.1	70.5	86.2	89.2	70.5	81.6	112.2	90.5	103.3
03	CONCRETE	103.4	106.4	104.9	98.7	110.6	104.6	103.1	95.4	99.3	99.1	80.0	89.7	91.1	80.0	85.6	106.6	92.5	99.6
04	MASONRY	119.7	102.5	109.0	108.3	104.0	105.6	114.3	94.6	102.0	97.5	84.7	89.5	96.9	84.7	89.3	110.9	104.2	106.7
05	METALS	96.9	103.3	98.9	104.5	108.6	105.8	98.4	105.1	100.5	92.8	90.1	92.0	92.8	90.1	92.0	98.8	93.8	97.3
06	WOOD, PLASTICS & COMPOSITES	91.3	106.8	100.0	89.8	107.9	99.9	85.9	94.4	90.6	87.4	90.2	89.0	93.6	90.2	91.7	95.4	92.4	93.7
07	THERMAL & MOISTURE PROTECTION	100.3	95.7	98.4	113.8	102.6	109.2	99.9	92.3	96.8	98.8	72.3	88.0	98.7	72.3	87.9	101.9	91.5	97.6
08	OPENINGS	100.8	98.8	100.3	99.9	115.6	103.5	98.7	102.3	99.6	102.0	76.4	96.0	103.3	76.4	97.0	104.9	87.2	100.8
0920	Plaster & Gypsum Board	101.0	107.0	105.1	97.5	108.1	104.7	92.0	94.1	93.4	100.0	89.3	92.8	101.0	89.3	93.1	103.9	91.6	95.6
0950, 0980	Ceilings & Acoustic Treatment	94.4	107.0	102.7	102.7	108.1	106.3	99.4	94.1	95.9	94.2	89.3	91.0	94.2	89.3	91.0	99.9	91.6	94.4
0960	Flooring	103.0	57.2	89.9	94.6	107.9	98.4	99.9	107.9	102.2	98.4	32.1	79.4	100.6	32.1	81.0	102.7	116.9	106.8
0970, 0990	Wall Finishes & Painting/Coating	102.6	121.0	113.7	105.4	121.0	114.8	102.6	80.8	89.4	96.4	45.7	65.8	96.4	46.0	65.9	95.8	95.4	95.6
09	FINISHES	96.8	98.9	98.0	95.8	110.4	103.9	95.6	94.3	94.9	95.3	69.7	81.2	95.6	69.8	81.4	97.4	97.9	97.7
COVERS	DIVS. 10 - 14, 25, 28, 41, 43, 44, 46	100.0	63.6	92.6	100.0	104.4	100.9	100.0	89.4	97.9	100.0	92.2	98.4	100.0	92.2	98.4	100.0	104.3	100.9
21, 22, 23	FIRE SUPPRESSION, PLUMBING & HVAC	97.7	104.5	100.5	100.1	104.4	101.8	100.0	80.1	92.0	95.3	39.6	72.8	95.3	39.7	72.9	99.9	84.4	93.7
26, 27, 3370	ELECTRICAL, COMMUNICATIONS & UTIL.	101.0	109.1	105.3	105.4	119.9	113.1	101.4	96.7	98.9	98.5	52.9	74.4	98.5	52.9	74.4	99.5	83.6	91.1
MF2014	WEIGHTED AVERAGE	99.4	102.2	100.6	100.9	108.7	104.3	99.6	93.3	96.8	96.7	69.5	84.8	95.9	69.5	84.4	101.1	92.5	97.4

NEW HAMPSHIRE / NEW JERSEY

DIVISION		KEENE 034			LITTLETON 035			MANCHESTER 031			NASHUA 030			PORTSMOUTH 038			ATLANTIC CITY 082,084		
		MAT.	INST.	TOTAL	MAT.	INST.	TOTAL	MAT.	INST.	TOTAL	MAT.	INST.	TOTAL	MAT.	INST.	TOTAL	MAT.	INST.	TOTAL
015433	CONTRACTOR EQUIPMENT		100.9	100.9		100.9	100.9		100.9	100.9		100.9	100.9		100.9	100.9		99.1	99.1
0241, 31 - 34	SITE & INFRASTRUCTURE, DEMOLITION	94.0	99.9	98.2	81.0	99.9	94.4	93.6	103.9	100.9	95.7	103.9	101.5	89.2	104.0	99.7	96.4	105.2	102.7
0310	Concrete Forming & Accessories	91.3	83.0	84.2	102.9	83.0	85.8	96.9	94.0	94.4	99.3	94.0	94.8	88.2	94.3	93.4	109.0	127.5	125.0
0320	Concrete Reinforcing	84.1	93.6	88.9	84.8	93.6	89.3	103.7	94.1	98.8	105.1	94.1	99.5	84.1	94.2	89.2	76.3	118.1	97.6
0330	Cast-in-Place Concrete	97.6	72.8	87.4	87.6	72.7	81.5	110.6	113.1	111.7	92.4	113.1	100.9	87.6	113.2	98.1	86.1	133.0	105.3
03	CONCRETE	98.7	81.6	90.3	90.5	81.6	86.1	107.2	100.4	103.9	98.9	100.4	99.7	90.6	100.6	95.5	93.1	126.5	109.5
04	MASONRY	101.3	88.3	93.2	107.9	88.3	95.7	102.9	104.2	103.7	101.4	104.2	103.1	97.2	104.2	101.5	111.8	132.0	124.4
05	METALS	93.5	90.7	92.6	93.5	90.7	92.6	101.3	94.3	99.2	98.8	94.3	97.5	95.0	94.9	94.9	95.6	106.1	98.8
06	WOOD, PLASTICS & COMPOSITES	92.0	90.2	91.0	103.5	90.2	96.0	95.3	92.4	93.7	101.9	92.4	96.6	88.7	92.4	90.8	109.6	127.2	119.4
07	THERMAL & MOISTURE PROTECTION	99.2	75.3	89.4	98.8	73.8	88.5	101.5	95.0	98.8	99.6	95.0	97.7	99.3	115.0	105.7	101.5	124.2	113.4
08	OPENINGS	100.4	81.2	96.0	104.4	77.1	98.0	105.2	87.2	101.0	105.4	87.2	101.2	106.2	84.0	101.0	101.7	124.4	107.0
0920	Plaster & Gypsum Board	100.3	89.3	92.9	114.6	89.3	97.5	104.7	91.6	95.8	110.0	91.6	97.5	100.0	91.6	94.3	105.4	127.5	120.3
0950, 0980	Ceilings & Acoustic Treatment	94.2	89.3	91.0	94.2	89.3	91.0	101.7	91.6	95.0	105.1	91.6	96.2	95.1	91.6	92.8	81.3	127.5	111.6
0960	Flooring	100.2	52.6	86.6	109.6	32.1	87.6	98.6	116.9	103.8	103.7	116.9	107.4	98.5	116.9	103.8	105.4	159.9	121.0
0970, 0990	Wall Finishes & Painting/Coating	96.4	45.7	65.8	96.4	59.9	74.4	100.7	95.4	97.5	96.4	95.4	95.8	96.4	95.4	95.8	93.9	130.2	115.8
09	FINISHES	97.0	74.5	84.6	100.2	72.2	84.8	98.6	97.9	98.2	101.7	97.9	99.6	96.2	97.9	97.1	96.0	134.5	117.2
COVERS	DIVS. 10 - 14, 25, 28, 41, 43, 44, 46	100.0	93.4	98.7	100.0	98.6	99.7	100.0	104.3	100.9	100.0	104.3	100.9	100.0	104.3	100.9	100.0	111.5	102.3
21, 22, 23	FIRE SUPPRESSION, PLUMBING & HVAC	95.3	43.3	74.3	95.3	63.8	82.6	99.9	84.4	93.7	100.1	84.4	93.8	100.1	84.4	93.8	99.7	123.6	109.3
26, 27, 3370	ELECTRICAL, COMMUNICATIONS & UTIL.	98.5	63.2	79.9	99.8	55.9	76.6	98.8	83.6	91.2	101.2	83.6	91.9	99.1	83.6	90.9	91.1	138.1	116.0
MF2014	WEIGHTED AVERAGE	97.2	73.3	86.7	97.1	76.3	88.0	101.4	93.8	98.1	100.5	93.8	97.6	97.9	94.3	96.4	98.2	124.8	109.8

NEW JERSEY

DIVISION		CAMDEN 081			DOVER 078			ELIZABETH 072			HACKENSACK 076			JERSEY CITY 073			LONG BRANCH 077		
		MAT.	INST.	TOTAL	MAT.	INST.	TOTAL	MAT.	INST.	TOTAL	MAT.	INST.	TOTAL	MAT.	INST.	TOTAL	MAT.	INST.	TOTAL
015433	CONTRACTOR EQUIPMENT		99.1	99.1		100.9	100.9		100.9	100.9		100.9	100.9		99.1	99.1		98.7	98.7
0241, 31 - 34	SITE & INFRASTRUCTURE, DEMOLITION	97.8	105.5	103.3	102.0	106.3	105.1	106.1	106.3	106.3	103.0	106.3	105.3	93.4	106.3	102.5	97.5	106.0	103.6
0310	Concrete Forming & Accessories	100.2	127.5	123.8	97.8	128.3	124.1	110.2	128.4	125.9	97.8	128.3	124.1	101.8	128.3	124.7	102.4	128.0	124.5
0320	Concrete Reinforcing	100.1	118.3	109.4	77.2	137.2	107.7	77.2	137.2	107.7	77.2	137.2	107.7	100.1	137.2	119.0	77.2	137.1	107.7
0330	Cast-in-Place Concrete	83.5	132.9	103.8	101.5	127.7	112.2	87.1	131.4	105.3	99.2	131.4	112.4	79.2	127.7	99.1	88.0	132.9	106.4
03	CONCRETE	94.0	126.5	110.0	98.8	128.7	113.5	94.5	130.1	112.0	97.0	130.0	113.2	92.1	128.6	110.0	96.2	130.2	112.9
04	MASONRY	101.3	132.0	120.4	93.0	132.5	117.6	108.7	132.5	123.5	96.9	132.5	119.1	87.0	132.5	115.4	101.3	132.0	120.5
05	METALS	101.1	106.1	102.6	93.3	116.2	100.4	94.8	116.2	101.4	93.4	116.1	100.4	98.9	113.9	103.5	93.4	113.7	99.7
06	WOOD, PLASTICS & COMPOSITES	98.4	127.2	114.5	96.2	127.2	113.5	111.7	127.2	120.4	96.2	127.2	113.5	97.1	127.2	114.0	98.2	127.1	114.4
07	THERMAL & MOISTURE PROTECTION	105.8	123.3	113.0	100.6	132.8	113.8	100.8	133.3	114.1	100.4	125.5	110.7	100.2	132.8	113.5	100.3	124.4	110.2
08	OPENINGS	104.1	124.5	108.8	105.5	127.7	110.6	103.6	127.7	109.2	102.9	127.7	108.7	101.6	127.7	107.7	97.8	127.6	104.7
0920	Plaster & Gypsum Board	101.5	127.5	119.0	100.9	127.5	118.9	107.7	127.5	121.0	100.9	127.5	118.9	104.3	127.5	120.0	102.7	127.5	119.4
0950, 0980	Ceilings & Acoustic Treatment	91.5	127.5	115.1	83.8	127.5	112.5	85.7	127.5	113.1	83.8	127.5	112.5	94.9	127.5	116.3	83.8	127.5	112.5
0960	Flooring	101.5	159.9	118.2	93.6	178.1	117.8	98.5	178.1	121.3	93.6	178.1	117.8	94.6	178.1	118.5	94.8	178.1	118.6
0970, 0990	Wall Finishes & Painting/Coating	93.9	130.2	115.8	98.4	132.5	119.0	98.4	132.5	119.0	98.4	132.5	119.0	98.5	132.5	119.0	98.5	130.2	117.6
09	FINISHES	96.4	134.5	117.4	93.7	136.9	117.5	96.9	136.9	118.9	93.5	136.9	117.4	96.6	136.9	118.8	94.5	137.5	118.2
COVERS	DIVS. 10 - 14, 25, 28, 41, 43, 44, 46	100.0	111.5	102.3	100.0	118.1	103.7	100.0	118.1	103.7	100.0	118.1	103.7	100.0	118.1	103.7	100.0	111.3	102.3
21, 22, 23	FIRE SUPPRESSION, PLUMBING & HVAC	100.0	123.6	109.5	99.7	127.4	110.9	100.1	125.5	110.3	99.7	127.4	110.9	100.1	127.4	111.1	99.7	127.1	110.8
26, 27, 3370	ELECTRICAL, COMMUNICATIONS & UTIL.	96.0	138.1	118.2	92.8	137.7	116.5	93.4	137.7	116.8	92.8	140.1	117.8	97.6	140.1	120.1	92.5	130.8	112.7
MF2014	WEIGHTED AVERAGE	99.5	124.8	110.5	97.9	127.8	110.9	98.8	127.6	111.3	97.6	128.1	110.9	97.9	127.9	111.0	97.1	126.3	109.8

For customer support on your Plumbing Cost Data, call 877.709.6968.

623

NEW JERSEY

DIVISION		NEW BRUNSWICK 088 - 089			NEWARK 070 - 071			PATERSON 074 - 075			POINT PLEASANT 087			SUMMIT 079			TRENTON 085 - 086		
		MAT.	INST.	TOTAL	MAT.	INST.	TOTAL	MAT.	INST.	TOTAL	MAT.	INST.	TOTAL	MAT.	INST.	TOTAL	MAT.	INST.	TOTAL
015433	CONTRACTOR EQUIPMENT		98.7	98.7		100.9	100.9		100.9	100.9		98.7	98.7		100.9	100.9		98.7	98.7
0241, 31 - 34	SITE & INFRASTRUCTURE, DEMOLITION	109.2	106.0	107.0	107.8	106.3	106.8	105.0	106.3	105.9	110.9	106.0	107.4	103.7	106.3	105.6	95.8	106.0	103.0
0310	Concrete Forming & Accessories	103.3	128.3	124.8	97.6	128.4	124.2	99.9	128.2	124.3	97.7	127.9	123.7	100.6	128.4	124.5	98.6	127.7	123.7
0320	Concrete Reinforcing	77.2	137.2	107.7	99.8	137.2	118.8	100.1	137.2	119.0	77.2	137.1	107.7	77.2	137.2	107.7	99.8	112.4	106.2
0330	Cast-in-Place Concrete	106.3	133.2	117.4	108.8	131.5	118.1	100.8	131.4	113.4	106.3	132.8	117.2	84.3	131.4	103.6	101.8	132.8	114.5
03	CONCRETE	110.4	130.5	120.3	105.8	130.1	117.7	102.2	130.0	115.9	110.1	130.1	119.9	91.5	130.1	110.5	102.5	125.5	113.8
04	MASONRY	109.3	132.5	123.7	99.6	132.5	120.1	93.6	132.5	117.9	97.0	132.0	118.8	95.5	132.5	118.6	103.9	132.0	121.4
05	METALS	95.6	113.8	101.2	100.8	116.3	105.5	94.2	116.1	100.9	95.6	113.5	101.1	93.3	116.2	100.4	100.8	105.2	102.1
06	WOOD, PLASTICS & COMPOSITES	103.0	127.1	116.5	93.9	127.2	112.5	98.8	127.2	114.7	95.9	127.1	113.4	100.1	127.2	115.3	95.6	127.1	113.3
07	THERMAL & MOISTURE PROTECTION	106.1	132.0	116.7	101.1	133.3	114.3	100.7	125.5	110.9	106.2	124.4	113.7	101.0	133.3	114.2	105.5	124.3	113.2
08	OPENINGS	96.1	127.6	103.5	104.4	127.7	109.8	108.3	127.7	112.8	98.2	129.2	105.4	110.2	127.7	114.2	106.1	122.2	109.8
0920	Plaster & Gypsum Board	102.6	127.5	119.4	101.1	127.5	118.9	104.3	127.5	120.0	98.3	127.5	118.0	102.7	127.5	119.4	98.7	127.5	118.1
0950, 0980	Ceilings & Acoustic Treatment	81.3	127.5	111.6	97.6	127.5	117.2	94.9	127.5	116.3	81.3	127.5	111.6	83.8	127.5	112.5	93.1	127.5	115.7
0960	Flooring	103.0	178.1	124.5	95.7	178.1	119.2	94.6	178.1	118.5	100.5	159.9	117.5	94.9	178.1	118.7	102.1	172.2	122.1
0970, 0990	Wall Finishes & Painting/Coating	93.9	132.5	117.2	99.8	132.5	119.5	98.4	132.5	119.0	93.9	130.2	115.8	98.4	132.5	119.0	99.1	130.2	117.8
09	FINISHES	96.0	136.8	118.5	95.6	136.9	118.3	96.7	136.9	118.9	94.7	134.5	116.6	94.6	136.9	117.9	96.6	136.5	118.6
COVERS	DIVS. 10 - 14, 25, 28, 41, 43, 44, 46	100.0	118.0	103.6	100.0	118.1	103.7	100.0	118.1	103.7	100.0	108.7	101.8	100.0	118.1	103.7	100.0	111.3	102.3
21, 22, 23	FIRE SUPPRESSION, PLUMBING & HVAC	99.7	127.4	110.8	100.1	127.4	111.1	100.1	127.4	111.1	99.7	127.0	110.7	99.7	125.5	110.1	100.1	126.9	110.9
26, 27, 3370	ELECTRICAL, COMMUNICATIONS & UTIL.	91.8	136.8	115.6	101.2	140.1	121.8	97.6	137.7	118.8	91.1	130.8	112.1	93.4	137.7	116.8	99.3	135.4	118.4
MF2014	WEIGHTED AVERAGE	99.7	127.6	111.9	101.2	128.4	113.0	99.6	127.8	111.9	99.2	125.9	110.8	97.9	127.6	110.8	100.9	125.1	111.5

DIVISION		NEW JERSEY VINELAND 080,083			NEW MEXICO ALBUQUERQUE 870 - 872			CARRIZOZO 883			CLOVIS 881			FARMINGTON 874			GALLUP 873		
		MAT.	INST.	TOTAL	MAT.	INST.	TOTAL	MAT.	INST.	TOTAL	MAT.	INST.	TOTAL	MAT.	INST.	TOTAL	MAT.	INST.	TOTAL
015433	CONTRACTOR EQUIPMENT		99.1	99.1		110.1	110.1		110.1	110.1		110.1	110.1		110.1	110.1		110.1	110.1
0241, 31 - 34	SITE & INFRASTRUCTURE, DEMOLITION	100.7	105.5	104.1	87.6	103.3	98.7	105.8	103.3	104.0	94.1	103.3	100.6	94.1	103.3	100.6	102.1	103.3	102.9
0310	Concrete Forming & Accessories	95.2	127.6	123.1	101.7	64.6	69.7	99.2	64.6	69.3	99.2	64.5	69.2	101.7	64.6	69.7	101.7	64.6	69.7
0320	Concrete Reinforcing	76.3	116.1	96.6	98.4	68.9	83.4	107.0	68.9	87.6	108.2	68.9	88.2	107.6	68.9	87.9	103.0	68.9	85.6
0330	Cast-in-Place Concrete	92.8	133.0	109.3	98.9	70.4	87.2	95.8	70.4	85.4	95.7	70.3	85.3	99.8	70.4	87.7	93.9	70.4	84.3
03	CONCRETE	97.9	126.2	111.8	100.2	68.5	84.6	117.0	68.5	93.2	105.8	68.4	87.4	103.9	68.5	86.5	110.0	68.5	89.6
04	MASONRY	99.4	132.0	119.7	105.0	58.7	76.1	103.4	58.7	75.5	103.4	58.7	75.5	113.7	58.7	79.4	99.4	58.7	74.0
05	METALS	95.5	105.5	98.6	105.2	86.8	99.5	99.5	86.8	95.6	99.2	86.7	95.3	102.9	86.8	97.9	102.0	86.8	97.3
06	WOOD, PLASTICS & COMPOSITES	92.8	127.2	112.1	97.5	65.5	79.6	93.1	65.5	77.6	93.1	65.5	77.6	97.7	65.5	79.7	97.7	65.5	79.7
07	THERMAL & MOISTURE PROTECTION	105.7	124.2	113.3	94.8	73.5	86.0	100.3	73.5	88.7	99.2	73.5	88.7	94.9	73.5	86.2	95.9	73.5	86.7
08	OPENINGS	97.7	124.1	103.8	101.2	67.7	93.4	98.6	67.7	91.4	98.7	67.7	91.5	103.6	67.7	95.3	103.6	67.7	95.3
0920	Plaster & Gypsum Board	96.9	127.5	117.5	95.8	64.2	74.4	78.6	64.2	68.8	78.6	64.2	68.8	89.4	64.2	72.3	89.4	64.2	72.3
0950, 0980	Ceilings & Acoustic Treatment	81.3	127.5	111.6	95.5	64.2	74.9	96.3	64.2	75.2	96.3	64.2	75.2	92.6	64.2	73.9	92.6	64.2	73.9
0960	Flooring	99.6	159.9	116.8	101.9	66.0	91.6	101.2	66.0	91.2	101.2	66.0	91.2	103.7	66.0	92.9	103.7	66.0	92.9
0970, 0990	Wall Finishes & Painting/Coating	93.9	130.2	115.8	113.2	66.6	85.1	103.3	66.6	81.1	103.3	66.6	81.1	107.1	66.6	82.7	107.1	66.6	82.7
09	FINISHES	93.5	134.5	116.1	95.9	64.8	78.8	96.6	64.8	79.1	95.4	64.8	78.5	94.8	64.8	78.3	96.0	64.8	78.8
COVERS	DIVS. 10 - 14, 25, 28, 41, 43, 44, 46	100.0	111.5	102.3	100.0	82.9	96.5	100.0	82.9	96.5	100.0	82.9	96.5	100.0	82.9	96.5	100.0	82.9	96.5
21, 22, 23	FIRE SUPPRESSION, PLUMBING & HVAC	99.7	123.6	109.3	100.1	69.0	87.6	97.2	69.0	85.8	97.2	68.7	85.7	100.0	69.0	87.5	97.1	69.0	85.8
26, 27, 3370	ELECTRICAL, COMMUNICATIONS & UTIL.	91.1	138.1	116.0	91.2	71.2	80.6	92.8	71.2	81.4	90.2	71.2	80.2	88.8	71.2	79.5	88.0	71.2	79.1
MF2014	WEIGHTED AVERAGE	97.5	124.7	109.4	99.6	72.6	87.2	100.3	72.6	88.2	98.3	72.5	87.1	100.1	72.6	88.1	99.5	72.6	87.8

NEW MEXICO

DIVISION		LAS CRUCES 880			LAS VEGAS 877			ROSWELL 882			SANTA FE 875			SOCORRO 878			TRUTH/CONSEQUENCES 879		
		MAT.	INST.	TOTAL	MAT.	INST.	TOTAL	MAT.	INST.	TOTAL	MAT.	INST.	TOTAL	MAT.	INST.	TOTAL	MAT.	INST.	TOTAL
015433	CONTRACTOR EQUIPMENT		86.0	86.0		110.1	110.1		110.1	110.1		110.1	110.1		110.1	110.1		86.0	86.0
0241, 31 - 34	SITE & INFRASTRUCTURE, DEMOLITION	94.0	82.9	86.1	93.6	103.3	100.5	96.3	103.3	101.3	98.8	103.3	102.0	89.8	103.3	99.4	108.3	82.9	90.3
0310	Concrete Forming & Accessories	95.7	63.4	67.8	101.7	64.6	69.7	99.2	64.6	69.3	100.4	64.6	69.5	101.7	64.6	69.7	99.3	63.4	68.3
0320	Concrete Reinforcing	104.6	68.7	86.4	104.7	68.9	86.5	108.2	68.9	88.2	103.8	68.9	86.1	106.8	68.9	87.5	100.8	68.7	84.5
0330	Cast-in-Place Concrete	90.4	62.7	79.0	97.1	70.4	86.1	95.8	70.4	85.3	105.3	70.4	91.0	95.1	70.4	85.0	104.2	62.7	87.1
03	CONCRETE	84.9	65.0	75.1	101.2	68.5	85.1	106.5	68.5	87.8	104.0	68.5	86.5	100.2	68.5	84.6	94.1	65.0	79.8
04	MASONRY	99.2	58.3	73.7	99.7	58.7	74.1	114.3	58.7	79.6	104.1	58.7	75.8	99.6	58.7	74.1	97.2	58.3	73.0
05	METALS	98.1	80.3	92.6	101.7	86.8	97.1	100.4	86.8	96.2	99.0	86.8	95.2	102.0	86.8	97.3	101.6	80.3	95.0
06	WOOD, PLASTICS & COMPOSITES	82.5	64.4	72.4	97.7	65.5	79.7	93.1	65.5	77.6	99.3	65.5	80.4	97.7	65.5	79.7	89.1	64.4	75.3
07	THERMAL & MOISTURE PROTECTION	86.0	68.6	78.8	94.5	73.5	85.9	99.3	73.5	88.7	96.9	73.5	87.3	94.5	73.5	85.9	83.3	68.6	77.2
08	OPENINGS	91.5	67.1	85.8	99.9	67.7	92.4	98.5	67.7	91.4	102.0	67.7	94.0	99.8	67.7	92.3	93.1	67.1	87.0
0920	Plaster & Gypsum Board	77.5	64.2	68.5	89.4	64.2	72.3	78.6	64.2	68.8	98.9	64.2	75.4	89.4	64.2	72.3	91.1	64.2	72.9
0950, 0980	Ceilings & Acoustic Treatment	84.2	64.2	71.0	92.6	64.2	73.9	96.3	64.2	75.2	93.0	64.2	74.0	92.6	64.2	73.9	82.4	64.2	70.4
0960	Flooring	131.2	66.0	112.5	103.7	66.0	92.9	101.2	66.0	91.2	101.8	66.0	92.9	103.7	66.0	92.9	135.3	66.0	115.4
0970, 0990	Wall Finishes & Painting/Coating	90.7	66.6	76.2	107.1	66.6	82.7	103.3	66.6	81.1	111.8	66.6	84.5	107.1	66.6	82.7	97.7	66.6	79.9
09	FINISHES	105.2	63.9	82.5	94.7	64.8	78.2	95.5	64.8	78.6	99.7	64.8	80.5	94.6	64.8	78.2	107.5	63.9	83.5
COVERS	DIVS. 10 - 14, 25, 28, 41, 43, 44, 46	100.0	80.2	96.0	100.0	82.9	96.5	100.0	82.9	96.5	100.0	82.9	96.5	100.0	82.9	96.5	100.0	80.2	96.0
21, 22, 23	FIRE SUPPRESSION, PLUMBING & HVAC	100.3	68.7	87.6	97.1	69.0	85.8	99.9	69.0	87.5	100.0	69.0	87.5	97.1	69.0	85.8	97.1	68.7	85.6
26, 27, 3370	ELECTRICAL, COMMUNICATIONS & UTIL.	92.1	71.2	81.0	90.6	71.2	80.4	91.7	71.2	80.9	102.9	71.2	86.2	88.5	71.2	79.4	92.3	71.2	81.2
MF2014	WEIGHTED AVERAGE	96.1	69.4	84.4	98.0	72.6	86.9	100.0	72.6	88.0	100.8	72.6	88.5	97.6	72.6	86.7	97.5	69.4	85.2

City Cost Indexes

DIVISION		NEW MEXICO TUCUMCARI 884			NEW YORK ALBANY 120-122			BINGHAMTON 137-139			BRONX 104			BROOKLYN 112			BUFFALO 140-142		
		MAT.	INST.	TOTAL	MAT.	INST.	TOTAL	MAT.	INST.	TOTAL	MAT.	INST.	TOTAL	MAT.	INST.	TOTAL	MAT.	INST.	TOTAL
015433	CONTRACTOR EQUIPMENT		110.1	110.1		112.9	112.9		114.1	114.1		110.6	110.6		113.3	113.3		97.1	97.1
0241, 31 - 34	SITE & INFRASTRUCTURE, DEMOLITION	93.8	103.3	100.5	83.8	106.3	99.8	95.9	94.1	94.6	108.8	120.8	117.3	120.3	127.6	125.5	98.3	98.2	98.2
0310	Concrete Forming & Accessories	99.2	64.5	69.2	100.1	100.6	100.5	100.8	88.5	90.2	98.4	175.3	164.8	107.2	182.9	172.5	97.3	116.5	113.9
0320	Concrete Reinforcing	106.0	68.9	87.1	104.0	103.8	103.9	93.7	97.8	95.8	103.9	184.4	144.9	95.1	205.9	151.5	97.8	102.1	100.0
0330	Cast-in-Place Concrete	95.7	70.3	85.3	91.9	111.3	99.9	102.7	127.9	113.1	95.9	173.4	127.7	104.8	172.1	132.4	106.7	119.7	112.0
03	CONCRETE	105.0	68.4	87.0	99.0	105.4	102.1	95.8	105.2	100.4	96.1	174.8	134.8	107.5	181.3	143.7	103.0	114.1	108.4
04	MASONRY	114.6	58.7	79.8	101.0	112.6	108.2	109.7	127.8	121.0	92.7	177.5	145.5	120.2	177.4	155.9	105.1	121.0	115.0
05	METALS	99.2	86.7	95.3	104.1	110.3	106.0	96.3	118.9	103.3	99.9	151.6	115.8	104.4	150.1	118.4	99.6	95.4	98.3
06	WOOD, PLASTICS & COMPOSITES	93.1	65.5	77.6	98.7	97.9	98.3	104.9	84.7	93.6	94.4	174.9	139.5	106.9	185.1	150.6	99.4	116.4	108.9
07	THERMAL & MOISTURE PROTECTION	99.1	73.5	88.6	105.9	105.8	105.9	107.3	103.6	105.8	108.5	163.2	131.0	108.4	164.1	131.2	101.8	110.6	105.4
08	OPENINGS	98.5	67.7	91.3	102.6	94.0	100.6	93.0	85.0	91.1	87.9	175.5	108.3	90.4	180.0	111.2	97.7	103.9	99.1
0920	Plaster & Gypsum Board	78.6	64.2	68.8	97.2	97.6	97.5	107.4	83.8	91.4	99.2	176.9	151.7	102.8	187.6	160.2	98.3	116.7	110.7
0950, 0980	Ceilings & Acoustic Treatment	96.3	64.2	75.2	92.4	97.6	95.8	91.1	83.8	86.3	83.8	176.9	145.0	87.1	187.6	153.2	102.0	116.7	111.6
0960	Flooring	101.2	66.0	91.2	97.1	113.7	101.9	106.8	103.6	105.9	98.3	186.8	123.6	113.9	186.8	134.8	101.3	121.2	107.0
0970, 0990	Wall Finishes & Painting/Coating	103.3	66.6	81.1	103.8	94.3	98.0	93.3	98.8	96.6	102.9	157.4	135.8	123.4	157.4	143.9	98.8	112.4	107.0
09	FINISHES	95.4	64.8	78.5	94.7	101.9	98.7	94.9	91.1	92.8	94.0	175.9	139.1	108.4	181.9	148.9	100.6	117.9	110.2
COVERS	DIVS. 10 - 14, 25, 28, 41, 43, 44, 46	100.0	82.9	96.5	100.0	99.0	99.8	100.0	96.2	99.2	100.0	135.0	107.1	100.0	135.4	107.1	100.0	105.6	101.1
21, 22, 23	FIRE SUPPRESSION, PLUMBING & HVAC	97.2	68.7	85.7	100.0	102.8	101.1	100.5	90.2	96.3	100.2	165.5	126.5	99.7	165.4	126.2	100.0	96.7	98.7
26, 27, 3370	ELECTRICAL, COMMUNICATIONS & UTIL.	92.8	71.2	81.4	98.7	104.1	101.6	99.9	105.4	102.8	97.0	181.9	141.9	99.7	181.9	143.1	100.2	102.7	101.5
MF2014	WEIGHTED AVERAGE	99.0	72.5	87.4	100.1	104.7	102.1	98.5	101.4	99.8	97.7	166.1	127.5	102.8	168.5	131.4	100.3	106.2	102.9

DIVISION		NEW YORK ELMIRA 148-149			FAR ROCKAWAY 116			FLUSHING 113			GLENS FALLS 128			HICKSVILLE 115,117,118			JAMAICA 114		
		MAT.	INST.	TOTAL	MAT.	INST.	TOTAL	MAT.	INST.	TOTAL	MAT.	INST.	TOTAL	MAT.	INST.	TOTAL	MAT.	INST.	TOTAL
015433	CONTRACTOR EQUIPMENT		116.0	116.0		113.3	113.3		113.3	113.3		112.9	112.9		113.3	113.3		113.3	113.3
0241, 31 - 34	SITE & INFRASTRUCTURE, DEMOLITION	97.1	94.1	95.0	123.4	127.6	126.4	123.4	127.6	126.4	73.6	105.8	96.5	113.3	126.3	122.6	117.7	127.6	124.7
0310	Concrete Forming & Accessories	81.3	93.2	91.6	93.5	175.1	163.9	97.3	175.1	164.4	85.7	90.3	89.7	90.0	154.5	145.6	97.3	175.1	164.4
0320	Concrete Reinforcing	97.4	95.7	96.5	95.1	205.9	151.5	95.1	205.9	152.3	95.4	94.4	94.9	95.1	210.2	153.7	95.1	205.9	151.5
0330	Cast-in-Place Concrete	93.7	103.2	97.6	113.3	172.1	137.5	113.3	172.1	137.5	85.3	106.7	94.1	96.3	165.5	124.7	104.8	172.1	132.4
03	CONCRETE	90.8	98.6	94.6	113.6	177.8	145.1	114.1	177.8	145.4	88.3	97.5	92.9	99.4	167.1	132.6	106.8	177.8	141.7
04	MASONRY	103.5	102.1	102.6	124.1	177.4	157.4	118.1	177.4	155.1	100.7	105.6	103.7	114.6	166.8	147.1	122.2	177.4	156.6
05	METALS	96.7	117.3	103.0	104.4	150.1	118.4	104.4	150.1	118.4	97.7	106.1	100.3	105.8	149.1	119.2	104.4	150.1	118.4
06	WOOD, PLASTICS & COMPOSITES	85.0	92.2	89.0	90.1	174.6	137.4	94.8	174.6	139.5	87.1	87.6	87.3	86.6	152.6	123.6	94.8	174.6	139.5
07	THERMAL & MOISTURE PROTECTION	103.9	94.1	99.9	108.3	163.0	130.7	108.3	163.0	130.7	98.8	97.2	98.2	107.9	156.3	127.8	108.1	163.0	130.6
08	OPENINGS	99.4	88.9	97.0	89.0	174.4	108.8	89.0	174.4	108.8	93.3	86.2	91.7	89.0	163.4	106.3	89.0	174.4	108.8
0920	Plaster & Gypsum Board	97.8	91.7	93.7	91.6	176.9	149.2	94.1	176.9	150.0	89.1	87.0	87.7	91.2	154.2	133.8	94.1	176.9	150.0
0950, 0980	Ceilings & Acoustic Treatment	95.4	91.7	93.0	76.2	176.9	142.4	76.2	176.9	142.4	82.1	87.0	85.3	75.2	154.2	127.2	76.2	176.9	142.4
0960	Flooring	94.6	103.6	97.1	109.2	186.8	131.4	110.6	186.8	132.4	86.2	111.3	93.3	108.3	185.1	130.2	110.6	186.8	132.4
0970, 0990	Wall Finishes & Painting/Coating	98.9	90.0	93.5	123.4	157.4	143.9	123.4	157.4	143.9	101.1	87.4	92.8	123.4	157.4	143.9	123.4	157.4	143.9
09	FINISHES	95.5	94.7	95.0	103.5	175.7	143.3	104.3	175.7	143.6	86.1	93.1	90.0	102.1	159.7	133.8	103.8	175.7	143.4
COVERS	DIVS. 10 - 14, 25, 28, 41, 43, 44, 46	100.0	99.4	99.9	100.0	134.2	106.9	100.0	134.2	106.9	100.0	96.2	99.2	100.0	128.2	105.7	100.0	134.2	106.9
21, 22, 23	FIRE SUPPRESSION, PLUMBING & HVAC	95.5	91.5	93.9	95.0	165.3	123.4	95.0	165.3	123.4	95.5	95.7	95.6	99.7	154.4	121.8	95.0	165.3	123.4
26, 27, 3370	ELECTRICAL, COMMUNICATIONS & UTIL.	96.5	96.2	96.3	107.0	181.9	146.6	107.0	181.9	146.6	93.6	100.5	97.3	99.0	143.4	122.5	97.9	181.9	142.3
MF2014	WEIGHTED AVERAGE	96.7	97.4	97.0	102.6	166.8	130.6	102.5	166.8	130.5	94.0	98.6	96.0	100.9	153.3	123.7	100.8	166.8	129.6

DIVISION		NEW YORK JAMESTOWN 147			KINGSTON 124			LONG ISLAND CITY 111			MONTICELLO 127			MOUNT VERNON 105			NEW ROCHELLE 108		
		MAT.	INST.	TOTAL	MAT.	INST.	TOTAL	MAT.	INST.	TOTAL	MAT.	INST.	TOTAL	MAT.	INST.	TOTAL	MAT.	INST.	TOTAL
015433	CONTRACTOR EQUIPMENT		93.9	93.9		113.3	113.3		113.3	113.3		113.3	113.3		110.6	110.6		110.6	110.6
0241, 31 - 34	SITE & INFRASTRUCTURE, DEMOLITION	98.4	94.2	95.4	140.6	122.9	128.0	121.3	127.6	125.8	135.5	122.7	126.4	115.2	118.5	117.6	114.6	118.5	117.4
0310	Concrete Forming & Accessories	81.3	86.9	86.1	87.1	104.4	102.0	101.6	175.1	165.0	94.7	102.9	101.8	88.8	139.1	132.1	103.6	139.0	134.2
0320	Concrete Reinforcing	97.6	98.9	98.2	95.8	140.6	118.6	95.1	205.9	151.5	95.1	140.1	118.0	102.9	183.3	143.8	103.0	183.3	143.9
0330	Cast-in-Place Concrete	97.2	102.1	99.2	115.2	135.1	123.4	108.1	172.1	134.4	107.8	124.8	114.8	107.1	141.3	121.1	107.0	141.3	121.1
03	CONCRETE	93.8	94.3	94.0	110.8	121.5	116.0	109.9	177.8	143.3	105.5	117.2	111.2	105.4	146.9	125.8	105.0	146.9	125.6
04	MASONRY	111.8	99.9	104.4	116.0	141.3	131.8	116.9	177.4	154.6	108.6	128.7	121.1	98.1	147.3	128.8	98.1	147.3	128.8
05	METALS	94.0	92.1	93.5	106.1	117.2	109.5	104.4	150.1	118.4	106.0	116.2	109.2	99.6	135.6	110.7	99.9	135.6	110.9
06	WOOD, PLASTICS & COMPOSITES	83.5	84.0	83.8	88.4	96.4	92.9	100.9	174.6	142.2	96.1	96.4	96.3	84.6	136.2	113.5	101.4	136.2	120.9
07	THERMAL & MOISTURE PROTECTION	103.4	93.5	99.3	122.3	137.2	128.4	108.3	163.0	130.7	122.0	132.1	126.1	109.4	144.1	123.6	109.5	144.1	123.7
08	OPENINGS	99.2	85.6	96.1	95.9	117.3	100.9	89.0	174.4	108.8	91.1	117.3	97.2	87.9	148.6	102.0	88.0	148.6	102.1
0920	Plaster & Gypsum Board	88.5	83.2	84.9	91.4	96.3	94.7	98.7	176.9	151.5	92.1	96.3	94.9	94.9	137.0	123.3	106.3	137.0	127.0
0950, 0980	Ceilings & Acoustic Treatment	92.0	83.2	86.2	72.9	96.3	88.3	76.2	176.9	142.4	72.9	96.3	88.3	82.1	137.0	118.2	82.1	137.0	118.2
0960	Flooring	97.7	103.6	99.4	103.8	72.5	94.9	112.1	186.8	133.5	106.1	72.5	96.5	90.2	186.8	117.8	97.1	186.8	122.8
0970, 0990	Wall Finishes & Painting/Coating	100.2	94.2	96.6	134.4	116.8	123.8	123.4	157.4	143.9	134.4	116.8	123.8	101.1	157.4	135.1	101.1	157.4	135.1
09	FINISHES	94.6	89.9	92.0	101.0	97.1	98.8	105.1	175.7	144.0	101.4	96.3	98.6	91.2	149.2	123.2	94.6	149.2	124.7
COVERS	DIVS. 10 - 14, 25, 28, 41, 43, 44, 46	100.0	98.6	99.7	100.0	111.9	102.4	100.0	134.2	106.9	100.0	110.8	102.2	100.0	125.2	105.1	100.0	121.9	104.4
21, 22, 23	FIRE SUPPRESSION, PLUMBING & HVAC	95.4	87.2	92.1	95.4	122.3	106.3	99.7	165.3	126.2	95.4	116.7	104.0	95.5	134.1	111.1	95.5	134.1	111.1
26, 27, 3370	ELECTRICAL, COMMUNICATIONS & UTIL.	95.4	96.3	95.9	95.1	111.0	103.5	98.5	181.9	142.6	95.1	111.0	103.5	95.2	155.2	126.9	95.2	155.2	126.9
MF2014	WEIGHTED AVERAGE	96.8	92.5	94.9	102.6	118.5	109.6	102.4	166.8	130.5	101.1	115.1	107.2	97.5	141.4	116.6	97.9	141.3	116.8

For customer support on your Plumbing Cost Data, call 877.709.6968.

625

NEW YORK

DIVISION		NEW YORK 100 - 102			NIAGARA FALLS 143			PLATTSBURGH 129			POUGHKEEPSIE 125 - 126			QUEENS 110			RIVERHEAD 119		
		MAT.	INST.	TOTAL	MAT.	INST.	TOTAL	MAT.	INST.	TOTAL	MAT.	INST.	TOTAL	MAT.	INST.	TOTAL	MAT.	INST.	TOTAL
015433	CONTRACTOR EQUIPMENT		111.1	111.1		93.9	93.9		98.9	98.9		113.3	113.3		113.3	113.3		113.3	113.3
0241, 31 - 34	SITE & INFRASTRUCTURE, DEMOLITION	117.3	121.5	120.2	100.5	95.5	97.0	106.1	103.0	103.9	136.5	123.5	127.2	116.6	127.6	124.4	114.5	125.9	122.6
0310	Concrete Forming & Accessories	102.6	183.3	172.2	81.3	114.4	109.8	91.1	95.4	94.9	87.1	165.0	154.3	90.1	175.1	163.4	94.6	153.4	145.3
0320	Concrete Reinforcing	109.9	210.4	161.0	96.3	102.5	99.5	99.9	102.7	101.3	95.8	141.0	118.8	96.7	205.9	152.3	96.9	184.2	141.3
0330	Cast-in-Place Concrete	107.8	177.7	136.5	100.6	120.3	108.7	104.4	104.4	104.4	111.5	137.9	122.4	99.6	172.1	129.4	97.9	164.5	125.3
03	CONCRETE	106.2	184.2	144.5	95.9	113.3	104.4	102.5	99.3	101.0	107.9	149.5	128.3	102.6	177.8	139.6	100.1	161.4	130.2
04	MASONRY	102.5	177.5	149.2	119.9	127.5	124.6	95.0	101.6	99.1	108.6	144.5	131.0	111.1	177.4	152.4	119.9	166.3	148.8
05	METALS	113.1	151.9	125.0	96.7	93.8	95.8	102.0	91.2	98.7	106.1	119.9	110.3	104.4	150.1	118.4	106.3	138.4	116.2
06	WOOD, PLASTICS & COMPOSITES	98.5	185.4	147.2	83.4	110.3	98.5	93.8	93.5	93.6	88.4	174.6	136.7	86.7	174.6	135.9	91.8	152.6	125.9
07	THERMAL & MOISTURE PROTECTION	108.6	164.5	131.5	103.5	112.9	107.4	116.3	100.5	109.8	122.3	146.8	132.3	107.9	163.0	130.5	108.8	156.1	128.2
08	OPENINGS	93.7	180.9	114.0	99.2	100.7	99.6	101.6	90.8	99.1	95.9	159.5	110.7	89.0	174.4	108.8	89.0	157.5	104.9
0920	Plaster & Gypsum Board	106.0	187.6	161.2	88.5	110.3	103.3	109.4	92.6	98.0	91.4	176.9	149.2	91.2	176.9	149.1	92.5	154.2	134.2
0950, 0980	Ceilings & Acoustic Treatment	102.2	187.6	158.4	92.0	110.3	104.1	98.1	92.6	94.5	72.9	176.9	141.2	76.2	176.9	142.4	76.1	154.2	127.5
0960	Flooring	99.5	186.8	124.5	97.7	121.2	104.4	108.9	113.7	110.3	103.8	167.2	122.0	108.3	186.8	130.7	109.2	142.7	118.8
0970, 0990	Wall Finishes & Painting/Coating	102.9	157.4	135.8	100.2	112.4	107.6	130.0	91.4	106.7	134.4	117.2	124.0	123.4	157.4	143.9	123.4	157.4	143.9
09	FINISHES	99.9	182.1	145.2	94.7	115.9	106.4	97.9	98.0	98.0	100.8	163.0	135.1	102.5	175.7	142.8	102.7	151.3	129.5
COVERS	DIVS. 10 - 14, 25, 28, 41, 43, 44, 46	100.0	136.1	107.3	100.0	107.1	101.4	100.0	99.9	99.6	100.0	121.6	104.4	100.0	134.2	106.9	100.0	127.6	105.6
21, 22, 23	FIRE SUPPRESSION, PLUMBING & HVAC	100.1	165.5	126.5	95.4	100.0	97.2	95.4	96.6	95.9	95.4	127.0	108.1	97.9	165.3	126.2	99.9	151.4	120.7
26, 27, 3370	ELECTRICAL, COMMUNICATIONS & UTIL.	104.7	181.9	145.5	94.0	100.3	97.3	91.6	90.3	90.9	95.1	119.9	108.2	99.0	181.9	142.8	100.6	133.6	118.1
MF2014	WEIGHTED AVERAGE	103.4	168.7	131.8	97.8	106.4	101.5	99.1	96.7	98.0	101.8	136.7	117.0	100.9	166.8	129.6	101.6	148.2	121.9

NEW YORK

DIVISION		ROCHESTER 144 - 146			SCHENECTADY 123			STATEN ISLAND 103			SUFFERN 109			SYRACUSE 130 - 132			UTICA 133 - 135		
		MAT.	INST.	TOTAL	MAT.	INST.	TOTAL	MAT.	INST.	TOTAL	MAT.	INST.	TOTAL	MAT.	INST.	TOTAL	MAT.	INST.	TOTAL
015433	CONTRACTOR EQUIPMENT		116.7	116.7		112.9	112.9		110.6	110.6		110.6	110.6		112.9	112.9		112.9	112.9
0241, 31 - 34	SITE & INFRASTRUCTURE, DEMOLITION	85.9	109.9	102.9	84.2	106.3	99.9	119.9	120.8	120.6	111.4	116.3	114.9	94.8	105.6	102.4	73.5	104.3	95.4
0310	Concrete Forming & Accessories	98.7	98.2	98.3	103.0	100.6	100.9	88.3	183.3	170.2	96.9	135.2	129.9	99.8	91.0	92.2	101.0	87.9	89.7
0320	Concrete Reinforcing	100.1	95.7	97.9	94.4	103.8	99.2	103.9	210.4	158.1	103.0	140.9	122.3	94.7	96.4	95.6	94.7	95.7	95.2
0330	Cast-in-Place Concrete	94.2	104.4	98.4	102.9	111.3	106.4	107.1	173.5	134.4	103.8	138.2	117.9	95.4	105.9	99.7	87.3	104.7	94.4
03	CONCRETE	99.3	100.6	99.9	102.9	105.4	104.1	107.1	182.7	144.2	102.0	136.3	118.9	98.9	97.8	98.4	96.9	95.9	96.4
04	MASONRY	107.0	104.9	105.7	97.7	112.6	107.0	104.7	177.5	150.1	97.6	142.6	125.7	101.5	105.2	103.8	93.1	103.9	99.8
05	METALS	104.1	107.3	105.1	101.9	110.3	104.5	97.8	151.8	114.4	97.8	119.4	104.5	99.8	105.6	101.6	97.7	105.1	100.0
06	WOOD, PLASTICS & COMPOSITES	97.6	97.6	97.6	107.4	97.9	102.1	83.3	185.4	140.4	94.0	136.2	117.6	101.3	88.5	94.2	101.3	84.7	92.0
07	THERMAL & MOISTURE PROTECTION	103.5	101.7	102.7	100.3	105.8	102.6	108.9	164.3	131.6	109.3	142.1	122.8	102.3	97.6	100.4	90.6	97.6	93.5
08	OPENINGS	105.3	92.2	102.2	99.4	94.0	98.1	87.9	181.2	109.6	88.0	139.0	99.8	95.0	85.6	92.8	98.0	83.4	94.6
0920	Plaster & Gypsum Board	107.2	97.4	100.6	98.8	97.6	98.0	95.0	187.6	157.6	98.5	137.0	124.5	98.8	88.0	91.2	98.8	84.0	88.5
0950, 0980	Ceilings & Acoustic Treatment	100.4	97.4	98.5	88.7	97.6	94.6	83.8	187.6	152.0	82.1	137.0	118.2	91.1	88.0	89.0	91.1	84.0	86.4
0960	Flooring	93.8	113.3	99.4	92.9	113.7	98.9	94.2	186.8	120.7	93.5	185.1	119.7	94.6	102.1	96.8	92.1	102.2	95.0
0970, 0990	Wall Finishes & Painting/Coating	100.3	99.1	99.6	101.1	94.3	97.0	102.9	157.4	135.8	101.1	124.9	115.5	98.5	99.8	99.3	91.3	99.8	96.4
09	FINISHES	100.4	101.3	100.9	91.6	101.9	97.3	93.2	182.1	142.2	92.3	140.0	118.6	94.0	93.5	93.7	92.1	91.1	91.6
COVERS	DIVS. 10 - 14, 25, 28, 41, 43, 44, 46	100.0	99.8	100.0	100.0	99.0	99.8	100.0	136.1	107.3	100.0	123.6	104.8	100.0	96.5	99.3	100.0	96.2	99.2
21, 22, 23	FIRE SUPPRESSION, PLUMBING & HVAC	100.1	90.3	96.1	100.2	102.8	101.2	100.2	165.5	126.5	95.5	123.3	106.7	100.2	91.9	96.9	100.2	92.2	97.0
26, 27, 3370	ELECTRICAL, COMMUNICATIONS & UTIL.	98.7	94.6	96.6	98.1	104.1	101.3	97.0	181.9	141.9	102.8	119.9	111.8	100.0	102.2	101.1	98.1	102.2	100.3
MF2014	WEIGHTED AVERAGE	101.1	99.1	100.2	99.3	104.7	101.7	99.2	168.4	129.4	97.6	129.0	111.3	98.9	98.0	98.5	97.1	97.1	97.1

NEW YORK / NORTH CAROLINA

DIVISION		WATERTOWN 136			WHITE PLAINS 106			YONKERS 107			ASHEVILLE 287 - 288			CHARLOTTE 281 - 282			DURHAM 277		
		MAT.	INST.	TOTAL	MAT.	INST.	TOTAL	MAT.	INST.	TOTAL	MAT.	INST.	TOTAL	MAT.	INST.	TOTAL	MAT.	INST.	TOTAL
015433	CONTRACTOR EQUIPMENT		112.9	112.9		110.6	110.6		110.6	110.6		96.3	96.3		96.3	96.3		101.7	101.7
0241, 31 - 34	SITE & INFRASTRUCTURE, DEMOLITION	80.9	105.7	98.5	108.5	118.5	115.6	116.7	118.5	118.0	102.4	76.9	84.3	105.6	76.9	85.2	100.4	85.7	89.9
0310	Concrete Forming & Accessories	85.9	94.3	93.1	101.9	139.1	134.0	102.2	143.9	138.2	96.0	41.3	48.8	99.7	42.8	50.6	100.2	44.3	52.0
0320	Concrete Reinforcing	95.3	96.5	95.9	103.0	183.3	143.9	107.1	183.3	145.9	93.0	63.0	77.7	98.8	58.5	78.3	91.2	57.9	74.2
0330	Cast-in-Place Concrete	101.5	107.6	104.1	95.1	141.3	114.1	106.3	141.4	120.7	115.4	51.3	89.1	120.2	49.8	91.3	101.6	47.1	79.2
03	CONCRETE	109.3	99.9	104.7	95.7	146.9	120.9	105.2	149.1	126.7	106.0	50.6	79.8	108.6	49.9	79.8	98.7	49.5	74.5
04	MASONRY	94.2	107.9	102.8	97.1	147.3	128.4	102.1	147.3	130.3	93.6	43.7	62.5	101.9	51.2	70.3	86.4	37.9	56.2
05	METALS	97.8	105.6	100.2	99.3	135.6	110.5	108.9	135.7	117.1	103.1	82.3	96.7	104.0	80.9	96.9	121.4	79.9	108.6
06	WOOD, PLASTICS & COMPOSITES	83.0	92.2	88.1	99.5	136.2	120.0	99.3	142.7	123.6	97.9	40.3	65.6	103.0	41.9	68.8	96.7	44.7	67.6
07	THERMAL & MOISTURE PROTECTION	90.9	100.1	94.7	109.2	144.1	123.5	109.5	144.8	124.0	106.7	43.7	80.8	100.8	46.6	78.6	106.9	45.7	81.8
08	OPENINGS	98.0	89.4	96.0	88.0	148.6	102.1	91.1	151.7	105.2	97.1	44.1	84.8	102.5	45.0	89.1	105.8	47.1	92.2
0920	Plaster & Gypsum Board	88.7	91.7	90.8	101.3	137.0	125.4	105.6	143.6	131.3	100.1	38.3	58.3	100.0	39.9	59.4	105.9	42.8	63.3
0950, 0980	Ceilings & Acoustic Treatment	91.1	91.7	91.5	82.1	137.0	118.2	100.5	143.6	128.9	85.7	38.3	54.5	88.8	39.9	56.7	88.1	42.8	58.3
0960	Flooring	86.0	102.2	90.6	95.6	186.8	121.7	95.2	186.8	121.4	102.2	42.7	85.2	101.6	43.4	84.9	103.9	42.7	86.4
0970, 0990	Wall Finishes & Painting/Coating	91.3	96.2	94.3	101.1	157.4	135.1	101.1	157.4	135.1	113.6	40.3	69.3	113.8	49.4	74.9	105.4	37.4	64.3
09	FINISHES	89.7	95.7	93.0	92.9	149.2	123.9	98.0	153.0	128.3	95.6	40.7	65.3	95.5	42.9	66.5	96.0	42.9	66.7
COVERS	DIVS. 10 - 14, 25, 28, 41, 43, 44, 46	100.0	97.5	99.5	100.0	125.2	105.1	100.0	126.3	105.3	100.0	77.7	95.5	100.0	78.1	95.6	100.0	72.1	94.4
21, 22, 23	FIRE SUPPRESSION, PLUMBING & HVAC	100.2	85.9	94.4	100.3	134.1	114.0	100.3	134.1	114.0	100.4	53.4	81.4	100.0	54.1	81.5	100.5	52.9	81.3
26, 27, 3370	ELECTRICAL, COMMUNICATIONS & UTIL.	100.0	90.4	94.9	95.2	155.2	126.9	102.9	163.8	135.1	102.2	55.7	77.6	101.2	58.5	78.6	96.1	55.6	74.7
MF2014	WEIGHTED AVERAGE	98.5	96.3	97.5	97.6	141.4	116.7	102.1	143.7	120.2	100.8	55.3	80.9	101.9	56.8	82.2	102.9	55.3	82.1

City Cost Indexes

NORTH CAROLINA

DIVISION		ELIZABETH CITY 279			FAYETTEVILLE 283			GASTONIA 280			GREENSBORO 270,272 - 274			HICKORY 286			KINSTON 285		
		MAT.	INST.	TOTAL	MAT.	INST.	TOTAL	MAT.	INST.	TOTAL	MAT.	INST.	TOTAL	MAT.	INST.	TOTAL	MAT.	INST.	TOTAL
015433	CONTRACTOR EQUIPMENT		106.0	106.0		101.7	101.7		96.3	96.3		101.7	101.7		101.7	101.7		101.7	101.7
0241, 31 - 34	SITE & INFRASTRUCTURE, DEMOLITION	104.6	87.4	92.4	101.7	85.6	90.3	102.3	76.9	84.2	100.3	85.7	89.9	101.2	85.5	90.1	100.3	85.6	89.8
0310	Concrete Forming & Accessories	85.3	42.5	48.4	95.6	60.3	65.2	103.1	38.7	47.5	99.9	44.4	52.1	92.0	37.1	44.6	88.2	42.1	48.4
0320	Concrete Reinforcing	89.2	45.9	67.2	96.8	58.0	77.0	93.5	56.5	74.7	90.1	58.0	73.8	93.0	56.1	74.2	92.5	45.8	68.7
0330	Cast-in-Place Concrete	101.8	47.0	79.3	121.0	48.3	91.2	112.8	52.7	88.1	100.8	47.6	79.0	115.4	48.1	87.7	111.4	44.8	84.0
03	CONCRETE	98.7	46.4	73.0	108.1	57.1	83.1	104.5	48.7	77.1	98.1	49.8	74.4	105.7	46.3	76.5	102.4	45.5	74.4
04	MASONRY	98.6	48.0	67.1	97.3	38.9	60.9	98.3	50.9	68.7	82.5	41.4	56.9	82.3	43.7	58.2	89.1	48.1	63.6
05	METALS	107.1	75.8	97.5	124.2	80.0	110.6	103.9	80.0	96.5	113.7	80.0	103.4	103.2	78.7	95.7	102.0	74.8	93.6
06	WOOD, PLASTICS & COMPOSITES	80.1	43.6	59.7	97.1	66.8	80.1	107.0	36.6	67.6	96.4	44.9	67.6	92.1	35.2	60.2	88.6	42.9	63.0
07	THERMAL & MOISTURE PROTECTION	106.2	43.3	80.4	106.3	45.9	81.5	106.9	45.2	81.6	106.7	42.8	80.5	107.0	41.7	80.2	106.8	42.8	80.6
08	OPENINGS	102.6	38.4	87.7	97.2	59.1	88.4	100.9	41.0	87.0	105.8	47.2	92.2	97.2	36.8	83.1	97.3	43.1	84.7
0920	Plaster & Gypsum Board	98.6	41.0	59.7	104.7	65.6	78.3	107.0	34.5	58.0	107.5	43.0	63.9	100.1	33.1	54.8	99.8	41.0	60.1
0950, 0980	Ceilings & Acoustic Treatment	88.1	41.0	57.1	86.5	65.6	72.7	89.0	34.5	53.2	88.1	43.0	58.4	85.7	33.1	51.1	89.0	41.0	57.4
0960	Flooring	95.7	23.2	75.0	102.4	42.7	85.3	105.4	42.7	87.4	103.9	39.7	85.6	102.1	32.8	82.3	99.6	22.6	77.6
0970, 0990	Wall Finishes & Painting/Coating	105.4	42.0	67.1	113.6	33.2	65.0	113.6	40.3	69.3	105.4	31.6	60.9	113.6	40.3	69.3	113.6	39.2	68.7
09	FINISHES	93.0	38.7	63.0	96.5	55.3	73.8	98.1	38.6	65.3	96.3	41.9	66.3	95.8	35.4	62.5	95.5	38.0	63.8
COVERS	DIVS. 10 - 14, 25, 28, 41, 43, 44, 46	100.0	79.4	95.8	100.0	74.3	94.8	100.0	77.3	95.4	100.0	75.9	95.1	100.0	77.1	95.4	100.0	71.6	94.3
21, 22, 23	FIRE SUPPRESSION, PLUMBING & HVAC	95.6	51.1	77.7	100.2	52.7	81.0	100.4	52.4	81.0	100.4	53.1	81.3	95.6	52.2	78.1	95.6	51.2	77.7
26, 27, 3370	ELECTRICAL, COMMUNICATIONS & UTIL.	95.9	34.5	63.4	101.4	50.7	74.6	101.6	57.4	78.2	95.2	55.7	74.3	99.6	57.4	77.3	99.4	46.1	71.2
MF2014	WEIGHTED AVERAGE	99.3	51.6	78.5	104.5	58.1	84.2	101.6	55.1	81.3	101.3	55.7	81.4	98.8	53.9	79.2	98.4	52.7	78.5

NORTH CAROLINA / NORTH DAKOTA

DIVISION		MURPHY 289			RALEIGH 275 - 276			ROCKY MOUNT 278			WILMINGTON 284			WINSTON-SALEM 271			BISMARCK 585		
		MAT.	INST.	TOTAL	MAT.	INST.	TOTAL	MAT.	INST.	TOTAL	MAT.	INST.	TOTAL	MAT.	INST.	TOTAL	MAT.	INST.	TOTAL
015433	CONTRACTOR EQUIPMENT		96.3	96.3		101.7	101.7		101.7	101.7		96.3	96.3		101.7	101.7		98.2	98.2
0241, 31 - 34	SITE & INFRASTRUCTURE, DEMOLITION	103.5	76.7	84.5	101.3	85.7	90.2	102.6	85.7	90.6	103.7	77.1	84.8	100.6	85.7	90.0	101.4	96.9	98.2
0310	Concrete Forming & Accessories	103.7	39.3	48.2	98.9	46.6	53.8	91.9	44.2	50.8	97.5	48.8	55.5	101.9	48.0	55.4	104.9	40.8	49.6
0320	Concrete Reinforcing	92.6	45.4	68.6	96.2	56.0	75.7	89.2	55.8	72.2	93.7	57.9	75.5	90.1	57.9	73.7	99.9	92.7	96.2
0330	Cast-in-Place Concrete	119.4	44.4	88.6	106.7	51.4	84.0	99.6	47.7	78.3	115.0	49.5	88.1	103.4	50.0	81.4	102.6	47.5	79.9
03	CONCRETE	109.2	44.1	77.2	101.7	51.7	77.1	99.5	49.3	74.9	105.9	52.4	79.6	99.4	52.2	76.2	101.5	54.2	78.3
04	MASONRY	85.3	41.2	57.8	86.6	42.2	58.9	76.7	39.5	53.5	82.8	41.7	57.2	82.7	38.6	55.2	114.6	56.4	78.3
05	METALS	100.9	74.2	92.7	105.7	79.2	97.5	106.3	78.4	97.7	102.7	79.9	95.7	110.8	79.9	101.3	103.4	88.4	98.8
06	WOOD, PLASTICS & COMPOSITES	107.8	39.7	69.7	95.1	47.5	68.5	87.4	44.7	63.5	100.1	49.0	71.5	96.4	49.4	70.1	95.3	34.7	61.4
07	THERMAL & MOISTURE PROTECTION	106.9	40.8	79.8	100.5	47.0	78.5	106.6	40.8	79.7	106.7	46.9	82.2	106.7	43.5	80.8	112.4	51.3	87.4
08	OPENINGS	97.1	38.9	83.5	104.8	47.4	91.4	101.8	41.9	87.9	97.3	51.0	86.5	105.8	49.7	92.8	108.0	50.1	94.5
0920	Plaster & Gypsum Board	106.1	37.6	59.8	99.8	45.7	63.3	100.5	42.8	61.5	102.4	47.2	65.1	107.5	47.6	67.0	98.8	33.0	54.3
0950, 0980	Ceilings & Acoustic Treatment	85.7	37.6	54.1	88.8	45.7	60.5	85.6	42.8	57.5	86.5	47.2	60.7	88.1	47.6	61.5	112.9	33.0	60.4
0960	Flooring	105.7	24.1	82.4	101.2	42.7	84.5	99.4	21.8	77.2	102.9	44.4	86.2	103.9	42.7	86.4	99.4	72.1	91.6
0970, 0990	Wall Finishes & Painting/Coating	113.6	39.4	68.8	104.2	36.9	63.5	105.4	38.5	65.0	113.6	38.7	68.4	105.4	36.7	64.0	101.9	29.4	58.1
09	FINISHES	97.7	36.1	63.7	95.5	44.7	67.5	93.9	39.3	63.8	96.3	46.8	69.0	96.3	45.7	68.4	102.7	42.8	69.7
COVERS	DIVS. 10 - 14, 25, 28, 41, 43, 44, 46	100.0	76.9	95.3	100.0	72.6	94.5	100.0	72.5	94.5	100.0	73.8	94.7	100.0	78.7	95.7	100.0	82.5	96.5
21, 22, 23	FIRE SUPPRESSION, PLUMBING & HVAC	95.6	50.9	77.6	100.0	52.5	80.8	95.6	52.7	78.3	100.4	54.9	82.0	100.4	53.3	81.4	100.1	71.9	88.7
26, 27, 3370	ELECTRICAL, COMMUNICATIONS & UTIL.	103.2	29.3	64.1	98.3	40.4	67.7	97.9	39.7	67.1	102.4	50.7	75.1	95.2	55.7	74.3	102.5	71.6	86.2
MF2014	WEIGHTED AVERAGE	99.6	48.3	77.2	100.5	54.1	80.3	98.5	52.3	78.3	100.3	55.9	81.0	101.0	56.5	81.6	103.1	66.2	87.0

NORTH DAKOTA

DIVISION		DEVILS LAKE 583			DICKINSON 586			FARGO 580 - 581			GRAND FORKS 582			JAMESTOWN 584			MINOT 587		
		MAT.	INST.	TOTAL	MAT.	INST.	TOTAL	MAT.	INST.	TOTAL	MAT.	INST.	TOTAL	MAT.	INST.	TOTAL	MAT.	INST.	TOTAL
015433	CONTRACTOR EQUIPMENT		98.2	98.2		98.2	98.2		98.2	98.2		98.2	98.2		98.2	98.2		98.2	98.2
0241, 31 - 34	SITE & INFRASTRUCTURE, DEMOLITION	105.7	94.4	97.7	113.6	92.9	98.9	102.6	96.9	98.5	109.6	92.9	97.7	104.7	92.9	96.3	107.1	96.9	99.8
0310	Concrete Forming & Accessories	101.2	35.4	44.4	90.8	34.9	42.5	99.3	41.5	49.4	94.5	34.2	42.5	92.3	34.1	42.1	90.5	67.0	70.2
0320	Concrete Reinforcing	100.2	93.1	96.6	101.1	92.6	96.6	101.1	91.3	95.9	98.7	92.9	95.7	100.1	93.2	97.6	100.1	93.2	97.6
0330	Cast-in-Place Concrete	126.3	46.0	93.3	114.3	44.4	85.6	113.6	49.2	87.2	114.3	44.1	85.5	124.7	44.1	91.6	114.3	46.5	86.4
03	CONCRETE	110.6	51.3	81.5	109.6	50.2	80.4	105.9	55.1	80.9	106.8	49.8	78.8	109.1	48.7	79.5	105.5	65.6	85.9
04	MASONRY	121.7	64.9	86.3	124.3	60.4	84.5	114.2	56.4	78.2	115.9	64.4	83.8	135.1	32.8	71.3	114.2	65.4	83.8
05	METALS	99.9	87.4	96.0	99.8	82.5	94.5	103.0	88.7	98.6	99.8	82.1	94.4	99.8	63.7	88.7	100.1	89.3	96.8
06	WOOD, PLASTICS & COMPOSITES	95.2	32.0	59.8	83.1	32.0	54.5	92.9	35.2	60.6	87.4	32.0	56.4	85.0	32.0	55.3	82.8	70.1	75.7
07	THERMAL & MOISTURE PROTECTION	107.6	49.8	83.9	108.1	48.5	83.7	107.4	51.6	84.5	107.8	49.8	84.0	107.4	41.5	80.4	107.5	57.4	87.0
08	OPENINGS	100.7	42.7	87.2	100.7	42.7	87.2	100.5	50.3	88.9	100.7	42.7	87.2	100.7	32.2	84.8	100.9	69.3	93.5
0920	Plaster & Gypsum Board	118.4	30.2	58.8	109.1	30.2	55.8	96.6	33.5	54.0	110.5	30.2	56.2	110.2	30.2	56.1	109.1	69.4	82.3
0950, 0980	Ceilings & Acoustic Treatment	114.3	30.2	59.0	114.3	30.2	59.0	112.1	33.5	60.5	114.3	30.2	59.0	114.3	30.2	59.0	114.3	69.4	84.8
0960	Flooring	107.9	35.9	87.3	101.4	35.9	82.7	102.8	72.1	94.0	103.3	35.9	84.0	102.1	35.9	83.2	101.1	89.3	97.8
0970, 0990	Wall Finishes & Painting/Coating	104.4	22.5	54.9	104.4	31.5	60.4	101.2	70.7	82.8	104.4	28.2	58.4	104.4	22.5	54.9	104.4	27.9	58.2
09	FINISHES	108.2	32.2	66.3	105.9	33.2	65.8	103.8	47.7	72.9	106.1	32.8	65.7	105.3	32.2	65.0	105.0	67.0	84.1
COVERS	DIVS. 10 - 14, 25, 28, 41, 43, 44, 46	100.0	32.4	86.3	100.0	32.6	86.4	100.0	82.6	96.5	100.0	32.5	86.4	100.0	80.6	96.1	100.0	86.4	97.3
21, 22, 23	FIRE SUPPRESSION, PLUMBING & HVAC	95.6	74.0	86.9	95.6	66.2	83.7	100.1	77.0	90.8	100.4	33.3	73.3	95.6	35.0	71.2	100.4	60.2	84.2
26, 27, 3370	ELECTRICAL, COMMUNICATIONS & UTIL.	98.3	35.8	65.3	107.7	72.5	89.1	102.4	68.4	84.4	102.1	53.5	76.4	98.3	35.7	65.2	105.5	75.7	89.8
MF2014	WEIGHTED AVERAGE	102.0	58.6	83.1	102.8	60.9	84.6	102.7	67.6	87.4	102.7	51.6	80.4	102.1	45.2	77.3	102.7	71.3	89.0

For customer support on your Plumbing Cost Data, call 877.709.6968.

NORTH DAKOTA / OHIO

DIVISION		WILLISTON 588 MAT.	INST.	TOTAL	AKRON 442-443 MAT.	INST.	TOTAL	ATHENS 457 MAT.	INST.	TOTAL	CANTON 446-447 MAT.	INST.	TOTAL	CHILLICOTHE 456 MAT.	INST.	TOTAL	CINCINNATI 451-452 MAT.	INST.	TOTAL
015433	CONTRACTOR EQUIPMENT		98.2	98.2		94.4	94.4		90.4	90.4		94.4	94.4		99.7	99.7		99.5	99.5
0241, 31 - 34	SITE & INFRASTRUCTURE, DEMOLITION	107.5	92.9	97.1	98.1	101.6	100.6	111.0	91.6	97.2	98.2	101.0	100.2	97.0	102.6	101.0	94.1	102.4	100.0
0310	Concrete Forming & Accessories	96.3	34.9	43.3	99.0	94.3	95.0	95.0	84.3	85.8	99.0	83.7	85.8	97.4	92.3	93.0	99.3	79.2	82.0
0320	Concrete Reinforcing	103.1	92.8	97.9	99.5	92.1	95.7	93.1	87.0	90.0	99.5	75.8	87.5	90.1	80.4	85.1	95.5	79.2	87.2
0330	Cast-in-Place Concrete	114.3	44.4	85.6	94.0	98.0	95.7	111.5	96.5	105.4	94.9	95.0	95.0	101.2	99.3	100.4	93.2	92.4	92.9
03	CONCRETE	106.9	50.2	79.1	97.3	94.4	95.9	107.6	88.6	98.3	97.7	85.7	91.8	101.9	92.4	97.2	96.3	84.0	90.3
04	MASONRY	108.8	60.4	78.7	93.3	96.2	95.1	86.4	90.4	88.9	95.4	84.3	87.9	94.0	100.5	98.1	93.7	85.0	88.2
05	METALS	100.0	82.5	94.6	95.4	81.6	91.1	103.0	80.0	95.9	95.4	74.5	88.9	94.9	85.7	92.1	97.2	84.6	93.3
06	WOOD, PLASTICS & COMPOSITES	88.8	32.0	57.0	99.1	93.6	96.0	85.7	84.9	85.3	99.5	82.8	90.2	98.4	89.6	93.5	100.9	76.8	87.4
07	THERMAL & MOISTURE PROTECTION	107.7	48.5	83.5	110.8	96.5	104.9	99.4	94.8	97.5	112.0	91.4	103.5	101.2	97.4	99.7	99.2	88.7	94.9
08	OPENINGS	100.8	42.7	87.3	110.7	93.3	106.7	101.7	82.3	97.2	104.3	77.9	98.2	93.8	83.5	91.4	102.0	77.4	96.3
0920	Plaster & Gypsum Board	110.5	30.2	56.2	96.9	93.2	94.4	91.4	84.1	86.5	98.0	82.1	87.2	93.3	89.5	90.7	94.9	76.3	82.3
0950, 0980	Ceilings & Acoustic Treatment	114.3	30.2	59.0	93.0	93.2	93.1	103.3	84.1	90.7	93.0	82.1	85.8	97.9	89.5	92.4	98.7	76.3	84.0
0960	Flooring	104.1	35.9	84.6	96.9	93.4	95.9	121.3	99.0	114.9	97.1	81.7	92.7	98.3	97.7	98.1	99.3	90.7	96.8
0970, 0990	Wall Finishes & Painting/Coating	104.4	31.5	60.4	96.3	106.3	102.3	100.5	99.1	99.6	96.3	83.2	88.4	97.8	92.9	94.9	97.8	83.9	89.4
09	FINISHES	106.2	33.2	66.0	98.0	95.3	96.5	101.8	89.0	94.8	98.2	82.9	89.8	98.8	93.3	95.8	99.2	81.1	89.3
COVERS	DIVS. 10 - 14, 25, 28, 41, 43, 44, 46	100.0	32.6	86.4	100.0	98.2	99.6	100.0	52.0	90.3	100.0	95.4	99.1	100.0	92.6	98.5	100.0	89.3	97.8
21, 22, 23	FIRE SUPPRESSION, PLUMBING & HVAC	95.6	66.2	83.7	100.1	94.6	97.9	95.3	51.3	77.6	100.1	81.4	92.5	95.8	95.2	95.6	100.0	83.3	93.3
26, 27, 3370	ELECTRICAL, COMMUNICATIONS & UTIL.	102.5	72.5	86.6	99.3	93.9	96.5	95.5	98.1	96.9	98.5	92.7	95.5	95.0	85.6	90.0	93.9	79.4	86.3
MF2014	WEIGHTED AVERAGE	101.2	60.9	83.7	99.9	94.2	97.4	99.6	80.4	91.2	99.3	85.6	93.3	96.8	93.0	95.1	98.3	84.5	92.2

OHIO

DIVISION		CLEVELAND 441 MAT.	INST.	TOTAL	COLUMBUS 430-432 MAT.	INST.	TOTAL	DAYTON 453-454 MAT.	INST.	TOTAL	HAMILTON 450 MAT.	INST.	TOTAL	LIMA 458 MAT.	INST.	TOTAL	LORAIN 440 MAT.	INST.	TOTAL
015433	CONTRACTOR EQUIPMENT		94.7	94.7		93.6	93.6		94.7	94.7		99.7	99.7		92.9	92.9		94.4	94.4
0241, 31 - 34	SITE & INFRASTRUCTURE, DEMOLITION	98.0	102.2	101.0	95.5	98.3	97.5	92.9	101.8	99.2	92.8	102.1	99.4	104.6	91.6	95.4	97.5	102.9	101.4
0310	Concrete Forming & Accessories	99.1	99.6	99.5	100.3	84.5	86.7	99.3	79.0	81.8	99.4	79.4	82.1	95.0	86.9	88.1	99.1	86.8	88.5
0320	Concrete Reinforcing	100.0	92.5	96.2	102.7	82.6	92.5	95.5	81.3	88.3	95.5	79.2	87.2	93.1	81.5	87.2	99.5	92.4	95.9
0330	Cast-in-Place Concrete	92.2	106.5	98.1	94.1	93.7	93.9	86.7	86.4	86.6	92.9	92.7	92.8	102.4	97.5	100.4	89.5	104.1	95.5
03	CONCRETE	96.5	99.8	98.2	97.9	87.1	92.6	93.3	81.8	87.6	96.2	84.2	90.3	100.4	89.3	94.9	95.2	93.2	94.2
04	MASONRY	97.8	105.6	102.7	96.6	94.0	95.0	93.2	83.5	87.1	93.5	85.5	88.5	117.7	86.4	98.2	89.9	104.6	99.1
05	METALS	96.9	84.4	93.1	96.9	80.7	91.9	96.4	78.6	91.0	96.5	84.5	92.8	103.0	81.5	96.4	96.0	82.8	91.9
06	WOOD, PLASTICS & COMPOSITES	98.2	97.1	97.6	97.1	82.3	88.8	102.2	77.1	88.1	100.9	76.8	87.4	85.6	86.3	86.0	99.1	80.9	88.9
07	THERMAL & MOISTURE PROTECTION	109.5	108.6	109.1	100.9	94.9	98.4	104.7	87.6	97.7	101.4	88.8	96.2	99.0	95.0	97.3	111.9	103.0	108.2
08	OPENINGS	100.6	95.2	99.3	102.4	80.0	97.2	101.3	78.1	95.9	98.9	77.4	93.9	101.7	79.8	96.6	104.3	86.4	100.2
0920	Plaster & Gypsum Board	96.2	96.8	96.6	93.5	81.8	85.6	94.9	76.6	82.6	94.9	76.3	82.3	91.4	85.5	87.4	96.9	80.1	85.5
0950, 0980	Ceilings & Acoustic Treatment	91.3	96.8	94.9	97.7	81.8	87.3	99.7	76.6	84.5	98.7	76.3	84.0	102.4	85.5	91.3	93.0	80.1	84.5
0960	Flooring	96.7	105.0	99.1	92.0	90.3	91.5	102.0	80.7	95.9	99.3	90.7	96.8	120.3	92.2	112.3	97.1	105.0	99.3
0970, 0990	Wall Finishes & Painting/Coating	96.3	105.1	101.6	95.5	92.9	94.0	97.8	85.0	90.1	97.8	84.7	89.9	100.5	82.7	89.8	96.3	105.1	101.6
09	FINISHES	97.5	100.9	99.4	94.1	85.9	89.6	100.2	79.3	88.7	99.1	81.4	89.3	100.8	87.2	93.3	98.0	91.1	94.2
COVERS	DIVS. 10 - 14, 25, 28, 41, 43, 44, 46	100.0	102.5	100.5	100.0	93.3	98.6	100.0	89.1	97.8	100.0	89.5	97.9	100.0	95.2	99.0	100.0	100.3	100.1
21, 22, 23	FIRE SUPPRESSION, PLUMBING & HVAC	100.0	100.5	100.2	100.1	92.0	96.8	100.9	86.6	95.2	100.6	83.6	93.7	95.3	89.0	92.8	100.1	92.1	96.9
26, 27, 3370	ELECTRICAL, COMMUNICATIONS & UTIL.	98.8	105.0	102.1	97.6	87.8	92.4	92.6	83.0	87.5	93.0	83.3	87.9	95.8	80.3	87.6	98.7	88.0	93.0
MF2014	WEIGHTED AVERAGE	99.0	100.3	99.6	98.6	89.2	94.5	98.0	84.4	92.1	97.9	85.1	92.3	100.0	86.8	94.3	98.9	93.0	96.3

OHIO

DIVISION		MANSFIELD 448-449 MAT.	INST.	TOTAL	MARION 433 MAT.	INST.	TOTAL	SPRINGFIELD 455 MAT.	INST.	TOTAL	STEUBENVILLE 439 MAT.	INST.	TOTAL	TOLEDO 434-436 MAT.	INST.	TOTAL	YOUNGSTOWN 444-445 MAT.	INST.	TOTAL
015433	CONTRACTOR EQUIPMENT		94.4	94.4		93.2	93.2		94.7	94.7		98.1	98.1		96.0	96.0		94.4	94.4
0241, 31 - 34	SITE & INFRASTRUCTURE, DEMOLITION	93.9	101.5	99.3	91.8	97.1	95.6	93.2	100.5	98.4	133.1	106.9	114.5	94.8	99.0	97.8	98.0	102.0	100.8
0310	Concrete Forming & Accessories	89.1	83.7	84.5	96.7	81.2	83.3	99.3	83.5	85.7	98.3	89.4	90.7	100.3	97.1	97.6	99.0	87.5	89.1
0320	Concrete Reinforcing	90.6	76.4	83.4	94.7	82.5	88.5	95.5	81.3	88.3	92.3	85.5	88.8	102.7	85.4	93.9	99.5	85.6	92.4
0330	Cast-in-Place Concrete	87.1	94.7	90.2	85.9	92.7	88.7	89.1	86.3	87.9	93.3	94.1	93.6	94.1	100.0	96.6	93.1	96.9	94.7
03	CONCRETE	89.7	85.7	87.7	89.9	85.1	87.6	94.4	83.8	89.2	93.8	89.7	91.8	97.9	95.5	96.7	96.9	89.8	93.4
04	MASONRY	92.4	97.0	95.3	98.8	95.5	96.7	93.4	83.5	87.2	86.2	94.2	91.2	104.8	99.9	101.7	93.6	93.6	93.6
05	METALS	96.2	75.5	89.8	96.0	78.7	90.6	96.4	78.4	90.9	92.5	79.5	88.5	96.7	85.5	93.3	95.4	78.7	90.3
06	WOOD, PLASTICS & COMPOSITES	86.7	80.9	83.4	92.8	79.8	85.5	103.6	83.6	92.4	88.4	88.3	88.3	97.1	96.9	97.0	99.1	86.1	91.8
07	THERMAL & MOISTURE PROTECTION	111.3	94.4	104.4	100.5	85.2	94.2	104.6	88.2	97.9	112.8	96.4	106.1	102.5	102.7	102.6	112.1	94.3	104.8
08	OPENINGS	104.9	76.0	98.2	96.6	76.5	91.9	99.3	79.0	94.6	97.0	83.7	94.0	99.7	91.3	97.7	104.3	85.1	99.8
0920	Plaster & Gypsum Board	90.7	80.1	83.5	91.4	79.3	83.2	94.9	83.3	87.1	90.7	87.5	88.6	93.5	96.9	95.8	96.9	85.5	89.2
0950, 0980	Ceilings & Acoustic Treatment	93.9	80.1	84.8	97.7	79.3	85.6	99.7	83.3	88.9	94.8	87.5	90.0	97.7	96.9	97.2	93.0	85.5	88.1
0960	Flooring	92.5	107.5	96.8	90.9	107.5	95.6	102.0	80.7	95.9	118.6	100.4	113.4	91.2	99.7	93.6	97.1	93.9	96.2
0970, 0990	Wall Finishes & Painting/Coating	96.3	88.5	91.6	95.5	47.5	66.5	97.8	85.0	90.1	107.8	103.1	104.9	95.6	102.0	99.4	96.3	92.6	94.1
09	FINISHES	95.8	88.2	91.6	93.1	82.6	87.3	100.2	83.2	90.8	110.0	92.6	100.4	93.8	98.1	96.2	98.1	88.8	93.0
COVERS	DIVS. 10 - 14, 25, 28, 41, 43, 44, 46	100.0	96.2	99.2	100.0	54.4	90.8	100.0	89.8	97.9	100.0	95.0	99.0	100.0	98.7	99.7	100.0	96.4	99.3
21, 22, 23	FIRE SUPPRESSION, PLUMBING & HVAC	95.3	88.9	92.7	95.3	91.0	93.6	100.9	81.3	93.0	95.7	92.0	94.2	100.1	100.4	100.2	100.1	87.1	94.8
26, 27, 3370	ELECTRICAL, COMMUNICATIONS & UTIL.	96.2	79.0	87.1	91.9	79.0	85.1	92.6	87.8	90.1	86.9	114.1	101.3	97.6	104.7	101.4	98.7	86.0	92.0
MF2014	WEIGHTED AVERAGE	96.7	87.3	92.6	95.1	85.3	90.8	98.0	84.7	92.2	96.8	94.9	96.0	98.7	98.1	98.4	99.2	89.0	94.7

For customer support on your Plumbing Cost Data, call 877.709.6968.

Table 1

DIVISION		OHIO ZANESVILLE 437-438 MAT.	INST.	TOTAL	OKLAHOMA ARDMORE 734 MAT.	INST.	TOTAL	CLINTON 736 MAT.	INST.	TOTAL	DURANT 747 MAT.	INST.	TOTAL	ENID 737 MAT.	INST.	TOTAL	GUYMON 739 MAT.	INST.	TOTAL
015433	CONTRACTOR EQUIPMENT		93.2	93.2		82.7	82.7		81.8	81.8		81.8	81.8		81.8	81.8		81.8	81.8
0241, 31 - 34	SITE & INFRASTRUCTURE, DEMOLITION	94.5	98.7	97.5	96.2	91.6	92.9	97.6	90.2	92.3	95.5	89.9	91.5	99.3	90.2	92.8	101.6	90.0	93.3
0310	Concrete Forming & Accessories	93.7	81.3	83.0	94.0	40.4	47.8	92.5	46.6	53.0	85.5	44.1	49.8	96.2	34.1	42.7	99.8	44.9	52.4
0320	Concrete Reinforcing	94.1	85.1	89.5	90.8	79.8	85.2	91.3	79.8	85.5	95.8	80.2	87.9	90.7	79.8	85.2	91.3	79.8	85.5
0330	Cast-in-Place Concrete	90.5	91.7	91.0	97.5	43.4	75.3	94.3	45.1	74.1	91.7	42.8	71.6	94.3	46.4	74.6	94.3	43.1	73.3
03	CONCRETE	93.5	85.3	89.5	95.2	49.3	72.7	94.7	52.7	74.1	92.3	50.8	71.9	93.5	47.6	71.8	98.2	51.2	75.1
04	MASONRY	95.5	85.9	89.5	103.4	57.1	74.5	129.7	57.1	84.4	95.6	62.3	74.9	110.1	57.1	77.0	105.9	53.7	73.3
05	METALS	97.4	80.4	92.1	102.5	70.1	92.5	102.6	70.1	92.6	93.5	70.7	86.5	104.0	70.0	93.6	103.2	69.8	92.9
06	WOOD, PLASTICS & COMPOSITES	88.5	79.8	83.6	97.4	37.9	64.0	96.4	46.2	68.3	88.1	43.0	62.9	100.1	29.3	60.5	104.0	46.0	71.5
07	THERMAL & MOISTURE PROTECTION	100.6	92.6	97.3	108.9	60.6	89.1	109.1	61.5	89.6	99.7	61.8	84.2	109.2	59.9	89.0	109.5	57.3	88.1
08	OPENINGS	96.6	79.3	92.5	102.9	49.0	90.3	102.9	53.3	91.4	96.3	52.0	86.0	102.9	43.9	89.1	103.0	49.2	90.5
0920	Plaster & Gypsum Board	88.2	79.3	82.2	89.8	36.6	53.8	89.4	45.2	59.6	79.4	41.9	54.1	90.5	27.8	48.1	90.7	44.9	59.8
0950, 0980	Ceilings & Acoustic Treatment	97.7	79.3	85.6	85.7	36.6	53.4	85.7	45.2	59.1	82.4	41.9	55.8	85.7	27.8	47.6	86.5	44.9	59.2
0960	Flooring	89.3	90.3	89.6	107.0	43.2	88.7	105.8	41.1	87.3	102.8	61.9	91.1	107.7	41.1	88.6	109.3	24.4	85.0
0970, 0990	Wall Finishes & Painting/Coating	95.5	92.9	94.0	104.7	51.3	72.5	104.7	51.3	72.5	103.9	51.3	72.1	104.7	51.3	72.5	104.7	33.0	61.5
09	FINISHES	92.4	83.6	87.5	92.9	40.1	63.8	92.7	44.7	66.3	90.2	46.6	66.2	93.5	34.6	61.1	94.5	38.9	63.9
COVERS	DIVS. 10 - 14, 25, 28, 41, 43, 44, 46	100.0	88.5	97.7	100.0	77.6	95.5	100.0	78.6	95.7	100.0	77.9	95.5	100.0	76.7	95.3	100.0	77.4	95.4
21, 22, 23	FIRE SUPPRESSION, PLUMBING & HVAC	95.3	89.9	93.1	95.6	65.8	83.6	95.6	65.9	83.6	95.5	65.4	83.4	100.3	65.8	86.4	95.6	63.8	82.7
26, 27, 3370	ELECTRICAL, COMMUNICATIONS & UTIL.	92.3	83.4	87.5	93.6	70.9	81.6	94.6	70.9	82.1	96.5	70.9	83.0	94.6	70.9	82.1	96.2	61.5	77.9
MF2014	WEIGHTED AVERAGE	95.6	86.5	91.6	98.3	61.9	82.4	99.6	63.1	83.7	95.1	63.5	81.3	100.2	60.5	82.9	99.4	59.7	82.1

Table 2 — OKLAHOMA

DIVISION		LAWTON 735 MAT.	INST.	TOTAL	MCALESTER 745 MAT.	INST.	TOTAL	MIAMI 743 MAT.	INST.	TOTAL	MUSKOGEE 744 MAT.	INST.	TOTAL	OKLAHOMA CITY 730-731 MAT.	INST.	TOTAL	PONCA CITY 746 MAT.	INST.	TOTAL
015433	CONTRACTOR EQUIPMENT		82.7	82.7		81.8	81.8		89.9	89.9		89.9	89.9		83.0	83.0		81.8	81.8
0241, 31 - 34	SITE & INFRASTRUCTURE, DEMOLITION	95.7	91.6	92.8	89.1	90.1	89.9	90.5	87.3	88.2	90.8	87.1	88.2	95.1	92.0	92.9	95.9	90.1	91.8
0310	Concrete Forming & Accessories	99.8	44.3	52.0	83.5	43.6	49.1	96.9	66.2	70.4	101.4	33.4	42.7	97.7	57.9	63.4	92.2	46.3	52.6
0320	Concrete Reinforcing	91.0	79.8	85.3	95.5	79.8	87.5	94.0	79.9	86.8	94.9	79.3	87.0	96.2	79.9	87.9	94.9	79.9	87.2
0330	Cast-in-Place Concrete	91.2	46.4	72.8	80.4	45.2	65.9	84.3	47.5	69.2	85.3	45.8	69.1	92.9	46.9	74.0	94.2	44.7	73.8
03	CONCRETE	91.5	52.1	72.1	82.9	51.3	67.4	87.6	63.0	75.5	89.4	47.7	68.9	94.9	58.3	76.9	94.4	52.4	73.8
04	MASONRY	105.8	57.1	75.4	114.3	56.1	78.1	98.4	56.3	72.1	116.9	45.2	72.2	109.8	56.4	76.5	90.9	56.1	69.2
05	METALS	108.0	70.1	96.4	93.4	70.0	86.2	93.4	82.1	89.9	94.8	80.4	90.4	100.7	70.2	91.3	93.4	70.3	86.3
06	WOOD, PLASTICS & COMPOSITES	103.0	43.1	69.5	85.6	43.0	61.7	101.0	72.9	85.3	105.6	30.7	63.6	97.0	61.8	77.3	96.3	46.0	68.1
07	THERMAL & MOISTURE PROTECTION	108.9	61.3	89.4	99.4	60.6	83.5	99.9	64.3	85.2	99.2	48.6	78.8	101.1	62.9	85.4	99.9	66.2	86.1
08	OPENINGS	104.6	51.4	92.2	96.2	51.8	85.9	96.2	68.3	89.7	96.2	42.4	83.7	104.0	61.6	94.2	96.2	53.1	86.2
0920	Plaster & Gypsum Board	92.4	42.0	58.3	78.4	41.9	53.7	84.8	72.6	76.5	86.9	29.0	47.8	98.0	61.2	73.1	83.7	44.9	57.5
0950, 0980	Ceilings & Acoustic Treatment	93.2	42.0	59.6	82.4	41.9	55.8	82.4	72.6	75.9	90.8	29.0	50.2	96.4	61.2	73.3	82.4	44.9	57.8
0960	Flooring	109.7	41.1	90.1	101.8	41.1	84.4	108.8	61.9	95.4	111.3	39.8	90.8	106.2	41.1	87.6	105.9	41.1	87.4
0970, 0990	Wall Finishes & Painting/Coating	104.7	51.3	72.5	103.9	37.8	64.0	103.9	77.5	87.9	103.9	34.2	61.8	106.4	51.3	73.2	103.9	51.3	72.1
09	FINISHES	95.5	42.8	66.5	89.2	41.0	62.6	92.0	67.3	78.4	94.9	33.0	60.8	97.6	53.7	73.4	91.8	44.3	65.6
COVERS	DIVS. 10 - 14, 25, 28, 41, 43, 44, 46	100.0	78.2	95.6	100.0	77.9	95.5	100.0	81.6	96.3	100.0	76.1	95.2	100.0	80.0	96.0	100.0	78.2	95.6
21, 22, 23	FIRE SUPPRESSION, PLUMBING & HVAC	100.3	65.9	86.4	95.5	61.6	81.8	95.5	61.9	81.9	100.3	60.6	84.3	100.1	65.5	86.2	95.5	61.8	81.9
26, 27, 3370	ELECTRICAL, COMMUNICATIONS & UTIL.	96.2	70.9	82.8	94.9	64.5	78.8	96.3	64.6	79.5	94.5	52.4	72.2	101.7	70.9	85.4	94.5	70.4	81.7
MF2014	WEIGHTED AVERAGE	100.7	62.8	84.2	94.5	60.4	79.7	94.8	67.5	82.9	97.4	55.7	79.2	100.3	65.7	85.2	95.1	62.1	80.7

Table 3 — OKLAHOMA / OREGON

DIVISION		POTEAU 749 MAT.	INST.	TOTAL	SHAWNEE 748 MAT.	INST.	TOTAL	TULSA 740-741 MAT.	INST.	TOTAL	WOODWARD 738 MAT.	INST.	TOTAL	BEND 977 MAT.	INST.	TOTAL	EUGENE 974 MAT.	INST.	TOTAL
015433	CONTRACTOR EQUIPMENT		89.1	89.1		81.8	81.8		89.9	89.9		81.8	81.8		98.8	98.8		98.8	98.8
0241, 31 - 34	SITE & INFRASTRUCTURE, DEMOLITION	77.2	85.8	83.3	99.0	90.1	92.7	97.1	87.5	90.3	97.9	90.2	92.4	105.4	102.9	103.6	96.1	102.9	100.9
0310	Concrete Forming & Accessories	90.1	41.0	47.8	85.4	43.7	49.5	101.5	40.3	48.7	92.6	46.8	53.1	110.1	98.8	100.4	106.5	98.7	99.8
0320	Concrete Reinforcing	95.9	79.9	87.8	94.9	79.8	87.2	95.1	79.8	87.3	90.7	79.9	85.2	93.3	99.6	96.5	97.4	99.6	98.5
0330	Cast-in-Place Concrete	84.3	44.9	68.1	97.2	42.9	74.9	92.9	46.2	73.7	94.3	45.3	74.2	104.4	102.7	103.7	101.1	102.6	101.7
03	CONCRETE	89.9	50.9	70.7	96.0	50.6	73.7	94.7	51.0	73.2	95.0	52.8	74.3	107.1	100.0	103.6	98.8	100.0	99.4
04	MASONRY	98.7	56.2	72.2	115.6	56.1	78.5	99.3	60.0	74.8	98.6	57.1	72.7	104.1	103.3	103.6	101.1	103.3	102.5
05	METALS	93.4	81.7	89.8	93.3	69.9	86.1	98.1	81.4	93.0	102.7	70.5	92.8	91.9	96.4	93.3	92.6	96.2	93.7
06	WOOD, PLASTICS & COMPOSITES	93.0	39.1	62.8	88.0	43.0	62.8	104.8	36.6	66.6	96.5	46.0	68.2	101.7	98.4	99.9	97.6	98.4	98.1
07	THERMAL & MOISTURE PROTECTION	99.9	60.5	83.7	99.9	59.3	83.2	99.8	62.2	84.4	109.1	66.6	91.7	107.9	94.9	102.6	107.2	91.6	100.8
08	OPENINGS	96.2	49.9	85.5	96.2	51.8	85.9	97.8	47.5	86.1	102.9	53.1	91.3	96.9	102.3	98.1	97.1	102.3	98.4
0920	Plaster & Gypsum Board	82.3	37.8	52.2	79.4	41.9	54.1	86.9	35.2	51.9	89.6	44.9	59.4	107.4	98.2	101.2	106.0	98.2	100.7
0950, 0980	Ceilings & Acoustic Treatment	82.4	37.8	53.1	82.4	41.9	55.8	90.8	35.2	54.3	86.5	44.9	59.2	91.3	98.2	95.8	92.3	98.2	96.1
0960	Flooring	105.2	61.9	92.8	102.8	32.7	82.8	110.0	42.8	90.8	105.8	43.2	87.9	110.5	103.4	108.5	108.9	103.4	107.3
0970, 0990	Wall Finishes & Painting/Coating	103.9	51.3	72.1	103.9	34.1	61.8	103.9	40.5	65.6	104.7	51.3	72.5	106.4	74.8	87.3	106.4	74.8	87.3
09	FINISHES	89.9	44.3	64.8	90.4	39.2	62.2	94.7	38.9	63.9	93.0	44.9	66.5	102.7	96.8	99.5	101.2	96.8	98.8
COVERS	DIVS. 10 - 14, 25, 28, 41, 43, 44, 46	100.0	77.7	95.5	100.0	77.9	95.5	100.0	78.8	95.7	100.0	78.5	95.7	100.0	99.6	99.9	100.0	99.6	99.9
21, 22, 23	FIRE SUPPRESSION, PLUMBING & HVAC	95.5	61.8	81.9	95.5	65.4	83.4	100.3	63.5	85.4	95.6	65.9	83.6	95.1	100.8	97.4	99.9	100.8	100.2
26, 27, 3370	ELECTRICAL, COMMUNICATIONS & UTIL.	94.6	64.6	78.7	96.6	70.9	83.0	96.5	64.6	79.6	96.1	70.9	82.8	101.6	96.1	98.7	100.1	96.1	98.0
MF2014	WEIGHTED AVERAGE	94.3	61.4	80.0	96.5	61.8	81.4	98.1	61.6	82.2	98.3	63.4	83.1	98.9	99.4	99.1	98.6	99.3	98.9

For customer support on your Plumbing Cost Data, call 877.709.6968.

629

OREGON

DIVISION		KLAMATH FALLS 976			MEDFORD 975			PENDLETON 978			PORTLAND 970 - 972			SALEM 973			VALE 979		
		MAT.	INST.	TOTAL	MAT.	INST.	TOTAL	MAT.	INST.	TOTAL	MAT.	INST.	TOTAL	MAT.	INST.	TOTAL	MAT.	INST.	TOTAL
015433	CONTRACTOR EQUIPMENT		98.8	98.8		98.8	98.8		96.2	96.2		98.8	98.8		98.8	98.8		96.2	96.2
0241, 31 - 34	SITE & INFRASTRUCTURE, DEMOLITION	109.2	102.9	104.7	103.4	102.9	103.0	102.5	96.4	98.2	98.5	102.9	101.6	91.9	102.9	99.7	90.5	96.4	94.7
0310	Concrete Forming & Accessories	102.9	98.6	99.2	102.0	98.6	99.1	103.3	99.0	99.6	107.7	98.9	100.1	106.0	98.8	99.8	109.9	98.7	100.2
0320	Concrete Reinforcing	93.3	99.6	96.5	94.9	99.6	97.3	92.6	99.7	96.2	98.1	99.7	98.9	103.6	99.6	101.6	90.4	99.6	95.1
0330	Cast-in-Place Concrete	104.4	102.6	103.7	104.4	102.6	103.7	105.2	103.9	104.7	103.9	102.7	103.4	97.9	102.7	99.9	82.9	103.8	91.5
03	CONCRETE	109.7	99.9	104.9	104.6	99.9	102.3	91.4	100.6	95.9	100.3	100.1	100.2	98.2	100.0	99.1	76.9	100.4	88.5
04	MASONRY	117.5	103.3	108.6	98.2	103.3	101.4	107.8	103.4	105.0	102.5	103.3	103.0	109.3	103.3	105.6	106.1	103.4	104.4
05	METALS	91.9	96.2	93.2	92.2	96.2	93.4	98.1	96.9	97.7	93.5	96.5	94.4	98.7	96.4	98.0	98.0	96.6	97.5
06	WOOD, PLASTICS & COMPOSITES	92.6	98.4	95.9	91.5	98.4	95.3	94.7	98.5	96.8	98.6	98.4	98.5	95.5	98.4	97.1	103.2	98.5	100.6
07	THERMAL & MOISTURE PROTECTION	108.2	92.9	101.9	107.8	92.9	101.7	101.1	93.5	98.0	107.1	96.7	102.9	104.9	94.9	100.8	100.5	94.4	98.0
08	OPENINGS	96.9	102.3	98.1	99.7	102.3	100.3	93.2	102.4	95.4	95.0	102.3	96.7	98.7	102.3	99.5	93.2	94.1	93.4
0920	Plaster & Gypsum Board	102.9	98.2	99.7	102.2	98.2	99.5	90.2	98.2	95.6	105.5	98.2	100.5	102.1	98.2	99.4	95.9	98.2	97.4
0950, 0980	Ceilings & Acoustic Treatment	98.9	98.2	98.4	105.4	98.2	100.7	64.4	98.2	86.6	94.2	98.2	96.8	99.3	98.2	98.5	64.4	98.2	86.6
0960	Flooring	107.6	103.4	106.4	107.1	103.4	106.1	74.7	103.4	82.9	106.4	103.4	105.5	107.2	103.4	106.1	76.7	103.4	84.4
0970, 0990	Wall Finishes & Painting/Coating	106.4	70.4	84.6	106.4	70.4	84.6	66.6	72.8	70.3	106.2	72.8	86.0	104.4	74.8	86.6	96.6	74.8	83.4
09	FINISHES	103.4	96.3	99.5	103.8	96.3	99.6	72.3	96.7	85.7	100.0	96.6	98.5	100.0	96.8	98.2	72.7	96.9	86.0
COVERS	DIVS. 10 - 14, 25, 28, 41, 43, 44, 46	100.0	99.5	99.9	100.0	99.5	99.9	100.0	90.1	98.0	100.0	99.6	99.9	100.0	99.6	99.9	100.0	99.9	100.0
21, 22, 23	FIRE SUPPRESSION, PLUMBING & HVAC	95.1	100.7	97.4	99.9	100.7	100.2	97.0	113.2	103.5	99.9	100.8	100.2	99.9	100.8	100.2	97.0	98.0	97.4
26, 27, 3370	ELECTRICAL, COMMUNICATIONS & UTIL.	100.2	80.8	89.9	103.9	80.8	91.6	92.3	97.1	94.8	100.4	103.2	101.9	108.0	96.1	101.7	92.3	97.0	94.8
MF2014	WEIGHTED AVERAGE	99.8	97.1	98.6	100.0	97.1	98.8	94.8	101.4	97.7	98.8	100.4	99.5	100.5	99.4	100.0	92.9	98.1	95.2

PENNSYLVANIA

DIVISION		ALLENTOWN 181			ALTOONA 166			BEDFORD 155			BRADFORD 167			BUTLER 160			CHAMBERSBURG 172		
		MAT.	INST.	TOTAL	MAT.	INST.	TOTAL	MAT.	INST.	TOTAL	MAT.	INST.	TOTAL	MAT.	INST.	TOTAL	MAT.	INST.	TOTAL
015433	CONTRACTOR EQUIPMENT		112.9	112.9		112.9	112.9		109.6	109.6		112.9	112.9		112.9	112.9		112.1	112.1
0241, 31 - 34	SITE & INFRASTRUCTURE, DEMOLITION	93.6	104.5	101.3	96.8	104.4	102.2	101.4	100.1	100.5	92.4	103.5	100.3	88.1	105.8	100.7	89.2	101.2	97.8
0310	Concrete Forming & Accessories	99.2	113.0	111.1	84.2	80.4	80.9	84.1	80.8	81.3	86.4	81.9	82.5	85.7	96.0	94.6	90.2	80.0	81.4
0320	Concrete Reinforcing	94.7	108.1	101.5	91.8	102.7	97.3	93.1	80.9	86.9	93.7	103.2	98.5	92.4	109.1	100.9	91.7	103.0	97.4
0330	Cast-in-Place Concrete	86.5	104.9	94.0	96.3	86.1	92.1	106.7	70.1	91.6	92.0	92.6	92.3	85.1	96.7	89.9	91.1	69.8	82.3
03	CONCRETE	93.6	110.1	101.7	89.1	88.1	88.6	102.1	78.5	90.5	95.4	91.0	93.2	81.3	99.9	90.4	101.9	82.2	92.3
04	MASONRY	97.4	101.0	99.7	100.5	64.5	78.0	114.1	87.0	97.2	97.6	88.4	91.9	102.6	98.9	100.3	102.6	87.4	93.1
05	METALS	100.0	123.0	107.1	93.9	116.8	101.0	97.6	104.0	99.6	97.9	116.5	103.6	93.6	121.9	102.3	97.5	114.6	102.7
06	WOOD, PLASTICS & COMPOSITES	100.7	115.8	109.1	79.2	83.2	81.4	84.0	80.7	82.2	85.5	80.1	82.5	80.6	95.5	89.0	88.0	80.8	84.0
07	THERMAL & MOISTURE PROTECTION	102.3	119.1	109.2	101.4	91.0	97.1	103.0	87.7	96.7	102.3	91.7	97.9	101.1	100.7	100.9	99.3	72.5	88.3
08	OPENINGS	95.0	115.0	99.6	88.7	89.3	88.8	98.0	81.6	94.2	95.2	91.0	94.2	88.7	104.8	92.4	94.1	82.8	91.4
0920	Plaster & Gypsum Board	95.9	116.0	109.5	87.7	82.5	84.2	92.3	80.0	84.0	88.5	79.2	82.2	87.7	95.2	92.7	96.7	80.0	85.4
0950, 0980	Ceilings & Acoustic Treatment	82.8	116.0	104.6	87.0	82.5	84.0	95.0	80.0	85.1	85.4	79.2	81.3	87.9	95.2	92.7	84.0	80.0	81.4
0960	Flooring	94.6	95.9	95.0	88.2	97.5	90.9	93.1	88.2	91.7	89.4	105.3	94.0	89.0	81.0	86.7	97.9	46.2	83.1
0970, 0990	Wall Finishes & Painting/Coating	98.5	70.5	81.6	94.0	109.2	103.2	100.2	81.6	89.0	98.5	94.6	96.1	94.0	109.2	103.2	105.9	81.6	91.2
09	FINISHES	91.8	105.5	99.4	90.0	85.8	87.7	95.0	80.8	87.2	89.9	85.5	87.5	89.9	93.6	91.9	92.1	73.7	82.0
COVERS	DIVS. 10 - 14, 25, 28, 41, 43, 44, 46	100.0	105.6	101.1	100.0	95.4	99.1	100.0	98.3	99.7	100.0	99.2	99.8	100.0	102.4	100.5	100.0	96.1	99.2
21, 22, 23	FIRE SUPPRESSION, PLUMBING & HVAC	100.2	112.1	105.0	99.7	82.5	92.8	95.2	87.3	92.0	95.5	90.7	93.6	95.0	93.9	94.6	95.5	87.1	92.1
26, 27, 3370	ELECTRICAL, COMMUNICATIONS & UTIL.	99.1	98.4	98.8	89.3	112.2	101.4	94.3	112.2	103.8	92.7	112.2	103.0	89.9	111.2	101.1	89.6	86.3	87.8
MF2014	WEIGHTED AVERAGE	97.9	108.6	102.5	94.6	91.9	93.4	98.2	91.3	95.2	95.7	96.4	96.0	92.5	102.0	96.6	96.2	88.0	92.6

PENNSYLVANIA

DIVISION		DOYLESTOWN 189			DUBOIS 158			ERIE 164 - 165			GREENSBURG 156			HARRISBURG 170 - 171			HAZLETON 182		
		MAT.	INST.	TOTAL	MAT.	INST.	TOTAL	MAT.	INST.	TOTAL	MAT.	INST.	TOTAL	MAT.	INST.	TOTAL	MAT.	INST.	TOTAL
015433	CONTRACTOR EQUIPMENT		93.9	93.9		109.6	109.6		112.9	112.9		109.6	109.6		112.1	112.1		112.9	112.9
0241, 31 - 34	SITE & INFRASTRUCTURE, DEMOLITION	106.9	90.2	95.0	106.1	100.5	102.1	93.7	105.2	101.9	97.8	102.9	101.4	89.6	103.5	99.5	86.9	104.7	99.6
0310	Concrete Forming & Accessories	83.2	128.2	122.0	83.5	84.6	84.4	98.5	88.1	89.5	90.7	95.9	95.2	100.4	87.9	89.6	80.8	88.7	87.6
0320	Concrete Reinforcing	91.6	130.9	111.6	92.6	103.5	98.1	93.7	103.3	98.6	92.6	109.1	101.0	101.4	106.2	103.9	91.9	107.1	99.6
0330	Cast-in-Place Concrete	81.7	86.7	83.8	102.9	92.9	98.8	94.7	81.8	89.4	99.0	96.3	97.9	93.2	95.8	94.3	81.7	93.1	86.4
03	CONCRETE	89.2	114.2	101.5	103.6	92.2	98.0	88.2	90.0	89.1	97.2	99.6	98.4	97.7	95.4	96.6	86.3	95.0	90.6
04	MASONRY	100.7	128.1	117.7	114.4	91.1	99.9	89.6	92.5	91.4	124.8	98.9	108.7	102.5	89.6	94.5	110.2	95.7	101.2
05	METALS	97.5	123.8	105.6	97.6	116.1	103.3	94.1	117.0	101.2	97.5	120.6	104.6	104.0	121.4	109.4	99.7	120.6	106.1
06	WOOD, PLASTICS & COMPOSITES	81.0	130.7	108.8	82.9	83.1	83.0	96.9	86.6	91.1	90.9	95.4	93.4	95.8	86.7	90.7	79.6	86.6	83.5
07	THERMAL & MOISTURE PROTECTION	100.1	131.4	112.9	103.2	96.5	100.5	101.9	92.6	98.1	102.8	100.7	102.0	102.5	110.2	105.7	101.8	106.6	103.8
08	OPENINGS	97.2	137.9	106.7	98.0	92.7	96.7	88.8	92.2	89.6	97.9	104.7	99.5	100.6	94.0	99.0	95.6	92.3	94.8
0920	Plaster & Gypsum Board	86.2	131.4	116.7	91.2	82.5	85.3	95.9	86.0	89.2	93.5	95.2	94.6	100.6	86.0	90.7	86.6	86.0	86.2
0950, 0980	Ceilings & Acoustic Treatment	82.0	131.4	114.5	95.0	82.5	86.7	82.8	86.0	84.9	94.2	95.2	94.8	92.6	86.0	88.3	83.7	86.0	85.2
0960	Flooring	79.4	134.3	95.1	92.9	105.3	96.4	93.2	91.8	92.8	96.2	68.0	88.2	103.2	91.7	99.8	97.4	94.6	96.6
0970, 0990	Wall Finishes & Painting/Coating	98.0	69.0	80.5	100.2	106.6	104.1	103.5	94.6	98.1	100.2	106.6	104.1	106.4	89.6	96.3	98.5	106.4	103.2
09	FINISHES	83.2	122.4	104.8	95.2	89.5	92.1	92.2	88.8	90.4	95.6	91.7	93.4	96.9	88.0	92.0	88.1	89.5	88.9
COVERS	DIVS. 10 - 14, 25, 28, 41, 43, 44, 46	100.0	70.6	94.1	100.0	99.1	99.8	100.0	100.8	100.2	100.0	102.2	100.4	100.0	97.4	99.5	100.0	101.0	100.2
21, 22, 23	FIRE SUPPRESSION, PLUMBING & HVAC	95.0	127.4	108.1	95.2	88.4	92.4	99.7	92.9	97.0	95.2	91.1	93.5	100.1	92.1	96.9	95.5	99.2	97.0
26, 27, 3370	ELECTRICAL, COMMUNICATIONS & UTIL.	92.1	128.1	111.1	94.9	112.2	104.0	91.1	96.7	94.0	94.9	112.2	104.1	96.8	88.3	92.3	93.6	91.7	92.6
MF2014	WEIGHTED AVERAGE	94.9	120.6	106.1	98.5	96.8	97.8	94.5	95.8	95.1	98.2	100.9	99.4	99.8	95.6	98.0	95.4	98.3	96.6

City Cost Indexes

PENNSYLVANIA

DIVISION		INDIANA 157 MAT.	INST.	TOTAL	JOHNSTOWN 159 MAT.	INST.	TOTAL	KITTANNING 162 MAT.	INST.	TOTAL	LANCASTER 175 - 176 MAT.	INST.	TOTAL	LEHIGH VALLEY 180 MAT.	INST.	TOTAL	MONTROSE 188 MAT.	INST.	TOTAL
015433	CONTRACTOR EQUIPMENT		109.6	109.6		109.6	109.6		112.9	112.9		112.1	112.1		112.9	112.9		112.9	112.9
0241, 31 - 34	SITE & INFRASTRUCTURE, DEMOLITION	95.9	101.2	99.7	101.9	102.2	102.1	90.7	105.7	101.4	81.6	103.5	97.2	90.7	104.3	100.4	89.4	102.1	98.4
0310	Concrete Forming & Accessories	84.7	86.3	86.1	83.5	84.8	84.6	85.7	95.9	94.5	92.3	87.5	88.2	92.8	112.7	110.0	81.8	88.5	87.6
0320	Concrete Reinforcing	91.8	109.2	100.6	93.1	108.9	101.2	92.4	109.2	101.0	91.4	106.1	98.9	91.9	108.1	100.1	96.3	106.3	101.4
0330	Cast-in-Place Concrete	97.1	95.9	96.6	107.6	92.4	101.4	88.4	96.5	91.7	77.5	97.9	85.9	88.4	104.7	95.1	86.7	90.9	88.4
03	CONCRETE	94.7	95.2	94.9	102.9	93.2	98.2	83.8	99.7	91.6	89.8	96.0	92.8	92.6	109.8	101.1	91.3	93.7	92.5
04	MASONRY	110.1	100.1	103.9	110.9	90.6	98.3	105.4	100.1	102.1	108.2	90.1	97.0	97.3	101.0	99.6	97.3	95.1	95.9
05	METALS	97.7	120.1	104.6	97.6	118.9	104.2	93.7	121.9	102.4	97.5	120.8	104.7	99.7	122.0	107.4	98.0	114.1	102.9
06	WOOD, PLASTICS & COMPOSITES	84.8	83.1	83.8	82.9	83.1	83.0	80.6	95.5	89.0	90.8	86.7	88.5	92.2	115.8	105.4	80.4	88.4	84.9
07	THERMAL & MOISTURE PROTECTION	102.7	98.2	100.9	103.0	95.1	99.7	101.1	101.0	101.1	98.8	97.1	98.1	102.2	102.8	102.4	101.8	91.7	97.7
08	OPENINGS	98.0	94.4	97.1	98.0	90.9	96.3	88.7	101.2	91.6	94.1	98.6	95.1	95.6	115.0	100.1	92.2	93.1	92.4
0920	Plaster & Gypsum Board	92.7	82.5	85.8	91.0	82.5	85.2	87.7	95.2	92.7	98.5	86.0	90.1	89.5	116.0	107.4	87.1	87.8	87.6
0950, 0980	Ceilings & Acoustic Treatment	95.0	82.5	86.7	94.2	82.5	86.5	87.9	95.2	92.7	84.0	86.0	85.3	83.7	116.0	105.0	85.4	87.8	87.0
0960	Flooring	93.7	105.3	97.0	92.9	98.2	94.4	89.0	105.3	93.6	98.8	91.7	96.8	91.9	95.9	93.0	87.6	57.9	79.1
0970, 0990	Wall Finishes & Painting/Coating	100.2	106.6	104.1	100.2	109.2	105.6	94.0	106.6	101.6	105.9	57.1	76.5	98.5	66.4	79.1	98.5	106.4	103.2
09	FINISHES	94.8	90.0	92.1	94.6	88.8	91.4	90.1	97.4	94.1	92.0	84.5	87.9	90.2	104.1	97.9	88.9	84.7	86.6
COVERS	DIVS. 10 - 14, 25, 28, 41, 43, 44, 46	100.0	100.6	100.1	100.0	99.3	99.9	100.0	100.2	100.5	100.0	97.4	99.5	100.0	100.6	101.2	100.0	101.2	100.2
21, 22, 23	FIRE SUPPRESSION, PLUMBING & HVAC	95.2	90.8	93.4	95.2	88.7	92.5	95.0	96.9	95.8	95.5	92.3	94.2	95.5	112.0	102.2	95.5	98.4	96.7
26, 27, 3370	ELECTRICAL, COMMUNICATIONS & UTIL.	94.9	112.2	104.1	94.9	112.2	104.0	89.3	112.2	101.4	91.0	44.0	66.2	93.6	143.4	119.9	92.7	97.6	95.3
MF2014	WEIGHTED AVERAGE	97.1	99.3	98.0	98.1	97.2	97.7	92.9	103.3	97.4	95.1	89.0	92.4	95.8	114.0	103.7	94.7	96.9	95.7

PENNSYLVANIA

DIVISION		NEW CASTLE 161 MAT.	INST.	TOTAL	NORRISTOWN 194 MAT.	INST.	TOTAL	OIL CITY 163 MAT.	INST.	TOTAL	PHILADELPHIA 190 - 191 MAT.	INST.	TOTAL	PITTSBURGH 150 - 152 MAT.	INST.	TOTAL	POTTSVILLE 179 MAT.	INST.	TOTAL
015433	CONTRACTOR EQUIPMENT		112.9	112.9		98.8	98.8		112.9	112.9		98.2	98.2	101.3	110.8	110.8		112.1	112.1
0241, 31 - 34	SITE & INFRASTRUCTURE, DEMOLITION	88.5	105.8	100.8	95.8	100.6	99.2	87.1	103.7	98.9	101.6	100.0	100.5	101.3	104.9	103.8	84.4	103.4	97.9
0310	Concrete Forming & Accessories	85.7	95.6	94.3	83.2	129.7	123.3	85.7	83.4	83.7	98.9	140.1	134.4	98.5	96.7	97.0	83.3	89.7	88.8
0320	Concrete Reinforcing	91.3	92.5	91.9	90.2	137.9	114.5	92.4	92.3	92.4	100.6	137.9	119.6	93.6	109.4	101.6	90.7	102.7	96.8
0330	Cast-in-Place Concrete	85.9	96.5	90.2	85.0	127.1	102.3	83.4	95.5	88.4	98.0	132.2	112.1	102.9	96.5	100.3	82.6	100.9	90.1
03	CONCRETE	81.6	96.4	88.9	89.2	130.1	109.3	80.1	90.5	85.2	99.4	136.4	117.6	100.5	100.1	100.3	93.3	97.3	95.3
04	MASONRY	101.9	97.7	99.3	110.4	124.9	119.4	101.8	95.8	98.0	96.7	130.7	117.9	103.0	102.1	102.4	102.1	92.2	95.9
05	METALS	93.7	113.6	99.8	99.2	130.0	108.7	93.7	111.5	99.2	102.0	130.2	110.7	99.0	121.3	105.9	97.7	118.8	104.2
06	WOOD, PLASTICS & COMPOSITES	80.6	95.9	89.2	79.9	130.6	108.3	80.6	80.1	80.3	98.6	141.9	122.9	100.6	95.8	97.9	80.2	88.1	84.6
07	THERMAL & MOISTURE PROTECTION	101.1	98.0	99.8	101.4	131.1	113.5	101.0	94.4	98.3	102.8	135.2	116.1	103.1	101.6	102.4	98.9	106.7	102.1
08	OPENINGS	88.7	96.6	90.5	86.9	140.1	99.2	88.7	80.3	86.7	98.6	146.2	109.7	101.8	105.0	102.5	94.1	91.8	93.6
0920	Plaster & Gypsum Board	87.7	95.6	93.0	85.6	131.4	116.5	87.7	79.2	82.0	96.6	143.1	128.0	99.8	95.6	96.9	93.5	87.5	89.5
0950, 0980	Ceilings & Acoustic Treatment	87.9	95.6	92.9	84.5	131.4	115.3	87.9	79.2	82.2	94.0	143.1	126.2	95.0	95.6	95.4	84.0	87.5	86.3
0960	Flooring	89.0	57.5	80.0	92.9	139.2	106.1	89.0	105.3	93.6	98.4	139.2	110.1	99.5	106.6	101.5	95.2	91.7	94.2
0970, 0990	Wall Finishes & Painting/Coating	94.0	109.2	103.2	98.6	147.8	128.3	94.0	106.6	101.6	99.8	154.7	133.0	100.2	119.2	111.7	105.9	106.4	106.2
09	FINISHES	90.0	89.6	89.8	90.7	133.0	114.0	89.8	87.8	88.7	98.9	141.8	122.5	97.7	100.2	99.1	90.5	91.3	90.9
COVERS	DIVS. 10 - 14, 25, 28, 41, 43, 44, 46	100.0	102.4	100.5	100.0	117.5	103.5	100.0	100.6	100.1	100.0	120.8	104.2	100.0	102.1	100.4	100.0	98.7	99.7
21, 22, 23	FIRE SUPPRESSION, PLUMBING & HVAC	95.0	93.6	94.4	95.2	127.3	108.2	95.0	92.9	94.2	100.0	131.8	112.9	99.9	100.4	100.1	95.5	98.6	96.8
26, 27, 3370	ELECTRICAL, COMMUNICATIONS & UTIL.	89.9	98.1	94.2	93.0	148.4	122.3	91.8	111.1	102.0	97.1	148.4	124.2	97.2	112.2	105.1	89.2	95.5	92.5
MF2014	WEIGHTED AVERAGE	92.5	97.9	94.8	95.0	129.5	110.0	92.4	96.9	94.4	99.7	133.6	114.5	99.9	104.6	102.0	94.9	98.5	96.5

PENNSYLVANIA

DIVISION		READING 195 - 196 MAT.	INST.	TOTAL	SCRANTON 184 - 185 MAT.	INST.	TOTAL	STATE COLLEGE 168 MAT.	INST.	TOTAL	STROUDSBURG 183 MAT.	INST.	TOTAL	SUNBURY 178 MAT.	INST.	TOTAL	UNIONTOWN 154 MAT.	INST.	TOTAL
015433	CONTRACTOR EQUIPMENT		118.0	118.0		112.9	112.9		112.1	112.1		112.9	112.9		112.9	112.9		109.6	109.6
0241, 31 - 34	SITE & INFRASTRUCTURE, DEMOLITION	100.3	112.7	109.1	94.1	104.8	101.7	84.4	103.1	97.7	88.5	102.2	98.2	96.6	104.5	102.2	96.5	102.9	101.0
0310	Concrete Forming & Accessories	98.8	90.2	91.4	99.3	88.6	90.0	84.4	80.3	80.8	87.2	89.4	89.1	96.1	88.5	89.6	77.6	96.0	93.5
0320	Concrete Reinforcing	91.5	104.7	98.2	94.7	106.6	100.8	93.0	103.4	98.3	95.0	111.6	103.4	93.2	106.2	99.8	92.6	109.1	101.0
0330	Cast-in-Place Concrete	76.2	97.3	84.9	90.3	93.1	91.5	87.1	65.6	78.2	85.1	72.6	80.0	90.2	95.5	92.4	97.1	96.4	96.8
03	CONCRETE	88.2	96.7	92.4	95.4	94.9	95.1	95.5	81.0	88.4	90.1	88.9	89.5	96.7	95.5	96.1	94.4	99.7	97.0
04	MASONRY	99.5	93.0	95.4	97.7	95.7	96.4	103.1	77.9	87.4	95.1	100.4	98.4	102.3	89.7	94.5	127.0	98.9	109.5
05	METALS	99.5	121.2	106.2	102.1	120.5	107.8	97.7	117.6	103.8	99.7	115.9	104.7	97.4	120.0	104.4	97.4	120.6	104.5
06	WOOD, PLASTICS & COMPOSITES	98.5	88.1	92.6	100.7	86.2	92.6	87.6	83.2	85.1	86.4	88.4	87.5	88.9	88.1	88.5	76.7	95.4	87.2
07	THERMAL & MOISTURE PROTECTION	101.8	112.1	106.0	102.2	95.6	99.5	101.5	90.6	97.0	102.0	85.3	95.2	100.0	104.9	102.0	102.6	100.7	101.8
08	OPENINGS	91.3	98.9	93.1	95.0	91.9	94.3	92.0	89.3	91.4	95.6	88.2	93.9	94.2	92.8	93.9	97.9	104.7	99.5
0920	Plaster & Gypsum Board	96.8	87.5	90.5	98.0	85.6	89.6	89.6	82.5	84.8	88.0	87.8	87.9	92.7	87.5	89.2	89.2	95.2	93.2
0950, 0980	Ceilings & Acoustic Treatment	76.9	87.5	83.9	91.1	85.6	87.5	82.9	82.5	82.6	82.0	87.8	85.8	80.7	87.5	85.2	94.2	95.2	94.8
0960	Flooring	96.8	91.7	95.4	94.6	107.0	98.2	92.3	97.5	93.8	89.9	51.3	78.9	95.9	87.7	93.5	90.4	105.3	94.7
0970, 0990	Wall Finishes & Painting/Coating	97.3	106.4	102.8	98.5	106.4	103.2	98.5	109.2	105.0	98.5	64.6	78.0	105.9	94.3	98.9	100.2	109.2	105.6
09	FINISHES	92.0	90.8	91.3	93.9	93.1	93.4	89.4	85.8	87.4	88.9	80.0	84.0	91.3	88.7	89.9	93.2	97.5	95.6
COVERS	DIVS. 10 - 14, 25, 28, 41, 43, 44, 46	100.0	100.3	100.1	100.0	101.0	100.2	100.0	93.1	98.6	100.0	62.4	92.4	100.0	97.9	99.6	100.0	102.2	100.4
21, 22, 23	FIRE SUPPRESSION, PLUMBING & HVAC	100.1	108.8	103.7	100.2	99.2	99.8	95.5	85.9	91.6	95.5	100.0	97.3	95.5	92.0	94.1	95.2	91.1	93.5
26, 27, 3370	ELECTRICAL, COMMUNICATIONS & UTIL.	99.6	95.5	97.5	99.2	97.7	98.4	91.9	112.2	102.6	93.6	143.3	119.9	89.5	90.8	90.2	92.0	112.2	102.7
MF2014	WEIGHTED AVERAGE	97.1	102.1	99.3	98.6	99.2	98.9	95.3	92.8	94.2	95.3	101.4	97.9	95.8	95.8	95.8	97.3	101.7	99.2

PENNSYLVANIA

DIVISION		WASHINGTON 153			WELLSBORO 169			WESTCHESTER 193			WILKES-BARRE 186 - 187			WILLIAMSPORT 177			YORK 173 - 174		
		MAT.	INST.	TOTAL	MAT.	INST.	TOTAL	MAT.	INST.	TOTAL	MAT.	INST.	TOTAL	MAT.	INST.	TOTAL	MAT.	INST.	TOTAL
015433	CONTRACTOR EQUIPMENT		109.6	109.6		112.9	112.9		98.8	98.8		112.9	112.9		112.9	112.9		112.1	112.1
0241, 31 - 34	SITE & INFRASTRUCTURE, DEMOLITION	96.6	102.9	101.1	95.9	102.0	100.3	101.6	97.8	98.9	86.5	104.7	99.5	87.9	103.4	98.9	85.2	103.5	98.2
0310	Concrete Forming & Accessories	84.8	96.2	94.6	85.8	86.7	86.6	89.8	128.2	122.9	89.9	88.9	89.1	92.5	61.0	65.3	86.9	88.1	87.9
0320	Concrete Reinforcing	92.6	109.3	101.1	93.0	106.2	99.7	89.3	113.6	101.7	93.7	107.1	100.6	92.5	64.6	78.3	93.2	106.2	99.8
0330	Cast-in-Place Concrete	97.1	96.4	96.8	91.2	88.9	90.3	94.1	126.4	107.4	81.7	93.0	86.4	76.4	74.2	75.5	83.0	98.1	89.2
03	CONCRETE	94.9	99.8	97.3	97.8	92.2	95.1	96.9	124.5	110.5	87.2	95.1	91.1	84.7	68.2	76.6	94.5	96.3	95.4
04	MASONRY	109.4	100.5	103.9	103.6	89.7	94.9	104.7	124.9	117.3	110.6	95.3	101.0	93.9	93.7	93.8	103.6	90.1	95.2
05	METALS	97.3	120.9	104.6	97.8	114.2	102.9	99.2	117.2	104.8	97.9	121.0	105.0	97.5	100.4	98.4	99.1	121.4	105.9
06	WOOD, PLASTICS & COMPOSITES	84.9	96.2	90.8	84.9	88.1	86.7	87.1	130.6	111.5	88.9	86.6	87.6	85.2	51.0	66.1	84.1	86.7	85.5
07	THERMAL & MOISTURE PROTECTION	102.7	101.1	102.1	102.5	89.6	97.2	101.7	129.9	113.3	101.8	106.4	103.7	99.4	102.4	100.6	99.0	110.4	103.7
08	OPENINGS	97.9	104.7	99.5	95.1	93.0	94.6	86.9	126.1	96.0	92.2	98.8	93.8	94.2	53.5	84.7	94.1	94.0	94.1
0920	Plaster & Gypsum Board	92.4	95.2	94.3	87.9	87.5	87.6	87.0	131.4	117.0	88.8	86.0	86.9	93.5	49.3	63.6	94.3	86.0	88.7
0950, 0980	Ceilings & Acoustic Treatment	94.2	95.2	94.8	82.9	87.5	85.9	84.5	131.4	115.3	85.4	86.0	85.8	84.0	49.3	61.2	83.1	86.0	85.0
0960	Flooring	93.8	105.3	97.1	89.1	50.8	78.2	95.7	139.2	108.1	90.7	94.6	91.8	94.8	49.7	81.9	96.4	91.7	95.1
0970, 0990	Wall Finishes & Painting/Coating	100.2	109.2	105.6	98.5	106.4	103.2	98.6	147.8	128.3	98.5	106.4	103.2	105.9	106.4	106.2	105.9	89.6	96.1
09	FINISHES	94.6	97.8	96.4	89.5	82.2	85.5	92.1	131.0	113.6	89.9	90.8	90.4	91.1	62.3	75.2	90.8	88.1	89.3
COVERS	DIVS. 10 - 14, 25, 28, 41, 43, 44, 46	100.0	102.2	100.4	100.0	97.9	99.6	100.0	117.3	103.5	100.0	100.8	100.2	100.0	94.6	98.9	100.0	97.6	99.5
21, 22, 23	FIRE SUPPRESSION, PLUMBING & HVAC	95.2	97.2	96.0	95.5	91.1	93.7	95.2	125.8	107.6	95.5	99.0	96.9	95.5	93.1	94.5	100.2	92.4	97.1
26, 27, 3370	ELECTRICAL, COMMUNICATIONS & UTIL.	94.3	112.2	103.8	92.7	86.5	89.5	92.9	114.9	104.5	93.6	91.7	92.6	90.0	74.6	81.8	91.0	88.3	89.5
MF2014	WEIGHTED AVERAGE	97.0	103.2	99.7	96.3	92.6	94.7	95.9	121.5	107.1	95.1	98.7	96.6	93.8	83.0	89.1	96.7	95.9	96.3

DIVISION		PUERTO RICO — SAN JUAN 009			RHODE ISLAND — NEWPORT 028			RHODE ISLAND — PROVIDENCE 029			SOUTH CAROLINA — AIKEN 298			SOUTH CAROLINA — BEAUFORT 299			SOUTH CAROLINA — CHARLESTON 294		
		MAT.	INST.	TOTAL	MAT.	INST.	TOTAL	MAT.	INST.	TOTAL	MAT.	INST.	TOTAL	MAT.	INST.	TOTAL	MAT.	INST.	TOTAL
015433	CONTRACTOR EQUIPMENT		90.5	90.5		102.4	102.4		102.4	102.4		101.3	101.3		101.3	101.3		101.3	101.3
0241, 31 - 34	SITE & INFRASTRUCTURE, DEMOLITION	133.8	91.2	103.5	89.2	105.0	100.4	91.4	105.0	101.1	118.9	87.4	96.5	114.3	85.6	93.9	99.7	86.3	90.2
0310	Concrete Forming & Accessories	92.4	17.8	28.1	101.5	122.4	119.5	99.8	122.4	119.3	97.5	67.4	71.5	96.4	37.9	46.0	95.4	63.0	67.5
0320	Concrete Reinforcing	188.3	12.7	98.9	106.4	149.2	128.2	102.3	149.2	126.2	93.7	65.2	79.2	92.8	25.8	58.7	92.7	59.3	75.7
0330	Cast-in-Place Concrete	103.8	30.4	73.7	79.5	124.1	97.8	95.5	124.1	107.2	79.2	69.8	75.4	79.2	47.2	66.1	92.9	49.5	75.1
03	CONCRETE	108.7	22.2	66.2	93.2	127.5	110.0	100.4	127.5	113.7	102.7	68.9	86.1	99.9	40.9	70.9	94.4	58.9	77.0
04	MASONRY	90.1	16.9	44.5	95.4	132.5	118.5	101.3	132.5	120.7	79.6	61.4	68.3	93.9	33.3	56.1	95.1	40.9	61.4
05	METALS	118.4	34.6	92.6	97.4	125.6	106.1	103.1	125.6	110.0	101.8	83.8	96.3	101.8	68.2	91.5	103.8	80.4	96.6
06	WOOD, PLASTICS & COMPOSITES	94.4	17.5	51.3	99.0	120.9	111.3	100.1	120.9	111.7	97.0	68.6	81.1	95.3	37.8	63.1	94.0	68.4	79.7
07	THERMAL & MOISTURE PROTECTION	129.1	21.6	85.0	100.6	121.5	109.2	100.5	121.5	109.1	102.7	66.4	87.8	102.4	38.8	76.3	101.6	47.3	79.3
08	OPENINGS	152.0	15.4	120.2	102.2	128.2	108.3	107.3	128.2	112.2	99.0	64.8	91.1	99.0	36.7	84.5	102.9	63.3	93.7
0920	Plaster & Gypsum Board	159.8	14.9	61.9	92.9	120.9	111.9	94.5	120.9	112.4	104.0	67.4	79.3	107.4	35.7	59.0	109.0	67.3	80.8
0950, 0980	Ceilings & Acoustic Treatment	225.7	14.9	87.2	93.2	120.9	111.4	90.1	120.9	110.4	86.4	67.4	73.9	89.8	35.7	54.3	89.8	67.3	75.0
0960	Flooring	224.3	16.6	164.9	97.7	141.5	110.4	98.0	141.5	110.4	105.4	68.4	94.8	106.8	51.2	90.9	106.5	58.2	92.7
0970, 0990	Wall Finishes & Painting/Coating	213.5	16.8	94.8	99.9	129.2	117.6	97.7	129.2	116.7	109.5	70.3	85.8	109.5	34.7	64.3	109.5	66.9	83.8
09	FINISHES	210.6	17.6	104.2	96.6	126.9	113.3	93.7	126.9	112.0	98.3	68.0	81.6	99.5	39.8	66.6	97.8	63.5	78.9
COVERS	DIVS. 10 - 14, 25, 28, 41, 43, 44, 46	100.0	17.5	83.3	100.0	108.6	101.7	100.0	108.6	101.7	100.0	71.8	94.3	100.0	70.4	94.0	100.0	69.0	93.7
21, 22, 23	FIRE SUPPRESSION, PLUMBING & HVAC	103.3	14.0	67.3	100.1	110.6	104.3	99.9	110.6	104.2	95.7	63.1	82.5	95.7	36.5	71.8	100.5	53.7	81.6
26, 27, 3370	ELECTRICAL, COMMUNICATIONS & UTIL.	125.9	13.2	66.3	100.0	99.0	99.5	99.6	99.0	99.2	96.9	65.9	80.5	100.7	33.6	65.2	99.0	88.8	93.6
MF2014	WEIGHTED AVERAGE	122.7	24.4	79.8	98.4	117.5	106.7	100.6	117.5	108.0	98.6	69.1	85.7	99.3	44.9	75.6	99.9	65.2	84.8

DIVISION		SOUTH CAROLINA — COLUMBIA 290 - 292			SOUTH CAROLINA — FLORENCE 295			SOUTH CAROLINA — GREENVILLE 296			SOUTH CAROLINA — ROCK HILL 297			SOUTH CAROLINA — SPARTANBURG 293			SOUTH DAKOTA — ABERDEEN 574		
		MAT.	INST.	TOTAL	MAT.	INST.	TOTAL	MAT.	INST.	TOTAL	MAT.	INST.	TOTAL	MAT.	INST.	TOTAL	MAT.	INST.	TOTAL
015433	CONTRACTOR EQUIPMENT		101.3	101.3		101.3	101.3		101.3	101.3		101.3	101.3		101.3	101.3		98.2	98.2
0241, 31 - 34	SITE & INFRASTRUCTURE, DEMOLITION	99.8	86.3	90.2	108.8	86.3	92.8	104.1	85.9	91.2	101.7	85.0	89.8	103.9	86.0	91.1	99.2	93.7	95.3
0310	Concrete Forming & Accessories	94.4	45.1	51.9	83.2	45.3	50.5	95.0	45.1	52.0	93.2	38.7	46.2	98.0	45.3	52.5	94.8	36.6	44.6
0320	Concrete Reinforcing	95.8	59.1	77.1	92.3	59.1	75.5	92.2	44.5	67.9	93.0	44.1	68.1	92.2	57.0	74.3	96.1	38.1	66.6
0330	Cast-in-Place Concrete	95.8	50.1	77.0	79.2	49.4	66.9	79.2	49.3	66.9	79.2	43.9	64.7	79.2	49.4	66.9	105.6	43.5	80.1
03	CONCRETE	96.2	51.1	74.1	94.2	51.0	73.0	93.0	48.2	71.0	90.9	43.3	67.5	93.2	50.6	72.3	102.8	40.7	72.3
04	MASONRY	91.5	37.6	57.9	79.8	40.9	55.6	77.4	40.9	54.7	101.6	34.6	59.8	101.8	40.9	55.6	115.9	55.4	78.2
05	METALS	100.9	79.6	94.3	102.6	80.0	95.6	102.6	74.5	93.9	101.8	71.2	92.4	102.6	79.1	95.4	103.6	63.6	86.0
06	WOOD, PLASTICS & COMPOSITES	98.0	44.8	68.2	79.7	44.8	60.2	93.7	44.8	66.4	92.0	38.9	62.3	98.0	44.8	68.2	99.6	36.0	64.0
07	THERMAL & MOISTURE PROTECTION	97.2	43.3	75.1	101.9	44.8	78.5	101.8	44.8	78.4	101.6	39.7	76.3	101.8	44.8	78.5	99.2	47.7	78.1
08	OPENINGS	104.6	50.5	92.0	99.1	50.5	87.8	99.0	47.0	86.9	99.0	39.4	85.2	99.0	50.0	87.6	99.1	36.7	84.6
0920	Plaster & Gypsum Board	102.3	42.9	62.2	97.1	42.9	60.5	102.6	42.9	62.3	101.9	36.8	57.9	105.5	42.9	63.2	103.7	34.3	56.8
0950, 0980	Ceilings & Acoustic Treatment	91.3	42.9	59.5	87.3	42.9	58.1	86.4	42.9	57.8	86.4	36.8	53.8	86.4	42.9	57.8	93.2	34.3	54.5
0960	Flooring	100.5	43.6	84.2	98.4	43.6	82.7	104.3	57.2	90.8	103.3	44.4	86.5	105.6	57.2	91.8	108.4	50.3	91.8
0970, 0990	Wall Finishes & Painting/Coating	106.2	66.9	82.5	109.5	66.9	83.8	109.5	66.9	83.8	109.5	39.2	67.0	109.5	66.9	83.8	102.6	37.1	63.1
09	FINISHES	95.9	46.6	68.7	94.4	47.1	68.3	96.3	49.3	70.4	95.7	39.4	64.6	97.1	49.3	70.8	100.3	38.8	66.4
COVERS	DIVS. 10 - 14, 25, 28, 41, 43, 44, 46	100.0	66.3	93.2	100.0	66.3	93.2	100.0	66.3	93.2	100.0	64.6	92.9	100.0	66.3	93.2	100.0	40.5	88.0
21, 22, 23	FIRE SUPPRESSION, PLUMBING & HVAC	100.0	53.0	81.1	100.5	53.1	81.4	100.5	53.0	81.3	95.7	44.2	74.9	100.5	53.1	81.3	100.1	39.8	75.8
26, 27, 3370	ELECTRICAL, COMMUNICATIONS & UTIL.	99.0	59.7	78.3	96.8	59.7	77.2	99.1	57.4	77.0	99.1	56.6	76.6	99.1	57.4	77.1	101.7	51.1	75.0
MF2014	WEIGHTED AVERAGE	99.3	56.5	80.6	98.2	56.9	80.2	98.3	55.8	79.8	97.8	50.3	77.1	98.6	56.7	80.3	100.5	49.6	78.3

For customer support on your Plumbing Cost Data, call 877.709.6968.

City Cost Indexes

SOUTH DAKOTA

DIVISION		MITCHELL 573			MOBRIDGE 576			PIERRE 575			RAPID CITY 577			SIOUX FALLS 570 - 571			WATERTOWN 572		
		MAT.	INST.	TOTAL	MAT.	INST.	TOTAL	MAT.	INST.	TOTAL	MAT.	INST.	TOTAL	MAT.	INST.	TOTAL	MAT.	INST.	TOTAL
015433	CONTRACTOR EQUIPMENT		98.2	98.2		98.2	98.2		98.2	98.2		98.2	98.2		99.2	99.2		98.2	98.2
0241, 31 - 34	SITE & INFRASTRUCTURE, DEMOLITION	96.0	93.7	94.3	95.9	93.7	94.3	100.5	93.7	95.7	97.7	93.8	94.9	94.3	95.4	95.0	95.8	93.7	94.3
0310	Concrete Forming & Accessories	94.0	37.0	44.8	84.9	36.8	43.4	97.7	38.4	46.6	102.2	36.7	45.7	98.8	40.5	48.5	81.6	36.6	42.8
0320	Concrete Reinforcing	95.5	46.3	70.5	98.0	38.1	67.5	99.0	71.1	84.8	89.9	71.3	80.4	98.0	71.2	84.3	92.9	38.2	65.1
0330	Cast-in-Place Concrete	102.6	42.3	77.8	102.6	43.5	78.3	98.5	42.2	75.4	101.8	42.8	77.6	92.3	42.9	72.0	102.6	45.9	79.3
03	CONCRETE	100.5	41.8	71.7	100.3	40.7	71.0	99.0	47.3	73.6	99.8	46.7	73.7	95.9	48.4	72.6	99.2	41.5	70.9
04	MASONRY	103.9	53.3	72.4	113.4	55.4	77.2	117.4	53.5	77.6	113.5	56.4	77.9	105.1	53.6	73.0	141.1	57.4	88.9
05	METALS	95.0	63.8	85.4	95.0	63.6	85.3	98.4	79.1	92.4	97.8	79.7	92.2	98.2	79.3	92.4	95.0	64.1	85.5
06	WOOD, PLASTICS & COMPOSITES	98.5	36.5	63.8	87.6	36.2	58.8	104.5	36.8	66.6	103.5	33.3	64.2	96.8	39.3	64.6	83.8	36.0	57.0
07	THERMAL & MOISTURE PROTECTION	98.9	45.6	77.1	99.0	47.8	78.0	101.8	45.9	78.9	99.6	47.8	78.3	101.9	48.2	79.9	98.8	48.3	78.1
08	OPENINGS	97.7	37.3	83.6	100.4	36.3	85.5	105.7	46.7	92.0	103.1	44.8	89.6	106.5	48.1	92.9	97.7	36.5	83.4
0920	Plaster & Gypsum Board	102.1	34.8	56.7	96.2	34.6	54.5	100.1	35.1	56.2	103.2	31.6	54.8	90.7	37.7	54.9	93.9	34.3	53.6
0950, 0980	Ceilings & Acoustic Treatment	89.9	34.8	53.7	93.2	34.6	54.7	93.3	35.1	55.0	95.7	31.6	53.6	90.6	37.7	55.8	89.9	34.3	53.3
0960	Flooring	107.9	50.3	91.4	103.5	50.3	88.3	107.4	35.8	86.9	107.6	78.6	99.3	103.8	74.9	95.5	102.1	50.3	87.3
0970, 0990	Wall Finishes & Painting/Coating	102.6	40.6	65.2	102.6	41.7	65.8	105.7	44.6	68.8	102.6	44.6	67.6	101.7	44.6	67.2	102.6	37.1	63.1
09	FINISHES	99.1	39.5	66.2	97.7	39.4	65.6	100.8	37.3	65.8	100.5	43.9	69.2	97.1	46.5	69.2	96.2	38.8	64.6
COVERS	DIVS. 10 - 14, 25, 28, 41, 43, 44, 46	100.0	37.6	87.4	100.0	40.5	88.0	100.0	75.8	95.1	100.0	75.7	95.1	100.0	76.2	95.2	100.0	40.5	88.0
21, 22, 23	FIRE SUPPRESSION, PLUMBING & HVAC	95.3	39.4	72.8	95.3	39.9	73.0	100.0	64.5	85.7	100.1	65.0	85.9	100.0	38.6	75.2	95.3	39.9	73.0
26, 27, 3370	ELECTRICAL, COMMUNICATIONS & UTIL.	99.9	44.0	70.3	101.7	44.8	71.7	104.8	51.8	76.8	98.0	51.8	73.6	101.3	75.1	87.5	99.0	44.8	70.4
MF2014	WEIGHTED AVERAGE	97.8	48.4	76.3	98.5	48.8	76.8	101.7	58.4	82.8	100.4	59.5	82.6	100.0	57.7	81.5	99.0	49.1	77.2

TENNESSEE

DIVISION		CHATTANOOGA 373 - 374			COLUMBIA 384			COOKEVILLE 385			JACKSON 383			JOHNSON CITY 376			KNOXVILLE 377 - 379		
		MAT.	INST.	TOTAL	MAT.	INST.	TOTAL	MAT.	INST.	TOTAL	MAT.	INST.	TOTAL	MAT.	INST.	TOTAL	MAT.	INST.	TOTAL
015433	CONTRACTOR EQUIPMENT		103.8	103.8		98.2	98.2		98.2	98.2		104.5	104.5		97.7	97.7		97.7	97.7
0241, 31 - 34	SITE & INFRASTRUCTURE, DEMOLITION	104.5	97.9	99.8	89.1	86.9	87.5	94.5	86.4	88.8	97.7	96.8	97.1	111.1	86.6	93.6	90.7	87.3	88.3
0310	Concrete Forming & Accessories	97.1	56.4	62.0	82.1	62.0	64.8	82.2	35.4	41.8	89.0	44.9	51.0	83.7	38.9	45.1	95.6	61.0	65.7
0320	Concrete Reinforcing	91.2	67.4	79.1	87.4	61.2	74.1	87.4	61.2	74.1	87.4	62.0	74.4	91.7	60.3	75.7	91.2	62.3	76.5
0330	Cast-in-Place Concrete	101.3	62.7	85.5	94.2	51.6	76.7	106.7	43.5	80.7	104.2	49.3	81.6	81.6	60.2	72.8	95.2	65.7	83.1
03	CONCRETE	96.0	62.3	79.4	96.7	59.8	78.6	106.9	45.2	76.6	97.7	51.6	75.1	103.3	52.3	78.3	93.4	64.4	79.1
04	MASONRY	109.7	51.7	73.5	127.7	54.6	82.1	121.9	43.8	73.2	127.9	47.2	77.6	125.7	44.8	75.3	86.6	55.3	67.1
05	METALS	98.7	89.2	95.8	96.4	86.0	93.2	96.5	85.9	93.2	98.8	85.9	94.8	95.9	85.2	92.6	99.3	87.0	95.5
06	WOOD, PLASTICS & COMPOSITES	104.0	57.3	77.8	70.1	64.4	66.9	70.3	33.7	49.8	85.4	45.4	63.0	76.9	36.9	54.5	91.0	60.6	73.9
07	THERMAL & MOISTURE PROTECTION	99.4	59.9	83.2	93.6	59.2	79.5	94.0	45.7	74.2	96.0	54.1	78.8	94.5	52.6	77.3	92.5	61.6	79.8
08	OPENINGS	99.7	57.6	89.9	92.6	57.9	84.6	92.7	42.2	80.9	100.3	51.6	89.0	96.2	45.1	84.3	93.3	58.6	85.2
0920	Plaster & Gypsum Board	82.8	56.5	65.0	87.7	63.7	71.5	87.7	32.1	50.1	89.3	44.1	58.8	102.4	35.5	57.2	109.0	59.8	75.8
0950, 0980	Ceilings & Acoustic Treatment	96.6	56.5	70.2	77.0	63.7	68.3	77.0	32.1	47.5	86.2	44.1	58.5	93.0	35.5	55.2	93.8	59.8	71.5
0960	Flooring	99.1	58.9	87.7	89.3	20.5	69.7	89.4	58.6	80.6	87.1	40.8	73.9	96.7	40.6	80.7	101.3	55.8	88.3
0970, 0990	Wall Finishes & Painting/Coating	102.5	70.7	83.3	94.0	33.4	57.4	94.0	36.5	59.3	95.8	49.6	67.9	99.6	42.1	64.9	99.6	81.2	88.5
09	FINISHES	97.2	58.2	75.7	89.8	52.0	69.0	90.3	38.4	61.7	89.7	43.7	64.4	101.9	38.1	66.7	94.5	62.0	76.6
COVERS	DIVS. 10 - 14, 25, 28, 41, 43, 44, 46	100.0	41.8	88.3	100.0	47.9	89.5	100.0	41.0	88.1	100.0	43.0	88.5	100.0	75.5	95.1	100.0	81.7	96.3
21, 22, 23	FIRE SUPPRESSION, PLUMBING & HVAC	100.2	62.6	85.0	97.4	78.1	89.6	97.4	72.6	87.4	100.1	68.0	87.1	99.9	59.1	83.4	99.9	67.9	87.0
26, 27, 3370	ELECTRICAL, COMMUNICATIONS & UTIL.	102.3	69.2	84.8	94.2	57.4	74.7	95.9	61.6	77.8	101.0	58.5	78.5	92.5	46.3	68.1	98.1	56.0	75.8
MF2014	WEIGHTED AVERAGE	100.0	66.2	85.2	96.9	66.1	83.5	98.1	59.1	81.1	99.9	61.2	83.0	99.8	56.5	80.9	96.6	66.9	83.7

TENNESSEE / TEXAS

DIVISION		MCKENZIE 382			MEMPHIS 375, 380 - 381			NASHVILLE 370 - 372			ABILENE 795 - 796			AMARILLO 790 - 791			AUSTIN 786 - 787		
		MAT.	INST.	TOTAL	MAT.	INST.	TOTAL	MAT.	INST.	TOTAL	MAT.	INST.	TOTAL	MAT.	INST.	TOTAL	MAT.	INST.	TOTAL
015433	CONTRACTOR EQUIPMENT		98.2	98.2		102.8	102.8		103.7	103.7		89.9	89.9		89.9	89.9		89.4	89.4
0241, 31 - 34	SITE & INFRASTRUCTURE, DEMOLITION	94.3	86.5	88.7	94.1	93.7	93.8	98.9	97.5	97.9	99.4	88.7	91.8	97.3	88.6	91.1	102.4	88.4	92.4
0310	Concrete Forming & Accessories	90.0	38.1	45.2	94.3	64.9	69.0	98.9	65.8	70.3	99.0	65.5	70.1	97.9	57.1	62.7	98.6	59.1	64.5
0320	Concrete Reinforcing	87.5	61.9	74.5	99.7	68.7	83.9	96.0	67.7	81.6	92.9	52.5	72.3	99.0	51.3	74.7	96.5	48.5	72.1
0330	Cast-in-Place Concrete	104.4	57.3	85.1	94.1	61.3	80.6	90.3	65.9	80.3	96.6	63.7	83.1	94.1	63.7	81.6	93.4	66.2	82.3
03	CONCRETE	105.5	51.3	78.9	94.1	65.8	80.2	92.4	67.6	80.2	95.0	63.0	79.3	96.7	59.0	78.2	95.7	60.2	78.3
04	MASONRY	126.2	46.2	76.3	99.8	62.7	76.7	93.8	59.0	72.1	104.2	63.6	78.9	109.7	63.6	81.0	107.6	56.5	75.7
05	METALS	96.5	86.1	93.3	100.1	90.6	97.2	100.4	89.9	97.2	105.3	69.8	94.4	101.6	69.2	91.7	100.5	66.3	90.0
06	WOOD, PLASTICS & COMPOSITES	79.1	36.3	55.1	95.4	67.0	79.5	101.6	66.6	82.0	101.1	68.5	82.8	100.8	57.1	76.4	94.4	59.9	75.0
07	THERMAL & MOISTURE PROTECTION	94.0	48.7	75.4	95.5	63.4	82.4	95.7	61.5	81.7	100.0	67.7	86.7	97.9	65.3	84.5	104.2	64.8	88.1
08	OPENINGS	92.7	44.1	81.4	100.1	66.0	92.2	98.5	65.9	90.9	97.1	63.3	89.2	101.1	56.8	90.8	104.3	55.6	92.9
0920	Plaster & Gypsum Board	90.5	34.8	52.8	95.6	66.3	75.8	95.8	66.0	75.6	84.0	68.0	73.2	92.1	56.3	67.9	86.7	59.2	68.1
0950, 0980	Ceilings & Acoustic Treatment	77.0	34.8	49.2	93.9	66.3	75.6	96.6	66.0	76.1	88.9	68.0	75.2	98.0	56.3	70.6	87.9	59.2	69.0
0960	Flooring	92.1	39.3	77.0	99.4	36.1	81.3	97.7	64.2	88.2	110.9	76.5	101.1	106.5	76.5	97.9	105.7	64.1	93.8
0970, 0990	Wall Finishes & Painting/Coating	94.0	49.6	67.2	97.0	55.7	72.1	100.5	72.0	83.3	106.6	56.1	76.1	100.8	56.1	73.8	104.3	48.8	70.8
09	FINISHES	91.3	38.2	62.1	97.3	58.5	75.9	100.2	66.0	81.4	93.7	67.1	79.0	96.8	60.3	76.7	95.6	58.7	75.2
COVERS	DIVS. 10 - 14, 25, 28, 41, 43, 44, 46	100.0	27.4	85.3	100.0	82.4	96.4	100.0	82.9	96.5	100.0	82.2	96.4	100.0	68.6	93.7	100.0	80.5	96.1
21, 22, 23	FIRE SUPPRESSION, PLUMBING & HVAC	97.4	68.3	85.6	100.0	73.9	89.5	100.0	84.0	93.5	100.3	49.0	79.6	100.0	54.9	81.8	100.1	60.1	84.0
26, 27, 3370	ELECTRICAL, COMMUNICATIONS & UTIL.	95.6	62.9	78.3	101.7	65.3	82.4	97.6	63.1	79.4	99.3	48.7	72.5	101.1	64.1	81.6	96.8	63.9	79.4
MF2014	WEIGHTED AVERAGE	98.2	59.2	81.2	99.0	71.3	86.9	98.4	74.1	87.8	99.7	62.2	83.3	100.2	63.2	84.1	100.0	63.6	84.1

For customer support on your Plumbing Cost Data, call 877.709.6968.

633

City Cost Indexes

TEXAS

DIVISION		BEAUMONT 776-777			BROWNWOOD 768			BRYAN 778			CHILDRESS 792			CORPUS CHRISTI 783-784			DALLAS 752-753		
		MAT.	INST.	TOTAL	MAT.	INST.	TOTAL	MAT.	INST.	TOTAL	MAT.	INST.	TOTAL	MAT.	INST.	TOTAL	MAT.	INST.	TOTAL
015433	CONTRACTOR EQUIPMENT		91.5	91.5		89.9	89.9		91.5	91.5		89.9	89.9		96.6	96.6		99.1	99.1
0241, 31-34	SITE & INFRASTRUCTURE, DEMOLITION	90.0	89.9	89.9	105.2	89.6	94.1	81.8	90.5	88.0	109.7	88.0	94.3	140.1	84.4	100.5	104.3	89.8	94.0
0310	Concrete Forming & Accessories	101.8	61.9	67.4	99.0	60.5	65.8	81.0	67.1	69.0	97.3	65.5	69.8	100.2	58.1	63.9	101.2	65.8	70.7
0320	Concrete Reinforcing	95.9	65.2	80.3	92.9	52.1	72.1	98.3	48.6	73.0	93.0	52.1	72.2	88.0	48.5	67.9	100.2	52.5	75.9
0330	Cast-in-Place Concrete	89.5	64.7	79.3	101.4	63.6	85.9	71.7	65.4	69.1	99.1	63.7	84.5	109.8	64.4	91.2	98.0	64.6	84.3
03	CONCRETE	96.3	64.2	80.5	101.2	60.7	81.3	80.9	63.7	72.5	104.0	62.9	83.8	100.1	59.9	80.4	99.0	64.1	81.9
04	MASONRY	106.8	63.9	80.0	141.8	56.6	88.7	146.3	60.4	92.7	108.5	56.6	76.2	91.0	56.4	69.6	102.4	56.7	73.9
05	METALS	95.2	75.5	89.1	99.9	69.1	90.5	94.8	70.1	87.2	102.7	69.2	92.4	94.7	77.9	89.5	100.4	80.0	94.1
06	WOOD, PLASTICS & COMPOSITES	106.1	61.8	81.3	99.5	61.9	78.4	73.5	68.5	70.7	100.4	68.5	82.6	116.0	58.4	83.7	104.1	68.6	84.2
07	THERMAL & MOISTURE PROTECTION	101.9	66.7	87.5	98.2	65.2	84.7	93.9	66.3	82.5	100.5	64.8	85.9	104.5	66.2	88.8	93.0	66.4	82.1
08	OPENINGS	94.0	62.3	86.7	94.7	59.7	86.6	97.2	61.6	88.9	94.2	63.3	87.1	109.1	54.7	96.5	99.5	63.5	91.1
0920	Plaster & Gypsum Board	102.3	61.2	74.5	87.8	61.2	69.8	90.3	68.0	75.2	83.6	68.0	73.1	98.4	57.5	70.7	96.7	68.0	77.3
0950, 0980	Ceilings & Acoustic Treatment	98.2	61.2	73.9	80.6	61.2	67.8	91.2	68.0	76.0	87.3	68.0	74.6	92.7	57.5	69.6	97.8	68.0	78.2
0960	Flooring	112.4	73.0	101.1	95.3	64.1	86.3	85.8	64.1	79.6	109.1	64.1	96.2	118.7	64.1	103.1	102.8	64.1	91.7
0970, 0990	Wall Finishes & Painting/Coating	95.1	59.4	73.6	104.2	56.1	75.1	92.5	62.2	74.2	106.6	56.1	76.1	120.1	56.5	81.7	107.0	56.1	76.3
09	FINISHES	93.7	63.4	77.0	87.7	60.7	72.8	82.3	66.2	73.4	94.0	64.6	77.8	105.9	57.8	79.9	100.0	64.7	80.5
COVERS	DIVS. 10-14, 25, 28, 41, 43, 44, 46	100.0	83.4	96.7	100.0	81.4	96.3	100.0	82.7	96.5	100.0	82.2	96.4	100.0	82.6	96.5	100.0	82.5	96.5
21, 22, 23	FIRE SUPPRESSION, PLUMBING & HVAC	100.2	64.1	85.6	95.4	50.0	77.1	95.4	66.5	83.8	95.5	54.9	79.1	100.2	58.9	83.5	100.0	62.4	84.9
26, 27, 3370	ELECTRICAL, COMMUNICATIONS & UTIL.	95.2	70.6	82.2	93.4	42.8	66.7	93.4	67.4	79.7	99.2	64.1	80.7	92.8	60.0	75.5	95.2	64.6	79.0
MF2014	WEIGHTED AVERAGE	97.4	68.6	84.9	98.9	59.4	81.7	94.9	68.2	83.3	99.3	64.3	84.0	100.9	63.5	84.6	99.5	67.4	85.5

TEXAS

DIVISION		DEL RIO 788			DENTON 762			EASTLAND 764			EL PASO 798-799,885			FORT WORTH 760-761			GALVESTON 775		
		MAT.	INST.	TOTAL	MAT.	INST.	TOTAL	MAT.	INST.	TOTAL	MAT.	INST.	TOTAL	MAT.	INST.	TOTAL	MAT.	INST.	TOTAL
015433	CONTRACTOR EQUIPMENT		89.4	89.4		95.8	95.8		89.9	89.9		89.9	89.9		89.9	89.9		100.5	100.5
0241, 31-34	SITE & INFRASTRUCTURE, DEMOLITION	121.8	88.4	98.0	105.5	81.3	88.3	108.0	87.6	93.5	101.7	87.6	91.7	102.1	88.7	92.6	107.2	88.9	93.8
0310	Concrete Forming & Accessories	96.5	57.8	63.1	106.2	65.3	70.9	99.8	65.3	70.0	97.8	63.4	68.1	98.0	65.6	70.1	89.8	64.6	68.1
0320	Concrete Reinforcing	88.7	48.4	68.2	94.3	52.1	72.8	93.1	50.9	71.6	99.2	52.0	75.2	98.4	52.4	75.0	97.8	62.4	79.8
0330	Cast-in-Place Concrete	118.2	63.2	95.6	78.5	64.6	72.8	107.2	63.6	89.3	88.4	63.7	78.3	95.0	63.7	82.1	95.2	66.4	83.4
03	CONCRETE	121.3	58.6	90.5	80.0	63.9	72.1	105.9	62.6	84.6	94.0	61.9	78.3	97.0	63.1	80.4	97.5	66.4	82.2
04	MASONRY	106.9	56.5	75.5	152.1	56.7	92.6	106.8	56.6	75.5	99.0	57.7	73.3	104.7	56.6	74.7	102.5	60.4	76.3
05	METALS	94.4	66.2	85.7	99.5	80.6	93.7	99.7	68.5	90.1	98.5	67.4	89.0	101.5	69.5	91.7	96.3	88.3	93.9
06	WOOD, PLASTICS & COMPOSITES	96.5	58.3	75.1	111.3	68.6	87.4	106.1	68.5	85.1	91.3	65.7	76.9	97.3	68.5	81.2	89.4	64.7	75.6
07	THERMAL & MOISTURE PROTECTION	101.3	65.3	86.5	96.0	66.7	84.0	98.6	65.9	85.1	99.8	64.4	85.3	96.0	65.9	83.6	93.0	67.1	82.4
08	OPENINGS	101.8	54.7	90.8	112.2	63.4	100.8	70.4	63.0	68.7	93.3	58.1	85.1	99.4	63.4	91.0	101.6	63.2	92.7
0920	Plaster & Gypsum Board	94.7	57.5	69.6	92.1	68.0	75.8	87.8	68.0	74.4	95.9	65.1	75.1	91.8	68.0	75.7	96.8	64.0	74.7
0950, 0980	Ceilings & Acoustic Treatment	89.1	57.5	68.3	84.7	68.0	73.8	80.6	68.0	72.3	91.4	65.1	74.1	90.5	68.0	75.7	94.6	64.0	74.5
0960	Flooring	100.7	64.1	90.2	90.4	64.1	82.9	121.4	64.1	105.0	110.4	64.1	97.1	117.9	64.1	102.5	100.1	64.1	89.8
0970, 0990	Wall Finishes & Painting/Coating	106.4	48.8	71.6	115.6	56.1	79.7	105.8	56.1	75.8	106.2	53.7	74.5	103.2	56.1	74.7	103.4	60.0	77.2
09	FINISHES	98.2	57.7	75.9	85.8	64.7	74.2	95.8	64.6	78.6	97.3	63.0	78.4	98.4	64.6	79.8	91.4	63.7	76.2
COVERS	DIVS. 10-14, 25, 28, 41, 43, 44, 46	100.0	80.5	96.1	100.0	82.5	96.5	100.0	82.2	96.4	100.0	81.0	96.2	100.0	82.3	96.4	100.0	84.5	96.9
21, 22, 23	FIRE SUPPRESSION, PLUMBING & HVAC	95.4	57.3	80.0	95.4	42.6	74.1	95.4	49.9	77.1	100.0	50.3	80.0	100.0	58.4	83.2	95.4	66.7	83.8
26, 27, 3370	ELECTRICAL, COMMUNICATIONS & UTIL.	94.9	68.8	81.1	95.9	64.6	79.3	93.3	64.5	78.1	99.1	57.4	77.0	94.7	64.5	78.8	95.0	67.9	80.7
MF2014	WEIGHTED AVERAGE	100.8	63.2	84.4	98.9	62.5	83.0	95.9	63.2	81.6	98.0	61.7	82.2	99.3	65.3	84.5	97.0	70.0	85.2

TEXAS

DIVISION		GIDDINGS 789			GREENVILLE 754			HOUSTON 770-772			HUNTSVILLE 773			LAREDO 780			LONGVIEW 756		
		MAT.	INST.	TOTAL	MAT.	INST.	TOTAL	MAT.	INST.	TOTAL	MAT.	INST.	TOTAL	MAT.	INST.	TOTAL	MAT.	INST.	TOTAL
015433	CONTRACTOR EQUIPMENT		89.4	89.4		96.7	96.7		100.4	100.4		91.5	91.5		89.4	89.4		91.4	91.4
0241, 31-34	SITE & INFRASTRUCTURE, DEMOLITION	107.8	88.4	94.0	97.9	85.4	89.0	106.0	88.2	93.4	96.6	89.9	91.8	102.1	88.4	92.3	95.8	92.3	93.3
0310	Concrete Forming & Accessories	94.1	57.9	62.9	92.3	65.3	69.0	91.8	64.6	68.4	87.8	61.6	65.2	96.6	58.0	63.3	88.4	64.9	68.1
0320	Concrete Reinforcing	89.2	45.9	67.2	100.7	52.1	76.0	97.5	61.4	79.1	98.5	50.6	74.1	88.7	48.8	68.4	99.6	52.0	75.4
0330	Cast-in-Place Concrete	100.2	63.2	85.0	89.3	63.9	78.9	92.3	66.4	81.7	98.6	64.6	84.6	84.6	63.3	75.8	103.8	62.8	86.9
03	CONCRETE	97.6	58.2	78.2	91.6	63.6	77.8	95.2	66.2	80.9	103.9	61.3	83.0	92.5	58.8	75.9	107.5	62.3	85.3
04	MASONRY	115.7	56.5	78.8	160.0	56.6	95.6	102.3	66.5	80.0	145.1	58.9	91.4	100.2	63.5	77.3	155.7	56.6	93.9
05	METALS	93.9	65.3	85.1	97.7	79.3	92.1	99.2	88.2	95.8	94.7	69.2	86.8	97.0	66.9	87.8	91.1	67.8	83.9
06	WOOD, PLASTICS & COMPOSITES	95.6	58.3	74.7	93.5	68.6	79.5	91.7	64.7	76.6	82.0	61.8	70.7	96.5	58.3	75.1	87.2	68.4	76.7
07	THERMAL & MOISTURE PROTECTION	102.0	64.5	86.6	93.0	66.0	81.9	92.6	67.8	82.4	94.9	65.9	83.0	100.2	66.5	86.4	94.5	65.2	82.5
08	OPENINGS	100.8	54.0	89.9	97.7	63.4	89.7	104.3	62.8	94.6	97.2	58.5	88.2	101.6	54.7	90.7	87.8	63.3	82.1
0920	Plaster & Gypsum Board	93.7	57.5	69.2	90.3	68.0	75.3	99.1	64.0	75.4	94.2	61.2	71.9	95.8	57.5	69.9	89.1	68.0	74.9
0950, 0980	Ceilings & Acoustic Treatment	89.1	57.5	68.3	93.6	68.0	76.8	99.6	64.0	76.2	91.2	61.2	71.5	93.5	57.5	69.8	90.3	68.0	75.7
0960	Flooring	101.1	64.1	90.5	98.6	64.1	88.7	101.2	64.1	90.6	89.4	64.1	82.1	100.5	64.1	90.1	103.5	64.1	92.2
0970, 0990	Wall Finishes & Painting/Coating	106.4	48.8	71.6	107.0	56.1	76.3	103.4	62.2	78.5	92.5	59.4	72.5	106.4	53.2	74.3	96.8	56.1	72.2
09	FINISHES	97.1	57.7	75.4	96.6	64.6	79.0	98.9	64.0	79.7	84.7	61.6	72.0	97.7	58.2	75.9	101.7	64.5	81.2
COVERS	DIVS. 10-14, 25, 28, 41, 43, 44, 46	100.0	80.8	96.1	100.0	82.3	96.4	100.0	84.5	96.9	100.0	81.0	96.2	100.0	80.5	96.1	100.0	82.0	96.4
21, 22, 23	FIRE SUPPRESSION, PLUMBING & HVAC	95.4	64.6	83.0	95.3	49.0	76.6	100.1	66.7	86.6	95.4	65.9	83.5	100.2	64.1	85.6	95.3	33.6	70.4
26, 27, 3370	ELECTRICAL, COMMUNICATIONS & UTIL.	91.8	63.9	77.1	92.0	64.6	77.5	96.9	67.5	81.4	93.4	68.0	80.0	95.1	60.9	77.0	92.2	55.6	72.9
MF2014	WEIGHTED AVERAGE	97.7	64.0	83.0	98.8	64.0	83.6	99.4	70.5	86.8	98.0	66.7	84.4	98.3	64.5	83.6	98.5	58.7	81.2

For customer support on your Plumbing Cost Data, call 877.709.6968.

City Cost Indexes

TEXAS

DIVISION		LUBBOCK 793-794 MAT.	INST.	TOTAL	LUFKIN 759 MAT.	INST.	TOTAL	MCALLEN 785 MAT.	INST.	TOTAL	MCKINNEY 750 MAT.	INST.	TOTAL	MIDLAND 797 MAT.	INST.	TOTAL	ODESSA 797 MAT.	INST.	TOTAL
015433	CONTRACTOR EQUIPMENT		98.5	98.5		91.4	91.4		96.8	96.8		96.7	96.7		98.5	98.5		89.9	89.9
0241, 31-34	SITE & INFRASTRUCTURE, DEMOLITION	123.8	86.1	97.0	91.2	94.0	93.1	144.1	84.4	101.7	94.6	85.3	88.0	126.7	86.1	97.8	99.7	88.0	91.4
0310	Concrete Forming & Accessories	97.9	57.2	62.8	91.7	61.7	65.9	101.2	57.5	63.5	91.3	65.3	68.8	101.9	65.4	70.5	98.9	65.4	70.0
0320	Concrete Reinforcing	94.1	52.5	72.9	101.3	64.9	82.8	88.2	48.4	67.9	100.7	50.9	75.3	95.1	52.1	73.2	92.9	52.1	72.1
0330	Cast-in-Place Concrete	96.8	64.7	83.6	92.9	64.0	81.0	119.3	64.1	96.6	83.8	63.8	75.6	102.9	64.6	87.2	96.6	63.6	83.1
03	CONCRETE	93.7	60.4	77.4	99.8	63.6	82.0	107.9	59.5	84.1	87.0	63.3	75.3	98.2	64.0	81.4	95.0	62.8	79.2
04	MASONRY	103.6	64.0	78.9	119.4	58.8	81.6	106.8	56.4	75.4	172.2	56.6	100.2	121.5	56.7	81.1	104.2	56.6	74.6
05	METALS	109.0	81.3	100.5	97.9	72.4	90.0	94.3	77.5	89.2	97.6	78.8	91.8	107.2	80.7	99.0	104.6	69.1	93.7
06	WOOD, PLASTICS & COMPOSITES	101.1	57.3	76.5	95.3	61.8	76.5	114.6	58.4	83.1	92.3	68.6	79.0	106.0	68.6	85.1	101.1	68.5	82.8
07	THERMAL & MOISTURE PROTECTION	89.3	66.2	79.8	94.3	65.6	82.5	104.8	59.8	86.3	92.8	66.0	81.8	89.5	65.4	79.7	100.0	65.9	86.0
08	OPENINGS	108.5	57.2	96.6	66.7	63.0	65.9	105.7	54.7	93.9	97.7	63.0	89.6	107.5	63.4	97.2	97.1	63.3	89.2
0920	Plaster & Gypsum Board	84.4	56.3	65.4	88.0	61.2	69.9	99.4	57.5	71.1	90.0	68.0	75.1	85.9	68.0	73.8	84.0	68.0	73.2
0950, 0980	Ceilings & Acoustic Treatment	90.6	56.3	68.1	84.4	61.2	69.1	93.3	57.5	69.8	93.6	68.0	76.8	88.1	68.0	74.9	88.9	68.0	75.2
0960	Flooring	104.4	64.1	92.8	138.2	64.1	117.0	118.1	63.8	102.6	98.2	64.1	88.4	105.7	64.1	93.8	110.9	64.1	97.5
0970, 0990	Wall Finishes & Painting/Coating	119.0	56.1	81.0	96.8	56.1	72.2	120.1	48.8	77.1	107.0	56.1	76.3	119.0	56.1	81.0	106.6	56.1	76.1
09	FINISHES	96.4	58.0	75.2	110.2	61.2	83.2	106.4	57.8	79.6	96.2	64.6	78.8	96.8	64.7	79.1	93.7	64.6	77.7
COVERS	DIVS. 10-14, 25, 28, 41, 43, 44, 46	100.0	81.3	96.2	100.0	80.9	96.1	100.0	80.7	96.1	100.0	82.3	96.4	100.0	81.2	96.2	100.0	80.9	96.1
21, 22, 23	FIRE SUPPRESSION, PLUMBING & HVAC	99.8	50.7	80.0	95.3	64.4	82.8	95.4	58.7	80.6	95.3	45.4	75.2	95.0	49.1	76.5	100.3	49.1	79.6
26, 27, 3370	ELECTRICAL, COMMUNICATIONS & UTIL.	97.9	64.2	80.1	93.5	70.6	81.4	92.6	36.3	62.8	92.1	64.6	77.5	97.9	64.2	80.1	99.4	64.1	80.7
MF2014	WEIGHTED AVERAGE	101.6	63.6	85.0	95.6	67.8	83.5	101.0	59.8	83.0	98.7	63.1	83.2	101.5	64.2	85.2	99.6	63.1	83.7

TEXAS

DIVISION		PALESTINE 758 MAT.	INST.	TOTAL	SAN ANGELO 769 MAT.	INST.	TOTAL	SAN ANTONIO 781-782 MAT.	INST.	TOTAL	TEMPLE 765 MAT.	INST.	TOTAL	TEXARKANA 755 MAT.	INST.	TOTAL	TYLER 757 MAT.	INST.	TOTAL
015433	CONTRACTOR EQUIPMENT		91.4	91.4		89.9	89.9		91.9	91.9		89.9	89.9		91.4	91.4		91.4	91.4
0241, 31-34	SITE & INFRASTRUCTURE, DEMOLITION	96.6	92.6	93.8	101.5	89.7	93.1	100.9	92.0	94.5	89.8	89.1	89.3	85.9	92.0	90.2	95.1	92.4	93.2
0310	Concrete Forming & Accessories	82.7	65.4	67.8	99.3	58.7	64.3	95.0	58.1	63.2	102.8	57.8	64.0	99.0	57.4	63.1	93.6	65.4	69.3
0320	Concrete Reinforcing	98.8	52.1	75.1	92.7	52.1	72.1	93.6	49.8	71.3	92.9	49.1	70.6	98.7	52.3	75.1	99.6	52.4	75.6
0330	Cast-in-Place Concrete	84.9	63.0	75.9	95.7	64.8	83.0	85.6	65.2	77.2	78.5	63.1	72.2	85.5	62.7	76.2	101.9	63.0	85.9
03	CONCRETE	101.6	62.6	82.4	96.7	60.3	78.8	91.7	59.7	76.0	82.9	58.7	71.0	92.8	59.0	76.2	107.1	62.7	85.3
04	MASONRY	113.8	56.6	78.2	138.0	58.7	88.5	96.8	65.3	77.2	151.0	56.4	92.1	176.5	56.6	101.8	166.0	56.6	97.8
05	METALS	97.6	68.3	88.6	100.1	69.1	90.6	96.9	68.6	88.2	99.8	66.5	89.6	91.0	67.8	83.9	97.5	68.4	88.5
06	WOOD, PLASTICS & COMPOSITES	85.3	68.4	75.8	99.8	58.3	76.5	94.7	57.4	73.8	108.9	58.3	80.6	99.7	58.2	76.5	97.0	68.4	81.0
07	THERMAL & MOISTURE PROTECTION	94.8	65.3	82.7	98.0	66.0	84.9	96.4	68.2	84.8	97.7	64.8	84.2	94.1	64.2	81.9	94.6	65.3	82.6
08	OPENINGS	66.7	63.3	65.9	94.7	57.8	86.1	102.7	54.4	91.5	67.1	57.0	64.7	87.7	57.8	80.8	66.6	63.4	65.9
0920	Plaster & Gypsum Board	85.2	68.0	73.6	87.8	57.5	67.3	90.0	56.5	70.0	87.8	57.5	67.3	93.4	57.5	69.1	88.0	68.0	74.5
0950, 0980	Ceilings & Acoustic Treatment	84.4	68.0	73.6	80.6	57.5	65.4	102.7	56.5	72.3	80.6	57.5	65.4	90.3	57.5	68.7	84.4	68.0	73.6
0960	Flooring	129.9	64.1	111.1	95.3	64.1	86.4	104.1	64.1	92.6	123.2	64.1	106.3	111.1	64.1	97.6	140.1	64.1	118.4
0970, 0990	Wall Finishes & Painting/Coating	96.8	56.1	72.2	104.2	56.1	75.1	106.3	53.2	74.2	105.8	48.8	71.4	96.8	56.1	72.2	96.8	56.1	72.2
09	FINISHES	108.0	64.5	84.0	87.5	59.1	71.8	103.4	58.1	78.4	95.1	57.7	74.5	104.0	58.5	78.9	111.1	64.5	85.4
COVERS	DIVS. 10-14, 25, 28, 41, 43, 44, 46	100.0	82.0	96.4	100.0	81.0	96.2	100.0	81.2	96.2	100.0	81.1	96.2	100.0	80.9	96.1	100.0	82.0	96.4
21, 22, 23	FIRE SUPPRESSION, PLUMBING & HVAC	95.3	59.6	80.9	95.4	50.1	77.1	100.0	63.2	85.1	95.4	54.7	79.0	95.3	37.0	71.8	95.3	62.4	82.0
26, 27, 3370	ELECTRICAL, COMMUNICATIONS & UTIL.	89.8	55.6	71.7	97.4	44.8	69.6	96.4	60.9	77.6	94.5	59.8	76.2	93.4	60.1	75.8	92.2	55.5	72.8
MF2014	WEIGHTED AVERAGE	95.0	64.4	81.7	98.5	59.6	81.5	98.5	65.1	84.0	94.7	61.7	80.3	98.0	58.4	80.7	98.5	65.0	83.9

TEXAS / UTAH

DIVISION		VICTORIA 779 MAT.	INST.	TOTAL	WACO 766-767 MAT.	INST.	TOTAL	WAXAHACHIE 751 MAT.	INST.	TOTAL	WHARTON 774 MAT.	INST.	TOTAL	WICHITA FALLS 763 MAT.	INST.	TOTAL	LOGAN 843 MAT.	INST.	TOTAL
015433	CONTRACTOR EQUIPMENT		99.3	99.3		89.9	89.9		96.7	96.7		100.5	100.5		89.9	89.9		97.5	97.5
0241, 31-34	SITE & INFRASTRUCTURE, DEMOLITION	111.8	85.7	93.2	98.6	88.7	91.6	96.0	85.4	88.4	117.1	87.8	96.2	99.3	88.7	91.8	96.0	95.7	95.8
0310	Concrete Forming & Accessories	90.0	59.0	63.3	101.2	65.5	70.4	91.3	65.5	69.0	84.9	61.7	64.9	101.2	65.5	70.4	104.7	58.1	64.5
0320	Concrete Reinforcing	93.9	48.5	70.8	92.5	48.5	70.1	100.7	50.9	75.4	97.7	50.6	73.7	92.5	51.3	71.5	99.6	81.3	90.3
0330	Cast-in-Place Concrete	106.9	65.6	89.9	84.9	66.8	77.4	88.4	63.9	78.3	109.8	65.5	91.6	90.7	63.7	79.6	87.3	73.4	81.6
03	CONCRETE	104.9	60.9	83.3	89.7	63.3	76.7	90.6	63.4	77.2	109.1	62.5	86.2	92.4	62.8	77.9	106.0	68.3	87.5
04	MASONRY	120.0	58.9	81.9	105.2	56.6	74.9	160.7	56.6	95.8	103.6	58.9	75.8	105.7	63.6	79.5	105.8	63.2	79.3
05	METALS	94.9	80.7	91.6	102.2	67.8	91.6	97.7	79.0	92.0	96.3	81.6	91.8	102.2	69.8	92.2	102.2	79.0	95.0
06	WOOD, PLASTICS & COMPOSITES	92.5	58.4	73.4	107.1	68.5	85.5	92.3	68.6	79.0	83.1	62.0	71.3	107.1	68.5	85.5	85.3	54.9	68.3
07	THERMAL & MOISTURE PROTECTION	96.8	66.9	84.6	98.5	65.8	85.1	92.9	66.0	81.9	93.4	66.6	82.4	98.5	67.7	85.9	97.4	67.9	85.3
08	OPENINGS	101.4	56.1	90.8	77.7	62.4	74.1	97.7	63.0	89.6	101.6	58.6	91.6	77.7	63.3	74.4	94.3	56.5	85.5
0920	Plaster & Gypsum Board	93.8	57.5	69.3	88.2	68.0	74.6	90.4	68.0	75.3	92.5	61.2	71.3	88.2	68.0	74.6	78.9	53.5	61.7
0950, 0980	Ceilings & Acoustic Treatment	95.4	57.5	70.5	82.2	68.0	72.9	95.3	68.0	77.4	94.6	61.2	72.6	82.2	68.0	72.9	97.2	53.5	68.5
0960	Flooring	99.5	64.1	89.4	122.4	64.1	105.8	98.2	64.1	88.4	97.7	64.1	88.1	123.4	76.5	110.0	101.2	47.7	85.9
0970, 0990	Wall Finishes & Painting/Coating	103.6	59.4	76.2	105.8	48.8	71.4	107.0	56.1	76.3	103.4	59.4	76.8	108.5	56.1	76.8	103.3	60.7	77.5
09	FINISHES	89.4	59.6	73.0	95.8	63.8	78.1	96.8	64.6	79.1	90.8	61.7	74.8	96.3	67.1	80.2	96.2	55.3	73.6
COVERS	DIVS. 10-14, 25, 28, 41, 43, 44, 46	100.0	80.8	96.1	100.0	82.3	96.4	100.0	82.3	96.4	100.0	81.3	96.2	100.0	69.8	93.9	100.0	85.1	97.0
21, 22, 23	FIRE SUPPRESSION, PLUMBING & HVAC	95.4	65.9	83.5	100.2	54.7	81.8	95.3	58.5	80.5	95.4	65.0	83.1	100.2	54.2	81.6	99.9	69.5	87.6
26, 27, 3370	ELECTRICAL, COMMUNICATIONS & UTIL.	99.8	57.6	77.5	97.6	63.6	79.6	92.1	64.6	77.5	98.6	68.0	82.4	99.3	60.1	78.6	97.5	73.0	84.6
MF2014	WEIGHTED AVERAGE	99.0	65.6	84.4	96.5	64.1	82.4	98.6	65.9	84.4	98.9	67.7	85.3	97.1	64.4	82.8	99.8	70.1	86.9

UTAH / VERMONT

DIVISION		OGDEN 842,844			PRICE 845			PROVO 846-847			SALT LAKE CITY 840-841			BELLOWS FALLS 051			BENNINGTON 052		
		MAT.	INST.	TOTAL	MAT.	INST.	TOTAL	MAT.	INST.	TOTAL	MAT.	INST.	TOTAL	MAT.	INST.	TOTAL	MAT.	INST.	TOTAL
015433	CONTRACTOR EQUIPMENT		97.5	97.5		96.6	96.6		96.6	96.6		97.5	97.5		100.9	100.9		100.9	100.9
0241, 31 - 34	SITE & INFRASTRUCTURE, DEMOLITION	84.9	95.7	92.6	93.3	93.8	93.7	92.3	94.1	93.6	84.5	95.6	92.4	88.2	100.5	96.9	87.5	100.5	96.7
0310	Concrete Forming & Accessories	104.7	58.1	64.5	107.1	48.3	56.4	106.3	58.0	64.7	107.1	58.0	64.8	98.3	85.5	87.3	95.9	106.6	105.2
0320	Concrete Reinforcing	99.2	81.3	90.1	106.8	81.1	93.7	107.7	81.3	94.3	101.5	81.3	91.2	83.2	85.7	84.5	83.2	85.7	84.5
0330	Cast-in-Place Concrete	88.7	73.4	82.4	87.4	59.5	75.9	87.4	73.3	81.6	96.9	73.3	87.2	87.0	110.7	96.7	87.0	110.7	96.7
03	CONCRETE	95.9	68.3	82.3	107.3	59.0	83.6	105.9	68.2	87.4	114.8	68.2	91.9	91.0	94.2	92.6	90.8	103.6	97.1
04	MASONRY	99.7	63.2	77.0	110.9	61.3	80.0	111.1	63.2	81.2	113.1	63.2	82.0	103.6	95.2	98.4	112.4	95.2	101.6
05	METALS	102.7	79.0	95.4	99.4	76.6	92.4	100.3	78.9	93.7	106.2	78.9	97.8	94.8	88.3	92.8	94.7	88.2	92.7
06	WOOD, PLASTICS & COMPOSITES	85.3	54.9	68.3	88.6	44.3	63.8	86.8	54.9	68.9	87.2	54.9	69.1	103.7	83.5	92.4	100.6	112.4	107.2
07	THERMAL & MOISTURE PROTECTION	96.4	67.9	84.7	99.0	59.9	82.9	99.0	67.9	86.2	103.4	67.9	88.9	97.0	82.7	91.1	97.0	81.1	90.4
08	OPENINGS	94.3	56.5	85.5	98.2	48.9	86.7	98.2	56.5	88.5	96.2	56.5	86.9	105.0	86.5	100.7	105.0	102.3	104.3
0920	Plaster & Gypsum Board	78.9	53.5	61.7	81.7	42.5	55.2	79.2	53.5	61.8	90.5	53.5	65.5	106.0	82.4	90.1	104.2	112.2	109.6
0950, 0980	Ceilings & Acoustic Treatment	97.2	53.5	68.5	97.2	42.5	61.3	97.2	53.5	68.5	91.5	53.5	66.5	91.5	82.4	85.6	91.5	112.2	105.1
0960	Flooring	99.0	47.7	84.3	102.3	35.5	83.2	102.0	47.7	86.5	103.2	47.7	87.3	100.6	104.6	101.8	99.7	104.6	101.1
0970, 0990	Wall Finishes & Painting/Coating	103.3	60.7	77.5	103.3	37.5	63.6	103.3	62.6	78.7	106.7	62.6	80.1	97.4	105.8	102.5	97.4	105.8	102.5
09	FINISHES	94.3	55.3	72.8	97.3	43.5	67.7	96.8	55.5	74.0	96.6	55.5	74.0	95.9	90.7	93.0	95.4	107.8	102.2
COVERS	DIVS. 10 - 14, 25, 28, 41, 43, 44, 46	100.0	85.1	97.0	100.0	45.8	89.0	100.0	85.0	97.0	100.0	85.0	97.0	100.0	97.6	99.5	100.0	100.7	100.1
21, 22, 23	FIRE SUPPRESSION, PLUMBING & HVAC	99.9	69.5	87.6	97.5	67.7	85.5	99.9	69.4	87.6	100.1	69.4	87.7	95.3	96.4	95.7	95.3	96.4	95.7
26, 27, 3370	ELECTRICAL, COMMUNICATIONS & UTIL.	97.9	73.0	84.7	103.5	73.0	87.4	98.4	59.3	78.0	100.5	59.3	79.0	101.5	89.3	95.1	101.5	61.2	80.2
MF2014	WEIGHTED AVERAGE	98.1	70.1	85.9	100.2	64.6	84.7	100.3	68.2	86.3	102.2	68.3	87.4	97.2	93.0	95.3	97.5	93.5	95.8

VERMONT

DIVISION		BRATTLEBORO 053			BURLINGTON 054			GUILDHALL 059			MONTPELIER 056			RUTLAND 057			ST. JOHNSBURY 058		
		MAT.	INST.	TOTAL	MAT.	INST.	TOTAL	MAT.	INST.	TOTAL	MAT.	INST.	TOTAL	MAT.	INST.	TOTAL	MAT.	INST.	TOTAL
015433	CONTRACTOR EQUIPMENT		100.9	100.9		100.9	100.9		100.9	100.9		100.9	100.9		100.9	100.9		100.9	100.9
0241, 31 - 34	SITE & INFRASTRUCTURE, DEMOLITION	88.9	100.5	97.1	93.0	100.4	98.2	87.2	99.0	95.6	91.5	100.1	97.6	91.7	100.1	97.7	87.3	99.0	95.6
0310	Concrete Forming & Accessories	98.6	85.4	87.2	98.0	87.6	89.0	96.1	78.3	80.7	97.9	84.1	86.0	98.9	87.6	89.1	94.4	78.3	80.5
0320	Concrete Reinforcing	82.3	85.7	84.0	101.1	85.6	93.2	83.9	85.6	84.8	94.1	85.6	89.8	103.9	85.6	94.6	82.3	85.6	84.0
0330	Cast-in-Place Concrete	89.7	110.7	98.3	100.9	109.6	104.5	84.3	101.0	91.2	100.0	109.6	104.5	85.2	109.6	95.2	84.3	101.0	91.2
03	CONCRETE	93.0	94.1	93.6	101.2	94.7	98.0	88.6	87.6	88.1	100.1	93.1	96.7	94.3	94.7	94.5	88.2	87.6	87.9
04	MASONRY	111.5	95.2	101.3	115.7	93.7	102.0	111.8	78.2	90.8	109.6	93.7	99.7	93.9	93.7	93.8	138.8	78.2	101.0
05	METALS	94.7	88.2	92.7	103.5	87.5	98.6	94.8	87.3	92.5	99.6	87.4	95.9	100.6	87.4	96.5	94.8	87.3	92.5
06	WOOD, PLASTICS & COMPOSITES	104.0	83.5	92.5	99.9	88.2	93.3	100.0	83.5	90.8	97.7	83.5	89.7	104.2	88.2	95.2	94.8	83.5	88.5
07	THERMAL & MOISTURE PROTECTION	97.1	78.0	89.3	103.8	80.4	94.2	96.8	70.7	86.1	103.3	79.9	93.7	97.2	80.4	90.3	96.7	70.7	86.1
08	OPENINGS	105.0	86.5	100.7	108.4	85.0	103.0	105.0	82.4	99.7	105.8	82.4	100.3	108.4	85.0	102.9	105.0	82.4	99.7
0920	Plaster & Gypsum Board	106.0	82.4	90.1	109.7	87.2	94.5	112.4	82.4	92.2	108.9	82.4	91.0	106.6	87.2	93.5	113.8	82.4	92.6
0950, 0980	Ceilings & Acoustic Treatment	91.5	82.4	85.6	101.0	87.2	91.9	91.5	82.4	85.6	95.1	82.4	86.8	95.9	87.2	90.2	91.5	82.4	85.6
0960	Flooring	100.7	104.6	101.8	102.4	104.6	103.1	103.6	104.6	103.9	103.2	104.6	103.6	100.6	104.6	101.8	106.9	104.6	106.3
0970, 0990	Wall Finishes & Painting/Coating	97.4	105.8	102.5	102.9	85.2	92.2	97.4	85.2	90.0	101.7	85.2	91.7	97.4	85.2	90.0	97.4	85.2	90.0
09	FINISHES	96.0	90.7	93.1	99.7	90.8	94.8	97.5	84.2	90.2	97.9	88.0	92.5	97.1	90.7	93.6	98.7	84.2	90.7
COVERS	DIVS. 10 - 14, 25, 28, 41, 43, 44, 46	100.0	97.6	99.5	100.0	97.5	99.5	100.0	91.8	98.3	100.0	97.0	99.4	100.0	97.5	99.5	100.0	91.8	98.3
21, 22, 23	FIRE SUPPRESSION, PLUMBING & HVAC	95.3	96.4	95.7	99.9	71.8	88.6	95.3	64.0	82.7	95.1	71.8	85.7	100.1	71.8	88.7	95.3	64.0	82.7
26, 27, 3370	ELECTRICAL, COMMUNICATIONS & UTIL.	101.5	89.3	95.1	102.2	61.2	80.5	101.5	61.2	80.2	99.7	61.2	79.3	101.6	61.2	80.2	101.5	61.2	80.2
MF2014	WEIGHTED AVERAGE	97.8	92.8	95.6	102.4	83.6	94.2	97.4	77.8	88.8	99.5	82.8	92.2	99.7	83.6	92.7	98.6	77.8	89.6

VERMONT / VIRGINIA

DIVISION		WHITE RIVER JCT. 050			ALEXANDRIA 223			ARLINGTON 222			BRISTOL 242			CHARLOTTESVILLE 229			CULPEPER 227		
		MAT.	INST.	TOTAL	MAT.	INST.	TOTAL	MAT.	INST.	TOTAL	MAT.	INST.	TOTAL	MAT.	INST.	TOTAL	MAT.	INST.	TOTAL
015433	CONTRACTOR EQUIPMENT		100.9	100.9		103.0	103.0		101.7	101.7		101.7	101.7		106.1	106.1		101.7	101.7
0241, 31 - 34	SITE & INFRASTRUCTURE, DEMOLITION	91.5	99.2	97.0	114.3	89.8	96.9	124.5	87.7	98.3	108.8	86.1	92.6	113.5	88.1	95.4	111.8	87.6	94.6
0310	Concrete Forming & Accessories	93.1	78.9	80.8	92.2	72.8	75.5	91.2	72.0	74.7	87.1	42.5	48.6	85.4	48.7	53.7	82.5	71.4	72.9
0320	Concrete Reinforcing	83.2	85.6	84.4	83.0	87.0	85.1	93.7	84.4	88.9	93.7	68.4	80.8	93.1	72.8	82.7	93.7	84.2	88.9
0330	Cast-in-Place Concrete	89.7	101.9	94.7	105.1	81.8	95.6	102.3	80.3	93.3	101.8	48.5	79.9	105.9	55.5	85.2	104.8	69.2	90.2
03	CONCRETE	94.7	88.2	91.5	102.0	79.6	91.0	106.7	78.3	92.8	102.9	51.2	77.5	103.5	57.2	80.8	100.6	74.1	87.6
04	MASONRY	124.5	79.7	96.6	91.3	74.2	79.8	106.3	70.7	84.1	95.4	51.4	68.0	120.3	53.2	78.5	107.5	70.7	84.6
05	METALS	94.8	87.3	92.5	105.7	97.0	103.0	104.3	97.2	102.1	103.1	85.7	97.8	103.4	90.2	99.3	103.5	95.3	101.0
06	WOOD, PLASTICS & COMPOSITES	97.4	83.5	89.6	95.0	71.0	81.6	91.5	71.0	80.0	83.7	40.0	59.2	82.2	45.6	61.7	80.8	71.0	75.3
07	THERMAL & MOISTURE PROTECTION	97.1	71.4	86.6	102.7	81.5	94.0	104.7	80.6	94.8	104.2	60.1	86.1	103.8	67.7	89.0	104.0	78.1	93.4
08	OPENINGS	105.0	82.4	99.7	100.9	75.2	94.9	99.0	75.2	93.5	102.0	46.5	89.1	100.2	52.0	89.0	100.5	75.2	94.6
0920	Plaster & Gypsum Board	102.8	82.4	89.0	107.4	69.9	82.1	104.2	69.9	81.0	99.7	38.0	58.0	99.7	43.0	61.4	99.9	69.9	79.6
0950, 0980	Ceilings & Acoustic Treatment	91.5	82.4	85.6	91.4	69.9	77.3	89.8	69.9	76.7	88.9	38.0	55.4	88.9	43.0	58.8	89.8	69.9	76.7
0960	Flooring	98.6	104.6	100.4	106.0	82.9	99.4	104.5	81.9	98.0	101.0	66.5	91.1	99.7	66.5	90.2	99.7	81.9	94.6
0970, 0990	Wall Finishes & Painting/Coating	97.4	85.2	90.0	124.0	80.0	97.5	124.0	80.0	97.5	109.6	54.3	76.2	109.6	76.1	89.3	124.0	76.1	95.1
09	FINISHES	95.2	84.6	89.3	100.1	74.6	86.0	100.0	73.8	86.5	96.5	47.2	69.3	96.2	53.5	72.6	96.9	73.2	83.9
COVERS	DIVS. 10 - 14, 25, 28, 41, 43, 44, 46	100.0	92.3	98.4	100.0	88.6	97.7	100.0	87.6	97.5	100.0	67.8	93.5	100.0	79.3	95.8	100.0	87.6	97.5
21, 22, 23	FIRE SUPPRESSION, PLUMBING & HVAC	95.3	64.8	83.0	100.3	86.9	94.9	100.3	85.4	94.3	95.6	50.5	77.4	95.6	69.3	85.0	95.6	72.9	86.4
26, 27, 3370	ELECTRICAL, COMMUNICATIONS & UTIL.	101.5	61.2	80.2	97.1	102.4	99.9	94.7	99.2	97.1	96.8	35.4	64.3	96.7	71.9	83.6	99.3	99.2	99.2
MF2014	WEIGHTED AVERAGE	98.5	78.3	89.7	101.1	85.5	94.3	101.9	84.0	94.1	99.4	54.9	80.0	100.5	67.1	86.0	99.9	80.5	91.4

VIRGINIA

DIVISION		FAIRFAX 220-221			FARMVILLE 239			FREDERICKSBURG 224-225			GRUNDY 246			HARRISONBURG 228			LYNCHBURG 245		
		MAT.	INST.	TOTAL	MAT.	INST.	TOTAL	MAT.	INST.	TOTAL	MAT.	INST.	TOTAL	MAT.	INST.	TOTAL	MAT.	INST.	TOTAL
015433	CONTRACTOR EQUIPMENT		101.7	101.7		106.1	106.1		101.7	101.7		101.7	101.7		101.7	101.7		101.7	101.7
0241, 31 - 34	SITE & INFRASTRUCTURE, DEMOLITION	123.1	87.8	98.0	109.2	87.6	93.8	111.4	87.7	94.5	106.6	84.8	91.1	120.1	86.3	96.1	107.5	86.3	92.4
0310	Concrete Forming & Accessories	85.4	72.0	73.9	98.0	43.7	51.2	85.4	69.4	71.6	90.3	36.4	43.8	81.4	42.6	47.9	87.1	60.3	64.0
0320	Concrete Reinforcing	93.7	87.0	90.3	90.9	49.9	70.0	94.4	87.0	90.6	92.4	50.4	71.0	93.7	64.9	79.0	93.1	68.5	80.5
0330	Cast-in-Place Concrete	102.3	80.6	93.4	103.1	53.5	82.7	103.9	69.3	89.7	101.8	47.5	79.5	102.3	58.0	84.1	101.8	57.7	83.7
03	CONCRETE	106.3	78.8	92.8	101.2	49.9	76.0	100.3	73.8	87.3	101.6	44.4	73.5	104.1	53.8	79.4	101.5	62.3	82.3
04	MASONRY	106.2	71.4	84.5	103.5	44.9	66.9	106.6	70.7	84.2	96.5	47.3	65.8	104.2	50.3	70.6	112.0	53.2	75.4
05	METALS	103.6	97.0	101.5	101.0	77.1	93.6	103.5	96.2	101.3	103.1	69.9	92.9	103.4	83.9	97.4	103.3	86.3	98.1
06	WOOD, PLASTICS & COMPOSITES	83.7	71.0	76.6	96.4	44.1	67.1	83.7	68.1	75.0	86.7	35.1	57.8	79.8	39.0	56.9	83.7	62.6	71.9
07	THERMAL & MOISTURE PROTECTION	104.6	73.7	91.9	104.3	49.3	81.8	104.0	78.5	93.6	104.2	45.3	80.0	104.4	63.9	87.8	104.0	63.9	87.6
08	OPENINGS	99.0	75.2	93.5	100.7	43.0	87.3	100.2	73.6	94.0	102.0	34.4	86.3	100.5	49.4	88.6	100.5	58.8	90.8
0920	Plaster & Gypsum Board	99.9	69.9	79.6	108.6	41.5	63.2	99.9	66.9	77.6	99.7	32.9	54.5	99.7	36.9	57.3	99.7	61.3	73.7
0950, 0980	Ceilings & Acoustic Treatment	89.8	69.9	76.7	84.8	41.5	56.3	89.8	66.9	74.8	88.9	32.9	52.1	88.9	36.9	54.8	88.9	61.3	70.8
0960	Flooring	101.4	81.9	95.8	106.5	60.3	93.3	101.4	81.9	95.8	102.3	33.7	82.6	99.4	74.0	92.1	101.0	66.5	91.1
0970, 0990	Wall Finishes & Painting/Coating	124.0	80.0	97.5	112.8	38.7	68.1	124.0	76.1	95.1	109.6	35.1	64.6	124.0	54.3	81.9	109.6	54.3	76.2
09	FINISHES	98.5	74.0	85.0	98.0	46.0	69.4	97.4	71.5	83.1	96.7	35.1	62.8	97.3	47.8	70.1	96.3	60.2	76.4
COVERS	DIVS. 10 - 14, 25, 28, 41, 43, 44, 46	100.0	87.8	97.5	100.0	45.7	89.0	100.0	83.4	96.6	100.0	43.7	88.6	100.0	67.6	93.5	100.0	70.9	94.1
21, 22, 23	FIRE SUPPRESSION, PLUMBING & HVAC	95.6	85.7	91.6	95.5	43.4	74.5	95.6	85.1	91.3	95.6	62.1	82.1	95.6	66.9	84.0	95.6	69.0	84.8
26, 27, 3370	ELECTRICAL, COMMUNICATIONS & UTIL.	97.9	99.2	98.6	90.1	48.2	68.0	94.9	99.2	97.2	96.8	44.6	69.2	96.9	96.4	96.7	97.9	52.7	74.0
MF2014	WEIGHTED AVERAGE	100.7	84.0	93.4	98.7	52.4	78.5	99.5	82.7	92.1	99.3	52.5	78.9	100.2	67.2	85.8	99.9	65.6	85.0

VIRGINIA

DIVISION		NEWPORT NEWS 236			NORFOLK 233 - 235			PETERSBURG 238			PORTSMOUTH 237			PULASKI 243			RICHMOND 230 - 232		
		MAT.	INST.	TOTAL	MAT.	INST.	TOTAL	MAT.	INST.	TOTAL	MAT.	INST.	TOTAL	MAT.	INST.	TOTAL	MAT.	INST.	TOTAL
015433	CONTRACTOR EQUIPMENT		106.1	106.1		106.7	106.7		106.1	106.1		106.0	106.0		101.7	101.7		106.1	106.1
0241, 31 - 34	SITE & INFRASTRUCTURE, DEMOLITION	108.2	88.8	94.4	108.9	89.8	95.3	111.7	89.0	95.6	106.7	88.6	93.8	105.9	85.5	91.4	103.6	89.0	93.2
0310	Concrete Forming & Accessories	97.3	64.1	68.7	97.9	64.3	68.9	90.6	57.2	61.8	86.4	53.1	57.7	90.3	38.7	45.8	97.6	57.2	62.8
0320	Concrete Reinforcing	90.6	72.0	81.2	94.7	72.1	83.2	90.3	73.0	81.5	90.3	72.0	81.0	92.4	64.6	78.3	97.5	73.0	85.0
0330	Cast-in-Place Concrete	100.2	66.1	86.2	105.2	66.1	89.1	106.3	53.7	84.7	99.2	66.0	85.6	101.8	48.2	79.8	98.1	53.7	79.8
03	CONCRETE	98.4	67.6	83.3	101.5	67.7	84.9	103.6	60.4	82.4	97.1	62.6	80.2	101.6	48.7	75.6	98.5	60.4	79.8
04	MASONRY	97.8	53.9	70.5	103.8	53.9	72.7	111.9	52.6	75.0	103.8	53.9	72.7	91.5	48.2	64.5	102.5	52.6	71.4
05	METALS	103.3	89.8	99.2	103.1	89.9	99.1	101.1	90.7	97.9	102.2	89.1	98.2	103.2	82.5	96.8	107.1	90.7	102.0
06	WOOD, PLASTICS & COMPOSITES	95.3	66.2	79.0	93.2	66.2	78.1	86.4	58.6	70.8	82.9	51.5	65.3	86.7	36.8	58.7	95.0	58.6	74.6
07	THERMAL & MOISTURE PROTECTION	104.3	68.4	89.6	100.9	68.4	87.5	104.3	68.3	89.5	104.3	66.8	88.9	104.2	51.1	82.4	104.8	68.3	88.3
08	OPENINGS	101.1	62.2	92.1	100.4	62.2	91.5	100.4	59.0	90.8	101.2	54.1	90.2	102.0	42.2	88.1	104.8	59.0	94.1
0920	Plaster & Gypsum Board	109.4	64.3	78.9	105.5	64.3	77.6	101.3	56.4	71.0	101.6	49.1	66.1	99.7	34.7	55.7	104.2	56.4	71.9
0950, 0980	Ceilings & Acoustic Treatment	88.2	64.3	72.5	88.9	64.3	72.7	85.7	56.4	66.4	88.2	49.1	62.5	88.9	34.7	53.3	90.6	56.4	68.1
0960	Flooring	106.5	66.5	95.1	105.6	66.5	94.4	102.5	71.2	93.6	99.5	66.5	90.0	102.3	66.5	92.0	103.4	71.2	94.2
0970, 0990	Wall Finishes & Painting/Coating	112.8	42.0	70.1	113.0	76.1	90.7	112.8	76.1	90.6	112.8	76.1	90.6	109.6	54.3	76.2	110.0	76.1	89.5
09	FINISHES	98.7	62.2	78.6	98.2	66.0	80.5	96.3	61.2	76.9	95.6	56.7	74.2	96.7	43.7	67.5	98.5	61.2	77.9
COVERS	DIVS. 10 - 14, 25, 28, 41, 43, 44, 46	100.0	82.2	96.4	100.0	82.2	96.4	100.0	79.6	95.9	100.0	72.0	94.3	100.0	44.0	88.7	100.0	79.6	95.9
21, 22, 23	FIRE SUPPRESSION, PLUMBING & HVAC	100.3	63.6	85.5	100.0	64.3	85.6	95.5	67.5	84.2	100.3	64.3	85.8	95.6	62.2	82.1	99.9	67.5	86.8
26, 27, 3370	ELECTRICAL, COMMUNICATIONS & UTIL.	92.8	63.5	77.3	95.5	58.3	75.8	93.0	71.9	81.8	91.2	58.3	73.8	96.8	54.4	74.4	96.8	71.9	83.6
MF2014	WEIGHTED AVERAGE	100.0	68.2	86.1	100.5	68.2	86.4	99.4	68.7	86.0	99.4	65.3	84.5	99.0	57.3	80.9	101.3	68.7	87.1

VIRGINIA / WASHINGTON

DIVISION		ROANOKE 240 - 241			STAUNTON 244			WINCHESTER 226			CLARKSTON 994			EVERETT 982			OLYMPIA 985		
		MAT.	INST.	TOTAL	MAT.	INST.	TOTAL	MAT.	INST.	TOTAL	MAT.	INST.	TOTAL	MAT.	INST.	TOTAL	MAT.	INST.	TOTAL
015433	CONTRACTOR EQUIPMENT		101.7	101.7		106.1	106.1		101.7	101.7		91.2	91.2		101.6	101.6		101.6	101.6
0241, 31 - 34	SITE & INFRASTRUCTURE, DEMOLITION	106.4	86.3	92.1	109.9	87.7	94.1	118.7	87.6	96.6	99.2	90.1	92.7	91.5	108.4	103.5	93.4	108.4	104.1
0310	Concrete Forming & Accessories	96.8	60.5	65.5	89.9	50.0	55.5	83.8	69.5	71.4	112.7	67.4	73.6	113.4	100.4	102.2	99.0	100.3	100.1
0320	Concrete Reinforcing	93.4	68.5	80.7	93.1	48.9	70.6	93.1	86.4	89.7	112.0	87.2	99.4	104.2	98.8	101.5	110.8	98.7	104.6
0330	Cast-in-Place Concrete	115.7	57.7	91.9	105.9	51.5	83.6	102.3	69.3	88.7	96.3	81.9	90.4	103.5	106.2	104.6	98.9	106.2	101.9
03	CONCRETE	105.9	62.4	84.5	102.9	52.0	77.9	103.5	73.7	88.9	107.2	76.3	92.0	98.7	101.6	100.1	98.9	101.6	100.2
04	MASONRY	97.7	53.2	70.0	106.9	51.4	72.3	101.7	59.7	75.5	102.3	81.5	89.3	110.6	100.8	104.5	103.0	99.1	100.6
05	METALS	105.5	86.5	99.7	103.4	80.5	96.3	103.5	94.7	100.8	88.3	81.6	86.2	101.8	92.4	98.9	101.2	92.1	98.4
06	WOOD, PLASTICS & COMPOSITES	95.9	62.6	77.3	86.7	49.9	66.1	82.2	68.1	74.3	113.2	63.8	85.5	111.7	100.0	105.2	91.1	100.0	96.1
07	THERMAL & MOISTURE PROTECTION	103.8	66.8	88.7	103.8	52.0	82.6	104.5	75.6	92.7	143.8	77.2	116.5	103.7	101.7	102.9	103.5	96.8	100.7
08	OPENINGS	100.9	58.8	91.1	100.5	48.1	88.3	102.1	73.7	95.5	119.0	68.7	107.3	101.3	99.8	101.0	106.9	99.9	105.3
0920	Plaster & Gypsum Board	107.4	61.3	76.2	99.7	47.5	64.4	99.9	66.9	77.6	145.7	62.5	89.5	109.9	99.9	103.0	101.3	99.9	100.4
0950, 0980	Ceilings & Acoustic Treatment	91.4	61.3	71.6	88.9	47.5	61.7	89.8	66.9	74.8	97.6	62.5	74.5	99.8	99.9	99.9	98.8	99.9	99.3
0960	Flooring	106.0	66.5	94.7	101.9	39.5	84.0	100.7	81.9	95.4	91.6	47.1	78.9	113.8	91.6	107.4	103.8	91.7	100.3
0970, 0990	Wall Finishes & Painting/Coating	109.6	54.3	76.2	109.6	34.5	64.3	124.0	87.2	101.8	100.0	58.5	75.0	95.4	91.5	93.0	99.0	91.5	94.5
09	FINISHES	99.0	61.1	78.1	96.6	46.6	69.0	97.9	72.7	84.0	110.3	61.3	83.3	104.7	97.6	100.8	98.9	97.6	98.2
COVERS	DIVS. 10 - 14, 25, 28, 41, 43, 44, 46	100.0	70.9	94.1	100.0	70.9	94.1	100.0	87.3	97.4	100.0	75.4	95.0	100.0	99.0	99.8	100.0	100.5	100.1
21, 22, 23	FIRE SUPPRESSION, PLUMBING & HVAC	100.3	64.8	86.0	95.6	58.9	80.8	95.6	87.7	92.4	95.6	83.4	90.7	100.1	99.1	99.7	100.0	99.1	99.7
26, 27, 3370	ELECTRICAL, COMMUNICATIONS & UTIL.	96.8	54.4	74.4	95.6	74.9	84.7	95.3	99.2	97.4	92.4	94.8	93.7	104.9	98.5	101.5	104.4	99.1	101.6
MF2014	WEIGHTED AVERAGE	101.5	65.1	85.6	99.7	61.8	83.2	100.0	82.2	92.2	101.5	80.2	92.2	101.6	99.6	100.7	101.2	99.4	100.4

637

WASHINGTON

DIVISION		RICHLAND 993			SEATTLE 980 - 981,987			SPOKANE 990 - 992			TACOMA 983 - 984			VANCOUVER 986			WENATCHEE 988		
		MAT.	INST.	TOTAL	MAT.	INST.	TOTAL	MAT.	INST.	TOTAL	MAT.	INST.	TOTAL	MAT.	INST.	TOTAL	MAT.	INST.	TOTAL
015433	CONTRACTOR EQUIPMENT		91.2	91.2		101.6	101.6		91.2	91.2		101.6	101.6		96.7	96.7		101.6	101.6
0241, 31 - 34	SITE & INFRASTRUCTURE, DEMOLITION	101.8	91.0	94.1	95.1	106.7	103.4	101.1	91.0	93.9	94.5	108.4	104.4	104.2	96.1	98.4	103.2	107.1	106.0
0310	Concrete Forming & Accessories	112.8	78.5	83.2	108.3	100.9	102.0	118.2	78.1	83.6	104.2	100.4	100.9	104.9	90.5	92.5	105.8	77.6	81.5
0320	Concrete Reinforcing	107.4	87.4	97.2	105.1	98.9	101.9	108.2	87.3	97.6	103.1	98.8	100.9	103.9	98.4	101.1	103.9	87.7	95.6
0330	Cast-in-Place Concrete	96.5	84.4	91.5	100.9	106.4	103.2	100.3	84.2	93.7	106.4	106.2	106.3	118.8	98.4	110.4	108.6	77.3	95.7
03	CONCRETE	106.8	82.2	94.7	99.6	102.1	100.8	109.1	82.0	95.8	100.4	101.6	101.0	110.7	94.6	102.8	108.4	79.6	94.2
04	MASONRY	104.3	84.3	91.8	108.2	100.9	103.6	104.9	84.3	92.0	106.2	100.9	102.9	106.7	95.6	99.8	108.6	91.4	97.9
05	METALS	88.7	84.9	87.5	102.5	94.2	99.9	90.9	84.5	89.0	103.6	92.3	100.1	101.0	91.8	98.2	101.0	86.2	96.4
06	WOOD, PLASTICS & COMPOSITES	113.4	76.1	92.5	104.9	100.0	102.2	122.4	76.1	96.5	101.0	100.0	100.4	92.9	90.2	91.4	102.5	75.8	87.5
07	THERMAL & MOISTURE PROTECTION	145.0	78.8	117.8	102.1	101.8	102.0	141.4	79.5	116.0	103.5	98.5	101.4	104.0	91.6	98.9	104.1	81.5	94.8
08	OPENINGS	121.4	71.3	109.7	100.1	99.9	100.0	121.9	71.8	110.3	102.0	99.9	101.5	98.3	92.8	97.0	101.5	71.0	94.4
0920	Plaster & Gypsum Board	145.7	75.2	98.1	103.6	99.9	101.1	137.7	75.2	95.5	107.8	99.9	102.5	105.7	90.1	95.1	110.2	74.9	86.4
0950, 0980	Ceilings & Acoustic Treatment	104.2	75.2	85.1	110.2	99.9	103.5	99.5	75.2	83.5	103.1	99.9	101.0	100.2	90.1	93.5	95.4	74.9	82.0
0960	Flooring	92.0	80.1	88.6	108.3	106.8	107.8	91.2	85.3	89.5	107.2	91.7	102.8	113.5	102.2	110.2	109.9	62.6	96.4
0970, 0990	Wall Finishes & Painting/Coating	100.0	67.6	80.4	96.3	91.5	93.4	100.2	70.9	82.5	95.4	91.5	93.0	97.6	70.8	81.4	95.4	72.6	81.6
09	FINISHES	112.0	76.7	92.6	105.0	100.6	102.6	109.8	78.1	92.3	103.4	97.6	100.2	101.5	90.6	95.5	103.8	73.5	87.1
COVERS	DIVS. 10 - 14, 25, 28, 41, 43, 44, 46	100.0	85.4	97.0	100.0	100.5	100.1	100.0	85.3	97.0	100.0	100.5	100.1	100.0	60.7	92.1	100.0	77.2	95.4
21, 22, 23	FIRE SUPPRESSION, PLUMBING & HVAC	100.5	107.6	103.3	100.0	115.1	106.1	100.4	84.3	93.9	100.2	99.2	99.8	100.3	87.0	94.9	95.4	86.7	91.9
26, 27, 3370	ELECTRICAL, COMMUNICATIONS & UTIL.	89.7	94.8	92.4	102.6	107.0	104.9	88.0	79.0	83.3	104.7	99.1	101.7	109.7	101.0	105.1	105.5	98.4	101.8
MF2014	WEIGHTED AVERAGE	103.1	89.3	97.1	101.4	104.7	102.8	103.4	82.3	94.2	101.8	99.7	100.9	102.7	92.1	98.1	101.6	86.6	95.0

DIVISION		WASHINGTON YAKIMA 989			WEST VIRGINIA BECKLEY 258 - 259			BLUEFIELD 247 - 248			BUCKHANNON 262			CHARLESTON 250 - 253			CLARKSBURG 263 - 264		
		MAT.	INST.	TOTAL	MAT.	INST.	TOTAL	MAT.	INST.	TOTAL	MAT.	INST.	TOTAL	MAT.	INST.	TOTAL	MAT.	INST.	TOTAL
015433	CONTRACTOR EQUIPMENT		101.6	101.6		101.7	101.7		101.7	101.7		101.7	101.7		101.7	101.7		101.7	101.7
0241, 31 - 34	SITE & INFRASTRUCTURE, DEMOLITION	96.8	107.8	104.7	100.5	89.9	92.9	100.6	89.9	92.9	106.8	89.9	94.8	101.5	90.7	93.8	107.4	89.9	95.0
0310	Concrete Forming & Accessories	104.6	95.8	97.0	83.5	93.6	92.2	87.0	93.6	92.7	86.4	94.3	93.2	96.7	94.1	94.4	83.8	94.2	92.8
0320	Concrete Reinforcing	103.5	87.4	95.3	92.7	89.0	90.8	92.0	84.1	88.0	92.6	84.2	88.3	101.0	89.1	94.9	92.6	84.1	88.3
0330	Cast-in-Place Concrete	113.6	83.9	101.4	98.1	103.0	100.1	99.6	102.9	101.0	99.3	101.8	100.3	96.8	103.0	99.3	108.8	100.0	105.2
03	CONCRETE	105.3	89.9	97.7	97.6	96.4	97.0	97.4	95.5	96.5	100.4	95.4	98.0	99.1	96.7	97.9	104.4	94.8	99.7
04	MASONRY	99.8	83.1	89.4	98.9	98.9	98.9	93.3	98.9	96.8	104.5	98.9	101.0	95.8	97.9	97.1	108.4	98.9	102.4
05	METALS	101.6	86.3	96.9	100.9	98.1	100.0	103.4	96.3	101.2	103.6	97.1	101.6	99.6	98.9	99.4	103.6	96.9	101.5
06	WOOD, PLASTICS & COMPOSITES	101.4	100.0	100.6	83.5	93.4	89.1	85.6	93.4	90.0	84.9	93.4	89.7	98.4	93.4	95.6	81.5	93.4	88.2
07	THERMAL & MOISTURE PROTECTION	103.6	82.7	95.0	102.9	90.6	97.9	104.0	90.6	98.5	104.2	91.7	99.1	99.4	90.4	95.7	104.1	90.6	98.6
08	OPENINGS	101.5	81.0	96.7	101.3	85.8	97.7	102.5	84.6	98.3	102.5	84.6	98.4	103.6	85.8	99.4	102.5	85.1	98.5
0920	Plaster & Gypsum Board	107.4	99.9	102.4	97.0	93.0	94.3	98.9	93.0	94.9	99.3	93.0	95.1	100.3	93.0	95.4	97.2	93.0	94.4
0950, 0980	Ceilings & Acoustic Treatment	97.3	99.9	99.0	83.9	93.0	89.9	87.3	93.0	91.1	88.9	93.0	91.6	92.2	93.0	92.7	88.9	93.0	91.6
0960	Flooring	108.1	58.0	93.8	102.4	118.0	106.8	98.9	118.0	104.3	98.6	118.0	104.1	106.4	118.0	109.7	97.6	118.0	103.4
0970, 0990	Wall Finishes & Painting/Coating	95.4	70.9	80.6	107.4	91.5	97.8	109.6	94.6	100.6	109.6	94.3	100.3	106.9	95.9	100.2	109.6	94.3	100.3
09	FINISHES	102.5	86.6	93.7	94.7	98.6	96.9	94.8	98.9	97.1	95.7	98.9	97.5	99.1	98.8	99.0	95.0	98.9	97.2
COVERS	DIVS. 10 - 14, 25, 28, 41, 43, 44, 46	100.0	97.6	99.5	100.0	56.9	91.3	100.0	56.9	91.3	100.0	101.5	100.3	100.0	101.2	100.2	100.0	101.5	100.3
21, 22, 23	FIRE SUPPRESSION, PLUMBING & HVAC	100.2	107.2	103.0	95.5	93.9	94.7	95.6	92.8	94.4	95.6	95.1	95.4	100.0	93.4	97.4	95.6	95.0	95.3
26, 27, 3370	ELECTRICAL, COMMUNICATIONS & UTIL.	107.8	94.9	101.0	93.9	91.1	92.4	95.7	91.1	93.3	97.1	96.1	96.6	99.1	91.1	94.8	97.1	96.1	96.6
MF2014	WEIGHTED AVERAGE	102.0	94.1	98.6	97.9	93.3	96.0	98.3	92.9	96.0	99.6	95.2	97.8	99.9	94.8	97.6	100.1	95.4	98.1

WEST VIRGINIA

DIVISION		GASSAWAY 266			HUNTINGTON 255 - 257			LEWISBURG 249			MARTINSBURG 254			MORGANTOWN 265			PARKERSBURG 261		
		MAT.	INST.	TOTAL	MAT.	INST.	TOTAL	MAT.	INST.	TOTAL	MAT.	INST.	TOTAL	MAT.	INST.	TOTAL	MAT.	INST.	TOTAL
015433	CONTRACTOR EQUIPMENT		101.7	101.7		101.7	101.7		101.7	101.7		101.7	101.7		101.7	101.7		101.7	101.7
0241, 31 - 34	SITE & INFRASTRUCTURE, DEMOLITION	104.2	89.9	94.0	105.1	90.9	95.0	116.3	89.8	97.5	104.2	90.2	94.3	101.6	90.5	93.7	109.8	90.6	96.2
0310	Concrete Forming & Accessories	85.8	92.2	91.4	95.3	95.9	95.8	84.2	93.5	92.2	83.6	82.8	82.9	84.2	93.9	92.6	88.6	90.3	90.1
0320	Concrete Reinforcing	92.6	84.0	88.3	94.1	92.9	93.5	92.6	83.9	88.2	92.7	80.6	86.6	92.6	84.1	88.3	92.0	83.9	87.9
0330	Cast-in-Place Concrete	104.0	103.5	103.7	106.9	102.5	105.1	99.7	102.9	101.0	102.8	96.7	100.3	99.3	101.6	100.2	101.5	96.4	99.4
03	CONCRETE	100.8	95.1	98.0	102.8	98.0	100.4	107.1	95.4	101.4	101.2	87.9	94.7	97.2	95.2	96.2	102.5	91.8	97.2
04	MASONRY	109.2	95.9	100.9	96.4	100.8	99.1	95.9	95.9	95.9	99.7	89.5	93.3	126.5	98.9	109.3	82.9	91.7	88.4
05	METALS	103.5	96.8	101.4	103.3	100.4	102.4	103.5	96.0	101.2	101.3	94.2	99.1	103.6	96.6	101.5	104.2	96.6	101.9
06	WOOD, PLASTICS & COMPOSITES	84.0	90.7	87.8	94.7	95.3	95.1	81.8	93.4	88.3	83.5	81.6	82.5	81.8	93.4	88.3	85.0	89.1	87.3
07	THERMAL & MOISTURE PROTECTION	104.0	90.6	98.5	103.0	90.8	98.0	104.8	89.8	98.7	103.1	79.3	93.4	104.0	90.6	98.5	103.9	89.0	97.8
08	OPENINGS	100.7	83.2	96.6	100.5	87.6	97.5	102.5	84.6	98.4	103.3	73.4	96.3	103.8	85.1	99.4	101.4	82.3	96.9
0920	Plaster & Gypsum Board	98.6	90.2	92.9	104.8	95.0	98.1	97.2	93.0	94.4	97.6	80.9	86.3	97.2	93.0	94.4	99.7	88.5	92.1
0950, 0980	Ceilings & Acoustic Treatment	88.9	90.2	89.8	86.4	95.0	92.0	88.9	93.0	91.6	86.4	80.9	82.8	88.9	93.0	91.6	88.9	88.5	88.7
0960	Flooring	98.4	118.0	104.0	109.8	121.7	113.2	97.7	118.0	103.5	102.4	118.0	106.8	97.7	118.0	103.5	101.5	118.0	106.2
0970, 0990	Wall Finishes & Painting/Coating	109.6	95.9	101.3	107.4	94.6	99.7	109.6	75.4	88.9	107.4	41.4	67.5	109.6	94.3	100.3	109.6	94.6	100.6
09	FINISHES	95.2	97.5	96.5	98.5	100.8	99.8	96.1	96.8	96.5	95.6	83.1	88.7	94.6	98.9	97.0	96.6	95.9	96.2
COVERS	DIVS. 10 - 14, 25, 28, 41, 43, 44, 46	100.0	101.2	100.2	100.0	101.8	100.4	100.0	56.9	91.3	100.0	97.7	99.5	100.0	65.8	93.1	100.0	100.5	100.1
21, 22, 23	FIRE SUPPRESSION, PLUMBING & HVAC	95.6	93.9	94.9	100.3	90.5	96.4	95.6	93.5	94.7	95.5	86.0	91.7	95.6	95.0	95.3	100.3	88.3	95.4
26, 27, 3370	ELECTRICAL, COMMUNICATIONS & UTIL.	97.1	91.1	93.9	97.8	95.3	96.5	93.0	91.1	92.0	100.1	81.8	90.4	97.3	96.1	96.7	97.2	92.7	94.8
MF2014	WEIGHTED AVERAGE	99.5	93.9	97.1	100.7	95.7	98.5	99.8	92.4	96.6	99.3	86.4	93.7	100.2	94.4	97.6	100.1	91.8	96.5

WEST VIRGINIA / WISCONSIN

DIVISION		PETERSBURG 268 MAT.	INST.	TOTAL	ROMNEY 267 MAT.	INST.	TOTAL	WHEELING 260 MAT.	INST.	TOTAL	BELOIT 535 MAT.	INST.	TOTAL	EAU CLAIRE 547 MAT.	INST.	TOTAL	GREEN BAY 541-543 MAT.	INST.	TOTAL
015433	CONTRACTOR EQUIPMENT		101.7	101.7		101.7	101.7		101.7	101.7		102.3	102.3		100.4	100.4		98.2	98.2
0241, 31 - 34	SITE & INFRASTRUCTURE, DEMOLITION	100.9	90.5	93.5	103.6	90.5	94.3	110.6	90.4	96.2	94.7	108.3	104.3	96.1	102.9	101.0	99.7	98.9	99.2
0310	Concrete Forming & Accessories	87.4	91.6	91.0	83.4	91.8	90.6	90.3	93.2	92.8	100.2	100.3	100.3	97.1	99.4	99.1	105.5	98.7	99.7
0320	Concrete Reinforcing	92.0	82.5	87.2	92.6	81.0	86.7	91.4	84.0	87.7	92.3	116.0	104.3	90.4	99.2	94.9	88.7	92.4	90.6
0330	Cast-in-Place Concrete	99.3	99.1	99.2	104.0	99.1	102.0	101.5	100.1	100.9	100.6	100.9	100.7	104.0	99.2	102.0	107.6	99.7	104.3
03	CONCRETE	97.3	93.0	95.2	100.6	92.8	96.8	102.5	94.4	98.5	98.8	103.5	101.1	99.7	99.4	99.6	102.8	98.0	100.5
04	MASONRY	99.5	98.9	99.1	96.2	98.9	97.9	107.6	93.9	99.1	100.6	104.3	102.9	94.7	102.1	99.3	126.7	101.4	110.9
05	METALS	103.7	95.8	101.2	103.7	95.5	101.2	104.4	96.8	102.1	93.1	107.7	97.6	95.0	101.4	97.0	97.4	98.2	97.6
06	WOOD, PLASTICS & COMPOSITES	85.9	90.7	88.6	80.9	90.7	86.4	86.7	93.4	90.5	99.9	99.2	99.5	105.1	99.3	101.9	110.0	99.3	104.0
07	THERMAL & MOISTURE PROTECTION	104.0	86.0	96.6	104.1	83.4	95.6	104.3	90.0	98.4	102.1	96.8	99.9	102.4	86.7	95.9	104.3	86.0	96.8
08	OPENINGS	103.8	82.8	98.9	103.7	82.4	98.8	102.2	85.1	98.2	102.5	108.8	103.9	103.7	96.2	102.0	100.1	96.0	99.2
0920	Plaster & Gypsum Board	99.3	90.2	93.2	96.8	90.2	92.4	99.7	93.0	95.2	94.5	99.6	97.9	102.9	99.6	100.6	98.7	99.6	99.3
0950, 0980	Ceilings & Acoustic Treatment	88.9	90.2	89.8	88.9	90.2	89.8	88.9	93.0	91.6	88.1	99.6	95.6	90.4	99.6	96.4	81.9	99.6	93.5
0960	Flooring	99.4	118.0	104.7	97.5	118.0	103.4	102.3	118.0	106.8	99.8	118.1	105.1	91.4	114.1	97.9	110.2	114.1	111.3
0970, 0990	Wall Finishes & Painting/Coating	109.6	94.3	100.3	109.6	94.3	100.3	109.6	94.3	100.3	98.3	97.2	97.7	94.5	86.4	89.6	105.4	83.2	92.0
09	FINISHES	95.4	97.3	96.4	94.7	97.3	96.1	96.9	98.2	97.6	97.4	103.1	100.5	94.6	101.2	98.3	99.0	100.7	99.9
COVERS	DIVS. 10 - 14, 25, 28, 41, 43, 44, 46	100.0	65.5	93.0	100.0	101.2	100.2	100.0	95.4	99.1	100.0	96.6	99.3	100.0	96.8	99.3	100.0	96.5	99.3
21, 22, 23	FIRE SUPPRESSION, PLUMBING & HVAC	95.6	93.7	94.8	95.6	93.4	94.7	100.3	93.6	97.6	100.1	98.6	99.5	100.1	87.6	95.0	100.3	85.1	94.2
26, 27, 3370	ELECTRICAL, COMMUNICATIONS & UTIL.	100.5	81.8	90.6	99.8	81.8	90.3	94.4	96.1	95.3	101.0	86.7	93.4	103.8	84.9	93.8	98.2	82.0	89.6
MF2014	WEIGHTED AVERAGE	99.3	91.3	95.8	99.4	92.2	96.3	101.2	94.3	98.2	98.9	100.7	99.7	99.3	95.3	97.5	101.2	93.4	97.8

WISCONSIN

DIVISION		KENOSHA 531 MAT.	INST.	TOTAL	LA CROSSE 546 MAT.	INST.	TOTAL	LANCASTER 538 MAT.	INST.	TOTAL	MADISON 537 MAT.	INST.	TOTAL	MILWAUKEE 530,532 MAT.	INST.	TOTAL	NEW RICHMOND 540 MAT.	INST.	TOTAL
015433	CONTRACTOR EQUIPMENT		100.2	100.2		100.4	100.4		102.3	102.3		102.3	102.3		90.5	90.5		100.8	100.8
0241, 31 - 34	SITE & INFRASTRUCTURE, DEMOLITION	100.3	105.2	103.8	90.1	102.8	99.1	93.9	108.2	104.1	92.4	108.3	103.7	94.6	98.5	97.4	94.4	103.3	100.8
0310	Concrete Forming & Accessories	108.0	100.5	101.6	84.1	99.1	97.1	99.5	97.9	98.2	102.0	100.1	100.4	104.1	118.0	116.1	92.4	100.9	99.7
0320	Concrete Reinforcing	92.1	97.9	95.0	90.1	91.5	90.9	93.5	91.3	92.3	97.9	91.6	94.7	95.7	98.2	97.0	87.8	99.0	93.5
0330	Cast-in-Place Concrete	109.4	107.3	108.5	93.5	95.9	94.5	100.0	100.9	100.4	91.1	100.8	95.1	99.8	109.6	103.8	108.3	82.5	97.7
03	CONCRETE	103.5	102.4	103.0	90.8	96.8	93.7	98.5	97.8	98.1	95.7	98.8	97.2	99.7	110.5	105.0	97.5	94.3	95.9
04	MASONRY	97.9	114.1	108.0	93.8	105.9	101.4	100.7	104.3	102.9	101.5	104.3	103.2	104.9	120.6	114.7	121.6	102.3	109.6
05	METALS	93.9	101.2	96.2	94.9	98.2	95.9	90.6	95.0	92.0	91.9	97.2	93.6	97.5	94.8	96.7	95.1	99.7	96.5
06	WOOD, PLASTICS & COMPOSITES	105.0	96.4	100.2	90.1	99.3	95.3	99.1	99.2	99.2	96.7	99.2	98.1	104.7	118.5	112.5	94.0	102.5	98.7
07	THERMAL & MOISTURE PROTECTION	102.2	108.7	104.9	101.8	90.2	97.1	101.9	82.5	93.9	102.8	105.1	103.8	100.8	114.0	106.2	103.6	92.2	98.9
08	OPENINGS	96.9	101.1	97.9	103.7	85.0	99.4	97.9	85.8	95.1	107.9	100.5	106.2	101.0	113.2	103.8	89.1	97.5	91.1
0920	Plaster & Gypsum Board	85.3	96.7	93.0	97.7	99.6	99.0	93.3	99.6	97.5	99.6	99.6	99.6	101.0	119.3	113.4	87.7	103.0	98.1
0950, 0980	Ceilings & Acoustic Treatment	88.1	96.7	93.7	89.5	99.6	96.1	83.1	99.6	93.9	94.0	99.6	97.6	95.9	119.3	111.3	58.5	103.0	87.8
0960	Flooring	118.8	110.4	116.4	85.6	114.1	93.7	99.5	110.5	102.7	103.2	110.5	105.3	98.2	114.0	102.7	103.2	114.1	106.3
0970, 0990	Wall Finishes & Painting/Coating	108.3	115.5	112.6	94.5	77.9	84.5	98.3	66.4	79.1	102.4	97.2	99.3	94.4	122.9	111.6	108.3	86.4	95.1
09	FINISHES	102.5	103.6	103.1	91.6	100.3	96.4	96.0	96.5	96.3	98.6	101.9	100.4	100.9	118.6	110.7	90.0	101.3	96.2
COVERS	DIVS. 10 - 14, 25, 28, 41, 43, 44, 46	100.0	101.1	100.2	100.0	96.7	99.3	100.0	49.7	89.8	100.0	99.7	99.9	100.0	104.6	100.9	100.0	96.1	99.2
21, 22, 23	FIRE SUPPRESSION, PLUMBING & HVAC	100.3	97.0	98.9	100.1	87.5	95.0	95.3	87.3	92.1	100.0	95.6	98.2	100.0	103.5	101.4	94.8	86.9	91.6
26, 27, 3370	ELECTRICAL, COMMUNICATIONS & UTIL.	101.5	98.7	100.1	104.2	85.1	94.1	100.8	85.1	92.5	102.1	92.1	96.8	100.8	99.5	100.1	101.9	85.1	93.0
MF2014	WEIGHTED AVERAGE	99.5	102.2	100.7	97.8	94.5	96.3	96.7	92.5	94.8	99.1	99.0	99.1	100.0	107.2	103.1	96.8	94.6	95.9

WISCONSIN

DIVISION		OSHKOSH 549 MAT.	INST.	TOTAL	PORTAGE 539 MAT.	INST.	TOTAL	RACINE 534 MAT.	INST.	TOTAL	RHINELANDER 545 MAT.	INST.	TOTAL	SUPERIOR 548 MAT.	INST.	TOTAL	WAUSAU 544 MAT.	INST.	TOTAL
015433	CONTRACTOR EQUIPMENT		98.2	98.2		102.3	102.3		102.3	102.3		98.2	98.2		100.8	100.8		98.2	98.2
0241, 31 - 34	SITE & INFRASTRUCTURE, DEMOLITION	91.5	98.9	96.8	85.0	108.0	101.4	94.5	109.0	104.8	103.3	98.8	100.1	91.3	103.3	99.9	87.4	98.9	95.6
0310	Concrete Forming & Accessories	88.6	98.1	96.8	91.7	98.9	97.9	100.5	100.5	100.5	86.2	97.8	96.2	90.5	91.9	91.7	88.1	98.7	97.2
0320	Concrete Reinforcing	88.8	92.3	90.6	93.5	91.4	92.5	92.3	97.9	95.1	89.0	90.9	90.0	87.8	91.4	89.6	89.0	91.3	90.2
0330	Cast-in-Place Concrete	99.7	99.4	99.6	85.9	101.8	92.4	98.7	107.0	102.1	113.1	99.3	107.4	102.0	97.9	100.3	93.0	99.6	95.7
03	CONCRETE	92.4	97.6	94.9	86.4	98.5	92.4	97.9	102.3	100.1	104.7	97.2	101.0	91.8	94.2	93.0	87.3	97.8	92.5
04	MASONRY	107.9	101.4	103.8	99.6	104.3	102.5	100.6	114.0	109.0	125.3	101.4	110.4	120.8	104.5	110.6	107.4	101.4	103.6
05	METALS	95.4	96.7	95.8	91.3	96.0	92.7	94.8	101.2	96.8	95.3	96.7	95.7	96.1	98.2	96.8	95.1	97.7	95.9
06	WOOD, PLASTICS & COMPOSITES	89.8	99.3	95.1	89.2	99.2	94.8	100.2	96.4	98.1	87.3	99.3	94.0	92.3	90.1	91.0	89.2	99.3	94.9
07	THERMAL & MOISTURE PROTECTION	103.4	85.7	96.1	101.4	92.6	97.7	102.1	108.3	104.7	104.2	83.5	95.7	103.3	92.2	98.7	103.3	84.4	95.5
08	OPENINGS	95.9	95.5	95.8	98.1	92.3	96.8	102.5	101.1	102.2	96.0	87.1	93.9	88.6	91.1	89.2	96.1	95.7	96.0
0920	Plaster & Gypsum Board	86.2	99.6	95.2	86.4	99.6	95.3	94.5	96.7	96.0	86.2	99.6	95.2	87.6	90.2	89.4	86.2	99.6	95.2
0950, 0980	Ceilings & Acoustic Treatment	81.9	99.6	93.5	85.6	99.6	94.8	88.1	96.7	93.7	81.9	99.6	93.5	59.3	90.2	79.6	81.9	99.6	93.5
0960	Flooring	101.6	114.1	105.2	96.2	118.1	102.5	99.8	110.4	102.8	100.9	114.1	104.7	104.4	122.4	109.6	101.5	114.1	105.1
0970, 0990	Wall Finishes & Painting/Coating	102.2	101.4	101.7	98.3	66.4	79.1	98.3	115.5	108.7	102.2	60.4	76.9	96.0	102.8	100.1	102.2	83.2	92.0
09	FINISHES	94.0	100.9	97.8	94.1	98.6	96.6	97.4	103.6	100.8	94.8	97.0	96.0	89.3	98.6	94.4	93.7	100.7	97.5
COVERS	DIVS. 10 - 14, 25, 28, 41, 43, 44, 46	100.0	64.6	92.8	100.0	63.4	92.6	100.0	101.1	100.2	100.0	55.0	90.9	100.0	94.1	98.8	100.0	96.5	99.3
21, 22, 23	FIRE SUPPRESSION, PLUMBING & HVAC	95.6	85.0	91.3	95.3	95.3	95.3	100.1	97.0	98.8	95.6	86.7	92.0	94.8	89.4	92.6	95.6	86.4	91.9
26, 27, 3370	ELECTRICAL, COMMUNICATIONS & UTIL.	102.5	80.4	90.9	104.8	92.1	98.1	100.8	98.6	99.6	101.8	81.8	91.3	107.1	99.0	102.8	103.7	81.8	92.1
MF2014	WEIGHTED AVERAGE	96.9	91.9	94.7	95.3	96.6	95.9	99.1	102.4	100.5	99.3	91.2	95.8	96.6	96.3	96.5	96.2	93.5	95.0

For customer support on your Plumbing Cost Data, call 877.709.6968.

639

City Cost Indexes

WYOMING

DIVISION		CASPER 826 MAT.	INST.	TOTAL	CHEYENNE 820 MAT.	INST.	TOTAL	NEWCASTLE 827 MAT.	INST.	TOTAL	RAWLINS 823 MAT.	INST.	TOTAL	RIVERTON 825 MAT.	INST.	TOTAL	ROCK SPRINGS 829-831 MAT.	INST.	TOTAL
015433	CONTRACTOR EQUIPMENT		98.2	98.2		98.2	98.2		98.2	98.2		98.2	98.2		98.2	98.2		98.2	98.2
0241, 31 - 34	SITE & INFRASTRUCTURE, DEMOLITION	98.6	94.4	95.6	92.5	94.4	93.8	85.1	94.0	91.5	98.4	94.0	95.3	92.1	94.0	93.5	89.2	94.0	92.6
0310	Concrete Forming & Accessories	102.5	50.1	57.3	104.4	53.5	60.5	94.8	62.2	66.7	99.2	62.2	67.3	93.7	62.1	66.5	101.5	62.1	67.5
0320	Concrete Reinforcing	105.7	70.6	87.8	98.0	70.6	84.1	105.5	70.8	87.9	105.2	70.8	87.7	106.1	70.6	88.0	106.1	70.6	88.0
0330	Cast-in-Place Concrete	102.8	72.0	90.2	95.0	72.1	85.6	95.9	61.6	81.8	96.0	61.6	81.9	96.0	60.0	81.2	95.9	59.9	81.1
03	CONCRETE	101.7	62.1	82.3	98.6	63.7	81.4	99.0	63.9	81.8	111.8	63.9	88.3	106.7	63.3	85.4	99.5	63.2	81.7
04	MASONRY	108.2	51.9	73.1	103.7	51.9	71.4	100.3	54.8	71.9	100.3	54.8	71.9	100.3	50.5	69.3	160.4	50.4	91.8
05	METALS	99.3	72.1	90.9	101.7	72.1	92.6	97.8	72.0	89.9	97.9	72.0	89.9	98.0	71.6	89.8	98.7	71.1	90.2
06	WOOD, PLASTICS & COMPOSITES	97.2	49.7	70.6	96.5	54.3	72.9	86.4	66.9	75.5	90.4	66.9	77.2	85.3	66.9	75.0	95.4	66.9	79.4
07	THERMAL & MOISTURE PROTECTION	105.9	55.1	85.1	100.1	55.6	81.8	101.8	55.8	82.9	103.2	55.8	83.7	102.7	59.9	85.1	101.9	59.9	84.7
08	OPENINGS	108.0	53.3	95.3	108.3	55.8	96.1	112.6	61.7	100.8	112.3	61.7	100.5	112.5	61.7	100.7	113.0	61.3	101.0
0920	Plaster & Gypsum Board	97.4	48.1	64.1	89.3	52.9	64.7	85.2	65.8	72.1	85.5	65.8	72.2	85.2	65.8	72.1	97.3	65.8	76.0
0950, 0980	Ceilings & Acoustic Treatment	99.2	48.1	65.6	91.0	52.9	66.0	92.9	65.8	75.1	92.9	65.8	75.1	92.9	65.8	75.1	92.9	65.8	75.1
0960	Flooring	112.0	56.4	96.1	109.9	56.4	94.6	103.5	56.4	90.0	106.2	56.4	92.0	103.0	56.4	89.7	108.4	56.4	93.5
0970, 0990	Wall Finishes & Painting/Coating	104.7	43.1	67.5	111.2	43.1	70.0	107.1	55.4	75.9	107.1	55.4	75.9	107.1	55.4	75.9	107.1	55.4	75.9
09	FINISHES	103.2	49.6	73.6	100.0	52.3	73.7	95.0	60.8	76.1	97.2	60.8	77.1	95.7	60.8	76.5	98.1	60.8	77.5
COVERS	DIVS. 10 - 14, 25, 28, 41, 43, 44, 46	100.0	87.6	97.5	100.0	88.2	97.6	100.0	89.1	97.8	100.0	89.1	97.8	100.0	81.4	96.2	100.0	80.8	96.1
21, 22, 23	FIRE SUPPRESSION, PLUMBING & HVAC	99.9	64.1	85.4	99.9	64.1	85.4	97.4	63.4	83.7	97.4	63.4	83.7	97.4	63.4	83.7	99.8	63.3	85.1
26, 27, 3370	ELECTRICAL, COMMUNICATIONS & UTIL.	103.3	64.1	82.6	105.8	68.6	86.1	104.5	69.0	85.7	104.5	69.0	85.7	104.5	62.9	82.5	102.4	62.9	81.5
MF2014	WEIGHTED AVERAGE	102.0	63.9	85.4	101.5	65.2	85.7	99.9	66.9	85.5	101.8	66.9	86.6	101.0	65.4	85.5	101.0	65.3	87.0

WYOMING / CANADA

DIVISION		SHERIDAN 828 MAT.	INST.	TOTAL	WHEATLAND 822 MAT.	INST.	TOTAL	WORLAND 824 MAT.	INST.	TOTAL	YELLOWSTONE NAT'L PA 821 MAT.	INST.	TOTAL	BARRIE, ONTARIO MAT.	INST.	TOTAL	BATHURST, NEW BRUNSWICK MAT.	INST.	TOTAL
015433	CONTRACTOR EQUIPMENT		98.2	98.2		98.2	98.2		98.2	98.2		98.2	98.2		102.1	102.1		102.2	102.2
0241, 31 - 34	SITE & INFRASTRUCTURE, DEMOLITION	92.4	94.4	93.8	89.6	94.0	92.8	87.1	94.0	92.0	87.2	94.2	92.2	118.5	100.4	105.6	103.4	96.6	98.6
0310	Concrete Forming & Accessories	102.2	50.6	57.7	96.8	52.7	58.8	96.9	62.0	66.8	96.9	63.1	67.8	124.4	89.4	94.2	104.4	63.7	69.3
0320	Concrete Reinforcing	106.1	70.9	88.2	105.5	70.8	87.8	106.1	70.6	88.0	107.9	70.6	88.9	173.2	83.8	127.7	137.0	56.5	96.0
0330	Cast-in-Place Concrete	99.2	72.0	88.1	100.2	61.5	84.3	95.9	59.9	81.1	95.9	61.6	81.8	165.6	89.0	134.2	131.6	61.9	103.0
03	CONCRETE	106.9	62.4	85.0	103.7	59.7	82.1	99.2	63.2	81.6	99.5	64.3	82.2	150.3	88.4	119.9	124.6	62.4	94.1
04	MASONRY	100.6	56.3	73.0	100.7	54.7	72.0	100.3	50.4	69.2	100.4	53.4	71.1	171.5	98.0	125.7	160.8	64.9	101.1
05	METALS	101.6	72.4	92.6	97.8	71.7	89.8	98.0	71.5	89.9	98.6	71.5	90.3	110.2	90.2	104.0	109.0	71.5	97.5
06	WOOD, PLASTICS & COMPOSITES	97.9	50.4	71.3	88.3	54.3	69.2	88.3	66.9	76.3	88.3	66.9	76.3	119.8	88.2	102.1	101.8	64.0	80.7
07	THERMAL & MOISTURE PROTECTION	102.9	56.3	83.8	102.1	55.8	83.1	101.9	57.5	83.7	101.4	58.8	83.9	111.8	89.1	102.5	106.0	62.2	88.1
08	OPENINGS	113.2	52.7	99.1	111.1	55.8	98.3	112.8	61.7	100.9	105.6	61.7	95.4	93.9	86.5	92.2	87.9	55.9	80.5
0920	Plaster & Gypsum Board	109.4	48.8	68.5	85.2	52.8	63.3	85.2	65.8	72.1	85.4	65.8	72.1	151.1	87.7	108.2	137.6	62.7	87.0
0950, 0980	Ceilings & Acoustic Treatment	95.7	48.8	64.9	92.9	52.8	66.6	92.9	65.8	75.1	93.7	65.8	75.4	91.5	87.7	89.0	106.4	62.7	77.7
0960	Flooring	107.5	56.4	92.9	105.0	56.4	91.1	105.0	56.4	91.1	105.0	56.4	91.1	127.2	98.9	119.1	107.9	47.2	90.6
0970, 0990	Wall Finishes & Painting/Coating	109.4	43.1	69.4	107.1	34.4	63.3	107.1	55.4	75.9	107.1	55.4	75.9	116.2	89.9	100.3	117.6	51.6	77.8
09	FINISHES	102.8	50.0	73.7	95.8	51.0	71.1	95.5	60.8	76.3	95.7	61.5	76.8	115.0	91.5	102.1	111.3	59.7	82.8
COVERS	DIVS. 10 - 14, 25, 28, 41, 43, 44, 46	100.0	87.7	97.5	100.0	80.3	96.0	100.0	81.4	96.2	100.0	82.4	96.4	139.2	74.4	126.1	131.1	65.3	117.8
21, 22, 23	FIRE SUPPRESSION, PLUMBING & HVAC	97.4	64.1	83.9	97.4	63.3	83.6	97.4	63.3	83.6	97.4	64.9	84.3	101.8	99.3	100.8	102.0	69.8	89.0
26, 27, 3370	ELECTRICAL, COMMUNICATIONS & UTIL.	107.4	63.0	83.9	104.5	68.6	85.5	104.5	62.9	82.5	103.2	62.9	81.9	116.8	90.5	102.9	112.7	61.8	85.8
MF2014	WEIGHTED AVERAGE	102.6	64.3	85.9	100.4	64.3	84.7	100.0	65.3	84.9	99.3	66.3	84.9	117.7	93.0	106.9	111.5	67.3	92.2

CANADA

DIVISION		BRANDON, MANITOBA MAT.	INST.	TOTAL	BRANTFORD, ONTARIO MAT.	INST.	TOTAL	BRIDGEWATER, NOVA SCOTIA MAT.	INST.	TOTAL	CALGARY, ALBERTA MAT.	INST.	TOTAL	CAP-DE-LA-MADELEINE, QUEBEC MAT.	INST.	TOTAL	CHARLESBOURG, QUEBEC MAT.	INST.	TOTAL
015433	CONTRACTOR EQUIPMENT		104.5	104.5		102.1	102.1		101.9	101.9		107.1	107.1		102.8	102.8		102.8	102.8
0241, 31 - 34	SITE & INFRASTRUCTURE, DEMOLITION	133.0	99.3	109.1	118.0	100.7	105.7	102.6	98.2	99.5	124.0	105.4	110.7	98.4	99.9	99.4	98.4	99.9	99.4
0310	Concrete Forming & Accessories	146.5	72.5	82.7	124.9	96.9	100.7	97.1	71.6	75.1	124.4	97.4	101.1	130.7	88.0	93.8	130.7	88.0	93.8
0320	Concrete Reinforcing	185.9	53.2	118.4	165.4	82.5	123.2	140.6	46.7	92.8	135.6	71.9	103.2	140.6	78.8	109.2	140.6	78.8	109.2
0330	Cast-in-Place Concrete	133.4	76.7	110.1	153.6	110.3	135.8	157.8	72.2	122.7	189.0	107.3	155.5	124.0	97.8	113.2	124.0	97.8	113.2
03	CONCRETE	145.6	71.0	109.0	141.8	98.8	120.7	139.4	67.9	104.3	159.1	96.1	128.2	124.3	89.8	107.4	124.3	89.8	107.4
04	MASONRY	233.5	67.6	130.1	168.0	102.6	127.2	163.4	71.9	106.4	196.3	91.4	130.9	163.9	87.1	116.0	163.9	87.1	116.0
05	METALS	125.6	76.8	110.6	109.1	90.9	103.5	108.2	74.2	97.7	137.0	89.3	122.3	107.4	87.6	101.3	107.4	87.6	101.3
06	WOOD, PLASTICS & COMPOSITES	154.2	73.5	109.0	123.7	95.7	108.1	92.9	71.1	80.7	102.3	97.2	99.4	135.3	88.0	108.8	135.3	88.0	108.8
07	THERMAL & MOISTURE PROTECTION	135.0	71.8	109.1	113.8	94.7	106.0	108.6	69.9	92.8	117.8	94.6	108.3	107.4	89.9	100.3	107.4	89.9	100.3
08	OPENINGS	105.5	64.8	96.1	92.4	92.7	92.5	86.1	64.9	81.1	89.9	86.2	89.1	93.6	81.1	90.7	93.6	81.1	90.7
0920	Plaster & Gypsum Board	119.3	72.2	87.4	130.1	95.5	106.7	130.7	70.0	89.7	139.4	96.6	110.5	157.0	87.4	109.9	157.0	87.4	109.9
0950, 0980	Ceilings & Acoustic Treatment	109.6	72.2	85.0	96.5	95.5	95.8	96.5	70.0	79.1	138.8	96.6	111.0	96.5	87.4	90.5	96.5	87.4	90.5
0960	Flooring	138.6	70.2	119.1	120.5	98.9	114.3	102.7	67.2	92.5	122.3	92.6	113.8	120.5	99.5	114.5	120.5	99.5	114.5
0970, 0990	Wall Finishes & Painting/Coating	126.5	58.7	85.6	118.3	98.6	106.4	118.3	64.1	85.6	127.4	112.4	118.3	118.3	91.7	102.2	118.3	91.7	102.2
09	FINISHES	122.5	71.0	94.1	112.4	97.9	104.4	106.6	70.3	86.6	125.5	98.7	110.7	115.3	90.8	101.8	115.3	90.8	101.8
COVERS	DIVS. 10 - 14, 25, 28, 41, 43, 44, 46	131.1	67.6	118.3	131.1	76.6	120.1	131.1	66.6	118.1	131.1	96.9	124.2	131.1	85.2	121.9	131.1	85.2	121.9
21, 22, 23	FIRE SUPPRESSION, PLUMBING & HVAC	102.1	84.1	94.9	102.0	102.3	102.1	102.0	84.4	94.9	101.2	92.9	97.8	102.3	91.0	97.8	102.3	91.0	97.8
26, 27, 3370	ELECTRICAL, COMMUNICATIONS & UTIL.	118.6	69.6	92.7	111.5	89.9	100.1	117.1	64.4	89.2	113.5	101.3	107.1	111.2	72.6	90.8	111.2	72.6	90.8
MF2014	WEIGHTED AVERAGE	125.2	75.8	103.7	114.9	96.9	107.1	112.9	74.6	96.2	123.5	95.7	111.4	112.4	87.6	101.6	112.4	87.6	101.6

City Cost Indexes

CANADA

DIVISION		CHARLOTTETOWN, PRINCE EDWARD ISLAND			CHICOUTIMI, QUEBEC			CORNER BROOK, NEWFOUNDLAND			CORNWALL, ONTARIO			DALHOUSIE, NEW BRUNSWICK			DARTMOUTH, NOVA SCOTIA		
		MAT.	INST.	TOTAL	MAT.	INST.	TOTAL	MAT.	INST.	TOTAL	MAT.	INST.	TOTAL	MAT.	INST.	TOTAL	MAT.	INST.	TOTAL
015433	CONTRACTOR EQUIPMENT		101.8	101.8		102.8	102.8		103.3	103.3		102.1	102.1		102.2	102.2		101.9	101.9
0241, 31 - 34	SITE & INFRASTRUCTURE, DEMOLITION	121.3	95.7	103.1	100.7	99.1	99.6	137.7	97.1	108.9	116.1	100.2	104.8	100.9	96.6	97.8	124.8	98.2	105.9
0310	Concrete Forming & Accessories	110.4	58.2	65.4	133.2	93.1	98.6	121.8	61.7	69.9	122.6	89.6	94.1	103.5	63.9	69.3	111.8	71.6	77.2
0320	Concrete Reinforcing	141.5	46.8	93.3	103.3	95.9	99.5	170.3	48.6	108.4	165.4	82.2	123.1	142.7	56.6	98.9	178.2	46.7	111.3
0330	Cast-in-Place Concrete	166.8	61.2	123.4	124.0	99.8	114.0	155.3	70.6	120.5	138.2	100.6	122.7	131.8	62.0	103.2	149.8	72.2	118.0
03	CONCRETE	147.1	57.9	103.3	114.6	95.9	105.4	182.2	63.2	123.7	134.3	92.2	113.6	132.5	62.5	98.1	162.1	67.9	115.8
04	MASONRY	178.4	62.6	106.3	162.7	95.0	120.5	229.3	64.1	126.4	166.8	94.0	121.4	160.2	64.9	100.8	244.8	71.9	137.1
05	METALS	133.9	67.6	113.5	108.8	92.2	103.7	126.0	72.9	109.7	109.1	89.6	103.1	104.4	71.8	94.3	125.9	74.2	110.0
06	WOOD, PLASTICS & COMPOSITES	97.8	57.8	75.4	137.4	93.6	112.9	128.3	60.8	90.5	122.1	89.0	103.6	100.2	64.0	80.0	116.1	71.1	90.9
07	THERMAL & MOISTURE PROTECTION	121.3	60.4	96.3	106.5	96.4	102.3	139.8	61.8	107.8	113.6	89.3	103.6	110.8	62.2	90.9	136.8	69.9	109.4
08	OPENINGS	88.2	50.0	79.3	92.2	80.6	89.5	112.4	57.1	99.5	93.6	86.2	91.9	89.1	55.9	81.4	95.3	64.9	88.2
0920	Plaster & Gypsum Board	131.6	56.3	80.7	159.4	93.2	114.6	149.0	59.3	88.4	189.4	88.6	121.3	141.2	62.7	88.2	144.6	70.0	94.2
0950, 0980	Ceilings & Acoustic Treatment	120.2	56.3	78.2	105.6	93.2	97.4	110.5	59.3	76.9	99.0	88.6	92.2	98.9	62.7	75.1	117.2	70.0	86.2
0960	Flooring	111.7	62.9	97.8	123.9	99.5	116.8	121.0	57.2	102.8	120.5	97.5	113.9	106.5	72.1	96.7	116.1	67.2	102.1
0970, 0990	Wall Finishes & Painting/Coating	125.6	42.9	75.7	117.6	106.4	110.8	126.4	61.8	87.4	118.3	91.9	102.4	120.7	51.6	79.0	126.4	64.1	88.8
09	FINISHES	117.1	57.8	84.4	118.4	96.5	106.4	122.0	60.9	88.4	121.0	91.6	104.8	110.6	64.5	85.2	120.1	70.3	92.6
COVERS	DIVS. 10 - 14, 25, 28, 41, 43, 44, 46	131.1	64.8	117.7	131.1	86.5	122.1	131.1	65.6	117.9	131.1	74.0	119.6	131.1	65.3	117.8	131.1	66.6	118.1
21, 22, 23	FIRE SUPPRESSION, PLUMBING & HVAC	102.3	63.8	86.8	102.0	92.3	98.1	102.1	71.8	89.9	102.3	100.1	101.4	102.0	69.8	89.0	102.1	84.4	95.0
26, 27, 3370	ELECTRICAL, COMMUNICATIONS & UTIL.	110.6	52.3	79.8	109.7	86.8	97.6	116.0	58.3	85.5	112.4	90.9	101.0	114.5	58.4	84.8	120.7	64.4	90.9
MF2014	WEIGHTED AVERAGE	120.0	62.8	95.1	111.4	92.9	103.3	129.6	67.6	102.6	115.0	93.3	105.5	111.9	67.5	92.5	126.1	74.6	103.7

CANADA

DIVISION		EDMONTON, ALBERTA			FORT MCMURRAY, ALBERTA			FREDERICTON, NEW BRUNSWICK			GATINEAU, QUEBEC			GRANBY, QUEBEC			HALIFAX, NOVA SCOTIA		
		MAT.	INST.	TOTAL	MAT.	INST.	TOTAL	MAT.	INST.	TOTAL	MAT.	INST.	TOTAL	MAT.	INST.	TOTAL	MAT.	INST.	TOTAL
015433	CONTRACTOR EQUIPMENT		107.1	107.1		104.4	104.4		102.2	102.2		102.8	102.8		102.8	102.8		101.9	101.9
0241, 31 - 34	SITE & INFRASTRUCTURE, DEMOLITION	144.1	105.4	116.6	123.1	101.3	107.6	105.0	96.6	99.1	98.2	99.8	99.4	98.7	99.8	99.5	109.2	98.3	101.5
0310	Concrete Forming & Accessories	123.2	97.4	100.9	122.8	92.3	96.5	120.3	64.2	71.9	130.7	87.8	93.7	130.7	87.8	93.7	108.4	78.8	82.9
0320	Concrete Reinforcing	134.2	71.9	102.5	152.9	71.8	111.7	132.8	56.7	94.1	148.7	78.8	113.2	148.7	78.8	113.2	144.3	62.7	102.7
0330	Cast-in-Place Concrete	183.9	107.3	152.4	203.9	105.1	163.3	128.3	62.0	101.1	122.3	97.8	112.2	126.4	97.7	114.6	149.4	81.9	121.7
03	CONCRETE	156.4	96.1	126.8	163.6	93.1	129.0	126.6	62.7	95.2	124.8	89.8	107.6	126.7	89.7	108.5	136.4	77.3	107.4
04	MASONRY	187.2	91.4	127.5	207.9	89.1	133.8	182.6	66.5	110.2	163.7	87.1	116.0	164.1	87.1	116.1	185.9	84.7	122.8
05	METALS	136.9	89.3	122.3	135.4	88.9	121.1	129.1	72.3	111.6	107.4	87.4	101.2	107.4	87.4	101.2	134.9	81.5	118.4
06	WOOD, PLASTICS & COMPOSITES	105.2	97.2	100.7	117.6	91.8	103.2	116.7	64.0	87.2	135.3	88.0	108.8	135.3	88.0	108.8	97.1	77.8	86.3
07	THERMAL & MOISTURE PROTECTION	122.7	94.6	111.2	121.9	91.9	109.6	116.0	63.2	94.4	107.4	89.9	100.3	107.4	88.3	99.6	119.0	79.4	102.8
08	OPENINGS	89.2	86.2	88.5	93.6	83.3	91.2	89.3	54.8	81.2	93.6	76.3	89.6	93.6	76.3	89.6	93.5	72.0	88.5
0920	Plaster & Gypsum Board	134.3	96.6	108.8	128.2	91.1	103.1	137.4	62.7	86.9	126.6	87.4	100.1	129.0	87.4	100.8	127.0	77.0	93.2
0950, 0980	Ceilings & Acoustic Treatment	139.5	96.6	111.3	104.9	91.1	95.8	116.1	62.7	81.0	96.5	87.4	90.5	96.5	87.4	90.5	118.7	77.0	91.3
0960	Flooring	122.5	92.6	114.0	120.5	92.6	112.5	116.8	75.3	104.9	120.5	99.5	114.5	120.5	99.5	114.5	109.2	88.3	103.3
0970, 0990	Wall Finishes & Painting/Coating	122.4	112.4	116.3	118.4	95.5	104.5	123.1	65.8	88.5	118.3	91.7	102.2	118.3	91.7	102.2	124.9	87.4	102.2
09	FINISHES	126.6	98.7	111.2	115.3	93.0	103.0	117.7	66.7	89.6	111.2	90.8	100.0	111.5	90.8	100.1	115.6	81.8	97.0
COVERS	DIVS. 10 - 14, 25, 28, 41, 43, 44, 46	131.1	96.9	124.2	131.1	95.2	123.9	131.1	65.3	117.8	131.1	85.2	121.9	131.1	85.2	121.9	131.1	68.5	118.5
21, 22, 23	FIRE SUPPRESSION, PLUMBING & HVAC	101.1	92.9	97.8	102.4	99.0	101.0	102.3	79.3	93.0	102.3	91.0	97.8	102.0	91.0	97.6	100.6	77.6	91.3
26, 27, 3370	ELECTRICAL, COMMUNICATIONS & UTIL.	110.6	101.3	105.7	105.8	85.4	95.1	116.1	77.1	95.5	111.2	72.6	90.8	111.8	72.6	91.1	117.0	81.2	98.1
MF2014	WEIGHTED AVERAGE	123.1	95.7	111.2	123.6	92.8	110.2	117.4	72.5	97.9	112.1	87.4	101.3	112.3	87.4	101.4	119.6	80.9	102.7

CANADA

DIVISION		HAMILTON, ONTARIO			HULL, QUEBEC			JOLIETTE, QUEBEC			KAMLOOPS, BRITISH COLUMBIA			KINGSTON, ONTARIO			KITCHENER, ONTARIO		
		MAT.	INST.	TOTAL	MAT.	INST.	TOTAL	MAT.	INST.	TOTAL	MAT.	INST.	TOTAL	MAT.	INST.	TOTAL	MAT.	INST.	TOTAL
015433	CONTRACTOR EQUIPMENT		108.9	108.9		102.8	102.8		102.8	102.8		106.1	106.1		104.4	104.4		104.3	104.3
0241, 31 - 34	SITE & INFRASTRUCTURE, DEMOLITION	115.5	112.4	113.3	98.2	99.8	99.4	98.8	99.9	99.6	120.5	103.3	108.3	116.1	104.0	107.5	99.6	104.8	103.3
0310	Concrete Forming & Accessories	120.9	94.7	98.3	130.7	87.8	93.7	130.7	88.0	93.8	123.4	90.8	95.3	122.8	89.6	94.2	113.0	87.2	90.8
0320	Concrete Reinforcing	146.4	92.0	118.7	148.7	78.8	113.2	140.6	78.8	109.2	110.3	76.8	93.2	165.4	82.2	123.1	100.8	91.9	96.3
0330	Cast-in-Place Concrete	143.5	101.0	126.1	122.3	97.8	112.2	127.4	97.8	115.2	110.3	101.6	106.7	138.2	100.6	122.7	129.0	93.6	114.4
03	CONCRETE	132.8	96.3	114.8	124.8	89.8	107.6	125.9	89.8	108.2	132.9	92.0	112.8	136.2	92.2	114.6	110.8	90.4	100.8
04	MASONRY	183.0	101.2	132.0	163.7	87.1	116.0	164.1	87.1	116.1	170.7	95.0	123.5	173.5	94.1	124.0	150.8	97.6	117.6
05	METALS	123.4	92.5	113.9	107.4	87.4	101.2	107.4	87.6	101.3	109.8	87.5	102.9	110.8	89.5	104.3	117.4	92.2	109.6
06	WOOD, PLASTICS & COMPOSITES	100.4	93.9	96.8	135.3	88.0	108.8	135.3	88.0	108.8	105.2	89.4	96.4	122.1	89.2	103.7	109.7	85.5	96.1
07	THERMAL & MOISTURE PROTECTION	118.2	95.7	109.0	107.4	89.9	100.2	107.4	89.9	100.2	123.3	86.8	108.3	113.6	90.4	104.1	108.6	92.8	102.1
08	OPENINGS	91.3	91.6	91.4	93.6	76.3	89.6	93.6	81.1	90.7	90.2	85.7	89.2	93.6	85.9	91.8	84.7	85.3	84.8
0920	Plaster & Gypsum Board	149.5	93.6	111.7	126.6	87.4	100.1	157.0	87.4	109.9	113.6	88.5	96.6	192.8	88.7	122.4	120.2	84.9	96.3
0950, 0980	Ceilings & Acoustic Treatment	122.1	93.6	103.4	96.5	87.4	90.5	96.5	87.4	90.5	96.5	88.5	91.2	112.4	88.7	96.8	98.6	84.9	89.6
0960	Flooring	123.1	102.3	117.2	120.5	99.5	114.5	120.5	99.5	114.5	119.6	55.6	101.3	120.5	97.5	113.9	110.5	102.3	108.1
0970, 0990	Wall Finishes & Painting/Coating	121.6	102.0	110.2	118.3	91.7	102.2	118.3	91.7	102.2	118.3	84.0	97.6	118.3	84.9	98.1	115.3	92.6	101.6
09	FINISHES	122.0	97.0	108.2	111.2	90.8	100.0	115.3	90.8	101.8	111.7	84.2	96.6	124.4	90.9	105.9	106.4	90.2	97.5
COVERS	DIVS. 10 - 14, 25, 28, 41, 43, 44, 46	131.1	98.8	124.6	131.1	85.2	121.9	131.1	85.2	121.9	131.1	95.6	124.0	131.1	74.0	119.6	131.1	97.0	124.2
21, 22, 23	FIRE SUPPRESSION, PLUMBING & HVAC	101.3	89.2	96.4	102.0	91.0	97.6	102.0	91.0	97.6	102.0	94.4	98.9	102.3	100.3	101.5	100.4	87.7	95.3
26, 27, 3370	ELECTRICAL, COMMUNICATIONS & UTIL.	108.3	100.5	104.2	112.9	72.6	91.6	111.8	72.6	91.1	115.1	81.9	97.6	112.4	89.6	100.3	111.5	98.0	104.4
MF2014	WEIGHTED AVERAGE	117.0	96.7	108.2	112.2	87.4	101.4	112.5	87.6	101.7	114.5	90.6	104.1	116.0	93.4	106.2	109.6	92.9	102.3

CANADA

DIVISION		LAVAL, QUEBEC			LETHBRIDGE, ALBERTA			LLOYDMINSTER, ALBERTA			LONDON, ONTARIO			MEDICINE HAT, ALBERTA			MONCTON, NEW BRUNSWICK		
		MAT.	INST.	TOTAL	MAT.	INST.	TOTAL	MAT.	INST.	TOTAL	MAT.	INST.	TOTAL	MAT.	INST.	TOTAL	MAT.	INST.	TOTAL
015433	CONTRACTOR EQUIPMENT		102.8	102.8		104.4	104.4		104.4	104.4		104.4	104.4		104.4	104.4		102.2	102.2
0241, 31 - 34	SITE & INFRASTRUCTURE, DEMOLITION	98.7	99.8	99.5	115.6	101.9	105.9	115.4	101.3	105.4	113.1	104.9	107.3	114.2	101.4	105.1	102.8	96.8	98.5
0310	Concrete Forming & Accessories	130.8	87.8	93.8	124.1	92.4	96.8	122.3	82.6	88.1	119.1	88.5	92.7	124.1	82.6	88.3	104.4	64.5	69.9
0320	Concrete Reinforcing	148.7	78.8	113.2	152.9	71.8	111.7	152.9	71.8	111.7	132.8	91.9	112.0	152.9	71.8	111.7	137.0	61.5	98.6
0330	Cast-in-Place Concrete	126.4	97.8	114.6	153.0	105.1	133.3	141.9	101.4	125.3	147.4	98.6	127.4	141.9	101.4	125.3	126.8	78.7	107.0
03	CONCRETE	126.7	89.8	108.6	139.5	93.2	116.7	134.1	87.5	111.3	132.4	92.7	112.9	134.3	87.5	111.3	122.3	69.7	96.5
04	MASONRY	164.0	87.1	116.1	181.7	89.1	124.0	163.4	82.5	113.0	182.6	98.7	130.3	163.4	82.5	113.0	160.5	65.0	101.0
05	METALS	107.4	87.4	101.3	128.8	88.9	116.5	110.0	88.8	103.4	124.9	91.8	114.7	110.0	88.7	103.4	109.0	80.4	100.2
06	WOOD, PLASTICS & COMPOSITES	135.5	88.0	108.9	121.3	91.8	104.8	117.6	82.0	97.7	108.2	86.6	96.1	121.3	82.0	99.3	101.8	64.0	80.7
07	THERMAL & MOISTURE PROTECTION	107.9	89.9	100.5	119.1	91.9	108.0	116.0	87.5	104.3	119.1	93.5	108.6	122.3	87.5	108.0	110.2	65.3	91.8
08	OPENINGS	93.6	76.3	89.6	93.6	83.3	91.2	93.6	77.9	89.9	86.4	86.6	86.4	93.6	77.9	89.9	87.9	61.0	81.6
0920	Plaster & Gypsum Board	129.3	87.4	100.9	119.1	91.1	100.2	114.4	81.0	91.8	152.4	86.1	107.6	117.2	81.0	92.8	137.6	62.7	87.0
0950, 0980	Ceilings & Acoustic Treatment	96.5	87.4	90.5	104.1	91.1	95.5	96.5	81.0	86.3	121.1	86.1	98.1	96.5	81.0	86.3	106.4	62.7	77.7
0960	Flooring	120.5	99.5	114.5	120.5	92.6	112.5	120.5	92.6	112.5	117.2	102.3	113.0	120.5	92.6	112.5	119.0	72.1	97.7
0970, 0990	Wall Finishes & Painting/Coating	118.3	91.7	102.2	118.2	104.2	109.8	118.4	81.2	96.0	122.0	99.4	108.3	118.2	81.2	95.9	117.6	51.6	77.8
09	FINISHES	111.5	90.8	100.1	112.9	94.0	102.5	110.7	84.0	96.0	120.5	91.9	104.7	110.9	84.0	96.1	111.3	64.5	85.5
COVERS	DIVS. 10 - 14, 25, 28, 41, 43, 44, 46	131.1	85.2	121.9	131.1	95.2	123.9	131.1	91.9	123.2	131.1	97.5	124.3	131.1	91.9	123.2	131.1	65.3	117.8
21, 22, 23	FIRE SUPPRESSION, PLUMBING & HVAC	100.4	91.0	96.6	102.2	95.6	99.6	102.3	95.6	99.6	101.4	86.6	95.4	102.0	92.3	98.1	102.0	70.1	89.1
26, 27, 3370	ELECTRICAL, COMMUNICATIONS & UTIL.	112.9	72.6	91.6	107.1	85.4	95.7	104.9	85.4	94.6	107.0	97.5	102.0	104.9	85.4	94.6	117.2	61.8	88.0
MF2014	WEIGHTED AVERAGE	112.1	87.4	101.3	118.4	92.2	107.0	113.4	88.9	102.8	116.4	93.3	106.3	113.6	88.2	102.5	111.8	70.1	93.6

CANADA

DIVISION		MONTREAL, QUEBEC			MOOSE JAW, SASKATCHEWAN			NEW GLASGOW, NOVA SCOTIA			NEWCASTLE, NEW BRUNSWICK			NORTH BAY, ONTARIO			OSHAWA, ONTARIO		
		MAT.	INST.	TOTAL	MAT.	INST.	TOTAL	MAT.	INST.	TOTAL	MAT.	INST.	TOTAL	MAT.	INST.	TOTAL	MAT.	INST.	TOTAL
015433	CONTRACTOR EQUIPMENT		104.5	104.5		100.6	100.6		101.9	101.9		102.2	102.2		102.1	102.1		104.3	104.3
0241, 31 - 34	SITE & INFRASTRUCTURE, DEMOLITION	112.1	99.4	103.1	115.5	95.9	101.6	118.5	98.2	104.1	103.4	96.6	98.6	133.9	99.8	109.7	110.7	103.9	105.9
0310	Concrete Forming & Accessories	124.3	93.4	97.7	107.1	60.5	66.9	111.7	71.6	77.1	104.4	63.9	69.5	147.4	87.0	95.3	118.4	89.1	93.1
0320	Concrete Reinforcing	134.0	96.0	114.7	107.9	61.9	84.5	170.3	46.7	107.4	137.0	56.6	96.1	201.7	81.8	140.7	159.6	84.4	121.3
0330	Cast-in-Place Concrete	161.7	101.2	136.8	137.7	71.1	110.4	149.8	72.2	118.0	131.6	62.0	103.0	144.3	87.1	120.8	149.1	87.9	124.0
03	CONCRETE	139.7	96.5	118.4	117.9	65.1	92.0	160.9	67.9	115.2	124.6	62.5	94.1	163.4	86.3	125.5	135.6	88.0	112.2
04	MASONRY	167.9	95.1	122.5	162.0	64.4	101.1	228.9	71.9	131.1	160.8	64.9	101.1	235.3	89.9	144.7	153.8	94.4	116.8
05	METALS	128.0	92.7	117.1	106.3	73.9	96.3	123.6	74.2	108.4	109.0	71.8	97.5	124.6	89.2	113.7	108.6	91.4	103.3
06	WOOD, PLASTICS & COMPOSITES	114.4	93.9	102.9	102.7	59.1	78.3	116.1	71.1	90.9	101.8	64.0	80.7	157.4	87.6	118.3	116.7	88.2	100.7
07	THERMAL & MOISTURE PROTECTION	116.5	96.9	108.5	106.8	64.1	89.3	136.8	69.9	109.4	110.2	62.2	90.5	143.5	85.6	119.8	109.4	87.4	100.4
08	OPENINGS	90.1	82.4	88.3	89.3	56.8	81.7	95.3	64.9	88.2	87.9	55.9	80.5	104.2	83.7	99.4	90.4	87.9	89.8
0920	Plaster & Gypsum Board	135.1	93.2	106.8	110.7	57.8	74.9	142.7	70.0	93.6	137.6	62.7	87.0	140.3	87.1	104.4	124.7	87.7	99.7
0950, 0980	Ceilings & Acoustic Treatment	126.1	93.2	104.4	96.5	57.8	71.0	109.6	70.0	83.6	106.4	62.7	77.7	109.6	87.1	94.8	95.3	87.7	90.3
0960	Flooring	122.0	99.5	115.6	110.6	63.3	97.1	116.1	67.2	102.1	107.9	72.1	97.7	138.6	97.5	126.9	113.7	104.9	111.2
0970, 0990	Wall Finishes & Painting/Coating	120.8	106.4	112.1	118.3	66.8	87.2	126.4	64.1	88.8	117.6	51.6	77.8	126.4	91.2	105.1	115.3	106.7	110.1
09	FINISHES	119.9	96.7	107.1	107.1	61.4	81.9	118.2	70.3	91.8	111.3	64.5	85.5	125.1	89.8	105.7	107.9	93.9	100.2
COVERS	DIVS. 10 - 14, 25, 28, 41, 43, 44, 46	131.1	87.1	122.3	131.1	64.6	117.7	131.1	66.6	118.1	131.1	65.3	117.8	131.1	72.7	119.3	131.1	97.2	124.3
21, 22, 23	FIRE SUPPRESSION, PLUMBING & HVAC	101.2	92.4	97.7	102.3	76.1	91.8	102.1	84.4	95.0	102.0	69.8	89.0	102.1	98.1	100.5	100.4	100.7	100.5
26, 27, 3370	ELECTRICAL, COMMUNICATIONS & UTIL.	112.6	86.8	99.0	113.9	62.8	86.9	116.3	64.4	88.9	112.0	61.8	85.5	116.4	90.8	102.9	112.6	90.5	100.9
MF2014	WEIGHTED AVERAGE	117.8	93.2	107.1	110.7	69.5	92.7	124.1	74.6	102.5	111.6	67.9	92.5	127.3	91.1	111.5	112.3	94.3	104.4

CANADA

DIVISION		OTTAWA, ONTARIO			OWEN SOUND, ONTARIO			PETERBOROUGH, ONTARIO			PORTAGE LA PRAIRIE, MANITOBA			PRINCE ALBERT, SASKATCHEWAN			PRINCE GEORGE, BRITISH COLUMBIA		
		MAT.	INST.	TOTAL	MAT.	INST.	TOTAL	MAT.	INST.	TOTAL	MAT.	INST.	TOTAL	MAT.	INST.	TOTAL	MAT.	INST.	TOTAL
015433	CONTRACTOR EQUIPMENT		104.3	104.3		102.1	102.1		102.1	102.1		104.5	104.5		100.6	100.6		106.1	106.1
0241, 31 - 34	SITE & INFRASTRUCTURE, DEMOLITION	108.2	104.6	105.7	118.5	100.3	105.5	118.0	100.2	105.3	116.6	99.3	104.3	110.9	96.1	100.3	124.0	103.3	109.3
0310	Concrete Forming & Accessories	121.2	91.4	95.5	124.4	85.5	90.9	124.9	88.1	93.1	124.3	72.0	79.2	107.1	60.3	66.8	112.7	85.5	89.3
0320	Concrete Reinforcing	138.9	91.9	115.0	173.2	83.8	127.7	165.4	82.3	123.1	152.9	53.2	102.2	112.7	61.8	86.8	110.3	76.8	93.2
0330	Cast-in-Place Concrete	145.6	99.9	126.8	165.6	83.1	131.7	153.6	88.7	127.0	141.9	76.1	114.9	124.8	71.0	102.7	138.2	101.6	123.2
03	CONCRETE	132.6	94.4	113.8	150.3	84.6	118.1	141.8	87.4	115.1	127.8	70.6	99.7	112.6	65.0	89.2	145.4	89.7	118.0
04	MASONRY	168.0	98.8	124.9	171.5	95.6	124.2	168.0	96.7	123.5	166.5	66.6	104.2	161.1	64.4	100.8	172.9	95.0	124.3
05	METALS	127.9	91.7	116.8	110.2	90.0	104.0	109.1	89.7	103.1	110.0	76.7	99.7	106.3	73.7	96.3	109.8	87.6	102.9
06	WOOD, PLASTICS & COMPOSITES	109.4	90.6	98.9	119.8	84.3	99.9	123.7	86.3	102.8	121.1	73.5	94.4	102.7	59.1	78.3	105.2	82.1	92.3
07	THERMAL & MOISTURE PROTECTION	124.4	93.8	111.8	111.8	86.4	101.4	113.8	91.2	104.5	107.3	71.3	92.5	106.7	63.0	88.8	117.2	86.1	104.4
08	OPENINGS	93.9	88.7	92.7	93.9	83.1	91.4	92.4	85.5	90.8	93.6	64.8	86.9	88.2	56.8	80.9	90.2	81.8	88.2
0920	Plaster & Gypsum Board	168.3	90.2	115.5	151.1	83.7	105.6	130.1	85.8	100.1	114.2	72.2	85.8	110.7	57.8	74.9	113.6	81.0	91.6
0950, 0980	Ceilings & Acoustic Treatment	125.6	90.2	102.4	91.5	83.7	86.4	96.5	85.8	89.4	96.5	72.2	80.5	96.5	57.8	71.0	96.5	81.0	86.3
0960	Flooring	112.3	97.5	108.1	127.2	98.9	119.1	120.5	97.5	113.9	120.5	70.2	106.1	110.6	63.3	97.1	116.0	76.2	104.6
0970, 0990	Wall Finishes & Painting/Coating	123.2	94.2	105.7	116.2	89.9	100.3	118.3	93.5	103.3	118.4	58.7	82.4	118.3	57.0	81.3	118.3	84.0	97.6
09	FINISHES	122.6	92.9	106.2	115.0	88.6	100.5	112.4	90.5	100.3	110.5	70.8	88.6	107.1	60.3	81.3	110.6	83.4	95.6
COVERS	DIVS. 10 - 14, 25, 28, 41, 43, 44, 46	131.1	95.7	124.0	139.2	73.2	125.9	131.1	74.2	119.6	131.1	67.2	118.2	131.1	64.6	117.7	131.1	94.8	123.8
21, 22, 23	FIRE SUPPRESSION, PLUMBING & HVAC	101.4	87.2	95.7	101.8	98.1	100.3	102.0	101.7	101.9	102.0	83.6	94.5	102.3	68.6	88.7	102.0	94.4	98.9
26, 27, 3370	ELECTRICAL, COMMUNICATIONS & UTIL.	107.3	98.7	102.8	118.2	89.6	103.1	111.5	90.4	100.4	113.6	60.1	85.3	113.9	62.8	86.9	111.7	81.9	96.0
MF2014	WEIGHTED AVERAGE	117.3	94.0	107.2	117.8	91.2	106.2	114.9	93.1	105.4	113.4	74.2	96.3	109.9	67.7	91.5	115.4	89.9	104.3

City Cost Indexes

CANADA

DIVISION		QUEBEC, QUEBEC			RED DEER, ALBERTA			REGINA, SASKATCHEWAN			RIMOUSKI, QUEBEC			ROUYN-NORANDA, QUEBEC			SAINT HYACINTHE, QUEBEC		
		MAT.	INST.	TOTAL	MAT.	INST.	TOTAL	MAT.	INST.	TOTAL	MAT.	INST.	TOTAL	MAT.	INST.	TOTAL	MAT.	INST.	TOTAL
015433	CONTRACTOR EQUIPMENT		104.9	104.9		104.4	104.4		100.6	100.6		102.8	102.8		102.8	102.8		102.8	102.8
0241, 31 - 34	SITE & INFRASTRUCTURE, DEMOLITION	110.8	99.5	102.8	114.2	101.4	105.1	124.4	98.0	105.6	98.7	99.1	99.0	98.2	99.8	99.4	98.7	99.8	99.5
0310	Concrete Forming & Accessories	125.6	93.7	98.1	139.6	82.6	90.4	117.6	90.9	94.6	130.7	93.1	98.3	130.7	87.8	93.7	130.7	87.8	93.7
0320	Concrete Reinforcing	138.3	96.0	116.8	152.9	71.8	111.7	138.0	84.8	111.0	106.0	95.9	100.9	148.7	78.8	113.2	148.7	78.8	113.2
0330	Cast-in-Place Concrete	141.8	101.6	125.2	141.9	101.4	125.3	171.5	95.7	140.4	128.6	99.8	116.8	122.3	97.8	112.2	126.4	97.8	114.6
03	CONCRETE	131.0	96.7	114.1	135.3	87.5	111.8	148.9	91.4	120.6	121.1	95.9	108.7	124.8	89.8	107.6	126.7	89.8	108.5
04	MASONRY	163.0	95.1	120.7	163.4	82.5	113.0	183.5	96.7	129.4	163.6	95.0	120.9	163.7	87.1	116.0	164.0	87.1	116.1
05	METALS	128.0	93.0	117.2	110.0	88.7	103.4	129.3	86.4	116.1	106.9	92.2	102.4	107.4	87.4	101.2	107.4	87.4	101.2
06	WOOD, PLASTICS & COMPOSITES	115.8	93.9	103.6	121.3	82.0	99.3	101.9	91.9	96.3	135.3	93.6	112.0	135.3	88.0	108.8	135.3	88.0	108.8
07	THERMAL & MOISTURE PROTECTION	118.4	97.0	109.6	132.2	87.5	113.9	123.2	86.0	107.9	107.4	96.4	102.9	107.4	89.9	100.2	107.7	89.9	100.4
08	OPENINGS	93.0	90.1	92.3	93.6	77.9	89.9	91.9	80.7	89.3	93.2	80.6	90.2	93.6	76.3	89.6	93.6	76.3	89.6
0920	Plaster & Gypsum Board	142.1	93.2	109.0	117.2	81.0	92.8	130.5	91.6	104.2	156.8	93.2	113.8	126.4	87.4	100.0	128.7	87.4	100.8
0950, 0980	Ceilings & Acoustic Treatment	126.1	93.2	104.4	96.5	81.0	86.3	125.4	91.6	103.2	95.7	93.2	94.0	95.7	87.4	90.2	95.7	87.4	90.2
0960	Flooring	125.8	99.5	118.3	123.0	92.6	114.3	120.0	65.5	104.4	120.5	99.5	114.5	120.5	99.5	114.5	120.5	99.5	114.5
0970, 0990	Wall Finishes & Painting/Coating	127.8	106.4	114.9	118.2	81.2	95.9	123.9	92.9	105.2	118.3	106.4	111.1	118.3	91.7	102.2	118.3	91.7	102.2
09	FINISHES	122.2	96.8	108.2	111.7	84.0	96.4	122.7	87.2	103.1	115.1	96.5	104.9	111.0	90.8	99.9	111.3	90.8	100.0
COVERS	DIVS. 10 - 14, 25, 28, 41, 43, 44, 46	131.1	87.3	122.3	131.1	91.9	123.2	131.1	72.4	119.3	131.1	86.5	122.1	131.1	85.2	121.9	131.1	85.2	121.9
21, 22, 23	FIRE SUPPRESSION, PLUMBING & HVAC	101.2	92.4	97.7	102.0	92.3	98.1	100.7	84.8	94.3	102.0	92.3	98.1	102.0	91.0	97.6	98.8	91.0	95.7
26, 27, 3370	ELECTRICAL, COMMUNICATIONS & UTIL.	111.4	86.8	98.4	104.9	85.4	94.6	114.4	92.4	102.8	111.8	86.8	98.6	111.8	72.6	91.1	112.5	72.6	91.4
MF2014	WEIGHTED AVERAGE	117.1	93.6	106.8	114.1	88.2	102.8	120.7	89.0	106.9	111.8	92.9	103.6	112.0	87.4	101.3	111.6	87.4	101.1

CANADA

DIVISION		SAINT JOHN, NEW BRUNSWICK			SARNIA, ONTARIO			SASKATOON, SASKATCHEWAN			SAULT STE MARIE, ONTARIO			SHERBROOKE, QUEBEC			SOREL, QUEBEC		
		MAT.	INST.	TOTAL	MAT.	INST.	TOTAL	MAT.	INST.	TOTAL	MAT.	INST.	TOTAL	MAT.	INST.	TOTAL	MAT.	INST.	TOTAL
015433	CONTRACTOR EQUIPMENT		102.2	102.2		102.1	102.1		100.6	100.6		102.1	102.1		102.8	102.8		102.8	102.8
0241, 31 - 34	SITE & INFRASTRUCTURE, DEMOLITION	103.5	98.0	99.6	116.5	100.3	105.0	111.7	98.0	102.0	106.7	99.8	101.8	98.7	99.8	99.5	98.8	99.9	99.6
0310	Concrete Forming & Accessories	124.6	67.0	74.9	123.6	95.3	99.1	107.4	90.9	93.2	112.8	87.9	91.3	130.7	87.8	93.7	130.7	88.0	93.8
0320	Concrete Reinforcing	137.0	62.0	98.8	117.4	83.6	100.2	114.5	84.8	99.4	106.1	82.1	93.9	148.7	78.8	113.2	140.6	78.8	109.2
0330	Cast-in-Place Concrete	129.7	80.1	109.3	141.7	102.1	125.5	135.4	95.7	119.1	127.4	87.2	110.9	126.4	97.8	114.6	127.4	97.8	115.2
03	CONCRETE	125.2	71.4	98.8	128.6	95.4	112.3	119.6	91.4	105.7	113.1	86.8	100.2	126.7	89.8	108.5	125.9	89.8	108.2
04	MASONRY	180.8	73.7	114.0	179.1	99.5	129.5	168.8	96.7	123.8	164.5	94.4	120.8	164.1	87.1	116.1	164.1	87.1	116.1
05	METALS	108.9	81.7	100.5	109.1	90.1	103.2	104.5	86.4	99.0	108.3	90.2	102.7	107.4	87.4	101.2	107.4	87.6	101.3
06	WOOD, PLASTICS & COMPOSITES	125.7	65.4	91.9	122.6	94.7	107.0	100.5	91.9	95.7	109.6	88.3	97.6	135.3	88.0	108.8	135.3	88.0	108.8
07	THERMAL & MOISTURE PROTECTION	110.5	70.9	94.3	113.9	94.7	106.0	107.1	86.0	98.4	112.7	88.4	102.7	107.4	89.9	100.3	107.4	89.9	100.2
08	OPENINGS	87.8	60.6	81.5	94.7	89.4	93.4	88.9	80.7	87.0	86.5	85.6	86.3	93.6	76.3	89.6	93.6	81.1	90.7
0920	Plaster & Gypsum Board	152.7	64.2	92.9	150.6	94.4	112.6	123.5	91.6	101.9	119.6	87.8	98.1	128.7	87.4	100.8	156.8	87.4	109.9
0950, 0980	Ceilings & Acoustic Treatment	111.5	64.2	80.4	100.7	94.4	96.6	117.4	91.6	100.4	96.5	87.8	90.8	95.7	87.4	90.2	95.7	87.4	90.2
0960	Flooring	119.5	72.1	105.9	120.5	106.6	116.5	111.0	65.5	98.0	113.4	100.8	109.8	120.5	99.5	114.5	120.5	99.5	114.5
0970, 0990	Wall Finishes & Painting/Coating	117.6	80.2	95.0	118.3	105.7	110.7	120.7	92.9	103.9	118.3	97.9	106.0	118.3	91.7	102.2	118.3	91.7	102.2
09	FINISHES	117.9	69.0	91.0	116.1	98.9	106.6	114.3	87.2	99.3	108.1	91.5	99.0	111.3	90.8	100.0	115.1	90.8	101.7
COVERS	DIVS. 10 - 14, 25, 28, 41, 43, 44, 46	131.1	66.3	118.0	131.1	75.7	119.9	131.1	74.1	119.6	131.1	96.4	124.1	131.1	85.2	121.9	131.1	85.2	121.9
21, 22, 23	FIRE SUPPRESSION, PLUMBING & HVAC	102.0	79.8	93.0	102.0	107.8	104.3	100.5	84.8	94.2	102.0	94.8	99.1	102.3	91.0	97.8	102.0	91.0	97.6
26, 27, 3370	ELECTRICAL, COMMUNICATIONS & UTIL.	120.3	88.2	103.3	114.2	92.7	102.8	116.2	92.4	103.6	113.0	90.8	101.3	111.8	72.6	91.1	111.8	72.6	91.1
MF2014	WEIGHTED AVERAGE	114.0	77.9	98.3	114.8	97.5	107.3	111.1	89.1	101.5	110.3	92.1	102.3	112.4	87.4	101.5	112.5	87.6	101.7

CANADA

DIVISION		ST CATHARINES, ONTARIO			ST JEROME, QUEBEC			ST JOHNS, NEWFOUNDLAND			SUDBURY, ONTARIO			SUMMERSIDE, PRINCE EDWARD ISLAND			SYDNEY, NOVA SCOTIA		
		MAT.	INST.	TOTAL	MAT.	INST.	TOTAL	MAT.	INST.	TOTAL	MAT.	INST.	TOTAL	MAT.	INST.	TOTAL	MAT.	INST.	TOTAL
015433	CONTRACTOR EQUIPMENT		102.1	102.1		102.8	102.8		103.3	103.3		102.1	102.1		101.8	101.8		101.9	101.9
0241, 31 - 34	SITE & INFRASTRUCTURE, DEMOLITION	100.0	101.4	101.0	98.2	99.8	99.4	123.4	99.9	106.7	100.1	100.9	100.7	128.4	95.7	105.2	114.3	98.2	102.8
0310	Concrete Forming & Accessories	110.9	94.2	96.5	130.7	87.8	93.7	123.9	84.8	90.1	107.0	89.0	91.5	112.0	58.3	65.7	111.7	71.6	77.1
0320	Concrete Reinforcing	101.7	92.0	96.7	148.7	78.8	113.2	158.9	75.5	116.4	102.5	91.3	96.8	168.1	46.8	106.4	170.3	46.7	107.4
0330	Cast-in-Place Concrete	123.2	100.6	113.9	122.3	97.8	112.2	167.5	96.4	138.3	124.2	97.0	113.0	142.3	61.2	109.0	115.6	72.2	97.8
03	CONCRETE	108.0	95.9	102.1	124.8	89.8	107.6	152.6	87.2	120.5	108.4	92.2	100.4	170.1	57.9	115.0	144.7	67.9	106.9
04	MASONRY	150.3	100.7	119.4	163.7	87.1	116.0	187.7	92.4	128.3	150.4	96.6	116.8	228.1	62.6	125.0	226.2	71.9	130.1
05	METALS	107.7	92.2	102.9	107.4	87.4	101.2	136.3	84.3	120.3	107.7	91.5	102.7	123.6	67.6	106.4	123.6	74.2	108.4
06	WOOD, PLASTICS & COMPOSITES	107.4	93.9	99.8	135.3	88.0	108.8	114.7	82.8	96.8	103.6	88.3	95.0	116.4	57.8	83.6	116.1	71.1	90.9
07	THERMAL & MOISTURE PROTECTION	108.6	96.5	103.6	107.4	89.9	100.2	127.1	91.4	112.5	108.0	91.7	101.3	136.1	61.1	105.4	136.8	69.9	109.4
08	OPENINGS	84.1	90.2	85.5	93.6	76.3	89.6	93.7	74.5	89.2	84.8	85.6	85.0	108.0	50.0	94.5	95.3	64.9	88.2
0920	Plaster & Gypsum Board	108.5	93.6	98.4	126.4	87.4	100.0	143.9	82.1	102.1	112.9	87.8	95.9	143.1	56.3	84.4	142.7	70.0	93.6
0950, 0980	Ceilings & Acoustic Treatment	95.3	93.6	94.1	95.7	87.4	90.2	120.6	82.1	95.3	90.3	87.8	88.6	109.6	56.3	74.6	109.6	70.0	83.6
0960	Flooring	109.3	98.9	106.3	120.5	99.5	114.5	116.0	59.3	99.8	107.3	100.8	105.5	116.2	62.9	100.9	116.1	67.2	102.1
0970, 0990	Wall Finishes & Painting/Coating	115.3	116.0	115.7	118.3	91.7	102.2	120.4	95.9	105.6	115.3	93.6	102.2	126.4	42.9	76.0	126.4	64.1	88.8
09	FINISHES	103.8	97.8	100.4	111.0	90.8	99.9	122.0	81.4	99.6	102.7	91.6	96.5	119.2	57.8	85.4	118.2	70.3	91.8
COVERS	DIVS. 10 - 14, 25, 28, 41, 43, 44, 46	131.1	75.2	119.8	131.1	85.2	121.9	131.1	72.6	119.3	131.1	97.1	124.3	131.1	64.8	117.7	131.1	66.6	118.1
21, 22, 23	FIRE SUPPRESSION, PLUMBING & HVAC	100.4	87.8	95.3	102.0	91.0	97.6	100.8	82.1	93.3	101.9	87.8	96.2	102.1	63.8	86.7	102.1	84.4	95.0
26, 27, 3370	ELECTRICAL, COMMUNICATIONS & UTIL.	112.9	98.8	105.5	112.4	72.6	91.3	114.4	83.2	97.9	110.3	99.2	104.4	115.4	52.3	82.0	116.3	64.4	88.9
MF2014	WEIGHTED AVERAGE	107.6	94.5	101.9	112.1	87.4	101.3	122.7	85.2	106.4	107.7	93.0	101.3	126.7	62.8	98.8	122.1	74.6	101.4

For customer support on your Plumbing Cost Data, call 877.709.6968.

643

City Cost Indexes

			CANADA															
	DIVISION	THUNDER BAY, ONTARIO			TIMMINS, ONTARIO			TORONTO, ONTARIO			TROIS RIVIERES, QUEBEC			TRURO, NOVA SCOTIA			VANCOUVER, BRITISH COLUMBIA		
		MAT.	INST.	TOTAL	MAT.	INST.	TOTAL	MAT.	INST.	TOTAL	MAT.	INST.	TOTAL	MAT.	INST.	TOTAL	MAT.	INST.	TOTAL
015433	CONTRACTOR EQUIPMENT		102.1	102.1		102.1	102.1		104.4	104.4		102.8	102.8		101.9	101.9		112.7	112.7
0241, 31 - 34	SITE & INFRASTRUCTURE, DEMOLITION	104.8	101.3	102.3	118.0	99.8	105.1	133.8	105.6	113.7	114.8	99.9	104.2	102.8	98.2	99.5	118.2	107.6	110.6
0310	Concrete Forming & Accessories	118.3	92.2	95.8	124.9	87.0	92.2	124.3	100.8	104.0	155.5	88.0	97.3	97.1	71.6	75.1	121.1	91.0	95.1
0320	Concrete Reinforcing	90.9	90.8	90.9	165.4	81.8	122.9	146.0	94.4	119.7	170.3	78.8	123.8	140.6	46.7	92.8	134.1	79.2	106.2
0330	Cast-in-Place Concrete	135.6	99.7	120.8	153.6	87.1	126.3	136.8	111.8	126.5	119.7	97.8	110.7	159.5	72.2	123.7	143.5	97.9	124.8
03	CONCRETE	115.8	94.5	105.3	141.8	86.3	114.5	129.7	103.1	116.7	147.4	89.8	119.1	140.2	67.9	104.7	137.1	91.4	114.6
04	MASONRY	151.0	100.6	119.6	168.0	89.9	119.3	182.0	110.9	137.7	230.7	87.1	141.2	163.6	71.9	106.5	166.7	91.4	119.7
05	METALS	107.5	91.1	102.5	109.1	89.2	103.0	132.5	93.9	120.7	122.7	87.6	111.9	108.2	74.2	97.7	139.3	90.3	124.2
06	WOOD, PLASTICS & COMPOSITES	116.7	90.9	102.2	123.7	87.6	103.5	115.1	98.9	106.0	173.0	88.0	125.4	92.9	71.1	80.7	104.8	91.5	97.4
07	THERMAL & MOISTURE PROTECTION	108.8	93.8	102.7	113.8	85.6	102.2	123.9	102.8	115.2	135.2	89.9	116.6	108.6	69.9	92.8	125.2	86.9	109.5
08	OPENINGS	83.3	88.0	84.4	92.4	83.7	90.4	90.5	96.8	92.0	105.5	81.1	99.8	86.1	64.9	81.1	88.9	86.9	88.4
0920	Plaster & Gypsum Board	139.8	90.5	106.5	130.1	87.1	101.1	133.0	98.8	109.9	173.4	87.4	115.3	130.7	70.0	89.7	129.7	90.6	103.3
0950, 0980	Ceilings & Acoustic Treatment	90.3	90.5	90.4	96.5	87.1	90.3	114.9	98.8	104.3	108.8	87.4	94.7	96.5	70.0	79.1	132.2	90.6	104.8
0960	Flooring	113.7	105.9	111.4	120.5	97.5	113.9	117.0	108.4	114.6	138.6	99.5	127.5	102.7	67.2	92.5	124.6	96.0	116.4
0970, 0990	Wall Finishes & Painting/Coating	115.3	94.6	102.8	118.3	91.2	101.9	120.4	106.7	112.1	126.4	91.7	105.4	118.3	64.1	85.6	121.7	95.6	106.0
09	FINISHES	108.5	95.1	101.1	112.4	89.8	99.9	116.9	103.0	109.2	128.7	90.8	107.8	106.6	70.3	86.6	123.8	92.7	106.6
COVERS	DIVS. 10 - 14, 25, 28, 41, 43, 44, 46	131.1	75.2	119.8	131.1	72.7	119.3	131.1	101.2	125.1	131.1	85.2	121.9	131.1	66.6	118.1	131.1	94.9	123.8
21, 22, 23	FIRE SUPPRESSION, PLUMBING & HVAC	100.4	88.0	95.4	102.0	98.1	100.4	101.0	97.3	99.5	102.1	91.0	97.7	102.0	84.4	94.9	101.2	80.5	92.8
26, 27, 3370	ELECTRICAL, COMMUNICATIONS & UTIL.	111.5	97.0	103.8	113.0	90.8	101.3	108.8	101.0	104.7	116.6	72.6	93.4	111.5	64.4	86.6	109.0	83.3	95.4
MF2014	WEIGHTED AVERAGE	108.8	93.4	102.1	115.1	91.1	104.6	118.3	101.3	110.9	124.8	87.6	108.6	112.5	74.6	96.0	119.5	89.1	106.2

			CANADA															
	DIVISION	VICTORIA, BRITISH COLUMBIA			WHITEHORSE, YUKON			WINDSOR, ONTARIO			WINNIPEG, MANITOBA			YARMOUTH, NOVA SCOTIA			YELLOWKNIFE, NWT		
		MAT.	INST.	TOTAL	MAT.	INST.	TOTAL	MAT.	INST.	TOTAL	MAT.	INST.	TOTAL	MAT.	INST.	TOTAL	MAT.	INST.	TOTAL
015433	CONTRACTOR EQUIPMENT		109.3	109.3		102.1	102.1		102.1	102.1		106.5	106.5		101.9	101.9		101.9	101.9
0241, 31 - 34	SITE & INFRASTRUCTURE, DEMOLITION	123.0	107.1	111.7	142.3	96.7	109.9	95.9	101.3	99.7	118.2	100.9	105.9	118.3	98.2	104.0	154.1	101.0	116.4
0310	Concrete Forming & Accessories	112.7	90.4	93.5	129.4	59.6	69.2	118.3	90.5	94.3	128.5	69.3	77.4	111.7	71.6	77.1	132.7	80.2	87.4
0320	Concrete Reinforcing	112.0	79.1	95.2	160.8	60.0	109.5	99.6	91.9	95.7	132.9	57.9	94.7	170.3	46.7	107.4	151.4	62.7	106.2
0330	Cast-in-Place Concrete	140.3	97.2	122.6	209.3	70.3	152.2	126.2	101.9	116.2	163.9	75.0	127.4	148.3	72.2	117.0	211.2	88.1	160.6
03	CONCRETE	150.1	90.7	120.9	182.4	64.1	124.3	109.6	94.7	102.3	144.4	69.7	107.7	160.1	67.9	114.8	182.0	80.0	131.9
04	MASONRY	171.7	91.3	121.6	260.3	62.8	137.2	150.5	99.9	119.0	176.6	70.8	110.7	228.8	71.9	131.0	260.3	74.8	144.7
05	METALS	107.0	86.3	100.6	135.3	73.6	116.3	107.6	91.7	102.7	141.7	75.0	121.2	123.6	74.2	108.4	137.9	78.7	119.7
06	WOOD, PLASTICS & COMPOSITES	104.4	91.4	97.1	117.6	58.3	84.4	116.7	89.0	101.2	113.6	70.2	89.3	116.1	71.1	90.9	124.0	81.9	100.4
07	THERMAL & MOISTURE PROTECTION	116.9	86.7	104.5	139.1	62.0	107.5	108.6	93.9	102.6	120.8	71.8	100.7	136.8	69.9	109.4	151.5	78.4	121.5
08	OPENINGS	90.5	83.2	88.8	107.6	55.5	95.4	83.1	87.5	84.1	85.5	62.6	80.2	95.3	64.9	88.2	102.2	68.9	94.4
0920	Plaster & Gypsum Board	115.8	90.6	98.8	150.5	56.8	87.2	125.7	88.6	100.6	129.5	68.6	88.3	142.7	70.0	93.6	163.7	81.1	107.9
0950, 0980	Ceilings & Acoustic Treatment	98.1	90.6	93.2	144.6	56.8	86.9	90.3	88.6	89.2	136.2	68.6	91.8	109.6	70.0	83.6	138.9	81.1	100.9
0960	Flooring	116.4	76.2	104.9	141.0	61.3	118.2	113.7	103.1	110.6	117.0	75.6	105.2	116.1	67.2	102.1	135.5	92.5	123.2
0970, 0990	Wall Finishes & Painting/Coating	120.7	95.6	105.5	143.1	56.0	90.5	115.3	95.5	103.3	124.5	56.5	83.4	126.4	64.1	88.8	135.3	80.3	102.1
09	FINISHES	112.1	89.3	99.5	140.9	59.3	95.9	106.3	93.4	99.2	122.3	69.5	93.2	118.2	70.3	91.8	137.5	82.0	106.9
COVERS	DIVS. 10 - 14, 25, 28, 41, 43, 44, 46	131.1	71.5	119.1	131.1	63.9	117.6	131.1	74.7	119.7	131.1	68.7	118.5	131.1	66.6	118.1	131.1	67.1	118.2
21, 22, 23	FIRE SUPPRESSION, PLUMBING & HVAC	102.0	80.5	93.3	102.5	74.8	91.3	100.4	88.0	95.4	101.1	66.9	87.3	102.1	84.4	95.0	102.7	94.0	99.2
26, 27, 3370	ELECTRICAL, COMMUNICATIONS & UTIL.	113.1	82.7	97.0	133.5	62.2	95.8	116.0	99.0	107.0	113.3	68.0	89.4	116.3	64.4	88.9	130.5	85.2	106.5
MF2014	WEIGHTED AVERAGE	115.7	87.2	103.3	135.3	68.5	106.2	108.2	93.4	101.7	121.0	71.8	99.5	124.0	74.6	102.5	135.3	84.2	113.0

For customer support on your Plumbing Cost Data, call 877.709.6968.

Location Factors

Costs shown in RSMeans cost data publications are based on national averages for materials and installation. To adjust these costs to a specific location, simply multiply the base cost by the factor and divide by 100 for that city. The data is arranged alphabetically by state and postal zip code numbers. For a city not listed, use the factor for a nearby city with similar economic characteristics.

STATE/ZIP	CITY	MAT.	INST.	TOTAL
ALABAMA				
350-352	Birmingham	100.6	76.7	90.2
354	Tuscaloosa	99.8	61.2	83.0
355	Jasper	100.2	57.5	81.6
356	Decatur	99.8	57.9	81.5
357-358	Huntsville	99.8	72.1	87.8
359	Gadsden	99.9	60.0	82.5
360-361	Montgomery	99.9	57.1	81.2
362	Anniston	98.6	61.8	82.6
363	Dothan	99.0	49.4	77.4
364	Evergreen	98.6	50.7	77.7
365-366	Mobile	99.9	65.4	84.9
367	Selma	98.7	49.9	77.4
368	Phenix City	99.5	54.6	79.9
369	Butler	98.9	49.4	77.3
ALASKA				
995-996	Anchorage	121.1	115.6	118.7
997	Fairbanks	121.0	116.4	119.0
998	Juneau	121.3	115.6	118.8
999	Ketchikan	132.0	115.6	124.8
ARIZONA				
850,853	Phoenix	99.2	73.8	88.1
851,852	Mesa/Tempe	98.6	72.8	87.4
855	Globe	98.8	70.0	86.3
856-857	Tucson	97.4	72.4	86.5
859	Show Low	98.9	72.5	87.4
860	Flagstaff	101.3	71.8	88.4
863	Prescott	99.2	68.9	86.0
864	Kingman	97.6	72.0	86.4
865	Chambers	97.6	71.5	86.2
ARKANSAS				
716	Pine Bluff	99.4	63.3	83.6
717	Camden	97.1	50.1	76.6
718	Texarkana	98.2	49.2	76.8
719	Hot Springs	96.4	51.8	76.9
720-722	Little Rock	98.5	64.0	83.4
723	West Memphis	96.1	62.4	81.4
724	Jonesboro	96.7	59.1	80.3
725	Batesville	94.6	54.6	77.1
726	Harrison	95.9	50.4	76.0
727	Fayetteville	93.4	52.0	75.4
728	Russellville	94.6	53.7	76.8
729	Fort Smith	97.2	61.6	81.7
CALIFORNIA				
900-902	Los Angeles	99.2	117.5	107.2
903-905	Inglewood	94.8	114.1	103.2
906-908	Long Beach	96.2	114.1	104.0
910-912	Pasadena	96.4	114.6	104.3
913-916	Van Nuys	99.1	114.6	105.9
917-918	Alhambra	98.3	114.8	105.5
919-921	San Diego	100.5	109.6	104.5
922	Palm Springs	97.2	112.0	103.6
923-924	San Bernardino	95.1	111.3	102.1
925	Riverside	99.4	114.6	106.0
926-927	Santa Ana	96.9	111.8	103.4
928	Anaheim	99.4	114.9	106.2
930	Oxnard	99.6	114.4	106.0
931	Santa Barbara	98.9	115.1	106.0
932-933	Bakersfield	100.7	113.6	106.3
934	San Luis Obispo	99.4	112.0	104.9
935	Mojave	96.8	109.8	102.5
936-938	Fresno	100.1	115.3	106.7
939	Salinas	100.2	121.6	109.5
940-941	San Francisco	105.2	145.4	122.7
942,956-958	Sacramento	100.5	119.3	108.7
943	Palo Alto	98.9	134.5	114.4
944	San Mateo	100.9	134.6	115.6
945	Vallejo	100.1	127.2	111.9
946	Oakland	103.1	134.3	116.7
947	Berkeley	102.7	135.1	116.8
948	Richmond	101.8	133.1	115.4
949	San Rafael	103.5	136.7	118.0
950	Santa Cruz	104.5	121.8	112.0

STATE/ZIP	CITY	MAT.	INST.	TOTAL
CALIFORNIA (CONT'D)				
951	San Jose	102.5	136.6	117.4
952	Stockton	100.9	117.2	108.0
953	Modesto	100.8	115.7	107.3
954	Santa Rosa	100.8	134.7	115.6
955	Eureka	102.1	116.4	108.4
959	Marysville	101.3	115.9	107.7
960	Redding	107.8	117.7	112.1
961	Susanville	106.8	118.1	111.7
COLORADO				
800-802	Denver	100.8	81.7	92.5
803	Boulder	97.8	81.3	90.6
804	Golden	99.8	78.9	90.7
805	Fort Collins	101.3	79.0	91.6
806	Greeley	98.9	75.5	88.7
807	Fort Morgan	98.3	79.0	89.9
808-809	Colorado Springs	100.4	84.0	93.2
810	Pueblo	101.1	80.3	92.0
811	Alamosa	102.2	73.7	89.8
812	Salida	101.9	75.3	90.3
813	Durango	102.7	75.7	90.9
814	Montrose	101.4	75.4	90.1
815	Grand Junction	104.8	75.3	91.9
816	Glenwood Springs	102.3	78.1	91.8
CONNECTICUT				
060	New Britain	100.1	120.6	109.1
061	Hartford	101.4	120.8	109.8
062	Willimantic	100.8	120.2	109.2
063	New London	97.1	120.7	107.4
064	Meriden	98.9	120.7	108.4
065	New Haven	101.6	120.7	109.9
066	Bridgeport	101.3	120.8	109.8
067	Waterbury	100.9	120.7	109.5
068	Norwalk	100.8	128.7	113.0
069	Stamford	100.9	128.7	113.0
D.C.				
200-205	Washington	101.1	91.9	97.1
DELAWARE				
197	Newark	98.8	109.2	103.3
198	Wilmington	99.3	109.2	103.6
199	Dover	99.3	109.2	103.6
FLORIDA				
320,322	Jacksonville	98.8	66.4	84.7
321	Daytona Beach	99.0	72.9	87.6
323	Tallahassee	100.8	56.9	81.7
324	Panama City	100.2	57.8	81.7
325	Pensacola	102.7	61.2	84.6
326,344	Gainesville	100.3	66.7	85.7
327-328,347	Orlando	100.5	70.0	87.2
329	Melbourne	101.7	76.3	90.6
330-332,340	Miami	99.8	73.1	88.2
333	Fort Lauderdale	98.6	72.6	87.2
334,349	West Palm Beach	97.4	71.1	85.9
335-336,346	Tampa	100.1	78.4	90.6
337	St. Petersburg	102.6	63.3	85.5
338	Lakeland	99.4	77.8	90.0
339,341	Fort Myers	98.6	72.6	87.3
342	Sarasota	100.8	74.3	89.2
GEORGIA				
300-303,399	Atlanta	97.6	74.4	87.5
304	Statesboro	97.3	55.8	79.2
305	Gainesville	96.0	61.6	81.0
306	Athens	95.5	63.9	81.7
307	Dalton	97.2	59.0	80.5
308-309	Augusta	96.3	63.6	82.1
310-312	Macon	97.2	64.7	83.0
313-314	Savannah	99.2	60.9	82.5
315	Waycross	98.3	59.4	81.4
316	Valdosta	98.5	61.4	82.3
317,398	Albany	98.5	61.3	82.3
318-319	Columbus	98.4	65.4	84.0

Location Factors

STATE/ZIP	CITY	MAT.	INST.	TOTAL	STATE/ZIP	CITY	MAT.	INST.	TOTAL
HAWAII					**KANSAS (CONT'D)**				
967	Hilo	114.8	120.2	117.2	678	Dodge City	102.7	66.0	86.7
968	Honolulu	119.5	120.2	119.8	679	Liberal	100.4	65.5	85.2
STATES & POSS.					**KENTUCKY**				
969	Guam	136.2	58.0	102.1	400-402	Louisville	96.6	83.3	90.8
					403-405	Lexington	96.6	81.4	90.0
IDAHO					406	Frankfort	98.1	80.7	90.5
832	Pocatello	101.1	77.4	90.8	407-409	Corbin	94.0	71.8	84.3
833	Twin Falls	102.2	67.3	87.0	410	Covington	95.9	91.4	93.9
834	Idaho Falls	99.8	74.7	88.9	411-412	Ashland	94.6	99.3	96.6
835	Lewiston	108.7	84.9	98.3	413-414	Campton	95.8	78.6	88.3
836-837	Boise	100.1	78.6	90.7	415-416	Pikeville	96.9	88.6	93.3
838	Coeur d'Alene	107.9	82.2	96.7	417-418	Hazard	95.2	78.7	88.0
					420	Paducah	93.9	84.5	89.8
ILLINOIS					421-422	Bowling Green	96.2	83.9	90.8
600-603	North Suburban	99.3	135.1	114.9	423	Owensboro	96.3	88.9	93.1
604	Joliet	99.2	138.4	116.3	424	Henderson	93.7	86.1	90.4
605	South Suburban	99.3	135.1	114.9	425-426	Somerset	93.2	79.9	87.4
606-608	Chicago	99.8	139.7	117.2	427	Elizabethtown	92.8	82.0	88.1
609	Kankakee	95.6	131.7	111.3					
610-611	Rockford	99.5	124.8	110.5	**LOUISIANA**				
612	Rock Island	97.2	100.5	98.6	700-701	New Orleans	101.4	69.5	87.5
613	La Salle	98.4	124.4	109.7	703	Thibodaux	98.2	65.4	83.9
614	Galesburg	98.2	106.7	101.9	704	Hammond	95.8	54.5	77.8
615-616	Peoria	100.5	109.3	104.4	705	Lafayette	97.9	63.0	82.7
617	Bloomington	97.5	109.4	102.7	706	Lake Charles	98.1	65.0	83.7
618-619	Champaign	101.0	107.6	103.9	707-708	Baton Rouge	99.1	63.7	83.7
620-622	East St. Louis	96.4	108.7	101.8	710-711	Shreveport	101.0	59.6	83.0
623	Quincy	98.1	101.0	99.3	712	Monroe	99.3	56.5	80.6
624	Effingham	97.3	106.0	101.1	713-714	Alexandria	99.4	57.2	81.0
625	Decatur	99.2	104.9	101.7					
626-627	Springfield	100.0	106.5	102.8	**MAINE**				
628	Centralia	95.0	108.5	100.9	039	Kittery	95.2	84.7	90.6
629	Carbondale	94.7	107.9	100.5	040-041	Portland	102.0	90.1	96.8
					042	Lewiston	99.6	90.1	95.5
INDIANA					043	Augusta	102.1	79.5	92.3
460	Anderson	97.0	83.8	91.3	044	Bangor	99.0	89.5	94.9
461-462	Indianapolis	99.1	84.8	92.9	045	Bath	97.5	81.9	90.7
463-464	Gary	98.3	110.4	103.6	046	Machias	97.1	85.5	92.0
465-466	South Bend	98.0	84.4	92.1	047	Houlton	97.2	85.6	92.1
467-468	Fort Wayne	97.6	79.1	89.5	048	Rockland	96.3	85.5	91.6
469	Kokomo	94.8	81.5	89.0	049	Waterville	97.5	79.5	89.7
470	Lawrenceburg	94.0	77.7	86.9					
471	New Albany	95.1	75.4	86.5	**MARYLAND**				
472	Columbus	97.2	80.9	90.1	206	Waldorf	97.1	81.7	90.4
473	Muncie	97.9	82.3	91.1	207-208	College Park	97.2	83.4	91.2
474	Bloomington	99.3	81.8	91.6	209	Silver Spring	96.5	83.2	90.7
475	Washington	95.8	81.8	89.7	210-212	Baltimore	100.8	81.9	92.6
476-477	Evansville	97.1	86.9	92.6	214	Annapolis	101.5	82.0	93.0
478	Terre Haute	97.9	86.5	92.9	215	Cumberland	96.8	81.8	90.2
479	Lafayette	96.9	82.7	90.7	216	Easton	98.3	68.6	85.3
					217	Hagerstown	97.6	83.3	91.4
IOWA					218	Salisbury	98.7	62.1	82.7
500-503,509	Des Moines	100.5	85.2	93.8	219	Elkton	95.9	81.9	89.8
504	Mason City	97.6	65.8	83.8					
505	Fort Dodge	97.9	58.2	80.6	**MASSACHUSETTS**				
506-507	Waterloo	99.5	71.8	87.4	010-011	Springfield	100.2	112.6	105.6
508	Creston	98.2	77.3	89.1	012	Pittsfield	99.7	111.3	104.8
510-511	Sioux City	99.7	72.6	87.9	013	Greenfield	97.7	112.3	104.1
512	Sibley	98.3	56.4	80.0	014	Fitchburg	96.5	126.5	109.6
513	Spencer	99.9	56.6	81.0	015-016	Worcester	100.2	126.4	111.6
514	Carroll	97.0	71.2	85.8	017	Framingham	95.7	135.1	112.9
515	Council Bluffs	100.7	78.1	90.9	018	Lowell	99.5	134.5	114.8
516	Shenandoah	97.6	71.2	86.1	019	Lawrence	100.3	134.6	115.3
520	Dubuque	99.1	80.2	90.9	020-022, 024	Boston	101.6	139.5	118.1
521	Decorah	98.0	64.2	83.3	023	Brockton	99.9	128.1	112.2
522-524	Cedar Rapids	100.0	83.2	92.7	025	Buzzards Bay	94.4	128.1	109.1
525	Ottumwa	97.9	73.7	87.3	026	Hyannis	97.0	128.1	110.6
526	Burlington	97.2	75.3	87.6	027	New Bedford	99.1	128.3	111.8
527-528	Davenport	99.0	93.1	96.4					
					MICHIGAN				
KANSAS					480,483	Royal Oak	94.9	106.9	100.1
660-662	Kansas City	99.5	94.8	97.5	481	Ann Arbor	97.1	107.5	101.7
664-666	Topeka	99.5	66.0	84.9	482	Detroit	97.8	109.6	102.9
667	Fort Scott	97.9	73.0	87.0	484-485	Flint	96.8	96.0	96.4
668	Emporia	97.9	69.2	85.4	486	Saginaw	96.4	91.8	94.4
669	Belleville	99.6	64.9	84.4	487	Bay City	96.6	91.7	94.5
670-672	Wichita	100.7	66.0	85.6	488-489	Lansing	99.3	94.7	97.3
673	Independence	100.9	69.2	87.1	490	Battle Creek	98.3	87.6	93.7
674	Salina	101.5	67.6	86.7	491	Kalamazoo	98.7	85.7	93.0
675	Hutchinson	96.5	65.0	82.8	492	Jackson	96.5	95.1	95.9
676	Hays	100.5	65.4	85.2	493,495	Grand Rapids	99.7	85.1	93.3
677	Colby	101.1	65.0	85.4	494	Muskegon	97.2	84.8	91.8

For customer support on your Plumbing Cost Data, call 877.709.6968.

Location Factors

STATE/ZIP	CITY	MAT.	INST.	TOTAL
MICHIGAN (CONT'D)				
496	Traverse City	96.3	80.1	89.3
497	Gaylord	97.2	83.6	91.3
498-499	Iron mountain	99.3	86.2	93.6
MINNESOTA				
550-551	Saint Paul	101.4	117.9	108.6
553-555	Minneapolis	100.4	120.0	108.9
556-558	Duluth	101.3	107.3	103.9
559	Rochester	99.9	104.7	102.0
560	Mankato	97.2	101.3	99.0
561	Windom	95.9	92.3	94.3
562	Willmar	95.5	101.3	98.0
563	St. Cloud	96.9	116.0	105.2
564	Brainerd	96.9	100.5	98.5
565	Detroit Lakes	98.9	95.1	97.3
566	Bemidji	98.2	98.2	98.2
567	Thief River Falls	97.8	92.4	95.4
MISSISSIPPI				
386	Clarksdale	97.3	52.6	77.8
387	Greenville	101.0	61.7	83.9
388	Tupelo	98.8	54.9	79.7
389	Greenwood	98.4	52.2	78.3
390-392	Jackson	100.7	62.2	83.9
393	Meridian	97.6	64.1	83.0
394	Laurel	98.9	55.8	80.1
395	Biloxi	99.2	58.1	81.3
396	Mccomb	97.4	51.7	77.5
397	Columbus	98.8	53.9	79.2
MISSOURI				
630-631	St. Louis	98.7	107.8	102.7
633	Bowling Green	97.3	94.4	96.0
634	Hannibal	96.2	92.4	94.6
635	Kirksville	99.5	89.9	95.3
636	Flat River	98.1	95.0	96.8
637	Cape Girardeau	98.2	92.4	95.7
638	Sikeston	96.7	92.2	94.8
639	Poplar Bluff	96.2	92.0	94.4
640-641	Kansas City	101.4	105.7	103.3
644-645	St. Joseph	100.6	93.5	97.5
646	Chillicothe	97.6	94.5	96.3
647	Harrisonville	97.1	101.5	99.1
648	Joplin	99.5	81.7	91.8
650-651	Jefferson City	99.9	93.7	97.2
652	Columbia	100.2	93.9	97.5
653	Sedalia	100.2	97.3	98.9
654-655	Rolla	97.9	94.6	96.5
656-658	Springfield	101.7	83.7	93.8
MONTANA				
590-591	Billings	102.7	77.1	91.5
592	Wolf Point	102.3	74.2	90.0
593	Miles City	100.1	74.0	88.7
594	Great Falls	104.1	76.2	91.9
595	Havre	101.2	73.1	89.0
596	Helena	102.4	73.8	89.9
597	Butte	102.6	77.3	91.6
598	Missoula	99.9	75.5	89.3
599	Kalispell	99.0	75.3	88.7
NEBRASKA	.			
680-681	Omaha	100.2	78.6	90.8
683-685	Lincoln	100.5	74.9	89.4
686	Columbus	98.4	77.1	89.1
687	Norfolk	100.0	76.7	89.8
688	Grand Island	100.0	77.1	90.0
689	Hastings	99.4	79.8	90.9
690	Mccook	98.7	71.1	86.7
691	North Platte	99.0	78.3	90.0
692	Valentine	101.2	72.5	88.7
693	Alliance	101.0	70.5	87.7
NEVADA				
889-891	Las Vegas	100.9	108.7	104.3
893	Ely	99.4	102.2	100.6
894-895	Reno	99.6	93.3	96.8
897	Carson City	99.7	93.5	97.0
898	Elko	98.3	84.9	92.4
NEW HAMPSHIRE				
030	Nashua	100.5	93.8	97.6
031	Manchester	101.4	93.8	98.1

STATE/ZIP	CITY	MAT.	INST.	TOTAL
NEW HAMPSHIRE (CONT'D)				
032-033	Concord	101.1	92.5	97.4
034	Keene	97.2	73.3	86.7
035	Littleton	97.1	76.3	88.0
036	Charleston	96.7	69.5	84.8
037	Claremont	95.9	69.5	84.4
038	Portsmouth	97.9	94.3	96.4
NEW JERSEY				
070-071	Newark	101.2	128.4	113.0
072	Elizabeth	98.8	127.6	111.3
073	Jersey City	97.9	127.9	111.0
074-075	Paterson	99.6	127.8	111.9
076	Hackensack	97.6	128.1	110.9
077	Long Branch	97.1	126.3	109.8
078	Dover	97.9	127.8	110.9
079	Summit	97.9	127.6	110.8
080,083	Vineland	97.5	124.7	109.4
081	Camden	99.5	124.8	110.5
082,084	Atlantic City	98.2	124.8	109.8
085-086	Trenton	100.9	125.1	111.5
087	Point Pleasant	99.2	125.9	110.8
088-089	New Brunswick	99.7	127.6	111.9
NEW MEXICO				
870-872	Albuquerque	99.6	72.6	87.8
873	Gallup	99.5	72.6	87.8
874	Farmington	100.1	72.6	88.1
875	Santa Fe	100.8	72.6	88.5
877	Las Vegas	98.0	72.6	86.9
878	Socorro	97.6	72.6	86.7
879	Truth/Consequences	97.5	69.4	85.2
880	Las Cruces	96.1	69.4	84.4
881	Clovis	98.3	72.5	87.1
882	Roswell	100.0	72.6	88.0
883	Carrizozo	100.3	72.6	88.2
884	Tucumcari	99.0	72.5	87.4
NEW YORK				
100-102	New York	103.4	168.7	131.8
103	Staten Island	99.2	168.4	129.4
104	Bronx	97.7	166.1	127.5
105	Mount Vernon	97.5	141.4	116.6
106	White Plains	97.6	141.4	116.7
107	Yonkers	102.1	143.7	120.2
108	New Rochelle	97.9	141.3	116.8
109	Suffern	97.6	129.0	111.3
110	Queens	100.9	166.8	129.6
111	Long Island City	102.4	166.8	130.5
112	Brooklyn	102.8	168.5	131.4
113	Flushing	102.5	166.8	130.5
114	Jamaica	100.8	166.8	129.6
115,117,118	Hicksville	100.9	153.3	123.7
116	Far Rockaway	102.6	166.8	130.6
119	Riverhead	101.6	148.2	121.9
120-122	Albany	100.1	104.7	102.1
123	Schenectady	99.3	104.7	101.7
124	Kingston	102.6	118.5	109.6
125-126	Poughkeepsie	101.8	136.7	117.0
127	Monticello	101.1	115.1	107.2
128	Glens Falls	94.0	98.6	96.0
129	Plattsburgh	99.1	96.7	98.0
130-132	Syracuse	98.9	98.0	98.5
133-135	Utica	97.1	97.1	97.1
136	Watertown	98.5	96.3	97.5
137-139	Binghamton	98.5	101.4	99.8
140-142	Buffalo	100.3	106.2	102.9
143	Niagara Falls	97.8	106.4	101.5
144-146	Rochester	101.1	99.1	100.2
147	Jamestown	96.8	92.5	94.9
148-149	Elmira	96.7	97.4	97.0
NORTH CAROLINA				
270,272-274	Greensboro	101.3	55.7	81.4
271	Winston-Salem	101.0	56.5	81.6
275-276	Raleigh	100.5	54.1	80.3
277	Durham	102.9	55.3	82.1
278	Rocky Mount	98.5	52.3	78.3
279	Elizabeth City	99.3	51.6	78.5
280	Gastonia	101.6	55.1	81.3
281-282	Charlotte	101.9	56.8	82.2
283	Fayetteville	104.5	58.1	84.2
284	Wilmington	100.3	55.9	81.0
285	Kinston	98.4	52.7	78.5

Location Factors

STATE/ZIP	CITY	MAT.	INST.	TOTAL
NORTH CAROLINA (CONT'D)				
286	Hickory	98.8	53.9	79.2
287-288	Asheville	100.8	55.3	80.9
289	Murphy	99.6	48.3	77.2
NORTH DAKOTA				
580-581	Fargo	102.7	67.6	87.4
582	Grand Forks	102.7	51.6	80.4
583	Devils Lake	102.0	58.6	83.1
584	Jamestown	102.1	45.2	77.3
585	Bismarck	103.1	66.2	87.0
586	Dickinson	102.8	60.9	84.6
587	Minot	102.7	71.3	89.0
588	Williston	101.2	60.9	83.7
OHIO				
430-432	Columbus	98.6	89.2	94.5
433	Marion	95.1	85.3	90.8
434-436	Toledo	98.7	98.1	98.4
437-438	Zanesville	95.6	86.5	91.6
439	Steubenville	96.8	94.9	96.0
440	Lorain	98.9	93.0	96.3
441	Cleveland	99.0	100.3	99.6
442-443	Akron	99.9	94.2	97.4
444-445	Youngstown	99.2	89.0	94.7
446-447	Canton	99.3	85.6	93.3
448-449	Mansfield	96.7	87.3	92.6
450	Hamilton	97.9	85.1	92.3
451-452	Cincinnati	98.3	84.5	92.2
453-454	Dayton	98.0	84.4	92.1
455	Springfield	98.0	84.7	92.2
456	Chillicothe	96.8	93.0	95.1
457	Athens	99.6	80.4	91.2
458	Lima	100.0	86.8	94.3
OKLAHOMA				
730-731	Oklahoma City	100.3	65.7	85.2
734	Ardmore	98.3	61.9	82.4
735	Lawton	100.7	62.8	84.2
736	Clinton	99.6	63.1	83.7
737	Enid	100.2	60.5	82.9
738	Woodward	98.3	63.4	83.1
739	Guymon	99.4	59.7	82.1
740-741	Tulsa	98.1	61.6	82.2
743	Miami	94.8	67.5	82.9
744	Muskogee	97.4	55.7	79.2
745	Mcalester	94.5	60.4	79.7
746	Ponca City	95.1	62.1	80.7
747	Durant	95.1	63.5	81.3
748	Shawnee	96.5	61.8	81.4
749	Poteau	94.3	61.4	80.0
OREGON				
970-972	Portland	98.8	100.4	99.5
973	Salem	100.5	99.4	100.0
974	Eugene	98.6	99.3	98.9
975	Medford	100.0	97.1	98.8
976	Klamath Falls	99.8	97.1	98.6
977	Bend	98.9	99.4	99.1
978	Pendleton	94.8	101.4	97.7
979	Vale	92.9	98.1	95.2
PENNSYLVANIA				
150-152	Pittsburgh	99.9	104.6	102.0
153	Washington	97.0	103.2	99.7
154	Uniontown	97.3	101.7	99.2
155	Bedford	98.2	91.3	95.2
156	Greensburg	98.2	100.9	99.4
157	Indiana	97.1	99.3	98.0
158	Dubois	98.5	96.8	97.8
159	Johnstown	98.1	97.2	97.7
160	Butler	92.5	102.0	96.6
161	New Castle	92.5	97.9	94.8
162	Kittanning	92.9	103.3	97.4
163	Oil City	92.4	96.9	94.4
164-165	Erie	94.5	95.8	95.1
166	Altoona	94.6	91.9	93.4
167	Bradford	95.7	96.4	96.0
168	State College	95.3	92.8	94.2
169	Wellsboro	96.3	92.6	94.7
170-171	Harrisburg	99.8	95.6	98.0
172	Chambersburg	96.2	88.0	92.6
173-174	York	96.7	95.9	96.3
175-176	Lancaster	95.1	89.0	92.4

STATE/ZIP	CITY	MAT.	INST.	TOTAL
PENNSYLVANIA (CONT'D)				
177	Williamsport	93.8	83.0	89.1
178	Sunbury	95.8	95.8	95.8
179	Pottsville	94.9	98.5	96.5
180	Lehigh Valley	95.8	114.0	103.7
181	Allentown	97.9	108.6	102.5
182	Hazleton	95.4	98.3	96.6
183	Stroudsburg	95.3	101.4	97.9
184-185	Scranton	98.6	99.2	98.9
186-187	Wilkes-Barre	95.1	98.7	96.6
188	Montrose	94.7	96.9	95.7
189	Doylestown	94.9	120.6	106.1
190-191	Philadelphia	99.7	133.6	114.5
193	Westchester	95.9	121.5	107.1
194	Norristown	95.0	129.5	110.0
195-196	Reading	97.1	102.1	99.3
PUERTO RICO				
009	San Juan	122.7	24.4	79.8
RHODE ISLAND				
028	Newport	98.4	117.5	106.7
029	Providence	100.6	117.5	108.0
SOUTH CAROLINA				
290-292	Columbia	99.3	56.5	80.6
293	Spartanburg	98.6	56.7	80.3
294	Charleston	99.9	65.2	84.8
295	Florence	98.2	56.9	80.2
296	Greenville	98.3	55.8	79.8
297	Rock Hill	97.8	50.3	77.1
298	Aiken	98.6	69.1	85.7
299	Beaufort	99.3	44.9	75.6
SOUTH DAKOTA				
570-571	Sioux Falls	100.0	57.7	81.5
572	Watertown	99.0	49.1	77.2
573	Mitchell	97.8	48.4	76.3
574	Aberdeen	100.5	49.6	78.3
575	Pierre	101.7	58.4	82.8
576	Mobridge	98.5	48.8	76.8
577	Rapid City	100.4	59.5	82.6
TENNESSEE				
370-372	Nashville	98.4	74.1	87.8
373-374	Chattanooga	100.0	66.2	85.2
375,380-381	Memphis	99.0	71.3	86.9
376	Johnson City	99.8	56.5	80.9
377-379	Knoxville	96.6	66.9	83.7
382	Mckenzie	98.2	59.2	81.2
383	Jackson	99.9	61.2	83.0
384	Columbia	96.9	66.1	83.5
385	Cookeville	98.1	59.1	81.1
TEXAS				
750	Mckinney	98.7	63.1	83.2
751	Waxahackie	98.6	65.9	84.4
752-753	Dallas	99.5	67.4	85.5
754	Greenville	98.8	64.0	83.6
755	Texarkana	98.0	58.4	80.7
756	Longview	98.5	58.7	81.2
757	Tyler	98.5	65.0	83.9
758	Palestine	95.0	64.4	81.7
759	Lufkin	95.6	67.8	83.5
760-761	Fort Worth	99.3	65.3	84.5
762	Denton	98.9	62.5	83.0
763	Wichita Falls	97.1	64.4	82.8
764	Eastland	95.9	63.2	81.6
765	Temple	94.7	61.7	80.3
766-767	Waco	96.5	64.1	82.4
768	Brownwood	98.9	59.4	81.7
769	San Angelo	98.5	59.6	81.5
770-772	Houston	99.4	70.5	86.8
773	Huntsville	98.0	66.7	84.4
774	Wharton	98.9	67.7	85.3
775	Galveston	97.0	70.0	85.2
776-777	Beaumont	97.4	68.6	84.9
778	Bryan	94.9	68.2	83.3
779	Victoria	99.0	65.6	84.4
780	Laredo	98.3	64.5	83.6
781-782	San Antonio	98.5	65.1	84.0
783-784	Corpus Christi	100.9	63.5	84.6
785	Mc Allen	101.0	59.8	83.0
786-787	Austin	100.0	63.6	84.1

For customer support on your Plumbing Cost Data, call 877.709.6968.

Location Factors

STATE/ZIP	CITY	MAT.	INST.	TOTAL	STATE/ZIP	CITY	MAT.	INST.	TOTAL
TEXAS (CONT'D)					**WISCONSIN (CONT'D)**				
788	Del Rio	100.8	63.2	84.4	538	Lancaster	96.7	92.5	94.8
789	Giddings	97.7	64.0	83.0	539	Portage	95.3	96.6	95.9
790-791	Amarillo	100.2	63.2	84.1	540	New Richmond	96.8	94.6	95.9
792	Childress	99.3	64.3	84.0	541-543	Green Bay	101.2	93.4	97.8
793-794	Lubbock	101.6	63.6	85.0	544	Wausau	96.2	93.5	95.0
795-796	Abilene	99.7	62.2	83.3	545	Rhinelander	99.3	91.2	95.8
797	Midland	101.5	64.2	85.2	546	La Crosse	97.8	94.5	96.3
798-799,885	El Paso	98.0	61.7	82.2	547	Eau Claire	99.3	95.3	97.5
					548	Superior	96.6	96.3	96.5
UTAH					549	Oshkosh	96.9	91.9	94.7
840-841	Salt Lake City	102.2	68.3	87.4					
842,844	Ogden	98.1	70.1	85.9	**WYOMING**				
843	Logan	99.8	70.1	86.9	820	Cheyenne	101.5	65.2	85.7
845	Price	100.2	64.6	84.7	821	Yellowstone Nat'l Park	99.3	66.3	84.9
846-847	Provo	100.3	68.2	86.3	822	Wheatland	100.4	64.3	84.7
					823	Rawlins	101.8	66.9	86.6
VERMONT					824	Worland	100.0	65.3	84.9
050	White River Jct.	98.5	78.3	89.7	825	Riverton	101.0	65.4	85.5
051	Bellows Falls	97.2	93.0	95.3	826	Casper	102.0	63.9	85.4
052	Bennington	97.5	93.5	95.8	827	Newcastle	99.9	66.9	85.5
053	Brattleboro	97.8	92.8	95.6	828	Sheridan	102.6	64.3	85.9
054	Burlington	102.4	83.6	94.2	829-831	Rock Springs	103.7	65.3	87.0
056	Montpelier	99.5	82.8	92.2					
057	Rutland	99.7	83.6	92.7	**CANADIAN FACTORS (reflect Canadian currency)**				
058	St. Johnsbury	98.6	77.8	89.6					
059	Guildhall	97.4	77.8	88.8	**ALBERTA**				
						Calgary	123.5	95.7	111.4
VIRGINIA						Edmonton	123.1	95.7	111.2
220-221	Fairfax	100.7	84.0	93.4		Fort McMurray	123.6	92.8	110.2
222	Arlington	101.9	84.0	94.1		Lethbridge	118.4	92.2	107.0
223	Alexandria	101.1	85.5	94.3		Lloydminster	113.4	88.9	102.8
224-225	Fredericksburg	99.5	82.7	92.1		Medicine Hat	113.6	88.2	102.5
226	Winchester	100.0	82.2	92.2		Red Deer	114.1	88.2	102.8
227	Culpeper	99.9	80.5	91.4					
228	Harrisonburg	100.2	67.2	85.8	**BRITISH COLUMBIA**				
229	Charlottesville	100.5	67.1	86.0		Kamloops	114.5	90.6	104.1
230-232	Richmond	101.3	68.7	87.1		Prince George	115.4	89.9	104.3
233-235	Norfolk	100.5	68.2	86.4		Vancouver	119.5	89.1	106.2
236	Newport News	100.0	68.2	86.1		Victoria	115.7	87.2	103.3
237	Portsmouth	99.4	65.3	84.5					
238	Petersburg	99.4	68.7	86.0	**MANITOBA**				
239	Farmville	98.7	52.4	78.5		Brandon	125.2	75.8	103.7
240-241	Roanoke	101.5	65.1	85.6		Portage la Prairie	113.4	74.2	96.3
242	Bristol	99.4	54.9	80.0		Winnipeg	121.0	71.8	99.5
243	Pulaski	99.0	57.3	80.9					
244	Staunton	99.7	61.8	83.2	**NEW BRUNSWICK**				
245	Lynchburg	99.9	65.6	85.0		Bathurst	111.5	67.3	92.2
246	Grundy	99.3	52.5	78.9		Dalhousie	111.9	67.5	92.5
						Fredericton	117.4	72.5	97.9
WASHINGTON						Moncton	111.8	70.1	93.6
980-981,987	Seattle	101.4	104.7	102.8		Newcastle	111.6	67.9	92.5
982	Everett	101.6	99.6	100.7		St. John	114.0	77.9	98.3
983-984	Tacoma	101.8	99.7	100.9					
985	Olympia	101.2	99.4	100.4	**NEWFOUNDLAND**				
986	Vancouver	102.7	92.1	98.1		Corner Brook	129.6	67.6	102.6
988	Wenatchee	101.6	86.6	95.0		St Johns	122.7	85.2	106.4
989	Yakima	102.0	94.1	98.6					
990-992	Spokane	103.4	82.3	94.2	**NORTHWEST TERRITORIES**				
993	Richland	103.1	89.3	97.1		Yellowknife	135.3	84.2	113.0
994	Clarkston	101.5	80.2	92.2					
					NOVA SCOTIA				
WEST VIRGINIA						Bridgewater	112.9	74.6	96.2
247-248	Bluefield	98.3	92.9	96.0		Dartmouth	126.1	74.6	103.7
249	Lewisburg	99.8	92.4	96.6		Halifax	119.6	80.9	102.7
250-253	Charleston	99.9	94.8	97.6		New Glasgow	124.1	74.6	102.5
254	Martinsburg	99.3	86.4	93.7		Sydney	122.1	74.6	101.4
255-257	Huntington	100.7	95.7	98.5		Truro	112.5	74.6	96.0
258-259	Beckley	97.9	93.3	95.9		Yarmouth	124.0	74.6	102.5
260	Wheeling	101.2	94.3	98.2					
261	Parkersburg	100.1	91.8	96.5	**ONTARIO**				
262	Buckhannon	99.6	95.5	97.8		Barrie	117.7	93.0	106.9
263-264	Clarksburg	100.1	95.4	98.1		Brantford	114.9	96.9	107.1
265	Morgantown	100.2	94.4	97.6		Cornwall	115.0	93.3	105.5
266	Gassaway	99.5	93.9	97.1		Hamilton	117.0	96.7	108.2
267	Romney	99.4	92.2	96.3		Kingston	116.0	93.4	106.2
268	Petersburg	99.3	91.3	95.8		Kitchener	109.6	92.9	102.3
						London	116.4	93.3	106.3
WISCONSIN						North Bay	127.3	91.1	111.5
530,532	Milwaukee	100.0	107.2	103.1		Oshawa	112.3	94.3	104.4
531	Kenosha	99.5	102.2	100.7		Ottawa	117.3	94.0	107.2
534	Racine	99.1	102.4	100.5		Owen Sound	117.8	91.2	106.2
535	Beloit	98.9	100.7	99.7		Peterborough	114.9	93.1	105.4
537	Madison	99.1	99.0	99.1		Sarnia	114.8	97.5	107.3
						Sault Ste Marie	110.3	92.1	102.3

Location Factors

STATE/ZIP	CITY	MAT.	INST.	TOTAL
ONTARIO (CONT'D)				
	St. Catharines	107.6	94.5	101.9
	Sudbury	107.7	93.0	101.3
	Thunder Bay	108.8	93.4	102.1
	Timmins	115.1	91.1	104.6
	Toronto	118.3	101.3	110.9
	Windsor	108.2	93.4	101.7
PRINCE EDWARD ISLAND				
	Charlottetown	120.0	62.8	95.1
	Summerside	126.7	62.8	98.8
QUEBEC				
	Cap-de-la-Madeleine	112.4	87.6	101.6
	Charlesbourg	112.4	87.6	101.6
	Chicoutimi	111.4	92.9	103.3
	Gatineau	112.1	87.4	101.3
	Granby	112.3	87.4	101.4
	Hull	112.2	87.4	101.4
	Joliette	112.5	87.6	101.7
	Laval	112.1	87.4	101.3
	Montreal	117.8	93.2	107.1
	Quebec	117.1	93.6	106.8
	Rimouski	111.8	92.9	103.6
	Rouyn-Noranda	112.0	87.4	101.3
	Saint Hyacinthe	111.6	87.4	101.1
	Sherbrooke	112.4	87.4	101.5
	Sorel	112.5	87.6	101.7
	St Jerome	112.1	87.4	101.3
	Trois Rivieres	124.8	87.6	108.6
SASKATCHEWAN				
	Moose Jaw	110.7	69.5	92.7
	Prince Albert	109.9	67.7	91.5
	Regina	120.7	89.0	106.9
	Saskatoon	111.1	89.1	101.5
YUKON				
	Whitehorse	135.3	68.5	106.2

R011105-05   Tips for Accurate Estimating

1. Use pre-printed or columnar forms for orderly sequence of dimensions and locations and for recording telephone quotations.

2. Use only the front side of each paper or form except for certain pre-printed summary forms.

3. Be consistent in listing dimensions: For example, length x width x height. This helps in rechecking to ensure that, the total length of partitions is appropriate for the building area.

4. Use printed (rather than measured) dimensions where given.

5. Add up multiple printed dimensions for a single entry where possible.

6. Measure all other dimensions carefully.

7. Use each set of dimensions to calculate multiple related quantities.

8. Convert foot and inch measurements to decimal feet when listing. Memorize decimal equivalents to .01 parts of a foot (1/8″ equals approximately .01″).

9. Do not "round off" quantities until the final summary.

10. Mark drawings with different colors as items are taken off.

11. Keep similar items together, different items separate.

12. Identify location and drawing numbers to aid in future checking for completeness.

13. Measure or list everything on the drawings or mentioned in the specifications.

14. It may be necessary to list items not called for to make the job complete.

15. Be alert for: Notes on plans such as N.T.S. (not to scale); changes in scale throughout the drawings; reduced size drawings; discrepancies between the specifications and the drawings.

16. Develop a consistent pattern of performing an estimate.
   For example:
   a. Start the quantity takeoff at the lower floor and move to the next higher floor.
   b. Proceed from the main section of the building to the wings.
   c. Proceed from south to north or vice versa, clockwise or counterclockwise.
   d. Take off floor plan quantities first, elevations next, then detail drawings.

17. List all gross dimensions that can be either used again for different quantities, or used as a rough check of other quantities for verification (exterior perimeter, gross floor area, individual floor areas, etc.).

18. Utilize design symmetry or repetition (repetitive floors, repetitive wings, symmetrical design around a center line, similar room layouts, etc.). Note: Extreme caution is needed here so as not to omit or duplicate an area.

19. Do not convert units until the final total is obtained. For instance, when estimating concrete work, keep all units to the nearest cubic foot, then summarize and convert to cubic yards.

20. When figuring alternatives, it is best to total all items involved in the basic system, then total all items involved in the alternates. Therefore you work with positive numbers in all cases. When adds and deducts are used, it is often confusing whether to add or subtract a portion of an item; especially on a complicated or involved alternate.

R011105-10   Unit Gross Area Requirements

The figures in the table below indicate typical ranges in square feet as a function of the "occupant" unit. This table is best used in the preliminary design stages to help determine the probable size requirement for the total project.

Building Type	Unit	Gross Area in S.F.		
		1/4	Median	3/4
Apartments	Unit	660	860	1,100
Auditorium & Play Theaters	Seat	18	25	38
Bowling Alleys	Lane		940	
Churches & Synagogues	Seat	20	28	39
Dormitories	Bed	200	230	275
Fraternity & Sorority Houses	Bed	220	315	370
Garages, Parking	Car	325	355	385
Hospitals	Bed	685	850	1,075
Hotels	Rental Unit	475	600	710
Housing for the elderly	Unit	515	635	755
Housing, Public	Unit	700	875	1,030
Ice Skating Rinks	Total	27,000	30,000	36,000
Motels	Rental Unit	360	465	620
Nursing Homes	Bed	290	350	450
Restaurants	Seat	23	29	39
Schools, Elementary	Pupil	65	77	90
Junior High & Middle		85	110	129
Senior High		102	130	145
Vocational		110	135	195
Shooting Ranges	Point		450	
Theaters & Movies	Seat		15	

R011105-20   Floor Area Ratios

Table below lists commonly used gross to net area and net to gross area
ratios expressed in % for various building types.

Building Type	Gross to Net Ratio	Net to Gross Ratio	Building Type	Gross to Net Ratio	Net to Gross Ratio
Apartment	156	64	School Buildings (campus type)		
Bank	140	72	Administrative	150	67
Church	142	70	Auditorium	142	70
Courthouse	162	61	Biology	161	62
Department Store	123	81	Chemistry	170	59
Garage	118	85	Classroom	152	66
Hospital	183	55	Dining Hall	138	72
Hotel	158	63	Dormitory	154	65
Laboratory	171	58	Engineering	164	61
Library	132	76	Fraternity	160	63
Office	135	75	Gymnasium	142	70
Restaurant	141	70	Science	167	60
Warehouse	108	93	Service	120	83
			Student Union	172	59

The gross area of a building is the total floor area based on outside
dimensions.

The net area of a building is the usable floor area for the function intended
and excludes such items as stairways, corridors and mechanical rooms. In
the case of a commercial building, it might be considered as the "leasable
area."

R011105-30   Occupancy Determinations

Function of Space	SF/Person Required
Accessory storage areas, mechanical equipment rooms	300
Agriculture Building	300
Aircraft Hangars	500
Airport Terminal	
Baggage claim	20
Baggage handling	300
Concourse	100
Waiting areas	15
Assembly	
Gaming floors (keno, slots, etc.)	11
Exhibit Gallery and Museum	30
Assembly w/ fixed seats	load determined by seat number
Assembly w/o fixed seats	
Concentrated (chairs only-not fixed)	7
Standing space	5
Unconcentrated (tables and chairs)	15
Bowling centers, allow 5 persons for each lane including 15 feet of runway, and for additional areas	7
Business areas	100
Courtrooms–other than fixed seating areas	40
Day care	35
Dormitories	50
Educational	
Classroom areas	20
Shops and other vocational room areas	50
Exercise rooms	50
Fabrication and Manufacturing areas where hazardous materials are used	200
Industrial areas	100
Institutional areas	
Inpatient treatment areas	240
Outpatient areas	100
Sleeping areas	120
Kitchens commercial	200
Library	
Reading rooms	50
Stack area	100
Mercantile	
Areas on other floors	60
Basement and grade floor areas	30
Storage, stock, shipping areas	300
Parking garages	200
Residential	200
Skating rinks, swimming pools	
Rink and pool	50
Decks	15
Stages and platforms	15
Warehouses	500

Excerpted from the 2012 *International Building Code*, Copyright 2011. Washington, D.C.: International Code Council. Reproduced with permission. All rights reserved. www.ICCSAFE.org

R011105-40   Weather Data and Design Conditions

City	Latitude (1)		Winter Temperatures (1) Med. of Annual Extremes	99%	97½%	Winter Degree Days (2)	Summer (Design Dry Bulb) Temperatures and Relative Humidity 1%	2½%	5%
	0	1'							
UNITED STATES									
Albuquerque, NM	35	0	5.1	12	16	4,400	96/61	94/61	92/61
Atlanta, GA	33	4	11.9	17	22	3,000	94/74	92/74	90/73
Baltimore, MD	39	2	7	14	17	4,600	94/75	91/75	89/74
Birmingham, AL	33	3	13	17	21	2,600	96/74	94/75	92/74
Bismarck, ND	46	5	-32	-23	-19	8,800	95/68	91/68	88/67
Boise, ID	43	3	1	3	10	5,800	96/65	94/64	91/64
Boston, MA	42	2	-1	6	9	5,600	91/73	88/71	85/70
Burlington, VT	44	3	-17	-12	-7	8,200	88/72	85/70	82/69
Charleston, WV	38	2	3	7	11	4,400	92/74	90/73	87/72
Charlotte, NC	35	1	13	18	22	3,200	95/74	93/74	91/74
Casper, WY	42	5	-21	-11	-5	7,400	92/58	90/57	87/57
Chicago, IL	41	5	-8	-3	2	6,600	94/75	91/74	88/73
Cincinnati, OH	39	1	0	1	6	4,400	92/73	90/72	88/72
Cleveland, OH	41	2	-3	1	5	6,400	91/73	88/72	86/71
Columbia, SC	34	0	16	20	24	2,400	97/76	95/75	93/75
Dallas, TX	32	5	14	18	22	2,400	102/75	100/75	97/75
Denver, CO	39	5	-10	-5	1	6,200	93/59	91/59	89/59
Des Moines, IA	41	3	-14	-10	-5	6,600	94/75	91/74	88/73
Detroit, MI	42	2	-3	3	6	6,200	91/73	88/72	86/71
Great Falls, MT	47	3	-25	-21	-15	7,800	91/60	88/60	85/59
Hartford, CT	41	5	-4	3	7	6,200	91/74	88/73	85/72
Houston, TX	29	5	24	28	33	1,400	97/77	95/77	93/77
Indianapolis, IN	39	4	-7	-2	2	5,600	92/74	90/74	87/73
Jackson, MS	32	2	16	21	25	2,200	97/76	95/76	93/76
Kansas City, MO	39	1	-4	2	6	4,800	99/75	96/74	93/74
Las Vegas, NV	36	1	18	25	28	2,800	108/66	106/65	104/65
Lexington, KY	38	0	-1	3	8	4,600	93/73	91/73	88/72
Little Rock, AR	34	4	11	15	20	3,200	99/76	96/77	94/77
Los Angeles, CA	34	0	36	41	43	2,000	93/70	89/70	86/69
Memphis, TN	35	0	10	13	18	3,200	98/77	95/76	93/76
Miami, FL	25	5	39	44	47	200	91/77	90/77	89/77
Milwaukee, WI	43	0	-11	-8	-4	7,600	90/74	87/73	84/71
Minneapolis, MN	44	5	-22	-16	-12	8,400	92/75	89/73	86/71
New Orleans, LA	30	0	28	29	33	1,400	93/78	92/77	90/77
New York, NY	40	5	6	11	15	5,000	92/74	89/73	87/72
Norfolk, VA	36	5	15	20	22	3,400	93/77	91/76	89/76
Oklahoma City, OK	35	2	4	9	13	3,200	100/74	97/74	95/73
Omaha, NE	41	2	-13	-8	-3	6,600	94/76	91/75	88/74
Philadelphia, PA	39	5	6	10	14	4,400	93/75	90/74	87/72
Phoenix, AZ	33	3	27	31	34	1,800	109/71	107/71	105/71
Pittsburgh, PA	40	3	-1	3	7	6,000	91/72	88/71	86/70
Portland, ME	43	4	-10	-6	-1	7,600	87/72	84/71	81/69
Portland, OR	45	4	18	17	23	4,600	89/68	85/67	81/65
Portsmouth, NH	43	1	-8	-2	2	7,200	89/73	85/71	83/70
Providence, RI	41	4	-1	5	9	6,000	89/73	86/72	83/70
Rochester, NY	43	1	-5	1	5	6,800	91/73	88/71	85/70
Salt Lake City, UT	40	5	0	3	8	6,000	97/62	95/62	92/61
San Francisco, CA	37	5	36	38	40	3,000	74/63	71/62	69/61
Seattle, WA	47	4	22	22	27	5,200	85/68	82/66	78/65
Sioux Falls, SD	43	4	-21	-15	-11	7,800	94/73	91/72	88/71
St. Louis, MO	38	4	-3	3	8	5,000	98/75	94/75	91/75
Tampa, FL	28	0	32	36	40	680	92/77	91/77	90/76
Trenton, NJ	40	1	4	11	14	5,000	91/75	88/74	85/73
Washington, DC	38	5	7	14	17	4,200	93/75	91/74	89/74
Wichita, KS	37	4	-3	3	7	4,600	101/72	98/73	96/73
Wilmington, DE	39	4	5	10	14	5,000	92/74	89/74	87/73
ALASKA									
Anchorage	61	1	-29	-23	-18	10,800	71/59	68/58	66/56
Fairbanks	64	5	-59	-51	-47	14,280	82/62	78/60	75/59
CANADA									
Edmonton, Alta.	53	3	-30	-29	-25	11,000	85/66	82/65	79/63
Halifax, N.S.	44	4	-4	1	5	8,000	79/66	76/65	74/64
Montreal, Que.	45	5	-20	-16	-10	9,000	88/73	85/72	83/71
Saskatoon, Sask.	52	1	-35	-35	-31	11,000	89/68	86/66	83/65
St. John's, N.F.	47	4	1	3	7	8,600	77/66	75/65	73/64
Saint John, N.B.	45	2	-15	-12	-8	8,200	80/67	77/65	75/64
Toronto, Ont.	43	4	-10	-5	-1	7,000	90/73	87/72	85/71
Vancouver, B.C.	49	1	13	15	19	6,000	79/67	77/66	74/65
Winnipeg, Man.	49	5	-31	-30	-27	10,800	89/73	86/71	84/70

(1) Handbook of Fundamentals, ASHRAE, Inc., NY 1989
(2) Local Climatological Annual Survey, USDC Env. Science Services
Administration, Asheville, NC

R011105-50 Metric Conversion Factors

Description: This table is primarily for converting customary U.S. units in the left hand column to SI metric units in the right hand column. In addition, conversion factors for some commonly encountered Canadian and non-SI metric units are included.

	If You Know		Multiply By		To Find
Length	Inches	x	25.4[a]	=	Millimeters
	Feet	x	0.3048[a]	=	Meters
	Yards	x	0.9144[a]	=	Meters
	Miles (statute)	x	1.609	=	Kilometers
Area	Square inches	x	645.2	=	Square millimeters
	Square feet	x	0.0929	=	Square meters
	Square yards	x	0.8361	=	Square meters
Volume (Capacity)	Cubic inches	x	16,387	=	Cubic millimeters
	Cubic feet	x	0.02832	=	Cubic meters
	Cubic yards	x	0.7646	=	Cubic meters
	Gallons (U.S. liquids)[b]	x	0.003785	=	Cubic meters[c]
	Gallons (Canadian liquid)[b]	x	0.004546	=	Cubic meters[c]
	Ounces (U.S. liquid)[b]	x	29.57	=	Milliliters[c, d]
	Quarts (U.S. liquid)[b]	x	0.9464	=	Liters[c, d]
	Gallons (U.S. liquid)[b]	x	3.785	=	Liters[c, d]
Force	Kilograms force[d]	x	9.807	=	Newtons
	Pounds force	x	4.448	=	Newtons
	Pounds force	x	0.4536	=	Kilograms force[d]
	Kips	x	4448	=	Newtons
	Kips	x	453.6	=	Kilograms force[d]
Pressure, Stress, Strength (Force per unit area)	Kilograms force per square centimeter[d]	x	0.09807	=	Megapascals
	Pounds force per square inch (psi)	x	0.006895	=	Megapascals
	Kips per square inch	x	6.895	=	Megapascals
	Pounds force per square inch (psi)	x	0.07031	=	Kilograms force per square centimeter[d]
	Pounds force per square foot	x	47.88	=	Pascals
	Pounds force per square foot	x	4.882	=	Kilograms force per square meter[d]
Bending Moment Or Torque	Inch-pounds force	x	0.01152	=	Meter-kilograms force[d]
	Inch-pounds force	x	0.1130	=	Newton-meters
	Foot-pounds force	x	0.1383	=	Meter-kilograms force[d]
	Foot-pounds force	x	1.356	=	Newton-meters
	Meter-kilograms force[d]	x	9.807	=	Newton-meters
Mass	Ounces (avoirdupois)	x	28.35	=	Grams
	Pounds (avoirdupois)	x	0.4536	=	Kilograms
	Tons (metric)	x	1000	=	Kilograms
	Tons, short (2000 pounds)	x	907.2	=	Kilograms
	Tons, short (2000 pounds)	x	0.9072	=	Megagrams[e]
Mass per Unit Volume	Pounds mass per cubic foot	x	16.02	=	Kilograms per cubic meter
	Pounds mass per cubic yard	x	0.5933	=	Kilograms per cubic meter
	Pounds mass per gallon (U.S. liquid)[b]	x	119.8	=	Kilograms per cubic meter
	Pounds mass per gallon (Canadian liquid)[b]	x	99.78	=	Kilograms per cubic meter
Temperature	Degrees Fahrenheit	(F-32)/1.8		=	Degrees Celsius
	Degrees Fahrenheit	(F+459.67)/1.8		=	Degrees Kelvin
	Degrees Celsius	C+273.15		=	Degrees Kelvin

[a]The factor given is exact
[b]One U.S. gallon = 0.8327 Canadian gallon
[c]1 liter = 1000 milliliters = 1000 cubic centimeters
   1 cubic decimeter = 0.001 cubic meter

[d]Metric but not SI unit
[e]Called "tonne" in England and "metric ton" in other metric countries

R011105-60   Weights and Measures

Measures of Length
1 Mile = 1760 Yards = 5280 Feet
1 Yard = 3 Feet = 36 inches
1 Foot = 12 Inches
1 Mil = 0.001 Inch
1 Fathom = 2 Yards = 6 Feet
1 Rod = 5.5 Yards = 16.5 Feet
1 Hand = 4 Inches
1 Span = 9 Inches
1 Micro-inch = One Millionth Inch or 0.000001 Inch
1 Micron = One Millionth Meter + 0.00003937 Inch

Surveyor's Measure
1 Mile = 8 Furlongs = 80 Chains
1 Furlong = 10 Chains = 220 Yards
1 Chain = 4 Rods = 22 Yards = 66 Feet = 100 Links
1 Link = 7.92 Inches

Square Measure
1 Square Mile = 640 Acres = 6400 Square Chains
1 Acre = 10 Square Chains = 4840 Square Yards =
43,560 Sq. Ft.
1 Square Chain = 16 Square Rods = 484 Square Yards =
4356 Sq. Ft.
1 Square Rod = 30.25 Square Yards = 272.25 Square Feet = 625 Square
Lines
1 Square Yard = 9 Square Feet
1 Square Foot = 144 Square Inches
An Acre equals a Square 208.7 Feet per Side

Cubic Measure
1 Cubic Yard = 27 Cubic Feet
1 Cubic Foot = 1728 Cubic Inches
1 Cord of Wood = 4 x 4 x 8 Feet = 128 Cubic Feet
1 Perch of Masonry = 16½ x 1½ x 1 Foot = 24.75 Cubic Feet

Avoirdupois or Commercial Weight
1 Gross or Long Ton = 2240 Pounds
1 Net or Short ton = 2000 Pounds
1 Pound = 16 Ounces = 7000 Grains
1 Ounce = 16 Drachms = 437.5 Grains
1 Stone = 14 Pounds

Shipping Measure
For Measuring Internal Capacity of a Vessel:
    1 Register Ton = 100 Cubic Feet

For Measurement of Cargo:
    Approximately 40 Cubic Feet of Merchandise is considered a Shipping
    Ton, unless that bulk would weigh more than 2000 Pounds, in which case
    Freight Charge may be based upon weight.

40 Cubic Feet = 32.143 U.S. Bushels = 31.16 Imp. Bushels

Liquid Measure
1 Imperial Gallon = 1.2009 U.S. Gallon = 277.42 Cu. In.
1 Cubic Foot = 7.48 U.S. Gallons

R011110-30 Engineering Fees

Typical **Structural Engineering Fees** based on type of construction and total project size. These fees are included in Architectural Fees.

Type of Construction	Total Project Size (in thousands of dollars)			
	$500	$500-$1,000	$1,000-$5,000	Over $5000
Industrial buildings, factories & warehouses	Technical payroll times 2.0 to 2.5	1.60%	1.25%	1.00%
Hotels, apartments, offices, dormitories, hospitals, public buildings, food stores		2.00%	1.70%	1.20%
Museums, banks, churches and cathedrals		2.00%	1.75%	1.25%
Thin shells, prestressed concrete, earthquake resistive		2.00%	1.75%	1.50%
Parking ramps, auditoriums, stadiums, convention halls, hangars & boiler houses		2.50%	2.00%	1.75%
Special buildings, major alterations, underpinning & future expansion	↓	Add to above 0.5%	Add to above 0.5%	Add to above 0.5%

For complex reinforced concrete or unusually complicated structures, add 20% to 50%.

Typical **Mechanical and Electrical Engineering Fees** are based on the size of the subcontract. The fee structure for both are shown below. These fees are included in Architectural Fees.

Type of Construction	Subcontract Size							
	$25,000	$50,000	$100,000	$225,000	$350,000	$500,000	$750,000	$1,000,000
Simple structures	6.4%	5.7%	4.8%	4.5%	4.4%	4.3%	4.2%	4.1%
Intermediate structures	8.0	7.3	6.5	5.6	5.1	5.0	4.9	4.8
Complex structures	10.1	9.0	9.0	8.0	7.5	7.5	7.0	7.0

For renovations, add 15% to 25% to applicable fee.

R012153-10 Repair and Remodeling

Cost figures are based on new construction utilizing the most cost-effective combination of labor, equipment and material with the work scheduled in proper sequence to allow the various trades to accomplish their work in an efficient manner.

The costs for repair and remodeling work must be modified due to the following factors that may be present in any given repair and remodeling project.

1. Equipment usage curtailment due to the physical limitations of the project, with only hand-operated equipment being used.

2. Increased requirement for shoring and bracing to hold up the building while structural changes are being made and to allow for temporary storage of construction materials on above-grade floors.

3. Material handling becomes more costly due to having to move within the confines of an enclosed building. For multi-story construction, low capacity elevators and stairwells may be the only access to the upper floors.

4. Large amount of cutting and patching and attempting to match the existing construction is required. It is often more economical to remove entire walls rather than create many new door and window openings. This sort of trade-off has to be carefully analyzed.

5. Cost of protection of completed work is increased since the usual sequence of construction usually cannot be accomplished.

6. Economies of scale usually associated with new construction may not be present. If small quantities of components must be custom fabricated due to job requirements, unit costs will naturally increase. Also, if only small work areas are available at a given time, job scheduling between trades becomes difficult and subcontractor quotations may reflect the excessive start-up and shut-down phases of the job.

7. Work may have to be done on other than normal shifts and may have to be done around an existing production facility which has to stay in production during the course of the repair and remodeling.

8. Dust and noise protection of adjoining non-construction areas can involve substantial special protection and alter usual construction methods.

9. Job may be delayed due to unexpected conditions discovered during demolition or removal. These delays ultimately increase construction costs.

10. Piping and ductwork runs may not be as simple as for new construction. Wiring may have to be snaked through walls and floors.

11. Matching "existing construction" may be impossible because materials may no longer be manufactured. Substitutions may be expensive.

12. Weather protection of existing structure requires additional temporary structures to protect building at openings.

13. On small projects, because of local conditions, it may be necessary to pay a tradesman for a minimum of four hours for a task that is completed in one hour.

All of the above areas can contribute to increased costs for a repair and remodeling project. Each of the above factors should be considered in the planning, bidding and construction stage in order to minimize the increased costs associated with repair and remodeling jobs.

R012909-80   Sales Tax by State

State sales tax on materials is tabulated below (5 states have no sales tax). Many states allow local jurisdictions, such as a county or city, to levy additional sales tax.

Some projects may be sales tax exempt, particularly those constructed with public funds.

State	Tax (%)	State	Tax (%)	State	Tax (%)	State	Tax (%)
Alabama	4	Illinois	6.25	Montana	0	Rhode Island	7
Alaska	0	Indiana	7	Nebraska	5.5	South Carolina	6
Arizona	5.6	Iowa	6	Nevada	6.85	South Dakota	4
Arkansas	6	Kansas	6.3	New Hampshire	0	Tennessee	7
California	7.5	Kentucky	6	New Jersey	7	Texas	6.25
Colorado	2.9	Louisiana	4	New Mexico	5.13	Utah	4.7
Connecticut	6.35	Maine	5	New York	4	Vermont	6
Delaware	0	Maryland	6	North Carolina	4.75	Virginia	5
District of Columbia	6	Massachusetts	6.25	North Dakota	5	Washington	6.5
Florida	6	Michigan	6	Ohio	5.5	West Virginia	6
Georgia	4	Minnesota	6.88	Oklahoma	4.5	Wisconsin	5
Hawaii	4	Mississippi	7	Oregon	0	Wyoming	4
Idaho	6	Missouri	4.23	Pennsylvania	6	Average	5.04 %

Sales Tax by Province (Canada)

GST - a value-added tax, which the government imposes on most goods and services provided in or imported into Canada. PST - a retail sales tax, which three of the provinces impose on the price of most goods and some services. QST - a value-added tax, similar to the federal GST, which Quebec imposes. HST - Five provinces have combined their retail sales tax with the federal GST into one harmonized tax.

Province	PST (%)	QST (%)	GST(%)	HST(%)
Alberta	0	0	5	0
British Columbia	7	0	5	0
Manitoba	8	0	5	0
New Brunswick	0	0	0	13
Newfoundland	0	0	0	13
Northwest Territories	0	0	5	0
Nova Scotia	0	0	0	15
Ontario	0	0	0	13
Prince Edward Island	0	0	0	14
Quebec	0	9.975	5	0
Saskatchewan	5	0	5	0
Yukon	0	0	5	0

R012909-85   Unemployment Taxes and Social Security Taxes

State unemployment tax rates vary not only from state to state, but also with the experience rating of the contractor. The federal unemployment tax rate is 6.0% of the first $7,000 of wages. This is reduced by a credit of up to 5.4% for timely payment to the state. The minimum federal unemployment tax is 0.6% after all credits.

Social security (FICA) for 2015 is estimated at time of publication to be 7.65% of wages up to $117,000.

R012909-90   Overtime

One way to improve the completion date of a project or eliminate negative float from a schedule is to compress activity duration times. This can be achieved by increasing the crew size or working overtime with the proposed crew.

To determine the costs of working overtime to compress activity duration times, consider the following examples. Below is an overtime efficiency and cost chart based on a five, six, or seven day week with an eight through twelve hour day. Payroll percentage increases for time and one half and double time are shown for the various working days.

| Days per Week | Hours per Day | Production Efficiency | | | | | Payroll Cost Factors | |
		1st Week	2nd Week	3rd Week	4th Week	Average 4 Weeks	@ 1-1/2 Times	@ 2 Times
5	8	100%	100%	100%	100%	100 %	1.000	1.000
	9	100	100	95	90	96	1.056	1.111
	10	100	95	90	85	93	1.100	1.200
	11	95	90	75	65	81	1.136	1.273
	12	90	85	70	60	76	1.167	1.333
6	8	100	100	95	90	96	1.083	1.167
	9	100	95	90	85	93	1.130	1.259
	10	95	90	85	80	88	1.167	1.333
	11	95	85	70	65	79	1.197	1.394
	12	90	80	65	60	74	1.222	1.444
7	8	100	95	85	75	89	1.143	1.286
	9	95	90	80	70	84	1.183	1.365
	10	90	85	75	65	79	1.214	1.429
	11	85	80	65	60	73	1.240	1.481
	12	85	75	60	55	69	1.262	1.524

General Requirements   R0131 Project Management & Coordination

R013113-40   Builder's Risk Insurance

Builder's Risk Insurance is insurance on a building during construction. Premiums are paid by the owner or the contractor. Blasting, collapse and underground insurance would raise total insurance costs above those listed. Floater policy for materials delivered to the job runs $.75 to $1.25 per $100 value. Contractor equipment insurance runs $.50 to $1.50 per $100 value. Insurance for miscellaneous tools to $1,500 value runs from $3.00 to $7.50 per $100 value.

Tabulated below are New England Builder's Risk insurance rates in dollars per $100 value for $1,000 deductible. For $25,000 deductible, rates can be reduced 13% to 34%. On contracts over $1,000,000, rates may be lower than those tabulated. Policies are written annually for the total completed value in place. For "all risk" insurance (excluding flood, earthquake and certain other perils) add $.025 to total rates below.

| Coverage | Frame Construction (Class 1) | | | Brick Construction (Class 4) | | | Fire Resistive (Class 6) | | |
	Range		Average	Range		Average	Range		Average
Fire Insurance	$.350 to	$ .850	$.600	$.158 to	$.189	$.174	$.052 to	$.080	$.070
Extended Coverage	.115 to	.200	.158	.080 to	.105	.101	.081 to	.105	.100
Vandalism	.012 to	.016	.014	.008 to	.011	.011	.008 to	.011	.010
Total Annual Rate	$.477 to	$1.066	$.772	$.246 to	$.305	$.286	$.141 to	$.196	$.180

659

R013113-50   General Contractor's Overhead

There are two distinct types of overhead on a construction project: Project Overhead and Main Office Overhead. Project Overhead includes those costs at a construction site not directly associated with the installation of construction materials. Examples of Project Overhead costs include the following:

1. Superintendent
2. Construction office and storage trailers
3. Temporary sanitary facilities
4. Temporary utilities
5. Security fencing
6. Photographs
7. Clean up
8. Performance and payment bonds

The above Project Overhead items are also referred to as General Requirements and therefore are estimated in Division 1. Division 1 is the first division listed in the CSI MasterFormat but it is usually the last division estimated. The sum of the costs in Divisions 1 through 49 is referred to as the sum of the direct costs.

All construction projects also include indirect costs. The primary components of indirect costs are the contractor's Main Office Overhead and profit. The amount of the Main Office Overhead expense varies depending on the the the following:

1. Owner's compensation
2. Project managers and estimator's wages
3. Clerical support wages
4. Office rent and utilities
5. Corporate legal and accounting costs
6. Advertising
7. Automobile expenses
8. Association dues
9. Travel and entertainment expenses

These costs are usually calculated as a percentage of annual sales volume. This percentage can range from 35% for a small contractor doing less than $500,000 to 5% for a large contractor with sales in excess of $100 million.

R013113-60   Workers' Compensation Insurance Rates by Trade

The table below tabulates the national averages for workers' compensation insurance rates by trade and type of building. The average "Insurance Rate" is multiplied by the "% of Building Cost" for each trade. This produces

the "Workers' Compensation" cost by % of total labor cost, to be added for each trade by building type to determine the weighted average workers' compensation rate for the building types analyzed.

Trade	Insurance Rate (% Labor Cost) Range	Insurance Rate (% Labor Cost) Average	% of Building Cost Office Bldgs.	% of Building Cost Schools & Apts.	% of Building Cost Mfg.	Workers' Compensation Office Bldgs.	Workers' Compensation Schools & Apts.	Workers' Compensation Mfg.
Excavation, Grading, etc.	3.4 % to 21.2%	9.7%	4.8%	4.9%	4.5%	0.47%	0.48%	0.44%
Piles & Foundations	6.1 to 28.3	14.3	7.1	5.2	8.7	1.02	0.74	1.24
Concrete	4.2 to 35.8	12.9	5.0	14.8	3.7	0.65	1.91	0.48
Masonry	4.9 to 36.4	13.7	6.9	7.5	1.9	0.95	1.03	0.26
Structural Steel	6.1 to 101.1	31.7	10.7	3.9	17.6	3.39	1.24	5.58
Miscellaneous & Ornamental Metals	4.3 to 31.5	12.2	2.8	4.0	3.6	0.34	0.49	0.44
Carpentry & Millwork	5.0 to 36.8	14.9	3.7	4.0	0.5	0.55	0.60	0.07
Metal or Composition Siding	6.1 to 69.2	18.4	2.3	0.3	4.3	0.42	0.06	0.79
Roofing	6.1 to 100.3	31.7	2.3	2.6	3.1	0.73	0.82	0.98
Doors & Hardware	4.1 to 36.8	11.3	0.9	1.4	0.4	0.10	0.16	0.05
Sash & Glazing	4.8 to 30.7	13.3	3.5	4.0	1.0	0.47	0.53	0.13
Lath & Plaster	3.1 to 36.2	10.9	3.3	6.9	0.8	0.36	0.75	0.09
Tile, Marble & Floors	3 to 24.8	8.9	2.6	3.0	0.5	0.23	0.27	0.04
Acoustical Ceilings	3.8 to 33.8	8.4	2.4	0.2	0.3	0.20	0.02	0.03
Painting	4.5 to 36.1	11.7	1.5	1.6	1.6	0.18	0.19	0.19
Interior Partitions	5.0 to 36.8	14.9	3.9	4.3	4.4	0.58	0.64	0.66
Miscellaneous Items	2.5 to 132.3	13.4	5.2	3.7	9.7	0.70	0.50	1.30
Elevators	1.3 to 12.8	5.3	2.1	1.1	2.2	0.11	0.06	0.12
Sprinklers	2.9 to 18.5	7.3	0.5	—	2.0	0.04	—	0.15
Plumbing	2.2 to 19.3	7.0	4.9	7.2	5.2	0.34	0.50	0.36
Heat., Vent., Air Conditioning	3.8 to 18.1	8.8	13.5	11.0	12.9	1.19	0.97	1.14
Electrical	2.4 to 13.5	5.8	10.1	8.4	11.1	0.59	0.49	0.64
Total	1.3 % to 132.3%	—	100.0%	100.0%	100.0%	13.61%	12.45%	15.18%
	Overall Weighted Average	13.75%						

Workers' Compensation Insurance Rates by States

The table below lists the weighted average workers' compensation base rate for each state with a factor comparing this with the national average of 13.5%.

State	Weighted Average	Factor	State	Weighted Average	Factor	State	Weighted Average	Factor
Alabama	20.6%	153	Kentucky	12.5%	93	North Dakota	8.1%	60
Alaska	12.5	93	Louisiana	19.6	145	Ohio	8.1	60
Arizona	13.8	102	Maine	11.4	84	Oklahoma	10.3	76
Arkansas	7.7	57	Maryland	13.5	100	Oregon	11.3	84
California	27.1	201	Massachusetts	11.5	85	Pennsylvania	18.3	136
Colorado	7.3	54	Michigan	13.5	100	Rhode Island	14.7	109
Connecticut	25.9	192	Minnesota	22.8	169	South Carolina	16.9	125
Delaware	14.0	104	Mississippi	14.0	104	South Dakota	14.1	104
District of Columbia	10.4	77	Missouri	13.9	103	Tennessee	12.2	90
Florida	10.8	80	Montana	8.3	61	Texas	9.3	69
Georgia	29.5	219	Nebraska	17.0	126	Utah	8.8	65
Hawaii	8.7	64	Nevada	9.2	68	Vermont	13.3	99
Idaho	10.5	78	New Hampshire	18.8	139	Virginia	8.7	64
Illinois	26.9	199	New Jersey	15.7	116	Washington	9.5	70
Indiana	5.0	37	New Mexico	15.3	113	West Virginia	7.4	55
Iowa	15.5	115	New York	19.0	141	Wisconsin	13.2	98
Kansas	9.2	68	North Carolina	17.3	128	Wyoming	6.5	48
			Weighted Average for U.S. is	13.7% of payroll = 100%				

The weighted average skilled worker rate for 35 trades is 13.5%. For bidding purposes, apply the full value of workers' compensation directly to total labor costs, or if labor is 38%, materials 42% and overhead and profit 20% of total cost, carry 38/80 × 13.5% =6.4% of cost (before overhead and profit)

into overhead. Rates vary not only from state to state but also with the experience rating of the contractor.

Rates are the most current available at the time of publication.

R013113-80   Performance Bond

This table shows the cost of a Performance Bond for a construction job scheduled to be completed in 12 months. Add 1% of the premium cost per month for jobs requiring more than 12 months to complete. The rates are "standard" rates offered to contractors that the bonding company considers financially sound and capable of doing the work. Preferred rates are offered by some bonding companies based upon financial strength of the contractor. Actual rates vary from contractor to contractor and from bonding company to bonding company. Contractors should prequalify through a bonding agency before submitting a bid on a contract that requires a bond.

Contract Amount	Building Construction Class B Projects			Highways & Bridges					
				Class A New Construction			Class A-1 Highway Resurfacing		
First $ 100,000 bid	$25.00 per M			$15.00 per M			$9.40 per M		
Next 400,000 bid	$ 2,500	plus $15.00	per M	$ 1,500	plus $10.00	per M	$ 940	plus $7.20	per M
Next 2,000,000 bid	8,500	plus 10.00	per M	5,500	plus 7.00	per M	3,820	plus 5.00	per M
Next 2,500,000 bid	28,500	plus 7.50	per M	19,500	plus 5.50	per M	15,820	plus 4.50	per M
Next 2,500,000 bid	47,250	plus 7.00	per M	33,250	plus 5.00	per M	28,320	plus 4.50	per M
Over 7,500,000 bid	64,750	plus 6.00	per M	45,750	plus 4.50	per M	39,570	plus 4.00	per M

R015423-10   Steel Tubular Scaffolding

On new construction, tubular scaffolding is efficient up to 60' high or five stories. Above this it is usually better to use a hung scaffolding if construction permits. Swing scaffolding operations may interfere with tenants. In this case, the tubular is more practical at all heights.

In repairing or cleaning the front of an existing building the cost of tubular scaffolding per S.F. of building front increases as the height increases above the first tier. The first tier cost is relatively high due to leveling and alignment.

The minimum efficient crew for erecting and dismantling is three workers. They can set up and remove 18 frame sections per day up to 5 stories high. For 6 to 12 stories high, a crew of four is most efficient. Use two or more on top and two on the bottom for handing up or hoisting. They can also set up and remove 18 frame sections per day. At 7' horizontal spacing, this will run about 800 S.F. per day of erecting and dismantling. Time for placing and removing planks must be added to the above. A crew of three can place and remove 72 planks per day up to 5 stories. For over 5 stories, a crew of four can place and remove 80 planks per day.

The table below shows the number of pieces required to erect tubular steel scaffolding for 1000 S.F. of building frontage. This area is made up of a scaffolding system that is 12 frames (11 bays) long by 2 frames high.

For jobs under twenty-five frames, add 50% to rental cost. Rental rates will be lower for jobs over three months duration. Large quantities for long periods can reduce rental rates by 20%.

Description of Component	Number of Pieces for 1000 S.F. of Building Front	Unit
5' Wide Standard Frame, 6'-4" High	24	Ea.
Leveling Jack & Plate	24	
Cross Brace	44	
Side Arm Bracket, 21"	12	
Guardrail Post	12	
Guardrail, 7' section	22	
Stairway Section	2	
Stairway Starter Bar	1	
Stairway Inside Handrail	2	
Stairway Outside Handrail	2	
Walk-Thru Frame Guardrail	2	

Scaffolding is often used as falsework over 15' high during construction of cast-in-place concrete beams and slabs. Two foot wide scaffolding is generally used for heavy beam construction. The span between frames depends upon the load to be carried with a maximum span of 5'.

Heavy duty shoring frames with a capacity of 10,000#/leg can be spaced up to 10' O.C. depending upon form support design and loading.

Scaffolding used as horizontal shoring requires less than half the material required with conventional shoring.

On new construction, erection is done by carpenters.

Rolling towers supporting horizontal shores can reduce labor and speed the job. For maintenance work, catwalks with spans up to 70' can be supported by the rolling towers.

R015433-10   Contractor Equipment

Rental Rates shown elsewhere in the book pertain to late model high quality machines in excellent working condition, rented from equipment dealers. Rental rates from contractors may be substantially lower than the rental rates from equipment dealers depending upon economic conditions; for older, less productive machines, reduce rates by a maximum of 15%. Any overtime must be added to the base rates. For shift work, rates are lower. Usual rule of thumb is 150% of one shift rate for two shifts; 200% for three shifts.

For periods of less than one week, operated equipment is usually more economical to rent than renting bare equipment and hiring an operator.

Costs to move equipment to a job site (mobilization) or from a job site (demobilization) are not included in rental rates, nor in any Equipment costs on any Unit Price line items or crew listings. These costs can be found elsewhere. If a piece of equipment is already at a job site, it is not appropriate to utilize mob/demob costs in an estimate again.

Rental rates vary throughout the country with larger cities generally having lower rates. Lease plans for new equipment are available for periods in excess of six months with a percentage of payments applying toward purchase.

Rental rates can also be treated as reimbursement costs for contractor-owned equipment. Owned equipment costs include depreciation, loan payments, interest, taxes, insurance, storage, and major repairs.

Monthly rental rates vary from 2% to 5% of the cost of the equipment depending on the anticipated life of the equipment and its wearing parts. Weekly rates are about 1/3 the monthly rates and daily rental rates about 1/3 the weekly rate.

The hourly operating costs for each piece of equipment include costs to the user such as fuel, oil, lubrication, normal expendables for the equipment, and a percentage of mechanic's wages chargeable to maintenance. The hourly operating costs listed do not include the operator's wages.

The daily cost for equipment used in the standard crews is figured by dividing the weekly rate by five, then adding eight times the hourly operating cost to give the total daily equipment cost, not including the operator. This figure is in the right hand column of the Equipment listings under Equipment Cost/Day.

Pile Driving rates shown for pile hammer and extractor do not include leads, crane, boiler or compressor. Vibratory pile driving requires an added field specialist during set-up and pile driving operation for the electric model. The hydraulic model requires a field specialist for set-up only. Up to 125 reuses of sheet piling are possible using vibratory drivers. For normal conditions, crane capacity for hammer type and size are as follows.

Crane Capacity	Hammer Type and Size		
	Air or Steam	Diesel	Vibratory
25 ton	to 8,750 ft.-lb.		70 H.P.
40 ton	15,000 ft.-lb.	to 32,000 ft.-lb.	170 H.P.
60 ton	25,000 ft.-lb.		300 H.P.
100 ton		112,000 ft.-lb.	

Cranes should be specified for the job by size, building and site characteristics, availability, performance characteristics, and duration of time required.

Backhoes & Shovels rent for about the same as equivalent size cranes but maintenance and operating expense is higher. Crane operators rate must be adjusted for high boom heights. Average adjustments: for 150' boom add 2% per hour; over 185', add 4% per hour; over 210', add 6% per hour; over 250', add 8% per hour and over 295', add 12% per hour.

Tower Cranes of the climbing or static type have jibs from 50' to 200' and capacities at maximum reach range from 4,000 to 14,000 pounds. Lifting capacities increase up to maximum load as the hook radius decreases.

Typical rental rates, based on purchase price are about 2% to 3% per month.

Erection and dismantling runs between 500 and 2000 labor hours. Climbing operation takes 10 labor hours per 20' climb. Crane dead time is about 5 hours per 40' climb. If crane is bolted to side of the building add cost of ties and extra mast sections. Climbing cranes have from 80' to 180' of mast while static cranes have 80' to 800' of mast.

Truck Cranes can be converted to tower cranes by using tower attachments. Mast heights over 400' have been used.

A single 100' high material **Hoist and Tower** can be erected and dismantled in about 400 labor hours; a double 100' high hoist and tower in about 600 labor hours. Erection times for additional heights are 3 and 4 labor hours

per vertical foot respectively up to 150', and 4 to 5 labor hours per vertical foot over 150' high. A 40' high portable Buck hoist takes about 160 labor hours to erect and dismantle. Additional heights take 2 labor hours per vertical foot to 80' and 3 labor hours per vertical foot for the next 100'. Most material hoists do not meet local code requirements for carrying personnel.

A 150' high **Personnel Hoist** requires about 500 to 800 labor hours to erect and dismantle. Budget erection time at 5 labor hours per vertical foot for all trades. Local code requirements or labor scarcity requiring overtime can add up to 50% to any of the above erection costs.

Earthmoving Equipment: The selection of earthmoving equipment depends upon the type and quantity of material, moisture content, haul distance, haul road, time available, and equipment available. Short haul cut and fill operations may require dozers only, while another operation may require excavators, a fleet of trucks, and spreading and compaction equipment. Stockpiled material and granular material are easily excavated with front end loaders. Scrapers are most economically used with hauls between 300' and 1-1/2 miles if adequate haul roads can be maintained. Shovels are often used for blasted rock and any material where a vertical face of 8' or more can be excavated. Special conditions may dictate the use of draglines, clamshells, or backhoes. Spreading and compaction equipment must be matched to the soil characteristics, the compaction required and the rate the fill is being supplied.

663

R015436-50 Mobilization

Costs to move rented construction equipment to a job site from an equipment dealer's or contractor's yard (mobilization), or to move the equipment off the job site (demobilization), are not included in the rental or operating rates, nor in the equipment cost on a unit price line or in a crew listing. These costs can be found consolidated in the Mobilization section of the data and elsewhere in particular site work sections. If a piece of equipment is already on the job site, it is not appropriate to include mob/demob costs in a new estimate that requires use of that equipment. The following table identifies approximate sizes of rented construction equipment that would be hauled on a towed trailer. Because this listing is not all-encompassing, the user can infer as to what size trailer might be required for a piece of equipment not listed.

3-ton Trailer	20-ton Trailer	40-ton Trailer	50-ton Trailer
20 H.P. Excavator	110 H.P. Excavator	200 H.P. Excavator	270 H.P. Excavator
50 H.P. Skid Steer	165 H.P. Dozer	300 H.P. Dozer	Small Crawler Crane
35 H.P. Roller	150 H.P. Roller	400 H.P. Scraper	500 H.P. Scraper
40 H.P. Trencher	Backhoe	450 H.P. Art. Dump Truck	500 H.P. Art. Dump Truck

R024119-10   Demolition Defined

Whole Building Demolition - Demolition of the whole building with no concern for any particular building element, component, or material type being demolished. This type of demolition is accomplished with large pieces of construction equipment that break up the structure, load it into trucks and haul it to a disposal site, but disposal or dump fees are not included. Demolition of below-grade foundation elements, such as footings, foundation walls, grade beams, slabs on grade, etc., is not included. Certain mechanical equipment containing flammable liquids or ozone-depleting refrigerants, electric lighting elements, communication equipment components, and other building elements may contain hazardous waste, and must be removed, either selectively or carefully, as hazardous waste before the building can be demolished.

Foundation Demolition - Demolition of below-grade foundation footings, foundation walls, grade beams, and slabs on grade. This type of demolition is accomplished by hand or pneumatic hand tools, and does not include saw cutting, or handling, loading, hauling, or disposal of the debris.

Gutting - Removal of building interior finishes and electrical/mechanical systems down to the load-bearing and sub-floor elements of the rough building frame, with no concern for any particular building element, component, or material type being demolished. This type of demolition is accomplished by hand or pneumatic hand tools, and includes loading into trucks, but not hauling, disposal or dump fees, scaffolding, or shoring. Certain mechanical equipment containing flammable liquids or ozone-depleting refrigerants, electric lighting elements, communication equipment components, and other building elements may contain hazardous waste, and must be removed, either selectively or carefully, as hazardous waste, before the building is gutted.

Selective Demolition - Demolition of a selected building element, component, or finish, with some concern for surrounding or adjacent elements, components, or finishes (see the first Subdivision (s) at the beginning of appropriate Divisions). This type of demolition is accomplished by hand or pneumatic hand tools, and does not include handling, loading, storing, hauling, or disposal of the debris, scaffolding, or shoring. "Gutting" methods may be used in order to save time, but damage that is caused to surrounding or adjacent elements, components, or finishes may have to be repaired at a later time.

Careful Removal - Removal of a piece of service equipment, building element or component, or material type, with great concern for both the removed item and surrounding or adjacent elements, components or finishes. The purpose of careful removal may be to protect the removed item for later re-use, preserve a higher salvage value of the removed item, or replace an item while taking care to protect surrounding or adjacent elements, components, connections, or finishes from cosmetic and/or structural damage. An approximation of the time required to perform this type of removal is 1/3 to 1/2 the time it would take to install a new item of like kind (see Reference Number R220105-10). This type of removal is accomplished by hand or pneumatic hand tools, and does not include loading, hauling, or storing the removed item, scaffolding, shoring, or lifting equipment.

Cutout Demolition - Demolition of a small quantity of floor, wall, roof, or other assembly, with concern for the appearance and structural integrity of the surrounding materials. This type of demolition is accomplished by hand or pneumatic hand tools, and does not include saw cutting, handling, loading, hauling, or disposal of debris, scaffolding, or shoring.

Rubbish Handling - Work activities that involve handling, loading or hauling of debris. Generally, the cost of rubbish handling must be added to the cost of all types of demolition, with the exception of whole building demolition.

Minor Site Demolition - Demolition of site elements outside the footprint of a building. This type of demolition is accomplished by hand or pneumatic hand tools, or with larger pieces of construction equipment, and may include loading a removed item onto a truck (check the Crew for equipment used). It does not include saw cutting, hauling or disposal of debris, and, sometimes, handling or loading.

R024119-20   Dumpsters

Dumpster rental costs on construction sites are presented in two ways.

The cost per week rental includes the delivery of the dumpster; its pulling or emptying once per week, and its final removal. The assumption is made that the dumpster contractor could choose to empty a dumpster by simply bringing in an empty unit and removing the full one. These costs also include the disposal of the materials in the dumpster.

The Alternate Pricing can be used when actual planned conditions are not approximated by the weekly numbers. For example, these lines can be used when a dumpster is needed for 4 weeks and will need to be emptied 2 or 3 times per week. Conversely the Alternate Pricing lines can be used when a dumpster will be rented for several weeks or months but needs to be emptied only a few times over this period.

Existing Conditions   R0265 Underground Storage Tank Removal

R026510-20   Underground Storage Tank Removal

Underground Storage Tank Removal can be divided into two categories: Non-Leaking and Leaking. Prior to removing an underground storage tank, tests should be made, with the proper authorities present, to determine whether a tank has been leaking or the surrounding soil has been contaminated.

To safely remove Liquid Underground Storage Tanks:
1. Excavate to the top of the tank.
2. Disconnect all piping.
3. Open all tank vents and access ports.
4. Remove all liquids and/or sludge.
5. Purge the tank with an inert gas.
6. Provide access to the inside of the tank and clean out the interior using proper personal protective equipment (PPE).
7. Excavate soil surrounding the tank using proper PPE for on-site personnel.
8. Pull and properly dispose of the tank.
9. Clean up the site of all contaminated material.
10. Install new tanks or close the excavation.

Existing Conditions | R0282 Asbestos Remediation

R028213-20 Asbestos Removal Process

Asbestos removal is accomplished by a specialty contractor who understands the federal and state regulations regarding the handling and disposal of the material. The process of asbestos removal is divided into many individual steps. An accurate estimate can be calculated only after all the steps have been priced.

The steps are generally as follows:
1. Obtain an asbestos abatement plan from an industrial hygienist.
2. Monitor the air quality in and around the removal area and along the path of travel between the removal area and transport area. This establishes the background contamination.
3. Construct a two part decontamination chamber at entrance to removal area.
4. Install a HEPA filter to create a negative pressure in the removal area.
5. Install wall, floor and ceiling protection as required by the plan, usually 2 layers of fireproof 6 mil polyethylene.

6. Industrial hygienist visually inspects work area to verify compliance with plan.
7. Provide temporary supports for conduit and piping affected by the removal process.
8. Proceed with asbestos removal and bagging process. Monitor air quality as described in Step #2. Discontinue operations when contaminate levels exceed applicable standards.
9. Document the legal disposal of materials in accordance with EPA standards.
10. Thoroughly clean removal area including all ledges, crevices and surfaces.
11. Post abatement inspection by industrial hygienist to verify plan compliance.
12. Provide a certificate from a licensed industrial hygienist attesting that contaminate levels are within acceptable standards before returning area to regular use.

Concrete | R0330 Cast-In-Place Concrete

R033053-60 Maximum Depth of Frost Penetration in Inches

THIS MAP IS REASONABLY ACCURATE FOR MOST PARTS OF THE UNITED STATES BUT IS NECESSARILY HIGHLY GENERALIZED, AND CONSEQUENTLY NOT TOO ACCURATE IN MOUNTAINOUS REGIONS, PARTICULARLY IN THE ROCKIES.

R050521-20   Welded Structural Steel

Usual weight reductions with welded design run 10% to 20% compared with bolted or riveted connections. This amounts to about the same total cost compared with bolted structures since field welding is more expensive than bolts. For normal spans of 18' to 24' figure 6 to 7 connections per ton.

Trusses — For welded trusses add 4% to weight of main members for connections. Up to 15% less steel can be expected in a welded truss compared to one that is shop bolted. Cost of erection is the same whether shop bolted or welded.

General — Typical electrodes for structural steel welding are E6010, E6011, E60T and E70T. Typical buildings vary between 2# to 8# of weld rod per ton of steel. Buildings utilizing continuous design require about three times as much welding as conventional welded structures. In estimating field erection by welding, it is best to use the average linear feet of weld per ton to arrive at the welding cost per ton. The type, size and position of the weld will have a direct bearing on the cost per linear foot. A typical field welder will deposit 1.8# to 2# of weld rod per hour manually. Using semiautomatic methods can increase production by as much as 50% to 75%.

R131113-20   Swimming Pools

Pool prices given per square foot of surface area include pool structure, filter and chlorination equipment, pumps, related piping, ladders/steps, maintenance kit, skimmer and vacuum system. Decks and electrical service to equipment are not included.

Residential in-ground pool construction can be divided into two categories: vinyl lined and gunite. Vinyl lined pool walls are constructed of different materials including wood, concrete, plastic or metal. The bottom is often graded with sand over which the vinyl liner is installed. Vermiculite or soil cement bottoms may be substituted for an added cost.

Gunite pool construction is used both in residential and municipal installations. These structures are steel reinforced for strength and finished with a white cement limestone plaster.

Municipal pools will have a higher cost because plumbing codes require more expensive materials, chlorination equipment and higher filtration rates.

Municipal pools greater than 1,800 S.F. require gutter systems to control waves. This gutter may be formed into the concrete wall. Often a vinyl/stainless steel gutter or gutter/wall system is specified, which will raise the pool cost.

Competition pools usually require tile bottoms and sides with contrasting lane striping, which will also raise the pool cost.

R211226-10   Standpipe Systems

The basis for standpipe system design is National Fire Protection Association NFPA 14, however, the authority having jurisdiction should be consulted for special conditions, local requirements and approval.

Standpipe systems, properly designed and maintained, are an effective and valuable time saving aid for extinquishing fires, especially in the upper stories of tall buildings, the interior of large commercial or industrial malls, or other areas where construction features or access make the laying of temporary hose lines time consuming and/or hazardous. Standpipes are frequently installed with automatic sprinkler systems for maximum protection.

There are three general classes of service for standpipe systems:
Class I for use by fire departments and personnel with special training for heavy streams (2-1/2" hose connections).
Class II for use by building occupants until the arrival of the fire department (1-1/2" hose connector with hose).

Class III for use by either fire departments and trained personnel or by the building occupants (both 2-1/2" and 1-1/2" hose connections or one 2-1/2" hose valve with an easily removable 2-1/2" by 1-1/2" adapter).

Standpipe systems are also classified by the way water is supplied to the system. The four basic types are:
Type 1: Wet standpipe system having supply valve open and water pressure maintained at all times.
Type 2: Standpipe system so arranged through the use of approved devices as to admit water to the system automatically by opening a hose valve.
Type 3: Standpipe system arranged to admit water to the system through manual operation of approved remote control devices located at each hose station.
Type 4: Dry standpipe having no permanent water supply.

Reprinted with permission from NFPA 14-2013, *Installation of Standpipe and Hose Systems*, Copyright © 2013, National Fire Protection Association, Quincy, MA. This reprinted material is not the complete and official position of the NFPA on the referenced subject, which is represented only by the standard in its entirety.

R211226-20   NFPA 14 Basic Standpipe Design

Class	Design-Use	Pipe Size Minimums	Water Supply Minimums
Class I	2 1/2" hose connection on each floor All areas within 150' of an exit in every exit stairway Fire Department Trained Personnel	Height to 100', 4" dia. Heights above 100', 6" dia. (275' max. except with pressure regulators 400' max.)	For each standpipe riser 500 GPM flow For common supply pipe allow 500 GPM for first standpipe plus 250 GPM for each additional standpipe (2500 GPM max. total) 30 min. duration 65 PSI at 500 GPM
Class II	1 1/2" hose connection with hose on each floor All areas within 130' of hose connection measured along path of hose travel Occupant personnel	Height to 50', 2" dia. Height above 50', 2 1/2" dia.	For each standpipe riser 100 GPM flow For multiple riser common supply pipe 100 GPM 300 min. duration, 65 PSI at 100 GPM
Class III	Both of above. Class I valved connections will meet Class III with additional 2 1/2" by 1 1/2" adapter and 1 1/2" hose.	Same as Class I	Same as Class I

*Note: Where 2 or more standpipes are installed in the same building or section of building they shall be interconnected at the bottom.

Combined Systems

Combined systems are systems where the risers supply both automatic sprinklers and 2-1/2" hose connection outlets for fire department use. In such a system the sprinkler spacing pattern shall be in accordance with NFPA 13 while the risers and supply piping will be sized in accordance with NFPA 14. When the building is completely sprinklered the risers may be sized by hydraulic calculation. The minimum size riser for buildings not completely sprinklered is 6".

The minimum water supply of a completely sprinklered, light hazard, high-rise occupancy building will be 500 GPM while the supply required for other types of completely sprinklered high-rise buildings is 1000 GPM.

General System Requirements

1. Approved valves will be provided at the riser for controlling branch lines to hose outlets.
2. A hose valve will be provided at each outlet for attachment of hose.
3. Where pressure at any standpipe outlet exceeds 100 PSI a pressure reducer must be installed to limit the pressure to 100 PSI. Note that the pressure head due to gravity in 100' of riser is 43.4 PSI. This must be overcome by city pressure, fire pumps, or gravity tanks to provide adequate pressure at the top of the riser.
4. Each hose valve on a wet system having linen hose shall have an automatic drip connection to prevent valve leakage from entering the hose.
5. Each riser will have a valve to isolate it from the rest of the system.
6. One or more fire department connections as an auxiliary supply shall be provided for each Class I or Class III standpipe system. In buildings having two or more zones, a connection will be provided for each zone.
7. There will be no shutoff valve in the fire department connection, but a check valve will be located in the line before it joins the system.
8. All hose connections street side will be identified on a cast plate or fitting as to purpose.

Reprinted with permission from NFPA 14-2013, *Installation of Standpipe and Hose Systems*, Copyright © 2013, National Fire Protection Association, Quincy, MA. This reprinted material is not the complete and official position of the NFPA on the referenced subject, which is represented only by the standard in its entirety.

R211313-10   Sprinkler Systems (Automatic)

Sprinkler systems may be classified by type as follows:

1. **Wet Pipe System.** A system employing automatic sprinklers attached to a piping system containing water and connected to a water supply so that water discharges immediately from sprinklers opened by a fire.

2. **Dry Pipe System.** A system employing automatic sprinklers attached to a piping system containing air under pressure, the release of which as from the opening of sprinklers permits the water pressure to open a valve known as a "dry pipe valve". The water then flows into the piping system and out the opened sprinklers.

3. **Pre-Action System.** A system employing automatic sprinklers attached to a piping system containing air that may or may not be under pressure, with a supplemental heat responsive system of generally more sensitive characteristics than the automatic sprinklers themselves, installed in the same areas as the sprinklers; actuation of the heat responsive system, as from a fire, opens a valve which permits water to flow into the sprinkler piping system and to be discharged from any sprinklers which may be open.

4. **Deluge System.** A system employing open sprinklers attached to a piping system connected to a water supply through a valve which is opened by the operation of a heat responsive system installed in the same areas as the sprinklers. When this valve opens, water flows into the piping system and discharges from all sprinklers attached thereto.

5. **Combined Dry Pipe and Pre-Action Sprinkler System.** A system employing automatic sprinklers attached to a piping system containing air under pressure with a supplemental heat responsive system of generally more sensitive characteristics than the automatic sprinklers themselves, installed in the same areas as the sprinklers; operation of the heat responsive system, as from a fire, actuates tripping devices which open dry pipe valves simultaneously and without loss of air pressure in the system. Operation of the heat responsive system also opens approved air exhaust valves at the end of the feed main which facilitates the filling of the system with water which usually precedes the opening of sprinklers. The heat responsive system also serves as an automatic fire alarm system.

6. **Limited Water Supply System.** A system employing automatic sprinklers and conforming to these standards but supplied by a pressure tank of limited capacity.

7. **Chemical Systems.** Systems using halon, carbon dioxide, dry chemical or high expansion foam as selected for special requirements. Agent may extinguish flames by chemically inhibiting flame propagation, suffocate flames by excluding oxygen, interrupting chemical action of oxygen uniting with fuel or sealing and cooling the combustion center.

8. **Firecycle System.** Firecycle is a fixed fire protection sprinkler system utilizing water as its extinguishing agent. It is a time delayed, recycling, preaction type which automatically shuts the water off when heat is reduced below the detector operating temperature and turns the water back on when that temperature is exceeded. The system senses a fire condition through a closed circuit electrical detector system which controls water flow to the fire automatically. Batteries supply up to 90 hour emergency power supply for system operation. The piping system is dry (until water is required) and is monitored with pressurized air. Should any leak in the system piping occur, an alarm will sound, but water will not enter the system until heat is sensed by a firecycle detector.

Area coverage sprinkler systems may be laid out and fed from the supply in any one of several patterns as shown below. It is desirable, if possible, to utilize a central feed and achieve a shorter flow path from the riser to the furthest sprinkler. This permits use of the smallest sizes of pipe possible with resulting savings.

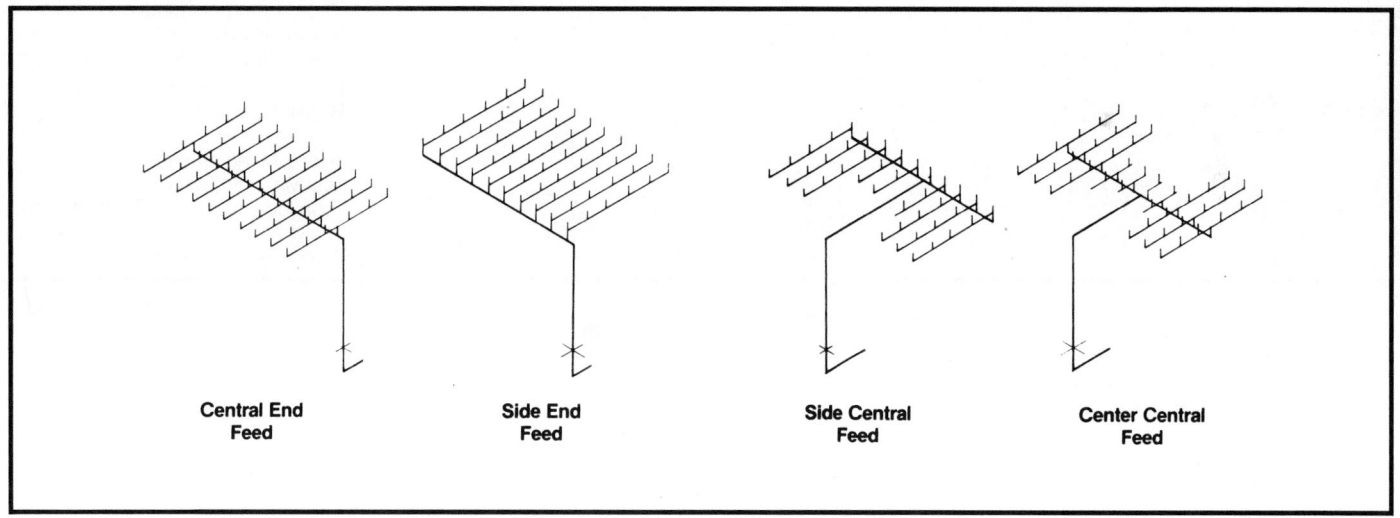

Central End Feed   Side End Feed   Side Central Feed   Center Central Feed

R211313-20 System Classification

System Classification
Rules for installation of sprinkler systems vary depending on the classification of occupancy falling into one of three categories as follows:

Light Hazard Occupancy
The protection area allotted per sprinkler should not exceed 225 S.F., with the maximum distance between lines and sprinklers on lines being 15'. The sprinklers do not need to be staggered. Branch lines should not exceed eight sprinklers on either side of a cross main. Each large area requiring more than 100 sprinklers and without a sub-dividing partition should be supplied by feed mains or risers sized for ordinary hazard occupancy.
Maximum system area = 52,000 S.F.

Included in this group are:

Churches	Nursing Homes
Clubs	Offices
Educational	Residential
Hospitals	Restaurants
Institutional	Theaters and Auditoriums
Libraries	(except stages and prosceniums)
(except large stack rooms)	Unused Attics
Museums	

Ordinary Hazard Occupancy
The protection area allotted per sprinkler shall not exceed 130 S.F. of noncombustible ceiling and 130 S.F. of combustible ceiling. The maximum allowable distance between sprinkler lines and sprinklers on line is 15'. Sprinklers shall be staggered if the distance between heads exceeds 12'. Branch lines should not exceed eight sprinklers on either side of a cross main.
Maximum system area = 52,000 S.F.

Included in this group are:

Group 1	Group 2
Automotive Parking and Showrooms	Cereal Mills
Bakeries	Chemical Plants—Ordinary
Beverage manufacturing	Confectionery Products
Canneries	Distilleries
Dairy Products Manufacturing/Processing	Dry Cleaners
Electronic Plans	Feed Mills
Glass and Glass Products Manufacturing	Horse Stables
Laundries	Leather Goods Manufacturing
Restaurant Service Areas	Libraries—Large Stack Room Areas
	Machine Shops
	Metal Working
	Mercantile
	Paper and Pulp Mills
	Paper Process Plants
	Piers and Wharves
	Post Offices
	Printing and Publishing
	Repair Garages
	Stages
	Textile Manufacturing
	Tire Manufacturing
	Tobacco Products Manufacturing
	Wood Machining
	Wood Product Assembly

Extra Hazard Occupancy
The protection area allotted per sprinkler shall not exceed 100 S.F. of noncombustible ceiling and 100 S.F. of combustible ceiling. The maximum allowable distance between lines and between sprinklers on lines is 12'. Sprinklers on alternate lines shall be staggered if the distance between sprinklers on lines exceeds 8'. Branch lines should not exceed six sprinklers on either side of a cross main.
Maximum system area:
   Design by pipe schedule = 25,000 S.F.
   Design by hydraulic calculation = 40,000 S.F.

Included in this group are:

Group 1	Group 2
Aircraft hangars	Asphalt Saturating
Combustible Hydraulic Fluid Use Area	Flammable Liquids Spraying
Die Casting	Flow Coating
Metal Extruding	Manufactured/Modular Home
Plywood/Particle Board Manufacturing	Building Assemblies (where
Printing (inks with flash points < 100	finished enclosure is present
degrees F	and has combustible interiors)
Rubber Reclaiming, Compounding,	Open Oil Quenching
Drying, Milling, Vulcanizing	Plastics Processing
Saw Mills	Solvent Cleaning
Textile Picking, Opening, Blending,	Varnish and Paint Dipping
Garnetting, Carding, Combing of	
Cotton, Synthetics, Wood Shoddy,	
or Burlap	
Upholstering with Plastic Foams	

R211313-40 Adjustment for Sprinkler/Standpipe Installations

Quality/Complexity Multiplier (For all installations)

Economy installation, add	0 to 5%
Good quality, medium complexity, add	5 to 15%
Above average quality and complexity, add	15 to 25%

R220102-20 Labor Adjustment Factors

Labor Adjustment Factors are provided for Divisions 21, 22, and 23 to assist the mechanical estimator account for the various complexities and special conditions of any particular project. While a single percentage has been entered on each line of Division 22 01 02.20, it should be understood that these are just suggested midpoints of ranges of values commonly used by mechanical estimators. They may be increased or decreased depending on the severity of the special conditions.

The group for "existing occupied buildings", has been the subject of requests for explanation. Actually there are two stages to this group: buildings that are existing and "finished" but unoccupied, and those that also are occupied. Buildings that are "finished" may result in higher labor costs due to the workers having to be more careful not to damage finished walls, ceilings, floors, etc. and may necessitate special protective coverings and barriers. Also corridor bends and doorways may not accommodate long pieces of pipe or larger pieces of equipment. Work above an already hung ceiling can be very time consuming. The addition of occupants may force the work to be done on premium time (nights and/or weekends), eliminate the possible use of some preferred tools such as pneumatic drivers, powder charged drivers etc. The estimator should evaluate the access to the work area and just how the work is going to be accomplished to arrive at an increase in labor costs over "normal" new construction productivity.

R220105-10 Demolition (Selective vs. Removal for Replacement)

Demolition can be divided into two basic categories.

One type of demolition involves the removal of material with no concern for its replacement. The labor-hours to estimate this work are found under "Selective Demolition" in the Fire Protection, Plumbing and HVAC Divisions. It is selective in that individual items or all the material installed as a system or trade grouping such as plumbing or heating systems are removed. This may be accomplished by the easiest way possible, such as sawing, torch cutting, or sledge hammer as well as simple unbolting.

The second type of demolition is the removal of some item for repair or replacement. This removal may involve careful draining, opening of unions, disconnecting and tagging of electrical connections, capping of pipes/ducts to prevent entry of debris or leakage of the material contained as well as transport of the item away from its in-place location to a truck/dumpster. An approximation of the time required to accomplish this type of demolition is to use half of the time indicated as necessary to install a new unit. For example; installation of a new pump might be listed as requiring 6 labor-hours so if we had to estimate the removal of the old pump we would allow an additional 3 hours for a total of 9 hours. That is, the complete replacement of a defective pump with a new pump would be estimated to take 9 labor-hours.

671

R220523-80   Valve Materials

VALVE MATERIALS

Bronze:
Bronze is one of the oldest materials used to make valves. It is most commonly used in hot and cold water systems and other non-corrosive services. It is often used as a seating surface in larger iron body valves to ensure tight closure.

Carbon Steel:
Carbon steel is a high strength material. Therefore, valves made from this metal are used in higher pressure services, such as steam lines up to 600 psi at 850°F. Many steel valves are available with butt-weld ends for economy and are generally used in high pressure steam service as well as other higher pressure non-corrosive services.

Forged Steel:
Valves from tough carbon steel are used in service up to 2000 psi and temperatures up to 1000°F in Gate, Globe and Check valves.

Iron:
Valves are normally used in medium to large pipe lines to control non-corrosive fluid and gases, where pressures do not exceed 250 psi at 450° or 500 psi cold water, oil or gas.

Stainless Steel:
Developed steel alloys can be used in over 90% corrosive services.

Plastic PVC:
This is used in a great variety of valves generally in high corrosive service with lower temperatures and pressures.

VALVE SERVICE PRESSURES

Pressure ratings on valves provide an indication of the safe operating pressure for a valve at some elevated temperature. This temperature is dependent upon the materials used and the fabrication of the valve. When specific data is not available, a good "rule-of-thumb" to follow is the temperature of saturated steam on the primary rating indicated on the valve body. Example: The valve has the number 150S printed on the side indicating 150 psi and hence, a maximum operating temperature of 367°F (temperature of saturated steam and 150 psi).

DEFINITIONS

1. "WOG" – Water, oil, gas (cold working pressures).
2. "SWP" – Steam working pressure.
3. 100% area (full port) – means the area through the valve is equal to or greater than the area of standard pipe.
4. "Standard Opening" – means that the area through the valve is less than the area of standard pipe and therefore these valves should be used only where restriction of flow is unimportant.
5. "Round Port" – means the valve has a full round opening through the plug and body, of the same size and area as standard pipe.
6. "Rectangular Port" – valves have rectangular shaped ports through the plug body. The area of the port is either equal to 100% of the area of standard pipe, or restricted (standard opening). In either case it is clearly marked.
7. "ANSI" – American National Standards Institute.

R220523-90   Valve Selection Considerations

INTRODUCTION: In any piping application, valve performance is critical. Valves should be selected to give the best performance at the lowest cost.

The following is a list of performance characteristics generally expected of valves.
1. Stopping flow or starting it.
2. Throttling flow (Modulation).
3. Flow direction changing.
4. Checking backflow (Permitting flow in only one direction).
5. Relieving or regulating pressure.

In order to properly select the right valve, some facts must be determined.

A. What liquid or gas will flow through the valve?

B. Does the fluid contain suspended particles?

C. Does the fluid remain in liquid form at all times?

D. Which metals does fluid corrode?

E. What are the pressure and temperature limits? (As temperature and pressure rise, so will the price of the valve.)

F. Is there constant line pressure?

G. Is the valve merely an on-off valve?

H. Will checking of backflow be required?

I. Will the valve operate frequently or infrequently?

Valves are classified by design type into such classifications as Gate, Angle, Check, Ball, Butterfly and Plug. They are also classified by end connection, stem, pressure restrictions and material such as bronze, cast iron, etc. Each valve has a specific use. A quality valve used correctly will provide a lifetime of trouble-free service, but a high quality valve installed in the wrong service may require frequent attention.

STEM TYPES
(OS & Y)—Rising Stem-Outside Screw and Yoke

Offers a visual indication of whether the valve is open or closed. Recommended where high temperatures, corrosives, and solids in the line might cause damage to inside-valve stem threads. The stem threads are engaged by the yoke bushing so the stem rises through the hand wheel as it is turned.

(R.S.)—Rising Stem-Inside Screw

Adequate clearance for operation must be provided because both the hand wheel and the stem rise.
The valve wedge position is indicated by the position of the stem and hand wheel.

(N.R.S.)—Non-Rising Stem-Inside Screw

A minimum clearance is required for operating this type of valve. Excessive wear or damage to stem threads inside the valve may be caused by heat, corrosion, and solids. Because the hand wheel and stem do not rise, wedge position cannot be visually determined.

VALVE TYPES
Gate Valves

Provide full flow, minute pressure drop, minimum turbulence and minimum fluid trapped in the line.
They are normally used where operation is infrequent.

Globe Valves

Globe valves are designed for throttling and/or frequent operation with positive shut-off. Particular attention must be paid to the several types of seating materials available to avoid unnecessary wear. The seats must be compatible with the fluid in service and may be composition or metal. The configuration of the Globe valve opening causes turbulence which results in increased resistance. Most bronze Globe valves are rising stem-inside screw, but they are also available on O.S. & Y.

Angle Valves

The fundamental difference between the Angle valve and the Globe valve is the fluid flow through the Angle valve. It makes a 90° turn and offers less resistance to flow than the Globe valve while replacing an elbow. An Angle valve thus reduces the number of joints and installation time.

Check Valves

Check valves are designed to prevent backflow by automatically seating when the direction of fluid is reversed.
Swing Check valves are generally installed with Gate-valves, as they provide comparable full flow. Usually recommended for lines where flow velocities are low and should not be used on lines with pulsating flow. Recommended for horizontal installation, or in vertical lines only where flow is upward.

Lift Check Valves

These are commonly used with Globe and Angle valves since they have similar diaphragm seating arrangements and are recommended for preventing backflow of steam, air, gas and water, and on vapor lines with high flow velocities. For horizontal lines, horizontal lift checks should be used and vertical lift checks for vertical lines.

Ball Valves

Ball valves are light and easily installed, yet because of modern elastomeric seats, provide tight closure. Flow is controlled by rotating up to 90° a drilled ball which fits tightly against resilient seals. This ball seats with flow in either direction, and valve handle indicates the degree of opening. Recommended for frequent operation readily adaptable to automation, ideal for installation where space is limited.

Butterfly Valves

Butterfly valves provide bubble-tight closure with excellent throttling characteristics. They can be used for full-open, closed and for throttling applications.

The Butterfly valve consists of a disc within the valve body which is controlled by a shaft. In its closed position, the valve disc seals against a resilient seat. The disc position throughout the full 90° rotation is visually indicated by the position of the operator.

A Butterfly valve is only a fraction of the weight of a Gate valve and requires no gaskets between flanges in most cases. Recommended for frequent operation and adaptable to automation where space is limited.

Wafer and Lug type bodies when installed between two pipe flanges, can be easily removed from the line. The pressure of the bolted flanges holds the valve in place.
Locating lugs makes installation easier.

Plug Valves

Lubricated plug valves, because of the wide range of service to which they are adapted, may be classified as all purpose valves. They can be safely used at all pressure and vacuums, and at all temperatures up to the limits of available lubricants. They are the most satisfactory valves for the handling of gritty suspensions and many other destructive, erosive, corrosive and chemical solutions.

R221113-40   Plumbing Approximations for Quick Estimating

Water Control

Water Meter; Backflow Preventer, ... 10 to 15% of Fixtures
Shock Absorbers; Vacuum Breakers;
Mixer.

Pipe And Fittings .... 30 to 60% of Fixtures

> **Note:** Lower percentage for compact buildings or larger buildings with plumbing in one area.
> Larger percentage for large buildings with plumbing spread out.
> In extreme cases pipe may be more than 100% of fixtures.
> Percentages **do not** include special purpose or process piping.

Plumbing Labor

1 & 2 Story Residential ....Rough-in Labor = 80% of Materials
Apartment Buildings .... Rough-in Labor = 90 to 100% of Materials
Labor for handling and placing fixtures is approximately 25 to 30% of fixtures

Quality/Complexity Multiplier (for all installations)

Economy installation, add.. 0 to 5%
Good quality, medium complexity, add .... 5 to 15%
Above average quality and complexity, add .... 15 to 25%

R221113-50   Pipe Material Considerations

1. Malleable fittings should be used for gas service.
2. Malleable fittings are used where there are stresses/strains due to expansion and vibration.
3. Cast fittings may be broken as an aid to disassembling of heating lines frozen by long use, temperature and minerals.
4. Cast iron pipe is extensively used for underground and submerged service.
5. Type M (light wall) copper tubing is available in hard temper only and is used for nonpressure and less severe applications than K and L.

6. Type L (medium wall) copper tubing, available hard or soft for interior service.
7. Type K (heavy wall) copper tubing, available in hard or soft temper for use where conditions are severe. For underground and interior service.
8. Hard drawn tubing requires fewer hangers or supports but should not be bent. Silver brazed fittings are recommended, however soft solder is normally used.
9. Type DMV (very light wall) copper tubing designed for drainage, waste and vent plus other non-critical pressure services.

Domestic/Imported Pipe and Fittings Cost

The prices shown in this publication for steel/cast iron pipe and steel, cast iron, malleable iron fittings are based on domestic production sold at the normal trade discounts. The above listed items of foreign manufacture may be available at prices of 1/3 to 1/2 those shown. Some imported items after minor machining or finishing operations are being sold as domestic to further complicate the system.

Caution: Most pipe prices in this book also include a coupling and pipe hangers which for the larger sizes can add significantly to the per foot cost and should be taken into account when comparing "book cost" with quoted supplier's cost.

R221113-70 Piping to 10' High

When taking off pipe, it is important to identify the different material types and joining procedures, as well as distances between supports and components required for proper support.

During the takeoff, measure through all fittings. Do not subtract the lengths of the fittings, valves, or strainers, etc. This added length plus the final rounding of the totals will compensate for nipples and waste.

When rounding off totals always increase the actual amount to correspond with manufacturer's shipping lengths.

A. Both red brass and yellow brass pipe are normally furnished in 12' lengths, plain end. The Unit Price section includes in the linear foot costs two field threads and one coupling per 10' length. A carbon steel clevis type hanger assembly every 10' is also prorated into the linear foot costs, including both material and labor.

B. Cast iron soil pipe is furnished in either 5' or 10' lengths. For pricing purposes, the Unit Price section features 10' lengths with a joint and a carbon steel clevis hanger assembly every 5' prorated into the per foot costs of both material and labor.

Three methods of joining are considered: lead and oakum poured joints or push-on gasket type joints for the bell and spigot pipe and a joint clamp for the no-hub soil pipe. The labor and material costs for each of these individual joining procedures are also prorated into the linear costs per foot.

C. Copper tubing covers types K, L, M, and DWV which are furnished in 20' lengths. Means pricing data is based on a tubing cut each length and a coupling and two soft soldered joints every 10'. A carbon steel, clevis type hanger assembly every 10' is also prorated into the per foot costs. The prices for refrigeration tubing are for materials only. Labor for full lengths may be based on the type L labor but short cut measures in tight areas can increase the installation labor-hours from 20 to 40%.

D. Corrosion-resistant piping does not lend itself to one particular standard of hanging or support assembly due to its diversity of application and placement. The several varieties of corrosion-resistant piping do not include any material or labor costs for hanger assemblies (See the Unit Price section for appropriate selection).

E. Glass pipe is furnished in standard lengths either 5' or 10' long, beaded on one end. Special orders for diverse lengths beaded on both ends are also available. For pricing purposes, R.S. Means features 10' lengths with a coupling and a carbon steel band hanger assembly every 10' prorated into the per foot linear costs.

Glass pipe is also available with conical ends and standard lengths ranging from 6" through 3' in 6" increments, then up to 10' in 12" increments. Special lengths can be customized for particular installation requirements.

For pricing purposes, Means has based the labor and material pricing on 10' lengths. Included in these costs per linear foot are the prorated costs for a flanged assembly every 10' consisting of two flanges, a gasket, two insertable seals, and the required number of bolts and nuts. A carbon steel band hanger assembly based on 10' center lines has also been prorated into the costs per foot for labor and materials.

F. Plastic pipe of several compositions and joining methods are considered. Fiberglass reinforced pipe (FRP) is priced, based on 10' lengths (20' lengths are also available), with coupling and epoxy joints every 10'. FRP is furnished in both "General Service" and "High Strength." A carbon steel clevis hanger assembly, 3 for every 10', is built into the prorated labor and material costs on a per foot basis.

The PVC and CPVC pipe schedules 40, 80 and 120, plus SDR ratings are all based on 20' lengths with a coupling installed every 10', as well as a carbon steel clevis hanger assembly every 3'. The PVC and

ABS type DWV piping is based on 10' lengths with solvent weld couplings every 10', and with carbon steel clevis hanger assemblies, 3 for every 10'. The rest of the plastic piping in this section is based on flexible 100' coils and does not include any coupling or supports.

This section ends with PVC drain and sewer piping based on 10' lengths with bell and spigot ends and 0-ring type, push-on joints.

G. Stainless steel piping includes both weld end and threaded piping, both in the type 304 and 316 specification and in the following schedules, 5, 10, 40, 80, and 160. Although this piping is usually furnished in 20' lengths, this cost grouping has a joint (either heli-arc butt-welded or threads and coupling) every 10'. A carbon steel clevis type hanger assembly is also included at 10' intervals and prorated into the linear foot costs.

H. Carbon steel pipe includes both black and galvanized. This section encompasses schedules 40 (standard) and 80 (extra heavy).

Several common methods of joining steel pipe — such as thread and coupled, butt welded, and flanged (150 lb. weld neck flanges) are also included.

For estimating purposes, it is assumed that the piping is purchased in 20' lengths and that a compatible joint is made up every 10'. These joints are prorated into the labor and material costs per linear foot. The following hanger and support assemblies every 10' are also included: carbon steel clevis for the T & C pipe, and single rod roll type for both the welded and flanged piping. All of these hangers are oversized to accommodate pipe insulation 3/4" thick through 5" pipe size and 1-1/2" thick from 6" through 12" pipe size.

I. Grooved joint steel pipe is priced both black and galvanized, in schedules 10, 40, and 80, furnished in 20' lengths. This section describes two joining methods: cut groove and roll groove. The schedule 10 piping is roll-grooved, while the heavier schedules are cut-grooved. The labor and material costs are prorated into per linear foot prices, including a coupled joint every 10', as well as a carbon steel clevis hanger assembly.

Notes:

The pipe hanger assemblies mentioned in the preceding paragraphs include the described hanger; appropriately sized steel, box-type insert and nut; plus 18" of threaded hanger rod.

C clamps are used when the pipe is to be supported from steel shapes rather than anchored in the slab. C clamps are slightly less costly than inserts. However, to save time in estimating, it is advisable to use the given line number cost, rather than substituting a C clamp for the insert.

Add to piping labor for elevated installation:

10' to 14.5' high	10%	30' to 34.5' high	40%
15' to 19.5' high	20%	35' to 39.5' high	50%
20' to 24.5' high	25%	40' and higher	55%
25' to 29.5' high	35%		

When using the percentage adds for elevated piping installations as shown above, bear in mind that the given heights are for the pipe supports, even though the insert, anchor, or clamp may be several feet higher than the pipe itself.

An allowance has been included in the piping installation time for testing and minor tightening of leaking joints, fittings, stuffing boxes, packing glands, etc. For extraordinary test requirements such as x-rays, prolonged pressure or demonstration tests, a percentage of the piping labor, based on the estimator's experience, must be added to the labor total. A testing service specializing in weld x-rays should be consulted for pricing if it is an estimate requirement. Equipment installation time includes start-up with associated adjustments.

R221316-10   Drainage Fixture Units for Fixtures and Groups

Fixture Type	Drainage Fixture Unit Value as Load Factors	Minimum Size of Trap (inches)
Automatic Clothes Washer, commercial (Note A)	3.0	2.0
Automatic Clothes Washer, residential	2.0	2.0
Full Bathroom Group w/bathtub or shower stall (1.6 gpf WC)	5.0	-
Full Bathroom Group w/bathtub or shower stall (WC flushing greater than 1.6 gpf) (Note D)	6.0	-
Bathtub (with or without overhead shower)	2.0	1.5
Bidet	1.0	1.25
Drinking Fountain	0.5	1025
Dishwasher, domestic	2.0	1.5
Emergency floor drains	0.0	2
Floor drains	2.0	2
Kitchen sinks, domestic	2.0	1.5
Lavatory	1.0	1.25
Shower (5.7 gpm or less flow rate)	2.0	1.5
Shower (5.7 gpm - 12.3 gpm flow rate)	3.0	2
Shower (12.3 gpm - 25.8 gpm flow rate)	5.0	3
Shower (25.8 gpm to 55.6 gpm flow rate)	6.0	4
Service Sink	2.0	1.5
Urinal	4.0	Note C
Urinal (1.0 gpf or less)	2.0	Note C
Urinal (nonwater supplied)	0.5	Note C
Water closet, flushometer tank, public or private	4.0	Note C
Water closet, private (1.6 gpf)	3.0	Note C
Water closet, private (flushing greater than 1.6 gpf)	4.0	Note C
Water closet, public (1.6 gpf)	4.0	Note C
Water closet, public (flushing greater than 1.6 gpf)	6.0	Note C

Notes:
A. A showerhead over a bathtub or whirlpool tub attachment does not increase the drainage fixture unit value.
B. Trap size shall be consistent with the fixture outlet size.
C. For fixtures added to a bathroom group, add the DFU value to those additional fixtures to the bathroom group fixture count.

Excerpted from the 2012 *International Plumbing Code*, Copyright 2011.
Washington, D.C.: International Code Council. Reproduced with permission.
All rights reserved. www.ICCSAFE.org

R221316-15   Drainage Fixture Units for Fixture Drains or Traps

Fixture Drain or Trap Size (inches)	Drainage Fixture Unit Value
1.25	1.0
1.5	2.0
2.0	3.0
2.5	4.0
3.0	5.0
4.0	6.0

Notes:
Drainage Fixture Unit values designate the relative load weight or different kinds of fixtures that shall be emplored in estimating the total load carried by a soil or waste pipe.

Excerpted from the 2012 *International Plumbing Code,* Copyright 2011.
Washington, D.C.: International Code Council. Reproduced with permission.
All rights reserved. www.ICCSAFE.org

R221316-20   Allowable Fixture Units (d.f.u.) for Branches and Stacks

Pipe Diam.	Horiz. Branch (not incl. drains)	Stack Size for 3 Stories or 3 Levels	Stack size for Over 3 levels	Maximum for 1 Story building Stack
1-1/2"	3	4	8	2
2"	6	10	24	6
2-1/2"	12	20	42	9
3"	20*	48*	72*	20*
4"	160	240	500	90
5"	360	540	1100	200
6"	620	960	1900	350
8"	1400	2200	3600	600
10"	2500	3800	5600	1000
12"	3900	6000	8400	1500
15"	7000			

*Not more than two water closets or bathroom groups within each branch interval nor more than six water closets or bathroom groups on the stack.

Stacks sized for the total may be reduced as load decreases at each story to a minimum diameter of 1/2 the maximum diameter.

R224000-10   Water Consumption Rates

Fixture Type	Water Supply Fixture Unit Value		
	Hot Water	Cold Water	Combined
Bathtub	1.0	1.0	1.4
Clothes Washer	1.0	1.0	1.4
Dishwasher	1.4	0.0	1.4
Kitchen Sink	1.0	1.0	1.4
Laundry Tub	1.0	1.0	1.4
Lavatory	0.5	0.5	0.7
Shower Stall	1.0	1.0	1.4
Water Closet (tank type)	0.0	2.2	2.2
Full Bath Group			
w/bathtub or shower stall	1.5	2.7	3.6
Half Bath Group			
w/W.C. and Lavatory	0.5	2.5	2.6
Kitchen Group			
w/Dishwasher and Sink	1.9	1.0	2.5
Laundry Group			
w/Clothes Washer and			
Laundry Tub	1.8	1.8	2.5
Hose bibb (sillcock)	0.0	2.5	2.5

Notes:
Typically, WSFU = 1GPM

Supply loads in the building water-distribution system shall be determined by total load on the pipe being sized, in terms of water supply fixture units (WSFU) and gallons per minute (GPM) flow rates. For fixtures not listed, choose a WSFU value of a fixture with similar flow characteristics. Water Fixture Supply Units determined the required water supply to fixtures and their service systems. Fixture units are equal to (1) cubic foot of water drained in a 1-1/4" pipe per minute. It is not a flow rate unit but a design factor.

Excerpted from the 2012 *International Plumbing Code*, Copyright 2011.
Washington, D.C.: International Code Council. Reproduced with permission.
All rights reserved. www.ICCSAFE.org

R224000-20   Fixture Demands in Gallons per Fixture per Hour

Table below is based on 140°F final temperature except for dishwashers in public places (*) where 180°F water is mandatory.

Supply Systems for Flush Tanks			Supply Systems for Flushometer Valves		
Load	Demand		Load	Demand	
WSFU	GPM	CU. FT.	WSFU	GPM	CU. FT.
1.0	3.0	0.041040	-	-	-
2.0	5.0	0.068400	-	-	-
3.0	6.5	0.868920	-	-	-
4.0	8.0	1.069440	-	-	-
5.0	9.4	1.256592	5.0	15	2.0052
6.0	10.7	1.430376	6.0	17.4	2.326032
7.0	11.8	1.577424	7.0	19.8	2.646364
8.0	12.8	1.711104	8.0	22.2	2.967696
9.0	13.7	1.831416	9.0	24.6	3.288528
10.0	14.6	1.951728	10.0	27	3.60936
11.0	15.4	2.058672	11.0	27.8	3.716304
12.0	16.0	2.138880	12.0	28.6	3.823248
13.0	16.5	2.205720	13.0	29.4	3.930192
14.0	17.0	2.272560	14.0	30.2	4.037136
15.0	17.5	2.339400	15.0	31	4.14408
16.0	18.0	2.906240	16.0	31.8	4.241024
17.0	18.4	2.459712	17.0	32.6	4.357968
18.0	18.8	2.513184	18.0	33.4	4.464912
19.0	19.2	2.566656	19.0	34.2	4.571856
20.0	19.6	2.620218	20.0	35.0	4.678800
25.0	21.5	2.874120	25.0	38.0	5.079840
30.0	23.3	3.114744	30.0	42.0	5.611356
35.0	24.9	3.328632	35.0	44.0	5.881920
40.0	26.3	3.515784	40.0	46.0	6.149280
45.0	27.7	3.702936	45.0	48.0	6.416640
50.0	29.1	3.890088	50.0	50.0	6.684000

Notes:
When designing a plumbing system that utilizes fixtures other than, or in addition to, water closets, use the data provided in the Supply Systems for Flush Tanks section of the above table.

To obtain the probable maximum demand, multiply the total demands for the fixtures (gal./fixture/hour) by the demand factor. The heater should have a heating capacity in gallons per hour equal to this maximum. The storage tank should have a capacity in gallons equal to the probable maximum demand multiplied by the storage capacity factor.

Excerpted from the 2012 *International Residential Code*, Copyright 2011. Washington, D.C.: International Code Council. Reproduced with permission. All rights reserved. www.ICCSAFE.org

R224000-30 Minimum Plumbing Fixture Requirements

Classification	Occupancy	Description	Water		Lavatories		Bathtubs/Showers	Drinking Fountains	Other
			Male	Female	Male	Female			
Assembly	A-1	Theaters and other buildings for the performing arts and motion pictures	1:125	1:65	1:200			1:500	1 Service Sink
	A-2	Nightclubs, bars, taverns dance halls	1:40		1:75			1:500	1 Service Sink
		Restaurants, banquet halls, food courts	1:75		1:200			1:500	1 Service Sink
	A-3	Auditorium w/o permanent seating, art galleries, exhibition halls, museums, lecture halls, libraries, arcades & gymnasiums	1:125	1:65	1:200			1:500	1 Service Sink
		Passenger terminals and transportation facilities	1:500		1:750			1:1000	1 Service Sink
		Places of worship and other religious services	1:150	1:75	1:200			1:1000	1 Service Sink
	A-4	Indoor sporting events and activities, coliseums, arenas, skating rinks, pools, and tennis courts	1:75 for the first 1500, then 1:120 for the remainder	1:40 for the first 1520, then 1:60 for the remainder	1:200	1:150		1:1000	1 Service Sink
	A-5	Outdoor sporting events and activities, stadiums, amusement parks, bleachers, grandstands	1:75 for the first 1500, then 1:120 for the remainder	1:40 for the first 1520, then 1:60 for the remainder	1:200	1:150		1:1000	1 Service Sink
Business	B	Buildings for the transaction of business, professional services, other services involving merchandise, office buildings, banks, light industrial	1:25 for the first 50, then 1:50 for the remainder		1:40 for the first 80, then 1:80 for the remainder			1:100	1 Service Sink
Educational	E	Educational Facilities	1:50		1:50			1:100	1 Service Sink
Factory and industrial	F-1 and F-2	Structures in which occupants are engaged in work fabricating, assembly or processing of products or materials	1:100		1:100		See International Plumbing Code	1:400	1 Service Sink
Institutional	I-1	Residential Care	1:10		1:10		1:8	1:100	1 Service Sink
	I-2	Hospitals, ambulatory nursing home care recipient	1 per room		1 per room		1:15	1:100	1 Service Sink
		Employees, other than residential care	1:25		1:35			1:100	
		Visitors, other than residential care	1:75		1:100			1:500	
	I-3	Prisons	1 per cell		1 per cell		1:15	1:100	1 Service Sink
		Reformatories, detention and correction centers	1:15		1:15		1:15	1:100	1 Service Sink
		Employees	1:25		1:35			1:100	
	I-4	Adult and child day care	1:15		1:15		1	1:100	1 Service Sink
Mercantile	M	Retail stores, service stations, shops, salesrooms, markets and shopping centers	1:500		1:750			1:1000	1 Service Sink
Residential	R-1	Hotels, Motels, boarding houses (transient)	1 per sleeping unit		1 per sleeping unit		1 per sleeping unit		1 Service Sink
	R-2	Dormitories, fraternities, sororities and boarding houses (not transient)	1:10		1:10		1:8	1:100	1 Service Sink
		Apartment House	1 per dwelling unit		1 per dwelling unit		1 per dwelling unit		1 Kitchen sink per dwelling; 1 clothes washer connection per 20 dwellings
	R-3	1 and 2 Family Dwellings	1 per dwelling unit		1:10		1 per dwelling unit		1 Kitchen sink per dwelling; 1 clothes washer connection per dwelling
	R-3	Congregate living facilities w/<16 people	1:10		1:10		1:8	1:100	1 Service Sink
	R-4	Congregate living facilities w/<16 people	1:10		1:10		1:8	1:100	1 Service Sink
Storage	S-1 and S-2	Structures for the storage of good, warehouses, storehouses and freight depots, low and moderate hazard	1:100		1:100		See International Plumbing Code	1;1000	1 Service Sink

Table 2902.1

Excerpted from the 2012 *International Building Code*, Copyright 2011. Washington, D.C.: International Code Council. Reproduced with permission. All rights reserved. www.ICCSAFE.org

R230500-10   Subcontractors

On the unit cost pages of the R.S. Means Cost Data books, the last column is entitled "Total Incl. O&P". This is normally the cost of the installing contractor. In the HVAC Division, this is the cost of the mechanical contractor. If the particular work being estimated is to be performed by a sub to the mechanical contractor, the mechanical's profit and handling charge (usually 10%) is added to the total of the last column.

R235616-60   Solar Heating (Space and Hot Water)

Collectors should face as close to due South as possible, however, variations of up to 20 degrees on either side of true South are acceptable. Local climate and collector type may influence the choice between east or west deviations. Obviously they should be located so they are not shaded from the sun's rays. Incline collectors at a slope of latitude minus 5 degrees for domestic hot water and latitude plus 15 degrees for space heating.

Flat plate collectors consist of a number of components as follows: Insulation to reduce heat loss through the bottom and sides of the collector. The enclosure which contains all the components in this assembly is usually weatherproof and prevents dust, wind and water from coming in contact with the absorber plate. The cover plate usually consists of one or more layers of a variety of glass or plastic and reduces the reradiation by creating an air space which traps the heat between the cover and the absorber plates.

The absorber plate must have a good thermal bond with the fluid passages. The absorber plate is usually metallic and treated with a surface coating which improves absorptivity. Black or dark paints or selective coatings are used for this purpose, and the design of this passage and plate combination helps determine a solar system's effectiveness.

Heat transfer fluid passage tubes are attached above and below or integral with an absorber plate for the purpose of transferring thermal energy from the absorber plate to a heat transfer medium. The heat exchanger is a device for transferring thermal energy from one fluid to another.

Piping and storage tanks should be well insulated to minimize heat losses.

Size domestic water heating storage tanks to hold 20 gallons of water per user, minimum, plus 10 gallons per dishwasher or washing machine. For domestic water heating an optimum collector size is approximately 3/4 square foot of area per gallon of water storage. For space heating of residences and small commercial applications the collector is commonly sized between 30% and 50% of the internal floor area. For space heating of large commercial applications, collector areas less than 30% of the internal floor area can still provide significant heat reductions.

A supplementary heat source is recommended for Northern states for December through February.

The solar energy transmission per square foot of collector surface varies greatly with the material used. Initial cost, heat transmittance and useful life are obviously interrelated.

Fig. R238313-11

R238313-10 Heat Trace Systems

Before you can determine the cost of a HEAT TRACE installation, the method of attachment must be established. There are (4) common methods:

1. Cable is simply attached to the pipe with polyester tape every 12'.
2. Cable is attached with a continuous cover of 2" wide aluminum tape.
3. Cable is attached with factory extruded heat transfer cement and covered with metallic raceway with clips every 10'.
4. Cable is attached between layers of pipe insulation using either clips or polyester tape.

Example: Components for method 3 must include:
A. Heat trace cable by voltage and watts per linear foot.
B. Heat transfer cement, 1 gallon per 60 linear feet of cover.
C. Metallic raceway by size and type.
D. Raceway clips by size of pipe.

When taking off linear foot lengths of cable add the following for each valve in the system. (E)

In all of the above methods each component of the system must be priced individually.

SCREWED OR WELDED VALVE:			FLANGED VALVE:			BUTTERFLY VALVES:		
1/2"	=	6"	1/2"	=	1' -0"	1/2"	=	0'
3/4"	=	9"	3/4"	=	1' -6"	3/4"	=	0'
1"	=	1' -0"	1"	=	2' -0"	1"	=	1' -0"
1-1/2"	=	1' -6"	1-1/2"	=	2' -6"	1-1/2"	=	1' -6"
2"	=	2'	2"	=	2' -6"	2"	=	2' -0"
2-1/2"	=	2' -6"	2-1/2"	=	3' -0"	2-1/2"	=	2' -6"
3"	=	2' -6"	3"	=	3' -6"	3"	=	2' -6"
4"	=	4' -0"	4"	=	4' -0"	4"	=	3' -0"
6"	=	7' -0"	6"	=	8' -0"	6"	=	3' -6"
8"	=	9' -6"	8"	=	11' -0"	8"	=	4' -0"
10"	=	12' -6"	10"	=	14' -0"	10"	=	4' -0"
12"	=	15' -0"	12"	=	16' -6"	12"	=	5' -0"
14"	=	18' -0"	14"	=	19' -6"	14"	=	5' -6"
16"	=	21' -6"	16"	=	23' -0"	16"	=	6' -0"
18"	=	25' -6"	18"	=	27' -0"	18"	=	6' -6"
20"	=	28' -6"	20"	=	30' -0"	20"	=	7' -0"
24"	=	34' -0"	24"	=	36' -0"	24"	=	8' -0"
30"	=	40' -0"	30"	=	42' -0"	30"	=	10' -0"

682

For customer support on your Plumbing Cost Data, call 877.709.6968.

R238313-10 Heat Trace Systems (cont.)

Add the following quantities of heat transfer cement to linear foot totals for each valve:

Nominal Valve Size	Gallons of Cement per Valve
1/2"	0.14
3/4"	0.21
1"	0.29
1-1/2"	0.36
2"	0.43
2-1/2"	0.70
3"	0.71
4"	1.00
6"	1.43
8"	1.48
10"	1.50
12"	1.60
14"	1.75
16"	2.00
18"	2.25
20"	2.50
24"	3.00
30"	3.75

The following must be added to the list of components to accurately price HEAT TRACE systems:

1. Expediter fitting and clamp fasteners (F)
2. Junction box and nipple connected to expediter fitting (G)
3. Field installed terminal blocks within junction box
4. Ground lugs
5. Piping from power source to expediter fitting
6. Controls
7. Thermostats
8. Branch wiring
9. Cable splices
10. End of cable terminations
11. Branch piping fittings and boxes

Deduct the following percentages from labor if cable lengths in the same area exceed:

150' to 250'	10%	351' to 500'	20%
251' to 350'	15%	Over 500'	25%

Add the following percentages to labor for elevated installations:

15' to 20' high	10%	31' to 35' high	40%
21' to 25' high	20%	36' to 40' high	50%
26' to 30' high	30%	Over 40' high	60%

R238313-20 Spiral-Wrapped Heat Trace Cable (Pitch Table)

In order to increase the amount of heat, occasionally heat trace cable is wrapped in a spiral fashion around a pipe; increasing the number of feet of heater cable per linear foot of pipe.

Engineers first determine the heat loss per foot of pipe (based on the insulating material, its thickness, and the temperature differential across it). A ratio is then calculated by the formula:

$$\text{Feet of Heat Trace per Foot of Pipe} = \frac{\text{Watts/Foot of Heat Loss}}{\text{Watts/Foot of the Cable}}$$

The linear distance between wraps (pitch) is then taken from a chart or table. Generally, the pitch is listed on a drawing leaving the estimator to calculate the total length of heat tape required. An approximation may be taken from this table.

	Feet of Heat Trace Per Foot of Pipe															
	Nominal Pipe Size in Inches															
Pitch In Inches	1	1¼	1½	2	2½	3	4	6	8	10	12	14	16	18	20	24
3.5	1.80															
4	1.65															
5	1.46	1.60	1.80													
6	1.34	1.45	1.55	1.75												
7	1.25	1.35	1.43	1.57	1.75											
8	1.20	1.28	1.34	1.45	1.60	1.80										
9	1.16	1.23	1.28	1.37	1.51	1.68										
10	1.13	1.19	1.24	1.32	1.44	1.57	1.82									
15	1.06	1.08	1.10	1.15	1.21	1.29	1.42	1.78								
20	1.04	1.05	1.06	1.08	1.13	1.17	1.25	1.49	1.73							
25		1.04	1.04	1.06	1.08	1.11	1.17	1.33	1.51	1.72						
30				1.04	1.05	1.07	1.12	1.24	1.37	1.54	1.70					
35						1.06	1.09	1.17	1.28	1.42	1.54	1.80				
40						1.05	1.07	1.14	1.22	1.33	1.44	1.52	1.64	1.75		
50							1.05	1.09	1.15	1.22	1.29	1.35	1.44	1.53	1.64	1.83
60								1.06	1.11	1.16	1.21	1.25	1.31	1.39	1.46	1.62
70								1.05	1.08	1.12	1.17	1.19	1.24	1.30	1.35	1.47
80									1.06	1.09	1.13	1.15	1.19	1.24	1.28	1.38
90									1.04	1.06	1.10	1.13	1.16	1.19	1.23	1.32
100										1.05	1.08	1.10	1.13	1.15	1.19	1.23

Note: Common practice would normally limit the lower end of the table to 5% of additional heat and above 80% an engineer would likely opt for two (2) parallel cables.

R312319-90   Wellpoints

A single stage wellpoint system is usually limited to dewatering an average 15' depth below normal ground water level. Multi-stage systems are employed for greater depth with the pumping equipment installed only at the lowest header level. Ejectors with unlimited lift capacity can be economical when two or more stages of wellpoints can be replaced or when horizontal clearance is restricted, such as in deep trenches or tunneling projects, and where low water flows are expected. Wellpoints are usually spaced on 2-1/2' to 10' centers along a header pipe. Wellpoint spacing, header size, and pump size are all determined by the expected flow as dictated by soil conditions.

In almost all soils encountered in wellpoint dewatering, the wellpoints may be jetted into place. Cemented soils and stiff clays may require sand wicks about 12" in diameter around each wellpoint to increase efficiency and eliminate weeping into the excavation. These sand wicks require 1/2 to 3 C.Y. of washed filter sand and are installed by using a 12" diameter steel casing and hole puncher jetted into the ground 2' deeper than the wellpoint. Rock may require predrilled holes.

Labor required for the complete installation and removal of a single stage wellpoint system is in the range of 3/4 to 2 labor-hours per linear foot of header, depending upon jetting conditions, wellpoint spacing, etc.

Continuous pumping is necessary except in some free draining soil where temporary flooding is permissible (as in trenches which are backfilled after each day's work). Good practice requires provision of a stand-by pump during the continuous pumping operation.

Systems for continuous trenching below the water table should be installed three to four times the length of expected daily progress to ensure uninterrupted digging, and header pipe size should not be changed during the job.

For pervious free draining soils, deep wells in place of wellpoints may be economical because of lower installation and maintenance costs. Daily production ranges between two to three wells per day, for 25' to 40' depths, to one well per day for depths over 50'.

Detailed analysis and estimating for any dewatering problem is available at no cost from wellpoint manufacturers. Major firms will quote "sufficient equipment" quotes or their affiliates offer lump sum proposals to cover complete dewatering responsibility.

Description for 200' System with 8" Header		Quantities
Equipment & Material	Wellpoints 25' long, 2" diameter @ 5' O.C.	40 Each
	Header pipe, 8" diameter	200 L.F.
	Discharge pipe, 8" diameter	100 L.F.
	8" valves	3 Each
	Combination jetting & wellpoint pump (standby)	1 Each
	Wellpoint pump, 8" diameter	1 Each
	Transportation to and from site	1 Day
	Fuel for 30 days x 60 gal./day	1800 Gallons
	Lubricants for 30 days x 16 lbs./day	480 Lbs.
	Sand for points	40 C.Y.
Labor	Technician to supervise installation	1 Week
	Labor for installation and removal of system	300 Labor-hours
	4 Operators straight time 40 hrs./wk. for 4.33 wks.	693 Hrs.
	4 Operators overtime 2 hrs./wk. for 4.33 wks.	35 Hrs.

R314116-40 Wood Sheet Piling

Wood sheet piling may be used for depths to 20' where there is no ground water. If moderate ground water is encountered Tongue & Groove sheeting will help to keep it out. When considerable ground water is present, steel sheeting must be used.

For estimating purposes on trench excavation, sizes are as follows:

Depth	Sheeting	Wales	Braces	B.F. per S.F.
To 8'	3 x 12's	6 x 8's, 2 line	6 x 8's, @ 10'	4.0 @ 8'
8' x 12'	3 x 12's	10 x 10's, 2 line	10 x 10's, @ 9'	5.0 average
12' to 20'	3 x 12's	12 x 12's, 3 line	12 x 12's, @ 8'	7.0 average

Sheeting to be toed in at least 2' depending upon soil conditions. A five person crew with an air compressor and sheeting driver can drive and brace 440 SF/day at 8' deep, 360 SF/day at 12' deep, and 320 SF/day at 16' deep.

For normal soils, piling can be pulled in 1/3 the time to install. Pulling difficulty increases with the time in the ground. Production can be increased by high pressure jetting.

R314116-45 Steel Sheet Piling

Limiting weights are 22 to 38#/S.F. of wall surface with 27#/S.F. average for usual types and sizes. (Weights of piles themselves are from 30.7#/L.F. to 57#/L.F. but they are 15" to 21" wide.) Lightweight sections 12" to 28" wide from 3 ga. to 12 ga. thick are also available for shallow excavations. Piles may be driven two at a time with an impact or vibratory hammer (use vibratory to pull) hung from a crane without leads. A reasonable estimate of the life of steel sheet piling is 10 uses with up to 125 uses possible if a vibratory hammer is used. Used piling costs from 50% to 80% of new piling depending on location and market conditions. Sheet piling and H piles can be rented for about 30% of the delivered mill price for the first month and 5% per month thereafter. Allow 1 labor-hour per pile for cleaning and trimming after driving. These costs increase with depth and hydrostatic head. Vibratory drivers are faster in wet granular soils and are excellent for pile extraction. Pulling difficulty increases with the time in the ground and may cost more than driving. It is often economical to abandon the sheet piling, especially if it can be used as the outer wall form. Allow about 1/3 additional length or more for toeing into ground. Add bracing, waler and strut costs. Waler costs can equal the cost per ton of sheeting.

R331113-80 Piping Designations

There are several systems currently in use to describe pipe and fittings. The following paragraphs will help to identify and clarify classifications of piping systems used for water distribution.

Piping may be classified by schedule. Piping schedules include 5S, 10S, 10, 20, 30, Standard, 40, 60, Extra Strong, 80, 100, 120, 140, 160 and Double Extra Strong. These schedules are dependent upon the pipe wall thickness. The wall thickness of a particular schedule may vary with pipe size.

Ductile iron pipe for water distribution is classified by Pressure Classes such as Class 150, 200, 250, 300 and 350. These classes are actually the rated water working pressure of the pipe in pounds per square inch (psi). The pipe in these pressure classes is designed to withstand the rated water working pressure plus a surge allowance of 100 psi.

The American Water Works Association (AWWA) provides standards for various types of **plastic pipe**. C-900 is the specification for polyvinyl chloride (PVC) piping used for water distribution in sizes ranging from 4" through 12". C-901 is the specification for polyethylene (PE) pressure pipe, tubing and fittings used for water distribution in sizes ranging from 1/2" through 3". C-905 is the specification for PVC piping sizes 14" and greater.

PVC pressure-rated pipe is identified using the standard dimensional ratio (SDR) method. This method is defined by the American Society for Testing and Materials (ASTM) Standard D 2241. This pipe is available in SDR numbers 64, 41, 32.5, 26, 21, 17, and 13.5. Pipe with an SDR of 64 will have the thinnest wall while pipe with an SDR of 13.5 will have the thickest wall. When the pressure rating (PR) of a pipe is given in psi, it is based on a line supplying water at 73 degrees F.

The National Sanitation Foundation (NSF) seal of approval is applied to products that can be used with potable water. These products have been tested to ANSI/NSF Standard 14.

Valves and strainers are classified by American National Standards Institute (ANSI) Classes. These Classes are 125, 150, 200, 250, 300, 400, 600, 900, 1500 and 2500. Within each class there is an operating pressure range dependent upon temperature. Design parameters should be compared to the appropriate material dependent, pressure-temperature rating chart for accurate valve selection.

Change Orders

Change Order Considerations

A change order is a written document, usually prepared by the design professional, and signed by the owner, the architect/engineer, and the contractor. A change order states the agreement of the parties to: an addition, deletion, or revision in the work; an adjustment in the contract sum, if any; or an adjustment in the contract time, if any. Change orders, or "extras" in the construction process occur after execution of the construction contract and impact architects/engineers, contractors, and owners.

Change orders that are properly recognized and managed can ensure orderly, professional, and profitable progress for all who are involved in the project. There are many causes for change orders and change order requests. In all cases, change orders or change order requests should be addressed promptly and in a precise and prescribed manner. The following paragraphs include information regarding change order pricing and procedures.

The Causes of Change Orders

Reasons for issuing change orders include:

- Unforeseen field conditions that require a change in the work
- Correction of design discrepancies, errors, or omissions in the contract documents
- Owner-requested changes, either by design criteria, scope of work, or project objectives
- Completion date changes for reasons unrelated to the construction process
- Changes in building code interpretations, or other public authority requirements that require a change in the work
- Changes in availability of existing or new materials and products

Procedures

Properly written contract documents must include the correct change order procedures for all parties—owners, design professionals and contractors—to follow in order to avoid costly delays and litigation.

Being "in the right" is not always a sufficient or acceptable defense. The contract provisions requiring notification and documentation must be adhered to within a defined or reasonable time frame.

The appropriate method of handling change orders is by a written proposal and acceptance by all parties involved. Prior to starting work on a project, all parties should identify their

authorized agents who may sign and accept change orders, as well as any limits placed on their authority.

Time may be a critical factor when the need for a change arises. For such cases, the contractor might be directed to proceed on a "time and materials" basis, rather than wait for all paperwork to be processed—a delay that could impede progress. In this situation, the contractor must still follow the prescribed change order procedures including, but not limited to, notification and documentation.

Lack of documentation can be very costly, especially if legal judgments are to be made, and if certain field personnel are no longer available. For time and material change orders, the contractor should keep accurate daily records of all labor and material allocated to the change.

Owners or awarding authorities who do considerable and continual building construction (such as the federal government) realize the inevitability of change orders for numerous reasons, both predictable and unpredictable. As a result, the federal government, the American Institute of Architects (AIA), the Engineers Joint Contract Documents Committee (EJCDC) and other contractor, legal, and technical organizations have developed standards and procedures to be followed by all parties to achieve contract continuance and timely completion, while being financially fair to all concerned.

Pricing Change Orders

When pricing change orders, regardless of their cause, the most significant factor is when the change occurs. The need for a change may be perceived in the field or requested by the architect/engineer *before* any of the actual installation has begun, or may evolve or appear *during* construction when the item of work in question is partially installed. In the latter cases, the original sequence of construction is disrupted, along with all contiguous and supporting systems. Change orders cause the greatest impact when they occur *after* the installation has been completed and must be uncovered, or even replaced. Post-completion changes may be caused by necessary design changes, product failure, or changes in the owner's requirements that are not discovered until the building or the systems begin to function.

Specified procedures of notification and record keeping must be adhered to and enforced regardless of the stage of construction: *before, during,* or *after* installation. Some bidding documents anticipate change orders by requiring that unit prices including overhead and profit percentages—for additional as well as deductible changes—be listed. Generally these unit prices do not fully take into account the ripple effect, or impact on other trades, and should be used for general guidance only.

When pricing change orders, it is important to classify the time frame in which the change occurs. There are two basic time frames for change orders: *pre-installation change orders*, which occur before the start of construction, and *post-installation change orders*, which involve reworking after the original installation. Change orders that occur between these stages may be priced according to the extent of work completed using a combination of techniques developed for pricing *pre-* and *post-installation* changes.

Factors To Consider When Pricing Change Orders

As an estimator begins to prepare a change order, the following questions should be reviewed to determine their impact on the final price.

General

- *Is the change order work* pre-installation *or* post-installation?

   Change order work costs vary according to how much of the installation has been completed. Once workers have the project scoped in their minds, even though they have not started, it can be difficult to refocus. Consequently they may spend more than the normal amount of time understanding the change. Also, modifications to work in place, such as trimming or refitting, usually take more time than was initially estimated. The greater the amount of work in place, the more reluctant workers are to change it. Psychologically they may resent the change and as a result the rework takes longer than normal. Post-installation change order estimates must include demolition of existing work as required to accomplish the change. If the work is performed at a later time, additional obstacles, such as building finishes, may be present which must be protected. Regardless of whether the change occurs pre-installation or post-installation, attempt to isolate the identifiable factors and price them separately. For example, add shipping costs that may be required pre-installation or any demolition required post-installation. Then analyze the potential impact on productivity of psychological and/or learning curve factors and adjust the output rates accordingly. One approach is to break down the typical workday into segments and quantify the impact on each segment.

Change Order Installation Efficiency

The labor-hours expressed (for new construction) are based on average installation time, using an efficiency level. For change order situations, adjustments to this efficiency level should reflect the daily labor-hour allocation for that particular occurrence.

- *Will the change substantially delay the original completion date?*

A significant change in the project may cause the original completion date to be extended. The extended schedule may subject the contractor to new wage rates dictated by relevant labor contracts. Project supervision and other project overhead must also be extended beyond the original completion date. The schedule extension may also put installation into a new weather season. For example, underground piping scheduled for October installation was delayed until January. As a result, frost penetrated the trench area, thereby changing the degree of difficulty of the task. Changes and delays may have a ripple effect throughout the project. This effect must be analyzed and negotiated with the owner.

- *What is the net effect of a deduct change order?*

In most cases, change orders resulting in a deduction or credit reflect only bare costs. The contractor may retain the overhead and profit based on the original bid.

Materials

- *Will you have to pay more or less for the new material, required by the change order, than you paid for the original purchase?*

The same material prices or discounts will usually apply to materials purchased for change orders as new construction. In some instances, however, the contractor may forfeit the advantages of competitive pricing for change orders. Consider the following example:

A contractor purchased over $20,000 worth of fan coil units for an installation, and obtained the maximum discount. Some time later it was determined the project required an additional matching unit. The contractor has to purchase this unit from the original supplier to ensure a match. The supplier at this time may not discount the unit because of the small quantity, and the fact that he is no longer in a competitive situation. The impact of quantity on purchase can add between 0% and 25% to material prices and/or subcontractor quotes.

- *If materials have been ordered or delivered to the job site, will they be subject to a cancellation charge or restocking fee?*

Check with the supplier to determine if ordered materials are subject to a cancellation charge. Delivered materials not used as result of a change order may be subject to a restocking fee if returned to the supplier. Common restocking charges run between 20% and 40%. Also, delivery charges to return the goods to the supplier must be added.

Labor

- *How efficient is the existing crew at the actual installation?*

Is the same crew that performed the initial work going to do the change order? Possibly the change consists of the installation of a unit identical to one already installed; therefore, the change should take less time. Be sure to consider this potential productivity increase and modify the productivity rates accordingly.

- *If the crew size is increased, what impact will that have on supervision requirements?*

Under most bargaining agreements or management practices, there is a point at which a working foreman is replaced by a nonworking foreman. This replacement increases project overhead by adding a nonproductive worker. If additional workers are added to accelerate the project or to perform changes while maintaining the schedule, be sure to add additional supervision time if warranted. Calculate the hours involved and the additional cost directly if possible.

- *What are the other impacts of increased crew size?*

The larger the crew, the greater the potential for productivity to decrease. Some of the factors that cause this productivity loss are: overcrowding (producing restrictive conditions in the working space), and possibly a shortage of any special tools and equipment required. Such factors affect not only the crew working on the elements directly involved in the change order, but other crews whose movement may also be hampered. As the crew increases, check its basic composition for changes by the addition or deletion of apprentices or nonworking foreman, and quantify the potential effects of equipment shortages or other logistical factors.

- *As new crews, unfamiliar with the project, are brought onto the site, how long will it take them to become oriented to the project requirements?*

The orientation time for a new crew to become 100% effective varies with the site and type of project. Orientation is easiest at a new construction site, and most difficult at existing, very restrictive renovation sites. The type of work also affects orientation time. When all elements of the work are exposed, such as concrete or masonry work, orientation is decreased. When the work is concealed or less visible, such as existing electrical systems, orientation takes longer. Usually orientation can be accomplished in one day or less. Costs for added orientation should be itemized and added to the total estimated cost.

- *How much actual production can be gained by working overtime?*

Short term overtime can be used effectively to accomplish more work in a day. However, as overtime is scheduled to run beyond several weeks, studies have shown marked decreases in output. The following chart shows the effect of long term overtime on worker efficiency. If the anticipated change requires extended overtime to keep the job on schedule, these factors can be used as a guide to predict the impact on time and cost. Add project overhead, particularly supervision, that may also be incurred.

Days per Week	Hours per Day	Production Efficiency					Payroll Cost Factors	
		1 Week	2 Weeks	3 Weeks	4 Weeks	Average 4 Weeks	@ 1-1/2 Times	@ 2 Times
5	8	100%	100%	100%	100%	100%	100%	100%
	9	100	100	95	90	96.25	105.6	111.1
	10	100	95	90	85	91.25	110.0	120.0
	11	95	90	75	65	81.25	113.6	127.3
	12	90	85	70	60	76.25	116.7	133.3
6	8	100	100	95	90	96.25	108.3	116.7
	9	100	95	90	85	92.50	113.0	125.9
	10	95	90	85	80	87.50	116.7	133.3
	11	95	85	70	65	78.75	119.7	139.4
	12	90	80	65	60	73.75	122.2	144.4
7	8	100	95	85	75	88.75	114.3	128.6
	9	95	90	80	70	83.75	118.3	136.5
	10	90	85	75	65	78.75	121.4	142.9
	11	85	80	65	60	72.50	124.0	148.1
	12	85	75	60	55	68.75	126.2	152.4

Effects of Overtime

Caution: Under many labor agreements, Sundays and holidays are paid at a higher premium than the normal overtime rate.

The use of long-term overtime is counterproductive on almost any construction job; that is, the longer the period of overtime, the lower the actual production rate. Numerous studies have been conducted, and while they have resulted in slightly different numbers, all reach the same conclusion. The figure above tabulates the effects of overtime work on efficiency.

As illustrated, there can be a difference between the *actual* payroll cost per hour and the *effective* cost per hour for overtime work. This is due to the reduced production efficiency with the increase in weekly hours beyond 40. This difference between actual and effective cost results from overtime work over a prolonged period. Short-term overtime work does not result in as great a reduction in efficiency and, in such cases, effective cost may not vary significantly from the actual payroll cost. As the total hours per week are increased on a regular basis, more time is lost due to fatigue, lowered morale, and an increased accident rate.

As an example, assume a project where workers are working 6 days a week, 10 hours per day. From the figure above (based on productivity studies), the average effective productive hours over a 4-week period are:

$$0.875 \times 60 = 52.5$$

Depending upon the locale and day of week, overtime hours may be paid at time and a half or double time. For time and a half, the overall (average) *actual* payroll cost (including regular and overtime hours) is determined as follows:

$$\frac{40 \text{ reg. hrs.} + (20 \text{ overtime hrs.} \times 1.5)}{60 \text{ hrs.}} = 1.167$$

Based on 60 hours, the payroll cost per hour will be 116.7% of the normal rate at 40 hours per week. However, because the effective production (efficiency) for 60 hours is reduced to the equivalent of 52.5 hours, the effective cost of overtime is calculated as follows:

For time and a half:

$$\frac{40 \text{ reg. hrs.} + (20 \text{ overtime hrs.} \times 1.5)}{52.5 \text{ hrs.}} = 1.33$$

Installed cost will be 133% of the normal rate (for labor).

Thus, when figuring overtime, the actual cost per unit of work will be higher than the apparent overtime payroll dollar increase, due to the reduced productivity of the longer workweek. These efficiency calculations are true only for those cost factors determined by hours worked. Costs that are applied weekly or monthly, such as equipment rentals, will not be similarly affected.

Equipment

- *What equipment is required to complete the change order?*

Change orders may require extending the rental period of equipment already on the job site, or the addition of special equipment brought in to accomplish the change work. In either case, the additional rental charges and operator labor charges must be added.

Summary

The preceding considerations and others you deem appropriate should be analyzed and applied to a change order estimate. The impact of each should be quantified and listed on the estimate to form an audit trail.

Change orders that are properly identified, documented, and managed help to ensure the orderly, professional and profitable progress of the work. They also minimize potential claims or disputes at the end of the project.

Estimating Tips

- The cost figures in this section were derived from approximately 11,000 projects contained in the RSMeans database of completed construction projects. They include the contractor's overhead and profit, but do not generally include architectural fees or land costs. The figures have been adjusted to January of the current year. New projects are added to our files each year, and outdated projects are discarded. For this reason, certain costs may not show a uniform annual progression. In no case are all subdivisions of a project listed.

- These projects were located throughout the U.S. and reflect a tremendous variation in square foot (S.F.) and cubic foot (C.F.) costs. This is due to differences, not only in labor and material costs, but also in individual owners' requirements. For instance, a bank in a large city would have different features than one in a rural area. This is true of all the different types of buildings analyzed. Therefore, caution should be exercised when using these square foot costs. For example, for courthouses, costs in the database are local courthouse costs and will not apply to the larger, more elaborate federal courthouses. As a general rule, the projects in the 1/4 column do not include any site work or equipment, while the projects in the 3/4 column may include both equipment and site work.

The median figures do not generally include site work.

- None of the figures "go with" any others. All individual cost items were computed and tabulated separately. Thus, the sum of the median figures for plumbing, HVAC, and electrical will not normally total up to the total mechanical and electrical costs arrived at by separate analysis and tabulation of the projects.

- Each building was analyzed as to total and component costs and percentages. The figures were arranged in ascending order with the results tabulated as shown. The 1/4 column shows that 25% of the projects had lower costs and 75% had higher. The 3/4 column shows that 75% of the projects had lower costs and 25% had higher. The median column shows that 50% of the projects had lower costs and 50% had higher.

- There are two times when square foot costs are useful. The first is in the conceptual stage when no details are available. Then, square foot costs make a useful starting point. The second is after the bids are in and the costs can be worked back into their appropriate categories for information purposes. As soon as details become available in the project design, the square foot approach should be discontinued and the project priced as to its particular components. When more precision is required, or

for estimating the replacement cost of specific buildings, the current edition of *RSMeans Square Foot Costs* should be used.

- In using the figures in this section, it is recommended that the median column be used for preliminary figures if no additional information is available. The median figures, when multiplied by the total city construction cost index figures (see City Cost Indexes) and then multiplied by the project size modifier at the end of this section, should present a fairly accurate base figure, which would then have to be adjusted in view of the estimator's experience, local economic conditions, code requirements, and the owner's particular requirements. There is no need to factor the percentage figures, as these should remain constant from city to city. All tabulations mentioning air conditioning had at least partial air conditioning.

- The editors of this book would greatly appreciate receiving cost figures on one or more of your recent projects, which would then be included in the averages for next year. All cost figures received will be kept confidential, except that they will be averaged with other similar projects to arrive at square foot cost figures for next year's book. See the last page of the book for details and the discount available for submitting one or more of your projects. ■

50 17 00 | S.F. Costs

		UNIT	UNIT COSTS			% OF TOTAL		
			1/4	MEDIAN	3/4	1/4	MEDIAN	3/4
01	0010	**APARTMENTS Low Rise (1 to 3 story)**	S.F.	74.50	94	125				**01**
	0020	Total project cost	C.F.	6.70	8.85	10.95			
	0100	Site work	S.F.	5.45	8.70	15.30	6.05%	10.55%	13.95%
	0500	Masonry		1.47	3.62	5.95	1.54%	3.92%	6.50%
	1500	Finishes		7.90	10.85	13.45	9.05%	10.75%	12.85%
	1800	Equipment		2.44	3.70	5.50	2.71%	3.99%	5.95%
	2720	Plumbing		5.80	7.45	9.50	6.65%	8.95%	10.05%
	2770	Heating, ventilating, air conditioning		3.70	4.56	6.70	4.20%	5.60%	7.60%
	2900	Electrical		4.33	5.75	7.80	5.20%	6.65%	8.35%
	3100	Total: Mechanical & Electrical	↓	15.40	19.95	24.50	16.05%	18.20%	23%
	9000	Per apartment unit, total cost	Apt.	69,500	106,000	156,500			
	9500	Total: Mechanical & Electrical	"	13,100	20,700	27,000			
02	0010	**APARTMENTS Mid Rise (4 to 7 story)**	S.F.	99	119	147				**02**
	0020	Total project costs	C.F.	7.70	10.65	14.55			
	0100	Site work	S.F.	3.95	8	15.45	5.25%	6.70%	9.20%
	0500	Masonry		4.13	9.05	12.40	5.10%	7.25%	10.50%
	1500	Finishes		12.95	17.35	20.50	10.70%	13.50%	17.70%
	1800	Equipment		2.77	4.30	5.65	2.54%	3.47%	4.31%
	2500	Conveying equipment		2.24	2.76	3.33	2.05%	2.27%	2.69%
	2720	Plumbing		5.80	9.30	9.85	5.70%	7.20%	8.95%
	2900	Electrical		6.50	8.85	10.75	6.35%	7.20%	8.95%
	3100	Total: Mechanical & Electrical	↓	21	26	31.50	18.25%	21%	23%
	9000	Per apartment unit, total cost	Apt.	112,000	132,000	218,500			
	9500	Total: Mechanical & Electrical	"	21,100	24,400	25,600			
03	0010	**APARTMENTS High Rise (8 to 24 story)**	S.F.	112	129	155				**03**
	0020	Total project costs	C.F.	10.90	12.65	14.95			
	0100	Site work	S.F.	4.07	6.60	9.20	2.58%	4.84%	6.15%
	0500	Masonry		6.50	11.80	14.65	4.74%	9.65%	11.05%
	1500	Finishes		12.45	15.55	18.35	9.75%	11.80%	13.70%
	1800	Equipment		3.61	4.44	5.85	2.78%	3.49%	4.35%
	2500	Conveying equipment		2.55	3.87	5.25	2.23%	2.78%	3.37%
	2720	Plumbing		7.15	9.75	11.95	6.80%	7.20%	10.45%
	2900	Electrical		7.70	9.75	13.15	6.45%	7.65%	8.80%
	3100	Total: Mechanical & Electrical	↓	23	29.50	35.50	17.95%	22.50%	24.50%
	9000	Per apartment unit, total cost	Apt.	116,500	128,500	178,000			
	9500	Total: Mechanical & Electrical	"	23,200	28,800	30,500			
04	0010	**AUDITORIUMS**	S.F.	117	161	234				**04**
	0020	Total project costs	C.F.	7.30	10.15	14.55			
	2720	Plumbing	S.F.	6.90	10.20	12.20	5.85%	7.20%	8.70%
	2900	Electrical		9.40	13.55	22.50	6.85%	9.50%	11.40%
	3100	Total: Mechanical & Electrical	↓	62.50	83	101	24.50%	27.50%	31%
05	0010	**AUTOMOTIVE SALES**	S.F.	86.50	119	146				**05**
	0020	Total project costs	C.F.	5.70	6.80	8.85			
	2720	Plumbing	S.F.	3.93	6.85	7.45	2.89%	6.05%	6.50%
	2770	Heating, ventilating, air conditioning		6.05	9.25	10	4.61%	10%	10.35%
	2900	Electrical		6.95	10.90	16	7.25%	8.80%	12.15%
	3100	Total: Mechanical & Electrical	↓	19.35	31	37.50	17.30%	20.50%	22%
06	0010	**BANKS**	S.F.	170	211	268				**06**
	0020	Total project costs	C.F.	12.10	16.45	21.50			
	0100	Site work	S.F.	19.40	30.50	43	7.90%	12.95%	17%
	0500	Masonry		7.70	16.50	29.50	3.36%	6.90%	10.05%
	1500	Finishes		15.15	23	28.50	5.85%	8.65%	11.70%
	1800	Equipment		6.35	14	28.50	1.34%	5.55%	10.50%
	2720	Plumbing		5.30	7.55	11.05	2.82%	3.90%	4.93%
	2770	Heating, ventilating, air conditioning		10.05	13.45	17.90	4.86%	7.15%	8.50%
	2900	Electrical		16.05	21.50	28	8.25%	10.20%	12.20%
	3100	Total: Mechanical & Electrical	↓	38.50	51.50	62	16.55%	19.45%	24%
	3500	See also division 11 22 00							

50 17 00	S.F. Costs		UNIT	UNIT COSTS			% OF TOTAL		
				1/4	MEDIAN	3/4	1/4	MEDIAN	3/4
13	0010	**CHURCHES**	S.F.	114	145	190			
	0020	Total project costs	C.F.	7.05	8.90	11.75			
	1800	Equipment	S.F.	1.22	3.19	6.80	.83%	2.04%	4.30%
	2720	Plumbing		4.43	6.20	9.15	3.51%	4.96%	6.25%
	2770	Heating, ventilating, air conditioning		10.35	13.50	19.15	7.50%	10%	12%
	2900	Electrical		9.60	13.20	18	7.30%	8.80%	10.95%
	3100	Total: Mechanical & Electrical	↓	29.50	39.50	53	18.30%	22%	25%
	3500	See also division 11 91 00							
15	0010	**CLUBS, COUNTRY**	S.F.	122	147	185			
	0020	Total project costs	C.F.	9.85	12	16.55			
	2720	Plumbing	S.F.	7.40	10.95	25	5.60%	7.90%	10%
	2900	Electrical		9.60	13.15	17.15	7%	8.95%	11%
	3100	Total: Mechanical & Electrical	↓	51	64	67.50	19%	26.50%	29.50%
17	0010	**CLUBS, SOCIAL Fraternal**	S.F.	103	141	188			
	0020	Total project costs	C.F.	6.10	9.25	11			
	2720	Plumbing	S.F.	6.15	7.65	11.55	5.60%	6.90%	8.55%
	2770	Heating, ventilating, air conditioning		7.15	10.70	13.75	8.20%	9.25%	14.40%
	2900	Electrical		7.15	12.05	13.75	5.95%	9.30%	10.55%
	3100	Total: Mechanical & Electrical	↓	38	41	52	21%	23%	23.50%
18	0010	**CLUBS, Y.M.C.A.**	S.F.	134	167	252			
	0020	Total project costs	C.F.	5.65	9.45	14.05			
	2720	Plumbing	S.F.	7.75	15.40	17.25	5.65%	7.60%	10.85%
	2900	Electrical		10.05	13	22.50	6.45%	7.95%	8.90%
	3100	Total: Mechanical & Electrical	↓	41	46.50	76	20.50%	21.50%	28.50%
19	0010	**COLLEGES Classrooms & Administration**	S.F.	118	167	216			
	0020	Total project costs	C.F.	7.10	12.50	20			
	0500	Masonry	S.F.	9.30	17.15	21	5.10%	8.05%	10.50%
	2720	Plumbing		5.30	12.90	24	5.10%	6.60%	8.95%
	2900	Electrical		10.05	15.85	22	7.70%	9.85%	12%
	3100	Total: Mechanical & Electrical	↓	42.50	58.50	77.50	23%	28%	31.50%
21	0010	**COLLEGES Science, Engineering, Laboratories**	S.F.	230	269	310			
	0020	Total project costs	C.F.	13.20	19.25	22			
	1800	Equipment	S.F.	6.25	29	31.50	2%	6.45%	12.65%
	2900	Electrical		18.95	27	41.50	7.10%	9.40%	12.10%
	3100	Total: Mechanical & Electrical	↓	70.50	83.50	129	28.50%	31.50%	41%
	3500	See also division 11 53 00							
23	0010	**COLLEGES Student Unions**	S.F.	147	199	241			
	0020	Total project costs	C.F.	8.20	10.70	13.25			
	3100	Total: Mechanical & Electrical	S.F.	55	59.50	70.50	23.50%	26%	29%
25	0010	**COMMUNITY CENTERS**	S.F.	122	150	203			
	0020	Total project costs	C.F.	7.90	11.30	14.65			
	1800	Equipment	S.F.	2.45	4.80	7.90	1.47%	3.01%	5.30%
	2720	Plumbing		5.75	10.05	13.70	4.85%	7%	8.95%
	2770	Heating, ventilating, air conditioning		8.65	13.35	19.15	6.80%	10.35%	12.90%
	2900	Electrical		10.25	13.55	19.90	7.15%	8.90%	10.40%
	3100	Total: Mechanical & Electrical	↓	34	43	61.50	18.80%	23%	30%
28	0010	**COURT HOUSES**	S.F.	175	206	284			
	0020	Total project costs	C.F.	13.40	16	20			
	2720	Plumbing	S.F.	8.30	11.60	13.15	5.95%	7.45%	8.20%
	2900	Electrical		18.55	21	30.50	9.05%	10.65%	12.15%
	3100	Total: Mechanical & Electrical	↓	52	68	74.50	22.50%	26.50%	30%
30	0010	**DEPARTMENT STORES**	S.F.	64.50	87.50	110			
	0020	Total project costs	C.F.	3.46	4.48	6.10			
	2720	Plumbing	S.F.	2.01	2.54	3.85	1.82%	4.21%	5.90%
	2770	Heating, ventilating, air conditioning	↓	4.88	9.05	13.65	8.20%	9.10%	14.80%

Row markers on right edge: 13, 15, 17, 18, 19, 21, 23, 25, 28, 30

For customer support on your Plumbing Cost Data, call 877.709.6968.

691

50 17 00 | S.F. Costs

			UNIT	UNIT COSTS			% OF TOTAL			
				1/4	MEDIAN	3/4	1/4	MEDIAN	3/4	
30	2900	Electrical	S.F.	7.40	10.15	12	9.05%	12.15%	14.95%	30
	3100	Total: Mechanical & Electrical	↓	13.05	16.65	29	13.20%	21.50%	50%	
31	0010	**DORMITORIES Low Rise (1 to 3 story)**	S.F.	122	170	212				31
	0020	Total project costs	C.F.	6.90	11.20	16.75				
	2720	Plumbing	S.F.	7.35	9.85	12.45	8.05%	9%	9.65%	
	2770	Heating, ventilating, air conditioning		7.80	9.35	12.45	4.61%	8.05%	10%	
	2900	Electrical		8.10	12.35	16.90	6.40%	8.65%	9.50%	
	3100	Total: Mechanical & Electrical	↓	42	45	70.50	22%	25%	27%	
	9000	Per bed, total cost	Bed	52,000	57,500	123,500				
32	0010	**DORMITORIES Mid Rise (4 to 8 story)**	S.F.	150	196	242				32
	0020	Total project costs	C.F.	16.55	18.20	22				
	2900	Electrical	S.F.	15.95	18.15	24.50	8.20%	10.20%	11.95%	
	3100	Total: Mechanical & Electrical	"	44.50	89	90.50	25%	30.50%	35.50%	
	9000	Per bed, total cost	Bed	21,400	48,800	282,500				
34	0010	**FACTORIES**	S.F.	57	85	131				34
	0020	Total project costs	C.F.	3.65	5.45	9.05				
	0100	Site work	S.F.	6.50	11.85	18.75	6.95%	11.45%	17.95%	
	2720	Plumbing		3.07	5.70	9.45	3.73%	6.05%	8.10%	
	2770	Heating, ventilating, air conditioning		5.95	8.55	11.55	5.25%	8.45%	11.35%	
	2900	Electrical		7.05	11.20	17.05	8.10%	10.50%	14.20%	
	3100	Total: Mechanical & Electrical	↓	20	32.50	41	21%	28.50%	35.50%	
36	0010	**FIRE STATIONS**	S.F.	113	156	213				36
	0020	Total project costs	C.F.	6.60	9.05	12.05				
	0500	Masonry	S.F.	15.40	30	38	7.50%	10.95%	15.55%	
	1140	Roofing		3.68	9.95	11.35	1.90%	4.94%	5.05%	
	1580	Painting		2.86	4.27	4.37	1.37%	1.57%	2.07%	
	1800	Equipment		1.40	2.69	4.97	.62%	1.63%	3.42%	
	2720	Plumbing		6.30	10.05	14.25	5.85%	7.35%	9.45%	
	2770	Heating, ventilating, air conditioning		6.25	10.15	15.65	5.15%	7.40%	9.40%	
	2900	Electrical		8.15	14.35	19.35	6.90%	8.60%	10.60%	
	3100	Total: Mechanical & Electrical	↓	41.50	53	60	18.40%	23%	27%	
37	0010	**FRATERNITY HOUSES & Sorority Houses**	S.F.	113	145	199				37
	0020	Total project costs	C.F.	11.25	11.70	14.10				
	2720	Plumbing	S.F.	8.55	9.75	17.90	6.80%	8%	10.85%	
	2900	Electrical	↓	7.45	16.10	19.70	6.60%	9.90%	10.65%	
38	0010	**FUNERAL HOMES**	S.F.	119	162	295				38
	0020	Total project costs	C.F.	12.15	13.55	26				
	2900	Electrical	S.F.	5.25	9.65	10.55	3.58%	4.44%	5.95%	
39	0010	**GARAGES, COMMERCIAL (Service)**	S.F.	67.50	104	144				39
	0020	Total project costs	C.F.	4.43	6.55	9.50				
	1800	Equipment	S.F.	3.80	8.55	13.30	2.21%	4.62%	6.80%	
	2720	Plumbing		4.67	7.20	13.10	5.45%	7.85%	10.65%	
	2730	Heating & ventilating		4.60	7.95	10.15	5.25%	6.85%	8.20%	
	2900	Electrical		6.40	9.75	14.10	7.15%	9.25%	10.85%	
	3100	Total: Mechanical & Electrical	↓	12.10	27	40	12.35%	17.40%	26%	
40	0010	**GARAGES, MUNICIPAL (Repair)**	S.F.	105	135	192				40
	0020	Total project costs	C.F.	6.20	7.85	13.50				
	0500	Masonry	S.F.	5.20	18.20	28	4.03%	9.15%	12.50%	
	2720	Plumbing		4.44	8.55	16.05	3.59%	6.70%	7.95%	
	2730	Heating & ventilating		7.60	11	21	6.15%	7.45%	13.50%	
	2900	Electrical		7.65	12.35	22.50	6.90%	9.40%	13%	
	3100	Total: Mechanical & Electrical	↓	34.50	56.50	68	21.50%	25.50%	35.50%	
41	0010	**GARAGES, PARKING**	S.F.	39.50	56	96.50				41
	0020	Total project costs	C.F.	3.62	4.92	7.15				

	50 17 00 | S.F. Costs		UNIT	UNIT COSTS			% OF TOTAL			
				1/4	MEDIAN	3/4	1/4	MEDIAN	3/4	
41	2720	Plumbing	S.F.	.66	1.69	2.61	1.72%	2.70%	3.85%	**41**
	2900	Electrical	↓	2.12	2.80	4.08	4.52%	5.40%	6.35%	
	3100	Total: Mechanical & Electrical	↓	4.11	6.05	7.50	7%	8.90%	11.05%	
	3200									
	9000	Per car, total cost	Car	16,300	20,400	26,000				
43	0010	**GYMNASIUMS**	S.F.	108	143	196				**43**
	0020	Total project costs	C.F.	5.35	7.25	8.90				
	1800	Equipment	S.F.	2.55	4.78	8.60	1.76%	3.26%	6.70%	
	2720	Plumbing		5.90	7.90	10.40	4.65%	6.40%	7.75%	
	2770	Heating, ventilating, air conditioning		6.40	11.10	22.50	5.15%	9.05%	11.10%	
	2900	Electrical		8.50	11.25	14.95	6.75%	8.50%	10.30%	
	3100	Total: Mechanical & Electrical	↓	29	40.50	49.50	19.75%	23.50%	29%	
	3500	See also division 11 66 00								
46	0010	**HOSPITALS**	S.F.	206	258	355				**46**
	0020	Total project costs	C.F.	15.60	19.40	28				
	1800	Equipment	S.F.	5.20	10.05	17.30	.80%	2.53%	4.80%	
	2720	Plumbing		17.70	25	32	7.60%	9.10%	10.85%	
	2770	Heating, ventilating, air conditioning		26	33.50	46.50	7.80%	12.95%	16.65%	
	2900	Electrical		22.50	30.50	45.50	10%	11.75%	14.10%	
	3100	Total: Mechanical & Electrical	↓	66.50	92.50	137	28%	33.50%	37%	
	9000	Per bed or person, total cost	Bed	238,000	328,500	378,500				
	9900	See also division 11 71 00								
48	0010	**HOUSING For the Elderly**	S.F.	101	128	157				**48**
	0020	Total project costs	C.F.	7.20	10	12.80				
	0100	Site work	S.F.	7.05	10.95	16.05	5.05%	7.90%	12.10%	
	0500	Masonry		1.91	11.55	16.85	1.30%	6.05%	11%	
	1800	Equipment		2.45	3.37	5.35	1.88%	3.23%	4.43%	
	2510	Conveying systems		2.29	3.31	4.49	1.78%	2.20%	2.81%	
	2720	Plumbing		7.50	9.60	12.10	8.15%	9.55%	10.50%	
	2730	Heating, ventilating, air conditioning		3.86	5.45	8.15	3.30%	5.60%	7.25%	
	2900	Electrical		7.55	10.25	13.10	7.30%	8.50%	10.25%	
	3100	Total: Mechanical & Electrical	↓	26	32	41	18.10%	22.50%	29%	
	9000	Per rental unit, total cost	Unit	94,000	110,000	122,500				
	9500	Total: Mechanical & Electrical	"	21,000	24,100	28,100				
50	0010	**HOUSING Public (Low Rise)**	S.F.	85	118	154				**50**
	0020	Total project costs	C.F.	6.75	9.45	11.75				
	0100	Site work	S.F.	10.85	15.60	25.50	8.35%	11.75%	16.50%	
	1800	Equipment		2.31	3.77	5.75	2.26%	3.03%	4.24%	
	2720	Plumbing		6.15	8.10	10.25	7.15%	9.05%	11.60%	
	2730	Heating, ventilating, air conditioning		3.08	6	6.55	4.26%	6.05%	6.45%	
	2900	Electrical		5.15	7.65	10.65	5.10%	6.55%	8.25%	
	3100	Total: Mechanical & Electrical	↓	24.50	31.50	35	14.50%	17.55%	26.50%	
	9000	Per apartment, total cost	Apt.	93,500	106,500	133,500				
	9500	Total: Mechanical & Electrical	"	19,900	24,600	27,200				
51	0010	**ICE SKATING RINKS**	S.F.	72.50	170	187				**51**
	0020	Total project costs	C.F.	5.35	5.45	6.30				
	2720	Plumbing	S.F.	2.72	5.10	5.20	3.12%	3.23%	5.65%	
	2900	Electrical	↓	7.80	11.95	12.65	6.30%	10.15%	15.05%	
52	0010	**JAILS**	S.F.	189	286	370				**52**
	0020	Total project costs	C.F.	19.95	28	32.50				
	1800	Equipment	S.F.	8.65	25.50	43.50	2.80%	6.95%	9.80%	
	2720	Plumbing		21.50	28.50	37.50	7%	8.90%	13.35%	
	2770	Heating, ventilating, air conditioning		19.95	26.50	51.50	7.50%	9.45%	17.75%	
	2900	Electrical		23.50	31.50	40	9.40%	11.70%	15.25%	
	3100	Total: Mechanical & Electrical	↓	62	110	131	28%	31%	36%	
53	0010	**LIBRARIES**	S.F.	143	190	248				**53**
	0020	Total project costs	C.F.	9.55	12	15.30				

For customer support on your Plumbing Cost Data, call 877.709.6968.

693

		50 17 00 | S.F. Costs		UNIT COSTS			% OF TOTAL			
			UNIT	1/4	MEDIAN	3/4	1/4	MEDIAN	3/4	
53	0500	Masonry	S.F.	9.50	19.35	32.50	5.60%	7.15%	10.95%	53
	1800	Equipment		1.91	5.15	7.75	.28%	1.39%	4.07%	
	2720	Plumbing		5.15	7.30	10.20	3.38%	4.60%	5.70%	
	2770	Heating, ventilating, air conditioning		11.45	19.35	23.50	7.80%	10.95%	12.80%	
	2900	Electrical		14.40	19	25.50	8.40%	10.55%	12%	
	3100	Total: Mechanical & Electrical	↓	47	57	67	21%	24%	28%	
54	0010	**LIVING, ASSISTED**	S.F.	129	154	180				54
	0020	Total project costs	C.F.	10.90	12.70	14.45				
	0500	Masonry	S.F.	3.75	4.53	5.55	2.36%	3.16%	3.86%	
	1800	Equipment		2.87	3.42	4.38	2.12%	2.45%	2.87%	
	2720	Plumbing		10.85	14.50	15.05	6.05%	8.15%	10.60%	
	2770	Heating, ventilating, air conditioning		12.85	13.45	14.75	7.95%	9.35%	9.70%	
	2900	Electrical		12.70	14.25	16.50	9%	9.95%	10.65%	
	3100	Total: Mechanical & Electrical	↓	35	41	48	24%	28.50%	31.50%	
55	0010	**MEDICAL CLINICS**	S.F.	132	162	207				55
	0020	Total project costs	C.F.	9.65	12.50	16.60				
	1800	Equipment	S.F.	1.68	7.50	11.65	1.05%	2.94%	6.35%	
	2720	Plumbing		8.75	12.30	16.50	6.15%	8.40%	10.10%	
	2770	Heating, ventilating, air conditioning		10.45	13.70	20	6.65%	8.85%	11.35%	
	2900	Electrical		11.30	16.10	21	8.10%	10%	12.20%	
	3100	Total: Mechanical & Electrical	↓	36	49	67.50	22%	27%	33.50%	
	3500	See also division 11 71 00								
57	0010	**MEDICAL OFFICES**	S.F.	124	154	190				57
	0020	Total project costs	C.F.	9.25	12.50	16.95				
	1800	Equipment	S.F.	4.09	8.10	11.35	.66%	4.87%	6.50%	
	2720	Plumbing		6.85	10.55	14.25	5.60%	6.80%	8.50%	
	2770	Heating, ventilating, air conditioning		8.25	11.95	15.75	6.10%	8%	9.70%	
	2900	Electrical		9.90	14.50	20.50	7.50%	9.80%	11.70%	
	3100	Total: Mechanical & Electrical	↓	27	39.50	56	19.30%	22.50%	27.50%	
59	0010	**MOTELS**	S.F.	78	113	148				59
	0020	Total project costs	C.F.	6.95	9.30	15.25				
	2720	Plumbing	S.F.	7.90	10.10	12.05	9.45%	10.60%	12.55%	
	2770	Heating, ventilating, air conditioning		4.83	7.20	12.90	5.60%	5.60%	10%	
	2900	Electrical		7.40	9.35	11.65	7.45%	9.05%	10.45%	
	3100	Total: Mechanical & Electrical	↓	22.50	31.50	54	18.50%	24%	25.50%	
	5000									
	9000	Per rental unit, total cost	Unit	39,800	75,500	82,000				
	9500	Total: Mechanical & Electrical	"	7,750	11,700	13,600				
60	0010	**NURSING HOMES**	S.F.	123	158	197				60
	0020	Total project costs	C.F.	9.65	12.05	16.45				
	1800	Equipment	S.F.	3.34	5.10	8.55	2%	3.62%	4.99%	
	2720	Plumbing		10.50	15.90	19.20	8.75%	10.10%	12.70%	
	2770	Heating, ventilating, air conditioning		11.05	16.80	22.50	9.70%	11.45%	11.80%	
	2900	Electrical		12.15	15.20	21	9.40%	10.60%	12.50%	
	3100	Total: Mechanical & Electrical	↓	29	40.50	68	26%	28%	30%	
	9000	Per bed or person, total cost	Bed	54,500	68,000	88,000				
61	0010	**OFFICES Low Rise (1 to 4 story)**	S.F.	103	135	175				61
	0020	Total project costs	C.F.	7.35	10.15	13.40				
	0100	Site work	S.F.	8.30	14.45	21.50	5.95%	9.70%	13.55%	
	0500	Masonry		4	8.05	14.70	2.61%	5.40%	8.45%	
	1800	Equipment		.93	2.15	5.85	.57%	1.50%	3.42%	
	2720	Plumbing		3.69	5.70	8.35	3.66%	4.50%	6.10%	
	2770	Heating, ventilating, air conditioning		8.15	11.40	16.65	7.20%	10.30%	11.70%	
	2900	Electrical		8.45	12.10	17.20	7.45%	9.65%	11.40%	
	3100	Total: Mechanical & Electrical	↓	23.50	32.50	48.50	18.20%	22%	27%	
62	0010	**OFFICES Mid Rise (5 to 10 story)**	S.F.	109	132	180				62
	0020	Total project costs	C.F.	7.75	9.90	14				

50 17 | Square Foot Costs

		50 17 00 | S.F. Costs	UNIT	UNIT COSTS			% OF TOTAL			
				1/4	MEDIAN	3/4	1/4	MEDIAN	3/4	
62	2720	Plumbing	S.F.	3.30	5.10	7.35	2.83%	3.74%	4.50%	62
	2770	Heating, ventilating, air conditioning		8.30	11.85	18.95	7.65%	9.40%	11%	
	2900	Electrical		8.10	10.40	15.70	6.35%	7.80%	10%	
	3100	Total: Mechanical & Electrical		21	27	51.50	18.95%	21%	27.50%	
63	0010	OFFICES High Rise (11 to 20 story)	S.F.	134	169	208				63
	0020	Total project costs	C.F.	9.40	11.75	16.85				
	2900	Electrical	S.F.	8.15	9.95	14.80	5.80%	7.85%	10.50%	
	3100	Total: Mechanical & Electrical		26.50	35.50	59.50	16.90%	23.50%	34%	
64	0010	POLICE STATIONS	S.F.	161	212	270				64
	0020	Total project costs	C.F.	12.85	15.70	21.50				
	0500	Masonry	S.F.	16.10	26.50	33.50	7.80%	9.10%	11.35%	
	1800	Equipment		2.31	11.25	17.80	.98%	3.35%	6.70%	
	2720	Plumbing		9	17.95	22.50	5.65%	6.90%	10.75%	
	2770	Heating, ventilating, air conditioning		13.20	18.70	26.50	5.85%	10.55%	11.70%	
	2900	Electrical		17.60	25.50	33	9.80%	11.70%	14.50%	
	3100	Total: Mechanical & Electrical		67	70.50	94	28.50%	31.50%	32.50%	
65	0010	POST OFFICES	S.F.	127	157	200				65
	0020	Total project costs	C.F.	7.65	9.70	11				
	2720	Plumbing	S.F.	5.15	7.10	8.95	4.24%	5.30%	5.60%	
	2770	Heating, ventilating, air conditioning		8	11.05	12.30	6.65%	7.15%	9.35%	
	2900	Electrical		10.50	14.75	17.50	7.25%	9%	11%	
	3100	Total: Mechanical & Electrical		30.50	39.50	45	16.25%	18.80%	22%	
66	0010	POWER PLANTS	S.F.	880	1,175	2,150				66
	0020	Total project costs	C.F.	24.50	53	113				
	2900	Electrical	S.F.	62.50	132	197	9.30%	12.75%	21.50%	
	8100	Total: Mechanical & Electrical		155	505	1,125	32.50%	32.50%	52.50%	
67	0010	RELIGIOUS EDUCATION	S.F.	103	137	170				67
	0020	Total project costs	C.F.	5.70	8.15	10.20				
	2720	Plumbing	S.F.	4.28	6.05	8.60	4.40%	5.30%	7.10%	
	2770	Heating, ventilating, air conditioning		10.80	12.25	17.30	10.05%	11.45%	12.35%	
	2900	Electrical		8.15	11.65	17.10	7.70%	9.05%	10.35%	
	3100	Total: Mechanical & Electrical		35.50	45	53	22%	23.50%	26%	
69	0010	RESEARCH Laboratories & Facilities	S.F.	157	222	325				69
	0020	Total project costs	C.F.	11.40	22.50	27				
	1800	Equipment	S.F.	6.80	13.40	32.50	1.22%	4.80%	9.35%	
	2720	Plumbing		11.20	19.70	31.50	6.15%	8.30%	10.80%	
	2770	Heating, ventilating, air conditioning		13.80	46.50	55	7.25%	16.50%	17.50%	
	2900	Electrical		18.45	29.50	48.50	9.45%	11.15%	14.60%	
	3100	Total: Mechanical & Electrical		59	104	147	29.50%	36%	41%	
70	0010	RESTAURANTS	S.F.	149	192	249				70
	0020	Total project costs	C.F.	12.50	16.40	21.50				
	1800	Equipment	S.F.	8.80	23.50	35.50	6.10%	13%	15.65%	
	2720	Plumbing		11.80	14.30	18.75	6.10%	8.15%	9%	
	2770	Heating, ventilating, air conditioning		14.95	21	25	9.20%	12%	12.40%	
	2900	Electrical		15.70	19.40	25	8.35%	10.55%	11.55%	
	3100	Total: Mechanical & Electrical		48.50	52	67.50	21%	25%	29.50%	
	9000	Per seat unit, total cost	Seat	5,450	7,275	8,600				
	9500	Total: Mechanical & Electrical	"	1,375	1,825	2,150				
72	0010	RETAIL STORES	S.F.	69.50	93.50	124				72
	0020	Total project costs	C.F.	4.71	6.70	9.35				
	2720	Plumbing	S.F.	2.52	4.20	7.15	3.26%	4.60%	6.80%	
	2770	Heating, ventilating, air conditioning		5.45	7.45	11.20	6.75%	8.75%	10.15%	
	2900	Electrical		6.25	8.55	12.35	7.25%	9.90%	11.60%	
	3100	Total: Mechanical & Electrical		16.65	21.50	28.50	17.05%	21%	23.50%	
74	0010	SCHOOLS Elementary	S.F.	114	141	172				74
	0020	Total project costs	C.F.	7.40	9.50	12.25				
	0500	Masonry	S.F.	8	17.30	25.50	4.89%	10.50%	14%	
	1800	Equipment		2.63	4.96	9.40	1.83%	3.13%	4.61%	

For customer support on your Plumbing Cost Data, call 877.709.6968.

695

50 17 00 | S.F. Costs			UNIT	UNIT COSTS			% OF TOTAL		
				1/4	MEDIAN	3/4	1/4	MEDIAN	3/4
74	2720	Plumbing	S.F.	6.50	9.20	12.30	5.70%	7.15%	9.35%	74
	2730	Heating, ventilating, air conditioning		9.80	15.55	22.50	8.15%	10.80%	14.90%	
	2900	Electrical		10.70	14.25	18.20	8.45%	10.05%	11.85%	
	3100	Total: Mechanical & Electrical	↓	39	47.50	60	25%	27.50%	30%	
	9000	Per pupil, total cost	Ea.	13,000	19,300	43,300				
	9500	Total: Mechanical & Electrical	"	3,675	4,650	11,700				
76	0010	**SCHOOLS Junior High & Middle**	S.F.	117	145	177				76
	0020	Total project costs	C.F.	7.40	9.60	10.75				
	0500	Masonry	S.F.	12.55	18.80	23	7.55%	11.10%	14.30%	
	1800	Equipment		3.20	6	8.80	1.80%	3.03%	4.80%	
	2720	Plumbing		6.80	8.40	10.40	5.30%	6.80%	7.25%	
	2770	Heating, ventilating, air conditioning		13.60	16.55	29	8.90%	11.55%	14.20%	
	2900	Electrical		11.90	14.70	18.60	8.05%	9.55%	10.60%	
	3100	Total: Mechanical & Electrical	↓	37.50	48	59.50	23%	25.50%	29.50%	
	9000	Per pupil, total cost	Ea.	14,800	19,400	26,100				
78	0010	**SCHOOLS Senior High**	S.F.	123	150	188				78
	0020	Total project costs	C.F.	7.35	10.30	17.40				
	1800	Equipment	S.F.	3.24	7.55	10.45	1.88%	2.67%	4.30%	
	2720	Plumbing		6.80	10.25	18.70	5.60%	6.90%	8.30%	
	2770	Heating, ventilating, air conditioning		11.75	15.95	24.50	8.95%	11.60%	15%	
	2900	Electrical		12.30	16.40	24	8.70%	10.35%	12.50%	
	3100	Total: Mechanical & Electrical	↓	40.50	48	77.50	24%	26.50%	28.50%	
	9000	Per pupil, total cost	Ea.	11,500	23,300	29,200				
80	0010	**SCHOOLS Vocational**	S.F.	98.50	143	177				80
	0020	Total project costs	C.F.	6.10	8.80	12.15				
	0500	Masonry	S.F.	4.33	14.25	22	3.20%	4.61%	10.95%	
	1800	Equipment		3.08	7.65	10.60	1.24%	3.10%	4.26%	
	2720	Plumbing		6.30	9.35	13.80	5.40%	6.90%	8.55%	
	2770	Heating, ventilating, air conditioning		8.80	16.40	27.50	8.60%	11.90%	14.65%	
	2900	Electrical		10.25	14	19.15	8.45%	11%	13.20%	
	3100	Total: Mechanical & Electrical	↓	38.50	67.50	86.50	27.50%	29.50%	31%	
	9000	Per pupil, total cost	Ea.	13,700	36,700	54,500				
83	0010	**SPORTS ARENAS**	S.F.	86	115	177				83
	0020	Total project costs	C.F.	4.67	8.35	10.80				
	2720	Plumbing	S.F.	4.31	7.55	15.95	4.35%	6.35%	9.40%	
	2770	Heating, ventilating, air conditioning		10.75	12.70	17.60	8.80%	10.20%	13.55%	
	2900	Electrical	↓	7.95	12.15	15.70	7.65%	9.75%	11.90%	
85	0010	**SUPERMARKETS**	S.F.	79.50	92	108				85
	0020	Total project costs	C.F.	4.42	5.35	8.10				
	2720	Plumbing	S.F.	4.44	5.60	6.50	5.40%	6%	7.45%	
	2770	Heating, ventilating, air conditioning		6.55	8.65	10.55	8.60%	8.65%	9.60%	
	2900	Electrical		9.30	11.35	13.50	10.40%	12.45%	13.60%	
	3100	Total: Mechanical & Electrical	↓	25.50	27	36.50	23.50%	27.50%	28.50%	
86	0010	**SWIMMING POOLS**	S.F.	129	300	460				86
	0020	Total project costs	C.F.	10.30	12.85	14				
	2720	Plumbing	S.F.	11.90	13.60	18.25	4.80%	9.70%	20.50%	
	2900	Electrical		9.75	15.65	34.50	6.05%	6.95%	7.60%	
	3100	Total: Mechanical & Electrical	↓	61	80.50	106	11.15%	14.10%	23.50%	
87	0010	**TELEPHONE EXCHANGES**	S.F.	160	243	315				87
	0020	Total project costs	C.F.	10.65	17.05	23.50				
	2720	Plumbing	S.F.	7.20	11.15	16.30	4.52%	5.80%	6.90%	
	2770	Heating, ventilating, air conditioning		16.75	33.50	42	11.80%	16.05%	18.40%	
	2900	Electrical		17.40	27.50	49	10.90%	14%	17.85%	
	3100	Total: Mechanical & Electrical		51.50	97.50	138	29.50%	33.50%	44.50%	
91	0010	**THEATERS**	S.F.	102	138	203				91
	0020	Total project costs	C.F.	4.96	7.35	10.80				

		50 17 00	S.F. Costs	UNIT	UNIT COSTS			% OF TOTAL		
				1/4	MEDIAN	3/4	1/4	MEDIAN	3/4
91	2720	Plumbing	S.F.	3.34	3.88	15.85	2.92%	4.70%	6.80%	**91**
	2770	Heating, ventilating, air conditioning		10.45	12.65	15.65	8%	12.25%	13.40%	
	2900	Electrical		9.40	12.70	26	8.05%	9.35%	12.25%	
	3100	Total: Mechanical & Electrical	↓	24	32.50	38.50	21.50%	25.50%	27.50%	
94	0010	**TOWN HALLS City Halls & Municipal Buildings**	S.F.	112	152	199				**94**
	0020	Total project costs	C.F.	8.10	12.60	18.45				
	2720	Plumbing	S.F.	5	9.35	16.45	4.31%	5.95%	7.95%	
	2770	Heating, ventilating, air conditioning		9.05	17.95	26.50	7.05%	9.05%	13.45%	
	2900	Electrical		11.40	15.70	21.50	8.05%	9.45%	11.65%	
	3100	Total: Mechanical & Electrical	↓	39.50	50	76.50	22%	26.50%	31%	
97	0010	**WAREHOUSES & Storage Buildings**	S.F.	44.50	63.50	95				**97**
	0020	Total project costs	C.F.	2.34	3.67	6.05				
	0100	Site work	S.F.	4.62	9.15	13.80	6.05%	12.95%	19.55%	
	0500	Masonry		2.54	6.35	13.70	3.60%	7.35%	12%	
	1800	Equipment		.68	1.55	8.70	.72%	1.69%	5.55%	
	2720	Plumbing		1.49	2.68	5	2.90%	4.80%	6.55%	
	2730	Heating, ventilating, air conditioning		1.70	4.80	6.45	2.41%	5%	8.90%	
	2900	Electrical		2.65	4.98	8.20	5.15%	7.20%	10.05%	
	3100	Total: Mechanical & Electrical	↓	7.40	11.35	22.50	13.30%	18.90%	26%	
99	0010	**WAREHOUSE & OFFICES Combination**	S.F.	55	73.50	101				**99**
	0020	Total project costs	C.F.	2.82	4.09	6.05				
	1800	Equipment	S.F.	.95	1.84	2.74	.42%	1.19%	2.40%	
	2720	Plumbing		2.13	3.70	5.50	3.74%	4.76%	6.30%	
	2770	Heating, ventilating, air conditioning		3.36	5.25	7.35	5%	5.65%	10.05%	
	2900	Electrical		3.75	5.55	8.65	5.85%	8%	10%	
	3100	Total: Mechanical & Electrical	↓	10.45	15.85	25	14.55%	19.95%	24.50%	

Square Foot Project Size Modifier

One factor that affects the S.F. cost of a particular building is the size. In general, for buildings built to the same specifications in the same locality, the larger building will have the lower S.F. cost. This is due mainly to the decreasing contribution of the exterior walls, plus the economy of scale usually achievable in larger buildings. The area conversion scale shown below will give a factor to convert costs for the typical size building to an adjusted cost for the particular project.

The square foot base size table lists the median costs, most typical project size in our accumulated data, and the range in size of the projects.

The size factor for your project is determined by dividing your project area in S.F. by the typical project size for the particular building type. With this factor, enter the area conversion scale at the appropriate size factor and determine the appropriate cost multiplier for your building size.

Example: Determine the cost per S.F. for a 100,000 S.F. Mid-rise apartment building.

$$\frac{\text{Proposed building area} = 100,000 \text{ S.F.}}{\text{Typical size from below} = 50,000 \text{ S.F.}} = 2.00$$

Enter area conversion scale at 2.0, intersect curve, read horizontally the appropriate cost multiplier of .94. Size adjusted cost becomes .94 × $119.00 = $112.00 based on national average costs.

Note: For size factors less than .50, the cost multiplier is 1.1
For size factors greater than 3.5, the cost multiplier is .90

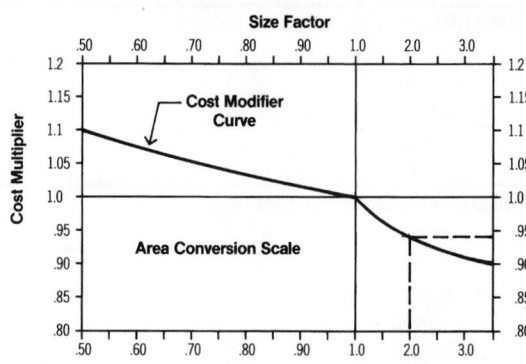

Square Foot Base Size							
Building Type	**Median Cost per S.F.**	**Typical Size Gross S.F.**	**Typical Range Gross S.F.**		**Building Type**	**Median Cost per S.F.**	**Typical Size Gross S.F.**	**Typical Range Gross S.F.**

Building Type	Median Cost per S.F.	Typical Size Gross S.F.	Typical Range Gross S.F.	Building Type	Median Cost per S.F.	Typical Size Gross S.F.	Typical Range Gross S.F.
Apartments, Low Rise	$ 94.00	21,000	9,700 - 37,200	Jails	$ 286.00	40,000	5,500 - 145,000
Apartments, Mid Rise	119.00	50,000	32,000 - 100,000	Libraries	190.00	12,000	7,000 - 31,000
Apartments, High Rise	129.00	145,000	95,000 - 600,000	Living, Assisted	154.00	32,300	23,500 - 50,300
Auditoriums	161.00	25,000	7,600 - 39,000	Medical Clinics	162.00	7,200	4,200 - 15,700
Auto Sales	119.00	20,000	10,800 - 28,600	Medical Offices	154.00	6,000	4,000 - 15,000
Banks	211.00	4,200	2,500 - 7,500	Motels	113.00	40,000	15,800 - 120,000
Churches	145.00	17,000	2,000 - 42,000	Nursing Homes	158.00	23,000	15,000 - 37,000
Clubs, Country	147.00	6,500	4,500 - 15,000	Offices, Low Rise	135.00	20,000	5,000 - 80,000
Clubs, Social	141.00	10,000	6,000 - 13,500	Offices, Mid Rise	132.00	120,000	20,000 - 300,000
Clubs, YMCA	167.00	28,300	12,800 - 39,400	Offices, High Rise	169.00	260,000	120,000 - 800,000
Colleges (Class)	167.00	50,000	15,000 - 150,000	Police Stations	212.00	10,500	4,000 - 19,000
Colleges (Science Lab)	269.00	45,600	16,600 - 80,000	Post Offices	157.00	12,400	6,800 - 30,000
College (Student Union)	199.00	33,400	16,000 - 85,000	Power Plants	1175.00	7,500	1,000 - 20,000
Community Center	150.00	9,400	5,300 - 16,700	Religious Education	137.00	9,000	6,000 - 12,000
Court Houses	206.00	32,400	17,800 - 106,000	Research	222.00	19,000	6,300 - 45,000
Dept. Stores	87.50	90,000	44,000 - 122,000	Restaurants	192.00	4,400	2,800 - 6,000
Dormitories, Low Rise	170.00	25,000	10,000 - 95,000	Retail Stores	93.50	7,200	4,000 - 17,600
Dormitories, Mid Rise	196.00	85,000	20,000 - 200,000	Schools, Elementary	141.00	41,000	24,500 - 55,000
Factories	85.00	26,400	12,900 - 50,000	Schools, Jr. High	145.00	92,000	52,000 - 119,000
Fire Stations	156.00	5,800	4,000 - 8,700	Schools, Sr. High	150.00	101,000	50,500 - 175,000
Fraternity Houses	145.00	12,500	8,200 - 14,800	Schools, Vocational	143.00	37,000	20,500 - 82,000
Funeral Homes	162.00	10,000	4,000 - 20,000	Sports Arenas	115.00	15,000	5,000 - 40,000
Garages, Commercial	104.00	9,300	5,000 - 13,600	Supermarkets	92.00	44,000	12,000 - 60,000
Garages, Municipal	135.00	8,300	4,500 - 12,600	Swimming Pools	300.00	20,000	10,000 - 32,000
Garages, Parking	56.00	163,000	76,400 - 225,300	Telephone Exchange	243.00	4,500	1,200 - 10,600
Gymnasiums	143.00	19,200	11,600 - 41,000	Theaters	138.00	10,500	8,800 - 17,500
Hospitals	258.00	55,000	27,200 - 125,000	Town Halls	152.00	10,800	4,800 - 23,400
House (Elderly)	128.00	37,000	21,000 - 66,000	Warehouses	63.50	25,000	8,000 - 72,000
Housing (Public)	118.00	36,000	14,400 - 74,400	Warehouse & Office	73.50	25,000	8,000 - 72,000
Ice Rinks	170.00	29,000	27,200 - 33,600				

Abbreviations

A	Area Square Feet; Ampere
AAFES	Army and Air Force Exchange Service
ABS	Acrylonitrile Butadiene Stryrene; Asbestos Bonded Steel
A.C., AC	Alternating Current; Air-Conditioning; Asbestos Cement; Plywood Grade A & C
ACI	American Concrete Institute
ACR	Air Conditioning Refrigeration
ADA	Americans with Disabilities Act
AD	Plywood, Grade A & D
Addit.	Additional
Adj.	Adjustable
af	Audio-frequency
AFUE	Annual Fuel Utilization Efficiency
AGA	American Gas Association
Agg.	Aggregate
A.H., Ah	Ampere Hours
A hr.	Ampere-hour
A.H.U., AHU	Air Handling Unit
A.I.A.	American Institute of Architects
AIC	Ampere Interrupting Capacity
Allow.	Allowance
alt., alt	Alternate
Alum.	Aluminum
a.m.	Ante Meridiem
Amp.	Ampere
Anod.	Anodized
ANSI	American National Standards Institute
APA	American Plywood Association
Approx.	Approximate
Apt.	Apartment
Asb.	Asbestos
A.S.B.C.	American Standard Building Code
Asbe.	Asbestos Worker
ASCE.	American Society of Civil Engineers
A.S.H.R.A.E.	American Society of Heating, Refrig. & AC Engineers
ASME	American Society of Mechanical Engineers
ASTM	American Society for Testing and Materials
Attchmt.	Attachment
Avg., Ave.	Average
AWG	American Wire Gauge
AWWA	American Water Works Assoc.
Bbl.	Barrel
B&B, BB	Grade B and Better; Balled & Burlapped
B&S	Bell and Spigot
B.&W.	Black and White
b.c.c.	Body-centered Cubic
B.C.Y.	Bank Cubic Yards
BE	Bevel End
B.F.	Board Feet
Bg. cem.	Bag of Cement
BHP	Boiler Horsepower; Brake Horsepower
B.I.	Black Iron
bidir.	bidirectional
Bit., Bitum.	Bituminous
Bit., Conc.	Bituminous Concrete
Bk.	Backed
Bkrs.	Breakers
Bldg., bldg	Building
Blk.	Block
Bm.	Beam
Boil.	Boilermaker
bpm	Blows per Minute
BR	Bedroom
Brg.	Bearing
Brhe.	Bricklayer Helper
Bric.	Bricklayer

Brk., brk	Brick
brkt	Bracket
Brng.	Bearing
Brs.	Brass
Brz.	Bronze
Bsn.	Basin
Btr.	Better
Btu	British Thermal Unit
BTUH	BTU per Hour
Bu.	bushels
BUR	Built-up Roofing
BX	Interlocked Armored Cable
°C	degree centegrade
c	Conductivity, Copper Sweat
C	Hundred; Centigrade
C/C	Center to Center, Cedar on Cedar
C-C	Center to Center
Cab	Cabinet
Cair.	Air Tool Laborer
Cal.	caliper
Calc	Calculated
Cap.	Capacity
Carp.	Carpenter
C.B.	Circuit Breaker
C.C.A.	Chromate Copper Arsenate
C.C.F.	Hundred Cubic Feet
cd	Candela
cd/sf	Candela per Square Foot
CD	Grade of Plywood Face & Back
CDX	Plywood, Grade C & D, exterior glue
Cefi.	Cement Finisher
Cem.	Cement
CF	Hundred Feet
C.F.	Cubic Feet
CFM	Cubic Feet per Minute
CFRP	Carbon Fiber Reinforced Plastic
c.g.	Center of Gravity
CHW	Chilled Water; Commercial Hot Water
C.I., CI	Cast Iron
C.I.P., CIP	Cast in Place
Circ.	Circuit
C.L.	Carload Lot
CL	Chain Link
Clab.	Common Laborer
Clam	Common maintenance laborer
C.L.F.	Hundred Linear Feet
CLF	Current Limiting Fuse
CLP	Cross Linked Polyethylene
cm	Centimeter
CMP	Corr. Metal Pipe
CMU	Concrete Masonry Unit
CN	Change Notice
Col.	Column
CO₂	Carbon Dioxide
Comb.	Combination
comm.	Commercial, Communication
Compr.	Compressor
Conc.	Concrete
Cont., cont	Continuous; Continued, Container
Corr.	Corrugated
Cos	Cosine
Cot	Cotangent
Cov.	Cover
C/P	Cedar on Paneling
CPA	Control Point Adjustment
Cplg.	Coupling
CPM	Critical Path Method
CPVC	Chlorinated Polyvinyl Chloride
C.Pr.	Hundred Pair
CRC	Cold Rolled Channel
Creos.	Creosote
Crpt.	Carpet & Linoleum Layer
CRT	Cathode-ray Tube
CS	Carbon Steel, Constant Shear Bar Joist

Csc	Cosecant
C.S.F.	Hundred Square Feet
CSI	Construction Specifications Institute
CT	Current Transformer
CTS	Copper Tube Size
Cu	Copper, Cubic
Cu. Ft.	Cubic Foot
cw	Continuous Wave
C.W.	Cool White; Cold Water
Cwt.	100 Pounds
C.W.X.	Cool White Deluxe
C.Y.	Cubic Yard (27 cubic feet)
C.Y./Hr.	Cubic Yard per Hour
Cyl.	Cylinder
d	Penny (nail size)
D	Deep; Depth; Discharge
Dis., Disch.	Discharge
Db	Decibel
Dbl.	Double
DC	Direct Current
DDC	Direct Digital Control
Demob.	Demobilization
d.f.t.	Dry Film Thickness
d.f.u.	Drainage Fixture Units
D.H.	Double Hung
DHW	Domestic Hot Water
DI	Ductile Iron
Diag.	Diagonal
Diam., Dia	Diameter
Distrib.	Distribution
Div.	Division
Dk.	Deck
D.L.	Dead Load; Diesel
DLH	Deep Long Span Bar Joist
dlx	Deluxe
Do.	Ditto
DOP	Dioctyl Phthalate Penetration Test (Air Filters)
Dp., dp	Depth
D.P.S.T.	Double Pole, Single Throw
Dr.	Drive
DR	Dimension Ratio
Drink.	Drinking
D.S.	Double Strength
D.S.A.	Double Strength A Grade
D.S.B.	Double Strength B Grade
Dty.	Duty
DWV	Drain Waste Vent
DX	Deluxe White, Direct Expansion
dyn	Dyne
e	Eccentricity
E	Equipment Only; East; emissivity
Ea.	Each
EB	Encased Burial
Econ.	Economy
E.C.Y	Embankment Cubic Yards
EDP	Electronic Data Processing
EIFS	Exterior Insulation Finish System
E.D.R.	Equiv. Direct Radiation
Eq.	Equation
EL	elevation
Elec.	Electrician; Electrical
Elev.	Elevator; Elevating
EMT	Electrical Metallic Conduit; Thin Wall Conduit
Eng.	Engine, Engineered
EPDM	Ethylene Propylene Diene Monomer
EPS	Expanded Polystyrene
Eqhv.	Equip. Oper., Heavy
Eqlt.	Equip. Oper., Light
Eqmd.	Equip. Oper., Medium
Eqmm.	Equip. Oper., Master Mechanic
Eqol.	Equip. Oper., Oilers
Equip.	Equipment
ERW	Electric Resistance Welded

Abbreviations

E.S.	Energy Saver	H	High Henry	Lath.	Lather
Est.	Estimated	HC	High Capacity	Lav.	Lavatory
esu	Electrostatic Units	H.D., HD	Heavy Duty; High Density	lb.; #	Pound
E.W.	Each Way	H.D.O.	High Density Overlaid	L.B., LB	Load Bearing; L Conduit Body
EWT	Entering Water Temperature	HDPE	High density polyethelene plastic	L. & E.	Labor & Equipment
Excav.	Excavation	Hdr.	Header	lb./hr.	Pounds per Hour
excl	Excluding	Hdwe.	Hardware	lb./L.F.	Pounds per Linear Foot
Exp., exp	Expansion, Exposure	H.I.D., HID	High Intensity Discharge	lbf/sq.in.	Pound-force per Square Inch
Ext., ext	Exterior; Extension	Help.	Helper Average	L.C.L.	Less than Carload Lot
Extru.	Extrusion	HEPA	High Efficiency Particulate Air Filter	L.C.Y.	Loose Cubic Yard
f.	Fiber stress			Ld.	Load
F	Fahrenheit; Female; Fill	Hg	Mercury	LE	Lead Equivalent
Fab., fab	Fabricated; fabric	HIC	High Interrupting Capacity	LED	Light Emitting Diode
FBGS	Fiberglass	HM	Hollow Metal	L.F.	Linear Foot
F.C.	Footcandles	HMWPE	high molecular weight polyethylene	L.F. Nose	Linear Foot of Stair Nosing
f.c.c.	Face-centered Cubic			L.F. Rsr	Linear Foot of Stair Riser
f'c.	Compressive Stress in Concrete; Extreme Compressive Stress	HO	High Output	Lg.	Long; Length; Large
		Horiz.	Horizontal	L & H	Light and Heat
F.E.	Front End	H.P., HP	Horsepower; High Pressure	LH	Long Span Bar Joist
FEP	Fluorinated Ethylene Propylene (Teflon)	H.P.F.	High Power Factor	L.H.	Labor Hours
		Hr.	Hour	L.L., LL	Live Load
F.G.	Flat Grain	Hrs./Day	Hours per Day	L.L.D.	Lamp Lumen Depreciation
F.H.A.	Federal Housing Administration	HSC	High Short Circuit	lm	Lumen
Fig.	Figure	Ht.	Height	lm/sf	Lumen per Square Foot
Fin.	Finished	Htg.	Heating	lm/W	Lumen per Watt
FIPS	Female Iron Pipe Size	Htrs.	Heaters	LOA	Length Over All
Fixt.	Fixture	HVAC	Heating, Ventilation & Air-Conditioning	log	Logarithm
FJP	Finger jointed and primed			L-O-L	Lateralolet
Fl. Oz.	Fluid Ounces	Hvy.	Heavy	long.	longitude
Flr.	Floor	HW	Hot Water	L.P., LP	Liquefied Petroleum; Low Pressure
FM	Frequency Modulation; Factory Mutual	Hyd.; Hydr.	Hydraulic	L.P.F.	Low Power Factor
		Hz	Hertz (cycles)	LR	Long Radius
Fmg.	Framing	I.	Moment of Inertia	L.S.	Lump Sum
FM/UL	Factory Mutual/Underwriters Labs	IBC	International Building Code	Lt.	Light
Fdn.	Foundation	I.C.	Interrupting Capacity	Lt. Ga.	Light Gauge
FNPT	Female National Pipe Thread	ID	Inside Diameter	L.T.L.	Less than Truckload Lot
Fori.	Foreman, Inside	I.D.	Inside Dimension; Identification	Lt. Wt.	Lightweight
Foro.	Foreman, Outside	I.F.	Inside Frosted	L.V.	Low Voltage
Fount.	Fountain	I.M.C.	Intermediate Metal Conduit	M	Thousand; Material; Male; Light Wall Copper Tubing
fpm	Feet per Minute	In.	Inch		
FPT	Female Pipe Thread	Incan.	Incandescent	M²CA	Meters Squared Contact Area
Fr	Frame	Incl.	Included; Including	m/hr.; M.H.	Man-hour
F.R.	Fire Rating	Int.	Interior	mA	Milliampere
FRK	Foil Reinforced Kraft	Inst.	Installation	Mach.	Machine
FSK	Foil/scrim/kraft	Insul., insul	Insulation/Insulated	Mag. Str.	Magnetic Starter
FRP	Fiberglass Reinforced Plastic	I.P.	Iron Pipe	Maint.	Maintenance
FS	Forged Steel	I.P.S., IPS	Iron Pipe Size	Marb.	Marble Setter
FSC	Cast Body; Cast Switch Box	IPT	Iron Pipe Threaded	Mat; Mat'l.	Material
Ft., ft	Foot; Feet	I.W.	Indirect Waste	Max.	Maximum
Ftng.	Fitting	J	Joule	MBF	Thousand Board Feet
Ftg.	Footing	J.I.C.	Joint Industrial Council	MBH	Thousand BTU's per hr.
Ft lb.	Foot Pound	K	Thousand; Thousand Pounds; Heavy Wall Copper Tubing, Kelvin	MC	Metal Clad Cable
Furn.	Furniture			MCC	Motor Control Center
FVNR	Full Voltage Non-Reversing	K.A.H.	Thousand Amp. Hours	M.C.F.	Thousand Cubic Feet
FVR	Full Voltage Reversing	kcmil	Thousand Circular Mils	MCFM	Thousand Cubic Feet per Minute
FXM	Female by Male	KD	Knock Down	M.C.M.	Thousand Circular Mils
Fy.	Minimum Yield Stress of Steel	K.D.A.T.	Kiln Dried After Treatment	MCP	Motor Circuit Protector
g	Gram	kg	Kilogram	MD	Medium Duty
G	Gauss	kG	Kilogauss	MDF	Medium-density fibreboard
Ga.	Gauge	kgf	Kilogram Force	M.D.O.	Medium Density Overlaid
Gal., gal.	Gallon	kHz	Kilohertz	Med.	Medium
gpm, GPM	Gallon per Minute	Kip	1000 Pounds	MF	Thousand Feet
Galv., galv	Galvanized	KJ	Kiljoule	M.F.B.M.	Thousand Feet Board Measure
GC/MS	Gas Chromatograph/Mass Spectrometer	K.L.	Effective Length Factor	Mfg.	Manufacturing
		K.L.F.	Kips per Linear Foot	Mfrs.	Manufacturers
Gen.	General	Km	Kilometer	mg	Milligram
GFI	Ground Fault Interrupter	KO	Knock Out	MGD	Million Gallons per Day
GFRC	Glass Fiber Reinforced Concrete	K.S.F.	Kips per Square Foot	MGPH	Thousand Gallons per Hour
Glaz.	Glazier	K.S.I.	Kips per Square Inch	MH, M.H.	Manhole; Metal Halide; Man-Hour
GPD	Gallons per Day	kV	Kilovolt	MHz	Megahertz
gpf	Gallon per flush	kVA	Kilovolt Ampere	Mi.	Mile
GPH	Gallons per Hour	kVAR	Kilovar (Reactance)	MI	Malleable Iron; Mineral Insulated
GPM	Gallons per Minute	KW	Kilowatt	MIPS	Male Iron Pipe Size
GR	Grade	KWh	Kilowatt-hour	mj	Mechanical Joint
Gran.	Granular	L	Labor Only; Length; Long; Medium Wall Copper Tubing	m	Meter
Grnd.	Ground			mm	Millimeter
GVW	Gross Vehicle Weight	Lab.	Labor	Mill.	Millwright
GWB	Gypsum wall board	lat	Latitude	Min., min.	Minimum, minute

700

For customer support on your Plumbing Cost Data, call 877.709.6968.

Misc.	Miscellaneous	PDCA	Painting and Decorating Contractors of America
ml	Milliliter, Mainline	P.E., PE	Professional Engineer; Porcelain Enamel; Polyethylene; Plain End
M.L.F.	Thousand Linear Feet		
Mo.	Month		
Mobil.	Mobilization	P.E.C.I.	Porcelain Enamel on Cast Iron
Mog.	Mogul Base	Perf.	Perforated
MPH	Miles per Hour	PEX	Cross linked polyethylene
MPT	Male Pipe Thread	Ph.	Phase
MRGWB	Moisture Resistant Gypsum Wallboard	P.I.	Pressure Injected
		Pile.	Pile Driver
MRT	Mile Round Trip	Pkg.	Package
ms	Millisecond	Pl.	Plate
M.S.F.	Thousand Square Feet	Plah.	Plasterer Helper
Mstz.	Mosaic & Terrazzo Worker	Plas.	Plasterer
M.S.Y.	Thousand Square Yards	plf	Pounds Per Linear Foot
Mtd., mtd., mtd	Mounted	Pluh.	Plumbers Helper
Mthe.	Mosaic & Terrazzo Helper	Plum.	Plumber
Mtng.	Mounting	Ply.	Plywood
Mult.	Multi; Multiply	p.m.	Post Meridiem
M.V.A.	Million Volt Amperes	Pntd.	Painted
M.V.A.R.	Million Volt Amperes Reactance	Pord.	Painter, Ordinary
MV	Megavolt	pp	Pages
MW	Megawatt	PP, PPL	Polypropylene
MXM	Male by Male	P.P.M.	Parts per Million
MYD	Thousand Yards	Pr.	Pair
N	Natural; North	P.E.S.B.	Pre-engineered Steel Building
nA	Nanoampere	Prefab.	Prefabricated
NA	Not Available; Not Applicable	Prefin.	Prefinished
N.B.C.	National Building Code	Prop.	Propelled
NC	Normally Closed	PSF, psf	Pounds per Square Foot
NEMA	National Electrical Manufacturers Assoc.	PSI, psi	Pounds per Square Inch
		PSIG	Pounds per Square Inch Gauge
NEHB	Bolted Circuit Breaker to 600V.	PSP	Plastic Sewer Pipe
NFPA	National Fire Protection Association	Pspr.	Painter, Spray
NLB	Non-Load-Bearing	Psst.	Painter, Structural Steel
NM	Non-Metallic Cable	P.T.	Potential Transformer
nm	Nanometer	P. & T.	Pressure & Temperature
No.	Number	Ptd.	Painted
NO	Normally Open	Ptns.	Partitions
N.O.C.	Not Otherwise Classified	Pu	Ultimate Load
Nose.	Nosing	PVC	Polyvinyl Chloride
NPT	National Pipe Thread	Pvmt.	Pavement
NQOD	Combination Plug-on/Bolt on Circuit Breaker to 240V.	PRV	Pressure Relief Valve
		Pwr.	Power
N.R.C., NRC	Noise Reduction Coefficient/ Nuclear Regulator Commission	Q	Quantity Heat Flow
		Qt.	Quart
N.R.S.	Non Rising Stem	Quan., Qty.	Quantity
ns	Nanosecond	Q.C.	Quick Coupling
nW	Nanowatt	r	Radius of Gyration
OB	Opposing Blade	R	Resistance
OC	On Center	R.C.P.	Reinforced Concrete Pipe
OD	Outside Diameter	Rect.	Rectangle
O.D.	Outside Dimension	recpt.	receptacle
ODS	Overhead Distribution System	Reg.	Regular
O.G.	Ogee	Reinf.	Reinforced
O.H.	Overhead	Req'd.	Required
O&P	Overhead and Profit	Res.	Resistant
Oper.	Operator	Resi.	Residential
Opng.	Opening	RF	Radio Frequency
Orna.	Ornamental	RFID	Radio-frequency identification
OSB	Oriented Strand Board	Rgh.	Rough
OS&Y	Outside Screw and Yoke	RGS	Rigid Galvanized Steel
OSHA	Occupational Safety and Health Act	RHW	Rubber, Heat & Water Resistant; Residential Hot Water
Ovhd.	Overhead	rms	Root Mean Square
OWG	Oil, Water or Gas	Rnd.	Round
Oz.	Ounce	Rodm.	Rodman
P.	Pole; Applied Load; Projection	Rofc.	Roofer, Composition
p.	Page	Rofp.	Roofer, Precast
Pape.	Paperhanger	Rohe.	Roofer Helpers (Composition)
P.A.P.R.	Powered Air Purifying Respirator	Rots.	Roofer, Tile & Slate
PAR	Parabolic Reflector	R.O.W.	Right of Way
P.B., PB	Push Button	RPM	Revolutions per Minute
Pc., Pcs.	Piece, Pieces	R.S.	Rapid Start
P.C.	Portland Cement; Power Connector	Rsr	Riser
P.C.F.	Pounds per Cubic Foot	RT	Round Trip
PCM	Phase Contrast Microscopy	S.	Suction; Single Entrance; South
		SBS	Styrene Butadiere Styrene

SC	Screw Cover		
SCFM	Standard Cubic Feet per Minute		
Scaf.	Scaffold		
Sch., Sched.	Schedule		
S.C.R.	Modular Brick		
S.D.	Sound Deadening		
SDR	Standard Dimension Ratio		
S.E.	Surfaced Edge		
Sel.	Select		
SER, SEU	Service Entrance Cable		
S.F.	Square Foot		
S.F.C.A.	Square Foot Contact Area		
S.F. Flr.	Square Foot of Floor		
S.F.G.	Square Foot of Ground		
S.F. Hor.	Square Foot Horizontal		
SFR	Square Feet of Radiation		
S.F. Shlf.	Square Foot of Shelf		
S4S	Surface 4 Sides		
Shee.	Sheet Metal Worker		
Sin.	Sine		
Skwk.	Skilled Worker		
SL	Saran Lined		
S.L.	Slimline		
Sldr.	Solder		
SLH	Super Long Span Bar Joist		
S.N.	Solid Neutral		
SO	Stranded with oil resistant inside insulation		
S-O-L	Socketolet		
sp	Standpipe		
S.P.	Static Pressure; Single Pole; Self-Propelled		
Spri.	Sprinkler Installer		
spwg	Static Pressure Water Gauge		
S.P.D.T.	Single Pole, Double Throw		
SPF	Spruce Pine Fir; Sprayed Polyurethane Foam		
S.P.S.T.	Single Pole, Single Throw		
SPT	Standard Pipe Thread		
Sq.	Square; 100 Square Feet		
Sq. Hd.	Square Head		
Sq. In.	Square Inch		
S.S.	Single Strength; Stainless Steel		
S.S.B.	Single Strength B Grade		
sst, ss	Stainless Steel		
Sswk.	Structural Steel Worker		
Sswl.	Structural Steel Welder		
St.; Stl.	Steel		
STC	Sound Transmission Coefficient		
Std.	Standard		
Stg.	Staging		
STK	Select Tight Knot		
STP	Standard Temperature & Pressure		
Stpi.	Steamfitter, Pipefitter		
Str.	Strength; Starter; Straight		
Strd.	Stranded		
Struct.	Structural		
Sty.	Story		
Subj.	Subject		
Subs.	Subcontractors		
Surf.	Surface		
Sw.	Switch		
Swbd.	Switchboard		
S.Y.	Square Yard		
Syn.	Synthetic		
S.Y.P.	Southern Yellow Pine		
Sys.	System		
t.	Thickness		
T	Temperature; Ton		
Tan	Tangent		
T.C.	Terra Cotta		
T & C	Threaded and Coupled		
T.D.	Temperature Difference		
Tdd	Telecommunications Device for the Deaf		
T.E.M.	Transmission Electron Microscopy		
temp	Temperature, Tempered, Temporary		
TFFN	Nylon Jacketed Wire		

701

Abbreviations

TFE	Tetrafluoroethylene (Teflon)	U.L., UL	Underwriters Laboratory	w/	With
T. & G.	Tongue & Groove;	Uld.	unloading	W.C., WC	Water Column; Water Closet
	Tar & Gravel	Unfin.	Unfinished	W.F.	Wide Flange
Th., Thk.	Thick	UPS	Uninterruptible Power Supply	W.G.	Water Gauge
Thn.	Thin	URD	Underground Residential	Wldg.	Welding
Thrded	Threaded		Distribution	W. Mile	Wire Mile
Tilf.	Tile Layer, Floor	US	United States	W-O-L	Weldolet
Tilh.	Tile Layer, Helper	USGBC	U.S. Green Building Council	W.R.	Water Resistant
THHN	Nylon Jacketed Wire	USP	United States Primed	Wrck.	Wrecker
THW.	Insulated Strand Wire	UTMCD	Uniform Traffic Manual For Control	W.S.P.	Water, Steam, Petroleum
THWN	Nylon Jacketed Wire		Devices	WT., Wt.	Weight
T.L., TL	Truckload	UTP	Unshielded Twisted Pair	WWF	Welded Wire Fabric
T.M.	Track Mounted	V	Volt	XFER	Transfer
Tot.	Total	VA	Volt Amperes	XFMR	Transformer
T-O-L	Threadolet	VAT	Vinyl Asbestos Tile	XHD	Extra Heavy Duty
tmpd	Tempered	V.C.T.	Vinyl Composition Tile	XHHW,	Cross-Linked Polyethylene Wire
TPO	Thermoplastic Polyolefin	VAV	Variable Air Volume	XLPE	Insulation
T.S.	Trigger Start	VC	Veneer Core	XLP	Cross-linked Polyethylene
Tr.	Trade	VDC	Volts Direct Current	Xport	Transport
Transf.	Transformer	Vent.	Ventilation	Y	Wye
Trhv.	Truck Driver, Heavy	Vert.	Vertical	yd	Yard
Trlr	Trailer	V.F.	Vinyl Faced	yr	Year
Trlt.	Truck Driver, Light	V.G.	Vertical Grain	Δ	Delta
TTY	Teletypewriter	VHF	Very High Frequency	%	Percent
TV	Television	VHO	Very High Output	~	Approximately
T.W.	Thermoplastic Water Resistant	Vib.	Vibrating	Ø	Phase; diameter
	Wire	VLF	Vertical Linear Foot	@	At
UCI	Uniform Construction Index	VOC	Volitile Organic Compound	#	Pound; Number
UF	Underground Feeder	Vol.	Volume	<	Less Than
UGND	Underground Feeder	VRP	Vinyl Reinforced Polyester	>	Greater Than
UHF	Ultra High Frequency	W	Wire; Watt; Wide; West	Z	zone
U.I.	United Inch				

Index

704

For customer support on your Plumbing Cost Data, call 877.709.6968.

706

Index

715

716

Division Notes

	CREW	DAILY OUTPUT	LABOR-HOURS	UNIT	BARE COSTS				TOTAL INCL O&P
					MAT.	LABOR	EQUIP.	TOTAL	

Division Notes

	CREW	DAILY OUTPUT	LABOR-HOURS	UNIT	BARE COSTS				TOTAL INCL O&P
					MAT.	LABOR	EQUIP.	TOTAL	

Division Notes

		CREW	DAILY OUTPUT	LABOR-HOURS	UNIT	BARE COSTS				TOTAL INCL O&P
						MAT.	LABOR	EQUIP.	TOTAL	

Division Notes

	CREW	DAILY OUTPUT	LABOR-HOURS	UNIT	BARE COSTS				TOTAL INCL O&P
					MAT.	LABOR	EQUIP.	TOTAL	

Division Notes

	CREW	DAILY OUTPUT	LABOR-HOURS	UNIT	BARE COSTS				TOTAL INCL O&P
					MAT.	LABOR	EQUIP.	TOTAL	

Division Notes

	CREW	DAILY OUTPUT	LABOR-HOURS	UNIT	BARE COSTS				TOTAL INCL O&P
					MAT.	LABOR	EQUIP.	TOTAL	

Other RSMeans Products & Services

RSMeans—
a tradition of excellence in construction cost information and services since 1942

Table Of Contents
Annual Cost Guides
RSMeans Online and *CostWorks* CD Comparison Matrix
Seminars
Reference Books
Order Form

For more information visit the RSMeans website at **www.rsmeans.com**

Unit prices according to the latest MasterFormat!

Book Selection Guide
The following table provides definitive information on the content of each cost data publication. The number of lines of data provided in each unit price or assemblies division, as well as the number of crews, is listed for each book. The presence of other elements such as reference tables, square foot models, equipment rental costs, historical cost indexes, and city cost indexes, is also indicated. You can use the table to help select the RSMeans book that has the quantity and type of information you most need in your work.

Unit Cost Divisions	Building Construction	Mechanical	Electrical	Commercial Renovation	Square Foot	Site Work Landsc.	Green Building	Interior	Concrete Masonry	Open Shop	Heavy Construction	Light Commercial	Facilities Construction	Plumbing	Residential
1	573	393	410	506		509	191	312	456	572	501	246	1042	403	178
2	777	279	85	733		993	178	399	213	776	734	481	1221	286	257
3	1690	340	230	1085		1471	987	355	2032	1689	1685	481	1788	316	388
4	957	21	0	922		720	179	612	1155	925	612	529	1172	0	441
5	1893	158	155	1090		841	1799	1095	720	1893	1037	979	1909	204	721
6	2452	18	18	2110		110	589	1528	281	2448	123	2140	2124	22	2660
7	1584	215	128	1623		581	757	532	524	1581	26	1317	1685	227	1037
8	2082	81	45	2609		265	1123	1753	105	2077	0	2193	2879	0	1526
9	1971	72	26	1767		313	449	2056	390	1911	15	1659	2220	54	1435
10	1026	17	10	642		215	27	842	158	1026	29	511	1118	235	224
11	1087	208	165	546		124	54	932	28	1071	0	230	1107	169	110
12	544	0	2	319		217	137	1654	14	532	0	352	1704	23	310
13	719	140	137	242		343	111	248	66	695	244	83	746	94	79
14	273	36	0	221		0	0	257	0	273	0	12	293	16	6
21	90	0	16	37		0	0	250	0	87	0	68	426	431	220
22	1157	7544	150	1190		1568	1067	838	20	1121	1677	857	7440	9347	712
23	1177	6971	580	928		158	902	776	38	1101	111	878	5187	1918	469
26	1354	454	10081	1021		793	611	1136	55	1349	562	1298	10030	399	617
27	72	0	297	34		13	0	71	0	72	39	52	279	0	22
28	96	58	136	71		0	21	78	0	99	0	40	151	44	25
31	1498	735	610	803		3217	289	7	1208	1443	3256	600	1560	660	611
32	822	53	8	886		4426	356	405	291	793	1841	420	1701	165	468
33	530	1079	538	252		2178	41	0	237	522	2159	128	1699	1286	154
34	107	0	47	4		190	0	0	31	62	193	0	128	0	0
35	18	0	0	0		327	0	0	0	18	442	0	84	0	0
41	60	0	0	32		7	0	22	0	60	30	0	67	14	0
44	75	79	0	0		0	0	0	0	0	0	0	75	75	0
46	23	16	0	0		274	261	0	0	23	264	0	33	33	0
48	12	0	25	0		0	25	0	0	12	17	12	12	0	12
Totals	24719	18967	13899	19673		19853	10154	16158	8022	24231	15597	15566	49880	16421	12682

Assem Div	Building Construction	Mechanical	Electrical	Commercial Renovation	Square Foot	Site Work Landscape	Assemblies	Green Building	Interior	Concrete Masonry	Heavy Construction	Light Commercial	Facilities Construction	Plumbing	Asm Div	Residential
A		15	0	188	150	577	598	0	0	536	571	154	24	0	1	376
B		0	0	848	2498	0	5658	56	329	1975	368	2089	174	0	2	211
C		0	0	647	926	0	1304	0	1629	146	0	816	250	0	3	588
D		1067	941	712	1859	72	2538	330	825	0	0	1345	1105	1088	4	851
E		0	0	86	260	0	300	0	5	0	0	0	257	5	0	5	392
F		0	0	0	114	0	114	0	0	0	0	0	114	3	0	6	357
G		527	447	318	249	3365	729	0	0	535	1350	205	293	677	7	307
															8	760
															9	80
															10	0
															11	0
															12	0
Totals		1609	1388	2799	6056	4014	11241	386	2788	3192	2289	4980	1854	1765		3922

Reference Section	Building Construction Costs	Mechanical	Electrical	Commercial Renovation	Square Foot	Site Work Landscape	Assem.	Green Building	Interior	Concrete Masonry	Open Shop	Heavy Construction	Light Commercial	Facilities Construction	Plumbing	Resi.
Reference Tables	yes	yes	yes	yes	no	yes	yes	yes	yes	yes	yes	yes	yes	yes	yes	yes
Models					111			25					50			28
Crews	571	571	571	550		571		571	571	571	548	571	548	550	571	548
Equipment Rental Costs	yes	yes	yes	yes		yes		yes	yes	yes	yes	yes	yes	yes	yes	yes
Historical Cost Indexes	yes	yes	yes	yes	yes	yes	yes	yes	yes	yes	yes	yes	yes	yes	yes	no
City Cost Indexes	yes	yes	yes	yes	yes	yes	yes	yes	yes	yes	yes	yes	yes	yes	yes	yes

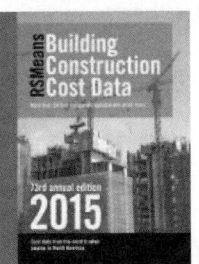

RSMeans Building Construction Cost Data 2015

Offers you unchallenged unit price reliability in an easy-to-use format. Whether used for verifying complete, finished estimates or for periodic checks, it supplies more cost facts better and faster than any comparable source. More than 24,700 unit prices have been updated for 2015. The City Cost Indexes and Location Factors cover more than 930 areas, for indexing to any project location in North America. Order and get *RSMeans Quarterly Update Service* FREE.

$206.95 | Available Sept. 2014 | Catalog no. 60015

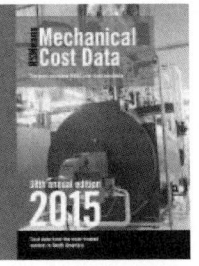

RSMeans Mechanical Cost Data 2015

Total unit and systems price guidance for mechanical construction . . . materials, parts, fittings, and complete labor cost information. Includes prices for piping, heating, air conditioning, ventilation, and all related construction.

Plus new 2015 unit costs for:

- Thousands of installed HVAC/controls, sub-assemblies and assemblies
- "On-site" Location Factors for more than 930 cities and towns in the U.S. and Canada
- Crews, labor, and equipment

$203.95 | Available Oct. 2014 | Catalog no. 60025

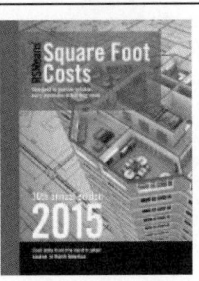

RSMeans Square Foot Costs 2015
Accurate and Easy-to-Use

- **Updated price information** based on nationwide figures from suppliers, estimators, labor experts, and contractors.
- Green building models
- Realistic graphics, offering true-to-life illustrations of building projects
- Extensive information on using square foot cost data, including sample estimates and alternate pricing methods

$218.95 | Available Oct. 2014 | Catalog no. 60055

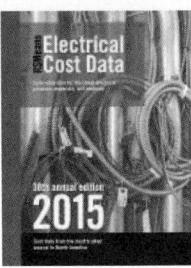

RSMeans Electrical Cost Data 2015

Pricing information for every part of electrical cost planning. More than 13,800 unit and systems costs with design tables; clear specifications and drawings; engineering guides; illustrated estimating procedures; complete labor-hour and materials costs for better scheduling and procurement; and the latest electrical products and construction methods.

- A variety of special electrical systems, including cathodic protection
- Costs for maintenance, demolition, HVAC/mechanical, specialties, equipment, and more

$208.95 | Available Oct. 2014 | Catalog no. 60035

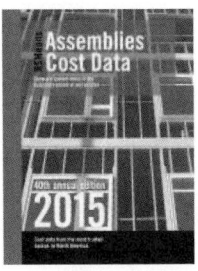

RSMeans Assemblies Cost Data 2015

RSMeans Assemblies Cost Data takes the guesswork out of preliminary or conceptual estimates. Now you don't have to try to calculate the assembled cost by working up individual component costs. We've done all the work for you.

Presents detailed illustrations, descriptions, specifications, and costs for every conceivable building assembly—over 350 types in all—arranged in the easy-to-use UNIFORMAT II system. Each illustrated "assembled" cost includes a complete grouping of materials and associated installation costs, including the installing contractor's overhead and profit.

$334.95 | Available Sept. 2014 | Catalog no. 60065

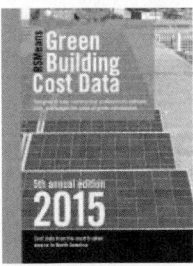

RSMeans Green Building Cost Data 2015

Estimate, plan, and budget the costs of green building for both new commercial construction and renovation work with this fifth edition of *RSMeans Green Building Cost Data*. More than 10,000 unit costs for a wide array of green building products plus assemblies costs. Easily identified cross references to LEED and Green Globes building rating systems criteria.

$170.95 | Available Nov. 2014 | Catalog no. 60555

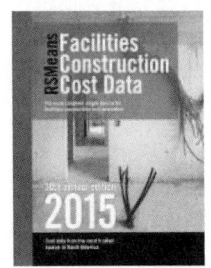

RSMeans Facilities Construction Cost Data 2015

For the maintenance and construction of commercial, industrial, municipal, and institutional properties. Costs are shown for new and remodeling construction and are broken down into materials, labor, equipment, and overhead and profit. Special emphasis is given to sections on mechanical, electrical, furnishings, site work, building maintenance, finish work, and demolition.

More than 49,800 unit costs, plus assemblies costs and a comprehensive Reference Section are included.

$524.95 | Available Nov. 2014 | Catalog no. 60205

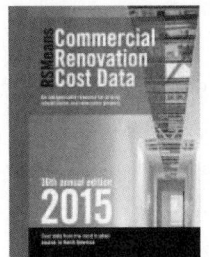

RSMeans Commercial Renovation Cost Data 2015
Commercial/Multi-family Residential

Use this valuable tool to estimate commercial and multi-family residential renovation and remodeling.

Includes: Updated costs for hundreds of unique methods, materials, and conditions that only come up in repair and remodeling, PLUS:

- Unit costs for more than 19,600 construction components
- Installed costs for more than 2,700 assemblies
- More than 930 "on-site" localization factors for the U.S. and Canada

$166.95 | Available Oct. 2014 | Catalog no. 60045

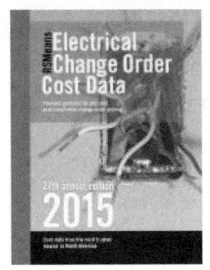

RSMeans Electrical Change Order Cost Data 2015

RSMeans Electrical Change Order Cost Data provides you with electrical unit prices exclusively for pricing change orders—based on the recent, direct experience of contractors and suppliers. Analyze and check your own change order estimates against the experience others have had doing the same work. It also covers productivity analysis, and change order cost justifications. With useful information for calculating the effects of change orders and dealing with their administration.

$199.95 | Available Dec. 2014 | Catalog no. 60235

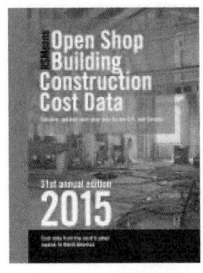

RSMeans Open Shop Building Construction Cost Data 2015

The latest costs for accurate budgeting and estimating of new commercial and residential construction . . . renovation work . . . change orders . . . cost engineering.

RSMeans Open Shop "BCCD" will assist you to:

- Develop benchmark prices for change orders.
- Plug gaps in preliminary estimates and budgets.
- Estimate complex projects.
- Substantiate invoices on contracts.
- Price ADA-related renovations.

$175.95 | Available Dec. 2014 | Catalog no. 60155

Annual Cost Guides

Unit prices according to the latest MasterFormat!

729

For more information visit the RSMeans website at **www.rsmeans.com**

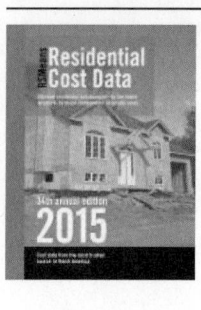

RSMeans Residential
Cost Data 2015

Contains square foot costs for 28 basic home models with the look of today, plus hundreds of custom additions and modifications you can quote right off the page. Includes more than 3,900 costs for 89 residential systems. Complete with blank estimating forms, sample estimates, and step-by-step instructions.

Contains line items for cultured stone and brick, PVC trim, lumber, and TPO roofing.

$148.95 | Available Oct. 2014 | Catalog no. 60175

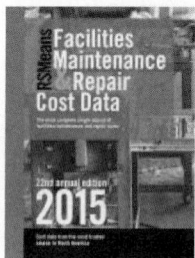

RSMeans Facilities Maintenance & Repair
Cost Data 2015

RSMeans Facilities Maintenance & Repair Cost Data gives you a complete system to manage and plan your facility repair and maintenance costs and budget efficiently. Guidelines for auditing a facility and developing an annual maintenance plan. Budgeting is included, along with reference tables on cost and management, and information on frequency and productivity of maintenance operations.

The only nationally recognized source of maintenance and repair costs. Developed in cooperation with the Civil Engineering Research Laboratory (CERL) of the Army Corps of Engineers.

$461.95 | Available Nov. 2014 | Catalog no. 60305

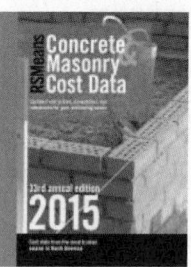

RSMeans Concrete & Masonry
Cost Data 2015

Provides you with cost facts for virtually all concrete/masonry estimating needs, from complicated formwork to various sizes and face finishes of brick and block—all in great detail. The comprehensive Unit Price Section contains more than 8,000 selected entries. Also contains an Assemblies [Cost] Section, and a detailed Reference Section that supplements the cost data.

$188.95 | Available Dec. 2014 | Catalog no. 60115

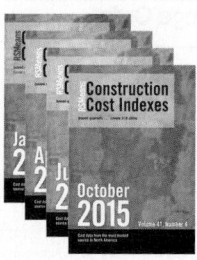

RSMeans Construction
Cost Indexes 2015

What materials and labor costs will change unexpectedly this year? By how much?

- Breakdowns for 318 major cities
- National averages for 30 key cities
- Expanded five major city indexes
- Historical construction cost indexes

$359.95 per year (subscription) | Catalog no. 50145
$94.95 individual quarters | Catalog no. 60145 A,B,C,D

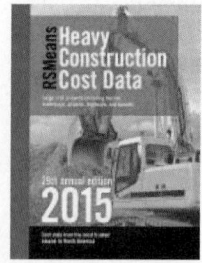

RSMeans Heavy Construction
Cost Data 2015

A comprehensive guide to heavy construction costs. Includes costs for highly specialized projects such as tunnels, dams, highways, airports, and waterways. Information on labor rates, equipment, and materials costs is included. Features unit price costs, systems costs, and numerous reference tables for costs and design.

$204.95 | Available Dec. 2014 | Catalog no. 60165

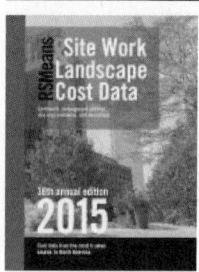

RSMeans Site Work & Landscape
Cost Data 2015

Includes unit and assemblies costs for earthwork, sewerage, piped utilities, site improvements, drainage, paving, trees and shrubs, street openings/repairs, underground tanks, and more. Contains more than 60 types of assemblies costs for accurate conceptual estimates.

Includes:

- Estimating for infrastructure improvements
- Environmentally-oriented construction
- ADA-mandated handicapped access
- Hazardous waste line items

$199.95 | Available Dec. 2014 | Catalog no. 60285

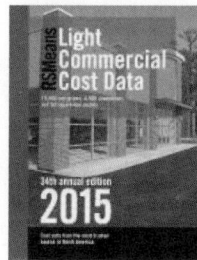

RSMeans Light Commercial
Cost Data 2015

Specifically addresses the light commercial market, which is a specialized niche in the construction industry. Aids you, the owner/designer/contractor, in preparing all types of estimates—from budgets to detailed bids. Includes new advances in methods and materials.

Assemblies Section allows you to evaluate alternatives in the early stages of design/planning.

More than 15,500 unit costs ensure that you have the prices you need, when you need them.

$153.95 | Available Nov. 2014 | Catalog no. 60185

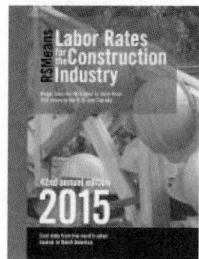

RSMeans Labor Rates for the
Construction Industry 2015

Complete information for estimating labor costs, making comparisons, and negotiating wage rates by trade for more than 300 U.S. and Canadian cities. With 46 construction trades in each city, and historical wage rates included for comparison. Each city chart lists the county and is alphabetically arranged with handy visual flip tabs for quick reference.

$450.95 | Available Dec. 2014 | Catalog no. 60125

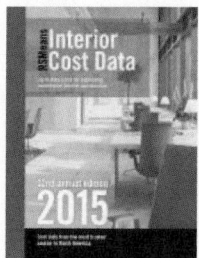

RSMeans Interior
Cost Data 2015

Provides you with prices and guidance needed to make accurate interior work estimates. Contains costs on materials, equipment, hardware, custom installations, furnishings, and labor for new and remodel commercial and industrial interior construction, including updated information on office furnishings, and reference information.

$208.95 | Available Nov. 2014 | Catalog no. 60095

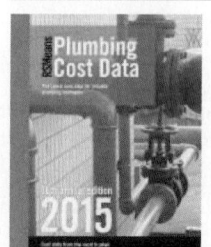

RSMeans Plumbing
Cost Data 2015

Comprehensive unit prices and assemblies for plumbing, irrigation systems, commercial and residential fire protection, point-of-use water heaters, and the latest approved materials. This publication and its companion, *RSMeans Mechanical Cost Data*, provide full-range cost estimating coverage for all the mechanical trades.

Contains updated costs for potable water, radiant heat systems, and high efficiency fixtures.

$206.95 | Available Oct. 2014 | Catalog no. 60215

For more information visit the RSMeans website at **www.rsmeans.com**

RSMeans Online and RSMeans CostWorks CD Comparison Matrix

Both the RSMeans CostWorks CD package and the RSMeans Online estimating tool provide the same comprehensive, up-to-date, and localized cost information to improve planning, estimating, and budgeting with RSMeans—the industry's most-trusted source of construction cost data.

In addition to key differences noted in the comparison matrix below, both products let you do the following:

• Quickly locate costs in the searchable RSMeans database.

• Instantly adjust the data to your specific location for accurate costs anywhere in the U.S. and Canada.

• Create reliable, updated cost estimates in minutes.

• Manage, share, and export your estimates to MS Excel® with ease.

• Access a wide range of reference materials, including crew rates and national average labor rates.

Both the CD and Online options are available in a convenient one-year subscription!

Use the handy comparison matrix below to help you decide the best product for your cost estimating needs!

Why us?

For more than 70 years, RSMeans has provided the quality cost data construction professionals everywhere depend on to estimate with confidence, save time, improve decision-making, and increase profits.

Improve planning and decision-making

Back your estimates with complete, accurate, and up-to-date cost data for informed business decisions.
• Verify construction costs from third parties.
• Check validity of subcontractor proposals.
• Evaluate material and assembly alternatives.

Increase profits

Use RSMeans to estimate projects quickly and accurately, so you can gain an edge over your competition.
• Create accurate and competitive bids.
• Minimize the risk of cost overruns.
• Reduce variability.
• Gain control over costs.

The following matrix summarizes the key comparison points between RSMeans' two electronic cost data products.

	Question	CostWorks CD	RSMeans Online.com
1	How is the data presented?	• The data is organized by book title and sold at comparable book cost. • Several packaged bundles of five titles are available at a cost savings. • Available in a convenient one-year subscription	• The data is organized by book title and sold at comparable book cost. • Several packaged bundles of five titles are available at a cost savings. • RSMeans Online also offers a Complete Library title with all online data at a cost savings. • Available in a convenient one-year subscription
2	Where is the data stored?	With a typical install, the data is read off of your hard drive or off of the CD.	All data is stored on our secure server.
3	How are updates delivered?	Updates are available on a quarterly basis for download from the Costworks HomePort website.	Data is updated automatically on a quarterly basis, and users are notified when updates occur.
4	Where are my projects and estimates stored?	Upon installation, a "My CostWorks Projects" folder is created within the "My Documents" directory. All projects are defaulted to save here unless otherwise specified.	Estimates are stored on our secure server.
5	Does my IT team need to assist me in installing and running the program?	Yes, depending on the level of your permission. See #11 for system requirements.	While the product is entirely Web-based, you may need IT support with permissions for certain tasks. See #11 for system requirements.
6	Can users collaborate on projects/estimates?	Yes, you can copy project folders between computers.	You can create shared folders that can be viewed, copied, and edited by anyone within your account workgroup.

Unit prices according to the latest MasterFormat!

RSMeans Online

For more information visit the RSMeans website at **www.rsmeans.com**

Comparison matrix continued

7	Can I create and use my own data with the RSMeans database?	Custom lines cannot be added to the CD database. With the Estimator tool, however, you can create and add custom cost lines to any estimate. You can also save custom lines as an Estimator template for use in creating new estimates.	You can create • a custom line within an estimate • a custom cost book consisting of custom unit (not assembly) cost lines, no more than once per quarter **Note** that custom data is searchable using the Include My Custom Data feature.
8	Can I enter my own trade/labor rates and markups?	Although you cannot adjust the trade labor rates, you can add custom markups for each report template using the Estimator tool.	For each quarterly update, you can create one custom unit cost book. You can edit the labor rates by individual trade and the crew equipment costs, and also edit the associated material and equipment markups.
9	What are my reporting options?	The Estimator tool lets you create editable custom reports and report templates. Reports can also be exported to MS Excel®.	You can generate pre-formatted estimate summaries and cost lists in both PDF and Excel formats.
10	How do I find specific cost data?	You can search by CSI line number and by keyword within a division.	You can search by CSI line number and by keyword within a division. There are several additional search options, including keyword and advanced search using special characters.
11	What are the system requirements?	• Operating system – Windows XP or later • PC requirements - 256 MB RAM (recommended), 1024 x 768 High Color Monitor, 800MHz Computer (recommended), 24X CD-ROM, 250 MB Hard Disk Space	• Resolution – Recommended resolution is 1024 x 768 or higher • Compatibility - Internet Explorer version 8.x, 9.x and Mozilla Firefox version 10 and higher • iPad - iOS 5 and higher

2015 RSMeans Seminar Schedule

Note: call for exact dates and details.

Location	Dates	Location	Dates
Seattle, WA	January and August	Jacksonville FL	September
Dallas/Ft. Worth, TX	January	Dallas, TX	September
Austin, TX	February	Charleston, SC	October
Anchorage, AK	March and September	Houston, TX	October
Las Vegas, NV	March and October	Atlanta, GA	November
Washington, DC	April and September	Baltimore, MD	November
Phoenix, AZ	April	Orlando, FL	November
Kansas City, MO	April	San Diego, CA	December
Toronto	May	San Antonio, TX	December
Denver, CO	May	Raleigh, NC	December
San Francisco, CA	June		
Bethesda, MD	June		
Columbus, GA	June	**1-800-334-3509, Press 1**	
El Segundo, CA	August		

Professional Development

For more information visit the RSMeans website at **www.rsmeans.com**

eLearning Training Sessions

Learn how to use *RSMeans Online®* or *RSMeans CostWorks®* CD from the convenience of your home or office. Our eLearning training sessions let you join a training conference call and share the instructors' desktops, so you can view the presentation and step-by-step instructions on your own computer screen. The live webinars are held from 9 a.m. to 4 p.m. eastern standard time, with a one-hour break for lunch.

For these sessions, you must have a computer with high speed Internet access and a compatible Web browser. Learn more at www.rsmeansonline.com or call for a schedule: 1-800-334-3509 and press 1. Webinars are generally held on selected Wednesdays each month.

RSMeans Online®
$299 per person

RSMeans CostWorks® CD
$299 per person

For more information visit the RSMeans website at **www.rsmeans.com**

RSMeans Online™ Training

Construction estimating is vital to the decision-making process at each state of every project. RSMeansOnline works the way you do. It's systematic, flexible and intuitive. In this one day class you will see how you can estimate any phase of any project faster and better.

Some of what you'll learn:
- Customizing RSMeansOnline
- Making the most of RSMeans "Circle Reference" numbers
- How to integrate your cost data
- Generate reports, exporting estimates to MS Excel, sharing, collaborating and more

Also available: RSMeans Online® training webinar

Facilities Construction Estimating

In this *two-day* course, professionals working in facilities management can get help with their daily challenges to establish budgets for all phases of a project.

Some of what you'll learn:
- Determining the full scope of a project
- Identifying the scope of risks and opportunities
- Creative solutions to estimating issues
- Organizing estimates for presentation and discussion
- Special techniques for repair/remodel and maintenance projects
- Negotiating project change orders

Who should attend: facility managers, engineers, contractors, facility tradespeople, planners, and project managers.

Construction Cost Estimating: Concepts and Practice

This *one-day* introductory course to improve estimating skills and effectiveness starts with the details of interpreting bid documents, ending with the summary of the estimate and bid submission.

Some of what you'll learn:
- Using the plans and specifications for creating estimates
- The takeoff process—deriving all tasks with correct quantities
- Developing pricing, using various sources; how subcontractor pricing fits in
- Summarizing the estimate to arrive at the final number
- Formulas for area and cubic measure, adding waste and adjusting productivity to specific projects
- Evaluating subcontractors' proposals and prices
- Adding Insurance and bonds
- Understanding how labor costs are calculated
- Submitting bids and proposals

Who should attend: project managers, architects, engineers, owner's representatives, contractors, and anyone who's responsible for budgeting or estimating construction projects.

Maintenance & Repair Estimating for Facilities

This *two-day* course teaches attendees how to plan, budget, and estimate the cost of ongoing and preventive maintenance and repair for existing buildings and grounds.

Some of what you'll learn:
- The most financially favorable maintenance, repair, and replacement scheduling and estimating
- Auditing and value engineering facilities
- Preventive planning and facilities upgrading
- Determining both in-house and contract-out service costs
- Annual, asset-protecting M&R plan

Who should attend: facility managers, maintenance supervisors, buildings and grounds superintendents, plant managers, planners, estimators, and others involved in facilities planning and budgeting.

Practical Project Management for Construction Professionals

In this *two-day* course, acquire the essential knowledge and develop the skills to effectively and efficiently execute the day-to-day responsibilities of the construction project manager.

Covers:
- General conditions of the construction contract
- Contract modifications: change orders and construction change directives
- Negotiations with subcontractors and vendors
- Effective writing: notification and communications
- Dispute resolution: claims and liens

Who should attend: architects, engineers, owner's representatives, project managers.

Mechanical & Electrical Estimating

This *two-day* course teaches attendees how to prepare more accurate and complete mechanical/electrical estimates, avoiding the pitfalls of omission and double-counting, while understanding the composition and rationale within the RSMeans mechanical/electrical database.

Some of what you'll learn:
- The unique way mechanical and electrical systems are interrelated
- M&E estimates—conceptual, planning, budgeting, and bidding stages
- Order of magnitude, square foot, assemblies, and unit price estimating
- Comparative cost analysis of equipment and design alternatives

Who should attend: architects, engineers, facilities managers, mechanical and electrical contractors, and others who need a highly reliable method for developing, understanding, and evaluating mechanical and electrical contracts.

Professional Development

Unit Price Estimating

This interactive *two-day* seminar teaches attendees how to interpret project information and process it into final, detailed estimates with the greatest accuracy level.

The most important credential an estimator can take to the job is the ability to visualize construction and estimate accurately.

Some of what you'll learn:

- Interpreting the design in terms of cost
- The most detailed, time-tested methodology for accurate pricing
- Key cost drivers—material, labor, equipment, staging, and subcontracts
- Understanding direct and indirect costs for accurate job cost accounting and change order management

Who should attend: corporate and government estimators and purchasers, architects, engineers, and others who need to produce accurate project estimates.

RSMeans CostWorks® CD Training

This one-day course helps users become more familiar with the functionality of *RSMeans CostWork*s program. Each menu, icon, screen, and function found in the program is explained in depth. Time is devoted to hands-on estimating exercises.

Some of what you'll learn:

- Searching the database using all navigation methods
- Exporting RSMeans data to your preferred spreadsheet format
- Viewing crews, assembly components, and much more
- Automatically regionalizing the database

This training session requires you to bring a laptop computer to class.

When you register for this course you will receive an outline for your laptop requirements.

Also offering web training for RSMeans CostWorks CD!

Facilities Estimating Using RSMeans CostWorks® CD

Combines hands-on skill building with best estimating practices and real-life problems. Brings you up-to-date with key concepts, and provides tips, pointers, and guidelines to save time and avoid cost oversights and errors.

Some of what you'll learn:

- Estimating process concepts
- Customizing and adapting RSMeans cost data
- Establishing scope of work to account for all known variables
- Budget estimating: when, why, and how
- Site visits: what to look for—what you can't afford to overlook
- How to estimate repair and remodeling variables

This training session requires you to bring a laptop computer to class.

Who should attend: facility managers, architects, engineers, contractors, facility tradespeople, planners, project managers and anyone involved with JOC, SABRE, or IDIQ.

Conceptual Estimating Using RSMeans CostWorks® CD

This *two-day* class uses the leading industry data and a powerful software package to develop highly accurate conceptual estimates for your construction projects. All attendees must bring a laptop computer loaded with the current year *Square Foot Costs* and the *Assemblies Cost Data* CostWorks titles.

Some of what you'll learn:

- Introduction to conceptual estimating
- Types of conceptual estimates
- Helpful hints
- Order of magnitude estimating
- Square foot estimating
- Assemblies estimating

Who should attend: architects, engineers, contractors, construction estimators, owner's representatives, and anyone looking for an electronic method for performing square foot estimating.

Assessing Scope of Work for Facility Construction Estimating

This *two-day* practical training program addresses the vital importance of understanding the SCOPE of projects in order to produce accurate cost estimates for facility repair and remodeling.

Some of what you'll learn:

- Discussions of site visits, plans/specs, record drawings of facilities, and site-specific lists
- Review of CSI divisions, including means, methods, materials, and the challenges of scoping each topic
- Exercises in SCOPE identification and SCOPE writing for accurate estimating of projects
- Hands-on exercises that require SCOPE, take-off, and pricing

Who should attend: corporate and government estimators, planners, facility managers, and others who need to produce accurate project estimates.

Unit Price Estimating Using RSMeans CostWorks® CD

Step-by-step instruction and practice problems to identify and track key cost drivers—material, labor, equipment, staging, and subcontractors—for each specific task. Learn the most detailed, time-tested methodology for accurately "pricing" these variables, their impact on each other and on total cost.

Some of what you'll learn:

- Unit price cost estimating
- Order of magnitude, square foot, and assemblies estimating
- Quantity takeoff
- Direct and indirect construction costs
- Development of contractor's bill rates
- How to use *RSMeans Building Construction Cost Data*

This training session requires you to bring a laptop computer to class.

Who should attend: architects, engineers, corporate and government estimators, facility managers, and government procurement staff.

Professional Development

Registration Information

Register early . . . Save up to $100! Register 30 days before the start date of a seminar and save $100 off your total fee. *Note: This discount can be applied only once per order. It cannot be applied to team discount registrations or any other special offer.*

How to register Register by phone today! The RSMeans toll-free number for making reservations is **1-800-334-3509, Press 1.**

Two-day seminar registration fee - $935. One-day *RSMeans CostWorks*® training registration fee - $375. To register by mail, complete the registration form and return, with your full fee, to: RSMeans Seminars, 700 Longwater Drive, Norwell, MA 02061.

Government pricing All federal government employees save off the regular seminar price. Other promotional discounts cannot be combined with the government discount.

Team discount program for two to four seminar registrations. Call for pricing: 1-800-334-3509, Press 1

Multiple course discounts When signing up for two or more courses, call for pricing.

Refund policy Cancellations will be accepted up to ten business days prior to the seminar start. There are no refunds for cancellations received later than ten working days prior to the first day of the seminar. A $150 processing fee will be applied for all cancellations. Written notice of cancellation is required. Substitutions can be made at any time before the session starts. **No-shows are subject to the full seminar fee.**

AACE approved courses Many seminars described and offered here have been approved for 14 hours (1.4 recertification credits) of credit by the AACE International Certification Board toward meeting the continuing education requirements for recertification as a Certified Cost Engineer/Certified Cost Consultant.

AIA Continuing Education We are registered with the AIA Continuing Education System (AIA/CES) and are committed to developing quality learning activities in accordance with the CES criteria. Many seminars meet the AIA/CES criteria for Quality Level 2. AIA members may receive (14) learning units (LUs) for each two-day RSMeans course.

Daily course schedule The first day of each seminar session begins at 8:30 a.m. and ends at 4:30 p.m. The second day begins at 8:00 a.m. and ends at 4:00 p.m. Participants are urged to bring a hand-held calculator, since many actual problems will be worked out in each session.

Continental breakfast Your registration includes the cost of a continental breakfast, and a morning and afternoon refreshment break. These informal segments allow you to discuss topics of mutual interest with other seminar attendees. (You are free to make your own lunch and dinner arrangements.)

Hotel/transportation arrangements RSMeans arranges to hold a block of rooms at most host hotels. To take advantage of special group rates when making your reservation, be sure to mention that you are attending the RSMeans seminar. You are, of course, free to stay at the lodging place of your choice. (**Hotel reservations and transportation arrangements should be made directly by seminar attendees.**)

Important Class sizes are limited, so please register as soon as possible.

Note: Pricing subject to change.

Registration Form

ADDS-1000

Call 1-800-334-3509, Press 1 to register or FAX this form 1-800-632-6732. Visit our website: www.rsmeans.com

Please register the following people for the RSMeans construction seminars as shown here. We understand that we must make our own hotel reservations if overnight stays are necessary.

☐ Full payment of $_____ enclosed.

☐ Bill me.

Please print name of registrant(s).

(To appear on certificate of completion)

P.O. #: _____
GOVERNMENT AGENCIES MUST SUPPLY PURCHASE ORDER NUMBER OR TRAINING FORM.

Please mail check to: RSMeans Seminars, 700 Longwater Drive, Norwell, MA 02061 USA

Firm name_____

Address_____

City/State/Zip_____

Telephone no._____ Fax no._____

E-mail address_____

Charge registration(s) to: ☐ MasterCard ☐ VISA ☐ American Express

Account no._____ Exp. date_____

Cardholder's signature_____

Seminar name_____

Seminar City_____

Professional Development

735

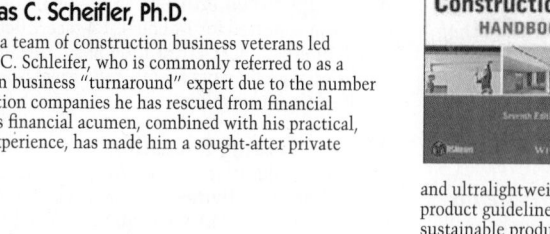

Managing the Profitable Construction Business: The Contractor's Guide to Success and Survival Strategies, 2nd Ed.

by Thomas C. Scheifler, Ph.D.

Learn from a team of construction business veterans led by Thomas C. Schleifer, who is commonly referred to as a construction business "turnaround" expert due to the number of construction companies he has rescued from financial distress. His financial acumen, combined with his practical, hands-on experience, has made him a sought-after private consultant.

$60.00 | 288 pages, hardover | Catalog no. 67370

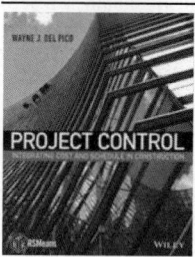

Construction Project Safety

by John Schaufelberger, Ken-Yu Lin

This essential introduction to construction safety for construction management personnel takes a project-based approach to present potential hazards in construction and their mitigation or prevention.

Beginning with an introduction to Accident Prevention Programs and OSHA compliance requirements, the book integrates safety instruction into the building process by following a building project from site construction through interior finish. Reinforcing this applied approach are photographs, drawings, contract documentation, and an online 3D BIM model to help visualize the on-site scenarios.

$85.00 | 320 pages, hardover | Catalog no. 67368

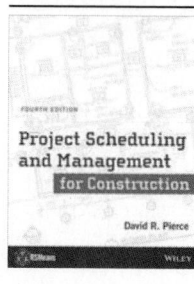

Project Control: Integrating Cost & Schedule in Construction

by Wayne J. Del Pico

Written by a seasoned professional in the field, *Project Control: Integrating Cost and Schedule in Construction* fills a void in the area of project control as applied in the construction industry today. It demonstrates how productivity models for an individual project are created, monitored, and controlled, and how corrective actions are implemented as deviations from the baseline occur.

$65.00 | 240 pages, softcover | Catalog no. 67366

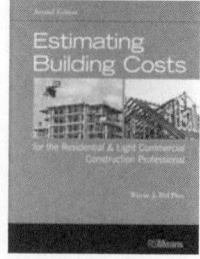

Project Scheduling and Management for Construction, 4th Edition

by David R. Pierce

First published in 1988 by RSMeans, the new edition of *Project Scheduling and Management for Construction* has been substantially revised for both professionals and students enrolled in construction management and civil engineering programs. While retaining its emphasis on developing practical, professional-level scheduling skills, the new edition is a relatable, real world case study.

$95.00 | 272 pages, softcover | Catalog no. 67367

Estimating Building Costs, 2nd Edition

For the Residential & Light Commercial Construction Professional

by Wayne J. Del Pico

Explains, in detail, how to put together a reliable estimate that can be used not only for budgeting, but also for developing a schedule, managing a project, dealing with contingencies, and, ultimately, making a profit.

Completely revised and updated to reflect the CSI MasterFormat 2010 system.

$75.00 | 528 pages, softcover | Catalog no. 67343A

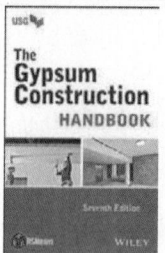

The Gypsum Construction Handbook, 7th Edition

by USG

The tried-and-true Gypsum Construction Handbook is a systematic guide to selecting and using gypsum drywall, veneer plaster, tile backers, ceilings, and conventional plaster building materials. A widely respected training text for aspiring architects and engineers, the book provides detailed product information and efficient installation methodology.

The 7th edition features updates in gypsum products, including ultralight panels, glass-mat panels, paper-faced plastic bead, and ultralightweight joint compound, and modern specialty acoustical and ceiling product guidelines. This comprehensive reference also incorporates the latest in sustainable products.

$40.00 | 576 pages, softcover | Catalog no. 67357B

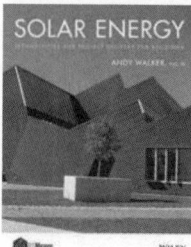

Solar Energy: Technologies & Project Delivery for Buildings

by Andy Walker

An authoritative reference on the design of solar energy systems in building projects, with applications, operating principles, and simple tools for the construction, engineering, and design professionals, the book simplifies the solar design and engineering process and provides sample documentation for the complete design of a solar energy system for buildings.

$85.00 | 320 pages, hardover | Catalog no. 67365

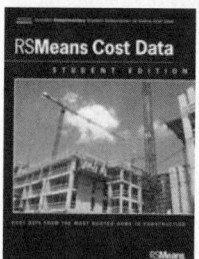

RSMeans Cost Data, Student Edition

Provides a thorough introduction to cost estimating in a print and online package. Clear explanations and example-driven. The ideal reference for students and new professionals. Features include:

- Commercial and residential construction cost data in print and online formats
- Complete how-to guidance on the essentials of cost estimating
- A supplemental website with plans, problem sets, and a full sample estimate

$110.00 | 512 pages, softcover | Catalog no. 67363

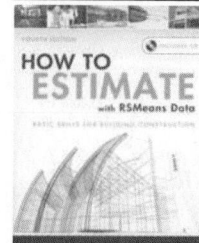

How to Estimate with Means Data & CostWorks, 4th Edition

by RSMeans and Saleh A. Mubarak, Ph.D.

This step-by-step guide takes you through all the major construction items with extensive coverage of site work, concrete and masonry, wood and metal framing, doors and windows, and other divisions. The only construction cost estimating handbook that uses the most popular source of construction data, RSMeans, this indispensable guide features access to the instructional version of *CostWorks* in electronic form, enabling you to practice techniques to solve real-world estimating problems.

$75.00 | 320 pages, softcover | Includes CostWorks CD | Catalog no. 67324C

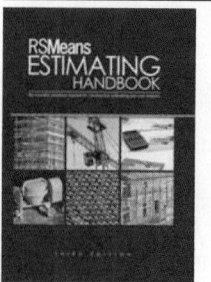

RSMeans Estimating Handbook, 3rd Edition

Widely used in the industry for tasks ranging from routine estimates to special cost analysis projects. This handbook will help construction professionals:

- Evaluate architectural plans and specifications
- Prepare accurate quantity takeoffs
- Compare design alternatives and costs
- Perform value engineering
- Double-check estimates and quotes
- Estimate change orders

$110.00 | Over 976 pages, hardcover | Catalog No. 67276B

Reference Books

For more information visit the RSMeans website at **www.rsmeans.com**

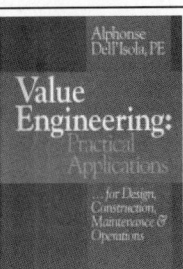

Value Engineering: Practical Applications
For Design, Construction, Maintenance & Operations

by Alphonse Dell'Isola, PE

A tool for immediate application—for engineers, architects, facility managers, owners, and contractors. Includes making the case for VE—the management briefing; integrating VE into planning, budgeting, and design; conducting life cycle costing; using VE methodology in design review and consultant selection; and case studies.

$90.00 | Over 427 pages, illustrated, softcover | Catalog no. 67319A

Risk Management for Design and Construction

Introduces risk as a central pillar of project management and shows how a project manager can prepare to deal with uncertainty. Includes:

- Integrated cost and schedule risk analysis
- An introduction to a ready-to-use system of analyzing a project's risks and tools to proactively manage risks
- A methodology that was developed and used by the Washington State Department of Transportation
- Case studies and examples on the proper application of principles
- Combining value analysis with risk analysis

$125.00 | Over 288 pages, softcover | Catalog no. 67359

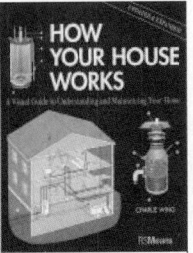

How Your House Works, 2nd Edition
by Charlie Wing

Knowledge of your home's systems helps you control repair and construction costs and makes sure the correct elements are being installed or replaced. This book uncovers the mysteries behind just about every major appliance and building element in your house. See-through, cross-section drawings in full color show you exactly how these things should be put together and how they function, including what to check if they don't work. It just might save you having to call in a professional.

$22.95 | 208 pages, softcover | Catalog no. 67351A

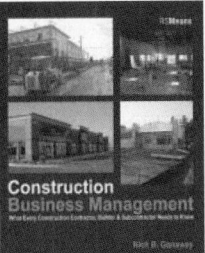

Construction Business Management
by Nick Ganaway

Only 43% of construction firms stay in business after four years. Make sure your company thrives with valuable guidance from a pro with 25 years of success as a commercial contractor. Find out what it takes to build all aspects of a business that is profitable, enjoyable, and enduring. With a bonus chapter on retail construction.

$60.00 | 201 pages, softcover | Catalog no. 67352

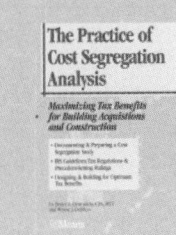

The Practice of Cost Segregation Analysis
by Bruce A. Desrosiers and Wayne J. Del Pico

This expert guide walks you through the practice of cost segregation analysis, which enables property owners to defer taxes and benefit from "accelerated cost recovery" through depreciation deductions on assets that are properly identified and classified.

With a glossary of terms, sample cost segregation estimates for various building types, key information resources, and updates via a dedicated website, this book is a critical resource for anyone involved in cost segregation analysis.

$105.00 | Over 240 pages | Catalog no. 67345

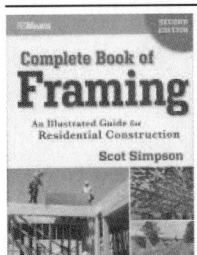

Complete Book of Framing, 2nd Edition
by Scot Simpson

This updated, easy-to-learn guide to rough carpentry and framing is written by an expert with more than thirty years of framing experience. Starting with the basics, this book begins with types of lumber, nails, and what tools are needed, followed by detailed, fully illustrated steps for framing each building element. Framer-Friendly Tips throughout the book show how to get a task done right—and more easily.

$29.95 | 368 pages, softcover | Catalog no. 67353A

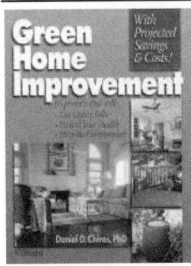

Green Home Improvement
by Daniel D. Chiras, PhD

With energy costs rising and environmental awareness increasing, people are looking to make their homes greener. This book, with 65 projects and actual costs and projected savings, helps homeowners prioritize their green improvements.

Projects range from simple water savers that cost only a few dollars, to bigger-ticket items such as HVAC systems. With color photos and cost estimates, each project compares options and describes the work involved, the benefits, and the savings.

$34.95 | 320 pages, illustrated, softcover | Catalog no. 67355

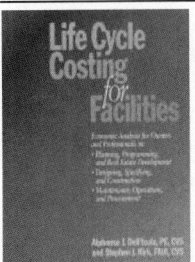

Life Cycle Costing for Facilities
by Alphonse Dell'Isola and Dr. Stephen Kirk

Guidance for achieving higher quality design and construction projects at lower costs! Cost-cutting efforts often sacrifice quality to yield the cheapest product. Life cycle costing enables building designers and owners to achieve both. The authors of this book show how LCC can work for a variety of projects—from roads to HVAC upgrades to different types of buildings.

$110.00 | 396 pages, hardcover | Catalog no. 67341

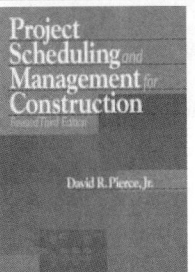

Project Scheduling and Management for Construction, 3rd Edition
by David R. Pierce, Jr.

A comprehensive, yet easy-to-follow guide to construction project scheduling and control—from vital project management principles through the latest scheduling, tracking, and controlling techniques. The author is a leading authority on scheduling, with years of field and teaching experience at leading academic institutions. Spend a few hours with this book and come away with a solid understanding of this essential management topic.

$64.95 | Over 286 pages, illustrated, hardcover | Catalog no. 67247B

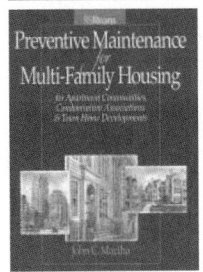

Preventive Maintenance for Multi-Family Housing
by John C. Maciha

Prepared by one of the nation's leading experts on multi-family housing.

This complete PM system for apartment and condominium communities features expert guidance, checklists for buildings and grounds maintenance tasks and their frequencies, and a dedicated website featuring customizable electronic forms. A must-have for anyone involved with multi-family housing maintenance and upkeep.

$95.00 | 290 pages | Catalog no. 67346

Reference Books

737

For more information visit the RSMeans website at **www.rsmeans.com**

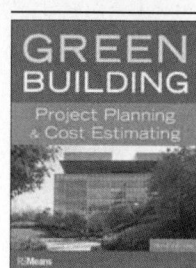

Green Building: Project Planning & Cost Estimating, 3rd Edition

Since the widely read first edition of this book, green building has gone from a growing trend to a major force in design and construction.

This new edition has been updated with the latest in green building technologies, design concepts, standards, and costs. Full-color with all new case studies—plus a new chapter on commercial real estate.

$110.00 | 480 pages, softcover | Catalog no. 67338B

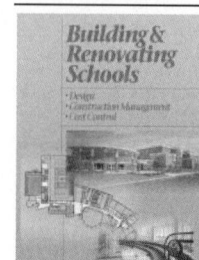

RSMeans Illustrated Construction Dictionary Unabridged 4th Edition, with CD-ROM

Long regarded as the industry's finest, *Means Illustrated Construction Dictionary* is now even better. With nearly 20,000 terms and more than 1,400 illustrations and photos, it is the clear choice for the most comprehensive and current information. The companion CD-ROM that comes with this new edition adds extra features, such as larger graphics and expanded definitions.

$110.00 | Over 880 pages, illust., hardcover | Catalog no. 67292B

Construction Supervision

Inspires excellence with proven tactics and techniques applied by thousands of construction supervisors. Leadership guidelines carve out a practical blueprint for motivating work performance and increasing productivity through effective communication. Features:

- A unique focus on field supervision and crew management
- Coverage of supervision from the foreman to the superintendent level
- An overview of technical skills whose mastery will build confidence and success for the supervisor
- A detailed view of "soft" management and communication skills

$95.00 | 464 pages, softcover | Catalog no. 67358

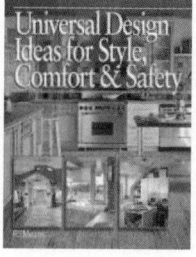

Building & Renovating Schools

This all-inclusive guide covers every step of the school construction process—from initial planning, needs assessment, and design, right through moving into the new facility. A must-have resource for anyone concerned with new school construction or renovation. With square foot cost models for elementary, middle, and high school facilities, and real-life case studies of recently completed school projects.

The contributors to this book—architects, construction project managers, contractors, and estimators who specialize in school construction—provide start-to-finish expert guidance on the process.

$110.00 | Over 412 pages, hardcover | Catalog no. 67342

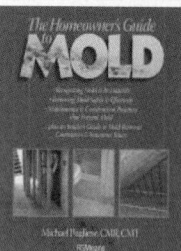

The Homeowner's Guide to Mold

By Michael Pugliese

Mold, whether caused by leaks, humidity or flooding, is a real health and financial issue—for homeowners and contractors. This full-color book explains:

- Construction and maintenance practices to prevent mold
- How to inspect for and remove mold
- Mold remediation procedures and costs
- What to do after a flood
- How to deal with insurance companies

$21.95 | 144 pages, softcover | Catalog no. 67344

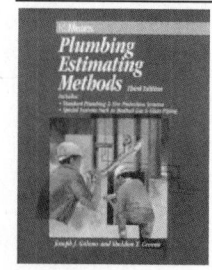

Universal Design Ideas for Style, Comfort & Safety

by RSMeans and Lexicon Consulting, Inc.

Incorporating universal design when building or remodeling helps people of any age and physical ability more fully and safely enjoy their living spaces. This book shows how universal design can be artfully blended into the most attractive homes. It discusses specialized products like adjustable countertops and chair lifts, as well as simple ways to enhance a home's safety and comfort. With color photos and expert guidance, every area of the home is covered. Includes budget estimates that give an idea how much projects will cost.

$21.95 | 160 pages, illustrated, softcover | Catalog no. 67354

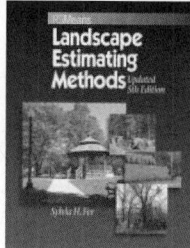

Landscape Estimating Methods, 5th Edition

Answers questions about preparing competitive landscape construction estimates, with up-to-date cost estimates and the new MasterFormat classification system. Expanded and revised to address the latest materials and methods, including new coverage on approaches to green building. Includes:

- Step-by-step explanation of the estimating process
- Sample forms and worksheets that save time and prevent errors

$75.00 | 336 pages, softcover | Catalog no. 67295C

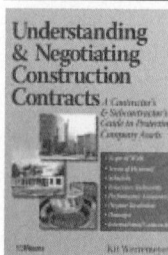

Plumbing Estimating Methods, 3rd Edition

by Joseph J. Galeno and Sheldon T. Greene

Updated and revised! This practical guide walks you through a plumbing estimate, from basic materials and installation methods through change order analysis. *Plumbing Estimating Methods* covers residential, commercial, industrial, and medical systems, and features sample takeoff and estimate forms and detailed illustrations of systems and components.

$65.00 | 381+ pages, softcover | Catalog no. 67283B

Understanding & Negotiating Construction Contracts

by Kit Werremeyer

Take advantage of the author's 30 years' experience in small-to-large (including international) construction projects. Learn how to identify, understand, and evaluate high risk terms and conditions typically found in all construction contracts—then, negotiate to lower or eliminate the risk, improve terms of payment, and reduce exposure to claims and disputes. The author avoids "legalese" and gives real-life examples from actual projects.

$80.00 | 320 pages, softcover | Catalog no. 67350

For more information visit the RSMeans website at **www.rsmeans.com**

Qty.	Book no.	COST ESTIMATING BOOKS	Unit Price	Total
	60065	Assemblies Cost Data 2015	$334.95	
	60015	Building Construction Cost Data 2015	206.95	
	60045	Commercial Renovaion Cost Data 2015	166.95	
	60115	Concrete & Masonry Cost Data 2015	188.95	
	50145	Construction Cost Indexes 2015 (subscription)	359.95	
	60145A	Construction Cost Index–January 2015	94.95	
	60145B	Construction Cost Index–April 2015	94.95	
	60145C	Construction Cost Index–July 2015	94.95	
	60145D	Construction Cost Index–October 2015	94.95	
	60345	Contr. Pricing Guide: Resid. R & R Costs 2015	39.95	
	60235	Electrical Change Order Cost Data 2015	199.95	
	60035	Electrical Cost Data 2015	208.95	
	60205	Facilities Construction Cost Data 2015	524.95	
	60305	Facilities Maintenance & Repair Cost Data 2015	461.95	
	60555	Green Building Cost Data 2015	170.95	
	60165	Heavy Construction Cost Data 2015	204.95	
	60095	Interior Cost Data 2015	208.95	
	60125	Labor Rates for the Const. Industry 2015	450.95	
	60185	Light Commercial Cost Data 2015	153.95	
	60025	Mechanical Cost Data 2015	203.95	
	60155	Open Shop Building Const. Cost Data 2015	175.95	
	60215	Plumbing Cost Data 2015	206.95	
	60175	Residential Cost Data 2015	148.95	
	60285	Site Work & Landscape Cost Data 2015	199.95	
	60055	Square Foot Costs 2015	218.95	
	62014	Yardsticks for Costing (2014)	189.95	
	62015	Yardsticks for Costing (2015)	201.95	
		REFERENCE BOOKS		
	67329	Bldrs Essentials: Best Bus. Practices for Bldrs	45.00	
	67307	Bldrs Essentials: Plan Reading & Takeoff	50.00	
	67342	Building & Renovating Schools	110.00	
	67261A	The Building Prof. Guide to Contract Documents	75.00	
	67353A	Complete Book of Framing, 2nd Ed.	29.95	
	67146	Concrete Repair & Maintenance Illustrated	75.00	
	67352	Construction Business Management	60.00	
	67364	Construction Law	115.00	
	67358	Construction Supervision	95.00	
	67363	Cost Data Student Edition	110.00	
	67314	Cost Planning & Est. for Facil. Maint.	95.00	
	67230B	Electrical Estimating Methods, 3rd Ed.	75.00	
	67343A	Est. Bldg. Costs for Resi. & Lt. Comm., 2nd Ed.	75.00	
	67276B	Estimating Handbook, 3rd Ed.	110.00	
	67318	Facilities Operations & Engineering Reference	115.00	
	67338B	Green Building: Proj. Planning & Cost Est., 3rd Ed.	110.00	
	67355	Green Home Improvement	34.95	
	67357B	The Gypsum Construction Handbook	40.00	
	67148	Heavy Construction Handbook	110.00	
	67308E	Home Improvement Costs–Int. Projects, 9th Ed.	24.95	
	67309E	Home Improvement Costs–Ext. Projects, 9th Ed.	24.95	
	67344	Homeowner's Guide to Mold	21.95	
	67324C	How to Est.w/Means Data & CostWorks, 4th Ed.	75.00	

Qty.	Book no.	REFERENCE BOOKS (Cont.)	Unit Price	Total
	67351A	How Your House Works, 2nd Ed.	22.95	
	67282A	Illustrated Const. Dictionary, Condensed, 2nd Ed.	65.00	
	67292B	Illustrated Const. Dictionary, w/CD-ROM, 4th Ed.	110.00	
	67362	Illustrated Const. Dictionary, Student Ed.	60.00	
	67348	Job Order Contracting	130.00	
	67295C	Landscape Estimating Methods, 5th Ed.	75.00	
	67341	Life Cycle Costing for Facilities	110.00	
	67370	Managing the Profitable Construction Business	60.00	
	67294B	Mechanical Estimating Methods, 4th Ed.	75.00	
	67345	Practice of Cost Segregation Analysis	105.00	
	67366	Project Control: Integrating Cost & Sched.	95.00	
	67346	Preventive Maint. for Multi-Family Housing	95.00	
	67367	Project Sched. and Manag. for Constr., 4th Ed.	95.00	
	67322B	Resi. & Light Commercial Const. Stds., 3rd Ed.	65.00	
	67359	Risk Management for Design and Construction	125.00	
	67365	Solar Energy: Technology & Project Delivery	85.00	
	67145B	Sq. Ft. & Assem. Estimating Methods, 3rd Ed.	75.00	
	67321	Total Productive Facilities Management	85.00	
	67350	Understanding and Negotiating Const. Contracts	80.00	
	67303B	Unit Price Estimating Methods, 4th Ed.	75.00	
	67354	Universal Design	21.95	
	67319A	Value Engineering: Practical Applications	90.00	

MA residents add 6.25% state sales tax	
Shipping & Handling**	
Total (U.S. Funds)*	

Prices are subject to change and are for U.S. delivery only. *Canadian customers may call for current prices. **Shipping & handling charges: Add 8% of total order for check and credit card payments. Add 8% of total order for invoiced orders.

Send order to: ADDV-1000

Name (please print) _____

Company _____

☐ Company

☐ Home Address _____

_____

City/State/Zip _____

Phone # _____ P.O. # _____

(Must accompany all orders being billed)

Mail to: RSMeans, 700 Longwater Drive, Norwell, MA 02061

RSMeans Project Cost Report

By filling out this report, your project data will contribute to the database that supports the RSMeans® Project Cost Square Foot Data. When you fill out this form, RSMeans will provide a $50 discount off one of the RSMeans products advertised in the preceding pages. Please complete the form including all items where you have cost data, and all the items marked (☑).

$50.00 Discount per product for each report you submit

Project Description (NEW construction only, please)

☑ Building Use (Office School . . .) _____

☑ Address (City, State)_____

_____

☑ Total Building Area (SF) _____

☑ Ground Floor (SF) _____

☑ Frame (Wood, Steel . . .) _____

☑ Exterior Wall (Brick, Tilt-up . . .) _____

☑ Basement: (check one) ☐ Full   ☐ Partial   ☐ None

☑ Number of Stories _____

☑ Floor-to-Floor Height _____

☑ Volume (C.F.) _____

% Air Conditioned _____ Tons _____

Total Project Cost $ _____

Owner _____

Architect_____

General Contractor _____

☑ Bid Date _____

☑ Typical Bay Size _____

☑ Occupant Capacity_____

☑ Labor Force: _____ % Union _____ % Non-Union

☑ Project Description (Circle one number in each line.)

1. Economy 2. Average 3. Custom 4. Luxury

1. Square 2. Rectangular 3. Irregular 4. Very Irregular

Comments _____

_____

A	☑	General Conditions	$
B	☑	Site Work	$
C	☑	Concrete	$
D	☑	Masonry	$
E	☑	Metals	$
F	☑	Wood & Plastics	$
G	☑	Thermal & Moisture Protection	$
GR		Roofing & Flashing	$
H	☑	Doors and Windows	$
J	☑	Finishes	$
JP		Painting & Wall Covering	$

K	☑	Specialties	$
L	☑	Equipment	$
M	☑	Furnishings	$
N	☑	Special Construction	$
P	☑	Conveying Systems	$
Q	☑	Mechanical	$
QP		Plumbing	$
QB		HVAC	$
R	☑	Electrical	$
S	☑	Mech./Elec. Combined	$

Please specify the RSMeans product you wish to receive.
Complete the address information.

Product Name _____

Product Number _____

Your Name _____

Title _____

Company_____

  ☐ Company
  ☐ Home   Street Address_____

City, State, Zip _____

Email Address _____

Return by mail or fax 888-492-6770.

Method of Payment:

Credit Card # _____

Expiration Date _____

Check_____

Purchase Order _____

Phone Number _____
(if you would prefer a sales call)

RSMeans
Square Foot Costs Department
700 Longwater Drive
Norwell, MA 02061
www.RSMeans.com